# In Praise of *Foundations of Analog and Digital Electronic Circuits*

*"This book, crafted and tested with MIT sophomores in electrical engineering and computer science over a period of more than six years, provides a comprehensive treatment of both circuit analysis and basic electronic circuits. Examples such as digital and analog circuit applications, field-effect transistors, and operational amplifiers provide the platform for modeling of active devices, including large-signal, small-signal (incremental), nonlinear and piecewise-linear models. The treatment of circuits with energy-storage elements in transient and sinusoidal-steady-state circumstances is thorough and accessible. Having taught from drafts of this book five times, I believe that it is an improvement over the traditional approach to circuits and electronics, in which the focus is on analog circuits alone."*

-PAUL E. GRAY, Massachusetts Institute of Technology

*"My overall reaction to this book is overwhelmingly favorable. Well-written and pedagogically sound, the book provides a good balance between theory and practical application. I think that combining circuits and electronics is a very good idea. Most introductory circuit theory texts focus primarily on the analysis of lumped element networks without putting these networks into a practical electronics context. However, it is becoming more critical for our electrical and computer engineering students to understand and appreciate the common ground from which both fields originate."*

-GARY MAY, Georgia Institute of Technology

*"Without a doubt, students in engineering today want to quickly relate what they learn from courses to what they experience in the electronics-filled world they live in. Understanding today's digital world requires a strong background in analog circuit principles as well as a keen intuition about their impact on electronics. In Foundations... Agarwal and Lang present a unique and powerful approach for an exciting first course introducing engineers to the world of analog and digital systems."*

-RAVI SUBRAMANIAN, Berkeley Design Automation

*"Finally, an introductory circuit analysis book has been written that truly unifies the treatment of traditional circuit analysis and electronics. Agarwal and Lang skillfully combine the fundamentals of circuit analysis with the fundamentals of modern analog and digital integrated circuits. I applaud their decision to eliminate from their book the usual mandatory chapter on Laplace transforms, a tool no longer in use by modern circuit designers. I expect this book to establish a new trend in the way introductory circuit analysis is taught to electrical and computer engineers."*

-TIM TRICK, University of Illinois at Urbana-Champaign

# Foundations of Analog and Digital Electronic Circuits

# ABOUT THE AUTHORS

Anant Agarwal is Professor of Electrical Engineering and Computer Science at the Massachusetts Institute of Technology. He joined the faculty in 1988, teaching courses in circuits and electronics, VLSI, digital logic and computer architecture. Between 1999 and 2003, he served as an associate director of the Laboratory for Computer Science. He holds a Ph.D. and an M.S. in Electrical Engineering from Stanford University, and a bachelor's degree in Electrical Engineering from IIT Madras. Agarwal led a group that developed Sparcle (1992), a multithreaded microprocessor, and the MIT Alewife (1994), a scalable shared-memory multiprocessor. He also led the VirtualWires project at MIT and was a founder of Virtual Machine Works, Inc., which took the VirtualWires logic emulation technology to market in 1993. Currently Agarwal leads the Raw project at MIT, which developed a new kind of reconfigurable computing chip. He and his team were awarded a Guinness world record in 2004 for LOUD, the largest microphone array in the world, which can pinpoint, track and amplify individual voices in a crowd. Co-founder of Engim, Inc., which develops multi-channel wireless mixed-signal chipsets, Agarwal also won the Maurice Wilkes prize for computer architecture in 2001, and the Presidential Young Investigator award in 1991.

Jeffrey H. Lang is Professor of Electrical Engineering and Computer Science at the Massachusetts Institute of Technology. He joined the faculty in 1980 after receiving his SB (1975), SM (1977) and Ph.D. (1980) degrees from the Department of Electrical Engineering and Computer Science. He served as the Associate Director of the MIT Laboratory for Electromagnetic and Electronic Systems between 1991 and 2003, and as an Associate Editor of "Sensors and Actuators" between 1991 and 1994. Professor Lang's research and teaching interests focus on the analysis, design and control of electromechanical systems with an emphasis on rotating machinery, micro-scale sensors and actuators, and flexible structures. He has also taught courses in circuits and electronics at MIT. He has written over 170 papers and holds 10 patents in the areas of electromechanics, power electronics and applied control, and has been awarded four best-paper prizes from IEEE societies. Professor Lang is a Fellow of the IEEE, and a former Hertz Foundation Fellow.

Agarwal and Lang have been working together for the past eight years on a fresh approach to teaching circuits. For several decades, MIT had offered a traditional course in circuits designed as the first core undergraduate course in EE. But by the mid-'90s, vast advances in semiconductor technology, coupled with dramatic changes in students' backgrounds evolving from a ham radio to computer culture, had rendered this traditional course poorly motivated, and many parts of it were virtually obsolete. Agarwal and Lang decided to revamp and broaden this first course for EE, ECE or EECS by establishing a strong connection between the contemporary worlds of digital and analog systems, and by unifying the treatment of circuits and basic MOS electronics. As they developed the course, they solicited comments and received guidance from a large number of colleagues from MIT and other universities, students, and alumni, as well as industry leaders.

Unable to find a suitable text for their new introductory course, Agarwal and Lang wrote this book to follow the lecture schedule used in their course. "Circuits and Electronics" is taught in both the spring and fall semesters at MIT, and serves as a prerequisite for courses in signals and systems, digital/computer design, and advanced electronics. The course material is available worldwide on MIT's OpenCourseWare website, http://ocw.mit.edu/OcwWeb/index.htm.

# Foundations of Analog and Digital Electronic Circuits

ANANT AGARWAL

*Department of Electrical Engineering and Computer Science,*
*Massachusetts Institute of Technology*

JEFFREY H. LANG

*Department of Electrical Engineering and Computer Science,*
*Massachusetts Institute of Technology*

AMSTERDAM • BOSTON • HEIDELBERG • LONDON
NEW YORK • OXFORD • PARIS • SAN DIEGO
SAN FRANCISCO • SINGAPORE • SYDNEY • TOKYO
MORGAN KAUFMANN PUBLISHERS IS AN IMPRINT OF ELSEVIER

Publisher: Denise E. M. Penrose
Publishing Services Manager: Simon Crump
Editorial Assistant: Valerie Witte
Cover Design: Frances Baca
Composition: Cepha Imaging Pvt. Ltd., India
Technical Illustration: Dartmouth Publishing, Inc.
Copyeditor: Eileen Kramer
Proofreader: Katherine Hasal
Indexer: Kevin Broccoli
Interior printer: King Printing Co., Inc.
Cover printer: King Printing Co., Inc.

Morgan Kaufmann Publishers is an imprint of Elsevier.
500 Sansome Street, Suite 400, San Francisco, CA 94111

This book is printed on acid-free paper.



**Library of Congress Cataloging-in-Publication Data**
ISBN: 1-55860-735-8

For information on all Morgan Kaufmann publications,
visit our Web site at www.mkp.com or www.books.elsevier.com

Printed in the United States of America
Last digit is the print number: 20 19 18 17 16 15

*To Anu, Akash, and Anisha*
—Anant Agarwal

*To Marija, Chris, John, Matt*
—Jeffrey Lang

# CONTENTS

Material marked with WWW appears on the Internet (please see Preface for details).

# PREFACE

## APPROACH

This book is designed to serve as a first course in an electrical engineering or an electrical engineering and computer science curriculum, providing students at the sophomore level a transition from the world of physics to the world of electronics and computation. The book attempts to satisfy two goals: Combine circuits and electronics into a single, unified treatment, and establish a strong connection with the contemporary worlds of both digital and analog systems.

These goals arise from the observation that the approach to introducing electrical engineering through a course in traditional circuit analysis is fast becoming obsolete. Our world has gone digital. A large fraction of the student population in electrical engineering is destined for industry or graduate study in digital electronics or computer systems. Even those students who remain in core electrical engineering are heavily influenced by the digital domain.

Because of this elevated focus on the digital domain, basic electrical engineering education must change in two ways: First, the traditional approach to teaching circuits and electronics without regard to the digital domain must be replaced by one that stresses the circuits foundations common to both the digital and analog domains. Because most of the fundamental concepts in circuits and electronics are equally applicable to both the digital and the analog domains, this means that, primarily, we must change the way in which we motivate circuits and electronics to emphasize their broader impact on digital systems. For example, although the traditional way of discussing the dynamics of first-order RC circuits appears unmotivated to the student headed into digital systems, the same pedagogy is exciting when motivated by the switching behavior of a switch and resistor inverter driving a non-ideal capacitive wire. Similarly, we motivate the study of the step response of a second-order RLC circuit by observing the behavior of a MOS inverter when pin parasitics are included.

Second, given the additional demands of computer engineering, many departments can ill-afford the luxury of separate courses on circuits and on electronics. Rather, they might be combined into one course.[1] Circuits courses

---

1. In his paper, "Teaching Circuits and Electronics to First-Year Students," in *Int. Symp. Circuits and Systems (ISCAS)*, 1998, Yannis Tsividis makes an excellent case for teaching an integrated course in circuits and electronics.

treat networks of passive elements such as resistors, sources, capacitors, and inductors. Electronics courses treat networks of both passive elements and active elements such as MOS transistors. Although this book offers a unified treatment for circuits and electronics, we have taken some pains to allow the crafting of a two-semester sequence — one focused on circuits and another on electronics — from the same basic content in the book.

Using the concept of "abstraction," the book attempts to form a bridge between the world of physics and the world of large computer systems. In particular, it attempts to unify electrical engineering and computer science as the art of creating and exploiting successive abstractions to manage the complexity of building useful electrical systems. Computer systems are simply one type of electrical system.

In crafting a single text for both circuits and electronics, the book takes the approach of covering a few important topics in depth, choosing more contemporary devices when possible. For example, it uses the MOSFET as the basic active device, and relegates discussions of other devices such as bipolar transistors to the exercises and examples. Furthermore, to allow students to understand basic circuit concepts without the trappings of specific devices, it introduces several abstract devices as examples and exercises. We believe this approach will allow students to tackle designs with many other extant devices and those that are yet to be invented.

Finally, the following are some additional differences from other books in this field:

- The book draws a clear connection between electrical engineering and physics by showing clearly how the lumped circuit abstraction directly derives from Maxwell's Equations and a set of simplifying assumptions.
- The concept of abstraction is used throughout the book to unify the set of engineering simplifications made in both analog and digital design.
- The book elevates the focus of the digital domain to that of analog. However, our treatment of digital systems emphasizes their analog aspects. We start with switches, sources, resistors, and MOSFETs, and apply KVL, KCL, and so on. The book shows that digital versus analog behavior is obtained by focusing on particular regions of device behavior.
- The MOSFET device is introduced using a progression of models of increased refinement — the S model, the SR model, the SCS model, and the SU model.
- The book shows how significant amounts of insight into the static and dynamic operation of digital circuits can be obtained with very simple models of MOSFETs.

- Various properties of devices, for example, the memory property of capacitors, or the gain property of amplifiers, are related to both their use in analog circuits and digital circuits.
- The state variable viewpoint of transient problems is emphasized for its intuitive appeal and since it motivates computer solutions of both linear or nonlinear network problems.
- Issues of energy and power are discussed in the context of both analog and digital circuits.
- A large number of examples are picked from the digital domain emphasizing VLSI concepts to emphasize the power and generality of traditional circuit analysis concepts.

With these features, we believe this book offers the needed foundation for students headed towards either the core electrical engineering majors — including digital and RF circuits, communication, controls, signal processing, devices, and fabrication — or the computer engineering majors — including digital design, architecture, operating systems, compilers, and languages.

MIT has a unified electrical engineering and computer science department. This book is being used in MIT's introductory course on circuits and electronics. This course is offered each semester and is taken by about 500 students a year.

## OVERVIEW

Chapter 1 discusses the concept of abstraction and introduces the lumped circuit abstraction. It discusses how the lumped circuit abstraction derives from Maxwell's Equations and provides the basic method by which electrical engineering simplifies the analysis of complicated systems. It then introduces several ideal, lumped elements including resistors, voltage sources, and current sources.

This chapter also discusses two major motivations of studying electronic circuits — modeling physical systems and information processing. It introduces the concept of a model and discusses how physical elements can be modeled using ideal resistors and sources. It also discusses information processing and signal representation.

Chapter 2 introduces KVL and KCL and discusses their relationship to Maxwell's Equations. It then uses KVL and KCL to analyze simple resistive networks. This chapter also introduces another useful element called the dependent source.

Chapter 3 presents more sophisticated methods for network analysis.

Chapter 4 introduces the analysis of simple, nonlinear circuits.

Chapter 5 introduces the digital abstraction, and discusses the second major simplification by which electrical engineers manage the complexity of building large systems.[2]

Chapter 6 introduces the switch element and describes how digital logic elements are constructed. It also describes the implementation of switches using MOS transistors. Chapter 6 introduces the S (switch) and the SR (switch-resistor) models of the MOSFET and analyzes simple switch circuits using the network analysis methods presented earlier. Chapter 6 also discusses the relationship between amplification and noise margins in digital systems.

Chapter 7 discusses the concept of amplification. It presents the SCS (switch-current-source) model of the MOSFET and builds a MOSFET amplifier.

Chapter 8 continues with small signal amplifiers.

Chapter 9 introduces storage elements, namely, capacitors and inductors, and discusses why the modeling of capacitances and inductances is necessary in high-speed design.

Chapter 10 discusses first order transients in networks. This chapter also introduces several major applications of first-order networks, including digital memory.

Chapter 11 discusses energy and power issues in digital systems and introduces CMOS logic.

Chapter 12 analyzes second order transients in networks. It also discusses the resonance properties of RLC circuits from a time-domain point of view.

Chapter 13 discusses sinusoidal steady state analysis as an alternative to the time-domain transient analysis. The chapter also introduces the concepts of impedance and frequency response. This chapter presents the design of filters as a major motivating application.

Chapter 14 analyzes resonant circuits from a frequency point of view.

Chapter 15 introduces the operational amplifier as a key example of the application of abstraction in analog design.

Chapter 16 discusses diodes and simple diode circuits.

The book also contains appendices on trignometric functions, complex numbers, and simultaneous linear equations to help readers who need a quick refresher on these topics or to enable a quick lookup of results.

---

2. The point at which to introduce the digital abstraction in this book and in a corresponding curriculum was arguably the topic over which we agonized the most. We believe that introducing the digital abstraction at this point in the course balances (a) the need for introducing digital systems as early as possible in the curriculum to excite and motivate students (especially with laboratory experiments), with (b) the need for providing students with enough of a toolchest to be able to analyze interesting digital building blocks such as combinational logic. Note that we recommend introduction of digital systems a lot sooner than suggested by Tsividis in his 1998 ISCAS paper, although we completely agree his position on the need to include some digital design.

## COURSE ORGANIZATION

The sequence of chapters has been organized to suit a one or two semester integrated course on circuits and electronics. First and second order circuits are introduced as late as possible to allow the students to attain a higher level of mathematical sophistication in situations in which they are taking a course on differential equations at the same time. The digital abstraction is introduced as early as possible to provide early motivation for the students.

Alternatively, the following chapter sequences can be selected to organize the course around a circuits sequence followed by an electronics sequence. The circuits sequence would include the following: Chapter 1 (lumped circuit abstraction), Chapter 2 (KVL and KCL), Chapter 3 (network analysis), Chapter 5 (digital abstraction), Chapter 6 (S and SR MOS models), Chapter 9 (capacitors and inductors), Chapter 10 (first-order transients), Chapter 11 (energy and power, and CMOS), Chapter 12 (second-order transients), Chapter 13 (sinusoidal steady state), Chapter 14 (frequency analysis of resonant circuits), and Chapter 15 (operational amplifier abstraction — optional).

The electronics sequence would include the following: Chapter 4 (nonlinear circuits), Chapter 7 (amplifiers, the SCS MOSFET model), Chapter 8 (small-signal amplifiers), Chapter 13 (sinusoidal steady state and filters), Chapter 15 (operational amplifier abstraction), and Chapter 16 (diodes and power circuits).

## WEB SUPPLEMENTS

We have gathered a great deal of material to help students and instructors using this book. This information can be accessed from the Morgan Kaufmann website:

**www.mkp.com/companions/1558607358**

The site contains:

- Supplementary sections and examples. We have used the icon WWW in the text to identify sections or examples.
- Instructor's manual
- A link to the MIT OpenCourseWare website for the authors' course, 6.002 Circuits and Electronics. On this site you will find:
    - Syllabus. A summary of the objectives and learning outcomes for course 6.002.
    - Readings. Reading assignments based on Foundations of Analog and Digital Electronic Circuits.
    - Lecture Notes. Complete set of lecture notes, accompanying video lectures, and descriptions of the demonstrations made by the instructor during class.

- Labs. A collection of four labs: Thevenin/Norton Equivalents and Logic Gates, MOSFET Inverting Amplifiers and First-Order Circuits, Second-Order Networks, and Audio Playback System. Includes an equipment handout and lab tutorial. Labs include pre-lab exercises, in-lab exercises, and post-lab exercises.
- Assignments. A collection of eleven weekly homework assignments.
- Exams. Two quizzes and a Final Exam.
- Related Resources. Online exercises in Circuits and Electronics for demonstration and self-study.

## ACKNOWLEDGMENTS

These notes evolved out of an initial set of notes written by Campbell Searle for 6.002 in 1991. The notes were also influenced by several who taught 6.002 at various times including Steve Senturia and Gerry Sussman. The notes have also benefited from the insights of Steve Ward, Tom Knight, Chris Terman, Ron Parker, Dimitri Antoniadis, Steve Umans, David Perreault, Karl Berggren, Gerry Wilson, Paul Gray, Keith Carver, Mark Horowitz, Yannis Tsividis, Cliff Pollock, Denise Penrose, Greg Schaffer, and Steve Senturia. We are also grateful to our reviewers including Timothy Trick, Barry Farbrother, John Pinkston, Stephane Lafortune, Gary May, Art Davis, Jeff Schowalter, John Uyemura, Mark Jupina, Barry Benedict, Barry Farbrother, and Ward Helms for their feedback. The help of Michael Zhang, Thit Minn, and Patrick Maurer in fleshing out problems and examples; that of Jose Oscar Mur-Miranda, Levente Jakab, Vishal Kapur, Matt Howland, Tom Kotwal, Michael Jura, Stephen Hou, Shelley Duvall, Amanda Wang, Ali Shoeb, Jason Kim, Charvak Karpe and Michael Jura in creating an answer key; that of Rob Geary, Yu Xinjie, Akash Agarwal, Chris Lang, and many of our students and colleagues in proofreading; and that of Anne McCarthy, Cornelia Colyer, and Jennifer Tucker in figure creation is also gratefully acknowledged. We gratefully acknowledge Maxim for their support of this book, and Ron Koo for making that support possible, as well as for capturing and providing us with numerous images of electronic components and chips. Ron Koo is also responsible for encouraging us to think about capturing and articulating the quick, intuitive process by which seasoned electrical engineers analyze circuits — our numerous sections on intuitive analysis are a direct result of his encouragement. We also thank Adam Brand and Intel Corp. for providing us with the images of the Pentium IV.

CHAPTER 1

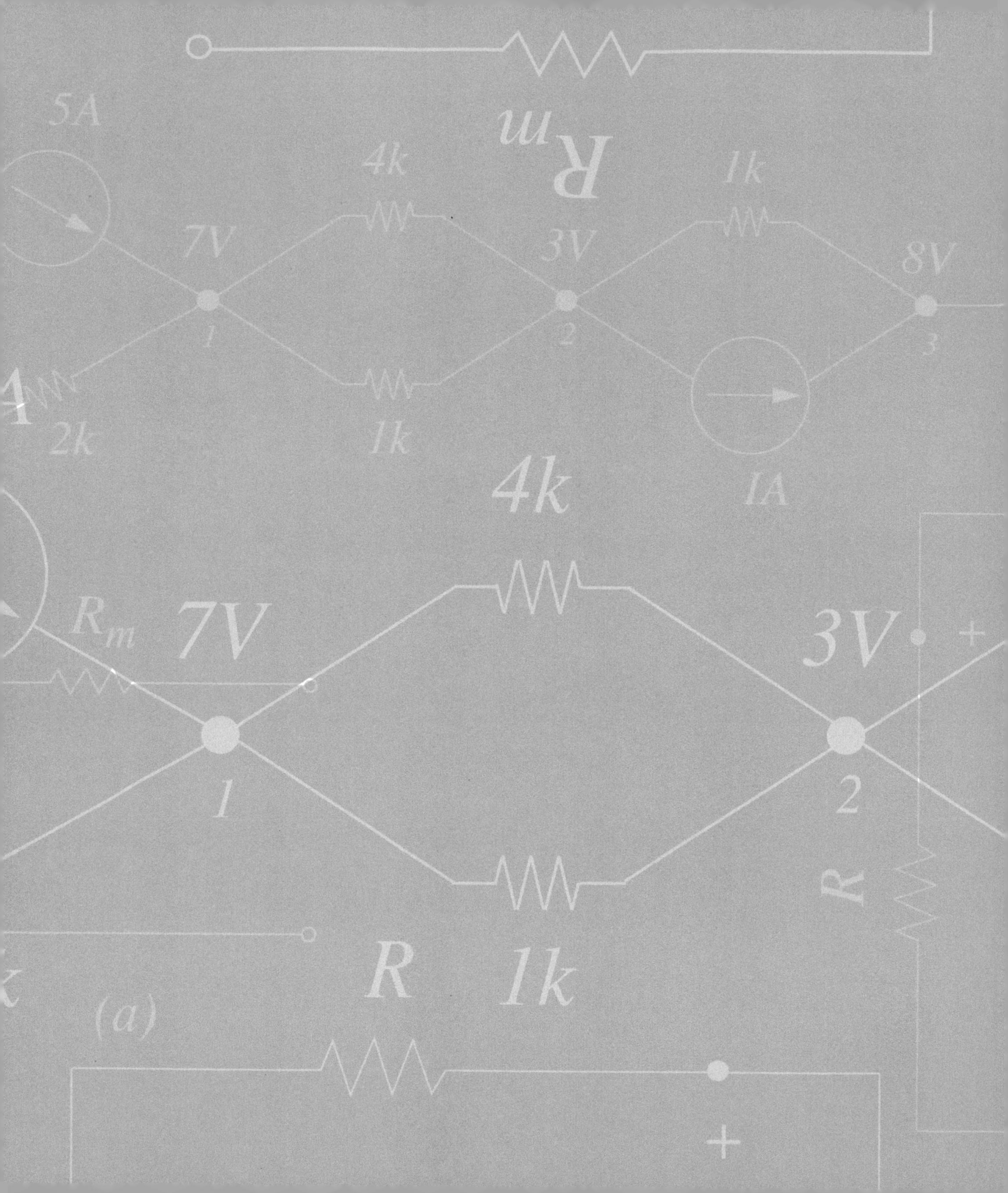

# 1 THE CIRCUIT ABSTRACTION

*"Engineering is the purposeful use of science."*

STEVE SENTURIA

## 1.1 THE POWER OF ABSTRACTION

Engineering is the purposeful use of science. Science provides an understanding of natural phenomena. Scientific study involves experiment, and scientific laws are concise statements or equations that explain the experimental data. The laws of physics can be viewed as a layer of abstraction between the experimental data and the practitioners who want to use specific phenomena to achieve their goals, without having to worry about the specifics of the experiments and the data that inspired the laws. Abstractions are constructed with a particular set of goals in mind, and they apply when appropriate constraints are met. For example, Newton's laws of motion are simple statements that relate the dynamics of rigid bodies to their masses and external forces. They apply under certain constraints, for example, when the velocities are much smaller than the speed of light. Scientific abstractions, or laws such as Newton's, are simple and easy to use, and enable us to harness and use the properties of nature.

Electrical engineering and computer science, or electrical engineering for short, is one of many engineering disciplines. Electrical engineering is the purposeful use of Maxwell's Equations (or Abstractions) for electromagnetic phenomena. To facilitate our use of electromagnetic phenomena, electrical engineering creates a new abstraction layer on top of Maxwell's Equations called the lumped circuit abstraction. By treating the lumped circuit abstraction layer, this book provides the connection between physics and electrical engineering. It unifies electrical engineering and computer science as the art of creating and exploiting successive abstractions to manage the complexity of building useful electrical systems. Computer systems are simply one type of electrical system.

The abstraction mechanism is very powerful because it can make the task of building complex systems tractable. As an example, consider the force equation:

$$F = ma. \tag{1.1}$$

The force equation enables us to calculate the acceleration of a particle with a given mass for an applied force. This simple force abstraction allows us to disregard many properties of objects such as their size, shape, density, and temperature, that are immaterial to the calculation of the object's acceleration. It also allows us to ignore the myriad details of the experiments and observations that led to the force equation, and accept it as a given. Thus, scientific laws and abstractions allow us to leverage and build upon past experience and work. (Without the force abstraction, consider the pain we would have to go through to perform experiments to achieve the same result.)

Over the past century, electrical engineering and computer science have developed a set of abstractions that enable us to transition from the physical sciences to engineering and thereby to build useful, complex systems.

The set of abstractions that transition from science to engineering and insulate the engineer from scientific minutiae are often derived through the *discretization discipline.* Discretization is also referred to as *lumping.* A discipline is a self-imposed constraint. The discipline of discretization states that we choose to deal with discrete elements or ranges and ascribe a single value to each discrete element or range. Consequently, the discretization discipline requires us to ignore the distribution of values within a discrete element. Of course, this discipline requires that systems built on this principle operate within appropriate constraints so that the single-value assumptions hold. As we will see shortly, the lumped circuit abstraction that is fundamental to electrical engineering and computer science is based on lumping or discretizing matter.[1] Digital systems use the digital abstraction, which is based on discretizing signal values. Clocked digital systems are based on discretizing both signals and time, and digital systolic arrays are based on discretizing signals, time *and* space.

Building upon the set of abstractions that define the transition from physics to electrical engineering, electrical engineering creates further abstractions to manage the complexity of building large systems. A lumped circuit element is often used as an abstract representation or a model of a piece of material with complicated internal behavior. Similarly, a circuit often serves as an abstract representation of interrelated physical phenomena. The operational amplifier composed of primitive discrete elements is a powerful abstraction that simplifies the building of bigger analog systems. The logic gate, the digital memory, the digital finite-state machine, and the microprocessor are themselves a succession of abstractions developed to facilitate building complex computer and control systems. Similarly, the art of computer programming involves the mastery of creating successively higher-level abstractions from lower-level primitives.

---

1. Notice that Newton's laws of physics are themselves based on discretizing matter. Newton's laws describe the dynamics of discrete bodies of matter by treating them as point masses. The spatial distribution of properties within the discrete elements are ignored.

Physics

Circuits and electronics

Digital logic

Computer architecture

Java programming

| Nature |
| --- |
| Laws of physics |
| Lumped circuit abstraction |
| Digital abstraction |
| Logic gate abstraction |
| Memory abstraction |
| Finite-state machine abstraction |
| Microprocessor abstraction |
| Assembly language abstraction |
| Programming language abstraction |
| Doom, mixed-signal chip |

**FIGURE 1.1** Sequence of courses and the abstraction layers introduced in a possible EECS course sequence that ultimately results in the ability to create the computer game "Doom," or a mixed-signal (containing both analog and digital components) microprocessor supervisory circuit such as that shown in Figure 1.2.

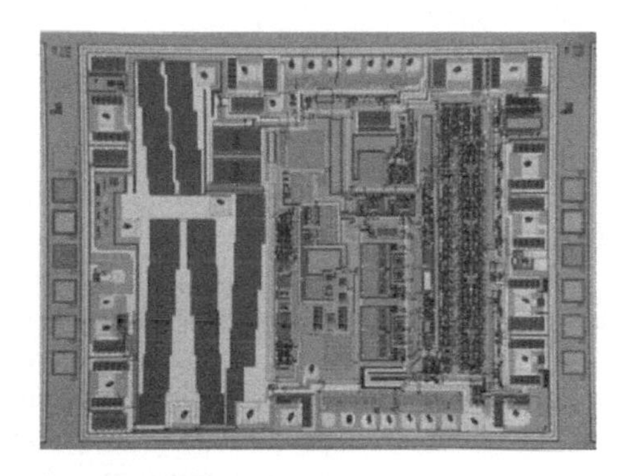

**FIGURE 1.2** A photograph of the MAX807L microprocessor supervisory circuit from Maxim Integrated Products. The chip is roughly 2.5 mm by 3 mm. Analog circuits are to the left and center of the chip, while digital circuits are to the right. (Photograph Courtesy of Maxim Integrated Products.)

Figures 1.1 and 1.3 show possible course sequences that students might encounter in an EECS (Electrical Engineering and Computer Science) or an EE (Electrical Engineering) curriculum, respectively, to illustrate how each of the courses introduces several abstraction layers to simplify the building of useful electronic systems. This sequence of courses also illustrates how a circuits and electronics course using this book might fit within a general EE or EECS course framework.

## 1.2 THE LUMPED CIRCUIT ABSTRACTION

Consider the familiar lightbulb. When it is connected by a pair of cables to a battery, as shown in Figure 1.4a, it lights up. Suppose we are interested in finding out the amount of current flowing through the bulb. We might go about this by employing Maxwell's equations and deriving the amount of current by

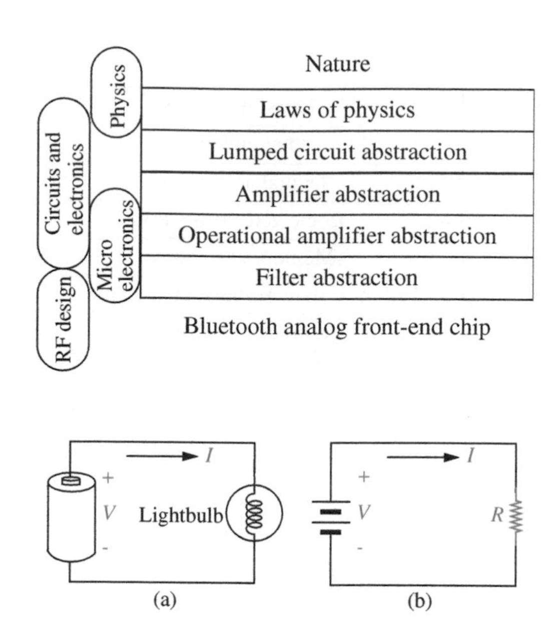


**FIGURE 1.3** Sequence of courses and the abstraction layers that they introduce in a possible EE course sequence that ultimately results in the ability to create a wireless Bluetooth analog front-end chip.

**FIGURE 1.4** (a) A simple lightbulb circuit. (b) The lumped circuit representation.

a careful analysis of the physical properties of the bulb, the battery, and the cables. This is a horrendously complicated process.

As electrical engineers we are often interested in such computations in order to design more complex circuits, perhaps involving multiple bulbs and batteries. So how do we simplify our task? We observe that if we discipline ourselves to asking only simple questions, such as what is the net current flowing through the bulb, we can ignore the internal properties of the bulb and represent the bulb as a discrete element. Further, for the purpose of computing the current, we can create a discrete element known as a resistor and replace the bulb with it.[2] We define the resistance of the bulb $R$ to be the ratio of the voltage applied to the bulb and the resulting current through it. In other words,

$$R = V/I.$$

Notice that the actual shape and physical properties of the bulb are irrelevant provided it offers the resistance $R$. We were able to ignore the internal properties and distribution of values inside the bulb simply by disciplining ourselves not to ask questions about those internal properties. In other words, when asking about the current, we were able to discretize the bulb into a single lumped element whose single relevant property was its resistance. This situation is

2. We note that the relationship between the voltage and the current for a bulb is generally much more complicated.

analogous to the point mass simplification that resulted in the force relation in Equation 1.1, where the single relevant property of the object is its mass.

As illustrated in Figure 1.5, a lumped element can be idealized to the point where it can be treated as a black box accessible through a few terminals. The behavior at the terminals is more important than the details of the behavior internal to the black box. That is, what happens at the terminals is more important than how it happens inside the black box. Said another way, the black box is a layer of abstraction between the user of the bulb and the internal structure of the bulb.

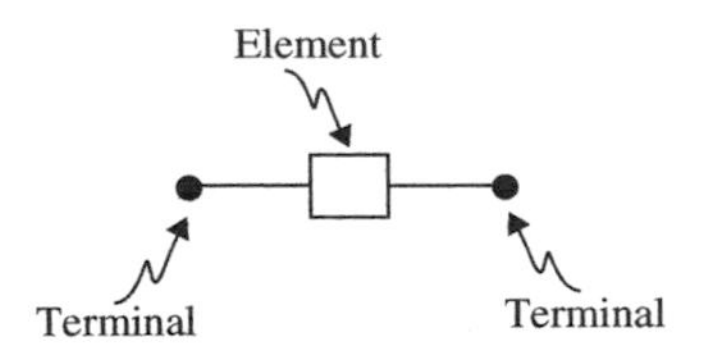


FIGURE 1.5 A lumped element.

The resistance is the property of the bulb of interest to us. Likewise, the voltage is the property of the battery that we most care about. Ignoring, for now, any internal resistance of the battery, we can lump the battery into a discrete element called by the same name supplying a constant voltage $V$, as shown in Figure 1.4b. Again, we can do this if we work within certain constraints to be discussed shortly, and provided we are not concerned with the internal properties of the battery, such as the distribution of the electrical field. In fact, the electric field within a real-life battery is horrendously difficult to chart accurately. Together, the collection of constraints that underlie the lumped circuit abstraction result in a marvelous simplification that allows us to focus on specifically those properties that are relevant to us.

Notice also that the orientation and shape of the wires are not relevant to our computation. We could even twist them or knot them in any way. Assuming for now that the wires are ideal conductors and offer zero resistance,[3] we can rewrite the bulb circuit as shown in Figure 1.4b using lumped circuit equivalents for the battery and the bulb resistance, which are connected by ideal wires. Accordingly, Figure 1.4b is called the lumped circuit abstraction of the lightbulb circuit. If the battery supplies a constant voltage $V$ and has zero internal resistance, and if the resistance of the bulb is $R$, we can use simple algebra to compute the current flowing through the bulb as

$$I = V/R.$$

Lumped elements in circuits must have a voltage $V$ and a current $I$ defined for their terminals.[4] In general, the ratio of $V$ and $I$ need not be a constant. The ratio is a constant (called the resistance $R$) only for lumped elements that

3. If the wires offer nonzero resistance, then, as described in Section 1.6, we can separate each wire into an ideal wire connected in series with a resistor.

4. In general, the voltage and current can be time varying and can be represented in a more general form as $V(t)$ and $I(t)$. For devices with more than two terminals, the voltages are defined for any terminal with respect to any other reference terminal, and the currents are defined flowing into each of the terminals.

obey Ohm's law.[5] The circuit comprising a set of lumped elements must also have a voltage defined between any pair of points, and a current defined into any terminal. Furthermore, the elements must not interact with each other except through their terminal currents and voltages. That is, the internal physical phenomena that make an element function must interact with external electrical phenomena only at the electrical terminals of that element. As we will see in Section 1.3, lumped elements and the circuits formed using these elements must adhere to a set of constraints for these definitions and terminal interactions to exist. We name this set of constraints *the lumped matter discipline.*

*The lumped circuit abstraction* Capped a set of lumped elements that obey the lumped matter discipline using ideal wires to form an assembly that performs a specific function results in the lumped circuit abstraction.

Notice that the lumped circuit simplification is analogous to the point-mass simplification in Newton's laws. The lumped circuit abstraction represents the relevant properties of lumped elements using algebraic symbols. For example, we use $R$ for the resistance of a resistor. Other values of interest, such as currents $I$ and voltages $V$, are related through simple functions. The ease of using algebraic equations in place of Maxwell's equations to design and analyze complicated circuits will become much clearer in the following chapters.

The process of discretization can also be viewed as a way of modeling physical systems. The resistor is a model for a lightbulb if we are interested in finding the current flowing through the lightbulb for a given applied voltage. It can even tell us the power consumed by the lightbulb. Similarly, as we will see in Section 1.6, a constant voltage source is a good model for the battery when its internal resistance is zero. Thus, Figure 1.4b is also called the lumped circuit model of the lightbulb circuit. Models must be used only in the domain in which they are applicable. For example, the resistor model for a lightbulb tells us nothing about its cost or its expected lifetime.

The primitive circuit elements, the means for combining them, and the means of abstraction form the graphical language of circuits. Circuit theory is a well established discipline. With maturity has come widespread utility. The language of circuits has become universal for problem-solving in many disciplines. Mechanical, chemical, metallurgical, biological, thermal, and even economic processes are often represented in circuit theory terms, because the mathematics for analysis of linear and nonlinear circuits is both powerful and well-developed. For this reason electronic circuit models are often used as analogs in the study of many physical processes. Readers whose main focus is on some area of electrical engineering other than electronics should therefore view the material in this

5. Observe that Ohm's law itself is an abstraction for the electrical behavior of resistive material that allows us to replace tables of experimental data relating $V$ and $I$ by a simple equation.

book from the broad perspective of an introduction to the modeling of dynamic systems.

## 1.3 THE LUMPED MATTER DISCIPLINE

*The scope of these equations is remarkable, including as it does the fundamental operating principles of all large-scale electromagnetic devices such as motors, cyclotrons, electronic computers, television, and microwave radar.*

—HALLIDAY AND RESNICK ON MAXWELL'S EQUATIONS

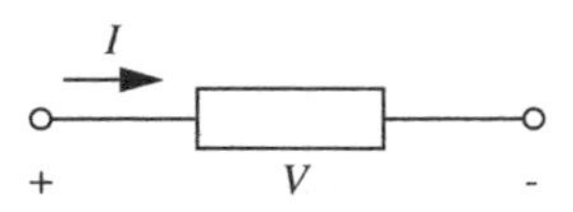


FIGURE 1.6 A lumped circuit element.

Lumped circuits comprise lumped elements (or discrete elements) connected by ideal wires. A lumped element has the property that a unique terminal voltage $V(t)$ and terminal current $I(t)$ can be defined for it. As depicted in Figure 1.6, for a two-terminal element, $V$ is the voltage across the terminals of the element,[6] and $I$ is the current through the element.[7] Furthermore, for lumped resistive elements, we can define a single property called the resistance $R$ that relates the voltage across the terminals to the current through the terminals.

The voltage, the current, and the resistance are defined for an element only under certain constraints that we collectively call the *lumped matter discipline* (LMD). Once we adhere to the lumped matter discipline, we can make several simplifications in our circuit analysis and work with the lumped circuit abstraction. Thus the lumped matter discipline provides the foundation for the lumped circuit abstraction, and is the fundamental mechanism by which we are able to move from the domain of physics to the domain of electrical engineering. We will simply state these constraints here, but relegate the development of the constraints of the lumped matter discipline to Section A.1 in Appendix A. Section A.2 further shows how the lumped matter discipline results in the simplification of Maxwell's equations into the algebraic equations of the lumped circuit abstraction.

The lumped matter discipline imposes three constraints on how we choose lumped circuit elements:

1. Choose lumped element boundaries such that the rate of change of magnetic flux linked with any closed loop outside an element must be zero for all time. In other words, choose element boundaries such that

$$\frac{\partial \Phi_B}{\partial t} = 0$$

through any closed path outside the element.

---

6. The *voltage* across the terminals of an element is defined as the work done in moving a unit charge (one coulomb) from one terminal to the other through the element against the electrical field. Voltages are measured *in volts* (V), where one volt is one joule per coulomb.

7. The *current* is defined as the rate of flow of charge from one terminal to the other through the element. Current is measured in *amperes* (A) , where one ampere is one coulomb per second.

2. Choose lumped element boundaries so that there is no total time varying charge within the element for all time. In other words, choose element boundaries such that

$$\frac{\partial q}{\partial t} = 0$$

where $q$ is the total charge within the element.

3. Operate in the regime in which signal timescales of interest are much larger than the propagation delay of electromagnetic waves across the lumped elements.

The intuition behind the first constraint is as follows. The definition of the voltage (or the potential difference) between a pair of points across an element is the work required to move a particle with unit charge from one point to the other *along some path* against the force due to the electrical field. For the lumped abstraction to hold, we require that this voltage be unique, and therefore the voltage value must not depend on the path taken. We can make this true by selecting element boundaries such that there is no time-varying magnetic flux outside the element.

If the first constraint allowed us to define a unique voltage across the terminals of an element, the second constraint results from our desire to define a unique value for the current entering and exiting the terminals of the element. A unique value for the current can be defined if we do not have charge buildup or depletion inside the element over time.

Under the first two constraints, elements do not interact with each other except through their terminal currents and voltages. Notice that the first two constraints require that the rate of change of magnetic flux outside the elements and net charge within the elements is zero *for all time.*[8] It directly follows that the magnetic flux and the electric fields outside the elements are also zero. Thus there are no fields related to one element that can exert influence on the other elements. This permits the behavior of each element to be analyzed independently.[9] The results of this analysis are then summarized by the

8. As discussed in Appendix A, assuming that the rate of change is zero for all time ensures that voltages and currents can be arbitrary functions of time.

9. The elements in most circuits will satisfy the restriction of non-interaction, but occasionally they will not. As will be seen later in this text, the magnetic fields from two inductors in close proximity might extend beyond the material boundaries of the respective inductors inducing significant electric fields in each other. In this case, the two inductors could not be treated as independent circuit elements. However, they could perhaps be treated together as a single element, called a transformer, if their distributed coupling could be modeled appropriately. A dependent source is yet another example of a circuit element that we will introduce later in this text in which interacting circuit elements are treated together as a single element.

relation between the terminal current and voltage of that element, for example, $V = IR$. More examples of such relations, or element laws, will be presented in Section 1.6.2. Further, when the restriction of non-interaction is satisfied, the focus of circuit operation becomes the terminal currents and voltages, and not the electromagnetic fields within the elements. Thus, these currents and voltages become the fundamental signals within the circuit. Such signals are discussed further in Section 1.8.

Let us dwell for a little longer on the third constraint. The lumped element approximation requires that we be able to define a voltage $V$ between a pair of element terminals (for example, the two ends of a bulb filament) and a current through the terminal pair. Defining a current through the element means that the current in must equal the current out. Now consider the following thought experiment. Apply a current pulse at one terminal of the filament at time instant $t$ and observe both the current into this terminal and the current out of the other terminal at a time instant $t + dt$ very close to $t$. If the filament were long enough, or if $dt$ were small enough, the finite speed of electromagnetic waves might result in our measuring different values for the current in and the current out.

We cannot make this problem go away by postulating constant currents and voltages, since we are very much interested in situations such as those depicted in Figure 1.7, in which a time-varying voltage source drives a circuit.

Instead, we fix the problem created by the finite propagation speeds of electromagnetic waves by adding the third constraint, namely, that the timescale of interest in our problem be much larger than electromagnetic propagation delays through our elements. Put another way, the size of our lumped elements must be much smaller than the wavelength associated with the $V$ and $I$ signals.[10]

Under these speed constraints, electromagnetic waves can be treated as if they propagated instantly through a lumped element. By neglecting propagation

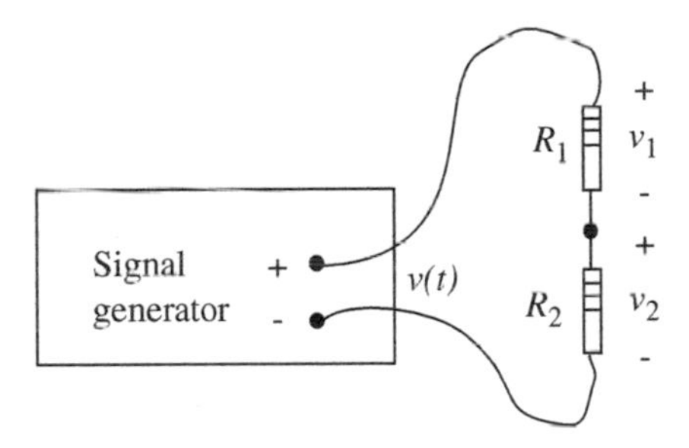


FIGURE 1.7 Resistor circuit connected to a signal generator.

10. More precisely, the wavelength that we are referring to is that wavelength of the electromagnetic wave launched by the signals.

effects, the lumped element approximation becomes analogous to the point-mass simplification, in which we are able to ignore many physical properties of elements such as their length, shape, size, and location.

Thus far, our discussion focused on the constraints that allowed us to treat individual elements as being lumped. We can now turn our attention to circuits. As defined earlier, circuits are sets of lumped elements connected by ideal wires. Currents outside the lumped elements are confined to the wires. An ideal wire does not develop a voltage across its terminals, irrespective of the amount of current it carries. Furthermore, we choose the wires such that they obey the lumped matter discipline, so the wires themselves are also lumped elements.

For their voltages and currents to be meaningful, the constraints that apply to lumped elements apply to entire circuits as well. In other words, for voltages between any pair of points in the circuit and for currents through wires to be defined, any segment of the circuit must obey a set of constraints similar to those imposed on each of the lumped elements.

Accordingly, the lumped matter discipline for circuits can be stated as

1. The rate of change of magnetic flux linked with any portion of the circuit must be zero for all time.

2. The rate of change of the charge at any node in the circuit must be zero for all time. A node is any point in the circuit at which two or more element terminals are connected using wires.

3. The signal timescales must be much larger than the propagation delay of electromagnetic waves through the circuit.

Notice that the first two constraints follow directly from the corresponding constraints applied to lumped elements. (Recall that wires are themselves lumped elements.) So, the first two constraints do not imply any new restrictions beyond those already assumed for lumped elements.[11]

The third constraint for circuits, however, imposes a stronger restriction on signal timescales than for elements, because a circuit can have a much larger physical extent than a single element. The third constraint says that the circuit must be much smaller in all its dimensions than the wavelength of light at the highest operating frequency of interest. If this requirement is satisfied, then wave phenomena are not important to the operation of the circuit. The circuit operates quasistatically, and information propagates instantaneously across it. For example, circuits operating in vacuum or air at 1 kHz, 1 MHz, and 1 GHz would have to be much smaller than 300 km, 300 m, and 300 mm, respectively.

---

11. As we shall see in Chapter 9, it turns out that voltages and currents in circuits result in electric and magnetic fields, thus appearing to violate the set of constraints to which we promised to adhere. In most cases these are negligible. However, when their effects cannot be ignored, we explicitly model them using elements called capacitors and inductors.

Most circuits satisfy such a restriction. But, interestingly, an uninterrupted 5000-km power grid operating at 60 Hz, and a 30-cm computer motherboard operating at 1 GHz, would not. Both systems are approximately one wavelength in size so wave phenomena are very important to their operation and they must be analyzed accordingly. Wave phenomena are now becoming important to microprocessors as well. We will address this issue in more detail in Section 1.4.

When a circuit meets these three constraints, the circuit can itself be abstracted as a lumped element with external terminals for which voltages and currents can be defined. Circuits that adhere to the lumped matter discipline yield additional simplifications in circuit analysis. Specifically, we will show in Chapter 2 that the voltages and currents across the collection of lumped circuits obey simple algebraic relationships stated as two laws: Kirchhoff's voltage law (KVL) and Kirchhoff's current law (KCL).

## 1.4 LIMITATIONS OF THE LUMPED CIRCUIT ABSTRACTION

We used the lumped circuit abstraction to represent the circuit pictured in Figure 1.4a by the schematic diagram of Figure 1.4b. We stated that it was permissible to ignore the physical extent and topology of the wires connecting the elements and define voltages and currents for the elements provided they met the lumped matter discipline.

The third postulate of the lumped matter discipline requires us to limit ourselves to signal speeds that are significantly lower than the speed of electromagnetic waves. As technology advances, propagation effects are becoming harder to ignore. In particular, as computer speeds pass the gigahertz range, increasing signal speeds and fixed system dimensions tend to break our abstractions, so that engineers working on the forefront of technology must constantly revisit the disciplines upon which abstractions are based and prepare to resort to fundamental physics if the constraints are violated.

As an example, let us work out the numbers for a microprocessor. In a microprocessor, the conductors are typically encased in insulators such as silicon dioxide. These insulators have dielectric constants nearly four times that of free space, and so electromagnetic waves travel only half as fast through them. Electromagnetic waves travel at the speed of approximately 1 foot or 30 cm per nanosecond in vacuum, so they travel at roughly 6 inches or 15 cm per nanosecond in the insulators. Since modern microprocessors (for example, the Alpha microprocessor from Digital/Compaq) can approach 2.5 cm in size, the propagation delay of electromagnetic waves across the chip is on the order of 1/6 ns. These microprocessors are approaching a clock rate of 2 GHz in 2001. Taking the reciprocal, this translates to a clock cycle time of 1/2 ns. Thus, the wave propagation delay across the chip is about 33% of a clock cycle. Although techniques such as pipelining attempt to reduce the number of elements (and therefore distance) a signal traverses in a clock cycle, certain clock or power lines in microprocessors can travel the full extent of the chip, and will suffer this large delay. Here, wave phenomena must be modeled explicitly.

In contrast, slower chips built in earlier times satisfied our lumped matter discipline more easily. For example, the MIPS microprocessor built in 1984 was implemented on a chip that was 1 cm on a side. It ran at a speed of 20 MHz, which translates to a cycle time of 50 ns. The wave propagation delay across the chip was 1/15 ns, which was significantly smaller then the chip cycle time.

As another example, a Pentium II chip built in 1998 clocked at 400 MHz, but used a chip size that was more or less the same as that of the MIPS chip — namely, about 1 cm on a side. As calculated earlier, the wave propagation delay across a 1-cm chip is about 1/15 ns. Clearly the 2.5-ns cycle time of the Pentium II chip is still significantly larger than the wave propagation delay across the chip.

Now consider a Pentium IV chip built in 2004 that clocked at 3.4 GHz, and was roughly 1 cm on a side. The 0.29-ns cycle time is only four times the wave propagation delay across the chip!

If we are interested in signal speeds that are comparable to the speed of electromagnetic waves, then the lumped matter discipline is violated, and therefore we cannot use the lumped circuit abstraction. Instead, we must resort to distributed circuit models based on elements such as transmission lines and waveguides.[12] In these distributed elements, the voltages and currents at any instant of time are a function of the location within the elements. The treatment of distributed elements are beyond the scope of this book.

The lumped circuit abstraction encounters other problems with time-varying signals even when signal frequencies are small enough that propagation effects can be neglected. Let us revisit the circuit pictured in Figure 1.7 in which a signal generator drives a resistor circuit. It turns out that under certain conditions the *frequency* of the oscillator and the lengths and layout of the wires may have a profound effect on the voltages. If the oscillator is generating a sine wave at some low frequency, such as 256 Hz (Middle C in musical terms), then the voltage divider relation developed in Chapter 2 (Equation 2.138) could be used to calculate with some accuracy the voltage across $R_2$. But if the frequency of the sine wave were 100 MHz ($1 \times 10^8$ Hertz), then we have a problem. As we will see later, capacitive and inductive effects in the resistors and the wires (resulting from electric fields and magnetic fluxes generated by the signal) will

---

12. In case you are wondering how the Pentium IV and similar chips get away with high clock speeds, the key lies in designing circuits and laying them out on the chip in a way that most signals traverse a relatively small fraction of the chip in a clock cycle. To enable succeeding generations of the chip to be clocked faster, signals must traverse progressively shorter distances. A technique called pipelining is the key enabling mechanism that accomplishes this. The few circuits in which signals travel the length of the chip must be designed with extreme care using transmission line analysis.

seriously affect the circuit behavior, and these are not currently represented in our model. In Chapter 9, we will separate these effects into new lumped elements called capacitors and inductors so our lumped circuit abstraction holds at high frequencies as well.

All circuit model discussions in this book are predicated on the assumption that the frequencies involved are low enough that the effects of the fields can be adequately modeled by lumped elements. In Chapters 1 through 8, we assume that the frequencies involved are even lower so we can ignore all capacitive and inductive effects as well.

Are there other additional practical considerations in addition to the constraints imposed by the lumped matter discipline? For example, are we justified in neglecting contact potentials, and lumping all battery effects in $V$? Can we neglect all resistance associated with the wires, and lump all the resistive effects in a series connected resistor? Does the voltage $V$ change when the resistors are connected and current flows? Some of these issues will be addressed in Sections 1.6 and 1.7.

## 1.5 PRACTICAL TWO-TERMINAL ELEMENTS

Resistors and batteries are two of our most familiar lumped elements. Such lumped elements are the primitive building blocks of electronic circuits. Electronic access to an element is made through its *terminals*. At times, terminals are paired together in a natural way to form *ports*. These ports offer an alternative view of how electronic access is made to an element. An example of an arbitrary element with two terminals and one port is shown in Figure 1.8. Other elements may have three or more terminals, and two or more ports.

Most circuit analyses are effectively carried out on circuits containing only two-terminal elements. This is due in part to the common use of two-terminal elements, and in part to the fact that most, if not all, elements having more than two terminals are usually modeled using combinations of two-terminal elements. Thus, two-terminal elements appear prominently in all electronic

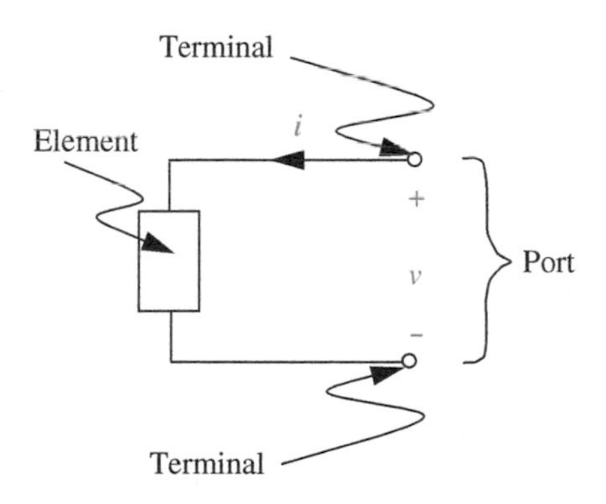


FIGURE 1.8 An arbitrary two-terminal circuit element.

circuit analyses. In this section, we discuss a couple of familiar examples of two-terminal elements — resistors and batteries.

### 1.5.1 BATTERIES

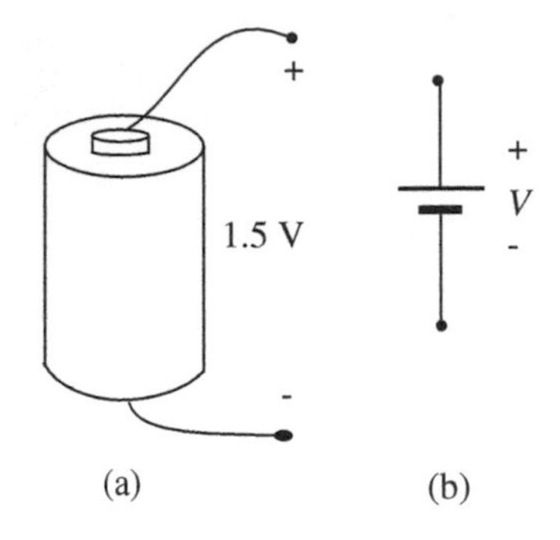


FIGURE 1.9 Symbol for battery.

Cell phone batteries, laptop batteries, flashlight batteries, watch batteries, car batteries, calculator batteries, are all common devices in our culture. All are sources of energy, derived in each case from an internal chemical reaction.

The important specifications for a battery are its nominal voltage, its total store of energy, and its internal resistance. In this section, we will assume that the internal resistance of a battery is zero. The voltage measured at the terminals of a single cell is fundamentally related to the chemical reaction releasing the energy. In a flashlight battery, for example, the carbon central rod is approximately 1.5 V positive with respect to the zinc case, as noted in Figure 1.9a. In a circuit diagram, such a single-cell battery is usually represented schematically by the symbol shown in Figure 1.9b. Of course, to obtain a larger voltage, several cells can be connected in *series*: the positive terminal of the first cell connected to the negative terminal of the second cell, and so forth, as suggested pictorially in Figure 1.10. Multiple-cell batteries are usually represented by the symbol in Figure 1.10b, (with no particular correspondence between the number of lines and the actual number of cells in series).

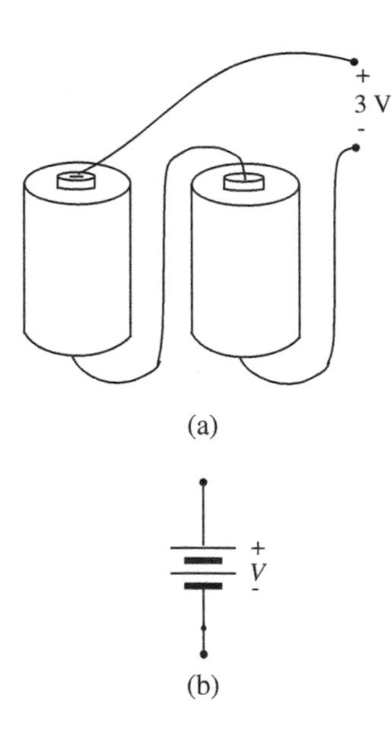


FIGURE 1.10 Cells in series.

The second important parameter of a battery is the total amount of energy it can store, often measured in *joules*. However, if you pick up a camcorder or flashlight battery, you might notice the ratings of *ampere-hours* or *watt-hours*. Let us reconcile these ratings. When a battery is connected across a resistive load in a circuit, it delivers power. The lightbulb in Figure 1.4a is an example of a resistive load.

The power delivered by the battery is the product of the voltage and the current:

$$p = VI. \tag{1.2}$$

Power is *delivered* by the battery when the current $I$ flowing out of the positive voltage terminal of the battery is positive. Power is measured in *watts*. A battery delivers one watt of power when $V$ is one volt and $I$ is one ampere.

Power is the rate of delivery of energy. Thus the amount of energy $w$ delivered by the battery is the time integral of the power.

If a constant amount of power $p$ is delivered over an interval $T$, the energy $w$ supplied is

$$w = pT. \tag{1.3}$$

The battery delivers one joule of energy if it supplies one watt of power over one second. Thus, joules and watt-seconds are equivalent units. Similarly, if a battery delivers one watt for an hour, then we say that it has supplied one watt-hour (3600 joules) of energy.

Assuming that the battery terminal voltage is constant at $V$, because the power delivered by the battery is the product of the voltage and the current, an equivalent indication of the power delivered is the amount of current being supplied. Similarly, the product of current and the length of time the battery will sustain that current is an indication of the energy capacity of the battery. A car battery, for instance, might be rated at 12 V and 50 A-hours. This means that the battery can provide a 1-A current for 50 hours, or a 100-A current for 30 minutes. The amount of energy stored in such a battery is

$$\text{Energy} = 12 \times 50 = 600 \text{ watt-hours} = 600 \times 3600 = 2.16 \times 10^6 \text{ joules.}$$

---

EXAMPLE 1.1 A LITHIUM-ION BATTERY A Lithium-Ion (Li-Ion) battery pack for a camcorder is rated as 7.2 V and 5 W-hours. What are its equivalent ratings in mA-hours and joules?

Since a joule(J) is equivalent to a W-second, 5 W-hours is the same as $5 \times 3600 = 18000$ J.

Since the battery has a voltage of 7.2 V, the battery rating in ampere-hours is $5/7.2 = 0.69$. Equivalently, its rating in mA-hours is 690.

---

EXAMPLE 1.2 ENERGY COMPARISON Does a Nickel-Cadmium (Ni-Cad) battery pack rated at 6 V and 950 mA-hours store more or less energy than a Li-Ion battery pack rated at 7.2 V and 900 mA-hours?

We can directly compare the two by converting their respective energies into joules. The Ni-Cad battery pack stores $6 \times 950 \times 3600/1000 = 20520$ J, while the Li-Ion battery pack stores $7.2 \times 900 \times 3600/1000 = 23328$ J. Thus the Li-Ion battery pack stores more energy.

---

When a battery is connected across a resistor, as illustrated in Figure 1.4, we saw that the battery delivers energy at some rate. The power was the rate of delivery of energy. Where does this energy go? Energy is dissipated by the resistor, through heat, and sometimes even light and sound if the resistor overheats and explodes! We will discuss resistors and power dissipation in Section 1.5.2.

If one wishes to increase the current capacity of a battery without increasing the voltage at the terminals, individual cells can be connected in *parallel*, as shown in Figure 1.11. It is important that cells to be connected in parallel be nearly identical in voltage to prevent one cell from destroying another. For example, a 2-V lead-acid cell connected in parallel with a 1.5 V flashlight cell will surely destroy the flashlight cell by driving a huge current through it.

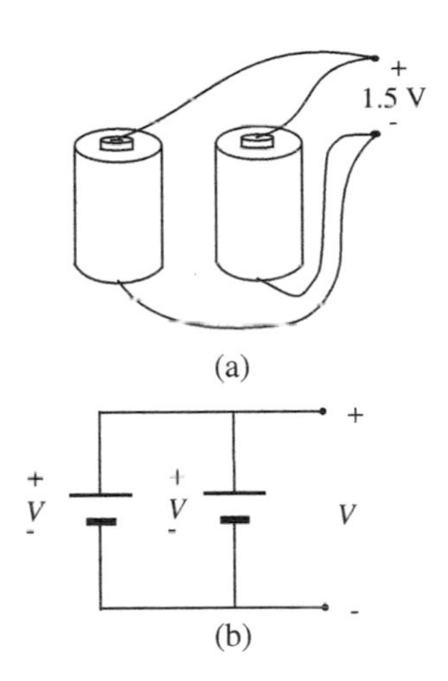


FIGURE 1.11 Cells in parallel.

The corresponding constraint for cells connected in series is that the nominal current capacity be nearly the same for all cells. The total energy stored in a multicell battery is the same for series, parallel, or series-parallel interconnections.

### 1.5.2 LINEAR RESISTORS

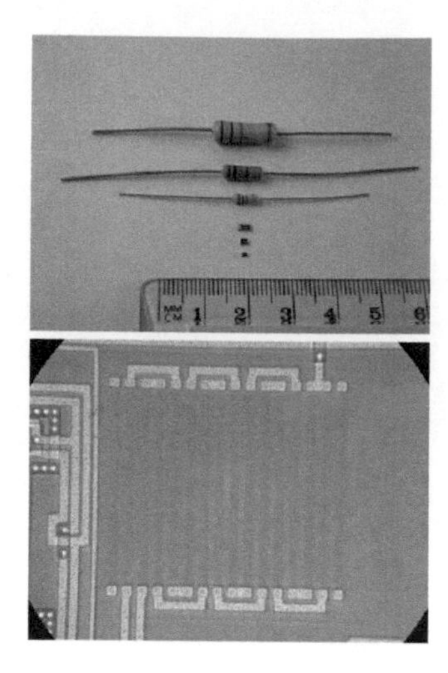

FIGURE 1.12 Discrete resistors (above), and Deposits integrated-circuit resistors (below). The image on the bottom shows a small region of the MAX807L microprocessor supervisory circuit from Maxim Integrated Products, and depicts an array of silicon-chromium thin-film resistors, each with 6 μm width and 217.5 μm length, and nominal resistance 50 kΩ. (Photograph Courtesy of Maxim Integrated Products.)

Resistors come in many forms (see Figure 1.12), ranging from lengths of nichrome wire used in toasters and electric stoves and planar layers of polysilicon in highly complex computer chips, to small rods of carbon particles encased in Bakelite commonly found in electronic equipment. The symbol for resistors in common usage is shown in Figure 1.13.

Over some limited range of voltage and current, carbon, wire and polysilicon resistors obey Ohm's law:

$$v = iR \tag{1.4}$$

that is, the voltage measured across the terminals of a resistor is linearly proportional to the current flowing through the resistor. The constant of proportionality is called the *resistance*. As we show shortly, the resistance of a piece of material is proportional to its length and inversely proportional to its cross-sectional area.

In our example of Figure 1.4b, suppose that the battery is rated at 1.5 V. Further assume that the resistance of the bulb is $R = 10\ \Omega$. Assume that the internal resistance of the battery is zero. Then, a current of $i = v/R = 150$ mA will flow through the bulb.

FIGURE 1.13 Symbol for resistor.

---

EXAMPLE 1.3 MORE ON RESISTANCE In the circuit in Figure 1.4b, suppose that the battery is rated at 1.5 V. Suppose we observe through some means a current of 500 mA through the resistor. What is the resistance of the resistor?

For a resistor, we know from Equation 1.4 that

$$R = \frac{v}{i}$$

Since the voltage $v$ across the resistor is 1.5 V and the current $i$ through the resistor is 500 mA, the resistance of the resistor is 3 Ω.

---

The resistance of a piece of material depends on its geometry. As illustrated in Figure 1.14, assume the resistor has a conducting channel with cross-sectional area $a$, length $l$, and resistivity $\rho$. This channel is terminated at its extremes by two conducting plates that extend to form the two terminals of the resistor. If this cylindrical piece of material satisfies the lumped matter

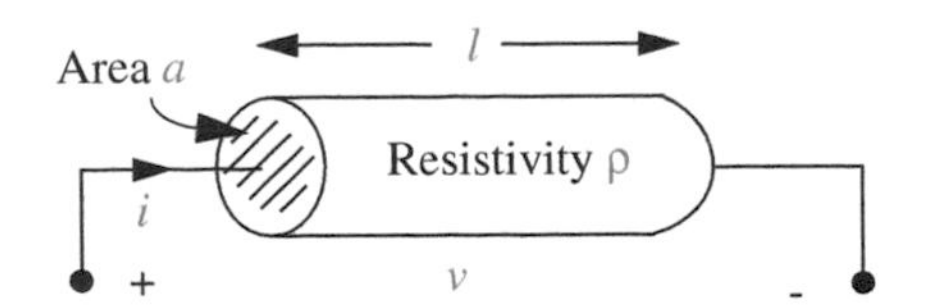


**FIGURE 1.14** A cylindrical-wire shaped resistor.

discipline and obey's Ohm's law, we can write[13]

$$R = \rho \frac{l}{a} \tag{1.5}$$

Equation 1.5 shows that the resistance of a piece of material is proportional to its length and inversely proportional to its cross-sectional area.

Similarly, the resistance of a cuboid shaped resistor with length $l$, width $w$, and height $h$ is given by

$$R = \rho \frac{l}{wh} \tag{1.6}$$

when the terminals are taken at the pair of surfaces with area $wh$.

---

EXAMPLE 1.4 RESISTANCE OF A CUBE Determine the resistance of a cube with sides of length 1 cm and resistivity 10 ohm-cms, when a pair of opposite surfaces are chosen as the terminals.

Substituting $\rho = 10$ Ω-cm, $l = 1$ cm, $w = 1$ cm, and $h = 1$ cm in Equation 1.6, we get $R = 10$ Ω.

---

EXAMPLE 1.5 RESISTANCE OF A CYLINDER By what factor is the resistance of a wire with cross-sectional radius $r$ greater than the resistance of a wire with cross-sectional radius $2r$?

A wire is cylindrical in shape. Equation 1.5 relates the resistance of a cylinder to its cross-sectional area. Rewriting Equation 1.5 in terms of the cross-sectional radius $r$ we have

$$R = \rho \frac{l}{\pi r^2}.$$

From this equation it is clear that the resistance of a wire with radius $r$ is four times greater than that of a wire with cross-sectional radius $2r$.

---

13. See Appendix A.3 for a derivation.

EXAMPLE 1.6 CARBON-CORE RESISTORS The resistance of small carbon-core resistors can range from 1 $\Omega$ to $10^6$ $\Omega$. Assuming that the core of these resistors is 1 mm in diameter and 5 mm long, what must be the range of resistivity of the carbon cores?

Given a 1-mm diameter, the cross-sectional area of the core is $A \approx 7.9 \times 10^{-7}$ m$^2$. Further, its length is $l = 5 \times 10^{-3}$m. Thus, $A/l \approx 1.6 \times 10^{-4}$ m.

Finally, using Equation 1.5, with $1\ \Omega \leq R \leq 10^6\ \Omega$, it follows that the approximate range of its resistivity is $1.6 \times 10^{-4}\ \Omega\text{m} \leq \rho \leq 1.6 \times 10^2\,\Omega\text{m}$.

EXAMPLE 1.7 POLY-CRYSTALLINE SILICON RESISTOR A thin poly-crystalline silicon resistor is 1 $\mu$m thick, 10 $\mu$m wide, and 100 $\mu$m long, where 1 $\mu$m is $10^{-6}$ m. If the resistivity of its poly-crystalline silicon ranges from $10^{-6}$ $\Omega$m to $10^2$ $\Omega$m, what is the range of its resistance?

The cross-sectional area of the resistor is $A = 10^{-11}$ m, and its length is $l = 10^{-4}$ m. Thus $l/A = 10^7$ m$^{-1}$. Using Equation 1.5, and the given range of resistivity, $\rho$, the resistance satisfies $10\ \Omega \leq R \leq 10^9\ \Omega$.

EXAMPLE 1.8 RESISTANCE OF PLANAR MATERIALS ON A CHIP Figure 1.15 shows several pieces of material with varying geometries. *Assume all the pieces have the same thickness.* In other words, the pieces of material are *planar.* Let us determine the resistance of these pieces between the pairs of terminals shown. For a given thickness, remember that the resistance of a piece of material in the shape of a cuboid is determined by the ratio of the length to the width of the piece of material (Equation 1.6). Assuming that $R_o$ is the resistance of a piece of planar material

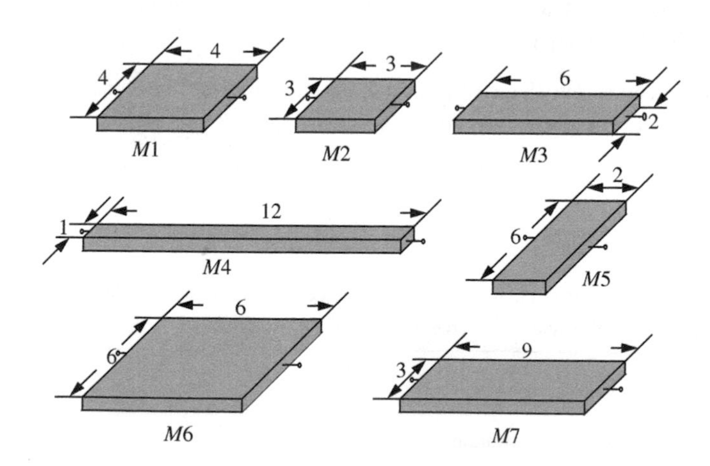


FIGURE 1.15 Resistors of various shapes.

with unit length and width, show that the resistance of a piece of planar material with length $L$ and width $W$ is $(L/W)R_o$.

From Equation 1.6, the resistance of a cuboid shaped material with length $L$, width $W$, height $H$, and resistivity $\rho$ is

$$R = \rho \frac{L}{WH}. \tag{1.7}$$

We are given that the resistance of a piece of the same material with $L = 1$ and $W = 1$ is $R_o$. In other words,

$$R_o = \rho \frac{1}{H}. \tag{1.8}$$

Substituting $R_o = \rho/H$ in Equation 1.7, we get

$$R = \frac{L}{W} R_o. \tag{1.9}$$

Now, assume $R_o = 2\ \text{k}\Omega$ for our material. Recall that Ohms are the unit of resistance and are written as $\Omega$. We denote a 1000-$\Omega$ value as kilo-$\Omega$ or k$\Omega$. Assuming that the dimensions of the pieces of material shown in Figure 1.15 are in $\mu$-m, or micrometers, what are their resistances?

First, observe that pieces $M1$, $M2$, and $M6$ must have the same resistance of 2 k$\Omega$ because they are squares (in Equation 1.9, notice that $L/W = 1$ for a square).

Second, $M3$ and $M7$ must have the same resistance because both have the same ratio $L/W = 3$. Therefore, both have a resistance of $3 \times 2 = 6$ k$\Omega$. Among them, $M4$ has the biggest $L/W$ ratio of 12. Therefore it has the largest resistance of 24 k$\Omega$. $M5$ has the smallest $L/W$ ratio of 1/3, and accordingly has the smallest resistance of 2/3 k$\Omega$.

Because all square pieces made out of a given material have the same resistance (provided, of course, the pieces have the same thickness), we often characterize the resistivity of planar material of a given thickness with

$$R_\square = R_o, \tag{1.10}$$

where $R_o$ is the resistance of a piece of the same material with unit length and width. Pronounced "R square," $R_\square$ is the resistance of a square piece of material.

---

EXAMPLE 1.9 MORE ON PLANAR RESISTANCES Referring back to Figure 1.15, suppose an error in the material fabrication process results in each dimension ($L$ and $W$) increasing by a fraction $e$. By what amount will the resistances of each of the pieces of material change?

Recall that the resistance $R$ of a planar rectangular piece of material is proportional to $L/W$. If each dimension increases by a fraction $e$, the new length becomes $L(1+e)$ and the new width becomes $W(1+e)$. Notice that the resistance given by

$$R = \frac{L(1+e)}{W(1+e)} R_o = \frac{L}{W} R_o$$

is unchanged.

---

EXAMPLE 1.10 RATIO OF RESISTANCES Referring again to Figure 1.15, suppose the material fabrication process undergoes a "shrink" to decrease each dimension (this time around, increasing the thickness $H$ in addition to $L$ and $W$) by a fraction $\alpha$ (e.g., $\alpha = 0.8$). Assume further, that the resistivity $\rho$ changes by some other fraction to $\rho'$. Now consider a pair of resistors with resistances $R_1$ and $R_2$, and original dimensions $L_1$, $W_1$ and $L_2$, $W_2$ respectively, and the same thickness $H$. By what fraction does the ratio of the resistance values change after the process shrink?

From Equation 1.7, the ratio of the original resistance values is given by

$$\frac{R_1}{R_2} = \frac{\rho L_1/(W_1 H)}{\rho L_2/(W_2 H)} = \frac{L_1/W_1}{L_2/W_2}.$$

Let the resistance values after the process shrink be $R_1'$ and $R_2'$. Since each dimension shrinks by the fraction $\alpha$, each new dimension will be $\alpha$ times the original value. Thus, for example, the length $L_1$ will change to $\alpha L_1$. Using Equation 1.7, the ratio of the new resistance values is given by

$$\frac{R_1'}{R_2'} = \frac{\rho' \alpha L_1/(\alpha W_1 \alpha H)}{\rho' \alpha L_2/(\alpha W_2 \alpha H)} = \frac{L_1/W_1}{L_2/W_2}$$

In other words, the ratio of the resistance values is unchanged by the process shrink.

The ratio property of planar resistance — that is that the ratio of the resistances of rectangular pieces of material with a given thickness and resistivity is independent of the actual values of the length and the width provided the ratio of the length and the width is fixed — enables us to perform process shrinks (for example, from a 0.25-μm process to a 0.18-μm process) without needing to change the chip layout. Process shrinks are performed by scaling the dimensions of the chip and its components by the same factor, thereby resulting in a smaller chip. The chip is designed such that relevant

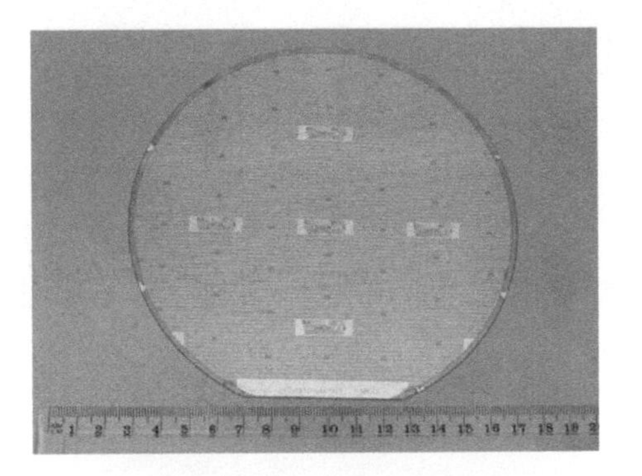

**FIGURE 1.16** A silicon wafer. (Photograph Courtesy of Maxim Integrated Products.)

**FIGURE 1.17** A chip photo of Intel's 2-GHz Pentium IV processor implemented in 0.18μm-technology. The chip is roughly 1 cm on a side. (Photograph courtesy of Intel Corp.)

signal values are derived as a function of resistance ratios,[14] thereby ensuring that the chip manufactured after a process shrink continues to function as before.

VLSI stands for "Very Large Scale Integration." Silicon-based VLSI is the technology behind most of today's computer chips. In this technology, lumped planar elements such as wires, resistors, and a host of others that we will soon encounter, are fabricated on the surface of a planar piece of silicon called a *wafer* (for example, see Figures 1.15 and 1.12). A wafer has roughly the shape and size of a Mexican tortilla or an Indian chapati (see Figure 1.16). The planar elements are connected together using planar wires to form circuits. After fabrication, each wafer is diced into several hundred chips or "dies," typically, each the size of a thumbnail. A Pentium chip, for example, contains hundreds of millions of planar elements (see Figure 1.17). Chips are attached, or bonded, to packages (for example, see Figure 12.3.4), which are in turn mounted on a printed-circuit board along with other discrete components such as resistors and capacitors (for example, see Figure 1.18) and wired together.

14. We will study many such examples in the ensuing sections, including the voltage divider in Section 2.3.4 and the inverter in Section 6.8.

**FIGURE 1.18** A printed-circuit board containing several interconnected chip packages and discrete components such as resistors (tiny box-like objects) and capacitors (tall cylindrical objects). (Photograph Courtesy of Anant Agarwal, the Raw Group.)

As better processes become available, VLSI fabrication processes undergo periodic shrinks to reduce the size of chips without needing significant design changes. The Pentium III, for example, initially appeared in the 0.25-μm process, and later in the 0.18-μm process. The Pentium IV chip shown in Figure 1.17 initially appeared in a 0.18-μm process in the year 2000, and later in 0.13-μm and 0.09-μm processes in 2001 and 2004, respectively.

---

There are two important limiting cases of the linear resistor: *open circuits* and *short circuits*. An open circuit is an element through which no current flows, regardless of its terminal voltage. It behaves like a linear resistor in the limit $R \to \infty$.

A short circuit is at the opposite extreme. It is an element across which no voltage can appear regardless of the current through it. It behaves like a linear resistor in the limit $R \to 0$. Observe that the short circuit element is the same as an ideal wire. Note that neither the open circuit nor the short circuit dissipate power since the product of their terminal variables ($v$ and $i$) is identically zero.

Most often, resistances are thought of as time-invariant parameters. But if the temperature of a resistor changes, then so too can its resistance. Thus, a linear resistor can be a time-varying element.

The linear resistor is but one example of a larger class of resistive elements. In particular, resistors need not be linear; they can be nonlinear as well. In general, *a two-terminal resistor is any two-terminal element that has an algebraic relation between its instantaneous terminal current and its instantaneous terminal voltage.* Such a resistor could be linear or nonlinear, time-invariant or time-varying. For example, elements characterized by the following element relationships are all general resistors:

$$\text{Linear resistor:} \quad v(t) = i(t)R(t)$$

$$\text{Linear, time-invariant resistor:} \quad v(t) = i(t)R$$

$$\text{Nonlinear resistor:} \quad v(t) = Ki(t)^3$$

However, as introduced in Chapter 9, elements characterized by these relationships are not general resistors:

$$v(t) = L\frac{di(t)}{dt}$$

$$v(t) = \frac{1}{C}\int_{-\infty}^{t} i(t')dt'$$

What is important about the general resistor is that its terminal current and voltage depend only on the instantaneous values of each other. For our convenience, however, an unqualified reference to a resistor in this book means a linear, time-invariant resistor.

### 1.5.3 ASSOCIATED VARIABLES CONVENTION

Equation 1.4 implies a specific relation between reference directions chosen for voltage and current. This relation is shown explicitly in Figure 1.19: the arrow that defines the positive flow of current (flow of positive charge) is directed *in* at the resistor terminal assigned to be positive in voltage. This convention, referred to as *associated variables*, is generalized to an arbitrary element in Figure 1.20 and will be followed whenever possible in this text. The variables $v$ and $i$ are called the *terminal variables* for the element. Note that the values of each of these variables may be positive or negative depending on the actual direction of current flow or the actual polarity of the voltage.

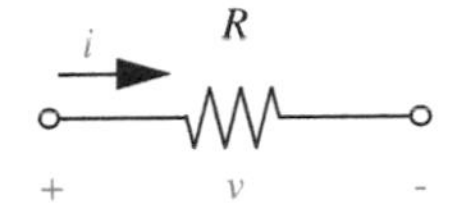


FIGURE 1.19 Definition of terminal variables $v$ and $i$ for the resistor.

*Associated Variables Convention* Define current to flow *in* at the device terminal assigned to be *positive* in voltage.

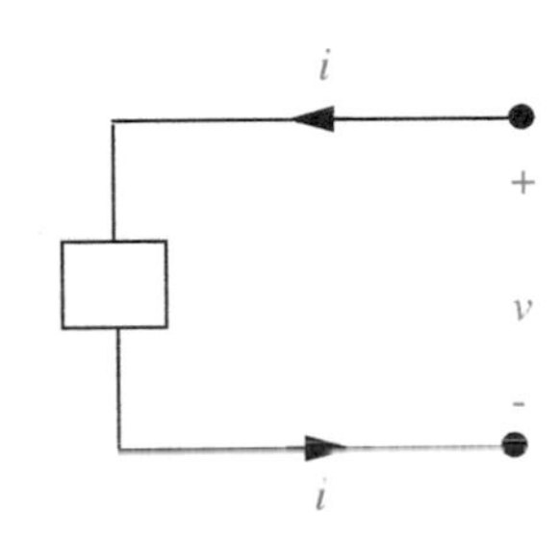


FIGURE 1.20 Definition of the terminal variables $v$ and $i$ for a two-terminal element under the associated variables convention.

When the voltage $v$ and current $i$ for an element are defined under the associated variables convention, the power *into* the element is positive when both $v$ and $i$ are positive. In other words, energy is pumped into an element when a positive current $i$ is directed *into* the voltage terminal marked positive. Depending on the type of element, the energy is either dissipated or stored. Conversely, power is supplied by an element when a positive current $i$ is directed *out* of the voltage terminal marked positive. When the terminal variables for a resistor are defined according to associated variables, the power dissipated in the resistor is a *positive* quantity, an intuitively satisfying result.

While Figure 1.20 is quite simple, it nonetheless makes several important points. First, the two terminals of the element in Figure 1.20 form a single port through which the element is addressed. Second, the current $i$ circulates through that port. That is, the current that enters one terminal is instantaneously equal to the current that exits the other terminal. Thus, according to the lumped matter discipline, net charge cannot accumulate within the element. Third, the voltage $v$ of the element is defined across the port. Thus, the element is assumed to respond only to the difference of the electrical potentials at its two terminals,

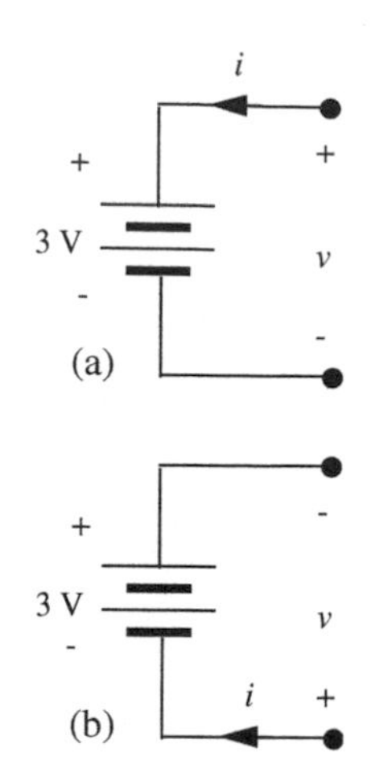


FIGURE 1.21 Terminal variable assignments for a battery.

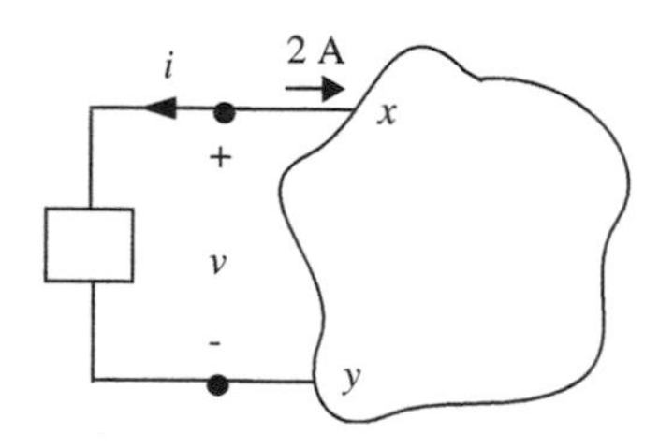


FIGURE 1.22 Terminal variable assignments for a two-terminal element.

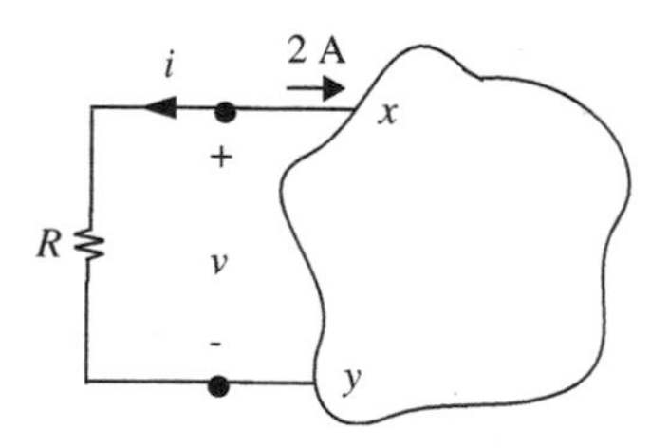


FIGURE 1.23 Terminal variable assignments for a resistor.

and not to the absolute electric potential at either terminal. Fourth, the current is defined to circulate positively through the port by entering the positive voltage terminal and exiting the negative voltage terminal. Which terminal is chosen as the positive voltage terminal is arbitrary, but the relation defined between the current and voltage is not. Lastly, for brevity, the current that exits the negative voltage terminal is usually never labeled, but it is always understood to be equal to the current that enters the positive voltage terminal.

---

EXAMPLE 1.11 TERMINAL VARIABLES VERSUS ELEMENT PROPERTIES Figures 1.21a and b shows two possible legal definitions for terminal variables for a 3 V battery. What is the value of terminal variable $v$ in each case?

For Figure 1.21a, we can see that terminal variable $v = 3$ V. For Figure 1.21b, however, $v = -3$ V.

This example highlights the distinction between a terminal variable and an element property. The battery voltage of 3 V is an element property, while $v$ is a terminal variable that we have defined. Element properties are usually written inside the element symbol, or if that is inconvenient, they are written next to the element (e.g., the battery voltage). Terminal polarities and terminal variables are written close to the terminals.

---

EXAMPLE 1.12 FUN WITH TERMINAL VARIABLES Figure 1.22 shows some two-terminal element connected to an arbitrary circuit at the points $x$ and $y$. The element terminal variables $v$ and $i$ are defined according to the associated variables convention. Suppose that a current of 2 A flows into the circuit terminal marked $x$. What is the value of terminal variable $i$?

Since the chosen direction of the terminal variable $i$ is opposite to that of the 2 A current, $i = -2$ A.

Now suppose that the two terminal element is a resistor (see Figure 1.23) with resistance $R = 10$ Ohms. Determine the value of $v$.

We know that under the associated variables convention the terminal variables for a resistor are related as

$$v = iR$$

Given that $R = 10\ \Omega$ and $i = -2$ A,

$$v = (-2)10 = -20 \text{ V}$$

Next, suppose that the two terminal element is a 3 V battery with the polarity shown in Figure 1.24a. Determine the values of terminal variables $v$ and $i$.

As determined earlier, $i = -2$ A. For the polarity of the battery shown in Figure 1.24a, $v = 3$ V.

Now, suppose the 3 V battery is connected with the polarity shown in Figure 1.24b. Determine that values of $v$ and $i$.

As before, $i = -2$ A. With the reversed battery connection shown in Figure 1.24b, $v = -3$ V.

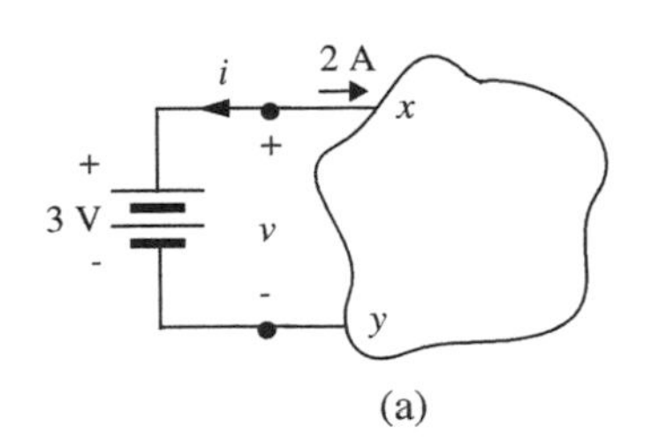


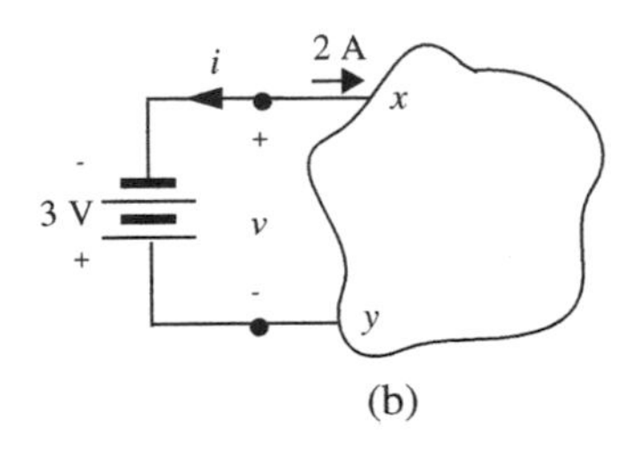


**FIGURE 1.24** The two-terminal element is a battery.

Under the associated variables convention, the instantaneous power $p$ supplied into an element is given by

$$p = vi \tag{1.11}$$

with units of watts (W).

Note that both $v$ and $i$, and therefore the instantaneous power $p$, can be functions of time. For a resistor, $p = vi$ represents the instantaneous power dissipated by the resistor.

Correspondingly, the amount of energy (in units of joules) supplied to an element during an interval of time between $t_1$ and $t_2$ under the associated variables convention is given by

$$w = \int_{t_1}^{t_2} vi\, dt. \tag{1.12}$$

For a resistor, by noting that $v = iR$ from Equation 1.4, the power relation for a two-terminal element (Equation 1.11) can be equivalently written as

$$p = i^2 R \tag{1.13}$$

or

$$p = \frac{v^2}{R}. \tag{1.14}$$

EXAMPLE 1.13 POWER INTO A RESISTOR Determine the power for the resistor in Figure 1.23. Confirm mathematically that the power is indeed supplied into the resistor.

We know that $i = -2$ A and $v = -20$ V. Therefore, the power is given by

$$p = vi = (-20\text{V})(-2\text{A}) = 40 \text{ W}$$

By our associated variables convention, the product $p = vi$ yields the power supplied *into* the element. Thus, we can confirm that 40 W of power is being supplied into the resistor. From the properties of a resistor, we also know that this power is dissipated in the form of heat.

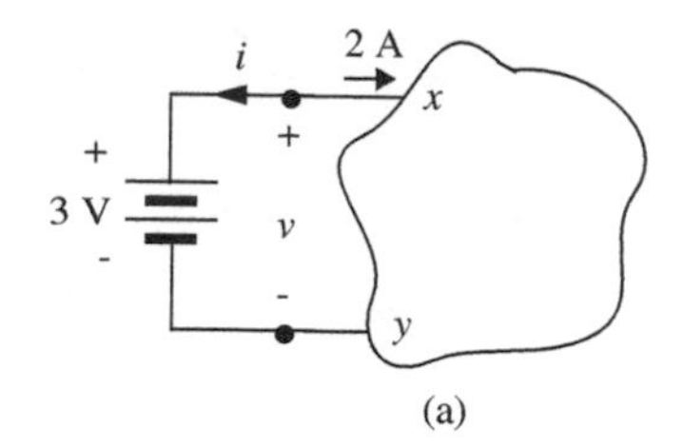


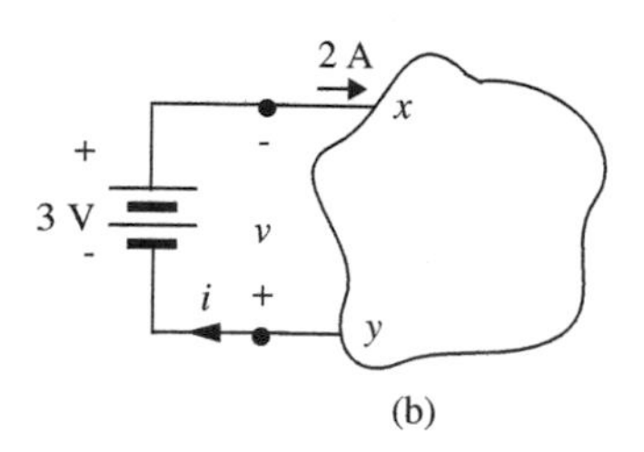


**FIGURE 1.25** Alternative assignments of terminal variables.

EXAMPLE 1.14 POWER SUPPLIED BY A BATTERY Determine the power for the battery using the two assignments of terminal variables in Figures 1.25a and 1.25b.

For the assignment of terminal variables in Figure 1.25a, $i = -2$ A and $v = 3$ V. Thus, by associated variables, power into the battery is given by

$$p = vi = (3\text{V})(-2\text{A}) = -6 \text{ W}$$

Since the power into the battery is negative, the power supplied by the battery is positive. Thus, in the circuit of Figure 1.25a, the battery is delivering power.

Next, let us analyze the same circuit with the assignment of terminal variables in Figure 1.25b. For this assignment, $i = 2$ A and $v = -3$ V. Thus, by associated variables, power into the battery is given by

$$p = vi = (-3\text{V})(2\text{A}) = -6 \text{ W}$$

In other words, the battery is delivering 6 watts of power. Since the circuit is the same, it is not surprising that our result has not changed when the terminal variable assignments are reversed.

EXAMPLE 1.15 POWER SUPPLIED VERSUS POWER ABSORBED BY A BATTERY In simple circuits, for example, circuits containing a single battery, we do not have to undergo the rigor of associated variables to determine whether power is being absorbed or supplied by an element. Let us work out such an example. In our lightbulb circuit of Figure 1.4b, suppose that the battery is rated at 1.5 V and 1500 J. Assume that the internal resistance of the battery is zero. Further assume that the resistance of the bulb is $R = 10\ \Omega$. What is the power dissipated in the resistor?

The power dissipated in the resistor is given by

$$p = VI = \frac{V^2}{R} = \frac{1.5^2}{10} = 0.225 \text{ W}$$

Since the entire circuit comprises a battery and a resistor, we can state without a lot of analysis that the resistor dissipates power and the battery supplies it. How much power does the battery provide when it is connected to the 10-Ω resistor? Suppose the battery supplies a current $I$. We can quickly compute the value of this current as:

$$I = \frac{V}{R} = \frac{1.5}{10} = 0.15 \text{ A}$$

Thus the power delivered by the battery is given by

$$p = VI = 1.5 \times 0.15 = 0.225 \text{ W}$$

Not surprisingly, the power delivered by the battery is the same as the power dissipated in the resistor. Note that since the circuit current $I$ has been defined to be directed *out* of the positive battery terminal in Figure 1.4b, and since the current is positive, the battery is supplying power.

How long will our battery last when it is connected to the 10-Ω resistor? Since the battery is supplying 0.225 W of power, and since a watt represents energy dissipation at the rate of one joule per second, the battery will last 1500/0.225 = 6667 s.

---

EXAMPLE 1.16 POWER RATING OF A RESISTOR In a circuit such as that shown in Figure 1.4b, the battery is rated at 7.2 V and 10000 J. Assume that the internal resistance of the battery is zero. Further assume that the resistance in the circuit is $R = 1$ kΩ. You are given that the resistor can dissipate a maximum of 0.5 W of power. (In other words, the resistor will overheat if the power dissipation is greater than 0.5 W.) Determine the current through the resistor. Further, determine whether the power dissipation in the resistor exceeds its maximum rating.

The current through the resistor is given by

$$I = \frac{V}{R} = \frac{7.2}{1000} = 7.2 \text{ mA}$$

The power dissipation in the resistor is given by

$$p = I^2R = (7.2 \times 10^{-3})^2 10^3 = 0.052 \text{ W}$$

Clearly, the power dissipation in the resistor is well within its capacity.

---

## 1.6 IDEAL TWO-TERMINAL ELEMENTS

As we saw previously, the process of discretization can be viewed as a way of *modeling physical systems.* For example, the resistor is a lumped model for a lightbulb. Modeling physical systems is a major motivation for studying electronic circuits. In our lightbulb circuit example, we used lumped electrical elements to model electrical components such as bulbs and batteries. In general, modeling physical systems involves representing real-world physical processes, whether they are electrical, chemical, or mechanical, in terms of a set of ideal electrical elements. This section introduces a set of ideal two-terminal elements including voltage and current sources, and ideal wires and resistors, which form our primitives in the vocabulary of circuits.

The same set of ideal two-terminal elements serve to build either *information processing* or *energy processing* systems as well. Information and energy processing includes the communication, storage, or transformation of information or energy, and is a second major motivation for studying electronic circuits. Whether we are interested in modeling systems or in information and energy processing, it is essential to be able to represent five basic processes in terms of our lumped circuit abstraction.

1. *Sources* of energy or information
2. *Flow* of energy or information in a system
3. *Loss* of energy or information in a system
4. *Control* of energy flow or information flow by some external force
5. *Storage* of energy or information

We will discuss ideal two-terminal elements that represent the first three of these in this section, deferring control and storage until Chapters 6 and 9, respectively.

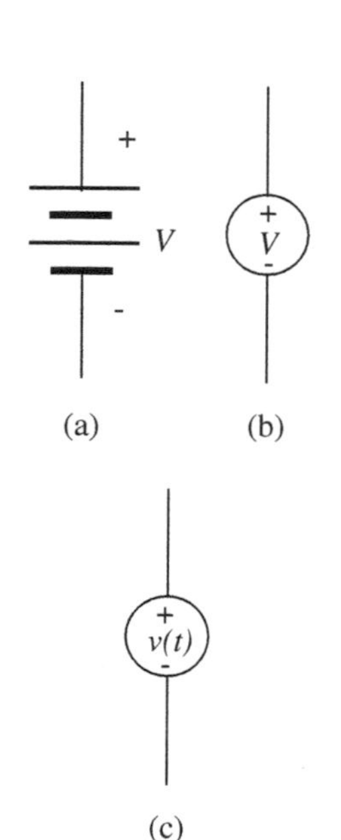


**FIGURE 1.26** Circuit symbol for a voltage source: (a) battery; (b) voltage source; (c) voltage source.

### 1.6.1 IDEAL VOLTAGE SOURCES, WIRES, AND RESISTORS

Familiar primary sources of energy in our daily lives are sunlight, oil, and coal. Secondary sources would be power plants, gasoline engines, home-heating furnaces, or flashlight batteries. In heating systems, energy flows through air ducts or heating pipes; in electrical systems the flow is through copper wires.[15] Similarly, information sources include speech, books, compact discs, videos, and the web (some of it, anyway!). Information flow in speech systems is through media such as air and water; in electronic systems, such as computers or phones, the flow relies on conducting wires. Sensors such as microphones, magnetic tape heads, and optical scanners convert information from various forms into an electrical representation. None of these elements is ideal, so our first task is to invent ideal energy or information sources and ideal conductors for energy or information flow.

Conceptually, it is relatively easy to extrapolate from known properties of a battery to postulate an *ideal voltage source* as a device that maintains a constant voltage at its terminals regardless of the amount of current drawn from those terminals. To distinguish such an ideal element from a battery[16] (see Figure 1.26a), we denote a voltage source by a single circle with polarity

15. Or, more accurately, in the fields between the wires.

16. In general, a physical battery has some internal resistance, which we ignored in our previous examples. A more precise relationship between the ideal voltage source and the battery is developed in Section 1.7.

markings inside it, as in Figure 1.26b. If the voltage source supplies a voltage $V$, then we also include the $V$ symbol inside the circle (or just outside the circle if there is not enough room to write the symbol inside). In the same manner, we might also represent an information source, such as a microphone or a sensor, as a voltage source providing a time-varying voltage $v(t)$ at its output (Figure 1.26c). We can assume that the voltage $v(t)$ depends solely on the microphone signal and is independent of the amount of current drawn from the terminals. (Note that $V$ and $v(t)$ in Figure 1.26 are element values and not terminal variables.)

We will see two types of voltage sources: *independent* and *dependent*. An independent voltage source supplies a voltage independent of the rest of the circuit. Accordingly, independent sources are a means through which inputs can be made to a circuit. Power supplies, signal generators, and microphones are examples of independent voltage sources. The circle symbol in Figure 1.26b represents an independent voltage source. In contrast to an independent voltage source, a dependent voltage source supplies a voltage as commanded by a signal from within the circuit of which the source is a part. Dependent sources are most commonly used to model elements having more than two terminals. They are represented with a diamond symbol; we shall see examples of these in future chapters.

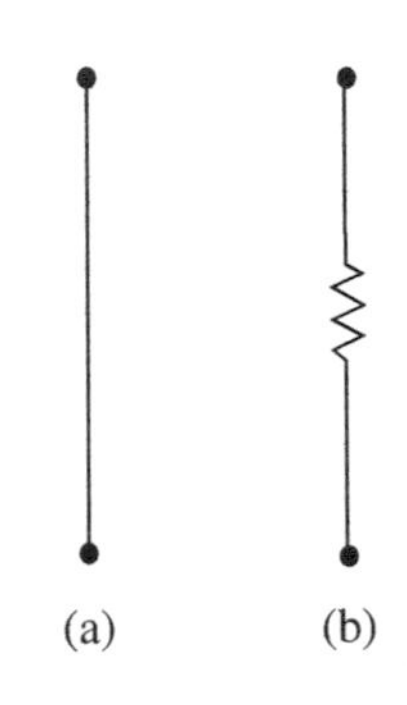


FIGURE 1.27 Circuit symbol for an ideal conductor: (a) perfect conductor; (b) wire with nonzero resistance.

In a manner similar to our invention of the ideal voltage source, we postulate an *ideal conductor* to be one in which any amount of current can flow without loss of voltage or power. The symbol for an ideal conductor is shown in Figure 1.27a. Notice that the symbol is just a line. The ideal conductor is no different from the ideal wire we saw earlier. Ideal conductors can be used to represent a channel for fluid flow in hydrodynamic systems.

Any physical length of wire will have some nonzero resistance. The resistance dissipates energy and represents a loss of energy from the system. If this resistance is important in a particular application, then we can model the wire as an ideal conductor in series with a resistor, as suggested in Figure 1.27b. To be consistent, we now state that the resistor symbol introduced in Figure 1.19 represents an *ideal linear resistor*, which by definition obeys Ohm's law

$$v = iR \tag{1.15}$$

for all values of voltage and current. Resistors can be used to model processes such as friction that result in energy loss in a system. Note that because this element law is symmetric, it is unchanged if the polarities of the current and voltage definitions are reversed. Sometimes it is convenient to work with reciprocal resistance, namely the conductance $G$ having the units of Siemens (S). In this case,

$$G = \frac{1}{R} \tag{1.16}$$

and

$$i = Gv. \tag{1.17}$$

Most often, resistances and conductances are thought of as time-invariant parameters. But if the temperature of a resistor changes, then so too can its resistance and conductance. Thus, a linear resistor can be a time-varying element.

### 1.6.2 ELEMENT LAWS

From the viewpoint of circuit analysis, the most important characteristic of a two-terminal element is the relation between the voltage across and the current through its terminals, or the $v$–$i$ relationship for short. This relation, called the *element law*, represents the lumped-parameter summary of the electronic behavior of the element. for example, as seen in Equation 1.15,

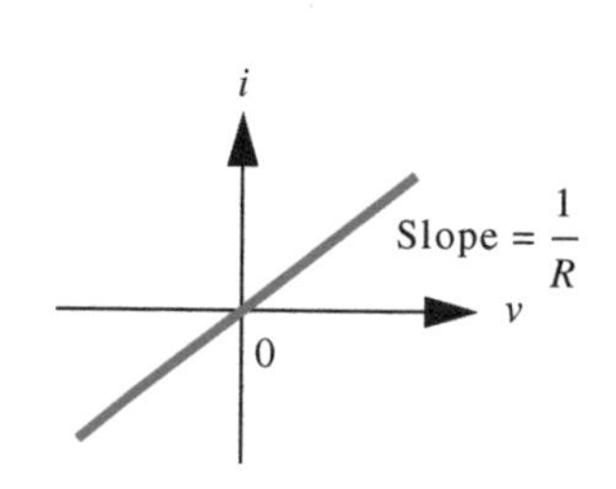


FIGURE 1.28 Plot of the $v$–$i$ relationship for a resistor.

$$v = iR$$

is the element law for the resistor. The element law is also referred to as the constituent relation, or the *element relation*. In order to standardize the manner in which element laws are expressed, the current and voltage for all two-terminal elements are defined according to the associated variables convention shown in Figure 1.19. Figure 1.28 shows a plot of the $v$–$i$ relationship for a resistor when $v$ and $i$ are defined according to the associated variables convention.

The constituent relation for the independent voltage source in Figure 1.26b supplying a voltage $V$ is given by

$$v = V \tag{1.18}$$

when its terminal variables are defined as in Figure 1.29a. A plot of the $v$–$i$ relationship is shown in Figure 1.29b. Observe the clear distinction between the element parameter $V$ and its terminal variables $v$ and $i$.

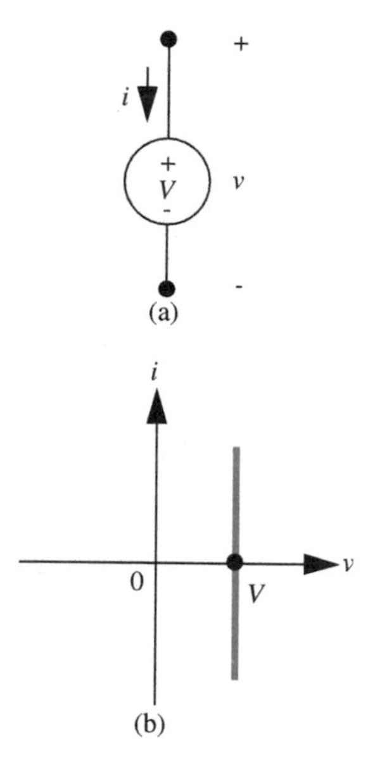


FIGURE 1.29 (a) Independent voltage source with assigned terminal variables, (b) $v$–$i$ relationship for the voltage source.

Similarly, the element law for the ideal wire (or a short circuit) is given by

$$v = 0. \tag{1.19}$$

Figure 1.30a shows the assignment of terminal variables and Figure 1.30b plots the $v$–$i$ relationship.

Finally, the element law for an open circuit is given by

$$i = 0. \tag{1.20}$$

Figure 1.31a shows the assignment of terminal variables and Figure 1.31b plots the $v - i$ relationship.

Comparing the $v$–$i$ relationship for the resistor in Figure 1.28 to those for a short circuit in Figure 1.30 and an open circuit in Figure 1.31, it is evident that the short circuit and open circuit are limiting cases for a resistor. The resistor approaches the short circuit case as its resistance approaches zero. The resistor approaches the open circuit case as its resistance approaches infinity.

EXAMPLE 1.17 MORE ON TERMINAL VARIABLES VERSUS ELEMENT PROPERTIES Figure 1.32 shows a 5-V voltage source connected to an arbitrary circuit at the points $x$ and $y$. Its terminal variables $v$ and $i$ are defined according to the associated variables convention as indicated in the figure. Suppose that a current of 2 A flows into the circuit terminal marked $x$. What are the values of $v$ and $i$?

For the assignment of terminal variables shown in Figure 1.32, $i = 2$ A and $v = -5$ V.

Notice the distinction between terminal variables and element properties in this example. The source voltage of 5 V is an element property, while $v$ is a terminal variable that we have defined. Similarly, the polarity markings inside the circle are a property of the source, while the polarity markings outside the circle representing the source relate to the terminal variable $v$. When possible, we attempt to write the element values inside the element symbol, while the terminal variables are written outside.

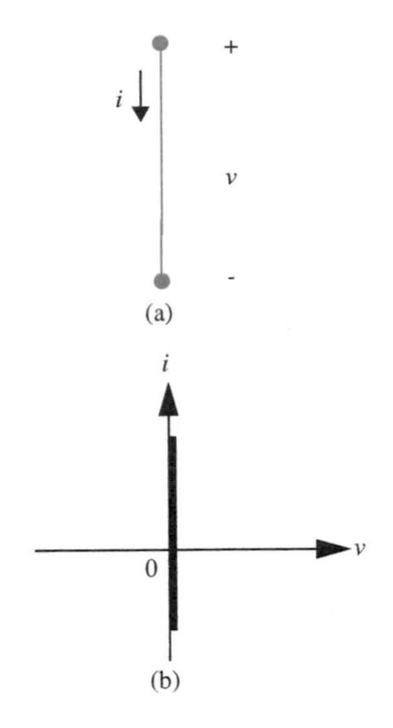


FIGURE 1.30 (a) Ideal wire with terminal variables, (b) $v$–$i$ relationship for the wire.

EXAMPLE 1.18 CHARTING $V$–$I$ RELATIONSHIPS An experimental way of charting the $v$–$i$ relationship for a two-terminal element is to connect an oscilloscope and an oscillator (or a signal generator set to produce an oscillatory output) in a *curve-plotter* configuration as suggested in Figure 1.33. The oscillator produces a voltage given by

$$v_i = V\cos(\omega t).$$

The basic concept is to use the oscillator to drive current into some arbitrary two-terminal device, and measure the resulting voltage $v_D$ and current $i_D$. Notice that the terminal variables for the two-terminal device, $v_D$ and $i_D$, are defined according to the associated variables convention. As can be seen from the circuit, the horizontal deflection will be proportional to $v_D$, and the vertical deflection will be proportional to $v_R$, and hence to $i_D$, assuming resistor $R$ obeys Ohm's law, and the horizontal and vertical amplifier inputs to the oscilloscope draw negligible current.

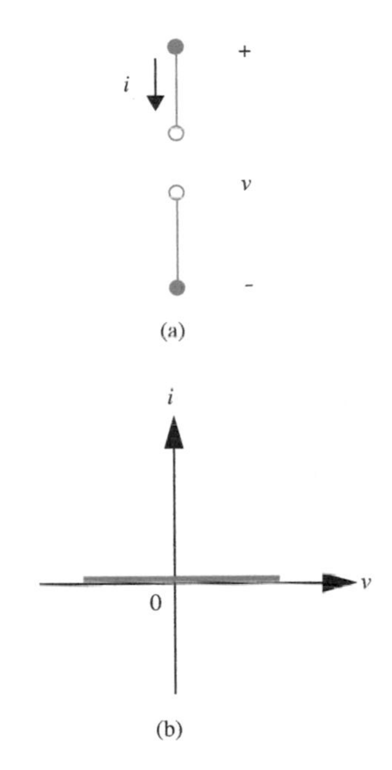


FIGURE 1.31 (a) An open circuit element with terminal variables, (b) $v$–$i$ relationship for the open circuit.

## 1.6.3 THE CURRENT SOURCE—ANOTHER IDEAL TWO-TERMINAL ELEMENT

In some fields of engineering, there are two obvious sources of power that appear to have dual properties. Think, for example, of air pumps. For an ordinary tire pump, the higher the air pressure, the harder the person at the pump-handle has to work. But with a household vacuum cleaner, also an air pump of sorts, you can hear the motor actually speed up if the air flow out of the machine is blocked, and a measurement of motor current would confirm that the power to the motor goes *down* under these conditions.

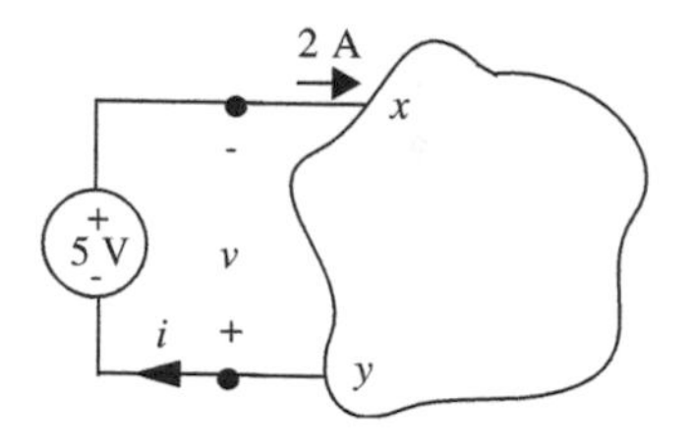


FIGURE 1.32 Terminal variables versus element properties.

It seems reasonable, then, to look for an electrical source that has characteristics that are the *dual* of those of the battery, in that the roles of current and voltage are interchanged. From the point of view of $v$–$i$ characteristics, this is a simple task. The ideal voltage source appears as a vertical line in $v$–$i$ space, so this other source, which we call an *ideal current source*, should appear as a horizontal line, as in Figure 1.34. Such a source maintains its output current at some constant value $I$ regardless of what voltage appears across the terminals.

The element law for a current source supplying a current $I$ is given by

$$i = I. \tag{1.21}$$

If the source were left with nothing connected across its terminals, then, in theory at least, the terminal voltage must rise to infinity because the constant current flowing through an infinite resistance gives infinite voltage. Recall the analogous problem with the ideal voltage source: If a short circuit is applied, the terminal current must become infinite.

It is difficult at first to have an intuitive grasp of the current source, principally because there is no familiar device available at the electronic parts counter that has these properties. However, one can still find special devices that deliver constant current to the arc lamps to illuminate the streets of Old Montreal, and we will show later that MOSFETs and Op Amps make excellent current sources. But these are not as familiar as the flashlight battery.

---

EXAMPLE 1.19 CURRENT SOURCE POWER Determine the power for the 3-A current source in Figure 1.35 if a measurement shows that $v = 5$ V.

For the assignment of terminal variables in Figure 1.35, $i = -3$ A. Further, we are given that $v = 5$V. Power *into* the current source is given by

$$p = vi = (5\text{V})(-3\text{A}) = -15 \text{ W}.$$

Since the power into the current source is negative, we determine that power is being supplied by the current source.

---

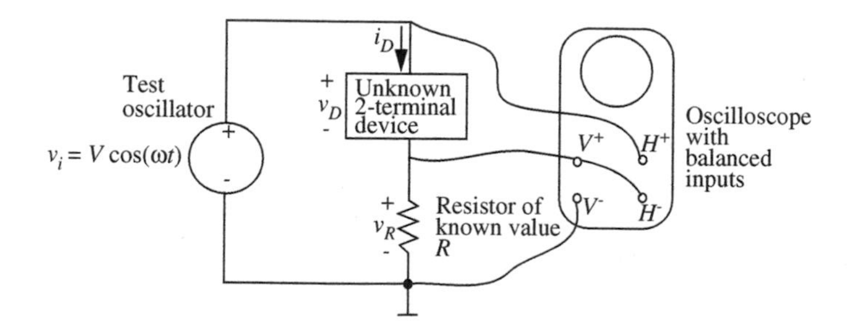


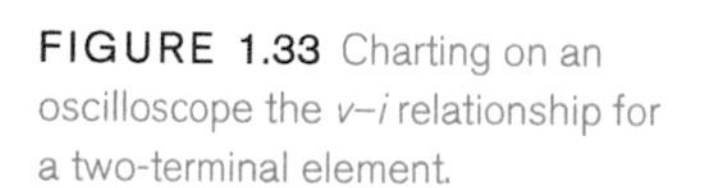
FIGURE 1.33 Charting on an oscilloscope the $v$–$i$ relationship for a two-terminal element.

Before proceeding further, it is important to distinguish between the model of a two-terminal element and the element itself. The models, or element laws, presented in this section are idealized. They describe a simplified behavior of the real elements (a voltage source for a battery, for instance). From this point forward, we will focus on circuits comprising only ideal elements, and make only occasional reference to reality. Nonetheless, it is important to realize in practice that the result of a circuit analysis is only as good as the models on which the analysis is based. Part of any discrepancy between theory and experiment may be a result of the fact that the elements do not really behave as the elements laws predict.

The $v$–$i$ relation is useful to describe other systems as well, not just primitive two-terminal elements such as sources and resistors. When creating circuit models for these systems, it is often the case that an electronic circuit can be abstracted as a black box accessible through a few terminals. As with any abstraction, the details of behavior at the interfaces (terminals, in our case) are more important than the details of behavior internal to the black box. That is, what happens at the terminals is more important than how it happens inside the black box. Furthermore, it is often the case that the terminals can be paired into ports in a natural way following the function of the circuit. For example, a complex amplifier or filter is often described by the relation between an input signal presented to the amplifier or filter at one pair of terminals or port, and an output signal presented by the amplifier or filter at a second port. In this case, the terminal pairs or ports take on special significance, and the voltage *across* the port and the current *through* the port become the port variables in terms of which the electronic circuit behavior is described.

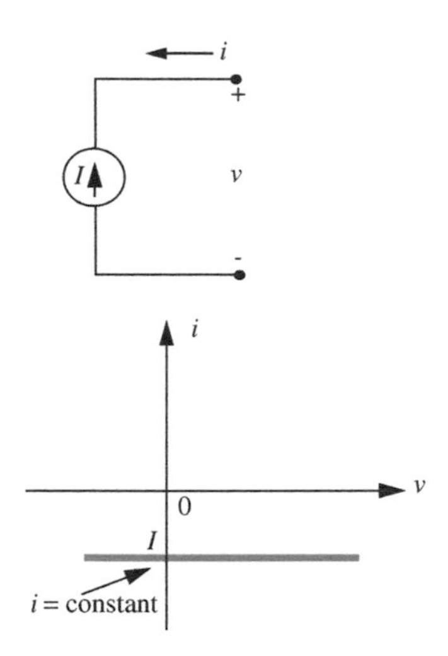


FIGURE 1.34 $v$–$i$ plot for current source.

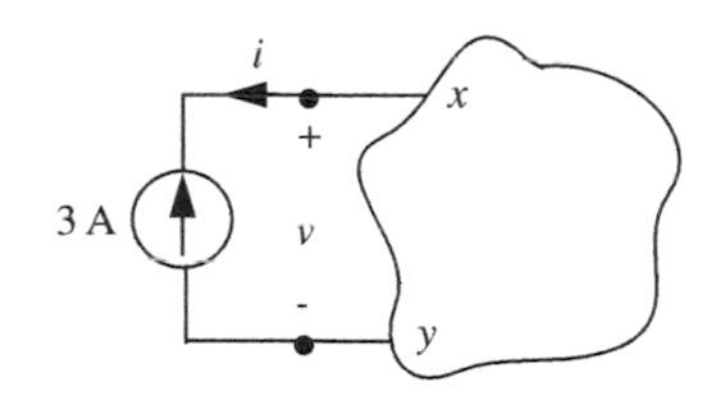


FIGURE 1.35 Power for the current source.

In principle, an electronic circuit can have one or more ports, although in practice it is common to define only a few ports to simplify matters. For example, an amplifier may be described in terms of its input port, its output port, and one or more ports for connection to power supplies. Even simple network elements such as sources, resistors, capacitors, and inductors can be thought of as one-port devices. Voltages are defined across the ports and currents through the ports as illustrated in Figure 1.36. Observe that the assignment of reference directions related to $v$ and $i$ follows the associated variables convention discussed in Section 1.5.2.

The notion of a port is much more general than its use in electronic circuit analysis would indicate. Many physical systems, such as mechanical, fluid, or thermal systems can be characterized by their behavior at a few ports. Furthermore, as depicted in Table 1.1 they have through and across parameters analogous to currents and voltages. Circuit models for these systems would use voltages and currents to model the corresponding through and across variables in those systems.

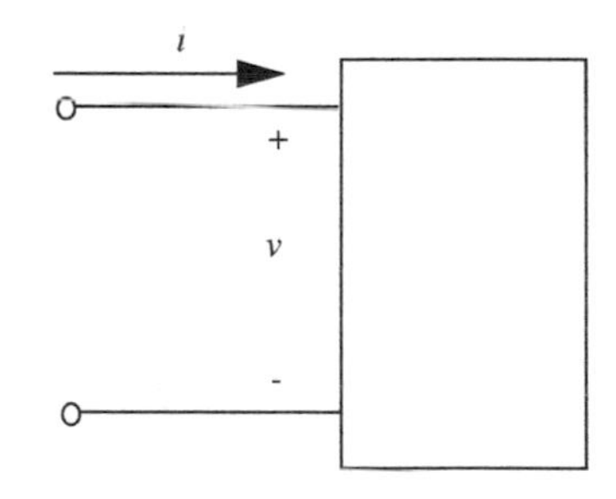


FIGURE 1.36 Definition of the voltage and current for a port.

TABLE 1.1 Through and across variables in physical and economic systems.

| THROUGH | ACROSS |
|---|---|
| Current | Voltage |
| Force | Motion |
| Flow | Pressure |
| Heat Flux | Temperature |
| Consumption | Wealth |

## 1.7 MODELING PHYSICAL ELEMENTS

Thus far, we have invented four ideal, primitive elements and studied their *v–i* characteristics. These ideal elements included the independent voltage source, the independent current source, the linear resistor, and the perfect conductor. Let us now return to building models for some of the physical elements we have seen thus far in terms of the four ideal elements.

Indeed, Figure 1.27b is one example of a model. We have modeled a physical device, namely, a length of copper wire, by a pair of ideal two-terminal elements: a perfect conductor and a linear resistance. Obviously this model is not exact. For example, if 1000 A of current flowed through a piece of 14-gauge copper wire (standard house wire designed to carry 15 A), the wire would become hot, glow brightly, and probably melt, thereby converting itself from a resistor with a very small resistance, for example, 0.001 Ω, to an infinite resistor. Our model, consisting of an ideal conductor in series with an ideal 0.001-Ω resistor, shows no such behavior: With 1000 A flowing, a one-volt drop would develop across the resistor, and one thousand watts of power would be dissipated, presumably in heat, as long as the current flowed. No smoke, no burnout.

In a similar way we can devise a model for a battery out of our ideal elements. When a flashlight bulb is connected to a new nominally 6-V battery, the voltage at the terminals of the battery (usually called the terminal voltage) drops from 6.2 V to perhaps 6.1 V. This drop results from the internal resistance of the battery. To represent this effect, we model the battery as an ideal voltage source in series with some small resistor $R$ as shown in Figure 1.37a. The drop in

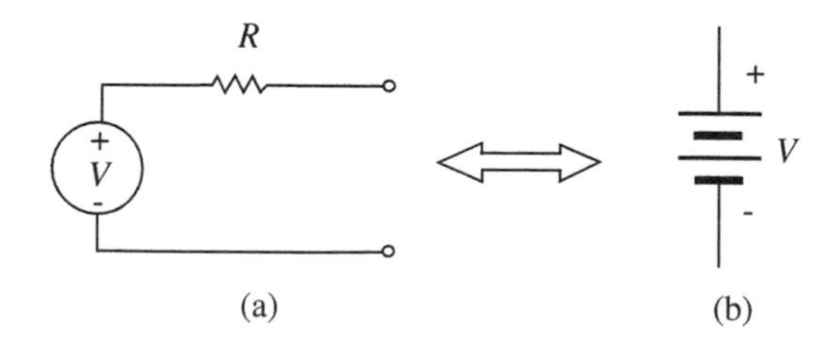


FIGURE 1.37 One model for a battery.

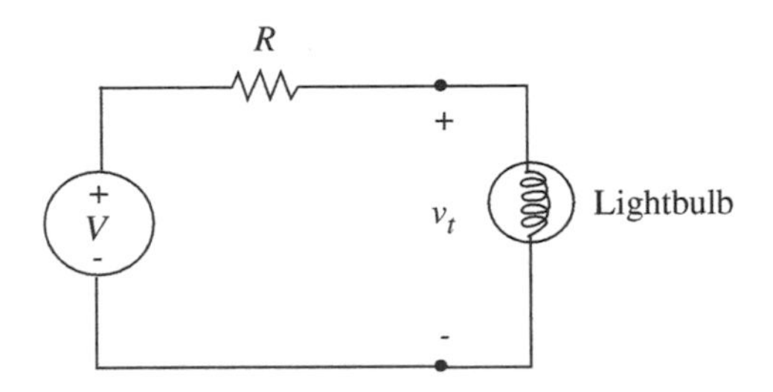


FIGURE 1.38 Battery model and lightbulb.

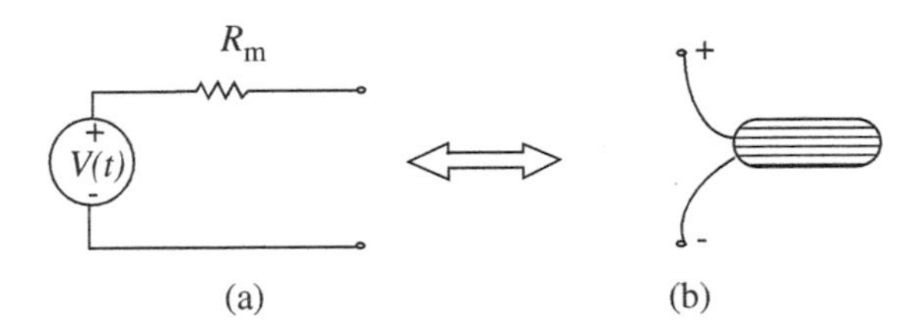


FIGURE 1.39 A microphone model.

terminal voltage when the bulb is initially connected will be properly represented in the circuit model of Figure 1.38 if the value of $R$ is appropriately chosen. However, the model is still not exact. For example, if a lightbulb is connected to a battery for some time, the battery voltage will slowly drop as the energy is drained from the battery. The model battery will not "run down," but will continue to light the bulb indefinitely.

A similar model might apply for a microphone. When an information processing system such as an amplifier is connected to the microphone, its output might drop from a 1-mV peak-to-peak signal to a 0.5-mV peak-to-peak signal due to the internal resistance of the microphone. As with the battery, we can model the microphone as a voltage source in series with a resistance $R_m$ as depicted in Figure 1.39. Although the output voltage of the microphone will not run down over time, its model is not exact for other reasons. For example, the voltage drop in the signal might be related to the signal frequency.

It is obvious that these "defects" in the models could be corrected by making the models more complicated. But the considerable increase in complexity might not be justified by the improvement in model accuracy. Unfortunately, it is not always obvious in a given problem how to find a reasonable balance between simplicity and accuracy. In this text we will always strive for simplicity on the following basis: Computer solutions for any of the problems we discuss are always available, and these can be structured to have great accuracy. So it makes sense in modeling with circuit elements, as opposed to computer modeling, to strive for insight rather than accuracy, for simplicity rather than complexity.

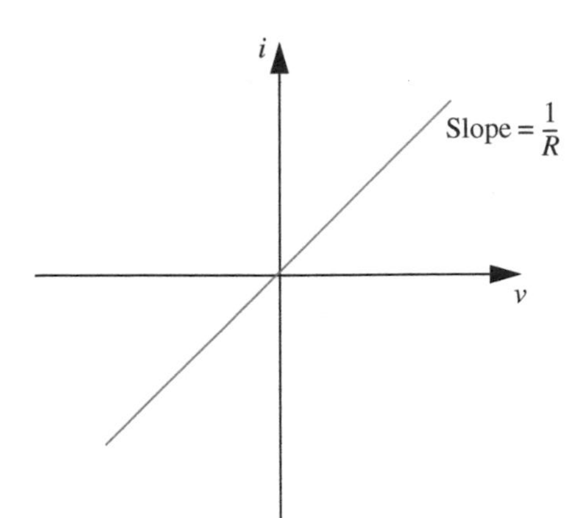


FIGURE 1.40 *i–v* plot for a resistor.

It is appropriate at this point to check experimentally the validity of the models developed here by plotting their *v–i* characteristics. The *v–i* characteristics can be plotted using the setup shown in Figure 1.33. First, use a 100-Ω 1/10-W resistor as the "unknown" two-terminal device. If the oscillator voltage is a few volts, a straight line passing through the origin with slope $1/R$ will appear on the screen (see Figure 1.40), showing that Ohm's law applies. However, if the voltage is increased so that $v_D$ is 5 or 10 V, then the 1/10-W resistor will heat up, and its value will change. If the oscillator is set to a very low frequency, say 1 Hz, the resistor heats up and cools down in the source of each cycle, so the trace is decidedly nonlinear. If the oscillator is in the mid-audio range, say 500 Hz, thermal inertia prevents the resistor from changing temperature rapidly, so some average temperature is reached. Thus the line will remain straight, but its slope will change as a function of the amplitude of the applied signal.

Resistor self-heating, with the associated change in value, is obviously undesirable in most circuit applications. For this reason manufacturers provide power ratings for resistors, to indicate maximum power dissipation ($p_{max}$) without significant value change or burnout. The power dissipated in a resistor is

$$p = vi, \tag{1.22}$$

which is the hyperbola in *v–i* space, as indicated in Figure 1.41. Our ideal-resistor model — ohmic with constant value — matches the actual resistor behavior only in the region between the hyperboli.

The plot on the oscilloscope face will also deviate from a straight line if the oscillator frequency is made high enough. Under this condition, capacitive and inductive effects in the circuit will generate an elliptical pattern. These will be discussed in later chapters.

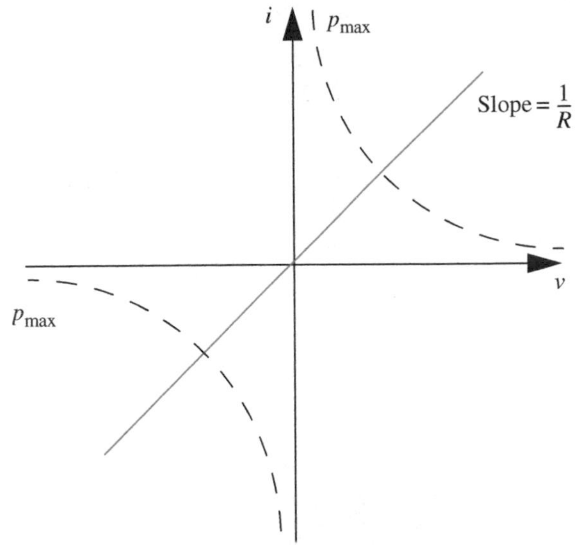

FIGURE 1.41 Power constraint for a resistor in the $i-v$ plane.

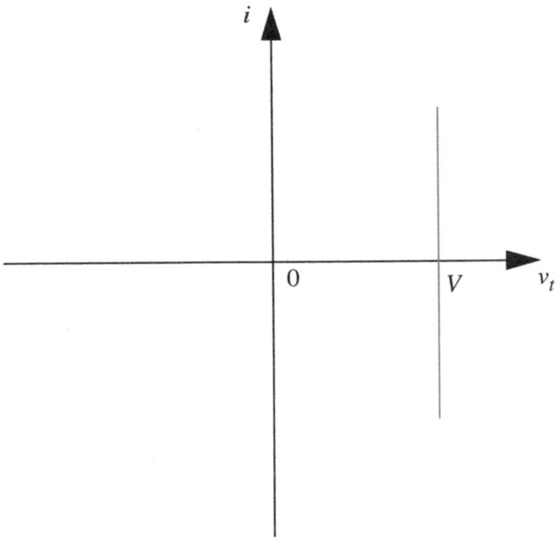

FIGURE 1.42 $i$–$v$ characteristic of a battery at low current levels.

Now plot the $i$–$v$ characteristic of a battery. At low current levels, the curve appears as a vertical line in $i$–$v$ space (see Figure 1.42). But if the oscillator amplitude is increased so that substantial currents are flowing, and we make an appropriate change in scope vertical sensitivity, the line remains straight, but now has a definite tip, as suggested by Figure 1.43a, indicating a nonzero series resistance. If the battery terminal voltage and current are defined as in

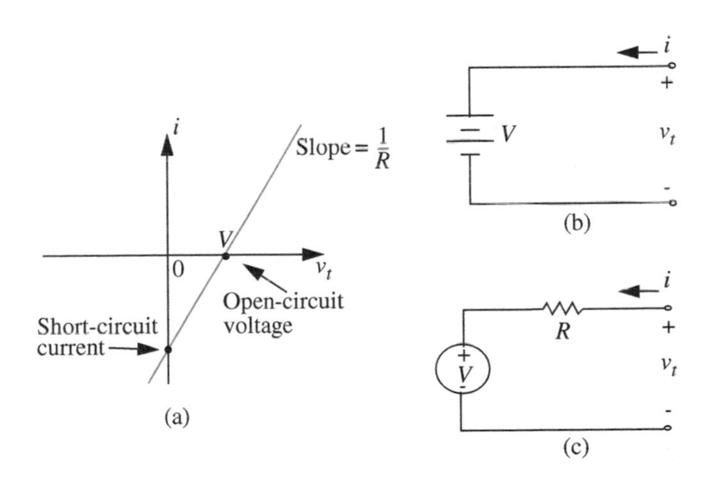


FIGURE 1.43 Battery characteristic, larger current scale.

Figure 1.43b, then from the model in Figure 1.43c, an appropriate expression for the terminal voltage is[17]

$$v_t = V + iR \tag{1.23}$$

Note that because of our choice of variables, in the first quadrant current is flowing *into* the positive terminal, that is, the battery is being charged, hence the terminal voltage is actually larger than the nominal voltage. The fact that the plot is almost a straight line validates our assumption that the battery can be modeled as a voltage source in series with a linear resistor. Note further that graphs such as Figure 1.43a can be characterized by only two numbers, a slope and an intercept. The slope is $1/R$ where $R$ is the series resistance in the model. The intercept can be specified either in terms of a voltage or a current. If we choose a voltage, then because the intercept is by definition at zero current, it is called the *open-circuit voltage*. If the intercept is specified in terms of current, it is called the *short-circuit current*, because by definition the voltage is zero at that point. These terms reappear in Chapter 3 from a very different perspective: Thevenin's Theorem.

This section described how we model physical elements such as batteries and wires in terms of ideal two-terminal elements such as independent voltage sources, resistors, and ideal wires. Our ideal circuit elements such as independent voltage sources and resistors also serve as models for physical entities such as water reservoirs and friction in water tubes, respectively. In the circuit model for a physical system, water pressure is naturally represented using a voltage, and water flow using a current. Water pressure and water flow, or the corresponding voltage and current, are fundamental quantities. In such systems, we will also be concerned with the amount of energy stored in the system, and the rate at which energy is being dissipated.

## 1.8 SIGNAL REPRESENTATION

The previous sections discussed how lumped circuit elements could serve as models for various physical systems or be used to process information. This section draws the correspondence between variables in physical systems and those in the electrical circuit model. It also discusses how electrical systems represent information and energy.

As discussed earlier, one of the motivations for building electronic circuits is to process information or energy. Processing includes communication, storage, and computation. Stereo amplifiers, computers, and radios are examples of commonplace electronic systems for processing information. Power supplies

17. For now, we simply state the equation, and postpone the derivation to Chapter 2 (see the example related to Figure 2.61) after we have mastered a few basic circuit analysis techniques.

and our familiar lightbulb circuit are examples of electronic circuits that process energy.

In both cases, the physical quantity of interest, either the information or the energy, is represented in the circuit by an electrical signal, namely a current or a voltage, and circuit networks are used to process these signals. Thus, the manner in which a circuit fulfills its purpose is effectively the manner in which it treats the signals that are its terminal currents and voltages.

### 1.8.1 ANALOG SIGNALS

Signals in the physical world are most commonly analog, that is, spanning a continuum of values. Sound pressure is such a signal. The electromagnetic signal picked up by a mobile phone antenna is another example of an analog signal. Not surprisingly, most circuits that interact with the physical world must be able to process analog signals.

Figure 1.44 shows several examples of analog signals. Figure 1.44b shows a DC current signal, while the remainder are various forms of voltage signals.

Figure 1.44a shows a sinusoidal signal with frequency 1 MHz and phase offset (or phase shift) $\pi/4$ rad. The same frequency can also be expressed as $10^6$ Hz, or $2\pi 10^6$ rad/s, and the phase offset as 45 degrees. The reciprocal of the frequency gives the period of oscillation or the cycle time, which is 1 $\mu$s for our sinusoid. Our sinusoid has an average value of zero. This signal can be described as a sinusoid with an *amplitude* (or magnitude, or maximum value) of 10 V, or equivalently as a sinusoid with a *peak-to-peak swing* of 20 V.

Sinusoids are an important class of signals that we will encounter frequently in this book. In general, a sinusoidal signal $v$ can be expressed as

$$v = A\sin(\omega t + \phi)$$

where $A$ is the amplitude, $\omega$ is the frequency in radians per second, $t$ is the time, and $\phi$ is the phase offset in radians.

Figures 1.44c and d show square wave signals. The square wave in Figure 1.44c has a peak-to-peak value of 10 V and an average value (or DC offset) of 5 V, while the square wave signal in Figure 1.44d has the same peak-to-peak value, but zero offset.

Information can be represented in one of many forms, for example, the amplitude, phase, or frequency. Figure 1.44e shows a signal (for example, from a microphone) that carries information in its amplitude, and Figure 1.44f shows a signal that is carrying information in its frequency.

To complete this section, we briefly touch on the concept of root mean square value to describe signals. Recall that our signal in Figure 1.44a was described as a sinusoid with an amplitude of 10 V. The same signal can be described as a sinusoid with *rms (root mean square)* value of $10/\sqrt{2}$. For a

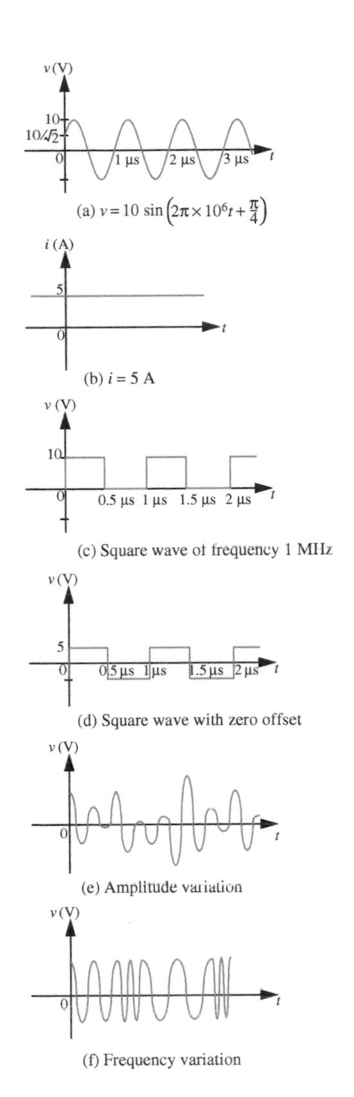


FIGURE 1.44 Several examples of analog signals: (a) a 1-MHz sinusoidal signal with amplitude 10 V and a phase offset of $\pi/4$; (b) a 5-A DC signal; (c) a 1-MHz square wave signal with a 5-V offset; (d) a 1-MHz square wave signal with zero offset; (e) a signal carrying information in its amplitude; (f) a signal carrying information in its frequency.

sinusoidal signal, or for that matter, any periodic signal $v$ with period $T$, the rms is computed as follows:

$$v_{\text{rms}} = \sqrt{\frac{1}{T}\int_{t_1}^{t_1+T} v^2(t)dt} \tag{1.24}$$

where the integration is performed over one cycle.

The significance of the rms value of a periodic signal can be seen by computing the average power $\bar{p}$ delivered to a resistance of value $R$ by a periodic voltage signal $v(t)$ with period $T$. For periodic signals, the average power can be obtained by integrating the power over one cycle and dividing by the cycle time:

$$\bar{p} = \frac{1}{T}\int_{t_1}^{t_1+T} \frac{v^2(t)}{R}dt. \tag{1.25}$$

For a linear, time-invariant resistor, we can pull $R$ out of the integral to write

$$\bar{p} = \frac{1}{R}\frac{1}{T}\int_{t_1}^{t_1+T} v^2(t)dt \tag{1.26}$$

By defining the rms value of a periodic signal as in Equation 1.24 we can rewrite Equation 1.26 as

$$\bar{p} = \frac{1}{R}v_{\text{rms}}^2 \tag{1.27}$$

By the artful definition of Equation 1.24, we have managed to obtain an expression for power resembling that due to a DC signal. In other words, the rms value of a periodic signal is the value of a DC signal that would have resulted in the same average power dissipation.[18] In like manner, the rms value of a DC signal is simply the constant value of the signal itself.

Thus, a sinusoidal voltage with rms value $v_{\text{rms}}$ applied across a resistor of value $R$ will result in an average power dissipation of $v_{\text{rms}}^2/R$.

For example, 120-V 60-Hz wall outlets in the United States are rated by their rms values. Thus, they supply a sinusoidal voltage with a peak amplitude of $120 \times \sqrt{2} = 170$ V.

### Native Signal Representation

Sometimes, circuit signals provide a native representation of physical quantities, as was the case with our lightbulb example in Figure 1.4. The circuit in Figure 1.4b was a model of the physical circuit in Figure 1.4a, which comprised

18. This new voltage unit, called the rms, was originally defined by the pioneers of the electric power industry to avoid (or possibly perpetuate) confusion between DC power and AC power.

a battery, wires, and a lightbulb. The purpose of the original circuit was to convert chemical energy stored in the battery into light. To do so, the battery converted the chemical energy to electrical energy, the wires then guided the electrical energy to the lightbulb, and the lightbulb converted at least some of this electrical energy to light. Thus, the circuit in Figure 1.4 performed a very primitive form of energy processing.

The circuit in Figure 1.4 was proposed to model the original circuit, and to help determine such quantities as the current flowing through the lightbulb and the power dissipated in it. In this case, the signal representations in the lumped-parameter circuit were chosen naturally. The quantities of interest in the physical circuit, namely its voltages and currents, were represented by the same voltages and currents in the circuit model. This is an example of native signal representation.

#### Non-Native Signal Representation

A more interesting occurrence is that of non-native signal representation. In this case, electrical signals are used to represent non-electrical quantities, which is common in electronic signal processing. For example, consider an electronic sound amplifier. Such a system might begin with a front-end transducer, such as a microphone, that converts sound into an electrical signal that represents the sound. This electrical signal is then amplified, and possibly filtered, to produce a signal representing the desired output sound. Finally, a back-end transducer, such as a speaker, converts the processed electrical signal back into sound. Because electrical signals can be transduced and processed with ease, electronic circuits provide an amazingly powerful means for information processing, and have all but replaced native processing. For example, electronic amplifiers have now replaced megaphones.

The choice of signal type, for example current or voltage, often depends on the availability of convenient transducers (elements that convert from one form of energy to another — for example, sound to electricity), power considerations, and the availability of appropriate circuit elements. Voltage is a popular representation and is used throughout this book. We will also see several situations later in which a voltage signal is converted to a current signal and vice versa as it is being processed in an electronic system.

### 1.8.2 DIGITAL SIGNALS—VALUE DISCRETIZATION

In contrast to the continuous representation of analog signals, we can quantize signals into discrete or lumped signal values. Value discretization forms the basis of the *digital abstraction*, which yields a number of advantages such as better noise immunity compared to an analog signal representation. Although most physical signals are analog in nature, it is worth noting that there are a few physical signals that are naturally quantized, and so would have

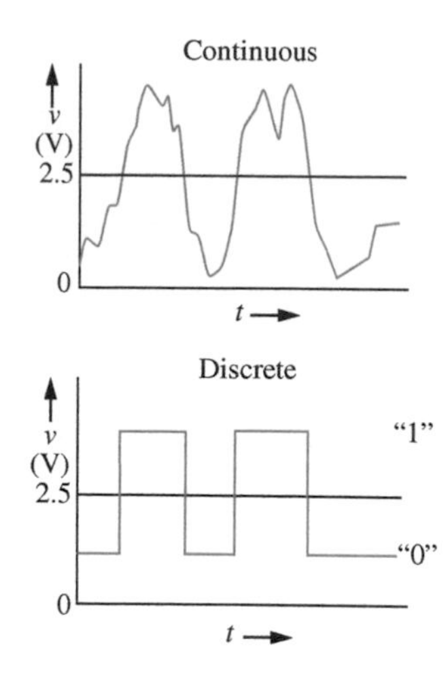


FIGURE 1.45 Voltage value discretization into two levels.

a naturally quantized signal representation. Wealth is an example of such a signal since monetary currencies are not generally considered to be infinitely divisible.[19]

To illustrate value discretization, consider the discretization of voltage as shown in Figure 1.45. Here, we discretize voltage into a finite number of information levels, for example, the two levels named "0" and "1." Under this quantization, if a voltage is observed to be below 2.5 V we interpret its value as representing the information level "0." If its value is above 2.5 V, we interpret it as representing the information level "1." Correspondingly, to produce the information level "0," we use any voltage less than 2.5 V. For example, we might use 1.25 V. Correspondingly, to produce the information level "1," we might use the voltage 3.75 V.

As discussed in Chapter 5, discrete signals offer better noise immunity than analog signals, but they do so at the expense of precision. If the noise that corrupts a discrete signal does not move its physical value past a discretization threshold, then the noise will be ignored. For example, suppose the information level "0" in Figure 1.45 is represented by a 1.25-V signal, and the information level "1" in Figure 1.45 is represented by a 3.75-V signal. Provided the voltage does not rise above 2.5 V for "0," or does not fall below 2.5 V for a "1," it will be interpreted correctly. Thus, this discrete signal representation is immune to $\pm 1.25$-V noise. Notice, however, the loss in precision — our coarse two-level representation is unable to distinguish between small changes in the voltage.[20]

In general, we can discretize values into any number of levels, for example, four. Thus the representation discussed thus far is a special case of value discretization called the *binary representation* where we discretize the voltage (or current for that matter) into two information levels: "0" and "1." Because systems using more than two levels are difficult to build, most digital systems in use today use the binary representation. Accordingly, the digital representation has become synonymous with the binary representation.

19. Notice that before the advent of currencies, the barter system prevailed, and wealth was indeed analog in nature, since a loaf of bread, or a plot of land for that matter, theoretically is infinitely divisible!

20. For applications that care only about whether a signal is above or below some threshold, the loss in precision is of no consequence, and a two-level representation is sufficient. However, for other applications that care about small changes in a signal, the basic two-level representation of a signal must be extended. We show in Chapter 5 that practical digital systems can offer both arbitrary degrees of precision and noise immunity through a process of discretization and coding. Briefly, to recover some precision while retaining noise immunity, digital systems quantize signals into a large number of levels — for example, 256 — and code these levels into a few binary digits — 8, in our example, where each binary digit can be represented as a two-level voltage on a single wire. This method converts an analog signal on a single wire into a binary encoded signal on several wires, where each wire carries a voltage that can vary between two levels.

### Native and Non-Native Signal Representation

As with analog signals, discrete signals can provided both native and non-native signal representations. The discrete binary values of 0 and 1 are a native representation for logic because they correspond naturally to the logical TRUE and FALSE values. Non-native signal representations can be derived from discrete signals by using sequences of digits having the value 0 or 1 to encode numbers whose values correspond to signal values of interest. Chapter 5 covers this topic in greater detail.

When designing a non-native information processing system, there are many choices for signal representation. For example, the use of voltage versus current, or analog versus discrete signals are two such choices. Each representation has its advantages and disadvantages, and facilitates a certain kind of processing. For example, digital representations offer noise immunity at the expense of precision. How these choices are made is usually application specific, and often depends on the availability of convenient transducers, power and noise considerations, and the availability of appropriate elements. The use of voltage to represent signals is probably most common, and is used routinely here. However, we will also encounter situations in which the signal representation switches from a voltage to a current and back again as the signal is processed.

## 1.9 SUMMARY

- The discretization of matter into lumped elements such as batteries and resistors that obey the lumped matter discipline and connecting them using ideal wires is the essence of the *lumped circuit abstraction.*

- The lumped matter discipline for lumped elements includes the following constraints:

  1. The boundaries of the discrete elements must be chosen so that

  $$\frac{\partial \Phi_B}{\partial t} = 0$$

     through any closed path outside the element for all time.

  2. The elements must not include any net time-varying charge for all time. In other words,

  $$\frac{\partial q}{\partial t} = 0$$

     where $q$ is the total charge within the element.

  3. We must operate in the regime in which timescales of interest are much larger than the propagation delay of electromagnetic waves through the elements.

- The lumped matter discipline for lumped circuits includes the following constraints:

  1. The rate of change of magnetic flux linked with any portion of the circuit must be zero for all time.

  2. The rate of change of the charge at any node in the circuit must be zero for all time.

  3. The signal timescales must be much larger than the propagation delay of electromagnetic waves through the circuit.

- The associated variables convention defines current to flow *in* at the device terminal assigned to be *positive* in voltage.

- The instantaneous power *consumed* by a device is given by $p(t) = v(t)i(t)$, where $v(t)$ and $i(t)$ are defined using the associated variables discipline. Similarly, the instantaneous power *delivered* by a device is given by $p(t) = -v(t)i(t)$. The unit of power is the watt.

- The amount of energy $w(t)$ *consumed* by a device over an interval of time $t_1 \rightarrow t_2$, is given by

$$w(t) = \int_{t_1}^{t_2} v(t)i(t)dt$$

where $v(t)$ and $i(t)$ are defined using the associated variables discipline. The unit of energy is the joule.

- Ohm's law states that resistors that obey Ohm's law satisfy the equation $v = iR$, where $R$ is constant. The resistance of a piece of homogeneous material is proportional to its length and inversely proportional to its cross-sectional area.

- The resistance of a planar piece of material with length $L$ and width $W$ is given by $\frac{L}{W} \times R_{\square}$, where $R_{\square}$ is the resistance of a square piece of material.

- The four ideal circuit elements are the ideal conductor, the ideal linear resistor, the voltage source, and the current source. The element law for the ideal conductor is

$$v = 0,$$

for the resistor with resistance $R$ is

$$v = iR,$$

for the voltage source supplying a voltage $V$ is

$$v = V,$$

and for the current source supplying a current $I$ is

$$i = I.$$

- The representation of parameters in physical systems by their equivalent electrical circuit parameters has been discussed.

- The representation of information in terms of analog and digital electrical signals has been discussed.

In the process of introducing the elements and their element laws, we defined the symbols and units for various physical quantities. These definitions are summarized in Table 1.2. The units can be further modified with engineering multipliers. Several common multipliers and their corresponding prefix symbols and values are given in Table 1.3.

**TABLE 1.2** Electrical engineering quantities, their units, and symbols for both.

| QUANTITY | SYMBOL | UNITS | SYMBOL |
|---|---|---|---|
| Time | $t$ | Second | s |
| Frequency | $f$ | Hertz | Hz |
| Current | $i$ | Ampere | A |
| Voltage | $v$ | Volt | V |
| Power | $p$ | Watt | W |
| Energy | $w$ | Joule | J |
| Resistance | $R$ | Ohm | Ω |
| Conductance | $G$ | Siemen | S |

**TABLE 1.3** Common engineering multipliers.

| MULTIPLIER | PREFIX | VALUE |
|---|---|---|
| peta | P | $10^{15}$ |
| tera | T | $10^{12}$ |
| giga | G | $10^{9}$ |
| mega | M | $10^{6}$ |
| kilo | k | $10^{3}$ |
| milli | m | $10^{-3}$ |
| micro | μ | $10^{-6}$ |
| nano | n | $10^{-9}$ |
| pico | p | $10^{-12}$ |
| femto | f | $10^{-15}$ |

## EXERCISES

EXERCISE 1.1 Quartz heaters are rated according to the average power drawn from a 120-V AC 60-Hz voltage source. Estimate the resistance (when operating) a 1200-W quartz heater.

NOTE: The voltage waveform for a 120-V AC 60-Hz waveform is

$$v(t) = \sqrt{2}\, 120 \cos(2\pi 60t).$$

The factor of $\sqrt{2}$ in the peak amplitude cancels when the average power is computed. One result is that the peak amplitude of the voltage from a 120-volt wall outlet is about 170 volts.

EXERCISE 1.2

a) The battery on your car has a rating stated in ampere-hours that permits you to estimate the length of time a fully charged battery could deliver any particular current before discharge. Approximately how much energy is stored by a 50 A-hour 12-V battery?

b) Assuming 100% efficient energy conversion, how much water stored behind a 30 m high hydroelectric dam would be required to charge the battery?

EXERCISE 1.3 In the circuit in Figure 1.46, $R$ is a linear resistor and $v = V_{DC}$ a constant (DC) voltage. What is the power dissipated in the resistor, in terms of $R$ and $V_{DC}$?

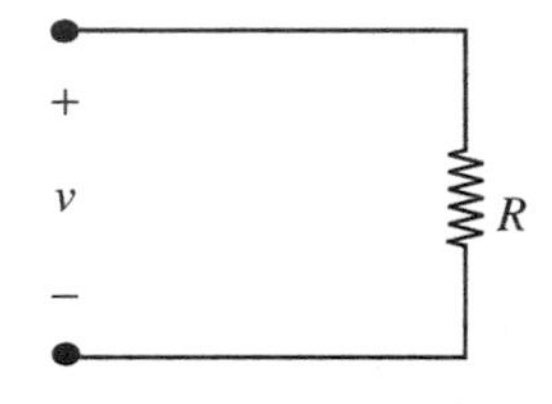


FIGURE 1.46

EXERCISE 1.4 In the circuit of the previous exercise (see Figure 1.46), $v = V_{AC} \cos\omega t$, a sinusoidal (AC) voltage with peak amplitude $V_{AC}$ and frequency $\omega$, in radians/sec.

a) What is the average power dissipated in R?

b) What is the relationship between $V_{DC}$ and $V_{AC}$ in Figure 1.46 when the average power in $R$ is the same for both waveforms?

## PROBLEMS

PROBLEM 1.1 Determine the resistance of a cube with sides of length $l$ cms and resistivity 10 Ω-cm, when a pair of opposite surfaces are chosen as the terminals.

PROBLEM 1.2 Sketch the $v$–$i$ characteristic of a battery rated at 10 V with an internal resistance of 10 Ohms.

PROBLEM 1.3 A battery rated at 7.2 V and 10000 J is connected across a lightbulb. Assume that the internal resistance of the battery is zero. Further assume that the resistance of the lightbulb is 100 Ω.

1. Draw the circuit containing the battery and the lightbulb and label the terminal variables for the battery and the lightbulb according to the associated variables discipline.

2. What is the power into the lightbulb?

3. Determine the power into the battery.

4. Show that the sum of the power into the battery and the power into the bulb is zero.

5. How long will the battery last in the circuit?

PROBLEM 1.4 A sinusoidal voltage source

$$v = 10\,\text{V}\,\sin(\omega t)$$

is connected across a 1 k$\Omega$ resistor.

1. Make a sketch of $p(t)$, the instantaneous power supplied by the source.
2. Determine the average power supplied by the source.
3. Now, suppose that a square wave generator is used as the source. If the square wave signal has a peak-to-peak of 20 V and a zero average value, determine the average power supplied by the source.
4. Next, if the square wave signal has a peak-to-peak of 20 V and a 10 V average value, determine the average power supplied by the source.

# CHAPTER 2

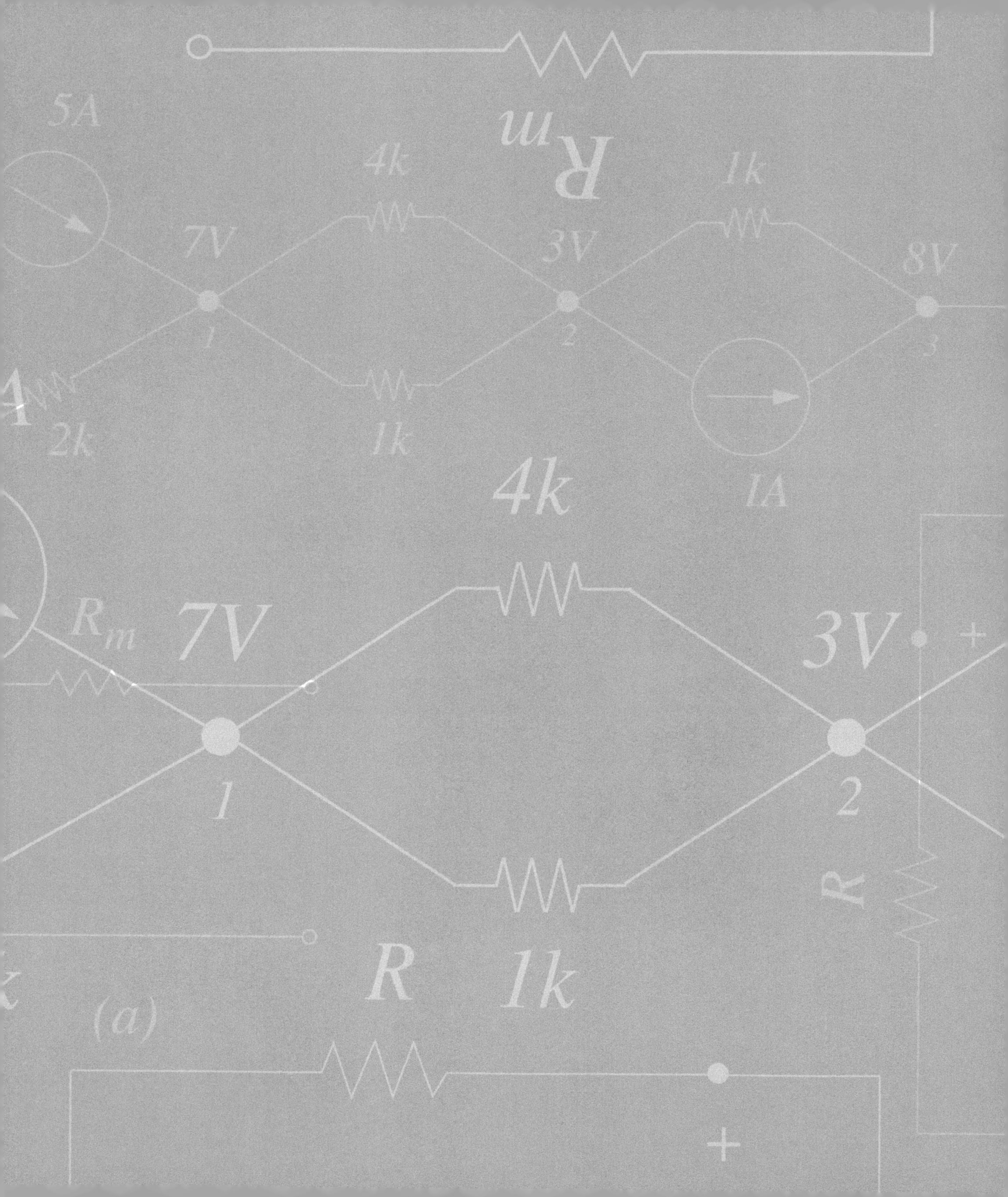

5A
4k
1k
7V
3V
8V
1
2
3
2k
1k
1A
4k
$R_m$
7V
3V
+
1
2
R
R
1k
(a)
+

# RESISTIVE NETWORKS

# 2

A simple electrical network made from a voltage source and four resistors is shown in Figure 2.1. This might be an abstract representation of some real electrical network, or a model of some other physical system, for example, a heat flow problem in a house. We wish to develop systematic general methods for analyzing circuits such as this, so that circuits of arbitrary complexity can be solved with dispatch. *Solving or analyzing a circuit generally involves finding the voltage across, and current through, each of the circuit elements.* Systematic general methods will also enable us to automate the solution techniques so that computers can be used to analyze circuits. Later on in this chapter and in the next, we will show how our problem formulation facilitates direct computer analysis.

To make the problem specific, suppose that we wish to find the current $i_4$ in Figure 2.1, given the values of the voltage source and the resistors. In general, we can resort to Maxwell's Equations to solve the circuit. But this approach is really impractical. Instead, when circuits obey the lumped matter discipline, Maxwell's Equations can be dramatically simplified into two algebraic relationships stated as Kirchhoff's voltage law (KVL) and Kirchhoff's current law (KCL). This chapter introduces these algebraic relationships and then uses them to develop a systematic approach to solving circuits, thereby finding the current $i_4$ in our specific example.

This chapter first reviews some terminology that will be useful in our discussions. We will then introduce Kirchhoff's laws and work out some examples to develop our facility with these laws. We will then introduce a systematic method for solving circuits based on Kirchhoff's laws using a very simple,

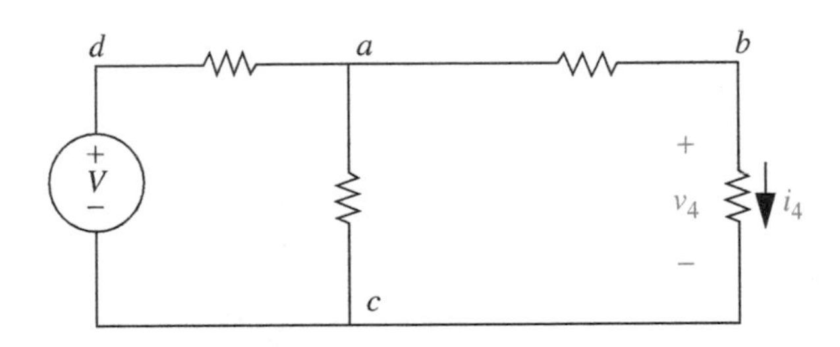


FIGURE 2.1 Simple resistive network.

illustrative circuit. We will then apply the same systematic method to solve more complicated examples, including the one shown in Figure 2.1.

## 2.1 TERMINOLOGY

*Lumped circuit elements* are the fundamental building blocks of electronic circuits. Virtually all of our analyses will be conducted on circuits containing two-terminal elements; multi-terminal elements will be modeled using combinations of two-terminal elements. We have already seen several two-terminal elements such as resistors, voltage sources, and current sources. Electronic access to an element is made through its *terminals*.

An electronic *circuit* is constructed by connecting together a collection of separate elements at their terminals, as shown in Figure 2.2. The junction points at which the terminals of two or more elements are connected are referred to as the *nodes* of a circuit. Similarly, the connections between the nodes are referred to as the edges or *branches* of a circuit. Note that each element in Figure 2.2 forms a single branch. Thus an element and a branch are the same for circuits comprising only two-terminal elements. Finally, circuit *loops* are defined to be closed paths through a circuit along its branches.

Several nodes, branches, and loops are identified in Figure 2.2. In the circuit in Figure 2.2, there are 10 branches (and thus, 10 elements) and 6 nodes.

As another example, $a$ is a node in the circuit depicted in Figure 2.1 at which three branches meet. Similarly, $b$ is a node at which two branches meet. $ab$ and $bc$ are examples of branches in the circuit. The circuit has five branches and four nodes.

Since we assume that the interconnections between the elements in a circuit are perfect (i.e., the wires are ideal), then it is not necessary for a set of elements to be joined together at a single point in space for their interconnection to be considered a single node. An example of this is shown in Figure 2.3. While the four elements in the figure are connected together, their connection does not occur at a single point in space. Rather, it is a distributed connection.

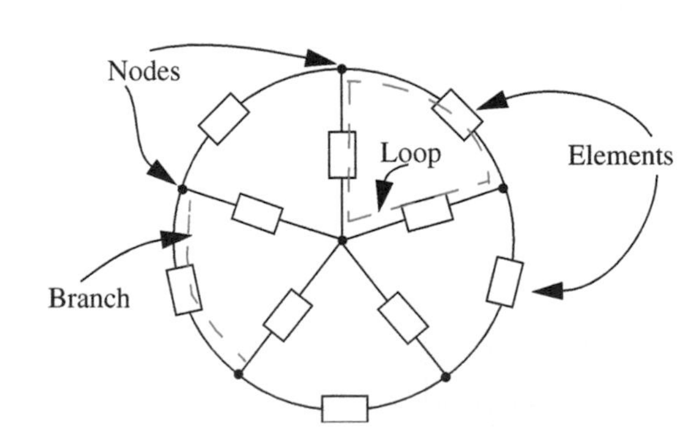


FIGURE 2.2 An arbitrary circuit.

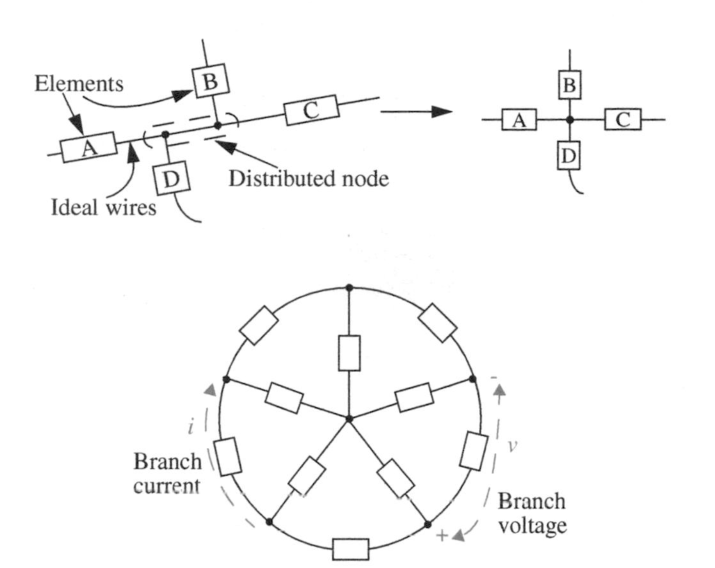


FIGURE 2.3 Distributed interconnections of four circuit elements that nonetheless occur at a single node.

FIGURE 2.4 Voltage and current definitions illustrated on a branch in a circuit.

Nonetheless, because the interconnections are perfect, the connection can be considered to be a single node, as indicated in the figure.

The primary signals within a circuit are its currents and voltages, which we denote by the symbols $i$ and $v$, respectively. We define a *branch current* as the current along a branch of the circuit (see Figure 2.4), and a *branch voltage* as the potential difference measured across a branch. Since elements and branches are the same for circuits formed of two-terminal elements, the branch voltages and currents are the same as the corresponding terminal variables for the elements forming the branches. Recall, as defined in Chapter 1, the terminal variables for an element are the voltage across and the current through the element.

As an example, $i_4$ is a branch current that flows through branch $bc$ in the circuit in Figure 2.1. Similarly, $v_4$ is the branch voltage for the branch $bc$.

## 2.2 KIRCHHOFF'S LAWS

Kirchhoff's current law and Kirchhoff's voltage law describe how lumped-parameter circuit elements couple at their terminals when they are assembled into a circuit. KCL and KVL are themselves lumped-parameter simplifications of Maxwell's Equations. This section defines KCL and KVL and justifies that they are reasonable using intuitive arguments.[1] These laws are employed in circuit analysis throughout this book.

1. The interested reader can refer to Section A.2 in Appendix A for a derivation of Kirchhoff's laws from Maxwell's Equations under the lumped matter discipline.

### 2.2.1 KCL

Let us start with Kirchhoff's current law (KCL).

*KCL* The current flowing out of any node in a circuit must equal the current flowing in. That is, the algebraic sum of all branch currents flowing into any node must be zero.

Put another way, KCL states that the net current that flows into a node through some of its branches must flow out from that node through its remaining branches.

Referring to Figure 2.5, if the currents through the three branches *into* node $a$ are $i_a$, $i_b$, and $i_c$, then KCL states that

$$i_a + i_b + i_c = 0.$$

Similarly, the currents into node $b$ must sum to zero. Accordingly, we must have

$$-i_b - i_4 = 0.$$

KCL has a simple intuitive justification. Referring to the closed box-like surface depicted in Figure 2.5, it is easy to see that the currents $i_a$, $i_b$, and $i_c$ must sum to zero, for otherwise, there would be a continuous charge buildup at node $a$. Thus, KCL is simply a statement of the conservation of charge.

Let us now illustrate the different interpretations of KCL with the help of Figure 2.6. Which interpretation you use depends upon convenience and the specific circuit you are trying to analyze. Figure 2.6 shows a node joining $N$ branches. Each of the branches contains some two-terminal element, the specifics of which are not relevant to our discussion. Note that all branch currents are defined to be positive into the node. Since KCL states that no net

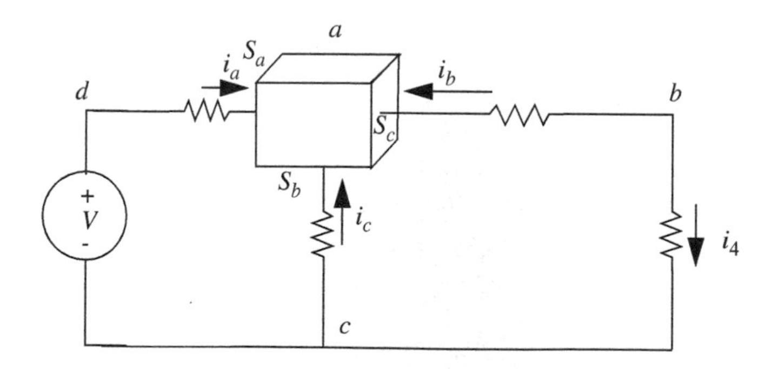


FIGURE 2.5 Currents into a node in the network.

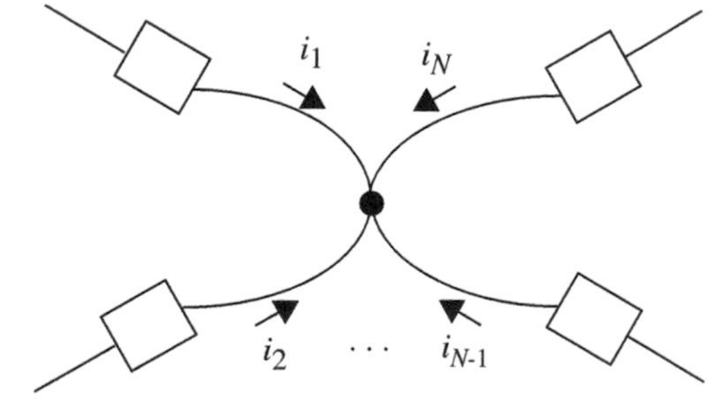


FIGURE 2.6 A node at which $N$ branches join.

FIGURE 2.7 Two series-connected circuit elements.

current can flow into a node, it follows for the node in Figure 2.6 that

$$\sum_{n=1}^{N} i_n = 0. \tag{2.1}$$

Next, by negating Equation 2.1, KCL becomes

$$\sum_{n=1}^{N} (-i_n) = 0. \tag{2.2}$$

Since $-i_n$ is a current defined to be positive out from the node in Figure 2.6, this second form of KCL states that no net current can flow out from a node. Finally, Equation 2.1 can be rearranged to take the form

$$\sum_{n=1}^{M} i_n = \sum_{n=M+1}^{N} (-i_n), \tag{2.3}$$

which demonstrates that the current flowing into a node through one set of branches must flow out from the node through the remaining branches.

An important simplification of KCL focuses on the two series-connected circuit elements shown in Figure 2.7. Taking KCL to state that no net current can flow into a node, the application of KCL at the node between the two elements yields

$$i_1 - i_2 = 0 \quad \Rightarrow \quad i_1 = i_2. \tag{2.4}$$

This result is important because it shows that the branch currents passing through two series-connected elements must be the same. That is, there is nowhere for the current $i_1$ to go as it enters the node connecting the two elements except to exit that node as $i_2$. In fact, with multiple applications of KCL, this observation is extendible to a longer string of series-connected elements. Such an extension would show that a common branch current passes through a string of series-connected elements.

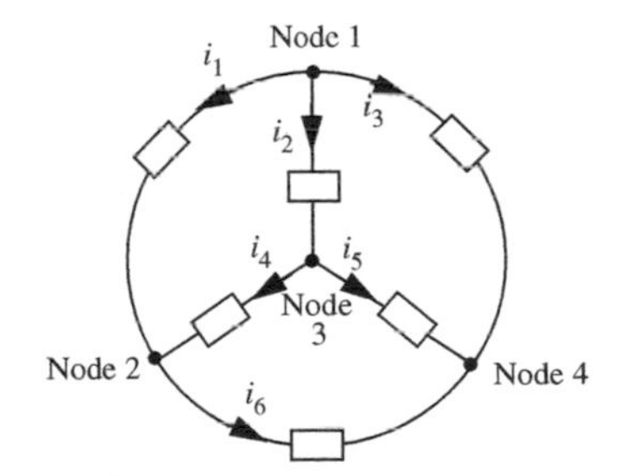


FIGURE 2.8 A circuit illustrating a more general use of KCL.

---

EXAMPLE 2.1 A MORE GENERAL USE OF KCL To illustrate the more general use of KCL consider the circuit in Figure 2.8, which has six branches connecting four nodes. Again, taking KCL to state that no net current can flow into a node, the application of KCL to the four nodes in the circuit yields

$$\text{Node 1:} \quad 0 = -i_1 - i_2 - i_3 \tag{2.5}$$

$$\text{Node 2:} \quad 0 = i_1 + i_4 - i_6 \tag{2.6}$$

$$\text{Node 3:} \quad 0 = i_2 - i_4 - i_5 \tag{2.7}$$

$$\text{Node 4:} \quad 0 = i_3 + i_5 + i_6. \tag{2.8}$$

Note that because each branch current flows into exactly one node and out from exactly one node, each branch current appears exactly once in Equations 2.5 through 2.8 positively, and exactly once negatively. This would also be true if Equations 2.5 through 2.8 were all written to state that no net current can flow out from a node. Such patterns can often be used to spot errors.

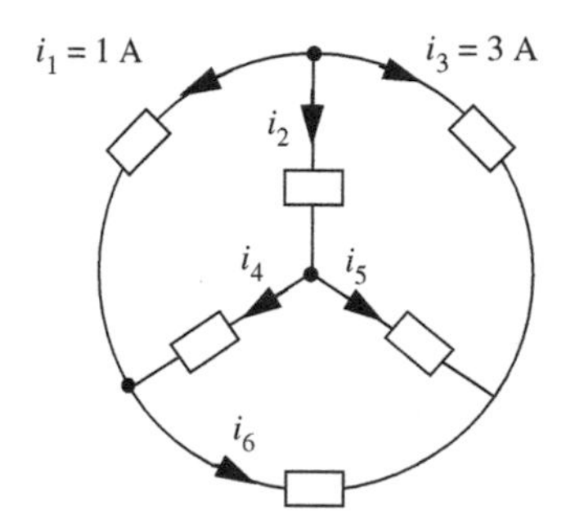


FIGURE 2.9 The circuit in Figure 2.8 with two branch currents numerically defined.

It is also because each branch current flows into exactly one node and out from exactly one node that summing Equations 2.5 through 2.8 yields $0 = 0$. This in turn shows that the four KCL equations are dependent. In fact, a circuit with $N$ nodes will have only $N - 1$ independent statements of KCL. Therefore, when fully analyzing a circuit it is both necessary and sufficient to apply KCL to all but one node.

If some of the branch currents in a circuit are known, then it is possible that KCL alone can be used to find other branch currents in the circuit. For example, consider the circuit in Figure 2.8 with $i_1 = 1$ A and $i_3 = 3$ A, as shown in Figure 2.9. Using Equation 2.5, namely KCL for Node 1, it can be seen that $i_2 = -4$ A. This is all that can be learned from KCL alone given the information in Figure 2.9.

But, if we further know that $i_5 = -2$ A, for example, we can learn from KCL applied to the other nodes that $i_4 = -2$ A and $i_6 = -1$ A.

---

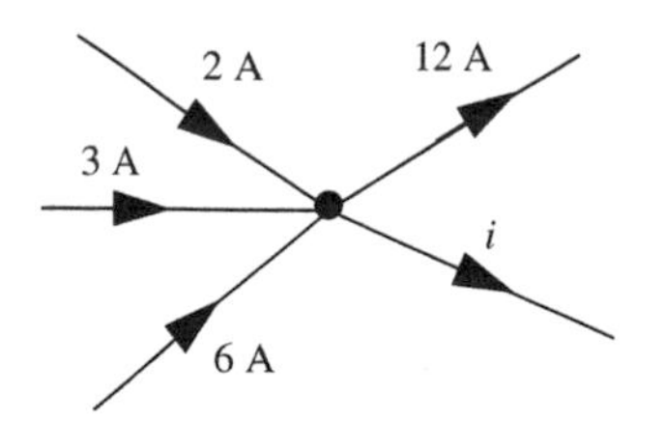


FIGURE 2.10 Five branches meeting at a node.

EXAMPLE 2.2 USING KCL TO DETERMINE AN UNKNOWN BRANCH CURRENT Figure 2.10 shows five branches meeting at a node in some circuit. As shown in the figure, four of the branch currents are given. Determine $i$.

By KCL, the sum of all the currents entering a node must equal the sum of all the currents exiting the node. In other words,

$$2\text{ A} + 3\text{ A} + 6\text{ A} = 12\text{ A} + i$$

Thus, $i = -1$A.

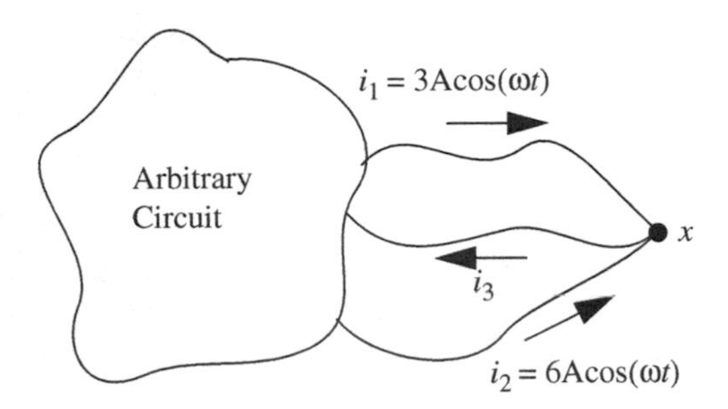


FIGURE 2.11 Node $x$ in a circuit pulled out for display.

EXAMPLE 2.3 KCL APPLIED TO AN ARBITRARY NODE IN A CIRCUIT Figure 2.11 shows an arbitrary circuit from which we have grabbed a node $x$ and pulled it out for display. The node is a junction point for three wires with currents $i_1$, $i_2$, and $i_3$. For the given values of $i_1$ and $i_2$, determine the value of $i_3$.

By KCL, the sum of all currents entering a node must be 0. Thus,

$$i_1 + i_2 - i_3 = 0$$

Note that $i_3$ is negated in this equation because it is defined to be positive for a current exiting the node. Thus $i_3$ is the sum of $i_1$ and $i_2$ and is given by

$$i_3 = i_1 + i_2 = 3\cos(\omega t) + 6\cos(\omega t) = 9\cos(\omega t)$$

EXAMPLE 2.4 EVEN MORE KCL Figure 2.12 shows a node connecting three branches. Two of the branches have current sources that supply the currents shown. Determine the value of $i$.

By KCL, the sum of all the currents entering a node must equal 0. Thus

$$2\text{ A} + 1\text{ A} + i = 0$$

and $i = -3$ A.

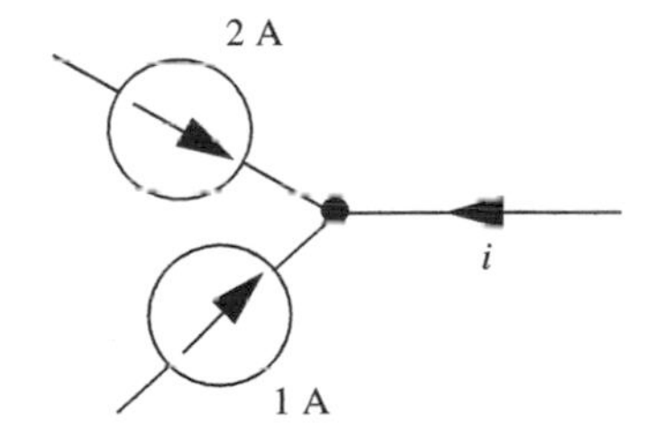


FIGURE 2.12 Node connecting three branches.

Finally, it is important to recognize that current sources can be used to construct circuits in which KCL is violated. Several examples of circuits constructed from current sources in which KCL is violated at every node are shown in Figure 2.13. We will not be concerned with such circuits here for two reasons. First, if KCL does not hold at a node, then electric charge must accumulate at that node. This is inconsistent with the constraint of the lumped matter discipline that $dq/dt$ be zero. Second, if a circuit were actually built to violate KCL, something would ultimately give. For example, the current sources might cease

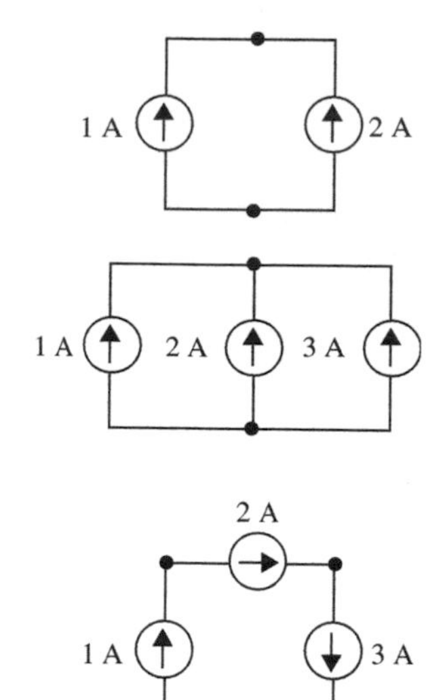


FIGURE 2.13 Circuits that violate KCL.

to function as ideal sources as they oppose one another. In any case, the behavior of the real circuit would not be well modeled by the type of circuit shown in Figure 2.13, and so there is no reason to study the latter.

### 2.2.2 KVL

Let us now turn our attention to Kirchhoff's voltage law (KVL). KVL is applied to circuit loops, that is, to interconnections of branches that form closed paths through a circuit. In a manner analogous to KCL, Kirchhoff's voltage law can be stated as:

*KVL* The algebraic sum of the branch voltages around any closed path in a network must be zero.

Alternatively, it states that the voltage between two nodes is independent of the path along which it is accumulated.

In Figure 2.14, the loop starting at node $a$, proceeding through nodes $b$ and $c$, and returning to $a$, is a closed path. In other words, the closed loop defined by the three circuit branches $a \to b$, $b \to c$, and $c \to a$ in Figure 2.14 is a closed path.

According to KVL, the sum of the branch voltages around this loop is zero. That is,

$$v_{ab} + v_{bc} + v_{ca} = 0$$

In other words,

$$v_1 + v_4 + v_3 = 0$$

where we have taken the positive sign for each voltage when going from the positive terminal to the negative terminal. It is important that we are consistent in how we assign polarities to voltages as we go around the loop.

A helpful mnemonic for writing KVL equations is to assign the polarity to a given voltage in accordance with the first sign encountered when traversing that voltage around the loop.

Like KCL, KVL has an intuitive justification as well. Recall that the definition of the voltage between a pair of nodes in a circuit is the potential

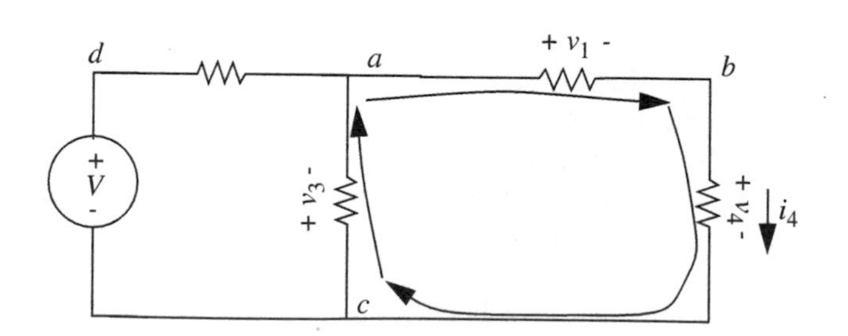


FIGURE 2.14 Voltages in a closed loop in the network.

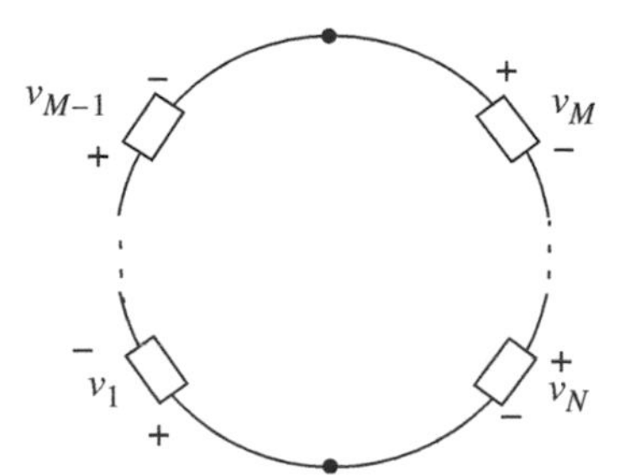


**FIGURE 2.15** A loop containing $N$ branches.

difference between the two nodes. The potential difference between two nodes is the sum of the potential differences for the set of branches along any path between the two nodes. For a loop, the start and end nodes are one and the same, and there cannot be a potential difference between a node and itself. Thus, since potential differences equate to voltages, the sum of branch voltages along a loop must equal zero. By the same reasoning, since the voltage between any pair of nodes must be unique, it must be independent of the path along which branch voltages are added. Notice from the definition of a voltage that KVL is simply an expression of the principle of conservation of energy.

The different interpretations of KVL are illustrated with the help of Figure 2.15, which shows a loop containing $N$ branches. Consider first the loop in Figure 2.15 in which all branch voltages decrease in the clockwise direction. Since KVL states that the sum of the branch voltages around a loop is zero, it follows for the loop in Figure 2.15 that

$$\sum_{n=1}^{N} v_n = 0. \tag{2.9}$$

Note that in summing voltages along a loop we have adopted the convention proposed earlier: A positive branch voltage is added to the sum if the path enters the positive end of a branch. Otherwise a negative branch voltage is added to the sum. Therefore, to arrive at Equation 2.9, we have traversed the loop in the clockwise direction. Next, by negating Equation 2.9, KVL becomes

$$\sum_{n=1}^{N} (-v_n) = 0. \tag{2.10}$$

Since $-v_n$ is a voltage defined to be positive in the opposite direction, this second form of KVL shows that KVL holds whether it is applied along the clockwise or counterclockwise path around the loop.

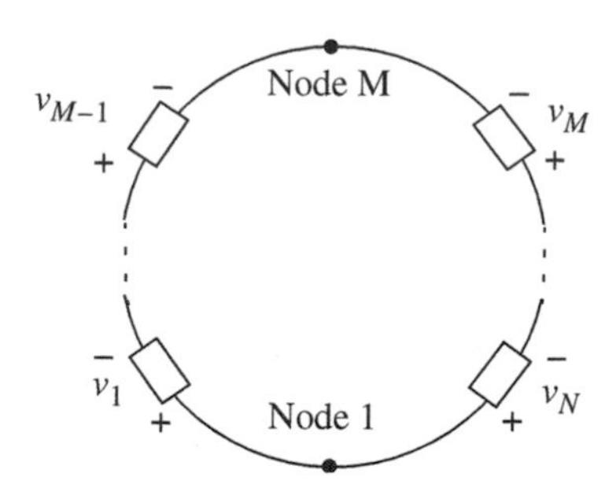


**FIGURE 2.16** A loop containing $N$ branches with some of the voltage definitions reversed.

---

EXAMPLE 2.5 PATH INDEPENDENCE OF KVL Consider the loop in Figure 2.16 in which some of the voltage definitions are reversed for convenience. Applying KVL to this loop yields

$$\sum_{n=1}^{M-1} v_n + \sum_{n=M}^{N} (-v_n) = 0 \quad \Rightarrow \quad \sum_{n=1}^{M-1} v_n = \sum_{n=M}^{N} v_n. \tag{2.11}$$

The second equality in Equation 2.11 demonstrates that the voltage between two nodes is independent of the path along which it accumulated. In this case, the second equality shows that the voltage between Nodes 1 and $M$ is the same whether accumulated along the path up the left side of the loop or the path up the right side of the loop.

---

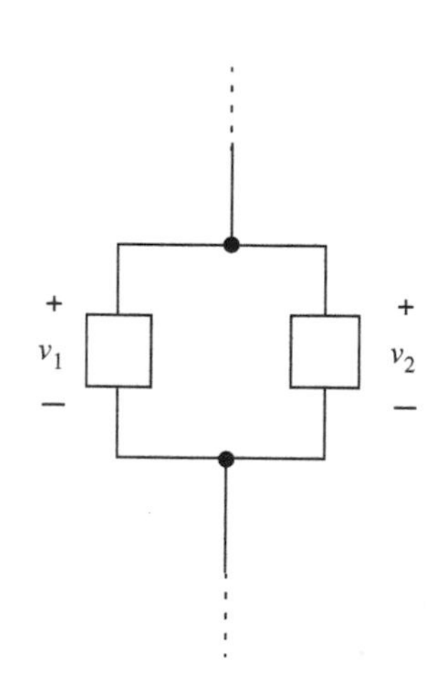


**FIGURE 2.17** Two parallel-connected circuit elements.

An important simplification of KVL focuses on the two parallel-connected circuit elements shown in Figure 2.17. Starting from the upper node and applying KVL in the counterclockwise direction around the loop between the two circuit elements yields

$$v_1 - v_2 = 0 \quad \Rightarrow \quad v_1 = v_2. \tag{2.12}$$

This result is important because it shows that the voltages across two parallel-connected elements must be the same. In fact, with multiple applications of KVL, this observation is extendible to a longer string of parallel-connected elements. Such an extension would show that a common voltage appears across all parallel-connected elements in the string.

---

EXAMPLE 2.6 A MORE GENERAL USE OF KVL To illustrate the more general use of KVL consider the circuit in Figure 2.18, which has six branches connecting four nodes. Four paths along the loops through the circuit are also defined in the figure; note that the external loop, Loop 4, is distinct from the other three. Applying

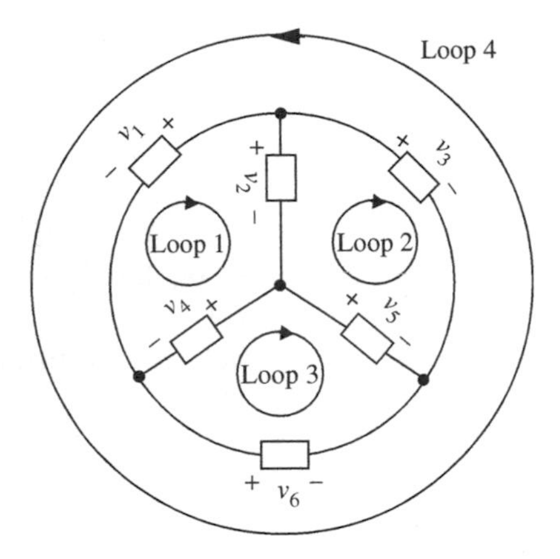


FIGURE 2.18 A circuit having four nodes and six branches.

KVL to the four loops yields

$$\text{Loop 1:}\quad 0 = -v_1 + v_2 + v_4 \tag{2.13}$$

$$\text{Loop 2:}\quad 0 = -v_2 + v_3 - v_5 \tag{2.14}$$

$$\text{Loop 3:}\quad 0 = -v_4 + v_5 - v_6 \tag{2.15}$$

$$\text{Loop 4:}\quad 0 = v_1 + v_6 - v_3 \tag{2.16}$$

Note that the paths along the loops have been defined so that each branch voltage is traversed positively around exactly one loop and negatively around exactly one loop. It is for this reason that each branch voltage appears exactly once in Equations 2.13 through 2.16 positively, and exactly once negatively. As with the application of KCL, such patterns can often be used to spot errors.

It is also because each branch voltage is traversed exactly once positively and once negatively that summing Equations 2.13 through 2.16 yields $0 = 0$. This in turn shows that the four KVL equations are dependent. In general, a circuit with $N$ nodes and $B$ branches will have $B - N + 1$ loops around which independent applications of KVL can be made. Therefore, while analyzing a circuit it is necessary to apply KVL only to these loops, which will in total, traverse each branch at least once in the process.

If some of the branch voltages in a circuit are known, then it is possible that KVL alone can be used to find other branch voltages in the circuit. For example, consider the circuit in Figure 2.18 with $v_1 = 1$ V and $v_3 = 3$ V, as shown in Figure 2.19. Using Equation 2.16, namely KVL for Loop 4, it can be seen that $v_6 = 2$ V. This is all that can be learned from KVL alone given the information in Figure 2.19. But, if we further know that $v_2 = 2$ V, for example, we can learn from KVL applied to the other loops that $v_4 = -1$ V and $v_5 = 1$ V.

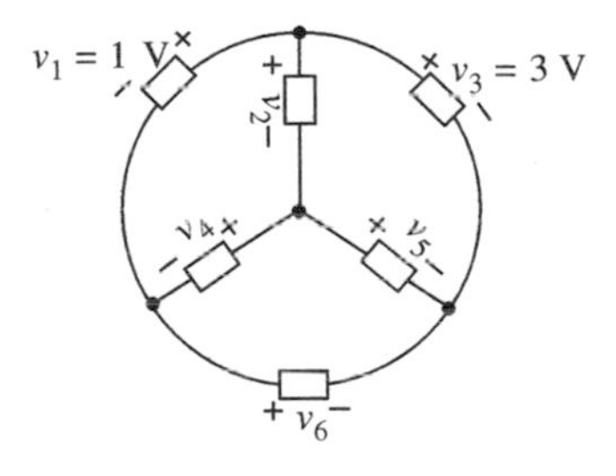


FIGURE 2.19 The circuit in Figure 2.18 with two branch voltages numerically defined.

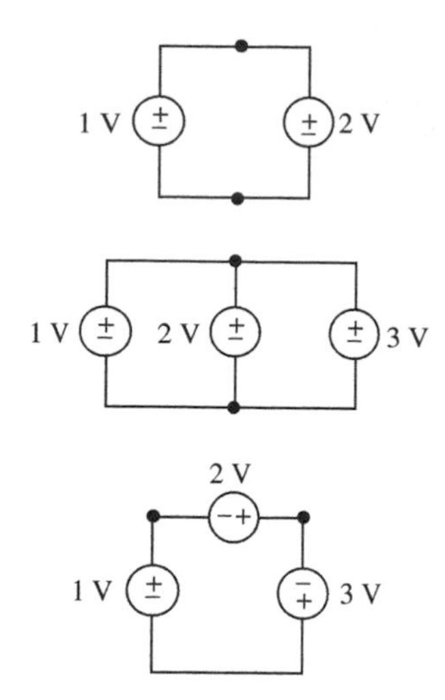


FIGURE 2.20 Circuits that violate KVL.

Finally, it is important to recognize that voltage sources can be used to construct circuits in which KVL is violated. Several examples of circuits constructed from voltage sources in which KVL is violated around every loop are shown in Figure 2.20. As with circuits that violate KCL, we will not be concerned with circuits that violate KVL, for two reasons. First, if KVL does not hold around a loop, then magnetic flux linkage will accumulate through that loop. This is inconsistent with the constraint of the lumped matter discipline that $d\Phi_B/dt = 0$ outside the elements. Second, if a circuit were actually built to violate KVL, something would ultimately give. For example, the voltage sources might cease to function as ideal sources as they oppose one another. Alternatively, the loop inductance might begin to accumulate flux linkage, leading to high currents that would damage the voltage sources or their interconnections. In any case, the behavior of the real circuit would not be well modeled by the type of circuit shown in Figure 2.20, and so there is no reason to study the latter.

---

EXAMPLE 2.7 VOLTAGE SOURCES IN SERIES Two 1.5-V voltage sources are connected in series as shown in Figure 2.21. What is the voltage $v$ at their terminals?

To determine $v$, employ, for example, a counterclockwise application of KVL around the circuit, treating the port formed by the two terminals as an element having voltage $v$. In this case, $1.5\ \text{V} + 1.5\ \text{V} - v = 0$, which has for its solution $v = 3$ V.

---

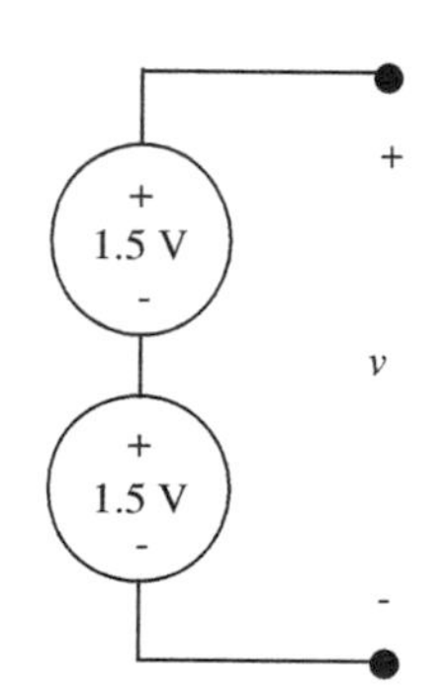


FIGURE 2.21 The series connection of two 1.5-V batteries.

---

EXAMPLE 2.8 KVL The voltages across two of the elements in the circuit in Figure 2.22 are measured as shown. What are the voltages, $v_1$ and $v_2$, across the other two elements?

Since element #1 is connected in parallel with element #4, the voltages across them must be the same. Thus, $v_1 = 5$ V. Similarly, the voltage across the series connection of elements #2 and #3 must also be 5 V, so $v_2 = 3$ V. This latter result can also be obtained through the counterclockwise application of KVL around the loop including elements #2, #3, and #4, for example. This yields, $v_2 + 2\ \text{V} - 5\ \text{V} = 0$. Again, $v_2 = 3$ V.

---

EXAMPLE 2.9 VERIFYING KVL FOR A CIRCUIT Verify that the branch voltages shown in Figure 2.23 satisfy KVL.

Summing the voltages in the loop $e, d, a, b, e$, we get

$$-3 - 1 + 3 + 1 = 0.$$

Similarly, summing the voltages in the loop $e, f, c, b, e$, we get

$$+1 - (-2) - 4 + 1 = 0.$$

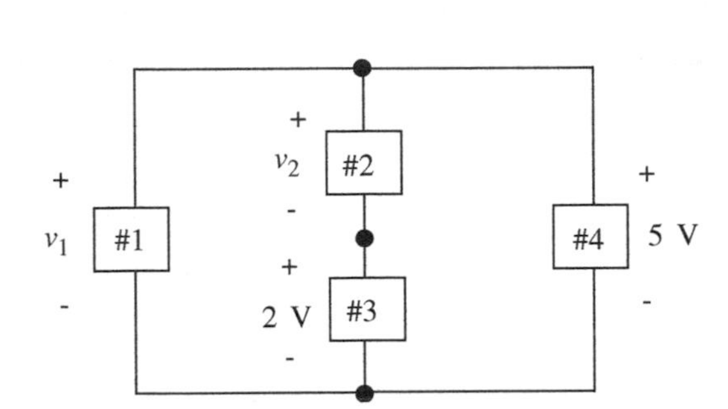


**FIGURE 2.22** A circuit with two measured and two unmeasured voltages.

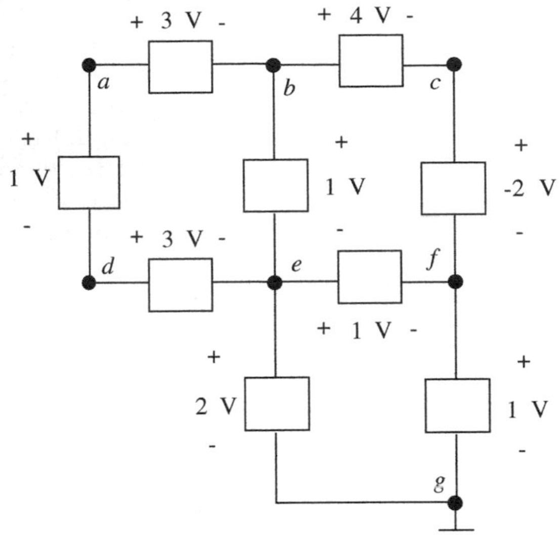

**FIGURE 2.23** A circuit with element voltages as shown.

Finally, summing the voltages in the loop *g*, *e*, *f*, *g*, we get

$$-2 + 1 + 1 = 0.$$

KVL is satisfied since the sum of the voltages around each of the three circuit loops is zero.

---

**EXAMPLE 2.10 SUMMING VOLTAGES ALONG DIFFERENT PATHS** Next, given the branch voltages shown in Figure 2.23, determine the voltage $v_{ga}$ between the nodes *g* and *a* by summing the branch voltages along the path *g*, *e*, *d*, *a*. Then, show that $v_{ga}$ is the same if path *g*, *f*, *c*, *b*, *a* is chosen.

Summing the voltage increases along the path *g*, *e*, *d*, *a*, we get

$$v_{ga} = -2\ \text{V} - 3\ \text{V} - 1\ \text{V} = -6\ \text{V}.$$

Similarly, summing the voltage increases along the path *g*, *f*, *c*, *b*, *a* we get

$$v_{ga} = -1\ \text{V} - (-2\ \text{V}) - 4\ \text{V} - 3\ \text{V} = -6\ \text{V}.$$

Clearly, both paths yield $-6$ V for $v_{ga}$.

Thus far, Chapters 1 and 2 have shown that the operation of a lumped system is described by two types of equations: equations that describe the behavior of its individual elements, or element laws (Chapter 1), and equations that describe how its elements interact when they are connected to form the system, or KCL and KVL (Chapter 2). For an electronic circuit, the element laws relate the branch currents to the branch voltages of the elements. The interactions between its elements are described by KCL and KVL, which are also expressed in terms of branch currents and voltages. Thus, branch currents and voltages become the fundamental signals within a lumped electronic circuit.

## 2.3 CIRCUIT ANALYSIS: BASIC METHOD

We are now ready to introduce a systematic method of solving circuits. It is framed in the context of a simple class of circuits, namely circuits containing only sources and linear resistors. Many of the important analysis issues can be understood through the study of these circuits. Solving a circuit involves determining all the branch currents and branch voltages in the circuit. In practice, some currents or voltages may be more important than others, but we will not make that distinction yet.

Before we return to the specific problem of analyzing the electrical network shown in Figure 2.1, let us first develop the systematic method using a few simpler circuits and build up our insight into the technique. We saw previously that under the lumped matter discipline, Maxwell's Equations reduce to the basic element laws and the algebraic KVL and KCL. Accordingly, a systematic solution of the network involves the assembly and subsequent joint solution of two sets of equations. The first set of equations comprise the constituent relations for the individual elements in the network. The second set of equations results from the application of Kirchhoff's current and voltage laws.

This basic method of circuit analysis, also called the *KVL and KCL method* or the *fundamental method*, is outlined by the following steps:

1. Define each branch current and voltage in the circuit in a consistent manner. The polarities of these definitions can be arbitrary from one branch to the next. However, for any given branch, follow the associated variables convention (see Section 1.5.3 in Chapter 1). In other words, the branch current should be defined as positive into the positive voltage terminal of the branch. By following the associated variables, element laws can be applied consistently, and the solutions will follow a much clearer pattern.

2. Assemble the element laws for the elements. These element laws will specify either the branch current or branch voltage in the case of an independent source, or specify the relation between the branch current

and voltage in the case of a resistor. Examples of these element laws were presented in Section 1.6.

3. Apply Kirchhoff's current and voltage laws as discussed in Section 2.2.

4. Jointly solve the equations assembled in Steps 2 and 3 for the branch variables defined in Step 1.

The remainder of this chapter is devoted to circuit analysis examples that rigorously follow these steps.

Once the two sets of equations are assembled, which is a relatively easy task, the analysis of a circuit essentially becomes a problem of mathematics, as indicated by Step 4. That is, the equations assembled earlier must be combined and used to solve for the branch currents and voltages of interest. However, because there is more than one way to approach this problem, our study of circuit analysis does not end with the direct approach outlined here. Considerable time can be saved, and considerable insight can be gained, by approaching circuit analyses in different ways. These gains are important subjects of this and future chapters.

### 2.3.1 SINGLE-RESISTOR CIRCUITS

To illustrate our basic approach to circuit analysis, consider the simple circuit shown in Figure 2.24. The circuit has one independent source and one resistor, and so has two branches, each with a current and a voltage. The goal of our circuit analysis is to find these branch variables.

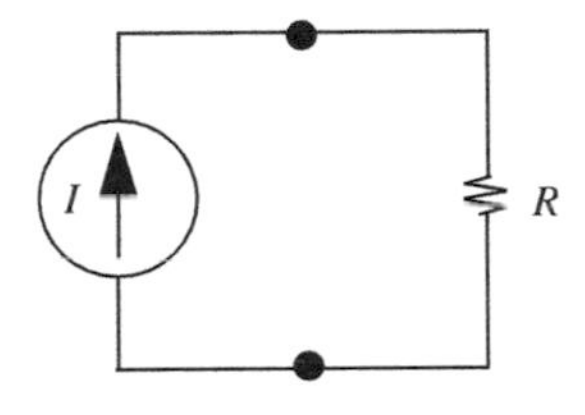


FIGURE 2.24 A circuit with only one independent current source and one resistor.

Step 1 in the analysis is to label the branch variables. We do so in Figure 2.25. Since there are two branches, there are two sets of variables. Notice that the branch variables for the current source branch and for the resistor branch each follow the associated variables convention.

Now, we proceed with Steps 2 through 4: assemble the element laws, apply KCL and KVL, and then simultaneously solve the two sets of equations to complete the analysis.

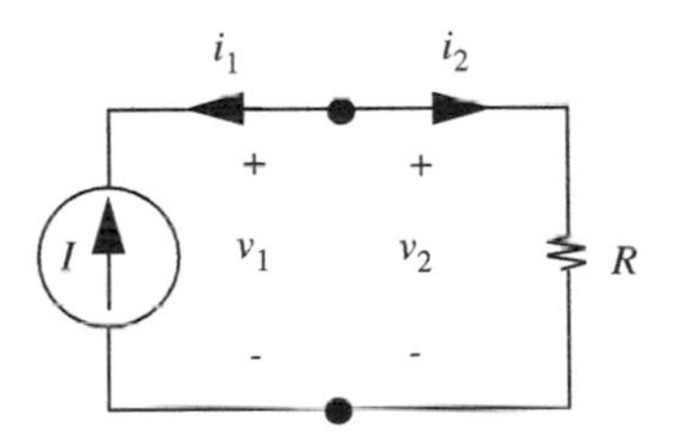


FIGURE 2.25 Assignment of branch variables.

The circuit has two elements. Following Step 2 we write the two element laws for these elements as

$$i_1 = -I, \tag{2.17}$$

$$v_2 = Ri_2, \tag{2.18}$$

respectively. Here, $v_1$, $i_1$, $v_2$, and $i_2$ are the branch variables. Note the distinction between the branch variable $i_1$ and the source amplitude $I$. Here, the independent source amplitude $I$ is assumed to be known.

Next, following Step 3, we apply KCL and KVL to the circuit. Since the circuit has two nodes, it is appropriate to write KCL for one node, as discussed in Section 2.2.1. The application of KCL at either node yields

$$i_1 + i_2 = 0. \tag{2.19}$$

The circuit also has two branches that form one loop. So, following the discussion in Section 2.2.2 it is appropriate to write KVL for one loop. Starting at the upper node and traversing the loop in a clockwise manner, the application of KVL yields

$$v_2 - v_1 = 0. \tag{2.20}$$

Notice we have used our mnemonic discussed in Section 2.2.2 for writing KVL equations. For example, in Equation 2.20, we have assigned a + polarity to $v_2$ since we first encounter the + sign when traversing the branch with variable $v_2$. Similarly, we have assigned a – polarity to $v_1$ since we first encounter the – sign when traversing the $v_1$ branch.

Finally, following Step 4, we combine Equations 2.17 through 2.20 and solve jointly to determine all four branch variables in Figure 2.25. This yields

$$-i_1 = i_2 = I, \tag{2.21}$$

$$v_1 = v_2 = RI, \tag{2.22}$$

and completes the analysis of the circuit in Figure 2.25.

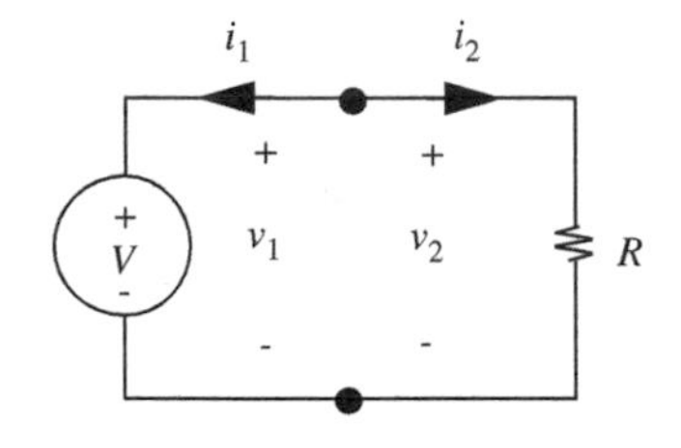


**FIGURE 2.26** A circuit with only one independent voltage source and one resistor.

---

EXAMPLE 2.11 SINGLE-RESISTOR CIRCUIT WITH ONE INDEPENDENT VOLTAGE SOURCE Now consider another simple circuit shown in Figure 2.26. This circuit can be analyzed in an identical manner. It too has two elements, namely a voltage source and a resistor. Figure 2.26 already shows the definitions of branch variables, and so accomplishes Step 1.

Next, following Step 2 we write the element laws for these elements as

$$v_1 = V \tag{2.23}$$

$$v_2 = Ri_2, \tag{2.24}$$

respectively. Here, the independent source amplitude $V$ is assumed to be known.

Next, following Step 3, we apply KCL and KVL to the circuit. Since the circuit has two nodes, it is again appropriate to write KCL for one node. The application of KCL at either node yields

$$i_1 + i_2 = 0. \tag{2.25}$$

The circuit also has two branches that form one loop, so it is again appropriate to write KVL for one loop. The application of KVL around the one loop in either direction yields

$$v_1 = v_2. \tag{2.26}$$

Finally, following Step 4, we combine Equations 2.23 through 2.26 to determine all four branch variables in Figure 2.26. This yields

$$-i_1 = i_2 = \frac{V}{R}, \tag{2.27}$$

$$v_1 = v_2 = V, \tag{2.28}$$

and completes the analysis of the circuit in Figure 2.26.

---

For the circuit in Figure 2.25, there are four equations to solve for four unknown branch variables. In general, a circuit having $B$ branches will have $2B$ unknown branch variables: $B$ branch currents and $B$ branch voltages. To find these variables, $2B$ independent equations are required, $B$ of which will come from element laws, and $B$ of which will come from the application of KVL and KCL. Moreover, if the circuit has $N$ nodes, then $N-1$ equations will come from the application of KCL and $B-N+1$ equations will come from the application of KVL.

While the two examples of circuit analysis presented here are admittedly very simple, they nonetheless illustrate the basic steps of circuit analysis: label the branch variables, assemble the element laws, apply KCL and KVL, and solve the resulting equations for the branch variables of interest. While we will not always follow these steps explicitly and in exactly the same order in future chapters, it is important to know that we will nonetheless process exactly the same information.

It is also important to realize that the physical results of the analysis of the circuit in Figure 2.25, and of any other circuit for that matter, cannot depend on the polarities of the definitions of the branch variables. We will work an example to illustrate this point.

---

EXAMPLE 2.12 POLARITIES OF BRANCH VARIABLES
Consider the analysis of the circuit in Figure 2.27, which is physically the same as the circuit in Figure 2.25. The only difference in the two figures is the reversal of the polarities of $i_2$ and $v_2$. The circuit in Figure 2.27 circuit has the same two elements, and their element laws are still

$$i_1 = -I \tag{2.29}$$

$$v_2 = Ri_2. \tag{2.30}$$

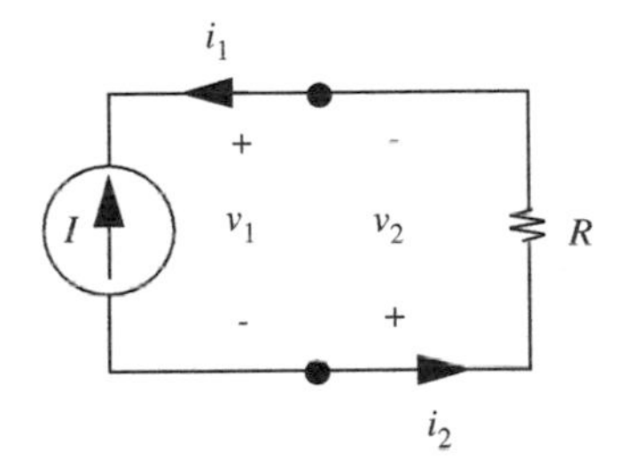


FIGURE 2.27 A circuit similar to the one shown in Figure 2.24.

Note that the polarity reversal of $i_2$ and $v_2$ has not changed the element law for the resistor from Equation 2.18 because the element law for a linear resistor is symmetric when the terminal variables are defined according to the associated variables convention. The circuit also has the same two nodes and the same loop. The application of KCL at either node now yields

$$i_1 - i_2 = 0, \tag{2.31}$$

and the application of KVL around the loop now yields

$$v_1 + v_2 = 0. \tag{2.32}$$

Note that Equations 2.31 and 2.32 differ from Equations 2.19 and 2.20 because of the polarity reversal of $i_2$ and $v_2$. Finally, combining Equations 2.29 through 2.32 yields

$$-i_1 = -i_2 = I \tag{2.33}$$

$$v_1 = -v_2 = RI, \tag{2.34}$$

which completes the analysis of the circuit in Figure 2.27.

Now compare Equations 2.33 and 2.34 to Equations 2.21 and 2.22. The important observation here is that they are the same except for the polarity reversal of the solutions for $i_2$ and $v_2$. This must be the case because the circuits in Figures 2.25 and 2.27 are physically the same, and so their branch variables must also be physically the same. Since we have chosen to define two of these branch variables with different polarities in the two figures, the signs of their values must differ accordingly so that they describe the same physical branch current and voltage.

### 2.3.2 QUICK INTUITIVE ANALYSIS OF SINGLE-RESISTOR CIRCUITS

Before moving on to more complex circuits, it is worthwhile to analyze the circuit in Figure 2.25 in a more intuitive and efficient manner. Here, the element law for the current source directly states that $i_1 = -I$. Next, the application of KCL to either node reveals that $i_2 = -i_1 = I$. In other words, the current from the source flows entirely through the resistor. Next, from the element law for the resistor, it follows that $v_2 = Ri_2 = RI$. Finally the application of KVL to the one loop yields $v_1 = v_2 = RI$ to complete the analysis.

**EXAMPLE 2.13 QUICK INTUITIVE ANALYSIS OF A SINGLE-RESISTOR CIRCUIT** This example considers the circuit in Figure 2.26. Here, the element law for the voltage source directly states that $v_1 = V$. Next, the application of KVL around the one loop reveals that $v_2 = v_1 = V$. In other words, the voltage from the source is applied directly across the resistor. Next, from the element law for the resistor, it follows that $i_2 = v_2/R = V/R$. Finally, the application of KCL to either node yields $i_1 = -i_2 = -V/R$ to complete the analysis. Notice that we had made use of a similar intuitive analysis in solving our battery and lightbulb example in Chapter 1.

The important message here is that it is not necessary to first assemble all the circuit equations, and then solve them all at once. Rather, using a little intuition, it is likely to be much faster to approach the analysis in a different manner. We will have more to say about this in Section 2.4 and in Chapter 3.

### 2.3.3 ENERGY CONSERVATION

Once the branch variables of a circuit have been determined, it is possible to examine the flow of energy through the circuit. This is often a very important part of circuit analysis. Among other things, such an examination should show that energy is conserved in the circuit. This is the case for the circuits in Figures 2.25 and 2.26. Using Equations 2.21 and 2.22 we see that the power into the current source in Figure 2.25 is

$$i_1 v_1 = -RI^2 \tag{2.35}$$

and that the power into the resistor is

$$i_2 v_2 = RI^2. \tag{2.36}$$

The negative sign in Equation 2.35 indicates that the current source actually supplies power.

Similarly, using Equations 2.27 and 2.28 we see that the power into the voltage source in Figure 2.26 is

$$i_1 v_1 = -\frac{V^2}{R} \tag{2.37}$$

and that the power into the resistor is

$$i_2 v_2 = \frac{V^2}{R}. \tag{2.38}$$

In both cases, the power generated by the source is equal to the power dissipated in the resistor. Thus, energy is conserved in both circuits.

Conservation of energy is itself an extremely powerful method for obtaining many types of results in circuits. It is particularly useful in dealing with complicated circuits that contain energy storage elements such as inductors and capacitors that we will introduce in later chapters. Energy methods can often allow us to obtain powerful results without a lot of mathematical grunge. We will use two energy-based approaches in this book.

1. One energy approach equates the energy supplied by a set of elements in a circuit to the energy absorbed by the remaining set of elements in a circuit. Usually, this method involves equating the power generated by the devices in a circuit to the power dissipated in the circuit.
2. Another energy approach equates the total amount of energy in a system at two different points in time (assuming that there are no dissipative elements in the circuit).

We will illustrate the use of the first method using a few examples in this section, and Section 9.5 in Chapter 9 will highlight examples using the second method.

---

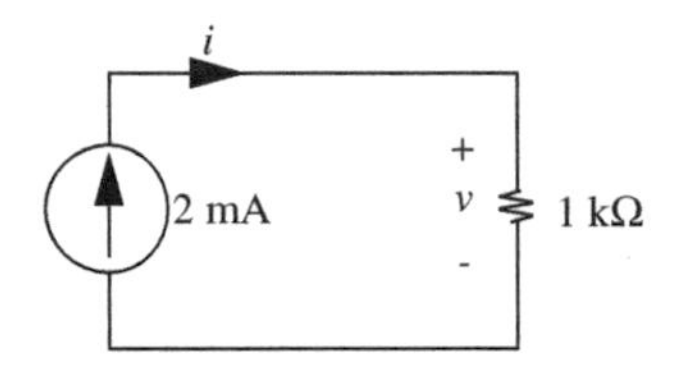


FIGURE 2.28 Energy conservation example.

EXAMPLE 2.14 ENERGY CONSERVATION Determine the value of $v$ in the circuit in Figure 2.28 using the method of energy conservation.

We will show that the mathematical grunge of the basic method can be eliminated using the energy method and some intuition. In Figure 2.28, the current source maintains a current $i = 0.002$ A through the circuit. To determine $v$, we equate the power supplied by the source to the power dissipated by the resistor. Since the current source and the resistor share terminals, the voltage $v$ appears across the current source as well. Thus, the power *into* the source is given by

$$v \times (-0.002) = -0.002v.$$

In other words, the power *supplied* by the source is $0.002v$.

Next, the power *into* the resistor is given by

$$\frac{v^2}{1\text{k}\Omega} = 0.001v^2.$$

Finally, equating the power supplied by the source to the power dissipated by the resistor, we have

$$0.002v = 0.001v^2.$$

In other words, $v = 0.5$ V.

---

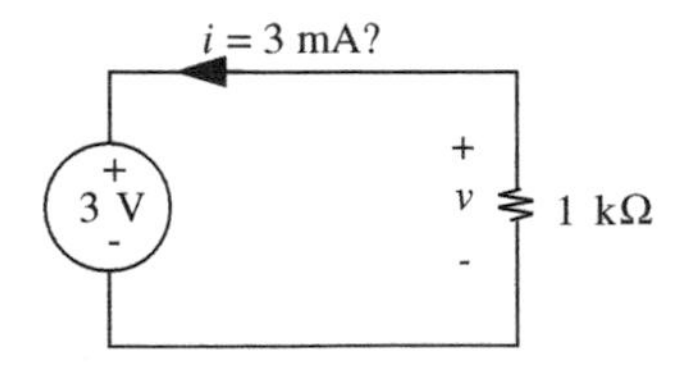


FIGURE 2.29 Another energy conservation example.

EXAMPLE 2.15 USING AN ENERGY-BASED APPROACH TO VERIFY A RESULT A student applies the basic method to the circuit in Figure 2.29 and obtains $i = 3$ mA. Determine whether this answer is correct by using the method of energy conservation.

By energy conservation, the power supplied by the source must be equal to the power dissipated by the resistor. Using the value of the current obtained by the student, the energy dissipated by the resistor is given by

$$i^2 \times 1K = 9 \text{ mW}.$$

The energy *into* the voltage source is given by

$$3 \text{ V} \times 3 \text{ mA} = 9 \text{ mW}.$$

In other words, the energy *supplied* by the source is given by −9 mW. Clearly the energy supplied by the source is not equal to the energy dissipated by the resistor, and so $i = 3$ mA is incorrect. Notice that if we reverse the polarity of $i$, energy will be conserved. Thus, $i = -3$ mA is the correct answer.

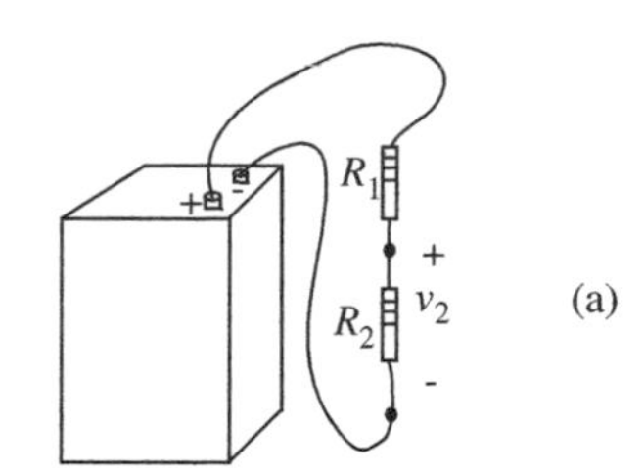


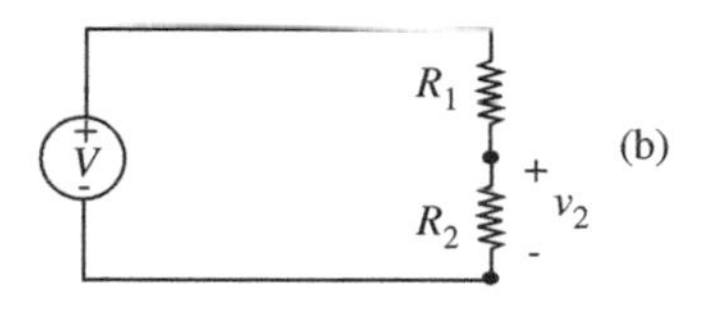


FIGURE 2.30 Voltage-divider circuit.

## 2.3.4 VOLTAGE AND CURRENT DIVIDERS

We will now tackle several circuits called *dividers* that are slightly more complex than the simplest single-loop, two-node, two-element circuits of the previous section. These circuits will comprise a single loop and three or more elements, or two nodes and three or more elements. Dividers produce fractions of input currents or voltages and will be encountered often in subsequent chapters. For the moment, however, they are good examples on which to practice circuit analysis, and we can use them to gain important insight into circuit behavior.

### Voltage Dividers

A voltage divider is an isolated loop that contains two or more resistors and a voltage source in series. A physical voltage divider circuit is illustrated pictorially in Figure 2.30a. We have connected two resistors in series, and connected the pair by some wires to a battery. Such a circuit is useful if we wish to obtain some arbitrary fraction, say 10%, of the battery voltage at the terminals marked $v_2$. To find the relation between $v_2$ and the battery voltage and resistor values, we draw the circuit in schematic form, as shown in Figure 2.30b. We then follow the basic four-step method outlined in Section 2.3 to solve the circuit.

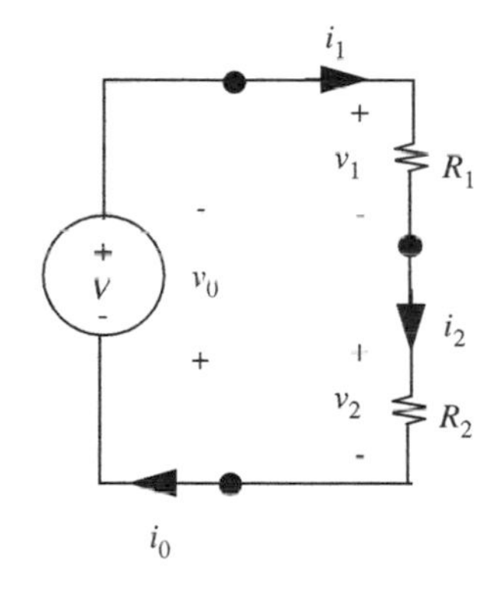


FIGURE 2.31 Assignment of branch variables to the voltage divider.

1. The circuit has three elements, or branches, and hence it will have six branch variables. Figure 2.31 shows one possible assignment of branch variables.

   To find these branch variables, we again assemble the element laws and the appropriate applications of KCL and KVL, and then simultaneously solve the resulting equations.

2. The three element laws are

$$v_0 = -V \tag{2.39}$$

$$v_1 = R_1 i_1 \tag{2.40}$$

$$v_2 = R_2 i_2. \tag{2.41}$$

3. Next, we apply KCL and KVL. The application of KCL to the two upper nodes yields

$$i_0 = i_1 \tag{2.42}$$

$$i_1 = i_2 \tag{2.43}$$

and the application of KVL to the one loop yields

$$v_0 + v_1 + v_2 = 0. \tag{2.44}$$

4. Finally, Equations 2.39 through 2.44 can be solved for the six unknown branch variables. This yields

$$i_0 = i_1 = i_2 = \frac{1}{R_1 + R_2} V \tag{2.45}$$

and

$$v_0 = -V \tag{2.46}$$

$$v_1 = \frac{R_1}{R_1 + R_2} V \tag{2.47}$$

$$v_2 = \frac{R_2}{R_1 + R_2} V. \tag{2.48}$$

This completes the analysis of the two-resistor voltage divider.

From the results of this analysis it should be apparent why the circuit in Figure 2.31 is called a voltage divider. Notice that $v_2$ is some fraction (specifically, $R_2/(R_1+R_2)$) of the source voltage $V$, as desired. The fraction is the ratio of the resistance about which the voltage is measured and the sum of the resistances. By adjusting the relative values of $R_1$ and $R_2$ we can make this fraction adjust anywhere from 0 to 1. If, for example, we wish $v_2$ to be one-tenth of $V$, as suggested at the start of this example, then $R_1$ should be nine times as big as $R_2$.

Notice also that $v_1 + v_2 = V$, and that the two resistors divide the voltage $V$ in proportion to their resistances since $v_1/v_2 = R_1/R_2$. For example, if $R_1$ is twice $R_2$ then $v_1$ is twice $v_2$.

The voltage-divider relationship in terms of conductance can be found from Equation 2.48 by substituting the conductances in place of the resistances:

$$v_2 = \frac{1/G_2}{1/G_1 + 1/G_2} V \tag{2.49}$$

$$= \frac{G_1}{G_1 + G_2} V. \tag{2.50}$$

Hence the voltage-divider relations expressed in terms of conductances involve the conductance *opposite* the desired voltage, divided by the sum of the two conductances.

The simple circuit topology of Figure 2.30 is so common that the *voltage-divider relation* given by Equation 2.48 will become a primitive in our circuit vocabulary. It is helpful to build up a set of such primitives, which are really solved simple cases, to speed up circuit analysis, and to facilitate intuition.

A simple mnemonic: For the voltage $v_2$, take the resistance associated with $v_2$ divided by the sum of the two resistances, multiplied by the voltage applied to the pair.

---

EXAMPLE 2.16 VOLTAGE DIVIDER A voltage divider circuit such as that in Figure 2.30 has $V = 10$ V and $R_2 = 1$ kΩ. Choose $R_1$ such that $v_2$ is 10% of $V$.

By the voltage divider relation of Equation 2.48, we have

$$v_2 = \frac{R_2}{R_1 + R_2} V.$$

For $v_2$ to be 10% of $V$ we must have

$$\frac{v_2}{V} = 0.1 = \frac{R_2}{R_1 + R_2}.$$

For $R_2 = 1$ kΩ, we must choose $R_1$ such that

$$0.1 = \frac{1\ \text{k}\Omega}{R_1 + 1\ \text{k}\Omega}$$

or $R_1 = 9$ kΩ.

---

EXAMPLE 2.17 TEMPERATURE VARIATION Consider the circuit in Figure 2.31 in which $V = 5$ V, $R_1 = 10^3$ Ω, and $R_2 = 10^3\ (1 + T/(500\,°\text{C}))$ Ω, where $T$ is the temperature of the second resistor. Over what range does $v_2$ vary if $T$ varies over the range $-100\,°\text{C} \leq T \leq 100\,°\text{C}$?

Given the temperature range, $R_2$ varies over the range:

$$0.8 \times 10^3 \ \Omega \leq R_2 \leq 1.2 \times 10^3 \ \Omega.$$

Therefore, following Equation 2.48, 2.2 V $\leq v_2 \leq$ 2.7 V, with the higher voltage occurring at the higher temperature.

---

Having determined its branch variables we can now examine the flow of energy through the two-resistor voltage divider. Using Equations 2.45 through 2.48 we see that the power into the source is

$$i_0 v_0 = -\frac{V^2}{R_1 + R_2} \tag{2.51}$$

and that the power into each resistor is

$$i_1 v_1 = \frac{R_1 V^2}{(R_1 + R_2)^2} \tag{2.52}$$

$$i_2 v_2 = \frac{R_2 V^2}{(R_1 + R_2)^2}. \tag{2.53}$$

Since the power into the voltage source is the opposite of the total power into the two resistors, energy is conserved in the two-resistor voltage divider. That is, the power generated by the voltage source is exactly dissipated in the two resistors.

## Resistors in Series

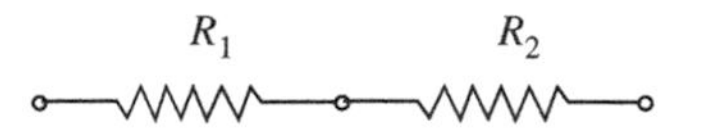


FIGURE 2.32 Resistors in series.

In electronic circuits one often finds resistors connected in series, as shown in Figures 2.31 and 2.32. For example, in our lightbulb example of Chapter 1, suppose the wire had a nonzero resistance, then the current through the wire would be related to the value of several resistances — including those of the wires and the bulb — in series. Our lumped circuit abstraction and the resulting Kirchhoff's laws allow us to calculate the *equivalent resistance* of such combinations using simple algebra.

Specifically, the analysis of the voltage divider shows that two resistors in series act as a single resistor having a resistance $R_S$ equal to the sum of the two individual resistances $R_1$ and $R_2$. In other words, series resistances add.

$$R_S = R_1 + R_2 \tag{2.54}$$

To see this, observe that the voltage source in Figure 2.31 applies the voltage $V$ to two series resistors $R_1$ and $R_2$, and that from Equation 2.43 these resistors respond with the common current $i_1 = i_2$ through their branches. Further, observe from Equation 2.45 that this common current, $i = i_1 = i_2$, is linearly proportional to the voltage from the source. Specifically, from Equation 2.45, the common current is given by

$$i = \frac{1}{R_1 + R_2} V. \tag{2.55}$$

By comparing Equation 2.55 to Equation 1.4, we conclude that for two resistors in series, the equivalent resistance of the pair when viewed from their outer terminals is the *sum* of the individual resistance values. Specifically, if $R_S$ is the resistance of the series resistor pair, then, from Equation 2.55, we find that

$$R_S = \frac{V}{i} = R_1 + R_2. \tag{2.56}$$

This is consistent with the physical derivation of resistance in Equation 1.6 since placing resistors in series essentially increases their combined length.

By substituting their conductances, we can also obtain the equivalent conductance of a pair of conductances in series as

$$\frac{1}{G_S} = \frac{1}{G_1} + \frac{1}{G_2}. \tag{2.57}$$

Simplifying,

$$G_S = \frac{G_1 G_2}{G_1 + G_2}. \tag{2.58}$$

As shown in the ensuing example, we can generalize our result for two series resistors to $N$ resistors in series as:

$$R_S = R_1 + R_2 + R_3 + \cdots R_N. \tag{2.59}$$

Remember this result as another common circuit primitive.

---

EXAMPLE 2.18 AN $N$-RESISTOR VOLTAGE DIVIDER Now consider a more general voltage divider having $N$ resistors, as shown in Figure 2.33. It can be analyzed in the same manner as the two-resistor voltage divider. The only difference is that there are now more unknowns to find, and hence more equations to work with. To begin, suppose we assign the branch variables as shown in Figure 2.33.

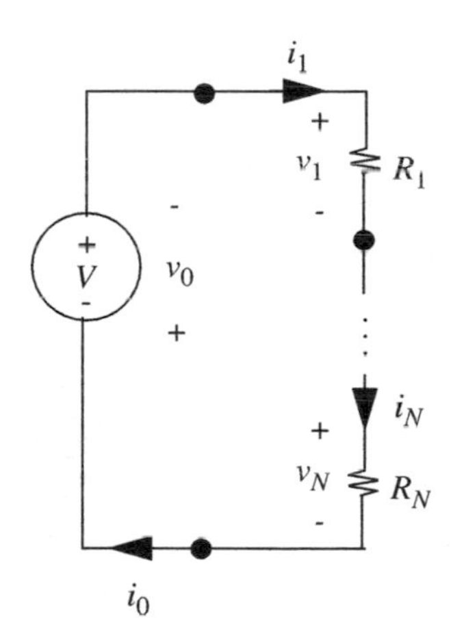

FIGURE 2.33 A voltage divider with $N$ resistors.

The element laws are

$$v_0 = -V \tag{2.60}$$

$$v_n = R_n i_n, \quad 1 \le n \le N. \tag{2.61}$$

Next, the application of KCL to the $N-1$ upper nodes yields

$$i_n = i_{n-1}, \quad 1 \leq n \leq N \tag{2.62}$$

and the application of KVL to the one loop yields

$$v_0 + v_1 + \cdots v_N = 0. \tag{2.63}$$

Finally, Equations 2.60 through 2.63 can be solved to yield

$$i_n = \frac{1}{R_1 + R_2 + \cdots R_N} V, \quad 0 \leq n \leq N \tag{2.64}$$

$$v_0 = -V \tag{2.65}$$

$$v_n = \frac{R_n}{R_1 + R_2 + \cdots R_N} V, \quad 1 \leq n \leq N. \tag{2.66}$$

This completes the analysis.

As was the case for the two-resistor voltage divider, the preceding analysis shows that series resistors divide voltage in proportion to their resistances. This follows from the $R_n$ in the numerator of the right-hand side of Equation 2.66.

Additionally, the analysis again shows that series resistances add. To see this, let $R_S$ be the equivalent resistance of the $N$ series resistors. Then, from Equation 2.64 we see that

$$R_S = \frac{V}{i_n} = R_1 + R_2 + \cdots R_N. \tag{2.67}$$

This result is summarized in Figure 2.34.

Finally, the two voltage-divider examples illustrate an important point, namely that series elements all carry the same branch current because the terminals from these elements are connected end-to-end without connection to additional branches through which the current can divert. This results in the KCL seen in Equations 2.42, 2.43, and 2.62, which state the equivalence of the branch currents.

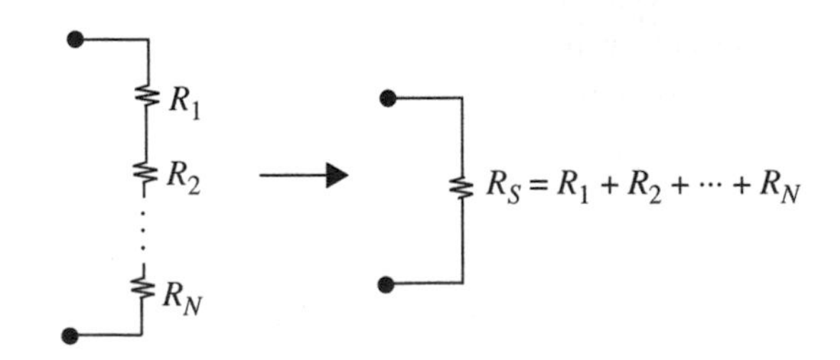


FIGURE 2.34 The equivalence of series resistors.

EXAMPLE 2.19 VOLTAGE-DIVIDER CIRCUIT Determine $v_1$ and $v_2$ for the voltage-divider circuit in Figure 2.35 with $R_1 = 10\ \Omega$, $R_2 = 20\ \Omega$, and $v(t) = 3$ V using (a) the basic method and (b) the results from voltage dividers.

(a) Let us first analyze the circuit using the basic method.

1. Assign variables as in Figure 2.36.
2. Write the constituent relations

$$v_0 = 3\ \text{V} \tag{2.68}$$

$$v_1 = 10 i_1 \tag{2.69}$$

$$v_2 = 20 i_2. \tag{2.70}$$

3. Write KCL

$$i_1 - i_2 = 0. \tag{2.71}$$

4. Write KVL

$$-v_0 + v_1 + v_2 = 0. \tag{2.72}$$

Now eliminate $i_1$ and $i_2$ from Equations 2.69, 2.70, and 2.71, to obtain

$$v_1 = \frac{v_2}{2}. \tag{2.73}$$

Substituting this result and $v_0 = 3$ V into Equation 2.72, we obtain

$$-3\ \text{V} + \frac{v_2}{2} + v_2 = 0. \tag{2.74}$$

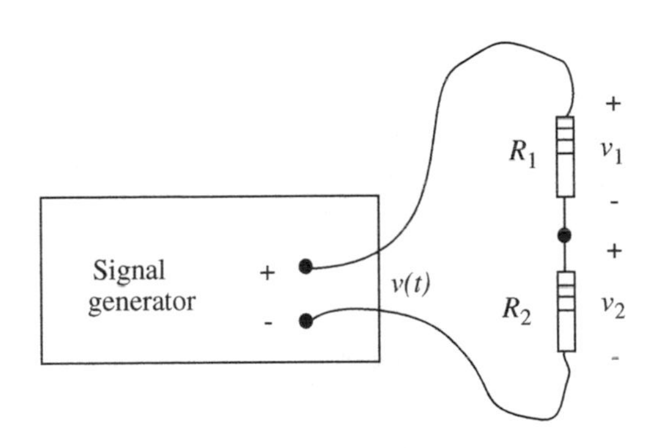


FIGURE 2.35 Voltage-divider circuit.

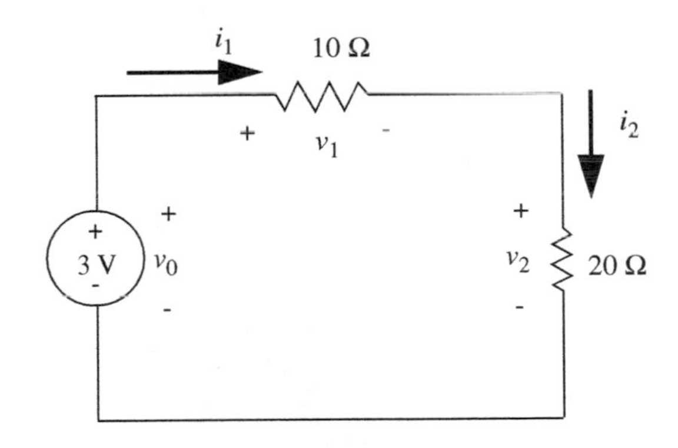


FIGURE 2.36 Voltage divider with variables assigned.

Hence,

$$v_2 = \frac{2}{3}3 \text{ V} = 2 \text{ V} \tag{2.75}$$

and from Equation 2.73, $v_1 = 1$ V.

(b) Using the voltage-divider relation, we can write by inspection the value $v_2$ as a function of the source voltage as follows:

$$v_2 = \frac{20}{10+20}3 \text{ V} = 2 \text{ V}.$$

Similarly,

$$v_1 = \frac{10}{10+20}3 \text{ V} = 1 \text{ V}.$$

## Current Dividers

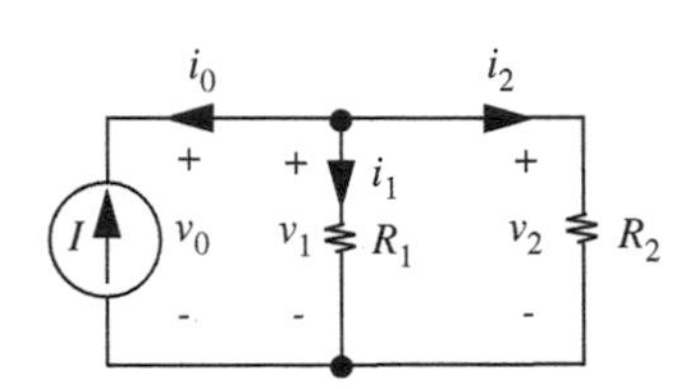


FIGURE 2.37 A current divider with two resistors.

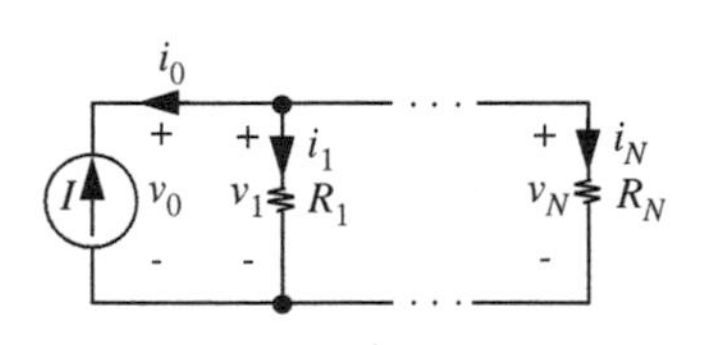


FIGURE 2.38 A current divider with $N$ resistors.

A current divider is a circuit with two nodes joining two or more parallel resistors and a current source. Two current dividers are shown in Figures 2.37 and 2.38, the first with two resistors and the second with $N$ resistors. In these circuits, the resistors share, or divide, the current from the source in proportion to their conductances. It turns out that the equations for voltage dividers comprising voltages and resistances, and those for current dividers comprising currents and conductances, are very similar. Therefore, to highlight the duality between these two types of circuits, we will attempt to mirror the steps from our voltage divider discussion.

Consider the two-resistor current divider shown in Figure 2.37. It has three elements, or branches, and hence six unknown branch variables. To find these branch variables we again assemble the element laws and the appropriate applications of KCL and KVL, and then simultaneously solve the resulting equations. First, the three element laws are

$$i_0 = -I \tag{2.76}$$

$$v_1 = R_1 i_1 \tag{2.77}$$

$$v_2 = R_2 i_2. \tag{2.78}$$

Next, the application of KCL to either node yields

$$i_0 + i_1 + i_2 = 0 \tag{2.79}$$

and the application of KVL to the two internal loops yields

$$v_0 = v_1 \tag{2.80}$$

$$v_1 = v_2. \tag{2.81}$$

Finally, Equations 2.76 through 2.81 can be solved for the six unknown branch variables. This yields

$$i_0 = -I \tag{2.82}$$

$$i_1 = \frac{R_2}{R_1 + R_2} I \tag{2.83}$$

$$i_2 = \frac{R_1}{R_1 + R_2} I \tag{2.84}$$

and

$$v_0 = v_1 = v_2 = \frac{R_1 R_2}{R_1 + R_2} I. \tag{2.85}$$

This completes the analysis of the two-resistor current divider.

The nature of the current division in Equations 2.83 and 2.84 is more obvious if they are expressed in terms of the conductances $G_1$ and $G_2$ where $G_1 \equiv 1/R_1$ and $G_2 \equiv 1/R_2$. With these definitions, $i_1$ and $i_2$ in Equations 2.83 and 2.84 become

$$i_1 = \frac{G_1}{G_1 + G_2} I \tag{2.86}$$

$$i_2 = \frac{G_2}{G_1 + G_2} I. \tag{2.87}$$

It is now apparent that $i_1 + i_2 = I$, and that the two resistors divide the current $I$ in proportion to their conductances since $i_1/i_2 = G_1/G_2$. For example, if $G_1$ is twice $G_2$ then $i_1$ is twice $i_2$.

To summarize our current divider discussion:

The current $i_2$ is equal to the input current $I$ multiplied by a factor, this time made up of the opposite resistor, $R_1$, divided by the sum of the two resistors (see Equation 2.84).

This relation will also become a useful primitive in our analysis vocabulary.

As we did with voltage dividers, we can now examine the flow of energy through the two-resistor current divider. Using Equations 2.82 through 2.85 we see that the power into the source is

$$i_0 v_0 = -\frac{R_1 R_2 I^2}{R_1 + R_2} \tag{2.88}$$

and that the power into each resistor is

$$i_1 v_1 = \frac{R_1 R_2^2 I^2}{(R_1 + R_2)^2} \tag{2.89}$$

$$i_2 v_2 = \frac{R_1^2 R_2 I^2}{(R_1 + R_2)^2}. \tag{2.90}$$

Since the power into the current source is the opposite of the total power into the two resistors, energy is conserved in the two-resistor current divider. That is, the power generated by the current source is exactly dissipated in the two resistors.

### Resistors in Parallel

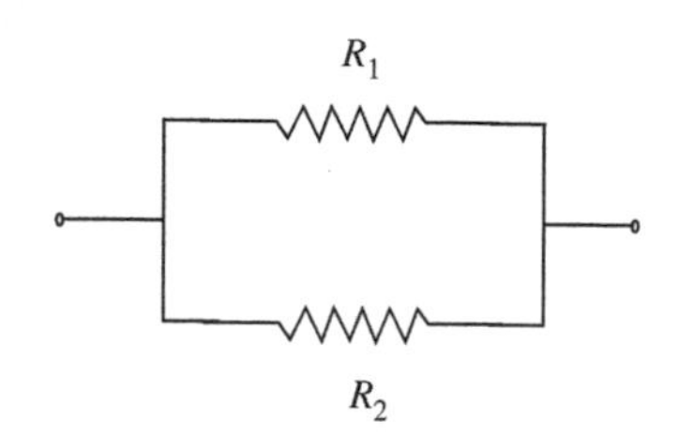


**FIGURE 2.39** Resistors in parallel.

Resistors in parallel occur as commonly as resistors in series. Two resistors in parallel are shown in Figures 2.37 and 2.39.

Our preceding analysis shows that the two resistors in parallel act as a single resistor $R_P$ having a conductance $G_P$ (where $G_P = 1/R_P$) equal to the sum of the two individual conductances. In other words, parallel conductances add:

$$G_P = G_1 + G_2 \tag{2.91}$$

To see this, observe that the current source in Figure 2.37 applies the current $I$ to two parallel resistors, and that from Equation 2.85 these resistors respond at their terminals with the common voltage $v = v_1 = v_2$ that is linearly proportional to the current from the source. Thus, the resistors behave together as a single resistor when viewed from their common terminals.

Let $G_P$ be the conductance of the parallel resistor pair. Then, from Equation 2.85, with the substitution of $G_1 \equiv 1/R_1$ and $G_2 \equiv 1/R_2$, and $v = v_1 = v_2$, we can write

$$v = \frac{R_1 R_2}{R_1 + R_2} I. \tag{2.92}$$

Or, in terms of conductances,

$$v = \frac{1}{G_1 + G_2} I. \tag{2.93}$$

In other words,

$$G_P = \frac{I}{v} = G_1 + G_2. \tag{2.94}$$

Hence, the equivalent conductance of the two parallel resistors is the sum of their individual conductances. This is consistent with the physical derivation of resistance in Equation 1.6 since placing resistors in parallel essentially increases their combined cross-sectional area.

In practice, it is more common to work with resistances than it is to work with conductances, although conductances are sometimes more convenient. For this reason, it is worthwhile to find the equivalent resistance of two parallel resistors in terms of the individual resistances. Let the equivalent resistance be $R_P$. Then, from Equation 2.94 it follows that

$$\frac{1}{R_P} = G_P = G_1 + G_2 = \frac{1}{R_1} + \frac{1}{R_2} \tag{2.95}$$

from which it follows that the equivalent resistance of two resistances in parallel is given by

$$R_P = \frac{R_1 R_2}{R_1 + R_2} \tag{2.96}$$

which is the product of the two resistor values divided by their sum. This relation can also be observed in 2.92, which has a form analogous to Ohm's law (Equation 1.4).

Parallel resistors occur frequently enough to merit a shorthand notation: the two resistor values separated by two parallel vertical lines

$$R_1 \| R_2 = \frac{R_1 R_2}{R_1 + R_2}. \tag{2.97}$$

As we show shortly, we can generalize this result to $N$ resistors connected in parallel. If the equivalent resistance for $N$ resistors connected in parallel is given by $R_P$, then the reciprocal of $R_P$ is given in terms of $R_1, R_2, R_3, \ldots R_N$ as

$$\frac{1}{R_P} = \frac{1}{R_1} + \frac{1}{R_2} + \frac{1}{R_3} + \cdots \frac{1}{R_n}. \tag{2.98}$$

The equivalent resistance of $N$ resistors in parallel is yet another example of a useful primitive that is worth remembering.

The shorthand notation for $N$ resistances in parallel is

$$R_P = R_1 \| R_2 \| R_3 \| \cdots R_N. \tag{2.99}$$

As an example, when $N$ resistors, each with resistance $R$, are connected in parallel, the effective resistance is simply

$$R_P = \frac{R}{N}. \tag{2.100}$$

WWW EXAMPLE 2.20 AN N-RESISTOR CURRENT DIVIDER

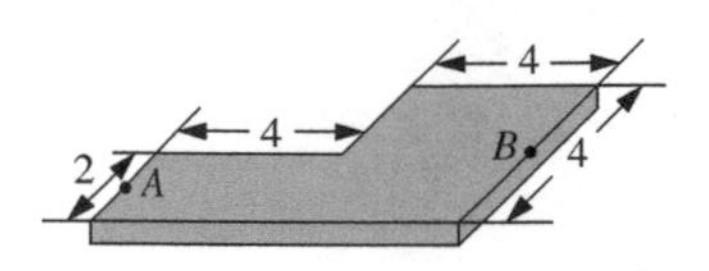


FIGURE 2.41 A VLSI resistor.

EXAMPLE 2.21 PLANAR RESISTOR Figure 2.41 depicts a planar resistor fabricated on a VLSI chip. Suppose $R_{\square} = 10\ \Omega$, find the effective resistance between terminals $A$ and $B$.

Recalling that the resistance of any square piece of the given material is $R_{\square}$, we can view the planar resistor as being composed of three series connected squares, each with resistance $R_{\square}$ as depicted in Figure 2.42.

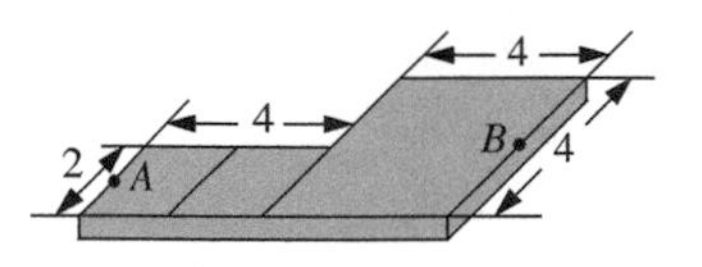


FIGURE 2.42 A VLSI resistor depicted as series connected squares.

Thus the effective resistance between $A$ and $B$ is simply $3R_{\square}$. In practice, however, the resistance of such a piece of material is likely to be larger than $3R_{\square}$ due to fringing effects. Section 1.4 discusses several such effects that limit the accuracy of our lumped circuit model.

EXAMPLE 2.22 EQUIVALENT RESISTANCE Compute the equivalent resistance of the resistor combination shown in Figure 2.43.

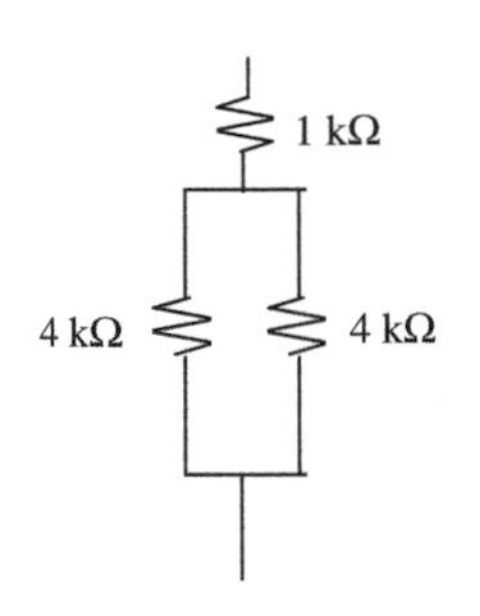


FIGURE 2.43 A series-parallel resistor combination.

Using the series-parallel simplification sequence shown in Figure 2.44, we find the equivalent resistance to be 3 kΩ.

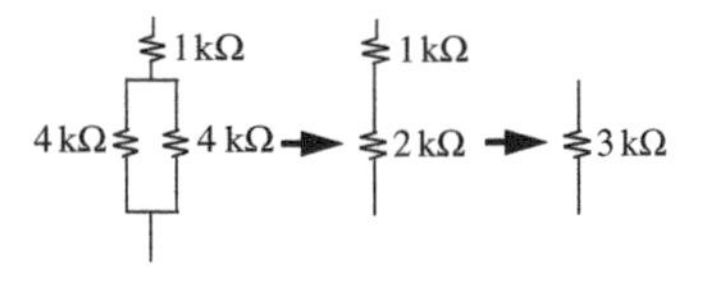


FIGURE 2.44 Equivalent resistance of a series-parallel resistor combination.

EXAMPLE 2.23 EQUIVALENT RESISTANCE COMBINATIONS What equivalent resistors can be made by combining up to three 1000-Ω resistors in series and/or in parallel?

Figure 2.45 shows the possible resistor combinations that use up to three resistors. To determine their equivalent resistance, use the parallel combination result from WWW Equation 2.109 and the series combination result from Equation 2.67. This yields equivalent resistances of: (A) 1000 Ω, (B) 500 Ω, (C) 2000 Ω, (D) 333 Ω, (E) 667 Ω, (F) 1500 Ω, and (G) 3000 Ω.

## 2.3.5 A MORE COMPLEX CIRCUIT

We are now ready to tackle more complex circuits, such as the electrical network shown in Figure 2.1. More specifically, let us suppose that the current $i_4$ is of particular interest to us. This circuit contains two loops and four nodes, and is amenable to our four-step solution procedure.

As our first step, we choose to assign the branch variables as shown in Figure 2.46. Recall that the assignment of voltage and current variables is still arbitrary (other than the constraint of associated variables), and that the solution is invariant under this choice.

As our second step, we write the element laws for each of the elements. The constituent relations for the resistors in this circuit are of the form $v = iR$, and the relation for the voltage source is $v_5 = V$. In terms of the variables defined in Figure 2.46, the constituent relations are

$$v_1 = i_1 R_1 \tag{2.110}$$

$$v_2 = i_2 R_2 \tag{2.111}$$

$$v_3 = i_3 R_3 \tag{2.112}$$

$$v_4 = i_4 R_4 \tag{2.113}$$

$$v_5 = V. \tag{2.114}$$

Our third step involves writing the KVL and KCL equations for the circuit. For KVL, one possible choice of closed paths is shown in Figure 2.47. If we assign the polarity to a voltage in accordance with the first sign encountered, we see that for Loop 1, $v_5$ and $v_2$ are negative, $v_1$ is positive. The corresponding KVL equations are

$$-v_5 + v_1 - v_2 = 0 \tag{2.115}$$

$$+v_2 + v_3 + v_4 = 0. \tag{2.116}$$

A different choice of paths is shown in Figure 2.48. The KVL equations for this choice are derivable from the set we already have; hence (1) they are equally valid, and (2) they contain no new information. It follows that adding

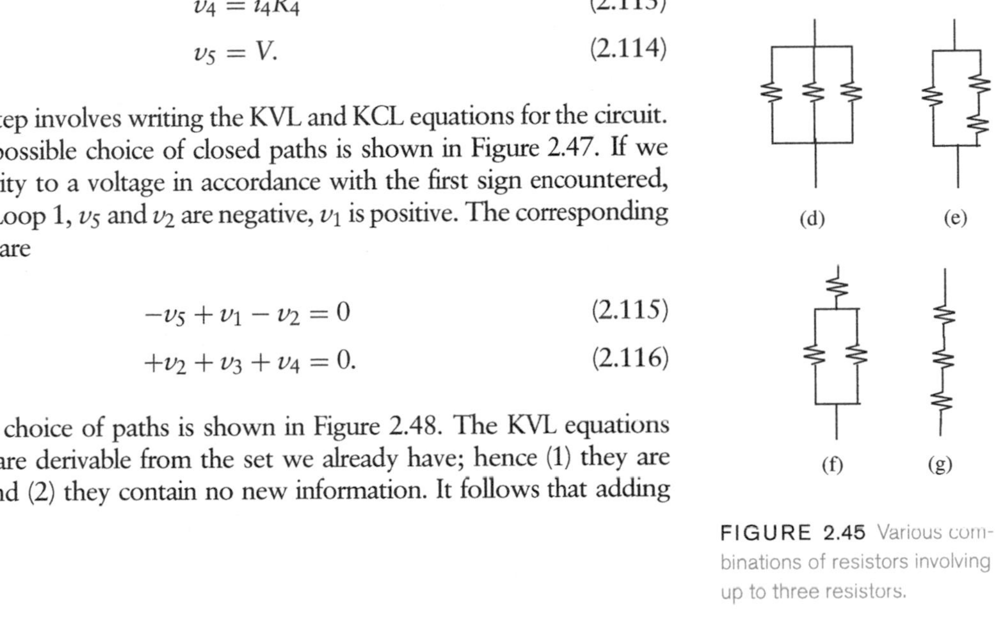


FIGURE 2.45 Various combinations of resistors involving up to three resistors.

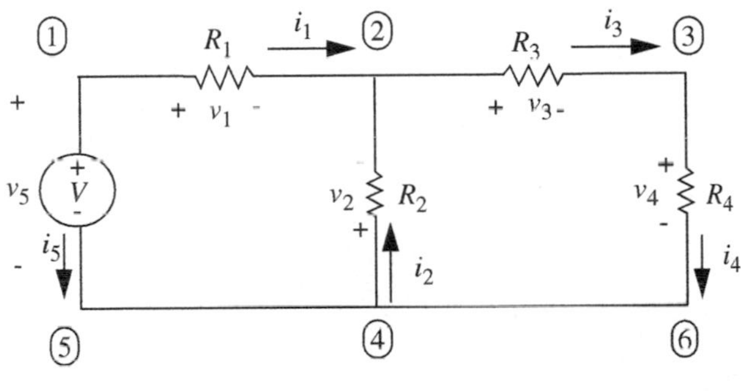


FIGURE 2.46 Assignment of branch variables.

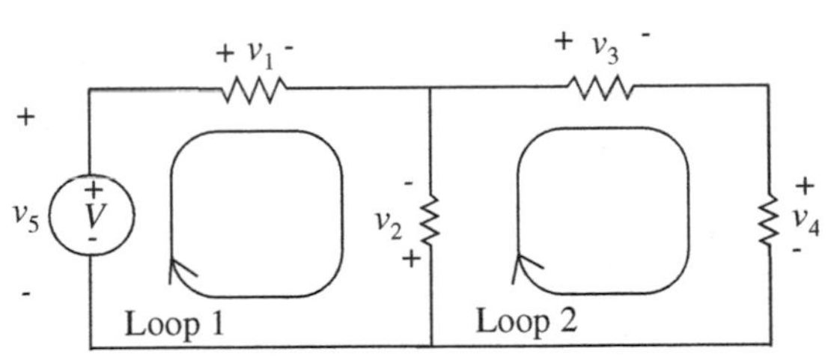


FIGURE 2.47 One choice of closed paths.

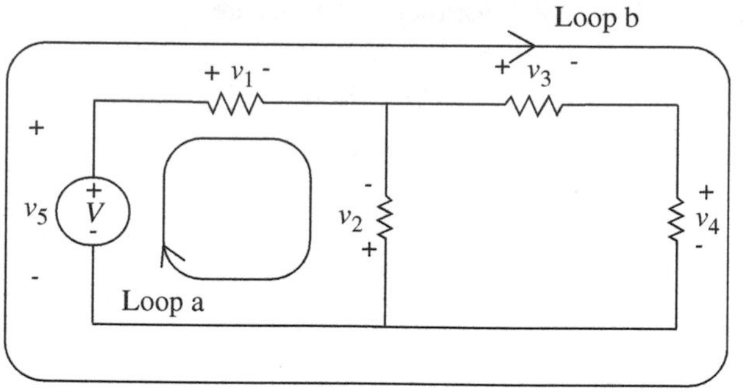

**FIGURE 2.48** Alternative choice of paths.

a third loop (loop b, Figure 2.48) to Figure 2.47 will not yield a KVL equation independent of Equations 2.115 and 2.116.[2]

We now write the KCL equations. From Figure 2.46, at Node 1, KCL yields

$$-i_5 - i_1 = 0 \tag{2.117}$$

and at Node 2

$$+i_1 + i_2 - i_3 = 0 \tag{2.118}$$

and at Node 3

$$i_3 - i_4 = 0. \tag{2.119}$$

As in the case of loops, it is possible to write KCL at Node 4, but the equation is not independent of those we already have.

One might be tempted to write node equations for the junctions labeled 5 and 6, but this doesn't make much sense. The branch between 4 and 6 is a perfect conductor, hence it is really just part of the copper lead attached to resistor $R_4$. For this reason we did not bother defining a separate current variable for this branch. A similar argument applies to branch 4-5.

Another way to emphasize that 5 and 6 are not true nodes is to redraw the circuit as shown in Figure 2.49. Clearly the circuit topology is unchanged in the sense that the interconnections among resistors and source are the same as before, but the false nodes have disappeared. We conclude that a node should be defined as a junction where two or more circuit elements, other than perfect conductors, join together. Whenever a number of circuit elements connect to one perfect conductor, (for example, 5, 4, 6 in Figure 2.46) only *one* node is created.

2. A detailed treatment of the topological issues underlying these rules is contained in Guillemin (*Introductory Circuit Theory*, Will, 1953) or Bose and Stevens (*Introductory Network Theory*, Harper and Row, 1965).

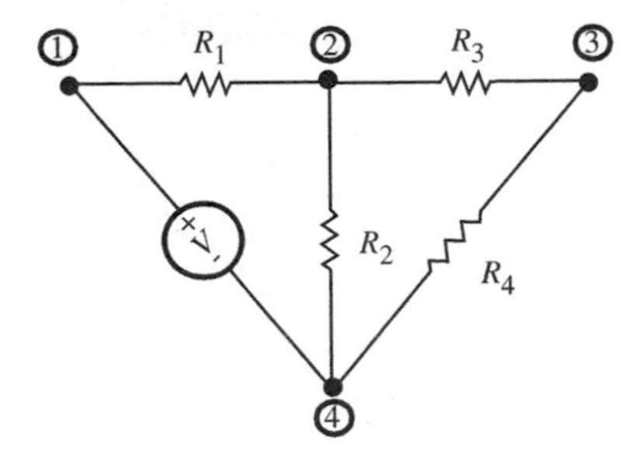


FIGURE 2.49 Circuit in Figure 2.46 redrawn.

We now have ten independent equations (Equations 2.110 through 2.119) and ten unknowns: five voltages and five currents. Thus the equations can be solved for any variable by simple algebra. To find $i_4$, for example, we can first substitute the constituent relations, Equations 2.110 through 2.114 into Equations 2.115 and 2.116:

$$-V + i_1 R_1 - i_2 R_2 = 0 \tag{2.120}$$

$$i_2 R_2 + i_3 R_3 + i_4 R_4 = 0. \tag{2.121}$$

Now eliminating $i_2$ and $i_3$ using Equations 2.118 and 2.119

$$-V + i_1 R_1 + (i_1 - i_4) R_2 = 0 \tag{2.122}$$

$$(-i_1 + i_4) R_2 + i_4 R_3 + i_4 R_4 = 0. \tag{2.123}$$

Rewriting to collect variables and place in the known voltages on the right-hand side of each equation, we obtain

$$i_1 (R_1 + R_2) - i_4 R_2 = V \tag{2.124}$$

$$-i_1 R_2 + i_4 (R_2 + R_3 + R_4) = 0, \tag{2.125}$$

which can be expressed in matrix form as

$$\begin{bmatrix} (R_1 + R_2) & -R_2 \\ -R_2 & (R_2 + R_3 + R_4) \end{bmatrix} \begin{bmatrix} i_1 \\ i_4 \end{bmatrix} = \begin{bmatrix} V \\ 0 \end{bmatrix}. \tag{2.126}$$

The matrix equation is in the form

$$Ax = b$$

where $x$ is a column vector of the unknowns ($i_1$ and $i_4$) and $b$ is the column vector of drive voltages and currents (in this case, just $V$). This vector of unknowns can be solved by using standard linear algebraic techniques.

For example, $i_4$ can be found by applying Cramer's Rule[3]

$$i_4 = \frac{VR_2}{(R_1 + R_2)(R_2 + R_3 + R_4) - R_2^2} \tag{2.127}$$

$$= \frac{VR_2}{R_1R_2 + R_1R_3 + R_1R_4 + R_2R_3 + R_2R_4}. \tag{2.128}$$

With some more effort, we can find the rest of the branch variables as given below

$$-i_5 = i_1 = \frac{R_2 + R_3 + R_4}{R_1(R_2 + R_3 + R_4) + R_2(R_3 + R_4)} V \tag{2.129}$$

$$i_2 = -\frac{R_3 + R_4}{R_1(R_2 + R_3 + R_4) + R_2(R_3 + R_4)} V \tag{2.130}$$

$$i_3 = i_4 = \frac{R_2}{R_1(R_2 + R_3 + R_4) + R_2(R_3 + R_4)} V \tag{2.131}$$

$$v_5 = V \tag{2.132}$$

$$v_1 = \frac{R_1(R_2 + R_3 + R_4)}{R_1(R_2 + R_3 + R_4) + R_2(R_3 + R_4)} V \tag{2.133}$$

$$v_2 = -\frac{R_2(R_3 + R_4)}{R_1(R_2 + R_3 + R_4) + R_2(R_3 + R_4)} V \tag{2.134}$$

$$v_3 = \frac{R_2R_3}{R_1(R_2 + R_3 + R_4) + R_2(R_3 + R_4)} V \tag{2.135}$$

$$v_4 = \frac{R_2R_4}{R_1(R_2 + R_3 + R_4) + R_2(R_3 + R_4)} V. \tag{2.136}$$

This completes our analysis.

Note that in Figure 2.46, resistors $R_1$ and $R_2$ alone do not form a simple voltage divider because of the presence of $R_3$ and $R_4$. It is true, however, that $R_3$ and $R_4$ form a voltage divider. Further, $R_1$ and the net resistance of $R_2, R_3$, and $R_4$ form a second voltage divider.

The analysis of the circuit in Figure 2.46 following the general approach developed in this chapter is both straightforward and tedious, with emphasis on

3. Cramer's Rule is a popular method for solving equations of the type $Ax = b$, where $x$ and $b$ are column vectors, and $A$ is a matrix. See Appendix D for more details.

tedious. Fortunately, as we shall see in Chapter 3, there are much less tedious approaches to this analysis. However, in advance of that we can still simplify the analysis by employing results taken solely from earlier sections of this chapter. Specifically, Section 2.4 shows that we can employ the equivalence of parallel and series resistors, and the behavior of current and voltage dividers, to develop an intuitive and simple approach to solving many types of circuits.

## 2.4 INTUITIVE METHOD OF CIRCUIT ANALYSIS: SERIES AND PARALLEL SIMPLIFICATION

To develop our intuition, let us first illustrate the method with the simple voltage divider in Figure 2.50a. Suppose we were interested in determining the voltage across resistor $R_2$. The figure shows a few important variables marked on it. An intuitive way of analyzing the circuit is to replace the two resistors by the series equivalent, as in Figure 2.50b, then find $i_1$ using Ohm's law. From Equation 2.56, the equivalent series resistance is given by

$$R_S = R_1 + R_2$$

and from Equation 1.4

$$i_1 = V/R_S.$$

Because $i_1$ must be the same in the two circuits (in Figures 2.50a and 2.50b), we can now find $v_2$ from Figure 2.50a

$$v_2 = i_1 R_2 \tag{2.137}$$

$$= \left(\frac{R_2}{R_1 + R_2}\right) V. \tag{2.138}$$

We have now determined the value of $v_2$ in a few simple steps using results from series resistances and Ohm's law. In the future, we will actually write down the result for the voltage divider in a single step by directly applying the voltage divider relation in Equation 2.138.

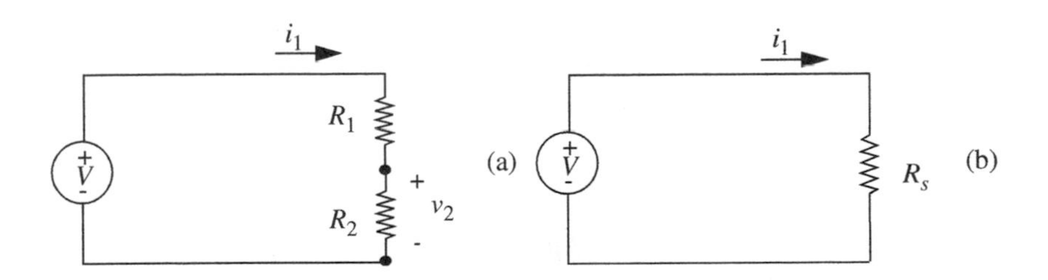


FIGURE 2.50 An intuitive way of analyzing the voltage divider circuit.

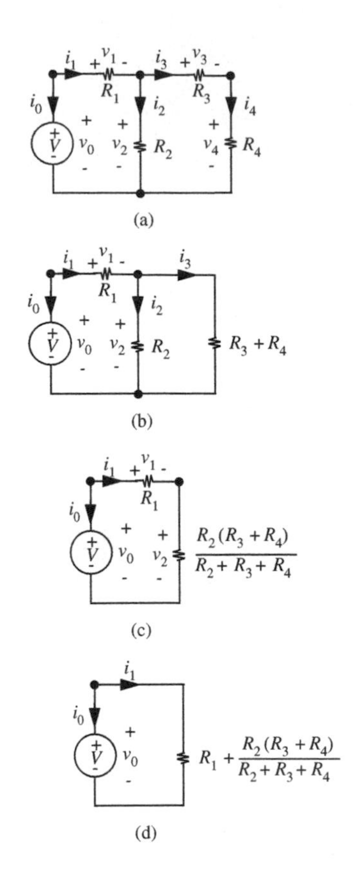


FIGURE 2.51 Collapsing the circuit.

It is worth dwelling on a couple of the "key moves" of our intuitive method. The basic approach is to first collapse, then expand. Notice that our first move was to *collapse* a set of resistances into a single equivalent resistance. Then, we found the current into the equivalent resistance. Finally, we took an *expanded view* of the two resistances to determine the specific voltage of interest.

Let us now use our intuition to develop an alternative method of analyzing the circuit in Figure 2.46 (repeated in Figure 2.51a for convenience). It will be obvious that the intuitive method is far less tedious than the rigorous application of the basic method in Section 2.3.

Our alternative analysis of the circuit in Figure 2.51a follows the two basic moves suggested earlier — first collapse, then expand. Accordingly, our analysis begins by collapsing the circuit using the equivalence of parallel and series resistors. This process is illustrated in Figure 2.51. Note that all branch variables that can be preserved during this collapse are shown in Figure 2.51. First, the series resistors $R_3$ and $R_4$ are combined to yield the circuit in Figure 2.51b. Next, $R_2$ is combined in parallel with the series equivalent of $R_3$ and $R_4$ to yield the circuit in Figure 2.51c. Finally, the two remaining series resistors are combined in series to yield the circuit in Figure 2.51d.

We now analyze our collapsed circuit in Figure 2.51d. Trivially, we know that

$$v_0 = V$$

and

$$i_0 = -i_1.$$

Now, following the results of Section 2.3.1, or equivalently by applying Ohm's law directly, we know that

$$i_1 = \frac{V}{\left(R_1 + \frac{R_2(R_3+R_4)}{R_2+R_3+R_4}\right)}.$$

Thus, at this point, $i_0$, $v_0$, and $i_1$ are known.

Our intuitive analysis concludes by expanding the circuit in Figure 2.51d progressively. As we expand, we determine the values of as many of the variables as we can in terms of previously computed variables. Following this process, first, the circuit in Figure 2.51c can be viewed as a voltage divider of $v_0$. In other words, $i_1$ can be multiplied by each of its two resistances to determine $v_1$ and $v_2$. Thus,

$$v_1 = V\frac{R_1}{\left(R_1 + \frac{R_2(R_3+R_4)}{R_2+R_3+R_4}\right)}$$

and

$$v_2 = V\frac{\frac{R_2(R_3+R_4)}{R_2+R_3+R_4}}{\left(R_1 + \frac{R_2(R_3+R_4)}{R_2+R_3+R_4}\right)}.$$

Next, since $v_2$ is now known, $R_2$ and the series equivalent of $R_3$ and $R_4$ in Figure 2.51b can each be divided into $v_2$ to determine $i_2$ and $i_3$. In other words,

$$i_2 = \frac{v_2}{R_2}$$

$$i_3 = \frac{v_2}{R_3 + R_4}.$$

Alternatively, $i_2$ and $i_3$ can be determined by viewing $R_2$ and the series equivalent of $R_3$ and $R_4$ as a current divider of $i_1$.

Finally, since $i_3$ is now known, $R_3$ and $R_4$ in Figure 2.51a can be viewed as a voltage divider of $v_2$, or they can be multiplied by $i_3 = i_4$, to determine $v_3$ and $v_4$. Doing so yields

$$v_3 = i_3R_3$$

and

$$v_4 = i_3R_4.$$

This completes the intuitive analysis of the circuit in Figure 2.51a. The important observation here is that the alternative approach to circuit analysis outlined in Figure 2.51 is considerably simpler than the direct approach.

---

EXAMPLE 2.24 CIRCUIT ANALYSIS SIMPLIFICATION As another example of circuit analysis simplification, consider the network of twelve resistors shown in Figure 2.52. Each resistor in the network has resistance $R$, and together the network outlines the shape of a cube. Additionally, the network has two terminals marked $A$ and $B$, which extend from opposite corners of the cube to form a port. We wish to determine the equivalent resistance of the network as viewed through this port.

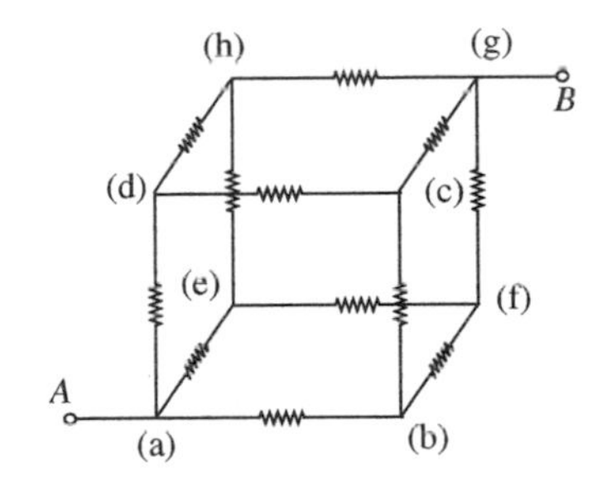


FIGURE 2.52 A cubic resistive network.

To determine the network resistance, we turn the network into a complete circuit by connecting a hypothetical current source to its terminals as shown in Figure 2.53. Note that the circuit in Figure 2.53 is now much the same as the circuit in Figure 2.25; the two circuits differ only in the complexity of the resistive network across the current source. Next, we compute the voltage across the port that appears in response to the application of the current source. The ratio of this voltage to the source current is then the equivalent resistance of the network as viewed through the port. Note that this procedure does work as desired for the circuit in Figure 2.25 since the division of Equation 2.22 by $I$ yields $v_2/I = R$.

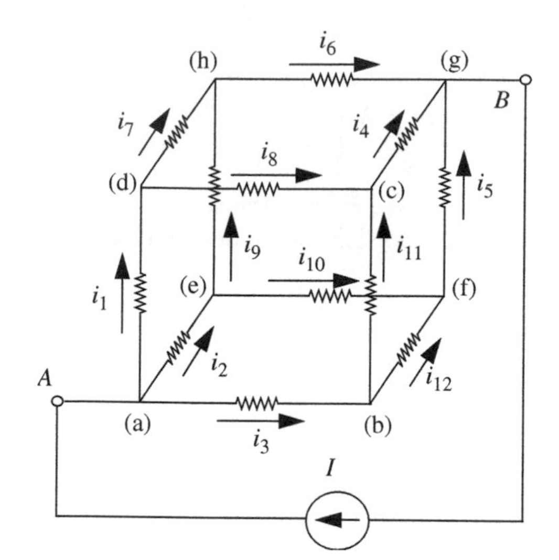


**FIGURE 2.53** Introducing a current into the $A$–$B$ terminal pair of the cubic network.

Turning now to the analysis of the circuit in Figure 2.53, we see that it involves the determination of 26 branch variables, which seems like a painful task. Fortunately, this analysis can be greatly simplified by taking advantage of the symmetry of the circuit, and that is the primary observation to be made here. As a consequence of the symmetry of the circuit, the three branch currents $i_1$, $i_2$, and $i_3$ are identical, as are the three branch currents $i_4$, $i_5$, and $i_6$. Further, KCL applied to the two nodes at the port terminals shows that the sum of each group of three branch currents is $I$, so all six branch currents equal $I/3$.

Next, again due to the symmetry of the circuit, the six branch currents $i_7$ through $i_{12}$ are all identical. Further, KCL applied to any interior node shows that these six branch currents all equal $I/6$. Now, all branch currents in the circuit are known.

So, through their element laws, the branch voltages across all twelve resistors are known, leaving the branch voltage across the current source as the only remaining unknown. Finally through the application of KVL around any loop that passes through the current source we see that its branch voltage is $5RI/6$. Dividing this voltage by $I$ yields $5R/6$ as the equivalent resistance of the cubic network of resistors.

While this solution yields an interesting result, the more important observation is the importance of simplifying a circuit, in this case through symmetry, before attempting its analysis.

---

EXAMPLE 2.25 RESISTANCE OF A CUBIC NETWORK An alternative method for determining the equivalent resistance of the cubic network in Figure 2.52 that uses series-parallel simplifications is now shown. Also suppose that each of the resistors in the network in Figure 2.52 has a resistance of 1 kΩ. Our goal

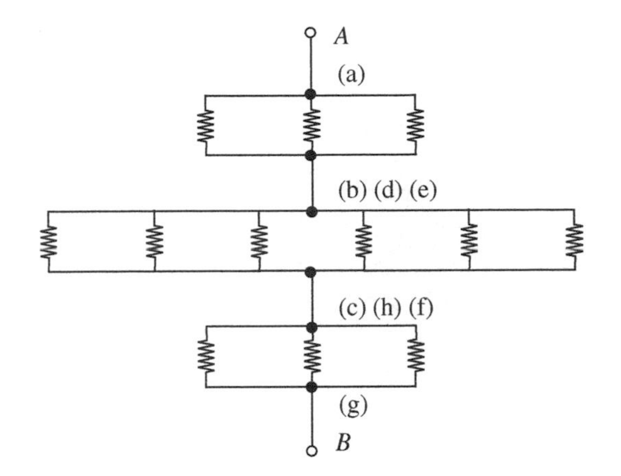


FIGURE 2.54 Simplified network.

is to find the equivalent resistance of this resistive network when looking into the $A$–$B$ terminal pair.

First, observe the symmetry property of this resistive network. From any of the eight vertices, the network looks identical, and therefore, the resistance between any pair of vertices connected by the solid diagonal (for example, (a)-(g), (b)-(h), (e)-(c), etc.) is the same. Furthermore, looking into $A$, the set of paths from $A$ to $B$ starting along the edge (a)-(d) are matched by a set of paths starting along the edge (a)-(b), or by a set of paths starting along the edge (a)-(e). Therefore, when we apply a current $I$ as shown in Figure 2.53, it must split evenly into $i_1$, $i_2$, and $i_3$. Likewise, it draws current off the network evenly, that is, $i_4$, $i_5$, and $i_6$ are the same. Since the same current and resistance causes the same voltage drop across the resistors, we conclude that nodes (b), (d), and (e) have the same voltage, and nodes (h), (c), and (f) also have the same voltage with respect to any reference node.

Notice that if we connect nodes with identical voltages by an ideal wire it does not draw any current and does not change the behavior of the circuit. Therefore, for the purpose of computing the resistance, we can connect all nodes with identical voltages, and simplify the network to the one shown in Figure 2.54.

We can now apply our series and parallel rules to determine the equivalent resistance as

$$1\text{ k}\Omega \| 1\text{ k}\Omega \| 1\text{ k}\Omega + 1\text{ k}\Omega \| 1\text{ k}\Omega \| 1\text{ k}\Omega \| 1\text{ k}\Omega \| 1\text{ k}\Omega \| 1\text{ k}\Omega + 1\text{ k}\Omega \| 1\text{ k}\Omega \| 1\text{ k}\Omega$$

which equals

$$\frac{1}{\frac{1}{1\text{ k}\Omega}+\frac{1}{1\text{ k}\Omega}+\frac{1}{1\text{ k}\Omega}} + \frac{1}{\frac{1}{1\text{ k}\Omega}+\frac{1}{1\text{ k}\Omega}+\frac{1}{1\text{ k}\Omega}+\frac{1}{1\text{ k}\Omega}+\frac{1}{1\text{ k}\Omega}+\frac{1}{1\text{ k}\Omega}} + \frac{1}{\frac{1}{1\text{ k}\Omega}+\frac{1}{1\text{ k}\Omega}+\frac{1}{1\text{ k}\Omega}} = \frac{5}{6}\text{ k}\Omega.$$

---

EXAMPLE 2.26 RESISTOR RATIOS Consider the more involved voltage-divider circuit in Figure 2.55a. The voltage source represents a battery that

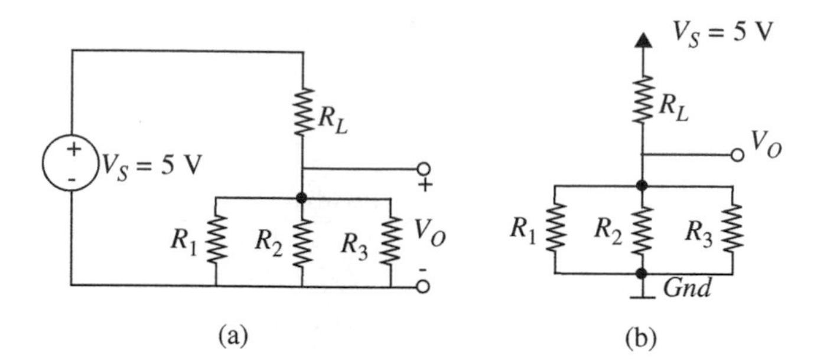


**FIGURE 2.55** Resistor circuit: (a) more complex voltage - divider circuit; (b) shorthand notation.

is supplying power to the rest of the circuit. Further, assume that the voltage $V_O$ is of interest to us. Notice also that the two voltages $V_S$ and $V_O$ share a common negative reference node. A power supply voltage source and a common voltage reference will be encountered so commonly in our circuit language that it is worth creating an idiomatic representation for them.

Figure 2.55b introduces our shorthand notation. First, batteries that serve as power supplies are often not shown explicitly and use the upwards pointing arrow notation instead. $V_S$ represents the power supply voltage. Often, we are also interested in measuring voltages with respect to a common reference node, termed the *ground* node. This node is represented with an upside-down "T" symbol as shown in the figure. The polarity symbols corresponding to voltages that are referenced from this node are not shown explicitly. Rather, the negative symbol is associated with the ground node and the plus symbol is associated with the node adjacent to which the voltage variable appears.[4]

Now, referring to Figure 2.55b, suppose $R_1 = R_2 = R_3 = 10$ kΩ, how do we choose $R_L$ such that $V_O < 1$ V?

The equivalent resistance of the three resistors in parallel is given by Equation 2.98. Thus,

$$R_{eq} = 10\text{k}\|10\text{k}\|10\text{k} = \frac{10}{3}\ \text{k}\Omega.$$

Using the voltage divider relationship, we require that

$$V_O = 5\frac{R_{eq}}{(R_L + R_{eq})} < 1,$$

which implies that $R_L$ has to be at least four times as large as $R_{eq}$. In other words

$$R_L > \frac{40}{3}\ \text{k}\Omega.$$

---

4. Chapter 3 will discuss the concepts of ground nodes as well as node voltages in more detail.

## 2.5 MORE CIRCUIT EXAMPLES

Let us now return to applying the basic method to several other circuits. Consider, for example, the circuit in Figure 2.56, which we will see again in Chapter 3. What is new about this circuit is that it contains two sources. It is not amenable to the intuitive method discussed in Section 2.4. Nonetheless, it can be analyzed by the basic approach presented in Section 2.3.

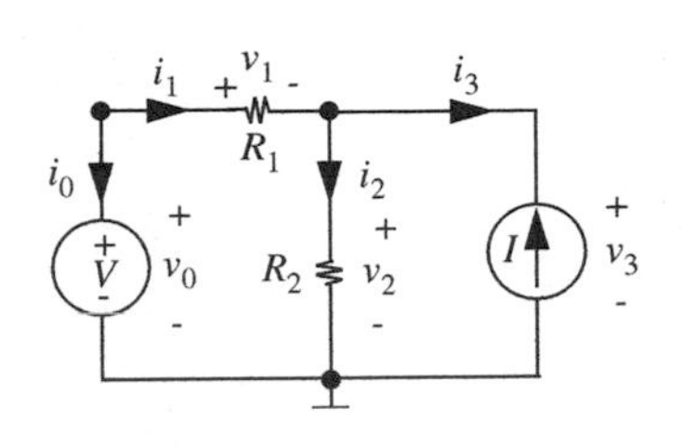


FIGURE 2.56 A circuit with two independent sources.

The element laws for this circuit are

$$v_0 = V \tag{2.139}$$

$$v_1 = R_1 i_1 \tag{2.140}$$

$$v_2 = R_2 i_2 \tag{2.141}$$

$$i_3 = -I. \tag{2.142}$$

Next, the application of KCL to the two upper nodes yields

$$i_0 = -i_1 \tag{2.143}$$

$$i_1 = i_2 + i_3 \tag{2.144}$$

and the application of KVL to the two internal loops yields

$$v_0 = v_1 + v_2 \tag{2.145}$$

$$v_2 = v_3. \tag{2.146}$$

Finally, Equations 2.139 through 2.146 can be solved to yield

$$-i_0 = i_1 = -\frac{R_2}{R_1 + R_2} I + \frac{1}{R_1 + R_2} V \tag{2.147}$$

$$i_2 = \frac{R_1}{R_1 + R_2} I + \frac{1}{R_1 + R_2} V \tag{2.148}$$

$$i_3 = -I \tag{2.149}$$

$$v_0 = V \tag{2.150}$$

$$v_1 = -\frac{R_1 R_2}{R_1 + R_2} I + \frac{R_1}{R_1 + R_2} V \tag{2.151}$$

$$v_2 = v_3 = \frac{R_1 R_2}{R_1 + R_2} I + \frac{R_2}{R_1 + R_2} V \tag{2.152}$$

to complete the analysis.

What is most interesting about the results of this analysis is that each branch variable in Equations 2.147 through 2.152 is a linear combination of a term proportional to $I$ and a term proportional to $V$. This suggests that we could analyze the circuit first with $V = 0$ and second with $I = 0$, and then combine the two analyses to obtain Equations 2.147 through 2.152. This is in fact possible, and it leads to yet further analysis simplifications as we shall see in Chapter 3.

Let us now practice the basic method on several other examples.

---

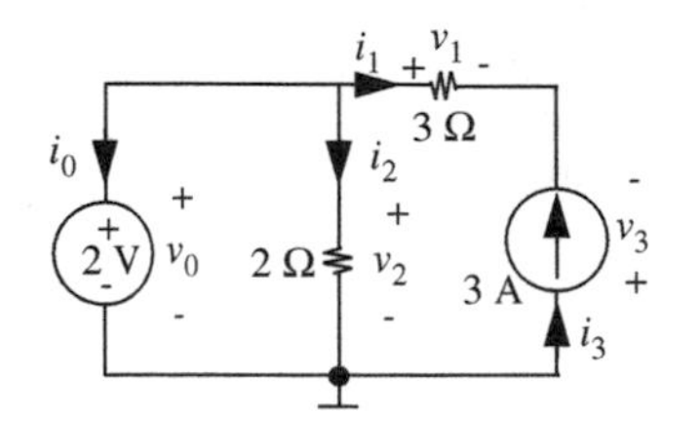


FIGURE 2.57 Another circuit with two independent sources.

EXAMPLE 2.27 CIRCUIT WITH TWO INDEPENDENT SOURCES Analyze the circuit in Figure 2.57 using the basic method. Further, show that energy is conserved in the circuit.

The branch variable assignments are shown in the figure. The element laws for this circuit are

$$v_0 = 2\text{ V}$$

$$v_1 = 3i_1$$

$$v_2 = 2i_2$$

$$i_3 = 3\text{ A}.$$

Applying KCL to the two upper nodes gives us

$$i_0 + i_1 + i_2 = 0$$

$$i_1 = -i_3.$$

Applying KVL to the two internal loops yields

$$v_0 = v_2$$

$$v_2 = -v_3 + v_1.$$

Solving the preceding eight equations, we get $v_0 = 2$ V, $v_1 = -9$ V, $v_2 = 2$ V, $v_3 = -11$ V, $i_0 = 2$ A, $i_1 = -3$ A, $i_2 = 1$ A, and $i_3 = 3$ A.

To show that energy is conserved, we need to compare the power dissipated by the resistors and the power generated by the sources. The power into the resistors is given by

$$(-9\text{ V}) \times (-3\text{ A}) + (2\text{ V}) \times (1\text{ A}) = 29\text{ W}.$$

The power into the sources is given by

$$(2\text{ V}) \times (2\text{ A}) + (-11) \times (3\text{ A}) = -29\text{ W}.$$

It is easy to see that the power dissipated by the resistors equals the power generated by the sources. Thus, energy is conserved.

---

**WWW EXAMPLE 2.28 BASIC CIRCUIT ANALYSIS METHOD**

**EXAMPLE 2.29 DETERMINING THE $I-V$ CHARACTERISTICS OF A CIRCUIT** Determine the $i$–$v$ relationship for the two-terminal device shown in Figure 2.61a. Make a sketch of the $i$–$v$ relationship for $R = 4\ \Omega$ and $V = 5$ V. As shown in the figure, assume that the internals of the device can be modeled as a voltage source in series with a resistor.

We will find the $i$–$v$ relationship of the device by applying some form of excitation to the device terminals and obtaining the relationship between the values of $i$ and $v$. One of the simplest inputs we can apply is a current source providing a current $i_{test}$, as illustrated in Figure 2.61b. The figure also shows the assignment of branch variables.

We will proceed by solving for the branch variables, $v_1$, $i_1$, $v_2$, $i_2$, $v_3$, and $i_3$, and then obtain the $i$–$v$ relationship by expressing $v$ and $i$ in terms of the expressions for the branch variables. Using the basic method, we first write the element laws

$$v_1 = V$$

$$v_2 = i_2 R$$

$$i_3 = -i_{test}.$$

Next, we apply KCL to the two upper nodes

$$i_1 = -i_2$$

$$i_2 = i_3$$

and KVL to the loop

$$v_1 - v_3 - v_2 = 0.$$

These six equations can be solved to yield

$$i_1 = -i_2 = -i_3 = i_{test}$$

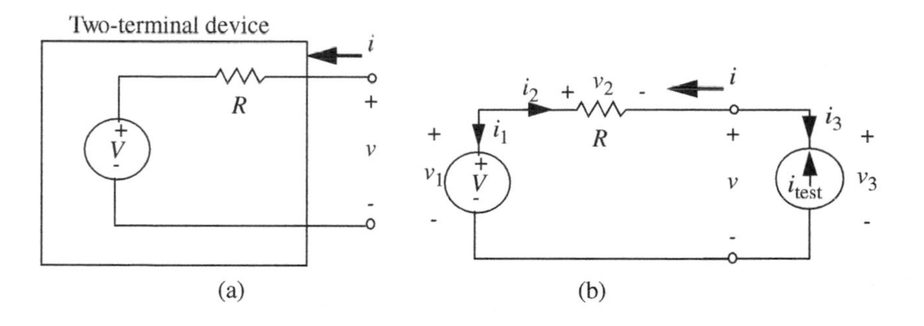


FIGURE 2.61 Determining the $i-v$ characteristics: (a) a two-terminal device; (b) assignment of branch variables to the circuit constructed to determine the $i-v$ characteristics of the device.

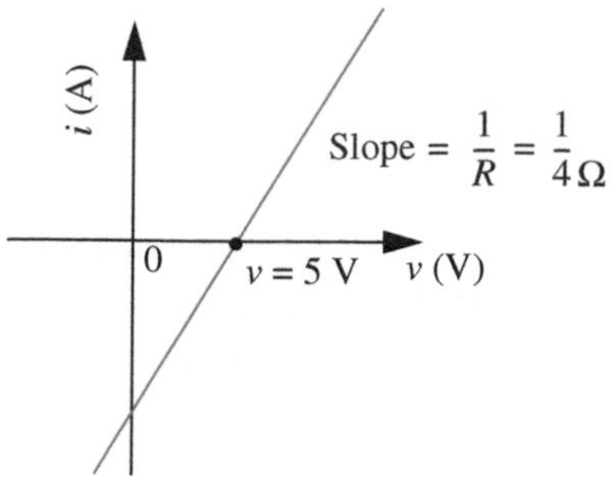

**FIGURE 2.62** A plot of the $i-v$ characteristics for the device.

and

$$v_1 = V, \quad v_2 = -i_{test}R, \quad \text{and} \quad v_3 = V + i_{test}R.$$

We can now write the expression for $v$ as

$$v = v_3 = V + i_{test}R$$

and substituting $i = i_{test}$, we obtain the relationship between $i$ and $v$ as

$$v = V + iR.$$

In other words, the $i$–$v$ relationship is given by

$$i = \frac{v - V}{R}.$$

Substituting $V = 5$ volts and $R = 4\ \Omega$, we get

$$i = \frac{v - 5}{4}.$$

This relationship is plotted in Figure 2.62.

## 2.6 DEPENDENT SOURCES AND THE CONTROL CONCEPT

Section 1.6 introduced the voltage source and the current source as ideal models for energy sources. We call these *independent sources* because their values are independent of circuit operation. But many sources have values that are dependent on, that is, *controlled by* some other parameters in the system. For example, the accelerator pedal in an automobile controls the power delivered by the engine; the handle on a sink faucet controls the flow of water; and room lights can be controlled by either a switch, a binary or two-state device, or a dimmer, a continuous control device. Chapter 6 will introduce another multi-terminal device called the MOSFET in which a control voltage between one pair of terminals of the device determines the MOSFET's behavior between another pair of terminals. Thus, when the multi-terminal dependent source is connected in a circuit, the behavior of the device can be controlled by a voltage or current in some other part of the circuit.

In the examples cited here, only a very small amount of power is needed to control large amounts of power at the output. In the car, for example, a trivial expenditure of energy controls hundreds of horsepower. To idealize, we assume that zero power is required to exercise control; we call this a *dependent source* or *controlled source*. The electrical forms of dependent sources are obvious extensions of the sources we have already seen: a dependent voltage source that can be controlled by some voltage or current, and a dependent current source which likewise has a value determined by some voltage or current. Dependent sources are most commonly used to model elements having more than two terminals.

Figure 2.63 shows an idealized voltage-controlled current source (VCCS). The device in the figure has four terminals. A pair of terminals serve as the control port and another pair of terminals are the output port. In many situations, the control port is also called the input port. Figure 2.63 shows a labeling of the branch variables at the output port $v_{OUT}$ and $i_{OUT}$, and the branch variables at the input port $v_{IN}$ and $i_{IN}$. The value of the voltage $v_{IN}$ across its control input port determines the value of the current $i_{OUT}$ through its output port. In principle, such a dependent source can provide power, but for simplicity the power terminals inherent to the source are not shown.

The diamond shape of the symbol indicates that the device is a dependent source, and the arrow inside indicates that it is a current source. The direction of the arrow indicates the direction of the sourced current and the label next to the symbol indicates the value of the sourced current. In the example in the figure, the sourced current is some function of the voltage $v_{IN}$

$$i_{OUT} = f(v_{IN}).$$

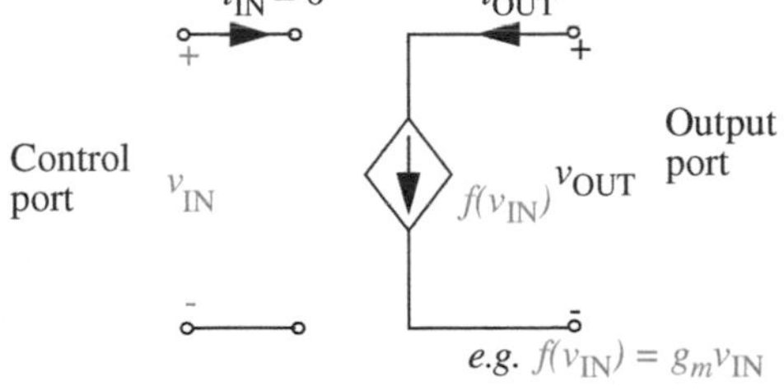

FIGURE 2.63 Voltage-controlled current source.

When the device is connected in a circuit, $v_{IN}$ might be another branch voltage in the circuit.

We often deal with linear dependent sources. A linear voltage-controlled current source is characterized by the equation:

$$i_{OUT} = g v_{IN} \tag{2.173}$$

where $g$ is a constant coefficient. When the dependent source is a voltage-controlled current source, the coefficient $g$ is called the *transconductance* with units of conductance. Notice that Equation 2.173 is the element law for our dependent source expressed as usual in terms of the branch variables. We also need to summarize the behavior of the input port to completely characterize the dependent source. Since our idealized VCCS does not require any power to be supplied at its input, the element law for the input port is simply

$$i_{IN} = 0 \tag{2.174}$$

which is simply the element law for an infinite resistance. For the ideal dependent sources considered in this book, we will assume that the control ports are ideal, that is, they draw zero power.

Figure 2.64 shows a circuit containing our dependent source. For clarity, the dependent source device is shown within the dashed box. In the figure, an independent voltage source (sourcing a voltage $V$) is connected to the control port and a resistor is connected to the output port. For the connection shown, because

$$v_{IN} = V$$

the output current $i_{OUT}$ will be $g$ times the input voltage $V$. We will complete the full analysis of the circuit shortly, and show that the presence of a dependent source does not alter the manner in which our approach to circuit analysis is applied.

Figure 2.65 shows another circuit containing our dependent source. In this circuit, the control port is connected across a resistor with resistance $R_I$. Accordingly, the voltage across $R_I$ becomes the guiding voltage for the dependent source.

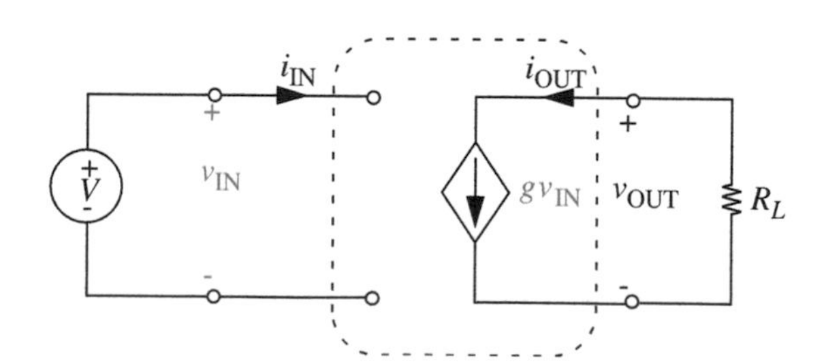


FIGURE 2.64 A circuit containing a voltage-controlled current source.

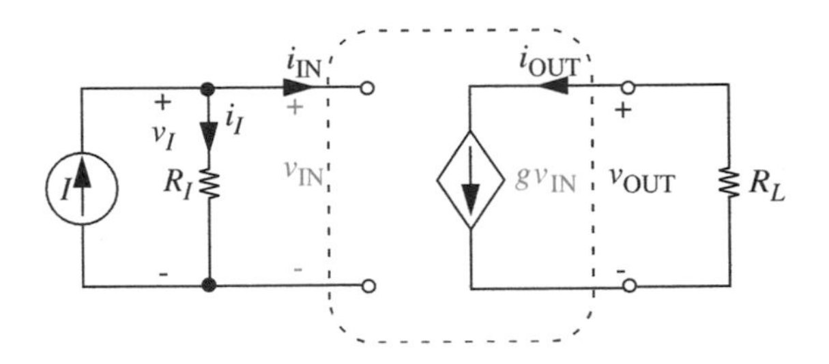


FIGURE 2.65 Another circuit containing a voltage-controlled current source.

Figure 2.66 illustrates the four types of linear dependent sources. Figure 2.66a depicts our now familiar voltage-controlled current source. Figure 2.66b depicts another type of dependent current source whose guiding variable is a branch current. This dependent source is called a current-controlled current source (CCCS).

The element law for the CCCS in Figure 2.66b is

$$i_{OUT} = \alpha i_{IN}. \tag{2.175}$$

The unitless coefficient $\alpha$ is referred to as a *current transfer ratio*. Furthermore, for a CCVS $v_{IN} = 0$.

Figures 2.66c and 2.66d depict the symbols for dependent voltage sources. A dependent voltage source supplies a branch voltage that is a function of some other signal within the circuit. Figure 2.66c shows a voltage-controlled voltage source (VCVS) and Figure 2.66d shows a current-controlled voltage source (CCVS). The guiding variable for a VCVS is a branch voltage, and that for a CCVS is a branch current. The diamond shape of their symbols again indicates that they are dependent sources, and the $\pm$ inside indicates that they are voltage sources. The polarity the $\pm$ indicates the polarity of the sourced voltage and the label next to the symbol indicates the value of the sourced voltage.

In the case of the VCVS in Figures 2.66c, the sourced voltage is equal to $\mu v_{IN}$, where $v_{IN}$ is a voltage across another branch of the circuit and $\mu$ is a unitless coefficient. Thus, the element law for the VCVS in Figure 2.66 is

$$v_{OUT} = \mu v_{IN}. \tag{2.176}$$

The coefficient $\mu$ is referred to as a *voltage transfer ratio*. Furthermore, for a VCVS $i_{IN} = 0$.

In the case of the CCVS in Figures 2.66d, the sourced voltage is equal to $ri_{IN}$, where $i_{IN}$ is the current through another branch of the circuit and $r$ is a coefficient having the units of resistance. Thus, the element law for the CCVS in the figure is

$$v_{OUT} = ri_{IN}. \tag{2.177}$$

The coefficient $r$ is referred to as a *transresistance*.

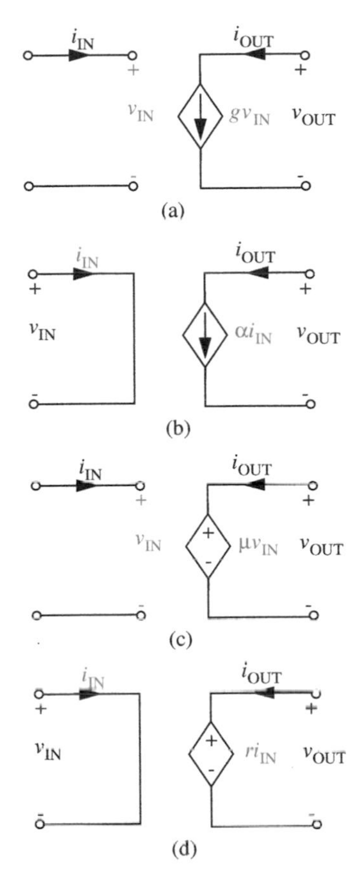


FIGURE 2.66 Four types of dependent sources: (a) VCCS (voltage-controlled current source); (b) CCCS (current-controlled current source); (c) VCVS (voltage-controlled voltage source); (d) CCVS (current-controlled voltage source).

Finally, for both the dependent current source and the dependent voltage source it is once again important to distinguish between the symbols that define them (e.g., $g$) and the branch variables that are defined (e.g., $v_{IN}$, $i_{IN}$, $v_{OUT}$, and $i_{OUT}$) in order to express their element laws. In particular, the branch variable definitions may be reversed for convenience, which will lead to a negation of the corresponding element laws.

## 2.6.1 CIRCUITS WITH DEPENDENT SOURCES

Let us now return to the analysis of our circuit in Figure 2.64, which contains a dependent voltage source. Nonetheless, the circuit can be analyzed by the basic approach presented in this chapter.

Figure 2.67 shows an assignment of the branch variables. The branch variables include $v_0$, $i_0$, $v_{IN}$, $i_{IN}$, $v_{OUT}$, $i_{OUT}$, $v_R$, and $i_R$.

The element laws for this circuit are

$$v_0 = V \tag{2.178}$$

$$i_{IN} = 0 \tag{2.179}$$

$$v_R = R_L i_R \tag{2.180}$$

$$i_{OUT} = g v_{IN}. \tag{2.181}$$

Next, the application of KCL to the two upper nodes yields

$$i_0 = -i_{IN} \tag{2.182}$$

$$i_{OUT} = -i_R \tag{2.183}$$

and the application of KVL to the two loops yields

$$v_0 = v_{IN} \tag{2.184}$$

$$v_R = v_{OUT}. \tag{2.185}$$

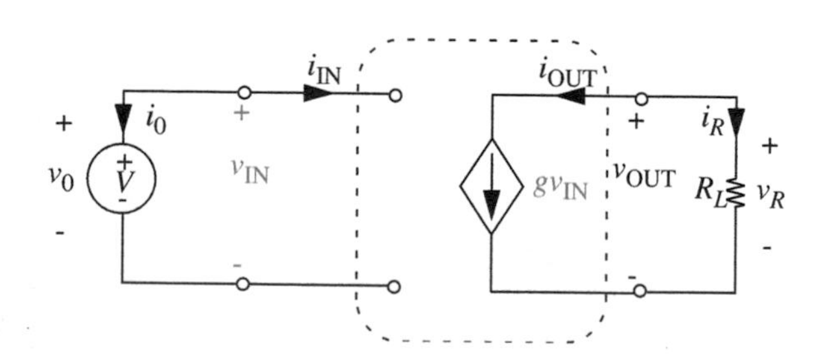


FIGURE 2.67 Assignment of branch variables.

Finally, Equations 2.178 through 2.185 can be solved for the branch variables to yield

$$i_0 = i_{IN} = 0 \tag{2.186}$$

$$v_0 = v_{IN} = V \tag{2.187}$$

$$i_{OUT} = -i_R = -gV \tag{2.188}$$

$$v_R = v_{OUT} = -gVR_L \tag{2.189}$$

to complete the analysis.

The presence of the dependent source in the circuit in Figure 2.67 does not alter the manner in which our approach to circuit analysis is applied. While this is an important observation, there is arguably a more important observation concerning the analysis of the dependent-source circuit, namely that it can proceed in stages. That is, it is possible to first analyze the operation of the "input side" of the circuit, that is, the independent voltage source and the input of the dependent source, and then separately analyze the operation of the "output side," that is, the dependent current source and the resistor $R_L$. We will term this approach the *sequential approach to circuit analysis.*

To see this, observe that the equations representing the input side of the circuit, namely, Equations 2.178, 2.179, 2.182, and 2.184 can be solved trivially by themselves to yield the values of $v_0$, $i_0$, $v_{IN}$, and $i_{IN}$ (see Equations 2.186 and 2.187).

Then, with $v_{IN}$ treated as a known signal, the equations representing the output side of the circuit, namely Equations 2.180, 2.181, 2.183, and 2.185, can be solved by themselves to yield the values of $v_{OUT}$, $i_{OUT}$, $v_R$, and $i_R$ (see Equations 2.188 and 2.189) — a result that is identical to that obtained for the circuit in Figure 2.25.

At this point you are probably wondering why it is that we were able to adopt such a sequential approach to analyzing the circuit in Figure 2.67. The same sequential approach does not work for the circuit in Figure 2.46. The intuition behind this useful property is that our idealized dependent source has decoupled the circuit into two parts — an input part and an output part. Because our dependent source model has an open circuit at its terminals marked by the branch voltage $v_{IN}$, the behavior of the input part is completely independent of the output part of the circuit. In other words, in determining the behavior of the input part, it is as if the output did not even exist. The output part, however, does depend on one of the input variables, namely, $v_{IN}$. However, once the value of the control input $v_{IN}$ is determined through an analysis of the input part, it fixes the value of the dependent source. Thus, the dependent source can be treated as an independent source for the purpose of analyzing the output part.

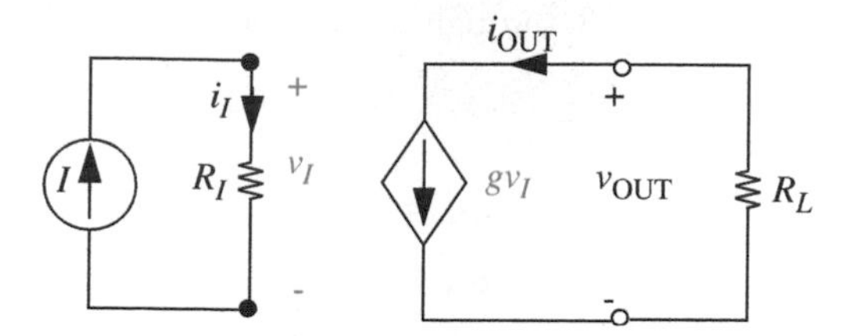


**FIGURE 2.68** The input port of an idealized dependent source is not shown explicitly.

Such a sequential approach to circuit analysis is commonly applied to circuits involving dependent sources, when the circuit does not introduce any external coupling between the control port and the output port of the dependent sources. We will use this approach to advantage in future chapters.

The analysis of circuits with idealized dependent sources admits one other simplification. In an idealized dependent source, the input port (or control port) is an open circuit if the guiding variable is a voltage. Similarly, the input is a short circuit if the guiding variable is a current. Thus, the presence of the input port does not really affect the behavior of the input part of the circuit. The idealized input port is simply present to sample the value of a branch current or voltage without changing the value of the existing branch variable. Therefore, we do not really need to show the input port of the dependent source explicitly, thereby reducing the number of branch variables that we have to deal with.

For example, the input port of the dependent source marked with the branch variables $v_{IN}$ and $i_{IN}$ in Figure 2.65 is an open circuit. Accordingly, $i_{IN} = 0$ and $v_{IN} = v_I$, the voltage across the resistor $R_I$. Therefore, we can equivalently use the circuit in Figure 2.68, where the input port of the dependent source is not shown explicitly, and the current sourced by the dependent source is specified directly in terms of $v_I$, the voltage across the resistor $R_I$. We have thus eliminated the branch variables $v_{IN}$ and $i_{IN}$ from our analysis.

As depicted in Figure 2.69, the same simplification can be made for a dependent source in which the guiding branch variable is a current. Figure 2.69a shows a circuit containing a current controlled current source with the control port marked and all branch variables labeled explicitly. Figure 2.69b shows the

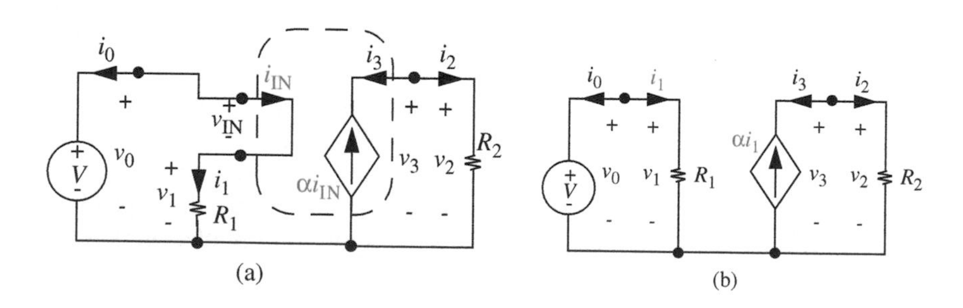


**FIGURE 2.69** Simplifying a circuit with a dependent source by not showing the control port explicitly: (a) with control port marked and branch variables labeled explicitly; (b) with simplification.

same circuit after making the simplification, where the sourced current is now specified in terms of $i_1$. Notice that there is a lot less clutter in the latter figure.

---

EXAMPLE 2.30 CURRENT-CONTROLLED CURRENT SOURCE Consider next the circuit shown in Figure 2.69b. This circuit contains a dependent current source. Notice that we have applied a simplification suggested earlier by not showing the control port of the dependent source explicitly. The current sourced by the dependent source is guided by the current $i_1$.

Let us now analyze this circuit. The branch variables are assigned as shown in Figure 2.69b.

The element laws for this circuit are

$$v_0 = V \tag{2.190}$$

$$v_1 = R_1 i_1 \tag{2.191}$$

$$v_2 = R_2 i_2 \tag{2.192}$$

$$i_3 = -\alpha i_1. \tag{2.193}$$

Next, the application of KCL to the two upper nodes yields

$$i_0 + i_1 = 0 \tag{2.194}$$

$$i_2 + i_3 = 0 \tag{2.195}$$

and the application of KVL to the two loops yields

$$v_0 = v_1 \tag{2.196}$$

$$v_2 = v_3. \tag{2.197}$$

Finally, Equations 2.190 through 2.197 can be solved to yield

$$-i_0 = i_1 = \frac{V}{R_1} \tag{2.198}$$

$$-i_3 = i_2 = \frac{\alpha V}{R_1} \tag{2.199}$$

$$v_0 = v_1 = V \tag{2.200}$$

$$v_2 = v_3 = \frac{\alpha R_2 V}{R_1} \tag{2.201}$$

to complete the analysis.

---

---

EXAMPLE 2.31 INTUITIVE SEQUENTIAL APPROACH FOR THE CCCS Alternatively, we can solve the circuit in Figure 2.69b in a few lines if we use the intuitive sequential approach. Assume that we are interested in finding out the branch variables related to $R_2$.

Using the sequential approach, first, let us tackle the input part of the circuit. Since the voltage $V$ appears across $R_1$, the current through $R_1$ is

$$i_1 = \frac{V}{R_1}.$$

Now let us tackle the output part of the circuit. The current through the current source is in the same direction as $i_2$, and so

$$i_2 = \alpha i_1 = \alpha \frac{V}{R_1}.$$

Applying Ohm's law, we get

$$v_2 = \alpha \frac{VR_2}{R_1}.$$

Not surprisingly, this result is the same as that in Equation 2.201.

---

EXAMPLE 2.32 BRANCH VARIABLES Analyze the circuit in Figure 2.70 and determine the values of all the branch variables. Further, show that energy is conserved in the circuit.

We will analyze the circuit intuitively, applying element laws, KVL and KCL, using the sequential approach. Looking at the input side, since the voltage source appears across an open circuit, it is easy to see that both $v_O$ and $v_{IN}$ are two volts. Similarly, both $i_O$ and $i_{IN}$ are zero. Thus, we have determined all the branch variables at the input side.

Next, let us analyze the output part of the circuit. Since we know the value of $v_{IN}$, the current through the current source is determined as

$$0.001 v_{IN} = 0.002 \text{ A}.$$

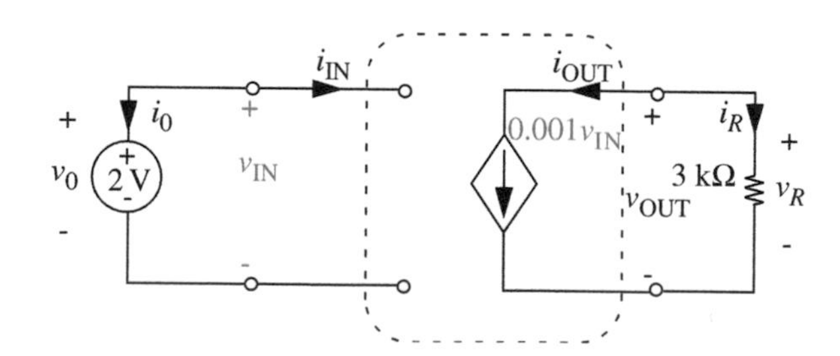


FIGURE 2.70 A circuit containing a voltage-controlled current source.

Since the current source current is in the same direction as $i_{OUT}$, and in the opposite direction as $i_R$, we obtain from KCL

$$i_{OUT} = 0.002 \text{ A}$$

and

$$i_R = -0.002 \text{ A}.$$

Finally, applying the element law for a resistor, we obtain

$$v_R = 3 \times 10^3 i_R = -6 \text{ V}$$

and from KVL, we obtain

$$v_{OUT} = v_R = -6 \text{ V}.$$

This completes our analysis, since all output side branch variables are also known.

To verify that energy is conserved in the circuit, we must show that the power dissipated by the elements is equal to the power supplied. Since the input side current is zero, there is no power dissipated or supplied at the input side. At the output side, the power dissipated in the 3-kΩ resistor is given by

$$3 \text{ k}\Omega \times i_R^2 = 0.012 \text{ W}.$$

The power into the dependent current source is given by

$$v_{OUT} \times i_{OUT} = -6 \times 0.002 = -0.012 \text{ W}.$$

In other words, the power *supplied by* the current source is 0.012 W. Since the power supplied is equal to the power dissipated, energy is conserved.

---

More examples containing dependent sources are given in Section 7.2.

---

WWW EXAMPLE 2.33 VOLTAGE-CONTROLLED RESISTOR

---

## WWW 2.7 A FORMULATION SUITABLE FOR A COMPUTER SOLUTION *

## 2.8 SUMMARY

- KCL is a law stating that the algebraic sum of the currents flowing into any node in a network must be zero.
- KVL is a law stating that the algebraic sum of the voltages around any closed path in a network must be zero.
  A helpful mnemonic for writing KVL equations is to assign the polarity to a given voltage in accordance with the first sign encountered when traversing that voltage around the loop.
- The following is the basic method (or fundamental method or KVL/KCL method) of solving networks:
  1. Define voltages and currents for each element.
  2. Write KVL.
  3. Write KCL.
  4. Write constituent relations.
  5. Solve.
- The series-parallel simplification method is an intuitive method of solving many types of circuits. This approach first collapses a set of resistances into a single equivalent resistance. Then, it successively expands the collapsed circuit and determines the values of all possible branch variables at each step.
- The equivalent resistance for two resistors in series is $R_S = R_1 + R_2$.
- The equivalent resistance of resistors in parallel is $R_P = R_1 \| R_2 = R_1 R_2/(R_1 + R_2)$.
- Voltage divider relation means that when two resistors with values $R_1$ and $R_2$ are connected in series across a voltage source with voltage $V$, the voltage across $R_2$ is given by $\left(R_2/(R_1 + R_2)\right) V$.
- Current divider relation means that when two resistors with values $R_1$ and $R_2$ are connected in parallel across a current source with current $I$, the current through $R_2$ is given by $\left(R_1/(R_1 + R_2)\right) I$.
- This chapter discussed four types of dependent sources: voltage-controlled current sources (VCCS), current-controlled current sources (CCCS), voltage-controlled voltage sources (VCVS), and current-controlled voltage sources (CCVS).
- The sequential method of circuit analysis is an intuitive approach that can often be applied to circuits containing dependent sources when the control port of the dependent source is ideal. This approach first analyzes the circuit on the input side of the dependent source, and then separately analyzes the operation of the output side of the dependent source.

- Conservation of energy is a powerful method for obtaining many types of results in circuits. Energy methods are intuitive and can often allow us to obtain powerful results without a lot of mathematical grunge. One energy approach equates the energy supplied by a set of elements in a circuit to the energy absorbed by the remaining set of elements in a circuit. Another energy approach equates the total amount of energy in a system at two different points in time (assuming that there are no dissipative elements in the circuit).

## EXERCISES

EXERCISE 2.1 Find the equivalent resistance from the indicated terminal pair of the networks in Figure 2.72.

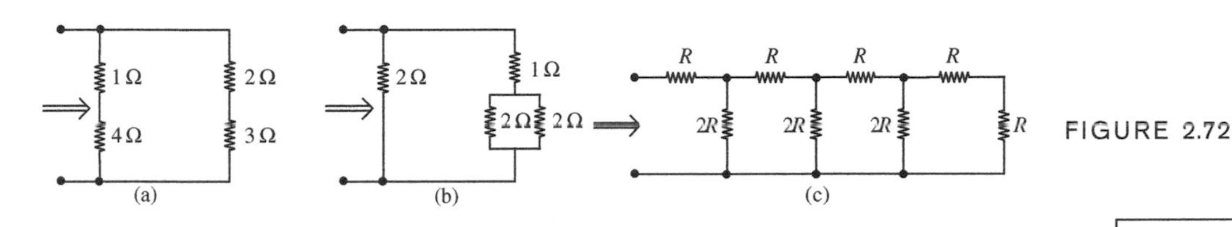


FIGURE 2.72

EXERCISE 2.2 Determine the voltages $v_A$ and $v_B$ (in terms of $v_S$) for the network shown in Figure 2.73.

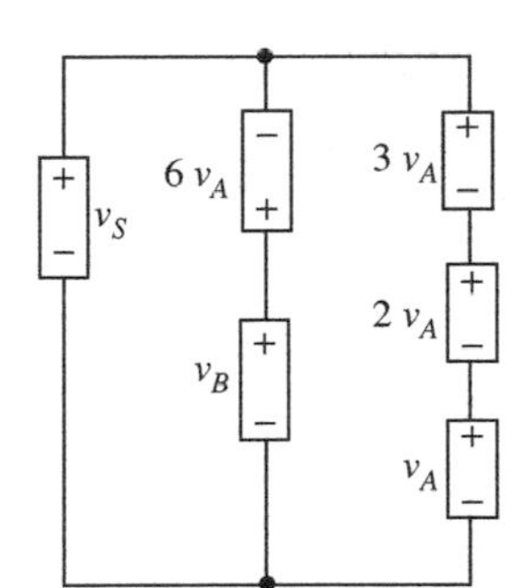


FIGURE 2.73

EXERCISE 2.3 Find the equivalent resistance between the indicated terminals (all resistances in ohms) in Figure 2.74.

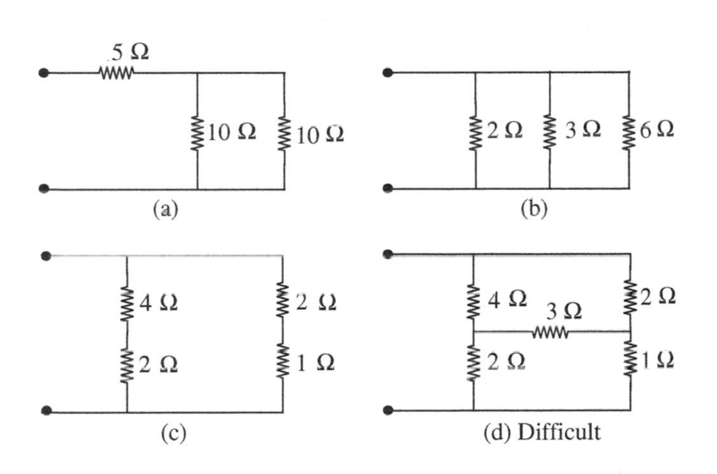


FIGURE 2.74

EXERCISE 2.4 Determine the indicated branch voltage or branch current in each network in Figure 2.75.

EXERCISE 2.5 Find the equivalent resistance at the indicated terminal pair for each of the networks shown in Figure 2.76.

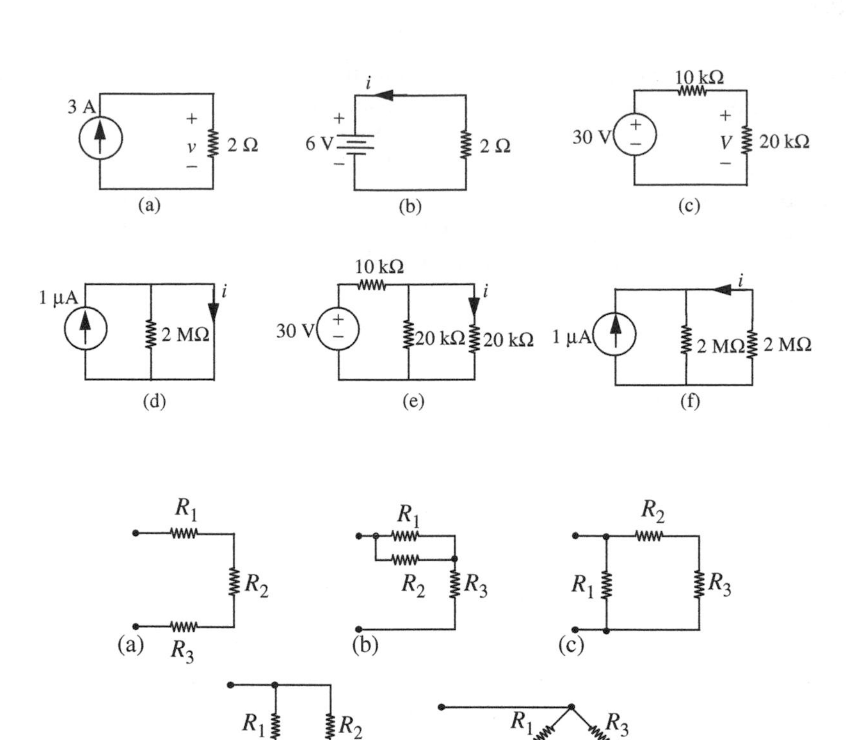


FIGURE 2.75

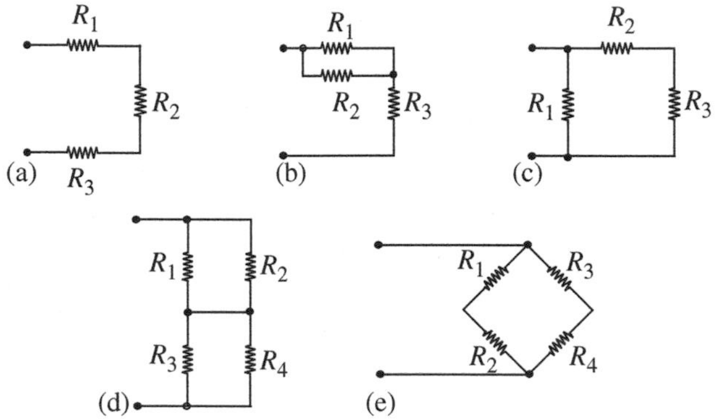

FIGURE 2.76

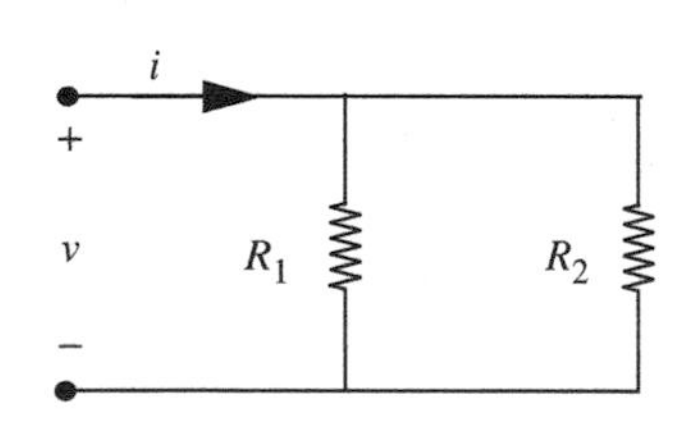


FIGURE 2.77

EXERCISE 2.6 In the circuit in Figure 2.77, $v$, $i$, and $R_1$ are known. Find $R_2$.

$$v = 5 \text{ V}$$

$$i = 40 \ \mu\text{A}$$

$$R_1 = 150 \text{ k}\Omega$$

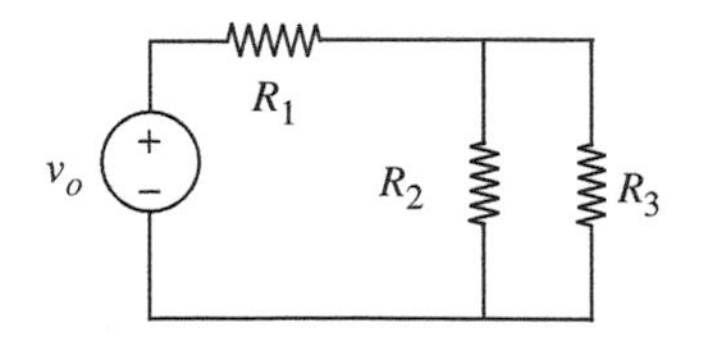


FIGURE 2.78

EXERCISE 2.7 In the circuit in Figure 2.78, $v_o = 6$ V, $R_1 = 100\ \Omega$, $R_2 = 25\ \Omega$, and $R_3 = 50\ \Omega$. Which of the resistors if any, are dissipating less than 1/4 watt?

EXERCISE 2.8 Sketch the $i - v$ characteristics for the networks in Figure 2.79. Label intercepts and slopes.

EXERCISE 2.9

a) Assign branch voltages and branch current variables to each element in the network in Figure 2.80. Use associated reference directions.

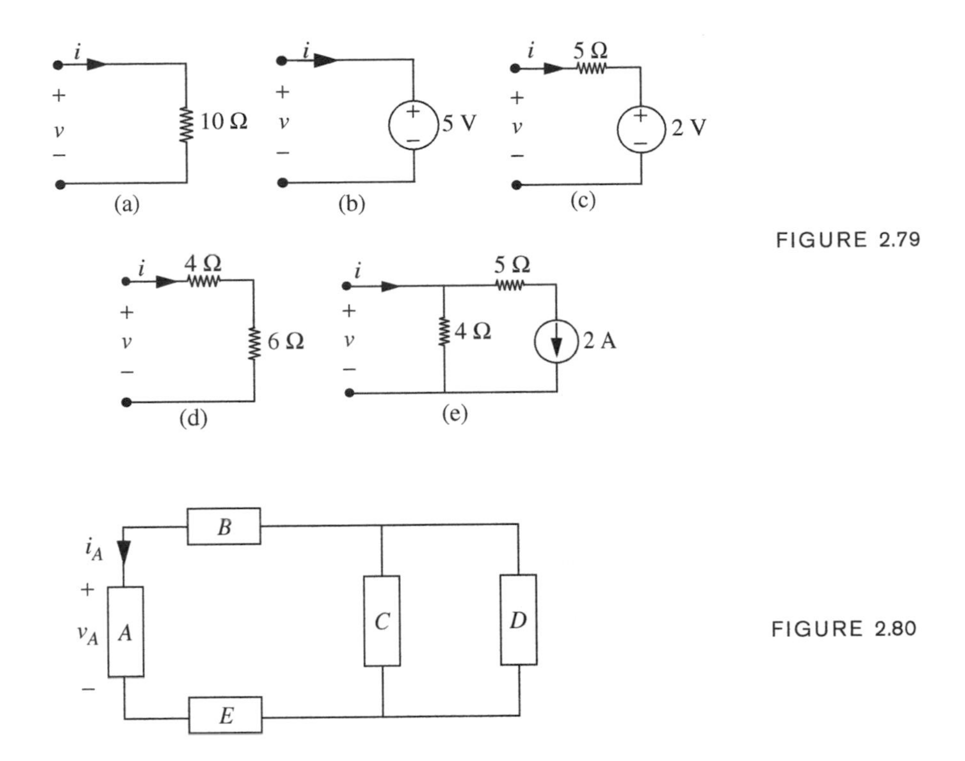


FIGURE 2.79

FIGURE 2.80

b) How many linearly independent KVL equations can be written for this network?

c) How many linearly independent KCL equations can be written for this network?

d) Formulate a set of KVL and KCL equations for the network.

e) Assign nonzero numbers to each branch current such that your KCL equations are satisfied.

f) Assign nonzero numbers to each branch voltage such that your KVL equations are satisfied.

g) As a check on your result, you can draw on the fact that power is conserved in a network that obeys KVL and KCL. Therefore calculate the quantity $\sum v_n i_n$. It should be zero.

EXERCISE 2.10 A portion of a larger network is shown in Figure 2.81. Show that the algebraic sum of the currents into this portion of the network must be zero.

## PROBLEMS

PROBLEM 2.1 A pictorial diagram for a flashlight is shown in Figure 2.82. The two batteries are identical, and each has an open-circuit voltage of 1.5 V. The lamp

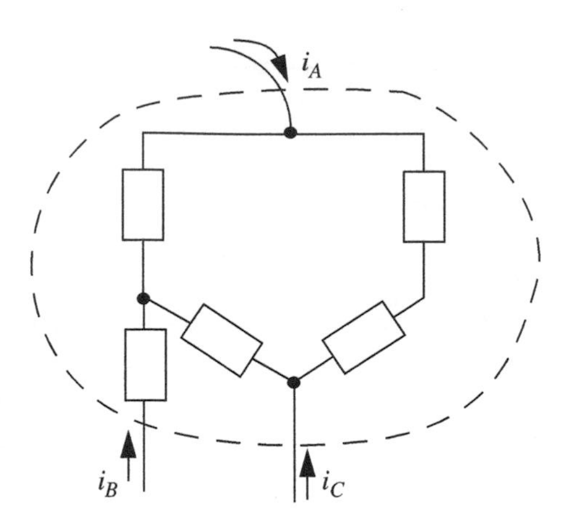


FIGURE 2.81

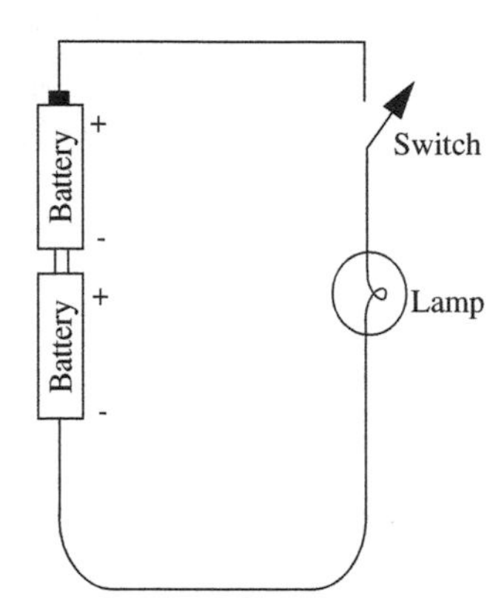


FIGURE 2.82

has a resistance of 5 Ω when lit. With the switch closed, 2.5 V is measured across the lamp. What is the internal resistance of each battery?

PROBLEM 2.2 Determine the current $i_O$ in the circuit in Figure 2.83 by working with resistors in series and parallel.

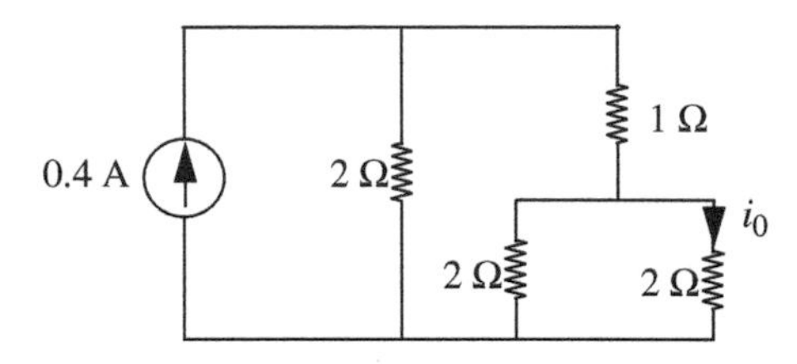


FIGURE 2.83

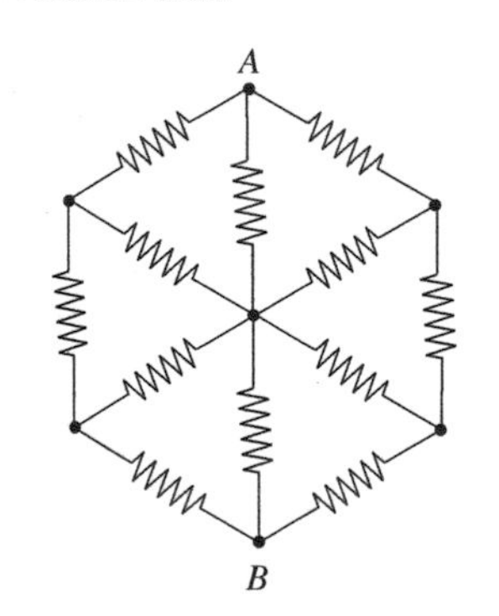


FIGURE 2.84

PROBLEM 2.3 Find the resistance between nodes $A$ and $B$ in Figure 2.84. All resistors equal 1 Ω.

PROBLEM 2.4 For the circuit in Figure 2.85, find values of $R_1$ to satisfy each of the following conditions:

a) v = 3 V

b) v = 0 V

c) i = 3 A

d) The power dissipated in $R_1$ is 12 W.

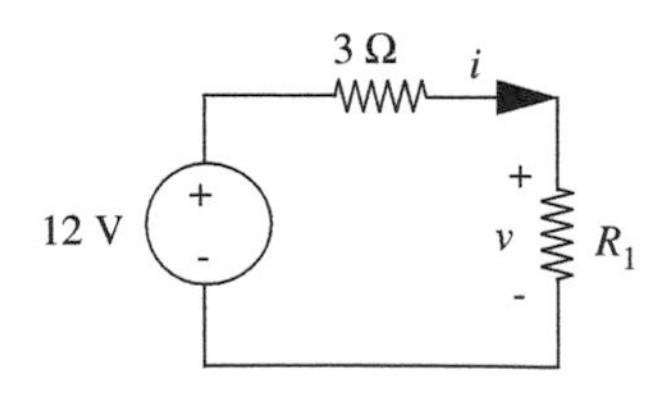


FIGURE 2.85

PROBLEM 2.5 Find the equivalent resistance $R_T$ at the indicated terminals for each of the networks in Figure 2.86.

PROBLEM 2.6 In each network in Figure 2.87, find the numerical values of the indicated variables (units are amperes, volts, and ohms).

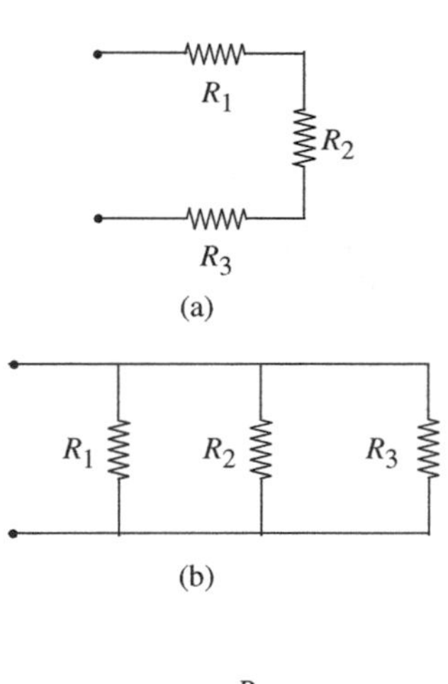


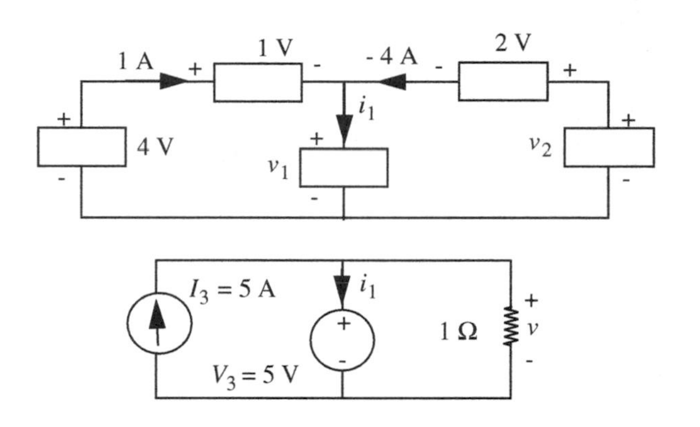


FIGURE 2.87

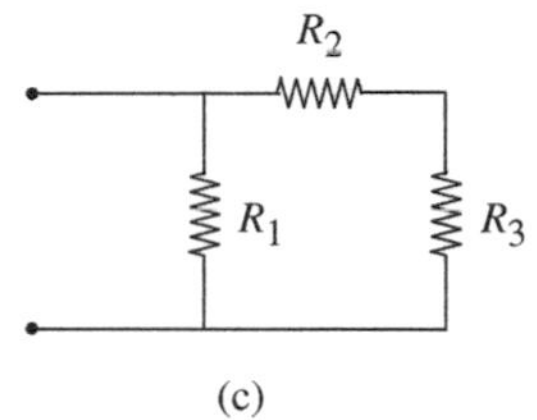


PROBLEM 2.7 For the circuit in Figure 2.88, determine the current $i_3$ explicitly in terms of all circuit parameters.

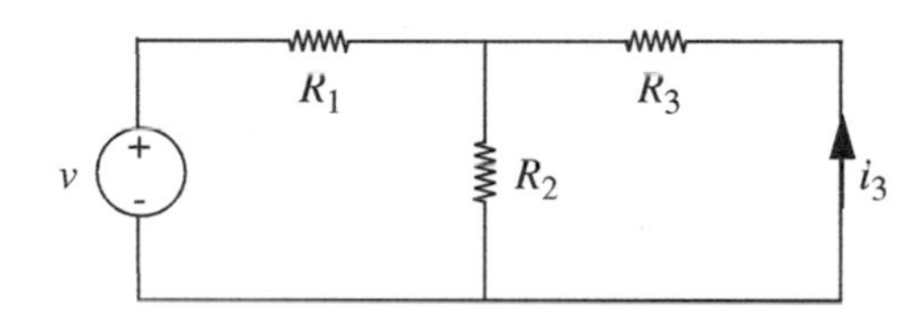


FIGURE 2.88

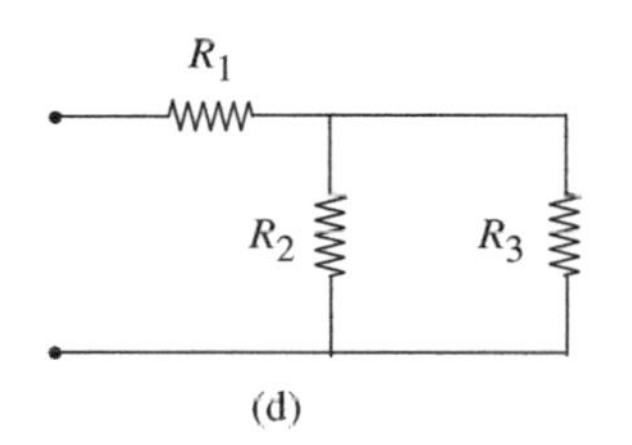


PROBLEM 2.8 Determine explicitly the voltage $v_3$ in the circuit in Figure 2.89.

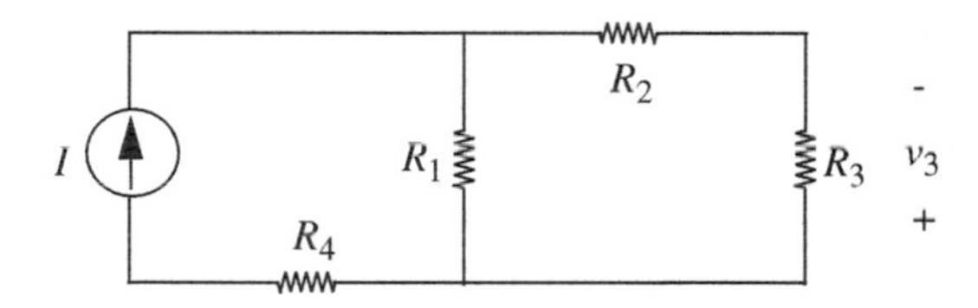


FIGURE 2.89

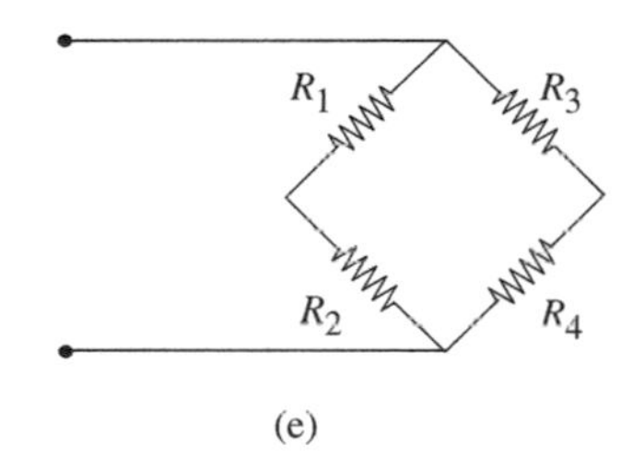


FIGURE 2.86

PROBLEM 2.9 Calculate the power dissipated in the resistor $R$ in Figure 2.90.

PROBLEM 2.10 Design a resistor attenuator to make $v_o = v_i/1000$, using the circuit configuration given in Figure 2.91, and resistor values available in your lab. This problem is underconstrained so it has many answers.

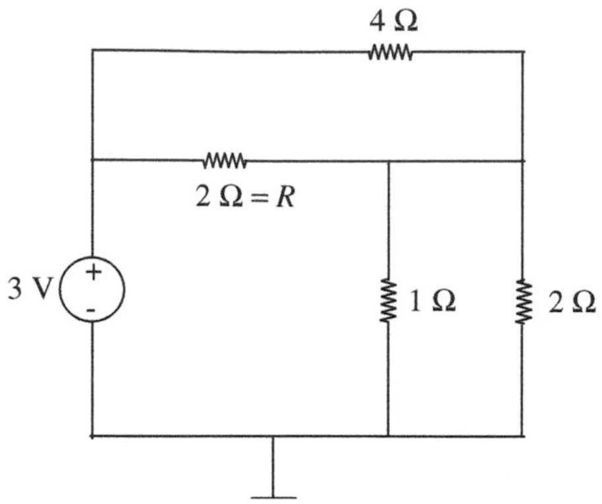

FIGURE 2.90

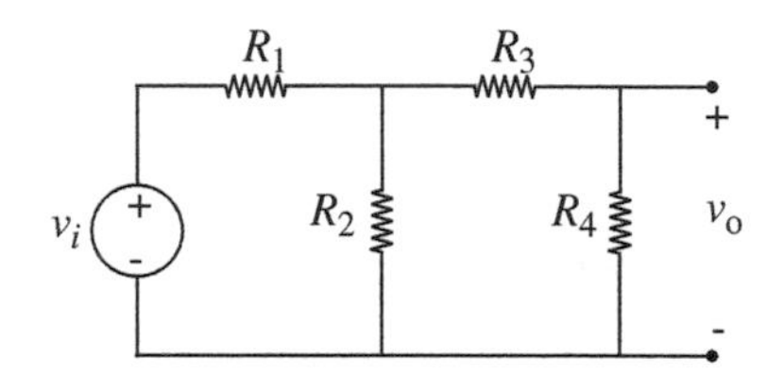


FIGURE 2.91

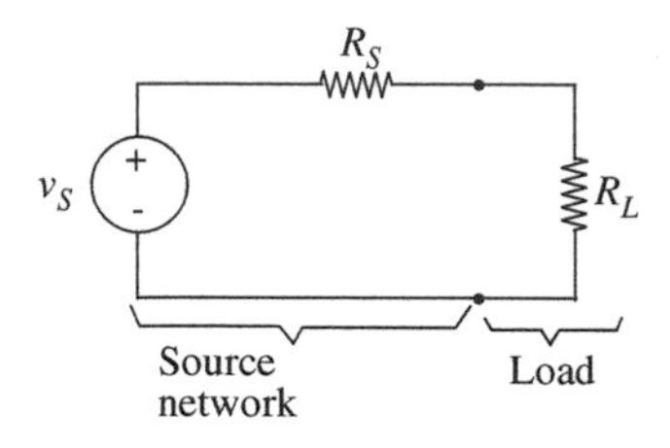


FIGURE 2.92

PROBLEM 2.11 Consider the network in Figure 2.92 in which a non-ideal battery drives a load resistor $R_L$. The battery is modeled as a voltage source $V_S$ in series with a resistor $R_S$. The following are some proofs about power transfer:

a) Prove that for $R_S$ variable and $R_L$ fixed, the power dissipated in $R_L$ is maximum when $R_S = 0$.

b) Prove that for $R_S$ fixed and $R_L$ variable, the power dissipated in $R_L$ is maximum when $R_S = R_L$ ("matched resistances").

c) Prove that for $R_S$ fixed and $R_L$ variable, the condition that maximizes the power delivered to the load $R_L$ requires that an equal amount of power be dissipated in the source resistance $R_S$.

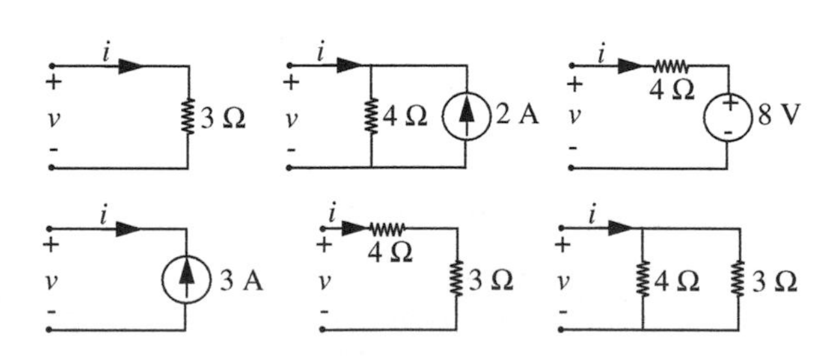


FIGURE 2.93

PROBLEM 2.12 Sketch the $v - i$ characteristics for the networks in Figure 2.93. Label intercepts and slopes.

PROBLEM 2.13

a) Find $i_1$, $i_2$, and $i_3$ in the network in Figure 2.94. (Note that $i_3$ does not obey the standard convention for current direction.)

b) Show that energy is conserved in this network.

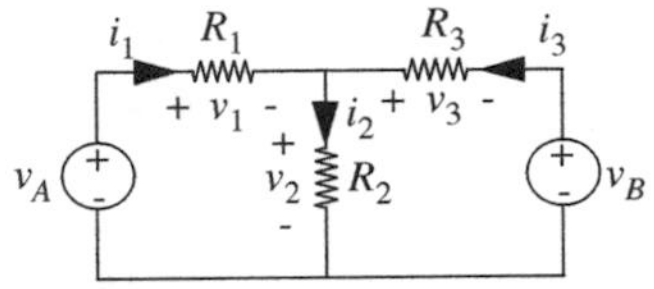

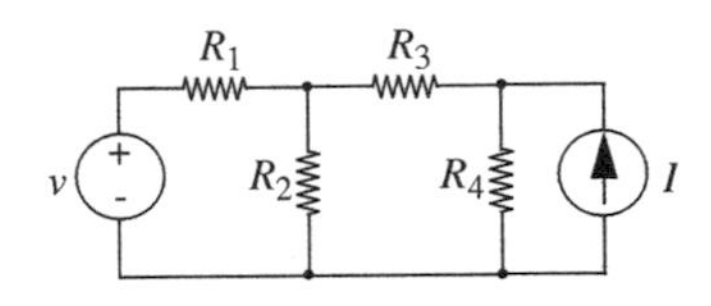


FIGURE 2.94

PROBLEM 2.14 Assume that you have an arbitrary network of passive two-terminal resistive elements in which the $i - v$ characteristic of each element does not touch either the $v$-axis or the $i$-axis, except that each $i - v$ characteristic passes through the origin. Prove that all branch currents and branch voltages in the network are zero.

PROBLEM 2.15 Solve for the voltage across resistor $R_4$ in the circuit in Figure 2.95 by assigning voltage and current variables for each resistor.

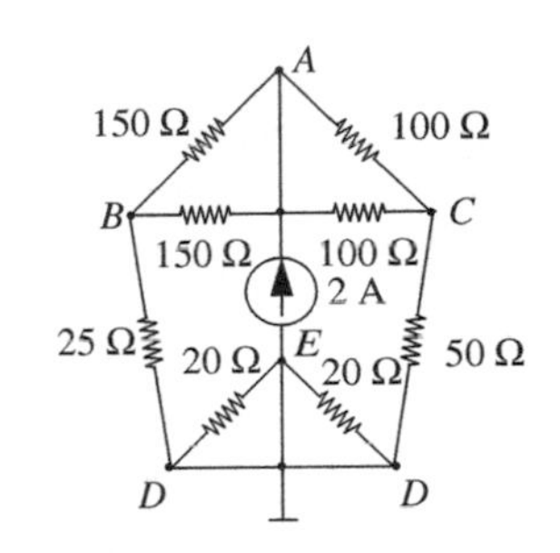


FIGURE 2.95

PROBLEM 2.16 Find the potential difference between each of the lettered nodes ($A$, $B$, $C$, and $D$) in Figure 2.96 and ground. All resistances are in ohms.

PROBLEM 2.17 Find the voltage between node $C$ and the ground node in Figure 2.97. All resistances are in ohms.

A
150 Ω
100 Ω
B
C
150 Ω
100 Ω
2 A
25 Ω
20 Ω
E
20 Ω
50 Ω
D
D

FIGURE 2.96

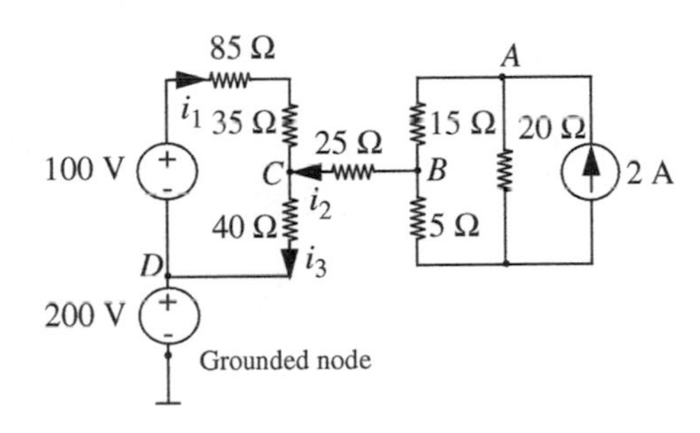


FIGURE 2.97

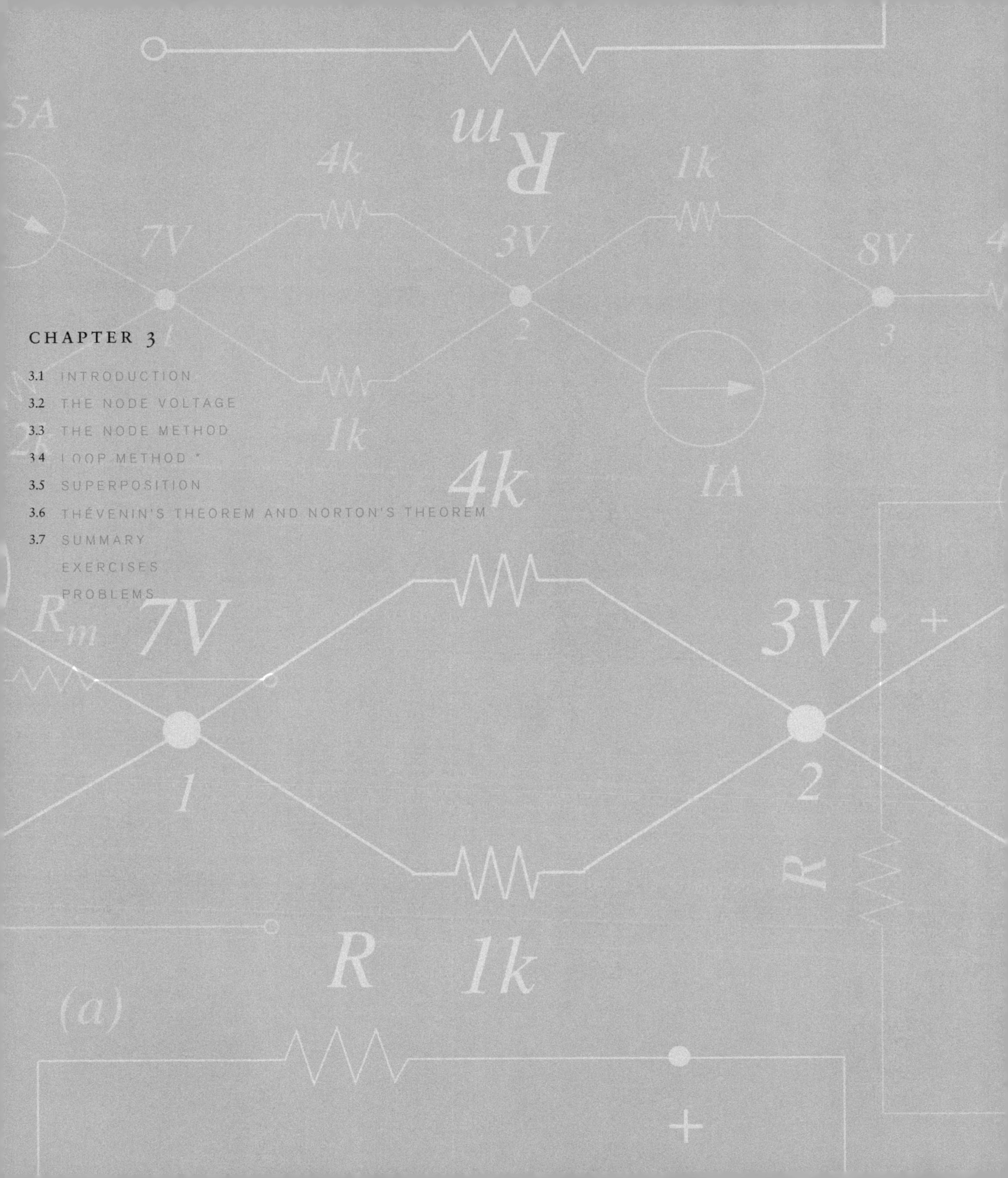

# CHAPTER 3

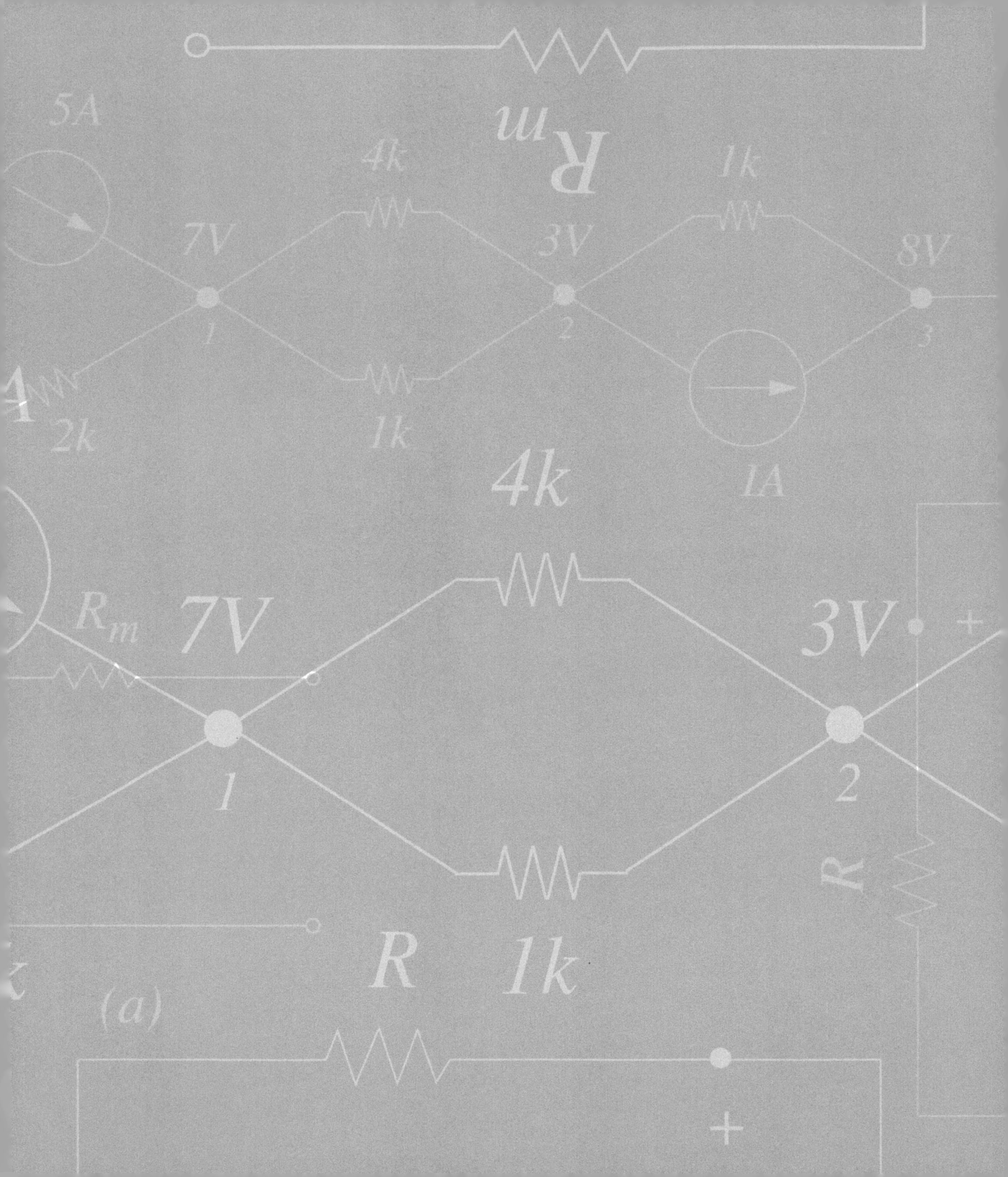

5A
4k
1k
7V
3V
8V
1
2
3
2k
1k
1A
4k
$R_m$
7V
3V
+
1
2
R
R
1k
(a)
+

# NETWORK THEOREMS

# 3

## 3.1 INTRODUCTION

The basic network analysis method introduced in Chapter 2 is fundamental but unfortunately often insufficient. The problem is that frequently we deal with complicated circuits in which we are interested in relating only one output variable to one input variable. For example, in analyzing a high-fidelity audio amplifier, we might wish to find only the relationship between the voltage at the output terminals and the voltage at the input terminals. The intermediate voltage and current variables might be of no direct interest to us, yet by the analysis method of Chapter 2, we are forced to define all such variables, and then systematically eliminate them. Even worse, a circuit with $N$ branches, each with its own voltage and current, will in general have $2N$ unknown branch variables. Thus, $2N$ equations must be solved simultaneously in order to complete the analysis. Even for a simple circuit, $2N$ can be an unwieldy number.

Fortunately, there exist better approaches to the organization of circuit analysis, and these approaches are the subject of this chapter. In this chapter, we develop a number of network theorems, all based on the fundamental methods of Chapter 2, which greatly simplify circuit analysis, and provide substantial insight about how circuits behave. These theorems also provide us with additional circuit vocabulary and a little more abstraction.

The first of these powerful techniques, called the *node method*, is fundamental and can be applied to any circuit, linear or nonlinear. The node method works with a set of variables called the *node voltages*. So, before we present the node method, let us discuss the concept of node voltages, and build up our facility to work with them.

## 3.2 THE NODE VOLTAGE

In Chapters 1 and 2 we worked with branch voltages. A branch voltage is the potential difference across the element in a branch. In like manner, we can define a *node voltage*.

A *node voltage* is the potential difference between the given node and some other node that has been chosen as a *reference node*. The reference node is called the *ground*.

Current flows from the node with the higher potential to the node with the lower potential.

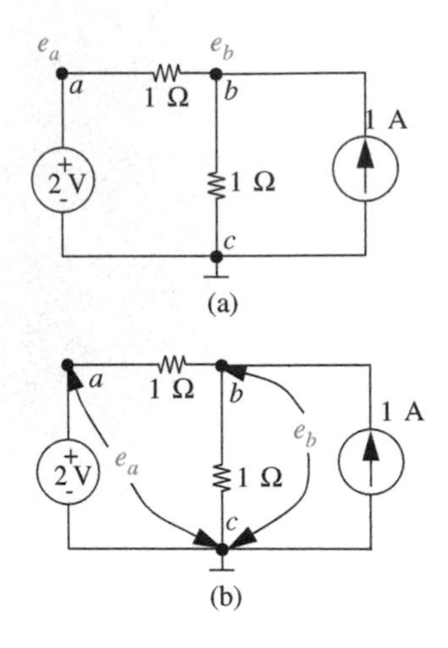


FIGURE 3.1 Ground node and node voltages.

Although the choice of reference node is in fact arbitrary, it is most convenient to choose the node that has the maximum number of circuit elements connected to it. The potential at this node is defined to be zero V, or ground-zero potential. In electrical and electronic circuits, this node will usually correspond to the "common ground" of the system, and is usually connected to the system chassis. Assigning zero potential to the ground node is permissible because elements respond only to their branch voltages and not to their absolute terminal voltages. Thus, an arbitrary constant potential may be uniformly added to all terminal voltages across the circuit thereby permitting any node to be selected as ground. A node will have a negative voltage if its potential is lower than that of the ground node.

Figure 3.1a shows a circuit that we saw earlier in Chapter 2, and illustrates some new notation. Node *c* has been chosen as ground. The upside down "*T*" symbol is the notation for the ground node. Nodes *a* and *b* are two other nodes of this circuit. Their node voltages $e_a$ and $e_b$ are marked. Figure 3.1b illustrates that the node voltages are measured with respect to the ground node.

Now, let us practice working with node voltages. Figure 3.2 shows our circuit from Figure 3.1 with a known set of branch voltages and currents. Let us determine the node voltages $e_a$ and $e_b$. The node voltage $e_a$ is the potential difference between node *a* and node *c*. To find the potential difference, let us start at node *c* and work our way to node *a* accumulating potential differences along the path $c \rightarrow a$. Thus, starting at node *c*, we count an *increase* in potential of 2 V as we traverse the voltage source and reach node *a*. Thus $e_a = 2$ V.

Similarly, $e_b$ is the potential difference between nodes *b* and *c*. Therefore, starting at node *c* and heading towards node *b* across the 1-Ω resistor, we notice a potential increase of 1.5 V. So $e_b = 1.5$ V.

Notice that from KVL, a given node's voltage should be the same irrespective of the path along which voltages are accumulated. Thus, let us confirm that the value of $e_b$ that is obtained by taking the path $c \rightarrow a \rightarrow b$ is the same as that obtained by taking the direct path $c \rightarrow b$. Starting at *c*, we first accumulate the voltage of 2 V as we cross the voltage source and reach node *a*. Then, proceeding towards node *b*, we notice a *0.5-V drop* across the 1-Ω resistor, resulting in a 1.5-V value for $e_b$, as seen earlier.

As we will see shortly, the node method will determine all the node voltages in a circuit. Once node voltages are known, we can readily determine all the

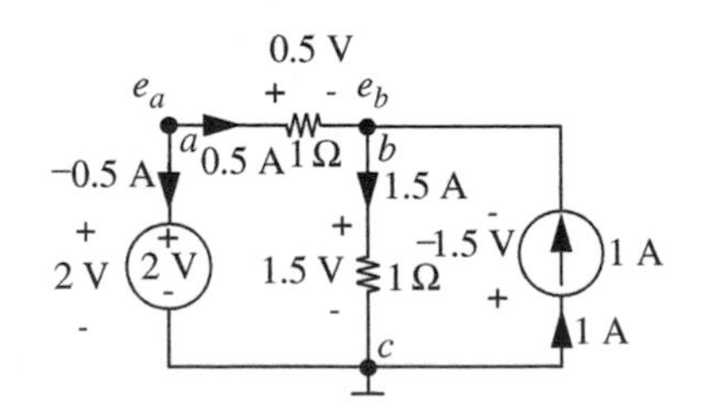


FIGURE 3.2 Determining the node voltages from the branch variables.

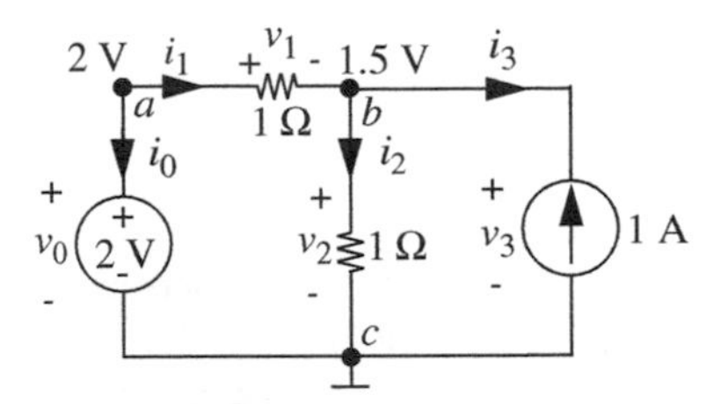


FIGURE 3.3 Determining the branch variable values from node voltages.

branch variables. As an example, Figure 3.3 shows our circuit from Figure 3.1 with a known set of node voltages. Let us determine the values of the branch variables.

Let us first determine the value of $v_1$. The branch voltage $v_1$ is the potential difference between the nodes $a$ and $b$. In other words,

$$v_1 = e_a - e_b = 2\text{ V} - 1.5\text{ V} = 0.5\text{ V}.$$

We need to be careful with voltage polarities as we obtain branch voltages by taking the difference of a pair of node voltages. As depicted in Figure 3.4, the relationship between the branch voltage $v_{ab}$ and node voltages $v_a$ and $v_b$ is given by

$$v_{ab} = v_a - v_b. \tag{3.1}$$

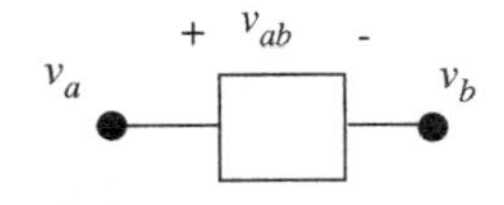


FIGURE 3.4 Branch and node voltages for the element are related as $v_{ab} = v_a - v_b$.

Intuitively, if $v_a > v_b$, then $v_{ab}$ is positive when its positive polarity coincides with the node with voltage $v_a$.

Similarly, noting that the potential of the ground node is taken as 0 V,

$$v_0 = e_a - e_c = 2\text{ V} \quad 0\text{ V} = 2\text{ V}$$

and

$$v_2 = v_3 = e_b - e_c = 1.5\text{ V} - 0\text{ V} = 1.5\text{ V}.$$

The branch currents are easily determined from the branch voltages and element laws as:

$$i_1 = \frac{v_1}{1\ \Omega} = 0.5\text{ A}$$

$$i_2 = \frac{v_2}{1\ \Omega} = 1.5\text{ A}$$

$$i_0 = -i_1 = -0.5\text{ A}$$

and

$$i_3 = -1 \text{ A}.$$

---

EXAMPLE 3.1 NODE VOLTAGES Determine the node voltages corresponding to nodes $c$ and $b$ for the circuit in Figure 3.5. Assume that $g$ is taken as the ground node.

Let $v_c$ and $v_b$ denote the voltages at nodes $c$ and $b$, respectively. To find $v_c$, let us follow the path $g \rightarrow f \rightarrow c$. Accordingly, there is a 1-V increase in potential from $g$ to $f$, and a further −2-V "increase" from $f$ to $c$ resulting in an accumulated potential of −1 V at c. Thus $v_c = -1$ V.

Similarly, because the potential at node $b$ is 4 V higher than that at node $c$, we get

$$v_b = 4\text{ V} + v_c = 4\text{ V} - 1\text{ V} = 3\text{ V}.$$

---

EXAMPLE 3.2 BRANCH VOLTAGES Determine all the branch voltages for the circuit in Figure 3.6 when the node voltages are measured with respect to node $e$.

We find each of the branch voltages by taking the difference of the appropriate node voltages. Let us denote the voltage of node $i$ as $v_i$:

$$v_1 = v_a - v_b = -1\text{ V}$$

$$v_2 = v_b - v_e = 2\text{ V}$$

$$v_3 = v_b - v_c = -1\text{ V}$$

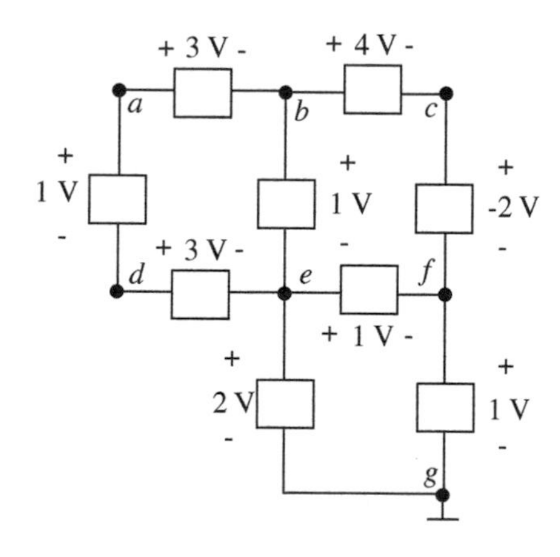


FIGURE 3.5 Circuit for determining node voltages.

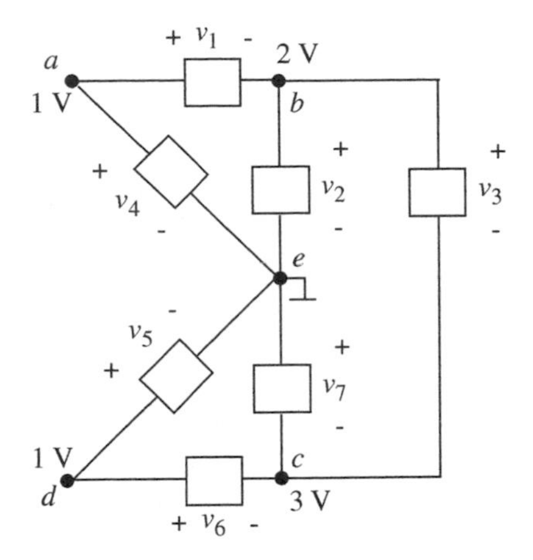


FIGURE 3.6 Circuit for determining branch voltages.

$$v_4 = v_a - v_e = 1 \text{ V}$$

$$v_5 = v_d - v_e = 1 \text{ V}$$

$$v_6 = v_d - v_c = -2 \text{ V}$$

$$v_7 = v_e - v_c = -3 \text{ V}.$$

Once all the branch voltages are known, the branch currents can readily be found from the branch voltages and the individual element laws. For example, if the element with the branch voltage $v_1$ is a resistor with resistance 1 kΩ, then its branch current $i_1$ defined according to associated variables is given by

$$i_1 = \frac{v_1}{1 \text{ k}\Omega} = -1 \text{ mA}.$$

---

Thus far, in this section, we have shown that once the node voltages for a circuit are known, we can readily determine all the branch voltages by applying KVL, and then the branch currents from the branch voltages and element laws. Since we can determine branch currents from node voltages and element laws, we can also write KCL for each of the nodes in a network in terms of node voltages and the element parameters. Although our doing so appears unmotivated at this point, we will make use of this fact in node analysis in Section 3.3.

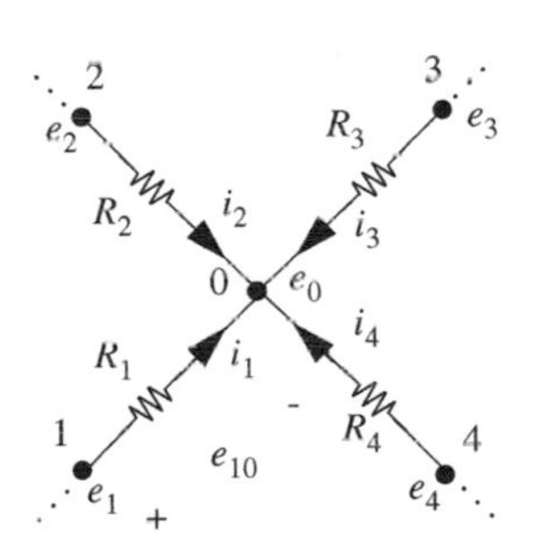


FIGURE 3.7 Circuit for writing KCL.

For example, consider the subcircuit shown in Figure 3.7. Let us write KCL for Node 0 directly in terms of the node voltages $e_0$, $e_1$, $e_2$, $e_3$, and $e_4$, (defined with respect to some ground).

Let us start by determining the current through the resistance $R_1$ into Node 0. The branch voltage across the resistance $R_1$ is given by applying

KVL as

$$e_{10} = e_1 - e_0$$

where the negative polarity of $e_{10}$ is defined to be at Node 0. Thus, the current $i_1$ through the resistance $R_1$ into Node 0 is given by using the element law for a resistor as

$$i_1 = \frac{e_{10}}{R_1}.$$

In terms of the node voltages,

$$i_1 = \frac{e_1 - e_0}{R_1}.$$

We can determine the currents into Node 0 through the other resistors in a similar manner:

$$i_2 = \frac{e_2 - e_0}{R_2}$$

$$i_3 = \frac{e_3 - e_0}{R_3}$$

$$i_4 = \frac{e_4 - e_0}{R_4}.$$

We can now write KCL for Node 0 in terms of node voltages and element values as

$$\frac{e_1 - e_0}{R_1} + \frac{e_2 - e_0}{R_2} + \frac{e_3 - e_0}{R_3} + \frac{e_4 - e_0}{R_4} = 0. \tag{3.2}$$

---

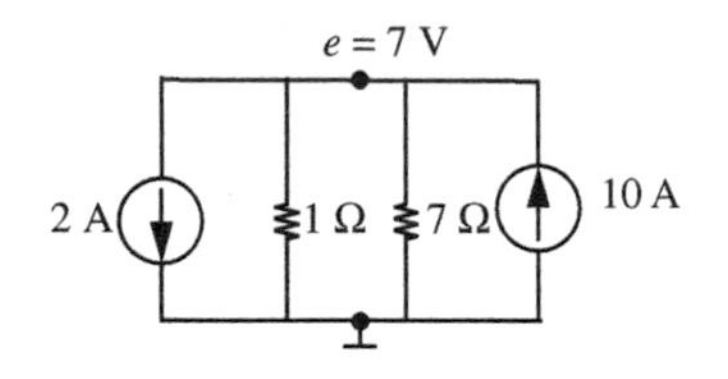


**FIGURE 3.8** Satisfying KCL.

EXAMPLE 3.3 KCL Show that the node with voltage $e = 7$ V in Figure 3.8 satisfies KCL.

For KCL to be satisfied at the node with node voltage $e$, the currents leaving the node must be zero. In other words

$$2\text{ A} + \frac{(7-0)\text{ V}}{1\ \Omega} + \frac{(7-0)\text{ V}}{7\ \Omega} - 10\text{ A}$$

must be 0. It is easy to see that this expression equals 0, and so KCL is satisfied.

---

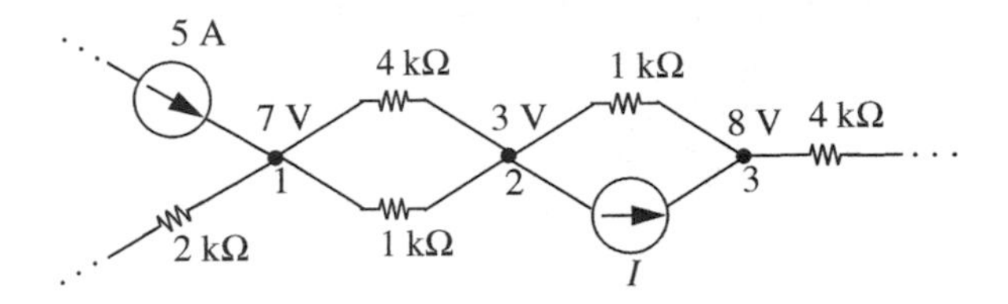


FIGURE 3.9 Portion of a circuit containing three nodes.

EXAMPLE 3.4 MORE KCL Figure 3.9 shows a portion of a circuit containing three nodes: 1, 2, and 3. The node voltages with respect to some ground are shown.

1. Write KCL for Node 2 in Figure 3.9 in terms of the node voltages and element values.
2. Determine the current $I$ through the current source.

KCL for Node 2 in terms of node voltages and element values is given by:

$$\frac{3\text{ V} - 7\text{ V}}{4\text{ k}\Omega} + \frac{3\text{ V} - 7\text{ V}}{1\text{ k}\Omega} + \frac{3\text{ V} - 8\text{ V}}{1\text{ k}\Omega} + I = 0.$$

Simplifying, we obtain $I = 10$ mA.

In summary, a voltage is always defined as the potential difference between a pair of points — the two branch terminals for a branch voltage, and two nodes for a node voltage. Accordingly, voltage measurement instruments have two leads — one to connect to the node in question and one to the reference node or ground. Thus, when we refer to a node voltage, we are also making implicit reference to a common ground node.

Interestingly, the significance of potential differences between pairs of nodes is easily illustrated with the example of a person hanging from a high voltage line. Although we do not recommend that you try this, a person hanging from a high voltage line is safe as long as no part of their body touches the ground. However, a deadly current would flow if the person were to touch the ground or another wire at a different potential.

Node voltages will be used in the next section as the variables in the node method. The node method will solve for the node voltages, which as we saw in this section, are sufficient to determine all the branch voltages and currents.

## 3.3 THE NODE METHOD

Perhaps the most powerful approach of circuit analysis is referred to as *node analysis*. Node analysis is based on the combination of element laws, KCL, and

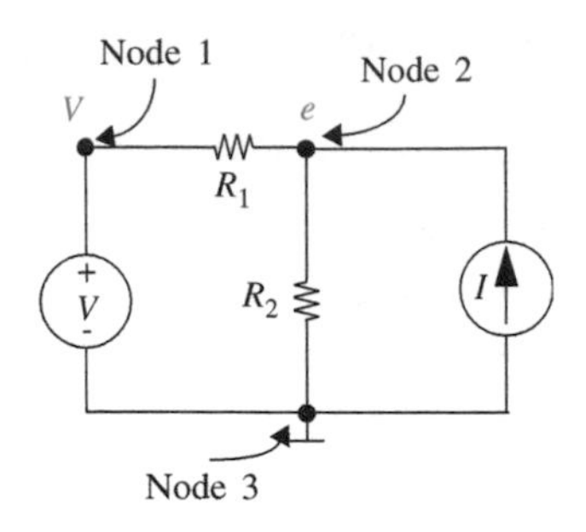


FIGURE 3.10 A resistive circuit.

KVL, just as was the basic approach presented in Chapter 2. Thus, it introduces no new physics, and it processes exactly the same information. However, node analysis organizes the analysis of a circuit in a manner that yields a relatively manageable problem, and this is what makes it particularly powerful.[1]

Let us illustrate the method with an example. Suppose we wish to find the voltage across and the current through resistor $R_1$ in the circuit shown in Figure 3.10. Notice that the circuit in the figure is identical to the one we analyzed in Figure 2.56 using the basic method, and therefore node analysis of it must yield the results in Equations 2.151 and 2.147 for the branch voltage and current corresponding to $R_1$. For node analysis, instead of defining voltage and current variables for each element in the network, we will choose node voltages as our variables.

As discussed in the previous section, since node voltages are defined with respect to a common reference, we first need to choose our reference ground node. While any node may be selected as the ground node, some nodes are more useful as ground nodes than others. Such useful nodes include those with the maximum number of circuit elements connected to it. Another useful ground node is one that connects to the maximum number of voltage sources. Sometimes the operation of a circuit may be more intuitively understood with a particular selection of the ground node. Alternatively, voltage measurements are often more easily or safely made with respect to a certain node and so that node might naturally be selected as the ground node.

One choice of ground node and a corresponding set of node voltages is defined in the figure. Node 3 is a good choice because it has three branches and it connects directly to the voltage source. Since the independent voltage source has a known voltage $V$, we can directly label the voltage of Node 1 as $V$ using the element law for an independent voltage source. Thus, we have one unknown node voltage $e$. Because *node voltages identically satisfy KVL*, it is not necessary to write KVL. To demonstrate this point, let us write KVL around the loops. Doing so, we find

$$-V + (V - e) + e = 0 \tag{3.3}$$

$$-e + e = 0. \tag{3.4}$$

Both of these equations are identically zero for all values of the node voltage variables: As promised, this choice of voltage variables automatically satisfies KVL. So to solve the circuit it is not necessary to write KVL. Instead, we

1. While node analysis is generally quite simple, it is complicated by the presence of floating independent voltage sources and by the presence of dependent sources. Note that a floating independent voltage source is a source that has neither terminal connected to ground, neither directly nor through one or more other independent voltage sources. Consequently, we first introduce node analysis without these complications, and then treat these complications in succession.

will directly proceed with writing KCL equations. Furthermore, to save time *the KCL equations can be written directly in terms of the node voltages and the resistors' values.* Since we have only one unknown, $e$, we need only one equation. Hence, at Node 2,

$$\frac{e-V}{R_1}+\frac{e}{R_2}-I=0. \tag{3.5}$$

Notice that the preceding step is actually two substeps bundled into one: (1) writing KCL in terms of currents and (2) substituting immediately node voltages and element parameters for the currents by using KVL and element laws. By doing these two substeps together, we have eliminated the need to define branch currents.

Note that in one step we have one unknown and one equation, whereas by the KVL and KCL method of Chapter 2 we would have written eight equations in eight unknowns. Further, note that both the device law for every resistor and all independent statements of KVL for the circuit have been used in writing Equation 3.5.

The voltage $e$ can now be determined easily as

$$e\left(\frac{1}{R_1}+\frac{1}{R_2}\right)=I+\frac{V}{R_1}. \tag{3.6}$$

It is wise to check dimensions at this point: Each term in this example should have the dimensions of current. Our equation can be somewhat simplified by rewriting in terms of *conductance* rather than resistance:

$$e(G_1+G_2)=I+VG_1 \tag{3.7}$$

where $G_1 = 1/R_1$ and $G_2 = 1/R_2$. Simplifying further,

$$e=\frac{1}{G_1+G_2}I+\frac{G_1}{G_1+G_2}V. \tag{3.8}$$

In terms of resistances,

$$e=\frac{R_1R_2}{R_1+R_2}I+\frac{R_2}{R_1+R_2}V. \tag{3.9}$$

Once we have determined the values of the node voltages, we can easily obtain the branch currents and voltages from the node voltages by using KVL and the constituent relations. For example, suppose we are interested only

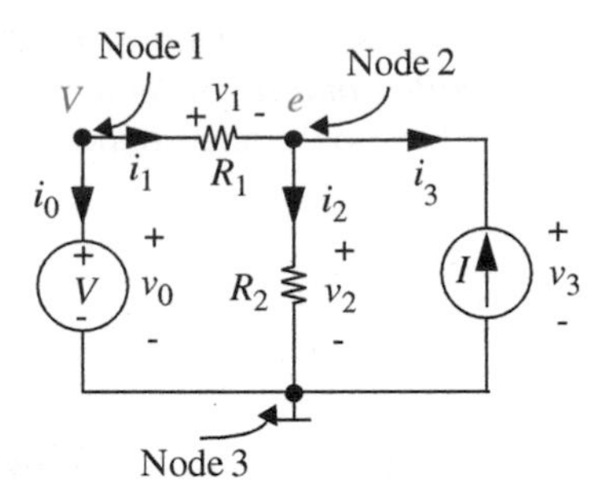


FIGURE 3.11 A resistive circuit.

in $v_1$, the voltage across $R_1$, and $i_1$, the current through $R_1$, as illustrated in Figure 3.11. Then

$$v_1 = V - e = -\frac{1}{G_1 + G_2} I + \frac{G_2}{G_1 + G_2} V \tag{3.10}$$

and

$$i_1 = (V - e)G_1 = -\frac{G_1}{G_1 + G_2} I + \frac{G_1 G_2}{G_1 + G_2} V. \tag{3.11}$$

In terms of resistances, $v_1$ and $i_1$ are given by

$$v_1 = -\frac{R_1 R_2}{R_1 + R_2} I + \frac{R_1}{R_1 + R_2} V$$

and

$$i_1 = -\frac{R_2}{R_1 + R_2} I + \frac{1}{R_1 + R_2} V.$$

For completeness, let us go ahead and determine the other branch voltages and currents as well:

$$v_0 = V \tag{3.12}$$

$$v_2 = v_3 = e = \frac{1}{G_1 + G_2} I + \frac{G_1}{G_1 + G_2} V \tag{3.13}$$

$$i_0 = -i_1 = \frac{G_1}{G_1 + G_2} I - \left( G_1 - \frac{G_1^2}{G_1 + G_2} \right) V \tag{3.14}$$

$$i_2 = eG_2 = \frac{G_2}{G_1 + G_2} I + \frac{G_1 G_2}{G_1 + G_2} V \tag{3.15}$$

$$i_3 = -I. \tag{3.16}$$

This completes the node analysis.

A comparison of the equations for the branch voltages and currents (Equations 3.10 through 3.16) with the corresponding Equations 2.147 through 2.152 in Chapter 2 shows that the node analysis has resulted in the same expressions for the branch variables as did the direct analysis presented. However, the node analysis obtained these results in a much simpler manner. The direct analysis of Chapter 2 involved the solution of eight simultaneous equations, namely Equations 2.139 through 2.146, while the node analysis involved the solution of only one equation, namely Equation 3.5, and the explicit back substitution of its solution.

In summary, the specific steps of the node method can be written as:

1. Select a reference node, called ground, from which all other voltages will be measured. Define its potential to be 0 V.
2. Label the potentials of the remaining nodes with respect to the ground node. Any node connected to the ground node through either an independent or a dependent voltage source should be labeled with the voltage of that source. The voltages of the remaining nodes are the primary unknowns and should be labeled accordingly. In this chapter we will denote the unknown node voltages by the symbol $e$. Since there are generally far fewer nodes than branches in a circuit, there will be far fewer primary unknowns to determine in a node analysis.
3. Write KCL for each of the nodes that has an unknown node voltage (in other words, the ground node and nodes with voltage sources connected to ground are excluded), using KVL and element laws to obtain the currents directly in terms of the node voltage differences and element parameters. Thus, one equation is written for each unknown node voltage.
4. Solve the equations resulting from Step 3 for the unknown node voltages. This is the most difficult step in the analysis.
5. Back-solve for the branch voltages and currents. More specifically, use node voltages and KVL to determine branch voltages as desired. Then, use the branch voltages, the element laws, and KCL to determine the branch currents, again as desired.

At this point, it is instructive to make some general comments about the equations produced by the node method. Although the actual collection of conductance terms in Equation 3.8 is not particularly educational in this somewhat contrived example, the general form of the equation is useful. The right-hand side has two terms, one for each source, and these source terms enter the equation as *sums*, and not products. Equations will always be of this form if the circuit is made up of linear elements. In fact, we use this property to define a linear network: *A network is linear if the response to an input $ax_1 + bx_2$ is the same as a times the response to $x_1$ alone plus b times the response to $x_2$ alone.* That is, if $f(x)$ is the response to some excitation $x$, then the system is linear if and only if

$$f(ax_1 + bx_2) = af(x_1) + bf(x_2). \tag{3.17}$$

### 3.3.1 NODE METHOD: A SECOND EXAMPLE

As a second, and slightly more complex, example of node analysis, consider the circuit shown in Figure 3.12, which is the same as that shown in Figure 2.46 except for the addition of an independent current source. Specifically, suppose we wish to find the voltage across and the current through resistor $R_3$.

The first two steps in its node analysis, namely the selection of a ground node and the labeling of its node voltages, are already complete. As shown in Figure 3.12, Node 4 is selected as the ground node, Node 3 is labeled with the known voltage $V$ of the independent source, and Nodes 1 and 2 are labeled with the unknown node voltages $e_1$ and $e_2$, respectively. Node 4 is a good choice for the ground node because it joins the largest number of branches and connects directly to the voltage source.

Next, following Step 3, we write KCL for Nodes 1 and 2 in terms of the unknown node voltages. This yields

$$\frac{(V - e_1)}{R_1} + \frac{(e_2 - e_1)}{R_3} - \frac{e_1}{R_2} = 0 \tag{3.18}$$

for Node 1, and

$$\frac{(e_1 - e_2)}{R_3} - \frac{e_2}{R_4} + I = 0 \tag{3.19}$$

for Node 2.

Note that in one step we have generated two equations and two unknowns, whereas by the KVL and KCL method of Chapter 2 we would have written twelve equations in twelve unknowns. The voltages $e_1$ and $e_2$ can now be

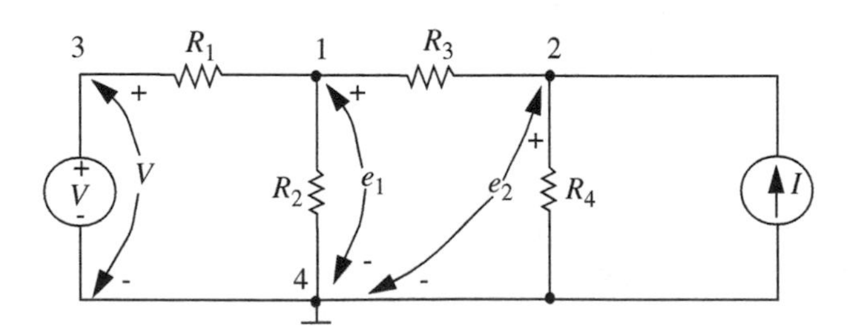


FIGURE 3.12 A resistive circuit.

determined by standard algebraic methods. First, rewrite the equations with the source terms on the left-hand side of the equations, and the dependent variables on the right:

$$\frac{V}{R_1} = e_1\left(\frac{1}{R_1} + \frac{1}{R_2} + \frac{1}{R_3}\right) - \frac{e_2}{R_3} \tag{3.20}$$

$$I = -\frac{e_1}{R_3} + e_2\left(\frac{1}{R_3} + \frac{1}{R_4}\right). \tag{3.21}$$

Rewriting in terms of conductance to simplify our calculations:

$$G_1 V = e_1(G_1 + G_2 + G_3) - e_2 G_3 \tag{3.22}$$

$$I = -e_1 G_3 + e_2(G_3 + G_4). \tag{3.23}$$

Application of Cramer's rule (see Appendix D), yields

$$e_1 = \frac{VG_1(G_3 + G_4) + IG_3}{(G_1 + G_2 + G_3)(G_3 + G_4) - G_3^2} \tag{3.24}$$

$$= \frac{V(G_1 G_3 + G_1 G_4) + IG_3}{G_1 G_3 + G_1 G_4 + G_2 G_3 + G_2 G_4 + G_3 G_4}. \tag{3.25}$$

Similarly, we can obtain $e_2$ as

$$e_2 = \frac{G_1 G_3 V + (G_1 + G_2 + G_3)I}{(G_1 + G_2 + G_3)(G_3 + G_4) - G_3^2}. \tag{3.26}$$

All node voltages are now known, and from these node voltages all branch variables in the circuit can be explicitly determined by using KVL and the constituent relations. For example, suppose the voltage across $R_3$ is $v_3$, and the current through $R_3$ is $i_3$, as illustrated in Figure 3.13. Then

$$v_3 = e_1 - e_2$$

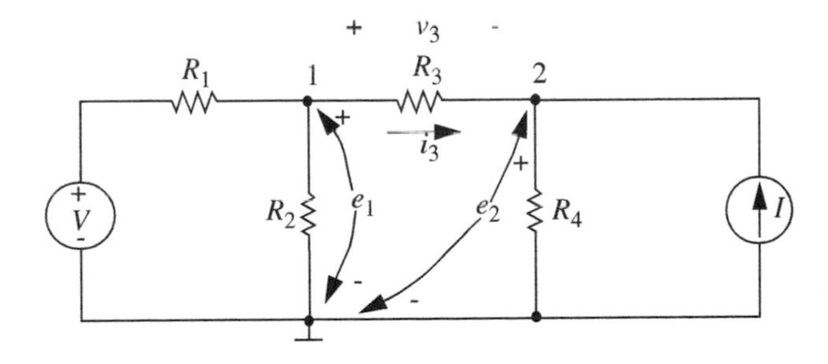


FIGURE 3.13 The resistive circuit.

and

$$i_3 = \frac{e_1 - e_2}{R_3}.$$

Since the circuit in Figure 3.12, with $I = 0$, is the same as the circuit in Figure 2.46, the analysis of the two circuits (with $I = 0$) using the basic and the node methods should yield the same results. Accordingly, the reader might want to compare the values for $v_3$ and $i_3$ obtained here, with those obtained in Equations 2.135 and 2.131.

This example illustrates an important circuit property: The structure of a node equation is closely related to the topology of the circuit. We will briefly introduce this relationship here, and spend some more time on this topic in Section 3.3.4. First, let us write our two node equations 3.22 and 3.23 in matrix form:

$$\begin{bmatrix} G_1 + G_2 + G_3 & -G_3 \\ -G_3 & G_3 + G_4 \end{bmatrix} \begin{bmatrix} e_1 \\ e_2 \end{bmatrix} = \begin{bmatrix} G_1 & 0 \\ 0 & 1 \end{bmatrix} \begin{bmatrix} V \\ I \end{bmatrix}. \tag{3.27}$$

The matrix equation is in the form

$$\bar{G}\,\bar{e} = \bar{S}\,\bar{s} \tag{3.28}$$

where $\bar{e}$ is a column vector of unknown voltages and $\bar{s}$ is the column vector of known source amplitudes. $\bar{G}$ is called the *conductance matrix* and $\bar{S}$ is called the *source matrix* for reasons that will be apparent shortly. In Equation 3.22, written at the $e_1$ node, we note from Figure 3.12 that the coefficient of the $e_1$ term (the first term in the first row of the $\bar{G}$ matrix) is the sum of the conductances connected to the $e_1$ node. Similarly in Equation 3.23, the coefficient of the $e_2$ term (the second term in the second row of the $\bar{G}$ matrix) is the sum of the conductances connected to the $e_2$ node. (These terms are often called the "self" conductances.) The off-diagonal coefficients represent conductances connected *between* the corresponding nodes, the "mutual" conductances. In Equation 3.22, for example, the coefficient of the $e_2$ term (the second term in the first row of the $\bar{G}$ matrix) is the mutual conductance between the $e_1$ node (because this is the $e_1$ equation) and $e_2$. For linear resistive circuits, the off-diagonal terms are negative, assuming that the equations have been structured to make the main-diagonal terms positive.

It is self-evident that with circuits made up of linear resistors, the mutual conductance $e_1$ to $e_2$ must be the same as the mutual conductance from $e_2$ to $e_1$. Hence the two off-diagonal coefficients in the node equations are identical. More generally, we expect node equation coefficients to exhibit mirror symmetry about the main diagonal for linear resistive circuits, as is evident from the $\bar{G}$ matrix. These helpful topological constraints are destroyed if we do not apply

KCL at the nodes defined by the node voltages. Such a procedure is mathematically correct (the new equations are derivable by algebraic manipulation of the original equations, Equations 3.22 and 3.23) but the symmetries are gone.

Interestingly, the SPICE software package uses the node method to solve circuits. The program takes as input a file containing a description of the circuit topology and by systematically following the node method produces a matrix equation such as that in Equation 3.27. It then solves for the vector of unknowns $\bar{e}$ using standard linear algebraic techniques.

---

EXAMPLE 3.5 NODE METHOD Determine the current $i$ through the 5-Ω resistor in the circuit in Figure 3.14.

Let us use the node method to solve the circuit. As Step 1 of node analysis, we will choose Node 1 as our ground node as depicted in Figure 3.14.

Step 2 labels the potentials of the remaining with respect to the ground node. Figure 3.14 shows such a labeling. Since Node 2 is connected to the ground node through an independent voltage source, it is labeled with the voltage of the source, namely 1 V. Node 3 is labeled with a node voltage $e_1$ and Node 4 is labeled with a node voltage $e_2$.

Next, following Step 3, we write KCL for Nodes 3 and 4. KCL for Node 3 is

$$\frac{e_1 - 1}{3} + \frac{e_1}{4} + \frac{e_1 - e_2}{2} + 2 = 0$$

and that for Node 4 is

$$-2 + \frac{e_2 - e_1}{2} + \frac{e_2}{5} - 1 = 0.$$

Following Step 4 we solve these equations to determine the unknown node voltages. This yields

$$e_1 = 0.65 \text{ V}$$

and

$$e_2 = 4.75 \text{ V}.$$

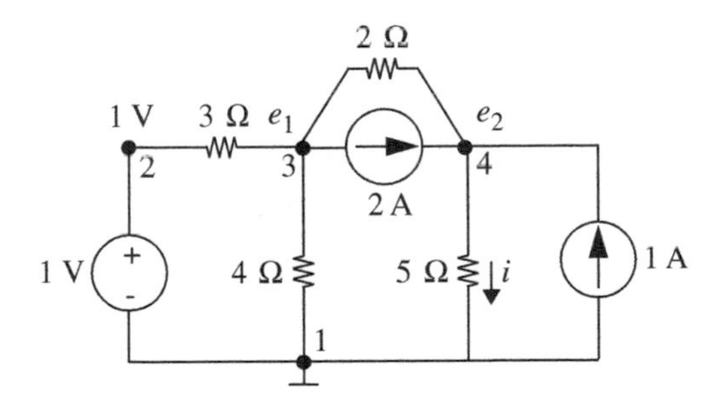


FIGURE 3.14 Determining the unknown current $i$.

We can now determine $i$ as

$$i = \frac{4.75}{5} = 0.95 \text{ A}.$$

---

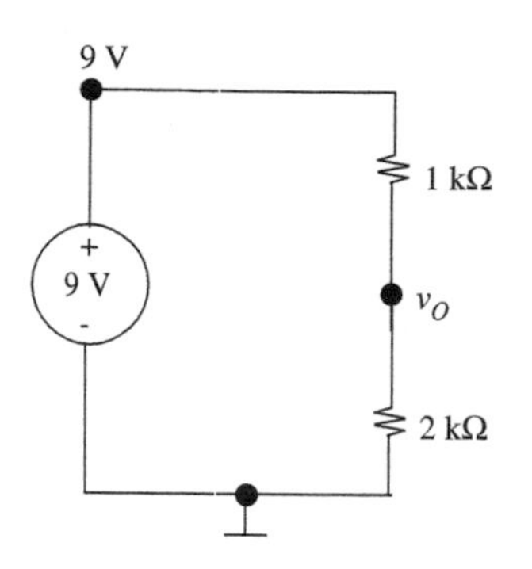


FIGURE 3.15 The voltage-divider circuit.

EXAMPLE 3.6 NODE METHOD SOLUTION OF THE VOLTAGE-DIVIDER CIRCUIT Lest you think the node method is applicable only to complex circuits with many nodes, let us apply the node method to the simple voltage-divider circuit in Figure 3.15 to obtain the voltage $v_O$.

The ground node is selected as shown in Figure 3.15. The circuit in Figure 3.15 has one unknown node voltage, $v_O$, also as marked in the figure. So, Steps 1 and 2 are complete.

Following Step 3, we write KCL for the node with the unknown node voltage:

$$\frac{v_O - 9}{1 \text{ k}\Omega} + \frac{v_O}{2 \text{ k}\Omega} = 0.$$

Multiplying throughout by $2k$ we obtain

$$2v_O - 18 + v_O = 0$$

which yields

$$v_O = 6 \text{ V}.$$

---

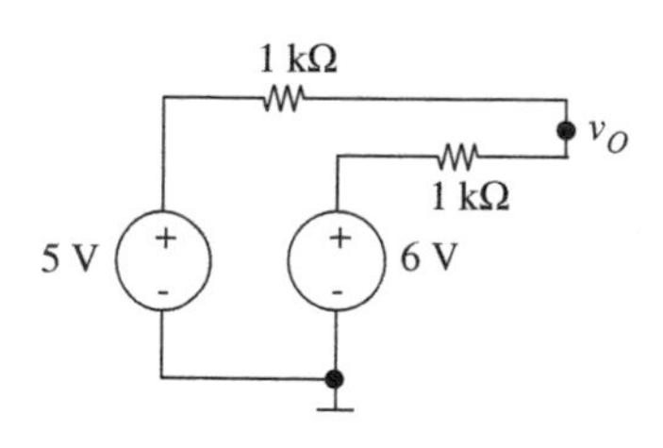


FIGURE 3.16 A summing circuit.

EXAMPLE 3.7 FIND NODE VOLTAGE USING THE NODE METHOD Determine the node voltage $v_O$ in the circuit shown in Figure 3.16 using the node method.

The circuit in Figure 3.16 has only one unknown node voltage, $v_O$, as marked in the figure. Figure 3.16 also shows a ground node, and so Steps 1 and 2 are complete.

Following Step 3, we write KCL for the node with the unknown node voltage:

$$\frac{v_O - 5}{1 \text{ k}\Omega} + \frac{v_O - 6}{1 \text{ k}\Omega} = 0.$$

Multiplying throughout by 1 kΩ we obtain

$$v_O - 5 + v_O - 6 = 0$$

which simplifies to

$$v_O = \frac{5\text{ V} + 6\text{ V}}{2}$$

or

$$v_O = 5.5\text{ V}.$$

The circuit in Figure 3.16 is called an *adder circuit* since $v_O$ is proportional to the sum of the input voltages.

---

EXAMPLE 3.8 MORE ON THE NODE METHOD Determine the node voltage $v$ in the circuit in Figure 3.17 using the node method.

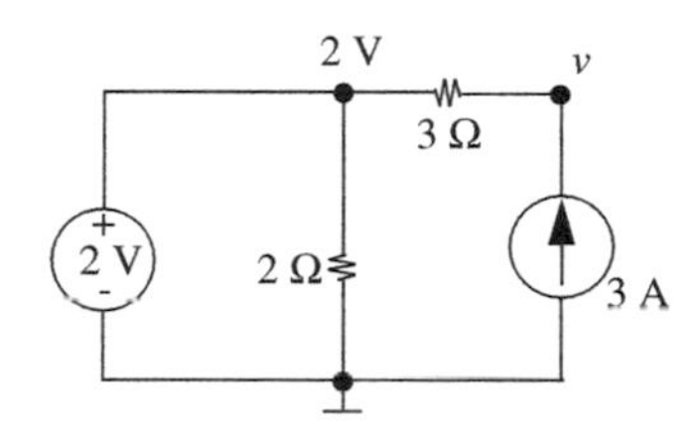


FIGURE 3.17 A circuit with two independent sources.

The ground node and unknown node variables are marked as shown in Figure 3.17. Next, following Step 3, we write KCL for the node with the unknown voltage.

Then, we write KCL for the node with the unknown node voltage:.

$$\frac{v - 2}{3} = 3\text{ V}.$$

Thus,

$$v = 11\text{ V}.$$

Compare the node analysis shown here with the basic method applied to the same circuit on page 190.

---

WWW EXAMPLE 3.9 EVEN MORE ON THE NODE METHOD

---

### 3.3.2 FLOATING INDEPENDENT VOLTAGE SOURCES

Node analysis as described here does not work for circuits that contain floating independent voltage sources such as the one shown in Figure 3.20. A floating independent voltage source is a voltage source that has neither terminal connected to ground, neither directly nor through one or more other independent voltage sources. The reason node analysis does not work is that the element law for an independent voltage source does not relate its branch current to its branch voltage. Therefore, it is not possible to complete Step 3 of node analysis if the circuit contains a floating independent voltage source. In this case, it is necessary to modify the node analysis slightly.

To apply node analysis to a circuit containing a floating voltage source we must realize that the node voltages at the terminals of the source are directly

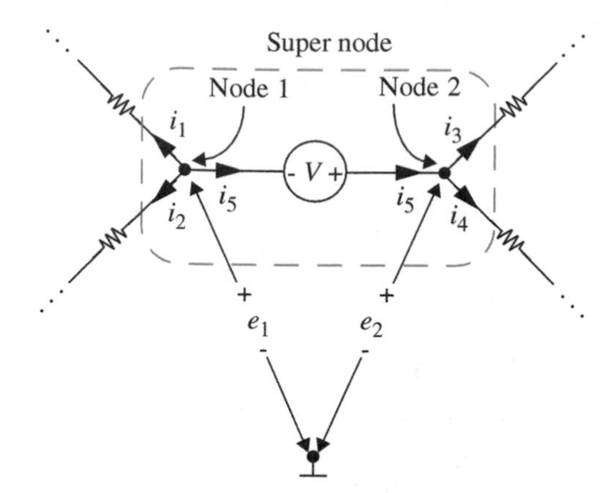


FIGURE 3.20 A floating independent voltage source and its treatment as a super node.

related by the element law for that source. For example, the application of KVL to the circuit in Figure 3.20 shows that

$$e_2 = V + e_1. \tag{3.32}$$

Because of this, the number of unknown node voltages in the circuit can be immediately reduced by one since $e_1$ and $e_2$ can be determined directly from each other using Equation 3.32. Consequently, the number of independent statements of KCL needed to determine the unknown node voltages can similarly be reduced by one. Thus, Nodes 1 and 2 in Figure 3.20 must together contribute one statement of KCL to the first part of Step 3 of the node analysis (namely, writing KCL for each of the nodes that has an unknown node voltage). Further, this single statement of KCL should not involve $i_5$ since $i_5$ cannot be determined from the element law of the voltage source in the second part of Step 3 (namely, using KVL and element laws to obtain the currents directly in terms of the node voltage differences and element parameters).

To derive the desired statement of KCL for Nodes 1 and 2, we draw a surface around both nodes, enclosing what is referred to as a *super node* in the process. Then, we write KCL for the super node. In the case of Figure 3.20, KCL applied to the super node yields

$$i_1 + i_2 + i_3 + i_4 = 0 \tag{3.33}$$

for the first part of Step 3. Note that this statement of KCL is nothing more than the sum of

$$i_1 + i_2 + i_5 = 0 \tag{3.34}$$

$$i_3 + i_4 - i_5 = 0, \tag{3.35}$$

which are the individual statements of KCL for Nodes 1 and 2. Following this, in the second part of Step 3, the currents are eliminated by substituting node voltages and element parameters in their place. In our example, $i_1$ and $i_2$ are determined using $e_1$ and the parameters of the elements through which $i_1$ and $i_2$ flow. Similarly $i_3$ and $i_4$ are determined using $e_1 + V$ and the parameters of the elements through which $i_3$ and $i_4$ flow, with $e_1$ serving as the one unknown node voltage.

Alternatively, $i_1$ and $i_2$ can be determined using $e_2 - V$, and $i_3$ and $i_4$ can be determined using $e_2$, with $e_2$ serving as the one unknown node voltage. Finally, it should be recognized that a floating string of independent voltage sources is handled in exactly the same manner as a floating isolated independent voltage source.

Let us illustrate node analysis applied to a circuit with a floating independent voltage source, and hence a super node, using the circuit shown in Figure 3.21. The circuit is the same as that shown in Figure 3.10 except that Node 2 is now selected as the ground node, and the node voltages for Nodes 1 and 3 are defined differently. The super node containing the floating voltage source is also marked in the figure.

The primary unknown in the circuit, $e$, is now the voltage at Node 3. Note also that the voltage at Node 1, the other node in the super node, is labeled in terms of $e$. By defining the ground node and labeling the node voltages, we have completed Steps 1 and 2 in the node analysis.

Next we perform Step 3 for the super node. This yields

$$\frac{e+V}{R_1} + \frac{e}{R_2} + I = 0. \tag{3.36}$$

In Equation 3.36, $(e + V)/R_1$ is the current (written in terms of node voltages and element parameters) out of the super node through the branch containing

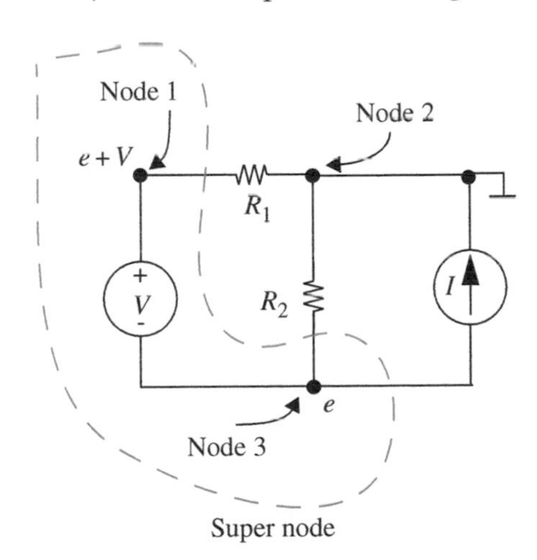


FIGURE 3.21 A circuit with a floating independent voltage source.

$R_1$. Similarly, $e/R_2$ is the current out of the super node through $R_2$, and $I$ is the current through the third branch from the super node.

Following Step 4, the solution of Equation 3.36 is

$$e = -\frac{R_1R_2}{R_1+R_2}I - \frac{R_2}{R_1+R_2}V. \tag{3.37}$$

Finally, to complete the node analysis, the solution for $e$ could be used in Step 5 of node analysis to determine the branch voltages and then the branch currents in the circuit. While we will not do this here, it is worthwhile to see that it will yield the same results as in Equations 3.10 through 3.16, providing that the branch currents and voltages are defined in the same manner. To see that this will be the case, observe that $e$ in Equation 3.37 is the same as in Equation 3.9 except for a minus sign, owing to the change in the sign of $e$ as defined in Figures 3.10 and 3.21.

---

EXAMPLE 3.10 FLOATING INDEPENDENT VOLTAGE SOURCE As another example of node analysis applied to a circuit with a floating independent voltage source, consider the circuit shown in Figure 3.22. In this circuit, the voltage source having value $V_3$ is the only floating independent voltage source. Because the source having value $V_1$ is connected to ground at Node 5 it is not a floating source, hence Node 1 is labeled with the node voltage $V_1$. Similarly, the source having value $V_2$ is not a floating source because it is connected to ground through the known voltage $V_1$, hence Node 2 is labeled with the known node voltage $V_1 + V_2$. Thus, only the voltages at Nodes 3 and 4 in the super node are unknown. In Figure 3.22, Node 3 is labeled with the unknown node voltage $e$, and so Node 4 is labeled with the node voltage $e + V_3$.

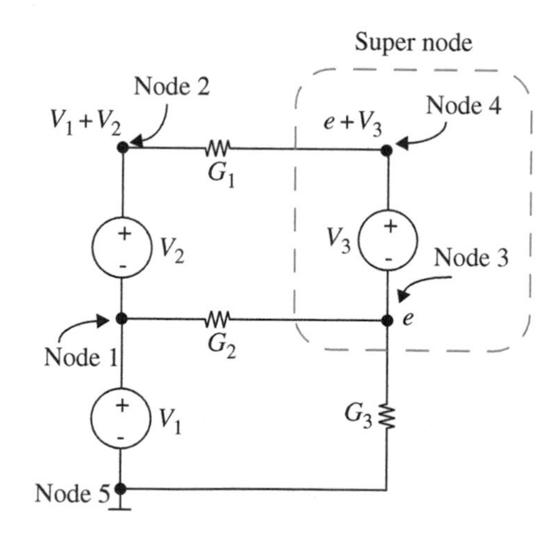


FIGURE 3.22 Another circuit with a floating independent voltage source.

To continue the node analysis of the circuit in Figure 3.22, we perform Step 3 for the super node. This yields

$$G_1\left[(e+V_3)-(V_1+V_2)\right]+G_2(e-V_1)+G_3e=0. \tag{3.38}$$

Here, conductances have been used for convenience. Following Step 4, the solution of Equation 3.38 is

$$e=\frac{(G_1+G_2)V_1+G_1V_2-G_1V_3}{G_1+G_2+G_3}. \tag{3.39}$$

Finally, to complete the node analysis, the solution for $e$ could be used in Step 5 to determine the branch voltages and then the branch currents in the circuit. We will not do this here.

---

### 3.3.3 DEPENDENT SOURCES AND THE NODE METHOD

A dependent source will also complicate the node analysis previously described when its element law does not easily relate its branch current to its branch voltage. In this case, it will again not be possible to complete Step 3, and so it is again necessary to modify the node analysis slightly. Since there are four types of dependent sources, and the branch currents and voltages that control them can appear through or across many different types of elements, it is impractical to treat each case separately in its most efficient manner. As a compromise, we present here a single method that treats all cases of dependent sources, and illustrate how this method can be made more efficient in a few illustrative cases. We will illustrate the method using the circuit in Figure 3.23, which contains a dependent current source, whose current is some function of a branch variable $i$ as shown in the figure.

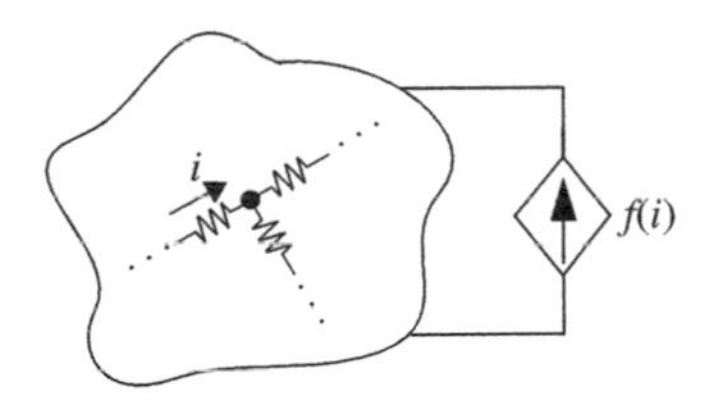


FIGURE 3.23 A circuit containing a dependent current source.

Our method of applying node analysis to a circuit containing dependent sources begins by assuming that we know the value of each dependent source. This assumption allows us to treat each dependent source as an independent source, and carry out a node analysis of the circuit as described in the previous subsections. For example, in the case of a dependent current source (see Figure 3.23), we replace the dependent source with an independent current source with some assumed current, say $I$ (see Figure 3.24), and carry out our usual five-step node analysis. As part of this analysis we solve for the branch variables that control the dependent sources in terms of the assumed source values.

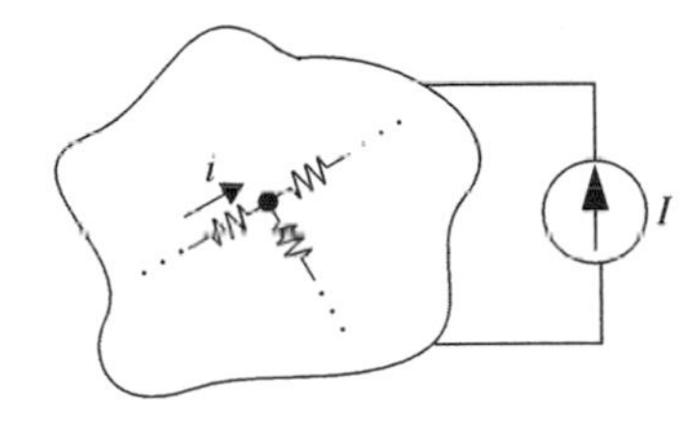


FIGURE 3.24 Replacing the dependent current source with an independent current source with an assumed current $I$.

Of immediate interest are the expressions for the branch variables that control dependent sources. In our example, this branch variable is $i$. Next, we substitute these expressions for the controlling variables into the element laws for the dependent sources, and self-consistently solve for the actual values of the dependent sources. Continuing with our dependent current source example of

Figure 3.23, suppose that the expression for $i$ is some function of the assumed current $I$ and is of the form

$$i = g(I). \tag{3.40}$$

We substitute this expression for the branch variable into the element law for the dependent current source as

$$I = f(i) = f(g(I)) \tag{3.41}$$

and solve for $I$. The solution for $I$ will not contain the variable $i$. Note that if the expression for $i$ shown in Equation 3.40 does not contain $I$, then no additional work needs to be done to solve for $I$, since $f(g(I))$ is itself a solution for $I$.

Finally, we back-substitute the actual values of the dependent sources — in other words, the solution for $I$ — into the original node analysis, thereby completing the analysis in total.

As a concrete example, suppose the dependence source function

$$f(i) = 10i.$$

Further, suppose we obtain the following expression for $i$ as a function of the assumed current $I$:

$$i = g(I) = \frac{I}{2} + 2\ \text{A}.$$

Then, according to Equation 3.41,

$$I = f(g(I)) = 10\left[\frac{I}{2} + 2\ \text{A}\right].$$

Solving, we get

$$I = -5\ \text{A}.$$

As expected, the solution for $I$ does not contain the variable $i$.

This modification to the original node analysis is not always the most efficient method of analysis, but it always works. However, when the element laws for the dependent sources can be easily expressed in terms of the node voltages, it is possible to take a more intuitive approach and apply the simple node analysis described in Section 3.3 without modification. In our example of Figure 3.23, suppose that the circuit on the left has the node voltages shown in Figure 3.25. In this case, it is easy to see that the element law for the current source can be easily written in terms of the node voltages as

$$f(i) = f\left(\frac{e_a - e_b}{R}\right)$$

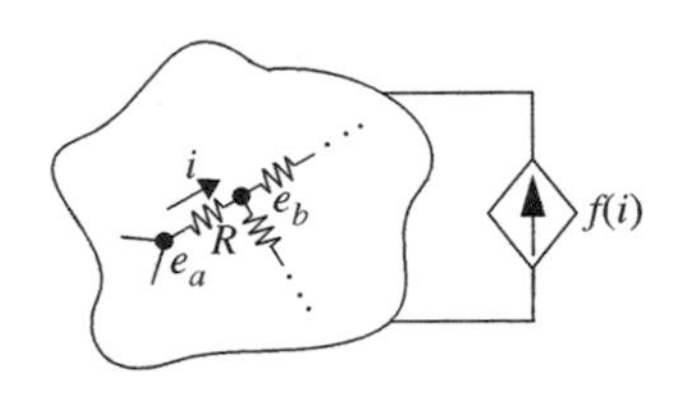


FIGURE 3.25 Node voltages.

and our simple node analysis can be applied without modification. We will do examples using both the modified and unmodified versions of the node method.

To illustrate our modified method of node analysis for a circuit containing a dependent source, consider the analysis of the circuit shown in Figure 3.26. This circuit has one dependent source, namely a CCCS. To analyze this circuit using the node method, we first replace its CCCS with an independent current source carrying a known current, say $I$, and analyze the resulting circuit. The resulting circuit, however, is exactly that shown in Figure 3.10, which we have already analyzed using the node method in Section 3.3. Note that the value $I$ in Figure 3.10 replaces the value $\alpha i_1$ in Figure 3.26. Thus, we are partially done with the analysis of the circuit in Figure 3.26.

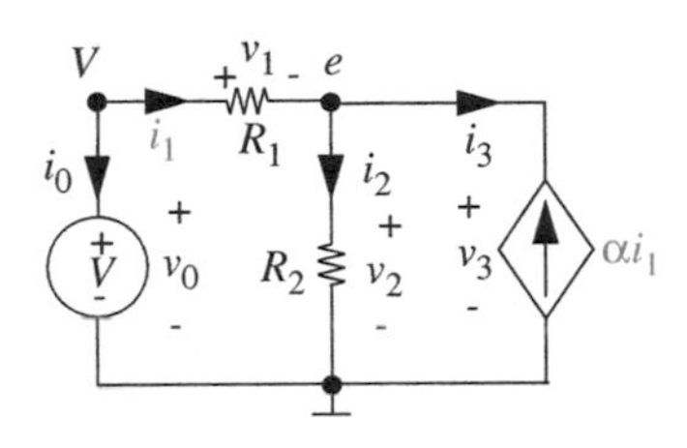


FIGURE 3.26 A circuit with a dependent source.

The results of our analysis of the circuit in Figure 3.10 appear in Equations 3.10 through 3.16. Let us copy them here for convenience after replacing conductances with resistances.

$$v_0 = V \tag{3.42}$$

$$i_0 = \frac{R_2}{R_1 + R_2} I - \frac{1}{R_1 + R_2} V \tag{3.43}$$

$$v_1 = -\frac{R_1 R_2}{R_1 + R_2} I + \frac{R_1}{R_1 + R_2} V \tag{3.44}$$

$$i_1 = -\frac{R_2}{R_1 + R_2} I + \frac{1}{R_1 + R_2} V \tag{3.45}$$

$$v_2 = v_3 = \frac{R_1 R_2}{R_1 + R_2} I + \frac{R_2}{R_1 + R_2} V \tag{3.46}$$

$$i_2 = \frac{R_1}{R_1 + R_2} I + \frac{1}{R_1 + R_2} V \tag{3.47}$$

$$i_3 = -I. \tag{3.48}$$

Of particular interest from that analysis is the value of $i_1$ because $i_1$ controls the CCCS in Figure 3.26. Using the result for $i_1$ from Equation 3.45 we next write

$$I = \alpha i_1 = \alpha \left[ -\frac{R_2}{R_1 + R_2} I + \frac{1}{R_1 + R_2} V \right]. \tag{3.49}$$

The first equality in Equation 3.49 expresses the equality of the CCCS in Figure 3.26 and its surrogate independent current source in Figure 3.10. The second equality follows from the substitution for $i_1$ using Equation 3.45 from the node analysis of the circuit in Figure 3.10. Since $i_1$ is determined in terms of $I$ during that analysis, Equation 3.49 becomes an implicit equation that must be solved for $I$. This solution yields

$$I = \frac{\alpha}{R_1 + (1+\alpha)R_2} V. \tag{3.50}$$

The actual value of the CCCS is now known.

Finally, we back-substitute Equation 3.50, namely the actual value of $I$, into Equations 3.42 through 3.48 to obtain

$$v_0 = V \tag{3.51}$$

$$i_0 = -\frac{1}{R_1 + (1+\alpha)R_2} V \tag{3.52}$$

$$v_1 = \frac{R_1}{R_1 + (1+\alpha)R_2} V \tag{3.53}$$

$$i_1 = \frac{1}{R_1 + (1+\alpha)R_2} V \tag{3.54}$$

$$v_2 = v_3 = \frac{(1+\alpha)R_2}{R_1 + (1+\alpha)R_2} V \tag{3.55}$$

$$i_2 = \frac{1+\alpha}{R_1 + (1+\alpha)R_2} V \tag{3.56}$$

$$i_3 = \frac{-\alpha}{R_1 + (1+\alpha)R_2} V. \tag{3.57}$$

This completes the analysis of the circuit in Figure 3.26.

While the preceding analysis is not terribly difficult, it can nonetheless be carried out more efficiently in many cases. As mentioned previously, commonly, it is possible to apply the simple node analysis described in Section 3.3 without modification because the element law for the CCCS can be easily expressed in terms of the node voltage $e$. To see this, we begin by performing Step 3 of node analysis to write

$$\frac{e - V}{R_1} + \frac{e}{R_2} - \alpha \frac{V - e}{R_1} = 0 \tag{3.58}$$

for the node at which $e$ is defined. Note that in the third term in Equation 3.58, $(V - e)/R_1$ has been substituted for $i_1$.

Next, following Step 4, we solve Equation 3.58 for $e$ to obtain

$$e = \frac{(1+\alpha)R_2}{R_1 + (1+\alpha)R_2} V. \tag{3.59}$$

This result is the same as expressed in Equation 3.55. The remainder of the node analysis, namely Step 5, then proceeds to yield Equations 3.51 through 3.57 directly. It is important to note, however, that the node analysis of circuits containing dependent sources cannot always be easily simplified in this manner.

---

EXAMPLE 3.11 DEPENDENT CURRENT SOURCE Now, let us analyze a slightly different circuit containing a dependent source as shown in Figure 3.27. The node voltages $v_O$ and $v_I$ are marked. The dependent current source supplies a current

$$i_O = f(x)$$

where we will consider two cases:

1. In the first case, $x$ is the voltage $v_I$, and the current

$$i_O = -G_m v_I.$$

2. In the second case, $x$ is the current $i_I$, and

$$i_O = -\beta i_I.$$

Let us suppose that we are specifically interested in determining $v_O$ as a function of $v_I$ in both cases.

Let us consider the first case in which

$$i_O = -G_m v_I.$$

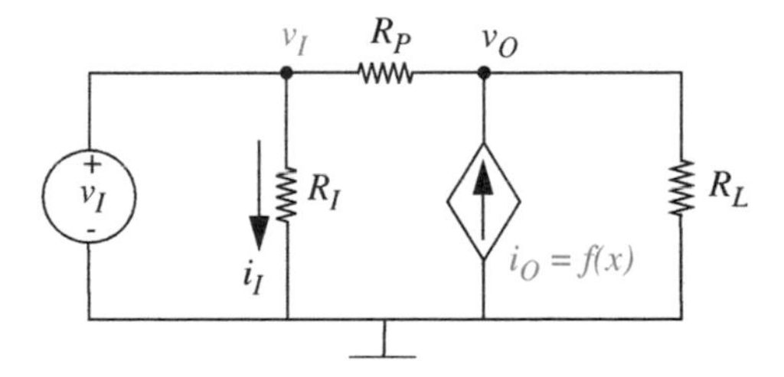


FIGURE 3.27 Another dependent current source circuit.

Notice that $i_O$ is directly expressed in terms of a node voltage, and so we can apply our simple node analysis technique without any modification, remembering, however, to substitute the element law for the dependent source current when writing KCL for the nodes with unknown voltages.

Since the ground and node voltages have been defined as shown in Figure 3.27, Steps 1 and 2 of node analysis are complete.

For Step 3, we write KCL at the node with the unknown voltage $v_O$ by summing the currents into the node as follows:

$$\frac{v_I - v_O}{R_P} + (-G_m v_I) = \frac{v_O}{R_L}. \tag{3.60}$$

Notice that we have used the element law for the dependent current source, namely,

$$i_O = -G_m v_I$$

to substitute for the current into the node from the dependent current source.

By simplifying Equation 3.60, we obtain:

$$v_O = \frac{(1 - G_m R_P)R_L}{R_P + R_L} v_I. \tag{3.61}$$

We have thus expressed $v_O$ as a function of $v_I$ when $i_O = -G_m v_I$.

Let us now consider the second case in which

$$i_O = -\beta i_I.$$

Although not directly expressed in terms of a node voltage, it easy to see that $i_O$ can be expressed in terms of a node voltage by substituting $i_I = v_I/R_I$ as follows:

$$i_O = -\beta \frac{v_I}{R_I}.$$

Thus, as in the first case, we can apply our simple node analysis technique without any modification. Going to Step 3 of node analysis, we write KCL at the node with the unknown voltage $v_O$ by summing the currents into the node as follows:

$$\frac{v_I - v_O}{R_P} + (-\beta \frac{v_I}{R_I}) = \frac{v_O}{R_L}. \tag{3.62}$$

Notice that we have used the element law for the dependent current source, namely,

$$i_O = -\beta \frac{v_I}{R_I}$$

to substitute for the current into the node from the dependent current source.

By simplifying Equation 3.62, we obtain:

$$v_O = \frac{\left(1 - \beta\frac{R_P}{R_I}\right) R_L}{R_P + R_L} v_I. \quad (3.63)$$

We have thus expressed $v_O$ as a function of $v_I$ when $i_O = -\beta i_I$.

---

WWW EXAMPLE 3.12 A MORE COMPLEX DEPENDENT-CURRENT SOURCE PROBLEM

---

### WWW 3.3.4 THE CONDUCTANCE AND SOURCE MATRICES *

## WWW 3.4 LOOP METHOD *

---

WWW EXAMPLE 3.13 LOOP METHOD

---

## 3.5 SUPERPOSITION

Suppose we make the circuit in Figure 3.12 one step more complicated by adding a third source, as shown in Figure 3.33. Straightforward node analysis following the procedure outlined by the node method yields

$$(V_1 - e_1)G_1 + (V_2 - e_1)G_2 + (e_2 - e_1)G_3 = 0 \quad (3.97)$$

$$(e_1 - e_2)G_3 - e_2G_4 + I = 0. \quad (3.98)$$

Collecting the source terms on the left side:

$$V_1G_1 + V_2G_2 = e_1(G_1 + G_2 + G_3) - e_2G_3 \quad (3.99)$$

$$I = -e_1G_3 + e_2(G_3 + G_4). \quad (3.100)$$

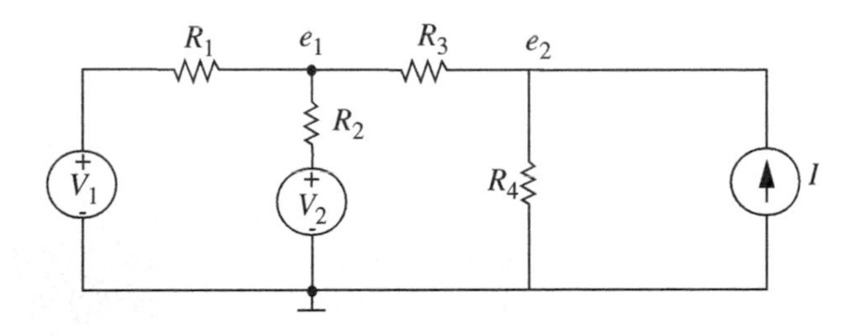


FIGURE 3.33 A network with three sources.

Let us again find $e_1$:

$$e_1 = \frac{(V_1G_1 + V_2G_2)(G_3 + G_4) + IG_3}{(G_1 + G_2 + G_3)(G_3 + G_4) - G_3^2} \tag{3.101}$$

$$= \frac{V_1G_1(G_3 + G_4) + V_2G_2(G_3 + G_4) + IG_3}{G_1G_3 + G_1G_4 + G_2G_3 + G_2G_4 + G_3G_4}. \tag{3.102}$$

Again note the structure of this expression:

- All denominator terms are of the same sign. Thus the denominator cannot be made zero for any nonzero values of conductances. (If the denominator could be made zero, we could get infinite $e_1$ for finite sources values, a violation of conservation of energy.)
- Each term on the right consists of one source term multiplied by a resistive (or conductive) factor. There are no products of source terms.

We now wish to translate these mathematical constraints to circuit constraints, to find simpler methods for analyzing multi-source networks. Specifically, we wish to find the terms in Equation 3.102, by inspection, from Figure 3.33. The mathematics says that, because of linearity, *the first term remains unchanged if the other two sources are set to zero.* We must now interpret this statement in circuit terms. Mathematically, we wish to set variable $V_2$ to zero, so in circuit terms we must set voltage source $V_2$ to zero. By definition, source $V_2$ must now be zero regardless of what current flows through it, that is, it must be a *short circuit.* So in general, setting a voltage source to zero is equivalent in circuit terms to replacing that source by a short circuit. Similarly, setting $I$ to zero means that no current can flow through that branch of the circuit regardless of the terminal voltage. Hence setting a current source to zero is equivalent in circuit terms to replacing that source by an *open circuit.* These are two additional important circuit primitives. Applying these two concepts to Figure 3.33, we can find the first term in Equation 3.102, that is, the part of $e_1$ arising from source $V_1$, by forming a subcircuit from Figure 3.33 with $V_2$ and $I$ set to zero as shown in Figure 3.34a. Thus, in Figure 3.34a, $e_{1A}$ is the voltage component of $e_1$ due to source $V_1$ acting alone. Now $e_{1A}$ can be found by inspection using the voltage-divider primitive:

$$e_{1A} = V_1 \frac{R_2 \| (R_3 + R_4)}{R_1 + R_2 \| (R_3 + R_4)} \tag{3.103}$$

where the two vertical lines are shorthand notation for "in parallel with." The numerator, for example, is $R_2$ in parallel with the sum of $R_3$ and $R_4$. The calculation is somewhat simplified if we use conductance instead of resistance.

Using the conductance form of the voltage-divider relation (Equation 2.50), we find

$$e_{1A} = V_1 \frac{G_1}{G_1 + G_2 + G_3G_4/(G_3 + G_4)} \tag{3.104}$$

where the two conductances in series, $G_3$ and $G_4$ are calculated using Equation 2.58. Both of these expressions are the same as the first term in Equation 3.102, after some manipulation. Note that the form of Equation 3.104 is much simpler and more insightful than the forms in Equations 3.101 and 3.102, because the derivation in terms of the voltage-divider primitive reveals the basic structure of the circuit. But the main point of this development is to show that the effect of source $V_1$ on the node voltage $e_1$ can be found very easily by forming a subcircuit in which $V_2$ and $I$ are set to zero.

By the same argument, the effect of $V_2$ and $I$ on $e_1$ can be calculated using the subcircuits shown in Figure 3.34b and 3.34c, respectively. For $V_2$, sources $V_1$ and $I$ are set to zero, as shown in Figure 3.34b. Clearly circuits 3.34a and 3.34b are identical in topology, so the effect of $V_2$ on $e_1$ can be written from Equation 3.103 by interchanging $R_1$ and $R_2$, or $G_1$ and $G_2$ in Equation 3.104. This will give us $e_{1B}$, the component of $e_1$ due to voltage source $V_2$.

To find the effect of $I$, it is necessary to set both $V_1$ and $V_2$ to zero, that is, replace each by a short circuit, as shown in Figure 3.34c. Now $e_{1C}$ can be

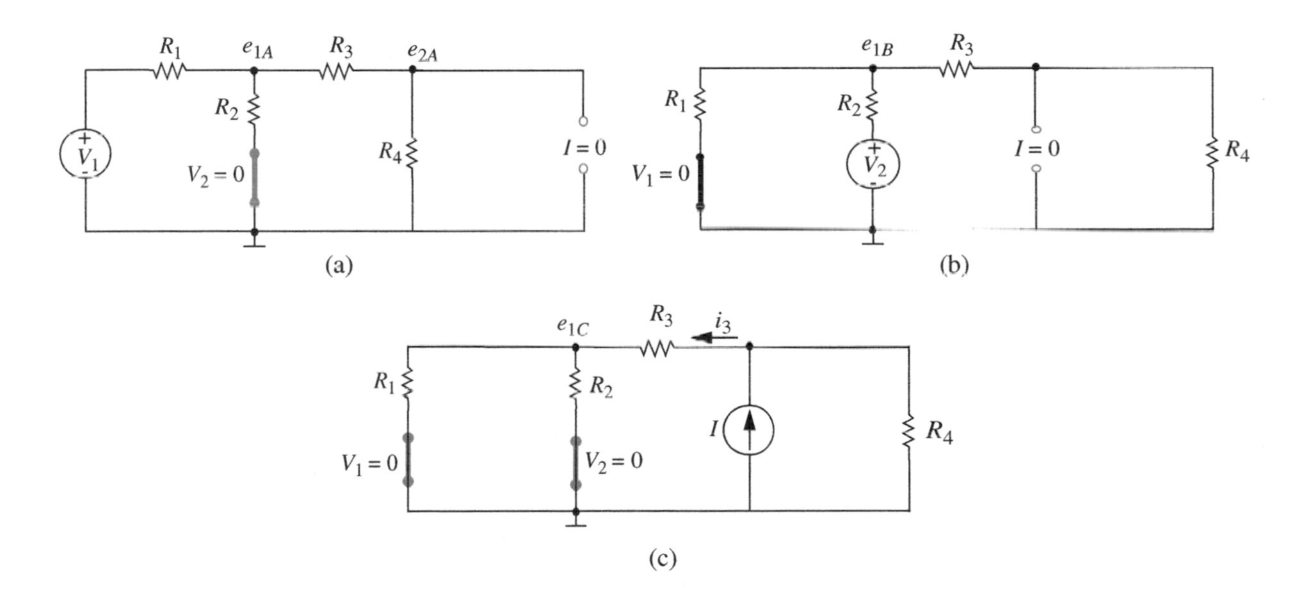


FIGURE 3.34 Subcircuits.

found by noting that the total conductance of the path to the left of the source is, from Equations 2.94 and 2.58,

$$G = \frac{(G_1 + G_2)G_3}{G_1 + G_2 + G_3}. \tag{3.105}$$

Hence, from the current divider relation, the current through $R_3$ is

$$i_{R_3} = \frac{GI}{G + G_4}. \tag{3.106}$$

Now $e_{1C}$ can be found from the relation

$$e_{1C} = \frac{i_{R_3}}{G_1 + G_2} \tag{3.107}$$

$$= \frac{GI}{(G + G_4)(G_1 + G_2)} \tag{3.108}$$

which, on substitution of Equation 3.105 and simplification, reduces to

$$e_{1C} = \frac{IG_3}{(G_1 + G_2)G_3 + G_4(G_1 + G_2) + G_3 G_4}. \tag{3.109}$$

This is equivalent to the third term in Equation 3.102.

This example illustrates both the use of superposition to solve a network with several sources, and also shows how primitives (elementary procedures) can be used to solve circuits by inspection. Generalizing, we note that any messy linear network — the one in Figure 3.35, for example — must somehow yield to straightforward network analysis and lead to a set of equations of the form

$$\begin{aligned} V_1 G_{1a} + V_2 G_{1b} + \cdots + I_1 + \cdots &= e_1 G_{11} + e_2 G_{12} + \cdots \\ V_1 G_{2a} + \cdots \qquad \cdots \qquad \cdots \qquad \cdots &= e_1 G_{21} + e_2 G_{22} + \cdots \\ V_1 G_{3a} + \cdots \qquad \cdots \qquad \cdots \qquad \cdots &= e_1 G_{31} + \cdots \end{aligned} \tag{3.110}$$

These have been written in the standard form, with source terms on the left in each equation. All of the unknown variables appear on the right side, each multiplied by conductances: the sum of the appropriate "self" conductances for terms along the main diagonal, and the sum of the appropriate "mutual" conductances elsewhere.

Further, the solution of such a set of linear simultaneous equations will always result in an expression of the general form of Equation 3.102, in which the voltage or current we are trying to evaluate will be equal to a *sum* of terms each involving only one source.

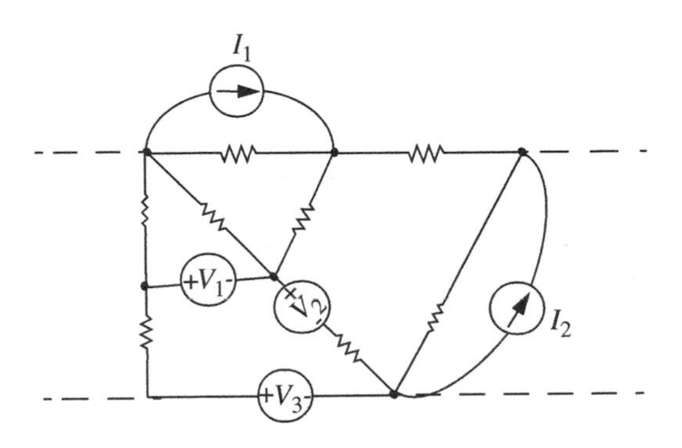


FIGURE 3.35 A resistive network.

The *superposition theorem* thus states that *in a linear network with a number of independent sources, the response can be found by summing the responses to each independent source acting alone, with all other independent sources set to zero*. These individual responses can be found very readily by forming subcircuits in which all independent sources except one are set to zero.

Accordingly, the superposition method for linear networks can be stated as follows:

*The Superposition Method*

1. For each independent source, form a subcircuit with all other independent sources set to zero. Setting a voltage source to zero implies replacing the voltage source with a short circuit, and setting a current source to zero implies replacing the current source with an open circuit.
2. From each subcircuit corresponding to a given independent source, find the response to that independent source acting alone. This step results in a set of individual responses.
3. Obtain the total response by summing together each of the individual responses.

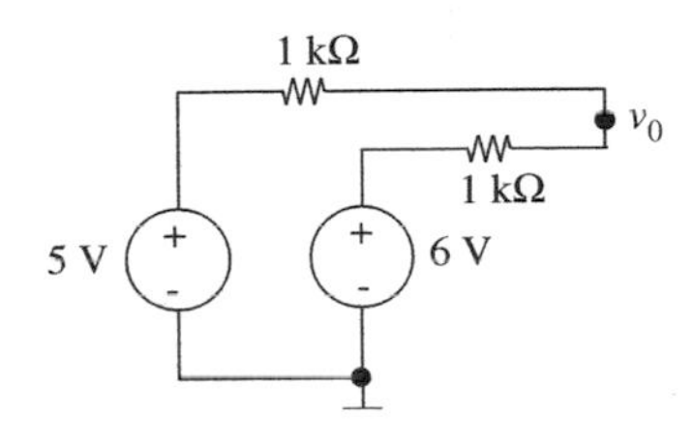


FIGURE 3.36 Circuit for performing superposition analysis.

---

EXAMPLE 3.14 SUPERPOSITION ANALYSIS OF AVERAGING CIRCUIT Show that the node voltage $v_0$ in the circuit shown in Figure 3.36 is the average of the two input voltages using the method of superposition.

By the method of superposition, the voltage $v_0$ can be determined by summing the responses of each of the sources acting alone. We will first obtain $v_{05}$, the response of the 5-V source acting alone. The subcircuit corresponding to the 5-V source acting alone is shown in Figure 3.37. Notice we have shorted the 6 V source.

By the voltage divider action, we can write

$$v_{05} = \frac{1\ \text{k}\Omega}{1\ \text{k}\Omega + 1\ \text{k}\Omega} 5\ \text{V} = \frac{5}{2}\ \text{V}.$$

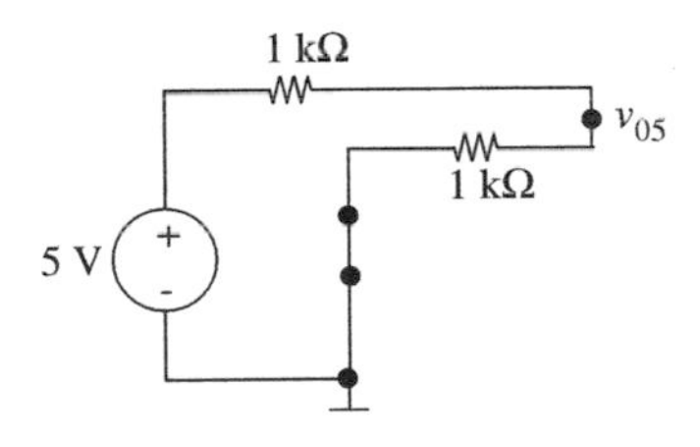


FIGURE 3.37 Circuit with 5-V source acting alone.

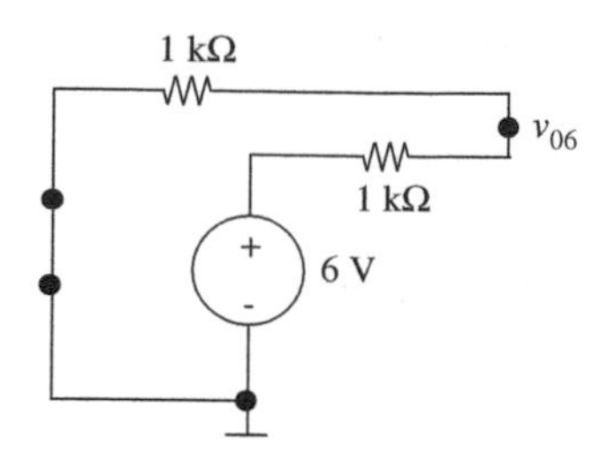


FIGURE 3.38 Circuit with 6-V source acting alone.

Next, we obtain $v_{06}$, the response of the 6-V source acting alone. The subcircuit corresponding to the 6-V source acting alone is shown in Figure 3.38. In this case, we have shorted the 5-V source.

Again, by the voltage divider action, we can write

$$v_{06} = \frac{1\ \text{k}\Omega}{1\ \text{k}\Omega + 1\ \text{k}\Omega} 6\ \text{V} = \frac{6}{2}\ \text{V}.$$

We now sum the two partial responses to obtain

$$v_0 = v_{05} + v_{06} = \frac{5+6}{2} = 5.5\ \text{V}.$$

It is easy to see that $v_0$ is the average of the two input voltages.

---

EXAMPLE 3.15 APPLYING THE METHOD OF SUPERPOSITION Figure 3.39 shows a circuit containing an independent voltage source and an independent current source. Determine the current $I$.

We will use the method of superposition to solve this circuit in two different ways. First, we will obtain the node voltage $e$ using superposition, and then, using the value of $e$, obtain the current $I$. Our second approach will directly determine $I$ using the method of superposition.

## First Method

Let us first determine the value of $e$ using superposition. By the method of superposition, the voltage $e$ can be determined by summing the responses of each of the sources acting alone. We will first obtain $e_v$, the response of the voltage source acting alone. The subcircuit corresponding to the voltage source acting alone is shown in Figure 3.40. Notice we have turned the current source off by open-circuiting it.

By the voltage divider action, we can write

$$e_v = 1\frac{2}{2+2} = \frac{1}{2}\ \text{V}.$$

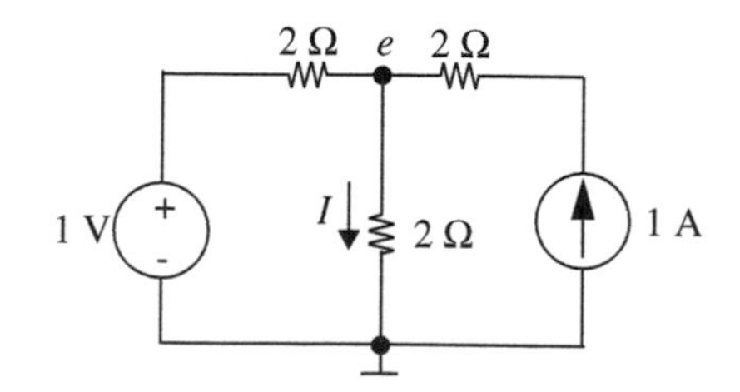


FIGURE 3.39 Circuit with two independent sources.

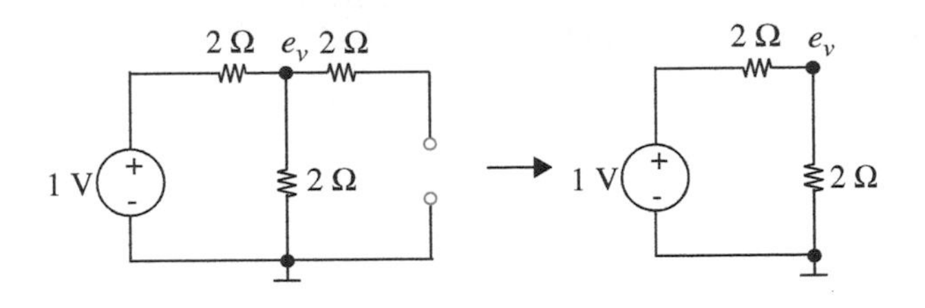


FIGURE 3.40 Subcircuit corresponding to the voltage source acting alone.

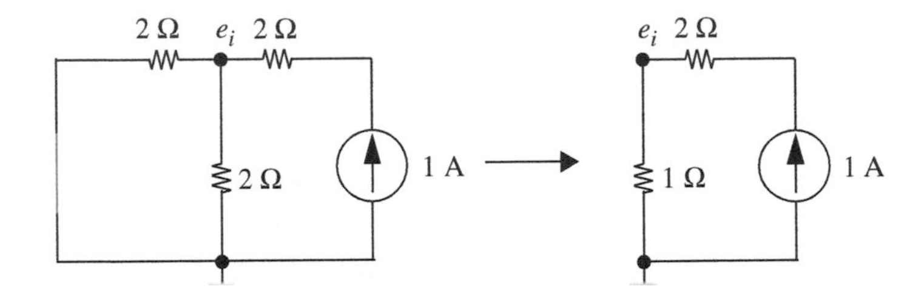


FIGURE 3.41 Subcircuit corresponding to the current source acting alone.

Next, we obtain $e_i$, the response of the current source acting alone. The subcircuit corresponding to the current source acting alone is shown in Figure 3.41. In this case, we have shorted the voltage source.

We first simplify the subcircuit by replacing the pair of 2-Ω resistors in parallel with an equivalent 1-Ω resistor as depicted in Figure 3.41. Then, since the 1-A current flows through each of the resistors, the voltage across the 1-Ω resistor is equal to $e_i$. In other words,

$$e_i = 1 \text{ A} \times 1 \ \Omega = 1 \text{ V}.$$

We now sum the two partial responses to obtain the total response $e$. That is,

$$e = e_v + e_i = \frac{1}{2} \text{ V} + 1 \text{ V} = 1.5 \text{ V}.$$

We can now determine $I$ as

$$I = \frac{e}{2 \ \Omega} = 0.75 \text{ A}.$$

## Second Method

Next, we will directly determine $I$ using superposition. Superposition says that $I$ can be determined by summing the currents generated by each of the sources acting alone. We will first obtain $I_v$, the current due to the voltage source acting alone. The subcircuit corresponding to the voltage source acting alone is shown in Figure 3.42.

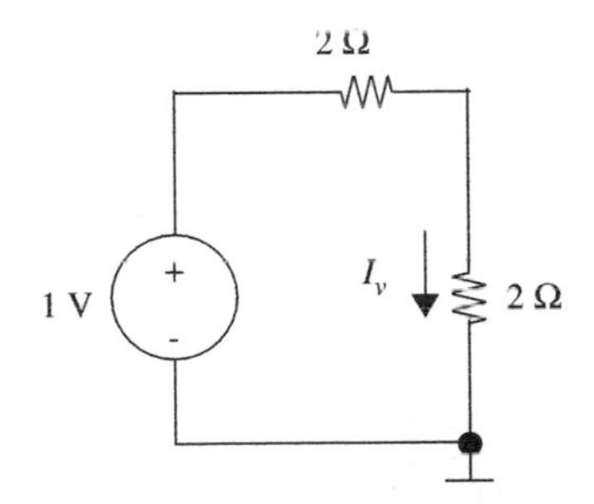


FIGURE 3.42 Subcircuit corresponding to the voltage source acting alone.

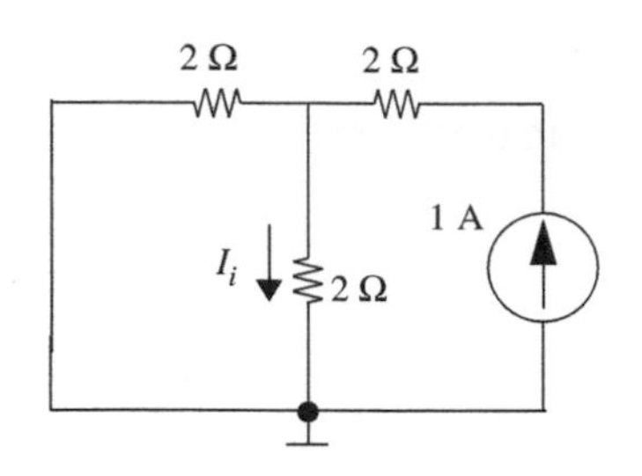

**FIGURE 3.43** Subcircuit corresponding to the current source acting alone.

The current in the subcircuit is given by the voltage divided by the sum of the resistors. In other words,

$$I_v = \frac{1 \text{ V}}{2\ \Omega + 2\ \Omega} = 0.25 \text{ A}.$$

Next, we obtain $I_i$, the response of the current source acting alone. The subcircuit corresponding to the current source acting alone is shown in Figure 3.43.

By the current divider relation, it is easy to see that $I_i = 0.5$ A, since the 1-A current supplied by the current source divides equally into the two branches of the subcircuit in Figure 3.43.

We now sum the two partial responses to obtain the total response $I$. That is,

$$I = I_v + I_i = 0.25 \text{ A} + 0.5 \text{ A} = 0.75 \text{ A}.$$

---

EXAMPLE 3.16 RESISTIVE ADDER CIRCUIT An elementary resistive adding circuit is shown in Figure 3.44a. This circuit might be used to add together a number of microphone signals before sending them to one amplifier. (Notice that this circuit is a generalization of the circuit in Figure 3.36.) We shall discover better ways of building such a circuit in later chapters, but the present form serves as a good illustration of the principle of superposition.

From the preceding discussion, the effect on the output voltage $V_o$ of the source $V_1$ acting alone can be found by forming a subcircuit in which all other independent sources are set to zero, which in this case means replacing $V_2$, $V_3$, and $V_4$ by short circuits, as shown in Figure 3.44b. Now $V_{oa}$, the response to $V_1$ alone, can be found by inspection

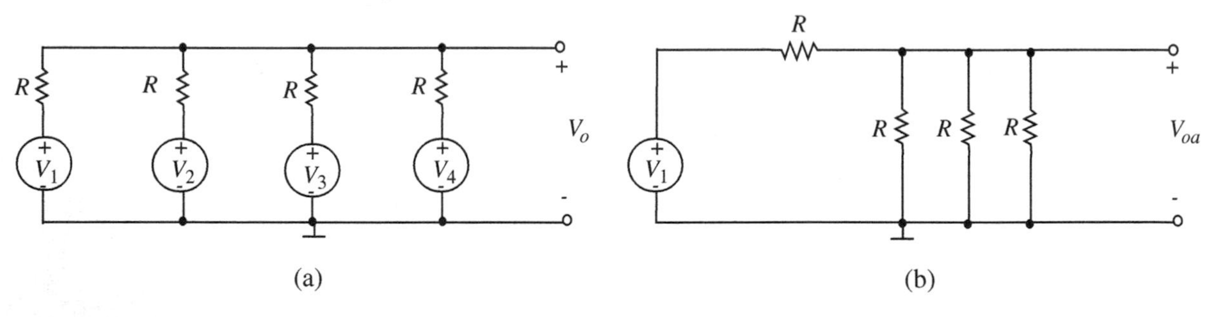

**FIGURE 3.44** Resistive adding circuit.

using the voltage-divider relation

$$V_{oa} = \frac{R/3}{R + R/3} V_1. \quad (3.111)$$

The complete response is the sum of four such terms, which in this special case all have the same coefficient

$$V_o = \frac{1}{4}(V_1 + V_2 + V_3 + V_4). \quad (3.112)$$

Note that there is no restriction on the nature of the sources (other than frequency limits, etc., as discussed in Chapter 1). The sources could be DC, sine waves or square waves, speech, or a mixture of these. Equation 3.112 states that the output will be the sum of these individual signals, each multiplied by a constant, a "scaling factor." If the inputs were four sine waves, each at a different frequency, then the output voltage would be the sum of these four sinusoids, appropriately scaled. No other frequencies would be present in the output signal. Thus a further consequence of linearity is that, whatever frequencies are present at the input or inputs of a linear system, these and only these frequencies will appear at the output.

---

WWW EXAMPLE 3.17 SUPERPOSITION APPLIED TO A BEEHIVE NETWORK

---

### 3.5.1 SUPERPOSITION RULES FOR DEPENDENT SOURCES

When the dependencies are linear, dependent sources are amenable to the set of analyses discussed earlier in Chapters 2 and 3. Care must be taken, however, in applying the superposition principle. Recall that the principle of superposition allows linear multisource networks to be solved for one source at a time by setting all other independent sources to zero. Setting a voltage source to zero means replacing it with a short circuit; a current source set to zero is an open circuit. The complete response is the sum of the responses to each individual source.

What do we do about dependent sources? A practical way is to leave all the *dependent* sources in the circuit. The network can then be solved for one *independent* source at a time by setting all other independent sources to zero, and summing the individual responses.

Alternatively, the dependent sources could be treated as independent sources, and in a final step of the analysis, their dependencies must be back-substituted in terms of other network parameters. However, this method tends to be impractical.

**EXAMPLE 3.18 A SINGLE DEPENDENT SOURCE AND SUPERPOSITION** Consider the circuit in Figure 3.49. It contains two independent sources and one dependent source. Using the superposition method, let us derive the output voltage $v_O$.

We will solve the circuit by leaving the dependent current source in the circuit and summing the responses of each of the independent sources acting alone.

### 1-V Source Acting Alone

Figure 3.50 shows the circuit corresponding to the 1-V source acting alone, where $v_{O1}$ is the corresponding response. Notice that the dependent current source has been left in the circuit, and the 2-V source has been shorted out.

By the voltage divider relation, we know that

$$v_1 = 0.5 \text{ V}.$$

Thus,

$$v_{O1} = \frac{1}{100} v_1 \times 1 \text{ k}\Omega = 5 \text{ V}.$$

### 2-V Source Acting Alone

Figure 3.51 shows the circuit corresponding to the 2-V source acting alone, where $v_{O2}$ is the corresponding response. By the voltage-divider relation, we know that

$$v_2 = 1 \text{ V}.$$

Thus,

$$v_{O2} = \frac{1}{100} v_2 \times 1 \text{ k}\Omega = 10 \text{ V}.$$

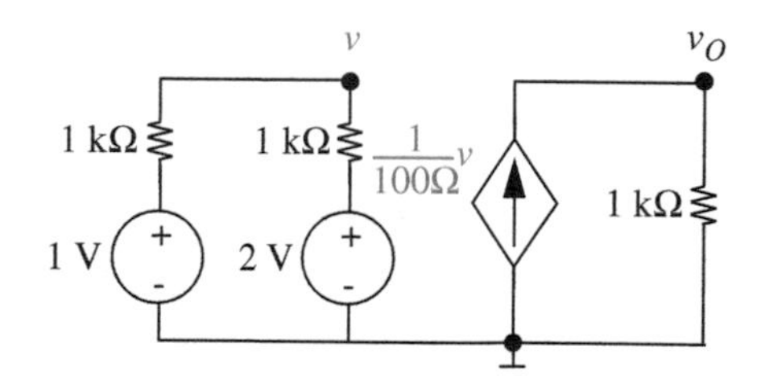


FIGURE 3.49 Circuit containing two independent sources and one dependent source.

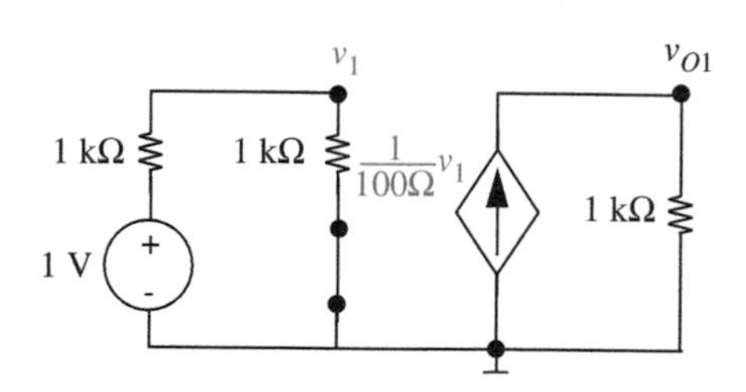


FIGURE 3.50 Subcircuit corresponding to the 1-V source acting alone.

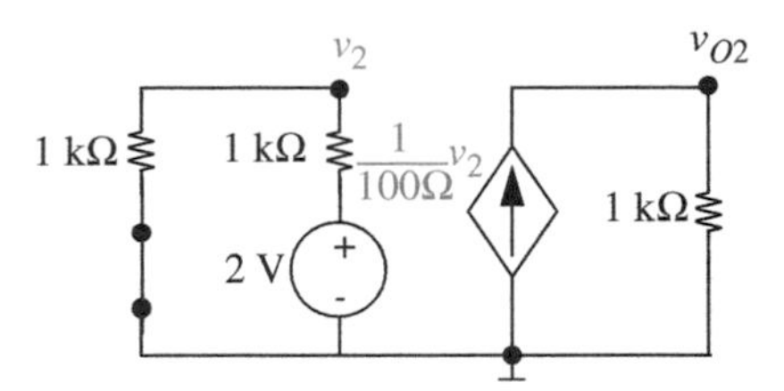


FIGURE 3.51 Subcircuit corresponding to the 2-V source acting alone.

Summing the two responses, we get the total response as

$$v_O = v_{O1} + v_{O2} = 15 \text{ V}.$$

---

EXAMPLE 3.19 MULTIPLE DEPENDENT SOURCES AND SUPERPOSITION As a more complicated example, consider the circuit in Figure 3.52. Using the superposition method, let us derive the output voltage $v_o$ as a function of $v_i$.

This circuit has two dependent current sources and two independent voltage sources ($v_1$ and $v_2$). We will solve this problem by leaving both the dependent current sources in the circuit and summing the responses of each of the independent sources acting alone. We also define two intermediate variables, the node voltages $v_a$ and $v_b$.

### $v_1$ Acting Alone

We will first obtain the response with $v_1$ acting alone. Figure 3.53 shows the circuit corresponding to $v_1$ acting alone. $v_{o1}$ is the corresponding response. Notice that the

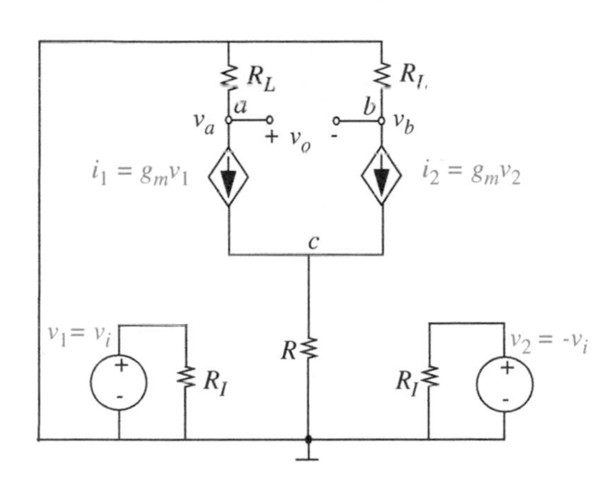


FIGURE 3.52 Circuit with multiple dependent sources.

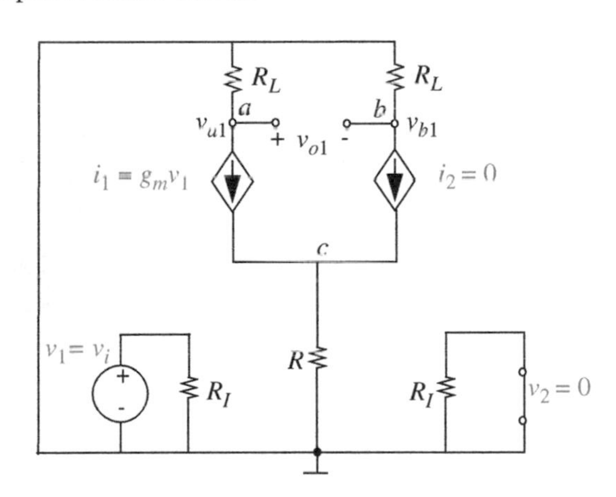


FIGURE 3.53 Subcircuit corresponding to $v_1$ acting alone.

dependent current sources have been left in the circuit, and $v_2$ has been shorted out. Since $v_2 = 0$, we find that $i_2 = 0$ (in other words, the dependent current source behaves like an open circuit).

We will first determine $v_{a1}$ and $v_{b1}$, the node voltages at the nodes $a$ and $b$ due to $v_1$ acting alone. We will then determine $v_{o1}$ as their difference.

Since $i_2 = 0$, there is no voltage drop across the resistor $R_L$ connected to node $b$. So, node $b$ will be at ground potential. In other words,

$$v_{b1} = 0.$$

We can obtain $v_{a1}$ by using KVL as

$$v_{a1} = 0 - i_1 R_L = -g_m v_1 R_L = -g_m v_i R_L.$$

Therefore

$$v_{o1} = v_{a1} - v_{b1} = -g_m v_i R_L.$$

## $v_2$ Acting Alone

Figure 3.54 shows the circuit corresponding to $v_2$ acting alone. $v_{o2}$ is the corresponding response. In this circuit, since $v_1 = 0$, we find that $i_1 = 0$.

Since $i_1 = 0$, there is no voltage drop across the resistor $R_L$ connected to node $a$, and so, this time around, node $a$ will be at ground potential. In other words,

$$v_{a2} = 0.$$

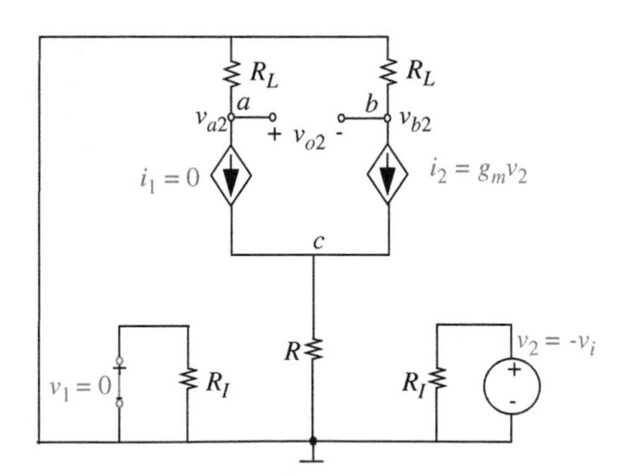


FIGURE 3.54 Subcircuit corresponding to $v_2$ acting alone.

We can obtain $v_{b2}$ by using KVL as

$$v_{b2} = 0 - i_2 R_L = -g_m v_2 R_L = -g_m(-v_i) R_L = g_m v_i R_L.$$

Therefore

$$v_{o2} = v_{a2} - v_{b2} = -g_m v_i R_L.$$

We can now obtain the total response by summing the responses to each of the independent sources acting alone. In other words,

$$v_o = v_{o1} + v_{o2} = -2 g_m v_i R_L.$$

---

## 3.6 THÉVENIN'S THEOREM AND NORTON'S THEOREM

### 3.6.1 THE THÉVENIN EQUIVALENT NETWORK

A simple extension of the concept of superposition yields two additional network theorems of great power, which allow us to suppress a lot of detail in circuit analysis and focus attention only on that part of a network we are really interested in. Consider, for example, a battery, or a high-fidelity power amplifier, or a wall outlet for 110-V AC power, or a power supply for a computer. What is the simplest way to describe the electrical properties of each of these systems at its output terminals? Is one parameter needed, or ten, or fifty? Clearly the voltage measured with a high-quality meter that draws negligible current is one important parameter (the *open-circuit voltage* mentioned in Section 1.7). Likewise we would want to know the frequency: zero frequency for the battery, 60 hertz (or 50 or even 25 in some countries) for the power line, etc. But we have already observed another effect that is important. When current is drawn from any of these systems, the voltage at the terminals drops. Depending on the quality of the wiring in a dormitory, the lights may dim noticeably when a toaster is plugged into the same circuit. Or the voltage of the flashlight battery will drop when a bulb is connected and current flows, as noted in Section 1.7. How can this effect be characterized? For the battery, is it necessary to make measurements at 100 current levels, and plot a curve of the characteristic?

If the system is linear, then the answer to this question is very simple. We will show that any collection of voltage sources, current sources, and resistors can be represented at any one pair of terminals by one voltage source and one resistor, or by one current source and one resistor. The graphical construction of Figure 1.43 already hinted at this fact, but we present here a more formal proof. We start with a general linear network containing sources and resistors, shown as an amorphous box in Figure 3.55a. We presume that the only two

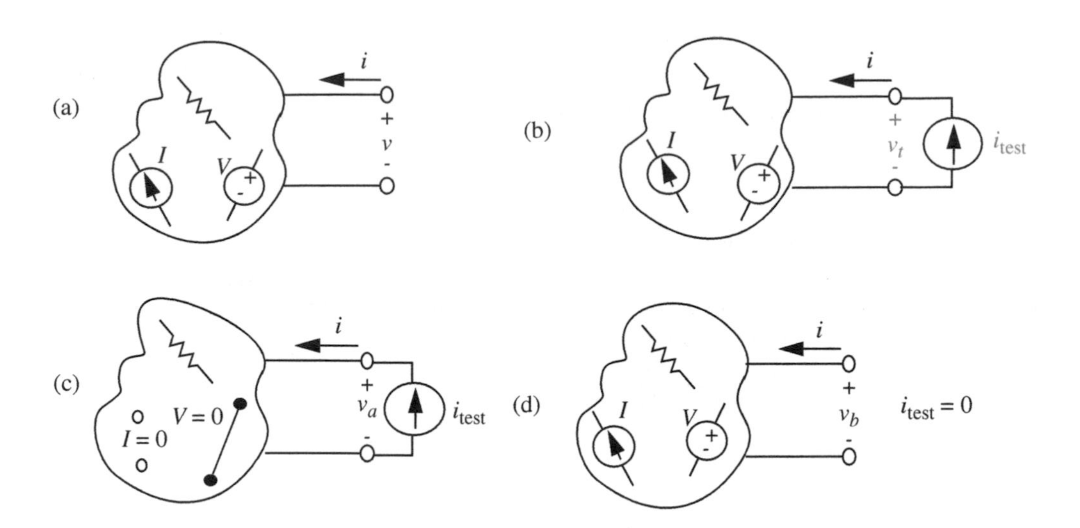


**FIGURE 3.55** Derivation of the Thévenin network.

terminals we are interested in are shown emerging on the right. We wish to find the relationship between $v$ and $i$ at these terminals.

To find $v$ in terms of $i$, we need to apply some form of excitation, and measure the response. The derivation is simplest if we use either a voltage source or a current source, rather than a complicated excitation network. In Figure 3.55b, we have chosen to apply a *test current source* to the terminals. To calculate the response $v_t$ by superposition, first set all the internal independent sources to zero, as in Figure 3.55c, and calculate the voltage $v_a$. As discussed in Section 3.5.1, dependent sources are left as is. Then set $i_{test}$ to zero, as in Figure 3.55d, and calculate $v_b$. The desired value of $v_t$ is the sum $v_a + v_b$. From Figure 3.55c,

$$v_a = i_{test} R_t \tag{3.113}$$

where $R_t$ is the net resistance measured between the two terminals when all internal independent sources are set to zero. Resistance $R_t$ is called the *Thévenin Equivalent Resistance*. From Figure 3.55d, $v_b$ is obviously just the voltage appearing at the terminals of the original network when no current is flowing; we call this the *open-circuit voltage*. That is,

$$v_b = v_{oc}. \tag{3.114}$$

Now by superposition,

$$v_t = v_a + v_b = v_{oc} + i_{test} R_t. \tag{3.115}$$

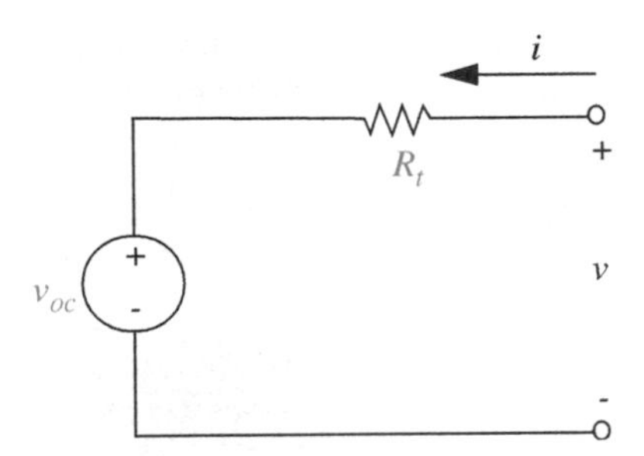


FIGURE 3.56 Thévenin equivalent.

This simple relation between voltage and current at a pair of terminals applies regardless of the complexity of the network, provided only that the network is linear. Thus, returning to the questions posed earlier, if we specify the *open-circuit voltage* and the *Thévenin equivalent resistance* of the battery, or the computer power supply, or the wall outlet, then to the extent that such systems can be considered to be linear, we have completely characterized the system as it appears at its terminals.

Equation 3.115 should be familiar from Section 1.7. It is the same as Equation 1.23, the volt-ampere relation for a voltage source in series with a resistor. In graphical terms it is the equation of a straight line in the $v$–$i$ plane with slope $1/R_t$ and voltage axis intercept $v_{OC}$. So the preceding calculation can be interpreted in terms of a circuit called the *Thévenin equivalent circuit* shown in Figure 3.56. If $v_{OC}$ and $R_t$ are calculated using the subcircuits in Figure 3.55c and 3.55d, then this circuit and the one in Figure 3.55a are equivalent, in the sense that any measurement at the indicated terminals are equivalent. In other words, any measurement at the indicated terminals of the two circuits will yield identical results.

Two independent measurements on a circuit are required to determine the parameters for the Thévenin model. One appropriate pair of measurements is as follows. The source parameter $v_{OC}$ is the voltage measured or calculated at the desired terminal pair when no current is flowing at these terminals:

$$v_{OC} = v_t \,\big|_{i_{\text{test}}=0} \,. \tag{3.116}$$

$R_t$ is the resistance measured or calculated at the desired terminal pair when all internal independent sources are set to zero:

$$R_t = \left. \frac{v_t}{i_{\text{test}}} \right|_{\text{internal source}=0} \,. \tag{3.117}$$

Summarizing, the Thévenin method allows us to abstract the behavior of a linear network at a given pair of terminals as a voltage source in series with a resistor. The voltage source in series with a resistor is called the Thévenin equivalent circuit of the network. The Thévenin equivalent circuit can be used to model the effect of the given network on other circuits external to the network.

*A Method for Determining the Thévenin Equivalent Circuit* The Thévenin equivalent circuit for any linear network at a given pair of terminals consists of a voltage source $v_{TH}$ in series with a resistor $R_{TH}$. The voltage $v_{TH}$ and resistance $R_{TH}$ can be obtained as follows:

1. $v_{TH}$ can be found by calculating or measuring the open-circuit voltage at the designated terminal pair on the original network.
2. $R_{TH}$ can be found by calculating or measuring the resistance of the open-circuit network seen from the designated terminal pair with all independent sources internal to the network set to zero. That is, with independent voltage sources replaced with short circuits, and independent current sources replaced with open circuits. (Dependent sources must be left intact, however.)

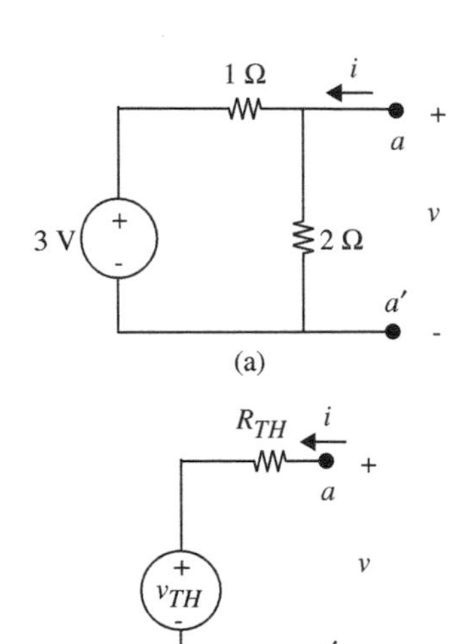


**FIGURE 3.57** Example circuit to illustrate the Thévenin method: (a) a network; (b) its Thévenin equivalent network.

---

EXAMPLE 3.20 THÉVENIN METHOD Let us now illustrate the method using a simple example. Figure 3.57a shows a network and Figure 3.57b shows its Thévenin equivalent network viewed from the network's *aa′* port. Determine the values of $v_{TH}$ and $R_{TH}$.

By the first step of the Thévenin method, the voltage $v_{TH}$ is given by the open-circuit voltage of the network at the *aa′* port. The open-circuit voltage is the voltage at the *aa′* network port when there is no *external* circuit element connected across the port. (Note that the 2-Ω resistor is internal to the network and should not be disconnected.) Figure 3.57a shows this situation. The open-circuit voltage that would be measured at the *aa′* port is given by the voltage-divider relation as

$$v_{TH} = 3\text{ V}\frac{2\ \Omega}{1\ \Omega + 2\ \Omega} = 2\text{ V}.$$

By the second step of the Thévenin method, the resistance $R_{TH}$ is found by measuring the resistance of the open-circuit network seen from the *aa′* port with the independent voltage source set to zero; that is, with the voltage source replaced with a short circuit. The network with the voltage source replaced with a short is shown in Figure 3.58.

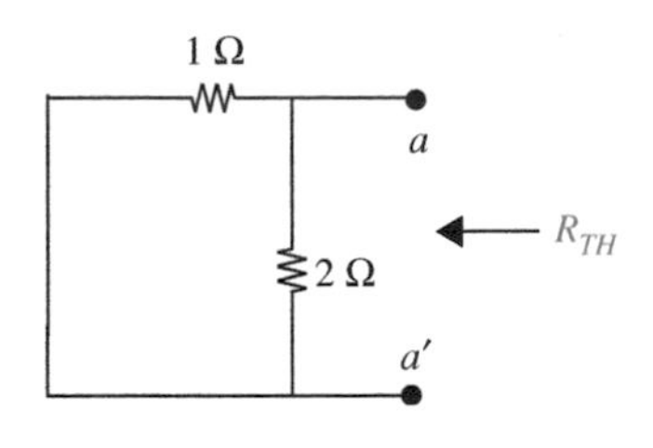


**FIGURE 3.58** Network with the voltage source replaced with a short.

The resistance viewed from the *aa′* port is given by

$$R_{TH} = 1\|2 = \frac{2}{3}\ \Omega.$$

The resulting Thévenin equivalent circuit is drawn in Figure 3.59.

---

EXAMPLE 3.21 MORE ON THE THÉVENIN METHOD Let us now work out a couple of related examples to illustrate the power of the Thévenin

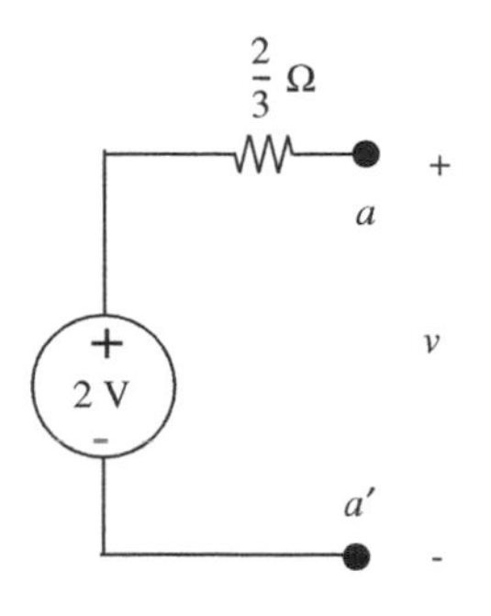


FIGURE 3.59 The resulting Thévenin equivalent circuit.

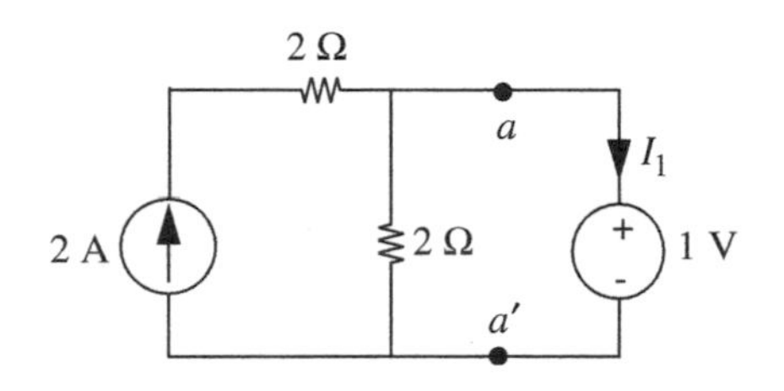


FIGURE 3.60 Circuit to illustrate the power of the Thévenin method.

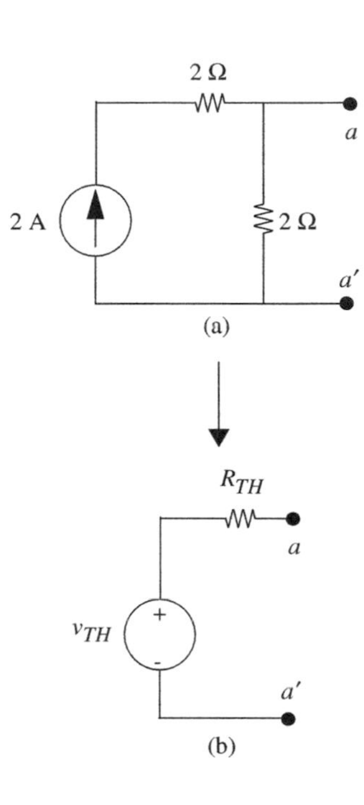


FIGURE 3.61 Thévenin equivalent network.

method. First, suppose we are asked to determine the current $I_1$ through the voltage source in the circuit in Figure 3.60.

Let us use the Thévenin method to obtain the desired current. To apply the Thévenin method, we will replace the network to the left of the voltage source (that is, to the left of the $aa'$ terminal pair, and depicted in Figure 3.61a) with its Thévenin equivalent network (depicted in Figure 3.61b). Once this replacement is made, as illustrated in Figure 3.62, then, the current $I_1$ can be written by inspection as

$$I_1 = \frac{v_{TH} - 1\text{ V}}{R_{TH}}. \tag{3.118}$$

$v_{TH}$ and $R_{TH}$ are the Thévenin equivalent parameters. The first step of the Thévenin method is to measure $v_{TH}$. As shown in Figure 3.63, $v_{TH}$ is the open-circuit voltage measured at the $aa'$ port.

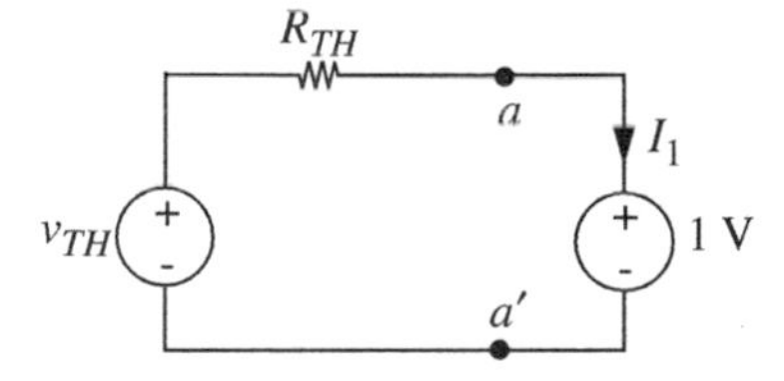


FIGURE 3.62 Circuit with network to the left of the $aa'$ terminal pair replaced with its Thévenin equivalent.

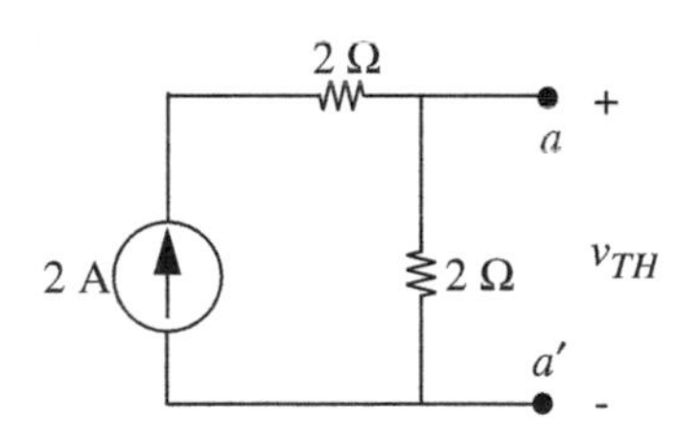


FIGURE 3.63 Open-circuit voltage.

Since the 2-A current flows through both the 2-Ω resistors in Figure 3.63, $v_{TH}$ can be written by inspection as

$$v_{TH} = 2\text{ A} \times 2\ \Omega = 4\text{ V}.$$

By the second step of the Thévenin method, the resistance $R_{TH}$ is found by measuring the resistance of the open-circuit network seen from the *aa′* port with the independent current source set to zero; that is, with the current source replaced with an open circuit as illustrated in Figure 3.64. It is easy to see that

$$R_{TH} = 2\ \Omega.$$

Having determined the Thévenin equivalent parameters $v_{TH}$ and $R_{TH}$, we can now obtain $I_1$ from Equation 3.118 as

$$I_1 = \frac{4\text{ V} - 1\text{ V}}{2\ \Omega} = \frac{3}{2}\text{ A}.$$

Notice that in this example the Thévenin method has allowed us to tackle a given problem (the circuit in Figure 3.60) by splitting it into three trivial subproblems, namely, the circuits in Figures 3.63, 3.64, and 3.62.

To further illustrate the power of the Thévenin method, suppose that the 1-*V* source in Figure 3.60 is replaced by a 10-Ω resistor as illustrated in Figure 3.65, and we are asked to find the current $I_2$ through the 10-Ω resistor.

We first notice that the network to the left of the terminal pair *aa′* in Figure 3.65 is unchanged from that in Figure 3.60. Thus, from the viewpoint of determining a parameter relating to the network on the right side of the *aa′* terminal pair, we can replace that network on the left with its Thévenin equivalent determined previously as illustrated in Figure 3.66.

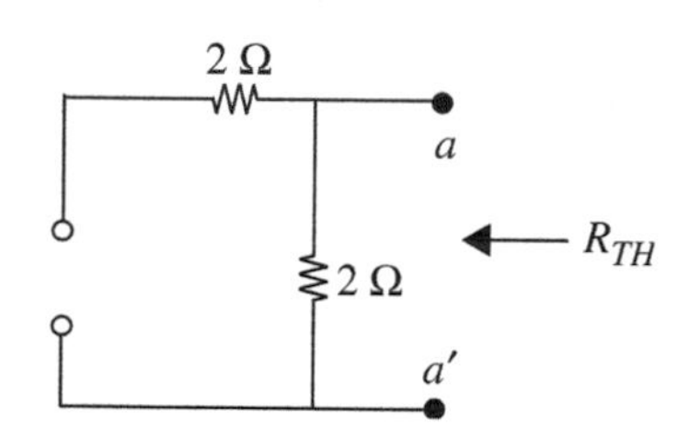


FIGURE 3.64 Measuring $R_{TH}$.

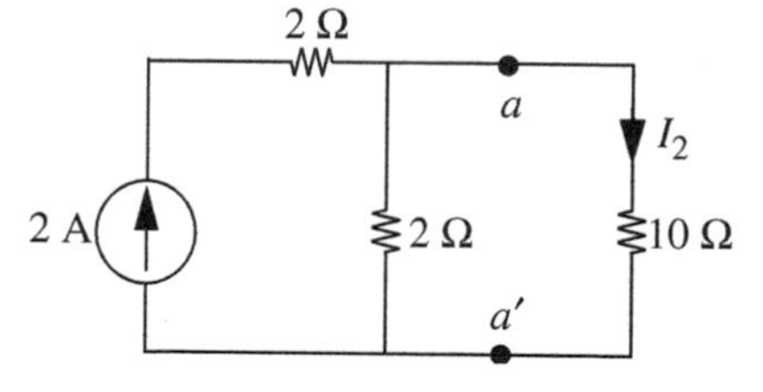


FIGURE 3.65 Circuit to further illustrate the power of the Thévenin method.

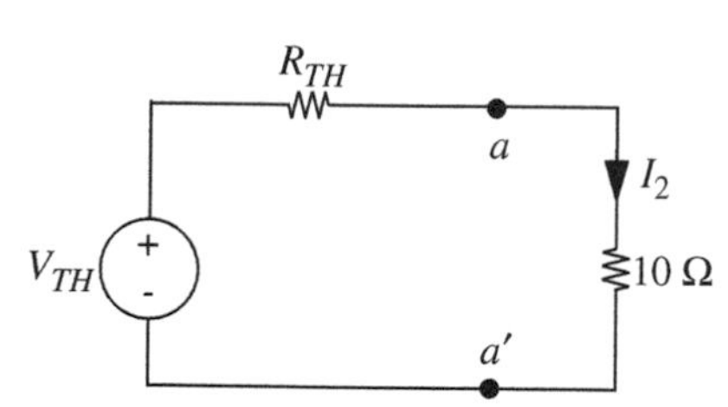


FIGURE 3.66 Circuit with network to the left of the *aa′* terminal pair replaced with its Thévenin equivalent.

The current $I_2$ can be quickly determined from the network in Figure 3.66 as

$$I_2 = \frac{v_{TH}}{R_{TH} + 10\ \Omega}.$$

We know that $v_{TH} = 4$ V and $R_{TH} = 2\ \Omega$, and so $I_2 = 1/3$ A.

---

EXAMPLE 3.22 BRIDGE CIRCUIT Determine the current $I$ in the branch $ab$ in the circuit in Figure 3.67.

There are many approaches that we can take to obtain the current $I$. For example, we could apply the node method and determine the node voltages at nodes $a$ and $b$ and thereby determine the current $I$. However, since we are interested only in the current $I$, a full blown node analysis is not necessary; rather we will find the Thévenin equivalent network for the subcircuit to the left of the $aa'$ terminal pair (Network A) and for the subcircuit to the right of the $bb'$ terminal pair (Network B), and then using these subcircuits solve for the current $I$.

Let us first find the Thévenin equivalent for Network A. This network is shown in Figure 3.68a. Let $v_{THA}$ and $R_{THA}$ be the Thévenin parameters for this network.

We can find $v_{THA}$ by measuring the open-circuit voltage at the $aa'$ port in the network in Figure 3.68b. We find by inspection that

$$v_{THA} = 1\ \text{V}$$

Notice that the 1-A current flows through each of the 1-Ω resistors in the loop containing the current source, and so $v_1$ is 1 V. Since there is no current in the resistor connected to the $a'$ terminal, the voltage $v_2$ across that resistor is 0. Thus $v_{THA} = v_1 + v_2 = 1$ V.

We find $R_{THA}$ by measuring the resistance looking into the $aa'$ port in the network in Figure 3.68c. The current source has been turned into an open circuit for the purpose

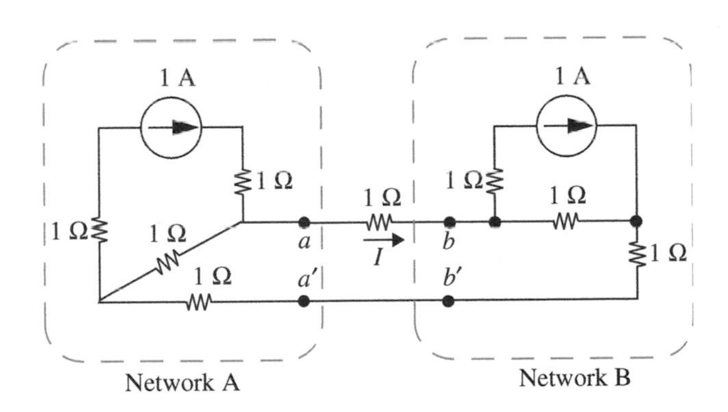


FIGURE 3.67 Determining the current in the branch $ab$.

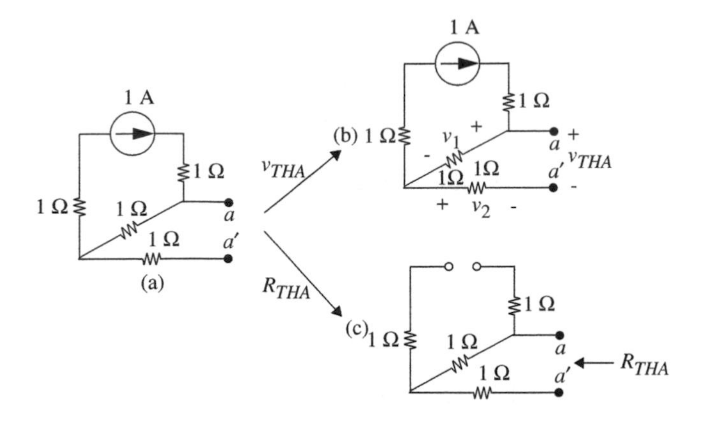


FIGURE 3.68 Finding the Thévenin equivalent for Network A.

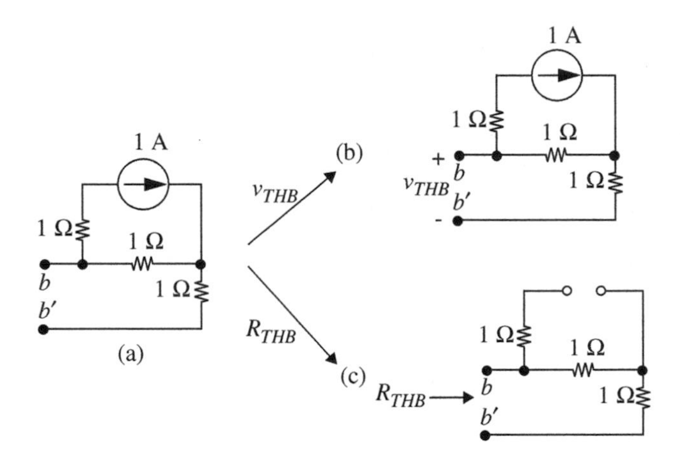


FIGURE 3.69 Finding the Thévenin equivalent for Network B.

of measuring $R_{THA}$. By inspection, we find that

$$R_{THA} = 2\ \Omega.$$

Let us now find the Thévenin equivalent for Network B shown in Figure 3.69a. Let $v_{THB}$ and $R_{THB}$ be the Thévenin parameters for this network.

$v_{THB}$ is the open-circuit voltage at the $bb'$ port in the network in Figure 3.69b. Using reasoning similar to that for $v_{THA}$ we find

$$v_{THB} = -1\ \mathrm{V}.$$

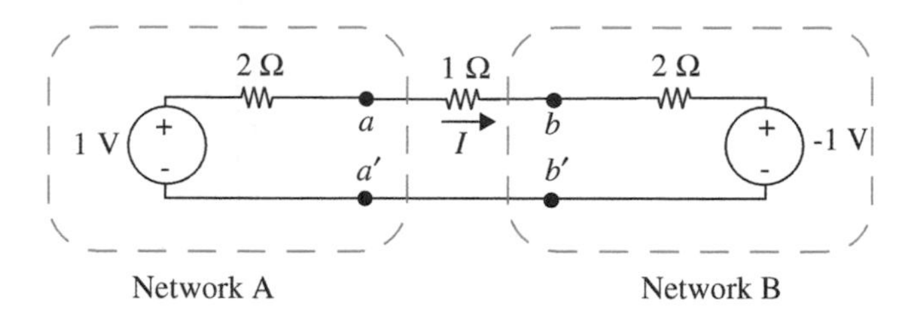


**FIGURE 3.70** Networks A and B replaced by their Thévenin equivalents.

$R_{THB}$ is the resistance looking into the $bb'$ port in the network in Figure 3.69c. By inspection,

$$R_{THB} = 2\ \Omega.$$

Replacing Network A and Network B with their Thévenin equivalents, we obtain the equivalent circuit in Figure 3.70.

The current $I$ is easily determined as

$$I = \frac{1\text{ V} - (-1\text{ V})}{2\ \Omega + 1\ \Omega + 2\ \Omega} = \frac{2}{5}\text{ A}.$$

Notice in this example we were able to solve a relatively complicated problem by composing the results of five subproblems (namely, the circuits in Figures 3.68b, 3.68c, 3.69b, 3.69c, and 3.70), each of which was solvable by inspection.

---

EXAMPLE 3.23 THÉVENIN ANALYSIS OF A CIRCUIT WITH A DEPENDENT SOURCE Find the Thévenin equivalent circuit for the network to the left of the $aa'$ terminal pair in Figure 3.71. Notice that this circuit contains a dependent source.

The network whose Thévenin equivalent is desired is shown in Figure 3.72. Let $v_{TH}$ and $R_{TH}$ be the Thévenin parameters for this network.

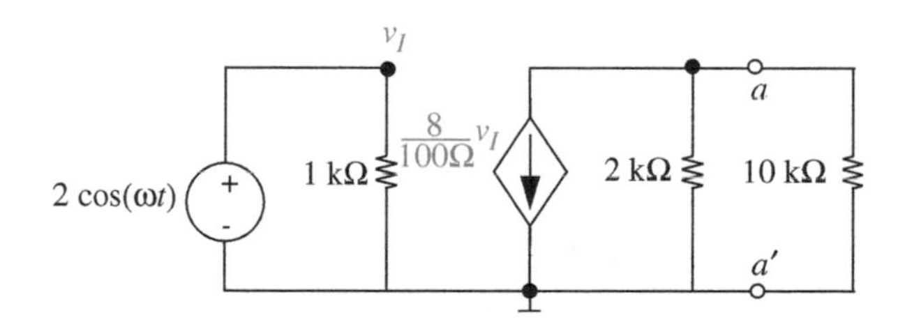


**FIGURE 3.71** Thévenin analysis of a circuit with a dependent source.

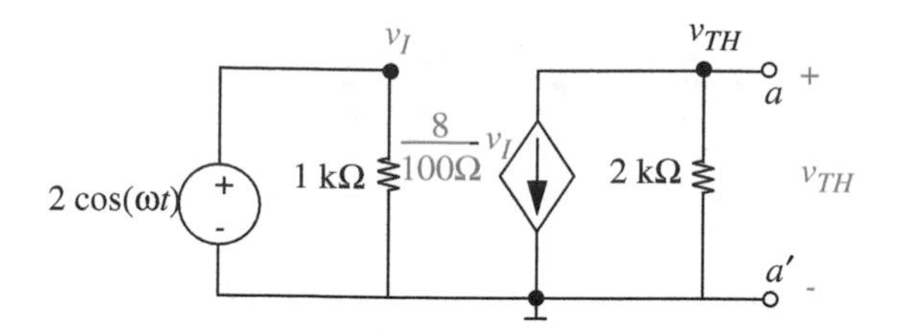


FIGURE 3.72 Network to be replaced by its Thévenin equivalent.

## Determining $v_{TH}$

We first find $v_{TH}$ by computing the open-circuit voltage at the *aa'* port of the circuit in Figure 3.72. We will find this voltage by applying the node method. Since the current of the dependent source is expressible directly in terms of a node voltage, we can apply the node method without modification.

Figure 3.72 shows the ground node, and the two other nodes labeled with the node voltages $v_I$ and $v_{TH}$. Notice that $v_I$ is already known to be

$$v_I = 2\cos(\omega t).$$

This completes Steps 1 and 2 of the node method.

Following Step 3 of the node method, we write KCL for Node *a*.

$$\frac{v_{TH}}{2\text{ k}\Omega} + \frac{8}{100\ S}v_I = 0.$$

Next, applying Step 4, we simplify the preceding equation to get

$$v_{TH} = -160v_I = -320\cos(\omega t).$$

Since we were interested only in the node voltage $v_{TH}$, we do not have to complete Step 5 of node analysis.

## Determining $R_{TH}$

We now find $R_{TH}$ by computing the resistance looking into the *aa'* port in the network in Figure 3.73. The independent voltage source has been turned into a short circuit for

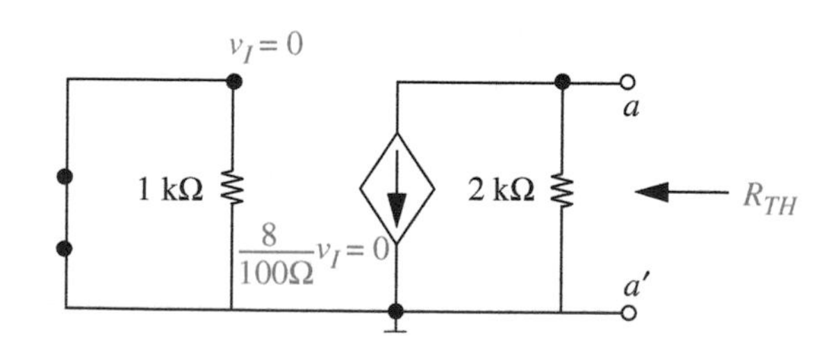


FIGURE 3.73 Determining $R_{TH}$.

the purpose of computing $R_{TH}$. The dependent source, however, is left in the circuit. Since $v_I = 0$, the current through the dependent current source is 0, and therefore, the dependent source behaves like an open circuit. Thus,

$$R_{TH} = 2\ \text{k}\Omega.$$

The resulting Thévenin circuit is shown in Figure 3.74.

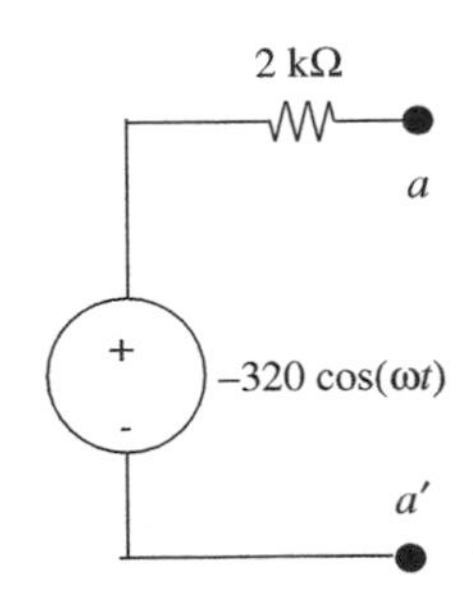


FIGURE 3.74 Resulting Thévenin circuit.

## 3.6.2 THE NORTON EQUIVALENT NETWORK

An analogous derivation to that in Section 3.6.1 gives rise to the *Norton equivalent network*. Recall that our goal is to find the $v$–$i$ relation for the network in Figure 3.75a so that we can replace the network with a simple equivalent circuit that yields the same $v$–$i$ relation as the original network. To find the $v$–$i$ relationship, this time we apply a test voltage $v_{test}$ to the circuit, as in Figure 3.75b, and find the resultant current $i_t$. Using superposition, the two subcircuits needed to find $i_t$ are shown in Figure 3.75c and 3.75d. In 3.75c, $v_{test}$ is set to zero and we measure $i_a$. In 3.75d, all independent sources are set to zero and we measure $i_b$. Then,

$$i_t = i_a + i_b.$$

From Figure 3.75c,

$$i_a = -i_{sc} \tag{3.119}$$

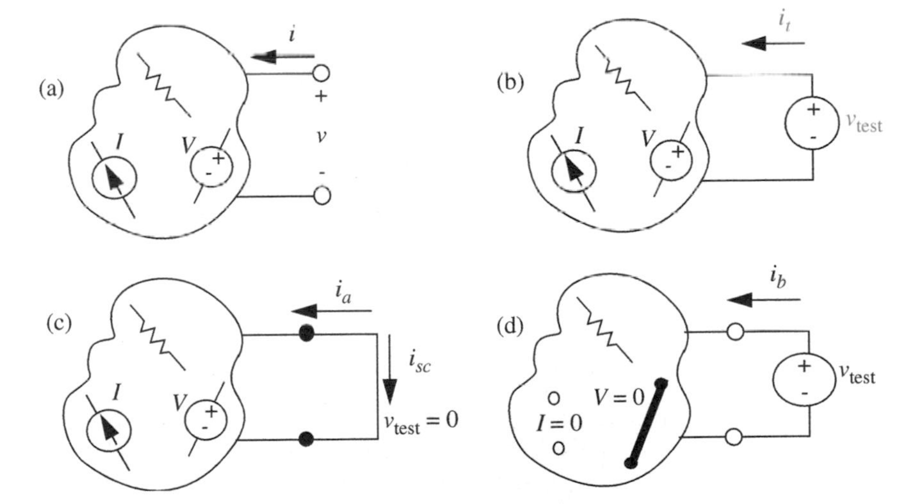


FIGURE 3.75 Derivation of Norton network.

where $i_{sc}$ is the current that flows in the short circuit across the network terminals in response to the internal sources, and thus is *the short circuit current.* From Figure 3.75d,

$$i_b = \frac{v_{\text{test}}}{R_t} \quad (3.120)$$

where $R_t$ is the net resistance measured between the terminals when all internal independent sources are set to zero. Because this calculation and the one in Figure 3.55c are identical (except for a change in excitation) the parameter $R_t$ is obviously the same in both calculations.

To complete the derivation, we find by superposition

$$i_t = i_a + i_b = -i_{sc} + \frac{v_{\text{test}}}{R_t}. \quad (3.121)$$

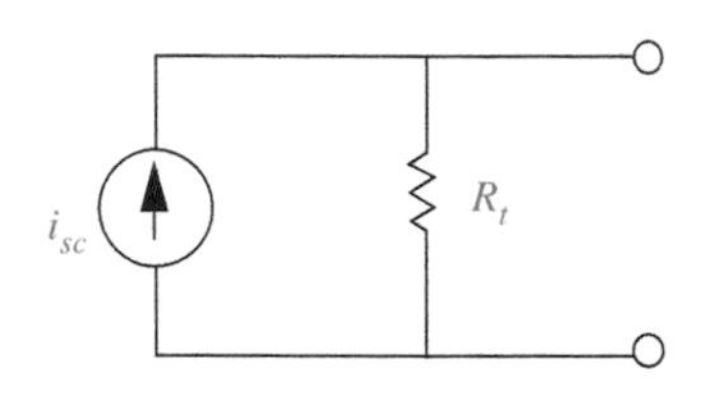


FIGURE 3.76 The Norton equivalent network.

As in the Thévenin derivation, this equation can be interpreted in terms of a circuit. It states that the terminal current is the sum of two components: a current source $i_{sc}$ and a resistor current $v_{\text{test}}/R_t$. Hence the *Norton equivalent* network, Figure 3.76, has a current source in parallel with a resistor. Examination of either the two equations, Equations 3.121 and 3.115, or the two figures, Figures 3.56 and 3.76 show that there is a simple relation between $v_{OC}$ and $i_{sc}$. Working from the figures, we can calculate the open-circuit voltage of each circuit to find

$$v_{OC} = i_{sc} R_t. \quad (3.122)$$

Thus it is a simple matter to change from one of these equivalent networks to the other.

To determine the Norton parameters for some circuit, again two independent measurements are required. The source parameter $i_{sc}$ could be found by applying a short to the circuit terminals and measuring the resultant current. The resistance parameter is measured as before in Equation 3.117. Note that the source parameters $i_{sc}$, $v_{OC}$ are related by Equation 3.122, so measuring or calculating any two of $v_{OC}$, $i_{sc}$ and $R_t$ is sufficient to characterize both the Norton and the Thévenin model. In particular, it is often convenient to find $R_t$ from two simple terminal measurements on the circuit

$$R_t = \frac{v_{OC}}{i_{sc}}. \quad (3.123)$$

In summary, the Norton method allows us to abstract the behavior of a linear network at a given pair of terminals as a current source in parallel with a resistor. The current source in parallel with the resistor is called the Norton equivalent circuit of the network. Like the Thévenin equivalent, the Norton equivalent circuit can also be used to model the effect of the given network on other circuits external to the network.

*A Method for Determining the Norton Equivalent Circuit* The Norton equivalent circuit for any linear network at a given pair of terminals consists of a current source $i_N$ in parallel with a resistor $R_N$. The current $i_N$ and resistance $R_N$ can be obtained as follows:

1. $i_N$ can be found by applying a short at the designated terminal pair on the original network and calculating or measuring the current through the short circuit.
2. $R_N$ can be found in the same manner as $R_{TH}$, that is, by calculating or measuring the resistance of the open-circuit network seen from the designated terminal pair with all independent sources internal to the network set to zero; that is, with voltage sources replaced with short circuits, and current sources replaced with open circuits.

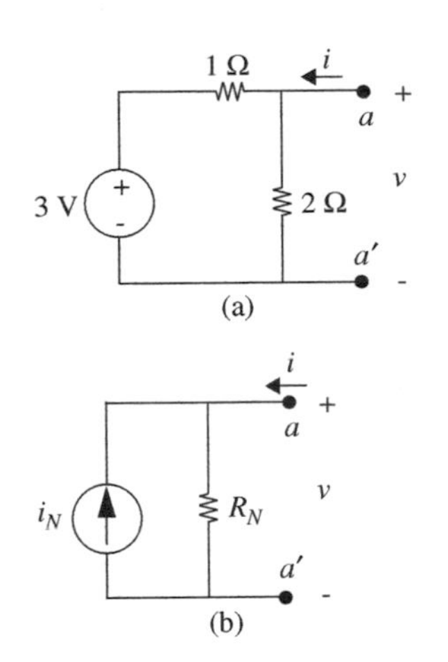


FIGURE 3.77 Norton equivalent network: (a) a network; (b) its Norton equivalent network.

---

EXAMPLE 3.24 NORTON EQUIVALENT Figure 3.77a shows a network and Figure 3.77b shows its Norton equivalent network viewed from the network's $aa'$ port. Determine the values of $i_N$ and $R_N$.

By the first step of the Norton method, the current $i_N$ is given by applying a short at the $aa'$ terminal pair and calculating the current through the short circuit. Figure 3.78 shows the network with a short at the $aa'$ terminal pair.

The current through the short at the $aa'$ terminal pair in Figure 3.78 is given by

$$i_N = \frac{3\text{ V}}{1\ \Omega} = 3\text{ A}.$$

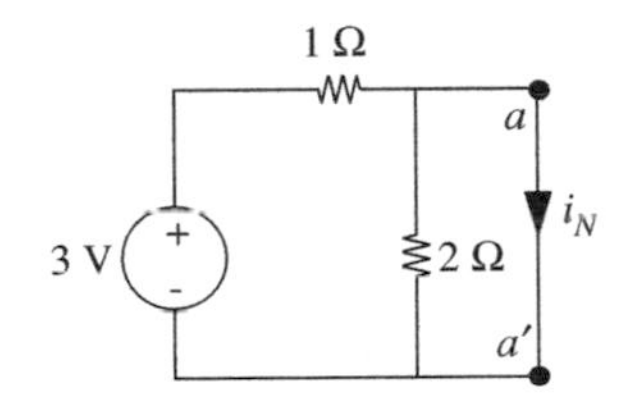


FIGURE 3.78 Determining $i_N$.

By the second step of the Norton method, the resistance $R_N$ is found by measuring the resistance of the open-circuit network seen from the $aa'$ port with the independent voltage source set to zero. The network with the voltage source replaced with a short is shown in Figure 3.79.

The resistance viewed from the $aa'$ port is given by

$$R_N = 1\ \Omega \| 2\ \Omega = \frac{2}{3}\ \Omega.$$

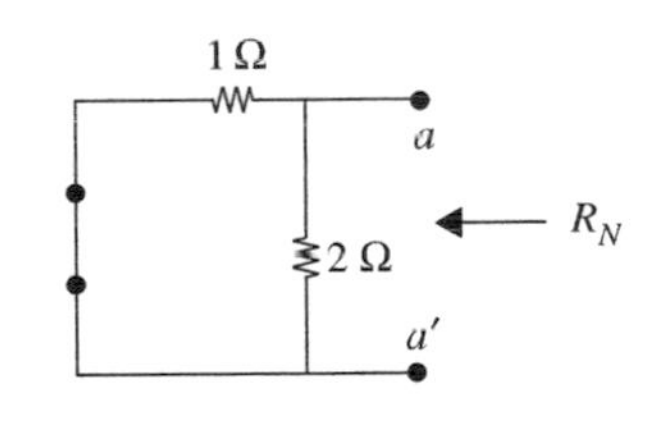


FIGURE 3.79 Determining $R_N$.

The resulting Norton equivalent circuit is drawn in Figure 3.80.

---

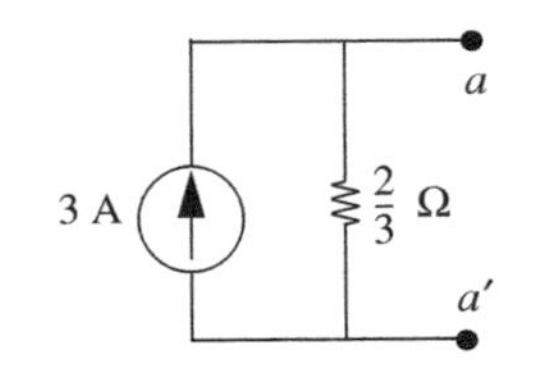


FIGURE 3.80 Resulting Norton equivalent circuit.

EXAMPLE 3.25 MORE ON THE NORTON METHOD Determine the current $I_1$ through the voltage source in the circuit in Figure 3.81 using the Norton method.

To apply the Norton method, we will replace the network to the left of the $aa'$ terminal pair with its Norton equivalent network comprising a current source with current $i_N$ in

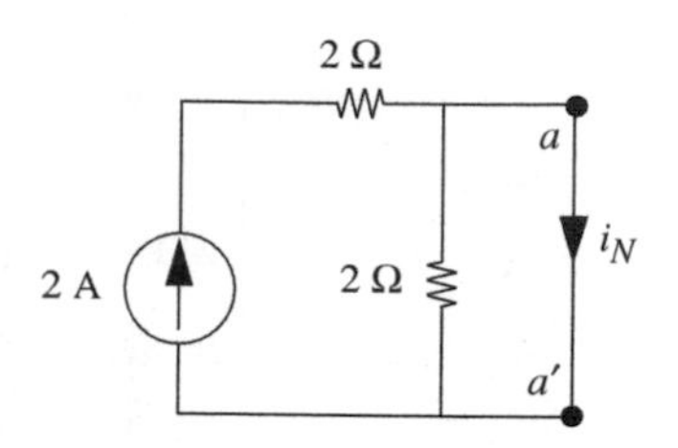


FIGURE 3.82 Determining $i_N$.

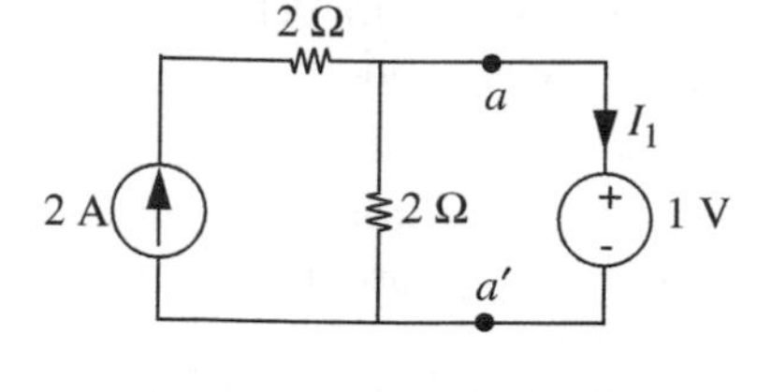


FIGURE 3.81 Circuit for applying the Norton method.

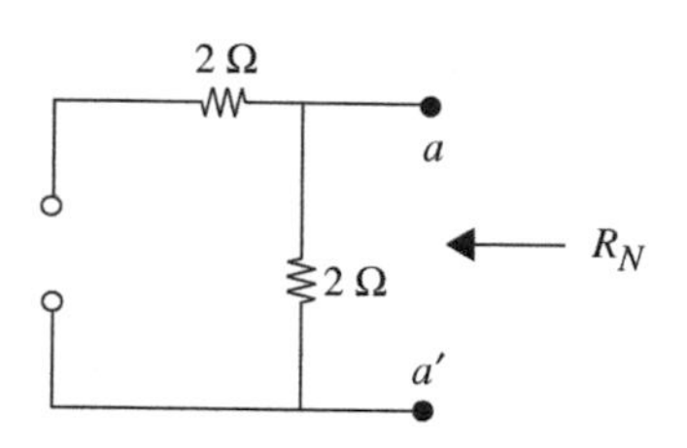


FIGURE 3.83 Determining $R_N$.

parallel with a resistance of value $R_N$. The first step of the Norton method is to measure $i_N$, which is the short-circuit current measured at a short circuit applied at the $aa'$ port as shown in Figure 3.82. Since all of the 2-A current flows through the short,

$$i_N = 2 \text{ A}.$$

By the second step of the Norton method, the resistance $R_N$ is found by measuring the resistance of the open-circuit network seen from the $aa'$ port with the current source replaced with an open circuit as illustrated in Figure 3.83. It is easy to see that

$$R_N = 2 \ \Omega.$$

The resulting Norton equivalent circuit is depicted in Figure 3.84.

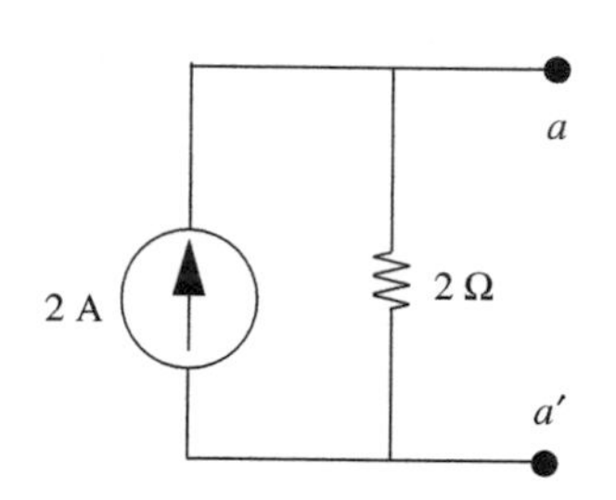


FIGURE 3.84 Resulting Norton equivalent circuit.

Having determined the Norton equivalent circuit, we can now obtain $I_1$ by connecting this equivalent circuit to the source on the right-hand side of the $aa'$ terminal pair as shown in Figure 3.85.

Since the voltage across the 2-Ω resistor is 1 V, the current through the 2-Ω resistor is 0.5 A. By applying KCL at Node $a$, we get

$$-2 \text{ A} + 0.5 \text{ A} + I_1 = 0.$$

Or, $I_1 = 1.5$ A.

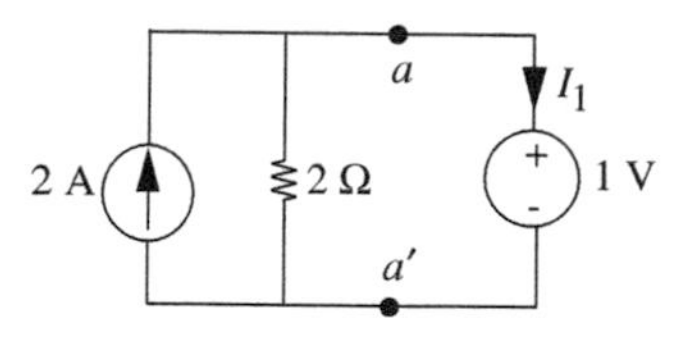


FIGURE 3.85 Connecting back the Norton equivalent circuit to determine $I_1$.

EXAMPLE 3.26 NORTON EQUIVALENT NETWORK Let us revisit the example in Figure 3.71 and this time around determine the Norton equivalent circuit for the network to the left of the $aa'$ terminal pair. Let $I_N$ and $R_N$ be the Norton parameters for this network.

## Determining $I_N$

We first find $I_N$ by computing the short-circuit current through the short placed at the $aa'$ terminal pair as depicted in Figure 3.86. $I_N$ can be determined by inspection as

$$I_N = -\frac{8}{100} v_I = -\frac{4}{25} \cos(\omega t).$$

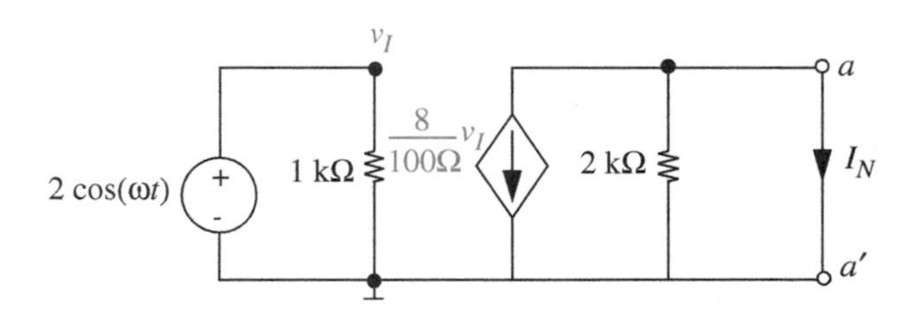


FIGURE 3.86 Determining $i_N$.

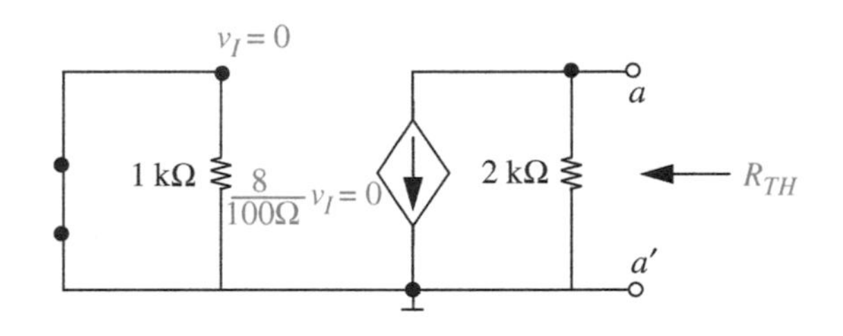


FIGURE 3.87 Determining $R_N$.

### Determining $R_N$

We now find $R_N$ by computing the resistance looking into the $aa'$ port in the network in Figure 3.87. As computed in the Thévenin version of this example,

$$R_N = 2\ \text{k}\Omega.$$

The resulting Norton circuit is shown in Figure 3.88.

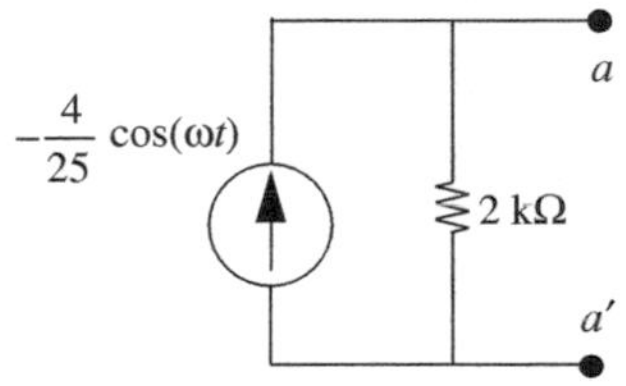

FIGURE 3.88 Resulting Norton equivalent circuit.

## 3.6.3 MORE EXAMPLES

Norton and Thévenin equivalents are particularly useful because often the two parameters are easy to find, as a consequence of the strong circuit constraints imposed as shown in Figures 3.55 and 3.75. This is best illustrated by an example. Suppose we are given the network in Figure 3.89a, and are asked to find the voltage across $R_3$ for a number of different values of $R_3$. We could just solve the whole network for each value of $R_3$, but a simpler approach is to find the Thévenin equivalent of the network driving $R_3$, that is, the network to the left of the points $x$–$x$. For clarity in this first example, we abstract this portion of the network in Figure 3.89b.

As we have noted, there are several ways to make the calculations, so it pays to examine the possibilities and choose the easiest route. The open-circuit voltage appears directly in the abstracted circuit, Figure 3.89b. The short-circuit current can be found from Figure 3.89c, and $R_t$ from 3.89d. By inspection from Figure 3.89d,

$$R_t = R_1 \| R_2. \tag{3.124}$$

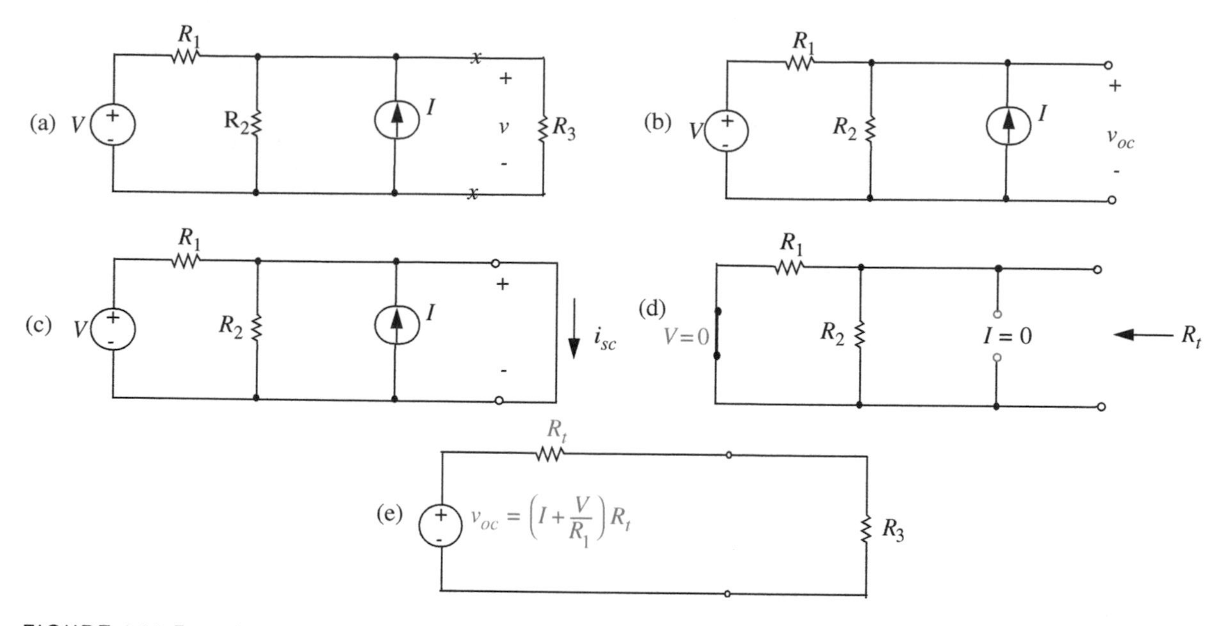


FIGURE 3.89 Example in which we are to find voltage across $R_3$ for several different values of $R_3$.

In Figure 3.89c, the short-circuit constraint makes the calculation of $i_{sc}$ for this particular topology easy. Because of the short circuit, $R_2$ can have no voltage across it, hence has no current flowing through it. Now by superposition,

$$i_{sc} = I + V/R_1. \tag{3.125}$$

The calculation of $v_{oc}$ from Figure 3.89b is straightforward, but a step more complicated than the preceding ones, so normally it would not be attempted. But for completeness, superposition of the two sources gives

$$v_{OC} = v\frac{R_2}{R_1 + R_2} + \frac{I(R_1 R_2)}{R_1 + R_2}. \tag{3.126}$$

It is clearly easier to find $v_{OC}$ from Equations 3.122, 3.124, and 3.125:

$$v_{OC} = (I + V/R_1)R_t. \tag{3.127}$$

Hence the complete circuit with the left half replaced by its Thévenin equivalent is as shown in Figure 3.89e. Now the voltage across $R_3$ for the various values of $R_3$ can be found by inspection. It should be noted that the Norton equivalent would have been just as effective in this problem. Also note that the circuit constraints imposed by the definitions of $R_t$, $i_{sc}$, and $v_{OC}$ often make the calculations of these parameters very easy, even in complicated networks.

EXAMPLE 3.27 BRIDGE CIRCUIT Another example is shown in Figure 3.90a. This is a *bridge circuit*, often used in the laboratory to measure values of unknown resistors by comparing against known standard resistors. We want to find the voltage across $R_5$, and then find the condition on the other resistor values that will make this voltage zero. Direct application of nodal analysis is quite messy, so we will seek an alternative method.

To solve, find the Thévenin equivalent of the circuit facing $R_5$, that is, the circuit shown in Figure 3.90b. The circuit now consists of two independent voltage dividers connected across a common voltage source $V$. The layout of the dividers is not quite as straightforward as in Figure 2.36 but, topologically, they are the same. Hence we can calculate the two voltage-dividers' voltages $v_a$ and $v_b$ by inspection; then subtract to find $v_{oc}$.

$$v_{oc} = v_a - v_b = V\left(\frac{R_3}{R_1 + R_3} - \frac{R_4}{R_2 + R_4}\right). \tag{3.128}$$

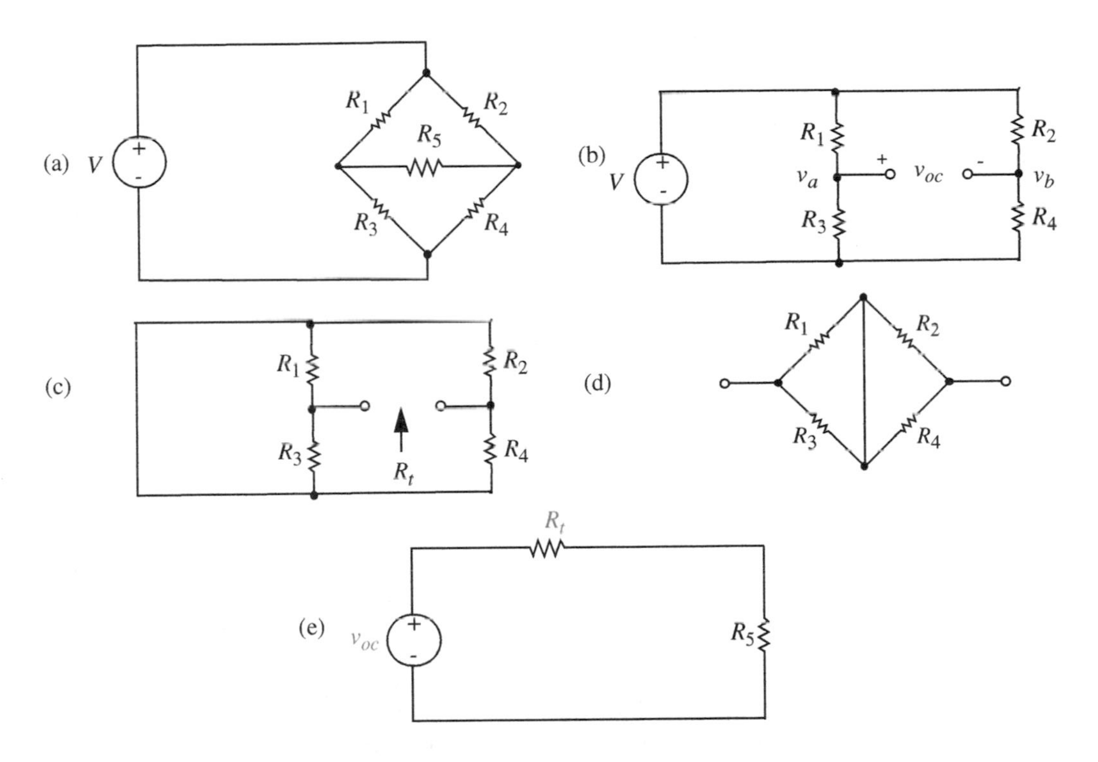


FIGURE 3.90 Example: A bridge circuit.

Now find the Thévenin equivalent resistance with $V$ set to zero, that is, for the circuit shown in Figure 3.90c. This is identical to the circuit in 3.90d,

$$R_t = (R_1 \| R_3) + (R_2 \| R_4). \tag{3.129}$$

The complete circuit can now be drawn as in Figure 3.90e. It is clear that the voltage across $R_5$ will be zero if $v_{oc}$ is zero, that is, if

$$\frac{R_3}{R_1 + R_3} = \frac{R_4}{R_2 + R_4} \tag{3.130}$$

or equivalently

$$\frac{R_3}{R_1} = \frac{R_4}{R_2}. \tag{3.131}$$

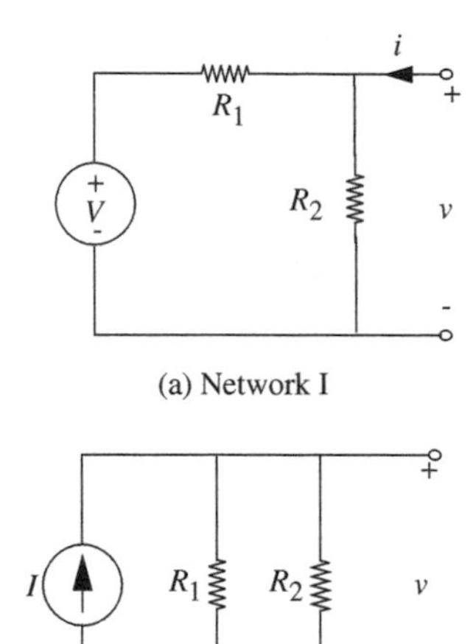


**FIGURE 3.91** Two simple networks: (a) Network I; (b) Network II.

Thus if $R_5$ is replaced by a voltmeter, the circuit can be used to find an unknown resistor, say $R_3$, in terms of three known resistors. Make one of the resistors, say $R_1$, a decade box with known resistance values and adjust until the voltmeter reads zero. The value of $R_3$ is then given by Equation 3.131.

Two closing comments: First, note that the identity of all voltages and currents inside the network that is replaced by the Thévenin or Norton circuit in general lose their identity; only the terminal voltage and current are preserved. Thus, for example, the current through $R_3$ in Figure 3.90a does not appear as any identifiable current flowing in the Thévenin circuit in Figure 3.90e. Second, if one wishes to measure the Thévenin or Norton parameters of a system in the laboratory, two independent measurements are required in order to specify the two parameters in the model. In addition, certain practical issues must be faced. For example, it is unwise, in fact dangerous, to apply a short circuit to a large battery such as an automobile storage battery in an attempt to measure the short-circuit current as suggested in Figure 3.75b. A better procedure is to first measure the open-circuit voltage, then measure the terminal voltage when some known resistor is connected to the battery. These two measurements can then be used to find $R_t$.

---

### EXAMPLE 3.28 NORTON AND THÉVENIN EQUIVALENTS

As another simple example, let us find the Norton and Thévenin equivalent networks and their $v$–$i$ characteristics for the two circuits shown in Figure 3.91.

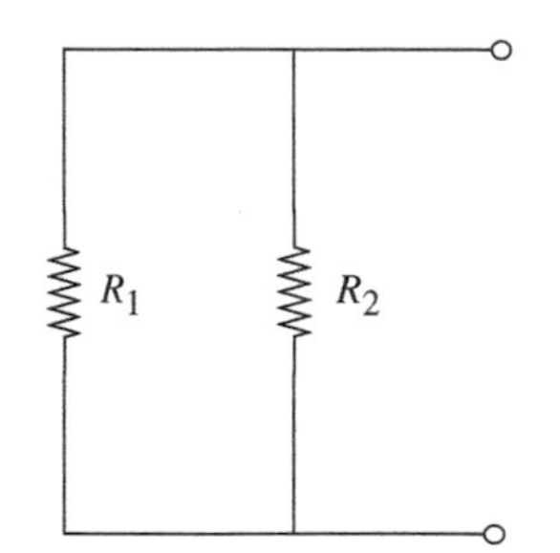


**FIGURE 3.92** Equivalent resistance of the network.

Let us start with Network A. First, let us find the Thévenin equivalent circuit. Shorting the voltage source results in the circuit shown in Figure 3.92. Therefore, $R_{TH} = R_N = R_1 \| R_2 = R_1 R_2/(R_1 + R_2)$, where $R_{TH}$ and $R_N$ are the Thévenin and Norton equivalent resistors, respectively. From the voltage-divider relationship, open circuit voltage $v_{oc}$ is $VR_2/(R_1 + R_2)$. This yields the Thévenin equivalent circuit on the left-hand side of Figure 3.93.

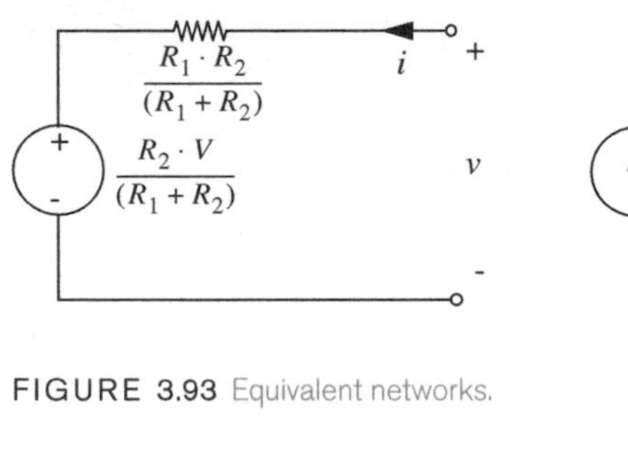


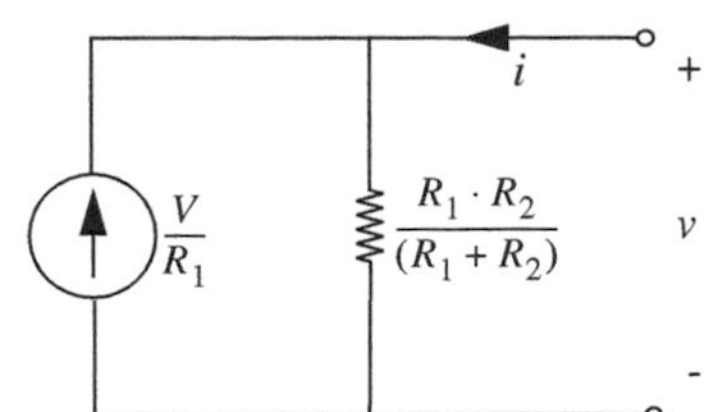


FIGURE 3.93 Equivalent networks.

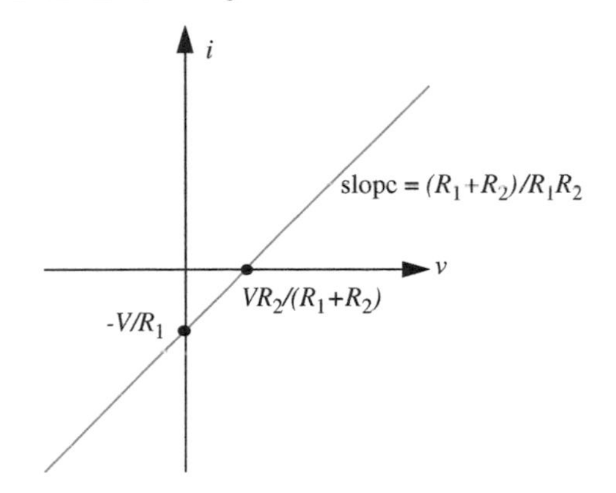


FIGURE 3.94 Short circuit current.

FIGURE 3.95 The $i$–$v$ characteristics of the network.

Now, let us find the Norton equivalent circuit for Network A. Referring to Figure 3.94, the short-circuit current $i_{sc}$ is $V/R_1$. Therefore, the Norton equivalent network is as shown on the right-hand side of Figure 3.93.

The $v$–$i$ curve of the circuit must pass through points $(v_{oc}, 0)$ and $(0, -i_{sc})$, as shown in Figure 3.95.

Let us now analyze Network B. Turning off the current source results in the circuit shown in Figure 3.92, which yields $R_1R_2/(R_1 + R_2)$ as the equivalent resistance for both the Thévenin and Norton equivalent networks. The open-circuit voltage is $IR_1 \| R_2$. Thus, $v_{oc} = R_1R_2I/(R_1 + R_2)$. As illustrated in Figure 3.96, all of the current will flow through the branch with zero resistance, that is $i_{sc} = I$.

The equivalent networks are shown in Figure 3.97, and the $v$–$i$ characteristics are shown in Figure 3.98.

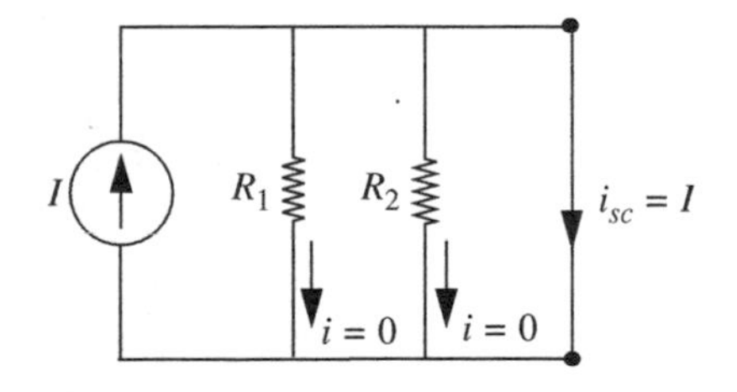


FIGURE 3.96 Short-circuit current.

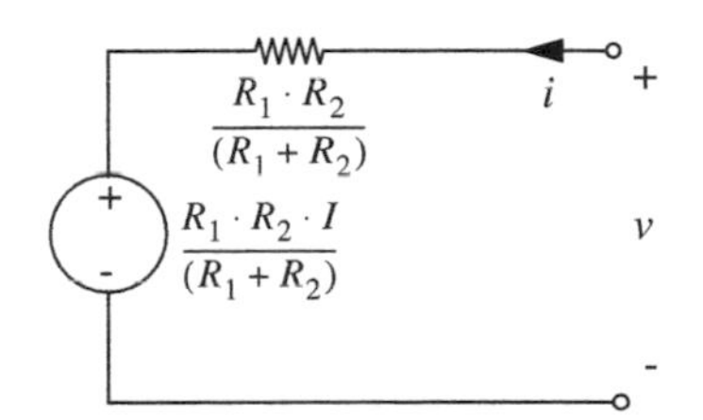


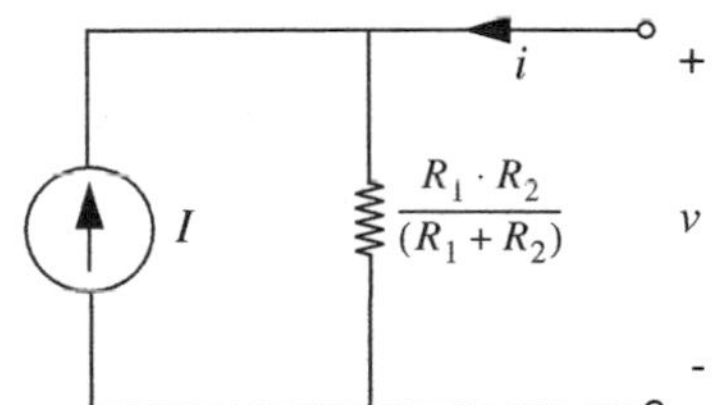


FIGURE 3.97 Equivalent networks.

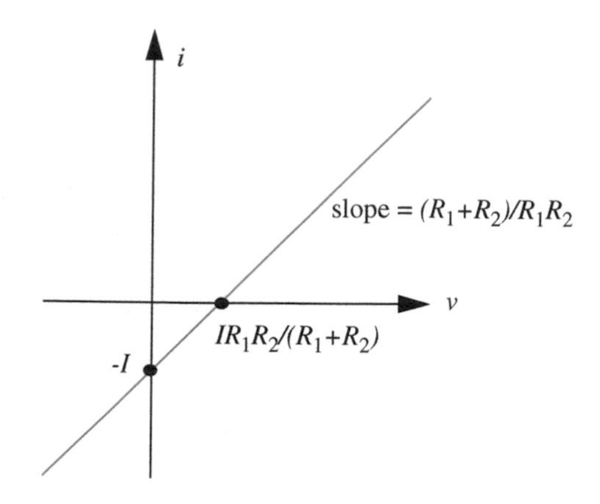

FIGURE 3.98 The v–i characteristics of the network.

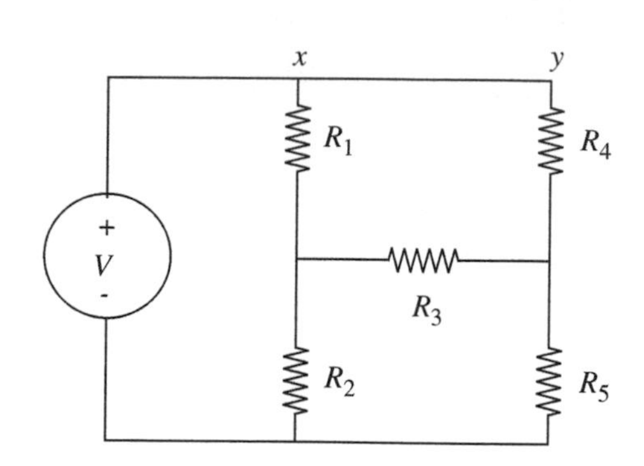


FIGURE 3.99 Resistive circuit.

---

EXAMPLE 3.29 A DIFFERENT APPROACH USING THE THÉVENIN METHOD The network shown in Figure 3.99 was solved earlier using the Thévenin method (see Figure 3.90). In this example, we will solve the same circuit using the Thévenin method, but with a slightly different approach.

Making the observation that the voltages at points $x$ and $y$ are the same, we can transform the circuit into the equivalent circuit shown in Figure 3.100. We can then transform the circuits in Figures 3.100a and 3.100b into their Thévenin equivalent networks. Figure 3.100a will have a source voltage of $VR_2/(R_1 + R_2)$ and an equivalent resistance of $R_1 \| R_2$. Figure 3.100b will have a source voltage of $VR_5/(R_4 + R_5)$ and an equivalent resistance of $R_4 \| R_5$.

The new circuit is shown in Figure 3.101. Notice that the new circuit is much easier to analyze. We leave the rest of the analysis as an exercise for you.

---

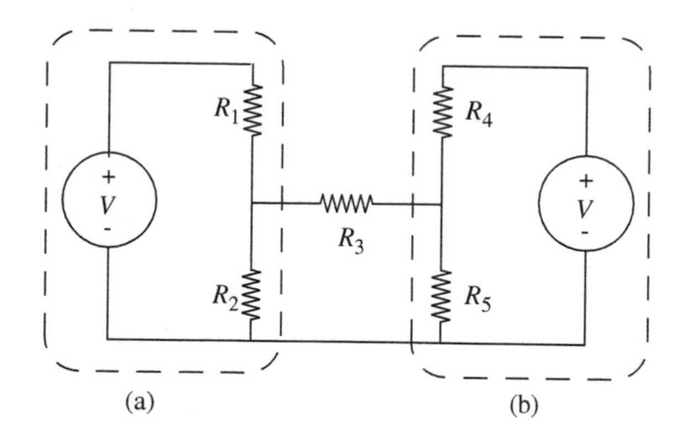


FIGURE 3.100 Equivalent circuit with two voltage sources.

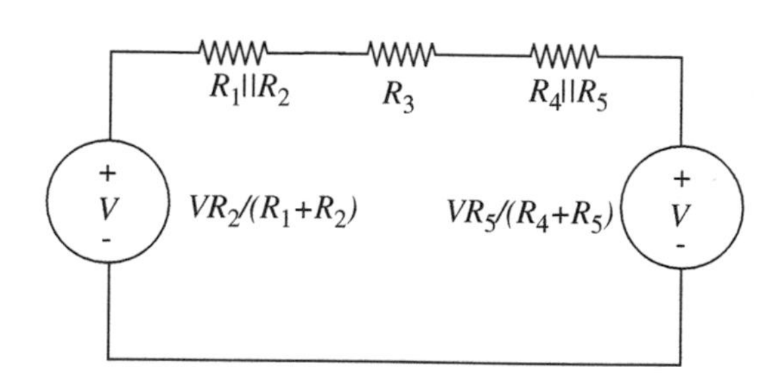


FIGURE 3.101 Equivalent Thévenin circuit.

## 3.7 SUMMARY

- In the node method one node is designated as a reference or ground node, and all other node voltages are measured with respect to that node. Only the KCL equations and the constituent relations need be written.
- In the loop method currents are defined to flow in loops. Loop currents are defined until all branches are traversed by at least one current. Only KVL equations need be written.
- Superposition means that if the circuit is linear, multisource networks can be solved for one source at a time by setting all other independent sources to zero. Setting a voltage source to zero means replacing it with a short circuit; a current source set to zero is an open circuit. The complete response is the sum of the responses to each individual source.

  For circuits with dependent sources, a practical solution is to leave all the dependent sources in the circuit. The network can then be solved for one independent source at a time by setting all other independent sources to zero, and summing the individual responses.
- The Thévenin equivalent circuit for any linear network at a given pair of terminals consists of a voltage source in series with a resistor. The element value for the Thévenin equivalent voltage source can be found by calculating or measuring at the designated terminal pair on the original network the open-circuit voltage. The equivalent resistance can be calculated or measured as the resistance of the network seen from the designated terminal pair with all independent sources internal to the network set to zero.
- The Norton equivalent circuit contains a current source in parallel with a resistor. The element value for the Norton equivalent current source can be found by calculating or measuring at the designated terminal pair on the original network the short circuit current. As with the Thévenin equivalent resistance, the Norton equivalent resistance can be calculated or measured as the resistance of the network seen from the designated terminal pair with all independent sources internal to the network set to zero. Note that the value of the equivalent resistance is the same for the Thévenin and Norton equivalent circuits, that is, $R_{TH} = R_N$.
- Since the Thévenin equivalent voltage $v_{TH}$, the Norton equivalent current $i_N$, and the equivalent resistance $R_{TH} = R_N$ are related as

  $$v_{TH} = i_N R_{TH},$$

  the element values for these equivalents can be found by calculating or measuring any two of the open-circuit voltage, the short-circuit current, or the resistance.

- Circuit analysis is often simplified by applying superposition or finding Thévenin or Norton equivalents, because complicated circuits are reduced to simpler circuits, for which the solution may already be known.

## EXERCISES

EXERCISE 3.1 Write node equations for the network in Figure 3.102. Solve for the node voltages, and use these voltages to find the branch current $i$. To minimize errors and facilitate answer-checking, it is helpful to obtain literal expressions before substituting numerical values for the parameters:

$$V = 2\text{ V} \quad R_3 = 3\ \Omega \quad R_1 = 2\ \Omega \quad R_4 = 2\ \Omega \quad R_2 = 4\ \Omega \quad R_5 = 1\ \Omega$$

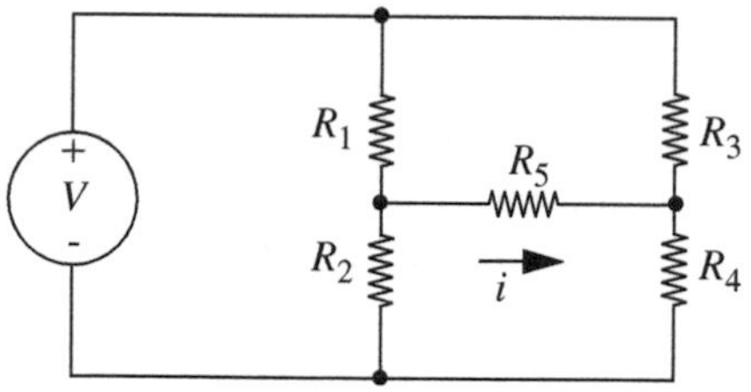

FIGURE 3.102

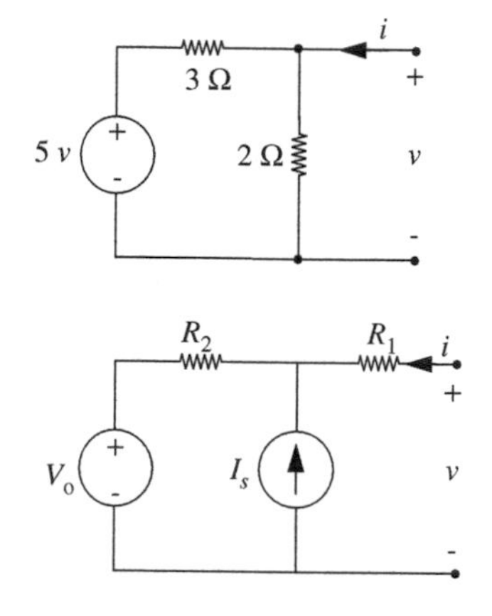


FIGURE 3.103

EXERCISE 3.2 Find the Norton equivalent at the indicated terminals for each network in Figure 3.103.

EXERCISE 3.3 Find the Thévenin equivalent for each network in Figure 3.104.

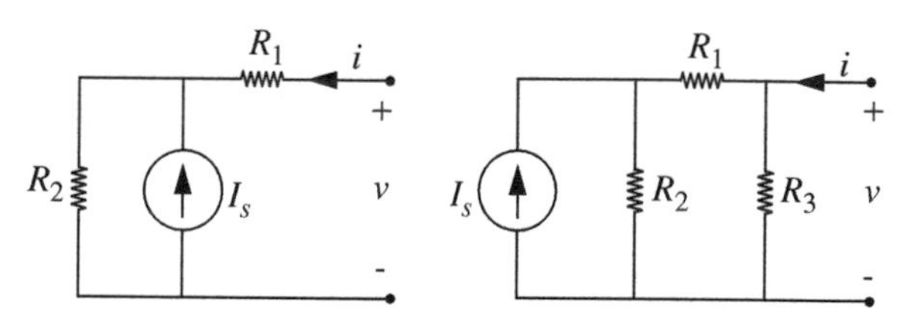


FIGURE 3.104

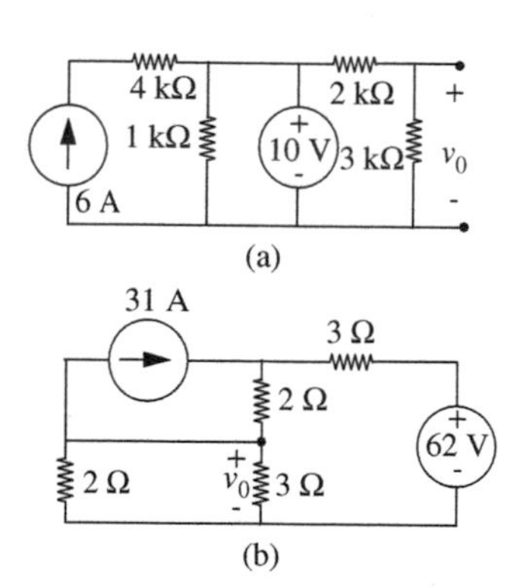


FIGURE 3.105

EXERCISE 3.4 Find $v_0$ in Figures 3.105a and 3.105b by superposition.

EXERCISE 3.5 Use superposition to find the voltage $v$ in the network in Figure 3.106.

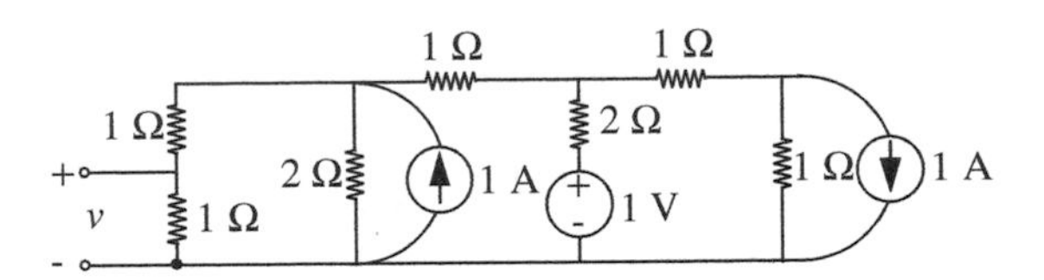


FIGURE 3.106

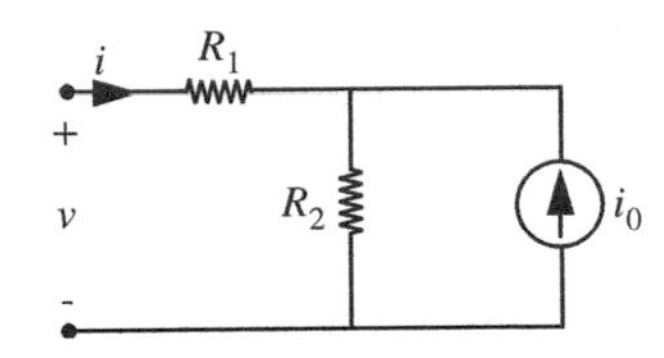


**EXERCISE 3.6** Determine (and label carefully) the Thévenin equivalent for the network in Figure 3.107:

$$R_1 = 2\ \text{k}\Omega \quad R_2 = 1\ \text{k}\Omega \quad i_0 = 3\ \text{mA}\ \cos(\omega t)$$

FIGURE 3.107

**EXERCISE 3.7** Determine and label carefully the Norton equivalent for the network in Figure 3.108.

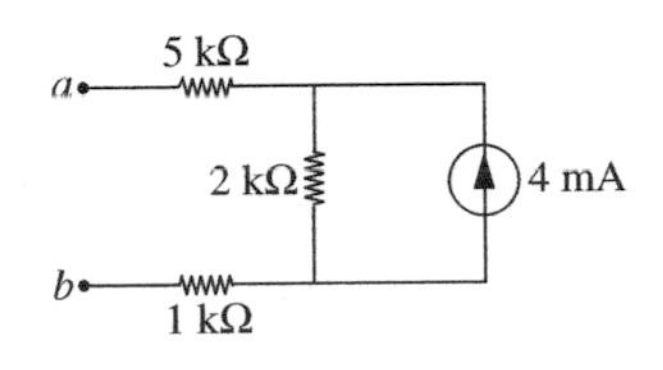


FIGURE 3.108

**EXERCISE 3.8** Find the Thévenin equivalent for the circuit at the terminals $AA'$ in Figure 3.109.

**EXERCISE 3.9** The resistive network shown in Figure 3.110 is excited by two voltage sources $v_1(t)$ and $v_2(t)$.

a) Express the current $i(t)$ through the 1-Ω resistor as a function of $v_1(t)$ and $v_2(t)$.

b) Determine the total energy dissipated in the 1-Ω resistor due to both $v_1(t)$ and $v_2(t)$ from time $T_1$ to time $T_2$.

c) Derive the constraint between $v_1(t)$ and $v_2(t)$ such that the value for (b) can be computed by adding the energies dissipated when each source acts alone (that is, by superposition).

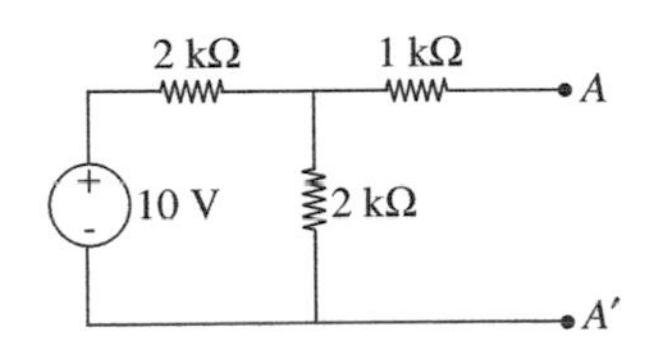


FIGURE 3.109

**EXERCISE 3.10** Find the Norton equivalent at the terminals marked $x$–$x$ in the circuit in Figure 3.111.

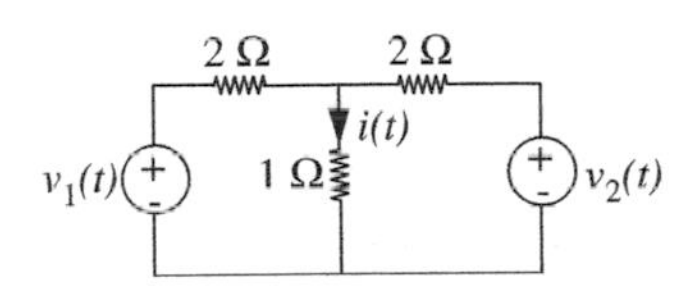

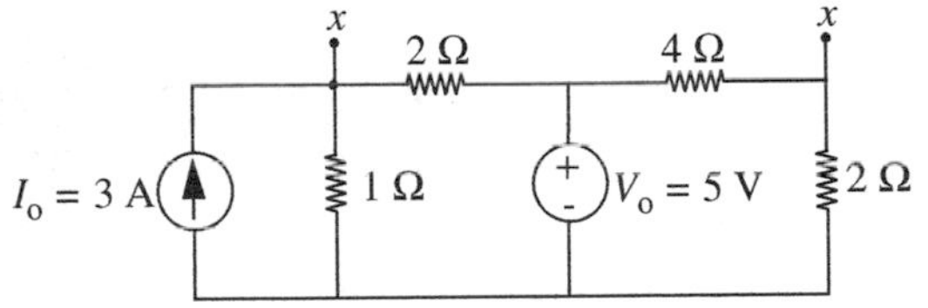

FIGURE 3.110

FIGURE 3.111

**EXERCISE 3.11** Find the Thévenin equivalent for the circuit in Figure 3.112 at the terminals $AA'$.

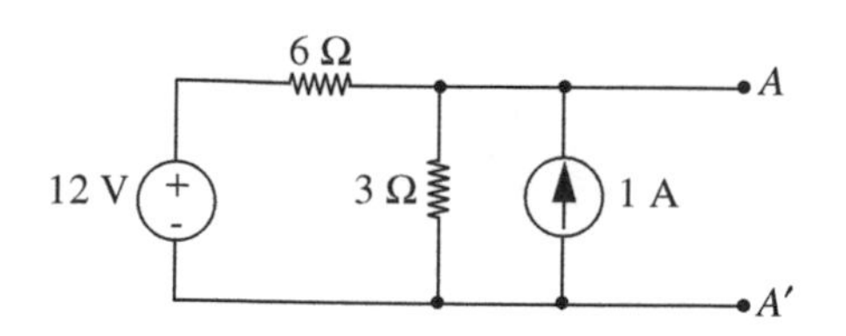


FIGURE 3.112

EXERCISE 3.12 In the network in Figure 3.113, find an expression for $v_2$.

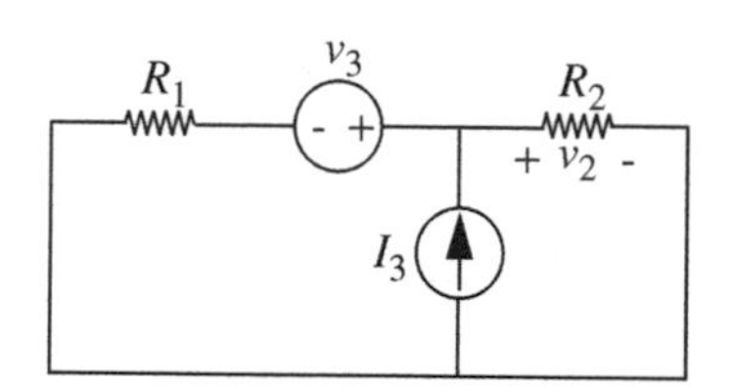


FIGURE 3.113

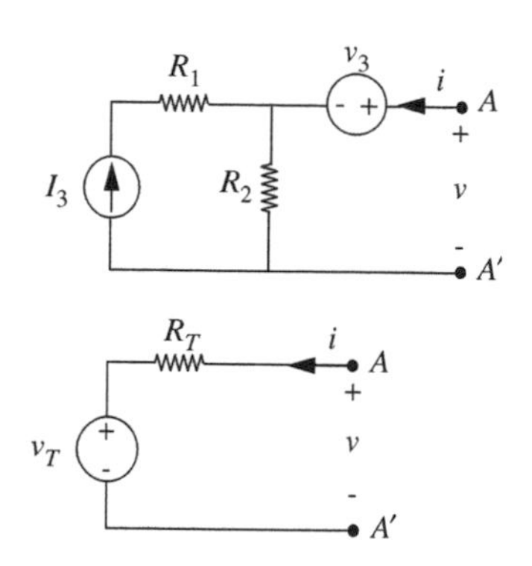


FIGURE 3.114

EXERCISE 3.13 The networks in Figure 3.114 are equivalent (that is, have the same $v$–$i$ relation) at terminals $A$–$A'$. Find $v_T$ and $R_T$.

EXERCISE 3.14 For each of the circuits in Figure 3.115 give the *number* of independent node variables needed for a solution of the problem by the node method.

EXERCISE 3.15 For the circuit shown in Figure 3.116, write a complete set of node equations for the voltages $v_a$, $v_b$, and $v_c$. Use conductance instead of resistance. Simplify the equations by collecting terms and arranging them in the "standard" form for n linear equations in n unknowns. (*Do not solve the equations.*)

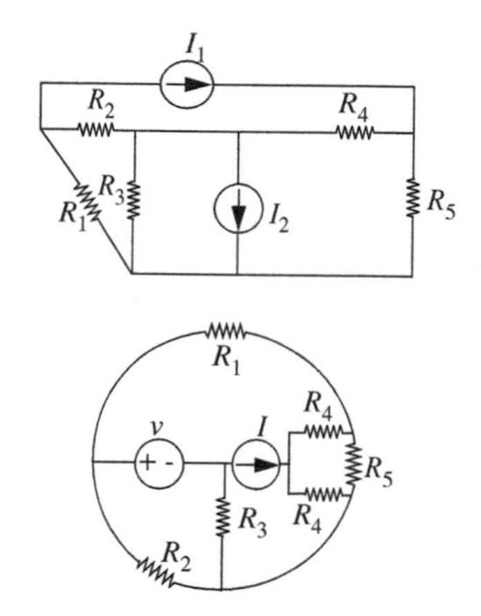


FIGURE 3.115

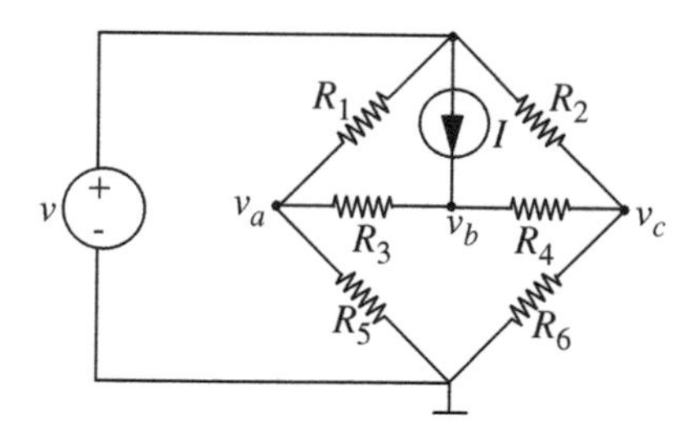


FIGURE 3.116

EXERCISE 3.16 For the circuit shown in Figure 3.117, use superposition to find $v$ in terms of the $R$'s and source amplitudes.

EXERCISE 3.17 Find the Thévenin equivalent of the circuit in Figure 3.118 at the terminals indicated.

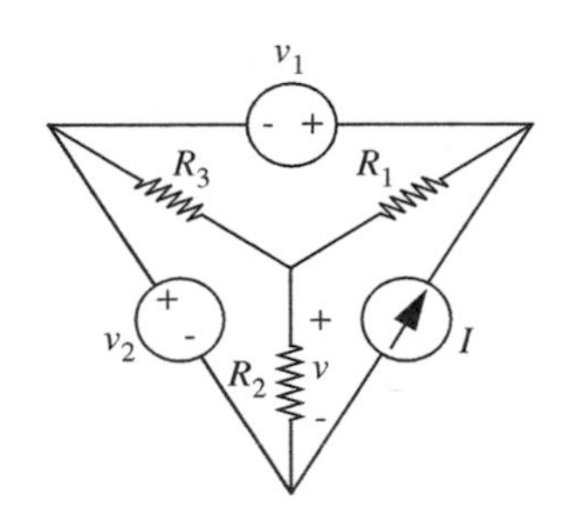

FIGURE 3.117

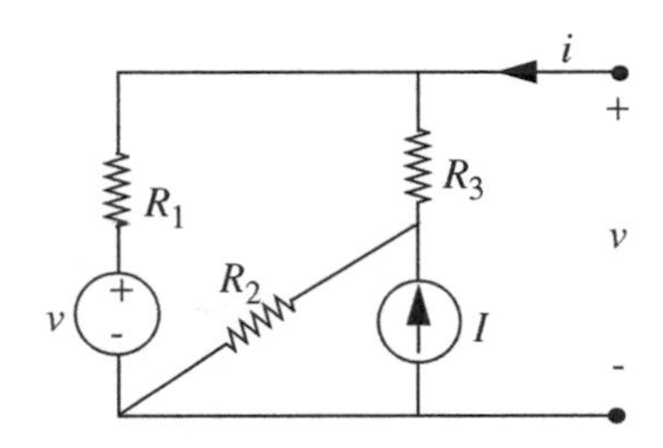

FIGURE 3.118

EXERCISE 3.18 In the circuit shown in Figure 3.119 there are five nodes, only three of which are independent. Take node $E$ as a reference node, and treat nodes $A$, $B$, and $D$ as the independent nodes.

a) Write an expression for $v_C$, the voltage on node $C$, in terms of $v_A$, $v_B$, $v_D$, and $v_1$.

b) Write a complete set of node equations that can be solved to find the unknown voltages in the circuit. (Do not solve the set of equations but do group them neatly.)

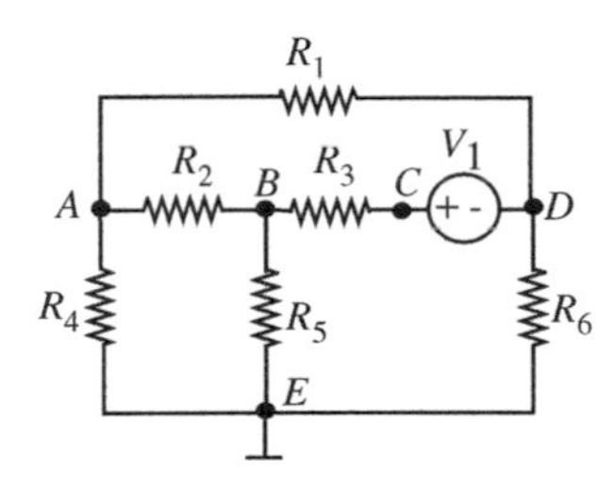

FIGURE 3.119

EXERCISE 3.19 Consider the circuit in Figure 3.120.

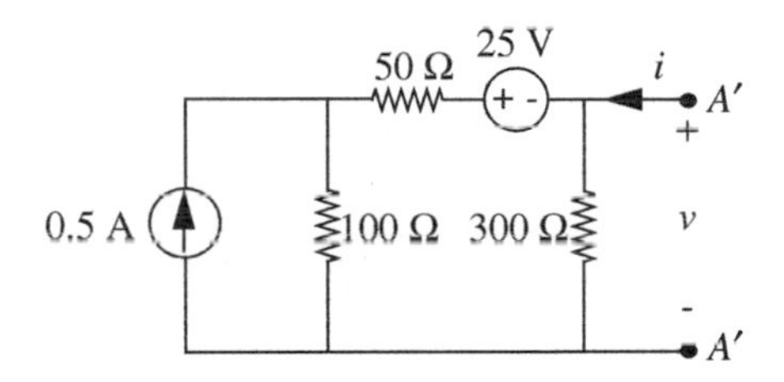

FIGURE 3.120

a) Find a Norton equivalent circuit for this circuit at terminals $A$–$A'$.

b) Find the Thévenin equivalent circuit corresponding to your answer in (a).

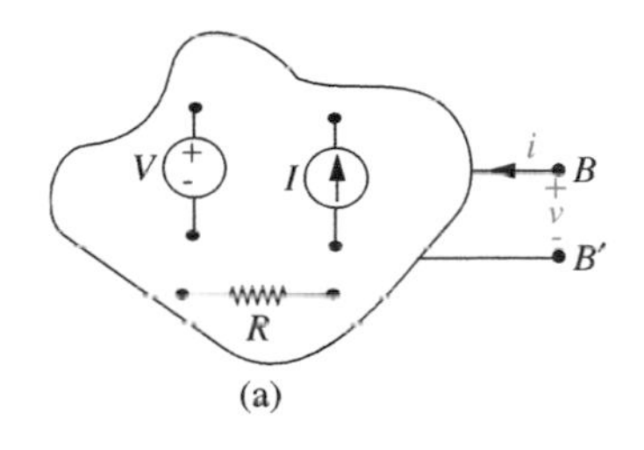


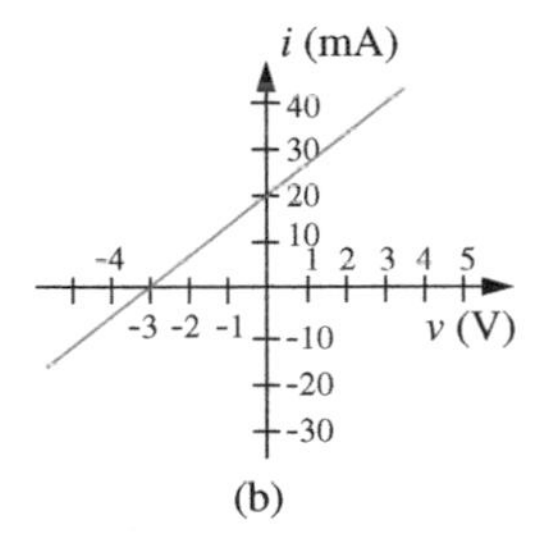

FIGURE 3.121

EXERCISE 3.20 Measurements made on terminals $B$–$B'$ of a linear circuit in Figure 3.121a, which is known to be made up only of independent voltage sources and current sources, and resistors, yield the current-voltage characteristics shown in Figure 3.121b.

a) Find the Thévenin equivalent of this circuit.

b) Over what portions, if any, of the $v$–$i$ characteristics does this circuit absorb power?

EXERCISE 3.21

a) Write in standard form the minimum number of node equations needed to analyze the circuit in Figure 3.122.

FIGURE 3.122

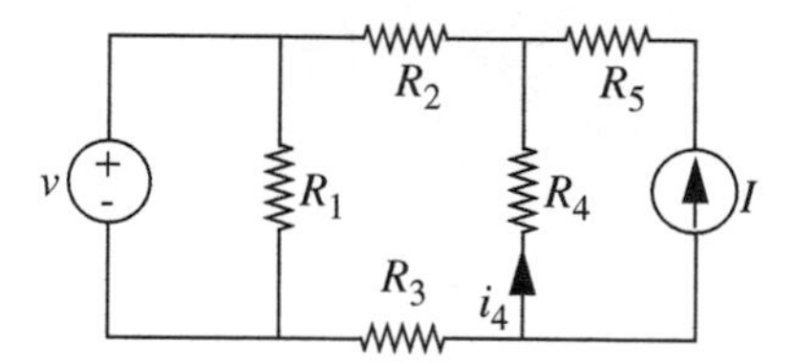


b) Determine explicitly the current $i_4$.

EXERCISE 3.22

a) Find the Thévenin equivalent of the circuit in Figure 3.123.

FIGURE 3.123

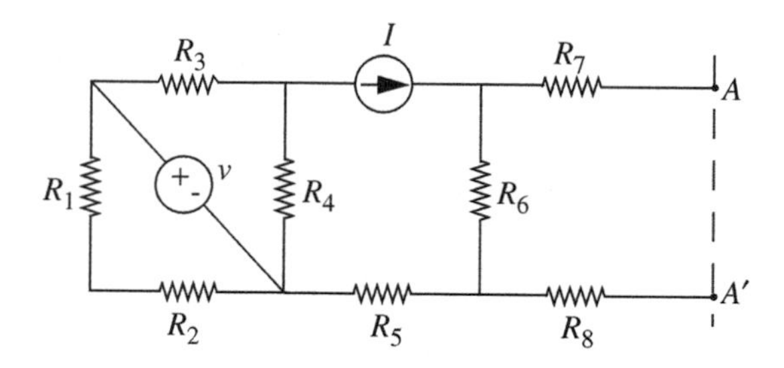


b) Find the Norton equivalent of the circuit in Figure 3.124.

FIGURE 3.124

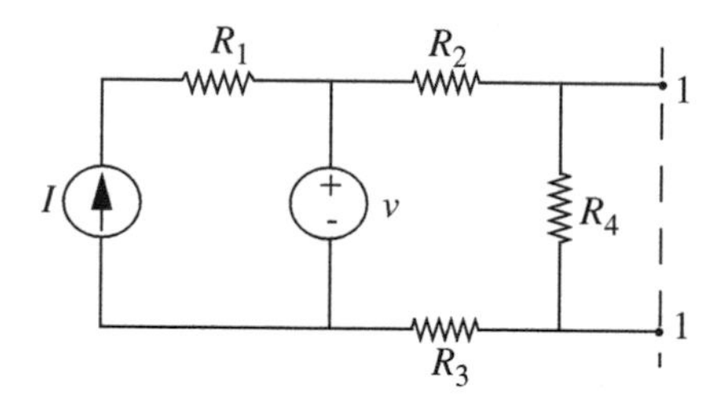


EXERCISE 3.23

a) Find the Norton equivalent of the circuit in Figure 3.125.

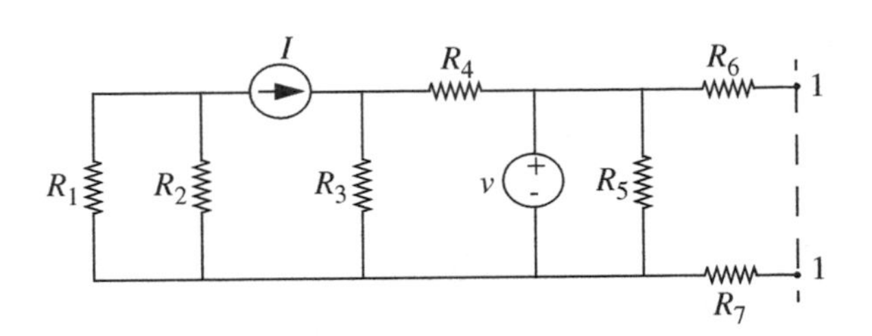


FIGURE 3.125

b) Find the Thévenin equivalent of the circuit in Figure 3.126.

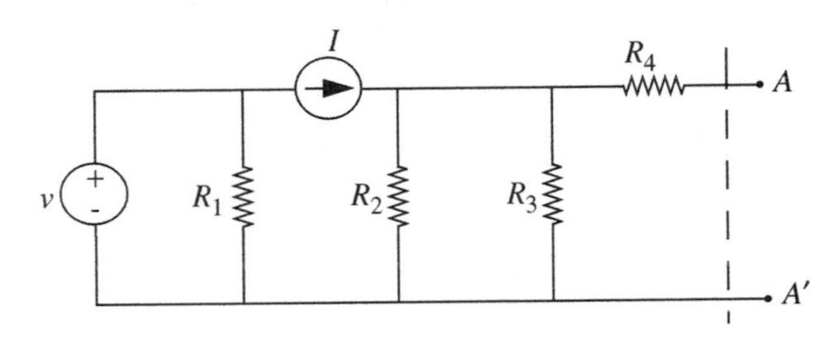


FIGURE 3.126

EXERCISE 3.24 Find the Thévenin equivalent circuit as seen from the terminals $a$–$b$ in Figure 3.127.

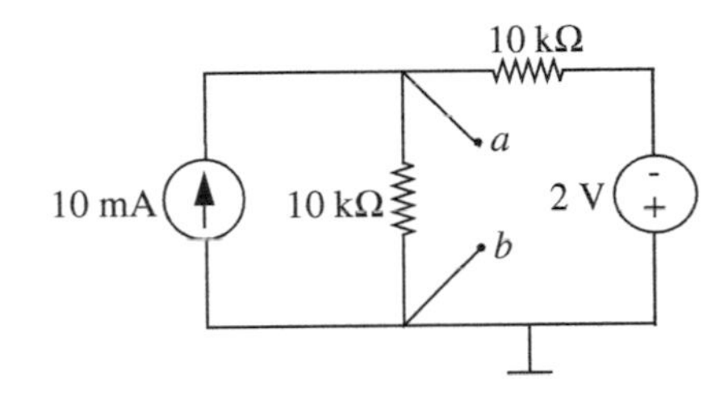


FIGURE 3.127

EXERCISE 3.25 Find the node potential $E$ in Figure 3.128.

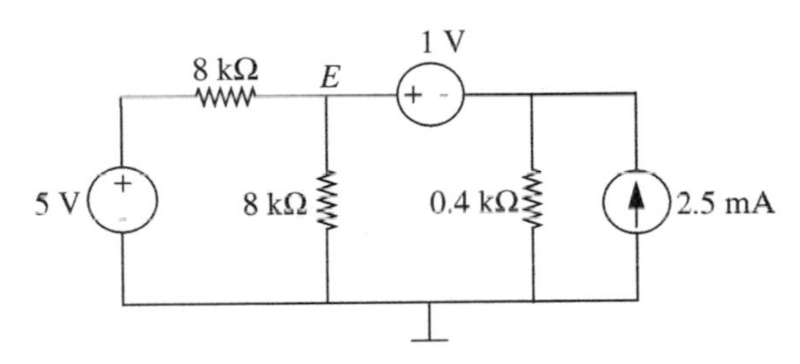


FIGURE 3.128

EXERCISE 3.26 For the circuit in Figure 3.129, write the node equations. Do not solve, but write in matrix form: source terms on the left, unknown variables on the right.

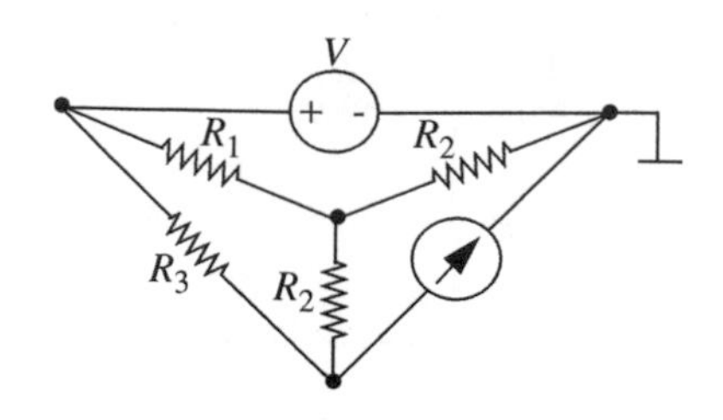


FIGURE 3.129

EXERCISE 3.27 Find $v_1$ by superposition for the circuit in Figure 3.130.

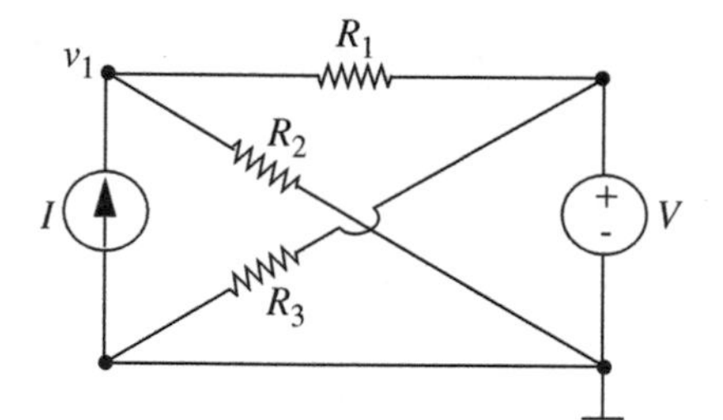


FIGURE 3.130

## PROBLEMS

PROBLEM 3.1 A fuse is a wire with a positive temperature coefficient of resistance (in other words, its resistance increases with temperature). When a current is passed through the fuse, power is dissipated in the fuse, which raises its temperature.

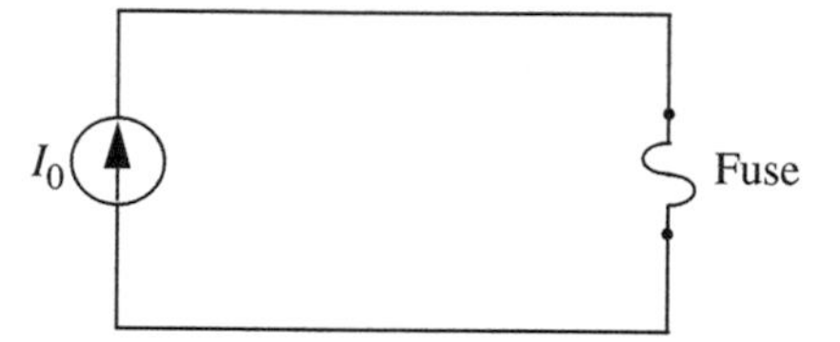


FIGURE 3.131

Use the following data to determine the current $I_0$ at which the fuse (in Figure 3.131) will blow (that is, its temperature goes up without limit).

Fuse Resistance:
$R = 1 + a\,T\;\Omega$
$a = 0.001\;\Omega/°C$
$T =$ *Temperature rise above ambient*

Temperature rise:
$T = \beta P$
$\beta = (1/.225)°C/W$
$P =$ *power dissipated in fuse*

PROBLEM 3.2

a) Prove, if possible, each of the following statements. If a proof is not possible, illustrate the failure with a counter-example and restate the theorem with a suitable restriction so it can be proved.

   i) In a network containing only linear resistors, every branch voltage and branch current must be zero.

   ii) The equivalent of a one-port network containing only linear resistors is a linear resistor.

b) To demonstrate that you understand superposition, construct an example that shows explicitly that a network containing a nonlinear resistor will not obey superposition. You may select any nonlinear element (provided you show that it is not linear) and any simple network containing that element.

PROBLEM 3.3 Find $V_0$ in Figure 3.132. Solve by (1) node method, (2) superposition. All resistances are in ohms.

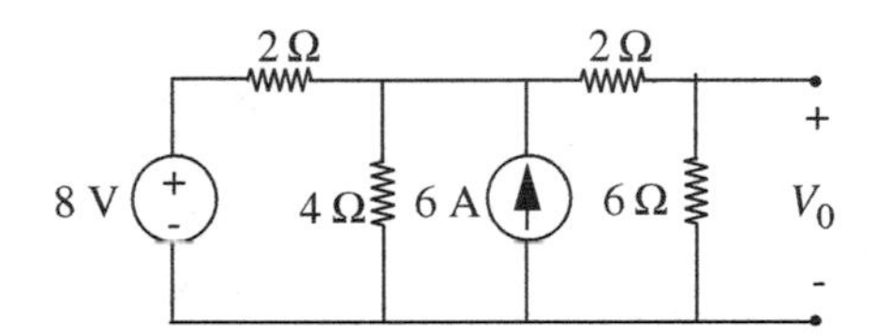


FIGURE 3.132

PROBLEM 3.4 Consider Figure 3.132. Find the Norton equivalent of the network as seen at the terminals on the right.

PROBLEM 3.5

a) Find $R_{eq}$, the equivalent resistance "looking into" the terminals on the right of the circuit in Figure 3.133.

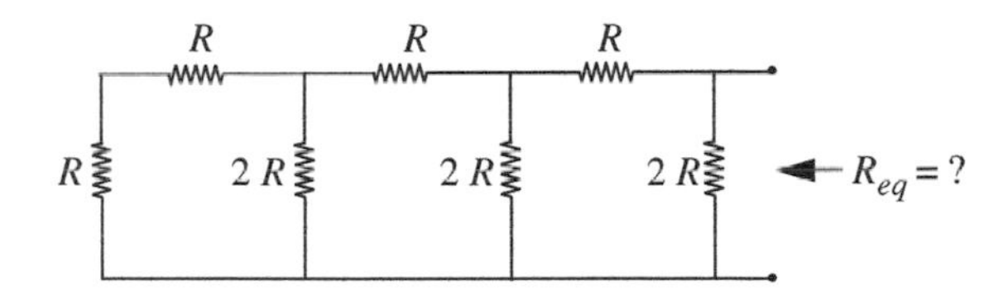


FIGURE 3.133

b) Find the Thévenin equivalent, looking into the terminals on the right of the circuit in Figure 3.134.

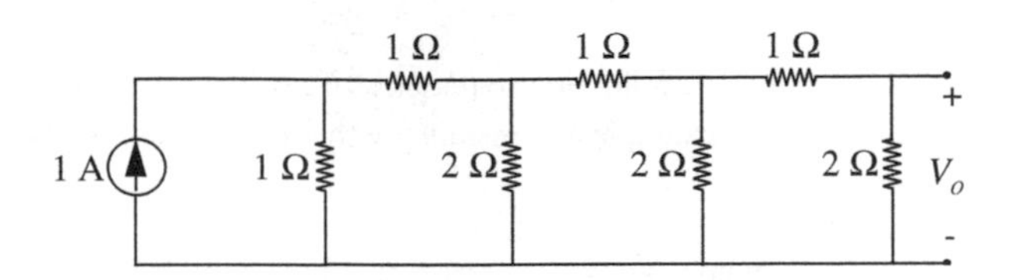


FIGURE 3.134

PROBLEM 3.6 Find $v_i$ for $I = 3$ A, $V = 2$ V in Figure 3.135. Strategy: To avoid numerical errors, derive expressions in literal form first, then check dimensions.

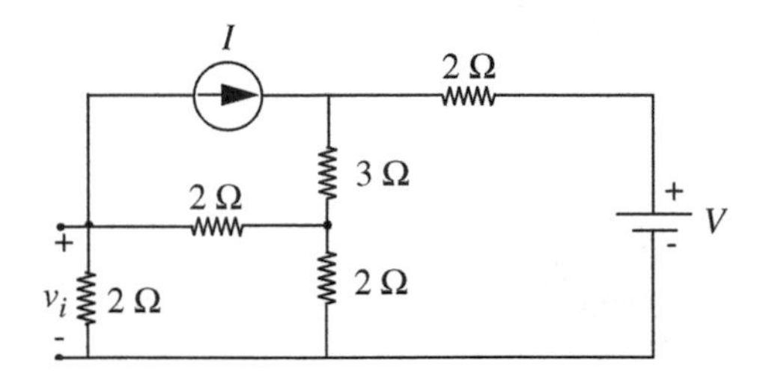


FIGURE 3.135

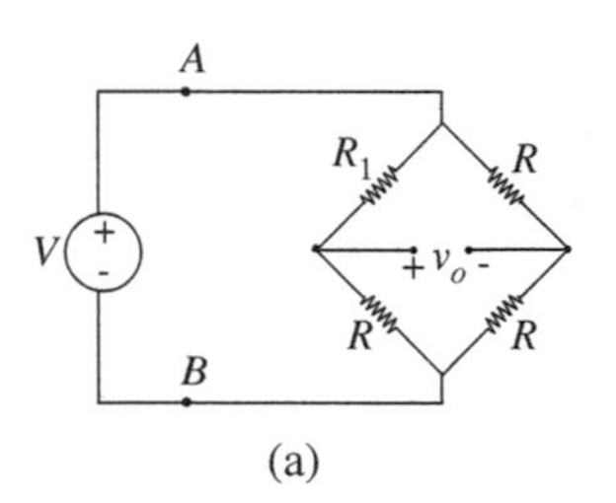


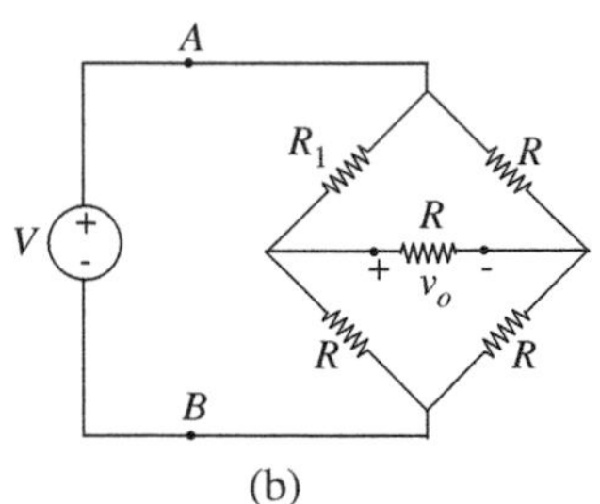


FIGURE 3.136

PROBLEM 3.7 For the circuits in Figures 3.136a and 3.136b:

a) Find $v_o$ for $R_1 = R$.

b) Find $v_o$ for $R_1 \neq R$.

c) Find the Thévenin equivalent for the network to the right of points $AB$, assuming $R_1 = R$.

PROBLEM 3.8

a) Determine the equation relating $i$ to $v$ in Figure 3.137.

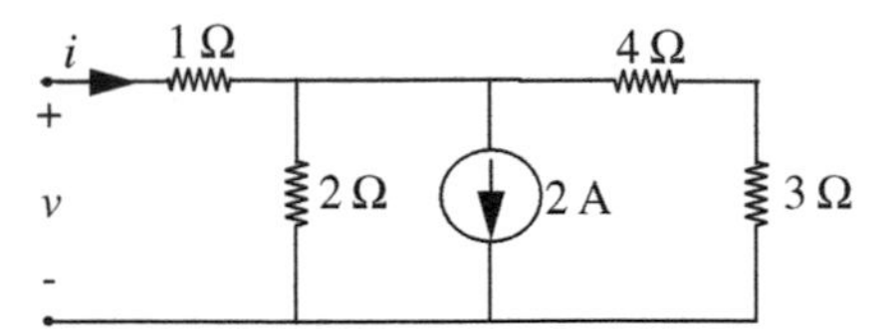


FIGURE 3.137

b) Plot the $v$–$i$ characteristics of the network.

c) Draw the Thévenin equivalent circuit.

d) Draw the Norton equivalent circuit.

PROBLEM 3.9 In Figure 3.138, find $v_O$ via (a) superposition, (b) the node method.

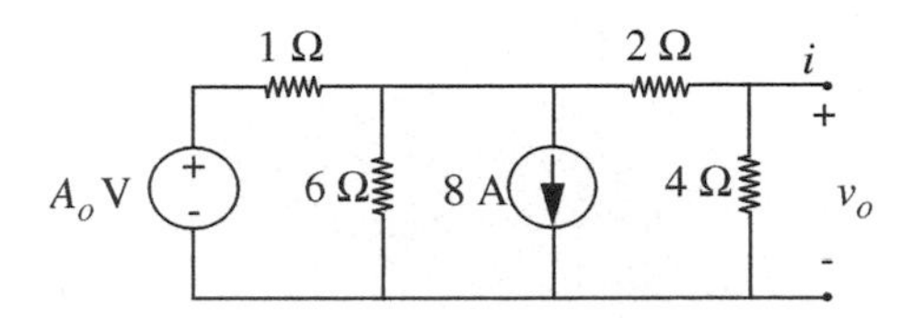


FIGURE 3.138

PROBLEM 3.10 Use the following three different methods to find $i$ in Figure 3.139:

1) Node method

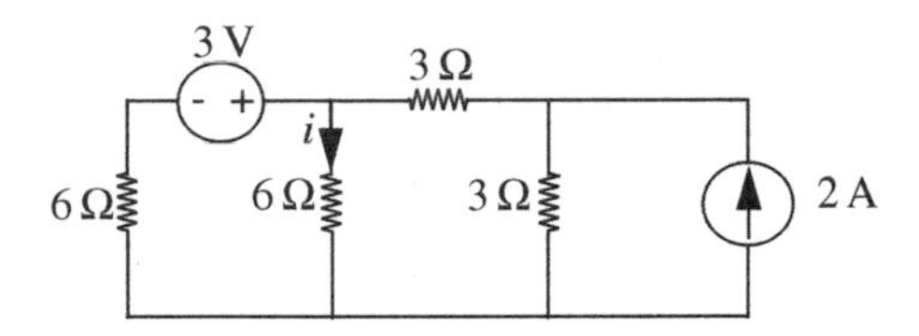


FIGURE 3.139

2) Superposition

3) Alternate Thévenin/Norton transformations

PROBLEM 3.11 A student is given an unknown resistive network as illustrated in Figure 3.140. She wishes to determine whether the network is linear, and if it is, what its Thévenin equivalent is.

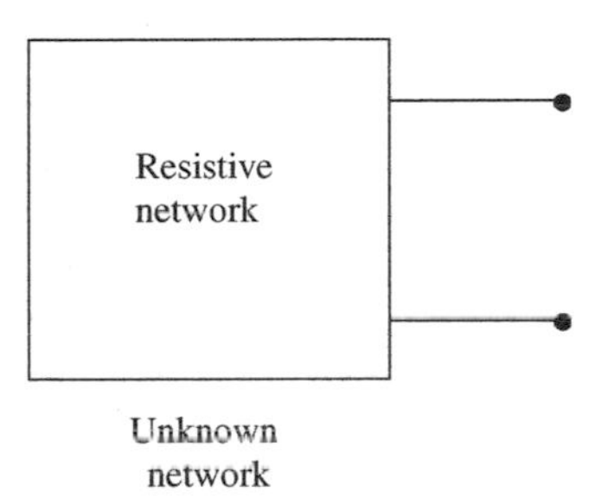


FIGURE 3.140

The only equipment available to the student is a voltmeter (assumed ideal), 100-kΩ and 1-MΩ test resistors that can be placed across the terminals during a measurement (see Figure 3.141).

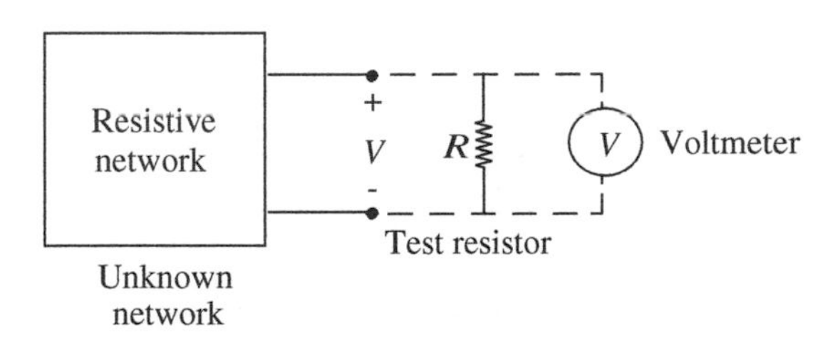


FIGURE 3.141

The following data were recorded:

| Test Resistor | Voltmeter Reading |
|---|---|
| Absent | 1.5 V |
| 100 kΩ | 0.25 V |
| 1 MΩ | 1.0 V |

What should the student conclude about the network from these results? Support your conclusion with plots of the network *v–i* characteristics.

PROBLEM 3.12

a) Devise an electrical circuit of voltage sources and resistors that will "calculate" the balance point (center of mass) of the massless bar shown in Figure 3.142, for three arbitrary masses hung at three arbitrary places along the bar. We want the circuit to generate a voltage that is proportional to the position of the balance point. Write the equation for your network, and show that it performs the required calculation. (Work with conductances and superposition for a simple solution.)

FIGURE 3.142

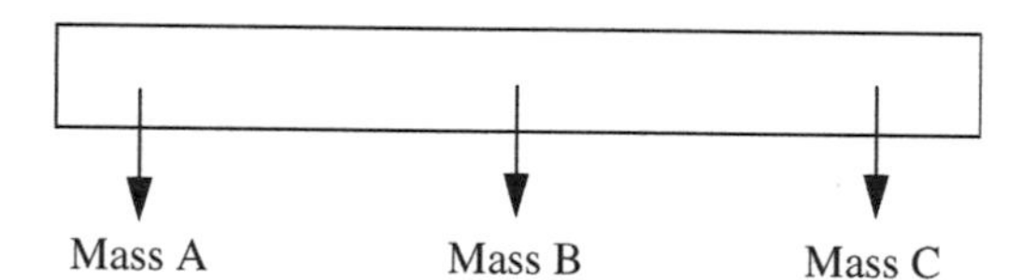


b) Extend your result in part (a) to two dimensions; that is, devise a new network (which will have more voltage sources and more resistors than above) that can find the center of mass of a triangle with arbitrary weights handing from its three corners. The network will now have to give you two voltages, one representing the x-coordinate and the other the y-coordinate of the center of mass. This system is a barycentric coordinate calculator, and can be used as the input for video games, or to simulate trichromatic color vision in the human eye.

PROBLEM 3.13

a) Find the Thévenin equivalent for the network in Figure 3.143 at the terminals *CB*. The current source is a *controlled source*. The current flowing through the current

FIGURE 3.143

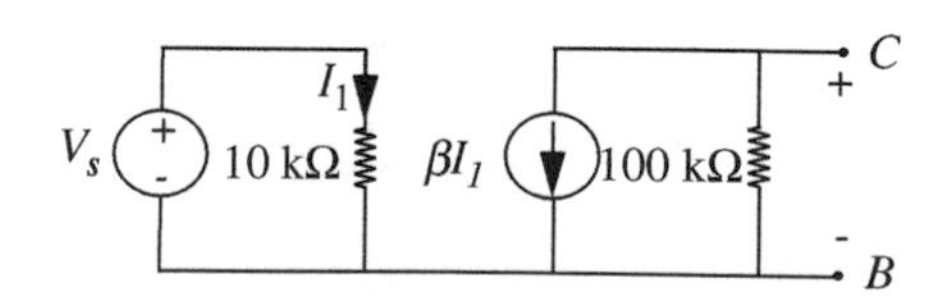

source is $\beta I_1$, where $\beta$ is some constant. (We will discuss controlled sources in more detail in the later chapters.)

b) Now suppose you connect a load resistor across the output of your equivalent circuit as shown in Figure 3.144. Find the value of $R_L$ which will provide the maximum power transfer to the load.

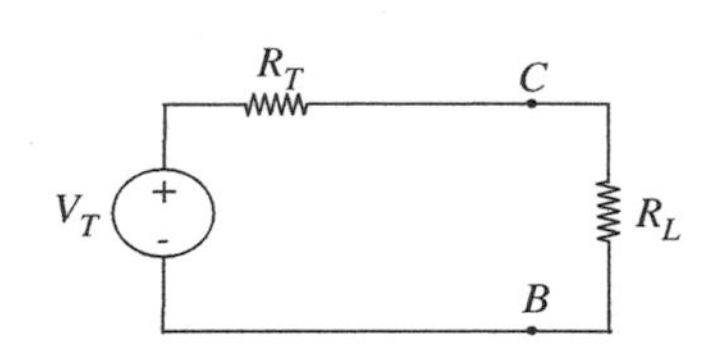


FIGURE 3.144

PROBLEM 3.14 You have been hired by the MITDAC Corporation to write a product description for a new 4-bit digital-to-analog-converter resistance ladder. Because of mask tolerances in VLSI chips, each resistor shown in Figure 3.145 is guaranteed to be only within 3% of its nominal value. That is, if $R_0$ is the nominal design resistance, then each resistance labeled R can have a resistance anywhere in the range $(1 \pm .03)R_0$ and each resistance labeled $2R$ can have a resistance anywhere in the range $(2 \pm .06)R_0$.

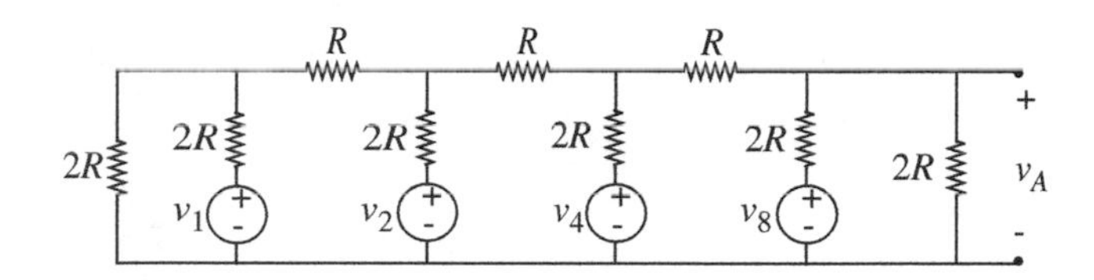


FIGURE 3.145

You are to write an *honest* description of the accuracy of this product. Remember that if you overstate the accuracy, your company will have many returns from dissatisfied customers, whereas if you understate the accuracy, your company won't have any customers.

Note: Part of this problem is to describe what the problem is. How should accuracy be specified? Is there an error level that is clearly unacceptable? Does your product avoid that error level? Is there an obvious "worst case" that can be easily analyzed? Have fun. And remember, common sense is an important ingredient of sound engineering.

PROBLEM 3.15 You have a 6-volt battery (assumed ideal) and a 1.5-volt flashlight bulb, which is known to draw 0.5 A when the bulb voltage is 1.5 V (in Figure 3.146). Design a network of resistors to go between the battery and the bulb to give $v_s = 1.5$ V when the bulb is connected, yet ensures that $v_s$ does not rise above 2 V when the bulb is disconnected.

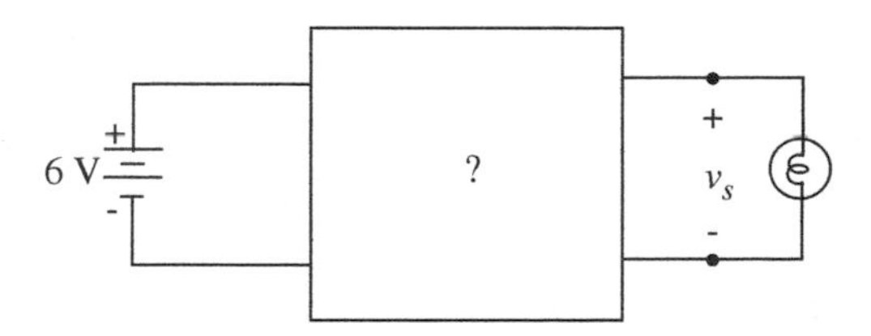


FIGURE 3.146

CHAPTER 4

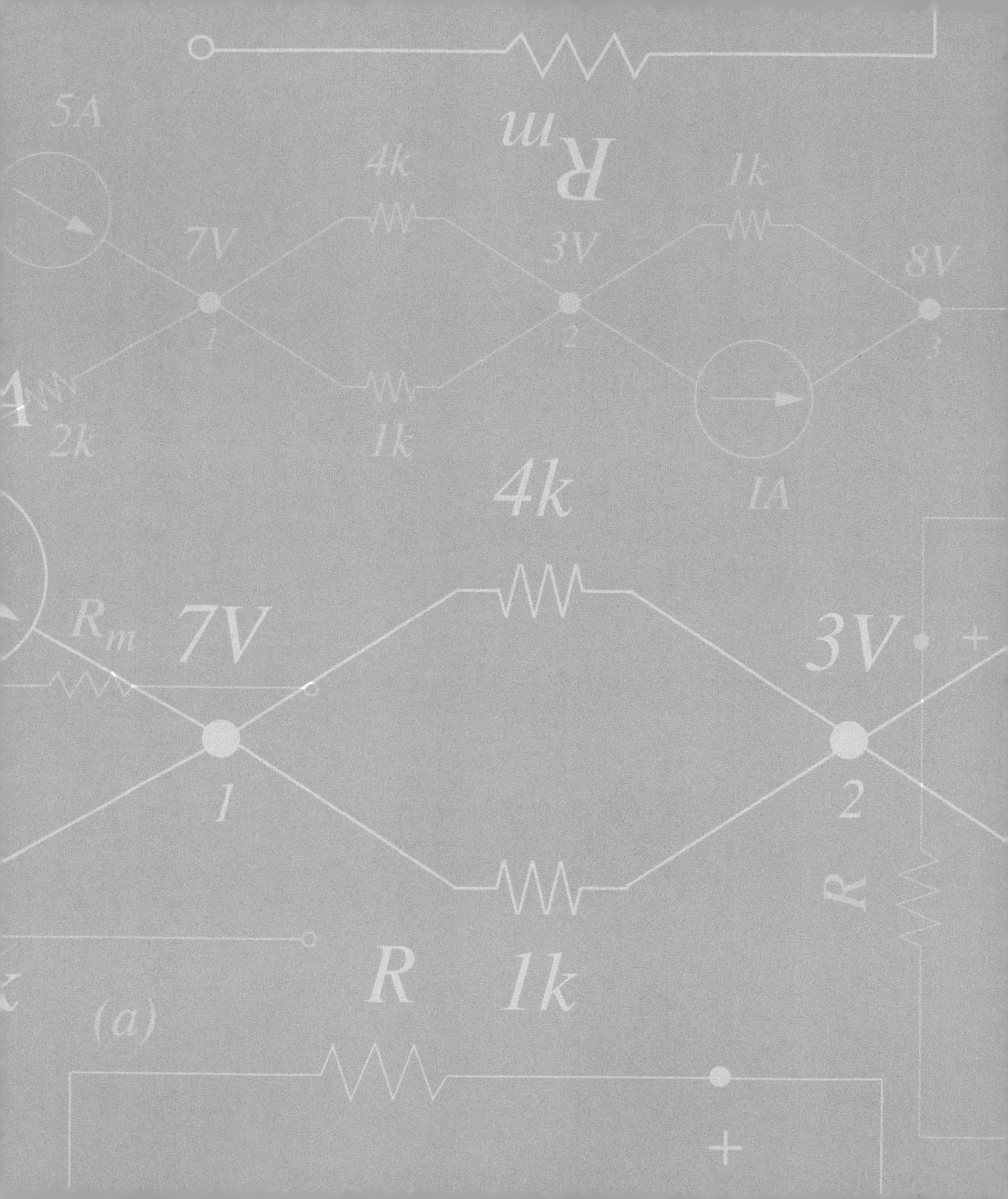
5A
4k
1k
7V
3V
8V
1
2
3
2k
1k
1A
4k
R_m
7V
3V
1
2
R
R
1k
(a)

# ANALYSIS OF NONLINEAR CIRCUITS

# 4

Thus far we have discussed a variety of circuits containing linear devices such as resistors and voltage sources. We have also discussed methods of analyzing linear circuits built out of these elements. In this chapter, we extend our repertoire of network elements and corresponding analysis techniques by introducing a nonlinear two-terminal device called a *nonlinear resistor*. Recall, from Section 1.5.2, a nonlinear resistor is an element that has a nonlinear, algebraic relation between its instantaneous terminal current and its instantaneous terminal voltage. A diode is an example of a device that behaves like a nonlinear resistor. In this chapter, we will introduce methods of analyzing general circuits containing nonlinear elements, trying whenever possible to use analysis methods already introduced in the preceding chapters. Chapter 7 will develop further the basic ideas on nonlinear analysis and Chapter 8 will expand on the concept of incremental analysis introduced in this chapter. Chapter 16 will elaborate on diodes.

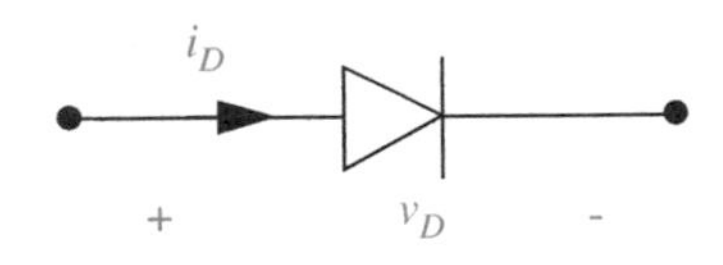


FIGURE 4.1 The symbol for a diode.

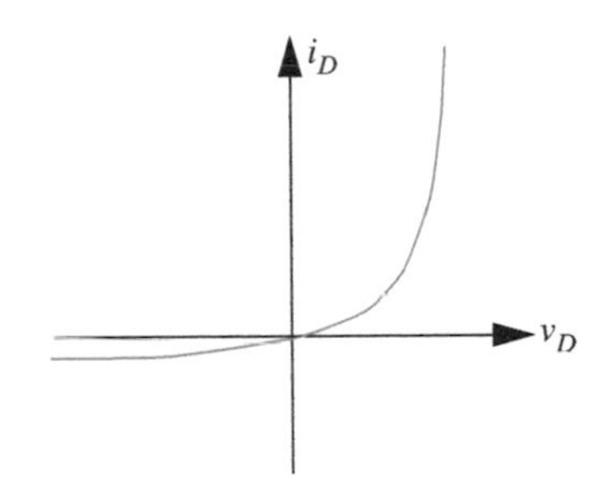


FIGURE 4.2 $v$–$i$ characteristics of a silicon diode.

## 4.1 INTRODUCTION TO NONLINEAR ELEMENTS

Before we begin our analysis of nonlinear resistors, we will describe as examples several nonlinear resistive devices, by their $v$–$i$ characteristics, just as we did for the resistor, the battery, etc. The first of the nonlinear devices that we discuss is the *diode*. Figure 4.1 shows the symbol for a diode. The diode is a two-terminal, nonlinear resistor whose current is exponentially related to the voltage across its terminals.

An analytical expression for the nonlinear relation between the voltage $v_D$ and the current $i_D$ for the diode is the following:

$$i_D = I_s(e^{v_D/V_{TH}} - 1). \tag{4.1}$$

For silicon diodes the constant $I_s$ is typically $10^{-12}$ A and the constant $V_{TH}$ is typically 0.025 V. This function is plotted in Figure 4.2.

An analytical expression for the relationship between voltage $v_H$ and current $i_H$ for another hypothetical nonlinear device is shown in Equation 4.2. In the equation, $I_K$ is a constant. The relationship is plotted in Figure 4.3.

$$i_H = I_K v_H^3. \tag{4.2}$$

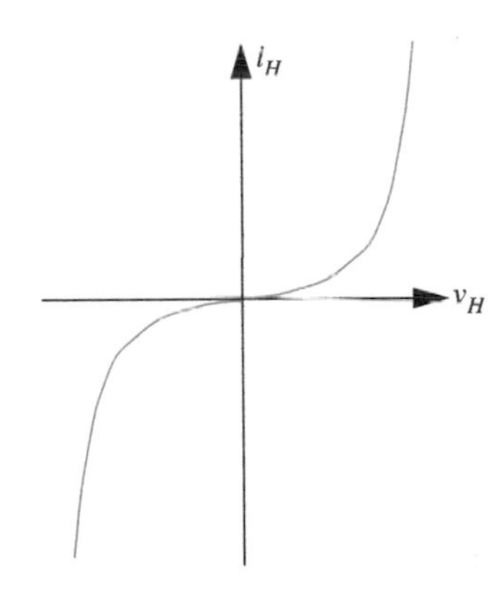


FIGURE 4.3 Another nonlinear $v$–$i$ characteristics.

The $v$–$i$ relationship for yet another two-terminal nonlinear device is shown in Equation 4.3. Figure 8.11 in Chapter 8 introduces such a nonlinear device. For this device the current is related to the square of the terminal voltage. In this equation, $K$ and $V_T$ are constants. The variables $i_{DS}$ and $v_{DS}$ are the terminal variables for the device. The relationship is plotted in Figure 4.4.

$$i_{DS} = \begin{cases} \dfrac{K(v_{DS} - V_T)^2}{2} & \text{for } v_{DS} \geq V_T \\ 0 & \text{for } v_{DS} < V_T \end{cases}. \tag{4.3}$$

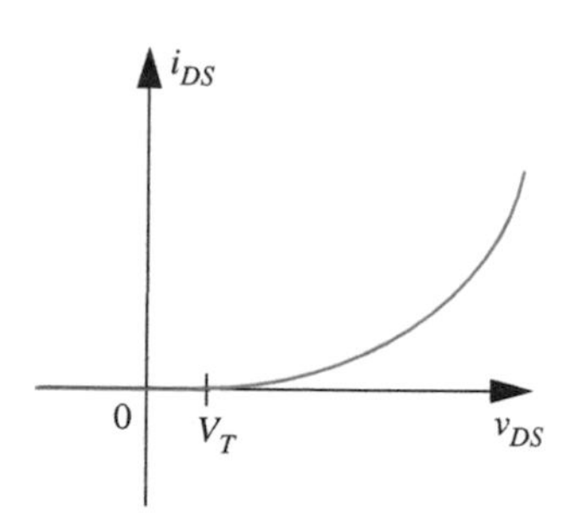


FIGURE 4.4 The $v$–$i$ characteristics for a square law device.

EXAMPLE 4.1 SQUARE LAW DEVICE For the nonlinear resistor device following the square law in Figure 4.4, determine the value of $i_{DS}$ for $v_{DS} = 2$ V. We are given that $V_T = 1$ V and $K = 4$ mA/V$^2$.

For the parameters that we have been given ($v_{DS} = 2$ V and $V_T = 1$ V), it is easy to see that

$$v_{DS} \geq V_T.$$

From Equation 4.3, the expression for $i_{DS}$ when $v_{DS} \geq V_T$ is

$$i_{DS} = \frac{K(v_{DS} - V_T)^2}{2}.$$

Substituting the known numerical values,

$$i_{DS} = \frac{4 \times 10^{-3}(2-1)^2}{2} = 2 \text{ mA}.$$

How does $i_{DS}$ change if $v_{DS}$ is doubled?

If $v_{DS}$ is doubled to 4 V,

$$i_{DS} = \frac{K(v_{DS} - V_T)^2}{2} = \frac{4 \times 10^{-3}(4-1)^2}{2} = 18 \text{ mA}$$

In other words, $i_{DS}$ increases to 18 mA when $v_{DS}$ is doubled.

What is the value of $i_{DS}$ if $v_{DS}$ is changed to 0.5 V?

For $v_{DS} = 0.5$ V and $V_T = 1$ V,

$$v_{DS} < V_T.$$

From Equation 4.3, we get

$$i_{DS} = 0.$$

When operating within some circuit, the current through our square law device is measured to be 4 mA. What must be the voltage across the device?

We are given that $i_{DS} = 4$ mA. Since there is a current through the device, the equation that applies is

$$i_{DS} = \frac{K(v_{DS} - V_T)^2}{2}.$$

Substituting known values,

$$8 \times 10^{-3} = \frac{4 \times 10^{-3}(v_{DS} - 1)^2}{2}.$$

Solving for $v_{DS}$, we get

$$v_{DS} = 3 \text{ V}.$$

---

EXAMPLE 4.2 DIODE EXAMPLE For the diode shown in Figure 4.1, determine the value of $i_D$ for $v_D = 0.5$ V, 0.6 V, and 0.7 V. We are given that $V_{TH} = 0.025$ V and $I_s = 1$ pA.

From the device law for a diode given in Equation 4.1, the expression for $i_D$ is

$$i_D = I_s(e^{v_D/V_{TH}} - 1).$$

Substituting the known numerical values for $v_D = 0.5$ V, we get

$$i_D = 1 \times 10^{-12}(e^{0.5/0.025} - 1) = 0.49 \text{ mA}.$$

Similarly, for $v_D = 0.6$ V, $i_D = 26$ mA, and for $v_D = 0.7$ V, $i_D = 1450$ mA. Notice the dramatic increase in current as $v_D$ increases beyond 0.6 V.

What is the value of $i_D$ if $v_D$ is $-0.2$ V?

$$i_D = I_s(e^{v_D/V_{TH}} - 1) = 1 \times 10^{-12}(e^{-0.2/0.025} - 1) = -0.9997 \times 10^{-12}\text{A}.$$

The negative sign for $i_D$ simply reflects the fact that when $v_D$ is negative, so is the current.

When operating within some circuit, the current through the diode is measured to be 8 mA. What must be the voltage across the diode?

We are given that $i_D = 8$ mA. Using the diode equation, we get

$$8 \times 10^{-3} = I_s(e^{v_D/V_{TH}} - 1) = 1 \times 10^{-12}(e^{v_D/0.025} - 1).$$

Simplifying, we get

$$e^{v_D/0.025} = 8 \times 10^9 + 1.$$

Taking logs on both sides, and solving for $v_D$, we get

$$v_D = 0.025 \ln(8 \times 10^9 + 1) = 0.57 \text{ V}.$$

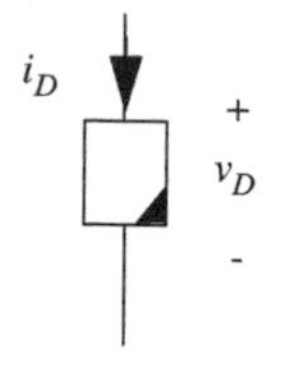


FIGURE 4.5 A nonlinear device.

EXAMPLE 4.3 ANOTHER SQUARE LAW DEVICE PROBLEM

The nonlinear device shown in Figure 4.5 is characterized by this device equation:

$$i_D = 0.1 v_D^2 \quad \text{for} \quad v_D \geq 0, \tag{4.4}$$

$i_D$ is given to be 0 for $v_D < 0$.

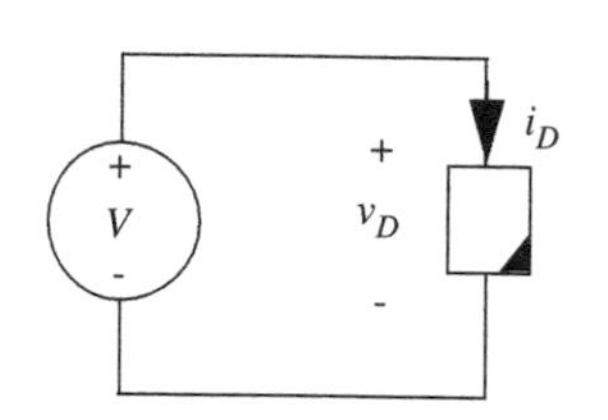


FIGURE 4.6 A circuit containing the nonlinear device.

Given that $V = 2$ V, determine $i_D$ for the circuit in Figure 4.6.

Using the device equation for $v_D \geq 0$,

$$i_D = 0.1 v_D^2 = 0.1 \times 2^2 = 0.4 \text{ A} \tag{4.5}$$

The nonlinear device is connected to some arbitrary circuit as shown in Figure 4.7. Following the associated variables discipline, the branch variables $v_B$ and $i_B$ for the device are defined as shown in the same figure. Suppose that a measurement reveals that $i_B = -1$ mA. What must be the value of $v_B$?

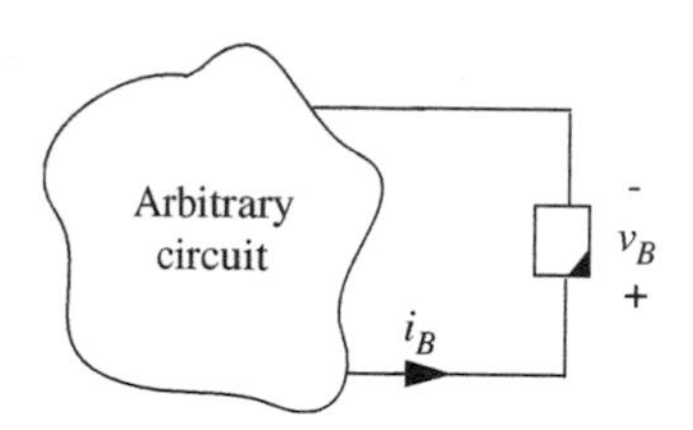


FIGURE 4.7 The nonlinear device connected to an arbitrary circuit.

Notice that the polarity of the branch variables has been reversed in Figure 4.7 from those in Figure 4.5. With this definition of the branch variables, the device equation becomes

$$-i_B = 0.1 v_B^2 \quad \text{for} \quad v_B \leq 0. \tag{4.6}$$

Furthermore, $i_B$ is 0 for $v_B > 0$.

Given that $i_B = -1$ mA, Equation 4.6 yields

$$-(-1 \times 10^{-3}) = 0.1 v_B^2 \quad \text{where} \quad v_B \leq 0.$$

In other words, $v_B = -0.1$ V.

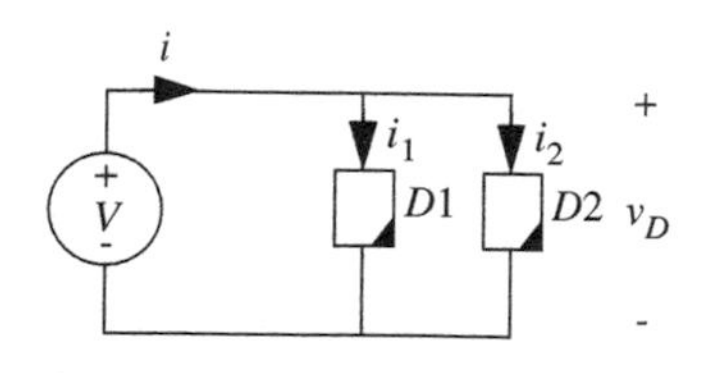


FIGURE 4.8 Nonlinear devices connected in parallel.

Given that $V = 2$ V, determine $i$ for the circuit in Figure 4.8.

Since the voltage across each of the nonlinear devices connected in parallel is $v_D = 2$ V, the current through each nonlinear device is the same as that calculated in Equation 4.5. In other words,

$$i_1 = i_2 = 0.4 \text{ A}.$$

Therefore, $i = i_1 + i_2 = 0.8$ A.

Given the analytical expression for the characteristic of a nonlinear device, such as that for the diode in Equation 4.1, how can we calculate the voltages and currents in a simple circuit such as Figure 4.9? In the following sections we will discuss four methods for solving such nonlinear circuits:

1. Analytical solutions
2. Graphical analysis
3. Piecewise linear analysis
4. Incremental or small signal analysis

## 4.2 ANALYTICAL SOLUTIONS

We first try to solve the simple nonlinear resistor circuit in Figure 4.9 by analytical methods. Assume that the hypothetical nonlinear resistor in the figure is characterized by the following $v$–$i$ relationship:

$$i_D = \begin{cases} Kv_D^2 & \text{for } v_D > 0 \\ 0 & \text{for } v_D \leq 0. \end{cases} \tag{4.7}$$

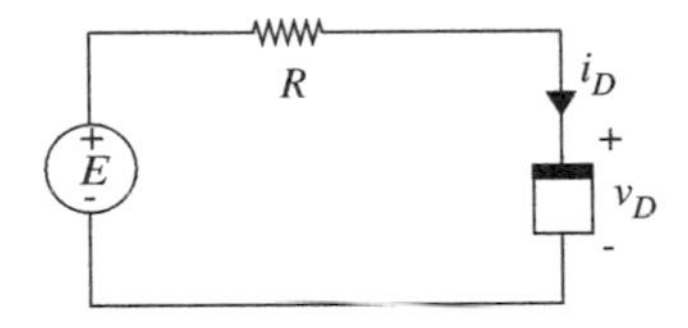


FIGURE 4.9 A simple circuit with a nonlinear resistor.

The constant $K$ is positive.

This circuit is amenable to a straightforward application of the node method. Recall that the node method and its foundational Kirchhoff's voltage and current laws are derived from Maxwell's Equations with no assumptions about linearity. (Note, however, that the superposition method, the Thévenin method, and the Norton method do require a linearity assumption.)

To apply the node method, we first choose a ground node and label the node voltages as illustrated in Figure 4.10. $v_D$ is our only unknown node voltage.

Next, following the node method, we write KCL for the node that has an unknown node voltage. As prescribed by the node method, we will use KVL and the device relation ($i_D = Kv_D^2$) to obtain the currents directly in terms of the node voltage differences and element parameters. For the node with voltage $v_D$,

$$\frac{v_D - E}{R} + i_D = 0 \tag{4.8}$$

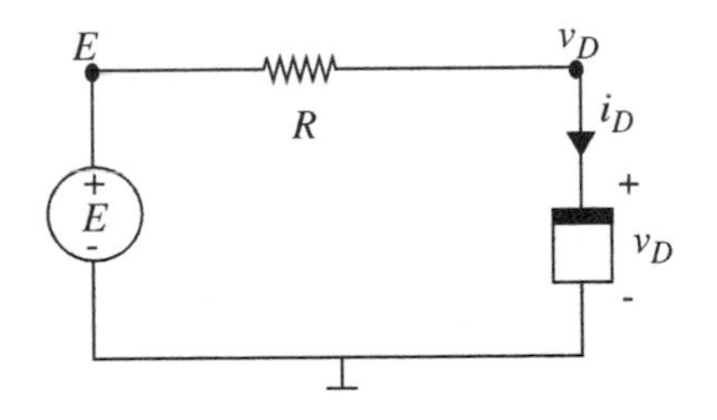


FIGURE 4.10 The nonlinear circuit with the ground node chosen and node voltages labeled.

Note that this is not quite our node equation, because of the presence of the $i_D$ term. To get the node equation we need to substitute for $i_D$ in terms of node voltages. Recall that the nonlinear device $v$–$i$ relation is

$$i_D = Kv_D^2. \tag{4.9}$$

Note that this device equation applies for positive $v_D$. We are given that $i_D = 0$ when $v_D \le 0$.

Substituting the nonlinear device $v$–$i$ relationship for $i_D$ in Equation 4.8, we get the required node equation in terms of the node voltages:

$$\frac{v_D - E}{R} + Kv_D^2 = 0. \tag{4.10}$$

For our device, note that Equation 4.9 holds only for $v_D > 0$. For $v_D \le 0$, $i_D$ is 0.

Simplifying Equation 4.10, we obtain the following quadratic equation.

$$RKv_D^2 + v_D - E = 0.$$

Solving for $v_D$ and choosing the positive solution

$$v_D = \frac{-1 + \sqrt{1 + 4RKE}}{2RK}. \tag{4.11}$$

The corresponding expression for $i_D$ can be obtained by substituting the previous expression for $v_D$ into Equation 4.9 as follows:

$$i_D = K\left[\frac{-1 + \sqrt{1 + 4RKE}}{2RK}\right]^2. \tag{4.12}$$

It is worth discussing why we ignored the negative solution. As shown in Figure 4.11, two *mathematical* solutions are possible when we solve Equations 4.10 and 4.9. However, the dotted curve in Figure 4.11 is part of

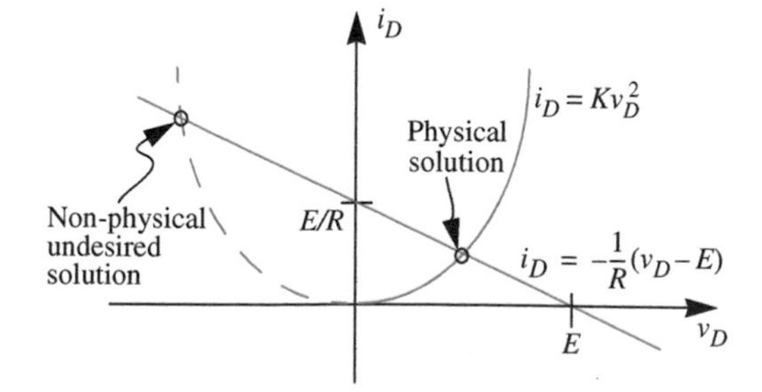


**FIGURE 4.11** Solutions to equations Equations 4.10 and 4.9.

Equation 4.9 but not the *physical* device. Because, recall, Equation 4.9 applies only for positive $v_D$. When $E$ is negative, $i_D$ will be equal to 0 and $v_D$ will be equal to $E$.

---

EXAMPLE 4.4 ONE NONLINEAR DEVICE, SEVERAL SOURCES, AND RESISTORS Shown in Figure 4.12 is a circuit of no obvious value, which we use to illustrate how to solve nonlinear circuits with more than one source present, using the nonlinear analysis method just discussed. Let us assume that we wish to calculate the nonlinear device current $i_D$.

Assume that the nonlinear device is characterized by the following $v$–$i$ relationship:

$$i_D = \begin{cases} Kv_D^2 & \text{for } v_D > 0 \\ 0 & \text{for } v_D \leq 0. \end{cases} \tag{4.13}$$

The terminal variables for the nonlinear device are defined as shown in Figure 4.9, and the constant $K$ is positive.

Linear analysis techniques such as superposition cannot be applied to the whole circuit because of the nonlinear element. But because there is only one nonlinear device, it is permissible to find the Thévenin (or Norton) equivalent circuit *faced by the nonlinear device* (see Figures 4.13a and b), because this part of the circuit is linear. Then we can compute easily the terminal voltage and current for the nonlinear device using the circuit in Figure 4.13b from Equations 4.11 and 4.12.

First, to find the open-circuit voltage, we draw the linear circuit as seen from the nonlinear device terminals in Figure 4.13c. Superposition or any other linear analysis method can now be used to calculate the open-circuit voltage:

$$V_{TH} = V\frac{R_2}{R_1 + R_2} - I_0 R_3. \tag{4.14}$$

The Thévenin equivalent resistance, $R_{TH}$, the resistance seen at the terminals in Figure 4.13d, with the sources set to zero is

$$R_{TH} = (R_1 || R_2) + R_3. \tag{4.15}$$

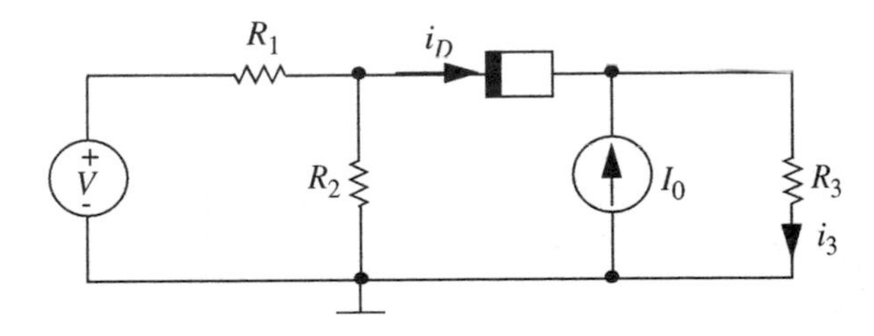


FIGURE 4.12 Circuit with several sources and resistors.

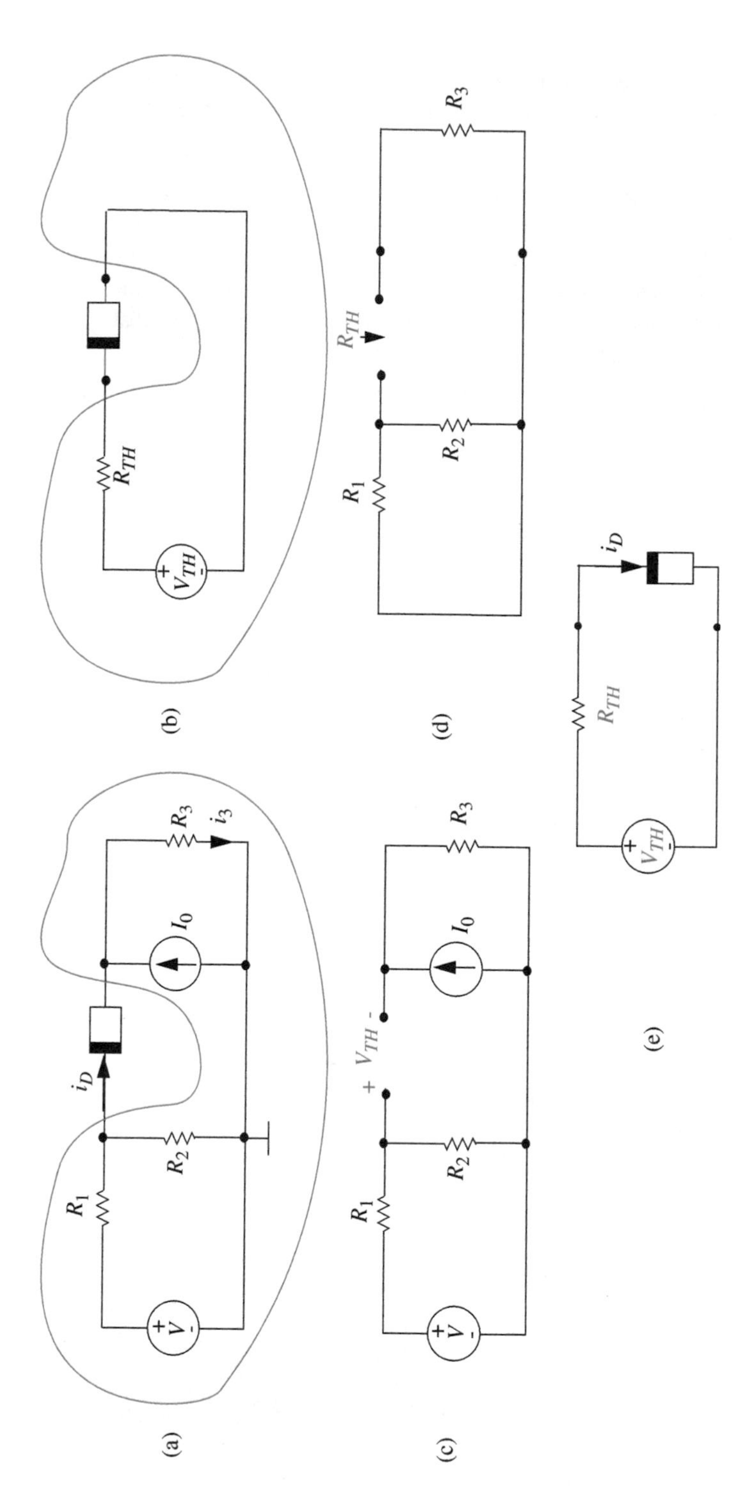


**FIGURE 4.13** Analysis using Thévenin's Theorem.

When we reconnect the nonlinear device to this Thévenin circuit, as in Figure 4.13e, we are back to a familiar example: one nonlinear device, one source, and one resistor. The desired device current $i_D$ can be found by a nonlinear analysis method, such as that used to solve the circuit in Figure 4.9.

One further comment: If in the problem statement we had been asked to find one of the resistor currents, say $i_3$, rather than $i_D$, then the Thévenin circuit, Figure 4.13e, would not give this current directly, because the identity of currents internal to the Thévenin network are in general lost, as noted in Chapter 3. Nonetheless, the Thévenin approach is probably the best, as it is a simple matter to work back through the linear part of the network to relate $i_3$ to $i_D$. In this case, once we have computed $i_D$, we can easily determine $i_3$ from Figure 4.13a using KCL,

$$i_3 = i_D + I_0. \tag{4.16}$$

---

WWW EXAMPLE 4.5 NODE METHOD

---

EXAMPLE 4.6 ANOTHER SIMPLE NONLINEAR CIRCUIT

Let us try to solve the nonlinear circuit containing a diode in Figure 4.16 by analytical methods. Following the node method, we first choose the ground node and label the node voltages as illustrated in Figure 4.17.

Next, we write KCL for the node with the unknown node voltage, and substitute for the diode current using the diode equation

$$\frac{v_D - E}{R} + i_D = 0 \tag{4.18}$$

$$i_D = I_s(e^{v_D/V_{TH}} - 1). \tag{4.19}$$

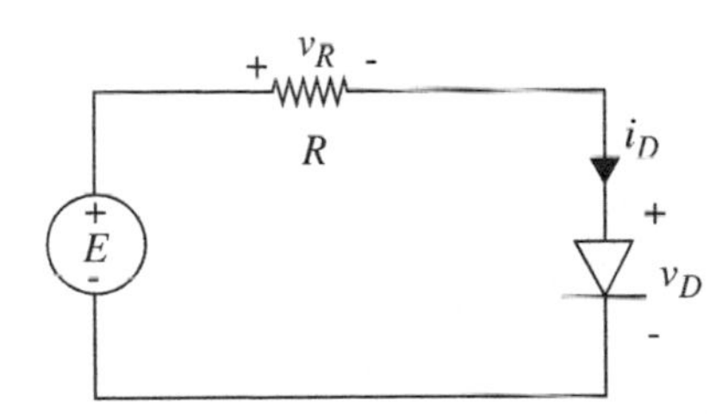


FIGURE 4.16 A simple non-linear circuit containing a diode.

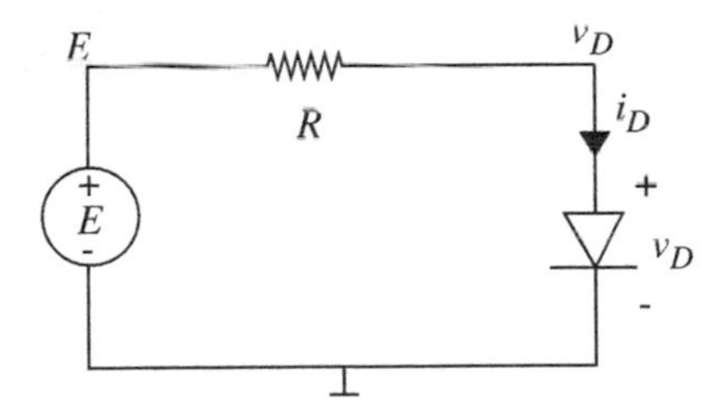


FIGURE 4.17 The circuit with the ground node and the node voltages marked.

If $i_D$ is eliminated by substituting Equation 4.19 into Equation 4.18, the following transcendental equation results:

$$\frac{v_D - E}{R} + I_s(e^{v_D/V_{TH}} - 1) = 0.$$

This equation must be solved by trial and error. Easy via computer, but not very insightful.

---

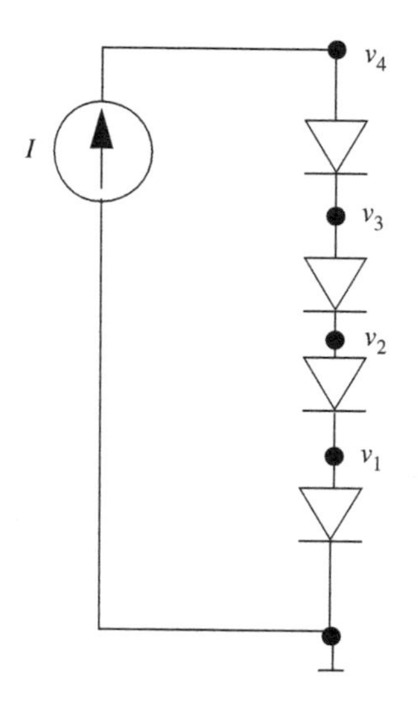


FIGURE 4.18 Series connected diodes.

EXAMPLE 4.7 SERIES-CONNECTED DIODES Referring to the series-connected diodes in Figure 4.18, determine $v_1$, $v_2$, $v_3$, and $v_4$, given that $I = 2$ A. The parameters in the diode relation are given to be $I_s = 10^{-12}$ A, $V_{TH} = 0.025$ V.

We will first use the node method to solve this problem. Figure 4.18 shows the ground node and the node voltages. There are four unknown node voltages. Next, we write KCL for each of the nodes. As prescribed by the node method, we will use KVL and the diode relation (Equation 4.1) to obtain the currents directly in terms of the node voltage differences and element parameters. For the node with voltage $v_1$,

$$10^{-12}(e^{v_1/0.025} - 1) = 10^{-12}(e^{(v_2-v_1)/0.025} - 1). \tag{4.20}$$

The term on the left-hand side is the current through the lowermost device expressed in terms of node voltages. Similarly, the term on the right-hand side is the current through the device that is second from the bottom.

Similarly, we can write the node equations for the nodes with voltages $v_2$, $v_3$, and $v_4$ as follows:

$$10^{-12}(e^{(v_2-v_1)/0.025} - 1) = 10^{-12}(e^{(v_3-v_2)/0.025} - 1) \tag{4.21}$$

$$10^{-12}(e^{(v_3-v_2)/0.025} - 1) = 10^{-12}(e^{(v_4-v_3)/0.025} - 1) \tag{4.22}$$

$$10^{-12}(e^{(v_4-v_3)/0.025} - 1) = I. \tag{4.23}$$

Simplifying, and taking the log on both sides of Equations 4.20 through 4.23, we get

$$v_1 = v_2 - v_1 \tag{4.24}$$

$$v_2 - v_1 = v_3 - v_2 \tag{4.25}$$

$$v_3 - v_2 = v_4 - v_3 \tag{4.26}$$

$$v_4 - v_3 = 0.025 \ln(10^{12} I + 1). \tag{4.27}$$

Given that $I = 2$ A, we can solve for $v_1$, $v_2$, $v_3$, and $v_4$, to get

$$v_1 = 0.025 \ln(10^{12} I + 1) = 0.025 \ln(10^{12} \times 2 + 1) = 0.71 \text{ V}$$

$$v_2 = 2v_1 = 1.42 \text{ V}$$

$$v_3 = 3v_1 = 2.13 \text{ V}$$

$$v_4 = 4v_1 = 2.84 \text{ V}.$$

Notice that we could have also solved the circuit intuitively by observing that the same 2-A current flows through each of the four identical diodes. Thus, the same voltage must drop across each of the diodes. In other words,

$$I = 10^{-12}(e^{v_1/0.025} - 1)$$

or,

$$v_1 = 0.025 \ln(10^{12} I + 1).$$

For $I = 2$ A,

$$v_1 = 0.025 \ln(10^{12} \times 2A + 1) = 0.71 \text{ V}.$$

Once the value of $v_1$ is known, we can easily compute the rest of the node voltages from

$$v_1 = v_2 - v_1 = v_3 - v_2 = v_4 - v_3.$$

WWW EXAMPLE 4.8 MAKING SIMPLIFYING ASSUMPTIONS

WWW EXAMPLE 4.9 VOLTAGE-CONTROLLED NONLINEAR RESISTOR

## 4.3 GRAPHICAL ANALYSIS

Unfortunately, the preceding examples are a rather special case. There are many nonlinear circuits that cannot be solved analytically. The simple circuit in Figure 4.16 is one such example. Usually we must resort to trial-and-error solutions on a computer. Such solutions provide answers, but usually give little insight about circuit performance and design. Graphical solutions, on the other hand, provide insight at the expense of accuracy. So let us re-examine the circuit in Figure 4.16 with a graphical solution in mind. For concreteness, we will assume that $E = 3$ V and $R = 500\ \Omega$, and that we are required to determine $v_D$, $i_D$, and $v_R$.

We have already found the two simultaneous equations, Equations 4.18 and 4.19, that describe the circuit. For convenience, let us rewrite these equations here after moving a few terms around:

$$i_D = -\frac{v_D - E}{R} \tag{4.31}$$

$$i_D = I_s(e^{v_D/V_{TH}} - 1). \tag{4.32}$$

To solve these expressions graphically, we plot both on the same coordinates and find the point of intersection. Because we are assuming that we have a graph of the nonlinear function, in this case Figure 4.2, the simplest course of action is to plot the linear expression, Equation 4.31, on this graph, as shown in Figure 4.20. The linear constraint of Equation 4.31 is usually called a "load line" for historical reasons arising from amplifier design (as we will see in Chapter 7).

Equation 4.31 plots as a straight line of slope $-1/R$ intersecting the $v_D$ axis, ($i_D = 0$) at $v_D = E$. (The negative sign may be a bit distressing, but does not represent a negative resistance, just the fact that $i_D$ and $v_D$ are *not associated variables for the resistors.*) For the particular values in this circuit, the graph indicates that $i_D$ must be about 5 mA, and $v_D$, about 0.6 V. Once we know that $i_D$ is 5 mA, it immediately follows that

$$v_R = i_D R = 5 \times 10^{-3} \times 500 = 2.5 \text{ V}.$$

It is easy to see from the construction that if $E$ were made three times as large, the voltage across the diode would increase by only a small amount, perhaps to about 0.65 V. This illustrates the kind of insight available from graphical analysis.

The graphical method described here is really more general than it might at first appear. For circuits containing many resistors and sources, but only one nonlinear element, the rest of the circuit, exclusive of the one nonlinear element, is by definition linear. Hence, as described previously in Example 4.4, regardless of circuit complexity we can reduce the circuit to the form in Figure 4.16 by

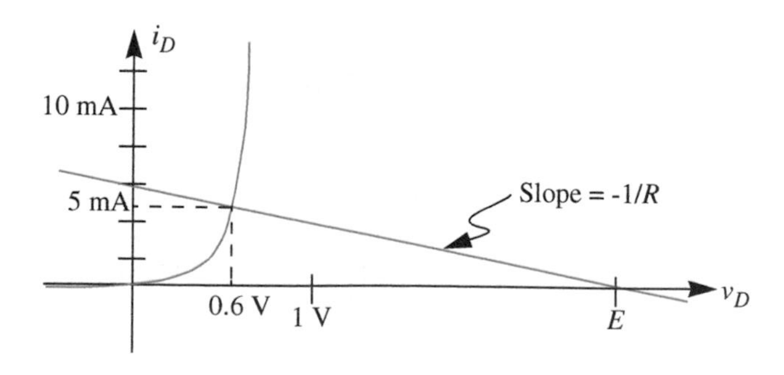


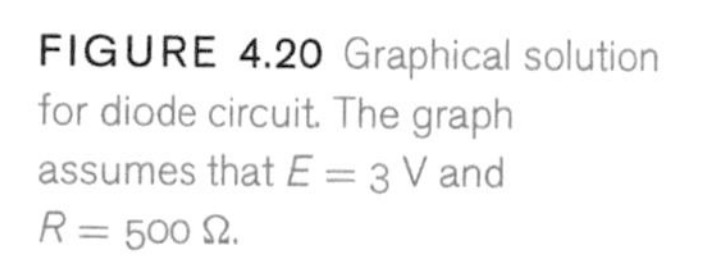
**FIGURE 4.20** Graphical solution for diode circuit. The graph assumes that $E = 3$ V and $R = 500\ \Omega$.

the application of Thévenin's Theorem *to the linear circuit facing the nonlinear element.*

For circuits with two nonlinear elements, the method is less useful, as it involves sketching one nonlinear characteristic on another. Nonetheless, crude sketches can still provide much insight.

---

EXAMPLE 4.10 HALF-WAVE RECTIFIER Let us carry the diode-resistor example of Figure 4.16 and Figure 4.20 a step further, and allow the driving voltage to be a sinusoid rather than DC. That is, let $v_I = E_o \cos(\omega t)$. Also, for reasons that will become evident, let us calculate the voltage across the resistor rather than the diode voltage. The graphical solution is no different than before, except that now we must solve for the voltage assuming a succession of values of $v_I$, and visualize how the resultant time waveform should appear.

The circuit now looks like Figure 4.21a. The diode characteristic with a number of different plots of Equation 4.31 (or load lines), corresponding to a representative set of values of $v_I$, is shown in Figure 4.21b. In Figures 4.21c and 4.21d we show the input sinusoid $v_I(t)$, and the corresponding succession of values of $v_O(t)$ derived from the graphical analysis in Figure 4.21b. Note from Figure 4.21a (or Equation 4.31) that

$$v_O = v_I - v_D \tag{4.33}$$

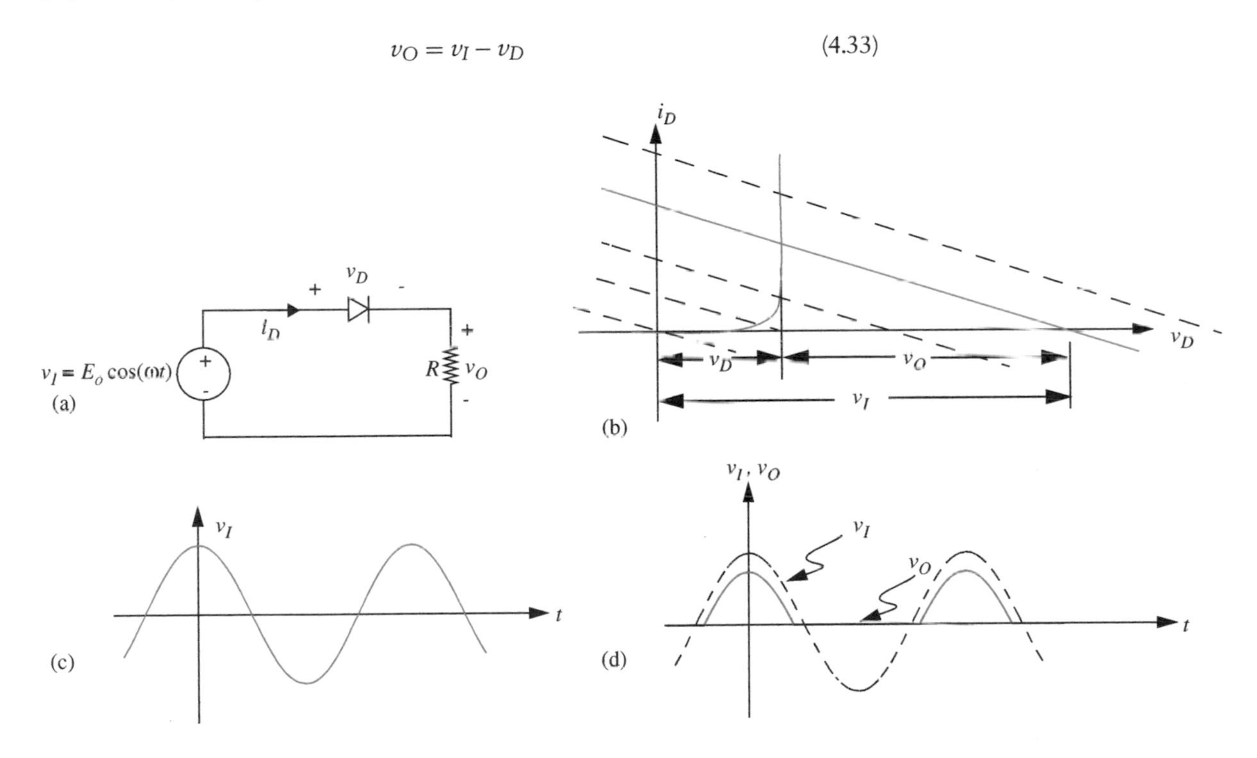


FIGURE 4.21 Half-wave rectifier.

and thus in the graph $v_O$ is the horizontal distance from the load line intersection on the $v_D$ axis to $v_I$.

A number of interesting conclusions can be drawn from this simple example. First, we really do not have to repeat the load line construction fifty times to visualize the output wave. It is clear from the graph that whenever the input voltage is negative, the diode current is so small that $v_O$ is almost zero. Also, for large positive values of $v_I$, the diode voltage stays relatively constant at about 0.6 volts (due to the nature of the exponential), so the voltage across the resistor will be approximately $v_I - 0.6$ V. This kind of insight is the principal value of the graphical method.

Second, in contrast to all previous examples, the output waveform in this circuit is a gross distortion of the input waveform. Note in particular that the input voltage waveform has no average value, (no DC value), whereas the output has a significant DC component, roughly 0.3 $E_o$. The DC motors in most toys, for example, will run nicely if connected across the resistor in the circuit of Figure 4.21a, whereas they will not run if driven directly by the sinusoid $v_I(t)$. This circuit is called a *half-wave rectifier*, because it reproduces only half of the input wave. Rectifiers are present in power supplies of most electronic equipment to generate DC from the 60-Hz "sinusoidal" wave from 110-V AC power line.

## 4.4 PIECEWISE LINEAR ANALYSIS

In the third of the four major methods of analysis for networks containing nonlinear elements, we represent the nonlinear $v$–$i$ characteristics of each nonlinear element by a succession of straight-line segments, then make calculations within each straight-line segment using the linear analysis tools already developed. This is called *piecewise linear analysis.* We will first illustrate piecewise linear analysis by using as an example a very simple piecewise linear model for the diode called *the ideal diode model.*

First, let us develop a simple piecewise linear model for the diode, and then use the piecewise linear method to analyze the circuit in Figure 4.16.

As can be seen from Figure 4.22a, the essential property of a diode is that for an applied positive voltage $v_D$ in excess of 0.6 volts, large amounts of current

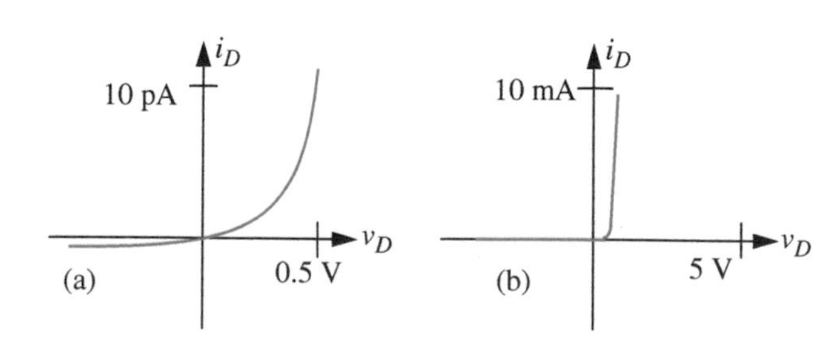


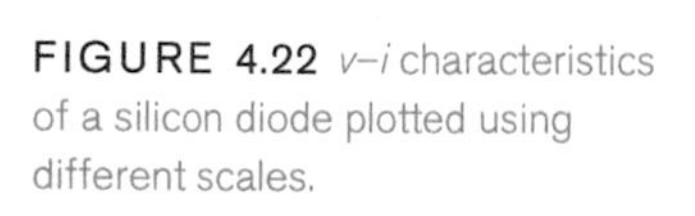
FIGURE 4.22 $v$–$i$ characteristics of a silicon diode plotted using different scales.

flow, whereas for negative voltages very small currents flow. Figure 4.22b draws the $v$–$i$ curve using a larger scale and highlights this dichotomy. The crudest approximation that preserves this dichotomy is the characteristic shown in Figure 4.23a: two linear segments intersecting at the origin, one of zero slope, indicating the behavior of an open circuit, the other infinite, indicating a short circuit. The abstraction is of sufficient use that we give it a special symbol, as shown in Figure 4.23b. This is yet another primitive in our vocabulary, called an *ideal diode*.

The behavior of this piecewise linear model can be summarized in two statements, one for each of the segments:

$$\text{Diode ON (short circuit): } v_D = 0 \quad \text{for all positive } i_D. \tag{4.34}$$

$$\text{Diode OFF (open circuit): } i_D = 0 \quad \text{for all negative } v_D. \tag{4.35}$$

We now use the diode model comprising two straight-line segments to illustrate the piecewise linear analysis method applied to the circuit in Figure 4.16 (also shown in Figure 4.24a). In particular, we will determine the voltage $v_R$ across the resistor and the current $i_D$ through the resistor for two values of the input voltage, $E = 3$ V and $E = -5$ V, and given that $R = 500\ \Omega$.

The piecewise linear analysis technique proceeds by focusing on one straight-line segment at a time, and using our previously developed *linear* analysis tools to make calculations within each segment. Notice that we are able to apply our linear analysis tools because the nonlinear device characteristics are approximated as linear within each segment. To facilitate our calculations, let us first draw the circuit that results for each of the straight-line segments comprising the ideal diode model.

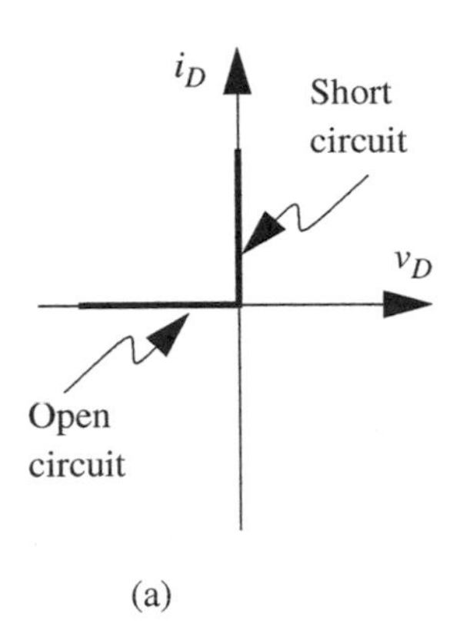


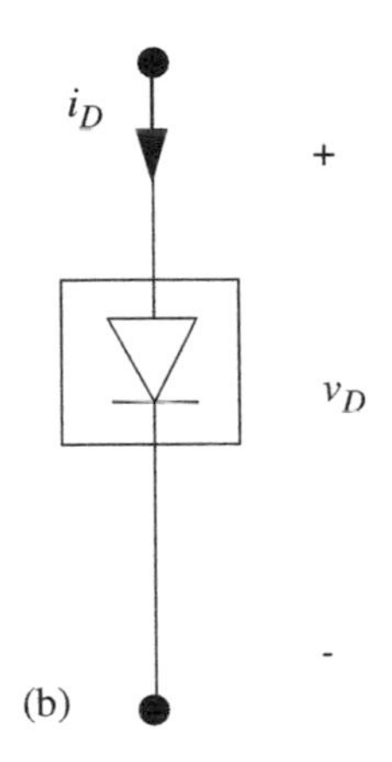


FIGURE 4.23 A piecewise linear approximation for the diode: the ideal diode model.

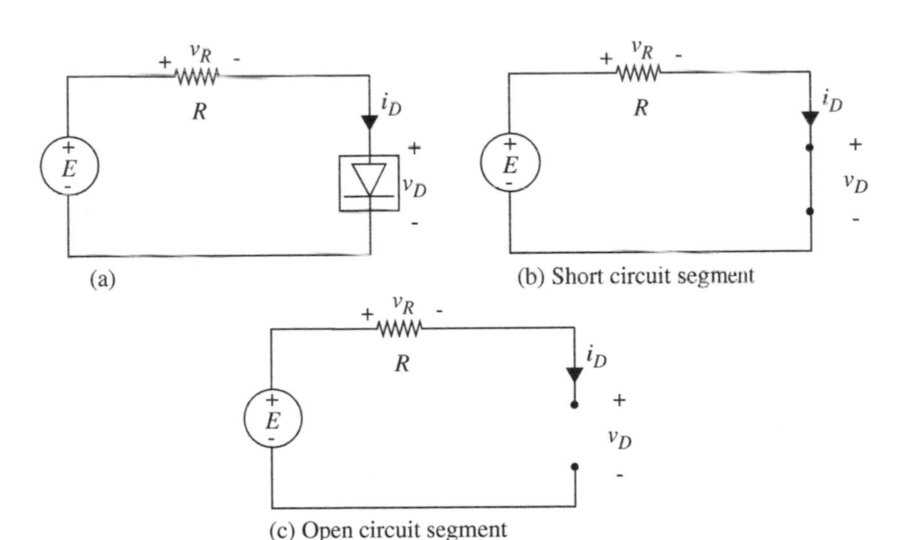


FIGURE 4.24 Piecewise linear analysis of a simple diode circuit.

*Short circuit segment:* Figure 4.24b shows the resulting circuit when the diode is operating as a short circuit. When $i_D$ and $v_D$ are in this straight-line segment of the characteristic, trivial calculations show that

$$i_D = \frac{E}{R} \tag{4.36}$$

and

$$v_R = i_D R = \frac{E}{R} R = E. \tag{4.37}$$

*Open circuit segment:* Figure 4.24c shows the corresponding circuit when the diode is operating as an open circuit. When $i_D$ and $v_D$ are in this part of the characteristic, it is clear that

$$i_D = 0 \tag{4.38}$$

and

$$v_R = 0. \tag{4.39}$$

*Combining the results:* All that remains now is to determine which of the two segments of operations apply when $E = 3$ V and when $E = -5$ V. A little bit of intuition tells us that when $E = 3$ V, the short circuit segment applies. Notice that both the resistor and the diode (a nonlinear resistor) do not produce power, and so the direction of the current must be such that the voltage source delivers power. In other words, when $E$ is positive, so must $i_D$. From Equation 4.34, when $i_D$ is positive, the diode is ON. In this segment, from Equations 4.36 and 4.37

$$i_D = \frac{E}{R} = \frac{3}{500} = 6 \text{ mA} \tag{4.40}$$

and

$$v_R = 3 \text{ V}.$$

Compared with the numbers obtained earlier in Section 4.3 using graphical analysis for $E = 3$ V, we see that piecewise linear analysis using an approximate model for the diode has yielded reasonably accurate results (6 mA versus 5 mA for $i_D$, and 3 V versus 2.5 V for $v_R$).

Intuition also tells us that when $E = -5$ V, the open circuit segment applies. For the negative input voltage, $v_D$ is negative. From Equation 4.35, when $v_D$ is negative, the diode is OFF. In this segment, from Equations 4.38 and 4.39, both $i_D$ and $v_R$ are 0.

Notice that the piecewise linear analysis method enabled us to break down a nonlinear analysis problem into multiple linear problems, each of which was very simple. However, an interesting aspect of the method is figuring out the segment of operation associated with each of the nonlinear devices. This was not too hard with a single nonlinear device such as an ideal diode, but can be challenging when there are a number of nonlinear devices. It turns out that the approach that we discussed in this example generalizes to the *method of assumed states*, which will be discussed in more detail in Chapter 16.

---

EXAMPLE 4.11 PIECEWISE LINEAR ANALYSIS OF A HYPOTHETICAL NONLINEAR DEVICE Figure 4.25a shows a circuit containing some hypothetical nonlinear device whose $v$–$i$ characteristics are approximated using the piecewise-linear graph shown in Figure 4.25b. The nonlinear device with its terminal voltage and current defined as shown in Figure 4.26a might have an actual $v$–$i$ curve as illustrated in Figure 4.26b. Figure 4.26c shows the correspondence between the device's actual $v$–$i$ curve and the piecewise linear model.

The behavior of the piecewise linear model for our nonlinear device can be summarized in two statements, one for each of the straight-line segments:

$$\text{Resistance } R_1 \text{ for all positive } i_D \tag{4.41}$$

$$\text{Resistance } R_2 \text{ for all negative } i_D \tag{4.42}$$

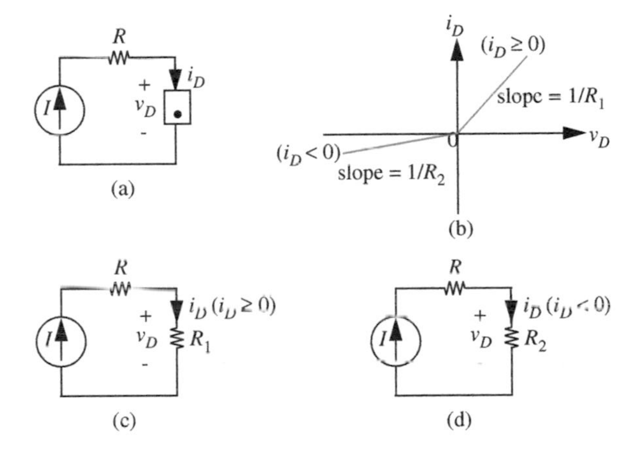


FIGURE 4.25 A circuit containing a nonlinear device whose characteristics are modeled using a piecewise linear approximation. In (b), $R_1 = 100\ \Omega$ and $R_2 = 10\ \text{k}\Omega$.

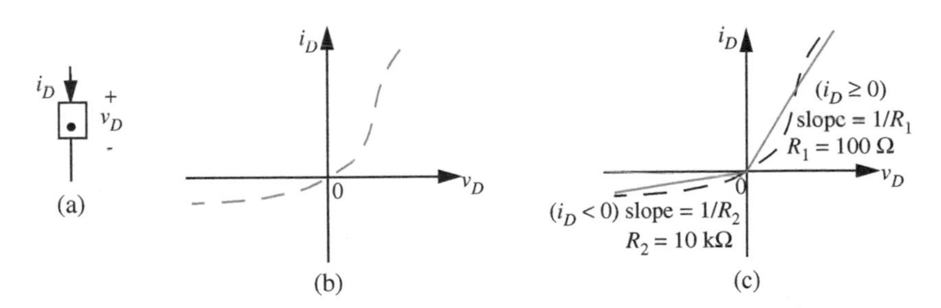


FIGURE 4.26 A hypothetical nonlinear device whose characteristics are modeled using a piecewise linear approximation.

Let us apply the piecewise linear analysis method to the circuit in Figure 4.25(a). Specifically, let us determine the voltage $v_D$ across the nonlinear device for various values of the current $I$ sourced by the independent current source. Specifically, we will determine $v_D$ for $I = 1$ mA, $I = -1$ mA, and when $I$ is a sinusoidal current of the form 0.002 A cos $(\omega t)$.

Following the piecewise linear analysis technique, let us focus on one straight-line segment at a time. Accordingly, we draw the circuit that results for each of the segments.

$R_1$ segment: Figure 4.25c shows the resulting circuit when the nonlinear device is operating in its $R_1$ segment. This segment applies when $i_D$ is positive. Since $i_D = I$, the $R_1$ segment applies when $I$ is positive. A simple application of Ohm's Law for the $R_1$ resistor yields

$$v_D = IR_1. \tag{4.43}$$

$R_2$ segment: Figure 4.25d shows the circuit when the nonlinear device is operating in its $R_2$ segment. This segment applies when $i_D$ is negative; in other words, when $I$ is negative. In this segment, we obtain

$$v_D = IR_2. \tag{4.44}$$

Summarizing,

$$\text{When } I \geq 0: \ v_D = IR_1 \tag{4.45}$$

$$\text{When } I < 0: \ v_D = IR_2. \tag{4.46}$$

Thus, for $I = 1$ mA, Equation 4.45 applies. Therefore

$$v_D = IR_1 = 0.001 \times 100 = 0.1 \text{ V}.$$

Similarly, for $I = -1$ mA, Equation 4.46 applies. Therefore

$$v_D = IR_2 = -0.001 \times 10000 = -10 \text{ V}.$$

Let us now determine $v_D$ for the cosine current input depicted in Figure 4.27a. When $I \geq 0$,

$$v_D = IR_1 = I \times 100$$

as shown in Figure 4.27b. Similarly, when $I < 0$,

$$v_D = IR_2 = I \times 10000$$

as shown in Figure 4.27c. Piecing together the two results for $I \geq 0$ and $I < 0$, we obtain the complete waveform for the output $v_D$ as shown in Figure 4.27d.

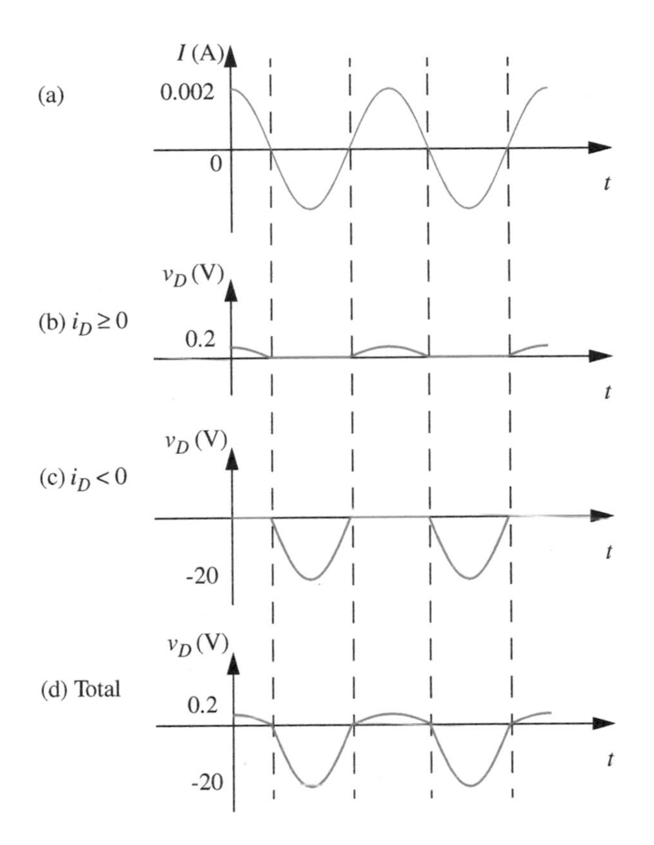


**FIGURE 4.27** Cosine input.

---

EXAMPLE 4.12 SUPERPOSITION APPLIED IN LINEAR SEGMENTS Although the previous examples illustrated the piecewise linear analysis method, they did not do full justice to the power of the technique, since the equivalent circuits within each of the linear segments were very simple (for example, the circuits in Figures 4.24b and 4.24c, or the circuits in Figures 4.25c and 4.25d, and did not require any of our powerful analysis techniques such as superposition that rely on linearity. We will now work a slightly more complicated example to illustrate the full power of the piecewise linear analysis method.

Consider the circuit in Figure 4.28 containing the hypothetical nonlinear device from Example 4.11 (shown in Figure 4.26a) and two independent sources. Suppose we are asked to determine the value of $v_B$. The presence of the nonlinear device does not allow the application of superposition, since superposition relies on the assumption of linearity.

Let us use the piecewise linear analysis method to solve this problem. The piecewise linear model for the device characteristics is shown in Figure 4.26c. Recall, the nonlinear device

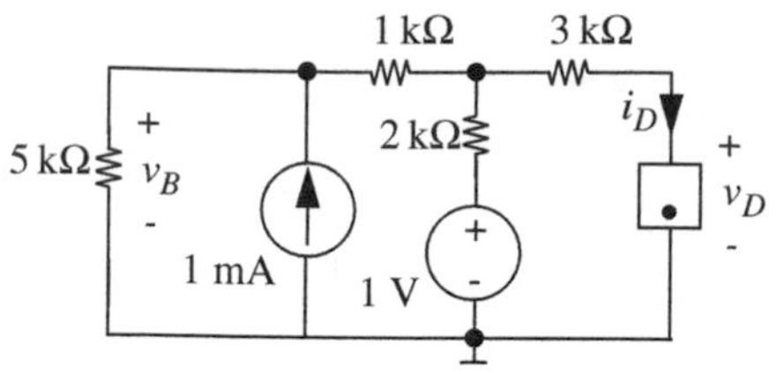

**FIGURE 4.28** A circuit containing a nonlinear device and multiple sources.

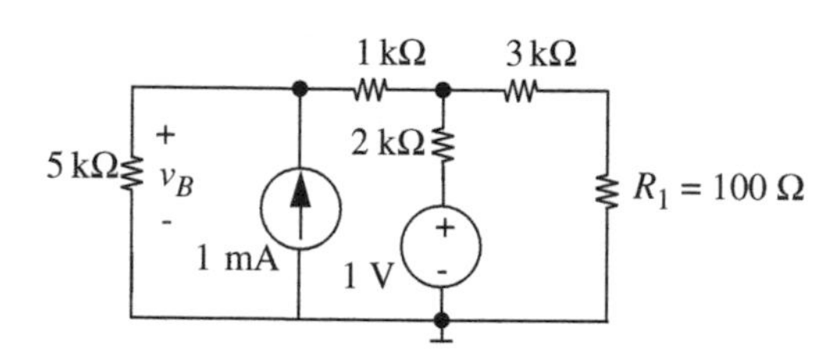


**FIGURE 4.29** Equivalent circuit in the linear segment with slope $1/R_1$.

behaves like a resistor of value $R_1$ when $i_D \geq 0$, and like a resistor of value $R_2$ when $i_D < 0$.

For the polarities of the current source and the voltage source shown in Figure 4.28, the current $i_D$ through the nonlinear device will be positive.[1] Accordingly, the $R_1$ segment of device operation applies and the equivalent circuit is as shown in Figure 4.29. In Figure 4.29, we have replaced the nonlinear device with a resistor of value $R_1$.

The circuit in Figure 4.29 is linear, so any of our linear techniques can be used. We will use superposition to solve this circuit. According to the first step of the superposition method, for each independent source, we must form a subcircuit with all other independent sources set to zero. Setting a voltage source to zero implies replacing the voltage source with a short circuit, and setting a current source to zero implies replacing the current source with an open circuit. Figure 4.30a shows the subcircuit with the voltage source set to zero, and Figure 4.30b shows the subcircuit with the current source set to zero.

Now, according to the second step of the superposition method, we must find the response of each independent source acting alone from the corresponding subcircuit. Let us denote the response of the current source acting alone as $v_{BI}$, and the response of the voltage source acting alone as $v_{BV}$.

---

1. In general, we can determine the polarity of $i_D$ by applying a Thévenin reduction on the circuit facing the nonlinear device. $i_D$ will be positive if the Thévenin voltage driving the device is also positive.

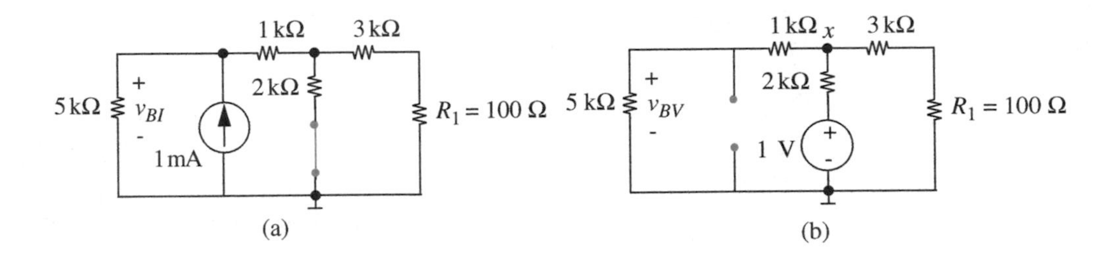


**FIGURE 4.30** Circuits with each of the sources acting alone.

$v_{BI}$: We will analyze the circuit in Figure 4.30a using the intuitive approach of series-parallel reductions discussed in Section 2.4 to obtain $v_{BI}$. In this approach, we will first collapse all the resistances into an equivalent resistance $R_{eq}$ seen by the current source and multiply that resistance by 1 mA. The equivalent resistance seen by the current source is given by

$$R_{eq} = (((3\text{ k}\Omega + 100\ \Omega) \| 2\text{ k}\Omega) + 1\text{ k}\Omega) \| 5\text{ k}\Omega.$$

Simplifying, we get

$$R_{eq} = 1.535\text{ k}\Omega.$$

Multiplying $R_{eq}$ by the current source current we get

$$v_{BI} = 1.535\text{ k} \times 1\ \text{ mA} = 1.535\text{ V}.$$

$v_{BV}$: We now analyze the circuit in Figure 4.30b to obtain $v_{BV}$. We will again use the intuitive approach suggested in Section 2.4 involving first collapsing, then expanding the circuit. Suppose we knew the voltage $v_x$ at node $x$, then we can easily obtain $v_{BV}$ by the voltage divider relation. We can obtain $v_x$ by first *collapsing* the circuit in Figure 4.30b into the equivalent circuit in Figure 4.31 and applying the voltage divider relation. $R_x$ in the circuit in Figure 4.31 is found by collapsing the 1-kΩ, 5-kΩ, 3-kΩ, and the 100-Ω resistances into an equivalent resistance as follows:

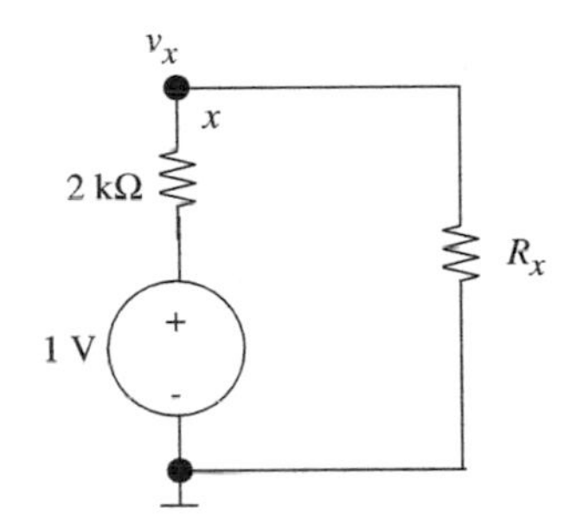


**FIGURE 4.31** Collapsed circuit.

$$R_x = (1\text{ k}\Omega + 5\text{ k}\Omega) \| (3\text{ k}\Omega + 100\ \Omega) = 2.05\text{ k}\Omega.$$

By the voltage divider relation

$$v_x = 1\text{ V}\frac{R_x}{2\text{ k}\Omega + R_x} = 1\text{ V}\frac{2.05\text{ k}\Omega}{2\text{ k}\Omega + 2.05\text{ k}\Omega} = 0.51\text{ V}.$$

We now obtain $v_{BV}$ by *expanding* the circuit in Figure 4.31 to the original circuit in Figure 4.30b and using the voltage divider relation as follows:

$$v_{BV} = v_x\frac{5\text{ k}\Omega}{1\text{ k}\Omega + 5\text{ k}\Omega} = 0.51\text{ V}\frac{5\text{ k}\Omega}{1\text{ k}\Omega + 5\text{ k}\Omega} = 0.425\text{ V}.$$

As the final step of the superposition method, we obtain the total response by summing together each of the individual responses:

$$v_B = v_{BI} + v_{BV} = 1.535\ \text{V} + 0.425\ \text{V} = 1.96\ \text{V}.$$

Thus, we have our desired answer. Notice that we were able to apply the powerful superposition method by focusing on a straight line segment of the nonlinear device.

---

WWW EXAMPLE 4.13 HALF-WAVE RECTIFIER RE-EXAMINED

---

### WWW 4.4.1 IMPROVED PIECEWISE LINEAR MODELS FOR NONLINEAR ELEMENTS *

---

WWW EXAMPLE 4.14 ANOTHER EXAMPLE USING PIECEWISE LINEAR MODELING

---

WWW EXAMPLE 4.15 THE DIODE RESISTANCE

---

WWW EXAMPLE 4.16 A MORE COMPLICATED PIECEWISE LINEAR MODEL

---

## 4.5 INCREMENTAL ANALYSIS

There are many applications in electronic circuits where nonlinear devices are operated only over a very *restricted* range of voltage or current, as in many sensor applications and most audio amplifiers, for example. In such cases, it makes sense to find a *piecewise linear* device model in a way that ensures maximum accuracy of fit over that *narrow operating range.* This process of linearizing device models over a very narrow operating range is called *incremental analysis* or *small-signal analysis.* The benefit of incremental analysis is that the incremental variables satisfy KVL and KCL, as well as linear $v$–$i$ relations over the narrow operating range.

We note, however, that this almost linear mode of operation of nonlinear devices over a narrow operating range is more common with MOSFET circuits (discussed in Chapter 8) than with nonlinear resistors. However, because of the simplicity of nonlinear resistor circuits, we introduce the concept of incremental analysis here, recognizing that the principal application will come later.

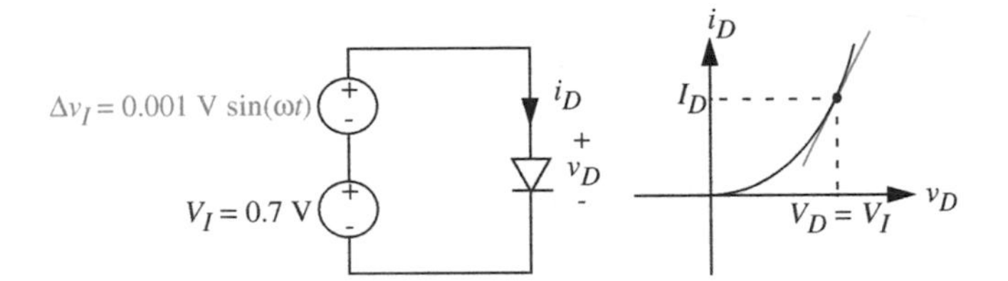


FIGURE 4.37 Incremental analysis.

We will discuss incremental analysis using the diode as an example. Suppose we wish to determine the value of the diode current $i_D$ in the circuit in Figure 4.37. Here we have a diode and a pair of voltage sources as in many previous examples, but in this case one source, $V_I$, is fixed at a value of, say, 0.7 V, and the other, $\Delta v_I$, is a 1-mV sinusoid. Inputs of this form — a DC value plus a small time-varying component — occur frequently in practice, and so it is important to find a simple way to solve for the circuit response for this type of input. We could, of course, take the obvious analytical approach and write

$$i_D = I_s\left(e^{(0.7\text{ V}+0.001\text{ V}\sin(\omega t))/V_{TH}} - 1\right) \tag{4.53}$$

but it leaves us with a complicated expression from which the form of the output is not readily apparent.

We will abandon the straightforward approach, and instead, cast off in a slightly different direction. Clearly, with the given values of the drive, this is a case where the diode is being operated only over a very restricted region of its nonlinear $v$–$i$ characteristics: the diode will always have a large positive DC offset voltage across it (given by $V_I$), and the diode current will vary only by a small amount around $I_D$ (as depicted in the graphical sketch in Figure 4.37) due to the small signal $\Delta v_I$ superimposed on the DC input voltage. Thus a sensible approach is to model the diode characteristic accurately in the vicinity of $I_D$ (as depicted by the small straight line segment tangent to the curve at the $V_I, I_D$ point, as shown in Figure 4.37) and disregard the rest of the curve. The Taylor Series expansion is the appropriate tool to employ for this task:

$$y = f(x) = f(X_o) + \left.\frac{df}{dx}\right|_{X_o}(x - X_o) + \frac{1}{2!}\left.\frac{d^2f}{dx^2}\right|_{X_o}(x - X_o)^2 + \cdots \tag{4.54}$$

This is an expansion of the $y$ versus $x$ relation about the point $f(X_o), X_o$. For our device $i_D$ versus $v_D$ relation,

$$i_D = f(v_D)$$

we need to develop the corresponding expansion about $f(V_D), V_D$, where $I_D = f(V_D)$.

For our example, the source voltages $V_I$ and $\Delta v_I$ are applied directly across the diode, so the corresponding diode voltages are given by $V_D = V_I$ and $\Delta v_D = \Delta v_I$.

Thus, in terms of diode parameters, the corresponding Taylor Series expansion of $i_D = f(v_D)$ about $f(V_D)$, $V_D$ is:

$$i_D = f(v_D) = f(V_D) + \left.\frac{df}{dv_D}\right|_{V_D} (v_D - V_D) + \frac{1}{2!}\left.\frac{d^2f}{dv_D^2}\right|_{V_D} (v_D - V_D)^2 + \cdots \tag{4.55}$$

For our diode example, mathematically we wish to expand the diode equation

$$i_D = I_s\left(e^{(V_D+\Delta v_D)/V_{TH}} - 1\right) \tag{4.56}$$

about the operating point $V_D$, $I_D$. In circuit terms we are calculating the response $i_D$ when a voltage $v_D = V_D + \Delta v_D$, is applied to a diode, as in Figure 4.37. The current $i_D$ will be of the form

$$i_D = I_D + \Delta i_D. \tag{4.57}$$

The Taylor series expansion of Equation 4.56 is

$$i_D = I_s\left(e^{V_D/V_{TH}} - 1\right) + \frac{1}{V_{TH}}\left(I_s e^{V_D/V_{TH}}\right)\Delta v_D + \frac{1}{2}\left(\frac{1}{V_{TH}}\right)^2\left(I_s e^{V_D/V_{TH}}\right)(\Delta v_D)^2 + \cdots \tag{4.58}$$

Simplifying,

$$i_D = I_s\left(e^{V_D/V_{TH}} - 1\right) + \left(I_s e^{V_D/V_{TH}}\right)\left[\frac{1}{V_{TH}}\Delta v_D + \frac{1}{2}\left(\frac{1}{V_{TH}}\right)^2(\Delta v_D)^2 + \cdots\right]. \tag{4.59}$$

Now if we are assured that the excursions away from the DC operating point $V_D$, $I_D$ are small, so that $\Delta v_D$ is very small compared to $V_{TH}$ (as in this case, since $V_{TH}$ is typically 0.025 V and we are given that $\Delta v_D = 0.001$ V) we can ignore the second and higher order terms in the expansion:

$$i_D = I_s\left(e^{V_D/V_{TH}} - 1\right) + \left(I_s e^{V_D/V_{TH}}\right)\left[\frac{1}{V_{TH}}\Delta v_D\right]. \tag{4.60}$$

We know that the output current is composed of a DC component $I_D$ and a small perturbation $\Delta i_D$. Thus, we can write

$$I_D + \Delta i_D = I_s\left(e^{V_D/V_{TH}} - 1\right) + \left(I_s e^{V_D/V_{TH}}\right)\left[\frac{1}{V_{TH}}\Delta v_D\right]. \tag{4.61}$$

Equating corresponding DC terms and corresponding incremental terms:

$$I_D = I_s\left(e^{V_D/V_{TH}} - 1\right) \tag{4.62}$$

$$\Delta i_D = \left(I_s e^{V_D/V_{TH}}\right)\frac{1}{V_{TH}}\Delta v_D. \tag{4.63}$$

Note that $I_D$ is simply the DC bias current related to the DC input voltage $V_D$. Accordingly, the DC terms relating $I_D$ to $V_D$ can be equated as in Equation 4.62 because the operating point values $I_D$, $V_D$ satisfy Equation 4.1, which is the diode equation. When the DC terms are eliminated from both sides of Equation 4.61, the incremental relation shown in Equation 4.63 results.

Thus the response current to an applied voltage $V_D + \Delta v_D$ contains two terms: a large DC current $I_D$ and a small current proportional to $\Delta v_D$, if we keep $\Delta v_D$ small enough.

A graphical interpretation of this result is often helpful. As shown in Figure 4.37, Equation 4.61 is the straight line passing through the DC operating point $V_D$, $I_D$ and tangent to the curve at that point. The higher order terms in Equation 4.58 that we neglected would add quadratic, cubic, etc., terms to the model, thereby improving the fit over a wider region.

For the particular case of incremental analysis with the diode equation, we commonly make the following approximation to Equation 4.63:

$$\Delta i_D = I_s\left(e^{V_D/V_{TH}} - 1\right)\frac{1}{V_{TH}}\Delta v_D \tag{4.64}$$

where the $-1$ term is artificially included because it is small in comparison to $e^{V_D/V_{TH}}$. With the inclusion of the $-1$ term, we can simplify further and write an approximate expression for the incremental diode current:

$$\Delta i_D = I_D\frac{1}{V_{TH}}\Delta v_D. \tag{4.65}$$

Figure 4.38 provides further insight into the result in Equations 4.62 and 4.63 (or its simplified form in Equation 4.65). Equation 4.62 establishes the $V_D, I_D$ operating point or the bias point of the diode. $I_D/V_{TH}$ is the slope of the $v$–$i$ curve at the point $V_D, I_D$. The product of the slope of the $v$–$i$ curve at $V_D, I_D$ (given by $I_D/V_{TH}$) and the small perturbation in applied diode voltage

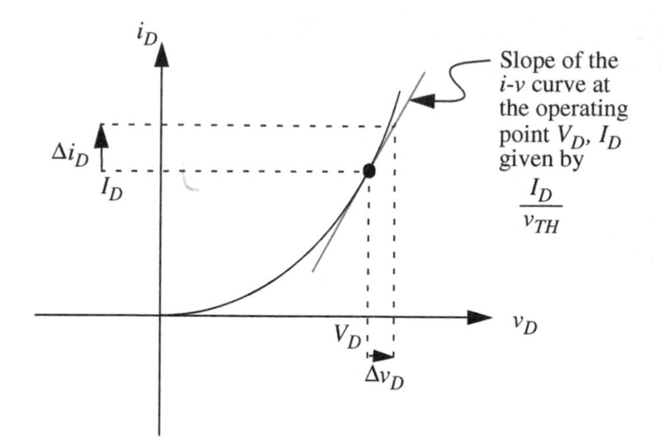


**FIGURE 4.38** Graphical interpretation of operating point and incremental signals.

(given by $\Delta v_D$) yields an approximation $\Delta i_D = I_D/V_{TH}\Delta v_D$ to the resulting perturbation in the diode current.

It is a simple matter to estimate the quality of fit. Taking the ratio of the third term to the second term on Equation 4.58, we obtain

$$\frac{\text{third term}}{\text{second term}} = \frac{1}{2}\frac{1}{V_{TH}}\Delta v_D. \tag{4.66}$$

At room temperature, $V_{TH}$ is roughly 25 mV. Thus, if we want the third term to be no more than 10% of the second, $\Delta v_D$ must be restricted to be less than 5 mV.

We do not have to go through the mechanics of a Taylor series expansion each time that we wish to find the relationship between the incremental parameters $\Delta i_D$ and $\Delta v_D$. Rather, we can find the relationship between the incremental parameters directly from the $i_D = f(v_D)$ relationship using

$$\Delta i_D = \left.\frac{df}{dv_D}\right|_{V_D} \Delta v_D \tag{4.67}$$

The relationship in Equation 4.67 is itself derived from the Taylor series expansion as follows. Recalling Equation 4.55,

$$i_D = f(v_D) = f(V_D) + \left.\frac{df}{dv_D}\right|_{V_D}(v_D - V_D) + \frac{1}{2!}\left.\frac{d^2f}{dv_D^2}\right|_{V_D}(v_D - V_D)^2 + \cdots \tag{4.68}$$

and replacing $i_D$ by its DC value plus an increment ($i_D = I_D + \Delta i_D$), the difference ($v_D - V_D$) by $\Delta v_D$, and

$$I_D = f(V_D) \tag{4.69}$$

we rewrite Equation 4.68 as

$$I_D + \Delta i_D = I_D + \left.\frac{df}{dv_D}\right|_{V_D} \Delta v_D + \frac{1}{2!}\left.\frac{d^2 f}{dv_D^2}\right|_{V_D} \Delta v_D^2 + \cdots \tag{4.70}$$

Now, deleting $I_D$ on both sides of the equation, and assuming $\Delta v_D$ is small enough that we can ignore second order terms in $\Delta v_D$ we get

$$\Delta i_D = \left.\frac{df}{dv_D}\right|_{V_D} \Delta v_D. \tag{4.71}$$

In words, the incremental change in the current is equal to $df/dv_D$ evaluated at $v_D = V_D$, multiplied by the incremental change in the voltage.

You can verify that applying Equation 4.71 to the diode equation

$$i_D = f(v_D) = I_s(e^{v_D/V_{TH}} - 1)$$

yields the same expression for $\Delta i_D$ as that in Equation 4.63.

The same result can be developed graphically from Figure 4.38. The incremental current $\Delta i_D$ is simply the product of $\Delta v_D$ and the slope of the $i_D$ versus $v_D$ curve at the point $I_D, V_D$. The slope of the $i_D$ versus $v_D$ curve at the point $I_D, V_D$ is given by:

$$\text{Slope of the } i_D \text{ versus } v_D \text{ curve} = \left.\frac{df(v_D)}{dv_D}\right|_{V_D}.$$

To wrap up our example of Figure 4.37, let us obtain the numerical value of $i_D$ for the given form of $v_D$. We are given that the input is of the form

$$v_I = V_I + \Delta v_I = 0.7\text{ V} + 0.001\text{ V}\sin(\omega t).$$

Since the input is applied directly across the diode, the corresponding relation in terms of diode voltages is

$$v_D = V_D + \Delta v_D = 0.7\text{ V} + 0.001\text{ V}\sin(\omega t).$$

When $\Delta v_D$ is small enough, $i_D$ can be written in the form

$$i_D = I_D + \Delta i_D.$$

From Equation 4.69,

$$I_D = f(V_D) = I_s\left(e^{0.7/V_{TH}} - 1\right)$$

and, from Equation 4.71,

$$\Delta i_D = \left.\frac{df}{dv_D}\right|_{V_D} \Delta v_D = I_s e^{0.7/V_{TH}} \frac{1}{V_{TH}} 0.001 \sin \omega t.$$

Substituting $I_s = 1$ pA and $V_{TH} = 0.025$ V (at room temperature), the diode parameters, we find that $I_D = 1.45$ A and $\Delta i_D = 0.058$ A $\sin(\omega t)$.

The values of $I_D$ and $\Delta i_D$ immediately confirm that $i_D$ is the sum of a DC term and a small time-varying sinusoidal term. Observe further the ease with which we obtained the form of $i_D$, and contrast with the uninsightful expression in Equation 4.53 that resulted from the brute-force analytical approach.

Although this process yielded fairly quickly the form of $i_D$, a bit of insight will simplify the process even further by enabling the use of *linear circuit techniques* to solve the problem as promised in the introduction of this section. We proceed by drawing attention to Equation 4.61. Equation 4.61 is certainly nonlinear. But an important interpretation central to all incremental arguments allows us to solve the problem by linear circuit methods. Note from Equation 4.62 that the first term in Equation 4.61, the DC current $I_D$, is *independent of* $\Delta v_D$. It depends only on the circuit parameters and the DC voltage $V_D$ which is the same as the DC source voltage $V_I$. Thus $I_D$ can be found with $\Delta v_D$, set to zero. On this basis, the second term in Equation 4.61 is *linear* in $\Delta v_D$, because we have shown that there is no hidden $\Delta v_D$ dependence in $I_D$. Hence the second term, *the change in the current i is linearly proportional to the change in v*, can be found from a linear circuit.

But what is the form of this linear circuit that can facilitate the computation of $\Delta i_D$? Observe that the constant of proportionality relating $\Delta i_D$ and $\Delta v_D$ is

$$\frac{\Delta i_D}{\Delta v_D} = g_d = \frac{1}{V_{TH}} I_D \tag{4.72}$$

or more generally, from Equation 4.71,

$$\frac{\Delta i_D}{\Delta v_D} = g_d = \left.\frac{df}{dv_D}\right|_{V_D}, \tag{4.73}$$

which can be interpreted as a linear conductance (the slope of the $v$–$i$ characteristic at $V_D$, $I_D$), or a linear resistance of value

$$r_d = \frac{V_{TH}}{I_D} \tag{4.74}$$

for a diode.

In general, the incremental behavior of a nonlinear device is that of a linear resistor, whose value $r_d$ is given by

$$r_d = \frac{1}{\left.\frac{df}{dv_D}\right|_{v_D=V_D}}. \tag{4.75}$$

For a diode, because $V_{TH}$ is about 25 mV at room temperature, for $I_D =$ 1 mA, the incremental resistance $r_d = 25\ \Omega$. Similarly, for $I_D = 1.45$ A, $r_d =$ 0.017 Ω. Note: The incremental resistance in general is not the same as the resistor $R_d$ used in the piecewise linear model of WWW Figure 4.33c. There we were trying for a fit over a large range of current, and hence would compromise on a different resistance value. The difference between $R_d$ and $r_d$ can be clearly understood by comparing the two graphical interpretations, WWW Figures 4.33d and 4.37.

In circuit terms, Equation 4.73 can be interpreted as depicted in Figure 4.39. $\Delta i_D$ can now be found trivially from the linear circuit in Figure 4.39, where $r_d = 1/g_d$. For $I_D = 1.45$ A, $r_d$ is 0.017 Ω at room temperature, and

$$\Delta i_D = \frac{\Delta v_D}{r_d} = 0.059 \text{ A } \sin(\omega t).$$

In summary, we began our analysis with the goal of determining the current ($i_D$) through the diode when an input voltage in the form of a DC value ($V_D$) plus a small time-varying component ($\Delta v_D$) is applied across it. Equation 4.61 shows that the resulting diode current is made up of two terms, a DC term, $I_D$, which depends only on the DC voltage applied $V_D$, and a small-signal or incremental term $\Delta i_D$, which depends on the small-signal voltage and also on the DC voltage $V_D$. But for fixed $V_D$, the incremental current $\Delta i_D$ is linearly related to $\Delta v_D$. The constant of proportionality is a conductance $g_d$ given by Equation 4.73. Because the incremental circuit model of Figure 4.39 correctly represents the relationship between $\Delta i_D$ and $\Delta v_D$, this linear circuit can be used to solve for $\Delta i_D$. In many situations, only the incremental change in the output

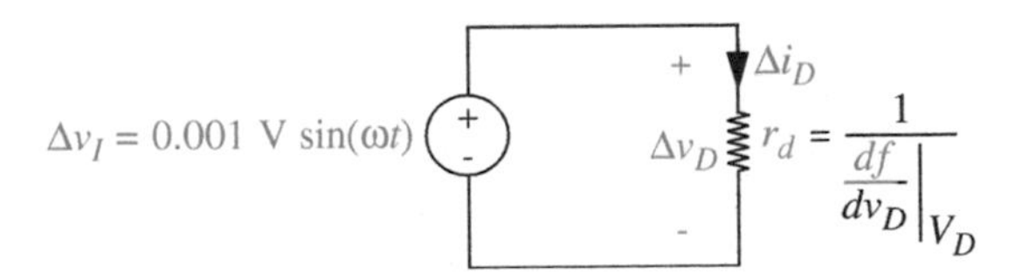


**FIGURE 4.39** Linear circuit for determining the value of $\Delta i_D$.

is of interest, and our analysis will end here. If the total value of the output ($i_D$) is desired, then it can be obtained by summing $\Delta i_D$ and the DC component $I_D$.

Thus, based on the preceding discussion, a systematic procedure for finding incremental voltages and currents for a circuit with a nonlinear device characterized by the $v$–$i$ relation $i_D = f(v_D)$ is as follows:

1. Find the DC operating variables, $I_D$ and $V_D$, using the subcircuit derived from the original circuit by setting all small-signal sources to zero. Any of the methods of analyzing nonlinear circuits discussed in the preceding sections — analytical, graphical, or piecewise linear — is appropriate.
2. Find the incremental output voltage and incremental nonlinear device current (the *change* away from the DC variables calculated in Step 1) by forming an incremental subcircuit in which the nonlinear device is replaced by a resistor of value $r_d$ (computed as shown in Equation 4.75), and all DC sources are set to zero. (That is, voltage sources are replaced by short circuits, and current sources by open circuits.) The incremental subcircuit is *linear*, so incremental voltages and currents can be calculated by any of the linear analysis techniques developed in Chapter 3, including superposition, Thévenin, etc.

One final note on notation before we work a few examples to illustrate the small signal approach. For convenience, we will introduce the following notation to distinguish between total variables, their DC operating or bias values, and their incremental excursions about the operating points. As illustrated in Figure 4.40, we will denote total variables with small letters and capital subscripts, DC operating point variables using all capitals, and incremental values using all small letters. Thus, $v_D$ denotes the total voltage across the device, $V_D$ the DC operating point, and $v_d = \Delta v_D$ the incremental component. Since the total variable is the sum of the two components, we have

$$v_D = V_D + v_d.$$

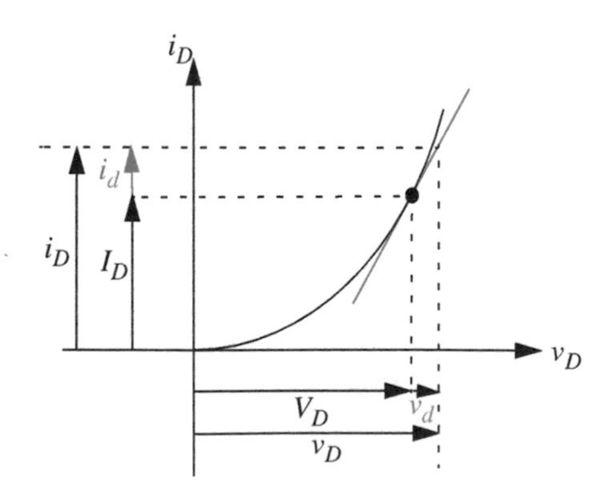


**FIGURE 4.40** Notation for operating point, small signal, and total variables.

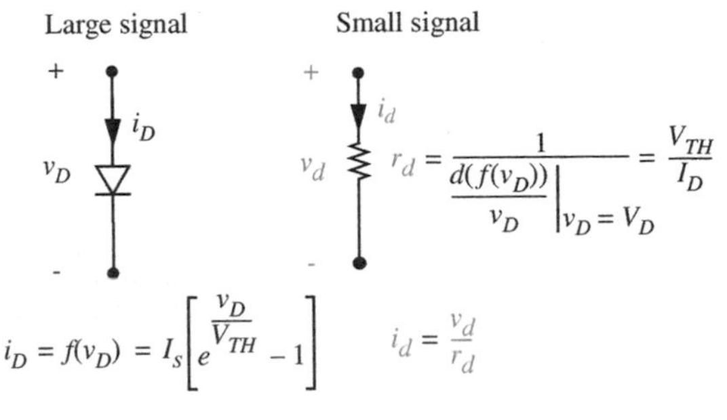

Large signal

+ $v_D$ $i_D$ -

$$i_D = f(v_D) = I_s\left[e^{\frac{v_D}{V_{TH}}} - 1\right]$$

Small signal

+ $v_d$ $i_d$ -

$$r_d = \frac{1}{\left.\frac{d(f(v_D))}{v_D}\right|_{v_D = V_D}} = \frac{V_{TH}}{I_D}$$

$$i_d = \frac{v_d}{r_d}$$

FIGURE 4.41 Large signal and small signal diode models.

Similarly, for the current

$$i_D = I_D + i_d.$$

Figure 4.41 summarizes the large and small signal models for the diode in terms of our new notation.

EXAMPLE 4.17 INCREMENTAL MODEL FOR SQUARE LAW DEVICE Derive an incremental model for the square law device shown in Figure 4.42a. Assume that the device is characterized by the following $v$–$i$ relationship:

$$i_D = Kv_D^2 \quad \text{for } v_D > 0$$
$$= 0 \quad \text{for } v_D \leq 0$$

where $K = 1$ mA/V$^2$, and that the operating point values for $V_D$ and $I_D$ are 1 V and 1 mA respectively.

We know from Equation 4.75 that the incremental model for a nonlinear device is a linear resistor of value $r_d$ as depicted in Figure 4.42. The value of the resistance is given by

$$r_d = \frac{1}{\left.\frac{df}{dv_D}\right|_{v_D = V_D}}.$$

Substituting, $f(v_D) = Kv_D^2$, we obtain

$$r_d = \frac{1}{2Kv_D|_{v_D=V_D=1\text{ V}}} = 500\ \Omega.$$

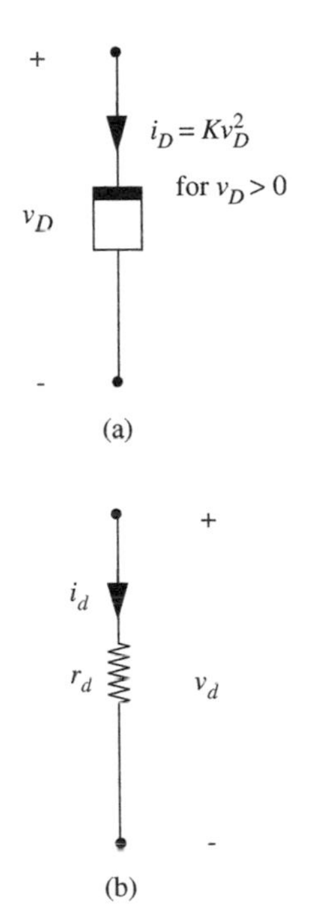


FIGURE 4.42 A square law device and its incremental model. (a) a square law device, (b) incremental model.

EXAMPLE 4.18 INCREMENTAL MODEL FOR A RESISTOR We will show that the incremental model for a linear resistor of value $R$ is also a resistor of value $R$. Intuitively, since the $v$–$i$ relation for a linear resistor is a straight line, the slope (given by $1/R$) is the same for all values of the resistor voltage and current. Further, since the incremental resistance $r$ is the reciprocal of the slope, it follows that $r = R$.

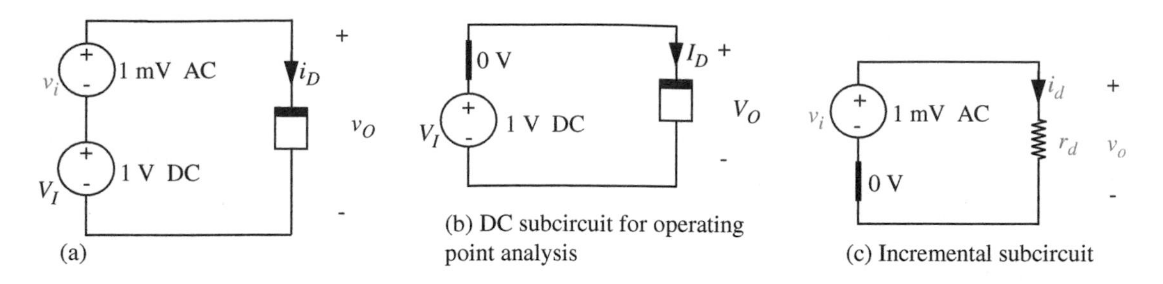


FIGURE 4.43 Incremental analysis of nonlinear resistor. Parts (b) and (c) show the DC and the incremental subcircuits, respectively.

We can also show the same result mathematically as follows. The $v$–$i$ relation for the resistor is given by

$$i = \frac{v}{R}.$$

We obtain the incremental resistance $r$ for resistor voltage and current $(V, I)$ using Equation 4.75:

$$r = \frac{1}{\left.\frac{d(v/R)}{dv}\right|_{v=V}} = R.$$

---

EXAMPLE 4.19 INCREMENTAL ANALYSIS OF SQUARE LAW DEVICE Suppose we are interested in finding the current $i_D$ through the square law device in Figure 4.43. The device is driven by a DC voltage in series with a small AC voltage. The square law device is characterized by the following $v$–$i$ relationship:

$$i_D = Kv_O^2 \qquad \text{for } v_O > 0. \tag{4.76}$$

The current $i_D$ is 0 for $v_O \leq 0$. Assume $K = 1\ \text{mA/V}^2$.

Since the input to the nonlinear device is the sum of a DC component and a relatively small AC component, incremental analysis is the appropriate tool for our task. Incremental analysis comprises the following steps:

1. Find the DC operating variables $I_D$ and $V_O$ by setting all small-signal sources to zero.

2. Find the incremental device current $i_d$ by forming an incremental subcircuit in which the nonlinear device is replaced by a linear resistor of value $r_d$ (from Equation 4.75), and all DC sources are set to zero.

Following the first step of incremental analysis, we draw the DC subcircuit Figure 4.43b and mark the operating-point variables $I_D, V_O$. The AC source is set to zero. By inspection from Figure 4.43b

$$V_O = V_I = 1\ \text{V}.$$

and

$$I_D = KV_O^2 = 1 \text{ mA}.$$

Next, following the second step of incremental analysis we draw the incremental subcircuit in Figure 4.43b. Here, we set the DC source to zero, and replace the nonlinear device with a linear resistor of value $r_d$, where

$$r_d = \frac{1}{\left.\frac{d(Kv_O^2)}{dv_O}\right|_{v_O=V_O}} = \frac{1}{2KV_O}.$$

Substituting $V_O = 1$ V, we obtain

$$r_d = 500\ \Omega.$$

Now that we know the value of $r_d$, we can obtain the small-signal component $i_d$ from the small signal circuit. Accordingly,

$$i_d = \frac{v_i}{r_d}.$$

Substituting numerical values, we find that $i_d$ is a 2-$\mu$A AC current. Thus, the total current $i_D$ is the sum of a 1-mA DC current and a 2-$\mu$A AC current. This completes our analysis.

---

EXAMPLE 4.20 VOLTAGE REGULATOR BASED ON A NONLINEAR RESISTOR To illustrate the use of incremental analysis, we examine the nonlinear device circuit shown in Figure 4.44a, a crude form of voltage regulator based on the hypothetical nonlinear resistor discussed earlier. Assume $R = 1k$ and that the nonlinear device is characterized by the following $v$–$i$ relationship:

$$i_D = Kv_O^2 \quad \text{for } v_O > 0. \tag{4.77}$$

The current $i_D$ is 0 for $v_O \leq 0$. Assume $K = 1 \text{ mA/V}^2$.

We assume that the supposedly DC source ($v_I$) supplying the circuit in reality has 5 volts of DC ($V_I$) with 50 mV of AC ($v_i$, also called a ripple) superimposed. The regulator is designed to reduce this unwanted AC component relative to the DC. Our task is to determine the magnitude of the ripple in the output, and the extent to which our regulator has been able to reduce the ripple amplitude relative to the DC voltage.

To understand how the circuit operates, we will perform an incremental analysis on the circuit by following these two steps:

1. Find the DC operating variables $I_D$ and $V_O$ by setting all small-signal sources to zero. This will require a nonlinear analysis using one of the nonlinear approaches

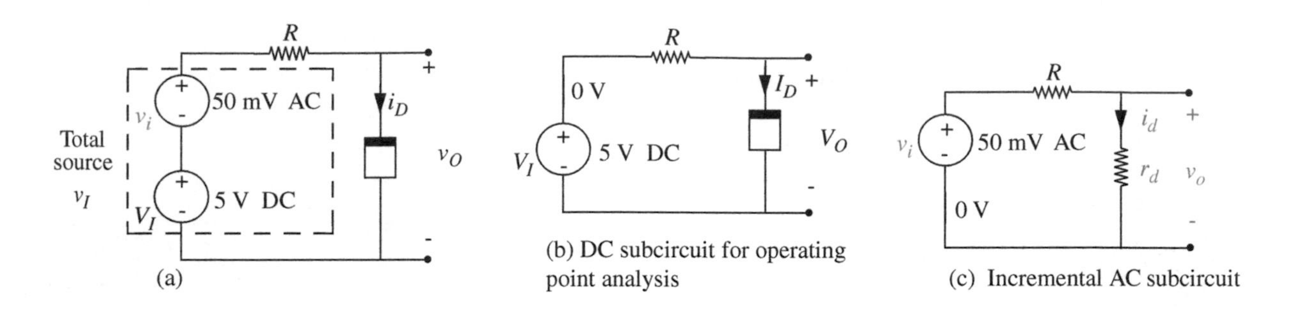


FIGURE 4.44 Nonlinear resistor voltage regulator. Parts (b) and (c) show the DC and the incremental subcircuits, respectively. The subcircuits are derived according to the discussion preceding this example. For more discussion of DC and incremental subcircuits, see Section 8.2.1 in Chapter 8.

previously discussed, for example, analytical (using the node method), graphical or piecewise linear.

2. Find the incremental output voltage $v_o$ and incremental nonlinear device current $i_d$ by forming an incremental subcircuit in which the nonlinear device is replaced by a linear resistor of value $r_d$ (from Equation 4.75), and all DC sources are set to zero. The incremental circuit will be linear, so any of our linear techniques will apply, for example, superposition, Thévenin.

Following the first step of incremental analysis, we draw the DC subcircuit Figure 4.44b and mark the operating-point variables $I_D$, $V_O$. Notice that we have set the small-signal source to zero.

We will now use the analytical analysis method to determine $I_D$ and $V_O$. By inspection from Figure 4.44b,

$$-V_I + I_D R + V_O = 0 \tag{4.78}$$

$$I_D = KV_O^2. \tag{4.79}$$

Eliminating $I_D$, we get

$$RKV_O^2 + V_O - V_I = 0.$$

Solving for $V_O$, we get:

$$V_O = \frac{-1 + \sqrt{1 + 4V_I RK}}{2RK}. \tag{4.80}$$

Substituting $K = 1 \text{ mA/V}^2$, $R = 1 \text{ k}\Omega$, and $V_I = 5$ V, we obtain the operating point values:

$$V_O = 1.8 \text{ V}$$

$$I_D = 3.24 \text{ mA}.$$

$V_O$ is the DC component of the output. This completes the first step of incremental analysis.

Next, following the second step of incremental analysis we draw the incremental subcircuit in Figure 4.44c. This time around, we set the DC source to zero, and replace the nonlinear resistor with a linear resistor of value $r_d$.

We can now find the incremental values $i_d$, $v_o$ from Figure 4.44c if we know the value of $r_d$. Accordingly, we first determine the value of $r_d$. We know from Equation 4.71 that

$$i_d = \left.\frac{d(Kv_O^2)}{dv_O}\right|_{v_O=V_O} v_o$$

so

$$r_d = \frac{1}{\left.\frac{d(Kv_O^2)}{dv_O}\right|_{v_O=V_O}}.$$

Simplifying,

$$r_d = \frac{1}{2KV_O}.$$

Substituting the numerical values, $r_d = 1/(2 \times 1 \times 10^{-3} \times 1.8) = 278\ \Omega$.

Now that we know the value of $r_d$, we can obtain the small-signal component of the output $v_o$ from the small signal AC circuit in Figure 4.44c. Notice that the circuit in Figure 4.44c is a voltage divider. Thus, the small signal AC output

$$v_o = v_i \frac{r_d}{R + r_d} \tag{4.81}$$

$$= 50 \times 10^{-3} \frac{278}{1000 + 278} = 10.9 \text{ mV}.$$

and

$$i_d = \frac{v_o}{r_d} = \frac{0.0109}{278} = 0.039 \text{ mA}.$$

This completes our analysis.[2]

Although both the DC and the AC components of the output voltage are smaller than the corresponding input components, the important parameter is the *fractional ripple*, the ratio of the ripple to the DC. At the input,

$$\text{fractional ripple} = \frac{50 \times 10^{-3}}{5} = 10^{-2} \tag{4.82}$$

---

2. As an interesting aside, we can alternatively obtain $v_o$ mathematically by starting from the equation relating $v_O$ to $v_I$:

$$v_O = \frac{-1 + \sqrt{1 + 4v_I RK}}{2RK}.$$

and at the output,

$$\text{fractional ripple} = \frac{10.9 \times 10^{-3}}{1.8} \simeq 0.6 \times 10^{-2}, \tag{4.83}$$

so the ripple has been reduced relative to the DC by a factor of about 1.7. This level of reduction is not particularly exciting. As can be seen from Equation 4.81, we can improve the level of reduction by reducing the value of $r_d$. One way to do so is to replace the nonlinear resistor of this example with one whose $v$–$i$ curve has a steeper slope as seen in the next example.

It is important to understand that the mathematical basis for incremental analysis and for the formation of the two subcircuits, as in Figure 4.44, is *not* superposition, but a particular interpretation of a Taylor series expansion. Even though we keep only the first two terms of the series, as in Equation 4.61, the relationship is still nonlinear, and hence superposition cannot apply.

---

### WWW EXAMPLE 4.21 DIODE REGULATOR

---

### WWW EXAMPLE 4.22 SMALL SIGNAL ANALYSIS USING A PIECEWISE LINEAR DIODE MODEL

---

Observing that the incremental change in $v_O$ is given by the product of the incremental change in $v_I$ and the slope of the $v_O$ versus $v_I$ curve evaluated at $V_I$, we can write

$$v_o = \left. \frac{d\left(\frac{-1+\sqrt{1+4v_I RK}}{2RK}\right)}{dv_I} \right|_{v_I = V_I} v_i.$$

Simplifying,

$$v_o = \frac{1}{\sqrt{1+4V_I RK}} v_i$$

$$= 10.9 \text{ mV}$$

which is the same as the value obtained by analyzing the small-signal circuit.

## 4.6 SUMMARY

- This chapter introduced nonlinear circuits and their analyses. Nonlinear circuits include one or more nonlinear devices, which display a nonlinear $v$–$i$ relationship. Nonlinear circuits obey KVL and KCL and can be solved using the basic KVL/KCL method or the node method. Note that the KVL/KCL method or the node method do not make any assumptions about linearity.
- We discussed four methods for solving nonlinear circuits including the analytical method, the graphical method, the piecewise linear method, and the small signal method (also known as the incremental method).

  The analytical method uses KVL/KCL or the node method to write the circuit equations and solves them directly. The graphical method uses a graph of the $v$–$i$ relation of the nonlinear device and the graph capturing the circuit constraint to solve for the operating point. The piecewise linear method represents the $v$–$i$ characteristics of a nonlinear element by a succession of straight-line segments, then makes calculations within each straight-line segment using linear analysis tools.

  The small signal method applies to circuits in which nonlinear devices are operated only over a very small range of voltage or current values. For small perturbations of voltages or currents about a nominal operating point, nonlinear device behavior can be approximated using a piecewise linear model that provides a good fit in the narrow operating range. Thus, incremental variables not only satisfy KVL and KCL, but also linear $v$–$i$ relations over the narrow operating range.
- We introduced the following notation to distinguish between total variables, DC operating values, and small signal variables:
  - We denote total variables with small letters and capital subscripts, e.g., $v_D$,
  - DC operating point variables using all capitals, e.g., $V_D$,
  - and incremental values using all small letters, e.g., $v_d$.
- A systematic procedure for finding incremental voltages and currents for a circuit with a nonlinear device characterized by the $v$–$i$ relation:

  $$i_D = f(v_D)$$

  is the following:

  1. Find the DC operating variables $I_D$ and $V_D$ using the subcircuit derived from the original circuit by setting all small-signal sources to zero. Any of the methods discussed in the preceding sections — analytical, graphical, or piecewise linear — is appropriate.

2. Find the incremental output voltage and incremental nonlinear device current by forming an incremental subcircuit in which the nonlinear device is replaced by a resistor of value $r_d$, where

$$r_d = \frac{1}{\left.\frac{df(v_D)}{dv_D}\right|_{v_D=V_D}},$$

other linear resistances are retained as is, and all DC sources are set to zero. The incremental subcircuit is linear, so incremental voltages and currents can be calculated by any of the linear analysis techniques.

## EXERCISES

**EXERCISE 4.1** Consider a two-terminal nonlinear device (see Figure 4.47) whose $v$–$i$ characteristic is given by:

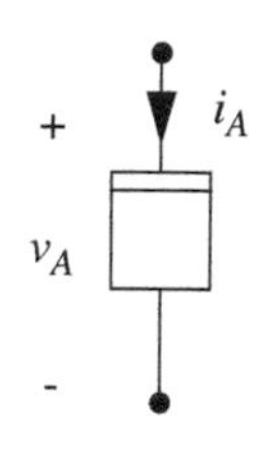

FIGURE 4.47

$$i_A = f(v_A) \tag{4.92}$$

Show that the incremental change in the current ($\Delta i_A = i_a$) for an incremental change in the voltage ($\Delta v_A = v_a$) at the DC operating point $V_a$, $I_A$ is given by:

$$i_a = \left.\frac{df(v_A)}{dv_A}\right|_{v_A=V_A} v_a.$$

(Hint: Substitute $i_A = I_A + i_a$ and $v_A = V_A + v_a$ in Equation 4.92, expand using Taylor series, ignore second order and higher terms in $v_a$, and equate corresponding DC and small signal terms.)

**EXERCISE 4.2** Suppose the two-terminal nonlinear device from the previous exercise (see Figure 4.47) has the following $v$–$i$ characteristic:

$$i_A = f(v_A) = c_X v_A^2 + c_Y v_A + c_Z \text{ for } v_A \geq 0, \text{ and } f(v_A) = 0 \text{ otherwise.}$$

a) Find the operating point current $I_A$ for an operating point voltage $V_A$, where $V_A > 0$.

b) Find the incremental change in the current $i_a$ for an incremental change in the voltage $v_a$ at the operating point $V_A$, $I_A$.

c) By what fraction does $i_a$ change for a $y$ percent change in $v_a$?

d) Suppose the nonlinear device is biased at $V'_A$ instead of $V_A$, where $V'_A$ is $y$ percent greater than $V_A$. Find the incremental change in the current ($i'_a$) for an incremental change in the voltage ($v_a$) at this new bias point. By what fraction is $i'_a$ different from the $i_a$ calculated in part (b).

e) Find the incremental change in the current $i_{acx}$ for an incremental change in the parameter $c_X$ (given by $\Delta c_X = c_x$) from its nominal value of $C_X$, assuming the operating point $v$–$i$ values are $V_A$, $I_A$.

Hint: Observe that if $i_A$ depends on the parameters $x_A$ and $y_B$, in other words,

$$i_A = f(x_A, y_B),$$

then the incremental change in $i_A$ for an incremental change in $y_B$ is given by

$$i_{ayb} = \left.\frac{\delta f(x_A, y_B)}{\delta y_B}\right|_{y_B=Y_B} y_b.$$

EXERCISE 4.3 The nonlinear device (NLD) in the circuit in Figure 4.48 has the $v$–$i$ characteristics shown. Find the operating point $i_D$ and $v_D$ for $R = 910\ \Omega$.

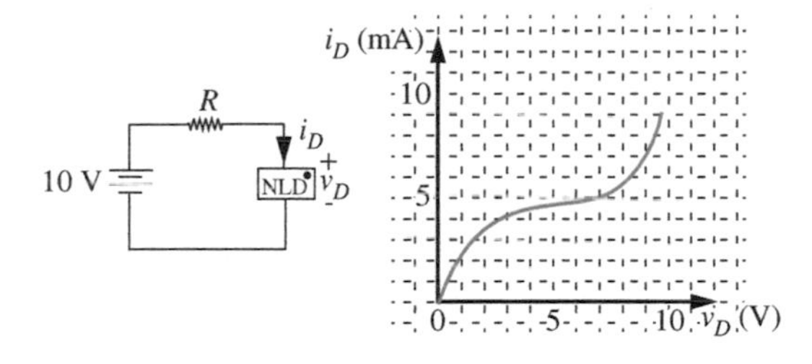


FIGURE 4.48

EXERCISE 4.4

a) Plot the $i_A$ vs. $v_A$ characteristics for the nonlinear network shown in Figure 4.49. Assume the diode is ideal.

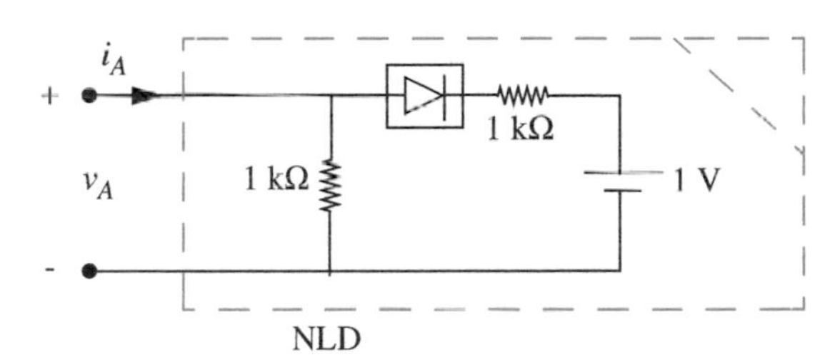


FIGURE 4.49

b) The nonlinear network from part (a) is connected as shown in Figure 4.50. Draw the load line on your $v$–$i$ characteristics from part (a), and find $i_T$.

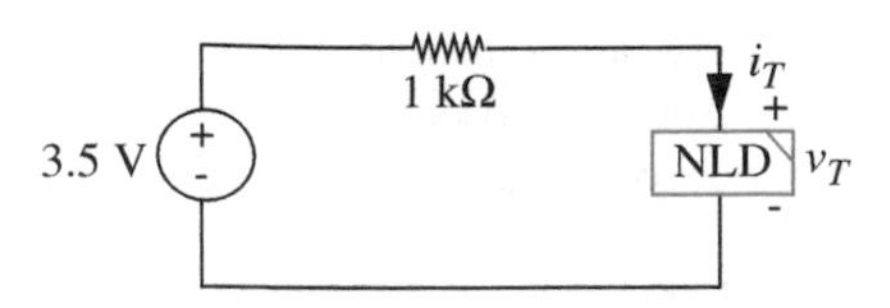


FIGURE 4.50

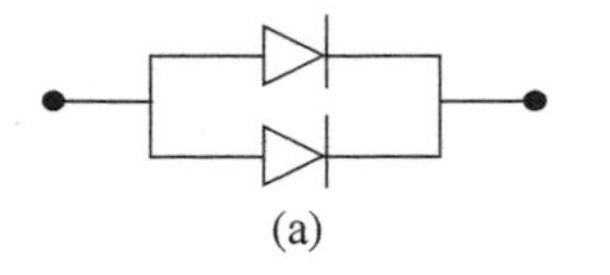


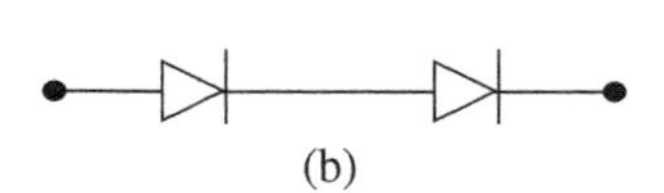


FIGURE 4.51

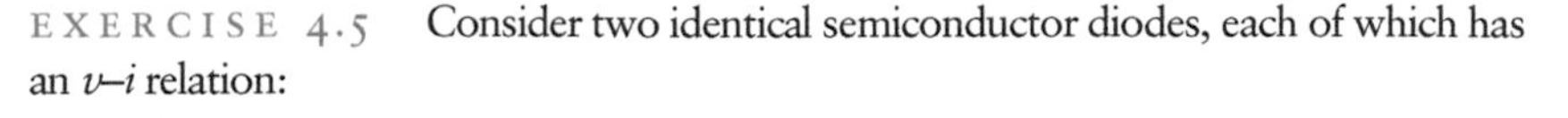

EXERCISE 4.5 Consider two identical semiconductor diodes, each of which has an $v$–$i$ relation:

$$i_D = I_S\left(e^{v_D/V_{TH}} - 1\right). \tag{4.93}$$

a) Find the relation of $i$ to $v$ for the pair connected in parallel as shown in Figure 4.51a.

b) Find the relation of $i$ to $v$ for the pair connected in series as shown in Figure 4.51b.

EXERCISE 4.6 For the circuit in Figure 4.52, find the input characteristic, $i$ versus $v$, and the transfer characteristic $i_2$ versus $v$. $I$ is fixed and positive. Express your results in graphs, labeling all slopes, intercepts, and coordinates of any break points.

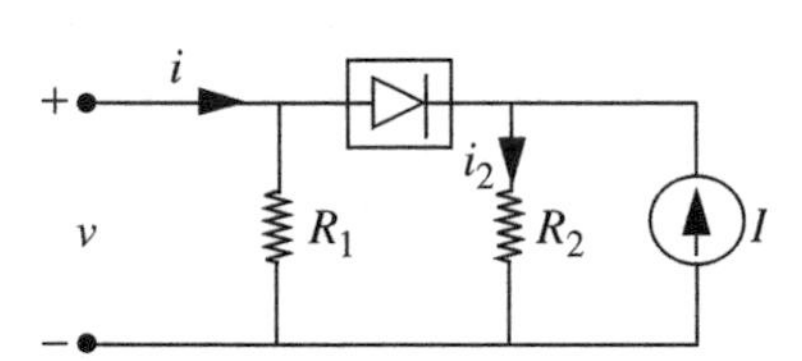


FIGURE 4.52

EXERCISE 4.7 For the circuit in Figure 4.53 and the values shown below, sketch the waveform of $i(t)$. On your sketch, show when the ideal diode is on and when it is off.

$$v_i = 10\sin(t) \quad V_0 = 5\text{ V} \quad R = 1\ \Omega.$$

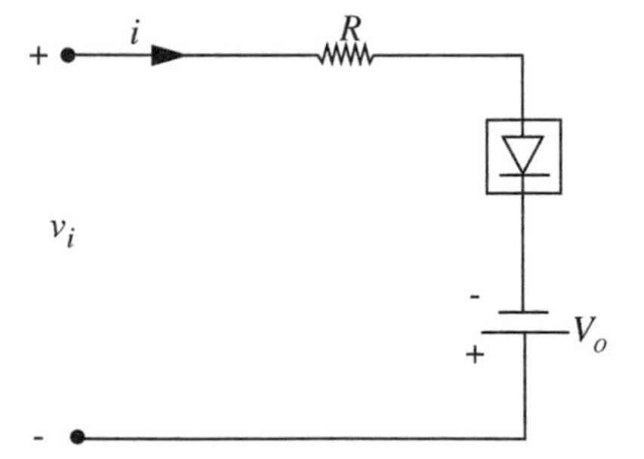


FIGURE 4.53

## PROBLEMS

PROBLEM 4.1 Consider the circuit containing a nonlinear element $N$ as shown in Figure 4.54. The $v$–$i$ relation for $N$ is given by:

$$i_A = c_2 v_A^2 + c_1 v_A + c_0 \text{ for } v_A \geq 0, \text{ and } i_A = 0 \text{ otherwise.}$$

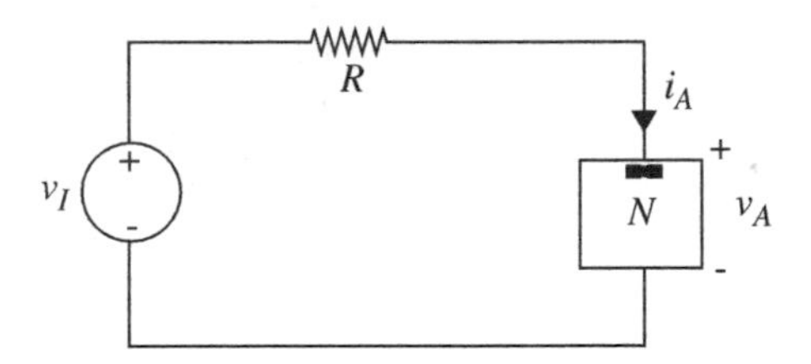


FIGURE 4.54

a) Solve for $i_A$ and $v_A$ using the analytical method.

b) Find the operating point values of the nonlinear element's voltage and current for $v_I = V_I$, where $V_I$ is positive.

c) Find the incremental change in $i_A$ (given by $i_a$) for an incremental change in $v_I$ (given by $v_i$).

d) Determine the incremental change in the voltage across the resistor $R$ for an incremental change in the input $v_I$ (given by $v_i$).

e) Find the incremental change in $i_A$ for a 2% increase in the value of $R$.

f) Find the incremental change in $i_A$ for an incremental change in $v_A$ at the bias point $V_A$, $I_A$.

g) Suppose we replace the source $v_I$ with a DC voltage $V_I$ in series with a small time-varying voltage $v_i = v_o \cos(\omega t)$. Determine the time varying component of $i_A$.

h) Suppose we now replace $v_I = V_I + v_i$, where $V_I = 10$ V and $v_i = 1$ V.

   i) Find the bias point DC current $I_A$ corresponding to $V_I = 10$ V.
   ii) Find the value of $i_a$ corresponding to $v_i = 1$ V using small signal analysis.
   iii) Find the value of $i_A$ using small signal analysis. (Use $i_A = I_A + i_a$.)
   iv) Find the value of $i_A$ using the analytical method for $v_I = V_I + v_i = 11$ V.
   v) Now, find the exact value of the $i_a$ using $i_a = i_A - I_A$.
   vi) What is the error in the value of $i_a$ computed using the small signal method?

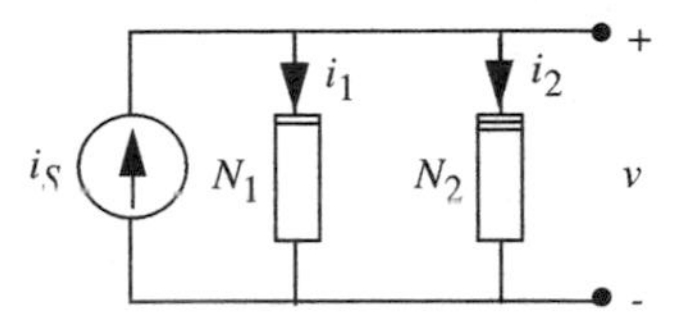


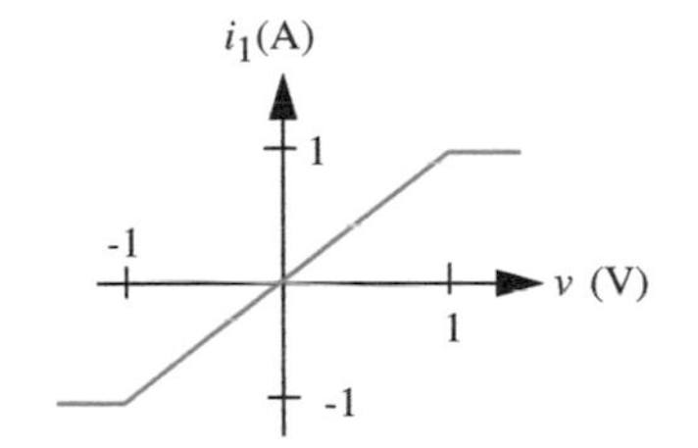


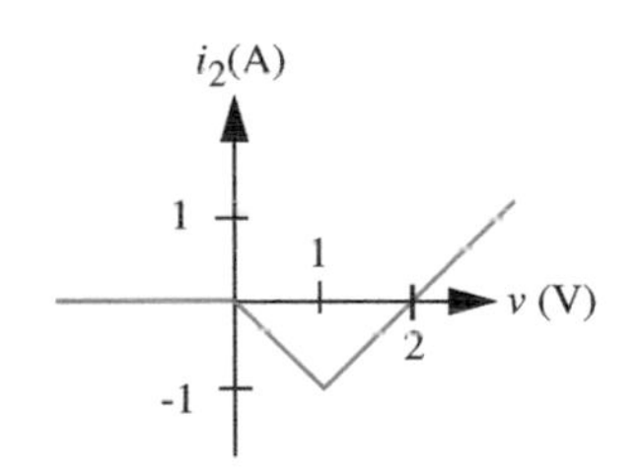


FIGURE 4.55

PROBLEM 4.2 The circuit shown in Figure 4.55 contains two nonlinear devices and a current source. The characteristics of the two devices are given. Determine the voltage, $v$, for (a) $i_S = 1$ A, (b) $i_S = 10$ A, (c) $i_S = 1 \cos(t)$.

PROBLEM 4.3 A plot (hypothetical) of the $v$–$i$ characteristics, (terminal voltage as a function of the current drawn *out*, and *not* its associated variables) for a battery is shown in Figure 4.56(a).

a) If a 2-Ω resistor is connected across the battery terminals, find the terminal voltage of the battery and the current through the resistor.

b) A lightbulb is a nonlinear resistance because of self-heating effects. A hypothetical $v$–$i$ plot is shown in Figure 4.56(b). Find the bulb current and bulb voltage if the lamp is connected to the battery.

c) Devise a piecewise-linear model for the battery which is reasonably accurate over the current range 0–2 A.

d) Use this piecewise-linear battery model to find the battery voltage and bulb current if the bulb and 2-Ω resistor are connected in series to the battery.

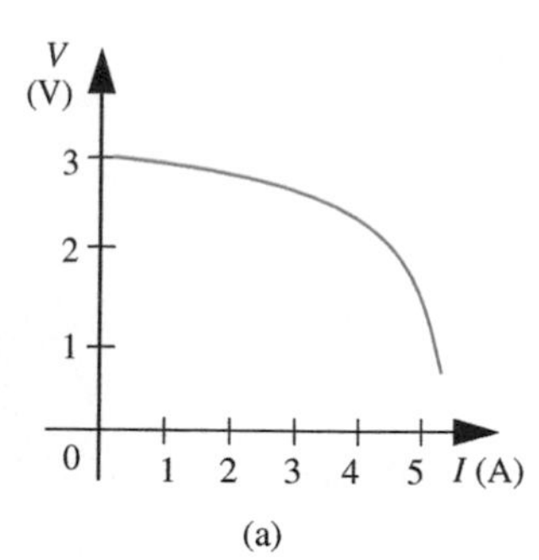


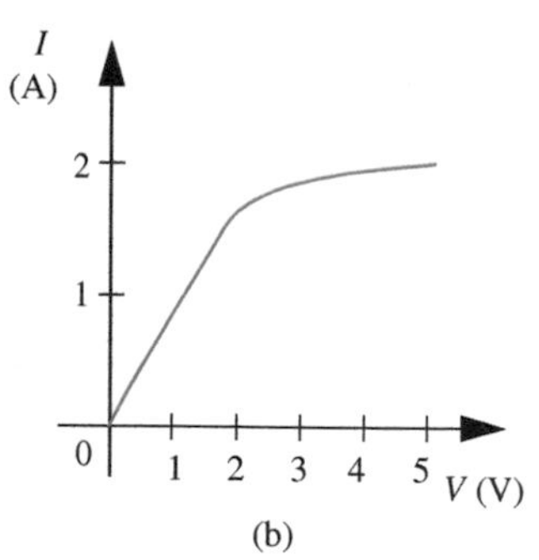


FIGURE 4.56

PROBLEM 4.4

a) Assuming the diode can be modeled as an ideal diode, and $R_1 = R_2$, plot the waveform $v_o(t)$ for the circuit in Figure 4.57, assuming a triangle wave input. Write an expression for $v_o(t)$ in terms of $v_i$, $R_1$, and $R_2$.

b) If the triangle wave has a peak amplitude of only 2 volts, and $R_1 = R_2$, a more accurate diode model must be used. Plot and write an expression for $v_o$ assuming that the diode is modeled using an ideal diode in series with a 0.6-volt source. Draw the transfer curve $v_o$ versus $v_i$.

PROBLEM 4.5 Figure 4.58 is an illustration of a crude Zener-diode regulator circuit.

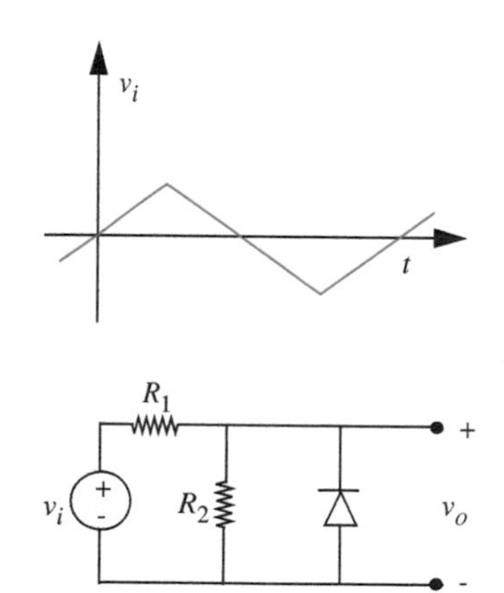


FIGURE 4.57

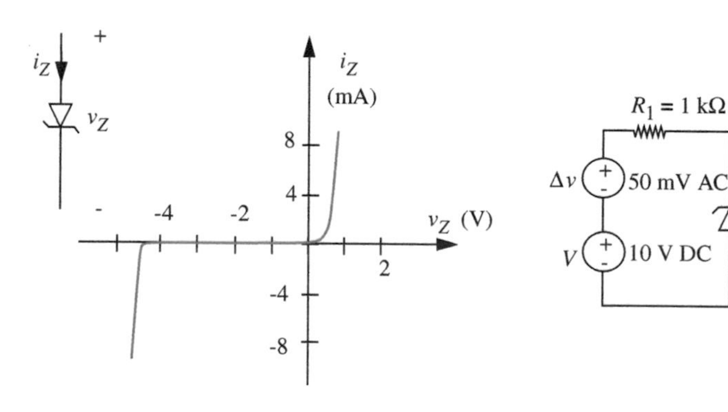

FIGURE 4.58

a) Using incremental analysis, estimate from the graph an analytical expression for $v_o$ in terms of $V$ and $\Delta v$.

b) Calculate the amount of DC and the amount of AC in the output voltage using the Zener-diode characteristic to find model values. (Numbers, please.)

c) What is the Thévenin output resistance of the power supply, that is, the Thévenin resistance seen looking in at the $v_o$ terminals?

PROBLEM 4.6 The terminal voltage-current characteristic of a *single* solar cell is shown in Figure 4.59a. Note that this is a sketch of the terminal voltage as a function of current drawn out (i.e., not the associated variable convention). An array is made by connecting a total of 100 such cells as follows: Ten solar cells are connected in series. Ten sets of these are made. These ten series strips are then connected in parallel (see Figure 4.59b).

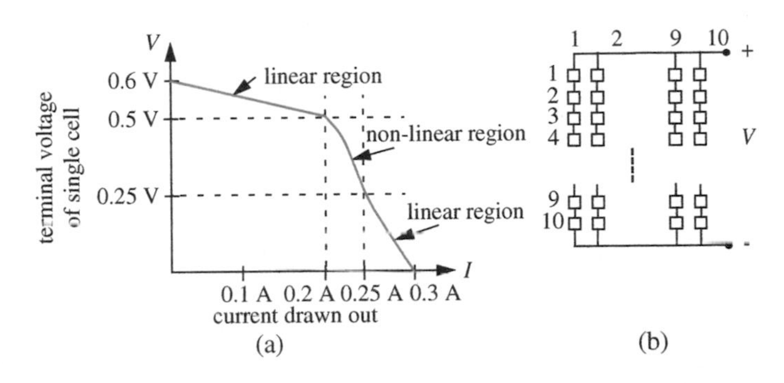


FIGURE 4.59

If a 3-Ω resistor is connected across this new two-terminal element (the 100-cell array), determine the terminal voltage across and the current through the resistor.

PROBLEM 4.7 The junction field-effect transistor (JFET) with the specific connection shown in Figure 4.60a (gate and source shorted together) behaves as a two-terminal device. The $v_D$–$i_D$ characteristics of the resulting two-terminal device shown in Figure 4.60b saturates at current $I_{DSS}$ for $v_D$ greater than a voltage $V_P$, called the pinch-off voltage. In the two-terminal configuration shown, the JFET characteristic is

$$i_D = I_{DSS}\left[2(v_D/V_P) - (v_D/V_P)^2\right] \quad \text{for } v_D \leq V_P$$

and

$$i_D = I_{DSS} \quad \text{for } v_D > V_P.$$

As illustrated in Figure 4.60c, this two-terminal device can be used to make a well-behaved DC current source, even starting with a ripple-containing power supply (depicted as $v_S$), as would be obtained from ordinary rectifier circuits. Suppose the voltage source $v_S$ has an average value $V_S$ and a 60-Hz "ripple component," $v_r = a\cos(\omega t)$ as shown in Figure 4.60d.

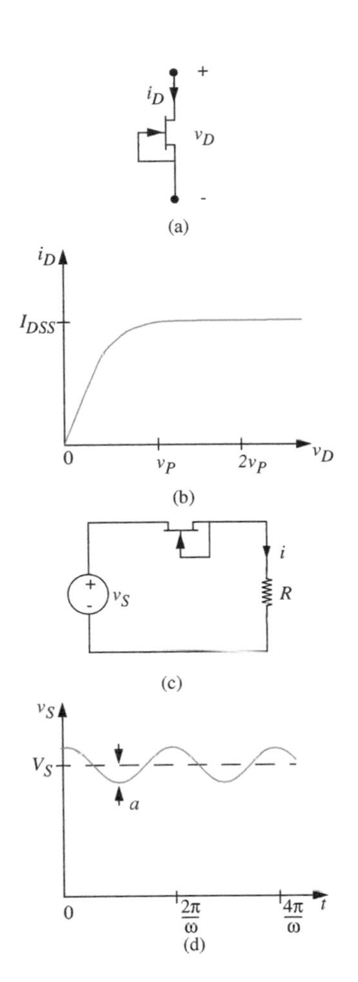


FIGURE 4.60

a) First assume that there is no ripple ($a = 0$). Find the current $i$ through the resistor $R$ as a function of $V_S$ for a value of $R$ = 1 kΩ. At what value of $V_S$ does the current stabilize at $I_{DSS}$? How would this value change if $R$ were doubled in value? Explain.

b) Now assume $a = 0.1$ V and $R = 1$ k$\Omega$. Make reasonable approximations to find the current waveform when $V_S = 5$ V, $V_S = 10$ V, and $V_S = 15$ V. Determine in each case the average value of the current $i$ and the magnitude and frequency of the largest sinusoidal component of the current.

PROBLEM 4.8 The current-voltage characteristic of a photovoltaic energy converter (solar cell) shown in Figure 4.61 can be approximated by

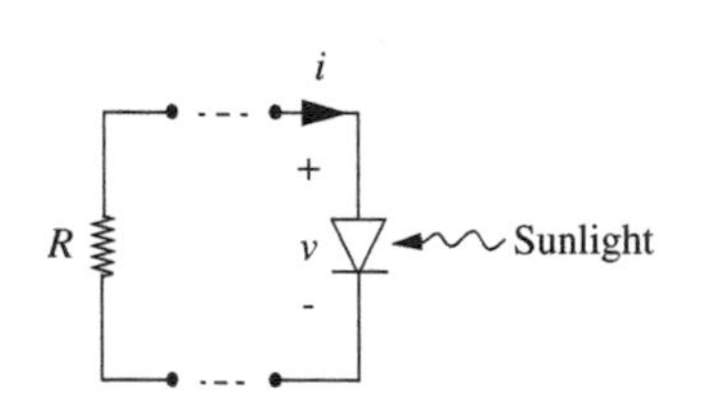


FIGURE 4.61

$$i = I_1(e^{v/V_{TH}} - 1) - I_2$$

where the first term characterizes the diode in the dark and $I_2$ is a term that depends on light intensity.

Assume $I_1 = 10^{-9}$ and assume light exposure such that $I_2 = 10^{-3}$ A.

a) Plot the $v$–$i$ characteristic of the solar cell. Be sure to note the values of open-circuit voltage and short-circuit current. (Note, however, that the characteristic is clearly nonlinear. Therefore, *Thévenin or Norton equivalents do not apply.*)

b) If it is desired to maximize the power that the solar cell can deliver to a resistive load, determine the optimum value of the resistor. How much power can this cell deliver?

PROBLEM 4.9

a) A nonlinear device has $v$–$i$ characteristics shown in Figure 4.62. Assuming that $S$ is an ideal voltage source, which connection, (a), (b), or (c) consumes most power? What if $S$ is an ideal current source?

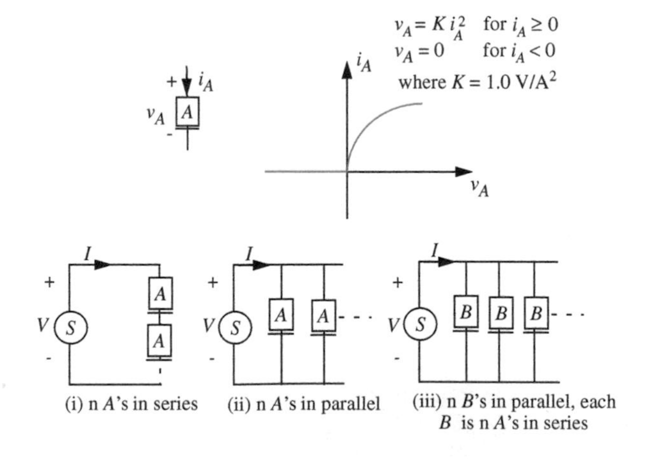


FIGURE 4.62

b) Another crazy device, C, with $v$–$i$ characteristics as shown in Figure 4.63, is introduced. If device $A$ and device $C$ are connected in series across an ideal voltage source of 6 volts, what is the current flow in the circuit? (You can solve it either analytically or graphically.)

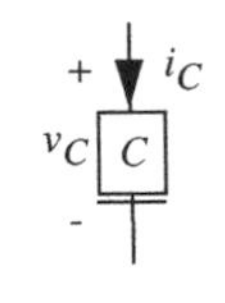


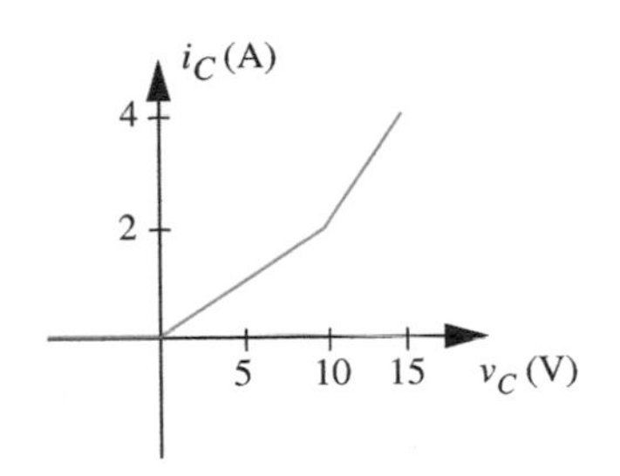


FIGURE 4.63

PROBLEM 4.10 In the circuit in Figure 4.64, assume $v_1 = 0.5$ V and $v_2 = A_2 \cos(\omega t)$, where $A_2 = 0.001$ V. Assume further that $V_{TH} = 25$ mV.

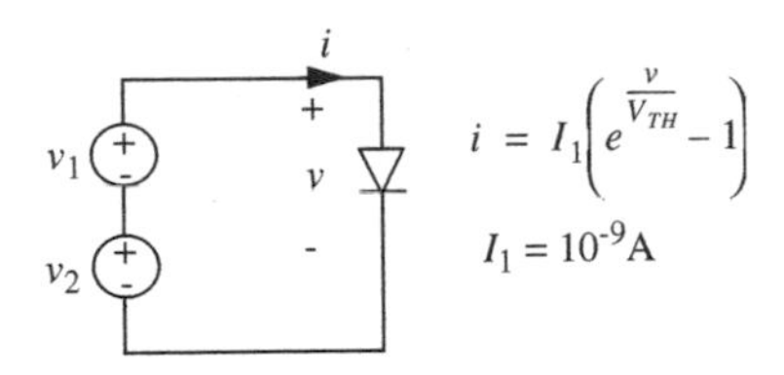


FIGURE 4.64

a) Find the current $i$ if only the $v_1$ source is connected (i.e., with the $v_2$ source shorted out).

b) Find the current $i$ if only the $v_2$ source is connected.

c) Find the current $i$ if both sources are connected as shown. Is superposition obeyed? Explain.

d) Based on your answer in (c) discuss the dependence of the amplitude of the *sinusoidal component* of the current on the amplitude $A_2$. How big can $A_2$ be before significant generation of harmonics will occur? (*HINT:* Taylor's theorem is relevant to this problem.)

PROBLEM 4.11 This problem concerns the circuit illustrated in Figure 4.65:

$R_1 = 1.0$ kΩ $R_2 = 1.0$ kΩ $R_3 = 0.5$ kΩ $R_4 = 1$ kΩ

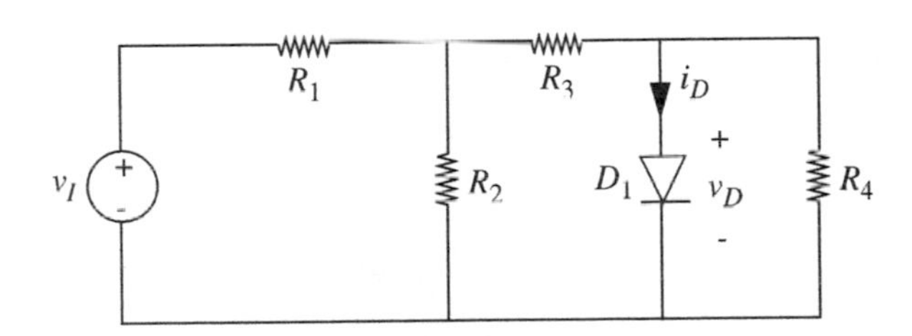


FIGURE 4.65

For $D_1 : i_D = I_S(e^{v_D/V_{TH}} - 1)$ with $I_S = 1 \times 10^{-9}$ A and $V_{TH} = 25$ mV.

a) Find the Thévenin equivalent circuit for the circuit connected to the diode.

b) Assume that for bias point determination the diode can be modeled by an ideal diode and a 0.6-volt battery. What are $v_D$ and $i_D$ when $v_1 = 4$ V?

c) Find a linear equivalent model for this diode valid for small signal incremental operation about the bias point determined from part b.

d) Use your model of part c to find $v_d(t)$ if $v_I = 4\text{ V} + 0.004\text{ V}\ \cos(\omega t)$ V.

PROBLEM 4.12 Consider the circuit in Figure 4.66. The voltage source and the current source are the sum of a DC-level and an AC-perturbation:

$$v = V + \Delta v$$

$$i = I + \Delta i$$

such that $V = 30$ V (DC), $I = 10$ A (DC), $\Delta v = 100$ mV (AC), $\Delta i = 50$ mA (AC).

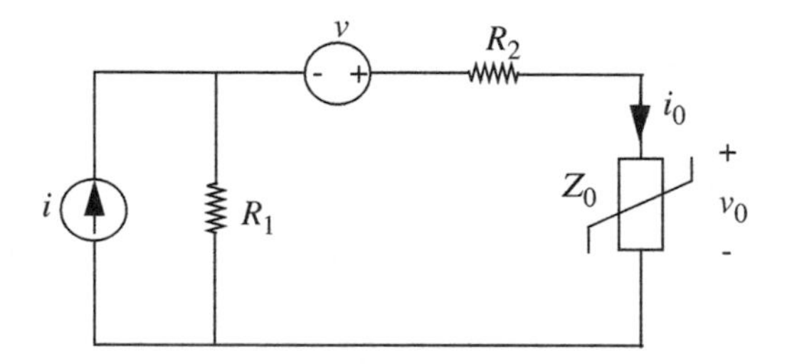


FIGURE 4.66

The resistors have the following values: $R_1 = R_2 = 1/2\ \Omega$. The nonlinear element $Z_O$ has the characteristic:

$$i_O = v_O + v_O^2.$$

Find, by incremental analysis, the DC and AC components of the output voltage $v_O$. (Remark: You can assume in your analysis that the nonlinear element is behaving as a passive element, i.e., is consuming power.)

PROBLEM 4.13 The circuit shown in Figure 4.67 contains a nonlinear element with the following properties:

$$i_N = 10^{-4} v_N^2 \quad \text{when } v_N > 0$$

$$i_N = 0 \quad \text{when } v_N < 0$$

where $i_N$ is in A and $v_N$ is in V.

The output voltage, $v_{OUT}$, may be written approximately as the sum of the two terms:

$$v_{OUT} \simeq V_{OUT} + v_{out}, \tag{4.94}$$

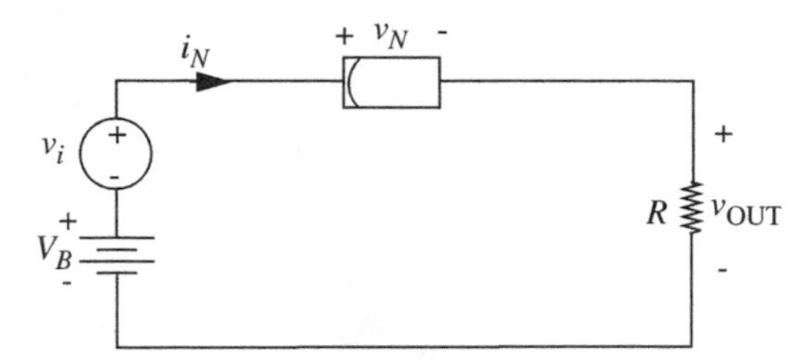


FIGURE 4.67

where $V_{OUT}$ is a DC voltage produced by $V_B$ and $v_{out}$ is the incremental voltage produced by the incremental voltage source $v_i$.

Assuming that $v_i = 10^{-3} \sin(\omega t)$ V and $V_B$ is such that the nonlinear element operates with $V_N = 10$ volts, determine the incremental output voltage $v_{out}$.

PROBLEM 4.14 Consider the diode network shown in Figure 4.68.

For purposes of this problem, the $i_D - v_D$ characteristics of all of the diodes can be accurately represented as

$$i_D = I_S e^{(v_D/25\text{ mV})} \quad \text{where } I_S = 1\text{ mA}/e^{25}.$$

Do not use a piecewise-linear model.

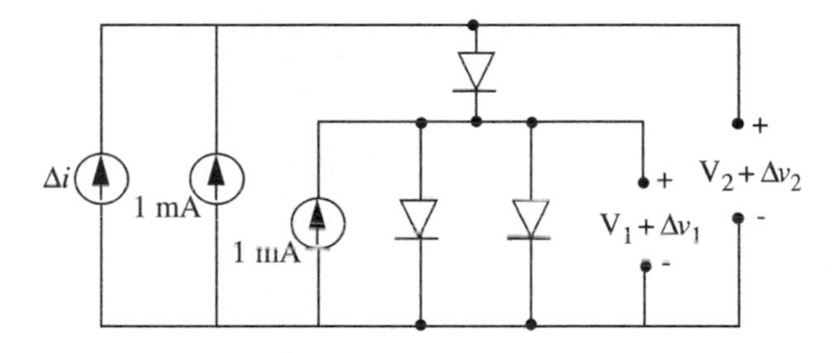


FIGURE 4.68

a) First, assume that $\Delta i = 0$. (Thus $\Delta v_i = \Delta v_2 = 0$.) What are the operating-point values of voltages $V_1$ and $V_2$?

b) Now assume that $\Delta i$ is nonzero, but small enough so that incremental analysis can be used to determine $\Delta v_1$ and $\Delta v_2$. What is the ratio $\Delta v_1 / \Delta v_2$ ?

# CHAPTER 5

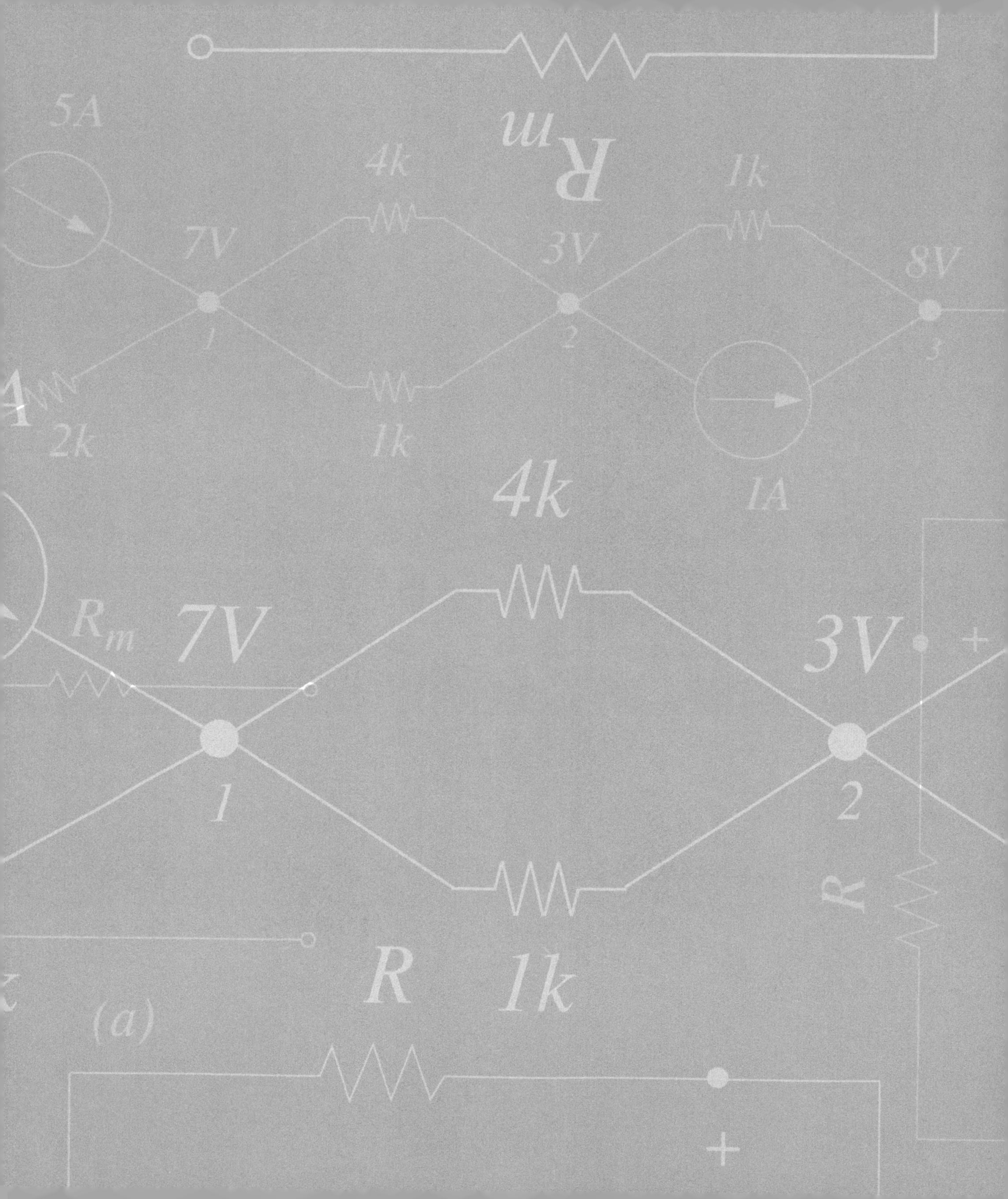

5A
4k
1k
7V
3V
8V
2k
1k
1A
4k
R
m
7V
3V
R
R
1k
(a)

# THE DIGITAL ABSTRACTION

# 5

Value discretization forms the basis of the *digital abstraction*. The idea is to lump signal values that fall within some interval into a single value. We saw an example of value discretization earlier in Figure 1.45 (repeated here for convenience as Figure 5.1) where a voltage signal was discretized into two levels. In this example, an observed voltage value between 0 volts and 2.5 volts is treated as a "0," and a value between 2.5 volts and 5 volts as a "1." Correspondingly, to transmit the logical value "0" over a wire, we place the nominal voltage level of 1.25 on the wire. Similarly, to transmit the logical "1," we place the nominal voltage level of 3.75 volts on the wire.[1] The discrete signal shown in Figure 5.1 comprises the sequence of values "0," "1," "0," "1," "0."

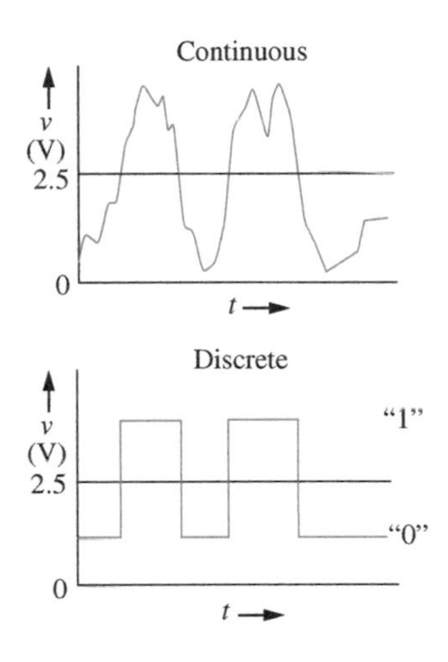


FIGURE 5.1 Value discretization into two levels.

Although the digital approach seems wasteful of signal dynamic range, it has a significant advantage over analog transmission in the presence of noise. Notice, this representation is immune to symmetric noise with a peak to peak value less than 2.5 V. To illustrate, consider the situation depicted in Figure 5.2 in which a sender desires to transmit a value $A$ to a receiver. The figure illustrates both an analog case and a digital case. In the analog case, let us suppose that the value $A$ is 2.4 V. The sender transmits $A$ by representing it as a voltage level of 2.4 V on a wire. Noise during transmission (represented as a 0.2-V noise voltage source in the figure) changes this voltage to 2.6 V at the receiver, resulting in the receiver interpreting the value incorrectly as 2.6.

In the digital case, suppose that the value $A$ is a logical "0." The sender transmits this value of $A$ by representing it as a voltage level of 1.25 V on the wire, which is received as a voltage level of 1.45 by the receiver because of the series noise source. In this situation, since the received voltage falls below the 2.5-V threshold, the receiver interprets it correctly as a logical "0." Thus, the sender and receiver were able to communicate without error in the digital case.

To illustrate further, consider the waveforms in Figure 5.3. Figure 5.3a shows a discretized signal waveform produced by a sender corresponding to

1. It turns out that the mapping of voltage ranges to logical values has a significant impact on the robustness of digital circuits, and a methodical way of selecting the mapping will be presented in Section 5.1. But for now, let us proceed with this rather arbitrarily chosen mapping, and continue to build our intuition.

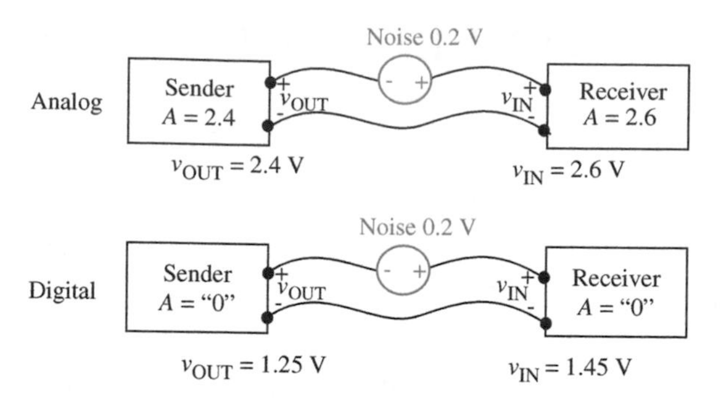


**FIGURE 5.2** Signal transmission in the presence of noise. The noise is represented as a series voltage source.

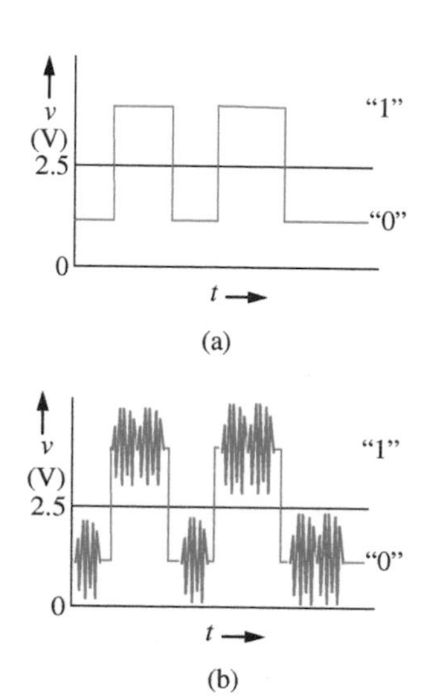


**FIGURE 5.3** Noise immunity for discretized signals: (a) a digital signal produced by a sender; (b) the signal received by a receiver following transmission through a noisy environment.

a "0," "1," "0," "1," "0" sequence. Figure 5.3b shows the same signal with the superposition of some amount of noise, possibly during transmission through a noisy environment. The receiver will be able to receive the sequence correctly provided the noise levels in Figure 5.3b are small enough that the voltages for a logical 0 signal do not exceed 2.5 V, and the voltages for a logical 1 signal do not fall below 2.5 V. Specifically, notice that the binary mapping we have chosen is immune to symmetric noise with a peak-to-peak value less than 2.5 V.

Of course, the discrete representation does not come for free. Considering our example in Figure 5.1, in the analog case, a single wire could carry any value, for example, 1.1, 2.9, or 0.9999999 V. However, in the digital case, a wire is restricted to one of only two values: "0" and "1," thereby losing precision significantly.

Two levels of signal precision are sufficient for many applications. As one example, logic computations involve signals that commonly take on one of two values: TRUE or FALSE. Indeed, most of this chapter (specifically, Sections 5.2, 5.3, and 5.5) deals exclusively with signals that can take on one of two values. Each of these two-level signals is communicated over a single wire. However, there are other applications that require more levels of precision. For example, a speech signal processing application might involve speech signals with 256 or more levels of precision. One approach to achieving more precision is to use coding to create multi-digit numbers. When each digit takes on one of two values, the digit is called a *binary digit,* or bit. Much as the familiar decimal system uses multiple digits to represent numbers other than 0 through 9, the binary system uses multiple bits to represent numbers other than 0 or 1. Multi-bit signals are commonly transmitted by allocating multiple wires — one for each bit, or occasionally, by time multiplexing multiple bits on a single wire. This approach of representing numbers in the binary system is discussed further in Section 5.6. For now, we return to our discussion of the two-level representation.

The two-level representation is commonly known as the *binary representation.* Virtually all digital circuits use the binary representation because two-level

| TRUE | FALSE |
|---|---|
| 0 V | 5 V |
| 5 V | 0 V |
| 2 V | 0 V |
| 0 V | 1 V |
| ON | OFF |
| 0 V < v < 2.5 V | 2.5 V < v < 5 V |
| 0 V < v < 1 V | 4 V < v < 5 V |
| 0 μA | 2 μA |

**TABLE 5.1** Binary signal representation. *v* represents the value of some parameter.

circuits are much easier to build than multilevel circuits. The two levels in the binary representation are variously called (a) TRUE or FALSE, (b) ON or OFF, (c) 1 or 0, (d) HIGH or LOW.

Digital signals are commonly implemented using voltage levels, for example, 0 V to represent FALSE, and 5 V to represent TRUE. We observe, however, that our choice of representing logical values with specific physical values (for example, representing a logical TRUE with 5 V and a logical FALSE with 0 V) is rather arbitrary. We can equivalently choose to represent a logical TRUE with 0 V and a logical FALSE with 5 V. Unless specifically mentioned otherwise, this book adopts the convention that TRUE and high correspond to the logical 1, and conversely, FALSE and low correspond to the logical 0. Table 5.1 depicts these and several other physical realizations of the binary signals, TRUE and FALSE.

## 5.1 VOLTAGE LEVELS AND THE STATIC DISCIPLINE

The previous section illustrated several ways to represent binary values. The representations differed not only in the signal type (for example, current versus voltage), but also in the signal values (for example, 5 V versus 4 V to represent a logical 1). Because we require that digital devices built by various manufacturers talk to each other, the devices must adhere to a common representation. The representation must allow for large enough design margins so that devices can be built out of a wide range of technologies. Furthermore, the representation should be such that the devices operate correctly even in the presence of some amount of noise.

The *static discipline* is a specification for digital devices. The static discipline requires devices to adhere to a common representation, and to guarantee that they interpret correctly inputs that are valid logical signals according to the common representation, and to produce outputs that are valid logical signals

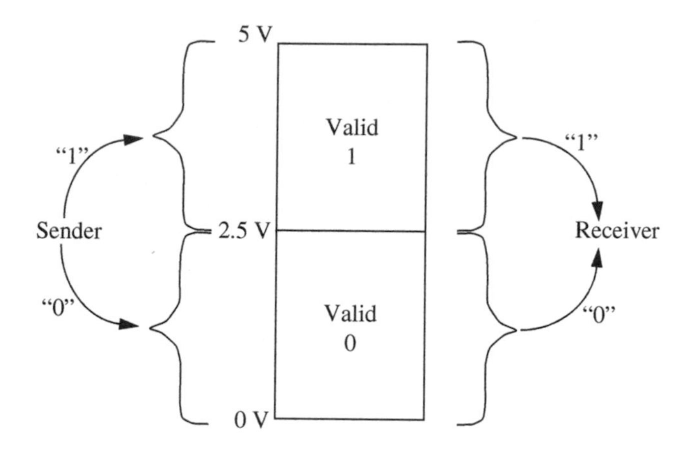


**FIGURE 5.4** Senders and receivers use an agreed-upon mapping between voltage levels and logical signals so that they can communicate with each other.

provided they receive valid logical inputs. By adhering to a common representation, digital devices based on different technologies or built by different manufacturers can communicate with each other.

We will begin with a simple representation, and then successively improve it until we have a representation that can serve as the basis for a static discipline. One of the representations we saw earlier divided a voltage range into two intervals and associated a logic value with each, namely,

$$\text{Logic } 0 : 0.0 \text{ V} \leq V < 2.5 \text{ V}. \tag{5.1}$$

$$\text{Logic } 1 : 2.5 \text{ V} \leq V \leq 5.0 \text{ V}. \tag{5.2}$$

This simple representation is illustrated in Figure 5.4. According to this representation, if a receiver saw 2 V on a wire it would interpret it as a 0. Similarly, a receiver would interpret 4 V on a wire as a 1. Assume, for now, that values outside this range are invalid.

What voltage level should a sender place on a wire? According to our representation, any value between 0 V and 2.5 V would suffice for a logical 0, and any value between 2.5 V and 5 V would work for a logical 1.

Devices that obey this representation would be able to communicate with each other successfully. In other words, as depicted in Figure 5.4, a sending device connected to a receiving device is allowed to output *any* value between 0 V and 2.5 V (for example, 0.5 V) for a logical 0, and *any* value between 2.5 V and 5 V (for example, 4 V) for a logical 1. Correspondingly, the receiving device must interpret *all* values between 0 V and 2.5 V as a logical 0, and *all* values between 2.5 V and 5 V as a logical 1. Thus, our simple representation allows a fair bit of flexibility because valid logical 1 signals and logical 0 signals can occupy a range of values.

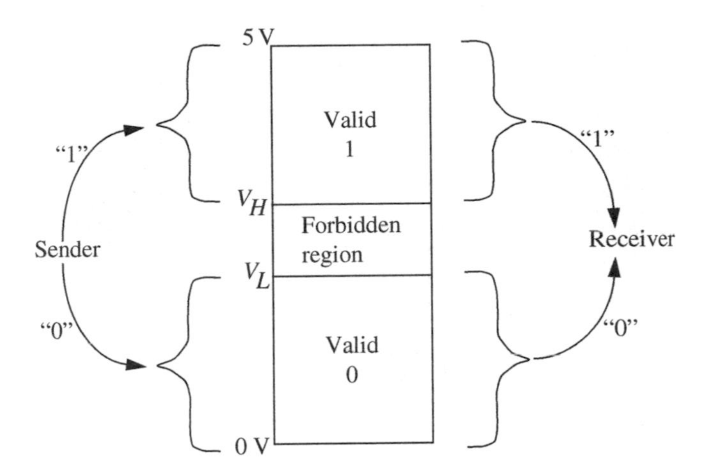


**FIGURE 5.5** A representation with forbidden regions showing the mapping of voltage levels and logical values.

There is one problem, however. What does the receiver do if it sees a voltage level of 2.5 V on the wire? Does it interpret this signal value as a logical 0 or as a logical 1? To eliminate such confusion, we further prescribe a *forbidden region* that separates the two valid regions. We further allow the behavior of the receiving device to be undefined if it sees a voltage in the forbidden region. Thus, the correspondence between voltage levels and logic signals from the viewpoint of a receiver might look like:

$$\text{Logic } 0 : 0 \text{ V} \leq V \leq 2 \text{ V}. \tag{5.3}$$

$$\text{Logic } 1 : 3 \text{ V} \leq V \leq 5 \text{ V}. \tag{5.4}$$

This representation using a forbidden region is illustrated in Figure 5.5. In this representation, a receiver interprets signals above 3 V as a logical 1 and voltages below 2 V as a logical 0. Signal voltages between 2 V and 3 V are invalid.

As marked in Figure 5.5, the largest voltage that a receiver will interpret as a valid logical 0 is termed the *low voltage threshold*, $V_L$, and the smallest voltage that a receiver will interpret as a valid logical 1 is termed the *high voltage threshold*, $V_H$.

In our representation with the forbidden region, a sender can output any voltage value between $V_H$ and 5 V for a logical 1, and any value between 0 V and $V_L$ for a logical 0. A sender must never output a value in the forbidden region. Correspondingly, as illustrated in Figure 5.5, a receiver must interpret any voltage value between $V_H$ and 5 V as a logical 1, and any value between 0 V and $V_L$ as a logical 0. The behavior of the receiver can be undefined if it sees a voltage value between $V_L$ and $V_H$ because these values are in the forbidden region.

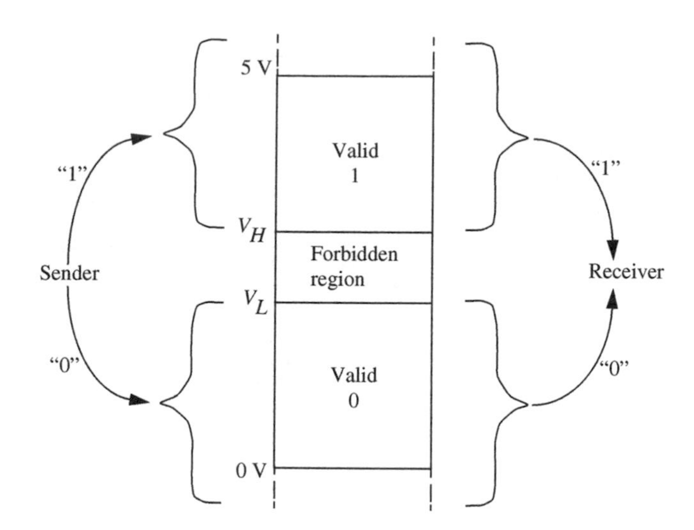


**FIGURE 5.6** For many practical devices, a sender can output any voltage value above $V_H$ for a logical 1, and any voltage value below $V_L$ for a logical 0.

It often turns out that practical circuits are able to correctly interpret values outside the extremum points (below 0 volts for a logical 0 and above 5 V for a logical 1), within certain limits, of course. When devices can make this interpretation, our representation with the forbidden region allows senders to output any voltage value above $V_H$ for a logical 1. Similarly, senders can output any value below $V_L$ for a logical 0. We will assume throughout this book that devices can make this interpretation safely. Figure 5.6 illustrates a simple and practical representation that uses this assumption.

There is one other problem with our representations illustrated in Figures 5.6 and 5.5: They do not offer any immunity to noise. To illustrate, consider our representation in Figure 5.5 with a high and low voltage threshold bounding a forbidden region. In that representation, recall that senders can output voltages above $V_H$ and below 5 V for logical 1's and voltages below $V_L$ and above 0 V for logical 0's. Receivers must correspondingly interpret output voltages above $V_H$ as logical 1's and voltages below $V_L$ as logical 0's.

A sender wishing to place a logical 0 on a wire can therefore output the voltage $V_L$, which falls within the valid range for a logical 0. Receivers observing the value $V_L$ transmitted on the wire will correctly interpret it as a logical 0. However, the presence of even the smallest amount of (positive) noise will force the voltage signal on the wire into the forbidden region, thereby causing the signal to become invalid. Thus, we say that the representation of Figure 5.5 offers no margin for noise.

Clearly we would like a representation that offers the maximum amount of noise immunity during transmission between the sender and the receiver. One way of achieving this is to place tighter restrictions on the values that senders can send. As an example, suppose that a receiver can interpret voltages that fall

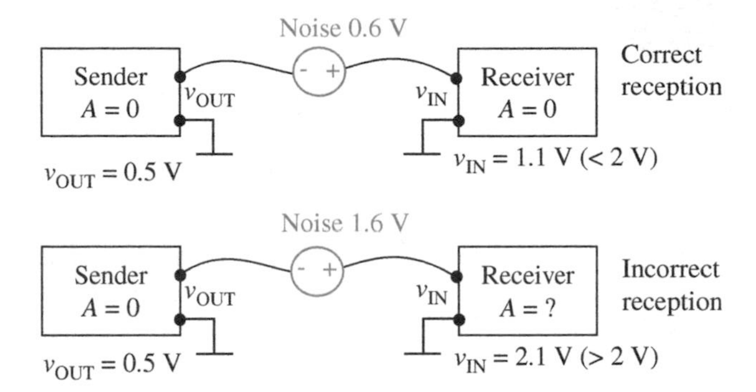


**FIGURE 5.7** Noise margins and signal transmission.

below 2 V as a logical 0. Further, suppose that senders are restricted to sending voltages lower than 0.5 V for a logical 0. Then, it takes at least 1.5 V of (positive) noise to force the sender's voltage signal on the wire into the forbidden region. We say that such a choice of voltage levels offers a noise immunity of 1.5 V for a logical 0.[2]

As an illustration of the notion of noise margins, consider the two situations in Figure 5.7.[3] In the first instance, the sender sends a 0 by placing $v_{OUT} = 0.5$ V (corresponding to the highest legal output voltage for a logic 0) on the wire. The receiver is able to interpret the value as a 0 because the received value is within the low input voltage threshold of 2 V.

In the second situation, however, the receiver is unable to interpret the signal correctly because the noise level of 1.6 V is higher than the noise margin of 1.5 V.

As another example concerning logical 1's, suppose that a receiver can interpret voltages that are above 3 V as a logical 1. Further, suppose that senders are restricted to sending voltages higher than 4.5 V for a logical 1. Then, it takes at least 1.5 V of noise to force the sender's voltage signal on the wire into the forbidden region. We say that such a choice of voltage levels offers a noise immunity of 1.5 volts for a logical 1.

2. As mentioned earlier, because most practical receiver circuits are able to correctly interpret values outside the extremum points (below 0 volts for a logical 0 and above 5 volts for a logical 1), we concern ourselves with providing for a noise margin only between the output value range and the forbidden region, and ignore the effect of noise that tends to push a value outside the extremum bounds.

3. Notice that the signal voltages in the figure are taken with respect to a ground node that is common to both the sender and the receiver. This ground node that is common to senders and receivers is often not shown explicitly, but it is always present! Many a novice designer has forgotten to connect together a common ground between subsystems and has found that the system does not work. Remember, currents flow in loops and the ground connection provides a return path for the current.

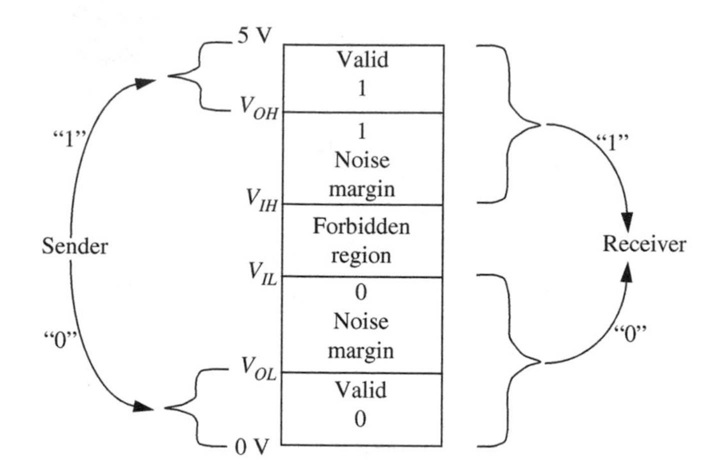


**FIGURE 5.8** A mapping between voltage levels and logical signals that provides noise margins. For a logical high, senders must output values in the $V_{OH}$ to 5-V range. For a logical low, senders must output values in the 0 V to $V_{OL}$ range. Receivers must correspondingly interpret values greater than $V_{IH}$ as a logical high, and output values lower than $V_{IL}$ as a logical low.

The tighter bounds on the voltage values for a sender compared to those for a receiver result in an asymmetry in input and output voltage thresholds. This asymmetry is reflected in Figure 5.8, which shows the correspondence between valid voltage levels and logic signals that is in common use in digital circuits.

To send a logical 0, the sender must produce an *output* voltage value that is less than $V_{OL}$. Correspondingly, the receiver must interpret *input* voltages below $V_{IL}$ as a logical 0.

To allow for a reasonable noise margin, $V_{IL}$ must be greater than $V_{OL}$.

Similarly, to send a logical 1, the sender must produce an *output* voltage value that is greater than $V_{OH}$. Further, the receiver must interpret voltages above $V_{IH}$ as a logical 1.

To allow for a reasonable noise margin, $V_{OH}$ must greater than $V_{IH}$.[4] We can define both a noise margin for transmitting logical 1's and for transmitting logical 0's.

*Noise Margin:* The absolute value of the difference between the prescribed output voltage for a given logical value and the corresponding forbidden region voltage threshold for the receiver is called the *noise margin* for that logical value.

---

4. The simple representation in Figure 5.5 can be viewed as one in which $V_{OH} = V_{IH} = V_H$ and $V_{OL} = V_{IL} = V_L$. Notice that the simple representation of Figure 5.5 offers zero noise margins.

As the name suggests, the noise margin allows the receiver to interpret a value correctly even if some amount of noise is imposed on a sender's signal. Figure 5.9a illustrates a scenario in which a sender outputs a 01010 sequence by producing the appropriate output voltage levels (between $V_{OH}$ and 5 V for a logical 1, and between 0 V and $V_{OL}$ for a logical 0). Provided that the noise does not exceed the noise margins (voltages for a logical 0 do not exceed $V_{IL}$ and voltages for a logical 1 do not fall below $V_{IH}$), a receiver is able to correctly interpret the signal as illustrated in Figure 5.9b.

As illustrated in Figure 5.8, the *noise margin for a logical 0* is given by

$$NM_0 = V_{IL} - V_{OL} \tag{5.5}$$

and the *noise margin for a logical 1* is given by

$$NM_1 = V_{OH} - V_{IH}. \tag{5.6}$$

The region between $V_{IL}$ and $V_{IH}$ is the forbidden region.

Devices that adhere to this discipline will be able to communicate with each other and be immune to noise levels that fall within the noise margins. When $NM_1$ and $NM_0$ are equal, we say that the noise margins are *symmetric.*

Relating the threshold voltage parameters to the numbers used in our example, $V_{OH}$ corresponds to 4.5 V, $V_{OL}$ corresponds to 0.5 V, $V_{IH}$ corresponds to 3 V, and $V_{IL}$ corresponds to 2 V. This mapping is illustrated in Figure 5.10. For our example, the noise margin for a logical 0, $NM_0$, is 1.5 V (2 V – 0.5 V), which is the difference between $V_{IL}$, the maximum input voltage recognized by a receiver as a logic 0, and $V_{OL}$, the highest legal output voltage for a logic 0. Similarly, the noise margin for a 1, $NM_1$, is also 1.5 V (4.5 V – 3 V), which is the difference between $V_{OH}$, the minimum legal output voltage for a logic 1, and $V_{IH}$, the minimum input voltage recognized by a receiver as a logic 1.

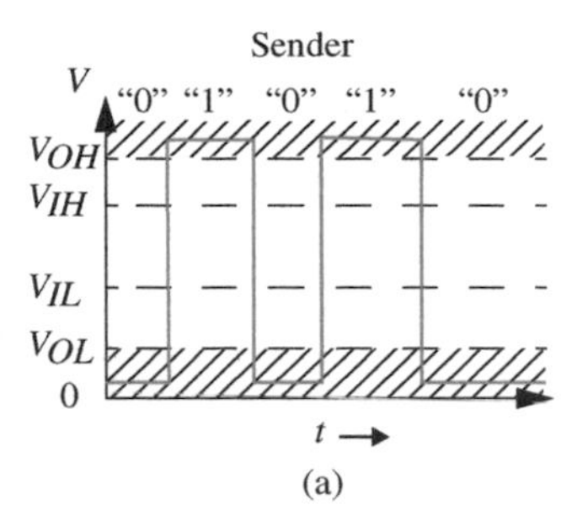


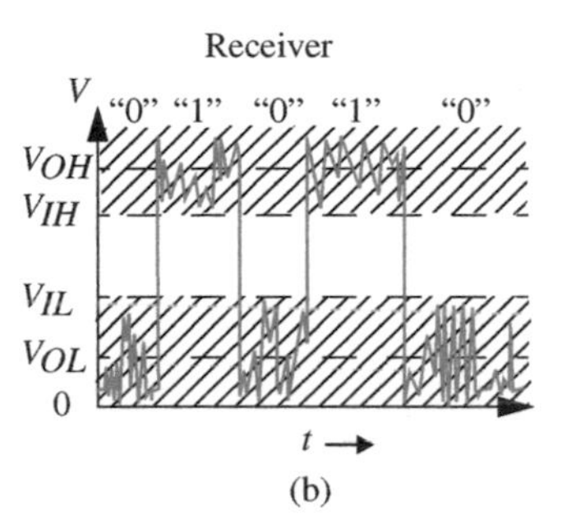


FIGURE 5.9 Senders must output voltages between $V_{OH}$ and 5 V to send a logical 1, and between 0 V and $V_{OL}$ for a logical 0. Correspondingly, receivers can interpret values greater than $V_{IH}$ as a logical high, and values lower than $V_{IL}$ as a logical low. The hashed regions are the valid ranges for senders and receivers.

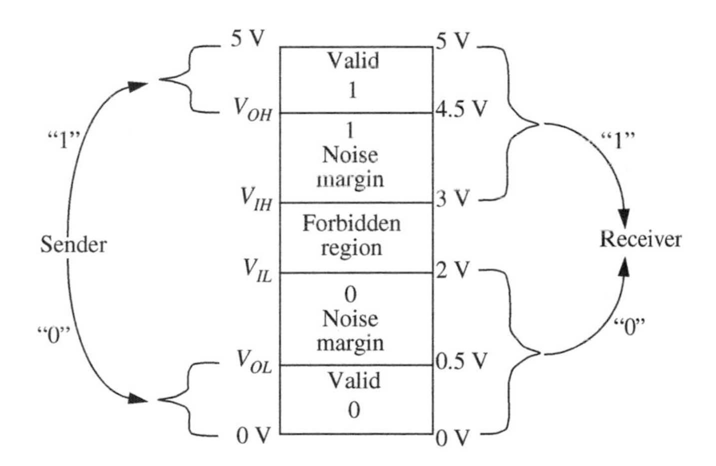


FIGURE 5.10 An example of a mapping between voltage levels and logical values.

*Static discipline* The *static discipline* is a specification for digital devices. The static discipline requires devices to interpret correctly voltages that fall within the input thresholds ($V_{IL}$ and $V_{IH}$). As long as valid inputs are provided to the devices, the discipline also requires the devices to produce valid output voltages that satisfy the output thresholds ($V_{OL}$ and $V_{OH}$).

When designing logic devices, we are often interested in maximizing the noise margins to achieve maximum noise immunity. Referring to Figure 5.8, the 0 noise margin, $NM_0 = V_{IL} - V_{OL}$, can be maximized by maximizing $V_{IL}$ and minimizing $V_{OL}$. Similarly, the 1 noise margin, $NM_1 = V_{OH} - V_{IH}$, can be maximized by maximizing $V_{OH}$ and minimizing $V_{IH}$. As we will see in Chapter 6, the maximum noise margins for devices are limited by the device characteristics or by considerations of symmetry between the low and high noise margins.

---

EXAMPLE 5.1 OBSERVING A STATIC DISCIPLINE The device company Yehaa Microelectronics, Inc. has developed a new process technology that is able to produce large quantities of a certain type of digital device known as an *adder* at a very low cost. For a logical 0, their adders produce a voltage level of 0.5 V at their outputs. Similarly, when outputting a logical 1, their adders produce the voltage level of 4.5 V. Furthermore, the Yehaa adders are able to interpret all signals between 0 V and 2 V at their inputs as a logical 0, and all signals between 3 V and 5 V as a logical 1.

Yehaa's sales team discovers that networking equipment company Disco Systems Inc. buys huge quantities of adder devices from a competitor Yikes Devices, Inc. Upon further research, the Yehaa sales team finds that the hardware systems in one of Disco's product lines operate under a static discipline with the following voltage thresholds:

$V_{IL} = 2$ V, $V_{IH} = 3.5$ V, $V_{OL} = 1.5$ V and $V_{OH} = 4$ V.

In other words, the mapping between voltage ranges and logical values in Disco's static discipline is as summarized in Figure 5.11.

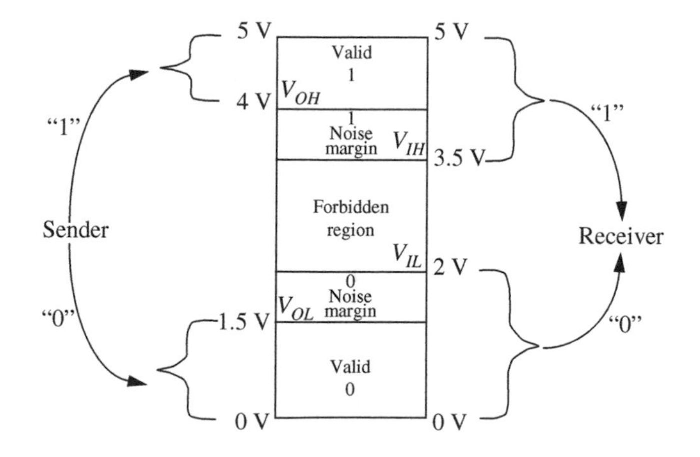


FIGURE 5.11 The mapping between voltage levels and logic values in the static discipline used by Disco Systems.

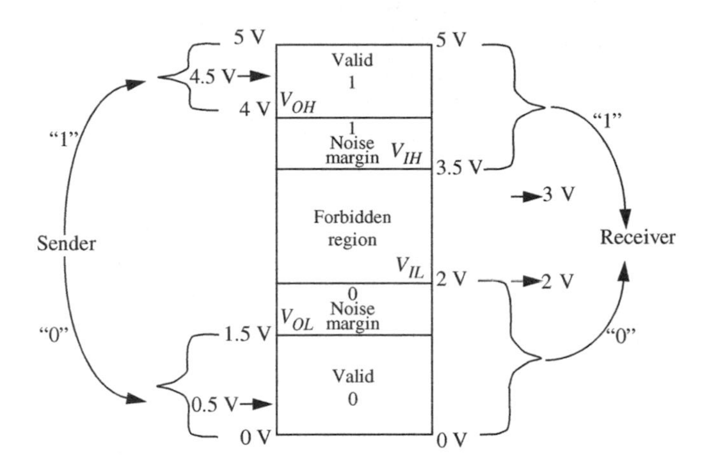


**FIGURE 5.12** Comparing the voltages produced by Yehaa against the voltage levels required by the static discipline used by Disco Systems.

Yehaa's sales team wishes to sell their adders to Disco at a lower cost than those from Yikes, but first, Yehaa must determine whether their adders can safely replace the adders from Yikes. The sales team asks their development engineers to determine whether Yehaa's adders satisfy the static discipline under which Disco's system operates.

The development team first looks at the output level for a logical 1 required by Disco's static discipline. The static discipline used by Disco requires devices to produce voltages between 4 V ($V_{OH}$) and 5 V for a logical 1. As illustrated in Figure 5.12, Yehaa's devices produce a voltage level of 4.5 V for a logical 1, which falls within the required range, and so they satisfy the $V_{OH}$ requirement.

Next, they look at the output voltage level for a logical 0. As depicted in Figure 5.12, Yehaa's devices produce a voltage level of 0.5 V for a logical 0, which falls within the 0 V to 1.5 V ($V_{OL}$) range required for logical 0's by Disco's static discipline. Thus, Yehaa's devices satisfy the $V_{OL}$ requirement.

The engineers now turn their attention to the input voltage levels required by Disco's static discipline. Yehaa's devices are able to interpret voltages as high as 2 V as a logical 0, so they can interpret any voltage between 0 V and 2 V ($V_{IL}$) as a logical 0, just as required by Disco's static discipline. Thus, the Yehaa devices satisfy the $V_{IL}$ requirement.

Similarly, Yehaa's devices are able to interpret voltages between 3.5 V ($V_{IH}$) and 5 V as a logical 1, which again satisfies the $V_{IH}$ requirement of Disco's static discipline. The fact that the Yehaa devices interpret certain voltages in Disco's forbidden region (specifically, those between 3 V and 3.5 V) as a logical 1 is irrelevant since devices are allowed arbitrary behavior for values in the forbidden region.

Thus, the development engineers are able to tell their sales team that Yehaa's adders satisfy Disco's static discipline and so they can be used as replacements for Disco's existing adders.

EXAMPLE 5.2 VIOLATING A STATIC DISCIPLINE Yikes discovers that Disco is considering switching to Yehaa adders because Yehaa's devices are cheaper than Yikes devices. The Yikes sales team goes over their own product list and notices that they do carry a new adder that they can sell to Disco at an even lower cost than the Yehaa adders. Overjoyed, the sales team asks their own development engineers to check whether these new adders satisfy Disco's static discipline.

The new adders of Yikes have the following properties: For a logical 0, the new adders produce a voltage level of 1.7 V at their outputs. Similarly, when outputting a logical 1, their adders produce the voltage level of 4.5 V. The new Yikes adders interpret all signals between 0 V and 1.5 V at their inputs as a logical 0, and all signals between 4 V and 5 V as a logical 1. Their behavior for input signals in the 1.5 V to 4 V range is undefined.

Furthermore, recall that Disco's systems operate under a static discipline with the following voltage thresholds: $V_{IL} = 2$ V, $V_{IH} = 3.5$ V, $V_{OL} = 1.5$ V, and $V_{OH} = 4$ V.

The Yikes development team first looks at the output voltage levels required by Disco's static discipline. They observe that the 4.5-V output produced by their new adders falls within Disco's legitimate range for a logical 1 (between $V_{OH} = 4$ V and 5 V), thus satisfying the $V_{OH}$ requirement.

Next, they turn their attention to the output voltage level required for a logical 0. To their disappointment, they discover that the 1.7-V output produced by their new adders for a logical 0 output is greater than the maximum value of $V_{OL} = 1.5$ V allowed by Disco. Thus their new adders violate the $V_{OL}$ requirement. At this point, the Yikes development team reluctantly concludes that their new adders cannot be sold to Disco.

As an exercise in futility, the development engineers further investigate the input voltage levels. Disco's static discipline requires that devices interpret any voltage between 0 V and 2 V ($V_{IL}$) as a logical 0. The new adders from Yikes fail this test because they are unable to interpret signals above 1.5 V as a logical 0.

Next, the engineers investigate the input high voltage level, but quickly discover that the situation is even worse. Disco's systems require that all devices interpret voltages in the range 3.5 V ($V_{IH}$) to 5 V as a logical 1. Unfortunately, their new adders can guarantee to interpret voltages only in the range 4 V to 5 V as a logical 1. Their behavior for input voltages in the range 3.5 V to 4 V is undefined. Since Disco's devices can legitimately produce voltages in this range for a logical 1, the new Yikes adders cannot co-exist with the existing Disco devices.

EXAMPLE 5.3 NOISE MARGINS Recall that the hardware systems in one of Disco's product lines operate under a static discipline with the following voltage thresholds: $V_{IL} = 2$ V, $V_{IH} = 3.5$ V, $V_{OL} = 1.5$ V, and $V_{OH} = 4$ V. Compute the noise margins.

From Equation 5.5, the noise margin for a logical 0 is given by

$$NM_0 = V_{IL} - V_{OL} = 2\ V - 1.5\ V = 0.5\ V.$$

Similarly, from Equation 5.6, the noise margin for a logical 1 is given by

$$NM_1 = V_{OH} - V_{IH} = 4\ V - 3.5\ V = 0.5\ V.$$

---

EXAMPLE 5.4 A STATIC DISCIPLINE WITH IMPROVED NOISE MARGINS Disco Systems Inc. has been having intermittent faults in its systems. Their system architects figure out that because the static discipline they have adopted does not provide a sufficient noise margin, their systems are susceptible to noise. To improve the noise immunity of their systems, they decide to upgrade their systems to a new static discipline in which the output high voltage threshold is increased by 0.5 V, and the output low voltage threshold is decreased by 0.5 V. Both the input voltage thresholds remain unchanged. In other words, the improved static discipline has the following voltage thresholds:

$$V_{IL} = 2\ V,\ V_{IH} = 3.5\ V,\ V_{OL} = 1\ V, \text{ and } V_{OH} = 4.5\ V.$$

This choice affords their system a symmetric noise margin of 1 V. In other words, the noise margins for a logical 0 and a logical 1 are equal and are given by

$$NM_0 = 2\ V - 1\ V = 1\ V$$

and

$$NM_1 = 4.5\ V - 3.5\ V = 1\ V$$

On hearing the upgrade announcement from Disco, the sales team of Yehaa claims that the adders they have sold Disco can be used under the upgraded static discipline. Let us determine whether this claim is true.

Recall, that Yehaa's adders behave as follows: For a logical 0, Yehaa's adders produce a voltage level of 0.5 V at their outputs. Similarly, when outputting a logical 1, their adders produce the voltage level of 4.5 V. Yehaa adders are able to interpret all signals between 0 V and 2 V at their inputs as a logical 0, and all signals between 3 V and 5 V as a logical 1.

To operate under Disco's upgraded static discipline, we know that the adders must operate correctly with the tighter bounds on the output thresholds:

- When outputting a logical 1, the voltage their outputs produce must be at least $V_{OH} = 4.5$ V. Since the Yehaa adders produce a 4.5 V output for a logical 1, they barely satisfy this condition.

- When outputting a logical 0, the voltage their outputs produce must be no greater than $V_{OL} = 1$ V. Since the Yehaa adders produce a 0.5-V output for a logical 0, they satisfy this condition easily.

Thus, we have shown that the claim made by the Yehaa sales team is true.

---

EXAMPLE 5.5 COMPARING NOISE MARGINS Which of the two static disciplines shown below offers better noise margins?

Static discipline A has the voltage thresholds given by:

$$V_{IL} = 1.5\text{ V},\quad V_{IH} = 3.5\text{ V},\quad V_{OL} = 1\text{ V},\quad \text{and}\quad V_{OH} = 4\text{ V}.$$

Static discipline B has the voltage thresholds given by:

$$V_{IL} = 1.5\text{ V},\quad V_{IH} = 3.5\text{ V},\quad V_{OL} = 0.5\text{ V},\quad \text{and}\quad V_{OH} = 4.5\text{ V}.$$

For static discipline A:

$$\text{NM}_0 = 1.5\text{ V} - 1\text{ V} = 0.5\text{ V}$$

and

$$\text{NM}_1 = 4\text{ V} - 3.5\text{ V} = 0.5\text{ V}.$$

For static discipline B:

$$\text{NM}_0 = 1.5\text{ V} - 0.5\text{ V} = 1\text{ V}$$

and

$$\text{NM}_1 = 4.5\text{ V} - 3.5\text{ V} = 1\text{ V}.$$

Thus, the voltage thresholds of static discipline B offer a better noise margin.

---

## 5.2 BOOLEAN LOGIC

The binary representation has a natural correspondence to logic, and therefore digital circuits are commonly used to implement logic procedures. For example, consider the logical "if" statement:

*If X is TRUE AND Y is TRUE then Z is TRUE else Z is FALSE.*

We can represent this statement using a boolean equation as:

$$Z = X\ \mathit{AND}\ Y.$$

In the previous equation, $Z$ is true only when both $X$ and $Y$ are *TRUE* and *FALSE* otherwise. For brevity we often represent the *AND* function using the "·" symbol as:

$$Z = X \cdot Y.$$

Just as we represent the algebraic expression $x \times y$ as $xy$, we often drop the *AND* symbol and write:

$$Z = X \cdot Y = XY.$$

The boolean equation for the statement:

*If* (*A is TRUE*) OR (*B is NOT TRUE*) *then* (*C is TRUE*) *else* (*C is FALSE*)

is

$$C = A + \overline{B}.$$

The preceding equation contains two other useful functions. The OR function is represented using "+" and the NOT function using the bar symbol as in "$\overline{X}$" or the $\sim$ symbol as in $\sim X$. For example, we represent the condition "$B$ is FALSE" as $\overline{B}$ or $\sim B$. We call $\overline{B}$ the complement of $B$. The logic operators that we have seen thus far are summarized in Table 5.2. For convenience, we will use 1, TRUE, and high interchangeably. Similarly, we will use 0, FALSE, and low interchangeably.

---

EXAMPLE 5.6 MOTION DETECTOR LOGIC Let us write the boolean expression for a motion detector that operates as follows: The circuit must produce a signal $L$ to turn on a set of lights when the signal $M$ from a motion sensor is high, provided it is not daytime. Assume that a light sensor produces a signal $D$ that is high during daytime.[5]

Notice that $L$ is nominally low. It must become high when $M$ is high and $D$ is low. Therefore, we can write

$$L = M\overline{D}.$$

---

*Truth table* We often find it convenient to use a *truth table* representation of boolean functions. A truth table enumerates all possible input value combinations and the corresponding output values.

For example, the truth table representation for $Z = X \cdot Y$ is shown in Table 5.3, that for $Z = X + Y$ is shown in Table 5.4, that for $Z = \overline{X}$ is shown in Table 5.5, and that for $C = A + \overline{B}$ is shown in Table 5.6. As discussed

5. Assume, of course, that the light sensor does not respond to the lights that are turned on by the motion detector.

| OPERATOR | SYMBOL |
|---|---|
| AND | · |
| OR | + |
| NOT | ~ |

TABLE 5.2 Some logic operations and their symbols.

| X | Y | Z |
|---|---|---|
| 0 | 0 | 0 |
| 0 | 1 | 0 |
| 1 | 0 | 0 |
| 1 | 1 | 1 |

TABLE 5.3 Truth table for $Z = X \cdot Y$.

| X | Y | Z |
|---|---|---|
| 0 | 0 | 0 |
| 0 | 1 | 1 |
| 1 | 0 | 1 |
| 1 | 1 | 1 |

TABLE 5.4 Truth table for $Z = X + Y$.

| X | Z |
|---|---|
| 0 | 1 |
| 1 | 0 |

TABLE 5.5 Truth table for $Z = \overline{X}$.

| A | B | C |
|---|---|---|
| 0 | 0 | 1 |
| 0 | 1 | 0 |
| 1 | 0 | 1 |
| 1 | 1 | 1 |

TABLE 5.6 Truth table for $C = A + \overline{B}$.

| A | B | C | D | OUTPUT |
|---|---|---|---|---|
| 0 | 0 | 0 | 0 | 1 |
| 0 | 0 | 0 | 1 | 0 |
| 0 | 0 | 1 | 0 | 0 |
| 0 | 0 | 1 | 1 | 0 |
| 0 | 1 | 0 | 0 | 1 |
| 0 | 1 | 0 | 1 | 0 |
| 0 | 1 | 1 | 0 | 0 |
| 0 | 1 | 1 | 1 | 0 |
| 1 | 0 | 0 | 0 | 1 |
| 1 | 0 | 0 | 1 | 0 |
| 1 | 0 | 1 | 0 | 0 |
| 1 | 0 | 1 | 1 | 0 |
| 1 | 1 | 0 | 0 | 0 |
| 1 | 1 | 0 | 1 | 0 |
| 1 | 1 | 1 | 0 | 0 |
| 1 | 1 | 1 | 1 | 0 |

**TABLE 5.7** Truth Table for $\overline{AB + C + D}$.

in Section 5.4, we can also go from a truth table representation to a logic expression.

---

EXAMPLE 5.7 TRUTH TABLE The truth table for the following logic expression is shown in Table 5.7.

$$\text{Output} = \overline{AB + C + D}.$$

---

## 5.3 COMBINATIONAL GATES

Yet another representation of boolean functions makes use of the *combinational gate* abstraction. We will see how gates are built out of primitive lumped circuit elements in Chapter 6. For now, let us focus on the gate-level abstraction. The digital gate notation for the boolean equation $Z = X$ AND $Y$ is shown in Figure 5.13.

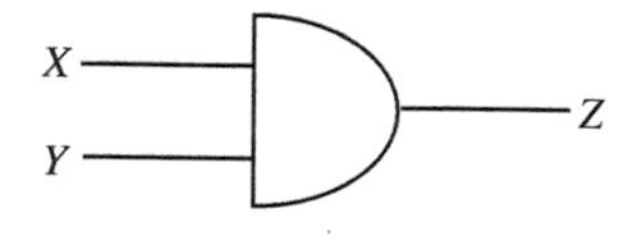


**FIGURE 5.13** The *AND* gate.

The output of combinational gates is purely a function of their inputs. Therefore, combinational functions can always be enumerated using truth tables. Combinational gates follow the static discipline. Provided they are given inputs that fall within valid input levels, they will produce outputs that satisfy valid output thresholds.

*Combinational gate abstraction* A combinational gate is an abstract representation of a circuit that satisfies two properties:

1. Its outputs are a function of its inputs alone.
2. It satisfies the static discipline.

Figure 5.14 shows several useful gate symbols. We have already seen the gate-level representation of the *AND* function. The *OR* gate performs the *OR* function of its inputs. The *NOT* gate takes the complement of its input. For convenience, we often denote the *NOT* function in logic circuits using the "o" symbol. The buffer gate or identity gate simply copies the input value to its output, that is, $A = A$. Its use will become apparent in Section 6.9.2. The *NAND* function is equivalent to the *AND* operation followed by the *NOT* operation. For example, $A = B$ NAND $C$ is equivalent to $A = \overline{B \text{ AND } C}$. It is also equivalent to the statement: $A$ is FALSE only if both $B$ and $C$ are TRUE. Similarly, the *NOR* operation is equivalent to the *OR* operation followed by the *NOT* operation.

A truth table illustrating several of these functions is shown in Table 5.8. Each output column in the truth table corresponds to the given boolean function.

Gates can have multiple inputs. For example, we can have a four-input *AND* gate that implements the function $E = A \cdot B \cdot C \cdot D$ as shown in Figure 5.15.

As shown in Figure 5.16, we can combine digital gates using wires to implement digital circuits, thereby creating more complicated boolean functions.

Figure 5.17 shows a graphical view of the inputs and output for the digital circuit in Figure 5.16. Notice that the output continues to be valid even when the input signal is noisy.

As we might expect, gates are themselves implemented using lumped circuit elements, such as resistors and current sources. In other words, a gate representing the function $F$ is simply an abstraction for a circuit that performs the function $F$. The circuit for an AND gate produces 5 V at its output when

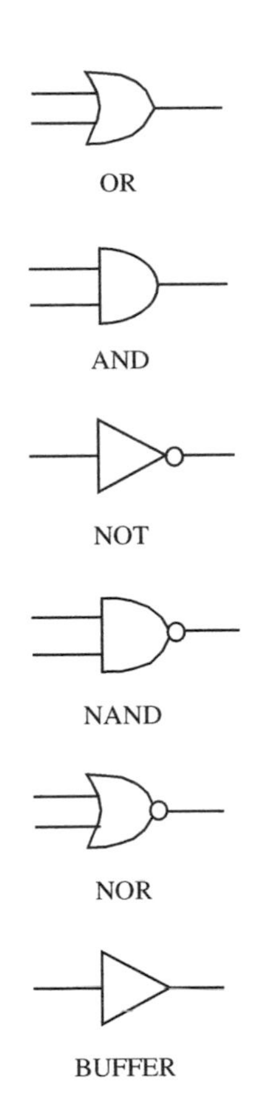


FIGURE 5.14 Gate symbols.

| INPUTS | | AND | OR | NAND | NOR |
|---|---|---|---|---|---|
| $B$ | $C$ | $B \cdot C$ | $B + C$ | $\overline{B \cdot C}$ | $\overline{B + C}$ |
| 0 | 0 | 0 | 0 | 1 | 1 |
| 0 | 1 | 0 | 1 | 1 | 0 |
| 1 | 0 | 0 | 1 | 1 | 0 |
| 1 | 1 | 1 | 1 | 0 | 0 |

TABLE 5.8 Truth table for several two-input functions.

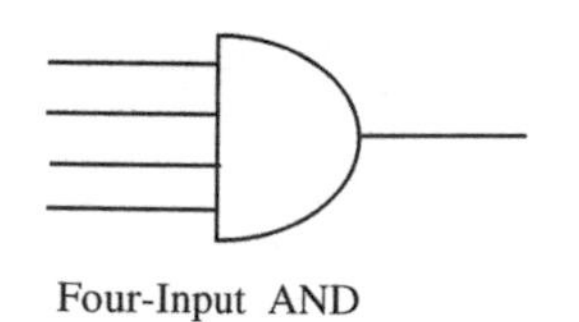


**FIGURE 5.15** A four-input *AND* gate.

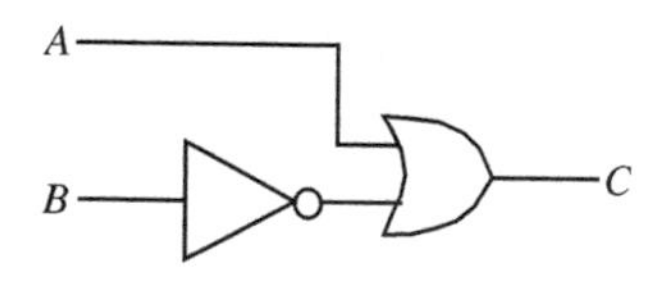


**FIGURE 5.16** The gate-level digital circuit for $C = A + \overline{B}$.

both its inputs are at 5 V, and produces 0 V otherwise. We will defer the actual implementation of digital gates to later chapters, and for now, use the convenient gate abstraction to build more complicated digital systems to process information.

---

EXAMPLE 5.8 GATE-LEVEL IMPLEMENTATION Let us implement the logic expression $Output = \overline{AB + C + D}$ using gates. Notice that we have a choice in implementing this expression. We can use one two-input *AND* gate, a three-input *OR* gate, and an inverter as shown in Figure 5.18. We can also replace the *OR* gate-inverter pair with a *NOR* gate. Alternatively, by rewriting the expression as

$$\text{Output} = \overline{((AB) + (C + D))}$$

we can implement the same circuit using an *AND* gate, an *OR* gate, and a *NOR* gate. We can also check each of the circuits against the truth table, and convince ourselves that they do work as desired.

---

EXAMPLE 5.9 MORE GATE-LEVEL IMPLEMENTATIONS Now, let us design a circuit for the expression: $\overline{(A + B)CD}$. We can rewrite the expression as

$$\overline{(A + B)CD} = \overline{((A + B)(CD))}.$$

The corresponding gate-level implementation is shown in Figure 5.19.

---

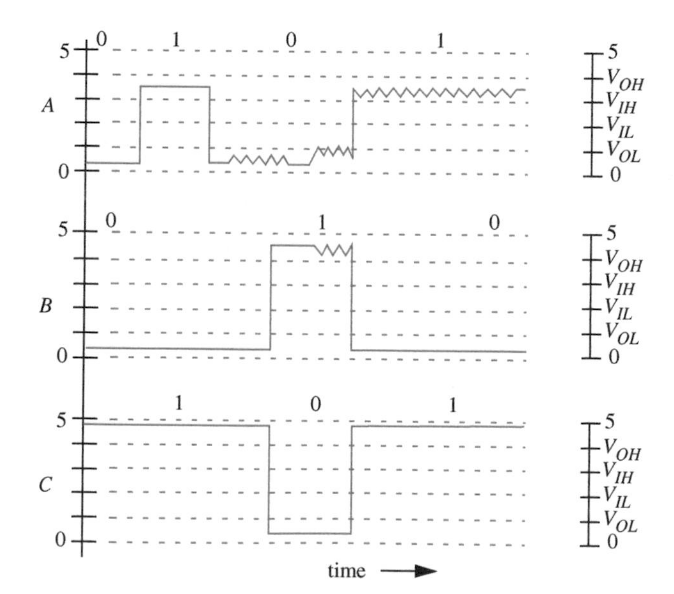


**FIGURE 5.17** A noisy signal input to a digital circuit.

EXAMPLE 5.10 YET ANOTHER GATE-LEVEL IMPLEMENTATION A circuit for the expression $A+B\overline{B}+C$ is shown in Figure 5.20. It requires three gates.

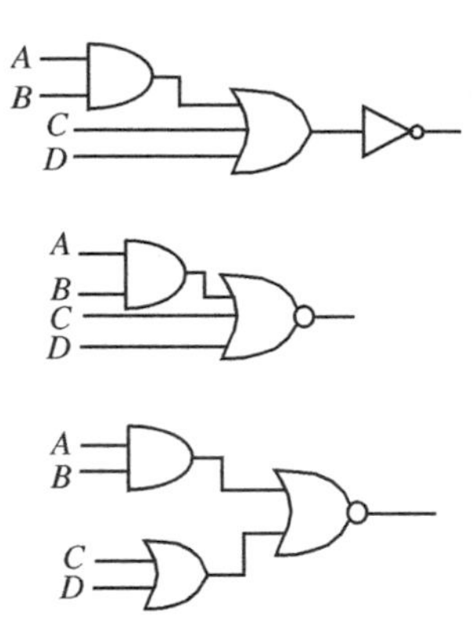

FIGURE 5.18 Implementations of $\overline{AB+C+D}$.

## 5.4 STANDARD SUM-OF-PRODUCTS REPRESENTATION

The previous two sections showed that logic expressions can be represented as truth tables or gate-level circuits. In this section, we will show the equivalence of the representations by discussing how we can derive automatically a logic expression from a truth table. Before we do so, it is useful to introduce a *standard or canonic form* of writing logic expressions called the *sum-of-products form.*

*Sum-of-products* As the name implies, logic expressions in the sum-of-products form are represented using two levels of operations as a set of product (*AND*) terms, each comprising one or more variables in their true forms (for example, $A$) or complement forms (e.g., $\overline{A}$), combined using the *OR* function.

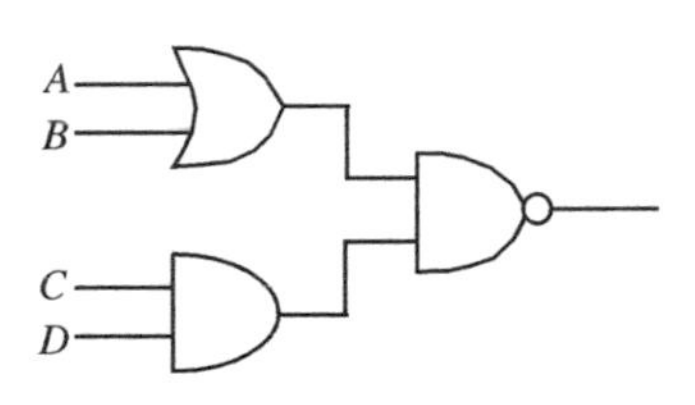

FIGURE 5.19 Implementation of $\overline{(A+B)CD}$.

For example, the logic expression $AD + A\overline{B}C + \overline{A}B\overline{C}$ is in a sum-of-products representation containing the sum of three product terms. The first term contains two variables, while the latter two terms contain three variables each. The expression $AB+C+\overline{D}+\overline{B}$ is also in a sum-of-products representation.

The expression $\overline{AB+C}$, however, is not in a sum of products representation, and neither is the expression $(A+B)(B+\overline{C})$. (Section 5.5 will discuss how we can convert such expressions to a sum-of-products representation.)

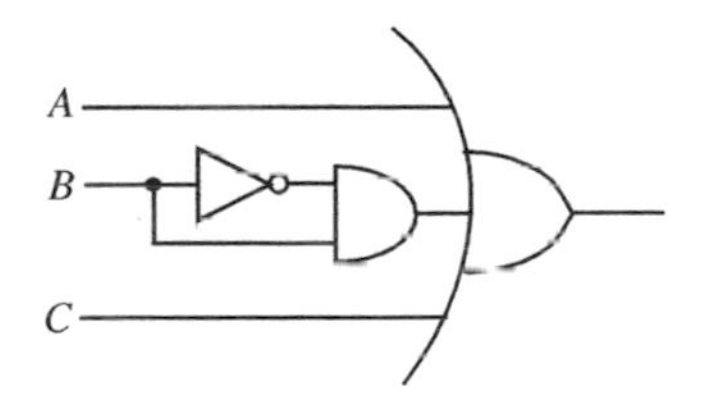

FIGURE 5.20 Implementation of $A+B\overline{B}+C$.

We can write a sum-of-products expression from a truth table representation by first writing a product term for each row in the truth table with a 1 in its output column, and then summing these product terms. Each product term comprises an *AND* function of all the input variables. A variable will appear in its true or complement form in a product term corresponding to a given row in the truth table depending on whether it appears as a 1 or a 0 in that row.

Thus, for example, a logic expression for the truth table in Figure 5.4 is

$$Z = \overline{X}Y + X\overline{Y} + XY. \tag{5.7}$$

By construction, this expression is in a sum-of-products form. It has three product terms corresponding to the three 1's in the output column of the truth table. Since $X$ and $Y$ appear as a 0 and a 1, respectively, in the first row with a 1 output, they contribute the product term $\overline{X}\,Y$ to the overall expression for $Z$. Similarly, the remaining two product terms come from the third and fourth rows of the truth table, respectively.

Notice, however, the expression for $Z$ in Equation 5.7 is quite a bit different from the expression for $Z$ (namely, $X + Y$) shown in Table 5.4. The two expressions are in fact equivalent. This should be self-evident since both expressions represent the same truth table, but this fact will become obvious after Section 5.5 (and specifically, Example 5.13) shows how such logic expressions can be simplified.

---

EXAMPLE 5.11 LOGIC EXPRESSION FROM A TRUTH TABLE Write a logic expression corresponding to the truth table in Table 5.7.

There are three 1's in the output column of the truth table in Table 5.7, and so we expect to see three product terms. The product term corresponding to the 1 in the first row is $\overline{A}\ \overline{B}\ \overline{C}\ \overline{D}$. Similarly, the next two products terms are $\overline{A}\ B\ \overline{C}\ \overline{D}$, and $A\ \overline{B}\ \overline{C}\ \overline{D}$. These three terms are combined with the *OR* function to yield the logic expression corresponding to the truth table as

$$\text{Output} = \overline{A}\,\overline{B}\,\overline{C}\,\overline{D} + \overline{A}\,B\,\overline{C}\,\overline{D} + A\,\overline{B}\,\overline{C}\,\overline{D}. \tag{5.8}$$

Example 5.14 will show that this sum-of-products expression is equivalent to the logic expression shown in the caption of Table 5.7.

---

## 5.5 SIMPLIFYING LOGIC EXPRESSIONS

We are often interested in simplifying logic expressions to minimize their implementation cost. For example, although the expression $A + B\overline{B} + C$ appears to require three gates,[6] simplification of the logic expression will result in a single gate. Notice that the expression $B\overline{B}$ always results in the answer 0 (a variable and its complement can never be TRUE at the same time). Furthermore, observe that $A + 0$ is always $A$. From these observations, we can simplify the expression as

$$A + B\overline{B} + C = A + 0 + C = A + C.$$

The reader can also verify that the expressions $A + B\overline{B} + C$ and $A + C$ are equivalent by developing the corresponding truth tables as illustrated in Table 5.9.

The following primitive rules come in handy for simplifying logic expressions:

$$A \cdot \overline{A} = 0 \tag{5.9}$$

$$A \cdot A = A \tag{5.10}$$

---

6. Assuming that $A$, $B$, $\overline{B}$, and $C$ are available as inputs.

$$A \cdot 0 = 0 \tag{5.11}$$

$$A \cdot 1 = A \tag{5.12}$$

$$A + \overline{A} = 1 \tag{5.13}$$

$$A + A = A \tag{5.14}$$

$$A + 0 = A \tag{5.15}$$

$$A + 1 = 1 \tag{5.16}$$

$$A + \overline{A}B = A + B \tag{5.17}$$

$$A(B + C) = AB + AC \tag{5.18}$$

$$AB = BA \tag{5.19}$$

$$A + B = B + A \tag{5.20}$$

$$(AB)C = A(BC) \tag{5.21}$$

$$(A + B) + C = A + (B + C) \tag{5.22}$$

You can verify these rules by comparing their truth tables. For example, the truth table comparing $A + \overline{A}B$ and $A + B$ is shown in Table 5.10.

The following are another set of useful equalities called De Morgan's laws:

$$\overline{A \cdot B} = \overline{A} + \overline{B} \tag{5.23}$$

$$\overline{A + B} = \overline{A} \cdot \overline{B}. \tag{5.24}$$

| $A$ | $B$ | $C$ | $A + B\overline{B} + C$ | $A + C$ |
|---|---|---|---|---|
| 0 | 0 | 0 | 0 | 0 |
| 0 | 0 | 1 | 1 | 1 |
| 0 | 1 | 0 | 0 | 0 |
| 0 | 1 | 1 | 1 | 1 |
| 1 | 0 | 0 | 1 | 1 |
| 1 | 0 | 1 | 1 | 1 |
| 1 | 1 | 0 | 1 | 1 |
| 1 | 1 | 1 | 1 | 1 |

**TABLE 5.9** Truth table for comparing the two expressions $A + B\overline{B} + C$ and $A + C$.

**TABLE 5.10** Truth table for comparing the two expressions $A + \overline{A}B$ and $A + B$. Notice that for convenience we have added an extra column for the intermediate expression $\overline{A}B$.

| $A$ | $B$ | $\overline{A}B$ | $A + \overline{A}B$ | $A + B$ |
|---|---|---|---|---|
| 0 | 0 | 0 | 0 | 0 |
| 0 | 1 | 1 | 1 | 1 |
| 1 | 0 | 0 | 1 | 1 |
| 1 | 1 | 0 | 1 | 1 |

De Morgan's laws can also be verified by developing truth tables as illustrated in Table 5.11. Notice that the columns for $\overline{A \cdot B}$ and $\overline{A} + \overline{B}$ are identical. Similarly, observe that the columns for $\overline{A + B}$ and $\overline{A} \cdot \overline{B}$ are identical, thereby verifying De Morgan's laws.

**TABLE 5.11** Truth table for verifying De Morgan's laws.

| $A$ | $B$ | $\overline{A}$ | $\overline{B}$ | $A \cdot B$ | $A + B$ | $\overline{A \cdot B}$ | $\overline{A + B}$ | $\overline{A} \cdot \overline{B}$ | $\overline{A} + \overline{B}$ |
|---|---|---|---|---|---|---|---|---|---|
| 0 | 0 | 1 | 1 | 0 | 0 | 1 | 1 | 1 | 1 |
| 0 | 1 | 1 | 0 | 0 | 1 | 1 | 0 | 0 | 1 |
| 1 | 0 | 0 | 1 | 0 | 1 | 1 | 0 | 0 | 1 |
| 1 | 1 | 0 | 0 | 1 | 1 | 0 | 0 | 0 | 0 |

De Morgan's laws can be expressed in terms of the gate notation as depicted in Figure 5.21. Consequently, the symbols on the right-hand side of the figure are often used in place of the corresponding *NAND* gate or the *NOR* gate.

These rules can be used to simplify logic expressions to reduce the number of gates required to implement them. For example, a direct implementation of the logic expression $AB\overline{B} + BC + \overline{C}$ appears to take five 2-input gates as seen from Figure 5.22. The implementation used in Figure 5.22 assumes that both TRUE and complement forms of each variable are available as inputs. In other words, for each variable $X$, we assume that both $X$ and $\overline{X}$ are available as inputs. Otherwise, we would need two additional inverters.

To reduce the number of gates required for its implementation, we can simplify the expression $AB\overline{B} + BC + \overline{C}$ as follows: We first collect terms as

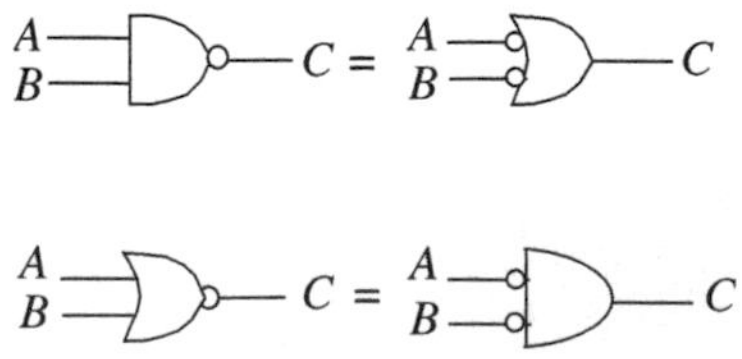

**FIGURE 5.21** Gate equivalences implied by De Morgan's laws.

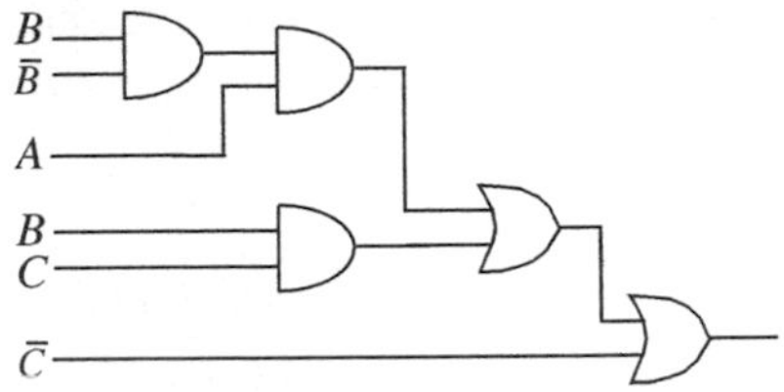

FIGURE 5.22 Direct implementation of $AB\overline{B} + BC + \overline{C}$.

shown below by applying the rule suggested by Equation 5.21:

$$AB\overline{B} + BC + \overline{C} = A(B\overline{B}) + BC + \overline{C}.$$

Then, we apply the simplification suggested by Equation 5.9 and obtain

$$A(B\overline{B}) + BC + \overline{C} = A0 + BC + \overline{C}.$$

Applying Equation 5.11 we get

$$A0 + BC + \overline{C} = 0 + BC + \overline{C}.$$

Grouping terms as suggested by Equation 5.22 we get

$$0 + BC + \overline{C} = (0 + BC) + \overline{C}.$$

Applying Equations 5.20 and 5.15 we get

$$(0 + BC) + \overline{C} = BC + \overline{C}.$$

Finally, applying Equation 5.17 after recognizing that both the AND and the OR operators are commutative (from Equations 5.19 and 5.20) we obtain the final simplified form

$$BC + \overline{C} = B + \overline{C}.$$

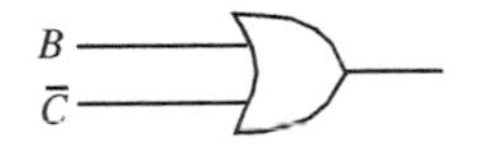

FIGURE 5.23 Implementation of $B + \overline{C}$, which results from simplifying $AB\overline{B} + BC + \overline{C}$.

The implementation of $B + \overline{C}$ takes just one gate and is shown in Figure 5.23. You might wish to work through some input values and verify that the circuits in Figures 5.22 and 5.23 are equivalent.

The preceding rules can also be used to simplify logic expressions into a standard or canonic form. A standard form of representation makes it easy to compare the costs of competing implementations. One canonic form that we have seen previously in Section 5.4 is the sum-of-products form. Recall that logic expressions in this form are represented using two levels of operations as a set of product (*AND*) terms combined using the *OR* function. For example, the expression $AB + C + D$ is in a sum-of-products representation. The expression $\overline{AB + C + D}$, however, is not. We can convert the latter expression into the sum-of-products form $\overline{A}\,\overline{C}\,\overline{D} + \overline{B}\,\overline{C}\,\overline{D}$ using the equivalence rules as follows:

$$\overline{AB + C + D} = \overline{(AB) + (C + D)} \tag{5.25}$$

$$= \overline{(AB)}\;\overline{(C + D)} \tag{5.26}$$

$$= (\overline{A} + \overline{B})(\overline{C}\,\overline{D}) \tag{5.27}$$

$$= (\overline{A})(\overline{C}\,\overline{D}) + (\overline{B})(\overline{C}\,\overline{D}) \tag{5.28}$$

$$= \overline{A}\,\overline{C}\,\overline{D} + \overline{B}\,\overline{C}\,\overline{D}. \tag{5.29}$$

We can also use the equalities to simplify expressions into their respective minimal forms. A commonly used form is called the *minimum sum-of-products form*. For example, $A + A + \overline{A}C + D$ is a valid sum-of-products representation. Since $A + A = A$, and $A + \overline{A}C = A + C$, its minimum sum-of-products form is simply $A + C + D$.

It turns out that the expression in our previous example $\overline{A}\,\overline{C}\,\overline{D} + \overline{B}\,\overline{C}\,\overline{D}$ is also the minimum sum-of-products representation.

---

EXAMPLE 5.12 MINIMUM SUM-OF-PRODUCTS FORM Find the minimum sum-of-products representation for the boolean function $A + \overline{\overline{A}C} + B$.

We first write the sum-of-products representation:

$$A + \overline{\overline{A}C} + B = A + (\overline{\overline{A}} + \overline{C}) + B$$
$$= A + (A + \overline{C}) + B$$
$$= A + A + \overline{C} + B$$
$$= A + \overline{C} + B.$$

Here, $A + A + \overline{C} + B$ is in a sum-of-products form. The minimum sum-of-products form, however, is $A + \overline{C} + B$.

---

EXAMPLE 5.13 SIMPLIFYING A LOGIC EXPRESSION Find the minimum sum-of-products representation for the boolean expression in Equation 5.7, namely

$$Z = \overline{X}Y + X\overline{Y} + XY.$$

The following sequence of simplifications show that this expression for $Z$ is equivalent to $X + Y$:

$$\begin{aligned} Z &= \overline{X}Y + X\overline{Y} + XY \\ &= \overline{X}Y + X(\overline{Y} + Y) \\ &= \overline{X}Y + X \cdot 1 \\ &= \overline{X}Y + X \\ &= Y + X. \end{aligned}$$

---

WWW EXAMPLE 5.14 SIMPLIFYING ANOTHER LOGIC EXPRESSION

---

EXAMPLE 5.15 IMPLEMENTATION USING NORS It turns out that certain types of gates take up less room or are easier to build in certain technologies than other types of gates. We can make use of the equivalence rules to convert a circuit from one form to another. Let us derive an implementation of the *AND* function based on two-input *NOR* gates. In other words, we wish to transform the expression $Z = A \cdot B$ into one that uses only *NOR* operators. The following steps show how we can transform the *AND* expression into one that uses three *NOR* operations:

$$A \cdot B = (A + A) \cdot (B + B) \tag{5.32}$$

$$= \overline{\overline{(A + A) \cdot (B + B)}} \tag{5.33}$$

$$= \overline{\overline{(A + A)} + \overline{(B + B)}}. \tag{5.34}$$

---

WWW EXAMPLE 5.16 YET ANOTHER IMPLEMENTATION USING NORS

---

## 5.6 NUMBER REPRESENTATION

As discussed earlier, the binary representation restricts a signal to either a high or low value. These two values can be used to represent two numbers: for example, 0 and 1. How do we represent other numbers? We briefly overview one alternative.[7] Just as a single decimal digit can represent one of ten

---

7. Number representation is a lengthy topic in itself, and the interested reader is referred to *Computation Structures*, by Ward and Halstead.

values (0, 1, 2, ..., 9), a single binary digit (termed a *bit*) represents one of two values (0, 1).

Bigger numbers are constructed by concatenating multiple digits. The multiple-digit decimal number *ijk* formed by concatenating the decimal digits *i*, *j*, and *k*, has the value

$$i \times 10^2 + j \times 10^1 + k \times 10^0.$$

Similarly, the multiple-digit binary number *lmn* formed by concatenating the binary digits *l*, *m*, and *n*, has the value

$$l \times 2^2 + m \times 2^1 + n \times 2^0.$$

In general, the value of the binary number $A_n A_{n-1} \ldots A_2 A_1 A_0$ is given by

$$\sum_{i=0}^{i=n} A_i 2^i. \tag{5.38}$$

Thus the binary number 10 corresponds to the decimal number 2, the binary number 11 corresponds to the decimal number 3, and the binary number 101 corresponds to the decimal number 5. To distinguish the binary number 10 from the decimal number 10, we denote the binary number as 0b10 when there is a possibility of confusion.

---

EXAMPLE 5.17 BINARY NUMBER REPRESENTATION What is the value of the binary number 1110?

The value of the binary number 1110 is given by Equation 5.38 as

$$1 \times 2^3 + 1 \times 2^2 + 1 \times 2^1 + 0 \times 2^0$$

which is 14 (decimal).

---

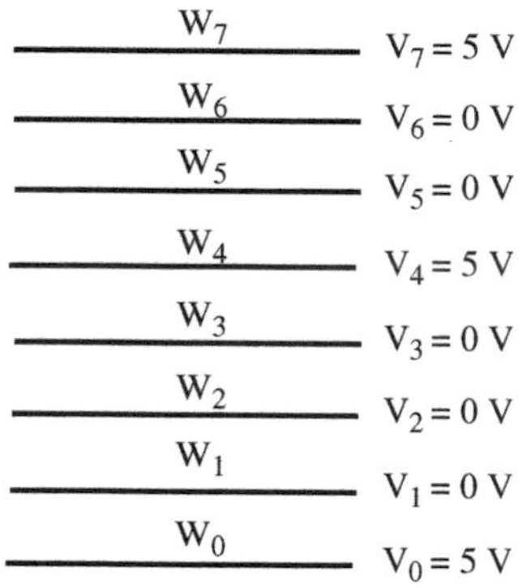

| Wire | Voltage |
|---|---|
| $W_7$ | $V_7 = 5$ V |
| $W_6$ | $V_6 = 0$ V |
| $W_5$ | $V_5 = 0$ V |
| $W_4$ | $V_4 = 5$ V |
| $W_3$ | $V_3 = 0$ V |
| $W_2$ | $V_2 = 0$ V |
| $W_1$ | $V_1 = 0$ V |
| $W_0$ | $V_0 = 5$ V |

FIGURE 5.26 Number representation.

How do we represent negative numbers? One simple alternative is to interpret the leading bit as a sign bit: a 0 denotes a positive number and a 1 denotes a negative number. Therefore, the number 110 represents −2, and the number 010 represents 2. When the interpretation of the leading bit (sign bit or value bit) is not clear from the context, to avoid confusion, it is important to indicate the numbering system being assumed when specifying a binary number.

---

EXAMPLE 5.18 NEGATIVE BINARY NUMBER REPRESENTATION Consider a bundle of 8 wires named $W_0$ through $W_7$ as shown in Figure 5.26. Let the value of the voltage on the wire $W_i$ be termed $V_i$. Let us use a voltage level of 0 V to denote a logical 0 and a voltage level of 5 V to denote a logical 1. Also, let us use the leading bit (value on $W_7$) to represent the sign of the number. What is the decimal representation of the number encoded in the wires?

Let the logic value on wire $W_i$ be $A_i$. Then, $A_7$ is the sign bit, and $A_6A_5A_4A_3A_2A_1A_0$ is the binary number. The decimal value of the number is given by the formula:

$$(-1)^{A_7} \sum_{i=0}^{i=6} A_i 2^i.$$

We are given, $V_7 = 5$ V, $V_6 = 0$ V, $V_5 = 0$ V, $V_4 = 5$ V, $V_3 = 0$ V, $V_2 = 0$ V, $V_1 = 0$ V, and $V_0 = 5$ V. Therefore, $A_7 = 1$, $A_6 = 0$, $A_5 = 0$, $A_4 = 1$, $A_3 = 0$, $A_2 = 0$, $A_1 = 0$, and $A_0 = 1$. In other words, the sign bit is 1, and the binary number is $A_6A_5A_4A_3A_2A_1A_0 = 0010001$. Thus, the corresponding decimal number is $-17$.

Operations on binary numbers can be performed in a manner analogous to operations on decimal numbers. To illustrate the correspondence, Figure 5.27a shows the addition of a pair of decimal numbers 26 and 87, and Figure 5.27b depicts the addition of a pair of positive binary numbers 11 and 11. In both the decimal and binary case, observe that the addition of the digits in a column generates a sum digit and a carry digit into the next higher digit column.[8] Observe further that adding a pair of two-digit numbers can sometimes result in a three-digit sum.

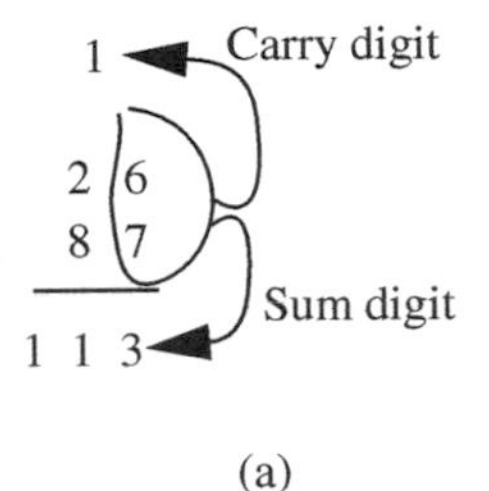


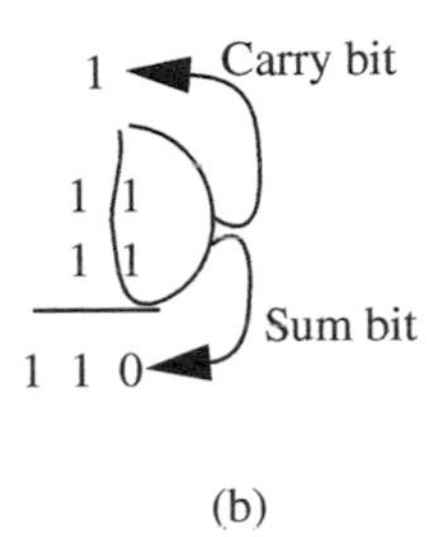


FIGURE 5.27 Addition of a pair of two-digit numbers.

---

EXAMPLE 5.19 ADDING A PAIR OF TWO-BIT POSITIVE INTEGERS Suppose we wish to add a pair of two-bit positive numbers $A : A_1A_0$ and $B : B_1B_0$. We will implement a two-bit adder using two techniques. The first method will write the truth table for the entire operation and implement it directly. The second method will first implement a one-bit adder using the truth table method, and then use the one-bit adder circuit to compose a two-bit adder. Let us denote the answer as $S : S_2S_1S_0$.

*First method* We first write the truth table for the two-bit adder as shown in Table 5.12. From the truth table, we obtain the sum-of-products representation for

8. Although a binary digit is called a *bit*, we use the term *digit* here since we are referring to both the decimal digit and the binary digit.

| $A_1$ | $A_0$ | $B_1$ | $B_0$ | $S_2$ | $S_1$ | $S_0$ |
|---|---|---|---|---|---|---|
| 0 | 0 | 0 | 0 | 0 | 0 | 0 |
| 0 | 0 | 0 | 1 | 0 | 0 | 1 |
| 0 | 0 | 1 | 0 | 0 | 1 | 0 |
| 0 | 0 | 1 | 1 | 0 | 1 | 1 |
| 0 | 1 | 0 | 0 | 0 | 0 | 1 |
| 0 | 1 | 0 | 1 | 0 | 1 | 0 |
| 0 | 1 | 1 | 0 | 0 | 1 | 1 |
| 0 | 1 | 1 | 1 | 1 | 0 | 0 |
| 1 | 0 | 0 | 0 | 0 | 1 | 0 |
| 1 | 0 | 0 | 1 | 0 | 1 | 1 |
| 1 | 0 | 1 | 0 | 1 | 0 | 0 |
| 1 | 0 | 1 | 1 | 1 | 0 | 1 |
| 1 | 1 | 0 | 0 | 0 | 1 | 1 |
| 1 | 1 | 0 | 1 | 1 | 0 | 0 |
| 1 | 1 | 1 | 0 | 1 | 0 | 1 |
| 1 | 1 | 1 | 1 | 1 | 1 | 0 |

**TABLE 5.12** Truth table for the two-bit adder.

each of $S_0$, $S_1$, and $S_2$ as follows:

$$\begin{aligned} S_0 = {} & \bar{A}_1\bar{A}_0\bar{B}_1B_0 + \bar{A}_1\bar{A}_0B_1B_0 \\ & + \bar{A}_1A_0\bar{B}_1\bar{B}_0 + \bar{A}_1A_0B_1\bar{B}_0 \\ & + A_1\bar{A}_0\bar{B}_1B_0 + A_1\bar{A}_0B_1B_0 \\ & + A_1A_0\bar{B}_1\bar{B}_0 + A_1A_0B_1\bar{B}_0 \qquad (5.39) \\ = {} & \bar{A}_0B_0 + A_0\bar{B}_0 \qquad (5.40) \end{aligned}$$

$$\begin{aligned} S_1 = {} & \bar{A}_1\bar{A}_0B_1\bar{B}_0 + \bar{A}_1\bar{A}_0B_1B_0 \\ & + \bar{A}_1A_0\bar{B}_1B_0 + \bar{A}_1A_0B_1\bar{B}_0 \\ & + A_1\bar{A}_0\bar{B}_1\bar{B}_0 + A_1\bar{A}_0\bar{B}_1B_0 \\ & + A_1A_0\bar{B}_1\bar{B}_0 + A_1A_0B_1B_0 \qquad (5.41) \end{aligned}$$

$$
\begin{aligned}
&= A_1A_0B_1B_0 + A_1\bar{B}_1\bar{B}_0 \\
&\quad + A_1\bar{A}_0\bar{B}_1 + \bar{A}_1B_1\bar{B}_0 \\
&\quad + \bar{A}_1A_0\bar{B}_1B_0 + \bar{A}_1\bar{A}_0B_1 \qquad (5.42)
\end{aligned}
$$

$$
\begin{aligned}
S_2 &= \bar{A}_1A_0B_1B_0 + A_1\bar{A}_0B_1\bar{B}_0 \\
&\quad + A_1\bar{A}_0B_1B_0 + A_1A_0\bar{B}_1B_0 \\
&\quad + A_1A_0B_1\bar{B}_0 + A_1A_0B_1B_0 \qquad (5.43) \\
&= A_1B_1 + A_1A_0B_0 + A_0B_1B_0 \qquad (5.44)
\end{aligned}
$$

Figure 5.28 displays a gate-level implementation.

*Second method* This implementation develops a two-bit adder circuit by composing two one-bit adders, and illustrates an important engineering technique called *divide-and-conquer*. The one-bit adders are called *full adders*. As illustrated in Figure 5.29, a full adder takes three inputs — two one-bit numbers to be added ($A_i$ and $B_i$), and one carry bit $C_i$ from a lower digit. The full adder produces two outputs: a sum bit $S_i$ and a carry bit $C_{i+1}$ to a higher digit.

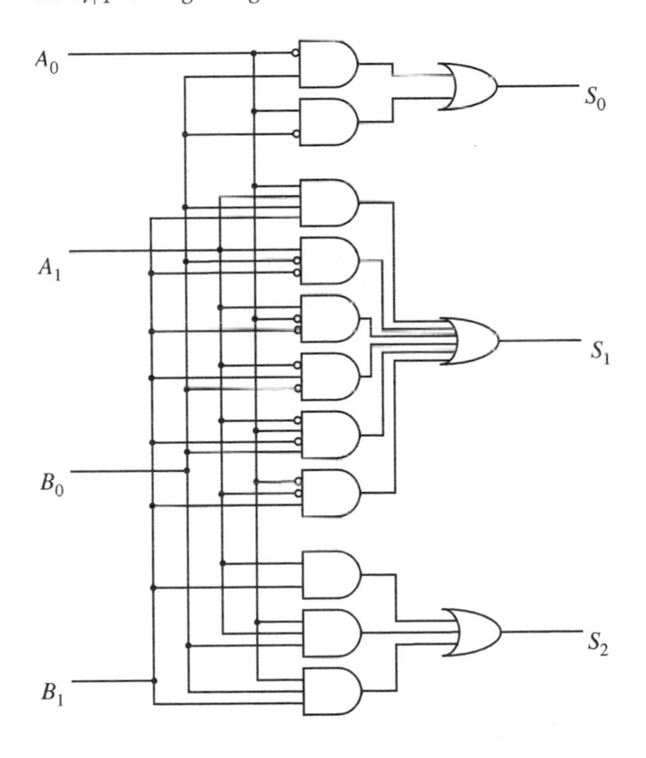


FIGURE 5.28 Direct implementation of two-bit adder.

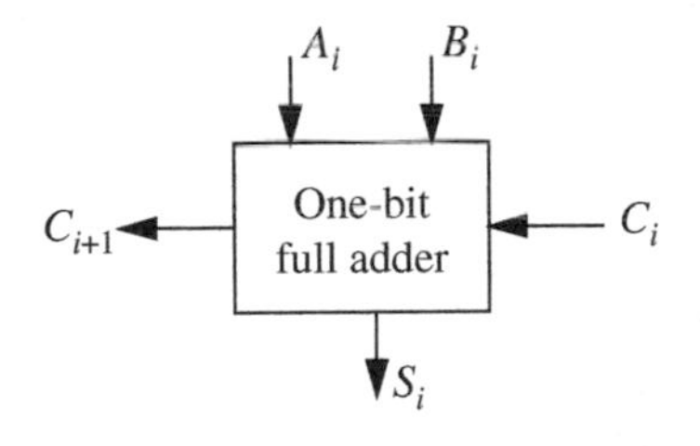


FIGURE 5.29 Straightforward two-bit adder implementation.

TABLE 5.13 Truth table for the one-bit full adder.

| $A_i$ | $B_i$ | $C_i$ | $C_{i+1}$ | $S_i$ |
|---|---|---|---|---|
| 0 | 0 | 0 | 0 | 0 |
| 0 | 0 | 1 | 0 | 1 |
| 0 | 1 | 0 | 0 | 1 |
| 0 | 1 | 1 | 1 | 0 |
| 1 | 0 | 0 | 0 | 1 |
| 1 | 0 | 1 | 1 | 0 |
| 1 | 1 | 0 | 1 | 0 |
| 1 | 1 | 1 | 1 | 1 |

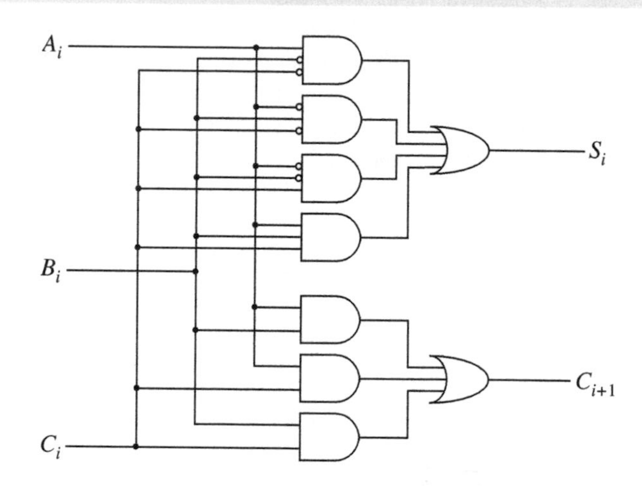


FIGURE 5.30 One-bit full adder implementation.

The truth table for the one-bit full adder is depicted in Table 5.13.

From the table, we derive the logic expression for the sum bit $S_i$ and the carry bit $C_{i+1}$ as follows:

$$S_i = \bar{A}_i\bar{B}_iC_i + \bar{A}_iB_i\bar{C}_i + A_i\bar{B}_i\bar{C}_i + A_iB_iC_i \tag{5.45}$$

$$C_{i+1} = \bar{A}_iB_iC_i + A_i\bar{B}_iC_i + A_iB_i\bar{C}_i + A_iB_iC_i. \tag{5.46}$$

Figure 5.30 shows a gate-level full adder circuit based on the logic expressions for $S_i$ and $C_{i+1}$.

We can create a two bit adder out of a pair of one-bit full adders by feeding the carry-out bit ($C_{i+1}$) of one adder to the carry-in bit ($C_i$) of another adder. The carry-in bit of the

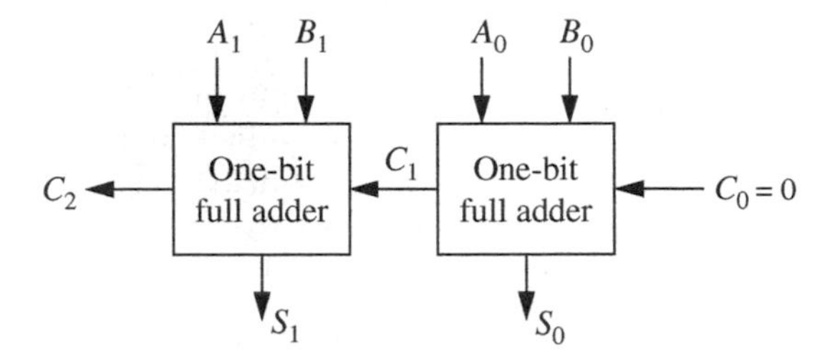


FIGURE 5.31 A two-bit ripple-carry adder using two one-bit full adders.

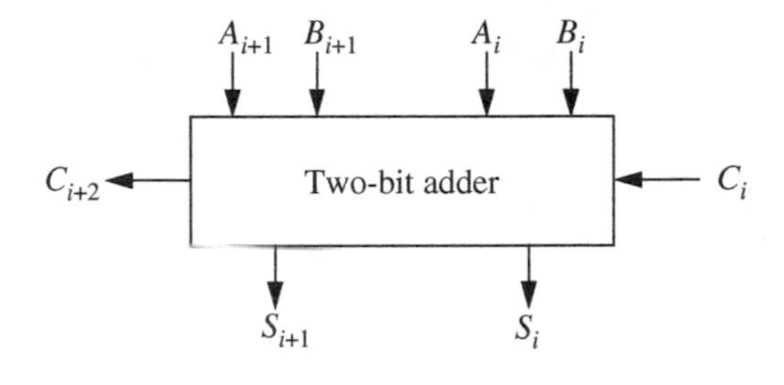


FIGURE 5.32 A two-bit adder block.

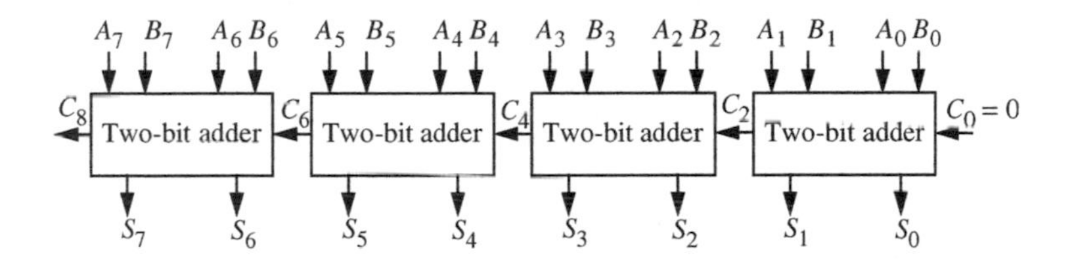


FIGURE 5.33 An eight-bit adder circuit.

low-digit adder is set to 0. The two bit-adder is shown in Figure 5.31. Because the carry bit ripples through the adders, this type of adder circuit is called a *ripple-carry adder*. Similarly, we can use $n$ one-bit full adders to construct an $n$-bit adder.[9]

---

EXAMPLE 5.20 BUILDING AN EIGHT-BIT ADDER Let us now build an adder that can add two eight-bit integers using the two-bit adder circuits from Figure 5.31 as the building blocks. First, for convenience, let us abstract the circuit from Figure 5.31 into an adder block for a pair of two-bit integers as shown in Figure 5.32.

Much as we ganged together the one-bit adder blocks, we can cascade together the two-bit adder blocks to form the eight-bit adder as shown in Figure 5.33.

---

9. As an exercise, you are encouraged to construct a one-bit full adder using two half adders. A half adder takes two bits $A_i$ and $B_i$ as inputs and produces as its output a sum bit $S_i$ and a carry-out bit $C_{i+1}$.

## 5.7 SUMMARY

- This chapter introduced the digital abstraction, which is based on the notion of lumping signal values into two levels — high and low. Digital circuits are designed to be more immune to noise than their analog counterparts. The degree of noise immunity of a digital circuit is governed by the voltage thresholds of the static discipline to which the circuit adheres.
- The static discipline requires digital devices to adhere to a common representation for their input and output voltages, and to guarantee that they interpret correctly inputs that are valid logical signals according to the common representation, and to produce outputs that are valid logical signals provided they receive valid logical inputs. By adhering to a common representation, digital devices based on different technologies or built by different manufacturers can communicate with each other. The common representation is specified in terms of four voltage thresholds:

  $V_{OH}$ The lowest output voltage value that a digital device can produce when it outputs a logical 1.

  $V_{OL}$ The highest output voltage value that a digital device can produce when it outputs a logical 0.

  $V_{IH}$ The lowest input voltage value that a digital device must recognize as a logical 1.

  $V_{IL}$ The highest input voltage value that a digital device must recognize as a logical 0.

- The voltage thresholds associated with a static discipline determine the noise margins. The 0 noise margin is given by

$$\mathrm{NM}_0 = V_{IL} - V_{OL}$$

  and the 1 noise margin is given by

$$\mathrm{NM}_1 = V_{OH} - V_{IH}.$$

- We also discussed several representations of digital logic including truth tables, which are a tabular representation; boolean expressions, which are akin to algebraic expressions; and combinational gates, which are a graphical circuit representation.

## EXERCISES

EXERCISE 5.1 Write a boolean expression for the following statement: "$Z$ is TRUE if either $X$ or $Y$ is FALSE, otherwise $Z$ is FALSE." Write a truth table for this expression.

EXERCISE 5.2 Write a boolean expression for the following statement: "$Z$ is FALSE if either $X$ or $Y$ is FALSE, otherwise $Z$ is TRUE." Write a truth table for this expression.

EXERCISE 5.3 Write a boolean expression for the following statement: "$Z$ is TRUE if no more than two of $W$, $X$, and $Y$ are TRUE, otherwise $Z$ is FALSE."

EXERCISE 5.4 Consider the statement: "$Z$ is TRUE if at least two of $W$, $X$, and $Y$ are TRUE, otherwise $Z$ is FALSE."

a) Write a boolean expression for this statement.

b) Write a truth table for the function $Z$.

c) Implement $Z$ using only AND, OR, and NOT gates. The inputs $W$, $X$, and $Y$ are available. Each gate may have an arbitrary number of inputs. (Hint: A sum-of-products representation of the boolean expression will facilitate this implementation.)

d) Implement $Z$ using only AND, OR, and NOT gates. Each gate may have no more than two inputs. As before, the inputs $W$, $X$, and $Y$ are available.

e) Implement $Z$ using only NAND and NOR gates. (Hint: a NAND gate or a NOR gate with its inputs tied together behaves like an inverter.)

f) Implement $Z$ using only NAND gates. (Hint: Use De Morgan's laws.)

g) Implement $Z$ using only NOR gates. (Hint: Use De Morgan's laws.)

h) Repeat part (d) and attempt to minimize the number of gates used.

i) Repeat part (d) and attempt to minimize the number of gates used, assuming that the inputs are available both in their true and complement forms. In other words, assume that in addition to $W$, $X$, and $Y$, the inputs $\overline{W}$, $\overline{X}$, and $\overline{Y}$, are also available.

EXERCISE 5.5 Represent the decimal number 4 as an unsigned, three-bit binary number and as an unsigned, four-bit binary number. Unsigned numbers do not include a sign bit. For example, 11110 is the unsigned, binary representation of the decimal number 30.

EXERCISE 5.6 Consider the functions $F(A, B, C)$ and $G(A, B, C)$ specified in the truth table given in Table 5.14.

a) Write a logic expression corresponding to the functions $F(A, B, C)$ and $G(A, B, C)$.

b) Implement $F(A, B, C)$ with logic gates.

**TABLE 5.14** Truth table for Exercise 5.6

| $A$ | $B$ | $C$ | $F(A, B, C)$ | $G(A, B, C)$ |
|---|---|---|---|---|
| 0 | 0 | 0 | 1 | 0 |
| 0 | 0 | 1 | 0 | 0 |
| 0 | 1 | 0 | 0 | 0 |
| 0 | 1 | 1 | 0 | 1 |
| 1 | 0 | 0 | 1 | 0 |
| 1 | 0 | 1 | 1 | 1 |
| 1 | 1 | 0 | 0 | 1 |
| 1 | 1 | 1 | 1 | 1 |

c) Implement $F(A, B, C)$ using only two-input gates.

d) Implement $F(A, B, C)$ using only two-input NAND gates. (Hint: Use De Morgan's laws.)

e) Repeat parts (b) through (d) for the function $G(A, B, C)$.

EXERCISE 5.7 Consider the following four logic expressions:

$$(A+\overline{B})(\overline{A}\cdot\overline{B}+C)+\overline{C\cdot D}$$

$$(A\cdot\overline{C}+\overline{B\cdot D})\overline{(D+\overline{B}+A}$$

$$A+\overline{\overline{B}\cdot D}+A\cdot C\cdot\overline{D}$$

$$\overline{(\overline{(A+\overline{C})}+B+\overline{D})+A\cdot\overline{C}\cdot D}$$

a) Give an implementation using gates for each of the four logic expressions.

b) Write the truth table for each of the four expressions.

c) Suppose you know that $A = 0$. Simplify the four expressions under this constraint.

d) Simplify the four expressions assuming that $A$ and $B$ are related as $A = \overline{B}$.

EXERCISE 5.8 A logic gate obeys a static discipline with the following voltage levels: $V_{IH} = 3.5$ V, $V_{OH} = 4.3$ V, $V_{IL} = 1.5$ V, and $V_{OL} = 0.9$ V. (a) What range of voltages will be treated as invalid under this discipline? (b) What are its noise margins?

EXERCISE 5.9 Consider a family of logic gates that operates under the static discipline with the following voltage thresholds: $V_{IL} = 1.5$ V, $V_{OL} = 0.5$ V, $V_{IH} = 3.5$ V, and $V_{OH} = 4.4$ V.

a) Graph an input-output voltage transfer function of a buffer satisfying the four voltage thresholds.

b) Graph an input-output voltage transfer function of an inverter satisfying the four voltage thresholds.

c) What is the highest voltage that can be output by an inverter for a logical 0 output?

d) What is the lowest voltage that can be output by an inverter for a logical 1 output?

e) What is the highest voltage that must be interpreted by a receiver as a logical 0?

f) What is the lowest voltage that must be interpreted by a receiver as a logical 1?

g) Does this choice of voltage thresholds offer any immunity to noise? If so, determine the noise margins.

EXERCISE 5.10 Consider a family of logic gates that operates under the static discipline with the following voltage thresholds: $V_{IL} = V_{OL} = 0.5$ V and $V_{IH} = V_{OH} = 4.4$ V.

a) Graph an input-output voltage transfer function of a buffer satisfying the two voltage thresholds.

b) Graph an input-output voltage transfer function of an inverter satisfying the two voltage thresholds.

c) What is the highest voltage that can be output by an inverter for a logical 0 output?

d) What is the lowest voltage that can be output by an inverter for a logical 1 output?

e) What is the highest voltage that must be interpreted by a receiver as a logical 0?

f) What is the lowest voltage that must be interpreted by a receiver as a logical 1?

g) Does this choice of voltage thresholds offer any immunity to noise?

# PROBLEMS

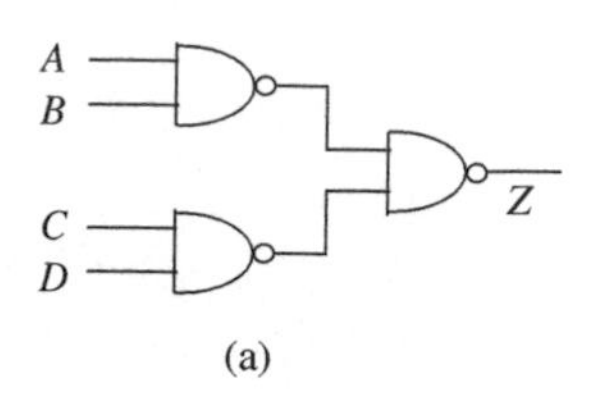

(a)

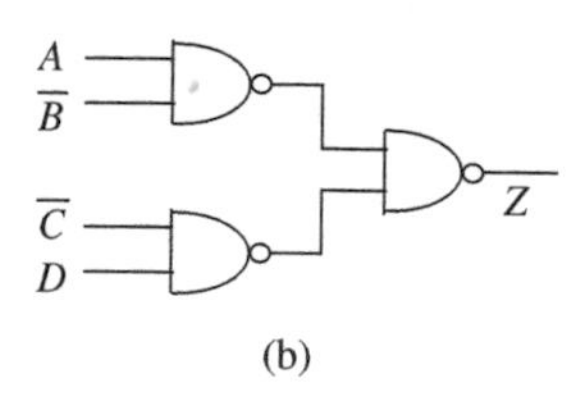

(b)

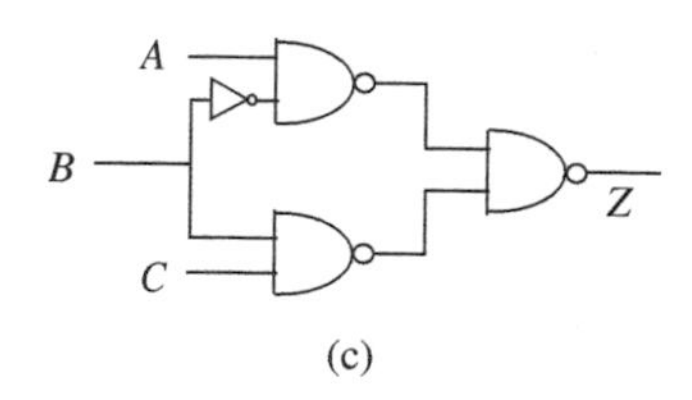

(c)

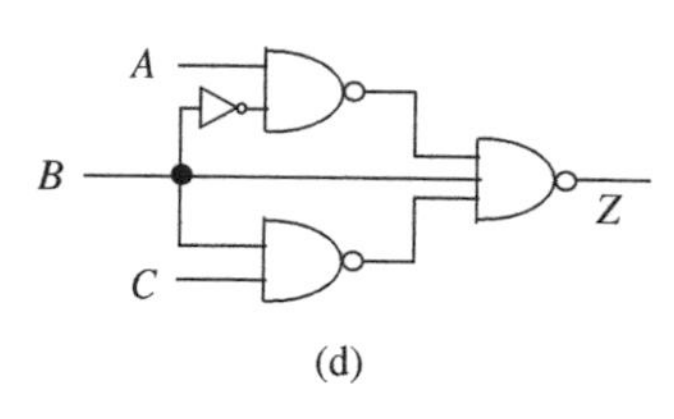

(d)

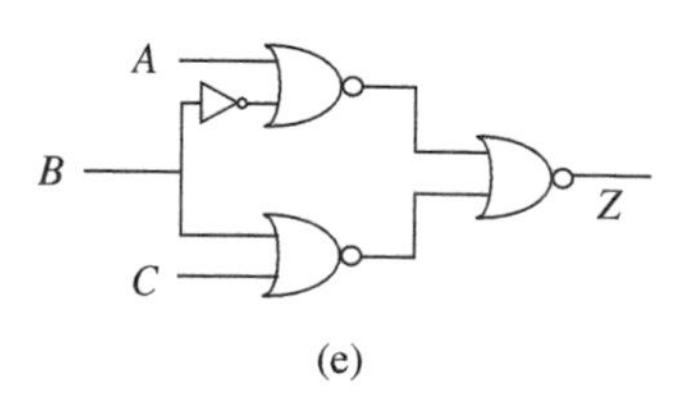

(e)

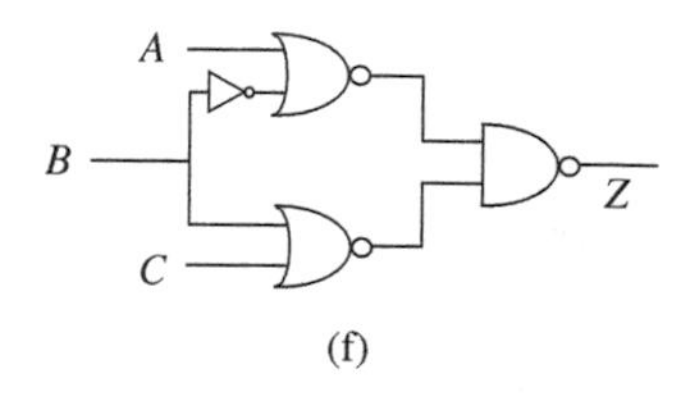

(f)

FIGURE 5.34

PROBLEM 5.1 Derive a truth table and a boolean expression that describes the operation of each digital circuit shown in Figure 5.34.

PROBLEM 5.2 Draw an output voltage waveform for the circuit in Figure 5.34c in response to the input voltage waveforms shown in Figure 5.35. Assume that the gates in the circuit obey the static discipline with $V_{OH} = 4$ V, $V_{IH} = 3$ V, $V_{OL} = 1$ V, and $V_{IL} = 2$ V.

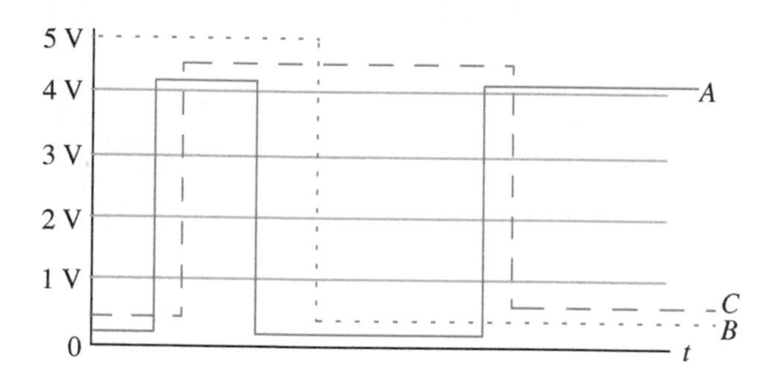


FIGURE 5.35

PROBLEM 5.3 The truth table for a "ones count" circuit is given in Table 5.15. This circuit has four inputs: *A*, *B*, *C*, and *D*, and three outputs $OUT_0$, $OUT_1$, and $OUT_2$. Together, the signals $OUT_0$, $OUT_1$, and $OUT_2$ represent a three-bit positive integer $OUT_2 OUT_1 OUT_0$. The output integer $OUT_2 OUT_1 OUT_0$ reflects the number of ones in the input. Using only NAND, NOR, and NOT gates, design an implementation for the circuit. Each gate may have an arbitrary number of inputs.

PROBLEM 5.4 A four-input multiplexer module is shown in Figure 5.36. The multiplexer has two select signals $S_1$ and $S_0$. The value on the select signals determines which of the inputs A, B, C, and D appears at the output. As illustrated in the figure, A is selected if $S_1 S_0$ is 00, B if $S_1 S_0$ is 01, C if $S_1 S_0$ is 10, and D if $S_1 S_0$ is 11. Write a boolean expression for Z in terms of $S_1 S_0$, A, B, C, and D. Implement the multiplexer using only NAND gates.

PROBLEM 5.5 A four-input demultiplexer module is shown in Figure 5.37. The demultiplexer has two select signals, $S_1$ and $S_0$. The select signals determine on which of the outputs (OUT0, OUT1, OUT2, or OUT3) the input IN appears. As illustrated in the figure, IN appears at output OUT0 if $S_1 S_0$ is 00, at OUT1 if $S_1 S_0$ is 01, at OUT2 if $S_1 S_0$ is 10, and at OUT3 if $S_1 S_0$ is 11. An output is 0 if it is not selected. Write a boolean expression for each of the outputs in terms of $S_1 S_0$ and IN. Implement the demultiplexer using only NAND gates.

PROBLEM 5.6 Implement the "greater-than" circuit depicted in Figure 5.38 using NAND gates. A and B represent one-bit positive integers. The output Z is 1 if A is greater than B, otherwise Z is 0.

| $A$ | $B$ | $C$ | $D$ | $OUT_2$ | $OUT_1$ | $OUT_0$ |
|---|---|---|---|---|---|---|
| 0 | 0 | 0 | 0 | 0 | 0 | 0 |
| 0 | 0 | 0 | 1 | 0 | 0 | 1 |
| 0 | 0 | 1 | 0 | 0 | 0 | 1 |
| 0 | 0 | 1 | 1 | 0 | 1 | 0 |
| 0 | 1 | 0 | 0 | 0 | 0 | 1 |
| 0 | 1 | 0 | 1 | 0 | 1 | 0 |
| 0 | 1 | 1 | 0 | 0 | 1 | 0 |
| 0 | 1 | 1 | 1 | 0 | 1 | 1 |
| 1 | 0 | 0 | 0 | 0 | 0 | 1 |
| 1 | 0 | 0 | 1 | 0 | 1 | 0 |
| 1 | 0 | 1 | 0 | 0 | 1 | 0 |
| 1 | 0 | 1 | 1 | 0 | 1 | 1 |
| 1 | 1 | 0 | 0 | 0 | 1 | 0 |
| 1 | 1 | 0 | 1 | 0 | 1 | 1 |
| 1 | 1 | 1 | 0 | 0 | 1 | 1 |
| 1 | 1 | 1 | 1 | 1 | 0 | 0 |

**TABLE 5.15** Truth table for a "ones count" circuit.

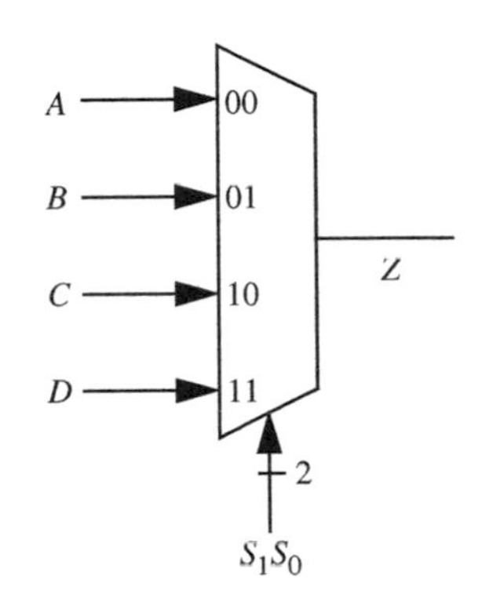


**FIGURE 5.36** A four-input multiplexer module. The "2" beside the wire corresponding to the select signals is a short-hand notation indicating there are two wires present.

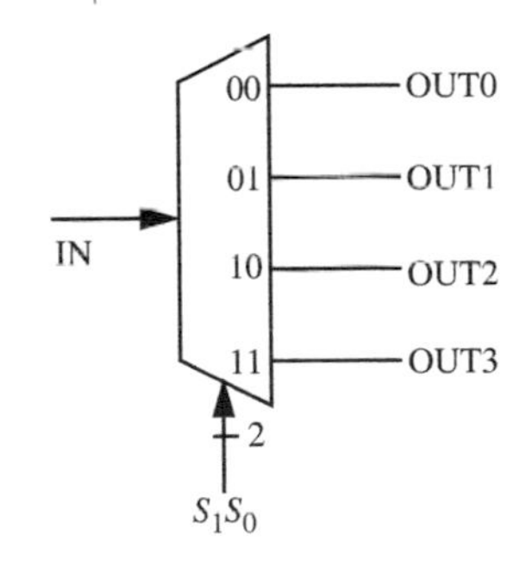


**FIGURE 5.37**

PROBLEM 5.7 Implement the four-input "odd" or "odd parity" circuit depicted in Figure 5.39 using NOR gates. In this circuit, the output Z is high if an odd number of the inputs are high, otherwise the output Z is low. How would you use the four-input "odd" circuit module shown in Figure 5.39 to implement a three-input "odd" circuit? If this cannot be done, discuss why not.

PROBLEM 5.8 Figure 5.40 depicts a four-input majority circuit module. The output Z of this circuit module is high if a majority of the inputs are high. Write a boolean expression for Z in terms of A0, A1, A2, and A3. How would you use the four-input majority circuit module shown in Figure 5.40 to implement a three-input majority circuit and a two-input majority circuit? If either of these cannot be done, discuss why not.

PROBLEM 5.9 Figure 5.41 illustrates a two-bit grey code converter. Its outputs OUT0, and OUT1, are equal to the inputs when the IN0, IN1 are 00 or 01. However, when the inputs IN0, IN1 are 10 and 11, the outputs OUT0 and OUT1 are 11 and 10, respectively. Implement the grey code converter using two-input NAND gates.

PROBLEM 5.10 Figure 5.42 illustrates input-output voltage transfer functions for several one-input one-output devices. For the voltage thresholds $V_{OL}$, $V_{IL}$, $V_{OH}$, and $V_{IH}$ as shown, which of the devices can serve as valid inverters?

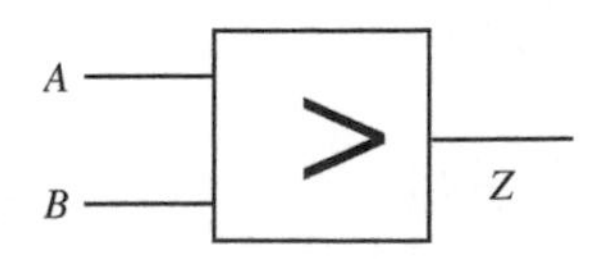

FIGURE 5.38

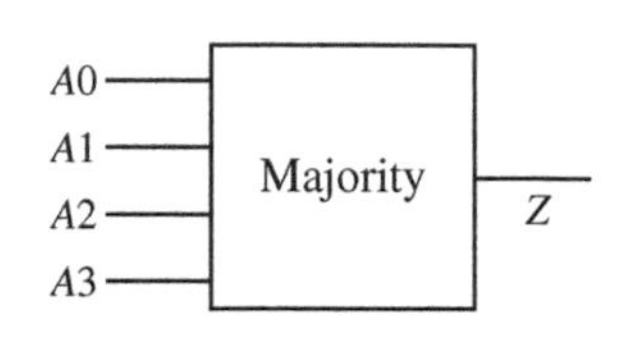

FIGURE 5.39

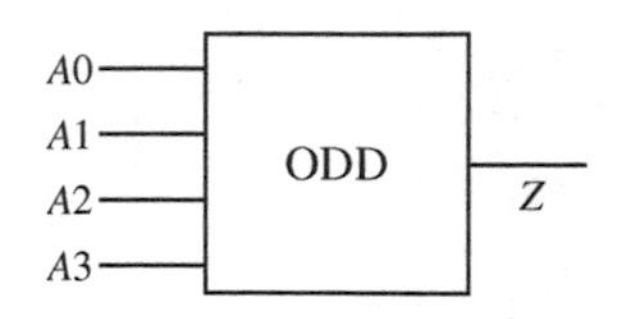

FIGURE 5.40

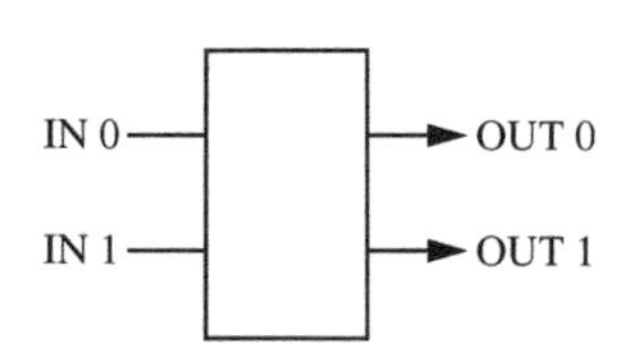

FIGURE 5.41

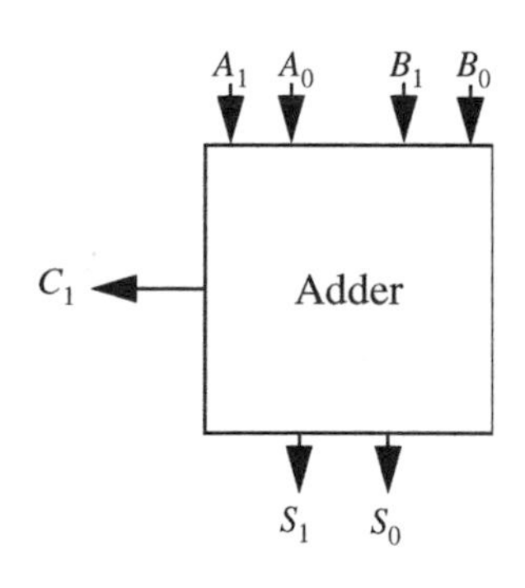

FIGURE 5.43

PROBLEM 5.11 Suppose we wish to build a two-bit adder circuit (see Figure 5.43) that takes as input a pair of two-bit positive integers $A_1A_0$ and $B_1B_0$ and produces a two-bit sum output $S_1S_0$ and a carry-out bit $C_1$. Write a truth table and a boolean expression for the carry-out bit in terms of the inputs.

Now, suppose we wish to build a two-bit adder circuit (see Figure 5.44) that takes as input a pair of two-bit positive integers $A_1A_0$ and $B_1B_0$, and a carry-in bit $C_0$, and produces a two-bit sum output $S_1S_0$ and a carry-out bit $C_1$. Write a truth table and a boolean expression for the carry-out bit in terms of the inputs.

PROBLEM 5.12 Suppose we have two logic families named NTL and YTL. The NTL family of logic gates operates under the static discipline with the following voltage thresholds: $V_{IL} = 1.5$ V, $V_{OL} = 1.0$ V, $V_{IH} = 3.5$ V, and $V_{OH} = 4$ V. The YTL family, on the other hand, is characterized by the voltage thresholds: $V_{IL} = 0.8$ V, $V_{OL} = 0.3$ V, $V_{IH} = 3.0$ V, and $V_{OH} = 4.5$ V. Will a YTL inverter driving the input of

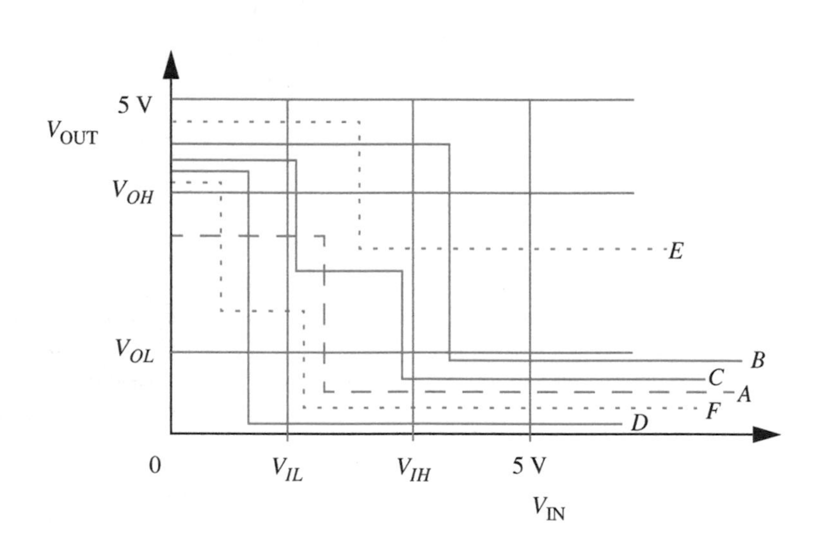

FIGURE 5.42

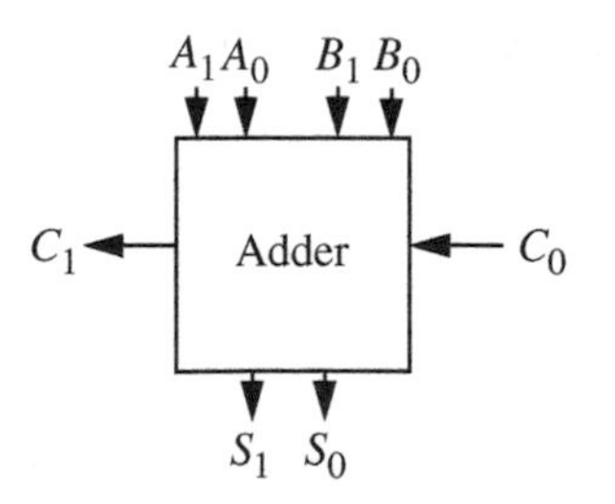

FIGURE 5.44

an NTL inverter operate correctly? Explain. Will an NTL inverter driving the input of an YTL inverter operate correctly? Explain.

PROBLEM 5.13 Consider a family of logic gates that operates under the static discipline with the following voltage thresholds: $V_{OL} = 0.5$ V, $V_{IL} = 1.6$ V, $V_{OH} = 4.4$ V, and $V_{IH} = 3.2$ V.

a) Graph an input-output voltage transfer function of a buffer satisfying the four voltage thresholds.

b) Graph an input-output voltage transfer function of an inverter satisfying the four voltage thresholds.

c) What is the highest voltage that can be output by an inverter for a logical 0 output?

d) What is the lowest voltage that can be output by an inverter for a logical 1 output?

e) What is the highest voltage that must be interpreted by a receiver as a logical 0?

f) What is the lowest voltage that must be interpreted by a receiver as a logical 1?

g) When transmitting information over a noisy wire, buffers can be used to minimize transmission errors by restoring signal values. Consider the transmission of data over a noisy wire that picks up a maximum of 80 mV symmetric peak-to-peak noise per centimeter. How many buffers are needed to transmit a signal over a distance of 2 meters in this noisy environment?

h) How large are the 0 and 1 noise margins for a buffer in this logic family? Now consider three buffers connected in series and behaving as a single buffer. What are the noise margins for this new buffer?

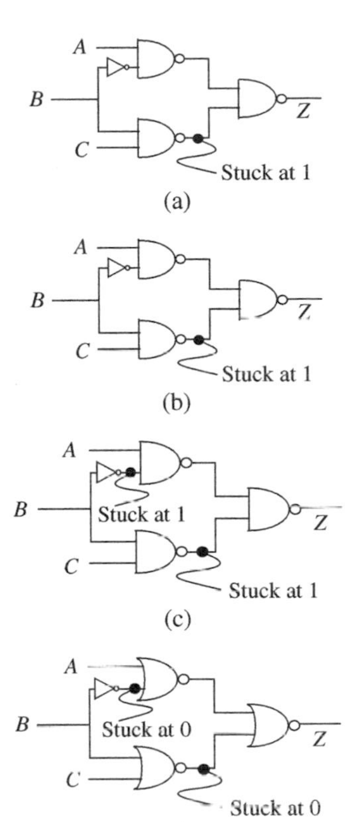


FIGURE 5.45

PROBLEM 5.14 Many manufacturing flaws in digital circuits can be modeled as *stuck-at faults*. The output of a gate is said to suffer from a *stuck-at-1* fault if the output is a 1 irrespective of its input values. Similarly, a *stuck-at-0 fault* at an output causes the output to produce a 0 at all times.

a) Consider the circuits shown in Figure 5.45 with one or more faults. Write an expression for each of the outputs in terms of the input variables for the given faults. (Hint: As an example, the output of the faulty circuit in Figure 5.45a will be independent of the input variable C).

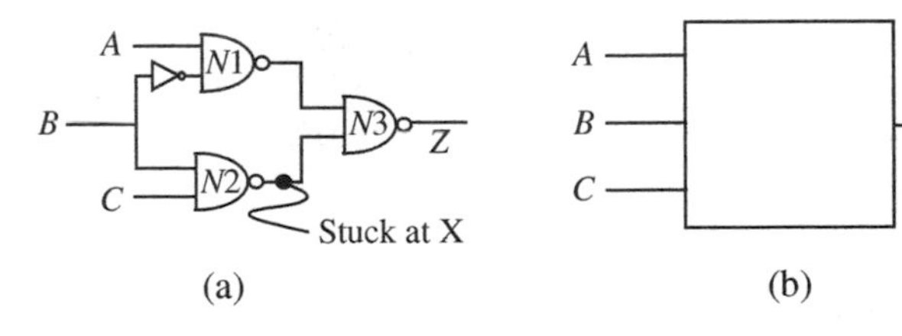


FIGURE 5.46

b) Suppose we are given the faulty circuit in Figure 5.46a where the output of NAND gate N2 is known to have a stuck-at fault. However, we do not know whether it is a stuck-at-1 fault or a stuck-at-0 fault. Further, as illustrated in Figure 5.46b, suppose that we have access only to the inputs A, B, and C, and the output Z. In other words, we are unable to directly observe the output X of the faulty NAND gate N2. How would you go about determining whether N2 suffers from a stuck-at-1 fault or a stuck-at-0 fault?

# CHAPTER 6

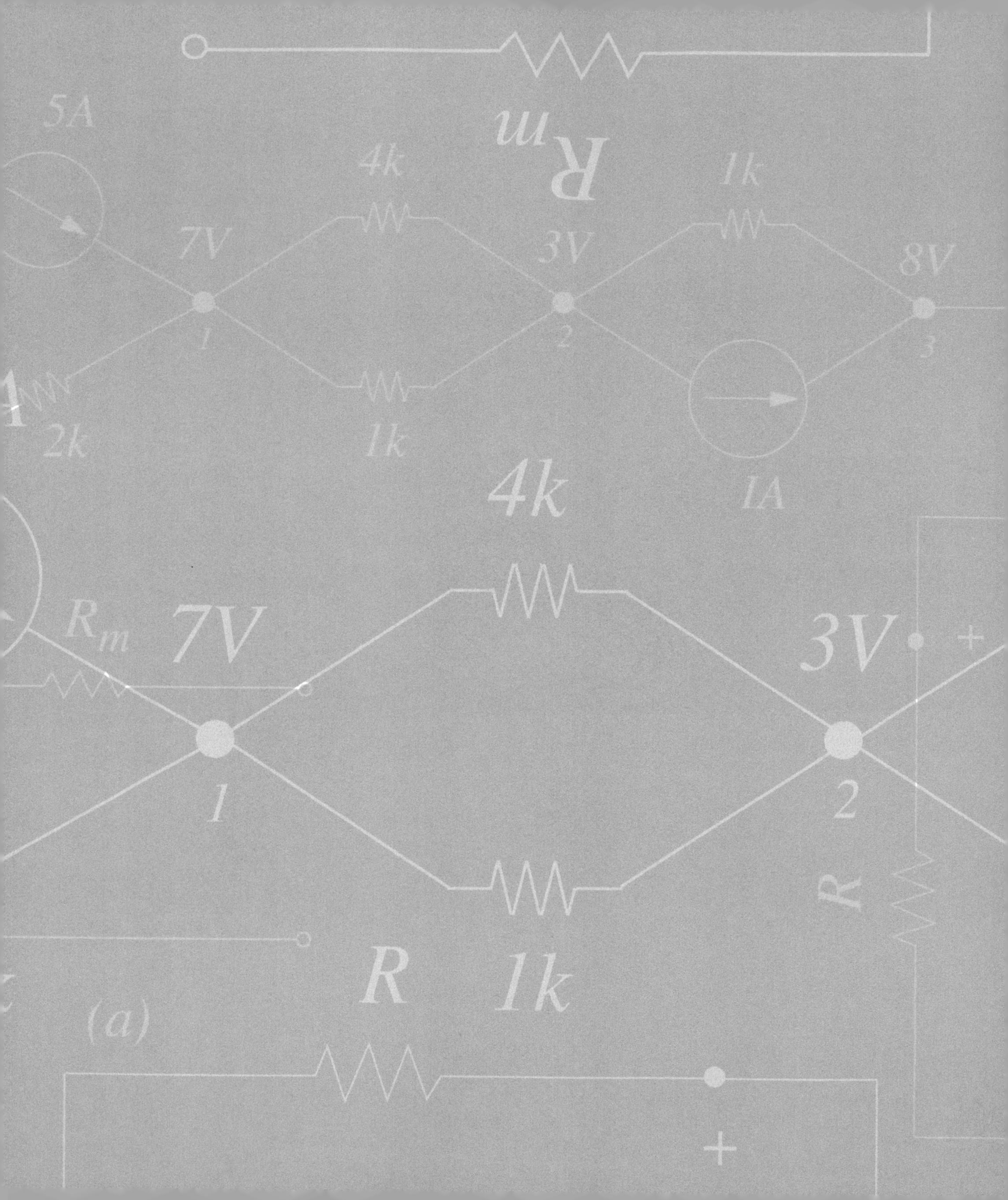
5A
4k
1k
7V
3V
8V
1
2
3
2k
1k
1A
4k
$R_m$
7V
3V
+
1
2
R
R
1k
(a)
+

# THE MOSFET SWITCH

# 6

This chapter introduces the switch circuit element and demonstrates how digital logic gates can be constructed using switches and other primitive circuit elements we have seen previously. This chapter also discusses a common implementation of the switch in VLSI technology using a device called a MOSFET (Metal Oxide Semiconductor Field-Effect Transistor).

## 6.1 THE SWITCH

Recall the electrical system and its lumped circuit model shown in Figure 1.4. As is commonly done in household electrical circuits, let us add a switch in the current path to turn the bulb on and off, as shown in Figure 6.1a. Figure 6.1b shows the corresponding lumped circuit model.

The switch is normally off and behaves like an open circuit. When pressure is applied to the switch, it closes and behaves like a wire and conducts current. Accordingly, the switch can be modeled as the three-terminal device shown in Figure 6.2. The three terminals include a control terminal, an input terminal, and an output terminal. The input and output terminals of a switch commonly exhibit symmetric properties. When the control terminal has a TRUE or a logical 1 signal on it, the input is connected to the output through a short circuit, and the switch is said to be in its ON state. Otherwise, there is an open

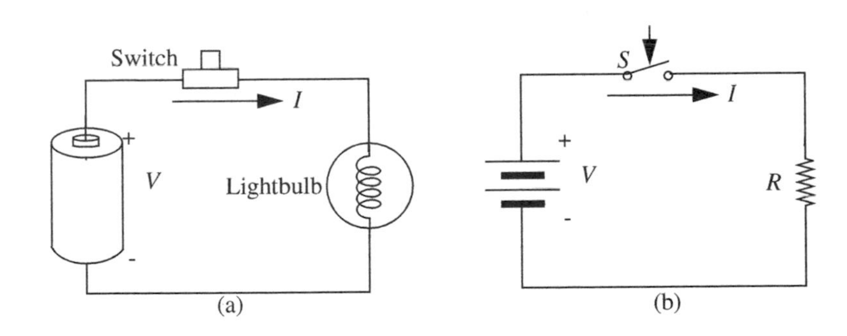


FIGURE 6.1 (a) The lightbulb circuit with a switch; (b) the lumped circuit representation.

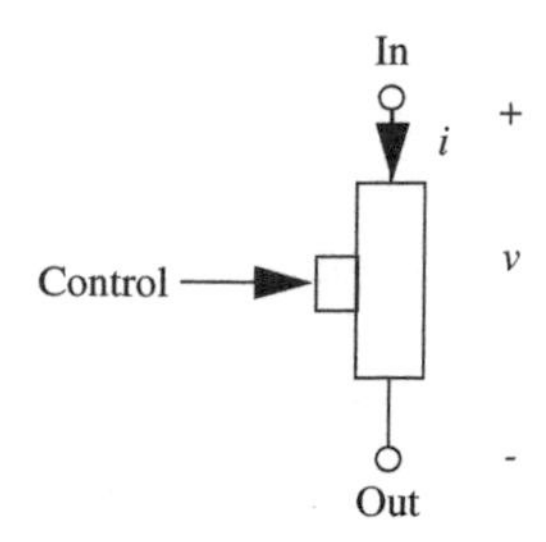


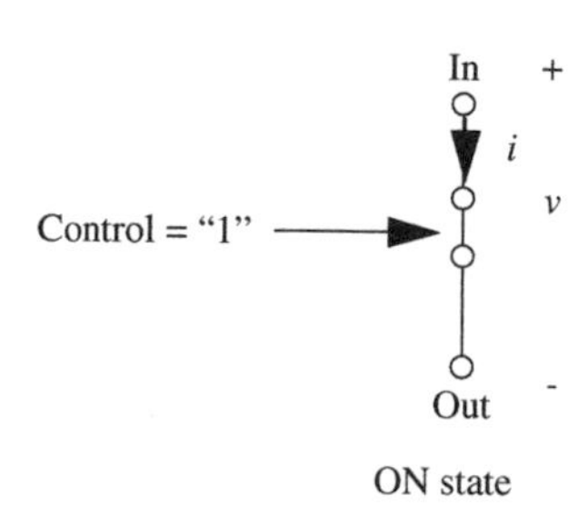


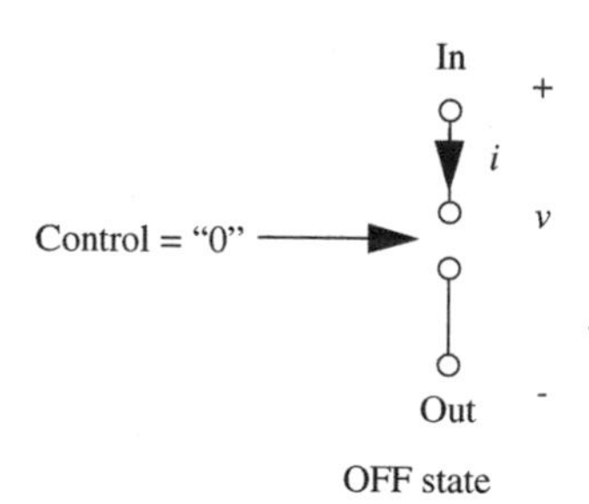


FIGURE 6.2 Three-terminal switch model.

circuit between the input and the output, and the switch is said to be in its OFF state.[1]

The $v$–$i$ curve for a switch is shown in Figure 6.3. Thus far, we drew the $v$–$i$ curves for two-terminal devices by plotting the relationship between the voltage and current for the two terminals. Likewise, for our three-terminal switch, we can draw the $v$–$i$ characteristics at the input-output terminal pair. The effect of the control terminal can be taken into account by drawing a different $v$–$i$ curve for each value at the control terminal. Thus, as illustrated in Figure 6.3, when the control input is a logical 0, the $v$–$i$ curve for the input-output terminal pair indicates that the current through the switch is 0, irrespective of the voltage applied. Conversely, the switch behaves like a short circuit between its input and output terminals when the control input is a logical 1. When behaving like a short circuit, the voltage across the input and output terminals is zero, and the current is unconstrained by the switch (rather, it is determined by constraints that are external to the switch).

The $v$–$i$ characteristics of a switch can also be expressed in algebraic form as:

$$\text{for Control} = \text{“0,”} \quad i = 0$$

and

$$\text{for Control} = \text{“1,”} \quad v = 0. \tag{6.1}$$

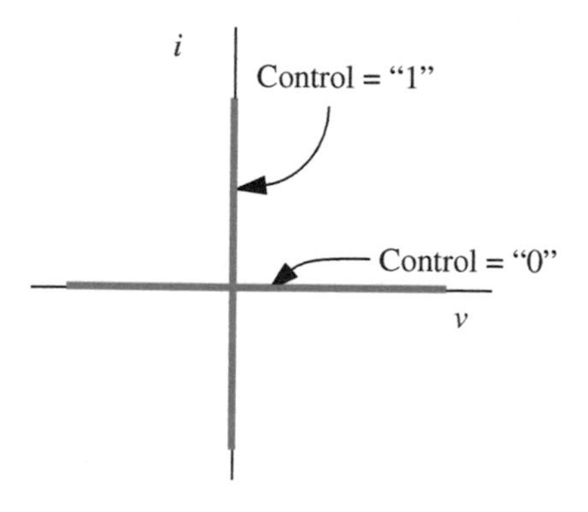


FIGURE 6.3 $v$–$i$ characteristics of a switch. $v$ is the voltage across the input and the output terminals of the switch and $i$ is the current through the same pair of terminals.

---

1. To build intuition, our switch example uses mechanical force to apply a logical 1 at its control input. However, there are other types of switches that work with electrical signals at their control terminals, and offer the same properties at their input and output terminals. We shall see an example of such a switch in Section 6.3.

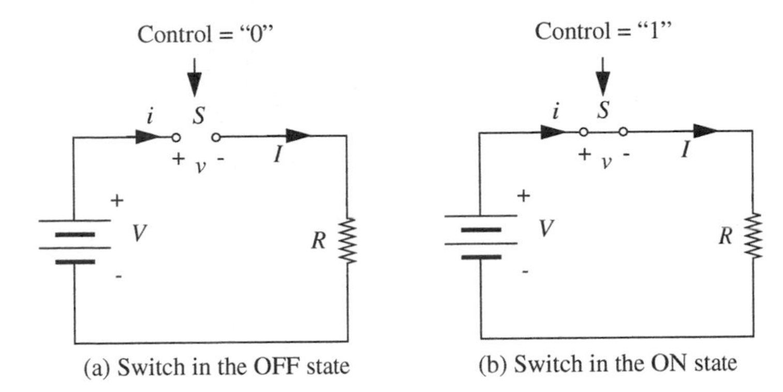


FIGURE 6.4 (a) Linear subcircuit formed when the switch is in the OFF state; (b) linear subcircuit formed when the switch is in the ON state.

Although the switch is a nonlinear device, circuits containing a switch and other linear devices can be analyzed by considering two linear subcircuits: one when the switch is in its ON state and one for the switch in its OFF state. Thus, standard linear techniques can be applied to each subcircuit. Figure 6.4 illustrates the two subcircuits for our lightbulb circuit example. Analyzing Figure 6.4a, it is easy to see that the current $I$ is zero when the switch is OFF. Similarly, analyzing Figure 6.4b, the current is given by $I = V/R$ when the switch is in the ON state.

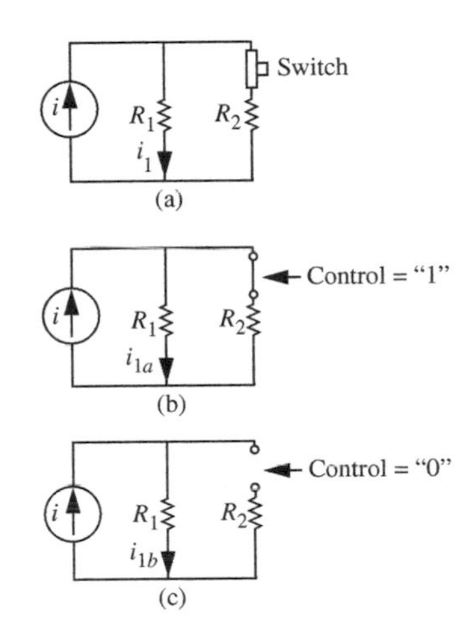


FIGURE 6.5 (a) Circuit containing a current source, two resistors, and a switch; (b) linear subcircuit formed when the switch is in the ON state; (c) linear subcircuit formed when the switch is in the OFF state.

---

EXAMPLE 6.1 CIRCUIT CONTAINING A SWITCH Determine the current through resistor $R_1$ in the circuit shown in Figure 6.5a.

The circuit in Figure 6.5a is nonlinear because it contains a switch. Since the only nonlinear element in the circuit is the switch, we can analyze this circuit by considering the two linear subcircuits formed for each of the two states of the switch.

Figure 6.5b shows the linear subcircuit formed when the switch is in the ON state. We can obtain the desired current $i_{1a}$ for the ON-state circuit by using the current divider relation from Equation 2.84. The current divider relation states that when two resistors are connected in parallel, the current through one of the resistors is equal to the total current through the two resistors multiplied by a factor, which is made up of the opposite resistance divided by the sum of the two resistances. Accordingly, (when the switch is in the ON state):

$$i_{1a} = i\frac{R_2}{R_1 + R_2}.$$

Figure 6.5c shows the linear subcircuit formed when the switch is in the OFF state. In this case, the entire current from the current source flows through the resistor $R_1$. Thus, (when the switch is in the OFF state):

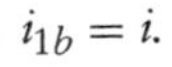

$$i_{1b} = i.$$

---

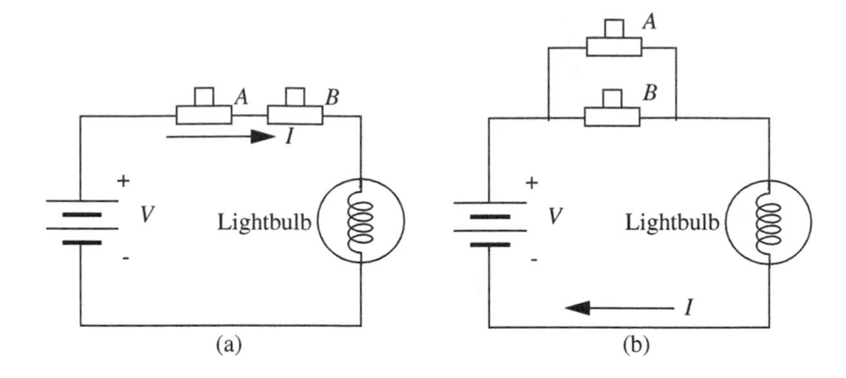


**FIGURE 6.6** (a) The lightbulb circuit with switch in an *AND* configuration; (b) the lightbulb circuit with switches in an *OR* configuration.

## 6.2 LOGIC FUNCTIONS USING SWITCHES

Next, consider the lightbulb circuit with a pair of switches connected in series as depicted in Figure 6.6a. The lightbulb can be turned on only by closing both the switches *A and B*. Similarly, Figure 6.6b shows a circuit with a pair of switches connected in an *OR* configuration. In the latter configuration, the bulb can be turned on by closing either switch *A or* switch *B*.

These circuits provide us with the insight into implementing logic functions using switches: Series connected switches implement the AND function and parallel connected switches implement the OR function. Switches can be combined in AND-OR configurations to implement more complicated functions. As shown previously, the switches implement a form of digital logic called *steering logic*. In this form, switches steer values (for example, a high voltage) along various paths. As we continue our discussion, we shall also see how switches can be used to build our familiar combinational logic gates.

One of the unappealing features of the mechanical switches in Figure 6.1 was that they responded only to mechanical pressure at their control terminal. The need for mechanical pressure would make it unacceptably hard to build logic circuits because the voltage at the output of a given switch would have to somehow be converted to a mechanical pressure to influence another switch. Preferably, a three-terminal switch device that responded to voltages would enable the construction of switching circuits using voltages alone. The MOSFET is one such device that can be implemented cheaply in VLSI technology.

## 6.3 THE MOSFET DEVICE AND ITS S MODEL

The MOSFET belongs to a class of devices called *transistors*. The MOSFET is a three-terminal device with a control terminal, an input terminal, and an output terminal (see Figure 6.7). We will discuss its physical structure in Section 6.7.

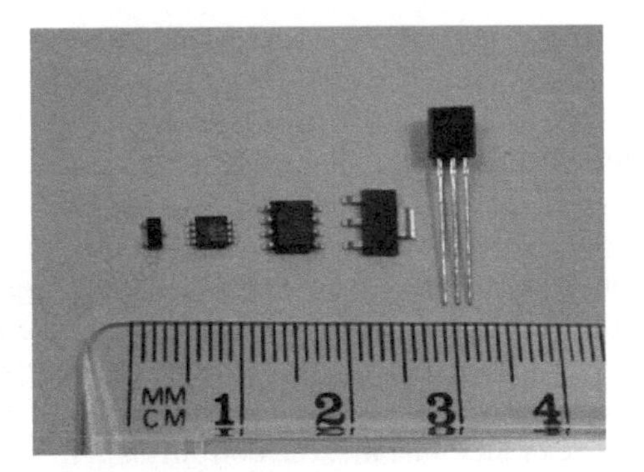


**FIGURE 6.7** Discrete MOSFETs. The rightmost device with three leads contains a single MOSFET, while the middle package contains multiple MOSFETs. (Photograph Courtesy of Maxim Integrated Products.)

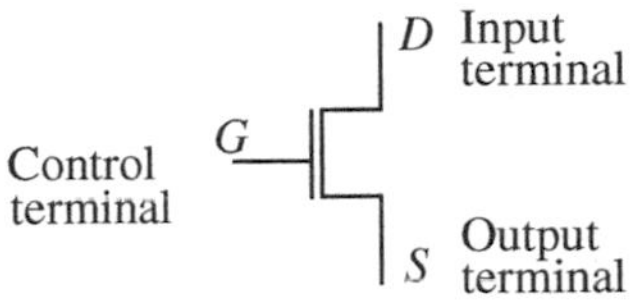

Its circuit symbol is shown in Figure 6.8. As shown in Figure 6.8, the control terminal of the MOSFET is called its *gate* G, the input terminal its *drain* D, and the output terminal its *source* S. For our purposes, we can treat the source and drain in a symmetric fashion. The name assignment is related to the direction of current flow. The terminals are labeled such that current flows from the drain to the source.[2] Equivalently, the channel terminal with the higher voltage is labeled as the drain.

D Input terminal
G Control terminal
S Output terminal

**FIGURE 6.8** The MOSFET circuit symbol.

As depicted in Figure 6.9, let the voltage across the gate and source of the MOSFET be $v_{GS}$, and the voltage across the drain and the source be $v_{DS}$. The current through G terminal is called $i_G$ and that through the D terminal is termed $i_{DS}$.

A simple circuit model for a specific type of MOSFET device called the n-channel MOSFET is depicted in Figure 6.10.[3] This model based on the simple switch is called the MOSFET's Switch Model, or S Model for short.[4] The device is in the ON state when $v_{GS}$ crosses a threshold voltage $V_T$, otherwise it is off. A typical value for $V_T$ for n-channel MOSFETs is 0.7 volts, but it can be varied by the manufacturing process.[5] In the ON state, the S model approximates the connection between the drain and the source as a short circuit. In practice, there is some nonzero resistance between the drain and the source,

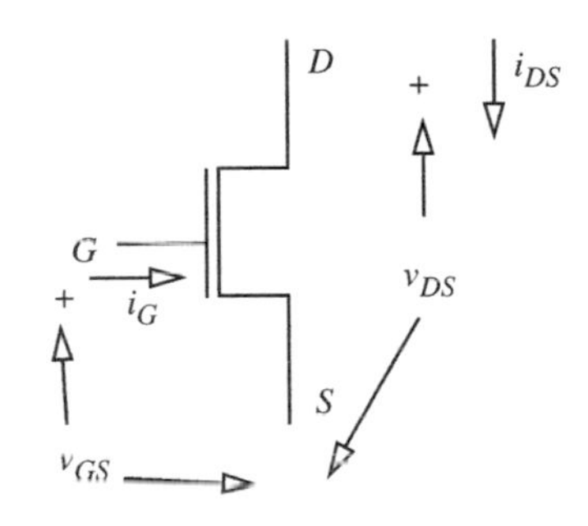


**FIGURE 6.9** Definitions of $v_{GS}$, $v_{DS}$, and $i_{DS}$.

2. You are is probably wondering why the nomenclature of the drain and the source seems reversed from the more logical choice relating the source to the input and the drain to the output. It turns out that the names stem from the internal conduction properties of the MOSFET. Electrons are the majority carriers in the *n*-channel MOSFET shown. *S* is the source of electrons and *D* is the drain of electrons.

3. We will see a complementary MOS transistor called the p-channel transistor later.

4. We will examine increasingly sophisticated models for the MOSFET device in the following chapters. These models will reflect aspects of the MOSFET's behavior that are not adequately captured by our simple switch model.

5. To simplify the math in our quantatitive examples on MOSFETs, this book commonly uses 1-volt thresholds.

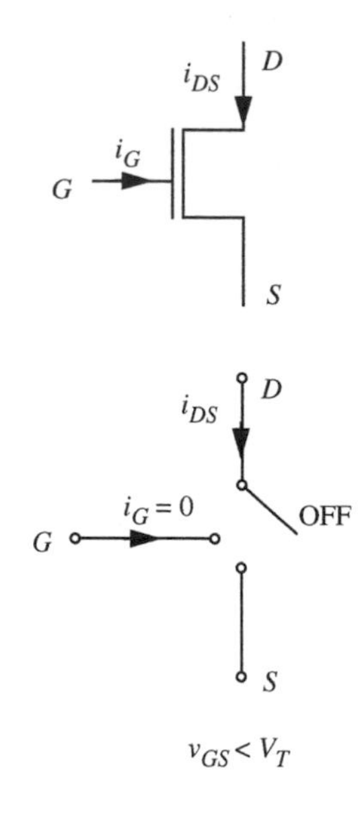


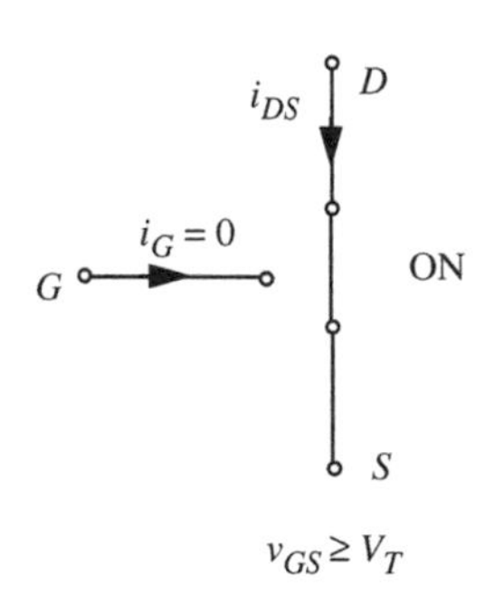


**FIGURE 6.10** The S model of the MOSFET.

but we ignore this resistance in the S model. Section 6.6 will discuss the switch-resistor model (or the SR model), which attempts to account for this resistance. In the OFF state, an open circuit exists between the drain and the source. As illustrated in Figure 6.10, an open circuit exists between the gate and the source, and between the gate and the drain at all times. Thus, $i_G = 0$ always.

Much as we did for the switch, we can plot the $v$–$i$ characteristics between the D and S terminals of the MOSFET for various gate-to-source voltages using the S model. Figure 6.11 shows this graph. Notice that the curves are shown only for the top-right quadrant. Because we define the drain to be the terminal with the higher voltage, by definition, $v_{DS}$ can never be negative.[6] Therefore, the left quadrants are irrelevant. Similarly, the bottom-right quadrant is not shown because $i_{DS}$ is positive when $v_{DS}$ is positive. Unlike the MOSFET, devices for which this is not true (for example, batteries) are capable of providing power. Like the switch with a zero on its control input, the connection between the D and S terminals looks like an open circuit ($i_{DS} = 0$) when $v_{GS} < V_T$. In contrast, the connection between D and S looks like a short circuit ($v_{DS} = 0$) when $v_{GS} \geq V_T$.

We can summarize the S model for the MOSFET in algebraic form by stating its $v$–$i$ characteristics as follows:

$$\text{for } v_{GS} < V_T, \quad i_{DS} = 0$$

and

$$\text{for } v_{GS} \geq V_T, \quad v_{DS} = 0 \tag{6.2}$$

Our discussion thus far treated a MOSFET as a three-terminal device. Notice, however, that the MOSFET is controlled by the voltage across a pair of terminals, namely, G and S. Similarly, we were interested in the voltage across, and current through, the D and S terminal pair. As discussed in Section 1.5, this natural pairing of terminals suggests an alternate MOSFET representation using ports. As shown in Figure 6.12, we can treat the G and S terminal pair as the input port or the control port and the D and S terminal pair as the output port of the MOSFET.

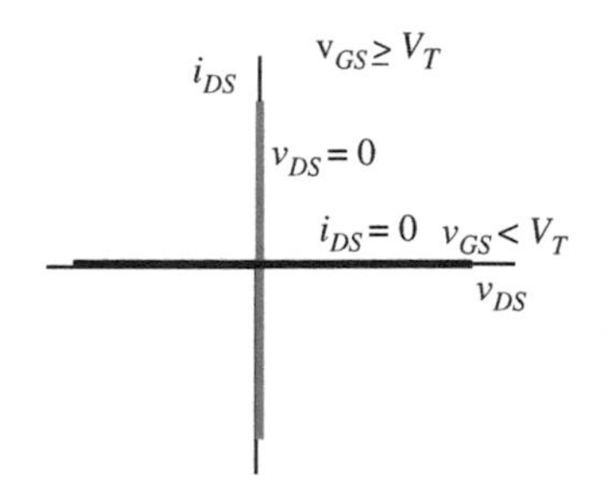


6. When fabricated on a VLSI chip, the physical MOSFET itself is symmetric with respect to the drain and source. (Note, however, that discrete MOSFET devices suitable for use on a breadboard are not symmetric.) So its drain and source labels can be interchanged without changing the device. Accordingly, if the drain and source nomenclature was unrelated to the potential difference between the two terminals, the MOSFET characteristic would look like as in the adjacent figure. As depicted, there is an open circuit between D and S when $v_{GS} < V_T$, and a short circuit when $v_{GS} \geq V_T$. When the open circuit exists, $v_{DS}$ can take on any value (positive or negative) as determined by external circuit constraints. Similarly, when the short circuit exists, $i_{DS}$ is unconstrained. Interestingly, notice here that the model we choose for an element depends on the way we use the element as much as on the physical construction of the element!

Notice further that we if choose our digital representation such that the logical 1 is represented with a value greater than 1 volt, then the MOSFET operates as a switch that turns on when its gate-to-source port has a logical 1 signal on it. Figure 6.13 shows the lightbulb circuit using MOSFETs to implement the *AND* switching function. In this circuit, the bulb turns on only when both A and B are 1.

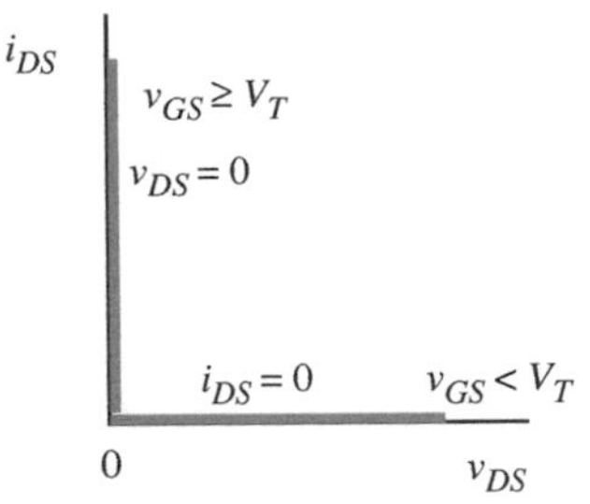

**FIGURE 6.11** MOSFET characteristics with the S model.

## 6.4 MOSFET SWITCH IMPLEMENTATION OF LOGIC GATES

Let us now build logic gates using MOSFETs. Consider the circuit shown in Figure 6.14, which comprises a MOSFET and a load resistor powered by a supply voltage $V_S$. The input to the circuit is connected to the gate of the MOSFET, the source is tied to ground, and drain is tied to $V_S$ in series with a load resister $R_L$. The same circuit is redrawn on the right-hand side using the shorthand notation for the power and ground terminals.

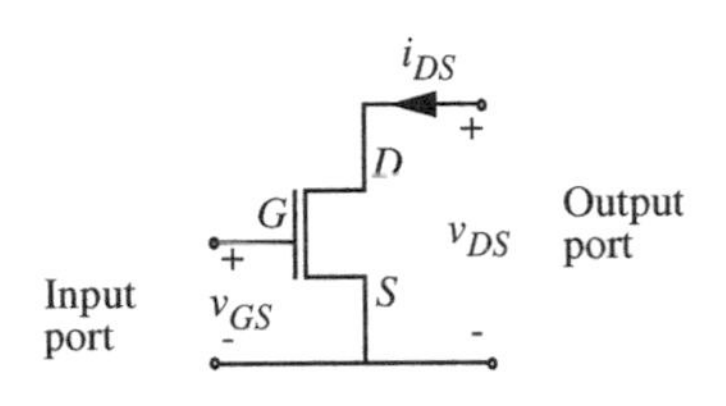


**FIGURE 6.12** Port representation of a MOSFET.

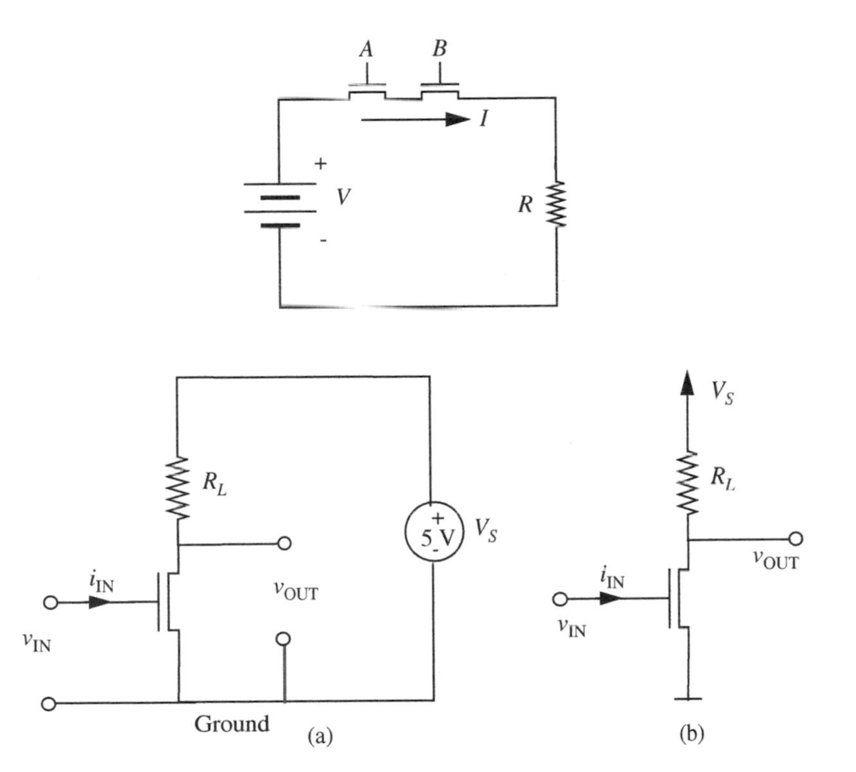


**FIGURE 6.13** The lightbulb circuit using MOSFETs.

**FIGURE 6.14** (a) The MOSFET inverter; (b) the same inverter circuit using the shorthand notation for power and ground.

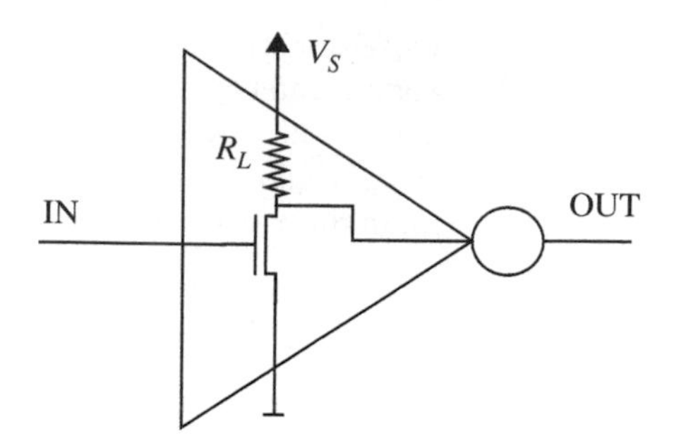


FIGURE 6.15 The inverter abstraction and its internal circuit. IN and OUT are the logical values represented by $v_{IN}$ and $v_{OUT}$.

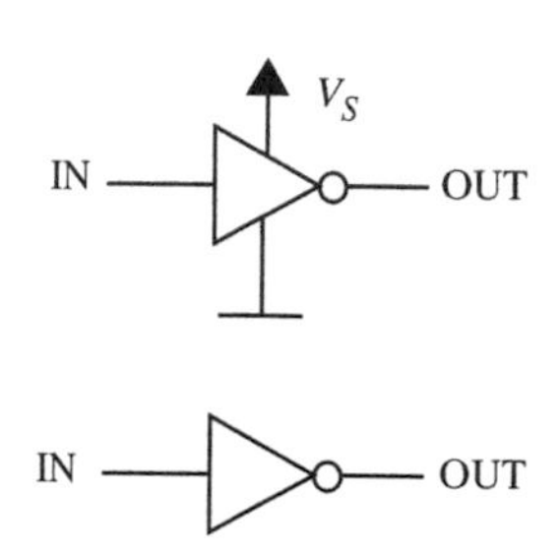


FIGURE 6.16 The inverter shown with explicit and implicit power connections.

Figure 6.15 shows the relationship between the terminals of the MOSFET-based inverter circuit and the abstract inverter gate.

The left-hand side of Figure 6.16 shows the corresponding abstract inverter with the power connections, and the right-hand side shows the inverter abstraction as we know it with implicit power connections. Notice that the inverter abstraction hides the internal circuit details and provides a simple usage model to the user of the inverter logic gate. The internal details are irrelevant to the gate-level logic designer.[7]

Let us analyze the behavior of the circuit by replacing the MOSFET with its equivalent S model. Figure 6.17 displays the equivalent model for the circuit shown in Figure 6.14. Let us assume that a logical high is represented using 5 V and a logical low using 0 V.

As shown in Figure 6.17, when the input $v_{IN}$ is high, the MOSFET is in the ON state (assuming that the high voltage level is above the threshold $V_T$), thereby pulling the output voltage to a low value.[8] In contrast, when the input is low, the MOSFET is off, and the output is raised to a high value by $R_L$. Here we see the purpose of the load resistor[9] $R_L$ — it provides a logical 1 output when the MOSFET is off. Furthermore, $R_L$ is usually chosen to be large so that the current is limited when the MOSFET is on.[10] Because the resistance

---

7. What if sophisticated logic designers want to optimize their design for certain parameters — such as speed or area — that are not captured by the gate-level abstraction? Later chapters will discuss how the gate-level abstraction is augmented with additional parameters — such as gate delay and size — derived from the internal circuit so that the logic designer can optimize their gate-level circuit without being forced to delve into the internal details of the gate.

8. Because the MOSFET in Figure 6.14 serves to pull the output voltage to a low value when it is its ON state, the MOSFET is sometimes referred to as the pulldown MOSFET.

9. Because it pulls the output to a high value, the load resistor is sometimes referred to as the pullup resistor.

10. We will see additional design constraints on $R_L$ as we progress to more accurate MOSFET models.

between the gate-to-source and the drain-to-source ports of the MOSFET is infinity in the S model, the current $i_{IN}$ is 0.

We can write the input-output relationships in a truth table as shown in Table 6.1. As is apparent from the table, the logical values IN and OUT represented by $v_{IN}$ and $v_{OUT}$ exhibit the behavior of an inverter.

Figure 6.18 shows a sample input waveform and the corresponding output waveform for our circuit. We can also plot the $v_{IN}$ versus $v_{OUT}$ voltage transfer curve for the inverter circuit as shown in Figure 6.19. The input-output transfer curve for the inverter — also called the inverter characteristic — allows us to determine whether the inverter satisfies a given static discipline. We will discuss the inverter characteristic and how it relates to the static discipline in Section 6.5.

We can also construct other gates in like manner. Figure 6.20 shows a NAND gate circuit and Figure 6.21 shows its equivalent S circuit model. It is

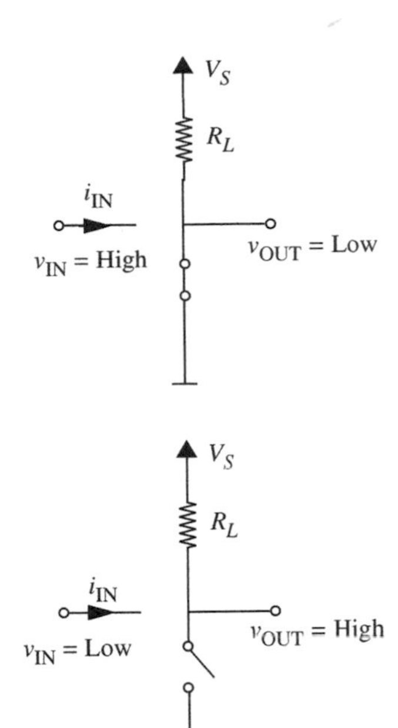


**FIGURE 6.17** The S circuit model of the n-channel MOSFET inverter.

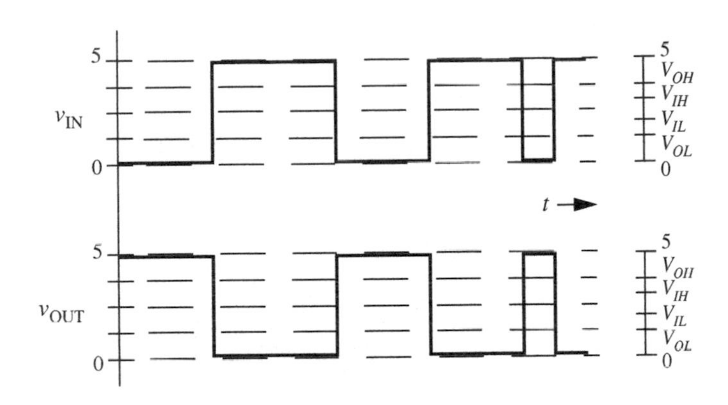


**FIGURE 6.18** Sample input-output waveforms for the inverter.

| IN | OUT |
|---|---|
| 0 | 1 |
| 1 | 0 |

**TABLE 6.1** Truth table for the MOSFET circuit.

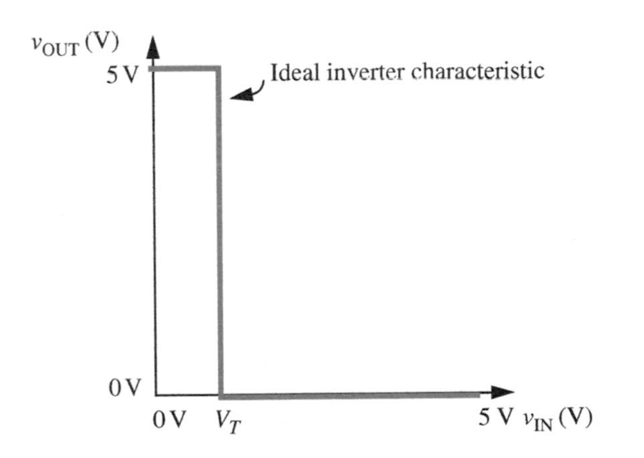


**FIGURE 6.19** The transfer characteristic of the inverter.

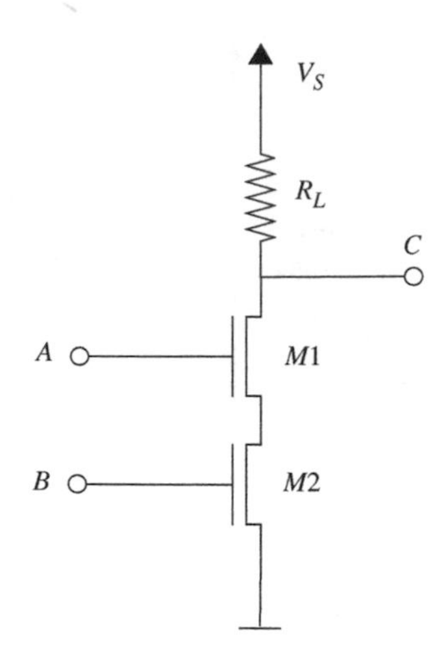


FIGURE 6.20 The circuit for a NAND gate.

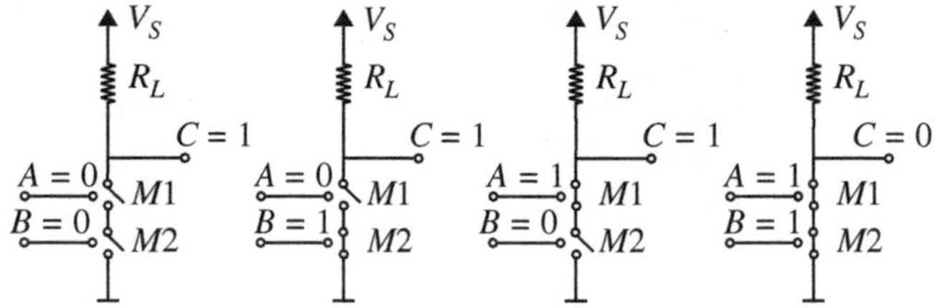

easy to see that the output is a 0 only when both inputs are high. The output is high otherwise.[11]

Using intuition from the two-input NAND circuit, we can build multiple-input NAND and NOR circuits. Figure 6.22a shows an $n$-input NOR gate and 6.22b shows an $n$-input NAND gate. In the multiple-input NOR gate, the output is pulled to ground when any of the inputs is high. Correspondingly, in the NAND gate, the output remains high if even one input is low.

FIGURE 6.21 The S circuit model for a NAND gate.

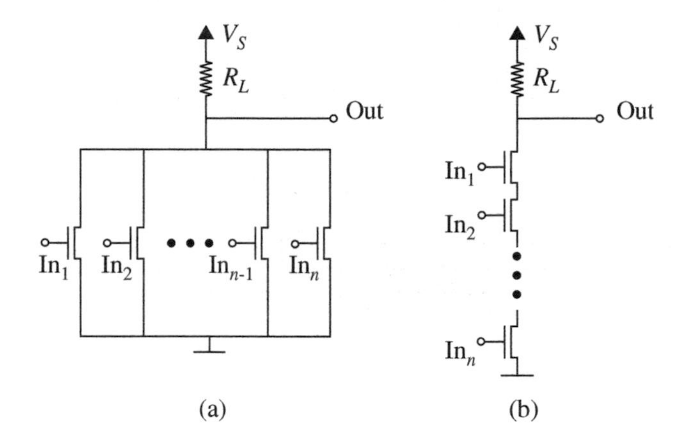


FIGURE 6.22 Multiple-input NOR and NAND gates.

---

EXAMPLE 6.2 COMBINATIONAL LOGIC USING MOSFETS

Recall the two combinational logic expressions whose gate-level implementations we had seen earlier:

$$\overline{AB + C + D}$$

$$\overline{(A + B)CD}$$

---

11. Notice that if both switches are off, the voltage at the node connecting M1 and M2 appears to be undefined. Therefore $v_{GS}$ for M1 also appears to be undefined. In practice, however, MOSFET switches are not perfect open circuits, rather they have a very high resistance between their drain and source in the off state. Thus, the voltage between M1 and M2 will be given by a voltage divider relationship. In any case, $v_{GS}$ for M1 will not impact the output voltage of the gate.

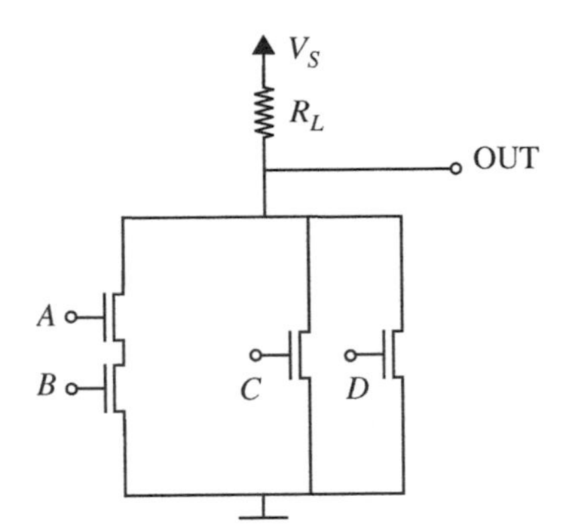


FIGURE 6.23 Transistor-level implementation of $\overline{AB + C + D}$.

In our earlier example, we had implemented these expressions using several abstract logic gate elements. Now that we understand how gates are constructed using MOSFETs and resistors, we can actually construct a single compound combinational logic gate using MOSFETs and resistors that implements each of these functions.[12]

Let us consider the first expression: $\overline{AB + C + D}$. Using the intuition that switches in series implement the AND property and switches in parallel implement the OR property, we can implement the first expression as shown in Figure 6.23. By checking the circuit against its truth table, we can convince ourselves that the circuit does work as desired.

Figure 6.24 shows the circuit for the second expression: $\overline{(A + B)CD}$.

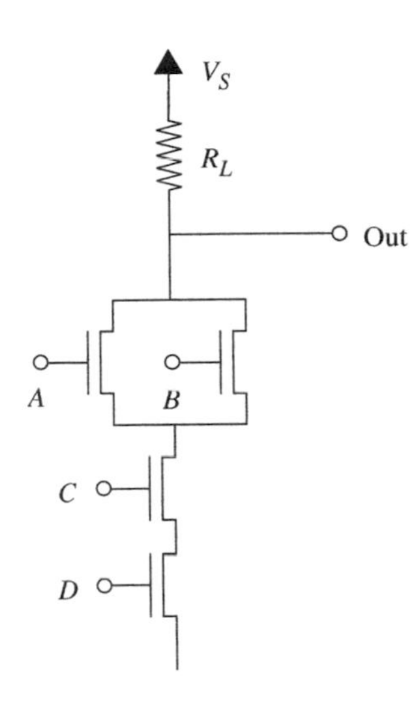


FIGURE 6.24 Transistor-level implementation of $\overline{(A + B)CD}$.

EXAMPLE 6.3 MORE COMBINATIONAL LOGIC USING MOSFETS Let us now construct the logic expression $(A + B)CD$ using MOSFETs. Since $(A + B)CD$ is simply the complement of $\overline{(A + B)CD}$, we can obtain $(A + B)CD$ by inverting the output from Figure 6.24, as illustrated in Figure 6.25.

It is important to point out two key properties of the MOSFET that make it an ideal component for building gates:

1. First, notice that we could compose multiple gate components into more complicated circuits without worrying about the internal circuit of

12. Direct transistor implementation potentially can yield savings in the number of transistors and resistors used. However, should you consider discarding the gate-level abstraction in favor of building logic functions directly out of transistors, remember that transistor-level implementation of complex logic functions can be a much more arduous task than a gate-level implementation. The difficulty will become apparent as we move to the more realistic switch-resistor MOSFET model discussed in Section 6.6. As a general guideline, the designer should use the highest level of abstraction with which to accomplish a design.

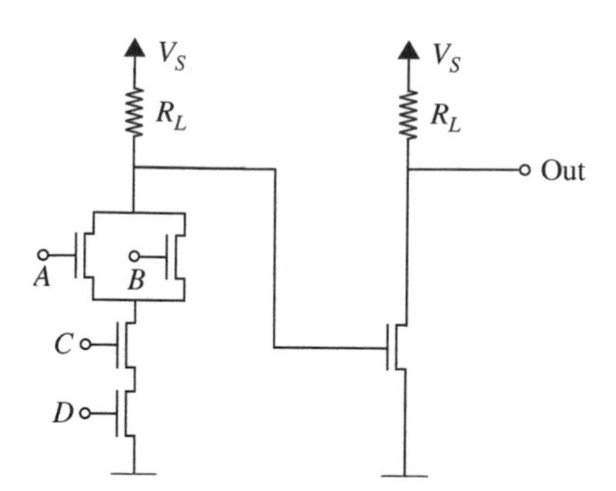


**FIGURE 6.25** Transistor-level implementation of $(A + B)CD$.

the gates. The reason we are able to do so is that the output of the MOSFET has no effect on its inputs. In other words, although the input voltage at G impacts the behavior of the MOSFET at its D and S terminals, the voltages or currents at its D and S terminals have no impact on G.

2. Second, the infinite resistance seen at the gate (G terminal) of a MOSFET makes it have no effect on the output of another gate driving its input. This feature of the MOSFET allows us to build systems containing many gates without worrying about how each gate affects the logical properties of other gates to which it is connected. This property of a gate is called *composability*. Imagine if the MOSFET input had zero resistance. In that case, we would not be able to connect the output of one inverter to the input of another and expect the first inverter to satisfy the static discipline.

We shall see later in Section 6.9.1 that the ability to build amplification into devices containing MOSFETs further facilitates composability.

## 6.5 STATIC ANALYSIS USING THE S MODEL

The input-output transfer curve for the inverter shown in Figure 6.19, or the inverter characteristic, contains all the information necessary to determine whether the inverter satisfies a given static discipline.

Recall that the *static discipline* for a logic gate guarantees that the outputs of the gate will meet the output constraints specified by the discipline, provided its inputs meet the input constraints. Recall, further, that a static discipline with its associated voltage thresholds is necessary to establish a standardized representation so that the logic devices from several vendors can operate with each other correctly. Similarly, a user who wants to build a system can select the best devices from several vendors provided they meet the voltage thresholds established by the static discipline adopted by the user.

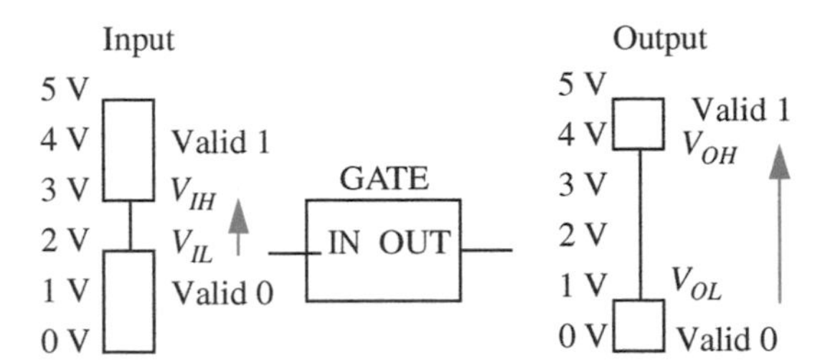


**FIGURE 6.26** Input and output voltage thresholds for a logic gate.

Output voltage levels are generally stricter than input voltage levels (higher than the corresponding input high and lower than the corresponding input low) to provide for noise margins. Figure 6.26 illustrates the asymmetry between inputs and outputs. At the input of the gate, any voltage level lower than $V_{IL}$ is recognized as a valid low, and any voltage higher than $V_{IH}$ is a valid high. At its output, the gate guarantees to produce a voltage level higher than $V_{OH}$ for a valid high, and a voltage level lower than $V_{OL}$ for a valid low.

Voltage levels between $V_{IL}$ and $V_{IH}$ are invalid at the input, and levels between $V_{OL}$ and $V_{OH}$ are invalid at the output. Because the output levels are stricter than the input thresholds, the static discipline provides for noise margins.

Based on the inverter characteristic (repeated here in Figure 6.27 for convenience), we can determine whether the inverter satisfies a given static discipline. As an example, let us determine whether the inverter satisfies a static discipline with the following voltage thresholds:

$$V_{OH} = 4.5\text{ V},\ V_{OL} = 0.5\text{ V},\ V_{IH} = 4\text{ V},\ \text{and } V_{IL} = 0.9\text{ V}.$$

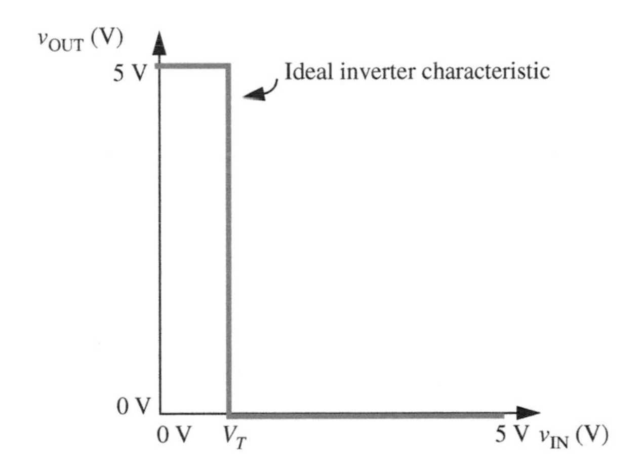


**FIGURE 6.27** The transfer characteristic of the inverter.

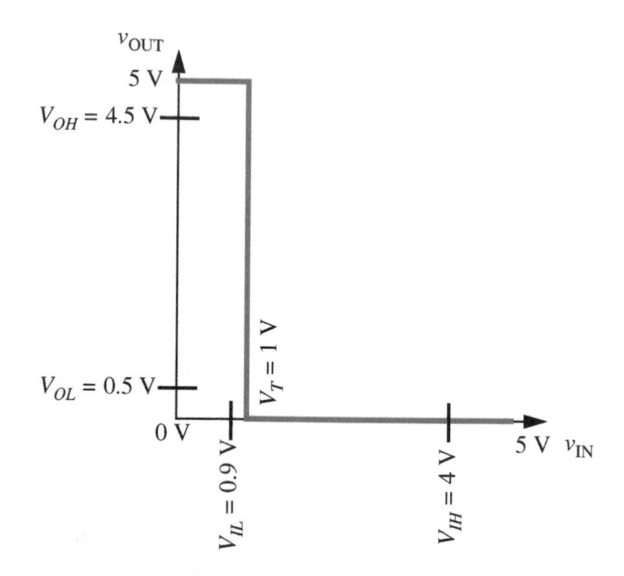


**FIGURE 6.28** A mapping between logic values and voltage levels corresponding to a static discipline appropriate for the inverter.

Figure 6.28 shows the voltage thresholds for the given static discipline superimposed on the inverter transfer function. Let us check each of the output and input thresholds.

$V_{OH}$: The inverter produces an output high of 5 V. Clearly, this output voltage level for a logical 1 is greater than the 4.5-V output-high threshold required by the static discipline.[13]

$V_{OL}$: The inverter produces an output low of 0 V. This output voltage is lower than the output-low threshold of 0.5 V required by the static discipline.[14]

$V_{IH}$: For our static discipline, $V_{IH} = 4$ V. To obey the static discipline the inverter must interpret any voltage above 4 V as a logical 1. This is certainly true for our inverter. Our inverter turns on when the input voltage is greater than $V_T = 1$ V and pulls the output to a valid low voltage. Thus it interprets any voltage above 1 V as a logical 1.[15]

---

13. It turns out that our inverter can satisfy a static discipline with a $V_{OH}$ as high as $5^-$ V. The notation $5^-$ V implies a voltage that is slightly below 5 V.

14. Notice that our inverter can satisfy a static discipline with a $V_{OL}$ as low as $0^+$ V. The notation $0^+$ V implies a voltage that is slightly above 0 V.

15. In fact, our inverter can satisfy a static discipline with a $V_{IH}$ as low as $1^+$ V, because the inverter produces a valid low output voltage for inputs greater than $V_T = 1$ V.

$V_{IL}$: For our static discipline, $V_{IL} = 0.9$ V. This means that to obey the static discipline the inverter must interpret any voltage below 0.9 V as a logical 0. This is true for our inverter. The inverter is off when its input voltage is below $V_T = 1$ V, and its output is at 5 V. Since the inverter produces a valid (output) high output voltage for input voltages below 0.9 V, it satisfies the static discipline.[16]

---

EXAMPLE 6.4 STATIC DISCIPLINE For fun, let us check to see whether our inverter satisfies a static discipline used by Disco Systems Inc. Assume that some of Disco's systems adhere to a static discipline with the following voltage thresholds:

$$V_{OH} = 4 \text{ V}, V_{OL} = 1 \text{ V}, V_{IH} = 3.5 \text{ V, and } V_{IL} = 1.5 \text{ V}.$$

To operate under this static discipline, we know that our inverters must operate as follows:

- When outputting a logical 1, the voltage their outputs produce must be at least $V_{OH} = 4$ V. Since our inverters produce a 5-V output for a logical 1, they satisfy this condition.
- When outputting a logical 0, the voltage their outputs produce must be no greater than $V_{OL} = 1$ V. Since our inverters produce a 0-V output for a logical 0, they satisfy this condition easily.
- At their inputs, the inverters must recognize voltages greater than $V_{IH} = 3.5$ V as a logical 1. Since our inverters recognize voltages above 1 V as a logical 1, they satisfy this condition as well.
- Finally, at their inputs, the inverters must recognize voltages less than $V_{IL} = 1.5$ V as a logical 0 if they are to satisfy Disco's static discipline. Unfortunately, our inverters can recognize voltages only below 1 V as a 0, and thus do not satisfy this condition.

Thus, our inverters cannot be used in Disco's systems.

---

We can also conduct a static analysis of other digital MOSFET circuits, such as those Section 6.4. When the S model for the MOSFET is used, the input and output voltage thresholds for the NAND and other digital circuits come out to be identical to those for the inverter. Thus, the results of static

16. In fact, our inverter can satisfy a static discipline with a $V_{IL}$ as high as $1^-$ V, because the inverter produces a valid high output voltage for inputs less than $V_T = 1$ V.

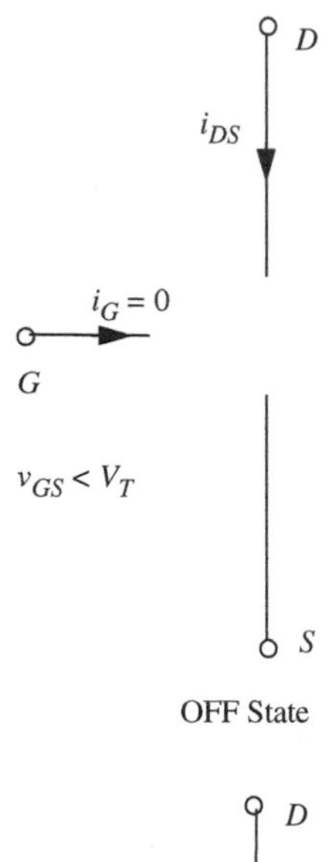


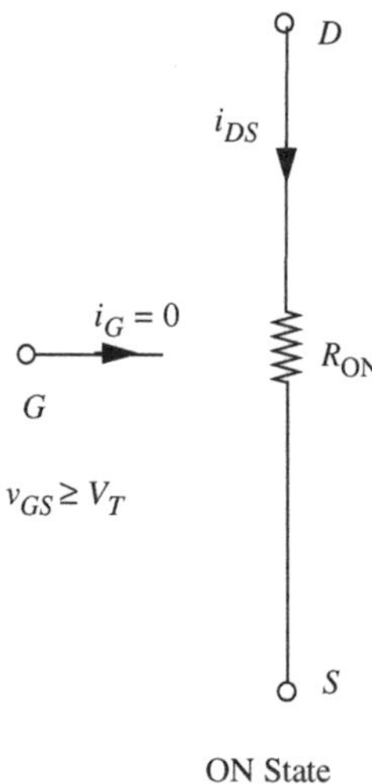


FIGURE 6.29 The switch-resistor model of the n-channel MOSFET.

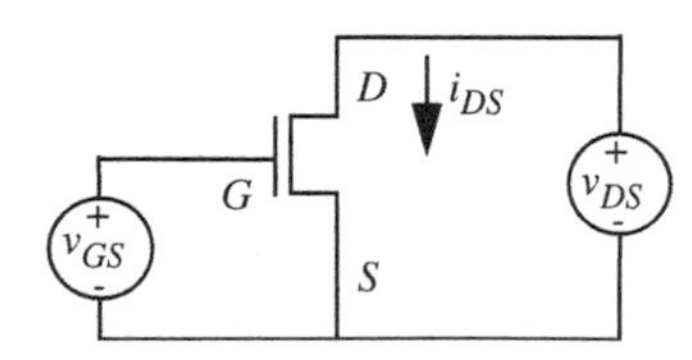


FIGURE 6.30 Setup for observing MOSFET characteristics.

analysis for these circuits are identical to those for the inverter. For example, like the inverter, the NAND circuit satisfies a static discipline with the following voltage thresholds: $V_{OH} = 4.5$ V, $V_{OL} = 0.5$ V, $V_{IH} = 4$ V, and $V_{IL} = 0.9$ V. Similarly, the NAND is not able to satisfy the static discipline with these voltage thresholds: $V_{OH} = 4$ V, $V_{OL} = 1$ V, $V_{IH} = 3.5$ V, and $V_{IL} = 1.5$ V.

## 6.6 THE SR MODEL OF THE MOSFET

The S model for the MOSFET discussed thus far is actually a gross simplification of the actual properties of the MOSFET. In particular, a practical MOSFET displays a non-zero resistance between its D and S terminals when it is on.[17] Accordingly, a slightly more accurate model for the MOSFET uses a resistance $R_{ON}$ in place of the short between D and S when the MOSFET is on. Figure 6.29 shows the Switch-Resistor model (or SR model) of the n-channel MOSFET.

When the MOSFET is off, there is no connection between the drain and the source. If the voltage $v_{GS}$ between the gate and source terminals is above $V_T$, the MOSFET turns on and displays a resistance $R_{ON}$ between its D and S terminals. As before, there is an open circuit between the gate and source terminals and the gate and drain terminals of the MOSFET, so $i_G = 0$.

The SR model is a better approximation of MOSFET behavior than the S model. In fact, it is easy to see that the SR model reduces to the S model if $R_{ON}$ is zero. However, the SR model still is a gross simplification of MOSFET behavior. In particular, although the MOSFET displays resistive behavior when $v_{DS} \ll v_{GS} - V_T$, the resistance $R_{ON}$ is not fixed. Rather, it is a function of $v_{GS}$. Furthermore, when $v_{DS}$ is comparable to, or greater than, $v_{GS} - V_T$, the drain-to-source behavior is not resistive at all, rather it is that of a current source. However, the fixed resistance model is much simpler and suffices for analyzing some aspects of digital circuits because the gate voltage is bimodal — low or high. When the voltage is low, the MOSFET turns off, and when the voltage is high, the drain-source connection offers a resistance $R_{ON}$ related to the gate voltage. Since there is only one value for the gate voltage when the input is high (for example $V_S$), and accordingly, one value for the resistance $R_{ON}$, we can use this value for $R_{ON}$ in our model. In summary, the SR model is valid only when $v_{DS} \ll v_{GS} - V_T$, and when there is only one value for the gate voltage when the input is high (for example $v_{GS} = V_S$). Chapter 7 will discuss more comprehensive models for the MOSFET that are valid across a wider range of values for $v_{GS}$ and $v_{DS}$.

The characteristics of the MOSFET according to the SR model are graphically displayed in Figure 6.31. These curves can be plotted by measuring the various voltages and currents from the setup in Figure 6.30.

17. And in fact, all practical switches display a nonzero resistance between their input and output terminals in the ON state.

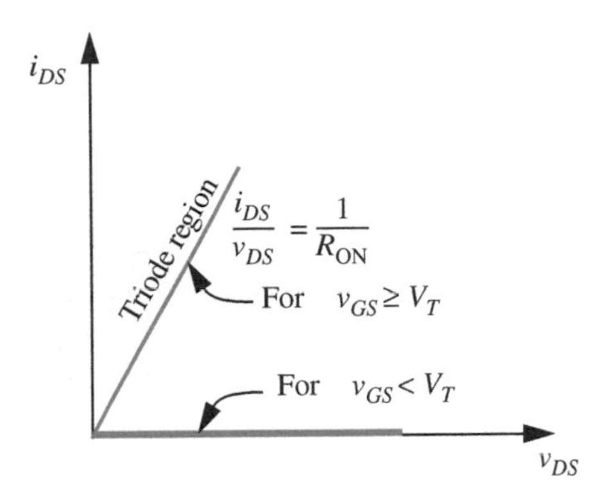


**FIGURE 6.31** Characteristics of the MOSFET according to the SR model. The top-left, bottom-left, and the bottom-right quadrants are not shown for the reasons given in Section 6.3. As discussed in more detail in Chapter 7, the region in which the MOSFET displays resistive behavior is within the triode region of MOSFET operation.

The SR model for the MOSFET can also be expressed in algebraic form as:

$$i_{DS} = \begin{cases} \dfrac{v_{DS}}{R_{ON}} & \text{for } v_{GS} \geq V_T \\ 0 & \text{for } v_{GS} < V_T \end{cases} . \tag{6.3}$$

The presence of the resistance $R_{ON}$ makes our analysis more realistic but complicates the design of logic gates somewhat. We shall discuss this further in Section 6.8.

## 6.7 PHYSICAL STRUCTURE OF THE MOSFET

The on-resistance of the MOSFET depends on several factors related to the physical properties of the MOSFET, such as its geometry. Typical values for the resistance range from a few milliohms for discrete MOSFET components to several kΩ for MOSFETs in VLSI technology. Let us take a quick look at the physical structure of MOSFETs to obtain some insight into the relationship between their on resistance and their geometry.

MOSFETs are constructed through several fabrication steps on the surface of a planar piece of single-crystal silicon called a wafer. Figure 6.32 shows the top view of several rectangular MOSFETs fabricated on a planar silicon surface. A wafer can be tens of centimeters in diameter, and typical MOSFETs might occupy an area that is less than a square micrometer. One micrometer is $10^{-6}$ meters. A micrometer ($\mu$m) is also referred to as a micron ($\mu$). The fabrication steps result in the construction of several planar layers sandwiched on the wafer surface. The layers might constitute insulating layers made up of silicon dioxide ($SiO_2$) created by oxidizing parts of the wafer surface, conducting layers comprising deposits of metals such as aluminum or copper, or polycrystalline silicon (poly), and semiconducting layers comprising silicon doped with materials with a high concentration of free electrons or holes. (A hole is

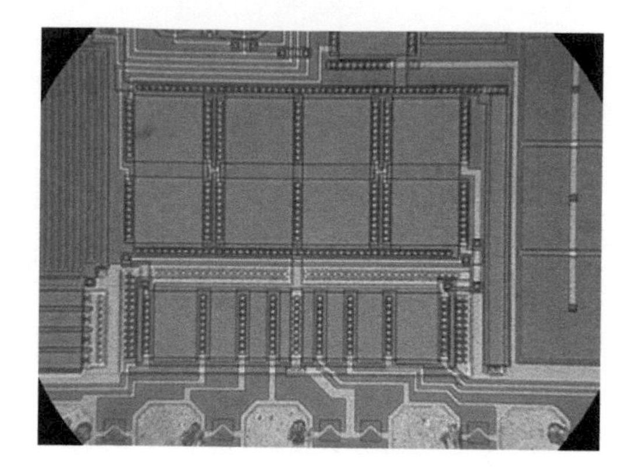

**FIGURE 6.32** Top view of several n-channel MOSFETs fabricated on a chip. The square MOSFETs in the center of the photograph have a width and length of 100 $\mu$m. (Photograph Courtesy of Maxim Integrated Products.)

a positively charged element, and comprises an atom with a missing electron.) Doping is accomplished by diffusion or by ion implantation. The conducting layers are separated by the insulating layers. Connections between layers separated by insulating material are created by etching contact holes in the insulating material and "pouring" metal through the contact holes.

Silicon doped with a material rich in electrons is called an *n-type semiconductor.* Similarly, silicon doped with a material rich in holes is called a *p-type semiconductor.* As the word "semiconductor" implies, silicon doped with either n-type or p-type material is a fairly good conductor of electricity. We use the notation $n^+$ or $p^+$ to refer to silicon that is heavily doped with n-type or p-type material, respectively. $n^+$ or $p^+$ type semiconductors are even better conductors. We commonly refer to n-type or p-type doped silicon areas as *diffusion regions.*

Figures 6.33 and 6.34 show sketches of two views of the physical structure of an n-channel MOSFET. An n-channel MOSFET is constructed on the surface of p-type silicon called the substrate. Two $n^+$ doped regions separated by a small distance (for example, 0.07 $\mu$ in digital devices, and quite a bit larger in

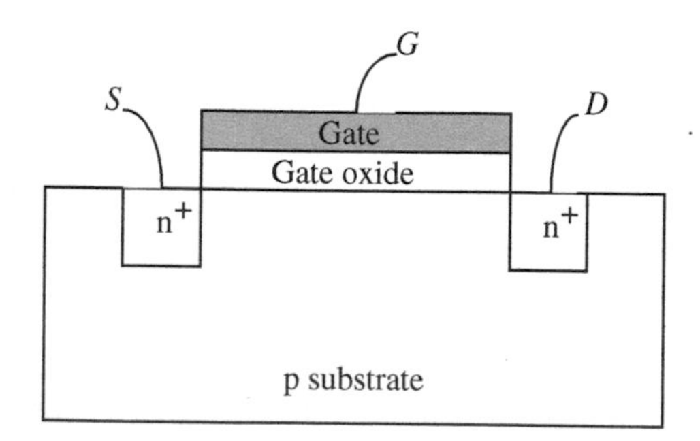


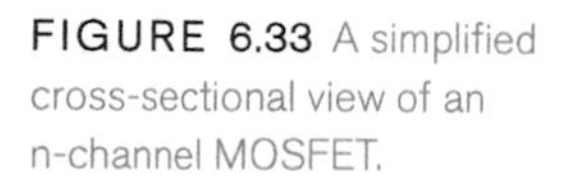

**FIGURE 6.33** A simplified cross-sectional view of an n-channel MOSFET.

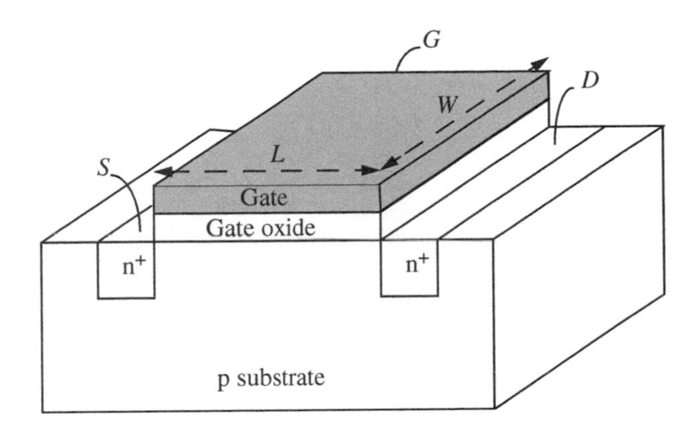


FIGURE 6.34 A three-dimensional view of an n-channel MOSFET.

analog applications) constitute the source and the drain. The region separating the source and the drain is called the channel region. The channel region is overlayed with a thin insulating layer (for example, 0.01 $\mu$m thick) made out of silicon dioxide (commonly called gate oxide). The gate oxide layer is in turn sandwiched between a layer of conducting polysilicon on top and the p-type substrate. The polysilicon layer top of the gate oxide forms the gate of the MOSFET.

Although the precise mechanics of how a MOSFET works is beyond the scope of this discussion, the following provides some intuition on its operation. First, recall that $n^+$-type silicon conducts through the motion of its free electrons. Let us consider the case where the gate and the source of the MOSFET are connected to ground, as illustrated in Figure 6.35. In this situation, $v_{GS} = 0$. Because the $n^+$ doped source and drain are separated by a p-type layer, they will not conduct any current when a voltage is applied across them ($v_{DS} > 0$).

However, when a positive voltage is applied at the gate of the device ($v_{GS} > 0$), negative charges are attracted to the surface from the nearby negative-charge-rich source region (as shown in Figure 6.36) and positive charges are repelled from the surface. Of course, no current flows between the gate and the substrate because of the insulating gate oxide layer. As the gate voltage

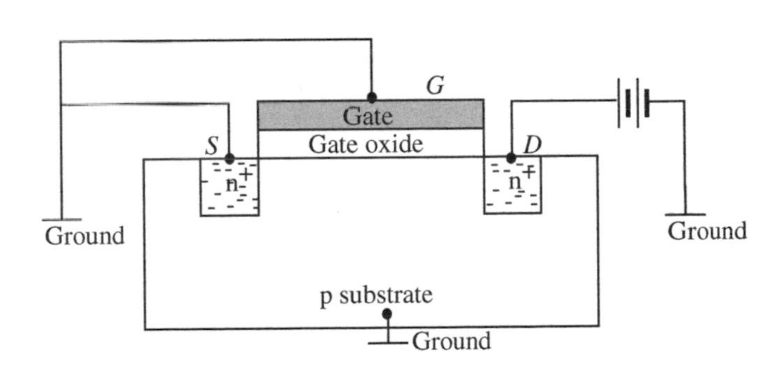


FIGURE 6.35 MOSFET operation when the gate is connected to ground.

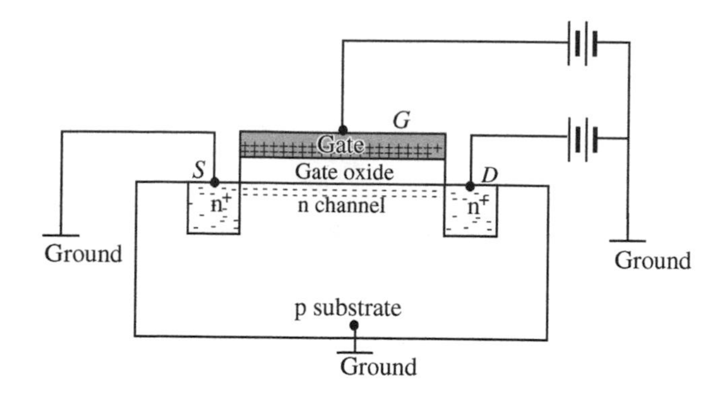


FIGURE 6.36 MOSFET operation when a positive gate voltage is applied.

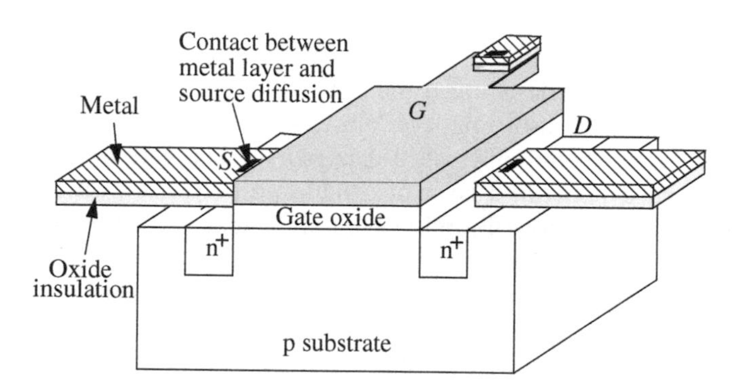


FIGURE 6.37 Connecting to a MOSFET.

increases, more negative charges are attracted to the surface until they form an n-type conducting channel that connects the source and the drain. The conducting channel forms when the gate voltage crosses a threshold voltage $V_T$ (in other words, $v_{GS} > V_T$). A current begins to flow between the drain and the source when a positive voltage is applied across the drain and the source ($v_{DS} > 0$). The MOSFET in our example is called an n-channel device because of the n-type channel that is formed.

It is easy to see that the MOSFET operates like a switch connected to the source and the drain that turns on when the gate voltage exceeds a threshold. Figure 6.37 shows how metal connections are made to the MOSFET terminals *G*, *S*, and *D* so the MOSFET can be coupled to other devices. As shown in the figure, the layers of metal are separated by layers of oxide (thickness is not to scale), so the metal does not inadvertently come in contact with other parts of the device.[18] Contact holes are etched between pairs of layers between

18. A metal-semiconductor connection behaves like a circuit element called a diode if the semiconductor is lightly doped, and like a short circuit if the semiconductor layer is heavily doped.

which connections are desired (much like a staple) and metal is allowed to flow through.

The conducting n-channel that is formed in the MOSFET discussed here is not an ideal conductor and has some resistance $R_{ON}$. Also notice that the resistance of the gate is related to the geometry of the channel. Let the channel length be $L$ and the channel width be $W$. Then, the resistance is proportional to $L/W$. If $R_n$ is resistance per square of the n-channel MOSFET in its on state, then the resistance of the channel is given by

$$R_{ON} = R_n \frac{L}{W}. \tag{6.4}$$

In any VLSI technology, there is a minimum fabricatable value for the MOSFET channel length. Clearly, smaller dimensions mean that a VLSI chip of a given size can hold more logic. As we shall see later, smaller dimensions also result in higher speeds of operation. VLSI technologies are characterized by this minimum channel length. For example, a 0.2-$\mu$m process yields gate lengths in the vicinity of $L = 0.2$ $\mu$m. Historically, technologists have been able to decrease gate lengths by about a factor of two every four years over the past two decades. See Table 6.2 for the scaling factors observed by the authors in projects in which they were involved, and Figure 6.38 for a cross-sectional view of Intel's 0.13 $\mu$m

| YEAR | DESIGN | MIN L |
|---|---|---|
| 1981 | Analog echo canceler | 8 $\mu$m |
| 1984 | Telecom bus controller | 4 $\mu$m |
| 1987 | RISC microprocessor | 2 $\mu$m |
| 1994 | Multiprocessor communications controller | 0.5 $\mu$m |
| 2002 | Raw microprocessor | 0.18 $\mu$m |

**TABLE 6.2** Historical gate-length scaling observations.

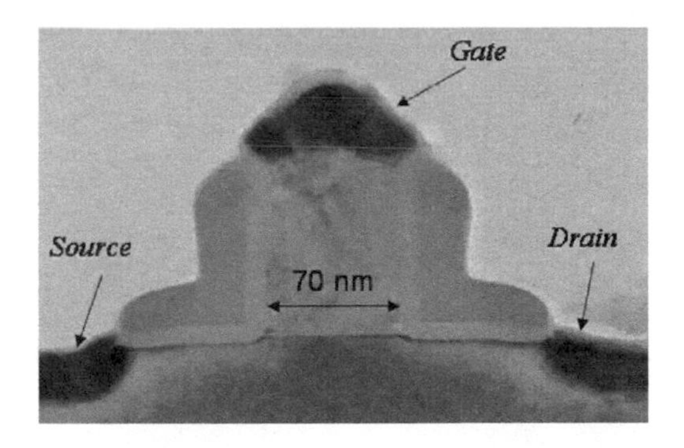


**FIGURE 6.38** A cross-sectional TEM (transmission electron microscope) picture of Intel's 0.13-$\mu$m generation logic transistor. (Photograph Courtesy of Intel Corporation.)

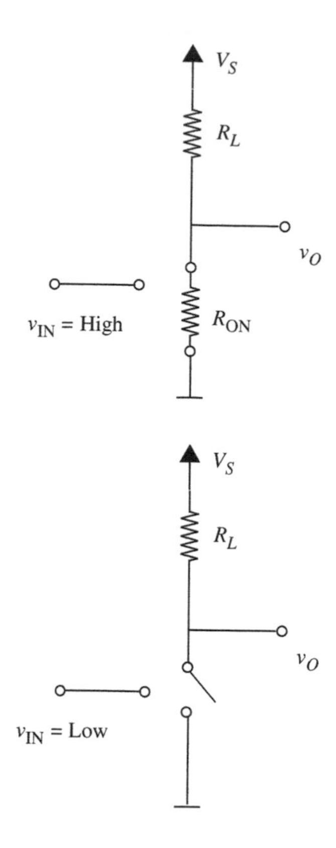


**FIGURE 6.39** Circuit model of the n-channel MOSFET inverter using the SR MOSFET model.

generation logic transistor. This torrid pace of technological development shows no signs of slowing down at the time of this writing.

## 6.8 STATIC ANALYSIS USING THE SR MODEL

The presence of the on resistance $R_{ON}$ complicates the design of logic gates slightly, but adds more realism to the model. Let us analyze our familiar inverter circuit shown in Figure 6.14 using the SR model of the MOSFET. In particular, let us derive its input-output transfer characteristic. Figure 6.39 shows the circuit model of the inverter using the SR model of the MOSFET.

As shown in Figure 6.39, when the input is low, the MOSFET is off, and the output is raised to a high value.

However, when the input $v_{IN}$ is high (and above the threshold $V_T$), the MOSFET is on and displays a resistance $R_{ON}$ between its D and S terminals, thereby pulling the output voltage lower. However, the output voltage is not 0 V as predicted by the simpler S model of the MOSFET. Instead, the value of the output voltage is given by the voltage-divider relationship:

$$v_{OUT} = V_S \frac{R_{ON}}{R_{ON} + R_L}. \tag{6.5}$$

The resulting inverter transfer characteristics, assuming $V_S = 5$ V, $V_T = 1$ V, $R_{ON} = 1\,\text{k}\Omega$, and $R_L = 14\,\text{k}\Omega$, are shown in Figure 6.40. Notice that the lowest output voltage of the inverter is no longer 0 V, rather it is

$$V_S \frac{R_{ON}}{R_{ON} + R_L} = 0.33 \text{ V}.$$

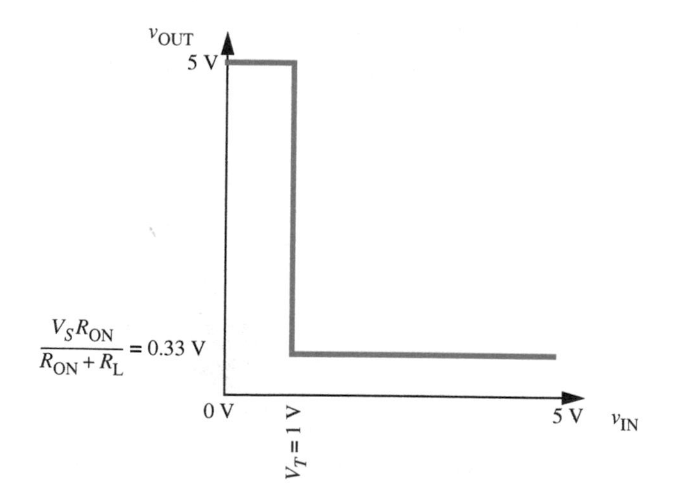


**FIGURE 6.40** Inverter transfer characteristics using the SR model.

Before embarking on a detailed static analysis of our inverter, let us conduct a simple electrical switching analysis to build intuition. In particular, we will analyze the relationship between $V_T$, $V_S$, $R_L$, and $R_{ON}$, and the switching behavior of the inverter. As a minimum, when a sending inverter drives a receiving inverter, the sender must be able to switch the MOSFET in the receiving inverter into its ON state when the sender produces a high voltage. Similarly, the sender must be able to switch the MOSFET in the receiving inverter into its OFF state when the sender produces a low voltage.

Our inverter produces a high output of $V_S$, so it is easy to see that the high output can turn the MOSFET in a receiving inverter into its ON state (provided, of course, that $V_S > V_T$). Because the inverter produces a nonzero low output, more care needs to be taken in the choice of resistance values and the MOSFET parameters. Specifically, for our inverter design, the voltage output for a logical 0 must be low enough that the MOSFET in the receiver stays OFF. Because the MOSFET turns ON for an input voltage greater than $V_T$, the following condition must be met for the low output of one inverter to be able to drive the MOSFET in another inverter into its OFF state:

$$V_S \frac{R_{ON}}{R_{ON} + R_L} < V_T. \tag{6.6}$$

Equation 6.6 specifies a key relation between the inverter device parameters for it to be usable as a switching device. Note that we do not distinguish between the MOSFET or resistance value parameters for the sender and the receiver, since the device must be able to serve as both a sender and a receiver.

---

EXAMPLE 6.5 SWITCHING ANALYSIS OF AN INVERTER
Assume the following values for the inverter circuit parameters: $V_S = 5$ V, $V_T = 1$ V, and $R_L = 10$ kΩ. Assume, further, that $R_n = 5$ kΩ for the MOSFET. Determine a $W/L$ sizing for the MOSFET so that the inverter gate output for a logical 0 is able to switch OFF the MOSFET of another inverter.

Equation 6.6 shows the condition that our inverter must satisfy for its logical 0 output to be able to turn off a MOSFET. From Equation 6.4, we know that the ON state resistance of the MOSFET is given by

$$R_{ON} = R_n \frac{L}{W}.$$

Substituting this relation into Equation 6.6, we get the following constraint on the $W/L$ ratio of the inverter MOSFET:

$$V_S \frac{R_n \frac{L}{W}}{R_n \frac{L}{W} + R_L} < V_T.$$

Simplifying, we obtain the following constraint on $W/L$:

$$\frac{W}{L} > \frac{R_n(V_S - V_T)}{V_T R_L}.$$

Substituting, $R_n = 5\ \text{k}\Omega$, $V_S = 5$ V, $V_T = 1$ V, and $R_L = 10\ \text{k}\Omega$, we obtain

$$\frac{W}{L} > 2.$$

For the parameter values that we have been given, this result indicates that our inverter MOSFET must be sized such that its $W/L$ ratio is greater than 2.

---

Commonly, it is not enough for the inverter to meet the preceding switching criteria. In real systems, however, it must also provide for adequate noise margins by satisfying a static discipline. The inverter characteristic shown in Figure 6.40 provides us with adequate information to determine whether the inverter satisfies a given static discipline. As an example, let us determine whether the inverter satisfies a static discipline with the following voltage thresholds:

$$V_{OH} = 4.5\text{ V},\ V_{OL} = 0.5\text{ V},\ V_{IH} = 4\text{ V, and } V_{IL} = 0.9\text{ V}.$$

Figure 6.41 shows the voltage thresholds for the given static discipline superimposed on the inverter transfer function. Let us check each of the output and input thresholds.

$V_{OH}$: The inverter produces an output high of 5 V. Clearly, this output voltage level for a logical 1 is greater than the 4.5-V output-high threshold required by the static discipline.[19]

$V_{OL}$: The inverter produces an output low of 0.33 V. This output voltage is lower than the output-low threshold of 0.5 V required by the static discipline.[20]

$V_{IH}$: For our static discipline, $V_{IH} = 4$ V. To obey the static discipline the inverter must interpret any voltage above 4 V as a logical 1. This is certainly true for our inverter.[21]

$V_{IL}$: For our static discipline, $V_{IL} = 0.9$ V. This means that to obey the static discipline the inverter must interpret any voltage below 0.9 V as a logical 0. This is also true for our inverter.[22]

---

19. In fact, our inverter can satisfy a static discipline with a $V_{OH}$ as high as $5^-$ V.

20. Notice that our inverter can satisfy a static discipline with a $V_{OL}$ as low as $0.33^+$ V.

21. In fact, our inverter can satisfy a static discipline with a $V_{IH}$ as low as $V_T^+$ V.

22. In fact, our inverter can satisfy a static discipline with a $V_{IL}$ as high as $V_T^-$ V.

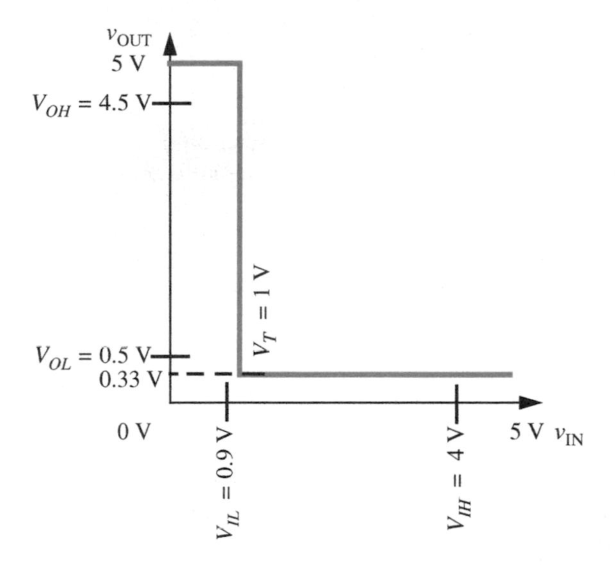


**FIGURE 6.41** A mapping between logic values and voltage levels corresponding to a static discipline appropriate for the inverter analyzed using the SR model.

Thus our inverter satisfies the static discipline with the voltage thresholds: $V_{OH} = 4.5$ V, $V_{OL} = 0.5$ V, $V_{IH} = 4$ V, and $V_{IL} = 0.9$ V, even when the SR model is used when $R_L = 14\ \mathrm{k\Omega}$ and $R_{ON} = 1\ \mathrm{k\Omega}$.

---

EXAMPLE 6.6 DESIGNING AN INVERTER TO MEET THE CONSTRAINTS OF A GIVEN STATIC DISCIPLINE Suppose we are given a static discipline with the following voltage thresholds: $V_{OH} = 4.5$ V, $V_{OL} = 0.2$ V, $V_{IH} = 4$ V, and $V_{IL} = 0.9$ V. Let us determine whether our inverter satisfies the constraints of this static discipline, and if it does not, let us redesign the inverter so that it does.

Let us begin by comparing the transfer characteristics of our inverter against the voltage thresholds of the given static discipline. As shown in Figure 6.40, recall that our inverter produces a high output of 5 V, and a low output of 0.33 V. It can interpret voltages below $V_T = 1$ V as a logical 0 and voltages above $V_T = 1$ V as a logical 1.

1. When outputting a logical 1, the voltage produced by our inverter must be greater than $V_{OH} = 4.5$ V. Since our inverters produce a 5-V output for a logical 1, they satisfy this condition.

2. When outputting a logical 0, the voltage must be no greater than $V_{OL} = 0.2$ V. Since our inverters produce a 0.33-V output for a logical 0, they violate this condition.

3. At their inputs, they must recognize voltages greater than $V_{IH} = 4$ V as a logical 1. Since our inverters recognize voltages above 1 V as a logical 1, they satisfy this condition as well.

4. Finally, at their inputs, the inverters must recognize voltages less than $V_{IL} = 0.9$ V as a logical 0 if they are to satisfy the static discipline. Our inverters satisfy this condition as well.

Since the output low voltage produced by our inverters is 0.33 V, which is higher than the required $V_{OL} = 0.2$ V, our inverters do not meet the constraints of the given static discipline.

How might we redesign our inverter to meet the given static discipline? Notice that according to Equation 6.5 the output voltage of the inverter for a high input is given by

$$v_{OUT} = V_S \frac{R_{ON}}{R_{ON} + R_L}.$$

The output $v_{OUT}$ is 0.33 V for $V_S = 5$ V, $R_{ON} = 1$ kΩ and $R_L = 14$ kΩ.

We need our inverter to produce an output lower than 0.2 V for a high input. In other words,

$$0.2\text{V} > V_S \frac{R_{ON}}{R_{ON} + R_L}. \tag{6.7}$$

We have three choices to reduce the output voltage: reduce $V_S$, reduce $R_{ON}$, or increase $R_L$. Reducing $V_S$ will also reduce the output high voltage, so that is not such a good strategy. Instead, we will look to working with the resistances.

First, let us try to increase $R_L$. Rearranging Equation 6.7, we get:

$$R_L > V_S \frac{R_{ON}}{0.2} - R_{ON}.$$

For $V_S = 5$ V and $R_{ON} = 1$ kΩ, we have

$$R_L > 24 \text{ k}\Omega.$$

In other words, we can choose $R_L > 24$ kΩ, which will result in a output voltage for a logical 0 that is lower than 0.2 V. However, it turns out that large values of resistance are hard to achieve in VLSI technology. Section 6.11 shows how another MOSFET can be used in place of the pullup resistor.

Alternatively, we can try to reduce $R_{ON}$ by increasing the $W/L$ ratio of the MOSFET. Let us determine the minimum $W/L$ ratio.

From Equation 6.7, we can find the constraint on $R_{ON}$ that allows the output low voltage to be less than 0.2 V as

$$R_{ON} < \frac{0.2 R_L}{V_S - 0.2}.$$

For $V_S = 5$ V and $R_L = 14$ kΩ, we have

$$R_{ON} < 0.58 \text{ k}.$$

Since $R_{ON} = R_n \frac{L}{W}$, (see Equation 6.4), and assuming $R_n = 5$ kΩ for our MOSFET, we get

$$5 \text{ k}\Omega \frac{L}{W} < 0.58 \text{ k}\Omega.$$

In other words, choosing a MOSFET with $W/L > 8.62$ will result in an output voltage for a logical 0 that is lower than 0.2 V.

### 6.8.1 STATIC ANALYSIS OF THE NAND GATE USING THE SR MODEL

We can also analyze other gates in like manner. Figure 6.42 shows the equivalent circuit for the NAND gate shown in Figure 6.20 based on the SR MOSFET model.

In this case, the output voltage when both inputs are high is given by

$$v_{OUT} = V_S \frac{2R_{ON}}{2R_{ON} + R_L}.$$

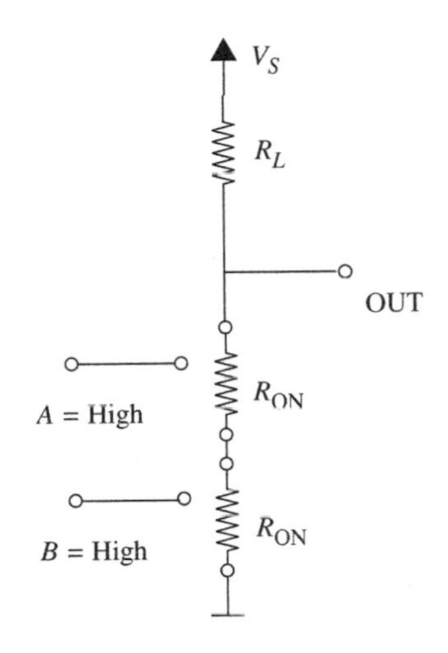


FIGURE 6.42 SR circuit model for NAND gate.

Let us determine whether our NAND gate with $V_S = 5$ V, $R_L = 14$ kΩ, and MOSFET properties $R_{ON} = 1$ kΩ and $V_T = 1$ V, satisfies a static discipline with the following voltage thresholds:

$$V_{OH} = 4.5 \text{ V}, V_{OL} = 0.5 \text{ V}, V_{IH} = 4 \text{ V, and } V_{IL} = 0.9 \text{ V}.$$

Recall that our inverter with characteristics shown in Figure 6.40 satisfied this static discipline. Now let's check out our NAND gate. Figure 6.43 shows the voltage thresholds for the given static discipline superimposed on the inverter transfer function. As before, let us check each of the output and input thresholds.

Like the inverter, the NAND gate produces an output high of 5 V, and therefore satisfies the output-high voltage threshold of 4.5 V. Similarly, the NAND gate satisfies both the $V_{IH} = 4$ V and the $V_{IL} = 0.9$ V thresholds since it interprets voltages above 4 V at its input as a logical 1 and voltages below 0.9 V at its inputs as a logical 0.

Let us now look at $V_{OL}$. When outputting a logical 0, the NAND gate produces a voltage

$$v_{OUT} = V_S \frac{2R_{ON}}{2R_{ON} + R_L}.$$

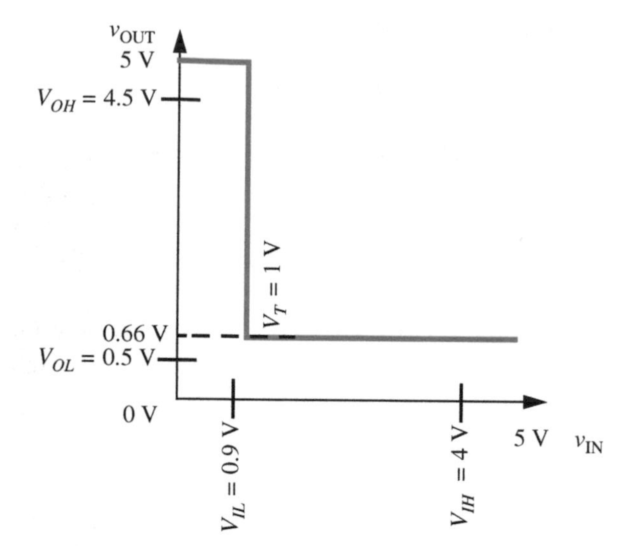


**FIGURE 6.43** The voltage levels corresponding to a static discipline superimposed on the NAND gate's transfer characteristics.

For $V_S = 5$ V, $R_L = 14\,\text{k}\Omega$, and $R_{ON} = 1\,\text{k}\Omega$, we get $v_{OUT} = 0.625$ V, which is nearly twice that produced by the inverter for a logical 0 output. This is not surprising since there are two MOSFETs in series in the pulldown network. Since this output voltage is greater than $V_{OL} = 0.5$ V, we conclude that our NAND does not satisfy the static discipline.

How might we redesign our NAND gate such that it satisfies the static discipline? One approach is to increase $R_L$ such that

$$0.5\text{ V} > V_S \frac{2R_{ON}}{2R_{ON} + R_L}.$$

In other words,

$$R_L > V_S \frac{2R_{ON}}{0.5} - 2R_{ON}.$$

For $V_S = 5$ V and $R_{ON} = 1$ k$\Omega$, we have

$$R_L > 18\text{ k}\Omega.$$

This means that we can choose $R_L > 18$ k$\Omega$ for the NAND gate, which will result in a output voltage for a logical 0 that is lower than 0.5 V, thereby satisfying the static discipline.

EXAMPLE 6.7 SWITCHING ANALYSIS OF A NAND GATE

Consider the NAND gate in Figure 6.42. Assume the following values for the circuit parameters: $V_S = 5$ V, $V_T = 1$ V, and $R_L = 10$ kΩ. Assume, further, that $R_n = 5$ kΩ for each of the MOSFETs. Determine a $W/L$ sizing for the MOSFETs so that the NAND gate output is able to switch ON or OFF the MOSFET of another gate such as an inverter.

Since the NAND gate produces a high output of 5 V, its output applied to the gate of another MOSFET (with a $V_T = 1$ V) can clearly drive the MOSFET into its ON state.

We now have to determine whether its low output can turn a MOSFET OFF. Recall that the NAND gate produces the following output voltage for a logical 0:

$$v_{OUT} = V_S \frac{2R_{ON}}{2R_{ON} + R_L}.$$

For a MOSFET driven by the output to remain OFF, we must have

$$v_{OUT} = V_S \frac{2R_{ON}}{2R_{ON} + R_L} < V_T.$$

From Equation 6.4, we know that the ON state resistance of the MOSFETs is given by

$$R_{ON} = R_n \frac{L}{W}.$$

Thus, we can write the following constraint on the $W/L$ ratio of the two NAND gate MOSFETs:

$$V_S \frac{2R_n \frac{L}{W}}{2R_n \frac{L}{W} + R_L} < V_T.$$

Simplifying, we obtain the following constraint on $W/L$:

$$\frac{W}{L} > \frac{2R_n(V_S - V_T)}{V_T R_L}.$$

Substituting, $R_n = 5$ kΩ, $V_S = 5$ V, $V_T = 1$ V, and $R_L = 10$ kΩ, we obtain

$$\frac{W}{L} > 4.$$

In other words, the two MOSFETs must be sized such that each of their $W/L$ ratios is greater than 4. Notice that as the number of MOSFETs connected in series increases, so must their sizes to ensure they produce a low enough output voltage that is able to switch OFF a MOSFET connected to the output.

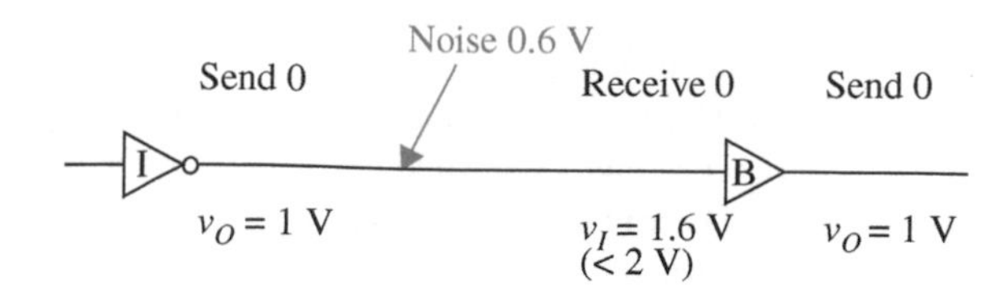


FIGURE 6.44 Noise margins and signal transmission.

## 6.9 SIGNAL RESTORATION, GAIN, AND NONLINEARITY

We saw in an earlier chapter (Figure 5.7) that the provision of noise margins enables error-free communication in the presence of noise. We will revisit the example in Figure 5.7 to demonstrate that logic devices must incorporate both *gain* and *nonlinearity* to provide nonzero noise margins.[23]

### 6.9.1 SIGNAL RESTORATION AND GAIN

Figure 6.44 shows a situation similar to that in Figure 5.7, but for concreteness, replaces the first logic gate with an inverter I and the second with a buffer B. Like the inverter, the buffer has a single input and a single output. It performs the identity function, that is it simply copies the input value to its output. This time around, we will focus on the conditions at the buffer. Assume that both our logic gates adhere to a static discipline with the following voltage levels:

$$V_{IL} = 2 \text{ V}$$

$$V_{IH} = 3 \text{ V}$$

$$V_{OL} = 1 \text{ V}$$

$$V_{OH} = 4 \text{ V}.$$

In our example, the inverter sends a 0 by placing $v_{OUT} = 1$ V (corresponding to $V_{OL}$) on the wire. Figure 6.44 shows 0.6 V of noise being added to the signal by the transmission channel. However, the buffer is able to correctly interpret the received value as a 0 because the received value of 1.6 V is within the low input voltage threshold of $V_{IL} = 2$ V. The buffer, in turn, performs the identity logical operation on the signal and produces a logical 0 at its output. According to the static discipline, the voltage level at the buffer's output is 1 V corresponding to $V_{OL}$.

Figure 6.45 shows the same situation replacing the actual voltage levels with the respective parameters for transmitting a logical 0 and a logical 1.

23. We will see the concept of gain showing up again in the context of analog design in Chapter 7.

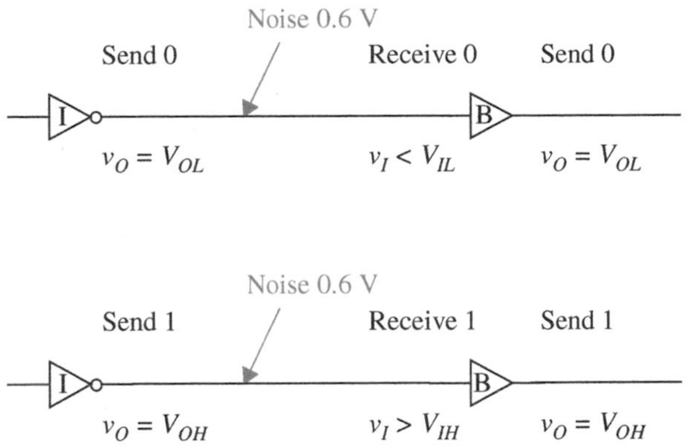

Noise 0.6 V
Send 0 Receive 0 Send 0
$v_O = V_{OL}$ $v_I < V_{IL}$ $v_O = V_{OL}$

Noise 0.6 V
Send 1 Receive 1 Send 1
$v_O = V_{OH}$ $v_I > V_{IH}$ $v_O = V_{OH}$

**FIGURE 6.45** Low and high thresholds for the input and output.

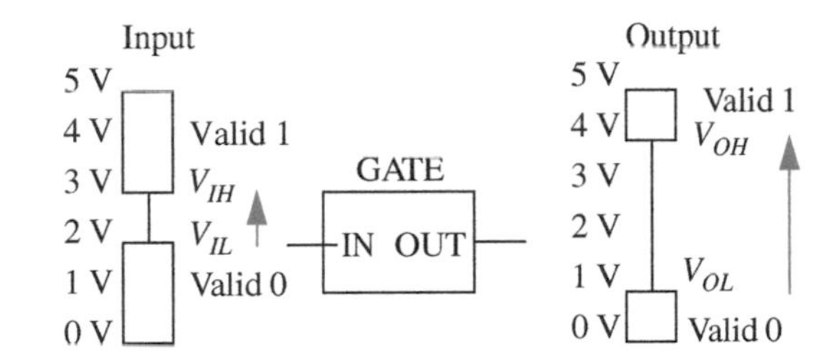


**FIGURE 6.46** Signal restoration and amplification.

In Figure 6.44, notice that to obey the static discipline, the buffer must convert the 1.6-V signal at its input to a 1-V value at its output. In fact, the buffer must *restore* any voltage up to 2 V at its input to voltage of 1 V or lower at its output. Similarly, corresponding to a logical high, it must restore any voltage above 3 V at its input to 4 V or higher at its output. This restoration property is key to our being able to *compose* multiple logic devices together. Because each level of logic restores or cleans up signals, we can decouple the noise introduced between each pair of levels. This noise decoupling benefit of restoring logic enables us to build complicated multistage logic systems.

As Figure 6.46 depicts, logic devices must restore input signals that lie in the range $0\text{ V} < v_I < V_{IL}$ for logical 0's and $V_{IH} < v_I < 5\text{ V}$ for logical 1's to output signals that are restricted to the range $0\text{ V} < v_O < V_{OL}$ for logical 0's and $V_{OH} < v_O < 5\text{ V}$ for logical 1's, respectively.

Observe further that the restrictions in Figure 6.46 imply that a non-inverting device such as a buffer or an AND gate must convert an input low to high transition of the form $V_{IL} \rightarrow V_{IH}$ to an output low to high transition of the form $V_{OL}$ (or lower) $\rightarrow$ $V_{OH}$ (or higher). This scenario is depicted by the arrows in Figure 6.46. The same situation is described using input and output waveforms in Figure 6.47. It is clear from the figures that a static discipline

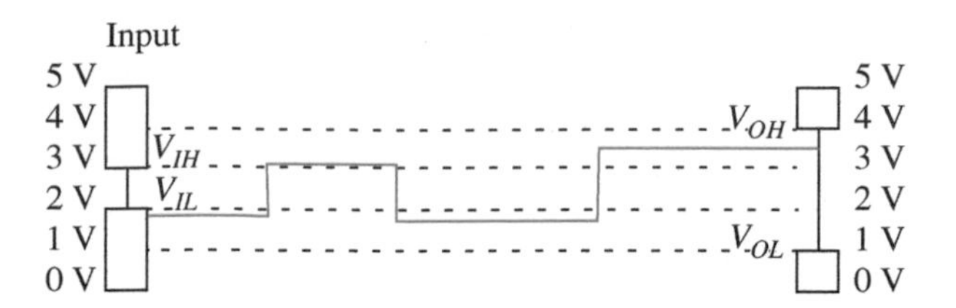


FIGURE 6.47 Input waveform and restored output waveform.

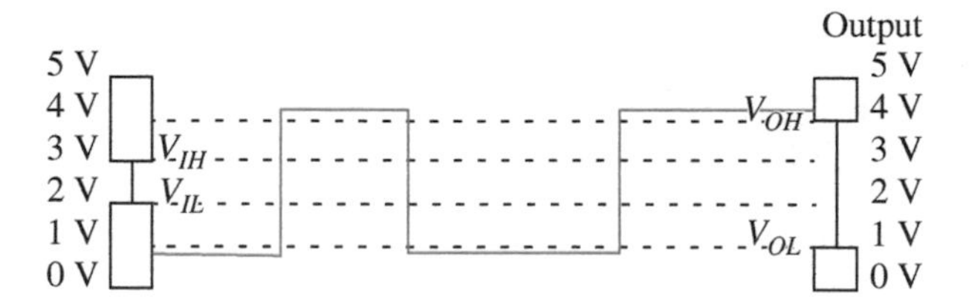


that provides for nonzero noise margins requires logic devices that provide a minimum gain.

Algebraically, nonzero noise margins require that

$$V_{IL} > V_{OL}. \tag{6.8}$$

and

$$V_{OH} > V_{IH}. \tag{6.9}$$

The magnitude of the change in the voltage for an input transition from $V_{IL} \rightarrow V_{IH}$ is given by

$$\Delta v_I = V_{IH} - V_{IL}.$$

The corresponding magnitude of the (minimum) change at the output is given by

$$\Delta v_O = V_{OH} - V_{OL}.$$

Therefore, the gain of a device that can convert a $V_{IL} \rightarrow V_{IH}$ transition at its input to a $V_{OL} \rightarrow V_{OH}$ transition at its output is given by

$$\text{Gain} = \frac{\Delta v_O}{\Delta v_I} = \frac{V_{OH} - V_{OL}}{V_{IH} - V_{IL}}.$$

From the noise-margin inequalities in Equations 6.8 and 6.9, we have

$$V_{OH} - V_{OL} > V_{IH} - V_{IL}.$$

Therefore, the magnitude of the gain for an input transition $V_{IL} \rightarrow V_{IH}$ must be greater than 1. In other words,

$$\text{Gain} = \frac{V_{OH} - V_{OL}}{V_{IH} - V_{IL}} > 1. \tag{6.10}$$

Similarly, inverting devices such as inverters or NAND gates must convert input low to high transitions of the form $V_{IL} \rightarrow V_{IH}$ to output high to low transitions of the form $V_{OH} \rightarrow V_{OL}$. Like the non-inverting case, the conditions on the magnitude of the gain for transitions from $V_{IL} \rightarrow V_{IH}$ remain unchanged.

Returning to our buffer example, the gain for the $V_{IL} \rightarrow V_{IH}$ transition is given by

$$\begin{aligned} \text{Gain} &= \frac{V_{OH} - V_{OL}}{V_{IH} - V_{IL}} \\ &= \frac{4V - 1V}{3V - 2V} \\ &= 3. \end{aligned}$$

Since the buffer and the inverter follow the same voltage thresholds, the magnitude of the gain for the $V_{IL} \rightarrow V_{IH}$ transition at the input of the inverter is also 3. Clearly, the greater the noise margins, the greater the required gain for the $V_{IL} \rightarrow V_{IH}$ transition.

### 6.9.2 SIGNAL RESTORATION AND NONLINEARITY

You might have realized that although logic devices must demonstrate a gain greater than unity when they transition from $V_{IL}$ to $V_{IH}$, they must also attenuate the signal at other times. For example, Figure 6.48 shows the signal from Figure 6.47 with some noise superimposed on it. It should be clear from Figure 6.48 that to obey the static discipline the buffer has reduced the 0-V to 2-V noise excursions at the input to 0-V to 1-V noise excursions at its output.

We can also verify this fact using the basic noise-margin inequalities in Equations 6.8 and 6.9. Equations 6.8 implies that any voltage between 0 and $V_{IL}$ at the input must be attenuated to a voltage between 0 and $V_{OL}$ at the output (see Figure 6.49). Since $V_{IL} > V_{OL}$ according to Equation 6.8, it follows that voltage transfer ratio must be less than unity. In other words,

$$\frac{V_{OL} - 0}{V_{IL} - 0} = \frac{V_{OL}}{V_{IL}} < 1.$$

The same reasoning applies to valid high voltages. Because $V_{IH} < V_{OH}$,

$$\frac{5 - V_{OH}}{5 - V_{IH}} < 1.$$

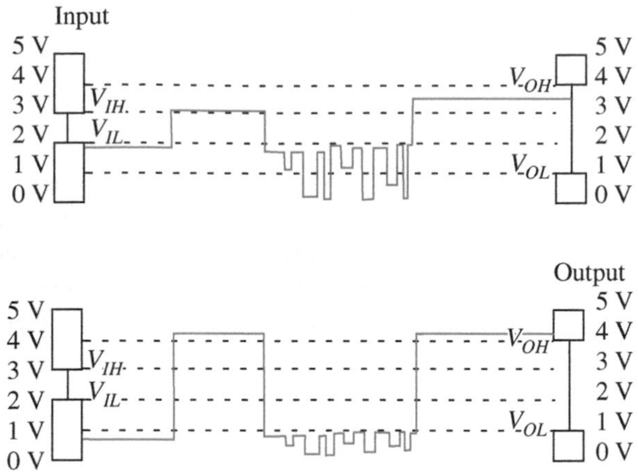

FIGURE 6.48 Input waveform and restored output waveform in the presence of noise.

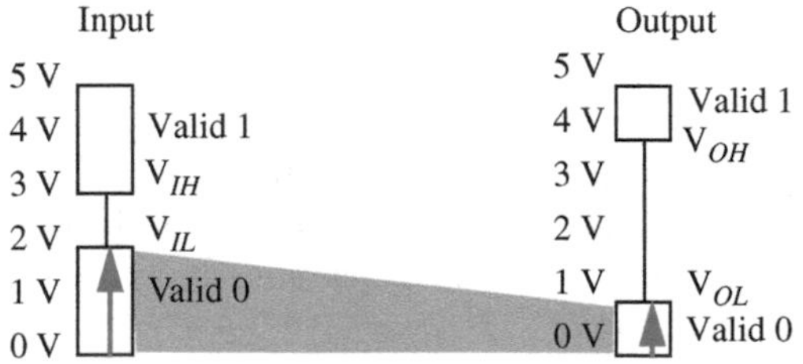

FIGURE 6.49 Signal restoration and attenuation.

The amplification requirement for low to high transitions of the form $V_{IL} \rightarrow V_{IH}$, and the attenuation requirement in other regions, mandates the use of nonlinear devices in logic gates.[24]

### 6.9.3 BUFFER TRANSFER CHARACTERISTICS AND THE STATIC DISCIPLINE

The presence of gain and nonlinearity in the buffer become abundantly clear if we look at its transfer characteristic. Figure 6.50 graphically plots the transfer characteristic of a logic device that can serve as a valid buffer. The shaded region depicts the valid region for the buffer transfer curve. The x-axis shows input voltages and the y-axis output voltages. We can make several interesting observations from this graph. Notice that valid input voltages result in valid output voltages. For example, input voltages less than $V_{IL}$ produce output voltages less than $V_{OL}$, and input voltages greater than $V_{IH}$ produce output voltages greater than $V_{OH}$. Also notice that the amplification occurs when the

24. As an exercise in futility, you might want to attempt building a simple logic gate, such as a buffer, with nonzero noise margins, using resistors alone.

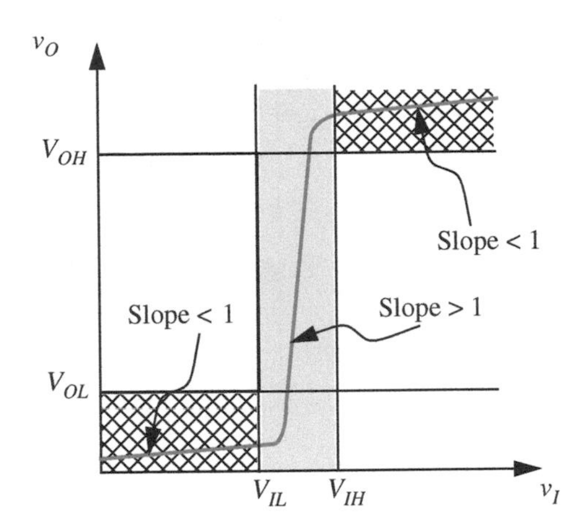


**FIGURE 6.50** The buffer characteristic.

curve is in the forbidden region. In other words, the slope of the transfer curve in the forbidden region is greater than one.

As discussed earlier, recall that it is not sufficient for a valid logic device to have gain in the forbidden region. The transfer characteristic for input values between 0 V and $V_{IL}$ must have an overall gain of less than unity. Accordingly, notice that the transfer curve shown in Figure 6.50 attenuates voltages that lie in the valid input low or valid input high intervals.

Observe further that the transfer curve for the buffer passes through the forbidden region. Doesn't this violate our initial premise that voltages in the forbidden region were disallowed? Recall that the static discipline requires the logic gate to guarantee valid outputs only for valid inputs. That the outputs are in the forbidden region when the inputs are invalid is of no consequence.

### 6.9.4 INVERTER TRANSFER CHARACTERISTICS AND THE STATIC DISCIPLINE

Let us now briefly examine a transfer curve for the hypothetical, but valid, inverter shown in Figure 6.51. Referring to Figure 6.51, provided the input voltage is lower than $V_{IL}$, an inverter satisfies the static discipline if it guarantees to provide an output voltage level greater than $V_{OH}$. Similarly, for an input voltage higher than $V_{IH}$, the inverter guarantees to provide an output that is below $V_{OL}$.

As discussed for the non-inverting buffer, the magnitude of the slope of the transfer curve in the forbidden region is greater than unity. Similarly, the magnitude of the slope of each of the curve segments in the valid regions is less than unity.

For maximum noise immunity, the separation between $V_{OL}$ and $V_{OH}$ should be as high as possible at output, and that between $V_{IL}$ and $V_{IH}$ to be as low as possible. This is equivalent to maximizing the area of the grey boxes

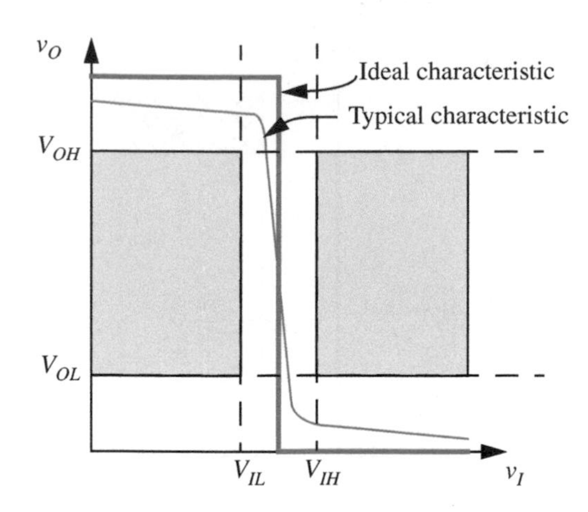


FIGURE 6.51 The inverter characteristic.

shown in Figure 6.51. An ideal inverter characteristic will look like the thick line shown in the figure.

## 6.10 POWER CONSUMPTION IN LOGIC GATES

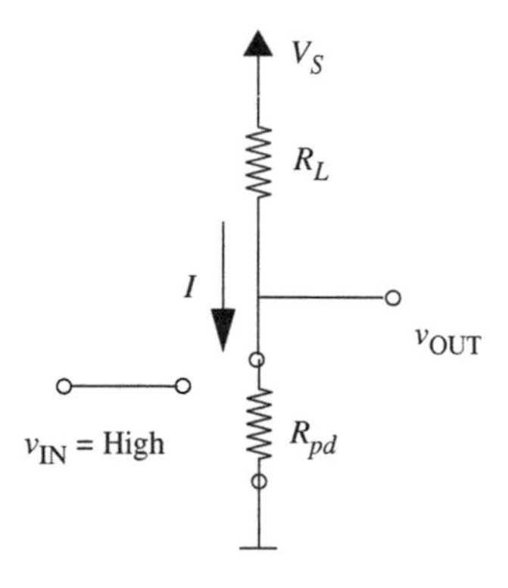


FIGURE 6.52 Power consumption in logic gates.

We can use the SR model to calculate the maximum power consumed by logic gates. We consider a simple case here, and postpone more discussion to Chapter 11. Referring to Figure 6.52, the power consumed by a logic gate is given by

$$\text{Power} = V_S I = \frac{V_S^2}{R_L + R_{pd}}. \tag{6.11}$$

The power consumed depends on the load resistance and the resistance of the pulldown network $R_{pd}$. For the inverter, the power consumed is zero when the input is low. The maximum power is consumed when the input is high and $R_{pd} = R_{ON}$.

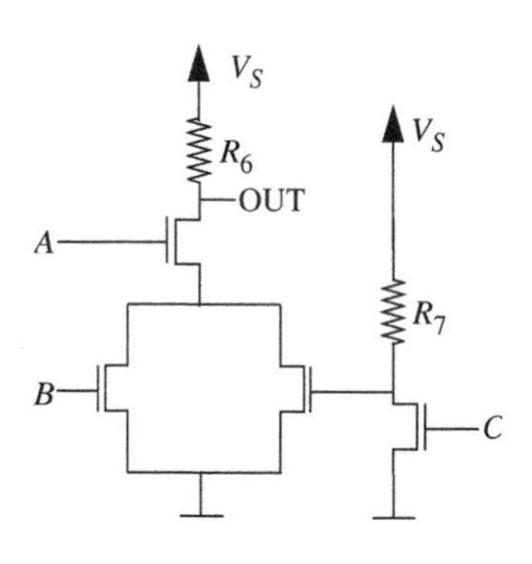


FIGURE 6.53 A logic circuit comprising MOSFET switches and resistors.

---

EXAMPLE 6.8 POWER IN LOGIC GATES Write a boolean equation for OUT in terms of the inputs for the circuit in Figure 6.53.

$$\text{OUT} = \overline{A(B + \bar{C})}.$$

Determine the power consumed by the circuit when $A = 1$, $B = 1$, and $C = 1$. Assume that the on-state resistance of the MOSFETs is $R_{ON}$.

When all the inputs are high, the relevant equivalent circuit is shown in Figure 6.54. The power is given by

$$P = V_S^2 \left( \frac{1}{2R_{ON} + R_6} + \frac{1}{R_{ON} + R_7} \right).$$

---

## WWW 6.11 ACTIVE PULLUPS

### WWW EXAMPLE 6.9 SIZING PULLUP DEVICES

### WWW EXAMPLE 6.10 COMBINATIONAL LOGIC USING MOSFET SWITCHES

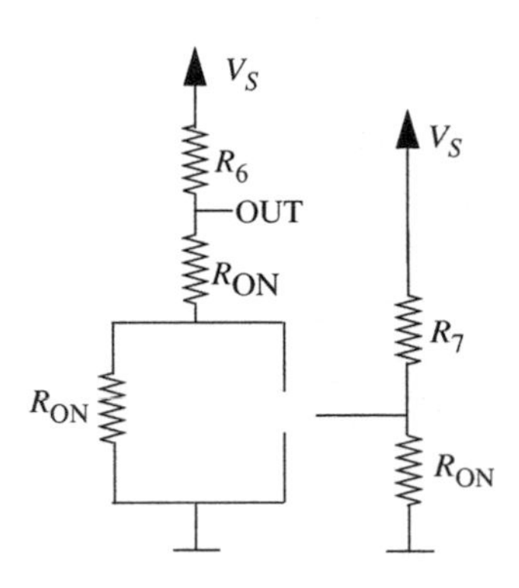


**FIGURE 6.54** Equivalent circuit when all inputs are high.

## 6.12 SUMMARY

- This chapter introduced our first three-terminal device, namely a switch. A common way of using three-terminal devices is to pair their terminals into a pair of ports called the control port and the output port.
- We also introduced the MOSFET device, which is a three-terminal circuit element. Although, as we will see in Chapter 7, the MOSFET has a very rich behavior, it can be grossly characterized as a switch. We developed the S and the SR models for the MOSFET, which capture its basic switch-like behavior.
- This chapter also showed how digital gates could be built using MOSFETs and resistors. We discussed how the digital circuits had to be designed so they met the $V_{IH}$, $V_{OH}$, $V_{IL}$, and $V_{OL}$ voltage thresholds specified by a given static discipline. We also estimated the power dissipation of logic gates using their circuit models.

## EXERCISES

EXERCISE 6.1 Give a resistor-MOSFET implementation of the following logic functions. Use the S model of the MOSFET for this exercise (in other words, you may assume that the on-state resistance of the MOSFETs is 0).

$$(A + B) \cdot (C + D)$$

$$\overline{A} \cdot \overline{B} \cdot C \cdot D$$

$$(\overline{Y \cdot W})(\overline{X \cdot W})(\overline{\overline{X} \cdot Y \cdot \overline{W}})$$

EXERCISE 6.2 Write a boolean expression that describes the function of each of the circuits in Figure 6.59.

EXERCISE 6.3 Figure 6.60 shows an inverter circuit using a MOSFET and a resistor. The MOSFET has a threshold voltage $V_T = 2$ V. Assume that $V_S = 5$ V and $R_L = 10$ kΩ. For this exercise, model the MOSFET using its switch model. In other words, assume that the on-state resistance of the MOSFET is 0.

a) Draw the input versus output voltage transfer curve for the inverter.

b) Does the inverter satisfy the static discipline for the voltage thresholds $V_{OL} = 1$ V, $V_{IL} = 1.5$ V, $V_{OH} = 4$ V and $V_{IH} = 3$ V? Explain. (Hint: To satisfy the static discipline, the inverter must interpret correctly input values that are valid logic signals. Furthermore, given valid logic inputs, the inverter must also output valid logic signals. Valid logic 0 input signals are represented by voltages less than $V_{IL}$, valid logic 1 input signals are represented by voltages greater than $V_{IH}$, valid logic 0 output signals are represented by voltages less than $V_{OL}$, and valid logic 1 output signals are represented by voltages greater than $V_{OH}$.)

c) Does the inverter satisfy the static discipline if the $V_{IL}$ specification was changed to $V_{IL} = 2.5$ V? Explain.

d) What is the maximum value of $V_{IL}$ for which the inverter will satisfy the static discipline?

e) What is the minimum value of $V_{IH}$ for which the inverter will satisfy the static discipline?

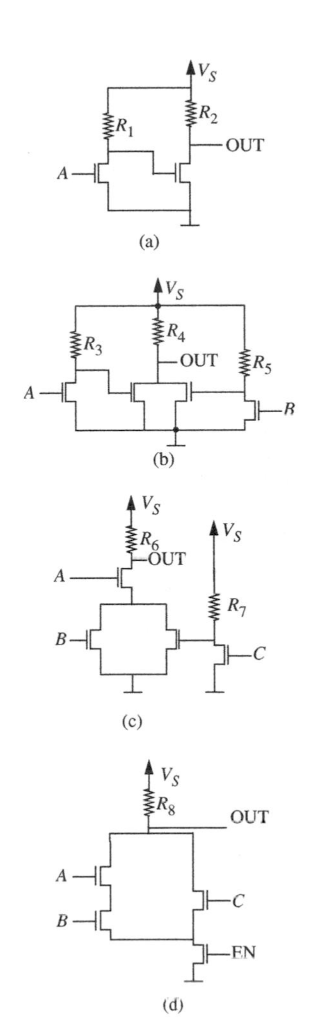


FIGURE 6.59

EXERCISE 6.4 Consider, again, the inverter circuit shown in Figure 6.60. The MOSFET has a threshold voltage $V_T = 2$ V. Assume that $V_S = 5$ V and $R_L = 10k$. For this exercise, model the MOSFET using its switch-resistor model. Assume that the on-state resistance of the MOSFET is $R_{ON} = 8$ kΩ.

a) Does the inverter satisfy the static discipline, which has voltage thresholds given by $V_{OL} = V_{IL} = 1$ V and $V_{OH} = V_{IH} = 4$ V? Explain.

b) Does the inverter satisfy the static discipline for the voltage thresholds $V_{OL} = V_{IL} = 2.5$ V and $V_{OH} = V_{IH} = 3$ V? Explain.

c) Draw the input versus output voltage transfer curve for the inverter.

d) Is there any value of $V_{IL}$ for which the inverter will satisfy the static discipline? Explain.

e) Now assume that $R_{ON} = 1k$ and repeat parts (a), (b), and (c).

EXERCISE 6.5 Compute the worst-case power consumed by the inverter shown in Figure 6.60. The MOSFET has a threshold voltage $V_T = 2$ V. Assume that $V_S = 5$ V and $R_L = 10$ kΩ. Model the MOSFET using its switch-resistor model, and assume that the on-state resistance of the MOSFET is $R_{ON} = 1$ kΩ.

EXERCISE 6.6 Consider again the circuits in Figure 6.59. Using the switch-resistor model of the MOSFET, choose minimum values for the various resistors in Figure 6.59 so each circuit satisfies the static discipline with voltage thresholds given by $V_{IL} = V_{OL} = V_S/10$ and $V_{IH} = V_{OH} = 4V_S/5$. Assume the on-state resistance of the MOSFET is $R_{ON}$ and that its turn-on threshold voltage $V_T = V_S/9$.

EXERCISE 6.7 Consider a family of logic gates that operates under the static discipline with the following voltage thresholds: $V_{OL} = 0.5$ V, $V_{IL} = 1.6$ V, $V_{OH} = 4.4$ V, and $V_{IH} = 3.2$ V.

a) Graph an input-output voltage transfer function of a buffer satisfying the four voltage thresholds.

b) What is the highest voltage that can be output by an inverter for a logical 0 output?

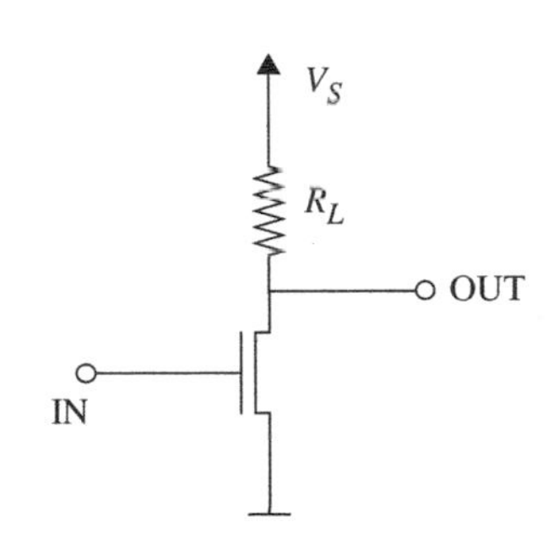


FIGURE 6.60

c) What is the lowest voltage that can be output by an inverter for a logical 1 output?

d) What is the highest voltage that must be interpreted by a receiver as a logical 0?

e) What is the lowest voltage that must be interpreted by a receiver as a logical 1?

f) What is the 0 noise margin provided by this logic family?

g) What is the 1 noise margin provided by this logic family?

h) What is the minimum voltage gain the buffer must provide in the forbidden region?

## PROBLEMS

### PROBLEM 6.1

a) Write a truth table and a boolean equation relating the output $Z$ to $A$, $\overline{A}$, $B$, and $C$, when these are input to the circuit shown in Figure 6.61.

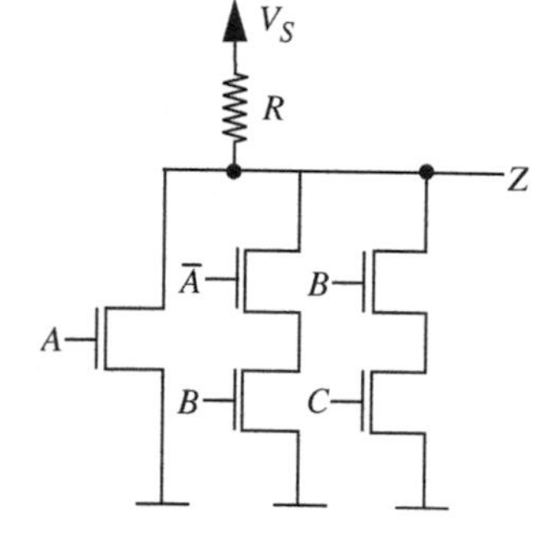


FIGURE 6.61

b) Suppose the circuit in Figure 6.61 suffers a manufacturing error that results in a short between the pair of wires depicted in Figure 6.62. Write a truth table and a boolean equation relating the output $Z$ to $A$, $\overline{A}$, $B$, and $C$, for the resulting circuit.

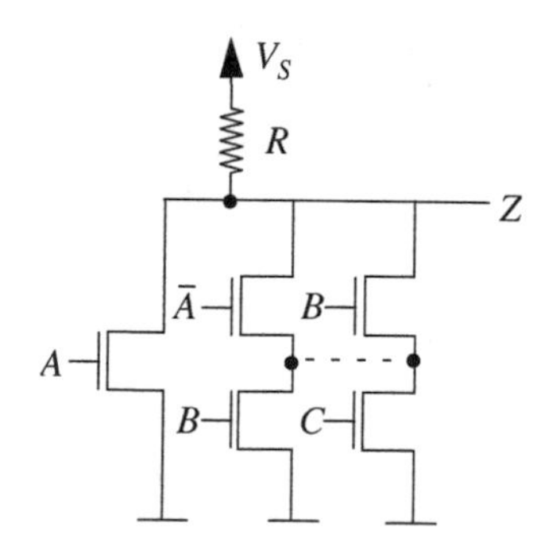


FIGURE 6.62

PROBLEM 6.2 A specific type of MOSFET has $V_T = -1$ V. The MOSFET is in the ON state (a short exists between its drain and source) when $v_{GS} \geq V_T$. The MOSFET is in the OFF state (an open circuit exists between its drain and source) when $v_{GS} < V_T$.

(a) Graph the $i_{DS}$ versus $v_{GS}$ characteristics of this MOSFET.

(b) Graph the $i_{DS}$ versus $v_{DS}$ characteristics this of the MOSFET for $v_{GS} \geq V_T$ and $v_{GS} < V_T$.

PROBLEM 6.3 Consider a family of logic gates that operate under the static discipline with the following voltage thresholds: $V_{OL} = 1$ V, $V_{IL} = 1.3$ V, $V_{OH} = 4$ V, and $V_{IH} = 3$ V. Consider the N-input NAND gate design shown in Figure 6.63. In the design $R = 100k$ and $R_{ON}$ for the MOSFETs is given to be $1k$. $V_T$ for the MOSFETs is 1.5 V. What is the maximum value of N for which the NAND gate will satisfy the static discipline? What is the maximum power dissipated by the NAND gate for this value of N?

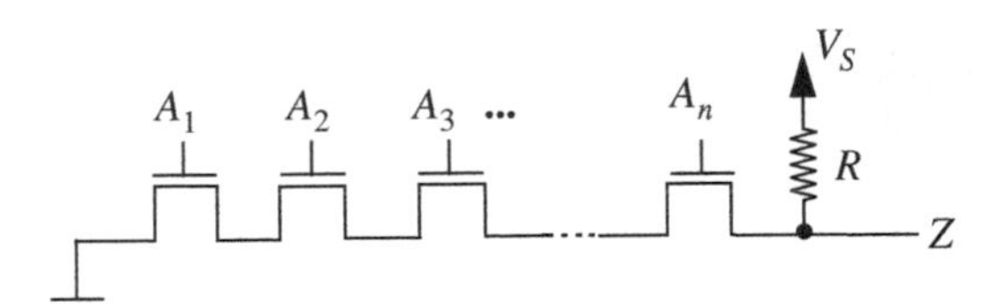


FIGURE 6.63

PROBLEM 6.4 Consider the N-input NOR gate shown in Figure 6.64. Assume that the on-state resistance of each of the MOSFETs is $R_{ON}$. For what set of inputs does this gate consume the maximum amount of power? Compute this worst-case power.

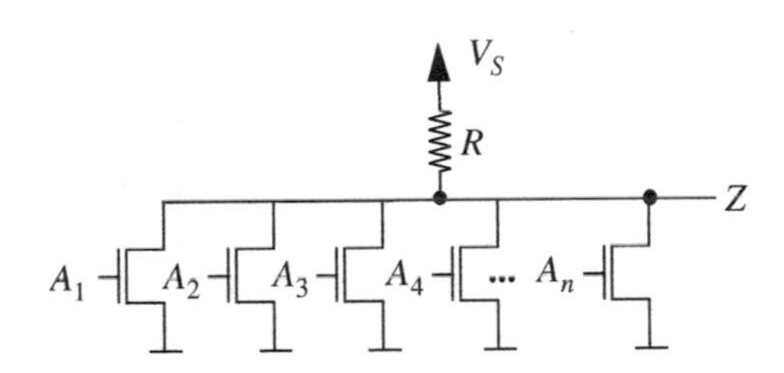


FIGURE 6.64

PROBLEM 6.5 Consider the circuit shown in Figure 6.65. We wish to design the circuit so it operates under a static discipline with voltage thresholds $V_{OL}$, $V_{IL}$, $V_{OH}$, and

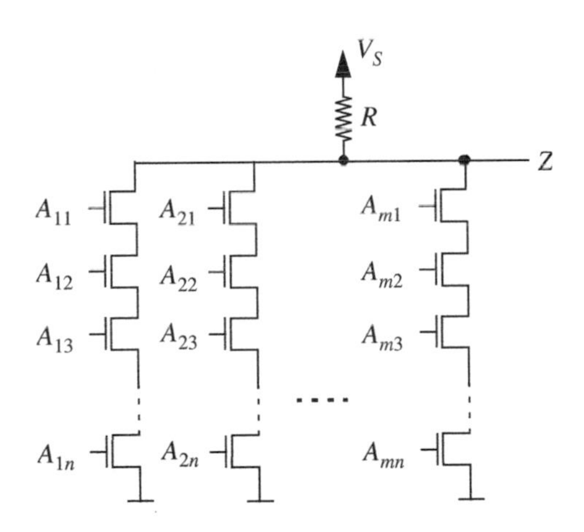


FIGURE 6.65

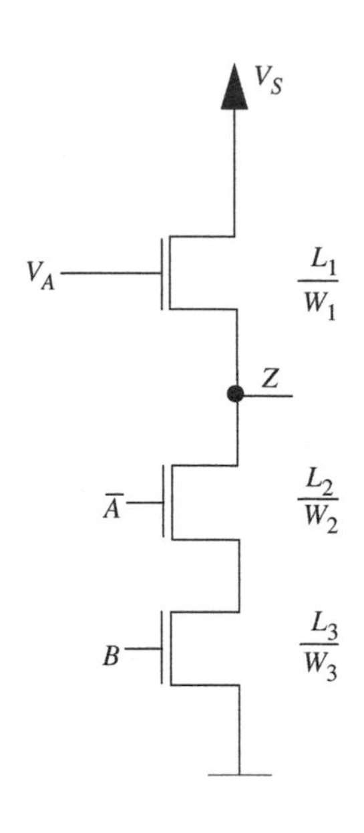


FIGURE 6.66

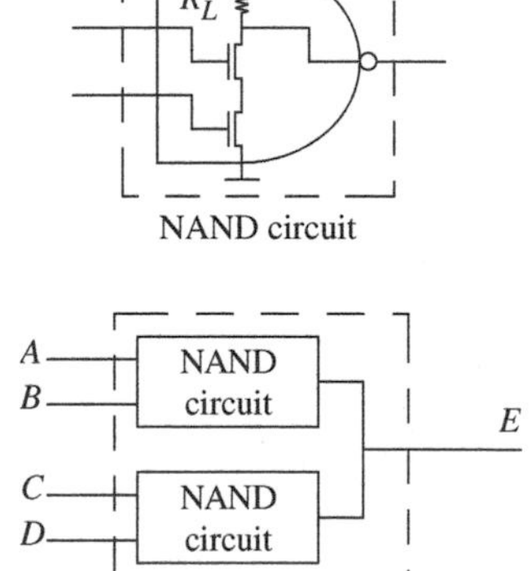


FIGURE 6.67

$V_{IH}$. Assume that the on-state resistance of each of the MOSFETs is $R_{ON}$ and that the MOSFET threshold voltage is $V_T$. Assume that the given values satisfy the constraints $V_S \geq V_{OH}$ and $V_{IL} < V_T$. For what values of $n$ and $m$ does this gate operate under the static discipline? What is the worst case power consumed by this circuit?

PROBLEM 6.6 Consider a family of logic gates that operate under the static discipline with the following voltage thresholds: $V_{OL} = 0.5$ V, $V_{IL} = 1$ V, $V_{OH} = 4.5$ V, and $V_{IH} = 4.0$ V.

a) Graph an input-output voltage transfer function of an inverter satisfying the four voltage thresholds.

b) Using the switch-resistor MOSFET model, design an inverter satisfying the static discipline for the four voltage thresholds using an n-channel MOSFET and a resistor. The MOSFET has $R_n = 1\,\text{k}\Omega$ and $V_T = 1.8$ V. Recall, $R_{ON} = R_n(L/W)$. Assume $V_S = 5$ V and $R_\square$ for a resistor is 500 Ω. Further assume that the area of the inverter is given by the sum of the areas of the MOSFET and the resistor. Assume that the area of a device is $L \times W$. The inverter should take as little area as possible with minimum size for $L$ or $W$ being 0.5 μm. Graph the input-output transfer function of the inverter. What is the total area of the inverter? What is its maximum static power dissipation?

PROBLEM 6.7 Consider a family of logic gates that operates under the static discipline with the following voltage thresholds: $V_{OL} = 0.5$ V, $V_{IL} = 0.9$ V, $V_{OH} = 4.5$ V, and $V_{IH} = 4$ V. Using the switch-resistor MOSFET model, design a 2-input NAND gate satisfying the static discipline for the four voltage thresholds using three n-channel MOSFETs as illustrated in Figure 6.66 (the MOSFET with its gate connected to a voltage $V_A$ and drain connected to the power supply $V_S$ serves as the pullup). $V_A$ is chosen such that $V_A > V_S + V_T$. The MOSFETs have $R_n = 1\,\text{k}\Omega$ and $V_T = 1.8$ V. Recall, $R_{ON} = R_n(L/W)$. Assume $V_S = 5$ V. Further assume that the area of the NAND gate is given by the sum of the areas of the three MOSFETs. Assume that the area of a device is $L \times W$. The NAND gate should take as little area as possible with minimum size for $L$ or $W$ being 0.5 μm. What is the total area of the NAND gate?

PROBLEM 6.8 Remember that a NAND gate can be implemented as a circuit with two n-channel MOSFETs and a pullup resistor $R_L$. Let us call it the NAND circuit shown in Figure 6.67. These NAND circuits are used by Penny-Wise Computer Corporation in their computer boards. In one ill-fated shipment of computer boards, the outputs of a pair of NAND circuits get shorted accidentally resulting in the effective Circuit $X$ shown in Figure 6.67.

a) What logic function does Circuit $X$ implement? Construct its truth table.

b) If we connect $n$ identical NAND circuits together in parallel forming Circuit $Y$ as shown in Figure 6.68, what is the general form of the logic function it implements?

c) If for each MOSFET, $R_{on} = 500\ \Omega$, $R_L = 100\,\text{k}\Omega$, and $V_T = 1.8$ V, how many NAND circuits can we connect in parallel and still satisfy the static discipline for the voltage thresholds given by: $V_{IL} = V_{OL} = 0.5$ V and $V_{IH} = V_{OH} = 4.5$ V?

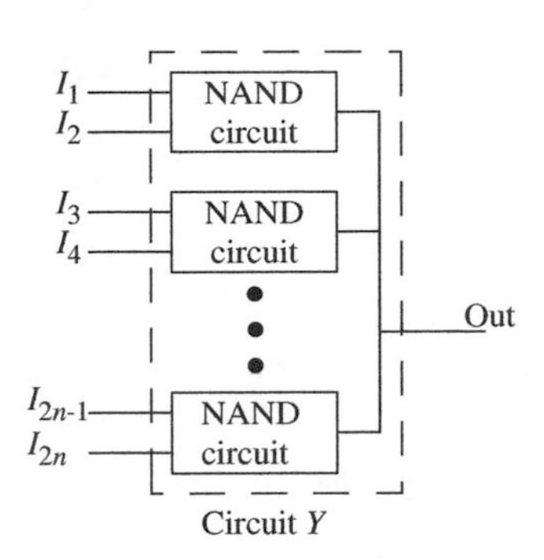


FIGURE 6.68

d) We now connect identical NAND circuits together and have the resulting Circuit $Y$ satisfy the static discipline for the voltage thresholds in part (c) with $R_L = 500\ \Omega$. Give specifications on the dimensions of the MOSFETs such that total MOSFET area is minimized. As before, assume that the area of a device is $L \times W$. Assume that $R_n = 1\,\text{k}\Omega$ and no resistor dimension or MOSFET gate dimension should be smaller than 0.5 µm. For what inputs does Circuit $Y$ dissipate maximum static power, and what is that power?

e) Now, choose a static discipline with voltage thresholds given by: $V_{OL} = 0.5$ V, $V_{IL} = 1.6$ V, $V_{OH} = 4.4$ V, and $V_{IH} = 3.2$ V. As before, each MOSFET has $R_{on} = 500\ \Omega$, $R_L = 100\ \text{k}\Omega$, and $V_T = 1.8$ V. How many NAND circuits can we connect in parallel and still satisfy this static discipline?

f) Repeat part (d) assuming the voltage thresholds given in part (e).

PROBLEM 6.9 Consider a family of logic gates that operates under the static discipline with the following voltage thresholds: $V_{OL} = 0.5$ V, $V_{IL} = 1.6$ V, $V_{OH} = 4.4$ V, and $V_{IH} = 3.2$ V.

a) Graph an input-output voltage transfer function of an inverter satisfying the four voltage thresholds.

b) Using the switch-resistor MOSFET model, design an inverter satisfying the static discipline for the four voltage thresholds using an n-channel MOSFET with $R_n = 1\ \text{k}\Omega$ and $V_T = 1.8$ V. Recall, $R_{on} = R_n(L/W)$. Assume $V_S = 5$ V and $R_\square$ for a resistor is 500 Ω. Further assume that the area of the inverter is given by the sum of the areas of the MOSFET and the resistor. Assume that the area of a device is $L \times W$. The inverter should take as little area as possible with minimum size for $L$ or $W$ being 0.5 µm. Graph the input-output transfer function of the inverter. What is the total area of the inverter? What is its static power dissipation?

# CHAPTER 7

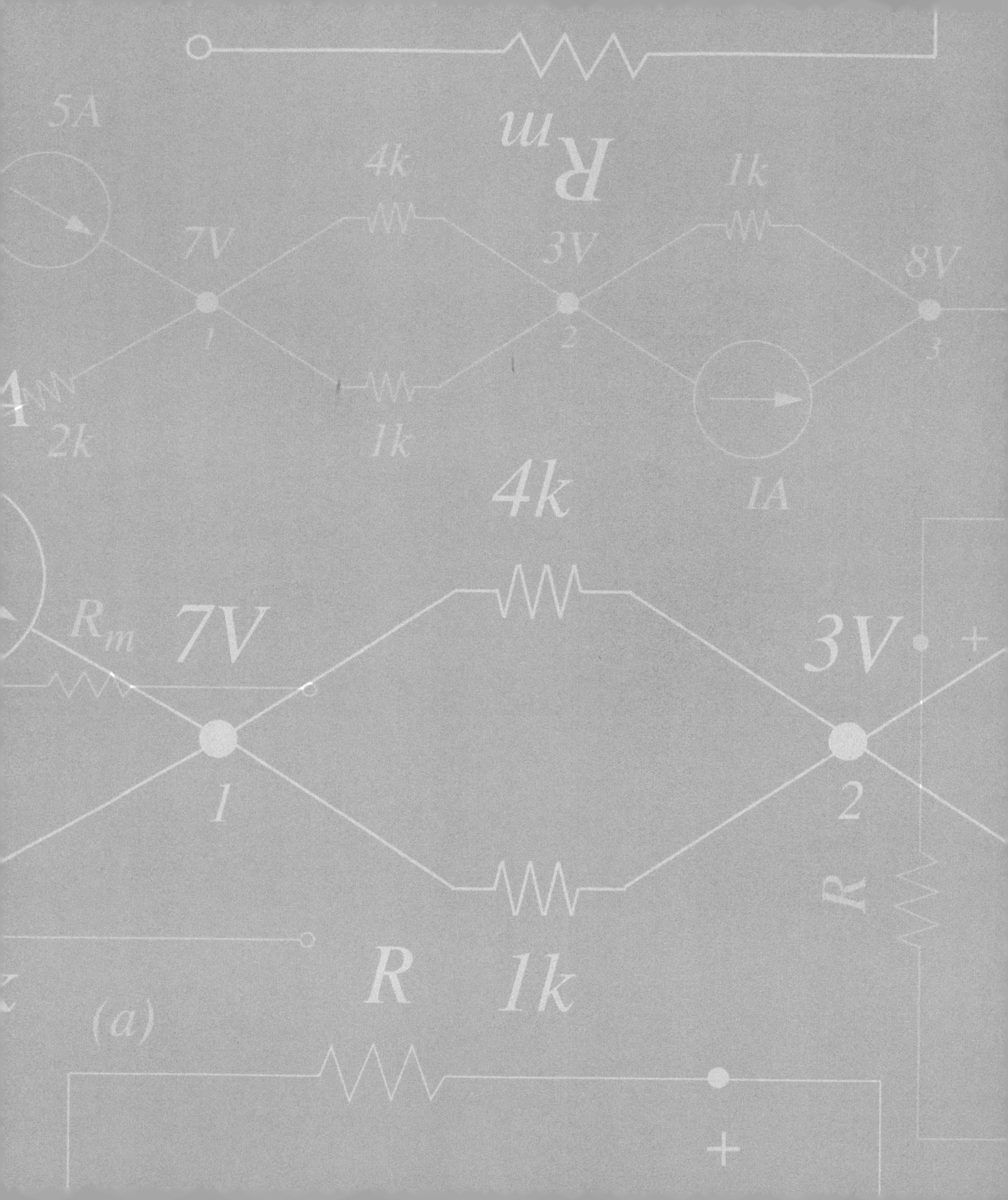

5A
4k
1k
7V
3V
8V
1
2
3
2k
1k
1A
4k
$R_m$
7V
3V
+
1
2
R
R
1k
(a)
+

# THE MOSFET AMPLIFIER

# 7

## 7.1 SIGNAL AMPLIFICATION

This chapter introduces the notion of amplification. Amplification, or gain, is key to both analog and digital processing of signals. Section 6.9.2 discussed how gain is employed in digital systems to achieve immunity to noise. This chapter will focus on the analog domain.

Amplifiers abound in the devices we use in our day to day life, such as stereos, loud speakers, and cell phones. Amplifiers can be represented as shown in Figure 7.1 as three-ported devices with a control input port, an output port, and a power port. Each port comprises two terminals. An input signal represented as a time-varying voltage or current is applied across or through the input terminals. An amplified version of the signal (either a voltage or a current) appears at the output. Depending on its internal structure, an amplifier can amplify the input current, the input voltage, or both. When the $V \times I$ product of the output exceeds that of the input, a power gain results. The power supply provides the necessary power for the resulting power amplification. The power supply also provides for the internal power consumption within the amplifier as well. A device must provide power gain to be called an amplifier.[1]

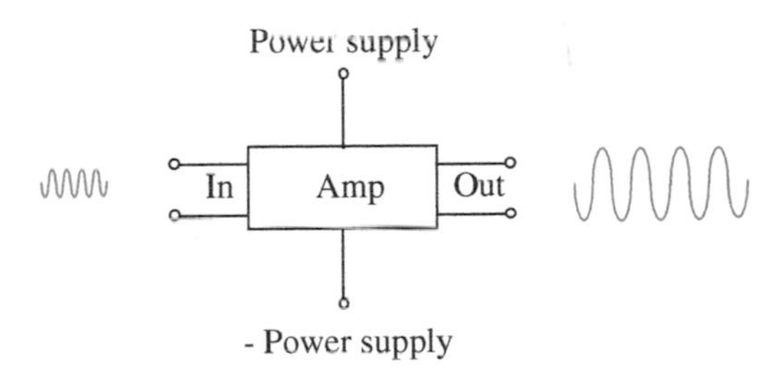


FIGURE 7.1 Signal amplification.

In practical amplifier designs, the input and the output signals commonly share a reference ground connection (see Figure 7.2). Correspondingly, one

1. We will see later a device called a transformer, which can provide a voltage gain but no power gain.

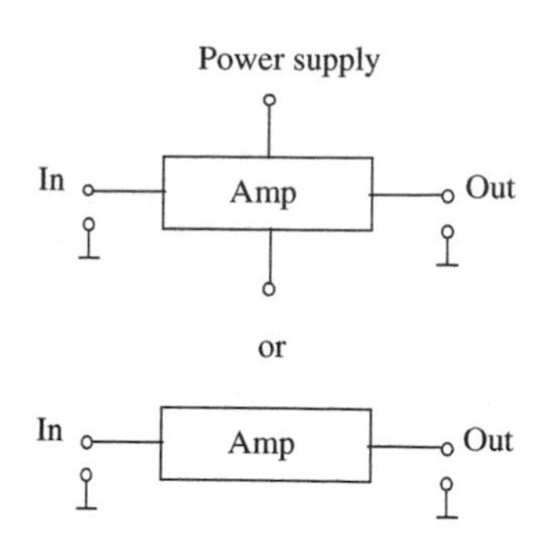


FIGURE 7.2 Reference ground and implicit power connections.

of the terminals of each of the ports is commonly tied to a reference ground. Furthermore, the power port is commonly not shown explicitly.

Besides their use in communication to overcome the dissipative effects of the communication medium (for example, in loud speakers or in wireless networking systems), amplifiers are useful for signal transmission in the presence of noise. Figures 7.3 and 7.4 show two signal transmission scenarios. In Figure 7.3, the signal transmitted in its native form is overwhelmed by noise at the receiver. In Figure 7.4, however, the amplified signal is seen to be much more tolerant to noise. (Contrast this with the application of amplification for noise immunity in digital systems as discussed in Section 6.9.2.)

A less obvious but equally important application of amplifiers is *buffering*. As the name implies, a buffer isolates one part of a system from another. Buffers allow us to compose complicated systems from smaller components by isolating the behavior of the individual components from each other. Many sensors, for example, produce a voltage signal, but cannot supply a large amount of current. (For instance, they might have a high Thévenin resistance.) However, later processing stages might require that the device supply a given amount of current. If this high current is drawn, a large voltage drop across the internal resistance of the sensor seriously attenuates the output voltage. In such situations, we might employ a buffer device that replicates the sensor's voltage signal at its output but can also provide a large amount of current. In such buffering applications, we shall often see amplifiers with less than unity voltage gain, but greater than unit current and power gain.

## 7.2 REVIEW OF DEPENDENT SOURCES

Before we get into the design and analysis of amplifiers, let us take a moment to review dependent sources. Because amplifiers are naturally modeled using

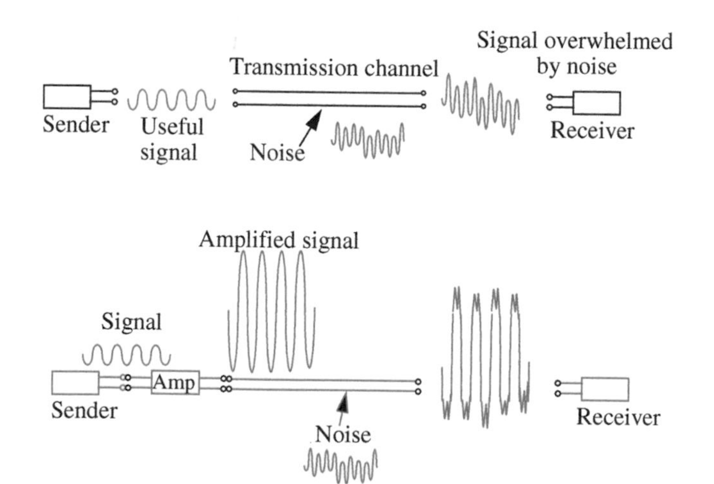


FIGURE 7.3 Signal transmission in the presence of noise.

FIGURE 7.4 Amplification provides noise tolerance.

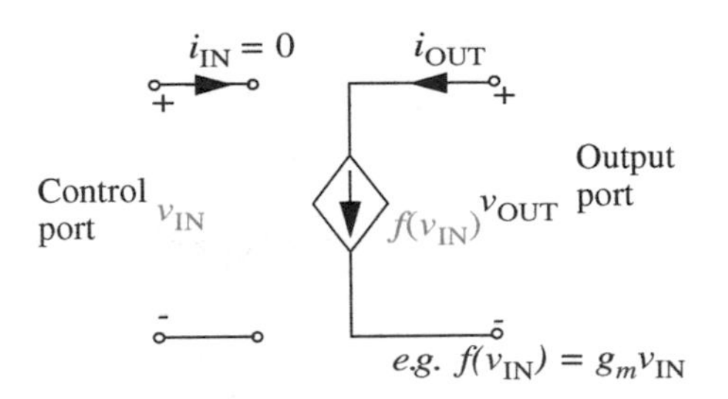


FIGURE 7.5 Voltage-controlled current source.

dependent sources, analysis of circuits with dependent sources will come in handy in their design.

Dependent sources serve to model control of energy or information flow. Recall that control of energy or information flow was one of five basic processes identified in Section 1.6 in Chapter 1. Figure 7.5 shows the familiar voltage controlled current source that we saw in Chapter 2. As we shall illustrate shortly, small amounts of energy at the control port of such dependent sources can control or steer huge amounts of energy at the output port.

---

EXAMPLE 7.1 VOLTAGE-CONTROLLED CURRENT SOURCE CIRCUIT Consider the circuit shown in Figure 7.6. Let $v_I$ be the voltage sourced by an independent voltage source. The current $i_O = f(x)$ produced by the dependent current source is a function of other values in the circuit. Let us first analyze the circuit when the output of the current source depends on a voltage:

$$i_O = f(v_I) = -g_m v_I.$$

*$v_O$ versus $v_I$* Let us attempt to determine $v_O$ as a function of $v_I$. Figure 7.6 shows our selection of a ground node and the labeling of nodes with their node voltage variables. $v_O$ is the only unknown node voltage. Writing the corresponding node equation, we get

$$\frac{v_O}{R_L} = i_O = f(v_I).$$

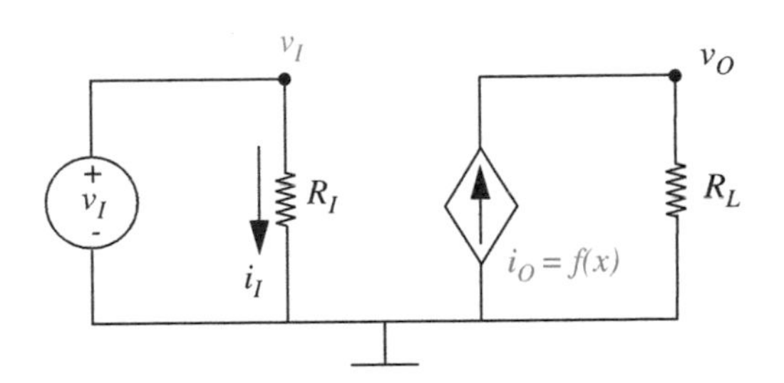


FIGURE 7.6 A circuit using a Voltage-controlled current source.

Since $f(v_I) = -g_m v_I$, we get our node equation

$$\frac{v_O}{R_L} = -g_m v_I. \tag{7.1}$$

Equation 7.1 shows the desired relationship between $v_O$ and $v_I$ and completes our solution. The voltage gain is given by

$$\frac{v_O}{v_I} = -g_m R_L.$$

Notice that we obtain a voltage gain greater than unity if $g_m R_L > 1$. Thus, the circuit in Figure 7.6 behaves as an amplifier for properly chosen values of $R_L$. In other words, the circuit produces an amplified version of $v_I$ at its output $v_O$. Section 7.4 will introduce a physical device that behaves as a voltage-controlled current source and develops an amplifier based on that device.

*$i_O$ versus $i_I$* Next, let us determine $i_O$ versus $i_I$. Substituting $v_I = i_I R_I$ and $v_O = i_O R_L$ in Equation 7.1, we can write

$$i_O R_L = -g_m R_L i_I R_I,$$

which simplifies to

$$i_O = -g_m R_I i_I. \tag{7.2}$$

Thus, the current gain is given by

$$\frac{i_O}{i_I} = -g_m R_I.$$

Notice that the dependent source provides a current *gain* greater than unity if $g_m R_I > 1$.

*$P_O$ versus $P_I$* Let us now determine the input power $P_I$ versus the output power $P_O$. By multiplying the left-hand sides and the right-hand sides of Equations 7.1 and 7.2, we can write

$$v_O i_O = g_m^2 R_L R_I v_I i_I. \tag{7.3}$$

In other words,

$$P_O = g_m^2 R_L R_I P_I.$$

The power gain is given by

$$\frac{P_O}{P_I} = g_m^2 R_L R_I.$$

Thus, the dependent source provides a power gain greater than unity when $g_m^2 R_L R_I > 1$.

---

EXAMPLE 7.2 CURRENT-CONTROLLED CURRENT SOURCE
Let us rework the circuit of Figure 7.6 and obtain its $v_O$ versus $v_I$ relation assuming that the output of the current source depends on a current:

$$i_O = f(i_I) = -\beta i_I$$

where $\beta$ is a constant. As before, let us attempt to determine $v_O$ as a function of $v_I$. Writing the node equation,

$$\frac{v_O}{R_L} = i_O = f(i_I)$$

or,

$$\frac{v_O}{R_L} = f(i_I).$$

Substituting $f(i_I) = -\beta i_I$, we get our desired node equation

$$\frac{v_O}{R_L} = -\beta i_I.$$

Since $i_I = v_I/R_I$, we get

$$v_O = -\beta \frac{R_L}{R_I} v_I. \tag{7.4}$$

Equation 7.4 gives the relationship between $v_O$ and $v_I$ and completes our solution.

---

## 7.3 ACTUAL MOSFET CHARACTERISTICS

Chapter 6 introduced the MOSFET and developed simple digital logic circuits using the device. That chapter also used the simplistic S model and the SR model of the MOSFET to analyze digital logic circuits. The SR model uses a fixed $R_{ON}$ between the D and S terminals of the MOSFET when $v_{GS} \geq V_T$. This model is a reasonable representation of MOSFET behavior only when the drain voltage is smaller than the gate voltage minus one threshold drop. In other words, when

$$v_{DS} < v_{GS} - V_T. \tag{7.5}$$

Accordingly, the SR model is useful to design and analyze digital circuit gates because a common mode of operation for the MOSFET within digital gates is one in which the gate voltage is high and the drain voltage is relatively low. For example, we might have $V_{OH} = 4$ V applied as a logical-high input to the gate of a MOSFET in an inverter (assume $V_T$ for the MOSFET is given to be 1 V), which might produce as the output a corresponding logical-low drain voltage $V_{OL} = 1$ V. With these values, $v_{DS} = 1$ V, $v_{GS} = 4$ V. Since $V_T = 1$ V, the constraint in Equation 7.5 is satisfied.

However, there are other situations demanding higher drain voltages in which we wish to use the MOSFET in an ON state. The SR model of the MOSFET is inappropriate in this region. This section will first show why the SR model is inadequate when $v_{DS} \geq v_{GS} - V_T$. We will then take a look at the actual MOSFET characteristics and then explore the possibility of creating a simple piecewise-linear model for the MOSFET in the region where $v_{DS} \geq v_{GS} - V_T$.

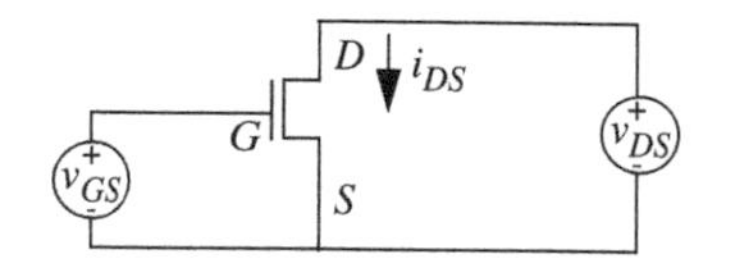


FIGURE 7.7 Setup for observing MOSFET characteristics.

We will use the setup shown in Figure 7.7 to observe the actual MOSFET characteristics. Let us start by applying a fixed, high gate-to-source voltage such that

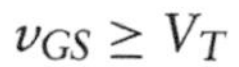

$$v_{GS} \geq V_T$$

and observing the value of $i_{DS}$ as the drain-to-source voltage $v_{DS}$ is increased. As illustrated in Figure 7.8, we observe that $i_{DS}$ increases more or less linearly as $v_{DS}$ is increased from 0 V. The approximately linear relationship between $i_{DS}$ and $v_{DS}$ exists for small values of $v_{DS}$, and

$$\frac{v_{DS}}{i_{DS}} = R_{\text{ON}}.$$

The linear relationship between $i_{DS}$ and $v_{DS}$ reflects resistive behavior for small $v_{DS}$, and is nicely captured by our SR model of the MOSFET.

Now, keeping the value of $v_{GS}$ at the same fixed value, we increase $v_{DS}$ further, and plot our observations in Figure 7.9. Notice that as $v_{DS}$ approaches the value of $v_{GS} - V_T$, the curve bends and begins to flatten out. In other words, the current $i_{DS}$ saturates as $v_{DS}$ begins to exceed $v_{GS} - V_T$. In fact,

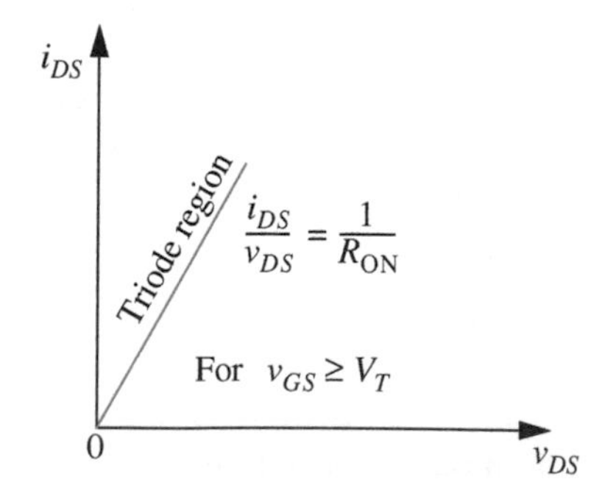


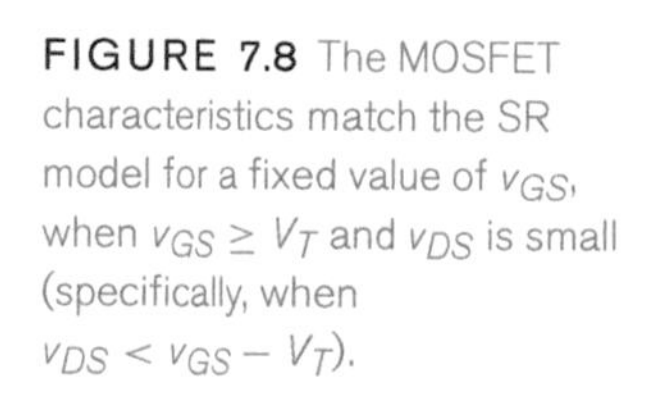

FIGURE 7.8 The MOSFET characteristics match the SR model for a fixed value of $v_{GS}$, when $v_{GS} \geq V_T$ and $v_{DS}$ is small (specifically, when $v_{DS} < v_{GS} - V_T$).

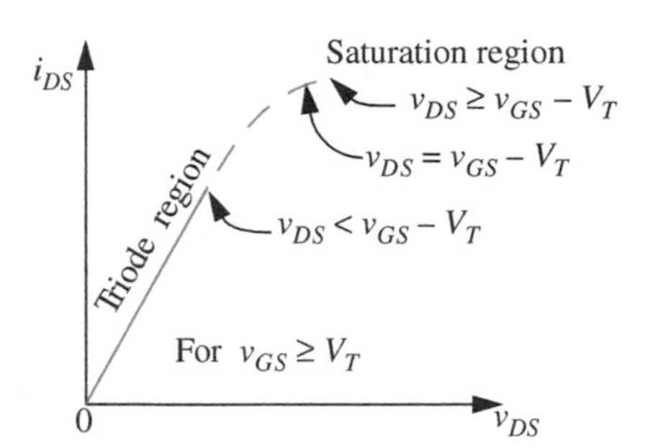


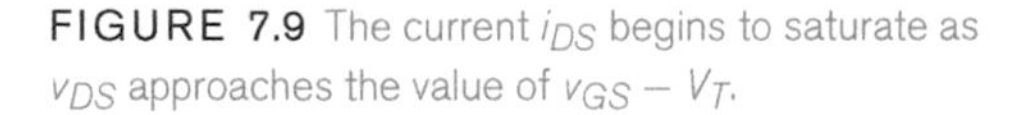

**FIGURE 7.9** The current $i_{DS}$ begins to saturate as $v_{DS}$ approaches the value of $v_{GS} - V_T$.

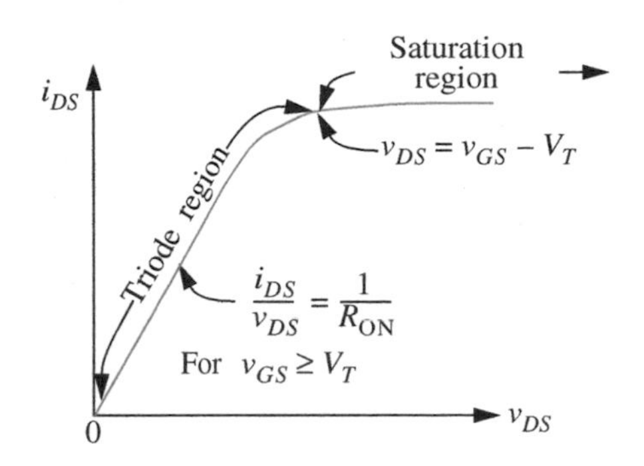


**FIGURE 7.10** The saturation region of MOSFET operation.

as illustrated in Figure 7.10, for a given value of $v_{GS}$, the $i_{DS}$ curve becomes virtually flat for large values of $v_{DS}$. Accordingly, the region where $v_{DS} \geq v_{GS} - V_T$ is called the *saturation region* of MOSFET operation. In contrast, the region where $v_{DS} < v_{GS} - V_T$ is called the *triode region.* Not surprisingly, the SR model applies with a fair degree of accuracy only in the triode region of MOSFET operation. In the saturation region, because $i_{DS}$ does not change as $v_{DS}$ increases, the MOSFET behaves like a current source. (Recall, from Figure 1.34, the $v$–$i$ curve for a current source is a horizontal line.)

The $i_{DS}$ curve in Figure 7.10 was measured keeping $v_{GS}$ constant at some value greater than $V_T$. It turns out that the $i_{DS}$ curves saturate at a different value for different values of $v_{GS}$. Thus, as illustrated in Figure 7.11, we get a different $i_{DS}$ versus $v_{DS}$ curve for each setting of $v_{GS}$ (for example, $v_{GS1}$, $v_{GS2}$, and so on), resulting in a family of $i_{DS}$ versus $v_{DS}$ curves. This family of curves represents the actual MOSFET characteristics. Notice that the slope of each of the curves in the triode region also varies somewhat with $v_{GS}$.

The actual MOSFET characteristics with the triode, saturation, and cutoff regions marked are shown in Figure 7.12. The dashed line represents the locus of the points for which

$$v_{DS} = v_{GS} - V_T.$$

The MOSFET is in cutoff for

$$v_{GS} < V_T.$$

The MOSFET operates in its triode region for points to the left of the dashed line, where

$$v_{DS} < v_{GS} - V_T \quad \text{and} \quad v_{GS} \geq V_T.$$

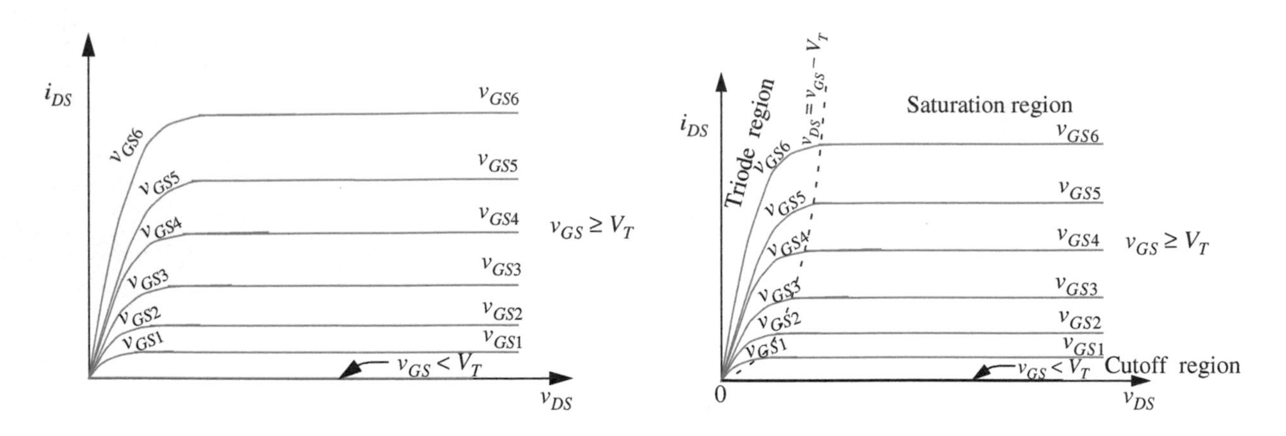


**FIGURE 7.11** Actual characteristics of the MOSFET. Each setting of $v_{GS}$ results in a separate $i_{DS}$ versus $v_{DS}$ curve.

**FIGURE 7.12** Actual characteristics of the MOSFET showing the triode, saturation, and cutoff regions.

The MOSFET operates in its saturation region for points to the right of the dashed line, where the following two conditions are met

$$v_{DS} \geq v_{GS} - V_T \quad \text{and} \quad v_{GS} \geq V_T.$$

*Saturation Region Operation of the MOSFET* A MOSFET operates in the saturation region when the following two conditions are met:

$$v_{GS} \geq V_T \tag{7.6}$$

and

$$v_{DS} \geq v_{GS} - V_T. \tag{7.7}$$

Given the more or less straight-line behavior of the $i_{DS}$ versus $v_{DS}$ curves to the left and to the right of the dashed line in Figure 7.12, it is natural to seek a piecewise-linear model for the MOSFET. Recall from Section 4.4, piecewise-linear modeling represents nonlinear $v$–$i$ characteristics by a succession of straight-line segments, and makes calculations within each straight-line segment using linear analysis tools. Figure 7.13 shows our choice of straight-line segments that model the actual MOSFET characteristics.

To the right of the $v_{DS} = v_{GS} - V_T$ boundary (represented by the dashed line in Figure 7.13) we have the saturation region, in which we use a set of horizontal straight-line segments (one for each value of $v_{GS}$) to represent the actual MOSFET characteristics. The straight-line segments representing the model are shown as thick grey lines. The circuit interpretation of each of the horizontal straight-line segments is a current source. Furthermore, because

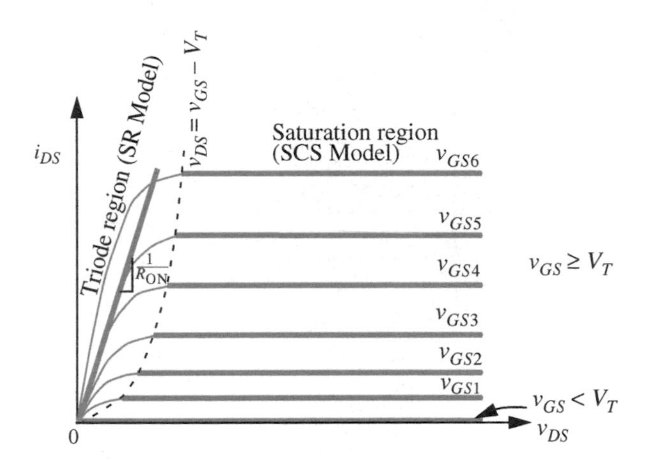


**FIGURE 7.13** SR and SCS models.

the value of the current depends on the value of $v_{GS}$, the behavior is that of a voltage-controlled current source. This behavior, captured by the *switch current source (SCS)* model of the MOSFET, applies only in the saturation region of the MOSFET. We will have a lot more to say about this saturation region model of the MOSFET in Section 7.4.

To the left of the $v_{DS} = v_{GS} - V_T$ boundary, we have the triode region, in which one possible modeling choice uses a single straight-line segment to approximate the $i_{DS}$ versus $v_{DS}$ curve for a given value of $v_{GS}$. Such a straight-line segment approximating the $i_{DS}$ versus $v_{DS}$ curve for a given $v_{GS}$ is shown as the thick grey line to the left of the $v_{DS} = v_{GS} - V_T$ boundary. You will notice that this choice of a single straight-line segment with a given slope $1/R_{ON}$ for a fixed value of $v_{GS}$ is our familiar SR model from Section 6.6. Intuitively, the single straight-line segment model suggests that the MOSFET behaves like a resistor with a fixed value $R_{ON}$ for a given value of $v_{GS}$, provided that $v_{GS} \geq V_T$ and $v_{DS} < v_{GS} - V_T$.

When the MOSFET curves are drawn using a compressed scale on the x-axis as in Figure 7.14, we see even the S model is not unreasonable in the triode region since it captures the gross characteristics of the MOSFET.

Of course, it is also possible to model the complete operation of the MOSFET (for any value of $v_{GS}$) using a more sophisticated nonlinear model. This results in the Switch Unified (SU) model, which is discussed further in Section 7.8. Although the SU model captures the complete characteristics of the MOSFET, for simplicity, we will focus on the SR and the SCS models. Accordingly, unless specifically mentioned otherwise, we will use the SR model for the triode region of the MOSFET when analyzing digital systems (since we work with a fixed high value of $v_{GS}$, where $v_{GS} \geq V_T$ and $v_{DS} < v_{GS} - V_T$), and the SCS model in the saturation region ($v_{GS} \geq V_T$ and $v_{DS} \geq v_{GS} - V_T$) for analog systems.

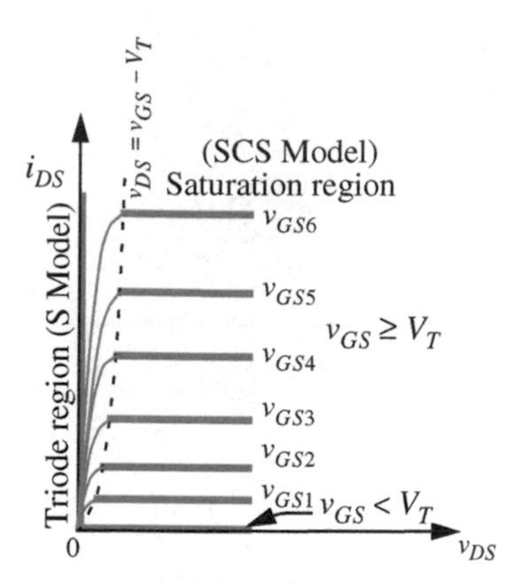


**FIGURE 7.14** S and SCS models.

As one final observation on the various models, notice in Figure 7.13 the discontinuity in the $i_{DS}$ versus $v_{DS}$ curve according to the SR model in the triode region, and the SCS model in the saturation region. In other words, if a MOSFET operates in a circuit such that $v_{DS} = v_{GS} - V_T$, the two models will give very different results. We can live with this discontinuity provided we do not attempt to reconcile the results from the two models in the same analysis.[2] You must choose between the two models depending on the particular situation. Specifically, use the SR model when operating with a fixed $v_{GS}$ in the triode region, and use the SCS model when operating in the saturation region. The SR model is appropriate for use in our digital circuits because of the inverting property of the type of digital circuits discussed here (for example, our familiar inverter). Since the drain voltage in our digital circuits is low when the gate voltage is high, the triode region of MOSFET operation applies, and therefore, the SR model is appropriate. Conversely, in the design of amplifiers, we will establish the saturation discipline, which will constrain amplifier designs to operate MOSFETs exclusively in their saturation region, thereby allowing the use of the SCS model.

## 7.4 THE SWITCH-CURRENT SOURCE (SCS) MOSFET MODEL

We saw in the previous section that when the gate voltage of the MOSFET is greater than the threshold voltage, and the drain voltage is greater than the gate voltage minus one threshold drop ($v_{DS} \geq v_{GS} - V_T$), a *voltage-controlled current source* model is appropriate for the MOSFET. The switch-current source

2. As discussed in Section 7.8, the SU model eliminates the discontinuity.

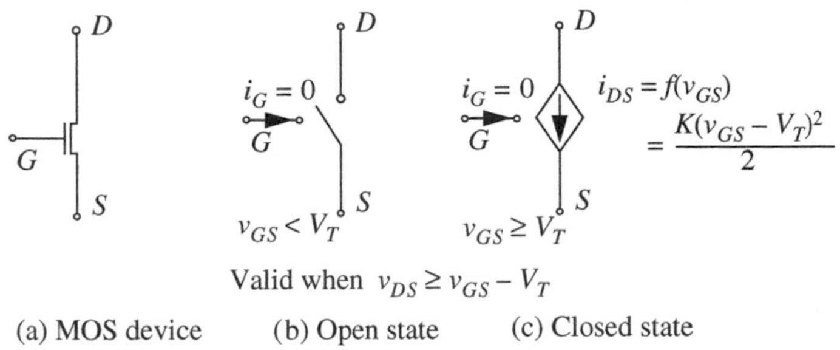

Valid when $v_{DS} \geq v_{GS} - V_T$

(a) MOS device (b) Open state (c) Closed state

**FIGURE 7.15** The switch-current source model of the MOSFET.

model (or SCS model) of the MOSFET captures this behavior and is depicted in Figure 7.15.

As depicted in Figure 7.15b, when $v_{GS} < V_T$, the MOSFET is OFF and an open circuit exists between the drain and the source. For the SCS model, the current $i_G$ into the gate terminal is zero.

When $v_{GS} \geq V_T$, and $v_{DS} \geq (v_{GS} - V_T)$, the amount of current provided by the source is given by

$$i_{DS} = \frac{K(v_{GS} - V_T)^2}{2} \tag{7.8}$$

where $K$ is a constant having units of $A/V^2$. The value of $K$ is related to the physical properties of the MOSFET.[3]

As in the OFF state, the current $i_G$ into the gate terminal is zero, reflecting an open circuit both between the gate and the source, and the gate and the drain.

As mentioned earlier, the region of operation in which $v_{DS} \geq (v_{GS} - V_T)$ is called the *saturation region*. The region in which $v_{DS} < (v_{GS} - V_T)$ is called the *triode region*. The characteristics of the MOSFET in the saturation region according to the SCS model are summarized graphically in Figure 7.16. Compare these characteristics with those for the SR model in the triode region displayed earlier in Figure 6.31.

The constraint curve separating the triode and saturation regions in Figure 7.16 given by

$$v_{DS} = v_{GS} - V_T \tag{7.10}$$

---

3. The parameter $K$ is related to the physical structure of the MOSFET as follows:

$$K = K_n \frac{W}{L}. \tag{7.9}$$

In Equation 7.9, $W$ is the MOSFET gate width and $L$ is the gate length. $K_n$ is a constant related to other MOSFET properties such as the thickness of its gate oxide.

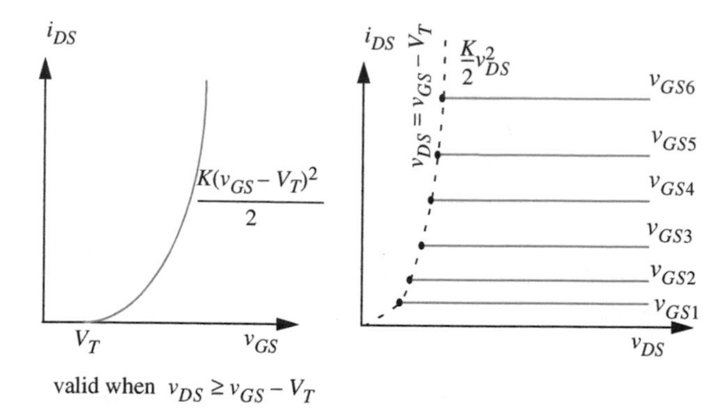


**FIGURE 7.16** Characteristics of the MOS device in the saturation region.

can also be rewritten in terms of $i_{DS}$ and $v_{DS}$ by substituting $v_{DS} = (v_{GS} - V_T)$ in Equation 7.8 as follows:

$$i_{DS} = \frac{K}{2} v_{DS}^2. \tag{7.11}$$

The following is a summary of the SCS model of the MOSFET in algebraic form. The model applies only in the saturation region of MOSFET operation, that is, when $v_{DS} \geq v_{GS} - V_T$.

$$i_{DS} = \begin{cases} \dfrac{K(v_{GS} - V_T)^2}{2} & \text{for } v_{GS} \geq V_T \text{ and } v_{DS} \geq v_{GS} - V_T \\ 0 & \text{for } v_{GS} < V_T. \end{cases} \tag{7.12}$$

---

EXAMPLE 7.3 A MOSFET CIRCUIT Determine the current $i_{DS}$ for the circuit in Figure 7.17. For the MOSFET, assume that $K = 1\ \text{mA}/\text{V}^2$ and $V_T = 1$ V.

It is easy to see that the MOSFET in Figure 7.17 is operating in its saturation region, since the drain-to-source voltage (5 V) is greater than $v_{GS} - V_T$ (2 V − 1 V = 1 V). Therefore, we can directly calculate the desired current using the MOSFET equation for saturation region operation

$$i_{DS} = \frac{K(v_{GS} - V_T)^2}{2}.$$

Substituting $v_{GS} = 2$ V, $K = 1\ \text{mA}/\text{V}^2$ and $V_T = 1$ V, we obtain $i_{DS} = 0.5$ mA.

---

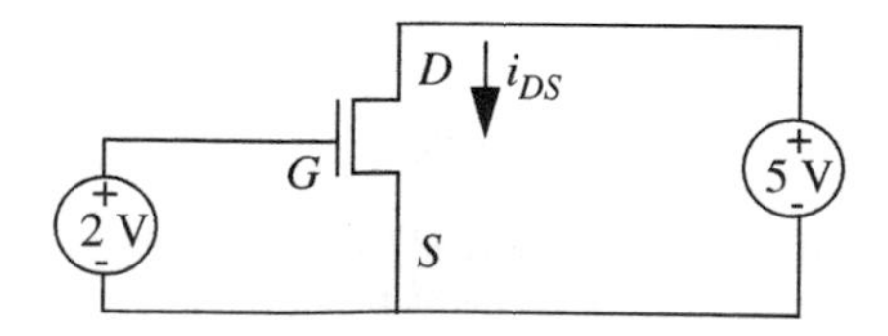


**FIGURE 7.17** A simple MOSFET circuit.

EXAMPLE 7.4 SATURATION REGION OPERATION Keeping the gate-to-source voltage for the MOSFET in the circuit shown in Figure 7.17 at 2 V, what is the minimum value of the drain-to-source voltage $v_{DS}$ for which the MOSFET will operate in saturation?

The MOSFET operates in saturation under the following constraints

$$v_{GS} \geq V_T$$

and

$$v_{DS} \geq v_{GS} - V_T.$$

Since $v_{GS}$ is given to be 2 V and $V_T$ is 1 V, the first constraint is satisfied. Substituting for $v_{GS}$ and $V_T$ in the second constraint, we obtain the following constraint on $v_{DS}$ for saturation region operation

$$v_{DS} \geq 1\text{ V}.$$

Thus the minimum value for $v_{DS}$ is 1 V.

EXAMPLE 7.5 SATURATION REGION OPERATION Next, keeping the drain-to-source voltage for the MOSFET in the circuit shown in Figure 7.17 at 5 V, what is the range of values for $v_{GS}$ for which the MOSFET will operate in saturation?

The lowest value for $v_{GS}$ is 1 V, since below that the MOSFET enters cutoff.

The highest value for $v_{GS}$ is determined by the constraint

$$v_{DS} \geq v_{GS} - V_T.$$

For $v_{DS} = 5$ V and $V_T = 1$ V, the highest value for $v_{GS}$ is 6 V. If $v_{GS}$ is increased beyond 6 V, the MOSFET enters the triode region.

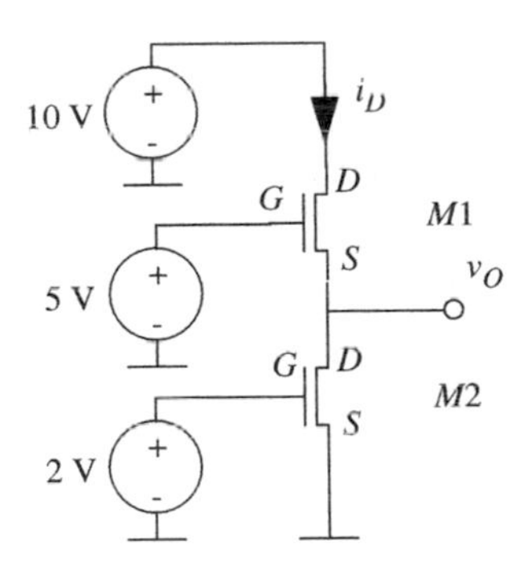


FIGURE 7.18 A circuit containing two MOSFETs. We are told that both MOSFETs operate in the saturation region.

EXAMPLE 7.6 A CIRCUIT CONTAINING TWO MOSFETS Determine the voltage $v_O$ for the MOSFET circuit shown in Figure 7.18. You are given that both MOSFETs operate in the saturation region. The MOSFETs are identical and are characterized by these parameter values: $K = 4\text{ mA}/\text{V}^2$ and $V_T = 1$ V.

Since we are told that both MOSFETs operate in the saturation region, and since $i_{DS}$ for both MOSFETs is the same, their respective gate-to-source voltages must also be equal. Recall that the drain-to-source current according to the SCS model is independent of $v_{DS}$, provided the MOSFET is in saturation. Thus, equating the gate-to-source voltages for MOSFETs $M1$ and $M2$ we have

$$5 - v_O = 2.$$

In other words, $v_O = 3$ V. It is easy to verify that $v_O = 3$ V implies that both MOSFETs are indeed in saturation.

Observe further that the drain-to-source voltages across the two MOSFETs operating as voltage-controlled voltage sources are not equal, even though the currents through the two devices are identical.

## 7.5 THE MOSFET AMPLIFIER

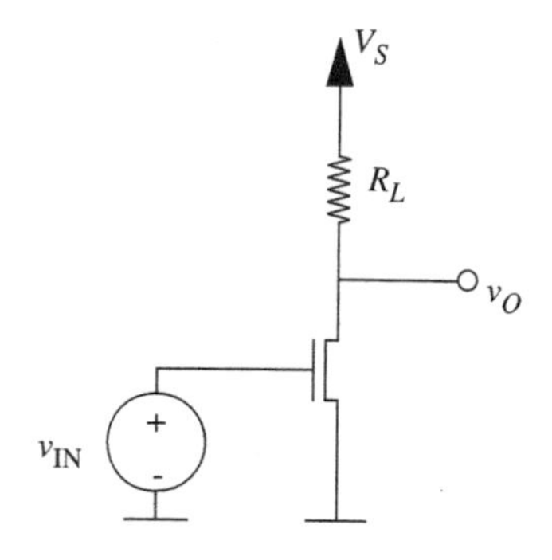


FIGURE 7.19 The MOSFET amplifier. The up-arrow labeled $V_S$ represents a connection through a power supply voltage source to ground.

A MOSFET amplifier circuit is shown in Figure 7.19. Remarkably, this circuit is identical to the inverter circuit we saw earlier! Unlike the inverter circuit, however, the input and output voltages of the MOSFET amplifier must be carefully chosen so that the MOSFET operates in its saturation region. In the saturation region of operation, the SCS model can be used to analyze the MOSFET amplifier. Constraining the inputs so that the MOSFET is always in saturation results in the desired amplifier behavior, and furthermore, it significantly simplifies our analysis. This constraint on how we use a MOSFET amplifier is yet another example of a discipline to which we adhere in circuit design and analysis. This discipline is called the *saturation discipline* and is discussed further in Section 7.5.2.

Let us examine the amplifier circuit in Figure 7.19. We will do so by replacing the MOSFET in Figure 7.19 with its SCS circuit model from Figure 7.15 as illustrated in Figure 7.20. As our first step, let us determine the conditions on the circuit such that the MOSFET is in saturation. When the MOSFET is connected in a circuit as shown, the following relationships between the MOSFET voltages and the circuit voltages apply:

$$v_{GS} = v_{IN}$$

$$v_{DS} = v_O$$

and

$$i_{DS} = i_D.$$

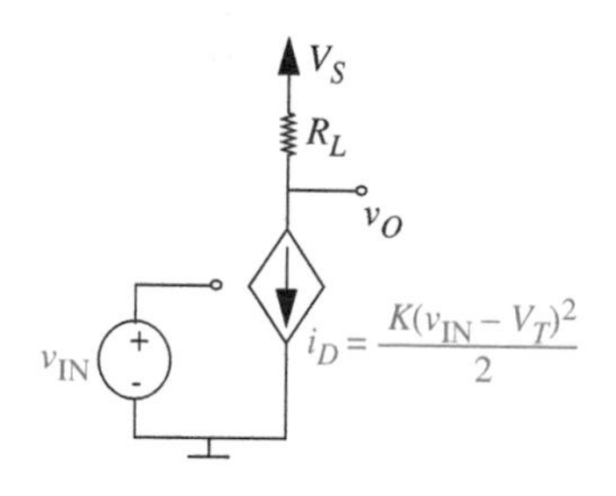


FIGURE 7.20 SCS circuit model of the MOSFET amplifier. $i_D$ is the MOSFET drain-to-source current.

Accordingly, the MOSFET is in saturation when the following constraints are met:

$$v_{IN} \geq V_T$$

and

$$v_O \geq v_{IN} - V_T.$$

In saturation, recall that the drain-to-source current of the MOSFET is given by Equation 7.12 in terms of the MOSFET parameters as

$$i_{DS} = \frac{K(v_{GS} - V_T)^2}{2}.$$

In terms of the amplifier circuit parameters, this equation becomes

$$i_D = \frac{K(v_{IN} - V_T)^2}{2}. \tag{7.13}$$

Next, we will attempt to answer the following question: What is the relationship between the amplifier output $v_O$ and its input $v_{IN}$? This relationship will describe the gain of the amplifier. Notice here an advantage of the saturation discipline — our constraint that the circuit inputs will be chosen so that the MOSFET is always in saturation allows us to focus on the saturation region of operation of the MOSFET and ignore its triode and cutoff region operation.

We will begin by formulating the output voltage $v_O$ as a function of the input voltage $v_{IN}$. Any of the methods described in Chapters 2 and 3 can be used to analyze this circuit. We will use the node method here. The ground node is marked in the circuit in Figure 7.20, and so are the node voltages $v_O$, $v_{IN}$, and $V_S$. Since the current into the MOSFET gate is zero, the node with voltage $v_O$ is the only interesting node in the circuit. Writing the node equation, we get

$$i_D = \frac{V_S - v_O}{R_L}.$$

Multiplying throughout by $R_L$ and rearranging terms, we get

$$v_O = V_S - i_D R_L.$$

In other words, $v_O$ is equal to the power supply voltage minus the voltage drop across $R_L$. When $v_{IN} \geq V_T$ and $v_O \geq v_{IN} - V_T$, we know that the MOSFET is in saturation and the SCS model for the MOSFET applies. Substituting for $i_D$ from Equation 7.13, we get the *transfer function* of the amplifier given by

$$v_O = V_S - K\frac{(v_{IN} - V_T)^2}{2}R_L. \tag{7.14}$$

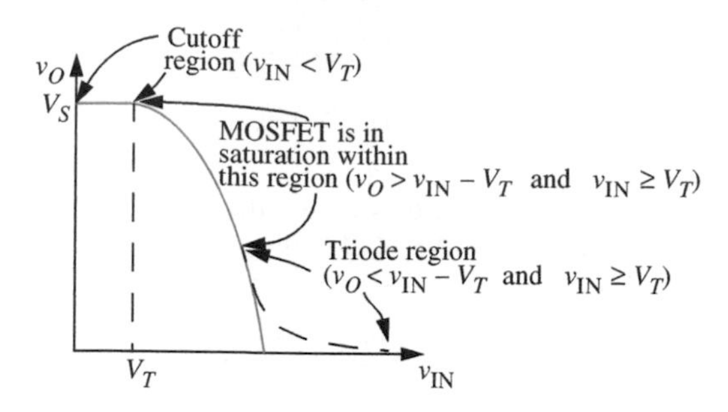


**FIGURE 7.21** $v_O$ versus $v_{IN}$ curve for the amplifier.

The transfer function relates the value of the output voltage to that of the input voltage. Accordingly, the gain of the amplifier is given by

$$\frac{v_O}{v_{IN}} = \frac{V_S - K\frac{(v_{IN}-V_T)^2}{2}R_L}{v_{IN}}. \tag{7.15}$$

Figure 7.21 plots $v_O$ versus $v_{IN}$ for the MOSFET amplifier. This decidedly nonlinear relationship is called the *transfer function* of the amplifier. When $v_{IN} < V_T$, the MOSFET is off and the output voltage is $V_S$. In other words, $i_D = 0$ when $v_{IN} < V_T$. As $v_{IN}$ increases beyond the threshold voltage $V_T$, so does the current sustained by the MOSFET. Therefore $v_O$ rapidly decreases as $v_{IN}$ increases. The MOSFET operates in the saturation region until the output voltage $v_O$ falls one threshold below the gate voltage, at which point the MOSFET enters the triode region (shown as a dashed line in Figure 7.21), and the saturation model and Equation 7.14 are no longer valid.

As shown in Figure 7.22, notice that the magnitude of the slope of certain regions of the curve is greater than one, thereby amplifying input signals that fall within this region. Shortly, we will take a more careful look at how we can connect an input signal to the amplifier so that it is amplified by leveraging the

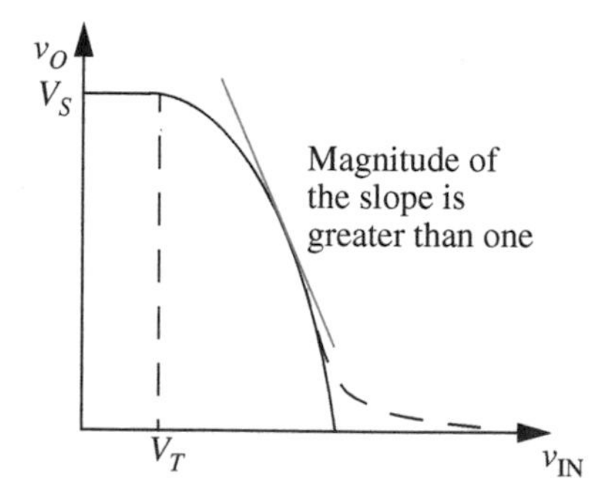


**FIGURE 7.22** In certain parts, the magnitude of the slope of the $v_O$ versus $v_{IN}$ curve is greater than one.

amplifier transfer function. But before we do so, let us examine the transfer function using numerical quantities to build up our insight.

We will examine the relationship between $v_{IN}$ and $v_{OUT}$ for the amplifier shown in Figure 7.20 for the following parameters:

$$V_S = 10 \text{ V}$$
$$K = 1 \text{ mA}/\text{V}^2$$
$$R_L = 10 \text{ k}\Omega$$
$$V_T = 1 \text{ V}.$$

Substituting in Equation 7.14, we get

$$v_O = V_S - K\frac{(v_{IN} - V_T)^2}{2}R_L \tag{7.16}$$

$$= 10 - (10^{-3})\left(\frac{(v_{IN} - 1)^2}{2}\right) 10 \times 10^3 \tag{7.17}$$

$$= 10 - 5(v_{IN} - 1)^2. \tag{7.18}$$

For example, substituting $v_{IN} = 2$ V in Equation 7.18, we obtain $v_O = 5$ V. We can tabulate the input-output voltage relationship for a larger number of quantities as shown in Table 7.1.

| $v_{IN}$ | $v_{OUT}$ |
|---|---|
| 1 | 10 |
| 1.4 | 9.2 |
| 1.5 | 8.8 |
| 1.8 | 6.8 |
| 1.9 | 6 |
| 2 | 5 |
| 2.1 | 4.0 |
| 2.2 | 2.8 |
| 2.3 | 1.6 |
| 2.32 | 1.3 |
| 2.35 | 0.9 |
| 2.4 | ~0 |

**TABLE 7.1** $v_{IN}$ versus $v_{OUT}$ for the MOSFET amplifier. All values are in volts. Observe that the MOSFET amplifier goes into the triode region for $v_{IN} > 2.3$ V and the SCS model for the MOSFET does not apply.

We can make a number of observations from Table 7.1. First, the amplifier clearly demonstrates voltage gain (*change* in the output voltage divided by the *change* in the input voltage) because the input ranging from 1 V to 2.4 V causes the output to change from 10 V to 0 V.

Second, the gain is nonlinear. From Table 7.1, when the input changes from 2 V to 2.1 V, the output changes from 5 V to 4 V, exhibiting a local voltage gain of 10. However, when the input changes from 1.4 V to 1.5 V, the output changes by merely 0.4 V, exhibiting a local voltage gain of 4. This fact is evident from the different slopes at various points in the transfer curve shown in Figure 7.22.

Third, the saturation discipline is met only for $v_{IN}$ values between 1 V and approximately 2.3 V. When the input $v_{IN}$ is less than 1 V, the MOSFET is in cutoff. Similarly, when $v_{IN}$ is greater than approximately 2.3 V, the output falls more than one threshold drop below the input. For instance, notice that when $v_{IN}$ is 2.32 V, the output is 1.3 V, which is more than one threshold drop below the input voltage.

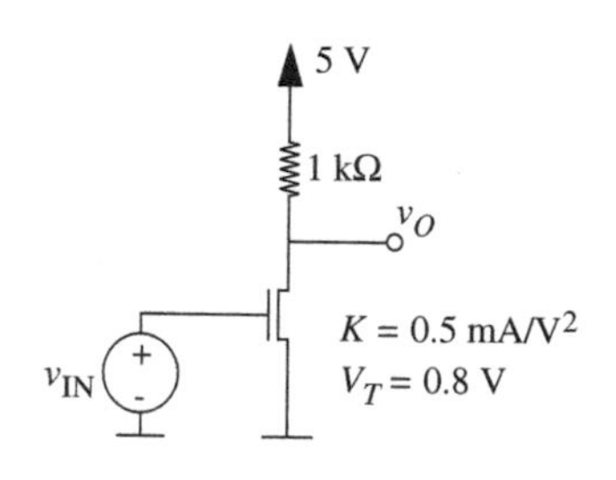


**FIGURE 7.23** A MOSFET amplifier example.

EXAMPLE 7.7 A MOSFET AMPLIFIER Consider the MOSFET amplifier shown in Figure 7.23. Assume that the MOSFET operates in the saturation region. For the parameters shown in the figure, determine the output voltage $v_O$ given that the input voltage $v_{IN} = 2.5$ V. From the value of $v_O$ verify that the MOSFET is indeed in saturation.

From Equation 7.14 we know that the relationship between the input and output voltages for a MOSFET amplifier under the saturation discipline is given by

$$v_O = V_S - K\frac{(v_{IN} - V_T)^2}{2}R_L.$$

Substituting for $K$, $V_S$, $V_T$, $R_L$, and $v_{IN}$, we obtain directly the value of $v_O$:

$$v_O = 5 - 0.5 \times 10^{-3}\frac{(2.5 - 0.8)^2}{2}1 \times 10^3$$

$$= 4.28 \text{ V}.$$

For the MOSFET to be in saturation, two conditions must be met:

$$v_{GS} \geq V_T$$

and

$$v_{DS} \geq v_{GS} - V_T.$$

Since $v_{GS} = v_{IN} = 2.5$ V, and $V_T = 0.8$ V, the first condition is met. Similarly, since $v_{DS} = v_O = 4.28$ V, and $v_{GS} - V_T = 1.7$ V, the second condition is also met. Thus the MOSFET is indeed in saturation.

---

EXAMPLE 7.8 A MOSFET SOURCE-FOLLOWER CIRCUIT Another useful MOSFET circuit is the *source follower* shown in Figure 7.24. For reasons that will be clear in our discussion of the source follower in Chapter 8, the source follower is also called a *buffer* circuit. Assuming that the MOSFET operates in the saturation region, determine the output voltage $v_{OUT}$ and the current $i_D$ given that the input voltage $v_{IN} = 2$ V, for the parameters indicated in the Figure 7.24.

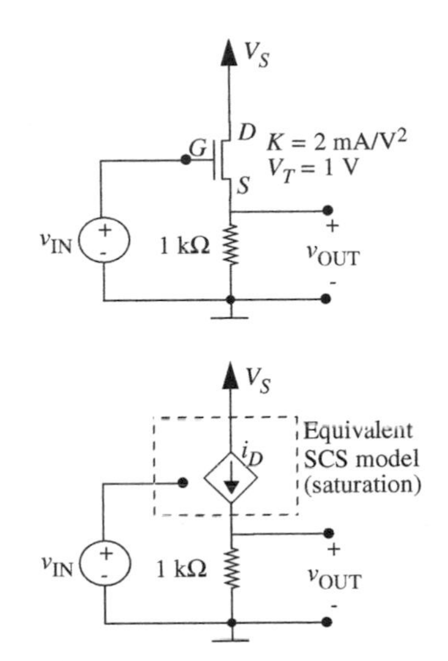


FIGURE 7.24 Source follower circuit.

We determine $v_{OUT}$ by writing the node equation for the output node:

$$i_D = \frac{v_{OUT}}{1 \times 10^3}. \tag{7.19}$$

Substituting for $i_D$ using the SCS model for the MOSFET we get

$$2 \times 10^{-3} \frac{(2\text{ V} - 1\text{ V} - v_{OUT})^2}{2} = \frac{v_{OUT}}{1 \times 10^3}.$$

Simplifying we get

$$v_{OUT}^2 - 3v_{OUT} + 1 = 0.$$

The two roots of the equation are 2.6 and 0.4. We pick the smaller of the two roots, since, for saturation operation, the solution must satisfy

$$v_{IN} - v_{OUT} \geq V_T.$$

In other words,

$$2.5\text{ V} - v_{OUT} \geq 1\text{ V}.$$

Thus, $v_{OUT} = 0.4$ *V*. Substituting into Equation 7.19

$$i_D = 0.4\text{ mA}.$$

---

### 7.5.1 BIASING THE MOSFET AMPLIFIER

Figure 7.21 showed that the MOSFET is in saturation only within a certain region of the amplifier transfer curve. The MOSFET circuit works as a reasonable amplifier only within this region, which, as shown in Table 7.1, ranges

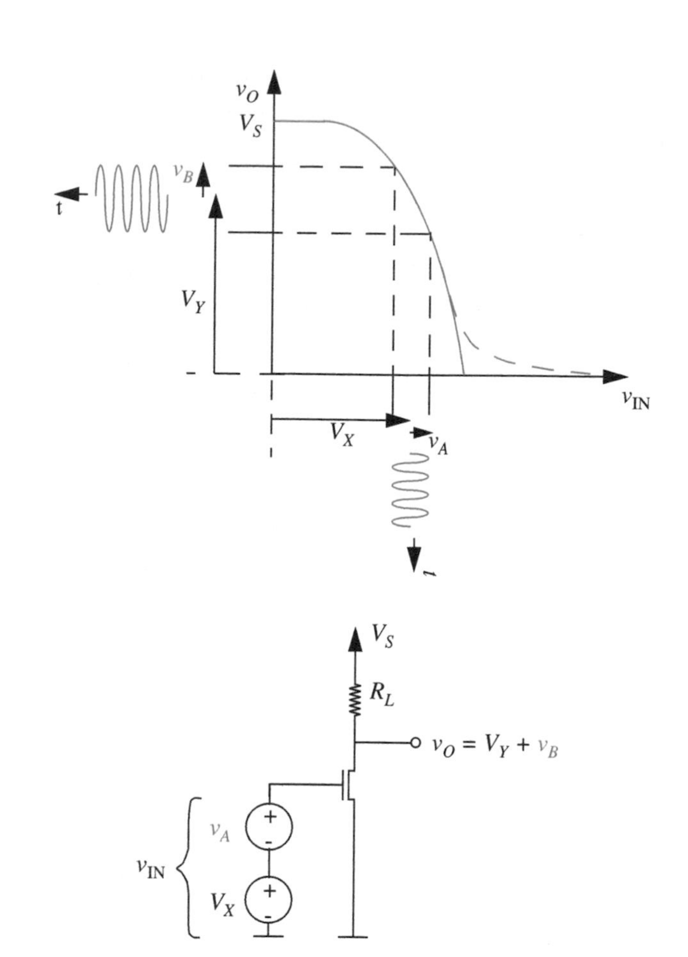


**FIGURE 7.25** Boosting the input signal of interest with a suitable DC offset so that the MOSFET operates in its saturation region for the entire range of input signal excursions.

**FIGURE 7.26** Circuit for boosting the input signal of interest ($v_A$) with a suitable DC offset ($V_X$) so that the MOSFET operates in its saturation region for the entire range of input signal excursions.

from an input of 1 volt to about 2.32 volts. In order to ensure that the amplifier operates within this region of the curve, we must transform the input voltage appropriately. As illustrated in Figure 7.25, one way of doing so is to boost the signal that we want to amplify (for example, $v_A$) with a DC offset (say, $V_X$) so that the amplifier operates in its saturation region even for negative excursions of the input signal. Figure 7.26 shows the corresponding circuit that adds an offset to the input signal by connecting a DC voltage source ($V_X$) in series with the input signal source ($v_A$). In other words, we have

$$v_{IN} = V_X + v_A$$

where $v_A$ is the desired input signal.

Notice in Figure 7.25 that the corresponding output voltage $v_O$ also contains a DC offset $V_Y$ added to the time varying output signal $v_B$. $v_B$ is an amplified version of the input signal $v_A$.

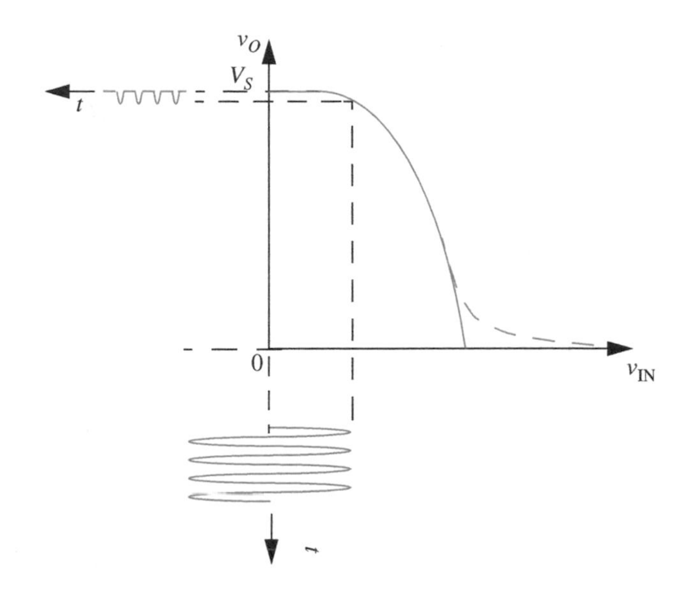


FIGURE 7.27 A sinusoidal input signal with zero offset results in a highly distorted output signal.

Contrast the amplifier behavior for the input signal source with a suitable DC offset voltage shown in Figure 7.25 with that for the input signal source applied directly Figure 7.27. When the input signal is applied to the amplifier without a DC offset, the MOSFET operates in its cutoff region for most of the input signal, and the output is highly distorted, bearing little resemblance to the input. The form of distortion suffered by the signal in the example in Figure 7.27 is called *clipping*.

The use of the amplifier with an input DC offset (and a resulting output offset) is important enough to merit some new terminology. The DC offset (for example, $V_X$) applied to the input of the amplifier is also called a *DC bias*. The use of the DC offset voltage at the input establishes an *operating point* for the amplifier. The operating point is sometimes referred to as the *bias point*. As an example, the operating point values of the input and output voltages for the amplifier in Figure 7.26 are $V_X$ and $V_Y$, respectively. We can select different operating points for the amplifier by applying different values of the input DC offset voltage. Section 7.7 discusses various methods of choosing an operating point.

We make one final observation about our amplifier. Although $v_B$ is an amplified version of the input signal $v_A$ when the input signal is boosted with a DC offset, $v_B$ is not linearly related to $v_A$. Notice from Equation 7.14 that our amplifier is nonlinear even when the MOSFET operates in the saturation region. Fortunately, the MOSFET amplifier behaves as an approximately linear amplifier for small signals; in other words, when the desired input signal $v_A$ is very small. However, we will postpone a more detailed analysis of

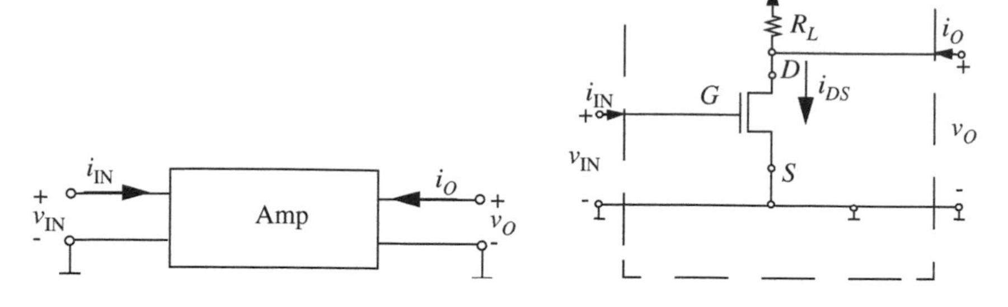


**FIGURE 7.28** The MOSFET amplifier abstraction.

the amplifier's small signal behavior to Chapter 8. Accordingly, for now, and in the rest of this chapter, we will *not* assume that the input is a small signal. Rather, we will assume that the input $v_{IN}$ that is fed into the amplifier comprises both the signal component of interest to this user (which may be a large valued signal), and a DC offset (or DC bias). For simplicity, all calculations will be performed on this boosted signal.

### 7.5.2 THE AMPLIFIER ABSTRACTION AND THE SATURATION DISCIPLINE

We would like the user of a MOSFET amplifier to be able to treat it as the abstract entity depicted in Figure 7.28, ignoring the internal details of the circuit. This abstract amplifier has $v_{IN}$ and $i_{IN}$ at its input port and $v_O$ and $i_O$ at its output port, and provides power gain. Details such as the power supply and the like are hidden from the user. The amplifier shown in Figure 7.28 uses ground as an implicit second terminal for both the input port and the output port. This form of amplifier is also called the *single-ended amplifier*.

Much like the gate abstraction went hand in hand with the static discipline—which dictated the valid range for applied inputs and expected outputs—the amplifier abstraction is associated with the saturation discipline, which prescribes constraints on the valid set of applied input signals and expected output signals. *The saturation discipline simply says that the amplifier be operated in the saturation region of the MOSFET.* As we shall see shortly, we choose this definition of the saturation discipline, because the amplifier provides a good amount of power gain in the saturation region, thereby operating well as an amplifier.

Specification of the saturation discipline serves two purposes: First, it prescribes constraints on how the device can be used; and second, it establishes a set of design criteria for the device. The amplifier abstraction and its associated usage discipline can be likened to procedural abstractions in software systems. Software procedures are an abstraction for the internal function they implement. Procedures are also associated with a usage discipline often articulated as comments at the head of the procedure. Section 7.6 will be concerned with identifying valid usage ranges under the saturation discipline.

## 7.6 LARGE-SIGNAL ANALYSIS OF THE MOSFET AMPLIFIER

Two forms of analysis come in handy for amplifiers: a large signal analysis and a small signal analysis. Large signal analysis deals with how the amplifier behaves for large changes in the input voltage, in other words, changes that are of the same magnitude as the operating parameters of the amplifier. Large signal analysis also determines the range of inputs for which the amplifier operates under the saturation discipline for the reasons discussed in Section 7.5.1. This section deals with large signal analysis. The next chapter deals with small signal analysis.

Large signal analysis attempts to answer the following specific questions related to the design of the amplifier:

1. What is the relationship between the amplifier output $v_O$ and its input $v_{IN}$ in the saturation region? Equation 7.14, developed using the analytical method, summarized the answer to this question. For variety, this section will use the graphical method to determine the same relationship.
2. What is the range of valid input values for the amplifier under the saturation discipline? What is the corresponding range of output values?

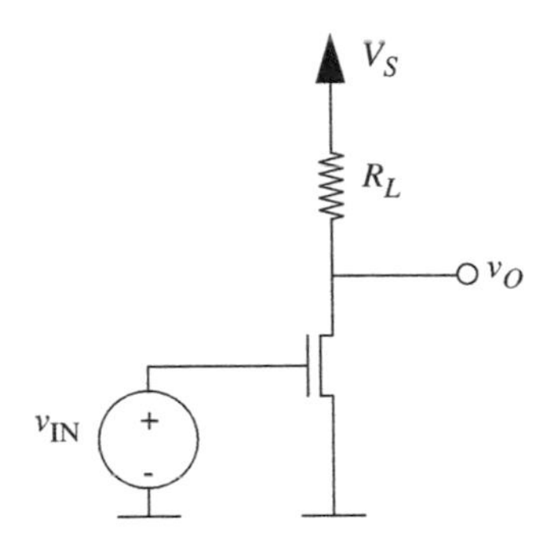


FIGURE 7.29 MOSFET amplifier circuit. For the parameters shown here, $v_{IN}$ is the same as $v_{GS}$ and $v_O$ is the same as $v_{DS}$. Similarly, $i_{DS}$ is the same as $i_D$.

Figure 7.29 shows the MOSFET amplifier, and Figure 7.30 replaces the MOSFET with its equivalent circuit model. In this section, we will use the graphical method of analyzing nonlinear circuits (introduced earlier in Section 4.3) to determine the answers to our questions.

Specifically, Section 7.6.1 will discuss the answer to the first question, and Section 7.6.2 will address the second question.

### 7.6.1 $v_{IN}$ VERSUS $v_{OUT}$ IN THE SATURATION REGION

Writing the node equation for the output node gives us the following relationship between $i_{DS}$ and $v_{DS}$:

$$v_{DS} = V_S - i_{DS}R_L. \tag{7.20}$$

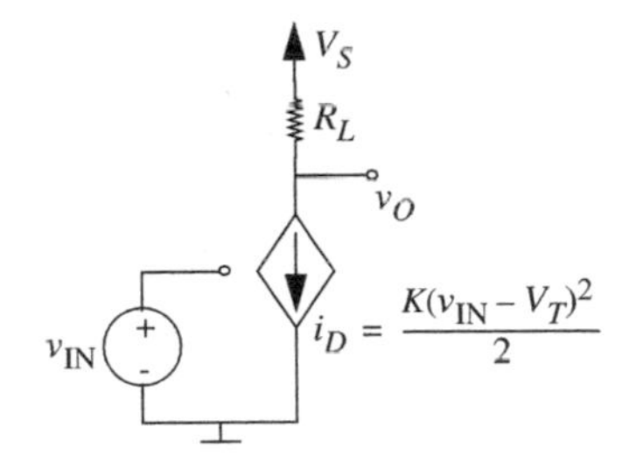


FIGURE 7.30 MOSFET amplifier — large signal model.

Recall that for our circuit $v_{IN}$ is the same as $v_{GS}$, $v_O$ is the same as $v_{DS}$, and $i_{DS}$ is the same as $i_D$, where $v_{IN}$, $v_O$, and $i_D$ are the amplifier circuit variables, and $v_{GS}$, $v_{DS}$, and $i_{DS}$ are the MOSFET variables.

Previously, using the analytical approach to solving our nonlinear amplifier problem, we substituted for $i_{DS}$ from Equation 7.12 into Equation 7.20 and obtained the input versus output voltage relationship shown in Equation 7.14.

This time around, we will use the graphical method to obtain the same relationship. We begin by rewriting Equation 7.20 as

$$i_{DS} = \frac{V_S}{R_L} - \frac{v_{DS}}{R_L}. \tag{7.21}$$

As Equation 7.21 demonstrates, the load resistor $R_L$ forces an affine relationship between $i_{DS}$ and $v_{DS}$. Figure 7.31 plots this affine relationship. The line representing the affine relationship between the output current and the voltage forced by the load resistor is called the *load line*. The slope of the line is inversely proportional to the load resistance.

Also, recall that the MOSFET SCS model forces the relationship captured by Equation 7.8, namely,

$$i_{DS} = \frac{K(v_{GS} - V_T)^2}{2}$$

and graphed in Figure 7.16 between the input voltage $v_{GS}$ and the MOSFET current $i_{DS}$. The output current and voltage must thus satisfy both the load-line constraint and the MOSFET $v_{DS}$ versus $i_{DS}$ relationships. We can graphically solve for the behavior of the output voltage by overlaying the load-line relationship on the $i_{DS}$ versus $v_{DS}$ characteristics of the MOSFET in the saturation region as depicted in Figure 7.32.

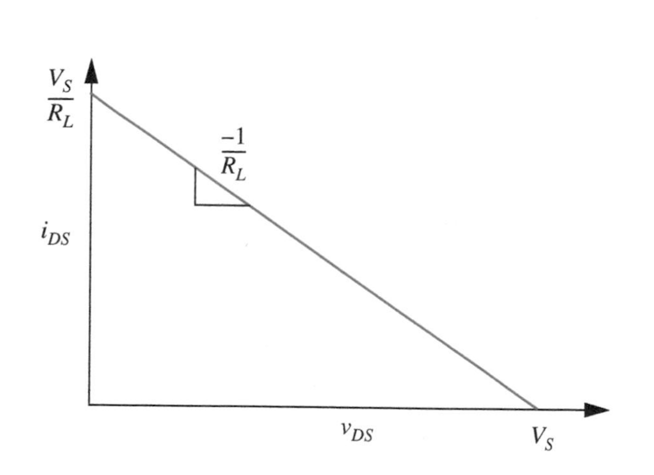


FIGURE 7.31 The load line for the MOSFET amplifier.

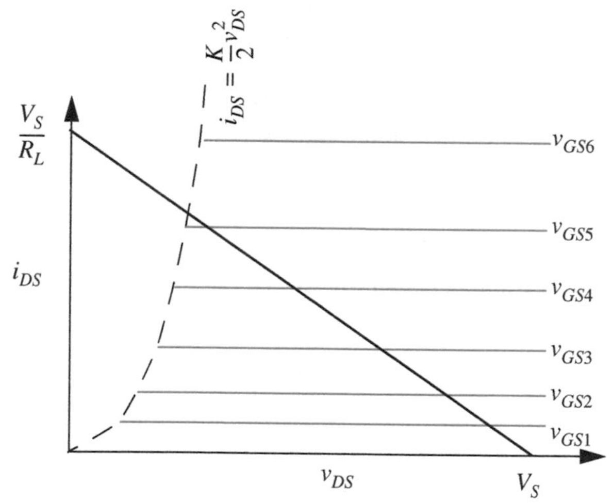


FIGURE 7.32 Load line super-imposed on the characteristic curves of the MOSFET.

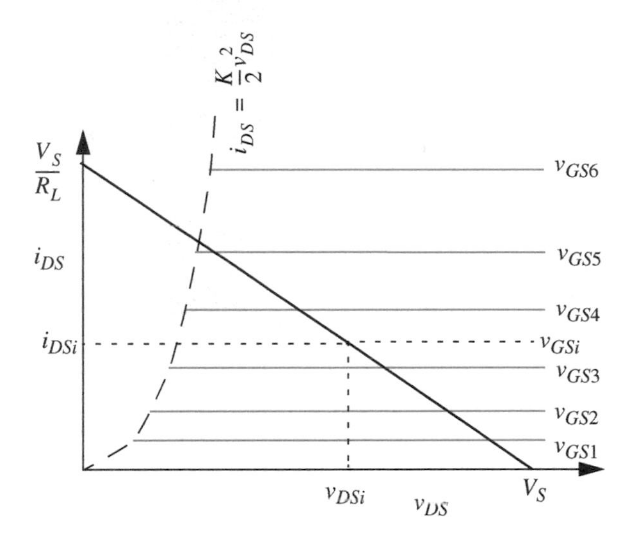


**FIGURE 7.33** Determining the transfer curve of the amplifier graphically.

Figure 7.33 illustrates how the amplifier transfer curve (that is, its $v_{IN}$ versus $v_O$ curve) can be determined. For some specific value of the input voltage, say $v_{IN} = v_{GSi}$, we can determine the output voltage $v_O = v_{DSi}$ by finding the intersection between the load line for $R_L$ and the output current $i_{DSi}$ for the given input voltage $v_{GSi}$. We can then plot these values to obtain the transfer function shown in Figure 7.21.

Figures 7.34 and 7.35 further show how an input sinusoid with a peak-to-peak voltage of 0.2 V with an offset of 1.5 V is amplified to an output peak-to-peak voltage of 1 V centered around 3.75 V for the following set

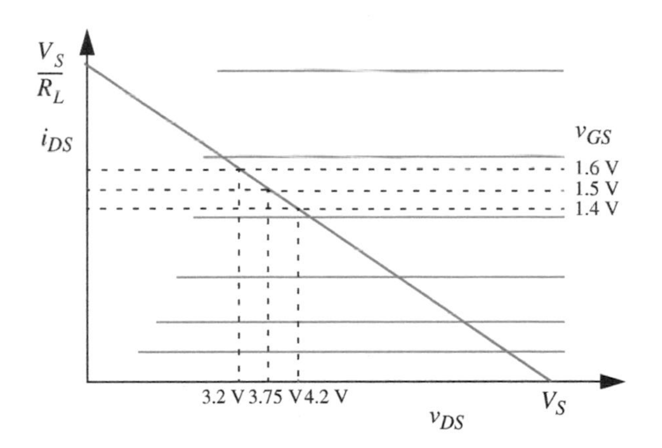


**FIGURE 7.34** Determining signal amplification graphically.

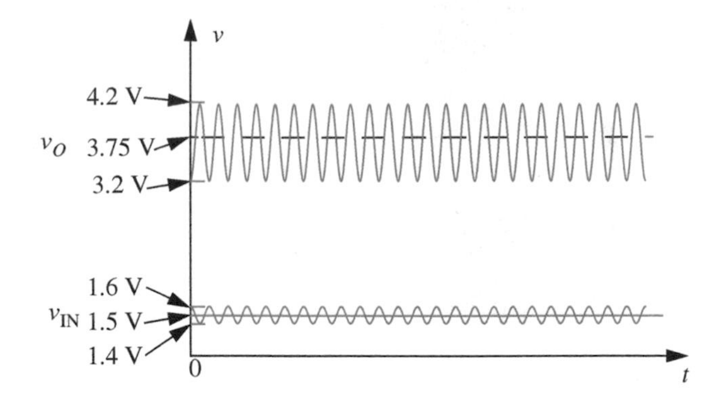


**FIGURE 7.35** Signal amplification.

of parameters:

$$R_L = 10\ \text{k}\Omega \tag{7.22}$$

$$K = 1\ \text{mA}/\text{V}^2 \tag{7.23}$$

$$V_S = 5\ \text{V} \tag{7.24}$$

$$V_T = 1\ \text{V}. \tag{7.25}$$

We note that the output will not be a perfect sinusoid like the input because the amplifier is non-linear.

This concludes our discussion of the first part of large signal analysis for a MOSFET based amplifier, namely, determining the relationship between the input and output voltage. Before we move on to the second part of large signal analysis, it is worth spending a few moments comparing the graphical method and the analytical method. To be sure, either method can be used in most situations. The analytical method is useful when simple expressions capture the behavior of the devices, as was the case for our MOSFET amplifier. The graphical method discussed in this section, however, is often more accurate when device characteristics measured from a physical device are available. The discrete devices you will come across in the laboratory, for example, will often come with data sheets containing their $v$–$i$ characteristics.

### 7.6.2 VALID INPUT AND OUTPUT VOLTAGE RANGES

Let us now answer the second question of large signal analysis, namely, what are valid input and output voltage ranges for the amplifier under the saturation discipline? These ranges will provide the outer voltage limits to input signals such as those in Figure 7.35. The limits will also provide insights into the nominal voltage about which the input signal should be centered, or, in other words, how to choose the operating point of the amplifier.

*Valid voltage range* The range of input voltages (and the resulting range of output voltages) for which the MOSFET (or MOSFETs) in the circuit operate in the saturation region.

The amplifier will amplify input signals when it is operating in this range without clipping the signal or introducing significant amounts of distortion. (Signal clipping occurs when the amplifier output cannot go beyond a certain voltage or current level.)

Let us begin by making some general observations about the current and voltage limits to build up our intuition. Observe that $i_{DS}$ can range only from $0 \rightarrow V_S/R_L$. The output voltage is $V_S$ when $i_{DS}$ is zero. $i_{DS}$ is zero for input voltages less than $V_T$. Similarly, the output voltage is 0 when the current is $V_S/R_L$, and the input voltage is at some high value greater than $V_T$. The limits of saturation region operation lie somewhere within the $i_{DS}$ current limits of 0 and $V_S/R_L$.

The valid range of input voltages has a lower limit and an upper limit. The lower limit on input voltages is easy to determine.

### Lowest Valid Input Voltage

Notice from Figure 7.36 that the input voltage must be greater than $V_T$ for the MOSFET to exit its cut off region. When the input voltage is $V_T$, the MOSFET exits its cutoff region and the output voltage of the amplifier is $V_S$. When the input voltage is equal to $V_T$, any positive value of $v_{DS}$ will cause the MOSFET to operate in its saturation region. Because we design the amplifier with $V_S > 0$, and since $v_{DS} = V_S$, the MOSFET will be in its saturation region. Since $V_T$ is the lowest voltage for which the MOSFET is in saturation, we get

$$\text{lowest valid input voltage} = V_T. \tag{7.26}$$

The corresponding value of the output voltage is $V_S$. The point labeled $(x)$ corresponding to the point $(V_T, V_S)$ on the amplifier $v_{IN}$ versus $v_O$ transfer curve in Figure 7.36 denotes the low end of the valid input voltage range.

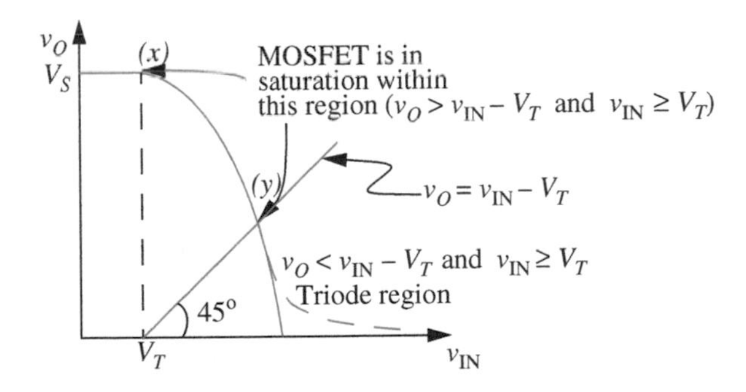


**FIGURE 7.36** The lowest valid input voltage under the saturation discipline is marked by the point $(x)$, and the highest valid input voltage under the saturation discipline is marked by the point $(y)$.

### Highest Valid Input Voltage

Next, we will determine the highest value of the input voltage for which the MOSFET satisfies the saturation discipline. Notice that the MOSFET goes into the triode region when the output voltage $v_O$ falls one threshold drop below the input voltage $v_{IN}$. In other words, when

$$v_O = v_{IN} - V_T.$$

Thus, the valid high input voltage is that value of $v_{IN}$ beyond which the MOSFET enters the triode region.

To build intuition, we first determine graphically the input voltage for which the output crosses into the triode region as follows. Referring to Figure 7.36, the straight line drawn at 45° to the $v_{IN}$ axis and intersecting it at $V_T$ reflects the set of points in the $v_{IN}$ versus $v_O$ plane for which

$$v_O = v_{IN} - V_T$$

assuming, of course, that $v_{IN}$ and $v_O$ use the same scale. Thus, the point (*y*) at which this 45° line intersects the $v_{IN}$ versus $v_O$ transfer curve marks the upper limit of the valid input range.

We can also determine analytically the value of this upper limit by solving for the intersection of the straight line in Figure 7.36 represented by

$$v_O = v_{IN} - V_T \tag{7.27}$$

and the transfer curve determined by Equation 7.14, which we rewrite here for convenience:

$$v_O = V_S - K\frac{(v_{IN} - V_T)^2}{2}R_L. \tag{7.28}$$

The intersection of these two curves is marked by the point (*y*) in Figure 7.36. Substituting the expression for $v_O$ from Equation 7.27 into Equation 7.28 we get

$$v_{IN} - V_T = V_S - K\frac{(v_{IN} - V_T)^2}{2}R_L. \tag{7.29}$$

Rearranging terms, we have

$$R_L\frac{K}{2}(v_{IN} - V_T)^2 + (v_{IN} - V_T) - V_S = 0. \tag{7.30}$$

The value of $v_{IN}$ that solves Equation 7.30 is the highest value of $v_{IN}$ for which the MOSFET operates in saturation.

Solving for $v_{IN} - V_T$, we get

$$v_{IN} - V_T = \frac{-1 + \sqrt{1 + 2V_S R_L K}}{R_L K}. \tag{7.31}$$

In other words,

$$v_{IN} = \frac{-1 + \sqrt{1 + 2V_S R_L K}}{R_L K} + V_T. \tag{7.32}$$

This value of $v_{IN}$ is the highest input voltage that satisfies the saturation discipline and corresponds to the point marked ($y$) in Figure 7.36.

Summarizing, the maximum valid input voltage range is

$$V_T \rightarrow \frac{-1 + \sqrt{1 + 2V_S R_L K}}{R_L K} + V_T$$

and the maximum valid output voltage range is

$$V_S \rightarrow \frac{-1 + \sqrt{1 + 2V_S R_L K}}{R_L K}.$$

As illustrated in Figure 7.37, the MOSFET enters its cutoff region for input voltages lower than $V_T$, and goes into the triode region for input voltages greater than $\left((-1 + \sqrt{1 + 2V_S R_L K})/R_L K\right) + V_T$. The corresponding drain current range is

$$0 \rightarrow \frac{K}{2}(v_{IN} - V_T)^2$$

where we substitute $\left((-1 + \sqrt{1 + 2V_S R_L K})/R_L K\right) + V_T$.

This completes the second step of large signal analysis.

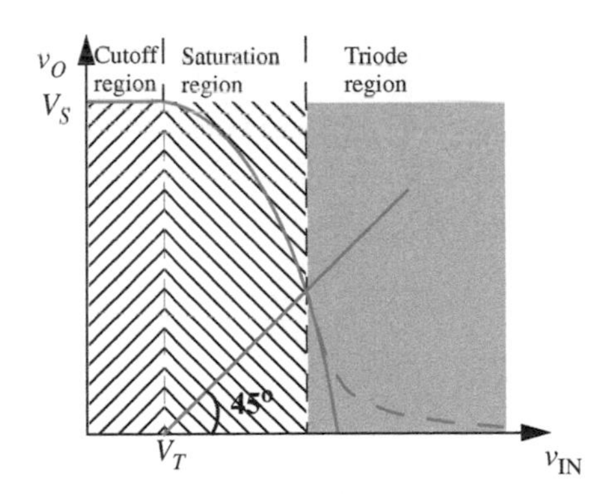


**FIGURE 7.37** Cutoff, saturation, and triode regions of operation of the MOSFET amplifier.

---

EXAMPLE 7.9 VALID INPUT AND OUTPUT RANGES FOR AMPLIFIER Let us now determine the valid input voltage range and the corresponding output voltage range for the amplifier given the following circuit parameters:

$$R_L = 10\ \text{k}\Omega \tag{7.33}$$

$$K = 1\ \text{mA/V}^2 \tag{7.34}$$

$$V_S = 5\ \text{V} \tag{7.35}$$

$$V_T = 1\ \text{V}. \tag{7.36}$$

From Equation 7.26, we know that $V_T = 1$ V is at the low end of the valid input range. The corresponding value of $v_O$ is $V_S = 5$ V and the current $i_D$ is 0.

Next, to obtain the highest value of the input voltage for saturation region operation of the MOSFET amplifier we substitute the values of these parameters in Equation 7.32.

$$\begin{aligned}\text{Highest valid input voltage} &= V_T + \frac{-1 + \sqrt{1 + 2V_S R_L K}}{R_L K} \\ &= 1 + \frac{-1 + \sqrt{1 + 2 \times 5 \times 10 \times 10^3 \times 10^{-3}}}{10 \times 10^3 \times 10^{-3}} \\ &\approx 1.9\ \text{V}.\end{aligned}$$

In other words, 1.9 V is the highest value of the input voltage that ensures saturation region operation of the amplifier. We can also solve for the corresponding values of $v_O$ and $i_D$ from Equations 7.27 and 7.8 as follows:

$$v_O = v_{IN} - V_T = 1.9 - 1 = 0.9\ \text{V}$$

$$i_D = \frac{K}{2}(v_{IN} - V_T)^2 = 0.41\ \text{mA}.$$

In summary, the maximum valid range for the input voltage is

$$1\ \text{V} \rightarrow 1.9\ \text{V}$$

and the maximum valid range for the output voltage is

$$5\ \text{V} \rightarrow 0.9\ \text{V}.$$

The corresponding drain current range is

$$0\ \text{mA} \rightarrow 0.41\ \text{mA}$$

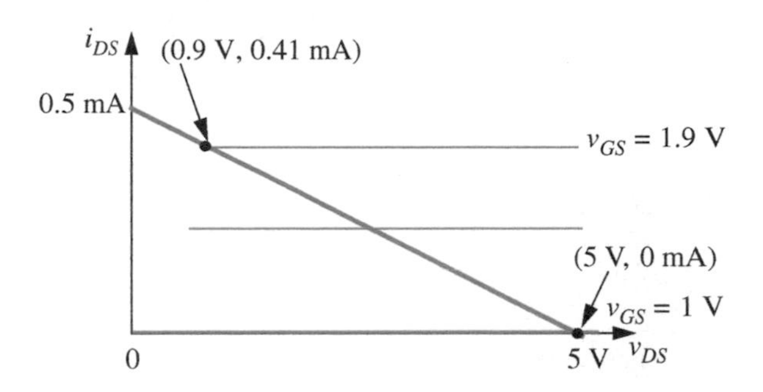


FIGURE 7.38 Valid input and output voltage ranges.

These values are plotted on a graph of the amplifier load line and the MOSFET device characteristics in Figure 7.38.

---

EXAMPLE 7.10 VALID RANGES FOR THE SOURCE FOLLOWER CIRCUIT Let us derive the valid operating ranges for the source follower circuit shown earlier in Figure 7.24, and repeated here in Figure 7.39 for convenience. Assume that $V_S = 10$ V.

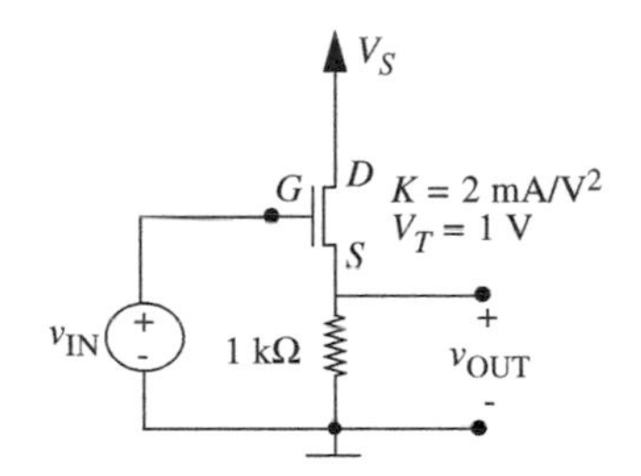


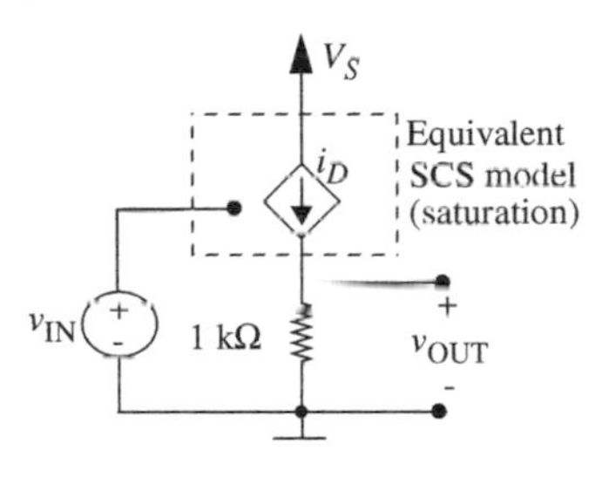


FIGURE 7.39 Source-follower circuit.

Recall that the valid input voltage range is defined as the range of input voltages for which the MOSFET operates under the saturation discipline. Two conditions must be met for the MOSFET to remain in saturation:

$$v_{GS} \geq V_T \tag{7.37}$$

and

$$v_{DS} \geq v_{GS} - V_T. \tag{7.38}$$

The first condition requires that $v_{IN} - v_{OUT} \geq V_T$, or,

$$v_{IN} \geq v_{OUT} + V_T.$$

Since the minimum value of $v_{OUT}$ is 0 V, the minimum value of $v_{IN}$ for saturation region MOSFET operation is given by

$$v_{IN} = V_T = 1 \text{ V}.$$

The second condition requires that

$$v_{DS} \geq v_{GS} - V_T,$$

which implies that

$$V_S - v_{OUT} \geq v_{IN} - v_{OUT} - V_T.$$

Rearranging terms and simplifying, we obtain

$$v_{IN} \leq V_S + V_T.$$

In other words, the maximum value of $v_{IN}$ is given by

$$v_{IN} = 10 \text{ V} + 1 \text{ V} = 11 \text{ V}.$$

Summarizing, the valid input range is given by

$$1 \text{ V} \rightarrow v_{IN} \rightarrow 11 \text{ V}.$$

The corresponding output voltage range is easily determined. At the low end of the valid range, we know that $v_{OUT} = 0$ for $v_{IN} = 1$ V. At the high end of the valid range, $v_{OUT}$ is determined by writing the node equation for the output node and substituting $v_{IN} = 11$ V:

$$i_D = \frac{v_{OUT}}{1 \times 10^3}.$$

Substituting for $i_D$ using the SCS model for the MOSFET we get

$$2 \times 10^{-3} \frac{(11 \text{ V} - 1 \text{ V} - v_{OUT})^2}{2} = \frac{v_{OUT}}{1 \times 10^3}.$$

Simplifying, we get

$$v_{OUT}^2 - 21 v_{OUT} + 100 = 0.$$

The two roots of the equation are 13.7 and 7.3. We pick the smaller of the two roots, since, for saturation operation, $v_{OUT}$ must be at least one $V_T$ below the input voltage. Thus,

$$v_{OUT} = 7.3 \text{ V}.$$

The valid output voltage range is given by

$$0 \text{ V} \rightarrow v_{OUT} \rightarrow 7.3 \text{ V}.$$

The corresponding valid current range is given by dividing the output voltage extremes by the resistance 1 kΩ:

$$0/10^3 \rightarrow i_D \rightarrow 7.3/10^3$$

or

$$0 \text{ mA} \rightarrow i_D \rightarrow 7.3 \text{ mA}.$$

---

### 7.6.3 ALTERNATIVE METHOD FOR VALID INPUT AND OUTPUT VOLTAGE RANGES

Section 7.6.2 showed that we could determine the valid range of amplifier operation under the saturation discipline using the transfer curve of the amplifier. Alternatively, we can solve for the same limits graphically from the load-line and the MOSFET device characteristics as illustrated in Figure 7.40.

Notice that under the saturation discipline the lowest valid value of the output voltage $v_O$ is identified by the point of intersection of the constraint curve separating the triode and saturation regions given by

$$i_{DS} = \frac{K}{2} v_{DS}^2 \tag{7.39}$$

and the load line given by

$$i_{DS} = \frac{V_S}{R_L} - \frac{v_{DS}}{R_L}. \tag{7.40}$$

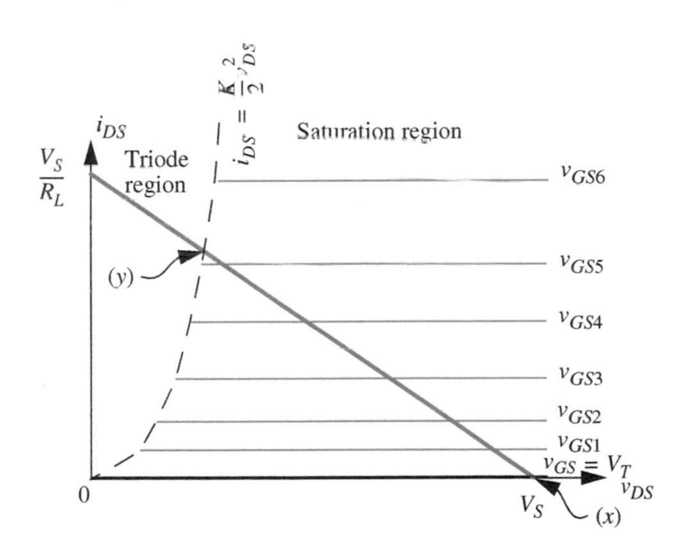


**FIGURE 7.40** Determining valid input and output voltage ranges from a graph of the amplifier load line and the MOSFET device characteristics.

The intersection point is marked (y) in Figure 7.40. Recall that the current $i_{DS}$ and the output voltage $v_O$ (which, in our amplifier circuit, is the same as $v_{DS}$) is constrained to the load line by KVL. Substituting for $i_{DS}$ from Equation 7.39 into 7.40, rearranging terms, and multiplying out by $R_L$, we get

$$R_L \frac{K}{2} v_{DS}^2 + v_{DS} - V_S = 0. \tag{7.41}$$

Observe that Equation 7.41 is the same as Equation 7.30 with $v_{DS}$ in place of $v_{IN} - V_T$. The two equations are consistent because

$$v_{DS} = v_{IN} - V_T \tag{7.42}$$

at the point where the load line intersects the boundary of the saturation region.

The positive solution to Equation 7.41 gives us the value of $v_{DS}$ at the point of intersection:

$$v_{DS} = \frac{-1 + \sqrt{1 + 2V_S R_L K}}{R_L K}. \tag{7.43}$$

This value of $v_{DS}$ is the desired lowest value of $v_O$ for saturation region operation of the MOSFET amplifier. The corresponding value of the highest valid input voltage can be obtained from Equation 7.42, and is given by $v_{IN} = v_O + V_T$. In other words, at the point (y) in Figure 7.40, $v_{IN}$ is given by $\left((-1 + \sqrt{1 + 2V_S R_L K})/R_L K\right) + V_T$, and $v_O$ is given by $\left(-1 + \sqrt{1 + 2V_S R_L K}\right)/R_L K$.

Next, we will determine the lowest value of the valid input voltage denoted by the point marked (x) in Figure 7.40. This point is the intersection of the load line and the $i_{DS}$ versus $v_{DS}$ line for which $v_{GS} = V_T$. At this point, $v_O = v_{DS} = V_S$ and $v_{IN} = v_{GS} = V_T$.

This completes our discussion of large signal analysis for the MOSFET amplifier. Large signal analysis determines the input-output transfer curve of the amplifier and the limits on the input voltage for which the amplifier operated under the saturation discipline. Specifically, the large signal analysis of an amplifier entails the following steps:

1. Derive the relationship between $v_{IN}$ and $v_O$ under the saturation discipline. Note that in general this might be a linear or fully nonlinear analysis.

2. Find the valid input voltage range and the valid output voltage range for saturation operation. The limits of the valid ranges occur when the MOSFET enters into a cutoff region or a triode region. In complicated circuits, this step may require numerical analysis.

Among other things, the limits determined in large signal analysis come in handy in determining a reasonable operating point for the amplifier. This will be the next topic of discussion.

## 7.7 OPERATING POINT SELECTION

We are often interested in amplifying time-varying signals. Because the amplifier turns off for input voltages less than $V_T$, it is important to add an appropriate DC offset voltage to the time-varying input signal so that the amplifier remains in the saturation region for the entire range of input voltage variation. This input DC offset voltage defines the operating point of the amplifier. The DC offset must be chosen carefully, for if it is too large, the amplifier will be pushed into the triode region, and if it is too low, the amplifier will slide into the cutoff region. How do we choose this operating point?

Time-varying signals such as those in Figure 7.35 are characterized by their peak-to-peak voltage and their DC offset. For example, the sinusoidal signal $v_{IN}$ in Figure 7.35 has a peak-to-peak value of 0.2 V and a DC offset of 1.5 V. Since the MOS amplifier is nonlinear, we define the output offset as the value of $v_O$ when the DC input offset voltage is the only signal applied at the input. Although the time-varying portion of the signal is of interest to us, as discussed in Section 7.5.1, the DC offset is provided simply to keep the amplifier operating in its saturation region.

The input offset voltage is also called the *input bias voltage* or the *input operating voltage*. The corresponding output voltage and the output current define the output operating point of the amplifier. Together, the input bias voltage, and the corresponding output voltage and the output current, define the operating point of the amplifier. We denote the operating point values of $v_{IN}$, $v_O$, and $i_D$ as $V_{IN}$, $V_O$, and $I_D$, respectively. As illustrated in Figure 7.41,

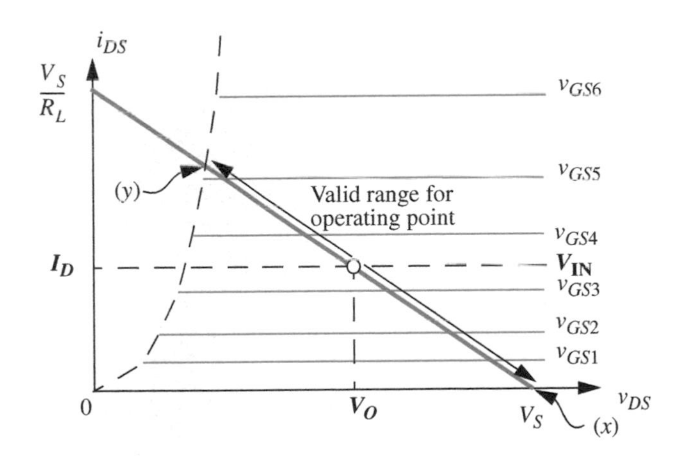


FIGURE 7.41 Valid range for operating point under the saturation discipline.

the operating point can be legally situated anywhere along the load line in the valid range between the points (*x*) and (*y*).

There are several factors that can govern our choice of the operating point. For example, the operating point dictates the maximum dynamic range of the input signal for both positive and negative excursions for which the MOSFET operates in saturation. As can be seen from Equation 7.15, the operating point value of the input voltage also governs the signal gain of the amplifier. This section will focus on selecting an operating point based on maximizing the useful input signal range. We will have more to say about the relationship between the gain of the amplifier and its operating point in Section 8.2.3 in Chapter 8.

Let us assume that the input signal has symmetric peak-to-peak swings about the DC offset. In other words, we will assume an equal magnitude for both the positive and negative excursions of the time-varying signal from the DC offset, as is the case for the input signal $v_{IN}$ in Figure 7.35 (but not for the output signal $v_O$). To obtain maximum useful input signal range, we might choose the input bias voltage $V_{IN}$ to be at the center of the valid range of input voltages for the amplifier, as illustrated in Figure 7.42.

Accordingly, for the amplifier parameters that we have been using thus far,

$$R_L = 10\ \text{k}\Omega \tag{7.44}$$

$$K = 1\ \text{mA}/\text{V}^2 \tag{7.45}$$

$$V_S = 5\ \text{V} \tag{7.46}$$

$$V_T = 1\ \text{V} \tag{7.47}$$

because our amplifier operates under the saturation discipline for input voltages in the range 1 V → 1.9 V, we might choose an input operating point voltage at the center of this range, namely $V_{IN} = 1.45$ V. This choice is illustrated in

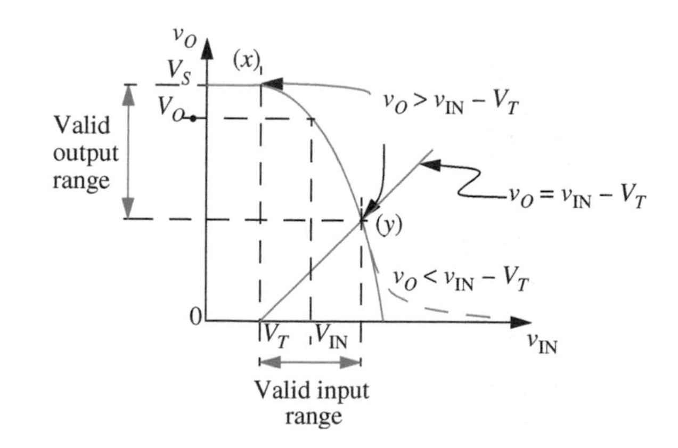


FIGURE 7.42 Selection of the input operating point.

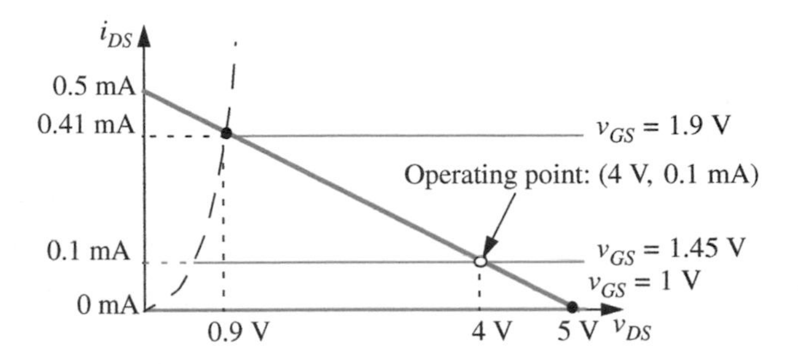


**FIGURE 7.43** Operating point and valid input and output voltage ranges.

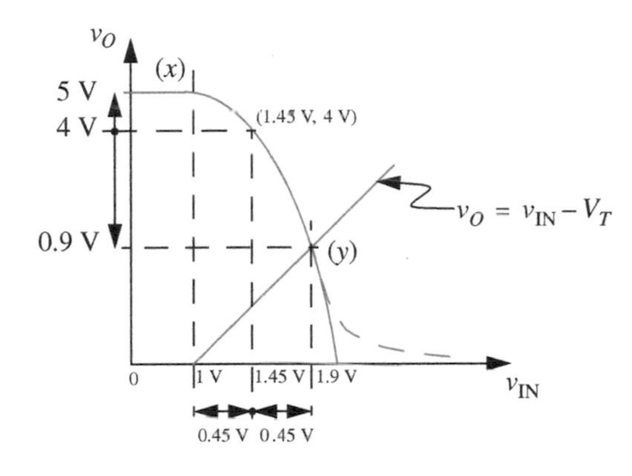


**FIGURE 7.44** Input operating point.

Figure 7.43, which is an $i_{DS}$ versus $v_{DS}$ graph, and in Figure 7.44, which shows the corresponding $v_{IN}$ versus $v_{OUT}$ graph. As we expect, the output will vary between 0.9 V and 5 V as the input varies between 1 V and 1.9 V.

Let us take a closer look at the behavior of the amplifier for the given input bias voltage by determining the corresponding output operating point. For a given input operating point voltage $V_{IN}$, we can determine the operating point output voltage $V_O$ from Equation 7.14, and the operating point output current $I_D$ from the MOSFET SCS model given in Equation 7.8. Substituting for the circuit parameters in Equation 7.14 we get

$$\begin{aligned} V_O &= V_S - K\frac{(V_{IN} - V_T)^2}{2}R_L \\ &= 5 - 10^{-3}\frac{(1.45 - 1)^2}{2}10^4 \\ &= 4 \text{ V}. \end{aligned}$$

From Equation 7.8 we get $I_D$ as

$$I_D = \frac{K(V_{IN} - V_T)^2}{2}$$
$$= \frac{10^{-3}(1.45 - 1)^2}{2}$$
$$= 0.1 \text{ mA.}$$

Thus the operating point for the amplifier is defined by

$$V_{IN} = 1.45 \text{ V}$$
$$V_O = 4 \text{ V}$$
$$I_D = 0.1 \text{ mA.}$$

This operating point maximizes the peak-to-peak input voltage swing for which the amplifier operates under the saturation discipline.

The operating point for our amplifier, along with the valid input and output voltage ranges, is shown in Figure 7.43. For this choice of the operating point, the maximum input voltage swing for positive excursions is 1.45 V $\rightarrow$ 1.9 V, and the maximum input voltage swing for negative excursions is 1.45 V $\rightarrow$ 1 V. The corresponding output voltage swings are 4 V $\rightarrow$ 0.9 V and 4 V $\rightarrow$ 5 V.

Although we chose the input operating point to be at the center of the valid input range, notice the asymmetry of the output voltage range about the output operating voltage. The asymmetry arises from the nonlinearity of the gain of the MOSFET amplifier. The next chapter will discuss an approach by which we can treat MOSFET amplifiers as linear amplifiers. Depending on our desired input and output voltage swings, and also amplifier gain, we can also choose other operating points for the amplifier. Other criteria for choosing the operating point might include concerns of stability and power dissipation, but these are beyond the scope of our discussion.

---

EXAMPLE 7.11 OPERATING POINT FOR THE MOSFET SOURCE FOLLOWER CIRCUIT Modify the source follower circuit from Figure 7.24 to include an input bias voltage that maximizes input voltage swing. Assume that $V_S = 10$ V.

Figure 7.45 shows the biased circuit, where $V_B$ is the bias voltage and $v_A$ is the input signal. The total signal, $v_{IN}$, is the sum of the offset voltage and the actual input. Recall that the input offset voltage ($V_B$) is applied to boost the input signal ($v_A$) in a way that the MOSFET remains in saturation for the maximum positive and negative excursions of the input signal.

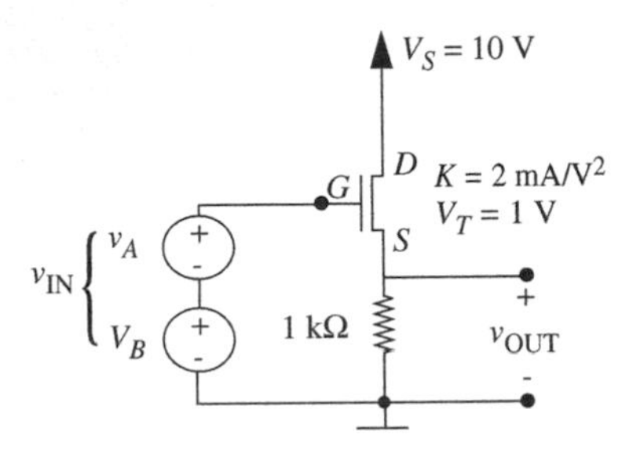


**FIGURE 7.45** Source-follower circuit with input bias.

From Example 7.10, we know that the valid range for the total input $v_{IN}$ is given by

$$1\text{ V} \rightarrow v_{IN} \rightarrow 11\text{ V}.$$

We can obtain the maximum input swing under saturation operation by biasing the input at the midpoint of the input valid range. In other words, we choose

$$V_B = 6\text{ V}.$$

This choice of input offset voltage allows a peak-to-peak swing of 10 V for the input signal $v_A$.

EXAMPLE 7.12 LARGE SIGNAL ANALYSIS OF ANOTHER MOSFET AMPLIFIER The circuit shown in Figure 7.46 is a MOSFET amplifier. We wish to determine the large-signal input-output behavior of this amplifier. We also wish to determine the range of $v_{IN}$ over which the MOSFET operation remains in the saturation region. In this example, we will assume that the MOSFET is characterized by $V_T = 1$ V and $K = 1$ mA/V$^2$.

Resistors $R_1$ and $R_2$ form a voltage divider from $V_S$ that establishes the constant bias voltage $V_B$ at the gate of the MOSFET. That bias voltage is $V_B = 1.6$ V.

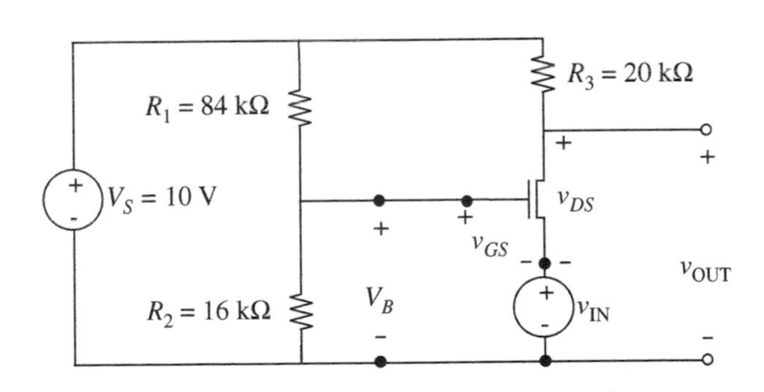


**FIGURE 7.46** Another MOSFET amplifier with the input connected to the source, and biasing provided by a voltage divider formed by resistors $R_1$ and $R_2$.

Next, applying KVL yields $v_{GS} = V_B - v_{IN}$. From this it follows that

$$v_{OUT} = V_S - \frac{R_3 K}{2}(V_B - v_{IN} - V_T)^2,$$

which evaluates to

$$v_{OUT} = 10 - 10 \times (0.6 - v_{IN})^2.$$

Thus, for example, for $v_{IN} = 0$ V, the output $v_{OUT} = 6.4$ V.

We now determine the range of $v_{IN}$ over which the MOSFET operation remains in the saturation region. To do so, the MOSFET voltages must satisfy $v_{DS} \geq v_{GS} - V_T \geq 0$. For the amplifier shown in Figure 7.46, this is equivalent to

$$v_{OUT} - v_{IN} \geq V_B - v_{IN} - V_T \geq 0.$$

By violating the first inequality, the MOSFET operation enters its triode region, and by violating the second inequality, the MOSFET operation enters its cutoff region. Numerically, this evaluates to

$$-0.3695 \text{ V} \leq v_{IN} \leq 0.6 \text{ V},$$

which corresponds to

$$0.6 \text{ V} \leq v_{OUT} \leq 10 \text{ V}.$$

Thus, note that the MOSFET operation can remain saturated for both positive and negative values of $v_{IN}$.

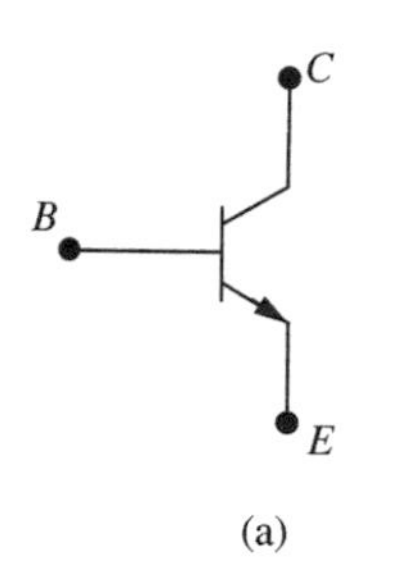


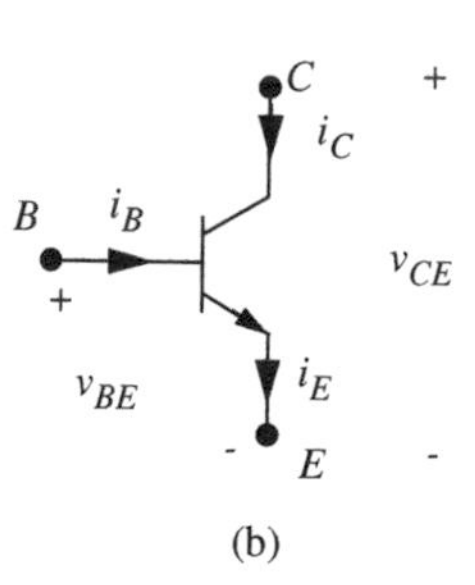


FIGURE 7.47 A bipolar junction transistor.

EXAMPLE 7.13 BIPOLAR JUNCTION TRANSISTOR (BJT) Figure 7.47a depicts another three-terminal device, called the bipolar junction transistor (BJT), that is in common use in VLSI circuits. A BJT has three terminals called the base (B), the collector (C), and the emitter (E). Figure 7.47b marks the device with its relevant voltage and current parameters.

In this example, we will compare the actual characteristics of the BJT with those predicted by a simple piecewise linear model. The actual characteristics of a BJT ($i_C$ versus $v_{CE}$ for various values of $i_B$) are shown in Figure 7.48. The horizontal nature of the $i_C$ versus $v_{CE}$ curves indicates that the device operates like a dependent current source when the base current $i_B > 0$ and the collector-to-emitter voltage ($v_{CE}$) is greater than approximately 0.2 V. The current supplied by the current source is typically about 100 times the base current. Although these curves are qualitatively similar to those of a MOSFET, there are also some differences. First, notice that we have chosen the BJT's base current $i_B$

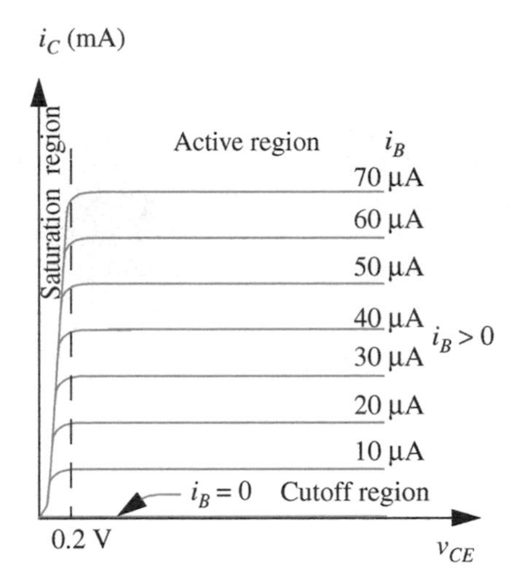


FIGURE 7.48 Actual characteristic curves for a bipolar junction transistor.

as our control parameter (the control parameter was the gate-to-source voltage for the MOSFET, and the gate current was zero). Second, the collector current is linearly related to the base current (when the MOSFET operated as a current source, its drain current was quadratically related to the gate-to-source voltage).

The BJT characteristics show three regions of operation:

1. When $i_B > 0$ and $v_{CE} > 0.2$ V, the BJT is said to be in the *active region* of operation. In this region, the horizontal collector current curves display a current-source-like behavior. As we shall see momentarily, the active region will be the predominant region of interest for analog circuit designs.
2. When $i_B = 0$, the BJT is said to be in the *cutoff region.*
3. Finally, when $i_B > 0$ and $v_{CE} \leq 0.2$ V, (that is, the region to the left of the vertical dashed line in Figure 7.48), the collector current drops sharply, and the BJT is said to be in the *saturation region.*[4]

Figure 7.49b shows a model for the BJT containing a current-controlled current source and a pair of diodes (a base-emitter diode and a base-collector diode). The current supplied by the dependent source is $\beta$ times $i_{B'}$. The parameter $\beta$ is a constant with

4. The saturation region in BJTs is completely unrelated to the saturation region in MOSFETs, and in fact, normal operation of BJTs attempts to avoid this region. This duplication of terms — one representing the favored region of operation in MOSFETs, and the other representing an avoided region of operation in BJTs — can be the source of confusion, but, unfortunately, has become the norm in circuit parlance.

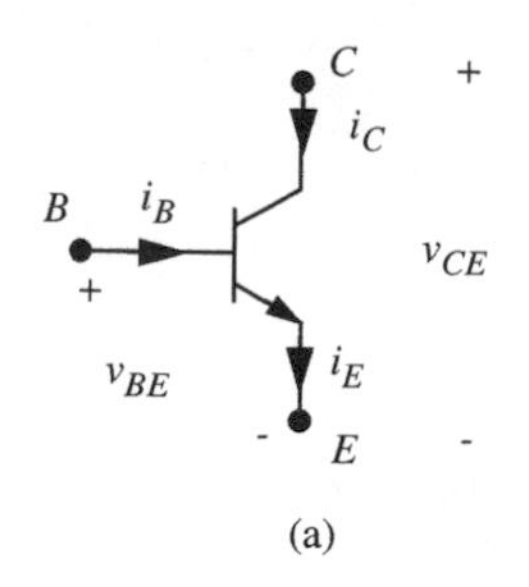


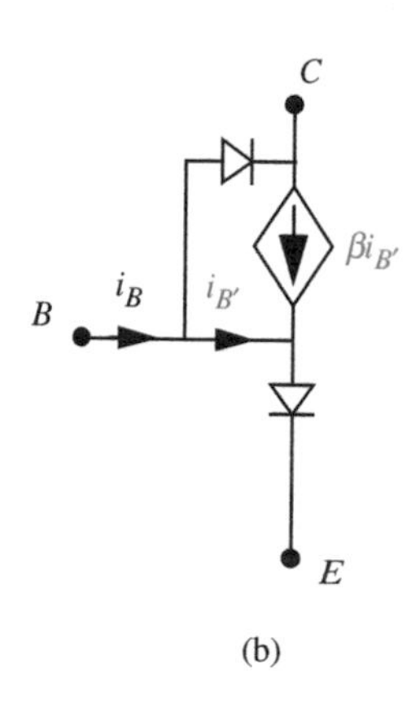


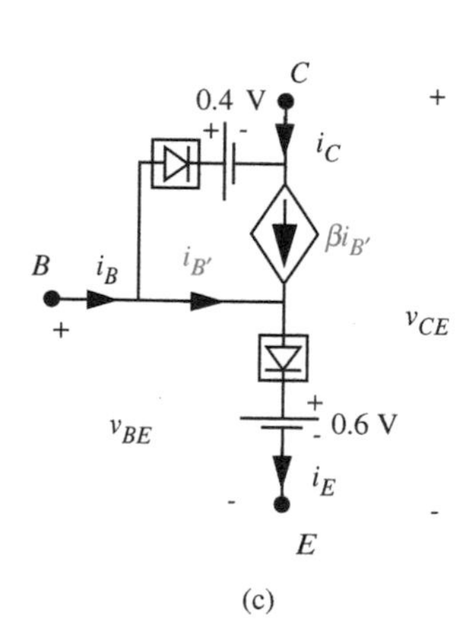


FIGURE 7.49 (a) A bipolar junction transistor; (b) a model for the BJT; (c) a piecewise-linear model for the BJT.

a typical value of around 100. (We show shortly that the base current $i_B = i_{B'}$ in the region of BJT operation that is of interest to us.)

Although we can analyze circuits directly with the model in Figure 7.49b, our analysis can be significantly simplified by using simple piecewise-linear models for the diodes. Figure 7.49c depicts such a piecewise-linear model for the BJT, in which we have replaced the diodes with simple piecewise-linear diode models comprising an ideal diode in series with a voltage source (from Figure 4.33a). In the model in Figure 7.49c, the dependent current source models the horizontal active region curves of the BJT.

The states of the two diodes (both ON, both OFF, and one OFF and one ON) result in distinct piecewise linear regions of BJT operation. Both diodes (in Figure 7.49c) in their OFF state model cutoff: when the base current $i_B$ is zero, both the diodes are OFF, and so is the current source. Figure 7.50a depicts the corresponding BJT model in the cutoff region. Observe that in the cutoff region

$$i_B = i_{B'}$$

because the base-to-collector diode is off.

When $i_B > 0$ and

$$v_{CE} > v_{BE} - 0.4 \text{ V} \tag{7.48}$$

the emitter diode is ON and the collector diode is OFF, and the active region results. In this region of operation, as illustrated in the active region BJT model in Figure 7.50b, the ideal diode between the base and emitter turns ON and appears as a short circuit. The 0.6-V source models the corresponding 0.6-V diode drop. Observe further that

$$i_B = i_{B'}$$

in the active region because the base-to-collector diode is off. In the active region, BJTs display a more-or-less constant voltage drop of about 0.6 V between their base and emitter terminals when the base current $i_B > 0$ (a fact not evident from the characteristic curves in Figure 7.48).

The condition $v_{CE} > v_{BE} - 0.4$ V ensures that the base-collector diode stays OFF. The condition states that the collector voltage must not fall below the base voltage by more than 0.4 V, because if it did, the base-collector diode would turn ON.[5] In the active

---

5. Although the constraint for active-region operation

$$v_{CE} > v_{BE} - 0.4 \text{ V}.$$

is equivalent to the simpler constraints

$$v_{BC} < 0.4 \text{ V} \quad \text{or} \quad v_{CE} > 0.2 \text{ V}.$$

region, the dependent current source amplifies the current supplied by the base by a factor $\beta$, so that the collector current becomes

$$i_C = \beta i_B$$

(recall, $i_B = i_{B'}$ in the active region), and the emitter current is

$$i_E = i_B(\beta + 1).$$

In the active region, the piecewise-linear model for the BJT can be summarized in words as

$$i_C = \begin{cases} \beta i_B & \text{for } i_B > 0 \text{ and } v_{CE} > v_{BE} - 0.4\text{ V} \\ 0 & \text{otherwise.} \end{cases} \quad (7.49)$$

The base-to-collector diode in Figure 7.49c helps model the onset of saturation. Specifically, saturation results when the both the base-to-collector and the base-to-emitter diodes are ON. When $i_B > 0$, and the condition implied by Equation 7.48 is violated, that is, if

$$v_{CE} = v_{BE} - 0.4\text{ V}$$

or equivalently, if

$$v_{BC} = 0.4\text{ V} \quad \text{or} \quad v_{CE} = 0.2\text{ V}$$

then the base-to-collector diode also turns ON, and the BJT saturation region results. The saturation region model for the BJT is shown in Figure 7.50c. In the BJT's saturation region, the BJT model stops looking like a current source, and instead displays a pair of very low resistance paths from the base into the collector and emitter (due to the pair of forward-biased diodes). Because of their low resistance, the path currents are determined by external circuit constraints. By summing voltages along the path $E$, $B$, $C$, we see that the collector-to-emitter voltage is pinned at 0.2 V, irrespective of the current $i_C$.

Our model is not yet complete. There is one additional state in which the emitter diode is OFF and the collector diode is ON, as can happen when the base-to-collector

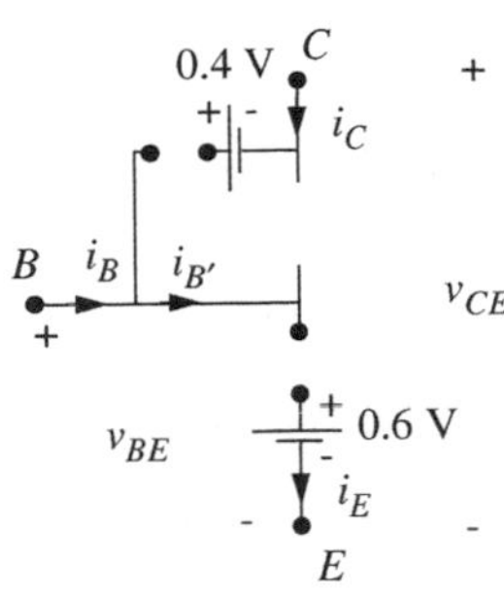

(a) Cutoff region
$i_B = 0$

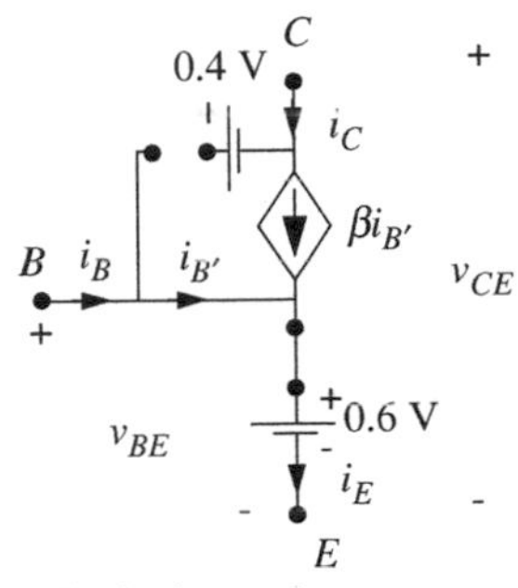

(b) Active region
$i_B > 0$
$v_{CE} > v_{BE} - 0.4$ (or, $v_{CE} > 0.2$)

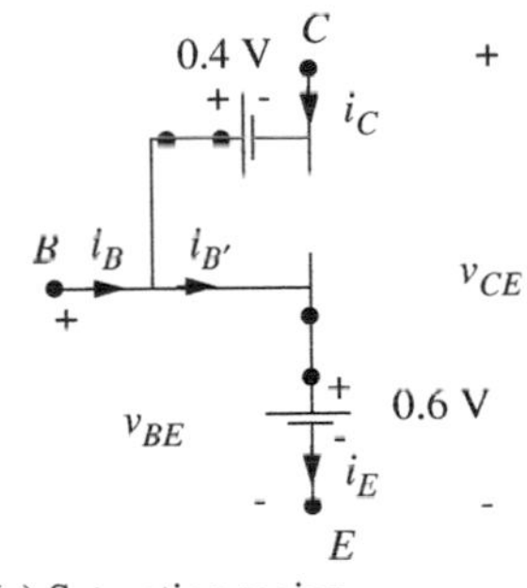

(c) Saturation region
$i_B > 0$
$v_{CE} = v_{BE} - 0.4$ (or, $v_{CE} = 0.2$)

**FIGURE 7.50** Bipolar junction transistor models in various regions of operation.

---

as can be seen by applying the voltage difference form of KVL to the model in Figure 7.49c, we use the former because, by a quirk of chance, it is reminiscent of our MOSFET drain-to-gate voltage constraint (namely,

$$v_{DS} > v_{GS} - V_T$$

for saturation operation of the MOSFET.)

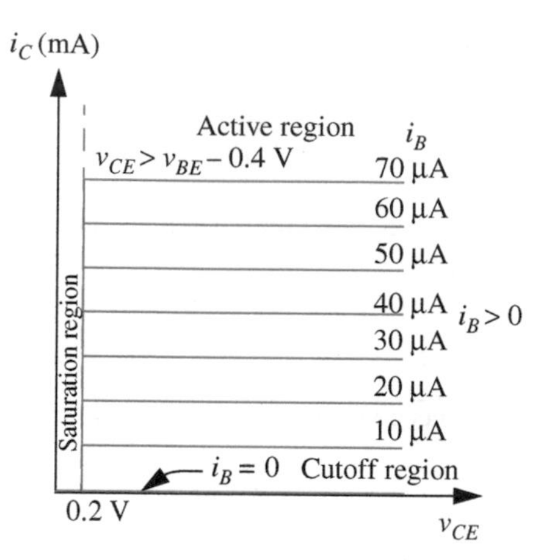


**FIGURE 7.51** Characteristic curves for a bipolar junction transistor as predicted by the piecewise linear model.

voltage is 0.4 V, and the base-to-emitter voltage is less than 0.6 V. This region of operation is called the *reverse injection region*. In this region, the behavior of the BJT is that of a forward biased diode between the base and the collector, and an open circuit at the emitter.

For simplicity, our introductory treatment will choose not to study both the reverse injection and the saturation regions. Accordingly, our BJT circuits will all be designed to avoid completely these regions of behavior.

In the rest of this example, we will discuss the piecewise-linear model for the BJT presented in Figure 7.49c and compare its predictions with the measured characteristics ($i_C$ versus $v_{CE}$ for various values of $i_B$) shown in Figure 7.48. We will plot the characteristics predicted by the piecewise-linear model assuming $\beta = 100$.

To plot the characteristics, we identify BJT behavior in the two piecewise-linear regions of operation that are of interest to us, and shown in Figure 7.50: cutoff and active. We first observe that $i_C$ is zero when the BJT is in cutoff, that is, when $i_B = 0$ (see Figure 7.50a). The curve labeled "Cutoff region" in Figure 7.51 depicts this situation.

Next, when $i_B > 0$ and $v_{CE} > v_{BE} - 0.4$ V (or, equivalently, $v_{CE} > 0.2$ V), the collector current is a constant at $\beta$ times the base current (Figure 7.50b). In the $i_C$ versus $v_{CE}$ plot in Figure 7.51, these constant current curves appear as horizontal lines. Because $\beta$ is a constant, the $i_C$ versus $i_B$ relationship is linear, and so the lines are equally spaced for equal increments in $i_B$.

Finally, when $i_B > 0$ and $v_{CE} = v_{BE} - 0.4$ V (or, equivalently, when $v_{CE} = 0.2$ V), the saturation region model applies (Figure 7.50c). $v_{CE}$ is correctly shown as being pinned at 0.2 V. The vertical line at $v_{CE} = 0.2$ V corresponding to $i_C$ indicates a short-circuit-like behavior in which the collector current is limited only by external circuit constraints.

The similarity of the curves in Figures 7.48 and 7.51 show that our simple piecewise-linear model is quite a good match for the behavior of the BJT.

As a final thought, although our piecewise linear model for the BJT seems a bit complicated at first glance, analog circuits are commonly designed such that BJT always operates in its active region, and the base-to-collector diode is always OFF.[6] We can achieve the desired effect by ensuring that the base-to-collector voltage never exceeds 0.4 V during normal operation (that is, $v_{BC} < 0.4$ V, or equivalently, $v_{CE} > v_{BE} - 0.4$ V). This assumption will be made in all the BJT circuits in this book, so the collector diode can be safely ignored. The resulting, simplified BJT model is depicted in Figure 7.52.

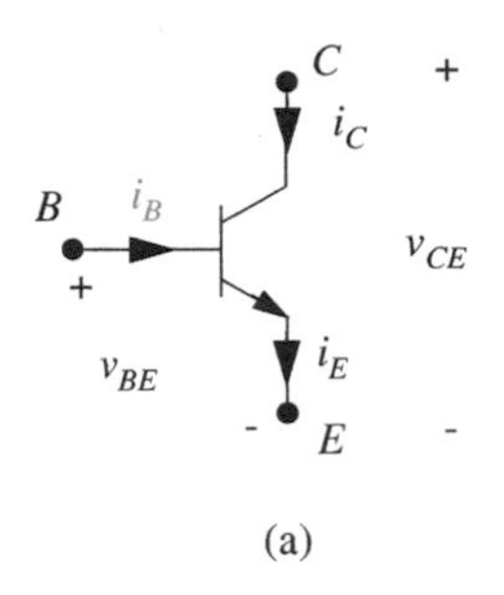


(a)

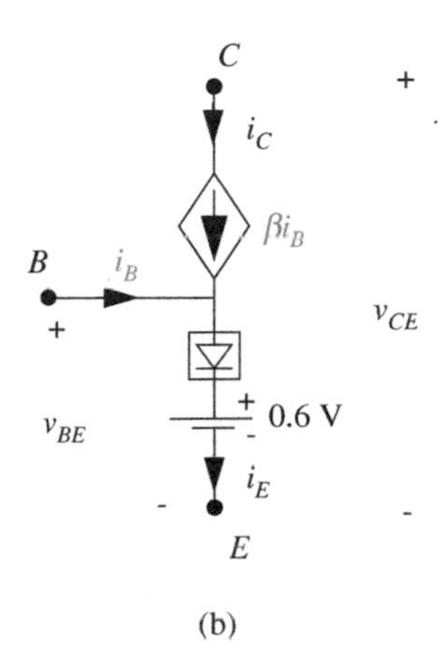


(b)

**FIGURE 7.52** A simpler BJT model suitable for the cutoff and active regions.

---

EXAMPLE 7.14 BJT CIRCUIT PARAMETERS Figure 7.53 shows measured values of $i_B$ and $v_{BE}$ for a BJT within a circuit. Find the corresponding values of $v_{BE}$, $i_C$, and $i_E$ using the BJT model containing two ideal diodes and a voltage source (Figure 7.49c).

Since $i_B > 0$ and $v_{CE} > 0.2$ V, it immediately follows that the BJT operates in its active region. In other words, the emitter diode in Figure 7.49c must be ON, and the collector diode must be OFF (see the active region BJT model in Figure 7.50b). Since the emitter diode is ON, it appears as a short circuit, and so

$$v_{BE} = 0.6 \text{ V}.$$

Based on the active region model in Figure 7.50b, since $i_B = 0.01$ mA,

$$i_C = \beta i_B = 1 \text{ mA}.$$

Summing the currents into the base and collector terminals, we get

$$i_E = i_B + i_C = 1.01 \text{ mA}.$$



**FIGURE 7.53** A bipolar junction transistor in a circuit.

---

6. This design choice is not unlike the one we made with MOSFETs, where circuit parameters were chosen so that the MOSFET always operated in saturation.

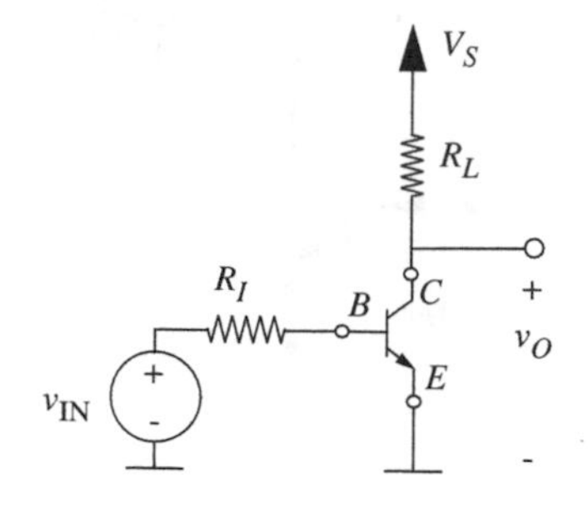


FIGURE 7.54 A BJT amplifier.

EXAMPLE 7.15 A BJT AMPLIFIER Figure 7.54 shows an amplifier circuit based on a BJT. This BJT amplifier configuration is called a *common emitter amplifier* since the emitter terminal of the BJT is common across the input and output ports. Using the piecewise-linear model for the BJT, determine the relationship between $v_O$ and $v_{IN}$, assuming that the BJT device is operating in its active region. Using this relation, determine the values of $v_O$ for $v_{IN} = 1$ V, 1.1 V, and 1.2 V, given that $R_I = 100$ kΩ, $R_L = 10$ kΩ, $\beta = 100$, and $V_S = 10$ V.

Figure 7.55 shows the equivalent circuit for the amplifier in which the BJT has been replaced with its piecewise-linear model. Notice we can safely ignore the collector diode and use the simple BJT model in Figure 7.52 since we are told that the BJT is operating in its active region. Figure 7.56 further shows the active region subcircuit for the amplifier.

The relationship between $v_O$ and $v_{IN}$ can be determined in a few short steps from the active region subcircuit. The current through $R_I$ is simply the voltage difference across the resistor divided by the resistance:

$$i_B = \frac{v_{IN} - 0.6}{R_I}. \tag{7.50}$$

Once $i_B$ is known, we can immediately determine the output voltage by writing the node equation for the node with voltage $v_O$ as follows:

$$\frac{V_S - v_O}{R_L} = \beta i_B.$$

Substituting for $i_B$ from Equation 7.50 and simplifying, we obtain the following relation between $v_O$ and $v_{IN}$:

$$v_O = V_S - \frac{(v_{IN} - 0.6)}{R_I} \beta R_L. \tag{7.51}$$

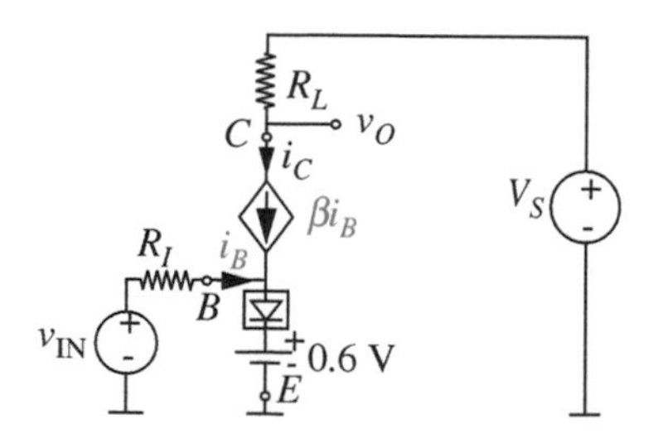


FIGURE 7.55 Equivalent circuit for the BJT amplifier.

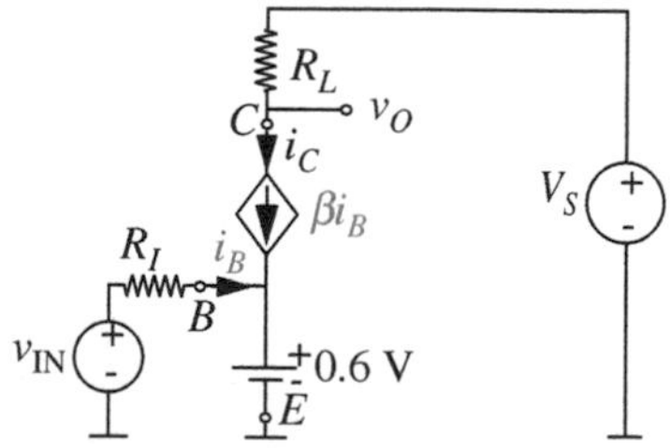

FIGURE 7.56 Active region subcircuit for the BJT amplifier.

Next, substituting $R_I = 100$ kΩ, $R_L = 10$ kΩ, $\beta = 100$, and $V_S = 10$ V, we obtain

$$v_O = 16 - 10v_{IN}.$$

For $v_{IN} = 1$ V, 1.1 V and 1.2 V, $v_O$ is 6 V, 5 V and 4 V, respectively.

As a further exercise, we will go ahead and confirm that the BJT is indeed in its active region for the highest input voltage applied. (Remember that the higher the base voltage the more likely it is that the collector diode in Figure 7.49c is turned ON. Thus, we need check only for the highest input voltage.)

The highest input voltage considered in this example is 1.2 V. For 1.2 V, the collector voltage $v_O = 4$ V. Since, in our circuit, $v_O = v_{CE} = 4$ V, the base-to-collector voltage is

$$v_{BC} = v_{BE} - v_{CE} = 0.6 - 4 = -3.4 \text{ V}.$$

Since the voltage $v_{BC}$ across the collector diode in Figure 7.49c is −3.4 V, which is less than 0.4 V, the collector diode is going to be OFF. (Equivalently, since $v_{CE} > 0.2$ V, we can directly say that that the collector diode is OFF.) We have thus confirmed that the BJT is in its active region.

---

EXAMPLE 7.16 LARGE SIGNAL ANALYSIS OF THE BJT AMPLIFIER Perform a large signal analysis of the BJT amplifier shown in Figure 7.54. Assume that $R_I = 100$ kΩ, $R_L = 10$ kΩ, $\beta = 100$, and $V_S = 10$ V.

For BJT circuits with input $v_{IN}$ and output $v_O$, large signal analysis attempts to answer the following questions:

1. What is the relationship between $v_O$ and $v_{IN}$ in the active region?
2. What is the range of valid input values for active region operation of the BJT? What is the corresponding range of output values?

From Equation 7.51 in Example 7.15, we know that the relation between $v_O$ and $v_{IN}$ for the BJT amplifier is

$$v_O = V_S - \frac{(v_{IN} - 0.6)}{R_I}\beta R_L,$$

thereby completing the first step of large signal analysis.

Next, let us determine the range of input values for which the BJT operates in its active region. To do so, we will first draw a graph of $v_O$ versus $v_{IN}$ to obtain insight into the behavior of the amplifier for various values of the input voltage. When $v_{IN} = 0$, we see

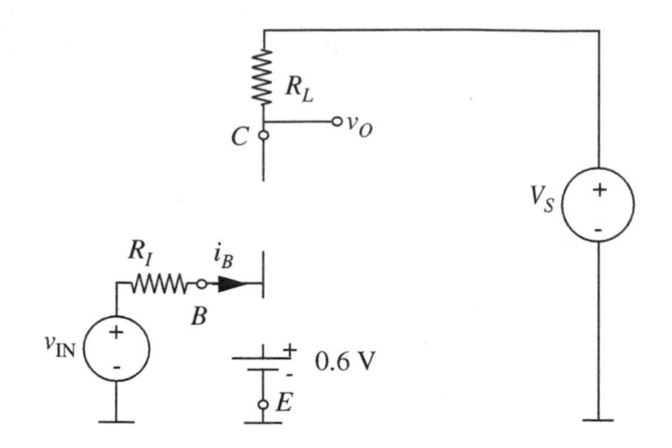


**FIGURE 7.57** Cutoff region subcircuit for the BJT amplifier.

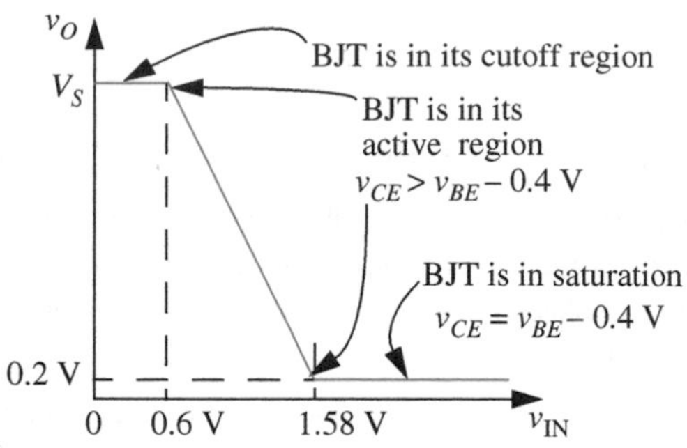

**FIGURE 7.58** $v_O$ versus $v_{IN}$ for the BJT amplifier.

that $i_B = 0$, and so the BJT is in cutoff. The cutoff region subcircuit for the BJT amplifier is shown in Figure 7.57. In cutoff, both the diode and current sources are replaced by open circuits. It is easy to see from the circuit in Figure 7.57 that

$$v_O = V_S.$$

Inspection of the amplifier equivalent circuit in Figure 7.55 indicates that the input current $i_B$ will be zero (and the ideal diode will remain OFF ) as long as $v_{IN} < 0.6$. Thus the output $v_O$ will remain at $V_S$ for $v_{IN} < 0.6$. This fact is graphed in Figure 7.58 as the horizontal straight line at voltage $V_S$ for $v_{IN} < 0.6$.

When $v_{IN}$ exceeds 0.6 V by a small amount,[7] the ideal diode turns ON, and current begins to flow through the resistor $R_I$. In this situation, the active region equivalent circuit in Figure 7.56 results. In the active region, $v_O$ is given by

$$v_O = V_S - \frac{(v_{IN} - 0.6)}{R_I}\beta R_L. \tag{7.52}$$

This relationship appears as a straight line with slope $-\beta R_L/R_I$ in the $v_O$ versus $v_{IN}$ graph and is plotted as such in Figure 7.58. Thus,

$$v_{IN} = 0.6 \text{ V}$$

7. If $v_{IN}$ exceeds 0.6 V by a large amount, the BJT might enter saturation. We will determine this saturation region boundary momentarily.

and

$$i_B = 0$$

are the input parameters at the lower boundary of the active region.

The $v_O$ versus $v_{IN}$ relationship in the active region shows that $v_O$ decreases linearly as $v_{IN}$ increases. The linear relationship applies as long as the BJT remains in its active region of operation. The upper boundary (with respect to the input voltage) of the active region is reached when $v_{IN}$ becomes large, and $v_O$ becomes small enough that the condition

$$v_{CE} > v_{BE} - 0.4 \text{ V}$$

is no longer met. Since $v_{CE} = v_O$ and $v_{BE}$ is pinned at 0.6 V (from the active region amplifier subcircuit in Figure 7.56), $v_O$ reaches the boundary point of the active region when

$$v_O = 0.6 - 0.4 = 0.2 \text{ V}.$$

The corresponding value of $i_C$ is given by

$$i_C = \frac{V_S - 0.2 \text{ V}}{R_L} = 980\ \mu\text{A}.$$

The input voltage corresponding to this output voltage can be found by solving for $v_{IN}$ from Equation 7.52 as follows

$$0.2 = 10 - \frac{v_{IN} - 0.6 \text{ V}}{100k} 100 \times 10 \text{ k}\Omega.$$

Solving, we get

$$v_{IN} = 1.58 \text{ V}.$$

This upper boundary of the active region (with respect to $v_{IN}$) is marked in Figure 7.58. The corresponding value of $i_B$ can be found from

$$i_B = \frac{v_{IN} - 0.6}{R_I} = 9.8\ \mu\text{A}.$$

Once the BJT exits the active region and enters the BJT saturation region (for $v_{IN} \geq 1.58$ V), the saturation model for the BJT in Figure 7.50c applies and the equivalent

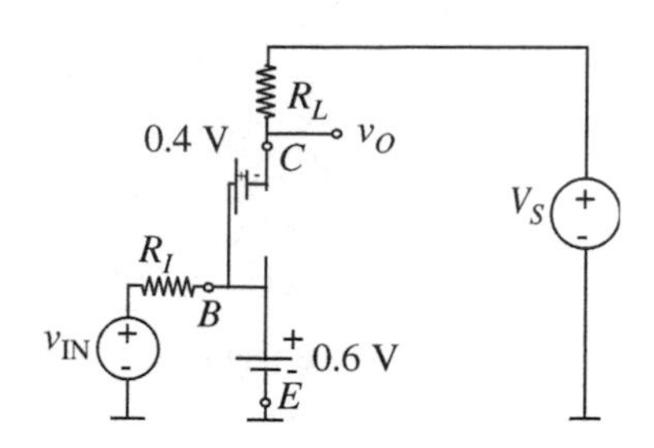


**FIGURE 7.59** Saturation region subcircuit for the BJT amplifier.

subcircuit in Figure 7.59 best models the saturation region operation of the amplifier. By a straightforward application of KVL, we find that $v_O$ is given by

$$v_O = 0.6 - 0.4 = 0.2 \text{ V}.$$

In other words, $v_O$ is pinned at 0.2 V by the BJT in its saturation region when the input voltage $v_{IN}$ exceeds 1.58 V. This fact is plotted as a horizontal line at $v_O = 0.2$ V in Figure 7.58.

To summarize, the limits on the inputs for active region operation are given by

$$0.6 \text{ V} < v_{IN} < 1.58 \text{ V}$$

and

$$0 < i_B < 9.8 \ \mu\text{A}.$$

The corresponding limits on the outputs are given by

$$10 \text{ V} > v_O > 0.2 \text{ V}$$

and using $i_C = \beta i_B$,

$$980 \ \mu\text{A} > i_C > 0 \text{ A}.$$

---

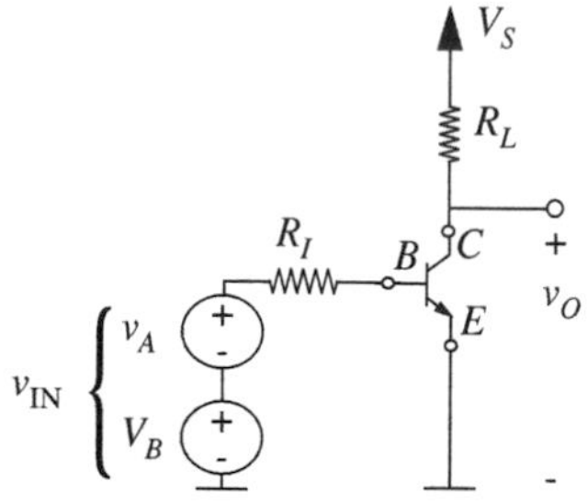

**FIGURE 7.60** BJT amplifier showing the input bias voltage explicitly.

---

**EXAMPLE 7.17 SELECTING AN OPERATING POINT FOR THE BJT AMPLIFIER** Choose an operating point for the amplifier analyzed in Example 7.16 to maximize the input voltage swing. What is the corresponding output operating point and the output voltage swing? Is the output swing symmetric about the output operating point?

The BJT amplifier circuit in Figure 7.54 is redrawn in Figure 7.60 to show explicitly that the input voltage $v_{IN}$ is the sum of a bias voltage $V_B$ and the signal $v_A$. Our first task is to find the input operating point $(V_B, I_B)$. We do so by reviewing the results from Example 7.16.

From Example 7.16, we know that the valid range for the total input voltage $v_{IN}$ to ensure active region operation is given by

$$0.6\ \text{V} < v_{IN} < 1.58\ \text{V}.$$

The corresponding range for the input current is

$$0 < i_B < 9.8\ \mu\text{A}.$$

We can obtain the maximum input swing for active region operation by biasing the input at the midpoint of the input valid range. In other words, we choose

$$V_B = 1.09\ \text{V}$$

and

$$I_B = 4.9\ \mu\text{A}.$$

The corresponding value for the output operating point voltage $V_O$ can be obtained from Equation 7.52 as

$$V_O = V_S - \frac{(V_B - 0.6\ \text{V})}{R_I}\beta R_L = 5.1\ \text{V}.$$

Similarly, the value of the output operating point current $I_C$ is given by

$$I_C = \beta I_B = 490\ \mu\text{A}.$$

We know from Example 7.16, that the output voltage swing for active region operation is given by

$$10\ \text{V} > v_O > 0.2\ \text{V}$$

Our output operating point of 5.1 V falls in the center of this range, and so the output swing is symmetric about the 5.1 V operating point. The symmetry results directly from the linearity of the BJT in its active region. Contrast this result with that for the MOSFET (Section 7.7), in which the output swing was asymmetric due to the MOSFET's nonlinear behavior in its saturation region.

---

WWW EXAMPLE 7.18 BETTER BJT MODELS

**EXAMPLE 7.19 LARGE SIGNAL ANALYSIS OF A DIFFERENTIAL AMPLIFIER** This example studies the differential amplifier shown in Figure 7.62. Differential amplifiers are widely used in analog signal processing, and are the heart of operational amplifiers. The applications of differential amplifiers are best discussed in the context of small signal analysis, and so we defer a detailed discussion of the applications until Chapter 8. Furthermore, a complete operational amplifier circuit will be studied in detail in Example 7.21, and operational amplifier applications will be discussed in Chapter 15. So for the present purposes, we will simply treat the amplifier in Figure 7.62 as yet another example of a MOSFET amplifier.

The amplifier in Figure 7.62 has two input voltages, $v_{IN1}$ and $v_{IN2}$; and one output voltage, $v_{OUT}$. The goal of this exercise is therefore to determine $v_{OUT}$ as a function of $v_{IN1}$ and $v_{IN2}$. The $V_S$ and $I_S$ sources serve only to bias the amplifier, and are assumed to be constant.

To begin the analysis, we assume that both MOSFETs are identical, and that both MOSFETs operate in their saturation regions. Therefore,

$$i_{D1} = \frac{K}{2}(v_{GS1} - V_T)^2 \tag{7.53}$$

$$i_{D2} = \frac{K}{2}(v_{GS2} - V_T)^2. \tag{7.54}$$

Further, from KCL applied to the node at which the two MOSFETs and the current source join,

$$i_{D1} + i_{D2} = I_S, \tag{7.55}$$

and from KVL applied to the loop around the two MOSFETs through ground,

$$v_{IN1} - v_{GS1} + v_{GS2} - v_{IN2} = 0. \tag{7.56}$$

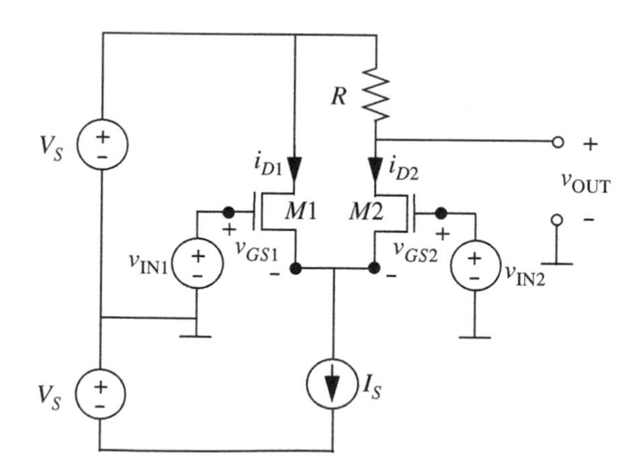


FIGURE 7.62 A differential amplifier.

Finally, at the output of the amplifier,

$$v_{OUT} = V_S - Ri_{D2}, \tag{7.57}$$

where it has been assumed that $i_{OUT} = 0$. Equations 7.53 through 7.57 may now be solved to determine $v_{OUT}$ as a function of $v_{IN1}$ and $v_{IN2}$. We will do so in two steps. First, Equations 7.53 through 7.56 will be solved to determine $i_{D2}$ as a function of $v_{IN1}$ and $v_{IN2}$. Then, Equation 7.57 will be used to determine $v_{OUT}$ from $i_{D2}$.

To determine $i_{D2}$, first substitute Equation 7.53 into Equation 7.55 to eliminate $i_{D1}$, then substitute Equation 7.56 into the result to eliminate $v_{GS1}$, and finally substitute Equation 7.54 into the result to eliminate $v_{GS2}$. This yields

$$I_S - i_{D2} + \frac{K}{2}\left(v_{IN1} - v_{IN2} + \sqrt{\frac{2i_{D2}}{K}}\right)^2. \tag{7.58}$$

Equation 7.58 is a quadratic equation in $\sqrt{2i_{D2}/K}$ and can be rewritten as

$$2\left(\sqrt{\frac{2i_{D2}}{K}}\right)^2 + 2(v_{IN1} - v_{IN2})\sqrt{\frac{2i_{D2}}{K}} + (v_{IN1} - v_{IN2})^2 - \frac{2I_S}{K}. \tag{7.59}$$

From Equations 7.58 and 7.59 it is apparent that $i_{D2}$ depends only on the difference voltage $v_{IN1} - v_{IN2}$ when both MOSFETs operate in their saturation regions. That is why the amplifier is referred to as a differential amplifier.

The solution to Equation 7.59 is

$$i_{D2} = \frac{K}{8}\left(\sqrt{\frac{4I_S}{K} - (v_{IN1} - v_{IN2})^2} - v_{IN1} + v_{IN2}\right)^2. \tag{7.60}$$

Note that the positive sign in the solution to Equation 7.59 is chosen in Equation 7.60 because $\sqrt{2i_{D2}/K}$ must be positive. Finally, Equation 7.60 can be substituted into Equation 7.57 to yield

$$v_{OUT} = V_S - \frac{RK}{8}\left(\sqrt{\frac{4I_S}{K} - (v_{IN1} - v_{IN2})^2} - v_{IN1} + v_{IN2}\right)^2. \tag{7.61}$$

From symmetry, $i_{D1}$ may also determined to be

$$i_{D1} = \frac{K}{8}\left(\sqrt{\frac{4I_S}{K} - (v_{IN2} - v_{IN1})^2} - v_{IN2} + v_{IN1}\right)^2. \tag{7.62}$$

From Equations 7.60 and 7.62 it is apparent that the amplifier functions such that the difference voltage $v_{IN1} - v_{IN2}$ steers the total current $I_S$ towards either $i_{D1}$ or $i_{D2}$, depending on its sign.

Equations 7.60 through 7.62 are valid only as long as both MOSFETs remain in their saturation region of operation. One requirement for saturation operation is that $|v_{IN1} - v_{IN2}|$ must not be so large that either MOSFET is cut off. Thus, $\sqrt{2i_{D2}/K}$, and similarly $\sqrt{2i_{D1}/K}$, must be positive. From Equations 7.53 and 7.54, this is equivalent to $v_{GS1} > V_T$ and $v_{GS2} > V_T$. From Equations 7.60 and 7.62, cutoff is therefore avoided as long as

$$\frac{2I_S}{K} > (v_{IN1} - v_{IN2})^2. \tag{7.63}$$

Additionally, neither MOSFET may be driven into its triode region. This may be avoided by using a sufficiently large value of $V_S$, or alternatively by further limiting the allowable range of $v_{IN1}$ and $v_{IN2}$.

---

EXAMPLE 7.20 MORE ON THE DIFFERENTIAL AMPLIFIER Next, we discuss a numerical example related to the differential amplifier of Example 7.19. For this amplifier, let $V_S = 10$ V, $I_S = 0.5$ mA, $K = 1$ mA/V$^2$, $V_T = 1$ V, and $R = 10$ kΩ.

Given these parameters, from Equation 7.61,

$$v_{OUT} = 10 \text{ V} - 1.25 \text{ V}^{-1} \left(\sqrt{2 \text{ V}^2 - (v_{IN1} - v_{IN2})^2} - v_{IN1} + v_{IN2}\right)^2. \tag{7.64}$$

Note that when $v_{IN1} = v_{IN2}$, $i_{D1} = i_{D2} = I_S/2 = 0.25$ mA, and so $v_{OUT} = 7.5$ V.

Further, from Equation 7.63,

$$|v_{IN1} - v_{IN2}| < 1 \text{ V} \tag{7.65}$$

to avoid cutoff. Correspondingly, $v_{OUT}$ will range from 10 V when MOSFET M2 is cut off by the application of $v_{IN1} - v_{IN2} = 1$ V, to 5 V when MOSFET M1 is cut off by the application of $v_{IN2} - v_{IN1} = 1$ V.

---

EXAMPLE 7.21 LARGE SIGNAL ANALYSIS OF AN OPERATIONAL AMPLIFIER CIRCUIT As will be clear in Chapter 15, the differential amplifier shown in Figure 7.62 does not quite fit our notion of an operational amplifier because $v_{OUT}$ is not zero for $v_{IN1} = v_{IN2}$. This can be remedied with the addition of a common-source stage built with a p-channel MOSFET, as shown in Figure 7.63. The common-source stage shifts the level of the output so that $v_{OUT}$ can be zero for $v_{IN1} = v_{IN2}$. It also provides additional voltage gain. Thus the circuit in Figure 7.63 serves as a simple operational amplifier.

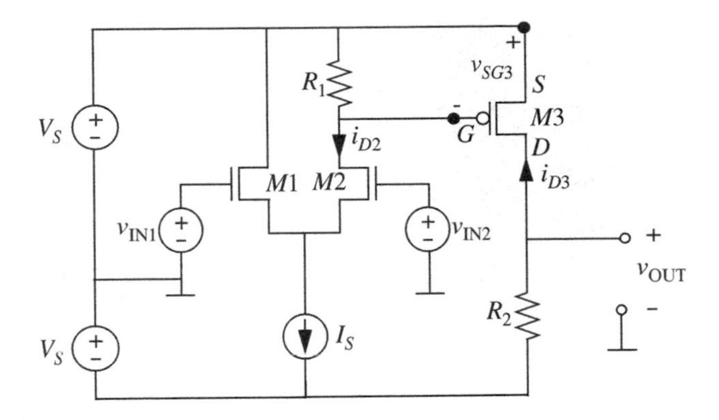


FIGURE 7.63 An operational amplifier built using a differential amplifier and a p-channel MOSFET amplifier.

The behavior of a p-channel MOSFET essentially mirrors that of the n-channel MOSFET. Correspondingly, $v_{GS}$, $v_{DS}$, and $i_D$ are all negative in the saturation region. Further, the threshold voltage $V_T$ is typically negative. Only the parameter $K$ is positive. Thus, for a p-channel MOSFET,

$$i_D = -\frac{K}{2}(v_{GS} - V_T)^2 \tag{7.66}$$

$$v_{DS} \le v_{GS} - V_T \le 0 \tag{7.67}$$

in the saturation region. Often, it is convenient to work with positive numbers. In this case, Equations 7.66 and 7.67 can be rewritten as

$$-(-i_D) = \frac{K}{2}(v_{SG} + V_T)^2 \tag{7.68}$$

$$v_{SD} \ge v_{SG} + V_T \ge 0. \tag{7.69}$$

In Equations 7.68 and 7.69, $v_{SG}$, $v_{SD}$, $-i_D$, and $K$ are all positive. Only $V_T$ is negative. We will use the latter formulation here.

To determine $v_{OUT}$ in the operational amplifier as a function of $v_{IN1}$ and $v_{IN2}$, we again assume that the two n-channel MOSFETs are identical, and that all three MOSFETs operate in the saturation region. To distinguish the n-channel MOSFETs from the p-channel MOSFET, denote the n-channel MOSFET parameters by $K_n$ and $V_{Tn}$, and the p-channel MOSFET parameters by $K_p$ and $V_{Tp}$. Again, all parameters are positive except for $V_{Tp}$.

The differential stage of the operation amplifier has already been analyzed in Example 7.19. In particular, from Equation 7.60 in Example 7.19 it was determined that

$$v_{SG3} = R_1 i_{D2} = \frac{R_1 K_n}{8}\left(\sqrt{\frac{4I_S}{K_n} - (v_{IN1} - v_{IN2})^2} - v_{IN1} + v_{IN2}\right)^2. \tag{7.70}$$

The common-source stage built with the p-channel MOSFET behaves according to

$$v_{OUT} = -V_S + R_2(-i_{D3}) = -V_S + \frac{R_2 K_p}{2}(v_{SG3} + V_{Tp})^2. \tag{7.71}$$

Combining Equations 7.70 and 7.71,

$$v_{OUT} = \frac{R_2 K_p}{2}\left(\frac{R_1 K_n}{8}\left(\sqrt{\frac{4I_S}{K_n} - (v_{IN1} - v_{IN2})^2} - v_{IN1} + v_{IN2}\right)^2 + V_{Tp}\right)^2 - V_S. \tag{7.72}$$

Finally, in order to meet the requirement that $v_{OUT} = 0$ when $v_{IN1} = v_{IN2}$, it must be the case that

$$V_S = \frac{R_2 K_p}{2}\left(\frac{R_1 I_S}{2} + V_{Tp}\right)^2. \tag{7.73}$$

In general, it is also necessary to derive the conditions under which all MOSFETs remain in their saturation region of operation. For brevity, we will not do that here.

---

EXAMPLE 7.22 NUMERICAL ANALYSIS OF OP AMP CIRCUIT Let us now conduct a numerical analysis of the operational amplifier of Example 7.21. Following Example 7.20, let $V_S = 10$ V, $I_S = 0.5$ mA, $K_n = 1$ mA/V$^2$, $V_{Tn} = 1$ V, and $R_1 = 10$ kΩ. Further, let $K_p = 1$ mA/V$^2$ and $V_{Tp} = -1.5$ V.

Then, from Equation 7.73, $R_2$ must be 20 kΩ in order for $v_{OUT}$ to be biased at 0 V when $v_{IN1} = v_{IN2}$. Given this design, Equation 7.72 yields

$$v_{OUT} = 10\ \text{V}^{-1}\left(1.25\ \text{V}^{-1}\left(\sqrt{2\ \text{V}^2 - (v_{IN1} - v_{IN2})^2} - v_{IN1} + v_{IN2}\right)^2 - 1.5\ \text{V}\right)^2 - 10\ \text{V} \tag{7.74}$$

as the unloaded input-output relation of the operational amplifier, assuming that all MOSFETs remain in their saturation region of operation. This is a complicated nonlinear equation, but as shown in the following chapter, it simplifies significantly and becomes linear for small signals.

---

## 7.8 SWITCH UNIFIED (SU) MOSFET MODEL

This section presents a more elaborate model of the MOSFET and can be skipped without loss of continuity.

The actual characteristics of the MOSFET shown in Figure 7.12 indicate that the MOSFET has very interesting behavior in the triode region. For a fixed

$v_{GS}$, we approximated the behavior as a linear resistor using the SR model. Clearly, the SR model does not capture MOSFET behavior if we vary $v_{GS}$. Worse yet, even for a given value of $v_{GS}$, the SR model becomes inaccurate as the value of $v_{DS}$ approaches $v_{GS} - V_T$. For more accuracy, we can develop a more elaborate model for the triode region operation of the MOSFET. Abandoning the piecewise-linear method, this more elaborate model characterizes the behavior of the MOSFET in the triode region as a *nonlinear* resistor, whose characteristics depend on $v_{GS}$. When combined with the SCS model for the saturation region, the nonlinear resistor model in the triode region results in a continuous set of MOSFET curves. The resulting combined model for the triode and saturation regions is called the *switch unified* model or the SU model of the MOSFET.

The SU model can be summarized as follows:

$$i_{DS} = \begin{cases} K\left[(v_{GS} - V_T)v_{DS} - \frac{v_{DS}^2}{2}\right] & \text{for } v_{GS} \geq V_T \text{ and } v_{DS} < v_{GS} - V_T \\ \frac{K(v_{GS}-V_T)^2}{2} & \text{for } v_{GS} \geq V_T \text{ and } v_{DS} \geq v_{GS} - V_T \\ 0 & \text{for } v_{GS} < V_T. \end{cases} \quad (7.75)$$

The characteristics of the MOSFET according to the SU model are plotted in Figure 7.64. As promised, notice that the curves in the triode and the saturation regions are continuous and provide a good match with actual MOSFET characteristics shown in Figure 7.11.

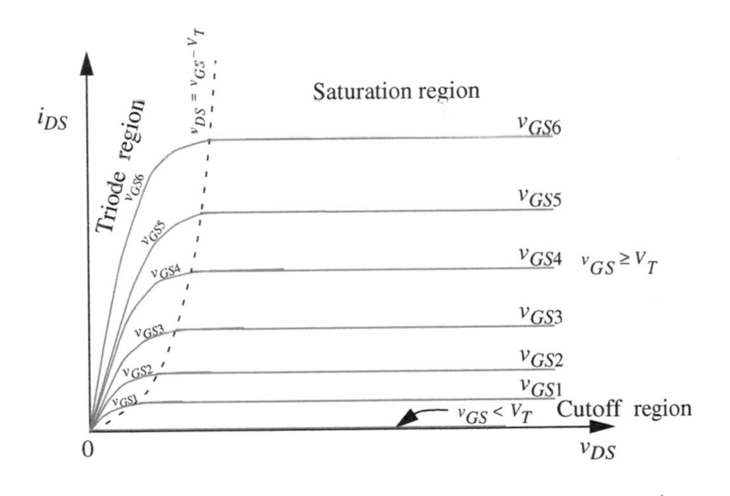


**FIGURE 7.64** Characteristics of the MOSFET device according to the SU model.

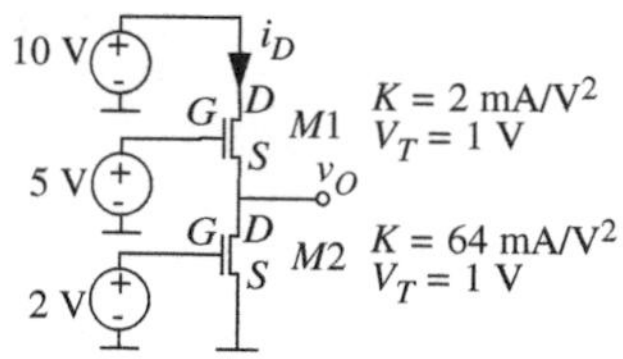

**FIGURE 7.65** A circuit containing two MOSFETs. We are told that *M*1 operates in the saturation region and that *M*2 operates in the triode region.

---

**EXAMPLE 7.23 ANALYSIS USING THE SU MODEL** Determine the voltage $v_O$ for the MOSFET circuit shown in Figure 7.65. You are given that MOSFET *M*1 operates in its saturation region, and that MOSFET *M*2 operates in the triode region. The MOSFET parameters are indicated in Figure 7.65.

For the MOSFET circuit shown in Figure 7.65, $i_{DS}$ for both MOSFETs is the same. Accordingly, we will write expressions for $i_{DS}$ for both MOSFETs, and equate them to obtain the voltage $v_O$. We are told that MOSFET *M*1 operates in its saturation region, and so the saturation region equation applies. Thus for MOSFET *M*1

$$i_D = K\frac{(v_{GS} - V_T)^2}{2}.$$

Substituting $v_{GS} = 5 - v_O$, $V_T = 1$ V, and $K = 2$ mA/V$^2$, we get

$$i_D = 10^{-3}(4 - v_O)^2. \tag{7.76}$$

Next, since we are given that *M*2 operates in the triode region, we can write

$$i_D = K\left[(v_{GS} - V_T)v_{DS} - \frac{v_{DS}^2}{2}\right].$$

Substituting $v_{GS} = 2\text{ V} - 1\text{ V}$, $V_T = 1$ V, $v_{DS} = v_O$, and $K = 64$ mA/V$^2$, and simplifying, we get

$$i_D = 64 \times 10^{-3}\left[v_O - \frac{v_O^2}{2}\right]. \tag{7.77}$$

Equating the right-hand sides of Equations 7.76 and 7.77, and simplifying, we get the following equation for $v_O$:

$$33v_O^2 - 72v_O + 16 = 0,$$

which yields

$$v_O = 0.25\text{ V}.$$

When $v_O$ is 0.25 V it is easy to see that *M*1 is indeed in saturation and *M*2 is in the triode region.

---

## 7.9 SUMMARY

- The last two chapters have discussed a set of progressively more elaborate models for the MOSFET. This section summarizes the models and discusses when it is appropriate to use each of the models.

- The simplest model for the MOSFET is the S model. This switch model models the on-off behavior of the MOSFET. Accordingly, the S model is appropriate when the designer cares only about the logical behavior of a circuit containing MOSFETs; in other words, where the voltage values of interest are only highs and lows. Thus, the S model is commonly used to arrive at the topology of a digital circuit to perform some given logical function. The S model is also useful in certain analog situations where the specific properties of the MOSFET beyond its on-off behavior have no effect on circuit behavior. Certain power circuits that use the MOSFET as a switch fall under this category.

- The SR model of the MOSFET characterizes the behavior of the MOSFET as a resistor when the MOSFET is in its ON state, and $v_{GS}$ is fixed. The SR model is appropriate for most types of simple analyses involving digital circuits, such as static discipline computations of voltage levels, simple power calculations, and, as will be discussed in later chapters, delay calculations. Although technically the SR model is valid only in the MOSFET's triode region (that is, when $v_{DS} < (v_{GS} - V_T)$), for simplicity, we ignore this limitation and apply it in digital circuit applications irrespective of the value of the drain voltage, since the model is such a gross simplification of the MOSFET's behavior in the first place.

- The SCS model characterizes the behavior of the MOSFET in its saturation region. By designing analog circuits to adhere to the saturation discipline, the SCS model is appropriate for most of our analog applications such as amplifiers and analog filters.

- The SU model provides accurate models of the MOSFET in both the triode and the saturation regions, but is more complicated. In its saturation region, it behaves as the SCS model. So, for analog circuits that are designed to adhere to the saturation discipline its use is no different than the use of the SCS model. Thus the SU model is useful when the designer wishes to conduct very accurate analyses of digital or analog circuits in which the MOSFETs are allowed to operate in both their triode and saturation regions. To analyze a circuit containing MOSFETs, the designer first makes an educated guess as to the region — triode, saturated, or cutoff — in which each of the MOSFETs operates. Then, the designer writes node equations for the circuit, selecting appropriate device equations for each of the MOSFETs. After solving the set of equations for the node voltages and edge currents, the designer must confirm that their initial guess as to the state of the MOSFET is consistent with the final node voltages. We leave a detailed treatment of the SU model for more advanced courses on circuits. In the rest of this book, we will focus on the S, the SR, and the SCS models.

- This chapter also introduced the MOSFET amplifier. The amplifier is an example of a nonlinear circuit. We chose to operate the amplifier under the saturation discipline so that it provided a voltage gain for an input signal and so that the MOSFET operated solely in its saturation region, where the SCS model applied. We also discussed the application of a DC offset voltage at the input of the amplifier to boost the signal of interest sufficiently so that the amplifier operated in saturation for the entire dynamic range of input signal variation. The application of a DC offset established a DC operating point for the amplifier.
- We introduced large signal analysis for the amplifier. Large signal analysis summarizes how the amplifier behaves for large swings in the input signal and involves answering the following questions:
  1. What is the relationship between the amplifier output $v_O$ and its input $v_{IN}$ in the saturation region?
  2. What is the range of valid input values for the amplifier under the saturation discipline? What is the corresponding range of valid output values?
- The next chapter will discuss a small signal analysis of the amplifier. Small signal analysis is appropriate when the input signal perturbations about the operating point are very small.

## EXERCISES

**EXERCISE 7.1** Determine the voltage $v_O$ across the voltage-dependent current source shown in the circuit in Figure 7.66 when

$$i = f(v) = \frac{K}{v^2}.$$

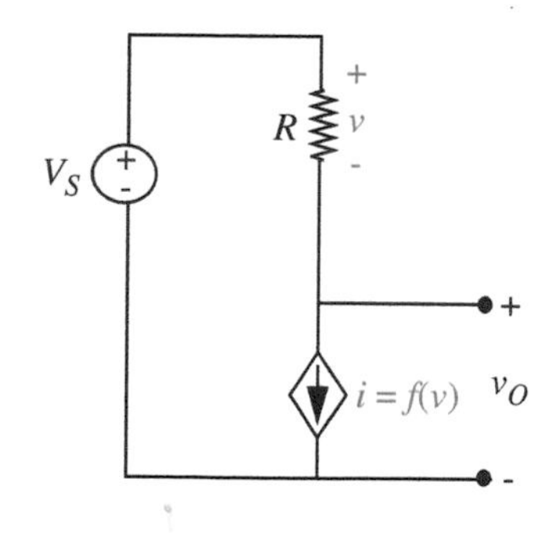


FIGURE 7.66

**EXERCISE 7.2** Consider the circuit containing the dependent current source shown in Figure 7.67.

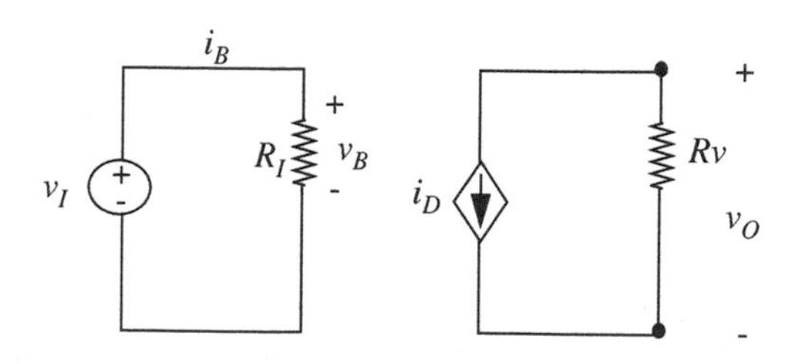


FIGURE 7.67

a) Determine $v_O$ in terms of $v_I$ if $i_D = K_1 v_B$. What are the units of $K_1$?

b) Determine $v_O$ in terms of $v_I$ if $i_D = K_2 i_B$. What are the units of $K_2$?

c) Determine $v_O$ in terms of $v_I$ if $i_D = K_3 v_B^2$. What are the units of $K_3$?

d) Determine $v_O$ in terms of $v_I$ if $i_D = K_4 i_B^2$. What are the units of $K_4$?

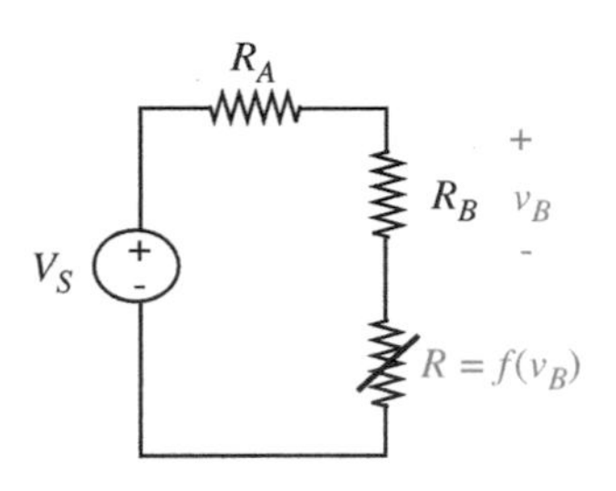


FIGURE 7.68

EXERCISE 7.3 The resistance $R$ in the circuit shown in Figure 7.68 depends on the voltage across resistor $R_B$. Determine $v_B$ if

$$R = \frac{K}{v_B}.$$

EXERCISE 7.4 A MOSFET is characterized by the following equation:

$$i_{DS} = \frac{K}{2}(v_{GS} - V_T)^2$$

in its saturation region. A MOSFET operates in the saturation region for

$$v_{DS} \geq v_{GS} - V_T \text{ and } v_{GS} \geq V_T.$$

Express the $v_{DS} \geq v_{GS} - V_T$ constraint in terms of $i_{DS}$ and $v_{DS}$.

EXERCISE 7.5 The MOSFET in Figure 7.69 is characterized by the equation:

$$i_{DS} = \frac{K}{2}(v_{GS} - V_T)^2$$

in its saturation region according to the SCS model. The MOSFET operates in the saturation region for

$$v_{DS} \geq v_{GS} - V_T \text{ and } v_{GS} \geq V_T.$$

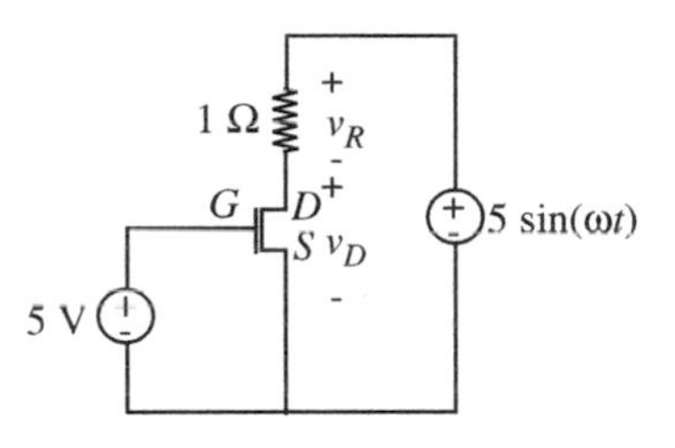


FIGURE 7.69

The MOSFET operates in its triode region for

$$v_{DS} < v_{GS} - V_T \text{ and } v_{GS} \geq V_T.$$

Suppose the MOSFET is characterized by the SR model in its triode region. In other words,

$$i_{DS} = \frac{v_{DS}}{R_{ON}}$$

in the triode region. Assume that $R_{ON}$ is a constant with respect to $i_{DS}$ and $v_{DS}$, but its value is some function of $v_{GS}$. Further suppose that $i_{DS} = 0$ when $v_{GS} < V_T$:

a) For $v_{GS} = 5$ V, what value of $R_{ON}$ makes the MOSFET $i_{DS}$ versus $v_{DS}$ characteristic continuous between its triode and saturation regions of operation?

b) Plot $v_R$ versus $v_D$ for the circuit shown in Figure 7.69. This circuit is useful in plotting the MOSFET characteristics. Assume that $K = 1\ \text{mA/V}^2$ and $V_T = 1$ V. Use the value of $R_{ON}$ calculated in (a). Use a volt scale for $V_D$ and a millivolt scale for $v_R$.

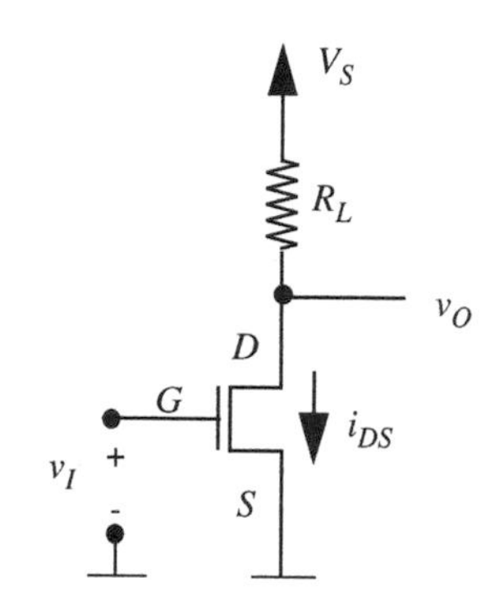


FIGURE 7.70

EXERCISE 7.6 Consider the MOSFET amplifier shown in Figure 7.70. Assume that the amplifier is operated under the saturation discipline. In its saturation region, the MOSFET is characterized by the equation:

$$i_{DS} = \frac{K}{2}(v_{GS} - V_T)^2$$

where $i_{DS}$ is the drain-to-source current when a voltage $v_{GS}$ is applied across its gate-to-source terminals.

a) Draw the equivalent circuit for the amplifier based on the SCS model of the MOSFET.

b) Write an expression relating $v_O$ to $i_{DS}$.

c) Write an expression relating $i_{DS}$ to $v_I$.

d) Write an expression relating $v_O$ to $v_I$.

e) Suppose that an input voltage $V_I$ results in an output voltage $V_O$. By what factor must $V_I$ be increased (or decreased) so that the output voltage is doubled?

f) Suppose, again, that an input voltage $V_I$ results in an output voltage $V_O$. Suppose, further, that we desire an output voltage that is $2V_O$. Assuming that both the input voltage and the MOSFET do not change, what are all the possible ways of accomplishing the desired doubling of the output voltage?

g) The power consumed by the MOSFET amplifier in Figure 7.70 is given by $V_S i_{DS}$, assuming that no current is draw out of the $v_O$ terminal. Which of the alternatives for doubling $V_O$ from parts (e) and (f) will result in the lowest power consumption?

EXERCISE 7.7 Consider, again, the MOSFET amplifier shown in Figure 7.70. Assume that the amplifier is operated under the saturation discipline. The MOSFET in doctored so its threshold voltage is 0. In other words, the saturation region of the

MOSFET is now characterized by the equation:

$$i_{DS} = \frac{K}{2} v_{GS}^2$$

where $i_{DS}$ is the drain-to-source current when a voltage $v_{GS}$ is applied across its gate-to-source terminals. The following questions relate to the large-signal analysis of the amplifier:

a) Derive the relationship between the output voltage $v_O$ and the input voltage $v_I$.

b) Derive the range of valid input voltages. Under the saturation discipline, valid input voltages are those that result in saturation region operation of the amplifier. Determine the corresponding range of output voltages ($v_O$) and output currents ($i_{DS}$).

c) Suppose we wish to amplify an AC input signal $v_i$. Assume that $v_i$ has a zero DC offset. Draw a circuit showing how a separate DC input voltage $V_I$ can be used to bias the amplifier in a region where saturation region operation is achieved for both positive and negative excursions of $v_i$. Assuming the $v_i$ has symmetric positive and negative swings, how would you choose the input operating point for the amplifier that allows a maximum peak-to-peak voltage range for $v_i$? What is the corresponding output operating point ($v_O$ and $i_{DS}$)?

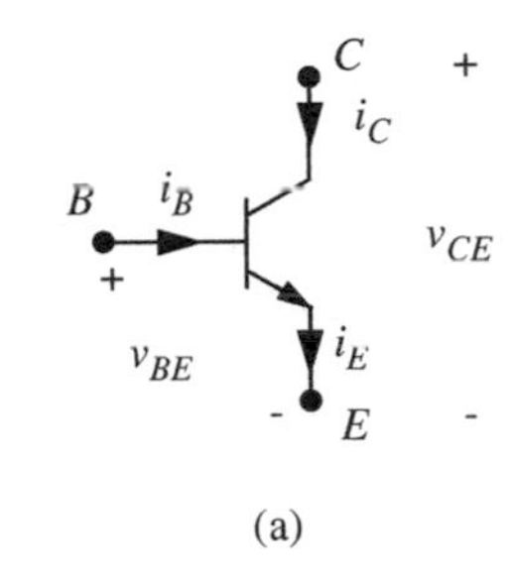


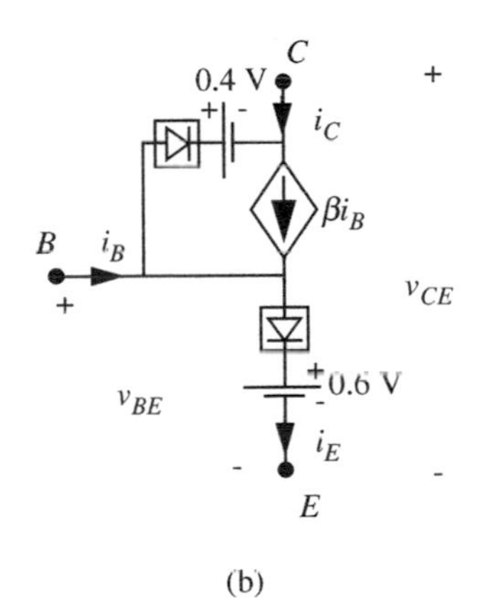


FIGURE 7.71 (a) A bipolar junction transistor. B stands for base, E for emitter, and C for collector; (b) a piecewise-linear model for the BJT.

EXERCISE 7.8 The three terminal device shown in Figure 7.71a is called a bipolar junction transistor (BJT). Figure 7.71b shows a piecewise-linear model for the device, in which the parameter $\beta$ is a constant. When

$$i_B > 0$$

and

$$v_{CE} > v_{BE} - 0.4 \text{ V},$$

the emitter diode behaves like a short circuit, the collector diode like an open circuit, and the collector current is given by:

$$i_C = \beta i_B.$$

Under the given constraints, the BJT is said to operate in its active region. For the rest of this exercise, assume that $\beta = 100$:

a) Determine the collector current $i_C$ for a base current $i_B = 1\ \mu\text{A}$ and $v_{CE} = 2$ V using the model in Figure 7.71b.

b) Sketch a graph of $i_C$ versus $v_{CE}$ for $i_B = 1\ \mu$A. Using the model in Figure 7.71b. In drawing this graph, assume that the current source turns off for

$$v_{CE} \leq v_{BE} - 0.4 \text{ V}.$$

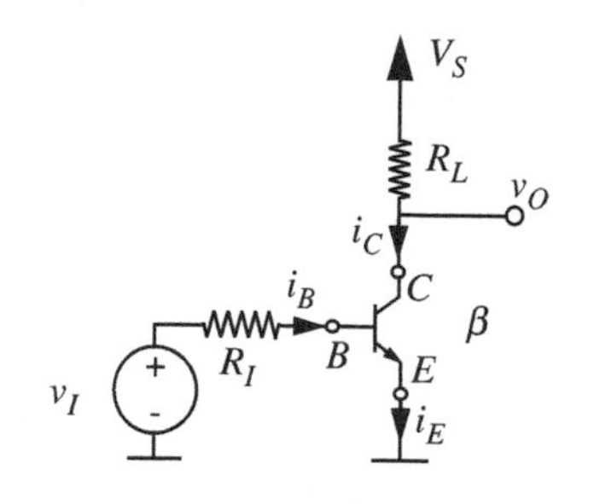


FIGURE 7.72

EXERCISE 7.9 Consider the bipolar junction transistor (BJT) amplifier shown in Figure 7.72. Assume that the BJT is characterized by the large signal model from Exercise 7.8, and that the BJT operates in its active region. Assume further that $V_S = 5$ V, $R_L = 10$ kΩ, $R_I = 500$ kΩ, and $\beta = 100$.

a) Draw the equivalent circuit for the BJT amplifier based on the large signal BJT model from Exercise 7.8.

b) Write an expression relating $v_O$ to $i_C$.

c) Write an expression relating $i_C$ to $v_I$.

d) Write an expression relating $i_E$ to $i_B$.

e) Write an expression relating $v_O$ to $v_I$.

f) What is the value of $v_O$ for an input voltage $v_I = 0.7$ V? What are the corresponding values of $i_B$, $i_C$, and $i_E$?

EXERCISE 7.10 In this exercise you will perform a large signal analysis of the BJT amplifier shown in Figure 7.72. Assume that the BJT is characterized by the large signal model from Exercise 7.8. Assume further that $V_S = 5$ V, $R_L = 10$ kΩ, $R_I = 500$ kΩ, and $\beta = 100$.

a) Write an expression relating $v_O$ to $v_I$.

b) What is the lowest value of the input voltage $v_I$ for which the BJT operates in its active region? What are the corresponding values of $i_B$, $i_C$, and $v_O$?

c) What is the highest value of the input voltage $v_I$ for which the BJT operates in its active region? What are the corresponding values of $i_B$, $i_C$, and $v_O$?

d) Sketch a graph of $v_O$ versus $v_I$ for the four parameter values given.

## PROBLEMS

PROBLEM 7.1 Consider the MOSFET voltage divider circuit shown in Figure 7.73. Assume that both MOSFETs operate in the saturation region. Determine the output voltage $V_O$ as a function of the supply voltage $V_S$, the gate voltages $V_A$ and $V_B$, and the MOSFET geometries $L_1$, $W_1$, and $L_2$, $W_2$. Assume that the MOSFET threshold voltage is $V_T$, and remember, $K = K_n W/L$.

PROBLEM 7.2 An inverting MOSFET amplifier is shown in Figure 7.74, together with an $i_{DS}$–$v_{DS}$ characteristic for the MOSFET. This characteristic is simpler than the SCS model presented in this chapter. The characteristic is simply the standard MOSFET characteristic with the triode region compressed onto the y-axis.

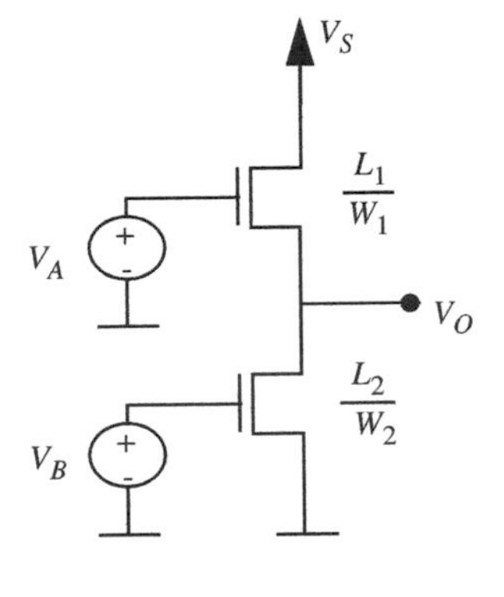


FIGURE 7.73

Alternatively, this characteristic can be viewed as describing ideal switch behavior that is extended to exhibit a saturating drain-source current. In other words, for $v_{GS} < V_T$, the MOSFET behaves like an open switch with $i_{DS} = 0$. For $v_{GS} \geq V_T$, the MOSFET behaves like a closed switch with $v_{DS} = 0$ provided that $i_{DS} < K/2(v_{GS} - V_T)^2$. However, once $i_{DS}$ reaches $K/2(v_{GS} - V_T)^2$, which is the maximum current the MOSFET can carry for a given $v_{GS}$, MOSFET operation enters a saturation region in which the MOSFET behaves as a current source of value $K/2(v_{GS} - V_T)^2$. Saturated operation is as described by the saturation model given in Figure 7.74.

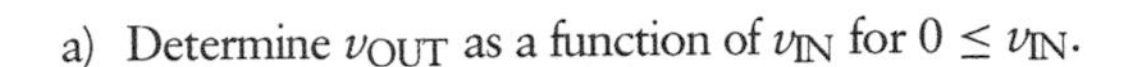

a) Determine $v_{OUT}$ as a function of $v_{IN}$ for $0 \leq v_{IN}$.

b) What is the lowest value of $v_{IN}$ for which $v_{OUT} = 0$?

c) Assume that $V_S = 15$ V, $R = 15$ kΩ, $V_T = 1$ V, and $K = 2$ mA/V$^2$. Graph $v_{OUT}$ versus $v_{IN}$ for $0$ V $\leq v_{IN} \leq 3$ V.

d) On the input-output graph, identify the regions over which the MOSFET behaves as an open circuit, behaves as a short circuit, and exhibits saturated behavior.

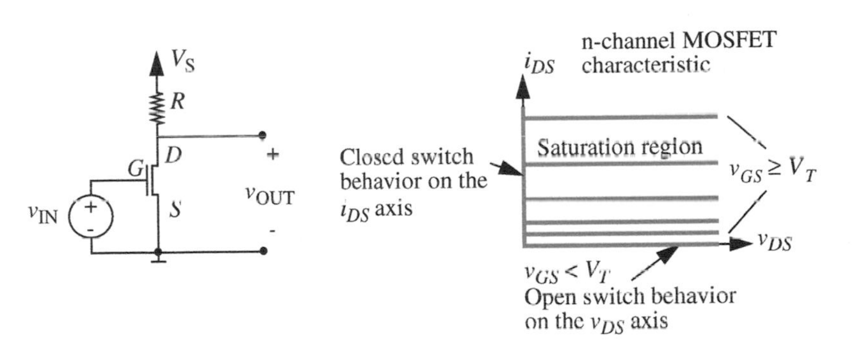


FIGURE 7.74

PROBLEM 7.3 A two-stage amplifier is shown in Figure 7.75. It is constructed by cascading two one-stage amplifiers of the type seen in Problem 7.2. In analyzing this amplifier, use the MOSFET model described in Problem 7.2 and illustrated in Figure 7.74.

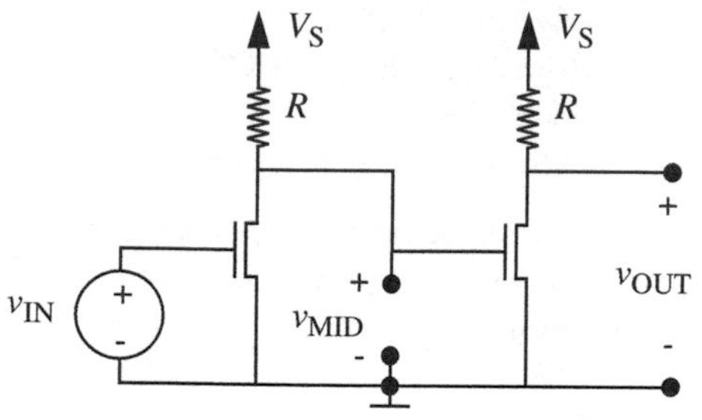

FIGURE 7.75

a) The fact that a second amplifier stage is connected to the first amplifier stage does not change the operation of the first stage. That is, the relation between $v_{MID}$ and $v_{IN}$ here is the same as the relation between $v_{OUT}$ and $v_{IN}$ in Problem 7.2. Why? What terminal characteristic of the second MOSFET must change in order for this not to be true?

b) Derive the relation between $v_{MID}$ and $v_{IN}$ for $0 \leq v_{IN}$, and the relation between $v_{OUT}$ and $v_{MID}$ for $0 \leq v_{MID} \leq V_S$. (Hint: see Problem 7.2.)

c) Derive the relation between $v_{OUT}$ and $v_{IN}$ for $0 \leq v_{IN}$.

d) Determine the range of input voltages for which both MOSFETs operate under the saturation discipline. What are the corresponding ranges for $v_{MID}$ and $v_{OUT}$?

e) Using the numerical parameters given in Problem 7.2, graph $v_{OUT}$ versus $v_{IN}$ for $v_{IN}$ for $0 \text{ V} \leq v_{IN} \leq 3 \text{ V}$. Compare this graph to the input-output graph found in Problem 7.2, and explain the differences.

PROBLEM 7.4 Consider again the two-stage amplifier shown in Figure 7.75. Suppose that the MOSFETs are characterized by the following equation in their saturation region:

$$i_{DS} = \frac{K}{2} v_{GS}^2.$$

In other words, the threshold voltage $V_T = 0$. Furthermore, the MOSFETs operate in their saturation region when

$$v_{DS} \geq v_{GS} \quad \text{and} \quad v_{GS} \geq 0.$$

Show that there is only one input voltage for which both stages simultaneously operate under the saturation discipline. What is that input voltage?

PROBLEM 7.5 Consider the "source-follower" or "buffer" circuit shown in Figure 7.76. Use the SCS MOSFET model (with parameters $V_T$ and $K$) to perform a large-signal analysis of this circuit according to the following steps:

a) Assuming that the MOSFET operates in its saturation region, show that $v_{OUT}$ is related to $v_{IN}$ according to

$$v_{OUT} = \left[\frac{\sqrt{(2/RK) + 4(v_{IN} - V_T)} - \sqrt{2/RK}}{2}\right]^2.$$

b) Determine the range of $v_{IN}$ over which the assumption of saturated MOSFET operation holds. What is the corresponding range for $v_{OUT}$?

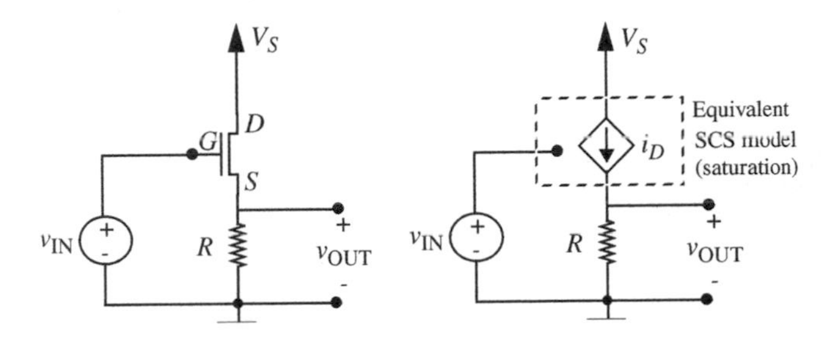


FIGURE 7.76

PROBLEM 7.6 This problem studies the use of a mythical MOSFET-like device called a ZFET to construct an amplifier as shown in Figure 7.77. The ZFET operates in its saturation region when $v_{GS} \geq 0$ and $v_{DS} > 0$. In this region, the drain-source terminal relation is $i_{DS} = Kv_{GS}^3$, where $K$ is a constant having units of $A/V^3$. When $v_{DS} = 0$, the ZFET exhibits a short circuit between its drain and source terminals, and is said to operate outside its saturation region. Similarly, the ZFET exhibits an open circuit for $v_{GS} < 0$ as it again operates outside its saturation region. Finally, the gate terminal always exhibits an open circuit. These characteristics are summarized in Figure 7.77, beneath the symbol for the ZFET.

a) Assuming saturated operation of the ZFET, determine $v_{OUT}$ as a function of $v_{IN}$.

b) Over what range of $v_{IN}$ will the ZFET operate in its saturation region?

c) Assume that $V_S = 10$ V, $R_L = 1$ kΩ, and $K = 0.001$ A/V$^3$. Sketch and clearly label $v_{OUT}$ as a function of $v_{IN}$ for $-1 \text{ V} \leq v_{IN} \leq 3$ V.

d) Given the parameters of part (c), can the amplifier be used as an inverter that provides a valid output high voltage threshold of $V_H = 7$ V? Why or why not? Assume that $V_L = 2$ V.

e) Given the parameters of part (c), can the amplifier can be used as an inverter that provides a valid output high voltage threshold of $V_H = 7$ V? Why or why not? This time around, assume that $V_L = 1$ V.

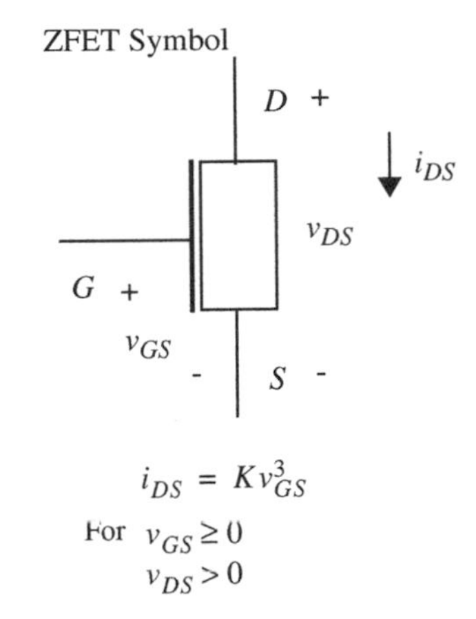


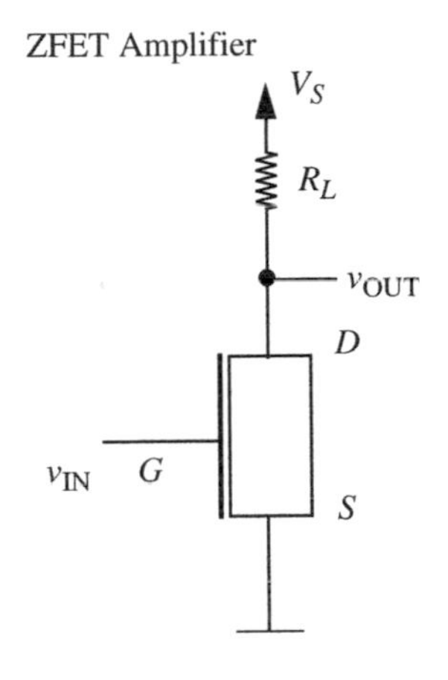


FIGURE 7.77

PROBLEM 7.7 Consider the difference amplifier circuit shown in Figure 7.78. Notice that the difference amplifier is powered by $+V_S$ and $-V_S$ power supplies. Assume that all MOSFETs operate under the saturation discipline, and, unless indicated otherwise, are characterized by the parameters $K$ and $V_T$.

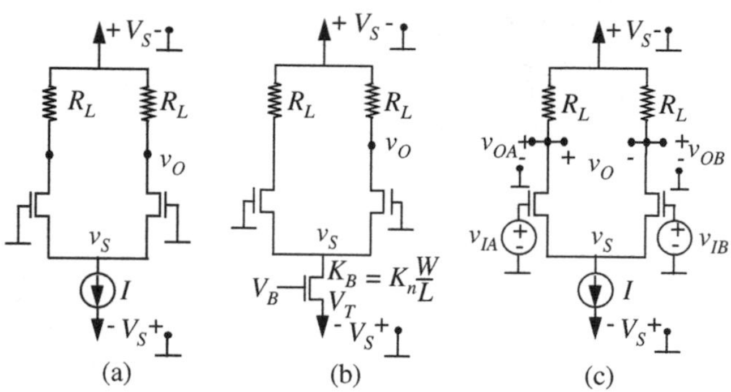


FIGURE 7.78

a) Determine $v_O$ and $v_S$ for the connection shown in Figure 7.78a. In this figure, the gates of the MOSFETs are connected to ground.

b) Consider the difference amplifier version shown in Figure 7.78b. In this figure, a MOSFET implementation of a current source replaces the abstract current source from Figure 7.78a. Determine values for $V_B$ and $W/L$ such that the circuit in (b) is equivalent to that in (a).

c) The difference amplifier in Figure 7.78c is driven by two input voltages $v_{IA}$ and $v_{IB}$ as shown. Assume that the input voltages satisfy the following constraint $v_{IA} = -v_{IB}$ at all times. Determine $v_{OA}$, $v_{OB}$, and $v_O$ as a function of $v_{IA}$.

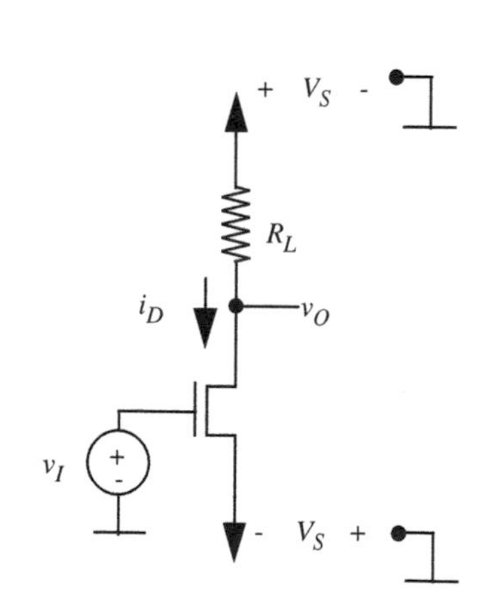


FIGURE 7.79

PROBLEM 7.8 Consider the amplifier circuit shown in Figure 7.79. The amplifier is powered by a $+V_S$ and a $-V_S$ power supply.

a) Determine $v_O$ and $i_D$ as a function of $v_I$ under the saturation discipline. Assume that the MOSFET parameters $K$ and $V_T$ are given.

b) Determine the range of valid input voltages for saturation region operation. Determine the corresponding valid range for $v_O$ and $i_D$.

c) Determine the output voltage when the input is grounded; in other words, for $v_I = 0$.

d) Determine the value of $v_I$ for which $v_I = v_O$ in terms of $V_S$, $R_L$, and the MOSFET parameters.

PROBLEM 7.9 Consider the current mirror circuit in Figure 7.80.

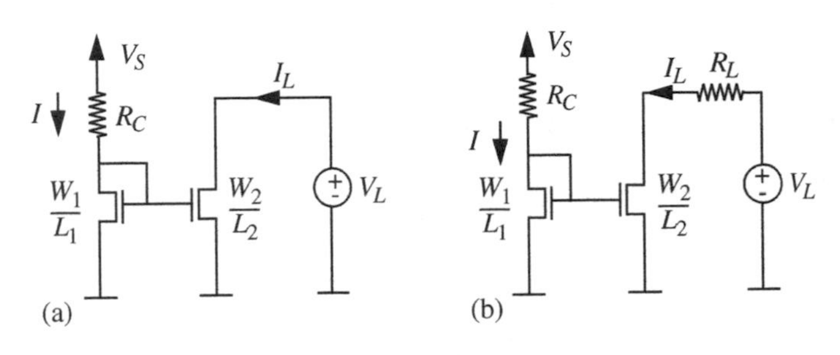


FIGURE 7.80

a) Referring to Figure 7.80a, determine $I_L$ as a function of $I$ assuming both MOSFETs operate under the saturation discipline. Both MOSFETs have the same values for $K_n$ and $V_T$. Does $I_L$ change if $V_L$ changes? What are the conditions under which $I_L = I$?

b) Now consider Figure 7.80b. The current $I$ can be increased either by increasing $V_S$ or decreasing $R_C$. Assuming that either $V_S$ or $R_C$ may be changed, and that $W_1/L_1 = W_2/L_2 = W/L$, determine the range of values of $I$ for which both MOSFETs operate under the saturation discipline. Assume both MOSFETs have the same values for $K_n$ and $V_T$.

PROBLEM 7.10 Consider the circuit shown in Figure 7.81. Assume that the MOSFET operates under the saturation discipline.

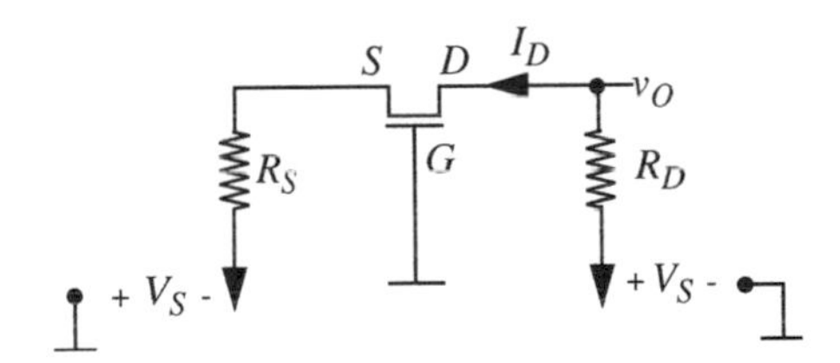


FIGURE 7.81

a) Draw the SCS equivalent circuit by replacing the MOSFET by its SCS model.

b) Determine $v_O$ and $i_D$ in terms of $R_D$, $R_S$, $V_S$, and the MOSFET parameters $K$ and $V_T$.

PROBLEM 7.11 Consider the "common-gate amplifier" circuit shown in Figure 7.82. Assume that the MOSFET operates under the saturation discipline.

a) Draw the SCS equivalent circuit by replacing the MOSFET by its SCS model.

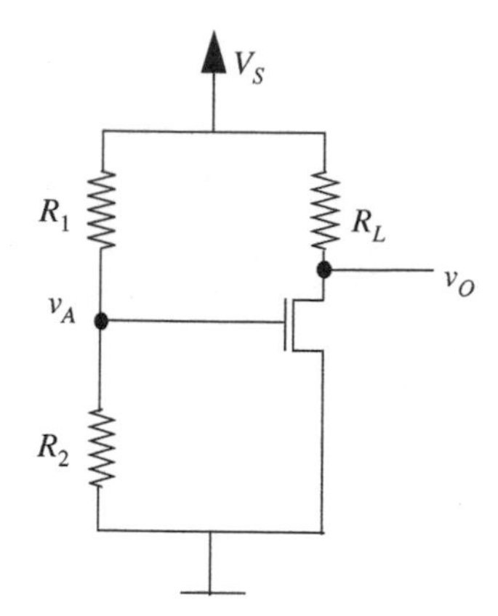


FIGURE 7.83

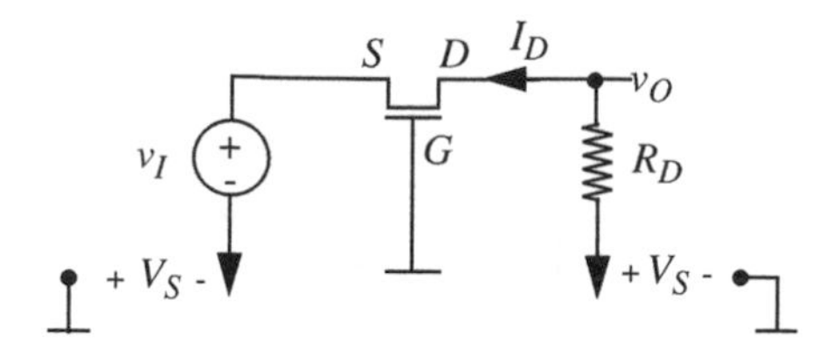


FIGURE 7.82

b) Determine $v_O$ and $i_D$ in terms of $v_I$, $R_D$, $V_S$, and the MOSFET parameters $K$ and $V_T$.

c) Determine the range of values of $v_I$ for which the MOSFET operates under the saturation discipline. What is the corresponding range of $v_O$?

PROBLEM 7.12 Consider the MOSFET circuit shown in Figure 7.83. Determine the value of $v_O$ in terms of the other circuit parameters. Assume the MOSFET is in saturation and is characterized by the parameters $K$ and $V_T$.

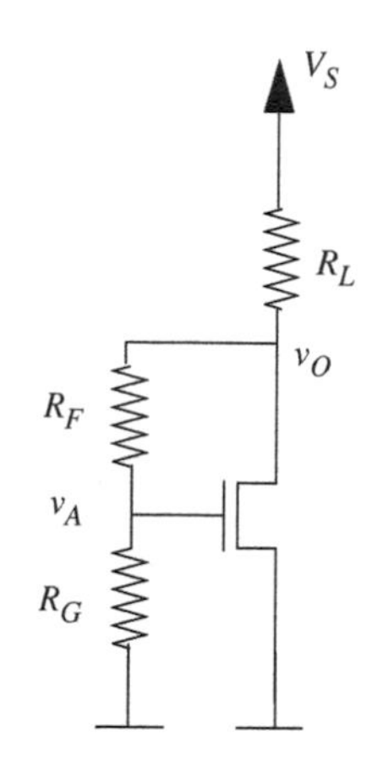


FIGURE 7.84

PROBLEM 7.13 Consider the MOSFET circuit shown in Figure 7.84. Determine the value of $v_O$ in terms of the other circuit parameters. Assume the MOSFET is in saturation and is characterized by the parameters $K$ and $V_T$.

PROBLEM 7.14 Figure 7.85 shows a MOSFET amplifier driving a load resistor $R_E$. The MOSFET operates in saturation and is characterized by parameters $K$ and $V_T$. Determine $v_{OUT}$ versus $v_{IN}$ for the circuit shown.

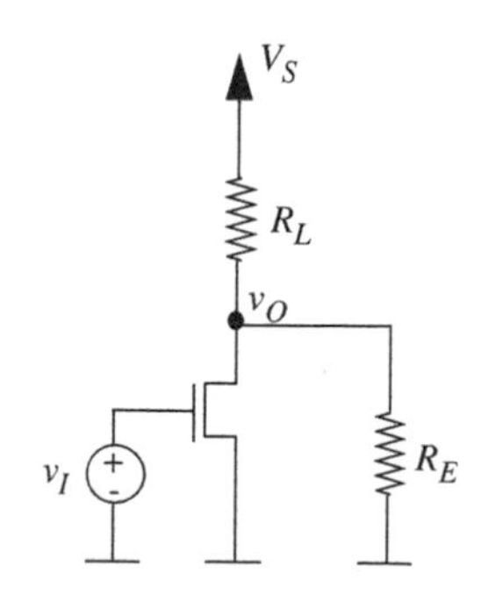


FIGURE 7.85

PROBLEM 7.15 Determine $v_{OUT}$ versus $v_{IN}$ for the circuit shown in Figure 7.86. Assume that the MOSFET operates in saturation and is characterized by the parameters $K$ and $V_T$. What is the value of $v_{OUT}$ when $v_{IN} = 0$?

PROBLEM 7.16 Determine $v_O$ versus $v_I$ for the circuit shown in Figure 7.87. Assume that the MOSFET operates in saturation and is characterized by the parameters $K$ and $V_T$. What is the value of $v_O$ when $v_I = 0$?

PROBLEM 7.17 Determine $v_O$ versus $v_I$ for the circuit shown in Figure 7.88. Assume that the MOSFET operates in saturation and is characterized by the parameters $K$ and $V_T$.

PROBLEM 7.18 Consider the BJT circuit called the "common-collector amplifier" shown in Figure 7.89. This BJT amplifier configuration is also called the source follower circuit. For this problem, use the piecewise-linear BJT model from Exercise 7.8. Assume that the BJT operates in its active region.

a) Draw the active-region equivalent circuit of the BJT source follower by replacing the BJT by its piecewise-linear model.

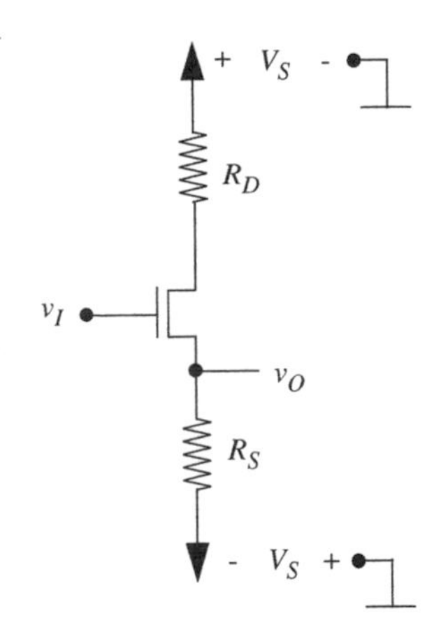


FIGURE 7.86

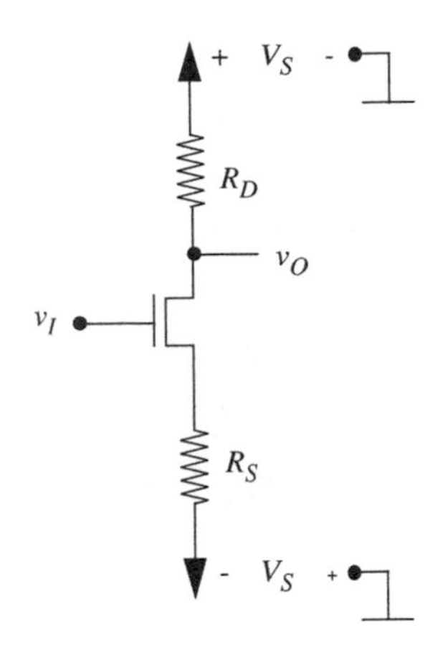


FIGURE 7.87

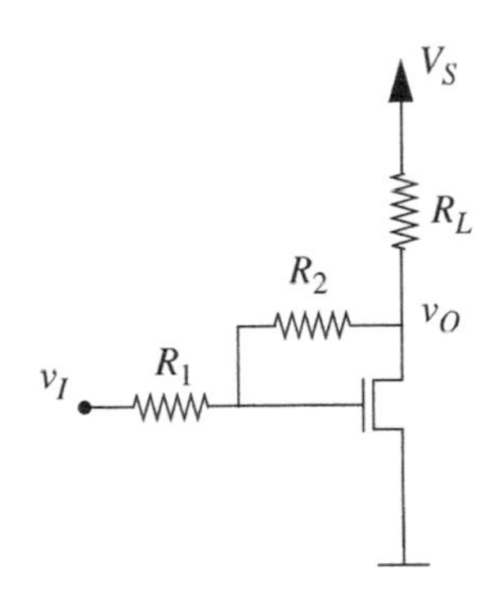


FIGURE 7.88

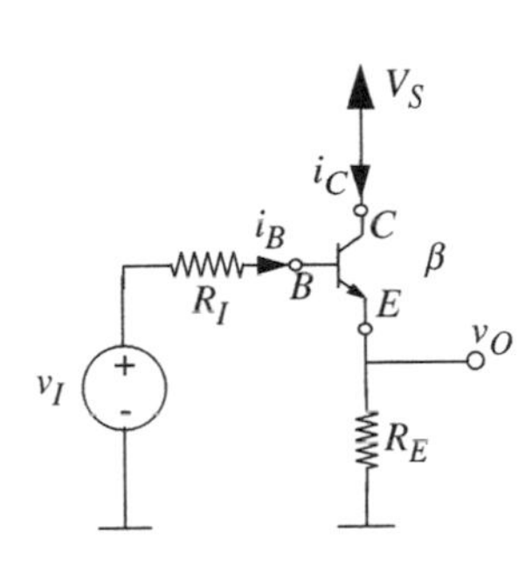


FIGURE 7.89

b) Assuming active region operation, determine $v_O$ in terms of $v_I$, $R_I$, $R_E$, and the BJT parameter $\beta$.

c) What is the value of $v_O$ when $\beta R_E >> R_I$?

d) Compute the value of $v_O$ given that $v_I = 3$ V, $R_I = 10$ kΩ, $R_E = 100$ kΩ, $\beta = 100$, and $V_S = 10$ V.

e) Determine the range of values of $v_I$ for which the BJT operates in its active region for the parameter values given in (d). What is the corresponding range of $v_O$?

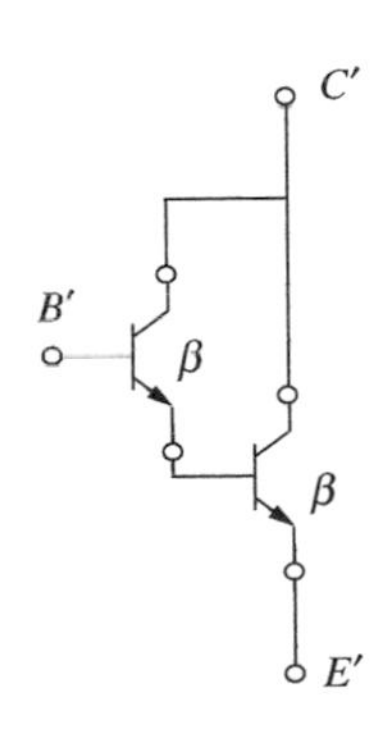


FIGURE 7.90

PROBLEM 7.19 Consider the compound three-terminal device formed by connecting two BJTs in the configuration shown in Figure 7.90. The three terminals are labeled $C'$, $B'$, and $E'$. The two BJTs are identical, each with $\beta = 100$. Assume that each of the BJTs operates in the active region.

a) Draw the active-region equivalent circuit of the compound BJT by replacing each of the BJTs by the piecewise-linear model shown in Exercise 7.8. Clearly label the $C'$, $B'$, and $E'$ terminals.

b) In the configuration shown, the compound device behaves like a BJT. Determine the value of the current gain $\beta'$ for this compound BJT.

c) When the base current $i_{B'} > 0$, determine the voltage between the $B'$ and $E'$ terminals.

# CHAPTER 8

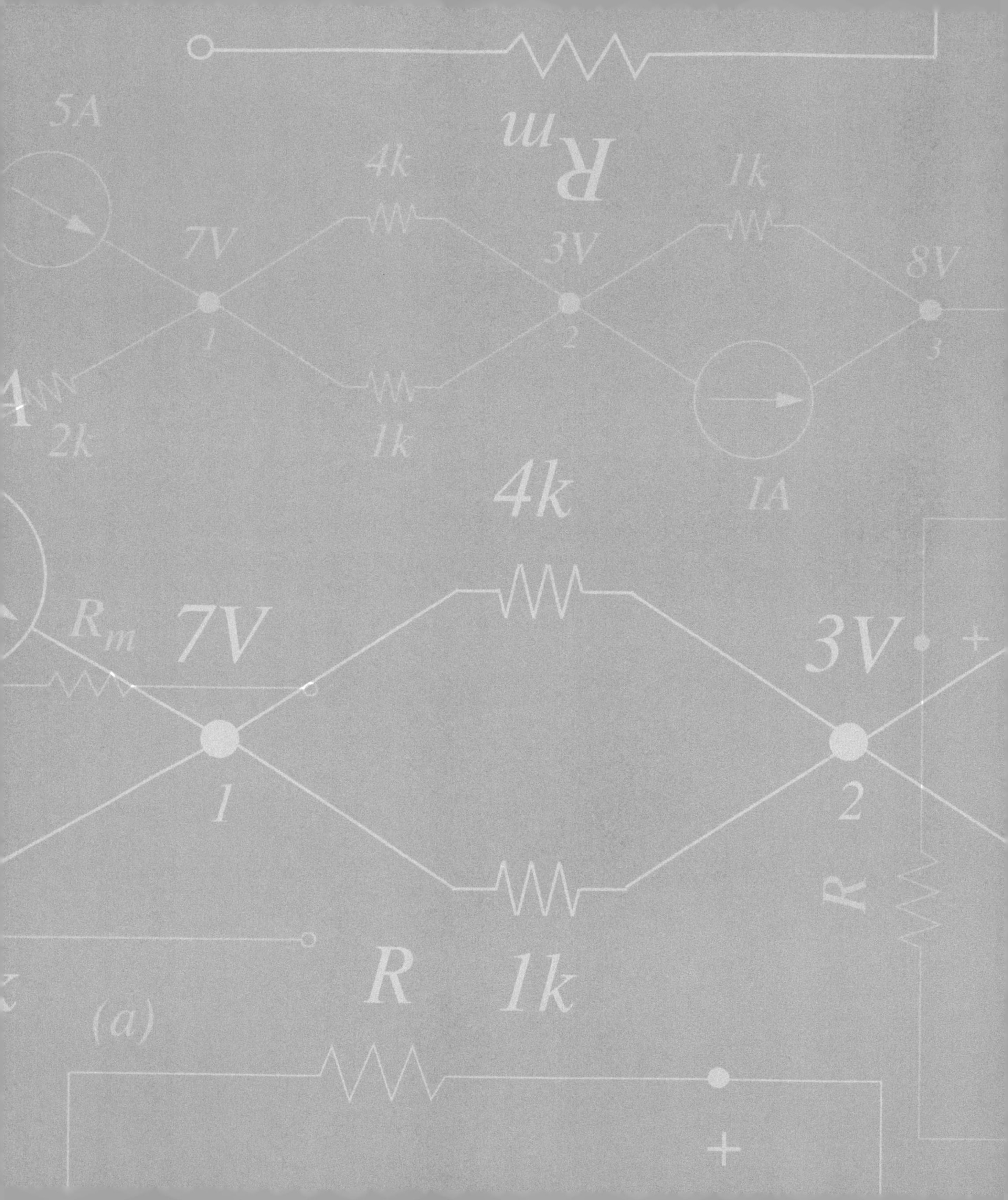
5A
4k
1k
7V
3V
8V
1
2
3
2k
1k
1A
4k
$R_m$
7V
3V
1
2
R
R
1k
(a)
+
+

# THE SMALL-SIGNAL MODEL

# 8

## 8.1 OVERVIEW OF THE NONLINEAR MOSFET AMPLIFIER

An unfortunate feature of the MOSFET amplifier discussed in Chapter 7 was its nonlinear input-output relationship. Shown in Figure 8.1, the MOSFET amplifier has the following input-output relationship:

$$v_O = V_S - i_D R_L. \tag{8.1}$$

Substituting for the current $i_D$ in terms of the MOSFET input voltage under the saturation discipline, we get the following nonlinear relationship between $v_I$ and $v_O$:

$$v_O = V_S - K\frac{(v_I - V_T)^2}{2}R_L. \tag{8.2}$$

The nonlinear relationship between the input and the output voltage is plotted in Figure 8.2. The nonlinear relationship makes it difficult for us to analyze and to build circuits using the amplifier.

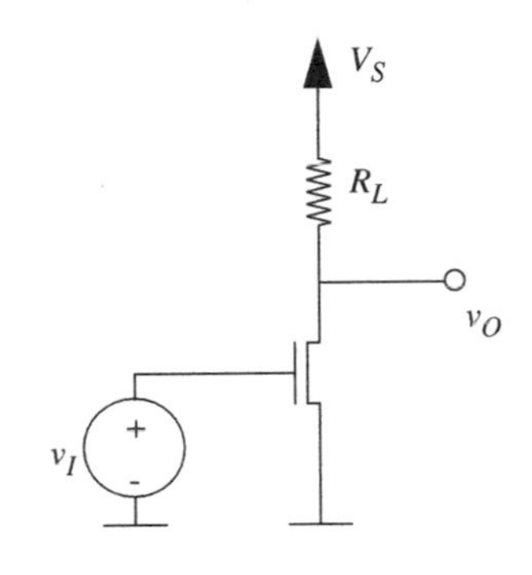


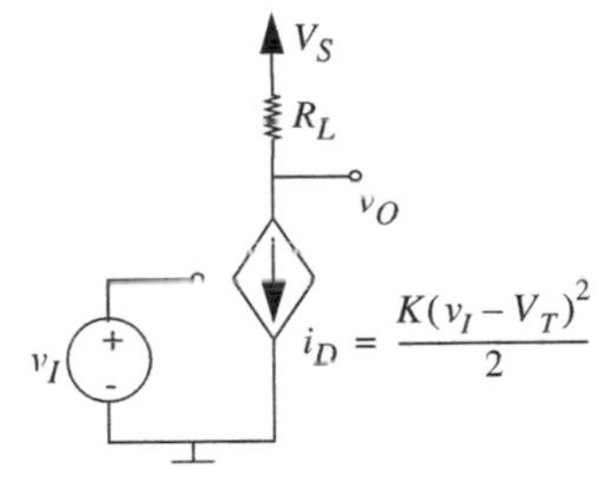


**FIGURE 8.1** The MOSFET amplifier and its SCS circuit model.

## 8.2 THE SMALL-SIGNAL MODEL

Many circuit applications, such as audio amplifiers, demand a linear amplifier of the form depicted in Figure 8.3. The amplifier shown in the figure has a constant gain $A$ that is independent of the input voltage. Does that mean we cannot use the MOSFET amplifier in these linear applications? It turns out that total variables representing signals such as those input to an audio amplifier commonly consist of two components: a DC offset (or an average value), plus a time-varying component with a zero average. We will show that if the time-varying component is small, then the incremental amplification provided by the MOSFET amplifier to the time-varying component about the operating point defined by the input DC offset will be approximately linear. As we saw in Section 4.5, this observation actually generalizes to arbitrary nonlinear circuits: The response of a circuit to small perturbations about an operating point will be linear. Thus, if the signals of interest to us can be represented as small perturbations about an operating point, then the response of arbitrary nonlinear

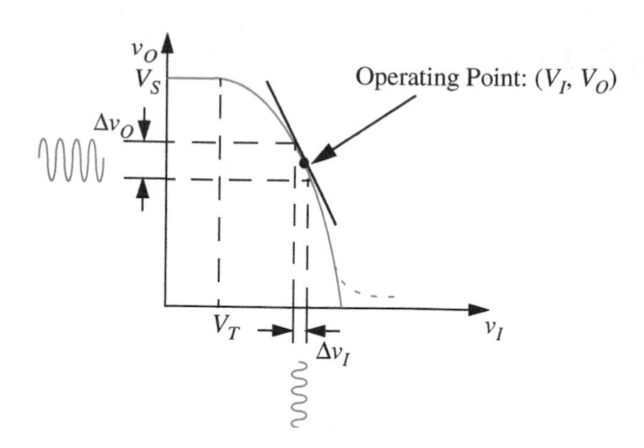


FIGURE 8.2 $v_O$ versus $v_I$ curve for the amplifier.

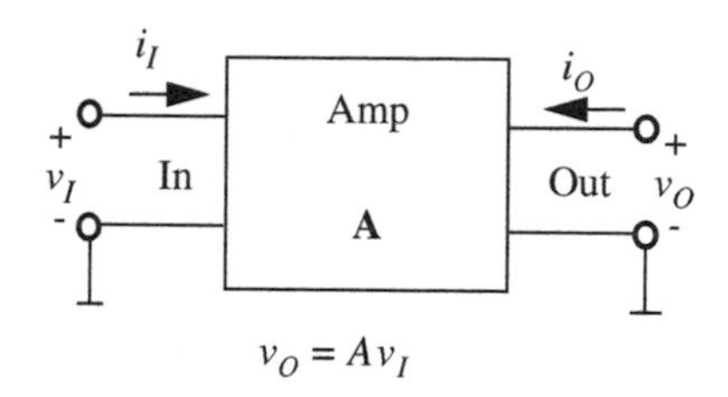


FIGURE 8.3 A linear amplifier abstraction with a constant gain $A$.

circuits to the small perturbations will be linear. As seen in Section 4.5, restricting signals to small perturbations about an operating point so the response of circuits to the perturbations is linear is a constrained way of using circuits that we call the *small-signal discipline.*

When the total variable comprises a DC operating value plus a small perturbation around the operating point, our models for the response of circuits to the perturbations will be linear and hence very simple. However, our incremental or small-signal models will apply only over a small range around the operating point. In contrast, our models of the previous chapter captured the behavior of the amplifier over a wide range of operation, but the models were complex. Separate models over different regions had to be spliced together to obtain the overall characteristics. Furthermore, the models were nonlinear. Such a tradeoff between complexity of the model and the range over which it is valid is not uncommon in modeling systems. In engineering practice, both extremes of models are useful: complex accurate models and simple approximate models. This chapter discusses small-signal models, which are simple models whose range of applicability is limited. Despite their limitations, the simple models are surprisingly useful engineering tools even when applied outside their strict range of validity.

Section 4.5 introduced the following notation to distinguish between total variables, their average DC values, and their incremental excursions about the average values. We will denote total variables with small letters and capital subscripts, average DC values using all capitals, and incremental values using all small letters. Thus, $v_I$ denotes the total input voltage, $V_I$ the DC offset, and $v_i$ the incremental component. Since the total variable is the sum of the two components, we have

$$v_I = V_I + v_i.$$

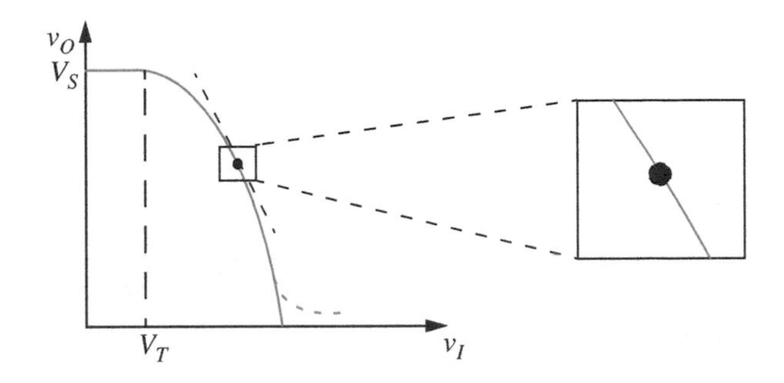


**FIGURE 8.4** A small segment of the $v_O$ versus $v_I$ curve.

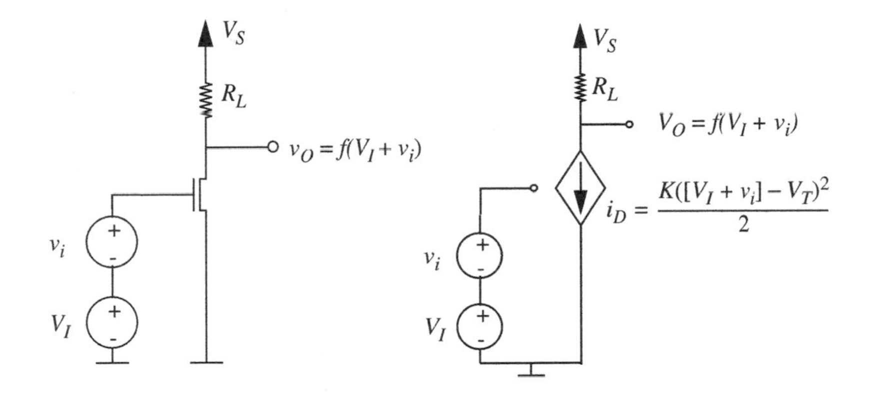


**FIGURE 8.5** Superimposing a small (possibly time-varying) signal on the DC bias voltage at the input of the MOSFET amplifier, and the corresponding SCS circuit model for the combined input signal.

Let us revisit the transfer curve of the amplifier shown in Figure 8.2. Consider a very small region of the transfer curve in the vicinity of the operating point $(V_I, V_O)$. The slope of the curve segment is depicted in the figure. As illustrated in Figure 8.4, if we focus our attention on the small curve segment shown, it looks more or less linear. We will use this intuition to develop an abstraction for amplifiers that appears linear for very small variations in the input voltage.

The basic idea is that the amplifier transfer function appears linear for small perturbations in the input voltage about a given bias point. We can arrive at the same result analytically. Suppose that the amplifier is biased at some bias point: $(V_I, V_O)$. Now suppose that we superimpose a small signal $\Delta v_I = v_i$ on $V_I$ as depicted in Figure 8.5. An example of a DC signal with a small superimposed time-varying signal is shown in Figure 8.6.

We know from the SCS model of the MOSFET (see Equation 7.8) that the current through the MOSFET is related to its gate voltage as:

$$i_{DS} = \frac{K(v_{GS} - V_T)^2}{2}. \tag{8.3}$$

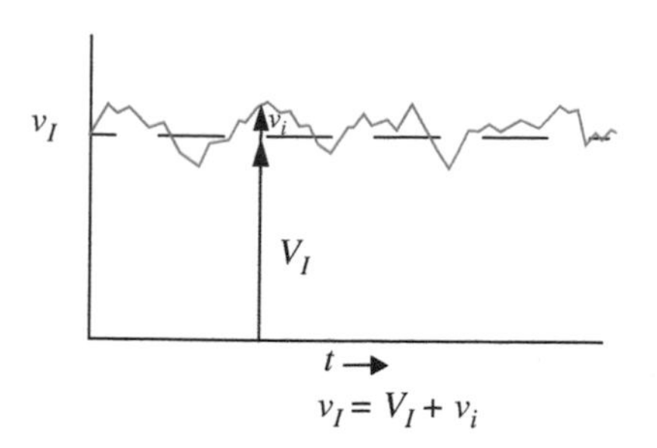


**FIGURE 8.6** A small time-varying signal combined with a DC offset voltage.

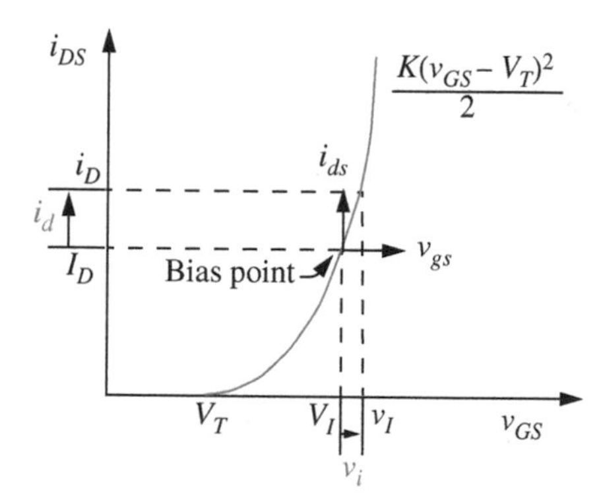


**FIGURE 8.7** Output current for the MOSFET for the combined input voltage.

For the combined input signal shown in Figure 8.5, the response current $i_D$ through the MOSFET is the sum of two components: a bias current $I_D$ and a change $i_d$ due to the incremental input signal $v_i$. As depicted in Figure 8.7, this combined current can be obtained by substituting for $v_{GS}$ as

$$i_D = f(V_I + v_i) = I_D + i_d = \frac{K([V_I + v_i] - V_T)^2}{2}. \tag{8.4}$$

Since we know that $v_i$ is small compared to $V_I$, we can adopt the following linearization technique to obtain the combined response: Model the MOSFET characteristic curve accurately only in the vicinity of the bias point $V_I$ and disregard the rest of the curve. The Taylor series expansion is the natural tool for this task.

The Taylor series expansion for the function $y = f(x)$ in the vicinity of $x = X_o$ is given by:

$$y = f(x) = f(X_o) + \left.\frac{df}{dx}\right|_{Xo} (x - X_o) + \frac{1}{2!} \left.\frac{d^2f}{dx^2}\right|_{Xo} (x - X_o)^2 + \cdots$$

Our goal is to use the Taylor series method to expand the MOSFET SCS equation for the combined input voltage given in Equation 8.4 about the bias voltage $V_I$. For our Taylor expansion, $V_I$ corresponds to $X_o$, $x$ corresponds to $V_I + v_i$, or $x - X_o$ corresponds to $v_i$, and $y$ corresponds to $i_D = I_D + i_d$. Applying the Taylor expansion to Equation 8.4 about $V_I$ we get

$$i_D = f(V_I + v_i) = \frac{K[(V_I + v_i) - V_T]^2}{2} \tag{8.5}$$

$$= \frac{K(V_I - V_T)^2}{2} + K(V_I - V_T)v_i + \frac{K}{2}v_i^2. \tag{8.6}$$

If the incremental signal $v_i$ is small enough to permit us to ignore the second order term (and higher terms, when they exist) in the Taylor series expansion, the following simplification results:

$$i_D \approx \frac{K(V_I - V_T)^2}{2} + K(V_I - V_T)v_i. \tag{8.7}$$

We know that the output current is composed of a DC component $I_D$ and a small perturbation $i_d$. Thus, we can write

$$I_D + i_d = \frac{K(V_I - V_T)^2}{2} + K(V_I - V_T)v_i. \tag{8.8}$$

Equating DC terms and corresponding incremental terms:

$$I_D = \frac{K(V_I - V_T)^2}{2} \tag{8.9}$$

$$i_d = K(V_I - V_T)v_i. \tag{8.10}$$

Note that $I_D$ is simply the DC bias current related to the DC input voltage $V_I$. Accordingly, the DC terms relating $I_D$ to $V_I$ can be equated as in Equation 8.9 because the operating point values $I_D, V_I$ satisfy Equation 8.3, which is the MOSFET equation. When the DC terms are eliminated from both sides of Equation 8.8, the incremental relation shown in Equation 8.10 results.

Notice that the change in the output current $i_d$ is linearly related to the change in the input voltage $v_i$ provided that $v_i$ is small compared to $V_I$. We note that Equation 8.9 is exact because the small-signal model goes through the exact model at the operating point. However, Equation 8.10 is approximate because of the linearization.

A graphical interpretation of this result provides additional intuition. As shown in Figure 8.8, Equation 8.8 is a straight line passing through the DC operating point $V_I, I_D$ and tangent to the curve at that point. Using the tangent

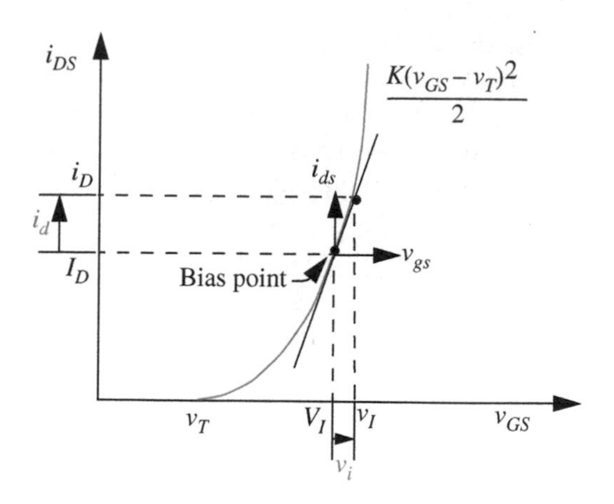


FIGURE 8.8 Incremental change in the output current for the MOSFET for a small change in the input voltage.

to compute the incremental change in the signal about the operating point is tantamount to replacing the actual curve with the tangent. Clearly, the tangent approximation is valid only for points that are close to the operating point. The higher-order term in Equation 8.6 that we neglected would add a quadratic term to the model, thereby making the fit exact for our model.

Let us return to the relationship between the incremental output current and the incremental input voltage for the MOSFET:

$$i_d = K(V_I - V_T)v_i. \tag{8.11}$$

The $K(V_I - V_T)$ term in Equation 8.11 relates the input voltage to the current through the MOSFET. Notice that for a given DC bias, the $K(V_I - V_T)$ term is a constant. Since the form of Equation 8.11 is similar to that for a conductance, the $K(V_I - V_T)$ term is called the *incremental transconductance* $g_m$ of the MOSFET. Accordingly, we can write

$$i_d = g_m v_i \tag{8.12}$$

where

$$g_m = K(V_{GS} - V_T). \tag{8.13}$$

In our example, $V_{GS} = V_I$.

Returning to our amplifier, we can express the total output voltage $v_O$ as the sum of the output operating voltage $V_O$ and the incremental change $v_o$ as

$$v_O = V_O + v_o.$$

From Equation 8.1 we know that

$$v_O = V_S - i_D R_L. \tag{8.14}$$

Replacing $v_O$ and $i_D$ with their corresponding DC and incremental components,

$$V_O + v_o = V_S - (I_D + i_d)R_L \tag{8.15}$$

$$= V_S - I_D R_L - i_d R_L. \tag{8.16}$$

Therefore,

$$V_O = V_S - I_D R_L \tag{8.17}$$

$$v_o = -i_d R_L \tag{8.18}$$

$$= -g_m v_i R_L. \tag{8.19}$$

In other words,

$$\text{Small signal gain} = \frac{v_o}{v_i} = -g_m R_L = A. \tag{8.20}$$

Notice from Equation 8.20 that the small signal gain is a constant $-g_m R_L$. Note, however, that $g_m$, and therefore the gain, depends on the choice of bias point for the amplifier. Equation 8.19 demonstrates that for small excursions from a DC operating point, a linear amplifier results! This result forms the basis of the *small-signal model.*

We can directly arrive at the small signal response — be it voltage or current — using basic calculus for circuit responses that are differentiable, which basically includes all physically realizable analog circuits. Recall that the derivative of a function $y = f(x)$ at the point $x_o$ is the slope of the function at that point, or $f'(x_o)$. As depicted in Figure 8.9, given a small change $\Delta x$ from the

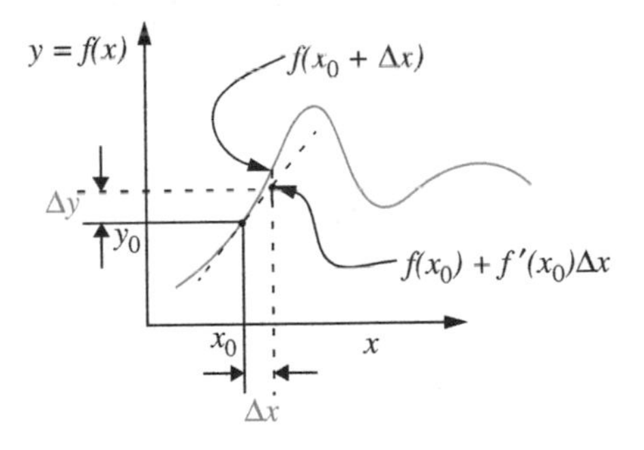


FIGURE 8.9 Incremental response.

point $x_o$, we can compute the response to the change as the product of the slope at that point and $\Delta x$. In other words,

$$f(x_o + \Delta x) = f(x_o) + \left.\frac{df(x)}{dx}\right|_{x_o} \Delta x.$$

Thus the incremental change in the output is given by

$$\Delta y = \left.\frac{dy}{dx}\right|_{x_o} \Delta x. \tag{8.21}$$

In particular, we can obtain the incremental voltage gain directly from the voltage transfer function, without first determining the incremental output current. The input-output voltage relationship for the MOSFET amplifier is given by

$$\begin{aligned} v_O &= f(v_I) \\ &= V_S - K\frac{(v_I - V_T)^2}{2}R_L. \end{aligned}$$

As before, let $v_i = \Delta v_I$ denote a small change in the input voltage, and let $v_o = \Delta v_O$ denote the corresponding change in the output voltage. Then,

$$\begin{aligned} v_o &= \left.\frac{df(v_I)}{dv_I}\right|_{v_I=V_I} v_i \\ &= -K(v_I - V_T)R_L\big|_{v_I=V_I}\, v_i \\ &= -K(V_I - V_T)R_L v_i \\ &= -g_m R_L v_i. \end{aligned}$$

Not surprisingly, this result is the same as the one we obtained earlier.

To summarize, the small-signal model is a statement of a particular type of linearized analysis of our circuits, which applies when the desired circuit responds to signals that can be represented as an incremental perturbation over a DC operating value. Put another way, it is a statement of a particular type of constraint on our use of circuits called the small-signal discipline that allows us to obtain linear behavior from nonlinear circuits over small ranges of operation.

*Small signal model* The responses of circuits to incremental changes from a known DC operating point will be linear to a good approximation.

A systematic procedure for finding incremental signal responses based on the preceding discussion involves two steps:

1. Find the DC operating point of the circuit using DC values and the complete characteristics of the devices. Determine the corresponding large-signal response (possibly nonlinear) to the desired input.
2. Apply the Taylor expansion method to the large-signal response to derive the small-signal response. Alternatively, as discussed in Section 8.2.1, replace the large-signal circuit with its equivalent small-signal model based on the Taylor expansion and obtain the small-signal response.

Small-signal analysis is an extremely useful technique that applies to all physical systems with differentiable characteristics. In essence, it says that if we operate within a small-signal discipline, the response of any physical system to small perturbation will be linear! In turn, the effectively linear system is amenable to linear analysis techniques, such as superposition.

For example, consider a two-terminal sensor $S$ that behaves like a temperature-dependent voltage source with the following nonlinear relationship between its terminal voltage $v_S$ and its temperature $t_S$:

$$v_S = Bt_S^3$$

where $B$ is some constant. If the ambient temperature is $T_S$ and the corresponding voltage is $V_S$, we can relate the incremental change in the terminal voltage $v_s$ to an incremental change in the temperature $t_s$ using Equation 8.21 as follows:

$$v_s = 3B\, t_S^2\Big|_{t_S=T_S} t_s.$$

In other words,

$$v_s = 3BT_S^2 t_s.$$

When operating at a given ambient temperature, $3BT_S^2$ is a constant. Therefore, the voltage response of the sensor to small changes in the temperature around an ambient will be linear.

### 8.2.1 SMALL-SIGNAL CIRCUIT REPRESENTATION

A model that involves only the small-signal variables of a circuit, and hence describes purely the small-signal behavior of that circuit, would greatly facilitate small-signal analysis. Fortunately, such a small-signal model is relatively straight forward to develop by executing the following procedure:

1. Set each source to its operating-point value, and determine the operating-point branch voltages and currents for each component in the circuit. This is most likely the longest step in the procedure.
2. Linearize the behavior of each circuit component about its operating point. That is, determine the linearized small-signal behavior of each component, and select a linear component to represent this behavior. The parameters of the small-signal components will commonly depend on the operating point voltages or currents.
3. Replace each original component in the circuit with its linearized equivalent and re-label the circuit with the small-signal branch variables. The resulting circuit is the desired small-signal model.

The circuit that is generated by this procedure is the desired small-signal circuit model, and is analogous to equating the small signal terms on both sides of Equation 8.8 yielding the equalities in Equation 8.10. Further, it is a linear circuit, and hence the analysis tools developed for linear circuits, such as superposition and the Thévenin equivalent model, may be applied to its analysis.

At this point, it is worth discussing why the procedure works. To begin, recognize that the operation of a circuit is described in total by two sets of equations: the circuit connection laws of KVL and KCL, and the constitutive laws that describe the behavior of the individual circuit components. With this recognition, the small-signal analysis of a circuit may also be described by the following more direct mathematical procedure:

1. Set each source to its operating-point value, and combine the equations to determine the operating point of the circuit. This is essentially the same step as in the previous procedure.
2. Return to the original set of equations. For each variable in every equation, substitute for the total variable the sum of its operating-point value and its small-signal value. Then, linearize the equations around the operating point assuming that the small-signal terms are small.
3. Cancel the operating-point variables from the linearized equations to yield a set of linear equations that relate the small signals to themselves. This cancellation must always be possible since the linearization is defined to pass through the operating-point. This cancellation is akin to separately equating the operating point variables and the incremental variables as we did in Equation 8.10.[1]

1. In other words, we start with a set of equalities defining the operating point using operating-point variables, for example,

$$V_O = AV_I.$$

4. Complete the small-signal analysis by combining the linearized equations to determine the desired small-signal variables in terms of the small-signal inputs at the sources.

Now, let us examine the last procedure more closely. Notice that in Step 2 it is actually necessary to linearize only the constitutive laws that describe the behavior of the individual circuit components because KVL and KCL are already linear equations. It is for this reason that the first procedure called for the linearization of only the constitutive laws. Further, because KVL and KCL constitute a linear set of equations, they are unchanged by the linearization step. This is important to recognize because KVL and KCL contain the information concerning the topology of the original circuit. That is, they state which branches are connected to which nodes, and which branches connect to form which loops. Since KVL and KCL are unaffected by the linearization step, the topological information is preserved during linearization. It is for this reason that the small-signal circuit model has the same topology as the original circuit. Thus, the linearized set of equations describing the behavior of the small-signal circuit variables that is generated by the more formal mathematical procedure comprises the original KVL and KCL equations, and linearized component constitutive laws. Thus, to develop a small-signal circuit model it is necessary to determine only equivalent linearized circuit components and substitute them into the circuit in place of their corresponding original circuit components.

Small-signal circuit models for various devices are summarized in Figure 8.10.

- The small-signal equivalent model for an independent DC voltage source is a short circuit because its output voltage does not change for any perturbation of the current through it. In particular, the power supply connection labeled $V_S$ in most of our circuits gets shorted to ground in the incremental circuit.
- The small-signal model for an independent DC current source is an open circuit.

---

We then linearize, and obtain a new set of equalities in operating-point variables and incremental variables, for example,

$$V_O + v_o = AV_I + Av_i.$$

The equalities that defined the operating point in the first place (namely, $V_O = AV_I$ in our example) may always be cancelled out of the linearized equations since they are only additively connected to the small-signal variables. For our example, we thus obtain

$$v_o = Av_i.$$

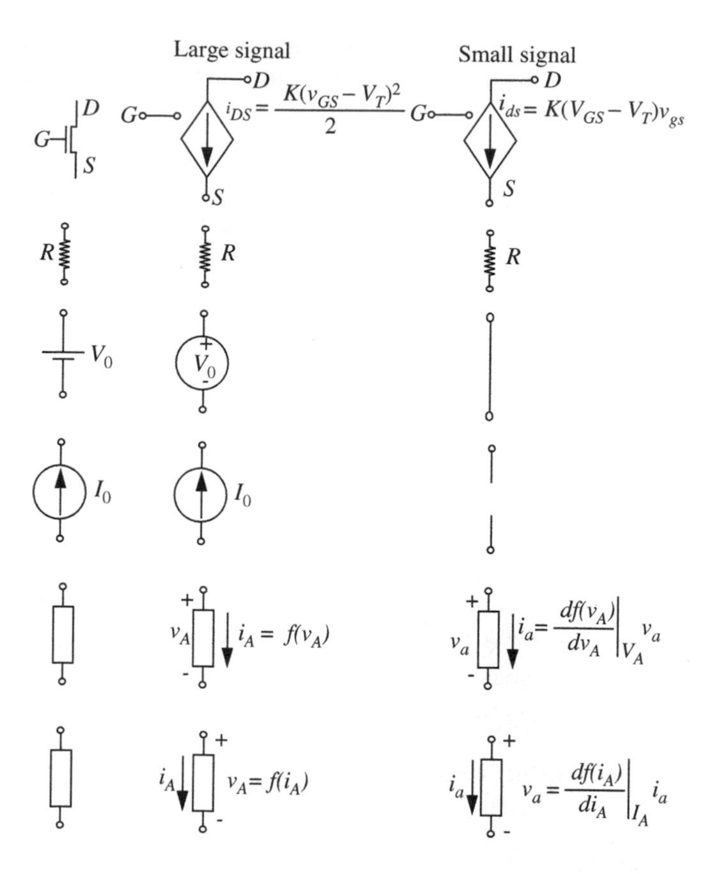


**FIGURE 8.10** Small-signal equivalent models.

- A resistor behaves identically for a large signal or a small-signal. Therefore its small-signal and large-signal models are the same.
- For a MOSFET, the derivation resulting in Equation 8.11 shows how to relate the incremental drain to source current $i_{ds}$ to the incremental gate to source voltage $v_{gs}$.
- By definition, an input signal $v_I$ has an incremental component $v_i$ and a DC component $V_I$.
- In general, if a device variable $x_B$ depends on some other variable $x_A$ as

$$x_B = f(x_A),$$

then the incremental change in $x_B$ due to an small change in $x_A$ is given by

$$x_b = \left.\frac{df(x_A)}{dx_A}\right|_{x_A = X_A} x_a \tag{8.22}$$

where $X_A$ is the operating point value of $x_A$.

EXAMPLE 8.1 A MOSFET WITH ITS GATE AND DRAIN TIED TOGETHER Let us derive the incremental model for a MOSFET that has its gate and drain terminal tied together as shown in Figure 8.11. When the G and D terminals of the MOSFET are tied together, we get an effective two-terminal device. Let us denote the two terminals as $D$ and $S$, respectively. Because the gate-to-source voltage of the device is the same as the drain-to-source voltage, the current $i_{DS}$ through the device is related to the voltage $v_{DS}$ across the device as

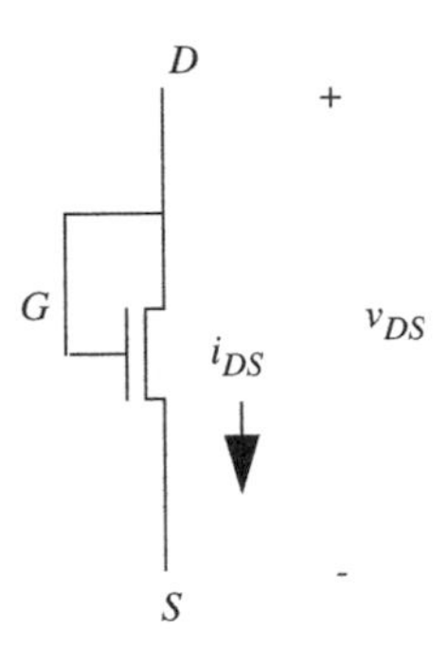


FIGURE 8.11 MOSFET with its G and D terminals connected together.

$$i_{DS} = K\frac{(v_{GS} - V_T)^2}{2}.$$

Since the gate and drain are connected, $v_{GS} = v_{DS}$. Therefore,

$$i_{DS} = K\frac{(v_{DS} - V_T)^2}{2}.$$

The large-signal model for the mosfet is shown in Figure 8.12.

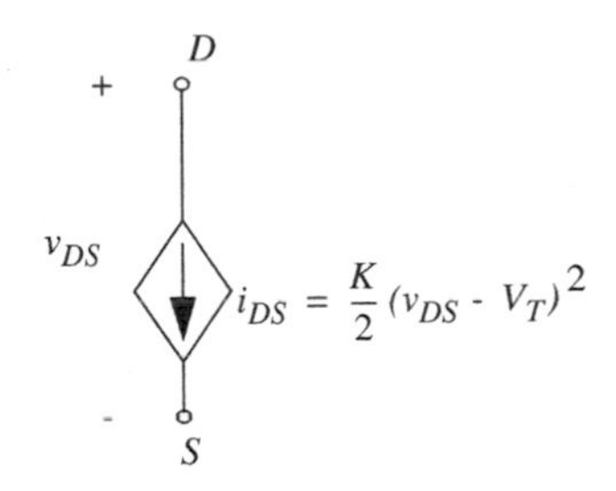


FIGURE 8.12 Large-signal model for a MOSFET with its G and D terminals connected together.

We can derive the change in $i_{DS}$ for a small change in $v_{DS}$ as follows. Let the DC value of $v_{DS}$ be $V_{DS}$ and let the change be denoted $v_{ds}$. Let the corresponding DC value of $i_{DS}$ be $I_{DS}$ and let its change be denoted $i_{ds}$. Then,

$$\begin{aligned} i_{ds} &= \left.\frac{di_{DS}}{dv_{DS}}\right|_{V_{DS}} v_{ds} \\ &= K(v_{DS} - V_T)|_{V_{DS}}\, v_{ds} \\ &= K(V_{DS} - V_T)v_{ds}. \end{aligned}$$

In other words,

$$v_{ds} = \frac{i_{ds}}{K(V_{DS} - V_T)}.$$

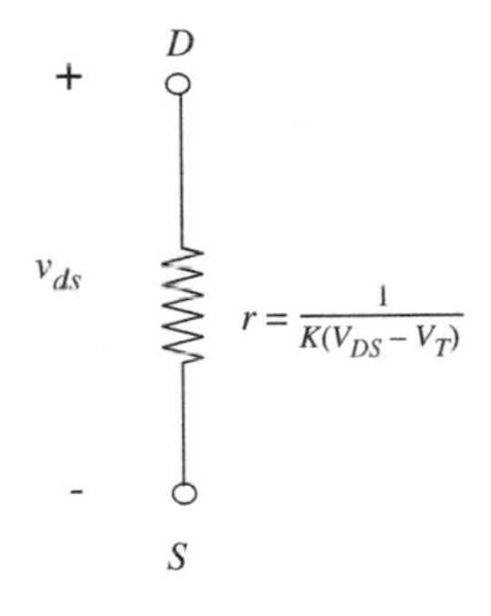


FIGURE 8.13 Small-signal model for a MOSFET with its G and D terminals connected together.

Notice that because $1/K(V_{GS} - V_T)$ is a constant, $v_{ds}$ is directly proportional to $i_{ds}$, which is a resistor relationship. Remarkably, a MOSFET with its gate and drain terminals connected behaves like a resistor with resistance $1/K(V_{GS} - V_T)$ to small signals.

The small-signal equivalent circuit for the preceding element is shown in Figure 8.13. Because of its resistive behavior for small signals, and because MOSFETs with a high resistance are easier to fabricate than resistors, MOSFETs are commonly used as the load resistor in amplifiers.

### 8.2.2 SMALL-SIGNAL CIRCUIT FOR THE MOSFET AMPLIFIER

Let us now develop the small-signal equivalent circuit for the MOS amplifier shown in Figure 8.14. Recall that developing the small-signal model involves the following steps:

1. Set each source to its operating-point value, and determine the operating-point branch voltages and currents for each component in the circuit.
2. Determine the linearized small-signal behavior of each component, and select a linear component to represent this behavior.
3. Replace each original component in the circuit with its linearized equivalent and re-label the circuit with the small-signal branch variables. The resulting circuit is the desired small-signal model.

As the first step, let us determine the operating point of the MOSFET amplifier for its bias voltages using the large-signal SCS circuit model depicted in Figure 8.15. Assuming that the input bias voltage is $V_I$, we can determine the output operating current $I_D$ and the output operating voltage $V_O$. We explicitly show the power supply voltage source $V_S$ to facilitate deriving the small-signal model.

$V_S$, $R_L$, $v_O$, $v_I$, $i_D = \frac{K(v_I - V_T)^2}{2}$

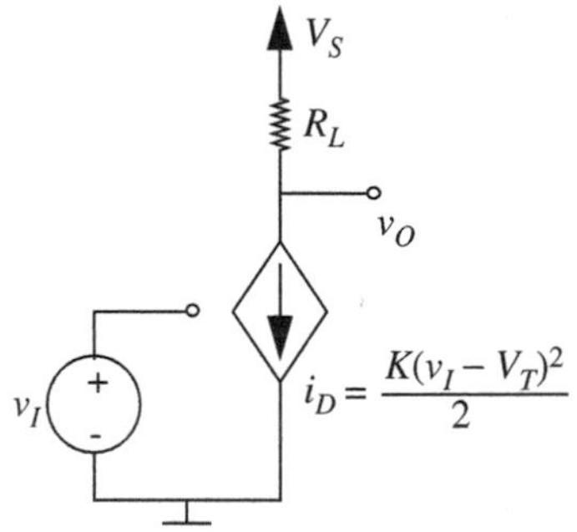

FIGURE 8.14 The MOSFET amplifier.

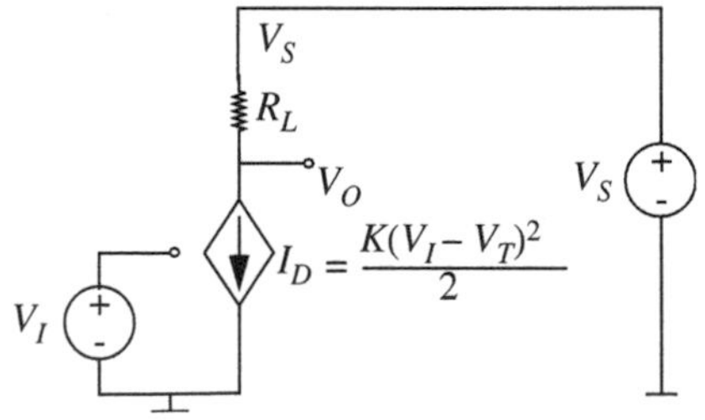

$V_S$, $R_L$, $V_O$, $V_S$, $V_I$, $I_D = \frac{K(V_I - V_T)^2}{2}$

FIGURE 8.15 Computing the operating point of the MOSFET amplifier based on the large-signal SCS model.

The output operating current $I_D$ is directly calculated from the MOSFET characteristic equation as:

$$I_D = \frac{K}{2}(V_I - V_T)^2.$$

The output operating voltage is obtained by applying KVL for the loop comprising the power supply, the MOSFET, and $R_L$ as follows:

$$V_O = V_S - I_D R_L \tag{8.23}$$

$$= V_S - \frac{K}{2}(V_I - V_T)^2 R_L. \tag{8.24}$$

As the second step, we determine the linearized small-signal models for each component. Referring to Figure 8.10, we see that the small-signal model for the DC power supply is a short. The small-signal model for the resistor is the same as its large-signal model. Finally, the linearized small-signal model for the MOSFET in saturation is a voltage-dependent current source whose small-signal current is linearly related to the small-signal gate-to-source voltage as:

$$i_{ds} = K(V_{GS} - V_T)v_{gs}.$$

Notice that the biasing of the large-signal circuit determines the parameters of the small-signal circuit (for example, the small-signal current source parameter $K(V_I - V_T)$ depends on the input bias voltage, $V_I$).

As the third step, we replace each original component in the circuit with its linearized equivalent and re-label the circuit with the small-signal branch variables $v_i$, $v_o$, and $i_d$ as depicted in Figure 8.16.

The small-signal circuit model can be analyzed to determine the circuit response to small signals. For example, we can use Figure 8.16 to determine the small-signal gain of the MOSFET amplifier. Applying KVL at the output, we get

$$v_o = -i_d R_L \tag{8.25}$$

$$= -K(V_I - V_T)v_i R_L \tag{8.26}$$

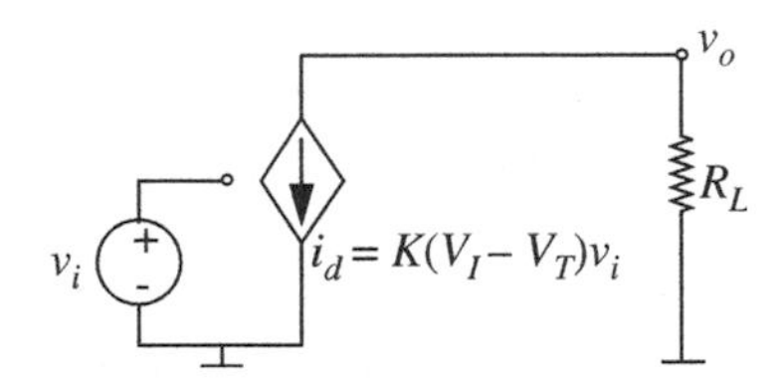


FIGURE 8.16 The small-signal SCS circuit model for the MOSFET amplifier.

Thus, the small-signal gain is given by

$$\frac{v_o}{v_i} = -K(V_I - V_T)R_L \tag{8.27}$$

$$= -g_m R_L \tag{8.28}$$

where

$$g_m = K(V_{GS} - V_T) \tag{8.29}$$

is the transconductance of the MOSFET.

As an example, let us compute the small-signal gain for the following amplifier parameters:

$$V_S = 10 \text{ V}$$

$$K = 1 \text{ mA/V}^2$$

$$R_L = 10 \text{ k}\Omega$$

$$V_T = 1 \text{ V}.$$

Also, suppose the input bias voltage is chosen to be $V_I = 2$ V. As determined earlier in Equation 8.24,

$$V_O = V_S - \frac{K}{2}(V_I - V_T)^2 R_L.$$

Substituting, the given parameters, we get $V_O = 5$ V.

We can now calculate the magnitude of the voltage gain as

$$\frac{v_o}{v_i} = K(V_I - V_T)R_L$$

$$= 10^{-3}(V_I - 1)10^4$$

$$= 10.$$

### 8.2.3 SELECTING AN OPERATING POINT

Small-signal operation requires that the total input signal appear as a *small* perturbation about a DC offset. The input DC offset establishes an operating point for the amplifier. Section 7.7 discussed the issue of operating points in the context of large signals, and proposed a method for selecting the operating point based on maximizing the dynamic input signal range. Specifically, Section 7.7 suggested that the operating point be chosen as the midpoint of the valid input voltage range of amplifier operation under the saturation discipline. This made

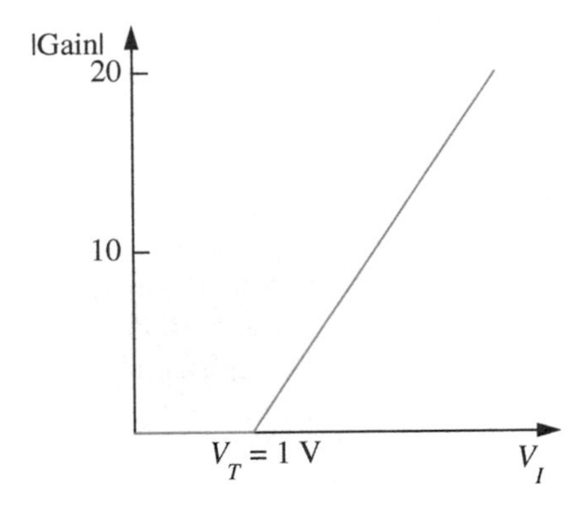


**FIGURE 8.17** Magnitude of the small-signal gain of the amplifier for various values of the input operating point voltage $V_I$.

sense, since the input signals were large and maximizing the input dynamic range enabled the amplifier to deal with the largest possible input signals.

When dealing with small signals, other criteria are often more important in selecting the operating point than just obtaining maximum dynamic range. One criterion is the small-signal gain of the amplifier. As evident from Equation 8.28, the small-signal gain of the amplifier is dependent on the input operating point voltage $V_I$. The magnitude of the small-signal gain is given by

$$\left|\frac{v_o}{v_i}\right| = K(V_I - V_T)R_L. \tag{8.30}$$

Figure 8.17 plots the magnitude of the gain for various values of $V_I$. The graph indicates that the amplifier gain increases with increasing $V_I$.

As an example, assuming these parameters for our amplifier,

$$V_S - 10 \text{ V}$$
$$K = 1 \text{ mA/V}^2$$
$$R_L = 10 \text{ k}\Omega$$
$$V_T = 1 \text{ V}$$

let us determine a value for the input operating-point voltage $V_I$ that will result in a gain of 12.

Substituting the required gain into Equation 8.30, we have

$$12 = 1 \times 10^{-3}(V_I - 1)10 \times 10^3.$$

Solving, we obtain $V_I = 2.2$ V. This means that an input DC offset of 2.2 V will result in a small-signal gain magnitude of 12.

Now, assuming that the input signal is a small-signal sinusoid superimposed on the DC offset of 2.2 V, let us determine the maximum valid peak-to-peak swing for the sinusoid. We refer back to Section 7.6.2 to answer this question. From Section 7.6.2, we know that under the saturation discipline, the maximum valid range for the input voltage is $V_T \rightarrow -1 + \sqrt{1 + 2V_S R_L K}/R_L K + V_T$.

For the given parameters, the valid range for input voltages is 1 V → 2.32 V. In other words, as discussed in Section 7.6.2, input voltages under 1 V will result in cutoff region operation of the MOSFET, while those over 2.32 V will result in triode region operation. Operation in either the cutoff region or the triode region will result in severe signal distortion.

Since the input offset is 2.2 V, and the maximum valid input voltage is 2.32 V, the maximum positive swing for saturation region operation of the MOSFET is given by 2.32 V – 2.2 V = 0.12 V. Thus, the maximum peak-to-peak swing for the input sinusoid is 2 × 0.12 V = 0.24 V. Notice the clear tradeoff we have made between gain and dynamic range. To increase the gain, we had to bias the amplifier with a high input bias voltage, which was close to the high end of the valid input signal range. However, the high input bias voltage limited the positive signal swing.

Another criterion that is often important is the output operating-point voltage. This is important when the amplifier must drive another circuit stage and the output operating-point voltage of the amplifier determines the input operating-point voltage of the next stage.

For example, consider the two-stage amplifier shown in Figure 8.18. In this circuit, $V_{IA}$ provides the DC bias for the first stage. Its output, in turn, $V_{OA}$ provides the DC bias for the second stage. Thus, $V_{OA} = V_{IB}$.

Assuming the following parameters for our amplifier,

$$V_S = 10 \text{ V}$$

$$K = 1 \text{ mA/V}^2$$

$$R = 10 \text{ k}\Omega$$

$$V_T = 1 \text{ V}$$

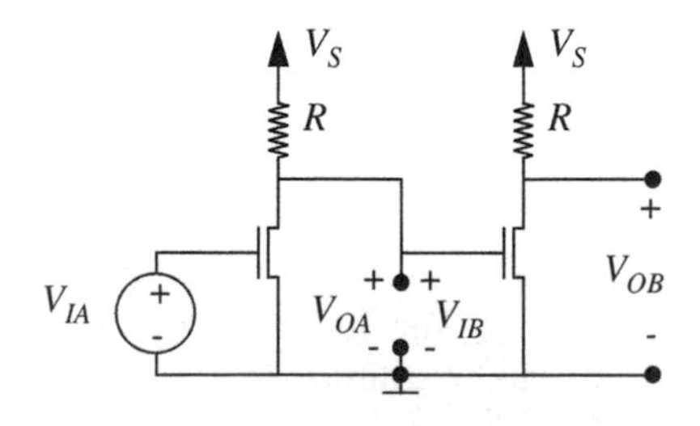


FIGURE 8.18 A two-stage amplifier.

suppose the first stage is biased at $V_{IA} = 2.2$ V to achieve a small-signal gain magnitude of 12. Let us determine whether the output operating-point voltage of the first stage can provide a valid input bias voltage for the second stage.

When the first stage is biased at $V_{IA} = 2.2$ V, the first stage operating-point output voltage $V_{OA}$ is given by Equation 8.24. Substituting the parameters for our circuit, we have

$$
\begin{aligned}
V_{OA} &= V_S - \frac{K}{2}(V_{IA} - V_T)^2 R \\
&= 10 - \frac{1 \times 10^{-3}}{2}(2.2 - 1)^2 10 \times 10^3 \\
&= 2.8 \text{ V}.
\end{aligned}
$$

From Section 7.6.2, we know that under the saturation discipline, the maximum valid range for the input voltage of the second stage is $V_T \rightarrow -1 + \sqrt{1 + 2V_S RK}/RK + V_T$. Substituting the circuit parameters, the valid input range for the second stage comes out to be 1 V $\rightarrow$ 2.32 V. Since $V_{OA}$ exceeds the upper bound (2.8 V > 2.32 V), we conclude that the first stage cannot provide a valid input bias voltage for the second stage when the first stage input bias is set at 2.2 V. We can correct this situation by increasing $V_{IA}$, or by increasing $R$ for the first stage.

### 8.2.4 INPUT AND OUTPUT RESISTANCE, CURRENT AND POWER GAIN

The small-signal equivalent circuit also allows us to determine other important circuit parameters, such as the small-signal input resistance, output resistance, current gain, and power gain. Since the amplifier behaves as a linear network for small signals, it can be characterized by a Thévenin equivalent when viewed from any given port. The input and output resistance come in handy in this Thévenin characterization. Let us determine these values for the MOSFET amplifier using its small-signal circuit in Figure 8.16. Since these parameters are externally observed quantities, they are defined with respect to the external ports of the amplifier abstraction. Thus, it is important that we define precisely what constitutes the input and output ports of the small-signal amplifier. Figure 8.19 shows the relationship between the external ports of the amplifier circuit and the small-signal model. Notice that we have internalized the input bias voltage into the small-signal amplifier abstraction so the user of the amplifier does not have to provide the appropriate input bias voltage. Instead, the user can simply provide a small input signal and observe the resulting signal output superimposed on the DC output offset.

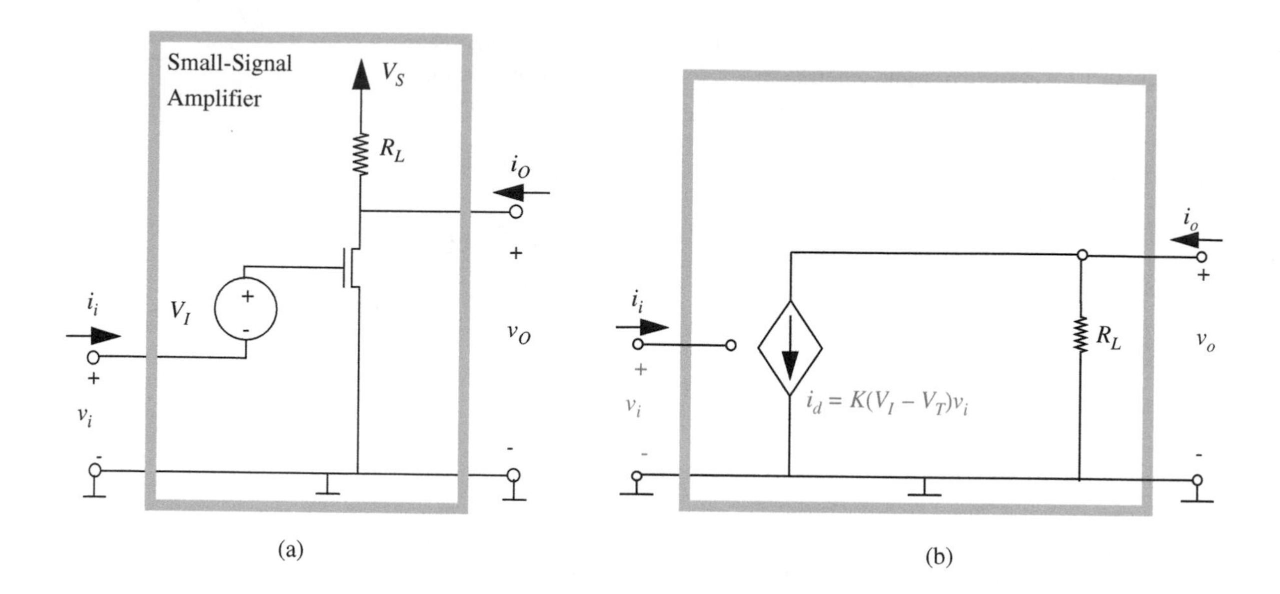


FIGURE 8.19 Amplifier input and output ports: (a) amplifier circuit; (b) small-signal model. As shown in the amplifier circuit, we have internalized the input bias voltage into the small-signal amplifier abstraction.

### Input Resistance $r_i$

Incremental input resistance The change in the input current for a small change in the input voltage.

Accordingly, as depicted in Figure 8.20 we compute it by applying a small test voltage $v_{test}$ at the input and measuring the corresponding current $i_{test}$. All other independent small-signal voltages or DC voltage sources are shorted. Similarly, all other independent small-signal or DC current sources are turned into open circuits.

The input resistance for the MOSFET amplifier is given by

$$r_i = \frac{v_{test}}{i_{test}} = \frac{v_{test}}{0} = \infty. \tag{8.31}$$

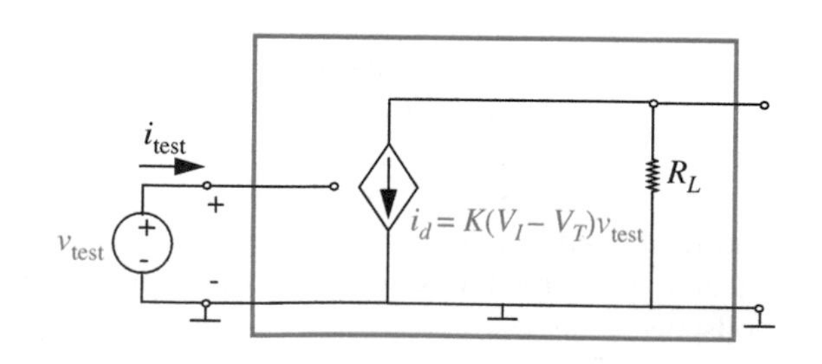


FIGURE 8.20 Input resistance measurement.

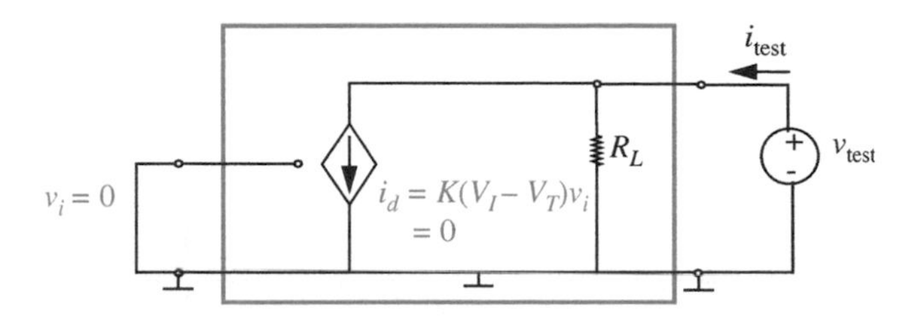


FIGURE 8.21 Output resistance measurement.

For the SCS MOSFET model, the gate does not draw any current ($i_{test} = 0$), so the input resistance is infinite.

### Output Resistance $r_{out}$

Incremental output resistance The change in the output current for a small change in the output voltage.

We must assume, of course, that the circuit is biased properly. As depicted in Figure 8.21, we compute the output resistance by applying a small test voltage $v_{test}$ at the output and measuring the corresponding current $i_{test}$. As before, all other independent small-signal voltages or DC voltage sources are shorted. Thus the small-signal input voltage $v_i$ is set to 0. Similarly, all other independent small-signal or DC current sources are turned into open circuits.

The output resistance is given by

$$r_{out} = \frac{v_{test}}{i_{test}} = R_L. \tag{8.32}$$

Because the input small-signal voltage is set to zero, the current through the MOSFET is 0. In other words, the MOSFET behaves like an open circuit. Thus the output resistance for small signals is $R_L$.

### Current Gain

Analogous to the voltage gain, we can define a current gain for an amplifier that supplies an external current.

Incremental current gain The change in the output current divided by the change in the input current, for a given external load resistance.

As depicted in Figure 8.22, we can compute the current gain by applying a small test voltage at the input and measuring both the input current $i_{test}$ and the output current $i_o$. The ratio $i_o/i_{test}$ is the current gain. Note that the output current is not the current that flows through the dependent current source, rather it is the current that is drawn by an external load resister $R_O$. Because it is dependent on the value of the load resistor, the current gain is defined

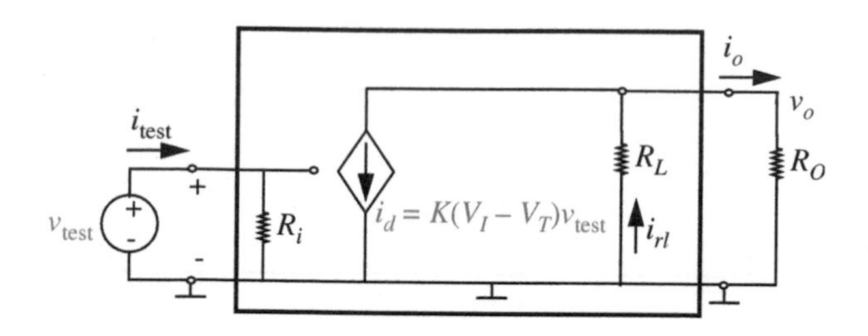


**FIGURE 8.22** Current gain measurement. As an exercise, we place a resistance $R_i$ between the input terminal and ground. For a MOSFET, $R_i = \infty$.

for a given load resistance. The introduction of an external load resistance also reduces the voltage gain of the amplifier because it appears in parallel with the internal load resistor $R_L$.

The current gain with an external load resistance $R_O$ is given by

$$\text{Current gain} = \frac{i_O}{i_{test}}. \tag{8.33}$$

Let us go through the exercise of determining the value of $i_o$ assuming there is some finite input resistance $R_i$ as shown in Figure 8.22. Substituting for $i_o$ and $i_{test}$ in terms of the respective voltages,

$$\text{Current gain} = \frac{\frac{v_O}{R_O}}{\frac{v_{test}}{R_i}} \tag{8.34}$$

$$= \frac{v_O}{v_{test}} \frac{R_i}{R_O}. \tag{8.35}$$

Equation 8.35 says that the current gain is proportional to the product of the voltage gain and the ratio of the input resistance and the output resistance.

We can determine the voltage gain $v_o/v_{test}$ by substituting for $v_o$ in terms of the current $i_d$ and the parallel resistance pair $R_L$ and $R_O$ as

$$\frac{v_O}{v_{test}} = -\frac{K(V_I - V_T)v_{test}(R_L \| R_O)}{v_{test}}.$$

In other words,

$$\frac{v_O}{v_{test}} = -K(V_I - V_T)(R_L \| R_O). \tag{8.36}$$

Notice that the voltage gain of the amplifier with an external load is lower than an unloaded amplifier. Substituting the expression for the voltage gain into

Equation 8.35, we get an expression for the current gain:

$$\text{Current gain} = -K(V_I - V_T)(R_L \| R_O)\frac{R_i}{R_O}. \tag{8.37}$$

Since $R_i = \infty$ for the MOSFET, the corresponding current gain is also infinite.

### Power Gain

Incremental power gain The ratio of the power supplied by the amplifier to an external load to that supplied to the amplifier by the input source.

Referring to Figure 8.22, we can compute the power gain as follows: We apply a small test voltage at the input and measure the input current $i_{test}$. We also measure the corresponding output voltage $v_o$ and output current $i_o$ supplied to the external load resistor. We compute the power supplied by the input source as $v_{test}i_{test}$. Similarly, we compute the power supplied to the external load as $v_o i_o$. As we did for the current gain, let us assume that the amplifier has an input resistance $R_i$. The power gain is given by

$$\text{Power gain} = \frac{v_o i_o}{v_{test} i_{test}} = \frac{v_o}{v_{test}} \frac{i_o}{i_{test}}. \tag{8.38}$$

We know both the voltage gain and the current gain from Equations 8.36 and 8.37, respectively. Substituting in the above equation we get,

$$\text{Power gain} = \frac{v_o}{v_{test}} \frac{i_o}{i_{test}} \tag{8.39}$$

$$= [-K(V_I - V_T)(R_L \| R_O)]\left[-K(V_I - V_T)(R_L \| R_O)\frac{R_i}{R_O}\right] \tag{8.40}$$

$$= [K(V_I - V_T)(R_L \| R_O)]^2 \frac{R_i}{R_O}. \tag{8.41}$$

Since $R_i = \infty$ for the MOSFET amplifier, the power gain is also infinite. In practical circuits, however, there is always some input resistance, so the power gain is finite.

---

EXAMPLE 8.2 VOLTAGE-CONTROLLED CURRENT SOURCE

Let us perform a small-signal analysis of the voltage-controlled current source circuit shown in Figure 8.23. Referring to Figure 8.23, the current $i_O$ depends on voltage source $v_I$ according to

$$i_O = \frac{1}{L(v_I - 1)}$$

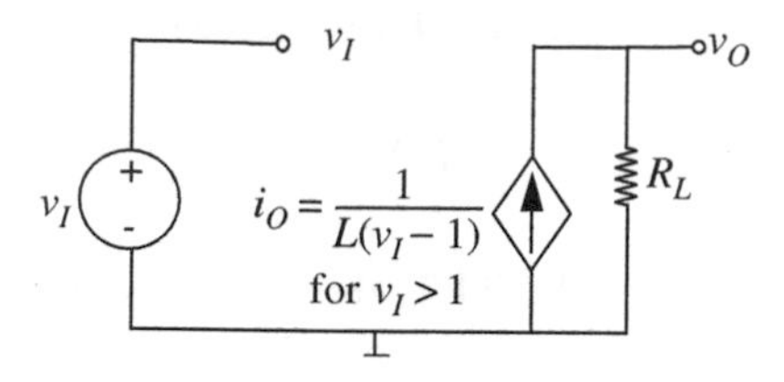


**FIGURE 8.23** Dependent current source circuit.

where $v_I > 1$ and $L$ is some constant. What is the change in $v_O$ for an incremental change in $v_I$, when the operating-point values of $v_I$ and $v_O$ are $V_I$ and $V_O$, respectively?

To find the incremental change in $v_I$, we follow the three-step process outlined in Section 8.2.1. We begin by writing the large-signal relationship between $v_O$ and $v_I$:

$$v_O = i_O R_L \tag{8.42}$$

$$= R_L \frac{1}{L(v_I - 1)}. \tag{8.43}$$

Substituting in the operating-point values, we get:

$$V_O = R_L \frac{1}{L(V_I - 1)}. \tag{8.44}$$

Next, we linearize the devices. The input voltage source with total voltage $v_I$ is replaced by its small-signal voltage $v_i$. The resistor remains unchanged. The small-signal equivalent of the dependent current source is derived using:

$$i_o = \left.\frac{di_O}{dv_I}\right|_{V_I} v_i$$

$$= -\frac{1}{L(V_I - 1)^2} v_i.$$

In the third step, we substitute in the small-signal models in place of the large-signal models for each of the devices. The corresponding small-signal circuit is shown in Figure 8.24.

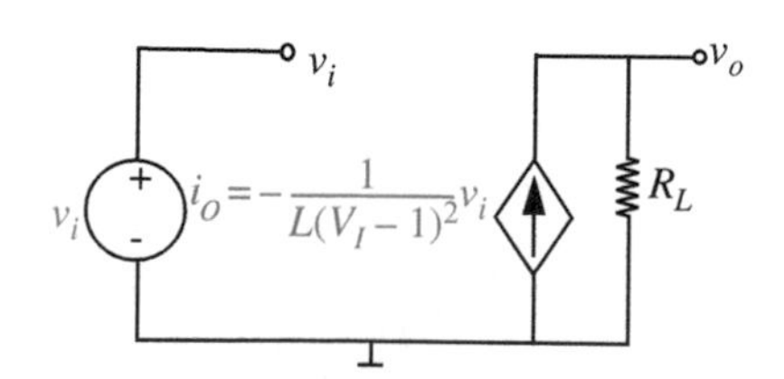


**FIGURE 8.24** Small-signal circuit model for the dependent current source.

We can now derive the change in the output voltage for a small change in the input voltage from the small-signal circuit by writing KVL for the output loop:

$$v_o = i_o R_L = -\frac{1}{L(V_I - 1)^2} v_i R_L.$$

We can also derive the change in the output voltage for a small change in the input voltage directly from the $v_O$ versus $v_I$ relationship given by Equation 8.43.

$$v_o = \left.\frac{dv_O}{dv_I}\right|_{V_I} v_i = -\frac{1}{L(V_I - 1)^2} v_i R_L.$$

---

---

EXAMPLE 8.3 SMALL-SIGNAL ANALYSIS OF A DIFFERENCE AMPLIFIER The *difference amplifier* is a building block for high-quality amplification and is useful for processing small signals. When a signal is noisy, straightforward use of an amplifier would amplify both the signal and the noise. However, under certain conditions that we will see shortly, a difference amplifier (also called a differential amplifier) can be used to amplify the signal by a much larger gain relative to the noise. Difference amplifiers are also used in building operational amplifiers, and suitable difference amplifier circuits are discussed in Examples 7.19 and 8.10.

Suppose the signal is available in differential form. In other words, suppose the signal is available as the relative voltage output ($v_A - v_B$) on a pair of terminals $A$ and $B$. For example, as the output of the tape-head in a tape-recorder, the output of an instrumentation device or a sensor. Such a sensor often resembles one of our primitive elements — for example, a variable resistor. The element might produce a voltage signal across its terminals related to some externally sensed parameter such as temperature, gas concentration, or magnetic field strength. Often, a pair of wires carrying the signal might travel through a noisy environment resulting in the coupling of more or less the same amount of noise ($v_n$) on each of the two wires, as depicted in Figure 8.25.[2] In other situations, the two wires might both carry a common DC bias. In such situations, a difference amplifier can help amplify just the differential signal component and discard the common noise component.

The difference amplifier abstraction is shown in Figure 8.26. It is a two-port device with one differential input port and one single-ended output port. The input port has two input terminals. The + input is called the *non-inverting input* and the − input is called the *inverting input*. It has an output port across which $v_O$ appears.

---

2. In fact, the wires are often twisted together to ensure that when there is noise, the same amount of noise infects both wires.

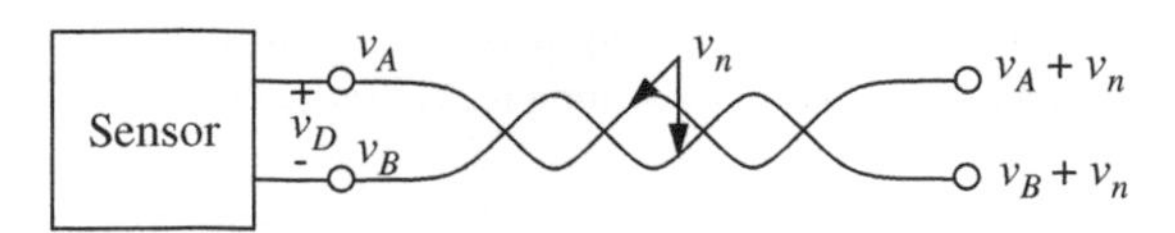


FIGURE 8.25 A differential signal.

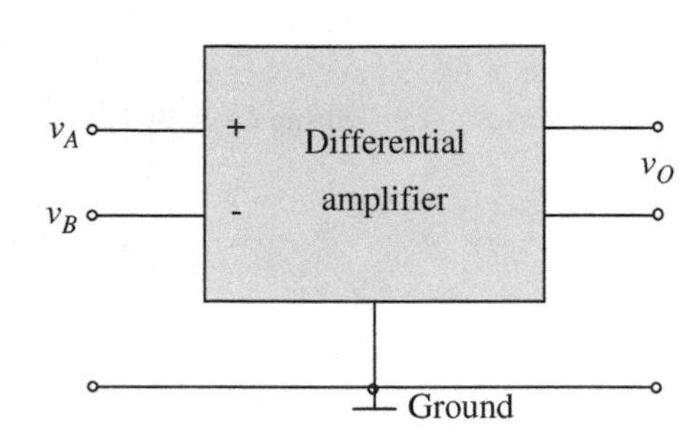


FIGURE 8.26 Difference amplifier black box representation.

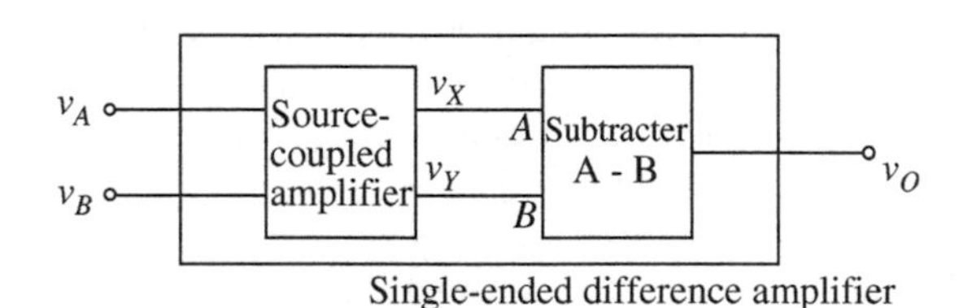


FIGURE 8.27 Single-ended difference amplifier structure.

We can also build a single-ended difference amplifier from a differential output difference amplifier as shown Figure 8.27.

The behavior of the difference amplifier is best explained by considering its effect on the following signals related to the two components, $v_A$ and $v_B$:

1. *A difference-mode component signal,*

$$v_D = v_A - v_B \tag{8.45}$$

2. *And a common-mode component signal,*

$$v_C = \frac{v_A + v_B}{2}. \tag{8.46}$$

The output of the difference amplifier is a function of these two components of the input,

$$v_O = A_D v_D + A_C v_C \tag{8.47}$$

where $A_D$ is called the *difference-mode gain* and $A_C$ is called the *common-mode gain.* The key in using a difference amplifier is to encode the useful signal in the difference-mode component and the noise in the common-mode component. Then if we make $A_D$ large and $A_C$ small, we achieve our goal of noise reduction. Usually we use the *common-mode rejection ratio* (CMRR) to describe the ability of the amplifier to reject the common-mode noise:

$$\text{CMRR} = \frac{A_D}{A_C}. \tag{8.48}$$

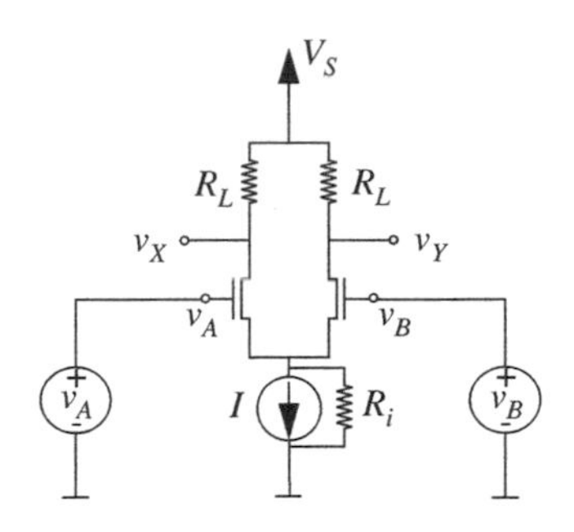


FIGURE 8.28 Source-coupled difference amplifier. All voltages are measured with respect to ground.

## MOSFET Implementation of the Difference Amplifier

Let us study a MOSFET version of the difference amplifier. The amplifier employs a pair of matching transistors called the *source-coupled pair.* The source-coupled amplifier is shown in Figure 8.28. $v_A$ and $v_B$ are the inputs, and $v_X$ and $v_Y$ are the outputs. Assume $v_A$ and $v_B$ are the input voltages measured with respect to ground. Also assume that $v_a$ and $v_b$ are small variations in the inputs, and that $v_x$ and $v_y$ are the corresponding small-signal variations in the output. The source-coupled pair is connected in series with a DC current source with a high internal resistance $R_i$. (We can implement the current source using a MOSFET biased to operate in its saturation region, but we do not show it here. For simplicity, we use an abstract non-ideal current source instead. In other words, the current source has a finite resistance, $R_i$.) Let the current provided by the DC current source be $I$.

Let us examine the difference amplifier using its small-signal model shown in Figure 8.29. Notice that an ideal current source acts like an open circuit, but a current source with an internal Norton equivalent resistance $R_i$ behaves like a resistor to incremental changes in its terminal variables. The MOSFETs are replaced by their small equivalent current sources. The voltages $v_{gs1}$ and $v_{gs2}$ are the small-signal voltages between the gate and source of the two input MOSFETs resulting from a small change in the input voltages $v_A$ and $v_B$.

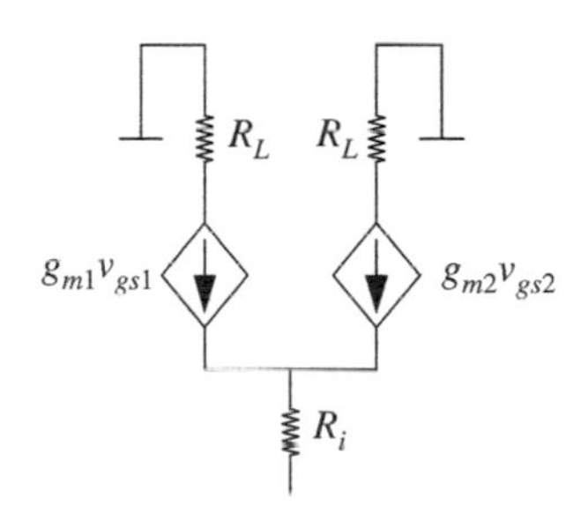


FIGURE 8.29 Source-coupled difference amplifier — small-signal model.

The gain parameters $g_{m1}$ and $g_{m2}$ for the MOSFETs depend on the operating-point values of the currents through them. Assuming that the current through $R_i$ is negligible, by symmetry, we find that the current $I$ divides equally between the two MOSFETs. Thus each has an operating-point current equal to $I/2$. From the SCS model for the MOSFETs, given $V_T$ and $K$, we can thus find the bias input voltages $V_{GS1}$ and $V_{GS2}$ in terms of $I$. In turn, the respective gains $g_{m1}$ and $g_{m2}$ can be determined in terms of $V_{GS1}$ and $V_{GS2}$, which are themselves functions of $I$.

Recall that

- the difference-mode component:

$$v_D = v_A - v_B$$

- and the common-mode component

$$v_C = \frac{v_A + v_B}{2}.$$

Therefore, we can decompose the inputs into their difference- and common-mode component as follows:

$$v_A = v_C + \frac{v_D}{2}$$
$$v_B = v_C - \frac{v_D}{2}.$$

We will discuss each mode separately, and then summarize the behavior of the entire amplifier.

## Difference-Mode Model

We first examine the circuit with the difference-mode part of the input only. Refer to Figure 8.30 for the circuit and its small-signal model. Assume that the two MOSFET's have identical characteristics, $g_{m1} = g_{m2} = g_m$. An application of KCL at the source node of the two MOSFETs (in other words, the node with the small-signal voltage $v_s$) yields

$$g_m v_{gs1} + g_m v_{gs2} = v_s/R_i. \tag{8.49}$$

From Figure 8.30 we can also write

$$\frac{v_d}{2} - v_{gs1} = v_s$$
$$\frac{-v_d}{2} - v_{gs2} = v_s.$$

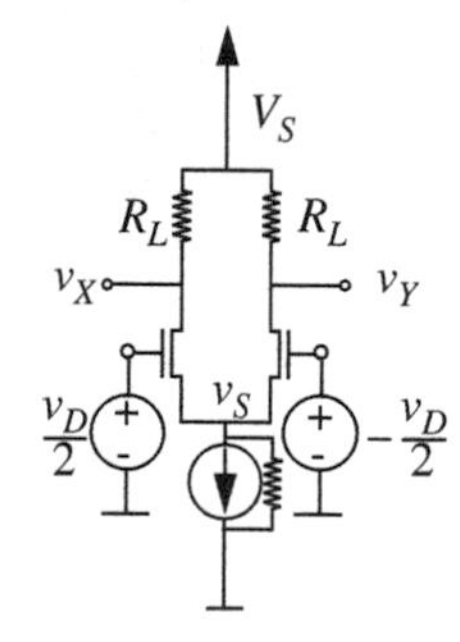


(a) Differential mode input only

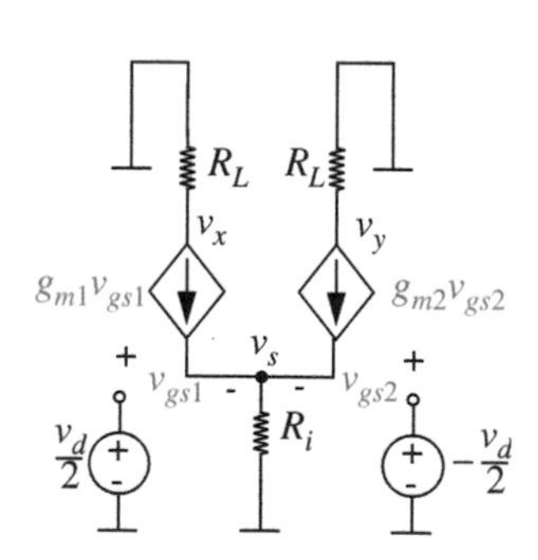


(b) Small-signal model

FIGURE 8.30 Difference-mode model. All voltages are measured with respect to ground.

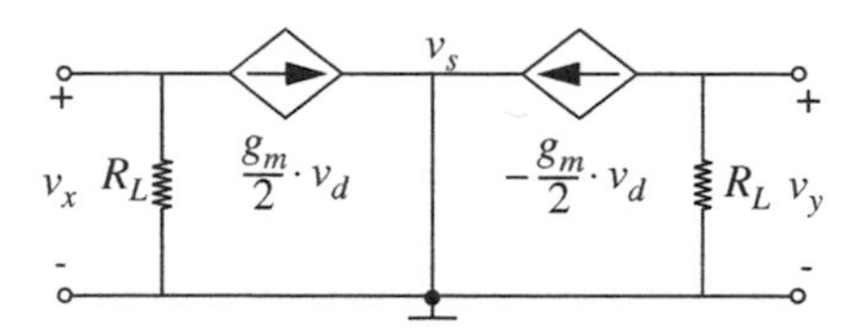


FIGURE 8.31 Difference-mode simplified model.

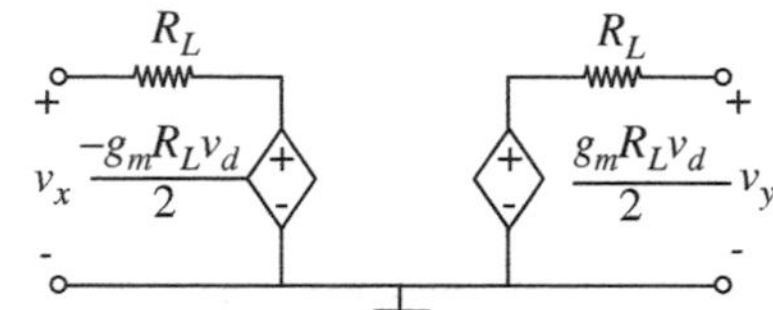


FIGURE 8.32 Difference-mode Thévenin equivalent circuit.

Substituting $v_{gs1}$ and $v_{gs2}$ in terms of $v_d$ into Equation 8.49, we obtain

$$g_m\left(\frac{v_d}{2} - v_s\right) + g_m\left(\frac{-v_d}{2} - v_s\right) = \frac{v_s}{R_i} \tag{8.50}$$

$$-2g_m v_s = \frac{v_s}{R_i}. \tag{8.51}$$

Since $g_m$ and $R_i$ are independent of each other, $v_s = 0$. This result greatly simplifies our circuit to the one in Figure 8.31. Converting it to the Thévenin equivalent model, we obtain the circuit shown in Figure 8.32.

We see that

$$v_x = -\frac{g_m R_L v_d}{2}$$

and

$$v_y = \frac{g_m R_L v_d}{2}.$$

Thus, the small-signal output voltage across the output terminal pair is given by

$$v_o = v_x - v_y = -g_m R_L v_d.$$

This yields a difference-mode small-signal gain

$$A_d = \frac{v_o}{v_d} = -g_m R_L.$$

## Common-Mode Model

We will now examine the behavior of the circuit for the common-mode input. The circuit and small-signal model is shown in Figure 8.33. The small-signal change in the

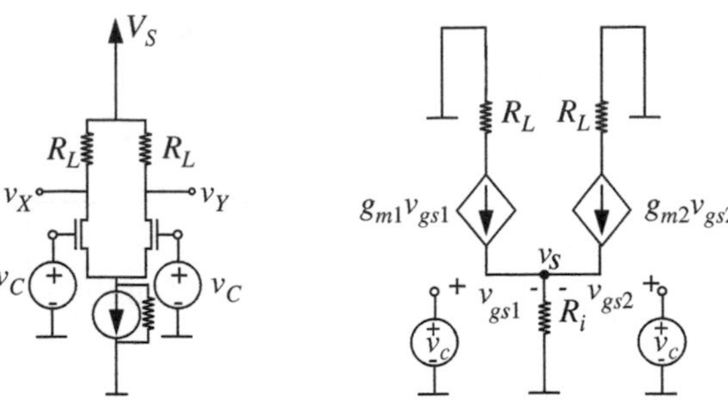


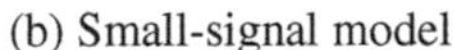

(a) Common-mode input only (b) Small-signal model

FIGURE 8.33 Common-mode model.

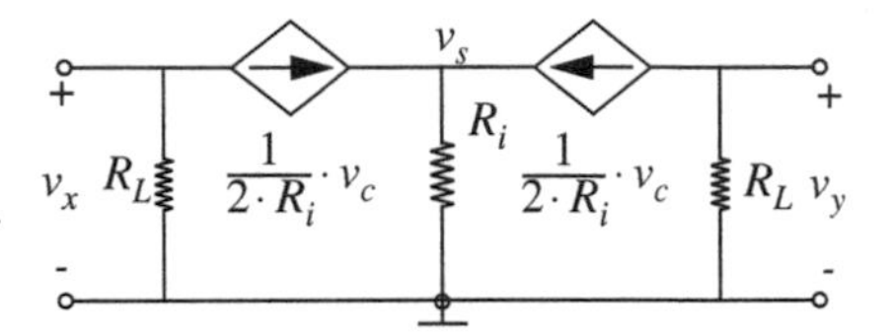


FIGURE 8.34 Common-mode Norton equivalent circuit.

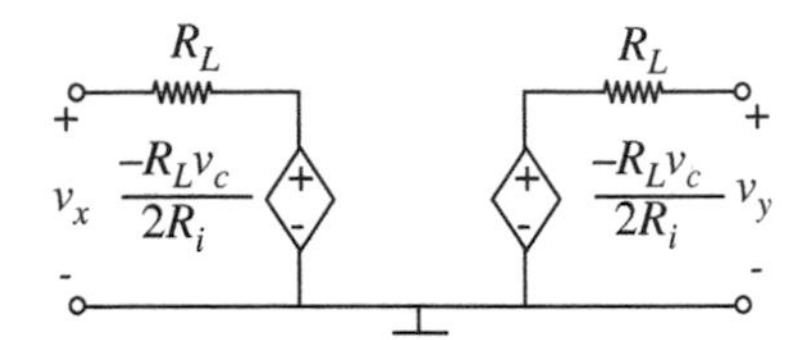


FIGURE 8.35 Common-mode Thévenin equivalent circuit.

common-mode input is denoted $v_c$. Observe that $v_{gs1} = v_{gs2} = v_{gs}$, and $v_{gs} = v_c - v_s$. Application of KCL at $v_s$ again yields

$$g_m v_{gs} + g_m v_{gs} = \frac{v_s}{R_i} \tag{8.52}$$

$$2g_m v_{gs} = \frac{v_c - v_{gs}}{R_i} \tag{8.53}$$

$$v_{gs} = \frac{1}{2g_m R_i + 1} v_c. \tag{8.54}$$

Assuming $R_i$ is large, so that $2g_m R_i \gg 1$, we can simplify Equation 8.54 to

$$v_{gs} \approx \frac{1}{2g_m R_i} v_c.$$

Therefore, the two dependent current sources will have value

$$\frac{1}{2R_i} v_c.$$

The simplified circuit is shown in Figure 8.34. Transforming the circuit into its Thévenin equivalent circuit gives the circuit shown in Figure 8.35.

From the Thévenin equivalent circuit, notice that

$$v_X = v_Y = -\frac{R_L V_C}{2R_i}.$$

Remarkably,

$$v_O = v_X - v_Y = 0$$

effectively yielding a common-mode small-signal gain of 0.

## Overall Behavior

Putting it all together, we combine the small-signal difference-mode circuit from Figures 8.32 with the small-signal common-mode circuit in 8.35 and obtain the circuit shown in Figure 8.36. Notice that we are able to do such a superposition because of the linearity property of our small-signal circuits. The output of the difference amplifier is the difference between $v_X$ and $v_Y$, which gives a difference-mode gain of $-g_m R_L$ and common-mode gain of 0.

## Input and Output Resistances

Computing the input and output resistances for the difference amplifier is fairly easy. When we apply the small input signals $v_a$ and $v_b$, there will not be any current flowing into the MOSFETs, so, we have infinite input resistance.

To compute the small-signal output resistance looking in from one of the terminals of the output port, we turn off all independent sources by setting $v_a = 0$ and $v_b = 0$, in effect, turning off $v_c$ and $v_d$. We introduce a test voltage at the desired output and short the other output to ground. Therefore, the overall circuit is transformed to the one shown in Figure 8.37. Thus the output resistance looking into ports $v_X$ or $v_Y$ and ground will be $R_L$.

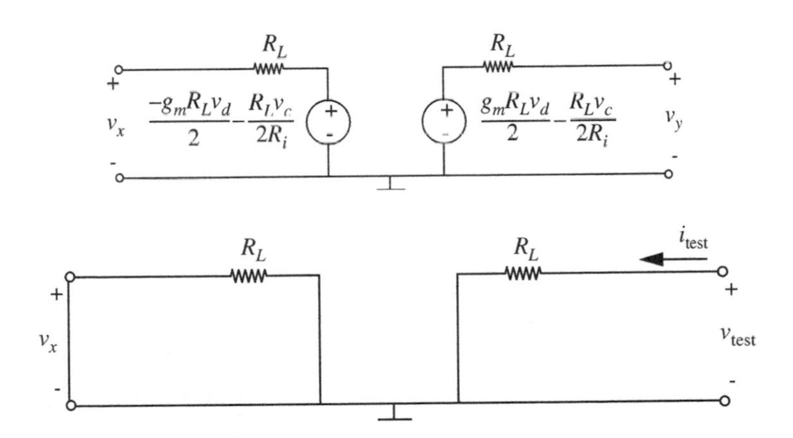


FIGURE 8.36 Difference amplifier Thévenin equivalent circuit.

FIGURE 8.37 Difference amplifier output resistance.

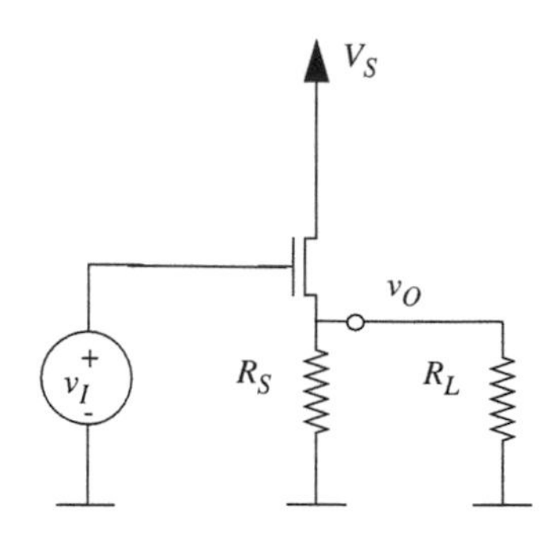


**FIGURE 8.38** Source-follower circuit.

**EXAMPLE 8.4 SOURCE FOLLOWER** A useful circuit we have seen before[3] is the *source follower* shown in Figure 8.38. The source follower in the figure is shown driving an external load resistor $R_L$. Assume that the total input voltage $v_I$ includes the appropriate DC bias voltage to meet the saturation discipline. The small-signal equivalent circuit for the source follower is shown in Figure 8.39. Let us analyze this circuit by computing its small-signal gain.

The small-signal output $v_o$ can be expressed in terms of the circuit parameters as

$$v_o = g_m v_{gs}(R_L \| R_S)$$

where $v_{gs}$ is the voltage between the gate and the source of the MOSFET. Using KVL, observe that $v_{gs} = v_i - v_o$. Therefore, we can write

$$v_o = g_m(v_i - v_o)(R_L \| R_S) \tag{8.55}$$

$$v_o\left(\frac{1}{R_L \| R_S} + g_m\right) = g_m v_i \tag{8.56}$$

$$v_o = \frac{R_L R_S g_m}{R_L + R_S + R_L R_S g_m} v_i \tag{8.57}$$

$$\frac{v_o}{v_i} = \frac{R_L R_S g_m}{R_L + R_S + R_L R_S g_m}. \tag{8.58}$$

Thus the gain is slightly less than 1. An important special case of Equation 8.59 is when $R_L$ is very large. Thus, when $R_L \to \infty$,

$$\frac{v_o}{v_i} = \frac{R_S g_m}{1 + R_S g_m}. \tag{8.59}$$

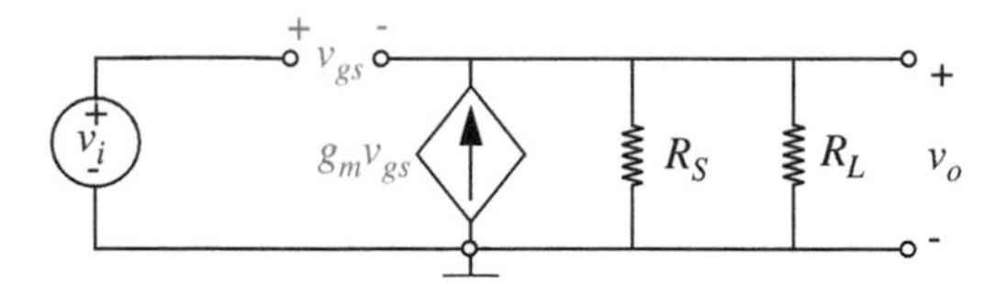


**FIGURE 8.39** Source-follower small-signal model. $g_m$, the transconductance of the MOSFET, is given by $K(V_{GS} - V_T)$, where $V_{GS}$ is the operating-point value of the gate-to-source voltage for the MOSFET. (See Example 7.8 or Problem 7.5 in Chapter 7 to see how the operating-point parameters of the source follower can be calculated.)

3. See Example 7.8 and Problem 7.5 in Chapter 7.

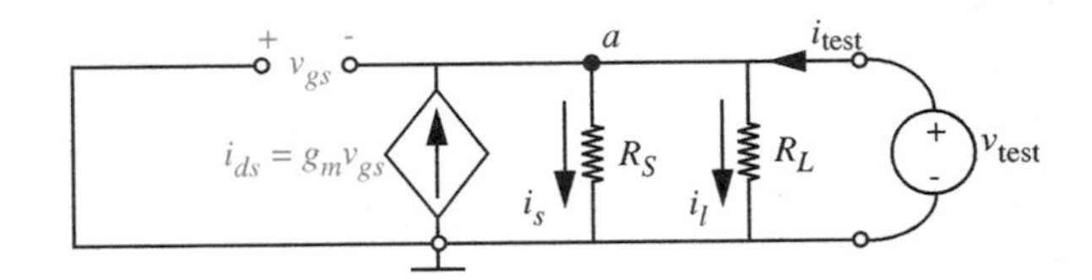


FIGURE 8.40 Source-follower output resistance.

When $g_m$ is large, irrespective of the values of $R_L$ and $R_S$, Equation 8.58 can be rewritten as

$$\frac{v_o}{v_i} \approx 1.$$

To find out why such a circuit is useful, let us compute the input and output resistances of the source-follower device.

## Small-Signal Input and Output Resistances

The input resistance $r_i$ is easily calculated. Since no current flows into the MOSFET, the input resistance is infinity.

Computing the output resistance needs more work. As depicted in Figure 8.40, let us turn off the independent sources, apply a small test voltage $v_{test}$ at the output terminal and measure the corresponding current $i_{test}$. The output resistance will be given by $r_{out} = v_{test}/i_{test}$.

In order to compute $r_{out}$, we apply KCL at node $a$ shown in Figure 8.40. The dependent source current $i_{ds}$ depends on $v_{gs}$, and $v_{gs}$ equals $-v_{test}$. Therefore, we have

$$i_{ds} + i_{test} = i_s + i_l \tag{8.60}$$

$$-g_m v_{test} + i_{test} = \frac{v_{test}}{R_L \| R_S}. \tag{8.61}$$

Rearranging the terms and simplifying the expression, we obtain

$$v_{test}\left(g_m + \frac{1}{R_L \| R_S}\right) = i_{test}$$

This leads to

$$r_{out} = \frac{v_{test}}{i_{test}} = \frac{R_L R_S}{g_m R_L R_S + R_L + R_S}.$$

When $g_m$, $R_L$, and $R_S$ are large, $R_L + R_S$ becomes insignificant compared to $g_m R_L R_S$. Therefore, we can simplify,

$$r_{out} \approx \frac{1}{g_m}.$$

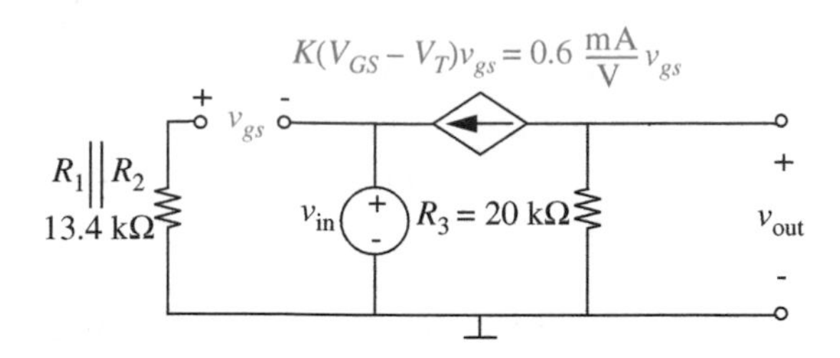


**FIGURE 8.41** Small-signal model of the MOSFET amplifier in Figure 7.46.

Since $g_m$ can be made very large, the output resistance can be made low. The low output resistance makes the source follower useful as a buffer device, which can provide a large amount of current gain.

---

EXAMPLE 8.5 SMALL-SIGNAL ANALYSIS OF ANOTHER MOSFET AMPLIFIER In this example, we examine the small-signal behavior of the MOSFET amplifier shown in Figure 7.46 and studied in Example 7.12. This amplifier works well for both positive and negative values of $v_{IN}$, and so we will choose the input bias voltage to be $V_{IN} = 0$ V for the small-signal analysis. Therefore

$$v_{IN} \equiv V_{IN} + v_{in} = v_{in}.$$

To determine the remaining bias voltages in the amplifier, we set $v_{in} = 0$ V, which results in $v_{IN} = 0$ V. From the results of Example 7.12, we can then determine the bias voltages $V_{OUT} = 6.4$ V and $V_{GS} = 1.6$ V.

Next, following the method of Section 8.2, we construct the small-signal circuit model shown in Figure 8.41. Analyzing the small-signal circuit model, we obtain

$$v_{out} = R_3 K(V_{GS} - V_T) v_{in} = 12\, v_{in}.$$

Therefore, the small-signal gain is 12 at the bias voltage $V_{IN} = 0$. The same result can be obtained by evaluating

$$dv_{OUT}/dv_{IN}|_{v_{IN}=0}$$

using the results of Example 7.12.

---

EXAMPLE 8.6 SMALL-SIGNAL MODEL FOR THE BJT In this example, we will develop the small-signal model for the BJT by linearizing the piecewise linear BJT model studied earlier in Figure 7.49c in Example 7.13. Figure 8.42b depicts the large-signal model (from Figure 7.49c) for the BJT under the constraint that the BJT operates in its active region. When operating in the active region, the base-to-collector diode shown in Figure 7.49c behaves like an open circuit, and so it can be safely ignored in our analysis.

Figure 8.42c depicts the small-signal model of the BJT based on the piecewise-linear model in Figure 8.42b. In the active region, the ideal diode in Figure 8.42b behaves like a short circuit. Furthermore, the 0.6-V voltage source appears as a short circuit for incremental changes. Finally, since the active region relationship between $i_B$ and $i_C$ is linear, and given by

$$i_C = \beta i_B,$$

the relationship between the incremental signals $i_c$ and $i_b$ is also the same:

$$i_c = \beta i_b.$$

Alternatively, we can derive the incremental change in the collector current for a small change in the base current mathematically from Equation 8.22 as follows:

$$\begin{aligned} i_c &= \left.\frac{di_C}{di_B}\right|_{i_B=I_B} i_b \\ &= \left.\frac{d\beta i_B}{di_B}\right|_{i_B=I_B} i_b \\ &= \beta i_b. \end{aligned}$$

Next, we will use the small-signal model for the BJT in a few examples.

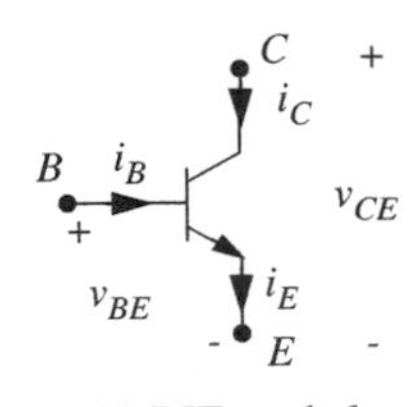


(a) BJT symbol

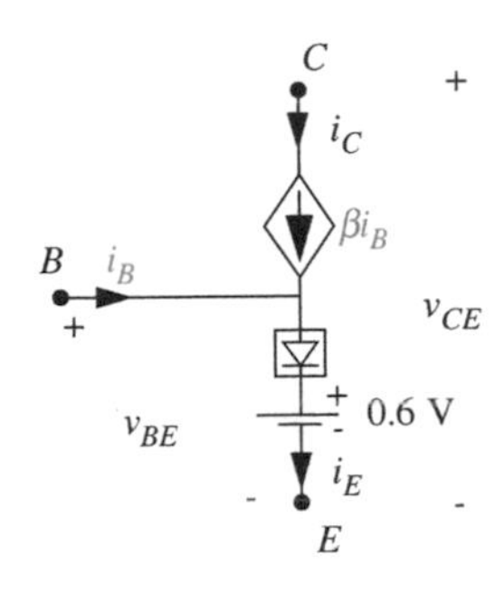


(b) BJT large-signal model assuming BJT is in active region

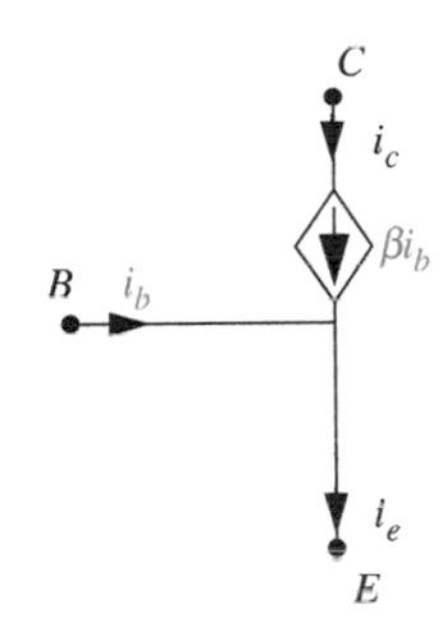


(c) BJT small-signal model

FIGURE 8.42 Small-signal model for the BJT.

EXAMPLE 8.7 SMALL-SIGNAL ANALYSIS OF THE BJT AMPLIFIER In this example, we will study the small-signal behavior of the common emitter BJT amplifier shown in Figure 7.54, which is redrawn here in Figure 8.43 to show that the total input $v_{IN}$ is the sum of a DC offset voltage $V_{IN}$ and a small-signal voltage $v_{in}$. In keeping with our usual small-signal notation, the total, operating point, and small-signal voltages at the output are given by $v_O$, $V_O$, and $v_o$ respectively. We will compute the small-signal gain of the amplifier assuming that the amplifier operates in its active region, and given that $R_I = 100$ kΩ, $R_L = 10$ kΩ, and $V_S = 10$ V. Assume that the current gain parameter $\beta$ for the BJT is 100, and that the input operating voltage is chosen to be $V_{IN} = 1$ V.

We now begin the small-signal analysis of our BJT amplifier. The first step of small-signal analysis is to determine the operating-point variables in the circuit. Although not strictly

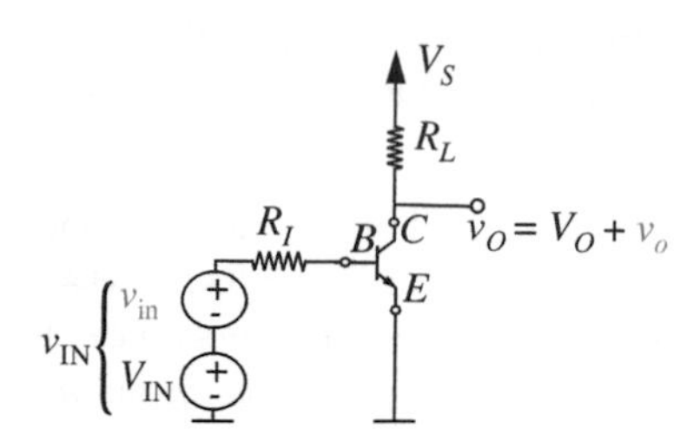


FIGURE 8.43 Our BJT amplifier showing the small-signal and bias input voltages.

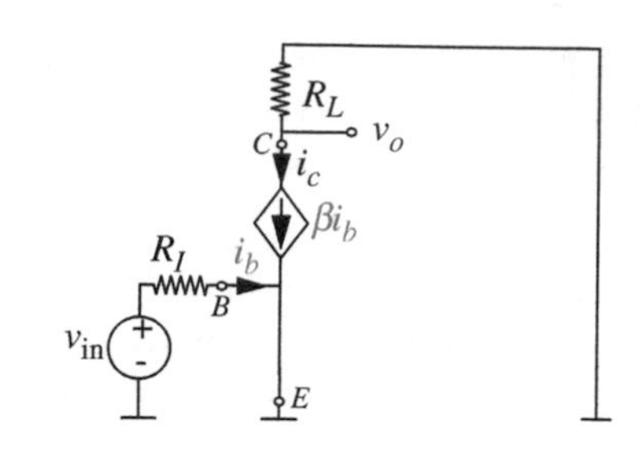


FIGURE 8.44 Small-signal circuit model for the BJT amplifier.

necessary,[4] we will go ahead with the operating-point analysis to verify that the BJT is indeed operating in its active region for the given parameters. From the transfer function relation in Equation 7.51, we know that

$$V_O = V_S - \frac{(V_{IN} - 0.6)}{R_I}\beta R_L.$$

Substituting our specific parameter values, we obtain

$$V_O = 6 \text{ V}.$$

Since, $V_{CE} = V_O = 6$ V and $V_{BE} = V_{IN} = 1$ V, it is easy to see that the BJT constraint for active region operation given by

$$V_{CE} > V_{BE} - 0.4 \text{ V}$$

is satisfied.

As the second step, we must determine linearized small-signal models for each of the circuit components. This step is trivial for our example, since all the elements are linear (including the BJT, since we are given that it always operates in its active region). The small-signal equivalents for the DC sources are short circuits, and those for the linear resistors are the resistors themselves. Finally, we will use the small-signal model for the BJT operating in its active region (developed in Example 8.6) illustrated in Figure 8.42c.

Proceeding with the third step of small-signal analysis, Figure 8.44 shows the small-signal circuit for the amplifier in which the components have been replaced by their respective

4. This step is not strictly necessary in our example because all the elements are linear (including the BJT, since we are given that it always operates in its active region). For linear elements, the small-signal model relationships are independent of their operating points. Compare, for example, the small-signal relations for the BJT and the MOSFET shown in Equations 8.62 and 8.10, respectively.

linearized equivalents, and in which small-signal branch variables have replaced the total variables.

The small-signal gain can now be determined by writing the node equation for the output node

$$\frac{v_O}{R_L} = -\beta i_b.$$

Substituting $i_b = v_{\rm in}/R_I$, we get

$$\frac{v_O}{R_L} = -\beta \frac{v_{\rm in}}{R_I}.$$

Simplifying, we obtain the small-signal gain of the BJT amplifier

$$\text{Small-signal gain} = \frac{v_O}{v_{\rm in}} = -\beta \frac{R_L}{R_I}. \tag{8.62}$$

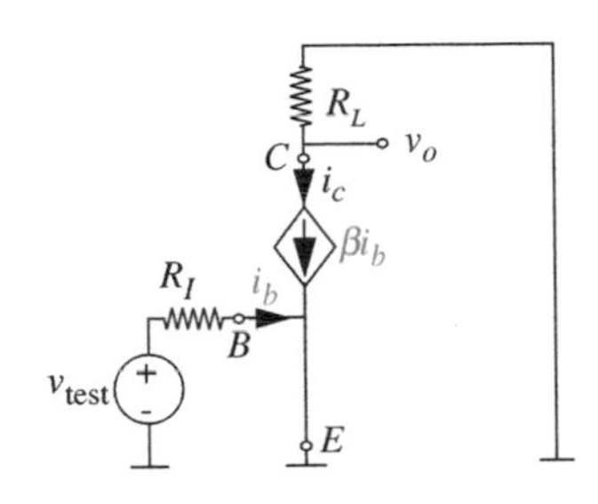


FIGURE 8.45 Applying a small-signal test voltage to the input port of the BJT amplifier to compute the small-signal input resistance.

Notice here that the gain of the BJT amplifier is independent of the operating point, provided the BJT operates in the active region. For a given BJT device (that is, a fixed value for $\beta$) the gain can be increased by increasing $R_L$ or decreasing $R_I$.

Finally, substituting $R_I = 100$ k$\Omega$, $R_L = 10$ k$\Omega$, $\beta = 100$, we obtain

$$\text{Small-signal gain} = -10.$$

This concludes our analysis.

---

EXAMPLE 8.8 SMALL-SIGNAL INPUT AND OUTPUT RESISTANCE OF THE BJT AMPLIFIER Let us first compute the small-signal input and output resistances of the common emitter BJT amplifier. The general approach to doing so is to turn off all independent sources and to apply a test voltage (or current) at the input or output port as appropriate and to measure the resulting current (or voltage). The ratio of the voltage to the current gives the resistance.

The input resistance $r_i$ is easily calculated. For an applied test voltage $v_{\rm in}$ (see Figure 8.45), the resulting current into the input B terminal $i_b$ is given by

$$i_b = \frac{v_{\rm test}}{R_I}.$$

Thus the input resistance $r_i$ is simply $R_I$.

As illustrated in Figure 8.46, we compute the output resistance by turning off all independent sources, and applying a small test voltage $v_{\rm test}$ at the output port and measuring the corresponding current $i_O$. The output resistance will be given by $r_{\rm out} = v_{\rm test}/i_O$.

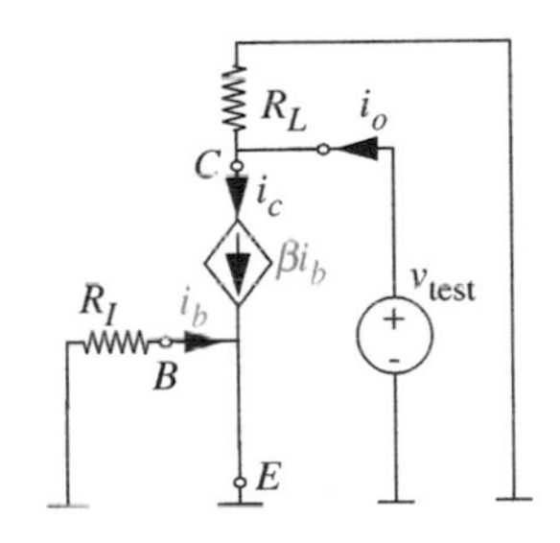


FIGURE 8.46 Applying a small-signal test voltage to the output port of the BJT amplifier to compute the small-signal output resistance.

In order to compute $r_{out}$, we apply KCL at the node labeled C shown in Figure 8.46. Summing all the currents going *into* node C, we get

$$i_o - \frac{v_{test}}{R_L} - \beta i_b = 0.$$

Since $i_b = 0$ (the voltage across $R_I$ is zero), we get

$$r_{out} = \frac{v_{test}}{i_o} = R_L.$$

---

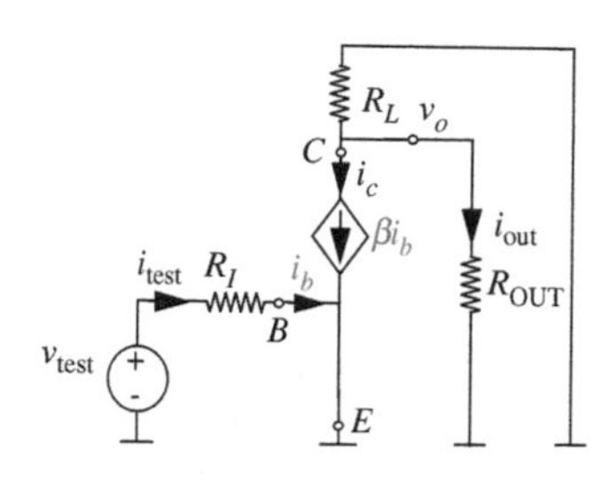


**FIGURE 8.47** Incremental circuit for the BJT amplifier including an external load resistor to facilitate current gain and power gain calculations.

EXAMPLE 8.9 SMALL-SIGNAL CURRENT GAIN AND POWER GAIN OF THE BJT AMPLIFIER In this example, let us compute the incremental current and power gain for the common emitter BJT amplifier. Both the current gain and the power gain are defined as the current or power supplied to an external load divided by the current or power supplied by an input source. Accordingly, as illustrated in Figure 8.47, let us add an external load resistance $R_{OUT}$ to our circuit to facilitate current and power gain measurements.

The incremental current gain is defined as the change in the output current ($i_{out}$) divided by the change in the input current ($i_{test}$), for a given external load resistance. We begin by writing the node equation for the node labeled C

$$i_{out} + i_c + \frac{v_o}{R_L} = 0. \tag{8.63}$$

We will obtain the desired relation between $i_{out}$ and $i_{test}$ if we can replace $i_c$ and $v_o$ in terms of $i_{test}$. From the BJT relation, we know that

$$i_c = \beta i_b = \beta i_{test}. \tag{8.64}$$

To determine $v_o$ in terms of $i_{test}$, observe that $v_o$ is the voltage drop across the parallel resistor pair comprising $R_L$ and $R_{OUT}$. In other words,

$$v_o = -i_c(R_L \| R_{OUT}).$$

Substituting for $i_c$, we get the desired relation between $v_o$ and $i_{test}$:

$$v_o = -\beta i_{test}(R_L \| R_{OUT}). \tag{8.65}$$

Substituting for $i_c$ and $v_o$ from Equations 8.64 and 8.65 into 8.63 we obtain

$$i_{out} + \beta i_{test} - \beta i_{test}\frac{(R_L \| R_{OUT})}{R_L} = 0.$$

Dividing throughout by $i_{test}$ and simplifying, we obtain the current gain as

$$\text{Current gain} = \frac{i_{out}}{i_{test}} = -\beta \frac{R_L}{R_L + R_{OUT}}. \quad (8.66)$$

Intuitively, we can also obtain the same current gain result in two short steps as follows: First, notice that the current $i_c$ is simply $i_{test}$ amplified by a factor $\beta$. Second, the fraction of the amplified current $\beta i_{test}$ that flows into $R_{OUT}$ is given by the current-divider relation from Equation 2.84 as the ratio of the opposite resistor $R_L$ divided by the sum of the two resistors $(R_L + R_{OUT})$.

Next, the incremental power gain is defined as the ratio of the power supplied into the output resistor $(v_o i_{out})$ and the power supplied by the input source $(v_{test} i_{test})$, for a given external load resistance. As suggested by Equation 8.38, the power gain is equivalent to the product of the current gain and the voltage gain for the BJT amplifier.

For the BJT amplifier that includes an output load resistance, the current gain is given by Equation 8.66. For reasons that will be obvious momentarily, we will rewrite the current gain in terms of the parallel combination of $R_L$ and $R_{OUT}$ as

$$\frac{i_{out}}{i_{test}} = -\beta \frac{(R_L \| R_{OUT})}{R_{OUT}}. \quad (8.67)$$

We can determine the voltage gain by including the effect of the output load resistance $R_{OUT}$ on the voltage gain equation of the BJT given by Equation 8.62. We do so by replacing the resistance $R_L$ in Equation 8.62 with the equivalent resistance of the parallel resistor pair $R_L$ and $R_{OUT}$ as

$$\frac{v_o}{v_{test}} = -\beta \frac{R_L \| R_{OUT}}{R_I}. \quad (8.68)$$

Taking the product of the current gain (Equation 8.67) and the voltage gain (Equation 8.68) and simplifying, we obtain

$$\text{Power gain} = \beta^2 \frac{(R_L \| R_{OUT})^2}{R_{OUT} R_I}.$$

---

EXAMPLE 8.10 SMALL SIGNAL OF THE OPERATIONAL AMPLIFIER CIRCUIT This example develops a small-signal model of the operational amplifier circuit shown in Figure 7.63 and previously discussed in Example 7.21. It then uses that model to determine the small-signal gain of the amplifier. The small-signal model and gain are determined for the bias conditions established by $V_{IN1} = V_{IN2} = 0$. Under these balanced bias conditions, $I_{D1} = I_{D2} = I/2$.

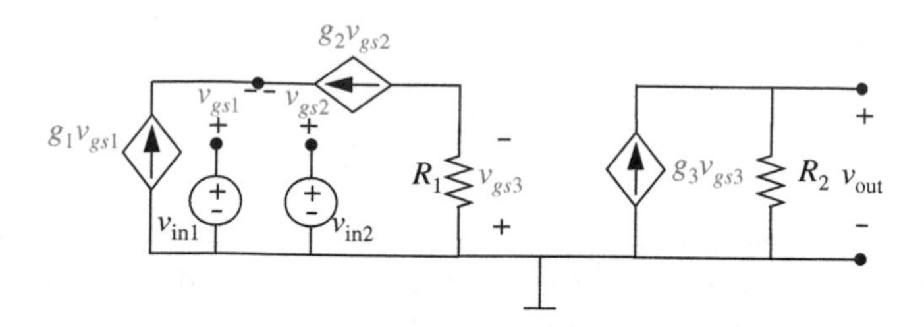


FIGURE 8.48 A small-signal model of the operational amplifier circuit.

Figure 8.48 shows a small-signal model of the operational amplifier shown in Figure 7.63. The three MOSFET transconductances $g_1$, $g_2$, and $g_3$ in Figure 8.48 are not yet determined.

Following the results summarized in Figure 8.10, the small-signal transconductances of the n-channel MOSFETs are given by

$$g_1 = K_n(V_{GS1} - V_T) \tag{8.69}$$

$$g_2 = K_n(V_{GS2} - V_T). \tag{8.70}$$

However, remember that we have chosen to bias the operational amplifier such that

$$I_{D1} = \frac{I}{2} = \frac{K_n}{2}(V_{GS1} - V_T)^2 \tag{8.71}$$

$$I_{D2} = \frac{I}{2} = \frac{K_n}{2}(V_{GS2} - V_T)^2. \tag{8.72}$$

Equations 8.71 and 8.72 can be substituted into Equations 8.69 and 8.70, respectively, to yield

$$g_1 = g_2 = \sqrt{K_n I}. \tag{8.73}$$

A similar small-signal model of the p-channel MOSFET can also be determined following the approach developed in Section 8.2. Specifically, taking the slope of Equation 8.67 at its bias point yields

$$- - i_d \approx K(V_{SG} + V_T)v_{sg} \tag{8.74}$$

and so the transconductance from $v_{sg}$ to $-i_d$ is in general given by

$$g = K(V_{SG} + V_T) = \sqrt{2K(-I_D)} \tag{8.75}$$

where the large-signal bias condition for the p-channel MOSFET has been used to derive the last equality. Applying this to the operational amplifier shown in Figure 8.48 yields

$$g_3 = \sqrt{2K_p(-I_{D3})}. \tag{8.76}$$

The small-signal model can now be used to determine the small-signal gain of the operational amplifier. Consider first the portion of the small-signal model that corresponds to the differential amplifier alone. KCL applied to the node between the two n-channel MOSFETs yields

$$i_{d1} + i_{d2} = g_1 v_{gs1} + g_2 v_{gs2} = 0. \tag{8.77}$$

Thus, an increase in one drain current in the differential amplifier is matched by an equal decrease in the other drain current since both drain currents must sum to $I$. Next, the application of KVL to the loop around the two MOSFETs through ground yields

$$v_{\text{in}1} - v_{\text{in}2} = v_{gs1} - v_{gs2}. \tag{8.78}$$

Finally, combining Equations 8.73, 8.77, and 8.78 with the observation from Figure 8.48 that $v_{sg3} = R_1 g_2 v_{gs2}$ yields

$$v_{sg3} = -\frac{R_1\sqrt{K_n I}}{2}(v_{\text{in}1} - v_{\text{in}2}) \tag{8.79}$$

as the small-signal gain of the differential amplifier.

Consider next the portion of the small-signal circuit that corresponds to the common-source stage built with the p-channel MOSFET. For this stage, the small-signal model shows that

$$v_{\text{out}} = R_2\sqrt{2K_p(-I_{D3})}v_{sg3} \tag{8.80}$$

where Equation 8.76 has been used to rewrite $g_3$. Note that the gain of this stage is positive because that gain is from $v_{sg3}$ to $v_{\text{out}}$.

Finally, Equations 8.79 and 8.80 can be combined to yield

$$v_{\text{out}} = \frac{R_1 R_2\sqrt{2K_n K_p I(-I_{D3})}}{2}(v_{\text{in}2} - v_{\text{in}1}) \tag{8.81}$$

as the small-signal gain of the unloaded operational amplifier.

In operational amplifier parlance (see Chapter 15), from Equation 8.81 we see that $v_{\text{IN}1}$ and $v_{\text{IN}2}$ play the roles of $v_-$ and $v_+$, respectively.

EXAMPLE 8.11 MORE ON THE SMALL-SIGNAL MODEL OF THE OPERATIONAL AMPLIFIER We will now work a numerical example related to the operational amplifier design described in Example 8.10, assuming that $-I_{D3} = 0.5$ mA.

Substitution of this value of $-I_{D3}$ and the parameters from Example 8.10, into Equation 8.81 yields

$$v_{out} = 50\sqrt{2}(v_{in2} - v_{in1}). \tag{8.82}$$

Thus, the small-signal gain of the operational amplifier is approximately 71.

## 8.3 SUMMARY

- This chapter expanded on our treatment of small-signal models, focusing on the model for three-terminal devices and amplifiers. As first introduced in Section 4.5, small-signal analysis applies when devices and circuits that are possibly nonlinear are operated over a very narrow range. Small-signal analysis finds a *piecewise linear* model that ensures maximum accuracy of fit over that *narrow operating range*. The principal benefit of small-signal models is that the small-signal variables display linear $v$–$i$ relations over the narrow operating range, thereby enabling the use of all of our linear analysis techniques such as superposition, Thévenin, and Norton.

- This chapter also introduced the small-signal circuit model. The small-signal circuit facilitates small-signal analysis by creating a circuit that is representative of the original large-signal circuit and involves only its small-signal variables. The small-signal circuit can be derived from the original circuit by executing the following procedure:

  1. Set each source to its operating-point value, and determine the operating-point branch voltages and currents for each component in the circuit. This step involves a large-signal analysis that is possibly nonlinear.

  2. Determine the linearized small-signal behavior of each component about its operating point, and select a linear component to represent this behavior.

  3. Replace each original component in the circuit with its linearized equivalent (also called the small-signal equivalent model) and re-label the circuit with the small-signal branch variables. The resulting circuit is the desired small-signal model.

- The small-signal equivalent model for an independent DC voltage source is a short circuit, while that for an independent DC current source is an open circuit. The small-signal equivalent model for a resistor is the resistor itself. The small-signal model for a MOSFET is shown in Figure 8.10.

## EXERCISES

EXERCISE 8.1 Consider the amplifier shown in Figure 8.49. The MOSFET operates in its saturation region and is characterized by the parameters $V_T$ and $K$. The input voltage $v_I$ comprises the sum of a DC bias voltage $V_I$ and a sinusoid of the form $v_i = A\sin(\omega t)$. Assume that $A$ is very small compared to $V_I$. Let the output voltage $v_O$ comprise a DC bias term $V_O$ and a small-signal response term $v_o$.

a) Determine the output operating point voltage $V_O$ for the input bias of $V_I$.

b) Determine the small-signal gain of the amplifier.

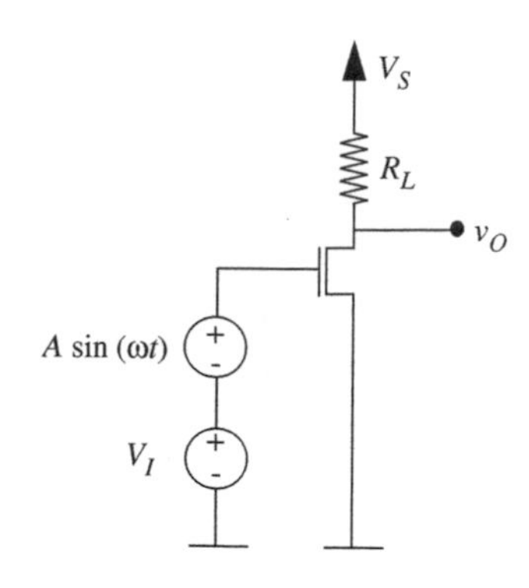


FIGURE 8.49

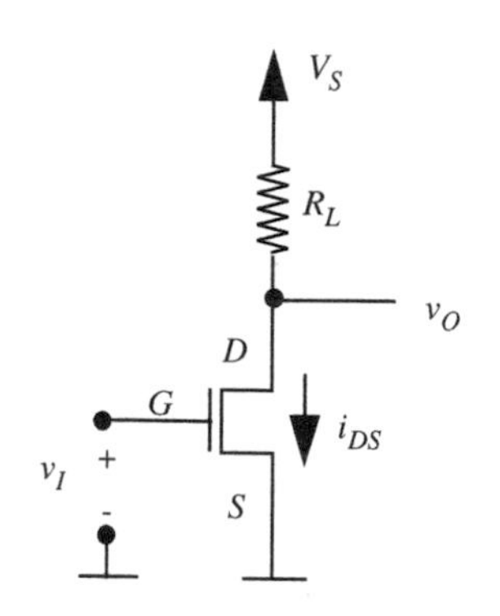


FIGURE 8.50

c) Draw the form of the input and output voltages as a function of time, clearly showing the DC and time-varying small-signal components.

EXERCISE 8.2 Develop the small-signal model for a two-terminal device formed by a MOSFET with its gate tied to its drain, operating under the saturation discipline, with parameters $V_T$ and $K$.

EXERCISE 8.3 Develop the small-signal model for a two-terminal device formed between the drain and source terminals of a MOSFET with a 2 volt DC source connected between its gate and source terminals ($V_{GS} = 2$ V). Assume the MOSFET operates under the saturation discipline. Assume further that $V_T = 1$ volt for the MOSFET.

EXERCISE 8.4 Consider the MOSFET amplifier shown in Figure 8.50. Assume that the amplifier is operated under the saturation discipline. In its saturation region, the MOSFET is characterized by the equation

$$i_{DS} = \frac{K}{2}(v_{GS} - V_T)^2$$

where $i_{DS}$ is the drain-to-source current when a voltage $v_{GS}$ is applied across its gate-to-source terminals.

a) Write an expression relating $v_O$ to $v_I$. What is its operating-point output voltage $V_O$, given an input operating-point voltage of $v_I$? What is the corresponding operating-point current $I_{DS}$?

b) Assuming an operating-point input voltage of $V_I$, derive the expression relating the small-signal output voltage $v_o$ to the small-signal input $v_i$ from the relationship between $v_O$ and $v_I$. What is the small-signal gain of the amplifier at the input operating point of $V_I$?

c) Draw the small-signal equivalent circuit for the amplifier based on the SCS model of the MOSFET assuming the operating-point input voltage is $V_I$.

d) Derive an expression for the small-signal gain of the amplifier from the small-signal equivalent circuit. Verify that the gain computed from the small-signal equivalent circuit is identical to the gain computed in part (b).

e) By what factor must $R_L$ change to double the small-signal gain of the amplifier? What is the corresponding change in the output bias voltage?

f) By what factor must $V_I$ change to double the small-signal gain of the amplifier? What is the corresponding change in the output bias voltage?

EXERCISE 8.5 Consider again the MOSFET amplifier shown in Figure 8.50. Assume as before that the MOSFET is operated under the saturation discipline, and that its parameters are $V_T$ and $K$.

a) What is the range of valid input voltages for the amplifier? What is the corresponding range of valid output voltages?

b) Assuming we desire to use voltages of the form $A\sin(\omega t)$ as AC inputs to the amplifier, determine the input bias point $V_I$ for the amplifier that will allow maximum input swing under the saturation discipline. What is the corresponding output bias point voltage $V_O$?

c) What is the largest value of $A$ that will allow saturation region operation for the bias point determined in (b)?

d) What is the small-signal gain of the amplifier for the bias point determined in (b)?

e) Suppose $A$ is small compared to $V_I$. Write an expression for the small-signal output voltage $v_o$ for the bias point determined in (b).

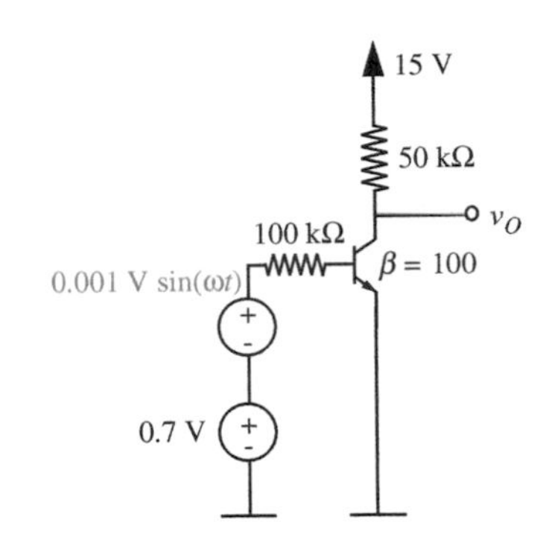


FIGURE 8.51

EXERCISE 8.6 Consider once more the MOSFET amplifier shown in Figure 8.50. Assume as before that the amplifier is operated under the saturation discipline, and that its parameters are $V_T$ and $K$.

a) Using the small-signal circuit model of the amplifier, and assuming an input bias voltage $V_I$, determine the small-signal output resistance of the amplifier. That is, determine the equivalent resistance of the amplifier at the output port of its small-signal model with $v_i \equiv 0$.

b) Develop a Thévenin equivalent model for the small-signal amplifier as observed at its output port.

c) What is its input resistance? That is, determine the equivalent resistance of the amplifier at the input port of its small-signal model.

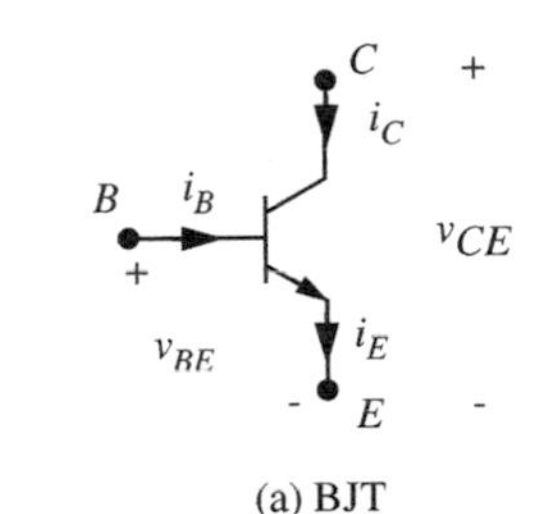


(a) BJT

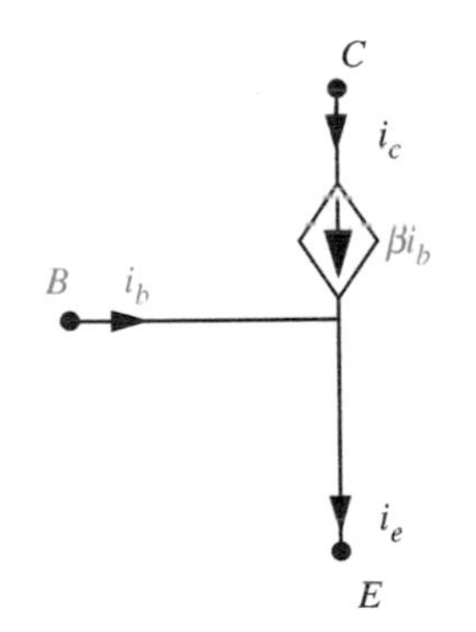


(b) BJT small-signal model

FIGURE 8.52

EXERCISE 8.7 Consider the common emitter BJT amplifier shown in Figure 8.51. The input voltage $v_I$ comprises the sum of a DC bias voltage $V_I = 0.7$ V and a sinusoid of the form $v_i = A\sin(\omega t)$, where $A = 0.001$ V. For the values shown, you may assume that $A$ is very small compared to $V_I$. You may further assume that the BJT always operates in its active region. Figure 8.52 shows a small-signal model for the BJT operating in its active region. Let the output voltage $v_O$ comprise a DC bias term $V_O$ and a small-signal response term $v_o$.

a) Determine the output operating-point voltage $V_O$ for the input bias of $V_I = 0.7$ V.

b) Draw the small-signal equivalent circuit for the amplifier.

c) Determine the small-signal gain of the amplifier.

d) What is the value of $v_o$, the small-signal component of the output, given the small-signal input shown in Figure 8.51?

e) Determine the small-signal input and output resistances of the amplifier.

f) Determine the small-signal current and power gain of the amplifier, assuming that the amplifier drives a load $R_O = 50\ \text{k}\Omega$ that is connected between the output node and ground.

## PROBLEMS

**PROBLEM 8.1** This problem studies the small-signal analysis of the MOSFET amplifier discussed in Problem 7.3 (Figure 7.75).

a) First, consider biasing the amplifier. Determine $V_{IN}$, the bias component of $v_{IN}$, so that $v_{OUT}$ is biased to $V_{OUT}$ where $0 < V_{OUT} < V_S$. Find $V_{MID}$, the bias component of $v_{MID}$ in the process.

b) Next, let $v_{IN} = V_{IN} + v_{in}$ where $v_{in}$ is considered to be a small perturbation of $v_{IN}$ around $V_{IN}$. Make the substitution for $v_{IN}$ and linearize the resulting expression for $v_{OUT}$. Your answer should take the form $v_{OUT} = V_{OUT} + v_{out}$, where $v_{out}$ takes the form $v_{out} = Gv_{in}$. Note that $v_{out}$ is the small-signal output and $G$ is the small-signal gain. Derive an expression for $G$.

c) For what value of $V_{IN}$ is $v_{OUT}$ biased to $V_{OUT} = V_S/2$? For this value of $V_{IN}$, evaluate $G_m$ using the numerical parameters given in Problem 7.2. You should find that this gain is the slope of the input-output graph from Problem 7.3 evaluated at the bias point.

**PROBLEM 8.2** Consider again the buffer described in Problem 7.5 (Figure 7.76). Perform a small-signal analysis of this circuit according to the following steps. Assume that the MOSFET operates in its saturation region and continue to use the SCS MOSFET model with parameters $V_T$ and $K$.

a) Draw the small-signal circuit model of the buffer.

b) Show that the small-signal transconductance $g_m$ of the MOSFET is given by

$$g_m = K(V_{IN} - V_{OUT} - V_T)$$

where $V_{IN}$ and $V_{OUT}$ are the bias, or operating-point, input and output voltages, respectively.

c) Determine the small-signal gain of the buffer. That is, determine the ratio $v_{out}/v_{in}$.

d) Determine the small-signal output resistance of the buffer. That is, determine the equivalent resistance of the buffer at the output port of its small-signal model with $v_{in} \equiv 0$.

e) Assume that $V_T = 1$ V, $K = 2$ mA/V$^2$, $R = 1\ \text{k}\Omega$, and $V_S = 10$ V. Under this assumption, design the input bias voltage to satisfy the following two objectives: First, MOSFET operation must remain within the saturation region for $|v_{in}| \leq 0.25$ V. Second, the output resistance of the small-signal model must be minimized.

f) Again assume that $V_T = 1$ V, $K = 2$ mA/V$^2$, $R = 1$ k$\Omega$, and $V_S = 10$ V. For $V_{IN} = 3$ V, compute the small-signal gain and output resistance.

g) Determine the small-signal input resistance of the buffer. That is, determine the equivalent resistance of the buffer at the input port of its small-signal model.

PROBLEM 8.3 This problem studies the small-signal analysis of the ZFET amplifier from Problem 7.6 (Figure 7.77). Assume that the amplifier is biased at an input voltage $V_{IN}$ such that the ZFET exhibits saturated operation; the corresponding bias output voltage is $V_{OUT}$. For this case, derive the small-signal voltage gain $v_{out}/v_{in}$ of the amplifier.

PROBLEM 8.4 The circuit shown in Figure 8.4 delivers a nearly constant current to its load despite the fact that the power supply is noisy. The noise is modeled by the small signal $v_s$ superimposed on the constant-supply voltage $V_S$. Thus, $V_S$ and $v_s$ are the large-signal and small-signal components of the total power supply voltage $v_S$, respectively. $I_L$ and $i_l$ are the large-signal and small-signal components of the load

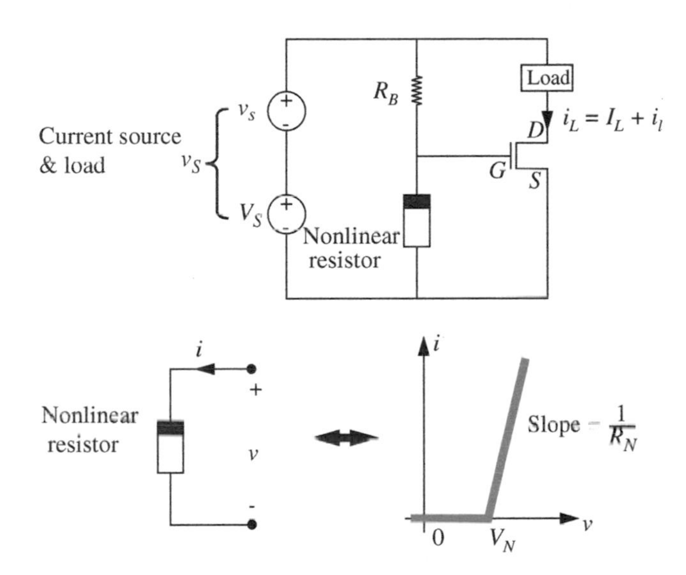


FIGURE 8.53

current $i_L$, respectively. The noise $v_s$ in the power supply voltage satisfies $v_s \ll V_S$, and is responsible for the presence of $i_l$ in $i_L$.

The current source contains a MOSFET which operates in its saturation region such that $i_{DS} = \frac{K}{2}(v_{GS} - V_T)^2$. The current source also contains a nonlinear resistor whose terminal characteristics are described graphically next. Assume that $V_S > V_N > V_T$.

a) Assume $v_s = 0$. Determine $V_{GS}$, the large-signal component of $v_{GS}$, in terms of $R_B$, $R_N$, $V_N$, and $V_S$.

b) Following the result of part (a), determine $I_L$ in terms of $R_B$, $R_N$, $V_N$, $V_S$, $K$, and $V_T$.

c) Now assume that $v_s \neq 0$. Draw a small-signal circuit model for the combined circuit comprising the power supply, current source and load, with which $i_l$ can be found from $v_s$. Clearly label the value of each component in the circuit model.

d) Using the small-signal model from part (c), determine the ratio $i_l/v_s$.

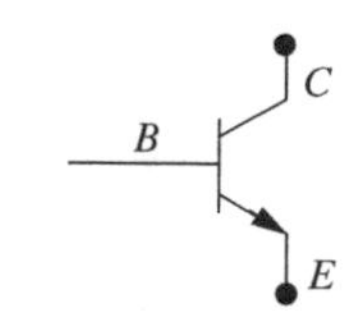


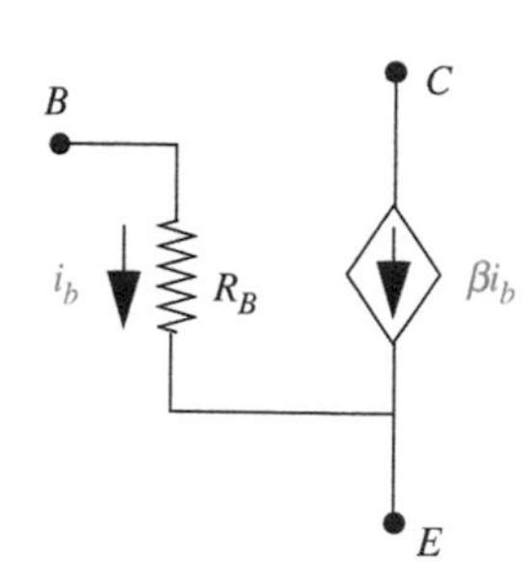


FIGURE 8.54

PROBLEM 8.5 Figure 8.54 depicts a bipolar junction transistor (BJT). Recall that a BJT has three terminals called the base (B), the collector (C), and the emitter (E). Figure 8.54 also shows an alternative small-signal model for the BJT operating in its active region. This model is slightly different from the small-signal BJT model discussed in this chapter in that it includes a base resistance $R_B$. In the model shown in the figure, $\beta$ is a constant.

a) Draw the small-signal equivalent circuit for the BJT amplifier shown in Figure 8.55. Use the small-signal equivalent circuit to derive the small-signal gain of the amplifier.

b) Draw the small-signal equivalent circuit for the BJT amplifier shown in Figure 8.56. Notice that the resistor divider provides the necessary bias voltage. Use the small-signal equivalent circuit to derive the small-signal gain of the amplifier.

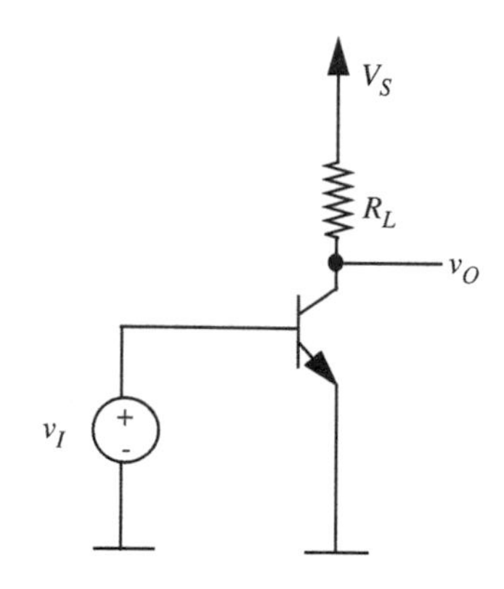


FIGURE 8.55

PROBLEM 8.6 Consider the MOSFET-based amplifier circuit discussed in Problem 7.8 (Figure 7.79). Assuming an input bias point voltage $V_I$, draw the small-signal circuit equivalent of the amplifier. Determine the small-signal gain of the amplifier. Assume throughout that the MOSFET operates in its saturation region.

PROBLEM 8.7 Consider again the amplifier circuit discussed in Problem 7.8 (Figure 7.79). Suppose that the amplifier is biased such that $v_I = v_O$ at the bias point. Draw the small-signal circuit equivalent of the amplifier assuming this bias point. Determine the small-signal gain of the amplifier at this bias point. Assume that the MOSFET operates in its saturation region.

PROBLEM 8.8 Consider the common gate amplifier circuit shown in Figure 7.82, and analyzed earlier in Problem 7.11. Assume that the MOSFET operates in its saturation region, and is characterized by the parameters $V_T$ and $K$.

a) Draw the SCS equivalent circuit by replacing the MOSFET by its SCS model.

b) Determine the output operating-point voltage $V_{OUT}$ and operating-point current $I_D$ in terms of an input operating-point voltage $V_{IN}$.

c) Assuming an input bias point voltage $V_{IN}$, draw the small-signal model of the amplifier.

d) Determine the small-signal gain $v_{out}/v_{in}$ of the amplifier.

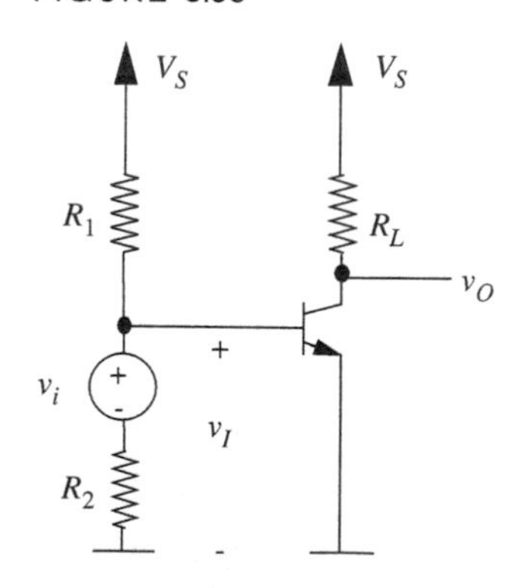


FIGURE 8.56

e) Determine the small-signal output resistance of the amplifier. That is, determine the equivalent resistance of the amplifier at the output port of its small-signal model with $v_i \equiv 0$. Is the small-signal output resistance greater than, less than, or equal to that of the "common source" amplifier shown in Figure 8.50?

f) Determine the small-signal input resistance of the amplifier. That is, determine the equivalent resistance of the amplifier at the input port of its small-signal model. Is the small-signal input resistance greater than, less than, or equal to that of the "common source" amplifier shown in Figure 8.50?

PROBLEM 8.9 Consider the circuit illustrated in Figure 7.86 and analyzed in Problem 7.15. Assume that the MOSFET operates in its saturation region, and is characterized by the parameters $V_T$ and $K$.

a) Draw the SCS equivalent circuit by replacing the MOSFET by its SCS model.

b) Determine the output operating-point voltage $V_O$ and operating-point current $I_D$ in terms of an input operating-point voltage $V_I$.

c) Assuming an input bias point voltage $V_I$, draw the small-signal model.

d) Determine the small-signal gain $v_o/v_i$.

e) Determine the small-signal output resistance.

f) Determine the small-signal input resistance.

PROBLEM 8.10 Consider the circuit illustrated in Figure 7.87 and analyzed in Problem 7.16. Assume that the MOSFET operates in its saturation region, and is characterized by the parameters $V_T$ and $K$.

a) Draw the SCS equivalent circuit by replacing the MOSFET by its SCS model.

b) Determine the output operating-point voltage $V_O$ and operating-point current $I_D$ in terms of an input operating-point voltage $V_I$.

c) Assuming an input bias point voltage $V_I$, draw the small-signal model.

d) Determine the small-signal gain $v_o/v_i$.

e) Determine the small-signal output resistance.

f) Determine the small-signal input resistance.

PROBLEM 8.11 This problem studies the small-signal analysis of the amplifier analyzed in Problem 7.14 (see Figure 7.85). Assume that the MOSFET operates in its saturation region, and is characterized by the parameters $V_T$ and $K$.

a) Draw the small-signal equivalent circuit of the amplifier driving the load resistor $R_E$, assuming an input bias voltage $V_I$.

b) Determine the small-signal gain of the amplifier when it is driving the load $R_E$.

PROBLEM 8.12 This problem studies the small-signal analysis of the circuit analyzed in Problem 7.17 (see Figure 7.88). Assume that the MOSFET operates in its saturation region, and is characterized by the parameters $V_T$ and $K$.

a) Draw the small-signal equivalent circuit assuming an input bias voltage $V_I$. What is the value of $g_m$ for the MOSFET under the given biasing conditions?

b) Determine the small-signal voltage gain $v_o/v_i$. What does the $v_o/v_i$ expression simplify to when each of $g_m R_1$, $g_m R_2$, and $g_m R_L$ is much greater than 1?

PROBLEM 8.13 This problem studies the small-signal analysis of the source follower (or common collector) BJT circuit analyzed in Problem 7.18 (see Figure 7.89). Assume that the BJT operates in its active region throughout this problem.

a) Determine the output operating-point voltage $V_O$ and operating-point current $I_E$ in terms of an input operating-point voltage $V_I$.

b) Assuming an input bias point voltage $V_I$, draw the small-signal model of the source-follower amplifier.

c) Determine the small-signal gain $v_o/v_i$ of the amplifier.

d) Determine the small-signal output resistance of the source follower amplifier. Is this resistance greater than, less than, or equal to that of the "common emitter" amplifier analyzed in Exercise 8.7 and shown in Figure 8.51?

e) Determine the small-signal input resistance of the amplifier. Is the input resistance greater than, less than, or equal to that of the "common emitter" amplifier shown in Figure 8.51?

f) Determine the small-signal current and power gain of the source follower amplifier. Assume for this part that the amplifier is driving an output load of $R_O$ connected between the output node and ground.

PROBLEM 8.14 Consider again the compound three-terminal device formed by connecting two BJTs in the configuration shown in Figure 7.90 (Problem 7.19). This problem relates to the small-signal analysis of this device. Assume that the two BJTs are identical, each with $\beta = 100$, and that each of the BJTs operates in the active region.

a) Draw the active-region equivalent circuit of the compound BJT by replacing each of the BJTs by the piecewise linear (large signal) model shown in Exercise 7.8. Clearly label the $C'$, $B'$, and $E'$ terminals.

b) Develop a small-signal model containing a single dependent current source for the compound device by linearizing the circuit model in (a) and simplifying suitably.

# CHAPTER 9

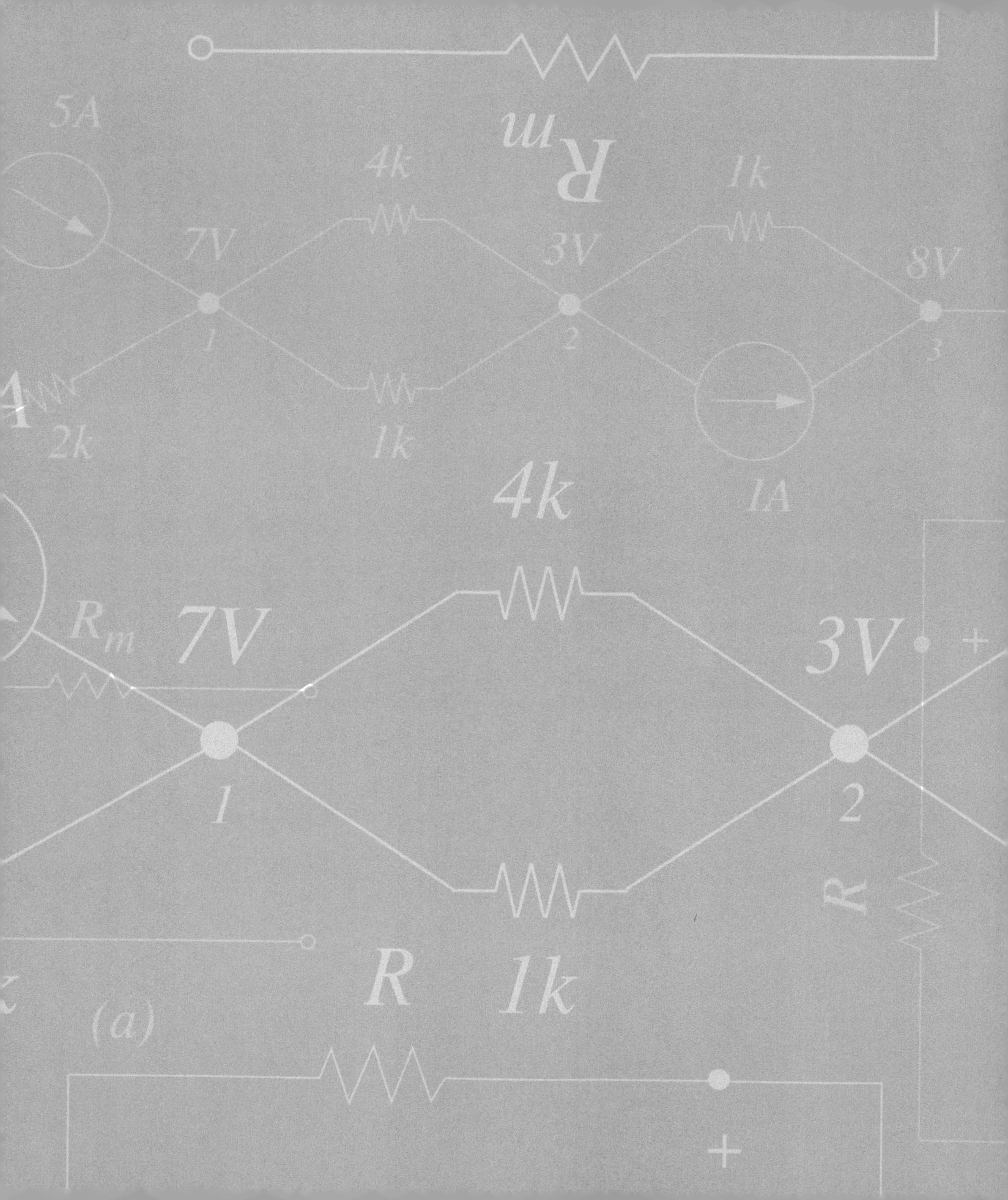
5A
4k
1k
7V
3V
8V
1
2
3
2k
1k
1A
4k
$R_m$
7V
3V
+
1
2
R
R
1k
(a)
+

# ENERGY STORAGE ELEMENTS

# 9

To this point in our study of electronic circuits, time has not been important. The analyses and designs we have performed so far have been static, and all circuit responses at a given time have depended only on the circuit inputs at that time. An important consequence of this is that our circuits have so far responded to input changes infinitely fast. This of course does not happen in reality. Circuits do take time to respond to their inputs, and this delay is often of significant importance.

As an example of circuit delays, and the importance of time in describing the response of a circuit, consider the two cascaded inverters shown in Figure 9.1. The ideal response of the first inverter, based on our analysis of electronic circuits to this point, is shown in Figure 9.2. A square-wave input yields an inverted square-wave output. However, in reality, the output shown in Figure 9.3 is more likely to occur, which is a much more complex function of time. This example is discussed in detail in Section 10.4, where we will show that the complex time behavior shown in Figure 9.3 directly relates to the speed at which circuits can operate. In this chapter, we will lay the foundation for that discussion.

In order to explain the temporal behavior of circuit responses such as that shown in Figure 9.3, we must introduce two new elements, namely *capacitors* and *inductors*. For example, we shall see that it is a capacitance internal to the MOSFET that is responsible for the non-ideal inverter response shown in Figure 9.3. For simplicity, we did not model that characteristic of the MOSFET in earlier chapters, but we will begin to do so now in Section 9.3.1.

There are other ways in which a capacitance or an inductance can inadvertently slow down a circuit. One way is shown in Figure 9.4. This figure

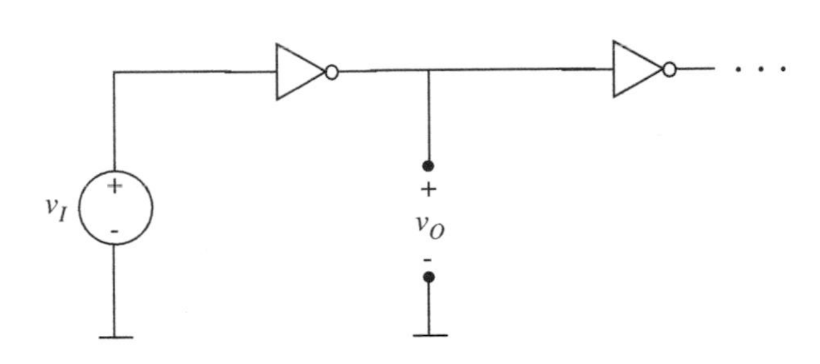


**FIGURE 9.1** Two cascaded inverters.

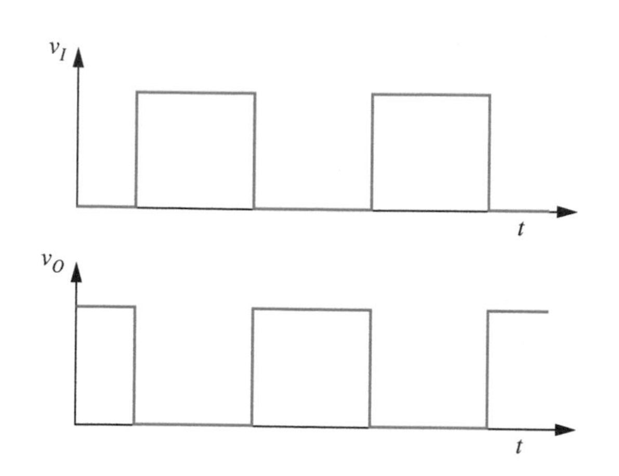


FIGURE 9.2 Ideal response of the first inverter to a square-wave input.

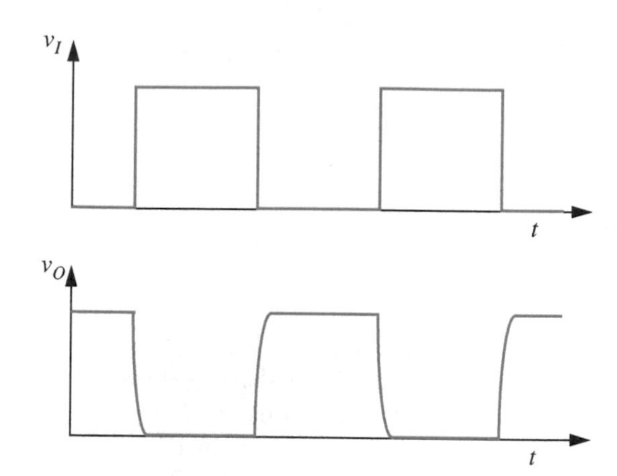


FIGURE 9.3 Observed response of the first inverter to a square-wave input.

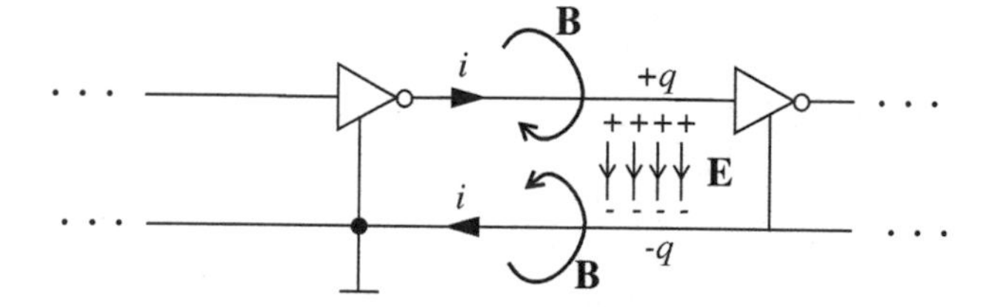


FIGURE 9.4 The behavior of a real interconnect between two inverters.

shows two inverters communicating over a long interconnect. As we discussed in Chapter 1, within our lumped-circuit abstraction, the interconnect is perfect. Specifically, by the definition of the lumped circuit abstraction (see Section 1.2), the wires interconnecting the elements have no resistance. Furthermore, by the lumped matter discipline which underlies the lumped circuit abstraction, the wires and other circuit elements store no electric charge and link no magnetic flux outside the elements. Reality, however, is different, and in some cases this difference is important. As Figure 9.4 shows, any interconnect having a potential difference with its surroundings actually stores an electric charge $q$ that sources an electric field **E** between that charge and its image. Furthermore, in order to supply the charge, a current $i$ must flow around the interconnect loop. This current in turn generates a magnetic flux density **B** that is linked by the loop. So, real interconnects do store electric charge and do link external magnetic flux, thereby appearing to violate the lumped matter discipline. They will also exhibit a nonzero resistance. These factors can all contribute to a reduction in the speed of the circuit as a whole, and at times it is important to study these effects.

Reality now presents us with a dilemma. On the one hand, we wish to work within the framework of the lumped circuit abstraction so that the circuits we study all fit within this easily-managed framework. On the other hand, we should not be forced to ignore circuit effects, in this case parasitic resistance, capacitance, and inductance, that significantly affect circuit performance. The resolution of this dilemma is the modeling compromise mentioned in Chapter 1. Figure 1.27 in Chapter 1 used an ideal wire in series with a lumped resistance to model a physical wire with some parasitic resistance. Similarly, we will introduce lumped capacitors and lumped inductors to model the effect of the charge and the flux. As illustrated in Figure 9.5, a capacitor comprising a pair of parallel plates collects the positive and negative charge on its plates and effectively models the distributed charge. Notice that because the capacitor contains equal positive and negative charges the net charge within the capacitor element is zero, thereby satisfying the lumped matter discipline. Thus, the capacitor can be viewed as a lumped element. In like manner, we will introduce a lumped inductor to model the effect of the flux linked with the wires as illustrated in Figure 9.6. The lumped matter discipline is satisfied because the flux is entirely contained inside the lumped inductor, and there is no net flux outside the element.

By using lumped resistors, capacitors, and inductors to model the effect of the resistance, charge, and flux associated with the physical wiring of the

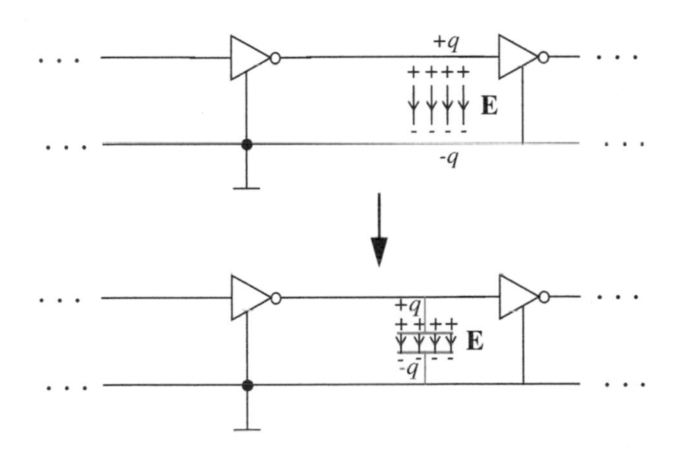


FIGURE 9.5 The capacitor models the effect of the distributed charge.

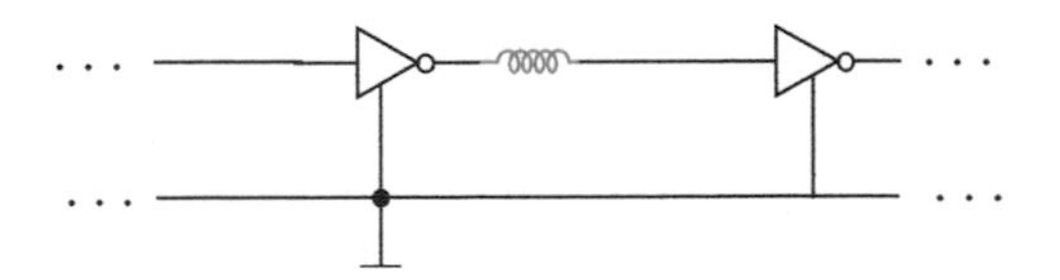

FIGURE 9.6 The inductor models the effect of the flux.

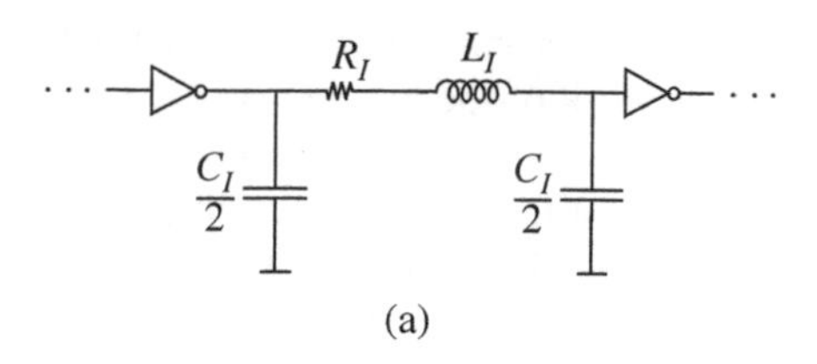


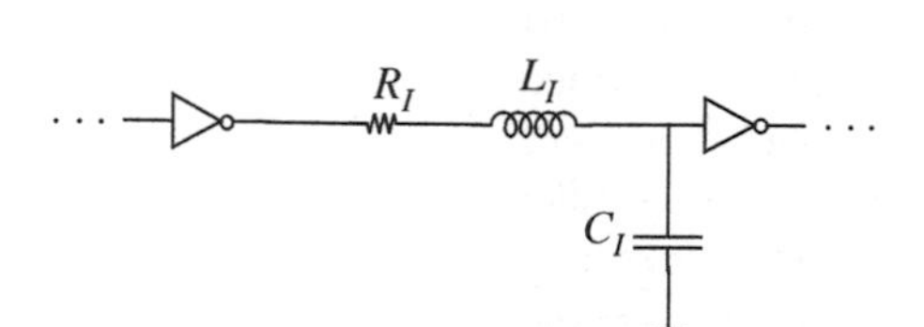


**FIGURE 9.7** Capturing the parasitic effects of resistance, charge, and flux through the use of resistors, capacitors, and inductors, respectively. Capacitors and inductors are formally introduced in Section 9.1.

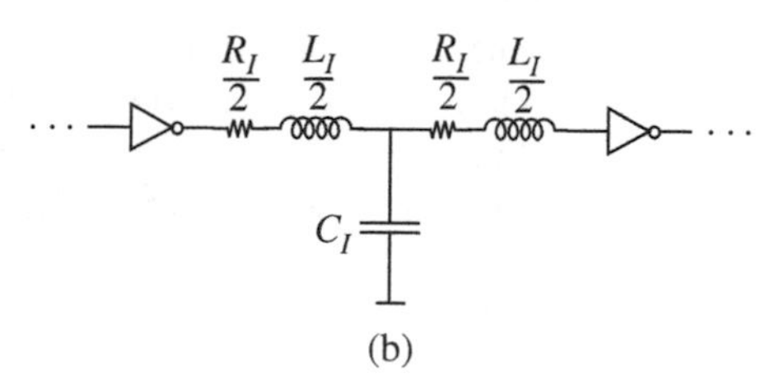


**FIGURE 9.8** Two different lumped models for an interconnect that account for interconnect resistance, capacitance, and inductance.

circuit, as shown in Figure 9.7, the wiring within the augmented circuit model remains perfect in keeping with the lumped circuit abstraction. In the figure, the interconnect resistance, capacitance, and inductance are $R_I$, $C_I$, and $L_I$ in total, respectively.

Figure 9.7 represents one of the simplest models used to model real interconnects. For more accuracy, since we can use as many additional lumped elements as we wish, we can arbitrarily approach the distributed modeling limit, although in general this is not necessary. For example, the two models shown in Figure 9.8 do a better job of modeling reality. The interconnect model in Figure 9.8a is a "Π" model in which the resistance and inductance is placed between the split capacitance. The interconnect model in Figure 9.8b is a "T" model in which the capacitance is placed between the split resistance and inductance. As discussed in Section 9.3.1, we will adopt a similar lumped modeling approach to the capacitances at work within the MOSFET.

From the preceding discussion it might appear that capacitors and inductors appear only as parasitics in circuits, causing undesirable delays. This is far from the truth. While they can and do act in that role, they are also often purposefully introduced into circuits, both as discrete devices on breadboards and printed-circuit boards, and as integrated-circuit components on a chip (see Figures 9.9 and 9.10 for examples of capacitors and inductors, respectively). For example, they are the cornerstones of memories, filters, samplers, and energy processing circuits. We shall see many examples of these in future chapters as well. Thus, we have many reasons to study capacitors and inductors.

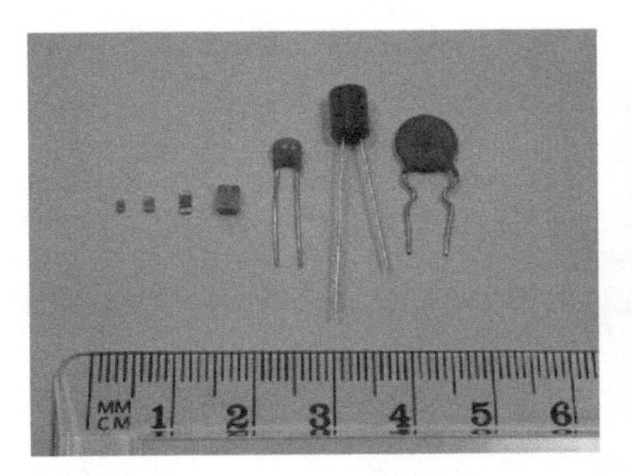


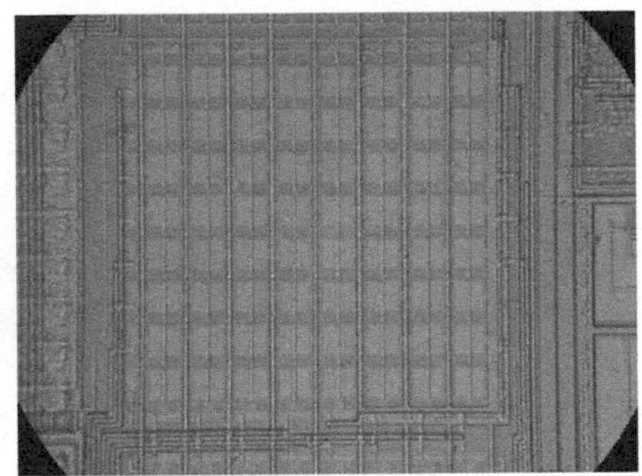

**FIGURE 9.9** Examples of discrete capacitors (left) and integrated-circuit capacitors (right). The image on the right shows a small region of the Maxim MAX1062 analog-to-digital converter chip and depicts an array of polysilicon-to-polysilicon capacitors, each measuring 15.9 $\mu$m by 15.9 $\mu$m. (Photograph Courtesy of Maxim Integrated Products.)

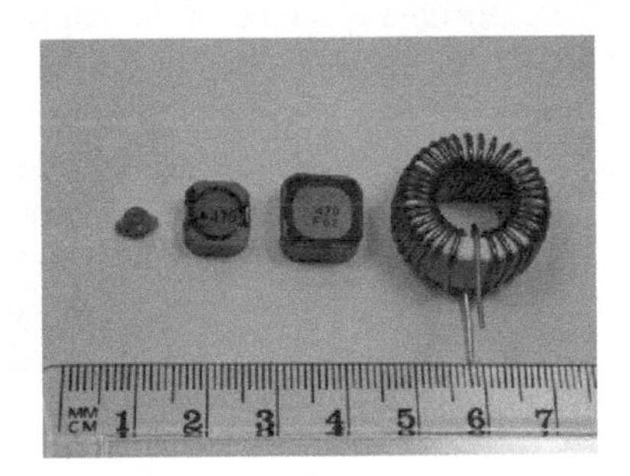


**FIGURE 9.10** Examples of discrete inductors. (Photograph Courtesy of Maxim Integrated Products.)

## 9.1 CONSTITUTIVE LAWS

In this section, we formally introduce the capacitor and inductor in the abstract, and develop the constitutive laws that relate their branch variables. Capacitors and inductors, which are the electric and magnetic duals of each other, differ from resistors in several significant ways. Most importantly, their branch variables do not depend algebraically upon each other. Rather, their relations involve temporal derivatives and integrals. Thus, the analysis of circuits containing capacitors and inductors involve differential equations in time. To emphasize this, we will explicitly show the time dependence of all variables in this chapter.

### 9.1.1 CAPACITORS

To understand the behavior of a capacitor, and to illustrate the manner in which a lumped model can be developed for it, consider the idealized two-terminal linear capacitor shown in Figure 9.11. In this capacitor each terminal is connected to a conducting plate. The two plates are parallel and are separated by a gap of length $l$. Their area of overlap is $A$. Note that these dimensions will be functions of time if the geometry of the capacitor varies. The gap is filled with an insulating linear dielectric having permittivity $\epsilon$.

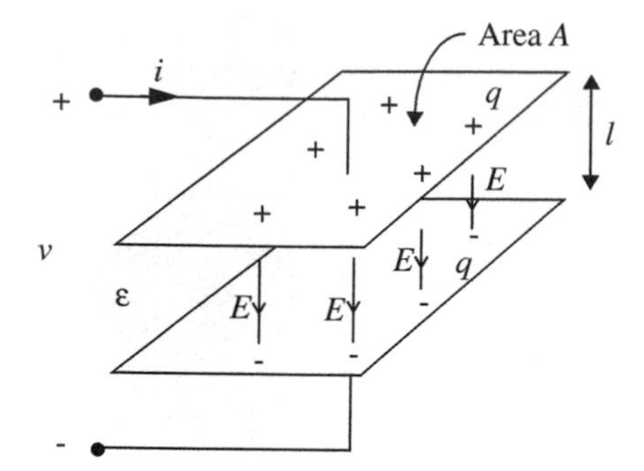


FIGURE 9.11 An idealized parallel-plate capacitor.

As current enters the positive terminal of the capacitor, it transports the electric charge $q$ onto the corresponding plate; the unit of charge is the Coulomb [C]. Simultaneously, an identical current exits the negative terminal and transports an equal charge off the other plate. Thus, although charge is separated within the capacitor, no net charge accumulates within it, as is required for lumped circuit elements by the lumped matter discipline discussed in Chapter 1.

The charge $q$ on the positive plate and its image charge $-q$ on the negative plate produce an electric field within the dielectric. It follows from Maxwell's Equations and the properties of linear dielectrics that the strength $E$ of this field is

$$E(t) = \frac{q(t)}{\epsilon A(t)}, \tag{9.1}$$

and its direction points from the positive plate to the negative plate. The electric field can then be integrated across the dielectric from the positive plate to the negative plate to yield

$$v(t) = l(t)E(t). \tag{9.2}$$

Combining Equations 9.1 and 9.2 then results in

$$q(t) = \frac{\epsilon A(t)}{l(t)} v(t). \tag{9.3}$$

We define

$$C(t) = \frac{\epsilon A(t)}{l(t)} \tag{9.4}$$

where $C$ is the capacitance of the capacitor having the units of Coulombs/Volt, or Farads [F]. Substituting for the capacitance in Equation 9.3, we get

$$q(t) = C(t)v(t). \tag{9.5}$$

In contrast to the resistor, which exhibits an algebraic relation between its branch current and voltage, the capacitor does not. Rather, it exhibits an algebraic relation between its branch voltage and its stored charge. Had the dielectric not been linear, this relation would have been nonlinear. While some capacitors exhibit such nonlinear behavior, we will focus only on linear capacitors.

The rate at which charge is transported onto the positive plate of the capacitor is

$$\frac{dq(t)}{dt} = i(t). \tag{9.6}$$

From Equation 9.6 we see that the Ampere is equivalent to a Coulomb/second. Equation 9.6 can be combined with Equation 9.5 to yield

$$i(t) = \frac{d(C(t)v(t))}{dt} \tag{9.7}$$

which is the element law for an ideal linear capacitor. Unless stated otherwise, we will assume in this text that capacitors are both linear and time-invariant. For linear, time-invariant capacitors, Equations 9.5 and 9.7 reduce to

$$q(t) = Cv(t) \tag{9.8}$$

$$i(t) = C\frac{dv(t)}{dt}, \tag{9.9}$$

respectively, *with the latter being the element law for a linear time-invariant capacitor.*[1]

The symbol for an ideal linear capacitor is shown in Figure 9.12. It is chosen to represent the parallel-plate capacitor shown in Figure 9.11. Also shown in Figure 9.12 is a graph of the relation between the branch voltage and stored charge of the capacitor.

One of the important properties of a capacitor is its memory property. In fact, it is this property that allows the capacitor to be the primary memory

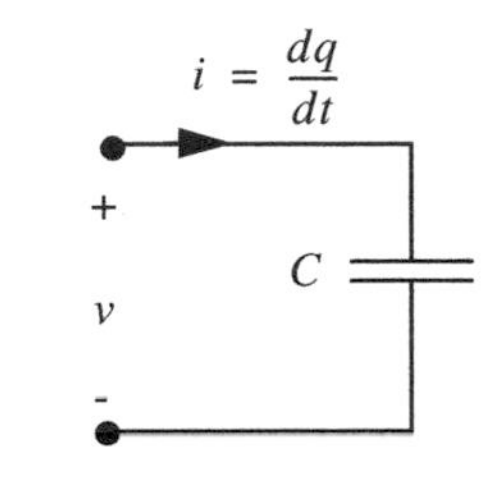


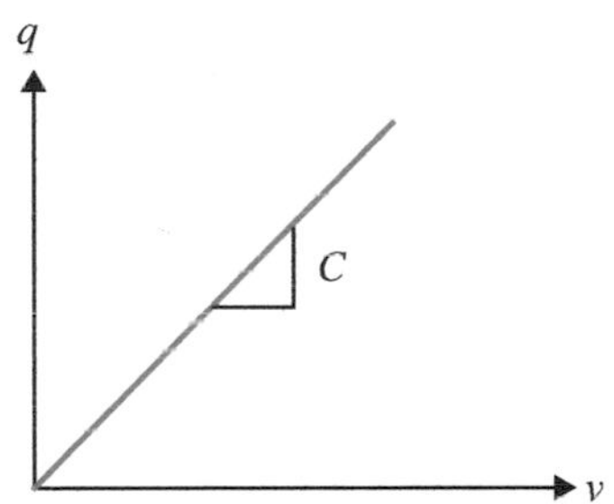


FIGURE 9.12 The symbol and voltage-charge relation for the ideal linear capacitor. The element law for the capacitor is $i = Cdv/dt$.

1. Although we will focus primarily on linear, time-invariant capacitors in this text, we note that some interesting transducers such as electric microphones and speakers, and other electric sensors and actuators, are appropriately modeled with time-varying capacitors. Similarly, most capacitors used in electronic equipment (paper, mica, ceramic, etc.) are linear, but often vary a small amount with temperature (a part of $10^4$ per degree centigrade). But many are nonlinear. The charge associated with a reverse-biased semiconductor diode, for example, varies as the 2/3 power of voltage, because the distance $d$, the effective width of the space-charge layer, is a function of voltage

$$q = K\left(\psi_O^{2/3} - (\psi_O - v)^{2/3}\right) \tag{9.10}$$

when $\psi_O$, the contact potential, is a few tenths of a volt. From the above we can determine that the capacitance of the reverse-biased diode varies as $v^{-1/3}$.

element in all integrated circuits. To see this property, we integrate Equation 9.6 to produce

$$q(t) = \int_{-\infty}^{t} i(t)dt \tag{9.11}$$

or, with the substitution of Equation 9.8, to produce

$$v(t) = \frac{1}{C}\int_{-\infty}^{t} i(t)dt. \tag{9.12}$$

Equation 9.12 shows that the branch voltage of a capacitor depends on the entire past history of its branch current, which is the essence of memory. This is in marked contrast to a resistor (either linear or nonlinear), which exhibits no such memory property.

At first glance, it might appear that it is necessary to know the entire history of the current $i$ in detail in order to carry out the integrals in Equations 9.11 and 9.12. This is actually not the case. For example, consider rewriting Equation 9.11 as

$$\begin{aligned} q(t_2) &= \int_{-\infty}^{t_2} i(t)dt \\ &= \int_{t_1}^{t_2} i(t)dt + \int_{-\infty}^{t_1} i(t)dt \\ &= \int_{t_1}^{t_2} i(t)dt + q(t_1). \end{aligned} \tag{9.13}$$

The latter equality shows that $q(t_1)$ perfectly summarizes, or memorizes, the entire accumulated history of $i(t)$ for $t \leq t_1$. Thus, if $q(t_1)$ is known, it is necessary and sufficient to know $i$ only over the interval $t_1 \leq t \leq t_2$ in order to determine $q(t_2)$. For this reason, $q$ is referred to as the *state* of the capacitor. For linear time-invariant capacitors, $v$ can also easily serve as a state because $v$ is proportionally related to $q$ through the constant C. Accordingly, we can rewrite Equation 9.12 as

$$\begin{aligned} v(t_2) &= \frac{1}{C}\int_{-\infty}^{t_2} i(t)dt \\ &= \frac{1}{C}\int_{t_1}^{t_2} i(t)dt + \frac{1}{C}\int_{-\infty}^{t_1} i(t)dt \\ &= \frac{1}{C}\int_{t_1}^{t_2} i(t)dt + v(t_1). \end{aligned} \tag{9.14}$$

Thus, we see that $v(t_1)$ also memorizes the entire accumulated history of $i(t)$ for $t \leq t_1$ and can serve as the state of the capacitor.

Associated with the ability to exhibit memory is the property of energy storage, which is often exploited by circuits that process energy. To determine the electric energy $w_E$ stored in a capacitor, we recognize that the power $iv$ is the rate at which energy is delivered to the capacitor through its port. Thus,

$$\frac{dw_E(t)}{dt} = i(t)v(t). \tag{9.15}$$

Next, substitute for $i$ using Equation 9.6, cancel the time differentials, and omit the parametric time dependence to obtain

$$dw_E = vdq. \tag{9.16}$$

Equation 9.16 is a statement of incremental energy storage within the capacitor. It states that the transport of the incremental charge $dq$ from the negative plate of the capacitor to the top plate across the electric potential difference $v$ stores the incremental energy $dw_E$ within the capacitor. To obtain the total stored electric energy, we must integrate Equation 9.16 with $v$ treated as a function of $q$. This yields

$$w_E = \int_0^q v(x)dx \tag{9.17}$$

where $x$ is a dummy variable of integration. Finally, substitution of Equation 9.8 and integration yields

$$\text{Stored energy} = w_E(t) = \frac{q^2(t)}{2C} = \frac{Cv(t)^2}{2} \tag{9.18}$$

as the electric energy stored in a capacitor. The units of energy is the Joule [J], or Watt-second. *Unlike a resistor, a capacitor stores energy rather than dissipates it.*

Capacitors come in an enormous range of values. For example, two pieces of insulated wire about an inch long, when twisted together, will have a capacitance of about 1 picofarad ($10^{-12}$ farads). A low-voltage power supply capacitor an inch in diameter and a few inches long could have a capacitance of 100,000 microfarads (0.1 farad; 1 microfarad, abbreviated as $\mu$F, is $10^{-6}$ F).

A real capacitor can exhibit richer behavior than that described here. For example, leakage current can flow through its dielectric. The practical significance of dielectric leakage is that eventually the charge stored on a capacitor can leak off. Thus, eventually a real capacitor will lose its memory. Fortunately, capacitors can be made with very low leakage (in other words, with very high resistance) in which case they are excellent long-term memory devices. However, if the dielectric leakage is large enough to be significant, then it can be modeled with a resistor in parallel with the capacitor.

Other non-idealities include the distributed series resistance, and even series inductance, that arises in foil-wound capacitors in particular. These characteristics limit the power-handling capability of a real capacitor, and the frequency range over which a real capacitor behaves like an ideal capacitor. They can often be explicitly modeled with a single series resistor and inductor, respectively.

---

EXAMPLE 9.1 PARALLEL PLATE CAPACITOR Suppose the parallel-plate capacitor in Figure 9.11 is 1 m square, has a gap separation of 1 $\mu$m, and is filled with a dielectric having permittivity of $2\epsilon_o$, where $\epsilon_o \approx 8.854 \times 10^{-12}$ F/m is the permittivity of free space. What is its capacitance? How much charge and energy does it store if its terminal voltage is 100 V?

The capacitance is determined from Equation 9.4 with $\epsilon = 1.8 \times 10^{-11}$ F/m, $A = 1$ m$^2$ and $l = 10^{-6}$ m. It is 18 $\mu$F. The charge is determined from Equation 9.8 with $v = 100$ V. It is 1.8 mC. Finally, the stored energy is determined from Equation 9.18. It is 90 mJ.

---

## 9.1.2 INDUCTORS

As we saw in Section 9.1.1, from the perspective of modeling electrical systems, the capacitor is a circuit element to model the effect of electric fields. Correspondingly, the *inductor* models the effect of magnetic fields. To understand the behavior of an inductor, and to illustrate the manner in which a lumped model can be developed for it, consider the idealized two-terminal linear inductor shown in Figure 9.13. In this inductor a coil with a terminal on each end is wound with $N$ turns around a toroidal core made from an insulator having magnetic permeability $\mu$. The length around the core is $l$ and its cross-sectional area is $A$. Note that these dimensions will be functions of time if the geometry of the inductor varies.

The current in the coil produces a magnetic flux in the inductor. Ideally, this magnetic flux does not stray significantly from the core, so that the flux outside

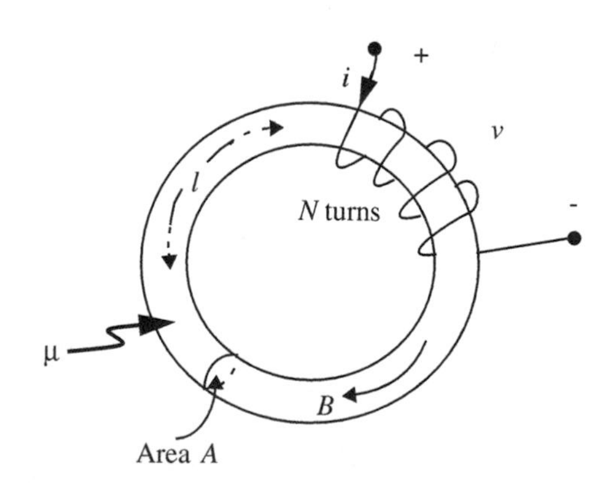


FIGURE 9.13 An idealized toroidal inductor.

the element is negligible. Thus the inductor can be treated as a lumped circuit element that satisfies the lumped matter discipline discussed in Chapter 1. From Maxwell's Equations and the properties of permeable materials, the density $B$ of the flux is

$$B(t) = \frac{\mu N i(t)}{l(t)}, \tag{9.19}$$

and its direction is around the core. The magnetic flux density can be integrated across the core to yield

$$\Phi(t) = A(t)B(t) \tag{9.20}$$

where $\Phi$ is the total flux passing through the core, and hence through one turn of the coil. Since the flux $\Phi$ is linked $N$ times by the $N$-turn coil, the total flux $\lambda$ linked by the coil is

$$\lambda(t) = N\Phi(t) = NA(t)B(t). \tag{9.21}$$

The units of flux linkage is the Weber [Wb]. Combining Equations 9.19 and 9.21 results in

$$\lambda(t) = \frac{\mu N^2 A(t)}{l(t)} i(t). \tag{9.22}$$

We define $L$, the inductance of the inductor, as

$$L(t) = \frac{\mu N^2 A(t)}{l(t)}. \tag{9.23}$$

$L$ has the units of Webers/Ampere, or Henrys [H]. That is, inductance is the number of flux linkages per ampere. Substituting for $L$ in Equation 9.22 we obtain the following relation for the total flux linked by the inductor

$$\lambda(t) = L(t)i(t). \tag{9.24}$$

In contrast to the resistor, which exhibits an algebraic relation between its branch current and voltage, the inductor does not. Rather, like the capacitor, it exhibits an algebraic relation between its branch current and its flux linkage. Had the core not been magnetically linear, this relation would have been nonlinear. While most inductors exhibit such nonlinear behavior for sufficiently high $B$, we will focus only on linear inductors.

Again from Maxwell's Equations, the rate at which flux linkage builds up in the inductor is

$$\frac{d\lambda(t)}{dt} = v(t). \tag{9.25}$$

From Equation 9.25 we see that the Volt is equivalent to a Weber/second. Equation 9.25 can be combined with Equation 9.24 to yield

$$v(t) = \frac{d(L(t)i(t))}{dt}, \tag{9.26}$$

which is the element law for an ideal linear inductor. For time-invariant inductors, Equations 9.24 and 9.26 reduce to

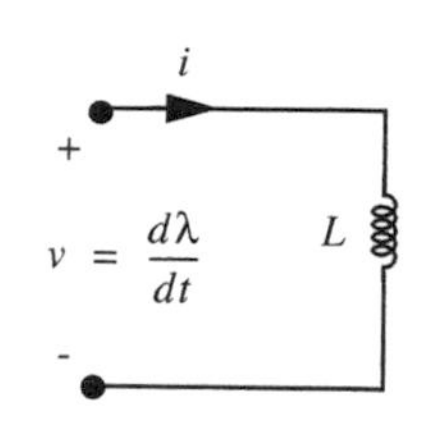


$$\lambda(t) = Li(t) \tag{9.27}$$

$$v(t) = L\frac{di(t)}{dt}, \tag{9.28}$$

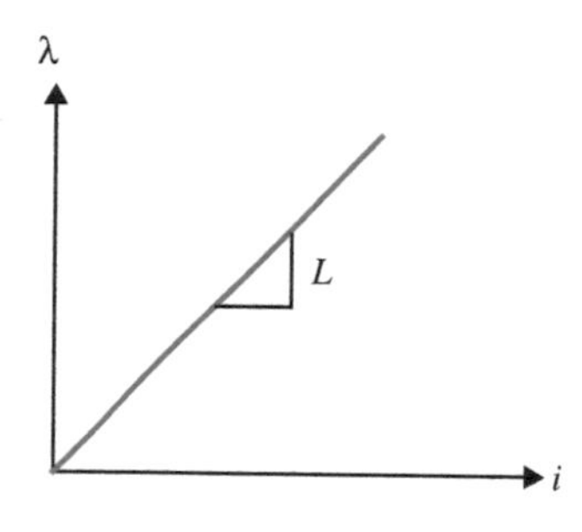


FIGURE 9.14 The symbol and current-flux-linkage relation for an ideal linear inductor. The element law for an inductor is $v = Ldi/dt$.

respectively, *with the latter being the element law for a linear time-invariant inductor.* This text will focus primarily on linear time-invariant inductors. Nonetheless, many interesting transducers such as motors, generators, and other magnetic sensors and actuators, are appropriately modeled with time-varying inductors.

The symbol for an ideal linear inductor is shown in Figure 9.14. It is chosen to represent the coil that winds the inductor shown in Figure 9.13. Also shown in Figure 9.14 is a graph of the relation between the branch current and flux linkage of the inductor.

One of the important properties of an inductor is its memory property. To see this property, we integrate Equation 9.25 to produce

$$\lambda(t) = \int_{-\infty}^{t} v(t)dt \tag{9.29}$$

or, with the substitution of Equation 9.27, to produce

$$i(t) = \frac{1}{L}\int_{-\infty}^{t} v(t)dt. \tag{9.30}$$

Equation 9.30 shows that the branch current of an inductor depends on the entire past history of its branch voltage, which is the essence of memory. As for the capacitor, this is in marked contrast to an ideal resistor, which exhibits no such memory property.

At first glance, it might appear that it is necessary to know the entire history of the voltage $v$ in detail in order to carry out the integrals in Equations 9.29 and 9.30. Again as for the capacitor, this is actually not the case. For example, consider rewriting Equation 9.29 as

$$\begin{aligned}\lambda(t_2) &= \int_{-\infty}^{t_2} v(t)dt \\ &= \int_{t_1}^{t_2} v(t)dt + \int_{-\infty}^{t_1} v(t)dt \\ &= \int_{t_1}^{t_2} v(t)dt + \lambda(t_1). \end{aligned} \tag{9.31}$$

The latter equality shows that $\lambda(t_1)$ perfectly summarizes, or memorizes, the entire accumulated history of $v(t)$ for $t \leq t_1$. Thus, if $\lambda(t_1)$ is known, it is necessary and sufficient to know $v$ only over the interval $t_1 \leq t \leq t_2$ in order to determine $\lambda(t_2)$. For this reason, $\lambda$, the total flux linked by the coil, is referred to as the *state* of the inductor. For linear time-invariant inductors, $i$ can also easily serve as a state because $i$ is proportionally related to $\lambda$ through the constant $L$. Accordingly, we can rewrite Equation 9.30 as

$$\begin{aligned}i(t_2) &= \frac{1}{L}\int_{-\infty}^{t_2} v(t)dt \\ &= \frac{1}{L}\int_{t_1}^{t_2} v(t)dt + \frac{1}{L}\int_{-\infty}^{t_1} v(t)dt \\ &= \frac{1}{L}\int_{t_1}^{t_2} v(t)dt + i(t_1). \end{aligned} \tag{9.32}$$

Equation 9.32 shows that $i$ can also serve as the state of an inductor.

As with the capacitor, associated with the ability to exhibit memory is the property of energy storage, which is often exploited by circuits that process energy. To determine the magnetic energy $w_M$ stored in an inductor, we recognize that the power $iv$ is the rate at which energy is delivered to the inductor through its port. Thus,

$$\frac{dw_M(t)}{dt} = i(t)v(t). \tag{9.33}$$

Next, substitute for $v$ using Equation 9.25, cancel the time differentials, and omit the parametric time dependence to obtain

$$dw_M = id\lambda. \tag{9.34}$$

Equation 9.34 is a statement of incremental energy storage within the inductor. To obtain the total stored magnetic energy, we must integrate Equation 9.34

with $i$ treated as a function of $\lambda$. This yields

$$w_M = \int_0^{\lambda} i(x)dx \tag{9.35}$$

where $x$ is a dummy variable of integration. Finally, substitution of Equation 9.27 and integration yields

$$\text{Stored energy} = w_M(t) = \frac{\lambda^2(t)}{2L} = \frac{Li(t)^2}{2} \tag{9.36}$$

as the magnetic energy stored in an inductor. Unlike a resistor, but like a capacitor, an inductor stores energy rather than dissipates it.

A real inductor exhibits richer behavior than that described here. For example, it can exhibit a significant coil resistance. The practical significance of this resistance is that it eventually dissipates any energy stored in the inductor. Unfortunately, this resistance is usually significant so that inductors make poor memory devices. When it is necessary to model this energy loss, the coil resistance can be modeled as a resistor in series with the ideal inductor.

Other non-idealities include core loss and inter-turn capacitance. These characteristics limit the power-handling efficiency of a real inductor, and the frequency range over which a real inductor behaves like an ideal inductor. They can often be modeled with a parallel resistor and capacitor, respectively.

---

EXAMPLE 9.2 TOROIDAL INDUCTOR Suppose the toroidal inductor in Figure 9.13 has a cross-sectional area of 1 $cm^2$, has a length around its toroid of 10 cm, has a coil with 100 turns, and is filled with free space having permeability $\mu_o = 4\pi \times 10^{-7}$ H/m. What is its inductance? How much flux does its coil link, and what energy does it store if its terminal current is 0.1 A?

The inductance is determined from Equation 9.23 with $\mu = 4\pi \times 10^{-7}$ H/m, $A = 10^{-4}$ $m^2$, $l = 0.1$ m and $N = 100$. It is 13 $\mu$H. The flux linkage is determined from Equation 9.24 with $i = 0.1$ A. It is 1.3 $\mu$Wb. Finally, the stored energy is determined from Equation 9.36. It is 0.063 $\mu$J.

---

## 9.2 SERIES AND PARALLEL CONNECTIONS

In Section 2.3.4, we saw that the resistances of resistors in series add, and that the conductances of resistors in parallel add. Thus, series and parallel resistors could be represented as a single resistor with an appropriate resistance. These addition rules later became useful as a means of simplifying circuits and their analyses. As we shall see in this section, similar rules may be derived for both capacitors and inductors, and these rules are equally useful.

### 9.2.1 CAPACITORS

Consider first the series combination of two capacitors as shown in Figure 9.15; we will assume here that the two capacitors were uncharged at the time of their connection. Since the two capacitors share a common current, it follows from Equation 9.11 that they store a common charge $q$, as shown in Figure 9.15. Thus, following Equation 9.8,

$$q(t) = C_1 v_1(t) = C_2 v_2(t). \tag{9.37}$$

Next, using KVL we observe that

$$v(t) = v_1(t) + v_2(t). \tag{9.38}$$

Finally, since the effective capacitance C of the two series capacitors is $q/v$, it follows that

$$\frac{1}{C} = \frac{v(t)}{q(t)} = \frac{1}{C_1} + \frac{1}{C_2},$$

or,

$$C = \frac{C_1 C_2}{C_1 + C_2} \tag{9.39}$$

where the second equality results from the substitution of Equation 9.38 and then Equation 9.37. Thus, we see that the reciprocal capacitances of capacitors in series add. This is consistent with the physical derivation of capacitance in Equation 9.4 since placing capacitors in series essentially increases their combined gap length.

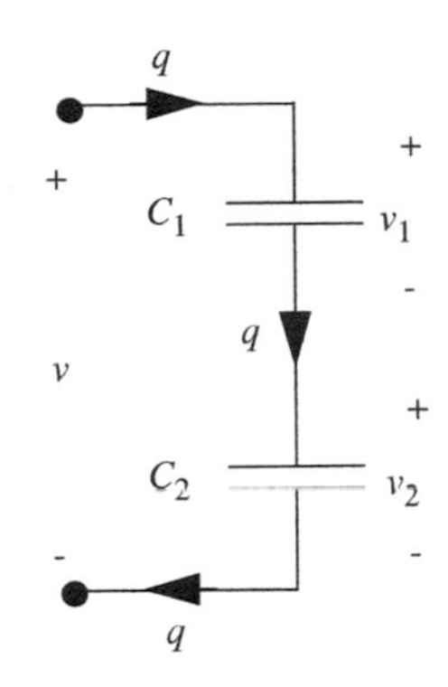


FIGURE 9.15 Two capacitors in series.

Now consider the parallel combination of two capacitors as shown in Figure 9.16. Since the two capacitors share a common voltage $v$, it follows from 9.8 that

$$v(t) = \frac{q_1(t)}{C_1} = \frac{q_2(t)}{C_2}. \tag{9.40}$$

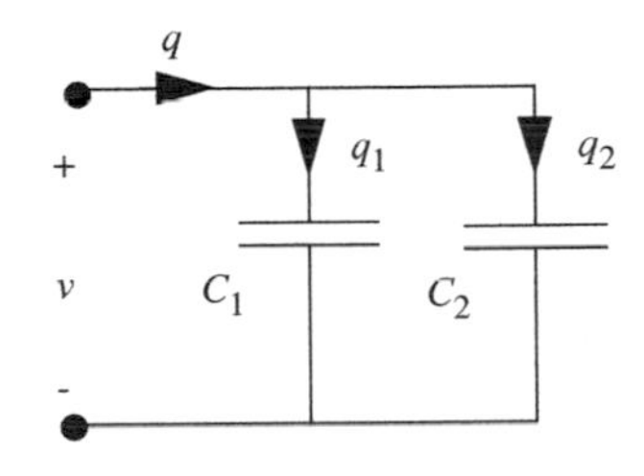


FIGURE 9.16 Two capacitors in parallel.

Next, using KCL and Equation 9.11 we observe that

$$q(t) = q_1(t) + q_2(t). \tag{9.41}$$

Finally, since the effective capacitance C of the two parallel capacitors is $q/v$, it follows that

$$C = \frac{q(t)}{v(t)} = C_1 + C_2 \tag{9.42}$$

where the second equality results from the substitution of Equation 9.41 and then Equation 9.40. Thus, we see that the capacitances of capacitors

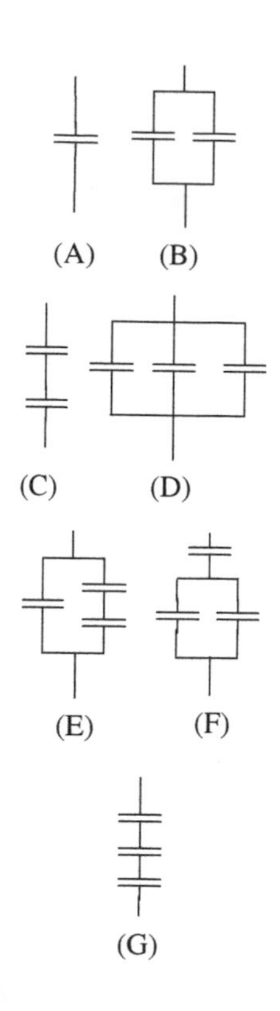


FIGURE 9.17 Various combinations of capacitors involving up to three capacitors.

in parallel add. This is consistent with the physical derivation of capacitance in Equation 9.4 since placing capacitors in parallel essentially increases their combined cross-sectional area.

---

EXAMPLE 9.3 CAPACITOR COMBINATIONS What equivalent capacitors can be made by combining up to three 1-$\mu$F capacitors in series and/or in parallel?

Figure 9.17 shows the possible capacitor combinations that use up to three capacitors. To determine their equivalent capacitances, use the series combination result from Equation 9.39 and/or the parallel combination result from Equation 9.42. This yields the equivalent capacitances of: (A) 1 $\mu$F, (B) 2 $\mu$F, (C) 0.5 $\mu$F, (D) 3 $\mu$F, (E) 1.5 $\mu$F, (F) 0.667 $\mu$F, and (G) 0.333 $\mu$F.

---

## 9.2.2 INDUCTORS

Consider the series combination of two inductors as shown in Figure 9.18; we will assume here that neither inductor carried a current at the time of their connection. Since the two inductors share a common current $i$, it follows from Equation 9.27 that

$$i(t) = \frac{\lambda_1(t)}{L_1} = \frac{\lambda_2(t)}{L_2}. \tag{9.43}$$

Next, using KVL and Equation 9.29 we observe that

$$\lambda(t) = \lambda_1(t) + \lambda_2(t). \tag{9.44}$$

Finally, since the effective inductance $L$ of the two series inductors is $\lambda/i$, it follows that

$$L = \frac{\lambda(t)}{i(t)} = L_1 + L_2 \tag{9.45}$$

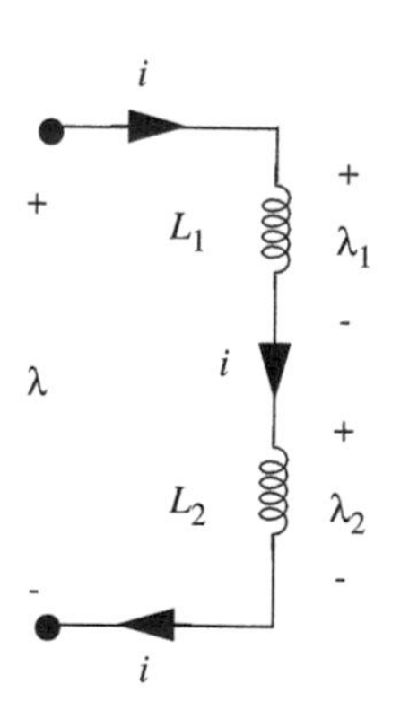


FIGURE 9.18 Two inductors in series.

where the second equality results from the substitution of Equation 9.44 and then Equation 9.43. Thus, we see that the inductances of inductors in series add. This is consistent with the physical derivation of inductance in Equation 9.23 since placing inductors in series essentially increases the total length of core around which the parallel turns are wound.

Now consider the parallel combination of two inductors as shown in Figure 9.19. Since the two inductors share a common voltage, it follows from Equation 9.29 that they share a common flux linkage $\lambda$, as shown in Figure 9.19. Thus, following Equation 9.27,

$$\lambda(t) = L_1 i_1(t) = L_2 i_2(t). \tag{9.46}$$

Next, using KCL we observe that

$$i(t) = i_1(t) + i_2(t). \tag{9.47}$$

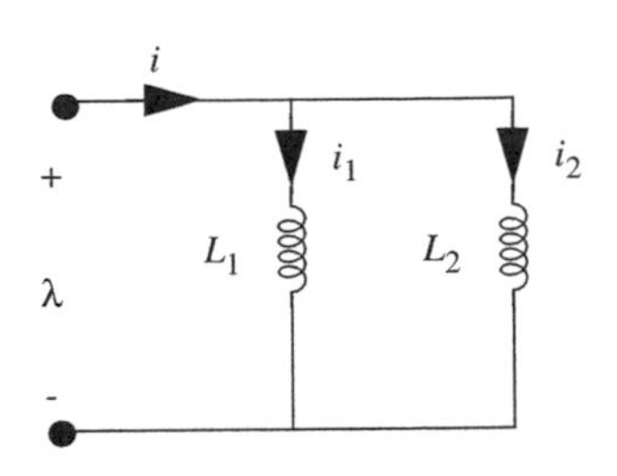


FIGURE 9.19 Two inductors in parallel.

Finally, since the effective inductance $L$ of the two parallel inductors is $\lambda/i$, it follows that

$$\frac{1}{L} = \frac{i(t)}{\lambda(t)} = \frac{1}{L_1} + \frac{1}{L_2},$$

or,

$$L = \frac{L_1 L_2}{L_1 + L_2} \tag{9.48}$$

where the second equality results from the substitution of Equation 9.47 and then Equation 9.46. Thus, we see that the reciprocal inductances of inductors in parallel add. This is consistent with the physical derivation of inductance in Equation 9.23 since placing inductors in parallel essentially increases the cross-sectional area of the core around which the turns are wound.

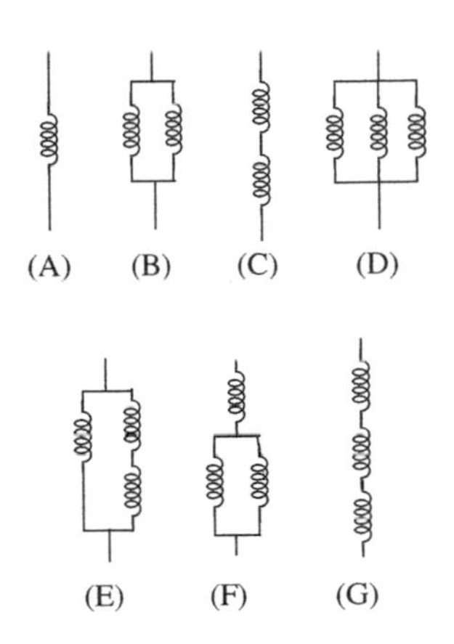


FIGURE 9.20 Various combinations of inductors involving up to three inductors.

EXAMPLE 9.4 INDUCTOR COMBINATIONS What equivalent inductors can be made by combining up to three 1-$\mu$H inductors in series and/or in parallel?

Figure 9.20 shows the possible inductor combinations that use up to three inductors. To determine their equivalent inductances, use the series combination result from Equation 9.45 and/or the parallel combination result from Equation 9.48. This yields the equivalent inductances of: (A) 1 $\mu$H, (B) 0.5 $\mu$H, (C) 2 $\mu$H, (D) 0.333 $\mu$H, (E) 0.667 $\mu$H, (F) 1.5 $\mu$H, and (G) 3 $\mu$H.

## 9.3 SPECIAL EXAMPLES

In this section, we examine several parasitic capacitances and inductances that are commonly encountered inside integrated circuits, and in external wiring connections to them and other circuit elements. There is again the danger that this discussion implies that capacitors and inductors appear most commonly as parasitics in circuits. This is certainly not the case. Rather, we examine the parasitics here primarily for interest sake, and because they will provide interesting and important circuit examples in future chapters.

### 9.3.1 MOSFET GATE CAPACITANCE

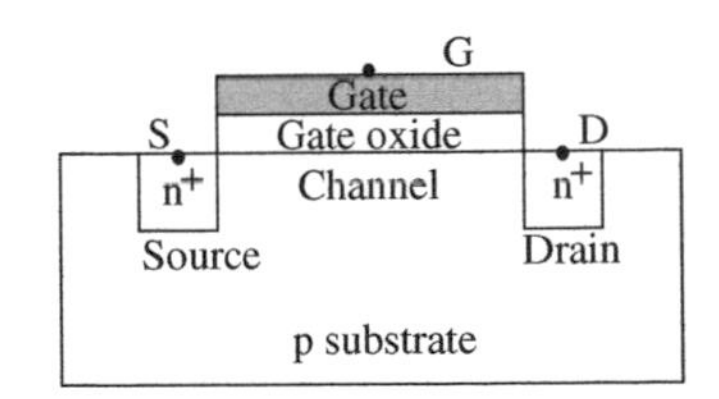


FIGURE 9.21 MOSFET structure.

Let us now take a closer look at the structure and operation of the MOSFET in order to better understand its dynamic behavior. Figure 9.21 reviews the

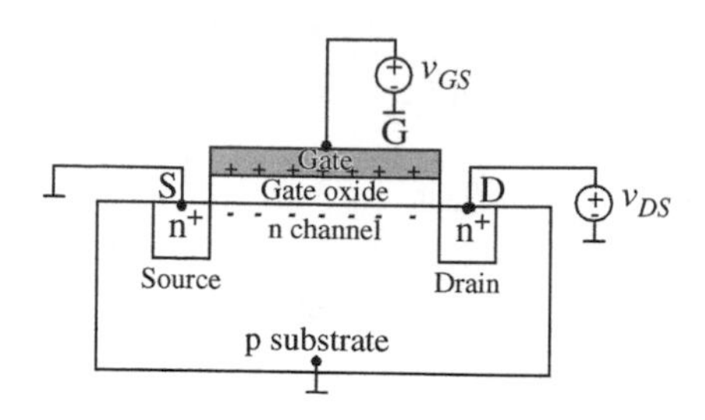


**FIGURE 9.22** MOSFET with a positive voltage applied at the gate relative to the source and substrate.

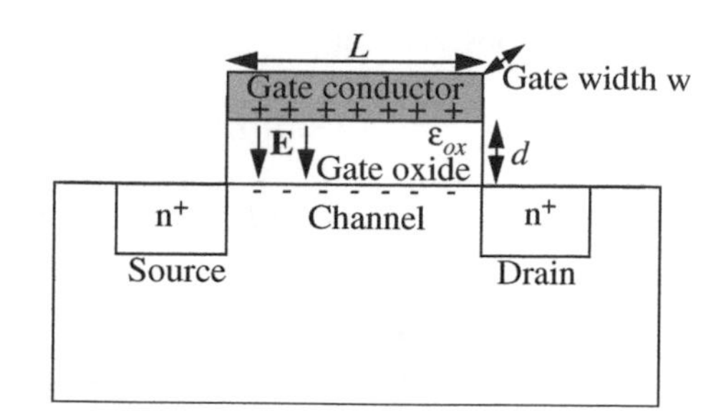


**FIGURE 9.23** Electric charge and field within the MOSFET with a positive voltage applied at the gate relative to the source and substrate.

structure of the n-channel MOSFET. The figure identifies its $n^+$-type source and drain, its p-type substrate, its channel region, its gate conductor, and the silicon dioxide dielectric that separates its gate and channel.

Figure 9.22 shows the same n-channel MOSFET with its source and substrate grounded, and positive voltages applied to its gate and drain. As the positive gate voltage is applied, electrons flow from the source into the channel and accumulate beneath the gate. When the gate voltage exceeds the threshold voltage of the MOSFET, the electron density beneath the gate becomes sufficiently high to invert the channel from p-type silicon to n-type silicon. Thus, a continuous n-type channel forms between the source and drain, thereby allowing electrons to flow from the source to the drain, and hence current to flow from the drain to the source, in the response to the positive drain voltage.

The important observation here from Figure 9.22 is that in the process of inverting its channel, and turning itself on, the MOSFET actually forms a parallel-plate capacitor between its gate and channel. This is emphasized in Figure 9.23, which shows the electric field **E** in the silicon dioxide emanating from the positive charge on the gate and terminating on the negative charge in the channel. Comparing this figure to Figure 9.11 leads to the realization (from Equation 9.4) that the gate-to-channel capacitance is approximately

$$\frac{\epsilon_{OX} LW}{d}$$

where $\epsilon_{OX} \approx 3.9\epsilon_o$ is the permittivity of the silicon dioxide, $d$ is the thickness of the silicon dioxide, $L$ is the channel length, and $W$ is the channel width. The product $LW$ is the gate area.

Since the electrons that fill the channel originate from the source, and since their image charges reside on the gate, the gate-to-channel capacitance that we identified in Figures 9.22 and 9.23 appears between the gate and source of the MOSFET when viewed from the MOSFET terminals. For this reason the capacitance is usually referred to as the gate-to-source capacitance of the MOSFET, or $C_{GS}$. In other words,

$$C_{GS} = \frac{\epsilon_{OX} LW}{d}. \tag{9.49}$$

Often, the ratio $\epsilon_{OX}/d$ is referred to as $C_{OX}$, the gate-to-channel capacitance per unit area of the MOSFET gate. In other words,

$$C_{OX} = \frac{\epsilon_{OX}}{d}$$

This realization also leads to the augmented switch-resistor-capacitor (SRC) model of the MOSFET shown in Figure 9.24. Here, a lumped capacitor is added to the SR model to account for the charge that must be supplied to the gate conductor and channel in order to turn on the MOSFET. Thus, we develop a model that describes the behavior of the MOSFET yet satisfies the lumped matter discipline.

Because the SRC model contains a capacitor between the gate and source terminals of the MOSFET, a current will flow into the gate terminal and out from the source terminal of that model as the gate-to-source voltage of the MOSFET varies. This current transports the charge that accumulates within the MOSFET as seen in Figures 9.22 and 9.23. Following Equation 9.9, the current is given by

$$i_G = C_{GS} \frac{dv_{GS}}{dt} \tag{9.50}$$

where

$$C_{GS} = C_{OX} LW. \tag{9.51}$$

From Equation 9.50 we can now begin to see the reason for the inverter behavior observed in Figure 9.3. It will take time for the gate current to transport charge onto the gate, and hence it will take time for the gate voltage to rise. Thus, it will take time for the inverter to pass a signal from its input to its output. We will have more to say about this in Section 10.4.

Finally, it is important realize that the dynamic behavior of a real MOSFET is actually much more complex than described here. In reality a MOSFET actually has many internal capacitances of importance, including capacitances between its gate and drain, its gate and source, its gate and substrate, its drain and source, its drain and substrate, and its source and substrate. Further, most of these capacitances are actually functions of $v_{GS}$ and $v_{DS}$. For our purposes, we will work primarily with $C_{GS}$ and assume that it is a constant capacitance.

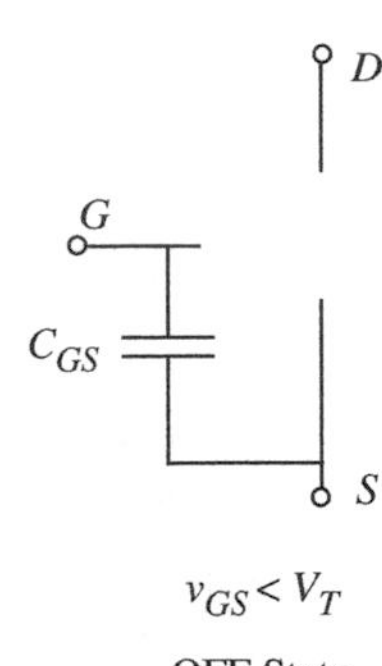


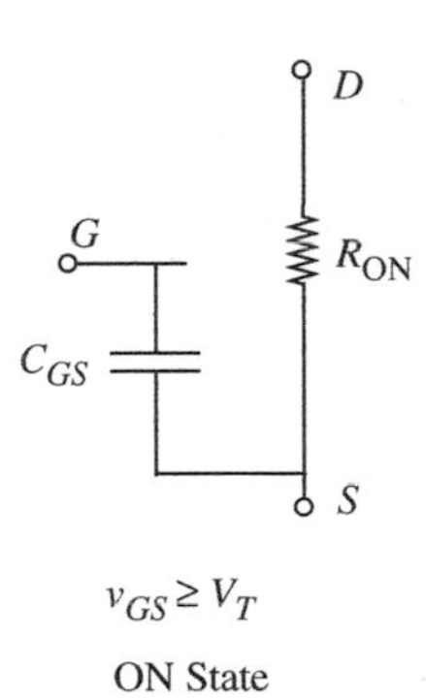


FIGURE 9.24 The switch-resistor-capacitor (SRC) model of the MOSFET.

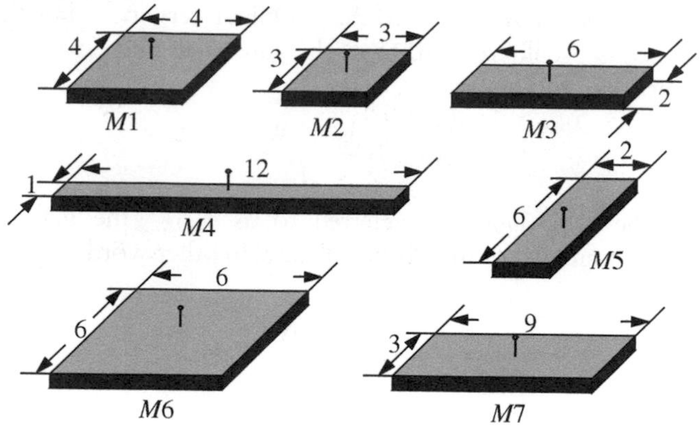

**FIGURE 9.25** MOSFET gates with different dimensions; all dimensions in the figure are in $\mu$m.

EXAMPLE 9.5 GATE CAPACITANCES OF MOSFETS Figure 9.25 shows the top view of several rectangular MOSFET gates fabricated within an integrated circuit. Let us assume that the silicon-dioxide dielectric is characterized by $C_{OX} \approx 4\ \text{fF}/\mu\text{m}^2$, and find the gate capacitances $C_{GS}$ for each MOSFET.

To do so, we use Equation 9.51. To begin, notice that MOSFETs M3, M4, and M5 must have the same capacitance because they have the same area of 12 $\mu\text{m}^2$. Their capacitance is therefore 48 fF. MOSFET M5 has the biggest area of 36 $\mu\text{m}^2$, and so it has the biggest capacitance of 144 fF, while MOSFET M2 has the smallest area of 9 $\mu\text{m}^2$, and so it has the smallest capacitance of 36 fF. MOSFETs M1 and M7 have capacitances 64 fF and 108 fF, respectively.

### 9.3.2 WIRING LOOP INDUCTANCE

The most common parasitic inductance is the inductance associated with a wiring loop. In the lumped circuit abstraction, this inductance is ignored unless it is explicitly modeled in a circuit using an additional lumped inductor. To estimate the inductance of a wiring loop, consider the circular loop of wire in free space shown in Figure 9.26. The loop has a loop radius $R$ and a wire radius $A$. Its inductance $L$ is given approximately by[2]

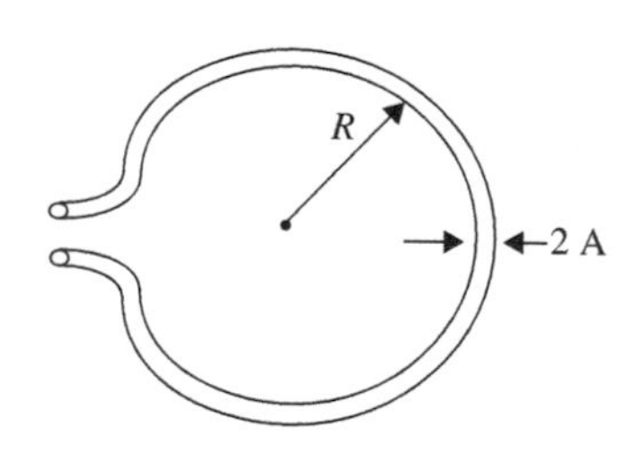


**FIGURE 9.26** A wiring loop.

$$L = \mu_o R \left( \ln\left(\frac{8R}{A}\right) - 2 \right). \tag{9.52}$$

This expression can also be used to successfully approximate the inductance of many noncircular wiring loops.

2. See Ramo, Whinnery, and Van Duzer, *Fields and Waves in Communication Electronics*, P. 311, John Wiley, 1965.

EXAMPLE 9.6 INDUCTANCE OF A WIRING LOOP Suppose a wiring loop in free space has a 5-mm diameter and is made from 200-$\mu$m-thick wire. What is its inductance?

Using Equation 9.52 with $R = 2.5 \times 10^{-3}$ m, $A = 10^{-4}$ m, and $\mu_o = 4\pi \times 10^{-7}$ H/m, the inductance is found to be 10 nH.

### 9.3.3 IC WIRING CAPACITANCE AND INDUCTANCE

Let us now return to Figure 9.4, and develop a model for the capacitance and inductance of the conductors inside an integrated circuit (IC) that are implied by the figure. Many conductors inside integrated circuits can be modeled as a flat conductor above a conducting substrate, or ground plane, as shown in Figure 9.27.

The conductor in the figure has a width $W$, and it is located the distance $G$ above the ground plane. Such conductors are typically surrounded by an insulating dielectric having a permittivity of $\epsilon > \epsilon_o$ and a permeability of $\mu_o$. Under the assumption that $W \gg G$, we can ignore the fringing electric and magnetic fields at the edges of the conductor. In this case, the capacitance $\tilde{C}$ and inductance $\tilde{L}$ of the conductor per unit length along its length is approximately

$$\tilde{C} = \frac{\epsilon W}{G} \tag{9.53}$$

$$\tilde{L} = \frac{\mu_o G}{W}. \tag{9.54}$$

In other cases, however, the width of the conductor is not large compared to its elevation above the ground plane. An example of this is a narrow printed circuit board trace. In such cases the conductor might alternatively be modeled as a cylindrical conductor above a ground plane as shown in Figure 9.28.

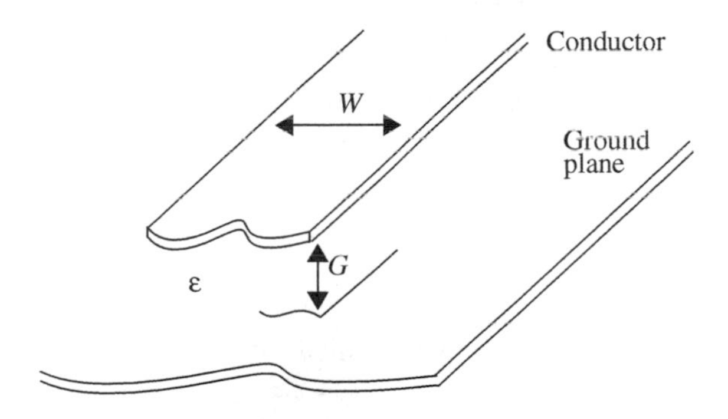


FIGURE 9.27 A flat conductor above a conducting ground plane.

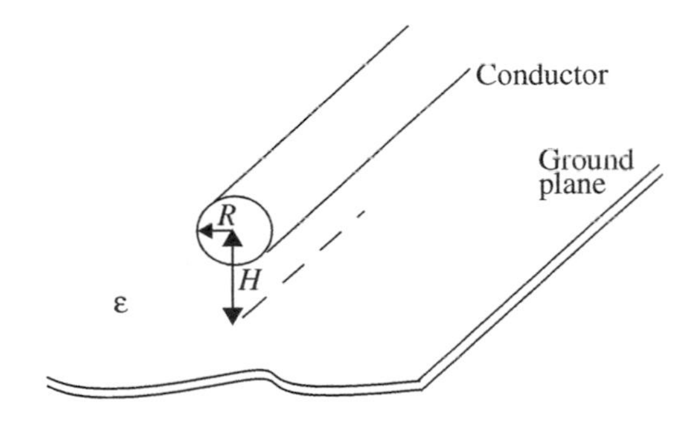


FIGURE 9.28 A cylindrical conductor above a conducting ground plane.

The conductor in Figure 9.28 has a radius $R$ and is centered the distance $H$ above the ground plane. It has a capacitance $\tilde{C}$ and inductance $\tilde{L}$ per unit length of approximately

$$\tilde{C} = \frac{2\pi\epsilon}{\ln\left(\frac{H}{R} + \sqrt{\frac{H^2}{R^2} - 1}\right)} \tag{9.55}$$

$$\tilde{L} = \frac{\mu_o}{2\pi} \ln\left(\frac{H}{R} + \sqrt{\frac{H^2}{R^2} - 1}\right) \tag{9.56}$$

along its length. Together, the conductors shown in Figures 9.27 and 9.28 can be used to model a wide variety of interconnects.

Finally, notice that for both interconnects,

$$\tilde{C}\tilde{L} = \epsilon\mu_o.$$

It follows from Maxwell's Equations that this will always be the case for any two-wire interconnect having constant cross section along its length. Thus, any effort to reduce either $\tilde{C}$ or $\tilde{L}$ will result in an increase of the other.

---

EXAMPLE 9.7 CAPACITANCE OF INTEGRATED-CIRCUIT INTERCONNECT Consider an integrated-circuit interconnect, such as the one shown in Figure 9.27, with $W = 2$ $\mu$m, $G = 0.1$ $\mu$m, and $\epsilon = 3.9\epsilon_o$. What is its capacitance and inductance per unit length?

Using Equations 9.53 and 9.54, $\tilde{C} = 690$ pF/m $= 0.69$ fF/$\mu$m, and $\tilde{L} = 63$ nH/m $=$ 63 fH/$\mu$m.

---

EXAMPLE 9.8 PRINTED-CIRCUIT-BOARD TRACE Consider modeling a printed-circuit-board trace as a cylindrical conductor above a ground plane, as shown in Figure 9.28. Let $R = 0.5$ mm, $H = 2$ mm, and $\epsilon = \epsilon_o$. What is its capacitance and inductance per unit length?

Using Equations 9.55 and 9.56, $\tilde{C} = 27$ pF/m, and $\tilde{L} = 410$ nH/m.

---

### 9.3.4 TRANSFORMERS

A transformer is a two-port device made by winding a second coil around the inductor, for example, that shown in Figure 9.13. Let the first (or primary) coil have $N_1$ turns and the second (or secondary) coil have $N_2$ turns. The symbol for an ideal transformer having this construction is shown in Figure 9.29. The two dots indicate the ends of the two coils that are wound in the same direction.

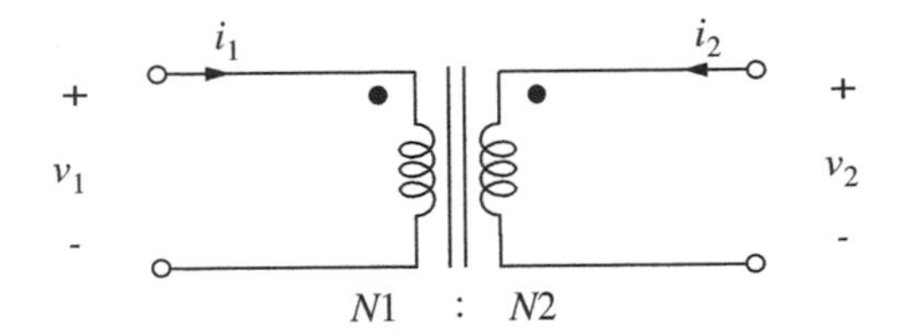


FIGURE 9.29 The symbol for an ideal transformer.

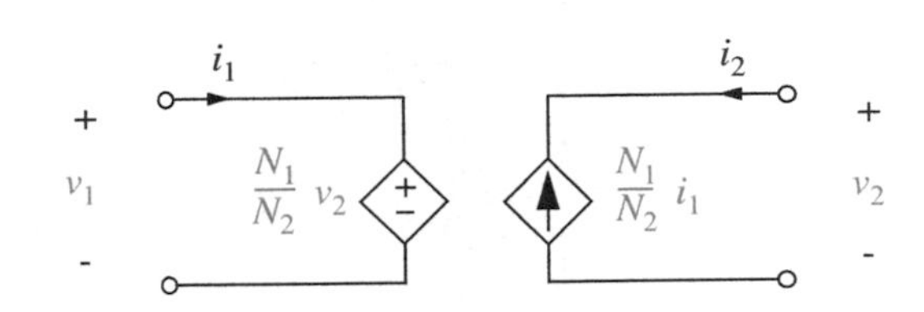


FIGURE 9.30 A useful model for an ideal transformer.

In an ideal transformer, the coils are wound so tightly against each other that each of their turns links the same flux $\Phi(t)$. It then follows from Equations 9.25 and 9.21 that

$$v_1 = N_1 \frac{d\Phi(t)}{dt} \tag{9.57}$$

$$v_2 = N_2 \frac{d\Phi(t)}{dt} \tag{9.58}$$

so that

$$\frac{v_1(t)}{N_1} = \frac{v_2(t)}{N_2}. \tag{9.59}$$

In an ideal transformer, the core is also infinitely permeable, that is, $\mu = \infty$. For a single-coil inductor carrying a finite flux $\Phi(t) = \lambda(t)/N$, Equation 9.22 shows that the total ampere-turns $Ni(t)$ flowing around the core through the coil must vanish as $\mu$ becomes infinite. In an ideal transformer, the total ampere turns must similarly vanish, and so

$$N_1 i_1(t) + N_2 i_2(t) = 0 \tag{9.60}$$

or

$$N_1 i_1(t) = -N_2 i_2(t). \tag{9.61}$$

Equations 9.59 and 9.61 are the constitutive equations for an ideal transformer.

By combining Equations 9.59 and 9.61, it can be observed that

$$v_1(t) i_1(t) = -v_2(t) i_2(t). \tag{9.62}$$

Thus, the power flowing into one port of an ideal transformer must instantaneously flow out from the second port. Said differently, an ideal transformer cannot store energy. This is consistent with having an infinitely permeable core.

A very useful model for an ideal transformer is shown in Figure 9.30. This model uses two dependent sources to enforce Equations 9.59 and 9.61. The voltage-dependent voltage source enforces Equation 9.59 and the current-dependent current source enforces Equation 9.61.

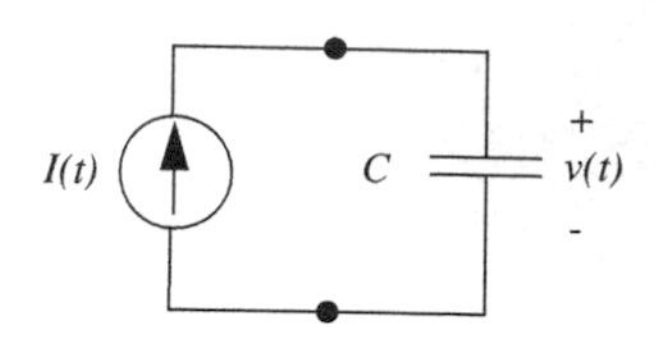

FIGURE 9.31 A current source driving a capacitor.

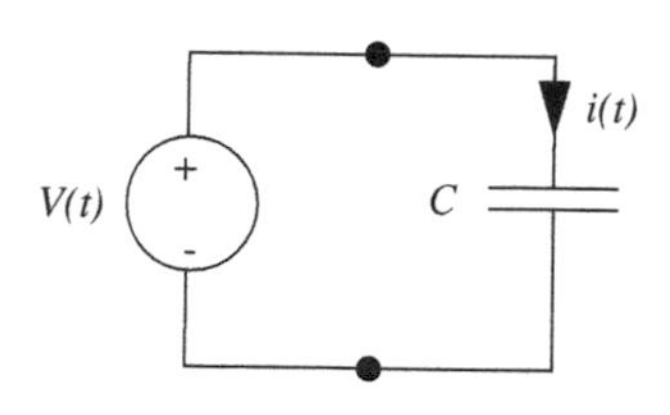

FIGURE 9.32 A voltage source driving a capacitor.

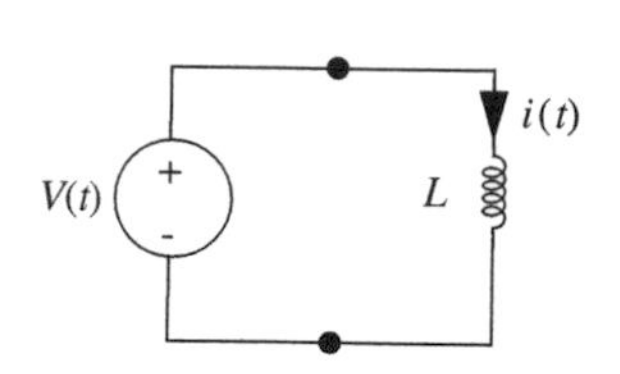

FIGURE 9.33 A voltage source driving an inductor.

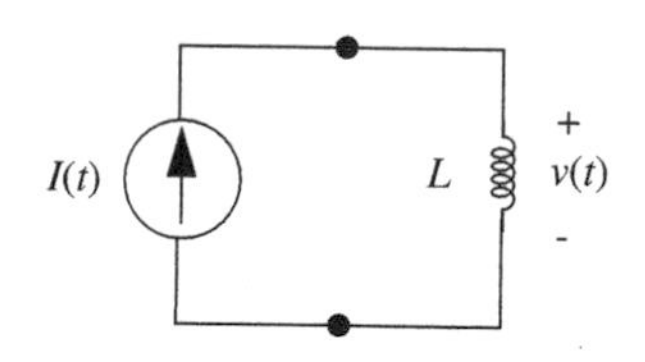

FIGURE 9.34 A current source driving an inductor.

EXAMPLE 9.9 A TRANSFORMER A transformer can be used to transform the 120-Volt rms utility voltage to a voltage that can power a 5-V DC load. To do so, what must be the approximate turns ratio of the transformer?

If the primary of the transformer is connected to the utility, then

$$v_1 = 120\sqrt{2}\sin(2\pi\ 60\ t),$$

where the frequency of the utility voltage is 60 Hz (or 60 cycles per second), or $2\pi$ 60 radians per second. Thus, the primary has a peak voltage of 170 V. At the secondary, it is desired that $v_2$ have a peak of 5 V, and so the turns ratio should be approximately

$$N_1/N_2 = 34.$$

A real transformer designed for this application would actually have a slightly smaller turns ratio so that $v_2$ would ideally be somewhat larger than 5 volts. This allows for voltage drops across coil resistances and leakage inductances found in practical devices.

## 9.4 SIMPLE CIRCUIT EXAMPLES

To complete our introduction to capacitors and inductors, let us now examine their behavior in the simple circuits shown in Figures 9.31 through 9.34. These circuits are the same as those shown in Figures 2.25 and 2.26, except for the replacement of the resistor in the latter figures by the capacitor or inductor in the former figures. Because the two sets of circuits are so similar, we could analyze the circuits shown in Figures 9.31 through 9.34 using the same approach applied in Chapter 2 to the circuits shown in Figures 2.25 and 2.26. Alternatively, we could carry out a node analysis as developed in Section 3.3. However, since the circuits here are simple, we will follow the more intuitive approach outlined at the end of Section 2.4, and save the formalities for the analysis of more complex circuits in future chapters.

Consider first the circuit shown in Figure 9.31. In this circuit, the current $I$ from the source must circulate through the capacitor. Thus, the current through both elements is known. Next, following Equation 9.12, the voltage $v$ across the capacitor, and hence across the current source, is given by

$$v(t) = \frac{1}{C}\int_{-\infty}^{t} I(t)dt. \tag{9.63}$$

All branch variables are now known.

Consider next the circuit shown in Figure 9.32. In this circuit, the voltage $V$ from the source must also appear across the capacitor. Thus, the voltage across both elements is known. Next, following Equation 9.9, the current $i$ circulating through both the capacitor and the voltage source is given by

$$i(t) = C\frac{dV(t)}{dt}. \tag{9.64}$$

Again, all branch variables are now known.

Now consider the circuit shown in Figure 9.33. In this circuit, the voltage $V$ from the source must also appear across the inductor, just as it appeared across the capacitor in Figure 9.32. Thus, following Equation 9.30, the current circulating through both the inductor and the voltage source is given by

$$i(t) = \frac{1}{L}\int_{-\infty}^{t} V(t)dt. \tag{9.65}$$

All branch variables are now known.

Finally, consider the circuit shown in Figure 9.34. In this circuit, the current $I$ from the source must circulate through the inductor, just as it did through the capacitor shown in Figure 9.31. Thus, following Equation 9.28, the voltage $v$ appearing across both the inductor and the current source is given by

$$v(t) = L\frac{dI(t)}{dt}. \tag{9.66}$$

Once again, all branch variables are now known.

In the following subsections, we will consider specific examples of the source current $I$ and source voltage $V$. However, before doing so, it is worth noting the similarity between the analyses of the four circuits we have just studied. Because capacitors and inductors are duals of each other, we find that the circuits are as well. For example, the circuits shown in Figures 9.31 and 9.33 are duals. Capacitance interchanges with inductance, and current interchanges with voltage, as can be seen by comparing Equations 9.63 and 9.65. Similarly, the circuits shown in Figures 9.32 and 9.34 are duals. Again, capacitance interchanges with inductance and current interchanges with voltage, as can be seen by comparing Equations 9.64 and 9.66.

It is also interesting to note that the circuits shown in Figures 9.31 through 9.34 perform either integration or differentiation of the source current or voltage as it produces the branch voltages or currents, respectively. Thus, if viewed in this way each circuit is an integrator or differentiator. We will make use of this capability of capacitors and inductors in future chapters as we build filters and other signal processing circuits.

### WWW 9.4.1 SINUSOIDAL INPUTS*

### 9.4.2 STEP INPUTS

Step functions, and their integrals and derivatives, constitute another important class of inputs to electronic circuits. So, as an example of a step input to the circuit shown in Figure 9.31 (redrawn here as Figure 9.36 for convenience), consider the source step function:

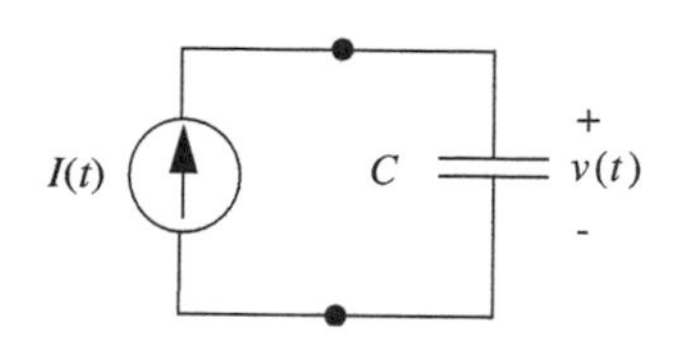


FIGURE 9.36 A current source driving a capacitor.

$$I(t) = \begin{cases} 0 & t \le 0 \\ I_o & t > 0. \end{cases} \tag{9.75}$$

Note that the source is zero for $t \le 0$, but nonzero for $t > 0$, so that it effectively turns on at $t = 0$. A sketch of the current step input is shown in Figure 9.37a.

To complete the analysis of the circuit, we substitute the corresponding source function from Equation 9.75 into Equation 9.63 and carry out the indicated integration.[3] The substitution of Equation 9.75 into Equation 9.63 yields

$$v(t) = \begin{cases} 0 & t \le 0 \\ \dfrac{I_o t}{C} & t > 0 \end{cases} \tag{9.76}$$

for the circuit shown in Figure 9.36. This result is shown in Figure 9.37b.

Let us now examine the operation of the circuit shown in Figure 9.36 more closely. Once the current source steps its value to $I_o$, it begins to deliver charge to the capacitor at that constant rate. The charge then accumulates linearly in the capacitor, much like water would accumulate in a glass from a faucet set to deliver that water at a constant rate. Since charge and voltage are proportional through the constant capacitance $C$ of the linear time-invariant capacitor, the voltage across the capacitor also increases linearly. This is as shown in Figure 9.37.

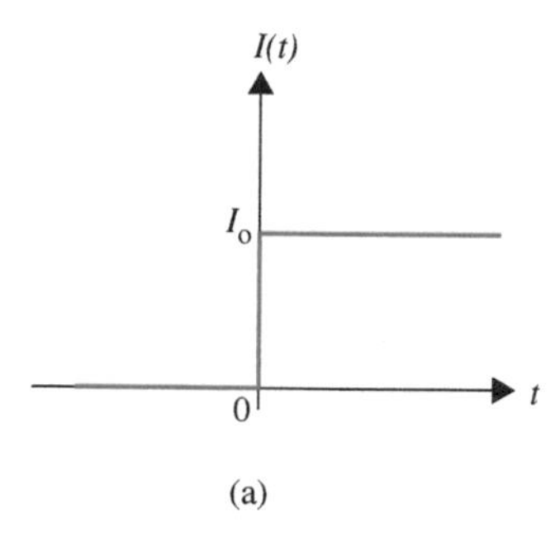


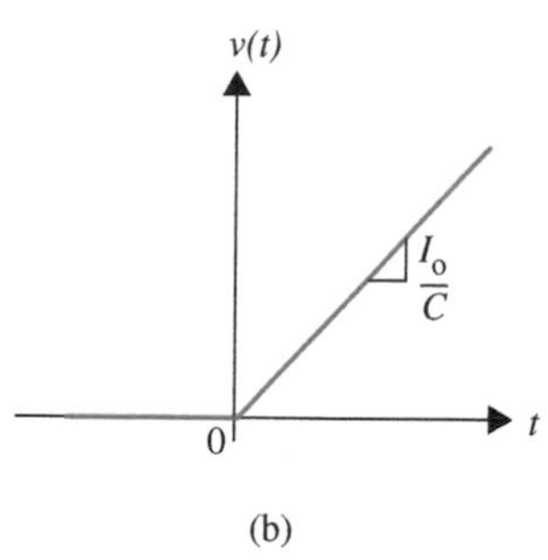


FIGURE 9.37 The current and voltage in the circuit shown in Figure 9.36.

Figure 9.37 also illustrates another very important point, namely that the charge stored in, and hence the voltage appearing across, a capacitor is a continuous function of time. Even though $I$ steps discontinuously at $t = 0$, $v$ does not; the state $q$, and hence $v$, is continuous. The only way for $v$ to take a discontinuous step would be for the current source to deliver a nonzero charge in zero time, which requires an infinite current. This is of course not a practical possibility, although we will see that such a mathematical construction can be used very effectively to model certain physical phenomena.

3. This is relatively easy for the integration, but as we will see shortly, it requires some thought when the circuit involves a differentiation.

The behavior seen in Figure 9.37 also begins to explain the delays seen in Figure 9.3, although as we shall see in Chapter 10 the details are slightly different. As we see in Figure 9.37, it takes time for a finite current source to charge a capacitor, and hence to raise its voltage. In the case of the two inverters shown in Figure 9.1, it takes time for the first inverter to change the voltage at the input to the second inverter. This is because it takes time for the first inverter to charge and discharge the gate-to-source capacitance of the MOSFET in the second inverter. The time it takes for this voltage to cross the threshold voltage of the MOSFET is also ultimately responsible for a delay at the output of the second inverter.

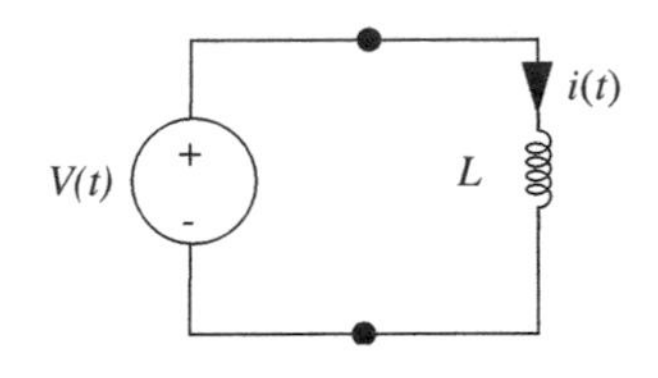


FIGURE 9.38 A voltage source driving an inductor.

EXAMPLE 9.10 MOSFET GATE-TO-SOURCE CAPACITANCE

Suppose that the gate-to-source capacitance $C_{GS}$ for a particular MOSFET is 100 fF. What constant gate current would be required to raise the gate-to-source voltage of that MOSFET from 0 V to 5 V in 10 ns?

This problem is well modeled by Figure 9.31 with $I$ being a current step. Hence, Equation 9.76 and Figure 9.37 apply. Since the voltage slope is to be 5 V in 10 ns, or $5 \times 10^8$ V/s, and since the capacitance is 100 fF, the current must be 50 $\mu$A.

A second way to solve this problem is to use Equation 9.8 with $C = 100$ fF and $v = 5$ V to determine that the gate current must deliver a charge of 500 fC. Since this charge flows at a constant rate over a 10-ns period, the current must be 50 $\mu$A.

EXAMPLE 9.11 VOLTAGE STEP INPUT TO AN INDUCTOR

As our next example, let us consider a voltage step input of the following form:

$$V(t) = \begin{cases} 0 & t \leq 0 \\ V_o & t > 0 \end{cases}. \tag{9.77}$$

to the circuit shown in Figure 9.33 (redrawn here as Figure 9.38). The voltage step input is sketched in Figure 9.39a.

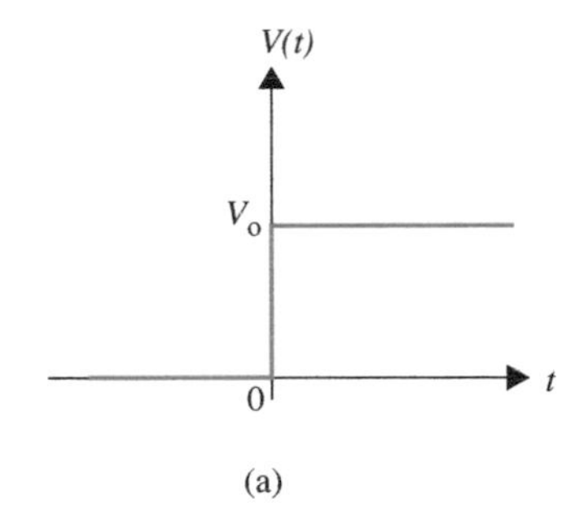


Since they are duals, the operation of the circuit in Figure 9.38 is much the same as that of the circuit in Figure 9.36. Once the voltage source steps its value to $V_o$ the current through the inductor begins to increase linearly. More formally, the substitution of Equation 9.77 into Equation 9.65 yields

$$i(t) = \begin{cases} 0 & t \leq 0 \\ \dfrac{V_o t}{L} & t > 0. \end{cases} \tag{9.78}$$

This is as shown in Figure 9.39b.

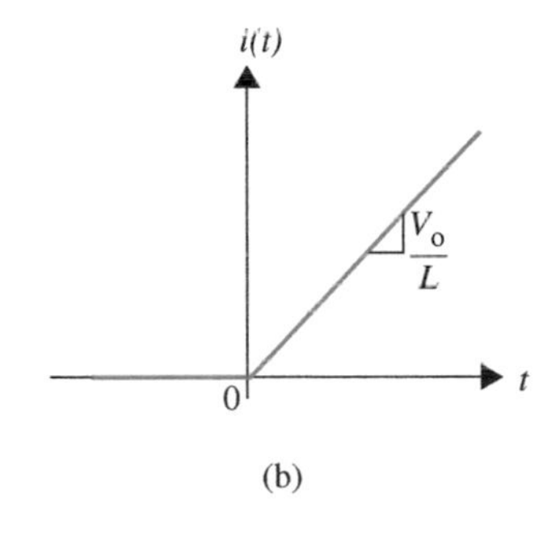


FIGURE 9.39 The current and voltage in the circuit shown in Figure 9.38.

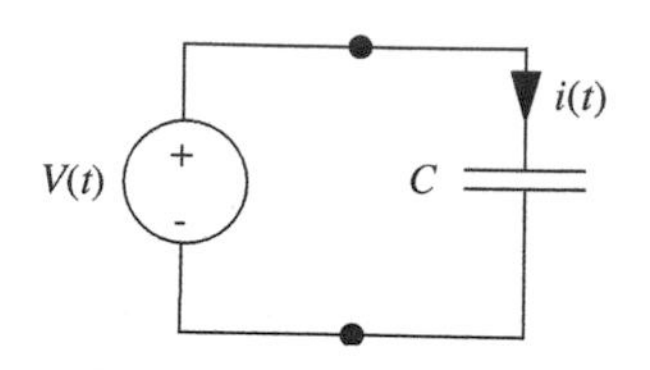


FIGURE 9.40 A voltage source driving a capacitor.

Figure 9.39 also illustrates an important point, namely that the flux linked by, and hence the current through, an inductor is a continuous function of time. Even though $V$ steps discontinuously at $t = 0$, $i$ does not; the state $\lambda$, and hence $i$, is continuous. The only way for $i$ to take a discontinuous step would be for the voltage source to deliver a nonzero flux linkage in zero time, which requires an infinite voltage. This is of course also not a practical possibility, although we will see that such a mathematical construction can also be used very effectively to model certain physical phenomena.

---

EXAMPLE 9.12 RELAY A relay is constructed as an electromagnet that opens and closes a mechanical switch. Suppose that the electromagnet can be modeled as an inductor having a 10-mH inductance. Suppose further that it will close the mechanical switch once its current reaches 10 mA. What voltage step must be applied to the electromagnet to close the switch in 100 $\mu$s?

This problem is well modeled by Figure 9.33 with $V$ being a voltage step. Hence, Equation 9.78 and Figure 9.39 apply. Since the current slope is to be 10 mA in 100 $\mu$s, or 100 A/s, and since the inductance is 10 mH, the voltage must be 1 V.

A second way to solve this problem is to use Equation 9.27 with $L = 10$ mH and $i = 10$ mA to determine that the voltage source must deliver a flux linkage of $10^{-4}$ Wb. Since this flux linkage is to be delivered at a constant rate over a 100-$\mu$s period, the voltage must be 1 V.

---

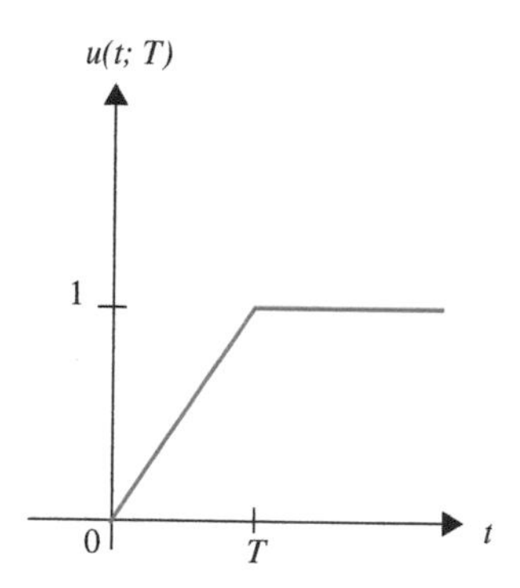


FIGURE 9.41 The ramping unit step function $u(t; T)$.

Let us now turn to the circuit shown in Figure 9.32 (redrawn here as Figure 9.40). Let us analyze its operation with a voltage source that takes a discontinuous step as expressed in the following equation

$$V(t) = \begin{cases} 0 & t \leq 0 \\ V_{\circ} & t > 0 \end{cases} . \tag{9.79}$$

and sketched in Figure 9.45a.

To analyze its operation with a source that takes a discontinuous step, we refer to Equation 9.64, and notice that we must contend with the differentiation of the step at $t = 0$. We can develop an understanding of this differentiation with the help of the ramping unit step function $u(t; T)$ defined in Figure 9.41. Here, $u(t; T)$ is a function of time $t$, having the ramp duration $T$ as a parameter.

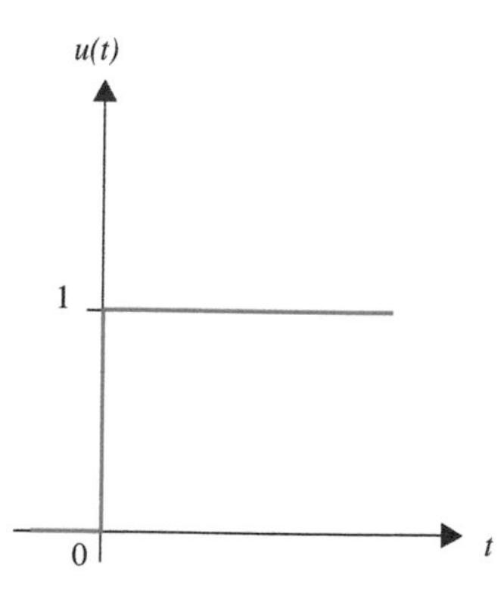


FIGURE 9.42 The unit step function $u(t)$.

Note that the ramp in $u(t; T)$, which occurs over the period $0 \leq t \leq T$, becomes increasingly steeper as the ramp width $T$ approaches 0. In fact, it is $u(t; T)$ in the limit $T \rightarrow 0$, or simply $u(t)$, as illustrated in Figure 9.42. Notice that the ideal unit step is the function at work in Equation 9.79. Recognizing this limiting behavior, our approach to dealing with the differentiation of a step will be to take a more roundabout, but easier route: We will compute the response of the circuit to a ramping unit step function, and then take the limit as $T \rightarrow 0$.

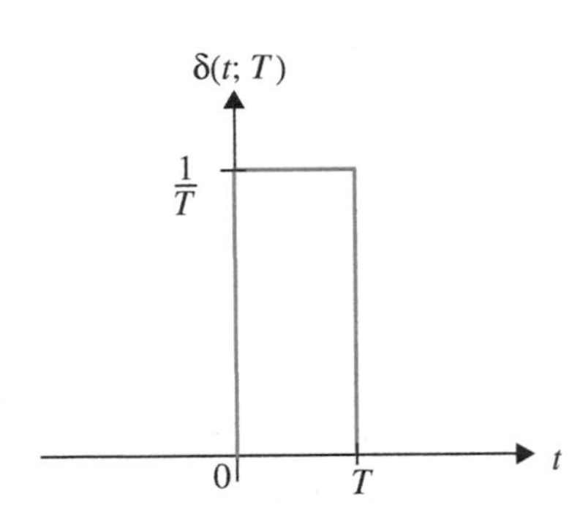


FIGURE 9.43 The unit-area pulse function $\delta(t; T)$ obtained by differentiating the ramping unit step function $u(t; T)$.

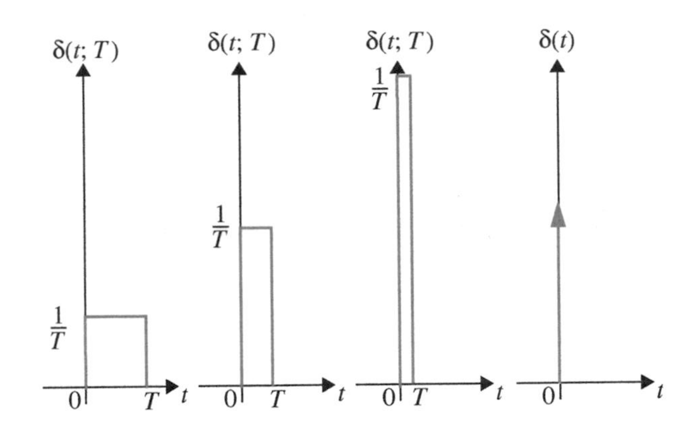


FIGURE 9.44 The unit-area pulse function becomes the unit impulse in the limit as $T \to 0$.

Thus, we can rewrite Equation 9.79 in terms of the unit step function as

$$V(t) = V_\circ \lim_{T \to 0} u(t; T) \equiv V_\circ u(t). \tag{9.80}$$

The ramping unit step function $u(t; T)$ can be differentiated to yield the unit-area pulse function $\delta(t; T)$ shown in Figure 9.43. This function becomes increasingly narrow and tall as $T$ approaches 0, but in doing so it maintains unit area, as depicted in Figure 9.44. In the limit $T \to 0$, $\delta(t; T)$ becomes the unit impulse (see the right-most graph in Figure 9.44, which we will simply denote by $\delta(t)$.

The unit impulse[4] has several important properties for our purposes. These properties are

$$\delta(t) = 0 \quad \text{for} \quad t \neq 0 \tag{9.81}$$

$$\int_{-\infty}^{t} \delta(t)dt = u(t) \quad \Leftrightarrow \quad \delta(t) = \frac{du(t)}{dt} \tag{9.82}$$

$$\int_{-\infty}^{\infty} \delta(t)dt = 1. \tag{9.83}$$

---

4. $u(t)$ and $\delta(t)$ are commonly used to represent the unit step function and the unit impulse function, respectively. Sometimes, the following notation is also used: $u_0(t)$ to represent a unit impulse at time $t = 0$. The notation $u_n(t)$ is used to represent the function that results from differentiating the impulse $n$ times, and the notation $u_{-n}(t)$ represents the function that results from integrating the impulse $n$ times. Thus $u_{-1}(t)$ represents the unit step at time $t = 0$, $u_{-2}(t)$ the ramp, and $u_1(t)$ the doublet at time $t = 0$.

Each of these properties can be deduced from the properties of $\delta(t; T)$ in the limit $T \rightarrow 0$. Finally, note that in accordance with Figure 9.43, the units of $\delta(t)$ are reciprocal time.

Now that we have the definitions and limiting interpretations of $u(t; T)$ and $\delta(t; T)$ in hand, we can complete the analysis of the circuit shown in Figure 9.40 with a source that steps discontinuously. Suppose that the voltage source in the figure produces the ramping voltage step given by

$$V(t) = V_o u(t; T). \tag{9.84}$$

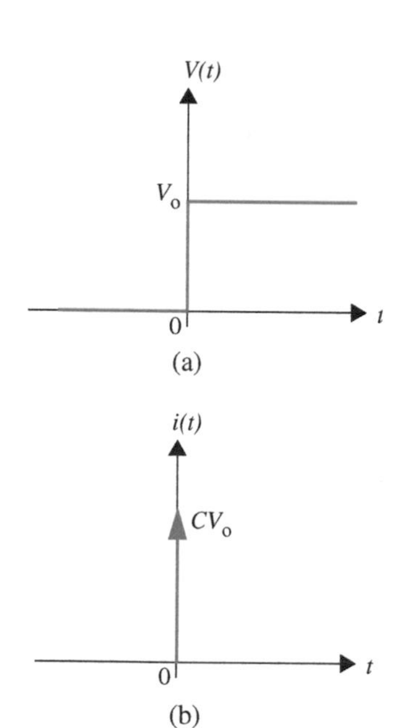


FIGURE 9.45 The voltage and current in the circuit shown in Figure 9.32 for a step voltage input.

Substitution of Equation 9.84 into Equation 9.64 then yields

$$i(t) = CV_o\delta(t; T). \tag{9.85}$$

Equation 9.85 shows that the voltage source supplies the current $CV_o/T$ during the period $0 \leq t \leq T$ as it ramps up the capacitor voltage from 0 V to $V_o$; note that $\delta(t; T) = 1/T$ during that period. In ramping up the capacitor voltage to $V_o$, the source delivers the charge $CV_o$ in accordance with Equation 9.8. This can be verified by integrating $i$ over $0 \leq t \leq T$.

Now consider the circuit behavior described by Equations 9.84 and 9.85 in the limit $T \rightarrow 0$. In this case, $V$ becomes the discontinuous voltage step described by Equations 9.79 and 9.80, and $i$ becomes

$$i(t) = CV_o\delta(t) \tag{9.86}$$

which is the desired response to the discontinuous step in the source voltage. The forms of the input voltage step and the current impulse response for the circuit in Figure 9.32 are depicted in Figure 9.45.

This response can also be obtained directly by substituting Equation 9.80 into Equation 9.64, and then making use of Equation 9.82. At first glance, it might not appear that $CV_o\delta(t)$ has the units of current, but it does because $CV_o$ has the units of charge, and $\delta(t)$ has the units of reciprocal time. In fact, $CV_o$ is the total charge delivered by the impulse current.

From our limiting interpretation of the impulse, we see that $i$ in Equation 9.86 is a current that instantaneously delivers the charge $CV_o$ to the capacitor at $t = 0$. Thus, the charge stored in the capacitor takes a step at $t = 0$, and so the voltage steps too as driven by the source. This illustrates an important point made earlier, namely that it takes an infinite current to cause the charge stored by, and hence the voltage appearing across, a capacitor to take a discontinuous step. Thus, except under unusual circumstances involving infinite currents, the state of a capacitor is a continuous function of time.

---

EXAMPLE 9.13 CURRENT STEP INPUT TO A INDUCTOR

Our next example considers a current step input of the following form (and sketched in Figure 9.47a):

$$I(t) = \begin{cases} 0 & t \leq 0 \\ I_o & t > 0 \end{cases} \tag{9.87}$$

to the circuit shown in Figure 9.34 (redrawn here as Figure 9.46).

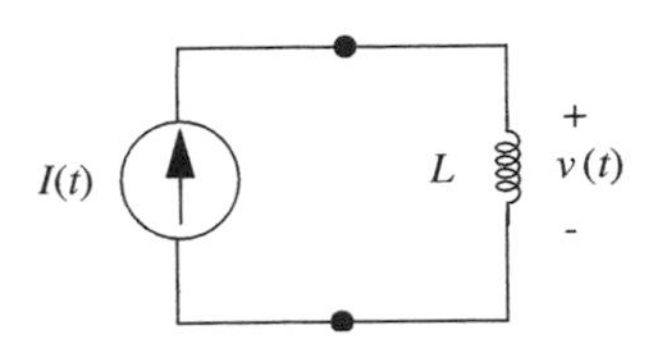


FIGURE 9.46 A current source driving an inductor.

Since they are duals, the behavior of the circuit shown in Figure 9.46 parallels that of the circuit shown in Figure 9.40. Suppose that the current source in the figure produces the ramping current step given by

$$I(t) = I_o u(t; T). \tag{9.88}$$

Substitution of Equation 9.88 into Equation 9.66 then yields

$$v(t) = LI_o \delta(t; T). \tag{9.89}$$

Equation 9.89 shows that the current source supplies the voltage $LI_o/T$ during the period $0 \leq t \leq T$ as it ramps up the inductor current from 0 A to $I_o$; note again that $\delta(t; T) = 1/T$ during that period. In ramping up the inductor current to $I_o$ the source delivers the flux linkage $LI_o$ in accordance with Equation 9.27. This can be verified by integrating $v$ over $0 \leq t \leq T$.

Now consider the circuit behavior described by Equations 9.88 and 9.89 in the limit $T \to 0$. In this case, $I$ becomes the discontinuous current step described by Equation 9.87 and

$$I(t) = I_o \lim_{T \to 0} u(t; T) \equiv I_o u(t) \tag{9.90}$$

and $v$ becomes

$$v(t) = LI_o \delta(t) \tag{9.91}$$

which is the desired response to the discontinuous step in the source current. The forms of the input current step and the voltage impulse response for the circuit in Figure 9.46 are depicted in Figure 9.47.

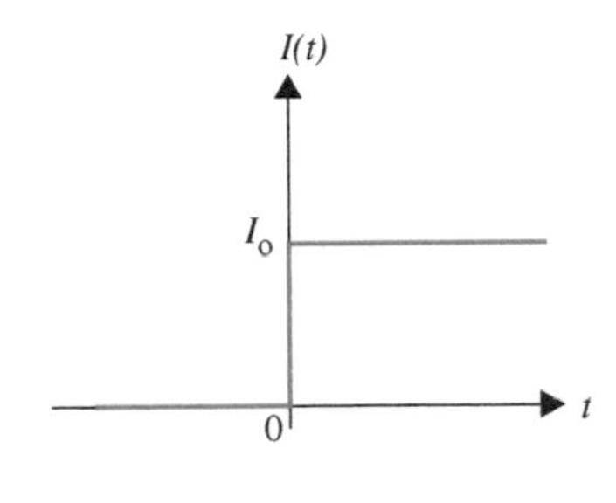


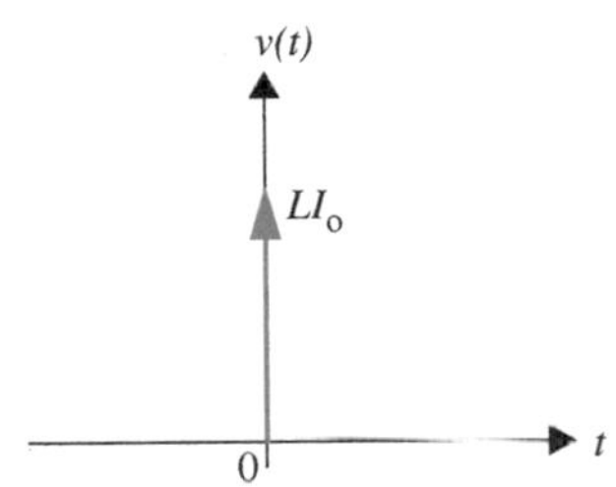


FIGURE 9.47 The current and voltage in the circuit shown in Figure 9.46 for a step current input.

This response can also be obtained directly by substituting Equation 9.90 into Equation 9.66, and then making use of Equation 9.82. At first glance it might not appear that $LI_o\delta(t)$ has the units of voltage, but it does because $LI_o$ has the units of flux linkage, and $\delta(t)$ has the units of reciprocal time. In fact, $LI_o$ is the total flux linkage delivered by the impulse voltage.

From our limiting interpretation of the impulse we now see that $v$ in Equation 9.91 is a voltage that instantaneously delivers the flux linkage $LI_o$ to the inductor at $t = 0$. Thus, the flux linked by the inductor takes a step at $t = 0$, and so the current steps too as driven by the source. This illustrates an important point made earlier, namely that it takes an infinite voltage to cause the flux linked by, and hence the current passing through, an inductor to take a discontinuous step. Thus, except under unusual circumstances involving infinite voltages, the state of an inductor is a continuous function of time.

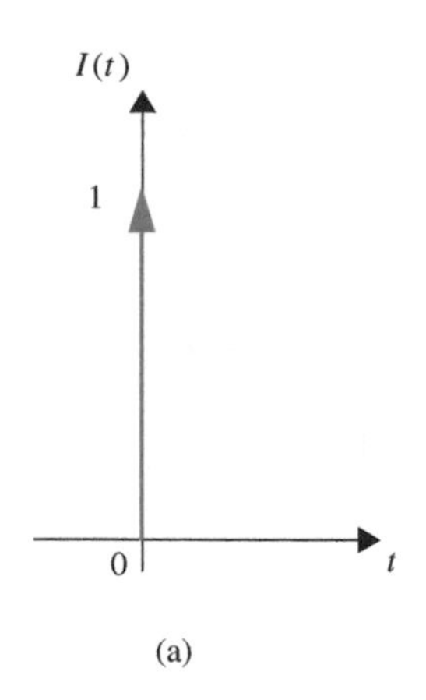


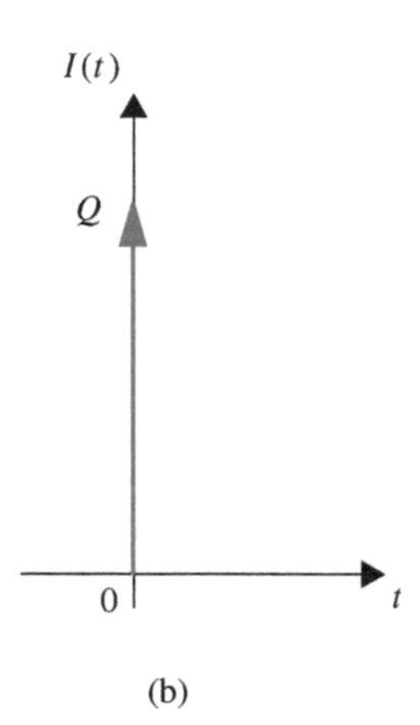


FIGURE 9.48 (a) The unit impulse $\delta(t)$; (b) an impulse with area Q.

### 9.4.3 IMPULSE INPUTS

We introduced impulse functions in the previous section. Recall that an unit impulse denoted by $\delta(t)$ has the following properties.

$$\delta(t) = 0 \quad \text{for} \quad t \neq 0 \tag{9.92}$$

$$\int_{-\infty}^{t} \delta(t)dt = u(t) \quad \Leftrightarrow \quad \delta(t) = \frac{du(t)}{dt} \tag{9.93}$$

$$\int_{-\infty}^{\infty} \delta(t)dt = 1. \tag{9.94}$$

In other words, the impulse $\delta(t)$ is nonzero only for $t = 0$. Its integral produces the unit step function, and the area under it is 1.

Recall further that the unit-area pulse function in Figure 9.44 becomes the unit impulse in the limit as $T \to 0$.

Figure 9.48a shows a unit current impulse and Figure 9.48b shows a current impulse with area under the impulse Q. In other words,

$$\int_{-\infty}^{\infty} i(t)dt = \int_{-\infty}^{\infty} \delta(t)dt = 1 \tag{9.95}$$

for the current in Figure 9.48a, and

$$\int_{-\infty}^{\infty} i(t)dt = \int_{-\infty}^{\infty} Q\delta(t)dt = Q \tag{9.96}$$

for the current in Figure 9.48b.

Let us analyze the circuit in Figure 9.31 (redrawn in Figure 9.49 for convenience) for an impulse input current with strength Q. In other words, an input current of the form:

$$I(t) = Q\delta(t).$$

Recall that solving the circuit implies finding the values of all the branch variables. The one branch variable that is unknown in the circuit is the voltage $v(t)$. We can obtain $v(t)$ by integrating the current through the capacitor as

$$\begin{aligned} v(t) &= \frac{1}{C}\int_{-\infty}^{t} I(t)dt \\ &= \frac{1}{C}\int_{-\infty}^{t} Q\delta(t)dt \\ &= \frac{1}{C}Qu(t). \end{aligned} \tag{9.97}$$

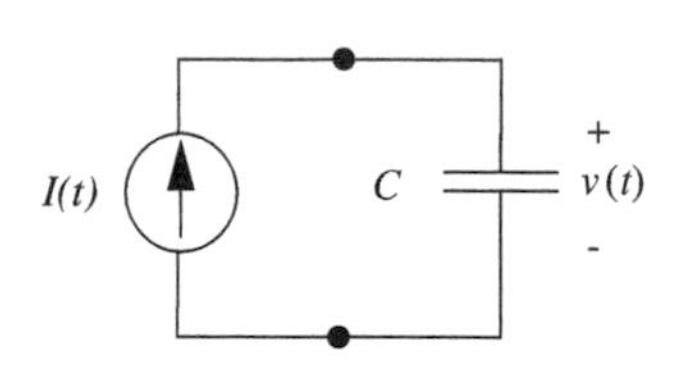


FIGURE 9.49 A current source driving a capacitor.

Thus, a current impulse of strength Q that occurs at time $t$ deposits a charge Q on the capacitor. This charge results in the capacitor voltage jumping to $\frac{1}{C}Q$ at time $t$ as illustrated in Figure 9.50.

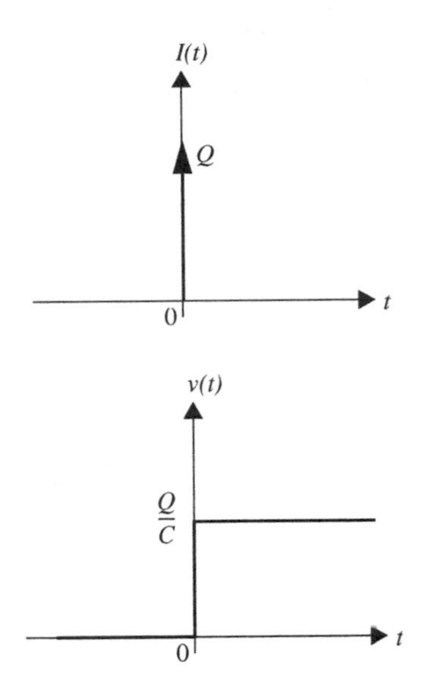


FIGURE 9.50 The current and voltage in the circuit shown in Figure 9.32 for an impulse current input.

Before ending this subsection, it is worthwhile to comment on impulse sources. Since impulses are a mathematical invention, and not a physical occurrence, it might appear that they have limited practical value. However, this is not the case. We often encounter sources that produce very narrow pulses of voltage or current. When these pulses are so narrow that we do not really care about the details of their shape, then we can model them very simply by an impulse with an equivalent area. From a mathematical viewpoint, this offers a significant savings since an impulse source is much easier to deal with than a source that produces a pulse with a complex shape. Thus, impulses are actually very useful modeling tools.

### WWW 9.4.4 ROLE REVERSAL*

## 9.5 ENERGY, CHARGE, AND FLUX CONSERVATION

In Section 9.2, we studied the parallel combination of capacitors that stored no charge at the time of their connection, and the series combination of inductors that linked no flux at the time of their connection. In this section, we extend that study to consider connections in the presence of initial charge and flux linkage.

Consider the parallel connection of the two initially-charged capacitors shown in Figure 9.51; the connection occurs when the switch closes. We wish to determine the state of the capacitors after the switch is closed. KCL applied to the bottom node of the circuit dictates that

$$\frac{dq_1(t)}{dt} + \frac{dq_2(t)}{dt} = \frac{d}{dt}(q_1(t) + q_2(t)) = 0. \tag{9.98}$$

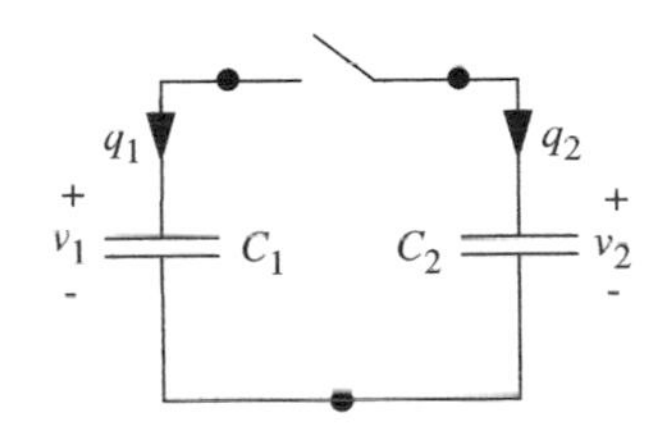


FIGURE 9.51 Two capacitors connected in parallel through a switch.

Equation 9.98 states that the total charge $q_1 + q_2$ on both capacitors is constant, and hence conserved for all time, even as the switch closes. Now, let the switch close at $t = 0$. After the switch closes, that is, for $t > 0$, KVL applied to the loop in Figure 9.51 dictates that

$$v_1(t) = v_2(t), \tag{9.99}$$

and, with the help of Equation 9.8, that

$$\frac{q_1(t)}{C_1} = \frac{q_2(t)}{C_2}. \tag{9.100}$$

We can now use Equations 9.98 and 9.100 to determine the capacitor charges, and, with the help of Equation 9.8, the common voltage after the switch is closed. To begin, let us denote the charges on the two capacitors prior to switch closure as $Q_1$ and $Q_2$. Then, from Equation 9.98,

$$q_1(t) + q_2(t) = Q_1 + Q_2. \tag{9.101}$$

Next, Equations 9.100 and 9.101 can be jointly solved to yield

$$q_1(t) = \frac{C_1}{C_1 + C_2}(Q_1 + Q_2) \tag{9.102}$$

$$q_2(t) = \frac{C_2}{C_1 + C_2}(Q_1 + Q_2) \tag{9.103}$$

for $t > 0$. Finally, Equations 9.102 and 9.103 can be substituted in Equation 9.8 to yield

$$v_1(t) = \frac{q_1(t)}{C_1} = \frac{Q_1 + Q_2}{C_1 + C_2} \tag{9.104}$$

$$v_2(t) = \frac{q_2(t)}{C_2} = \frac{Q_1 + Q_2}{C_1 + C_2}, \tag{9.105}$$

again for $t > 0$. According to Equations 9.102 and 9.103, the capacitors share the total charge in proportion to their capacitance.

While charge is conserved during the closure of the switch, it is interesting to note that energy is not. Using Equation 9.18, the total energy stored between the two capacitors before the switch is closed is found to be

$$w_E(t < 0) = \frac{Q_1^2}{2C_1} + \frac{Q_2^2}{2C_2}. \tag{9.106}$$

Using Equations 9.102, 9.103, and 9.18, the total energy stored after the switch is closed is found to be

$$w_E(t > 0) = \left(\frac{C_1(Q_1 + Q_2)}{C_1 + C_2}\right)^2 \frac{1}{2C_1} + \left(\frac{C_2(Q_1 + Q_2)}{C_1 + C_2}\right)^2 \frac{1}{2C_2}$$

$$= \frac{(Q_1 + Q_2)^2}{2(C_1 + C_2)}. \tag{9.107}$$

The two energies are not equal. Further, with a little algebraic manipulation, it can be seen that

$$w_E(t < 0) - w_E(t > 0) = \frac{1}{2}\frac{C_1 C_2}{C_1 + C_2}\left(\frac{Q_1}{C_1} - \frac{Q_2}{C_2}\right)^2 \geq 0, \qquad (9.108)$$

and so energy is always lost during the closure of the switch, except for the special case in which $v_1$ and $v_2$ are equal before the switch closes. Where the lost energy goes is not apparent from Figure 9.51 because it is an idealized figure. Perhaps the energy is dissipated in the wires used to connect the two capacitors and the switch. Or, perhaps it is dissipated in a spark as the switch closes. It might even be lost as radiated electromagnetic energy. In any case, it is lost. One of the problems at the end of this chapter explores this loss further.

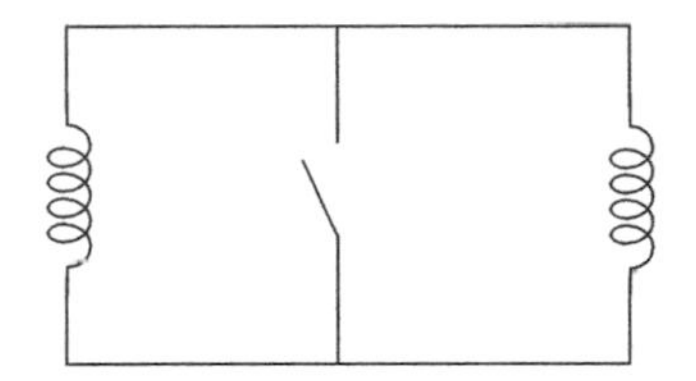

FIGURE 9.52 Two inductors connected in series through a switch.

Finally, we note that arguments similar to those presented here can be made for inductors connected in series. The corresponding circuit is shown in Figure 9.52. Initially, the switch is closed, allowing each inductor to carry an arbitrary current. Then, the switch opens. In this way, two inductors with different initial currents are connected in series. Using arguments similar to those presented earlier it is possible to determine the inductor currents and flux linkages after the switch opens. Then, it is straightforward to show that energy is lost during the opening of the switch.

---

EXAMPLE 9.14 CHARGE SHARING IN CAPACITORS In Figure 9.51, suppose that $C_1 = 1\ \mu F$ and $C_2 = 10\ \mu F$. Further, suppose that $v_1 = 10$ V and $v_2 = 1$ V before the switch is closed. How much energy is stored in the two capacitors before the switch is closed? Also, how much energy is lost during the switch closure, and what is the common voltage across the capacitors after the switch closes?

To begin, we find the charge on each capacitor before the switch closes. Using Equation 9.8, $Q_1 = 10\ \mu C$ and $Q_2 = 10\ \mu C$. Next, use Equations 9.102 and 9.103 to determine that $q_1 = 20/11\ \mu C$ and $q_2 = 200/11\ \mu C$ after the switch closes. Equation 9.103 or 9.104 can then be used to determine that $v_1 = v_2 = 20/11$ V after the switch closes. Finally, Equations 9.106 and 9.108 are used to determine that the initial stored energy is 55 $\mu J$, and that the energy lost is approximately 36.8 $\mu J$.

---

## 9.6 SUMMARY

- In this chapter, we introduced capacitors and inductors, and derived their lumped element laws from more fundamental distributed physics. Since those element laws involved time derivatives, or alternatively time integrals, time became an important variable in this chapter. This was not the case in previous chapters because the circuits we studied there contained only sources, resistors, and idealized MOSFETs, all of which have purely algebraic element laws. Because time was an important variable in this chapter, we also began to consider a richer class of functions for the sources that would input signals into a circuit. These functions included sinusoids, steps, and impulses. The circuit responses to these inputs became our first examples of transients in electronic circuits.

- When we first introduced capacitors and inductors, we did so through the effects that parasitic capacitance and inductance can have on the performance of an electronic circuit. This in turn caused us to re-evaluate our lumped matter discipline, under which such parasitics do not, by definition, exist. In the end we made a modeling compromise to preserve the lumped matter discipline while admitting the existence of important parasitics. That compromise was to augment an original circuit with lumped elements to model the important parasitics, with the understanding that the augmented model obeys the lumped matter discipline. While this is certainly an important issue, it is also important to realize that capacitors and inductors are useful well beyond the modeling of parasitics. As we shall see in future chapters they are frequently used on purpose.

- Through our analysis of capacitors and inductors, and several simple circuits that contained them, we have seen that these elements exhibit memory and are capable of reversible energy storage. A simple experiment will illustrate. As illustrated in Figure 9.53, charge up a capacitor by connecting it to a power supply (position 1), then disconnect the supply (position 2). The capacitor will "remember" the voltage of the supply for hours if a high-quality capacitor is used. A similar experiment performed with a resistor produces no memory; when the power supply is disconnected, the resistor voltage instantly falls to zero.

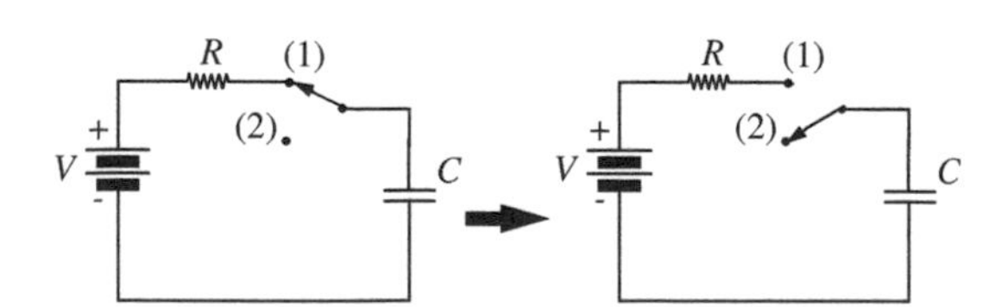


FIGURE 9.53 The capacitor holds its voltage for a long period of time.

- The corresponding experiment on an inductor yields less exciting results, however. Establish an inductor current by the circuit in Figure 9.54, with the switch in position 1. Then move the switch to position 2 (make before break). The current will decay to zero in fractions of a second, because the energy stored in the magnetic field is rapidly dissipated in the internal resistance of the coil.

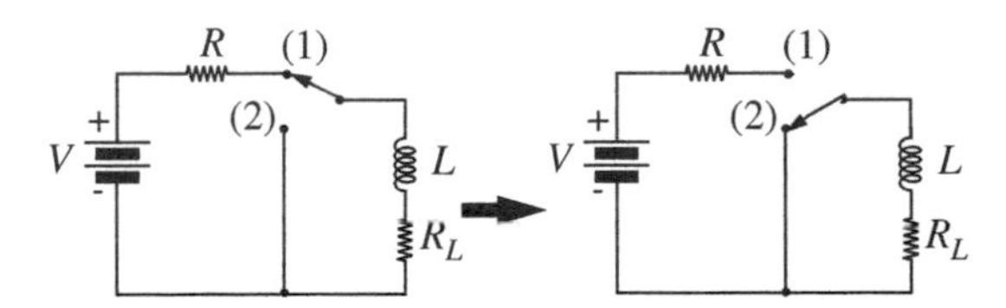


FIGURE 9.54 The inductor holds its current for a very short period of time.

- Memory and reversible energy storage are characteristics associated with the state of the elements: charge in the case of the capacitor and flux linkage in the case of the inductor. This behavior is quite different from the behavior of ideal resistors. Ideal resistors do not exhibit memory, and they irreversibly dissipate energy.
- The element law for a capacitor is

$$i = C\frac{dv}{dt}$$

and that for an inductor is

$$v = L\frac{di}{dt}.$$

- The energy stored in a capacitor is

$$w_{\mathrm{E}}(t) = \frac{q^2(t)}{2C} = \frac{Cv(t)^2}{2}.$$

- The energy stored in an inductor is

$$w_{\mathrm{M}}(t) = \frac{\lambda^2(t)}{2L} = \frac{Li(t)^2}{2}.$$

- Finally, in the process of introducing capacitors and inductors, we defined the symbols and units for various physical quantities. These definitions are summarized in Table 9.1. The units can be further modified with the engineering multipliers listed in Table 1.3.

**TABLE 9.1** Electrical engineering quantities, their units, and symbols for both.

| QUANTITY | SYMBOL | UNITS | SYMBOL |
|---|---|---|---|
| Time | $t$ | Second | s |
| Charge | $q$ | Coulomb | C |
| Capacitance | $C$ | Farad | F |
| Flux Linkage | $\lambda$ | Weber | Wb |
| Inductance | $L$ | Henry | H |
| Energy | $w$ | Joule | $J$ |

## EXERCISES

**EXERCISE 9.1** Find the equivalent capacitance between the two terminals in each of the networks in Figure 9.55.

(a)

1 μF 3 μF

(b)

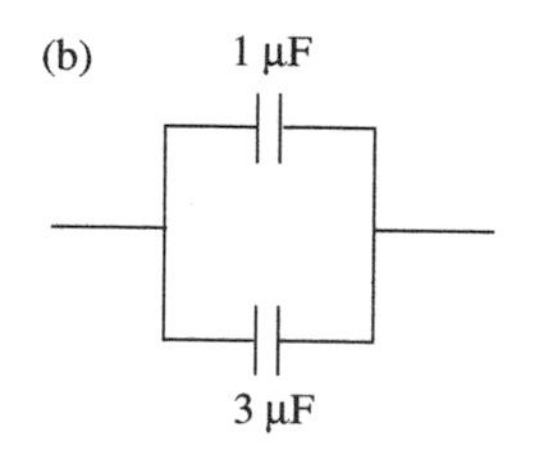


(c)

3 μF
2 μF
1 μF

FIGURE 9.55

**EXERCISE 9.2** Find the equivalent capacitance or inductance for each case in Figure 9.56.

**EXERCISE 9.3** Consider a power line on a computer backplane that is 2.5 mm wide, and separated from its underlying ground plane by 25 $\mu$m. Let the permittivity and permeability of the separating insulator be $2\epsilon_o$ and $\mu_o$, respectively. What is the capacitance and inductance of the line per 10 cm of length?

If the voltage on the line is 5 V, how much energy is stored in its capacitance per 10 cm of length? If the current through the line is 1 A, how much energy is stored in its inductance per 10 cm of length?

**EXERCISE 9.4** A current source drives a capacitor as shown in Figure 9.57. The source current is as shown in Figure 9.58 for $0 \leq t \leq T$. If the capacitor voltage is $V_o$ at $t = T$, what was it at $t = 0$?

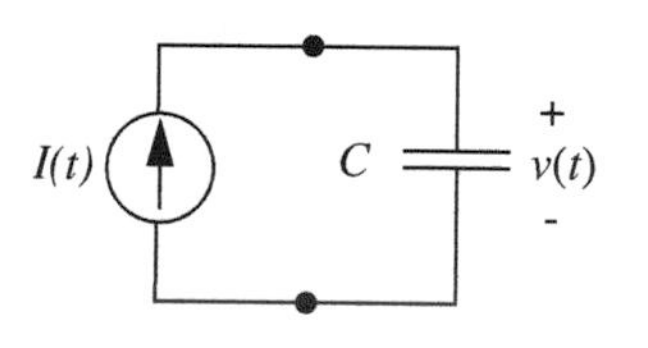


**FIGURE 9.57** A current source driving a capacitor.

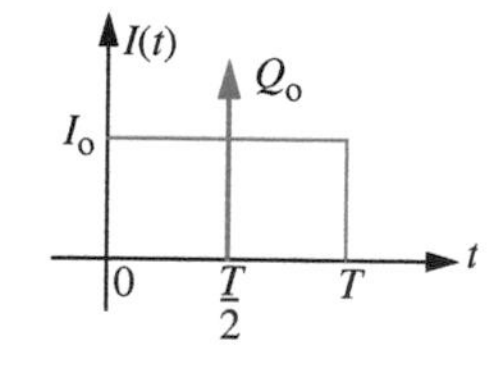


**FIGURE 9.58** Source current.

**EXERCISE 9.5** A voltage source drives an inductor as shown in Figure 9.59. The source voltage is as shown in Figure 9.60 for $0 \leq t \leq T$. If the inductor current is $I_o$ at $t = T$, what was it at $t = 0$?

**EXERCISE 9.6** Figure 9.61 shows four circuits, labeled "1" through "4," together with the waveform for the source in each circuit. The figure also shows four branch-variable waveforms, labeled "a" through "d," that could correspond to the branch current $i$ or branch voltages $v$ labeled in the circuits. Match the branch variable waveforms to the appropriate circuit and source waveform.

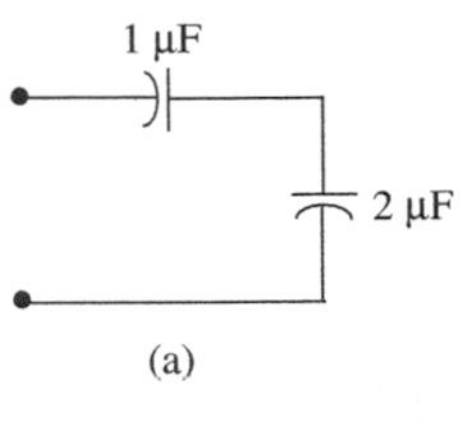


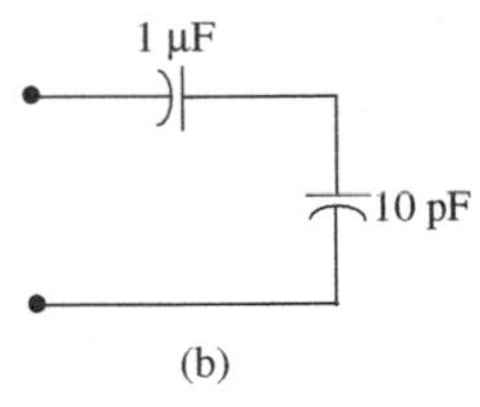


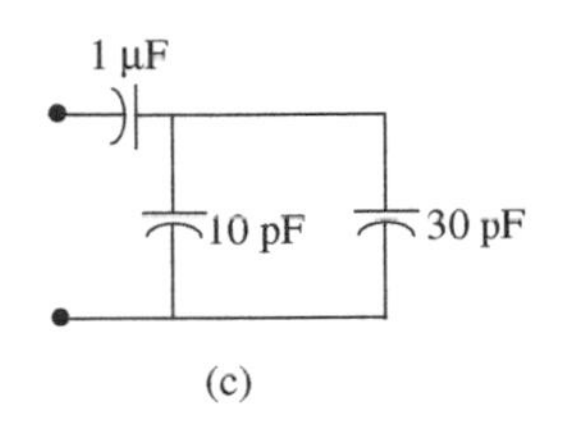


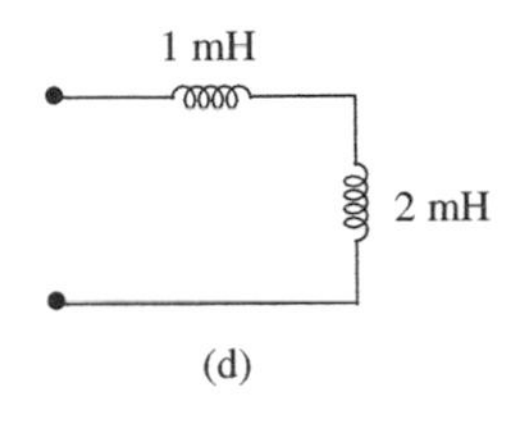


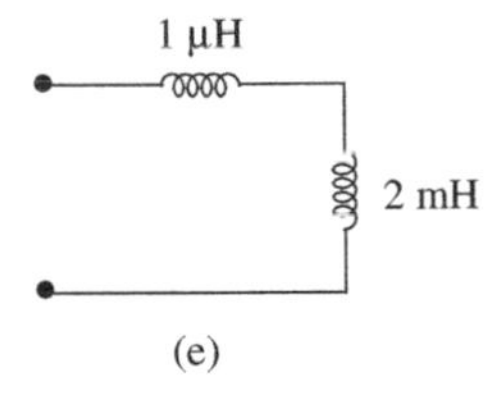


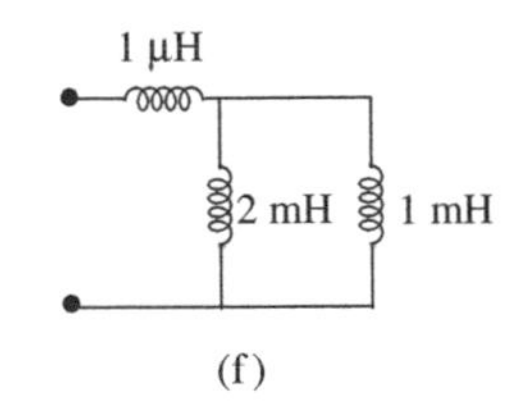


FIGURE 9.56

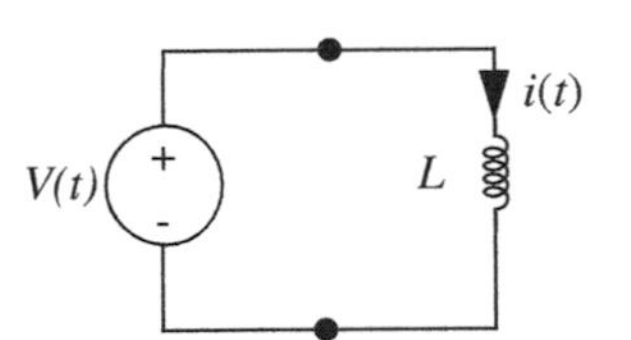


FIGURE 9.59 A current source driving an inductor.

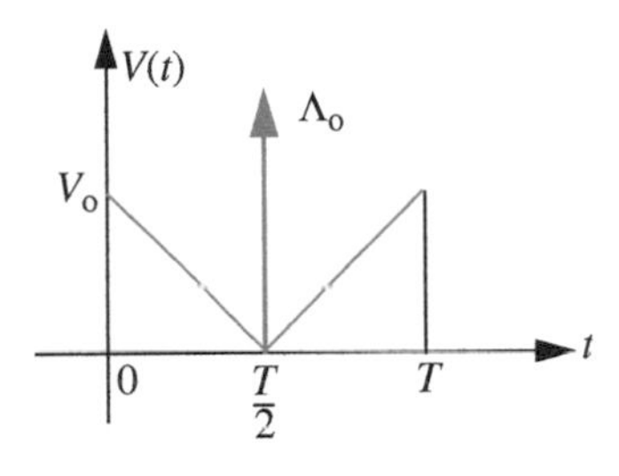


FIGURE 9.60 Source current.

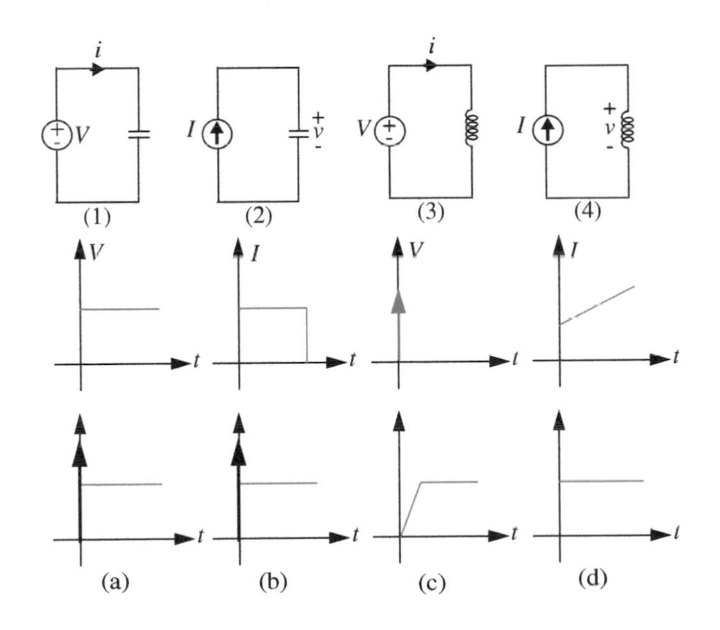


FIGURE 9.61

# PROBLEMS

**PROBLEM 9.1** A voltage source is connected in series with two capacitors as shown in Figure 9.62. The source voltage is $V(t) = 5\text{ V } u(t)$, as shown. If the current $i$ and voltage $v$ are given by $i(t) = 4\ \mu\text{C } \delta(t)$ and $v(t) = 1\text{ V } u(t)$, again as shown, what are $C_1$ and $C_2$?

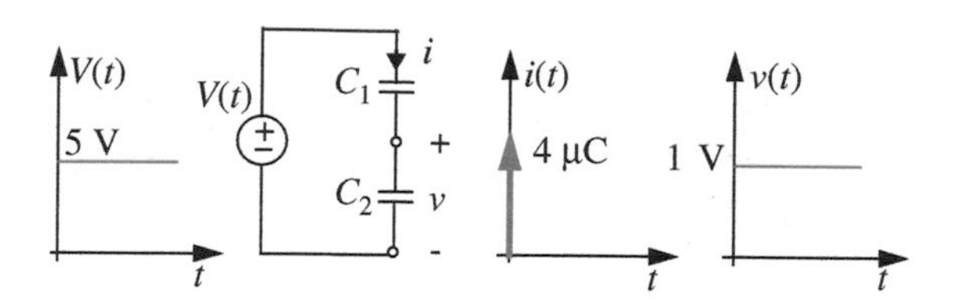


FIGURE 9.62

**PROBLEM 9.2** A current source is connected in parallel with two inductors as shown in Figure 9.63. The source current is $i(t) = 400\text{ A/s } u(t)$, as shown. If the current $i$ and voltage $v$ are given by $i(t) = 100\text{ A/s } u(t)$ and $v(t) = 0.3\text{ V } u(t)$, again as shown, what are $L_1$ and $L_2$?

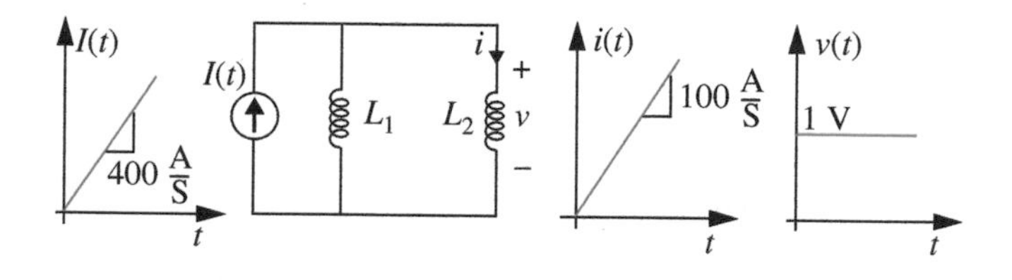


FIGURE 9.63

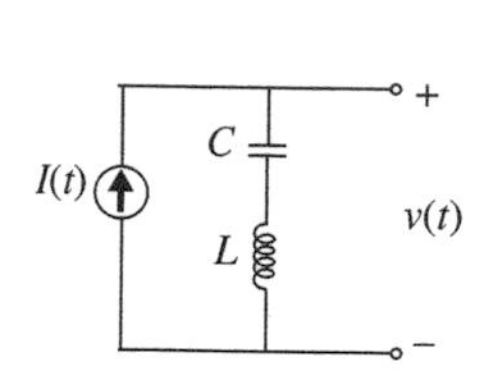


FIGURE 9.64

**PROBLEM 9.3** A current source drives a series-connected capacitor and inductor as shown in Figure 9.64. Let $I(t) = I_\circ \sin(\omega t)u(t)$, and assume that the inductor and capacitor both stored no energy prior to $t = 0$.

Determine the voltage $v$ for $t \geq 0$.

Is there any relation between $I_\circ$, $\omega$, $C$, and $L$ for which $v$ is constant for $t \geq 0$? If so, state the relation and determine $v$.

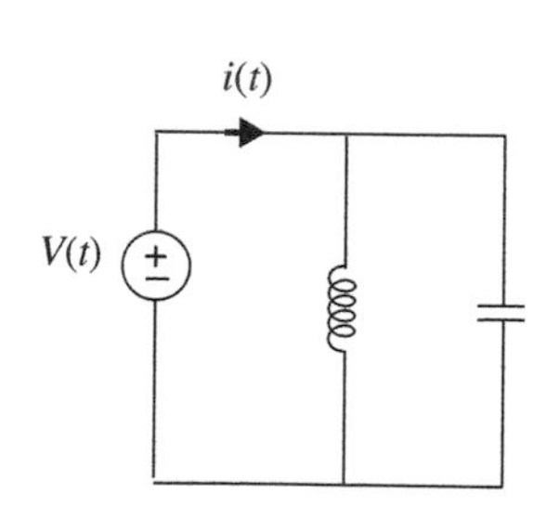


FIGURE 9.65

**PROBLEM 9.4** A voltage source drives a parallel-connected capacitor and inductor as shown in Figure 9.65. Let $V(t) = V_\circ \sin(\omega t)u(t)$, and assume that the inductor and capacitor both stored no energy prior to $t = 0$.

Determine the current $i$ for $t \geq 0$.

Is there any relation between $V_\circ$, $\omega$, $C$, and $L$ for which $i$ is constant for $t \geq 0$? If so, state the relation and determine $i$.

PROBLEM 9.5 A constant voltage source having value $V$ drives a time-varying capacitor as shown in Figure 9.66. The time-varying capacitance is given by $C(t) = C_0 + C_1 \sin(\omega t)$. Determine the capacitor current $i(t)$.

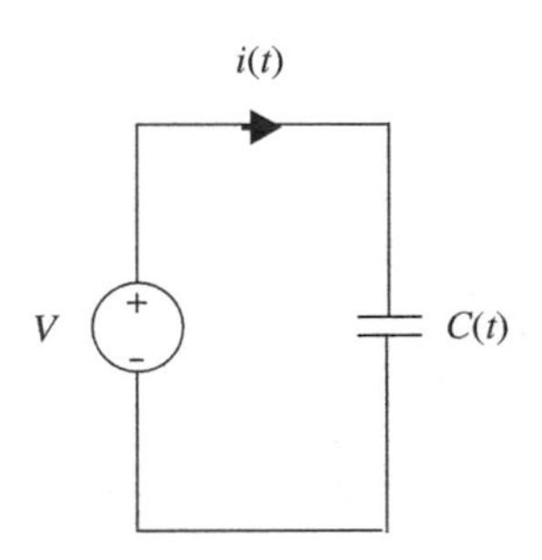


FIGURE 9.66

PROBLEM 9.6 A constant current source having value $I$ drives a time-varying inductor as shown in Figure 9.67. The time-varying inductance is given by $L(t) = L_0 + L_1 \sin(\omega t)$. Determine the inductor voltage $v(t)$.

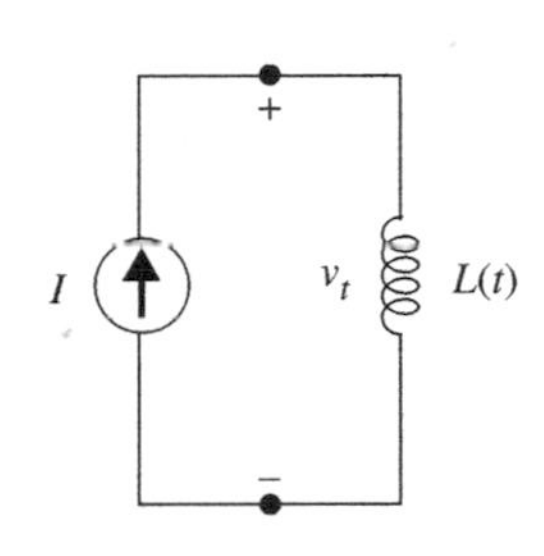


FIGURE 9.67

PROBLEM 9.7 Consider the parallel plate capacitor shown in Figure 9.68. Assume that the dielectric is free space so that $\epsilon = \epsilon_o$.

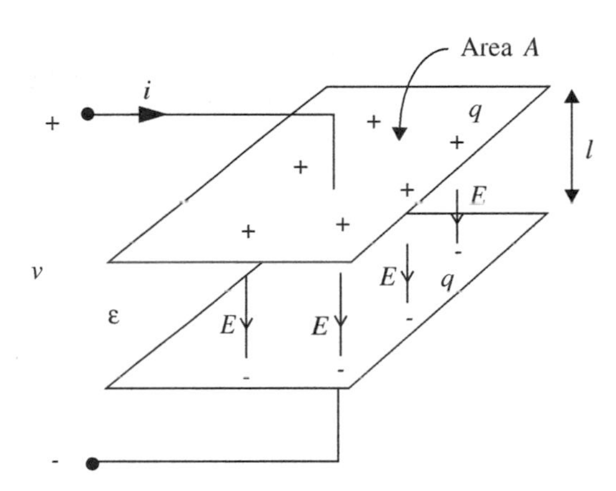


FIGURE 9.68

Suppose the capacitor is charged to the voltage $V$. Determine the charge and the electric energy stored in the capacitor in this case.

The capacitor is disconnected from the charging source so that its stored charge remains constant. Following that, its plates are pulled apart so as to double the distance between them; that is, the gap separation is now $2l$. For this new configuration, determine the voltage across the terminals of the capacitor and the energy stored in the capacitor. Explain how the stored energy changes.

PROBLEM 9.8 Figure 9.69 shows two capacitive two-port networks. One is a "Π" network, and one is a "T" network. For the Π network, find $i_{1P}$ and $i_{2P}$ as functions of $v_{1P}$ and $v_{2P}$. For the T network, find $i_{1T}$ and $i_{2T}$ as functions of $v_{1T}$ and $v_{2T}$.

How must $C_{1P}$, $C_{2P}$, and $C_{3P}$ be related to $C_{1T}$, $C_{2T}$, and $C_{3T}$ for both networks to have the same terminal relations?

PROBLEM 9.9 Figure 9.70 shows two inductive two-port networks. One is a "Π" network, and one is a "T" network. For the Π network, find $v_{1P}$ and $v_{2P}$ as functions of $i_{1P}$ and $i_{2P}$. For the T network, find $v_{1T}$ and $v_{2T}$ as functions of $i_{1T}$ and $i_{2T}$.

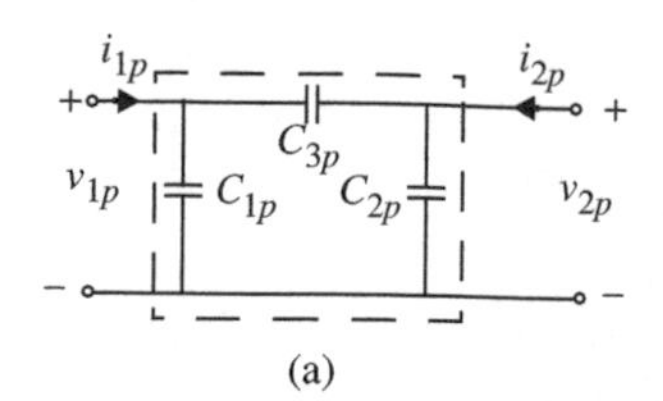


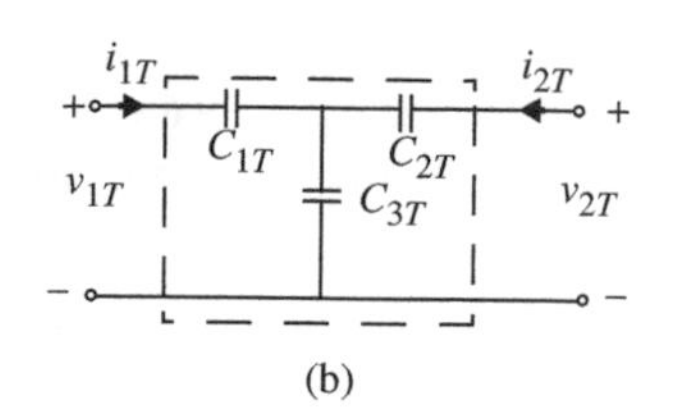


**FIGURE 9.69** (a) A capacitive T two-port network; and (b) a capacitive Π two-port network.

How must $L_{1P}$, $L_{2P}$, and $L_{3P}$ be related to $L_{1T}$, $L_{2T}$, and $L_{3T}$ for both networks to have the same terminal relations?

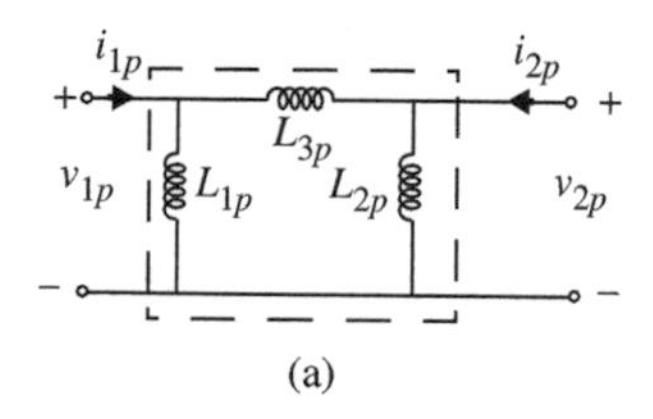


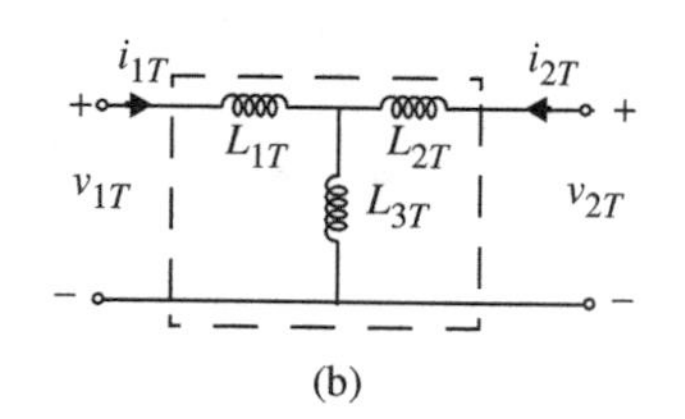


**FIGURE 9.70** (a) An inductive T two-port network; and (b) an inductive Π two-port network.

**FIGURE 9.71**

PROBLEM 9.10 This problem examines in more detail why energy is lost when the switch in Figure 9.71 closes. To do so, we examine the transient that occurs during the closure of the switch. In preparation for this, let $t = 0$ be the time at which the switch first begins to close, and let $t = T$ be the time at which the circuit reaches steady state. The charges on the two capacitors prior to switch closure are given to be $Q_1$ and $Q_2$.

Further, let $q_1(t)$ be any function defined over the interval $0 \leq t \leq T$ such that

$$q_1(0) = Q_1$$

and $q_1(T)$ is the steady state charge on the capacitor given by

$$q_1(T) = \frac{C_1}{C_1 + C_2}(Q_1 + Q_2).$$

In this way, the function $q_1$ is an arbitrary transient connecting the initial and final charge during the switch closure.

(a) Use the charge conservation relation:

$$q_1(t) + q_2(t) = Q_1 + Q_2$$

to find $q_2$ in terms of $q_1$ for $0 \leq t \leq T$. Then, use the equation:

$$\frac{dq(t)}{dt} = i(t)$$

to determine $i_1$ and $i_2$, again in terms of $q_1$ for $0 \leq t \leq T$. Finally, use the equation:

$$q(t) = Cv(t)$$

to find $v_1$ and $v_2$, also in terms of $q_1$ for $0 \leq t \leq T$. The entire transient is now described in terms of the arbitrary function $q_1$.

(b) During the transient, the difference between $v_1$ and $v_2$ must appear across some element or elements within the circuit. KVL requires this. For example, it could appear across the wiring resistance or the switch, or a combination of both. In any case, energy is lost as a current passes through this voltage difference. If we consider the voltage difference to be $(v_1 - v_2)$, as opposed to its opposite, then it is $i_2$ that passes into the positive terminal of this difference. Why?

(c) The product $i_2(v_1 - v_2)$ is the power dissipated during the transient. Determine this power in terms of $q_1$ for $0 \leq t \leq T$.

(d) Integrate the power found in the part (c) over the interval $0 \leq t \leq T$ to find the energy lost during the transient. Also, show that the energy lost is equal to the energy difference in

$$w_E(t < 0) - w_E(t > 0) = \frac{1}{2}\frac{C_1 C_2}{C_1 + C_2}\left(\frac{Q_1}{C_1} - \frac{Q_2}{C_2}\right)^2.$$

Remarkably, the energy lost is independent of the interior details of the function chosen for $q_1$. Since these details are equivalent to the details of the loss mechanism, it is apparent that the amount of energy lost is independent of how it is lost.

# CHAPTER 10

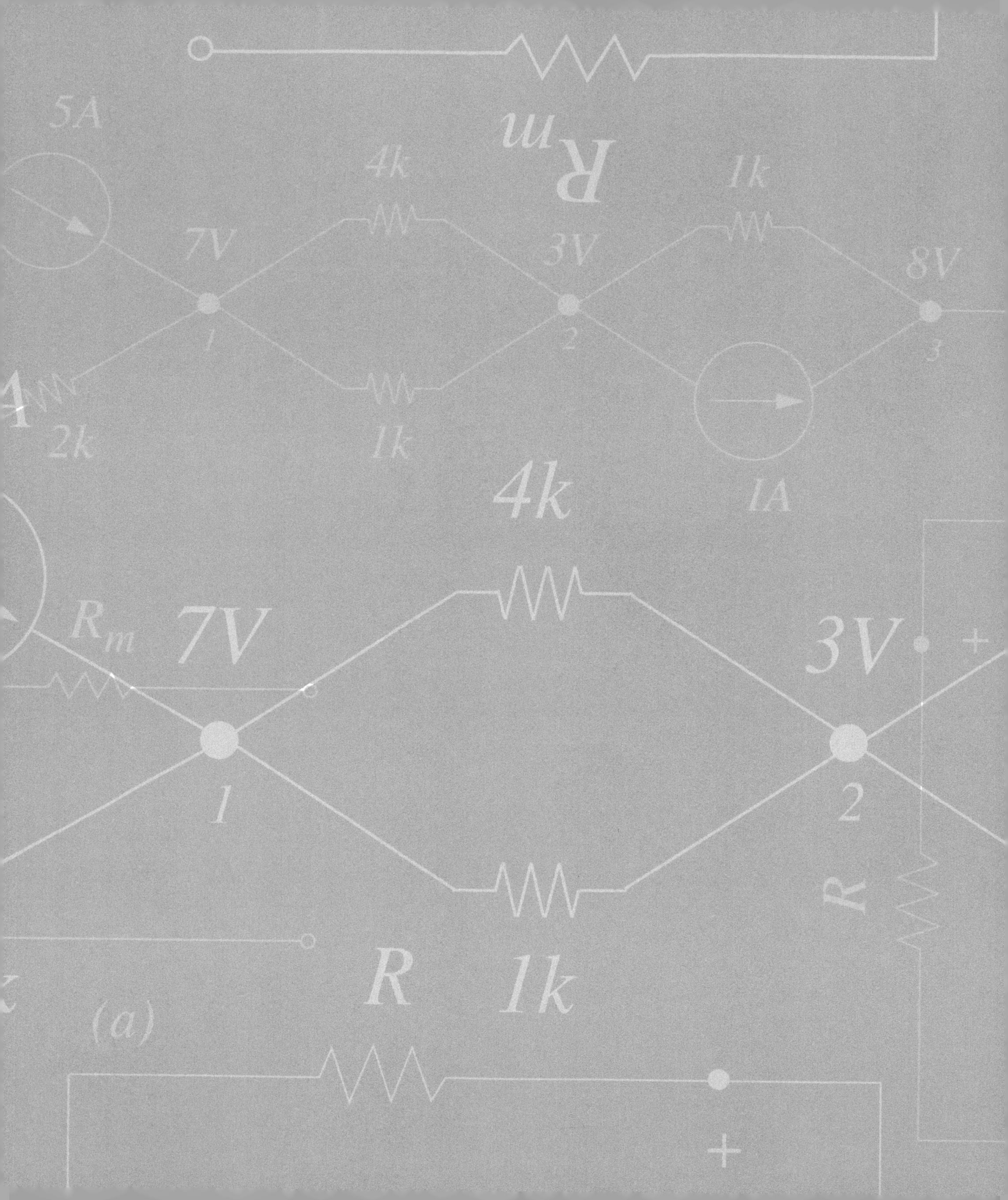

5A
4k
R
m
1k
7V
3V
8V
1
2
3
2k
1k
1A
4k
$R_m$
7V
3V
+
1
2
R
R
1k
(a)
+

# FIRST-ORDER TRANSIENTS IN LINEAR ELECTRICAL NETWORKS

# 10

As illustrated in Chapter 9, capacitances and inductances impact circuit behavior. The effect of capacitances and inductances is so acute in high-speed digital circuits, for example, that our simple digital abstractions developed in Chapter 6 based on a static discipline become insufficient for signals that undergo transitions. Therefore, understanding the behavior of circuits containing capacitors and inductors is important. In particular, this chapter will augment our digital abstraction with the concept of delay to include the effects of capacitors and inductors.

Looked at positively, because they can store energy, capacitors and inductors display the memory property, and offer signal-processing possibilities not available in circuits containing only resistors. Apply a square-wave voltage to a multi-resistor linear circuit, and all of the voltages and currents in the network will have the same square-wave shape. But include one capacitor in the circuit and very different waveforms will appear — sections of exponentials, spikes, and sawtooth waves. Figure 10.1 shows an example of such waveforms for the two-inverter system of Figure 9.1 in Chapter 9. The linear analysis techniques already developed — node equations, superposition, etc. — are adequate for finding appropriate network equations to analyze these kinds of circuits. However, the formulations turn out to be *differential equations* rather than algebraic equations, so additional skills are needed to complete the analyses.

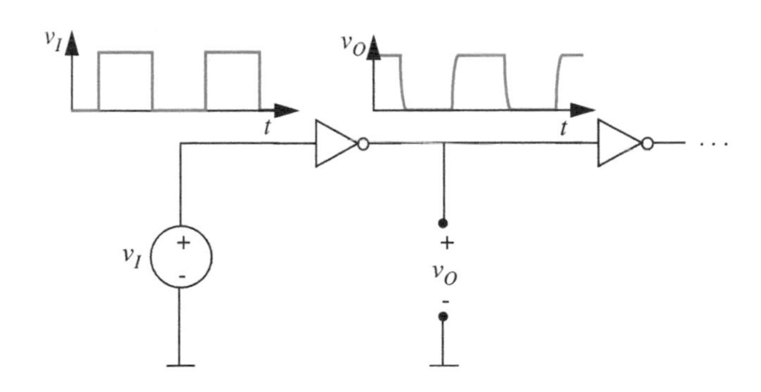


**FIGURE 10.1** Observed response of the first inverter to a square-wave input.

This chapter will discuss systems containing a single storage element, namely, a single capacitor or a single inductor. Such systems are described by simple, first-order differential equations. Chapter 12 will discuss systems containing two storage elements. Systems with two storage elements are described by second-order differential equations.[1] Higher-order systems are also possible, and are discussed briefly in Chapter 12.

This chapter will start by analyzing simple circuits containing one capacitor, one resistor, and possibly a source. We will then analyze circuits containing one inductor and one resistor. The two-inverter circuit of Figure 10.1 is examined in detail in Section 10.4.

## 10.1 ANALYSIS OF RC CIRCUITS

Let us illustrate first-order systems with a few primitive examples containing a resistor, a capacitor, and a source. We first analyze a current source driving the so-called parallel RC circuit.

### 10.1.1 PARALLEL RC CIRCUIT, STEP INPUT

Shown in Figure 10.2a is a simple source-resistor-capacitor circuit. On the basis of the Thévenin and Norton equivalence discussion in Section 3.6.1, this circuit could result from a Norton transformation applied to a more complicated

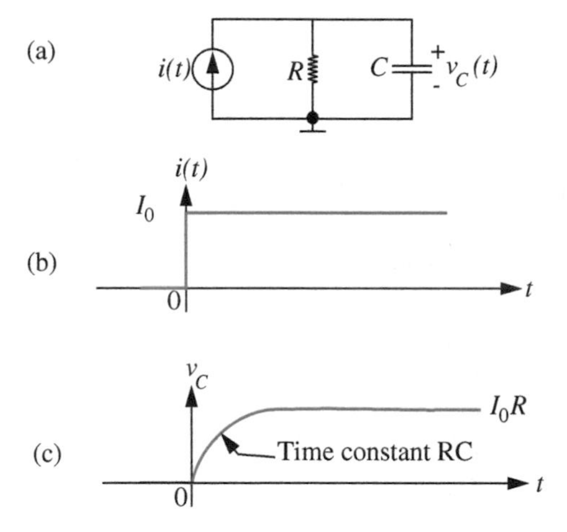


FIGURE 10.2 Capacitor charging transient.

1. However, a circuit with two storage elements that can be replaced by a single equivalent storage element remains a first-order circuit. For example, a pair of capacitors in parallel can be replaced with a single capacitor whose capacitance is the sum of the two capacitances.

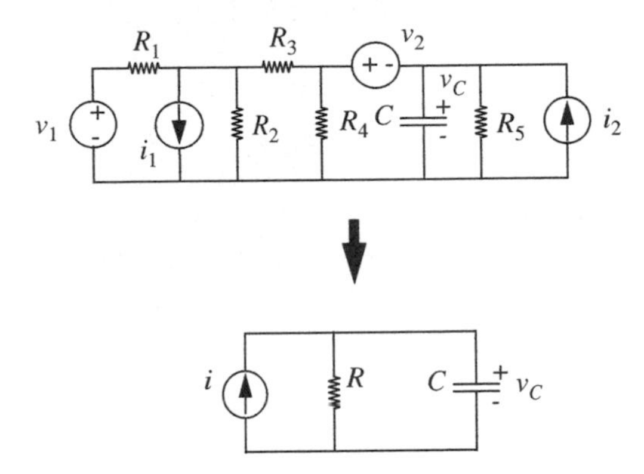


FIGURE 10.3 A more complicated circuit that can be transformed into the simpler circuit in Figure 10.2a by using Thévenin and Norton transformations.

circuit containing many sources and resistors, and one capacitor, as suggested in Figure 10.3. Let us assume we wish to find the capacitor voltage $v_C$. We will use the node method described in Chapter 3 to do so. As shown in Figure 10.2a, we take the bottom node as ground, which leaves us with one unknown node voltage corresponding to the top node. The voltage at the top node is the same as the voltage across the capacitor, and so we will proceed to work with $v_C$ as our unknown. Next, according to Step 3 of the node method, we write KCL for the top node in Figure 10.2a, substituting the constituent relation for a capacitor from Equation 9.9,

$$i(t) = \frac{v_C}{R} + C\frac{dv_C}{dt}. \tag{10.1}$$

Or, rewriting,

$$\frac{dv_C}{dt} + \frac{v_C}{RC} = \frac{i(t)}{C}. \tag{10.2}$$

As promised, the problem can be formulated in one line. But to find $v_C(t)$, we must solve a nonhomogeneous, linear first-order ordinary differential equation with constant coefficients. This is not a difficult task, but one that must be done systematically using any method of solving differential equations.

To solve this equation, we will use the *method of homogeneous and particular solutions* because this method can be readily extended to higher-order equations. As a review, the method of homogeneous and particular solutions arises from a fundamental theorem of differential equations. The method states that the solution to the nonhomogeneous differential equation can be obtained by summing together the homogeneous solution and the particular solution. More specifically, let $v_{CH}(t)$ be any solution to the homogeneous differential equation

$$\frac{dv_C}{dt} + \frac{v_C}{RC} = 0 \tag{10.3}$$

associated with our nonhomogeneous differential equation 10.2. The homogeneous equation is derived from the original nonhomogeneous equation by setting the driving function, $i(t)$ in this case, to zero. Further, let $v_{CP}(t)$ be *any* solution to Equation 10.2. Then, the sum of the two solutions,

$$v_C(t) = v_{CH}(t) + v_{CP}(t)$$

is a general solution or a total solution to Equation 10.2. $v_{CH}(t)$ is called the *homogeneous solution* and $v_{CP}(t)$ is called the *particular solution*. When dealing with circuit responses, the homogeneous solution is also called the *natural response* of the circuit because it depends only on the internal energy storage properties of the circuit and not on external inputs. The particular solution is also called the *forced response* or the *forced solution* because it depends on the external inputs to the circuit.

Let us now return to the business of solving Equation 10.2. To make the problem specific, assume that the current source $i(t)$ is a step function

$$i(t) = I_0 \qquad t > 0 \tag{10.4}$$

as shown in Figure 10.2b. Further, we assume for now that the voltage on the capacitor was zero before the current step was applied. In mathematical terms, this is an *initial condition*

$$v_C = 0 \qquad t < 0. \tag{10.5}$$

The method of homogeneous and particular solutions proceeds in three steps:

1. Find the homogeneous solution $v_{CH}$.
2. Find the particular solution $v_{CP}$.
3. The total solution is then the sum of the homogeneous solution and the particular solution. Use the initial conditions to solve for the remaining constants.

The first step is to solve the homogeneous equation, formed by setting the driving function in the original differential equation to zero. Then, any method of solving homogeneous equations can be used. In this case the homogeneous equation is

$$\frac{dv_{CH}}{dt} + \frac{v_{CH}}{RC} = 0. \tag{10.6}$$

We assume a solution of the form

$$v_{CH} = Ae^{st} \tag{10.7}$$

because the homogeneous solution for any linear constant-coefficient ordinary differential equation is always of this form. Now we must find values for the constants $A$ and $s$. Substitution into Equation 10.6 yields

$$Ase^{st} + \frac{Ae^{st}}{RC} = 0. \tag{10.8}$$

The value for $A$ cannot be determined from this equation, but discarding the trivial solution of $A = 0$, we find

$$s + \frac{1}{RC} = 0 \tag{10.9}$$

because $e^{st}$ is never zero for finite $s$ and $t$, so can be factored out. Hence

$$s = -\frac{1}{RC}. \tag{10.10}$$

Equation 10.9 is called the characteristic equation of the system, and $s = -1/RC$ is a root of this characteristic equation. The characteristic equation summarizes the fundamental dynamic properties of a circuit, and we will have much more to say about it later chapters. For reasons that will become clear in Chapter 12, the root of the characteristic equation, $s$, is also called the *natural frequency* of the system.

We now know that the homogeneous solution is of the form

$$v_{CH} = Ae^{-t/RC}. \tag{10.11}$$

The product RC has the dimensions of time and is called the *time constant* of the circuit.

The second step is to find a particular solution, that is, to find any solution $v_{CP}$ that satisfies the original differential equation; it need not satisfy the initial conditions. That is, we are looking for any solution to the equation

$$I_0 = \frac{v_{CP}}{R} + C\frac{dv_{CP}}{dt}. \tag{10.12}$$

Since the drive $I_0$ is constant in time for $t > 0$, one acceptable particular solution is also a constant:

$$v_{CP} = K. \tag{10.13}$$

To verify this, we substitute into Equation 10.12

$$I_0 = \frac{K}{R} + 0 \tag{10.14}$$

$$K = I_0 R. \tag{10.15}$$

Because Equation 10.14 can be solved for $K$, we are assured that our "guess" about the form of the particular solution, that is, Equation 10.13, was correct.[2] Hence the particular solution is

$$v_{CP} = I_0 R. \tag{10.16}$$

The total solution is the sum of the homogeneous solution (Equation 10.11) and the particular solution (Equation 10.16)

$$v_C = Ae^{-t/RC} + I_0 R. \tag{10.17}$$

The only remaining unevaluated constant is $A$, and we can solve for this by applying the initial condition. Equation 10.5 applies for $t$ less than zero, and our solution, Equation 10.17 is valid for $t$ greater than zero. These two parts of the solution are patched together by a *continuity condition* derived from Equation 9.9: An instantaneous jump in capacitor voltage requires an infinite spike in current, so *for finite current, the capacitor voltage must be continuous.* This circuit cannot support infinite capacitor current (because $i(t)$ is finite, the infinite current would have to come from the resistor, and this is impossible). Thus we are justified in assuming continuity of $v_C$, hence can equate the solutions for negative time and positive time by solving at $t = 0$

$$0 = A + I_0 R. \tag{10.18}$$

Thus

$$A = -I_0 R \tag{10.19}$$

and the complete solution for $t > 0$ is

$$v_C = -I_0 Re^{-t/RC} + I_0 R$$

---

2. Alternatively, a guess of

$$v_{CP} = Kt,$$

where $K$ is a constant independent of $t$, would not be correct, since substituting into Equation 10.12 yields

$$I_0 = \frac{Kt}{R} + CK$$

which cannot be solved for a time-independent $K$.

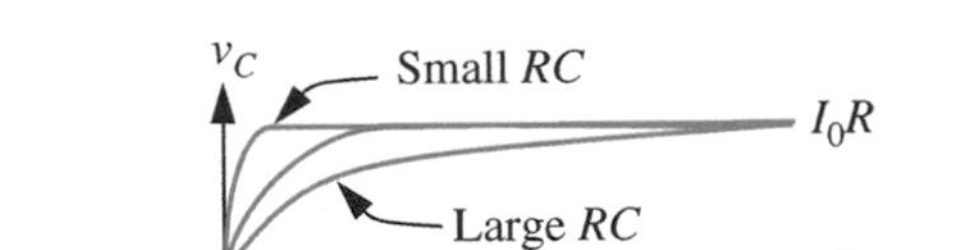


**FIGURE 10.4** Significance of the $RC$ time constant.

or

$$v_C = I_0R(1 - e^{-t/RC}). \qquad (10.20)$$

This is plotted in Figure 10.2c.

Some comments at this point help to give perspective. First, notice that capacitor voltage starts from a zero value at $t = 0$ and reaches its final value of $I_0R$ for large $t$. The increase from 0 to $I_0R$ has a time constant $RC$. The final value of $I_0R$ for the capacitor voltage implies that all of the current from the current source flows through the resistor, and the capacitor behaves like an *open circuit* (for large $t$).

Second, the initial value of 0 for the capacitor voltage implies that at $t = 0$ all of the current from the current source must be flowing through the capacitor, and none through the resistor. Thus the capacitor behaves like an *instantaneous short circuit* at $t = 0$.

Third, the physical significance of the time constant $RC$ can now be seen. Illustrated in Figure 10.4, it is the temporal scale factor that determines how rapidly the transient goes to completion.

Finally, it may seem that the solution to such a simple problem can't possibly be as involved as this appears Correct. This problem and most first-order systems with step excitation can be solved by inspection (see Section 10.3). But here we are trying to establish general methods, and have chosen the simplest example to illustrate the method.

### 10.1.2 RC DISCHARGE TRANSIENT

With the capacitor now charged, assume that the current source is suddenly set to zero as suggested in Figure 10.5a, where for convenience, the time axis is redefined so that the turn-off occurs at $t = 0$. The relevant circuit to analyze the RC turn-off or discharge transient now contains just a resistor and a capacitor as indicated in Figure 10.5c. The voltage on the capacitor at the start of the experiment is represented by the initial condition

$$v_C = I_0R \qquad t < 0. \qquad (10.21)$$

This RC discharge scenario is identical to that of a circuit containing a resistor and a capacitor, where there is an initial voltage $v_C(0) = I_0R$ on the capacitor.

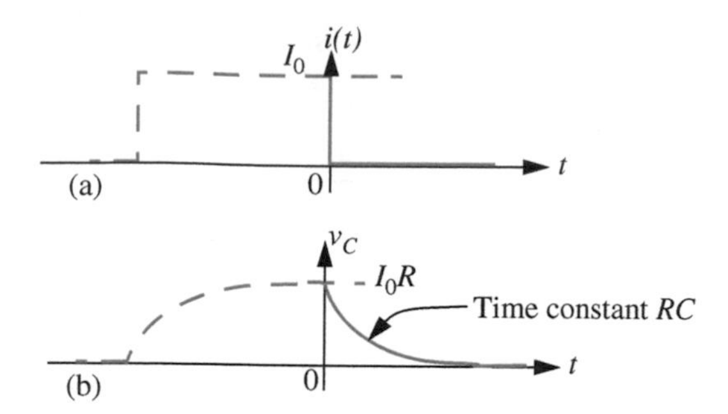


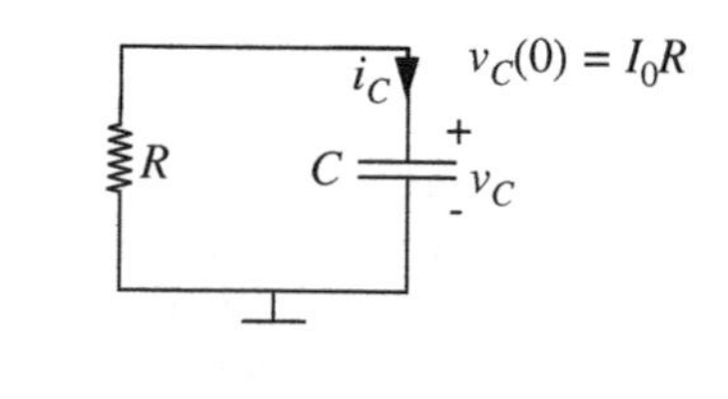


**FIGURE 10.5** RC discharge transient.

Because the drive current is zero, the differential equation for $t$ greater than zero is now

$$0 = \frac{v_C}{R} + \frac{Cdv_C}{dt}. \tag{10.22}$$

As before, the homogeneous solution is

$$v_{CH} = Ae^{-t/RC} \tag{10.23}$$

but now the particular solution is zero, since there is no forcing input, so Equation 10.23 is the total solution. In other words,

$$v_C = v_{CH} = Ae^{-t/RC}.$$

Equating Equations 10.21 and 10.23 at $t = 0$, we find

$$I_0R = A \tag{10.24}$$

so the capacitor voltage waveform for $t > 0$ is

$$v_C = I_0Re^{-t/RC}. \tag{10.25}$$

This solution is sketched in Figure 10.5b.

In general, for a resistor and capacitor circuit with an initial voltage $v_C(0)$ on the capacitor, the capacitor voltage waveform for $t > 0$ is

$$v_C = v_C(0)e^{-t/RC}. \tag{10.26}$$

### Properties of Exponentials

Because decaying exponentials occur so frequently in solutions to simple RC and RL transient problems, it is helpful at this point to discuss some of the properties of these functions as an aid to sketching waveforms.

- For a general exponential function of the form

$$x = Ae^{-t/\tau} \tag{10.27}$$

  the initial slope of the exponential is

$$\left.\frac{dx}{dt}\right|_{t=0} = \frac{-A}{\tau}.$$

  Hence the initial slope of the curve, projected to the time axis, intercepts the time axis at $t = \tau$, irrespective of the value of $A$, as shown in Figure 10.6a.

- Furthermore, notice that when $t = \tau$, the function in Equation 10.27 becomes

$$x(t = \tau) = \frac{A}{e}.$$

  In other words, the function reaches $1/e$ of its initial value irrespective of the value of $A$. Figure 10.6b depicts this point in the exponential curve.

- Because $e^{-5} = 0.0067$, it is common to assume for the $t$ greater than five time constants, that is,

$$t > 5\tau$$

  the function is essentially zero (see Figure 10.6a). That is, we assume the transient has gone to completion.

We will see later that these properties of the time constant $\tau$ make it useful in obtaining rough estimates for time durations associated with rising or falling exponentials.

### 10.1.3 SERIES RC CIRCUIT, STEP INPUT

Let us now convert the Norton source in Figure 10.2 to a Thévenin source in Figure 10.7 and determine the capacitor voltage as a function of time. The input waveform $v_S$ is assumed to be a voltage step of magnitude $V$ applied at $t = 0$,

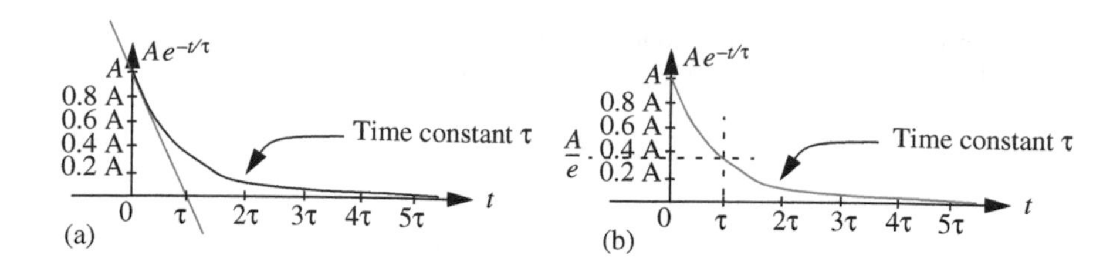


FIGURE 10.6 Properties of exponentials.

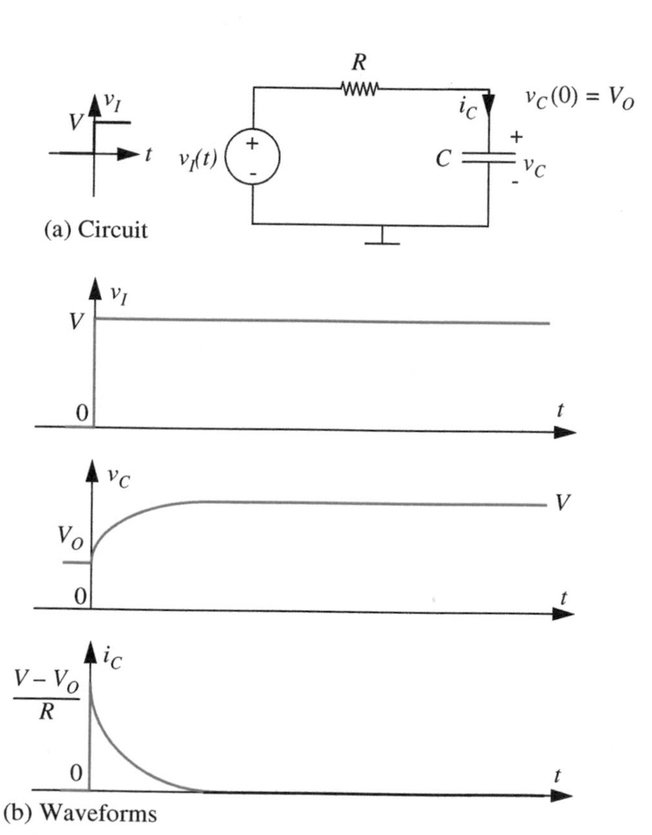


FIGURE 10.7 Series RC circuit with step input.

but this time around, we assume the capacitor voltage is $V_O$ just before the step.[3] That is, the initial condition on the circuit is

$$v_C = V_O \qquad t < 0. \tag{10.28}$$

3. For the purpose of determining the response for $t \geq 0$, it does not really matter to us how the capacitor voltage became $V_O$ for $t = 0$, or the value of the capacitor voltage for $t < 0$. Nevertheless,

The differential equation can be found by using the node method. Applying KCL at the node with voltage $v_C$, we get

$$\frac{v_C - v_I}{R} + C\frac{dv_C}{dt} = 0.$$

Dividing by $C$ and rearranging terms,

$$\frac{dv_C}{dt} + \frac{v_C}{RC} = \frac{v_I}{RC}. \tag{10.29}$$

The homogeneous equation is

$$\frac{dv_{CH}}{dt} + \frac{v_{CH}}{RC} = 0 \tag{10.30}$$

which, as expected, is the same as that in Equation 10.6 for the Norton circuit, since the Norton and Thévenin circuits are equivalent. Borrowing the homogeneous solution to Equation 10.6, we have

$$v_{CH} = Ae^{-t/RC} \tag{10.31}$$

where $RC$ is the time constant of the circuit.

Let us now find the particular solution. Since the input drive is a step of magnitude $V$, the particular solution is any solution to

$$\frac{dv_{CP}}{dt} + \frac{v_{CP}}{RC} = \frac{V}{RC}. \tag{10.32}$$

---

the following is one possible circuit that will realize the given initial condition on the capacitor and the effect of a step input:

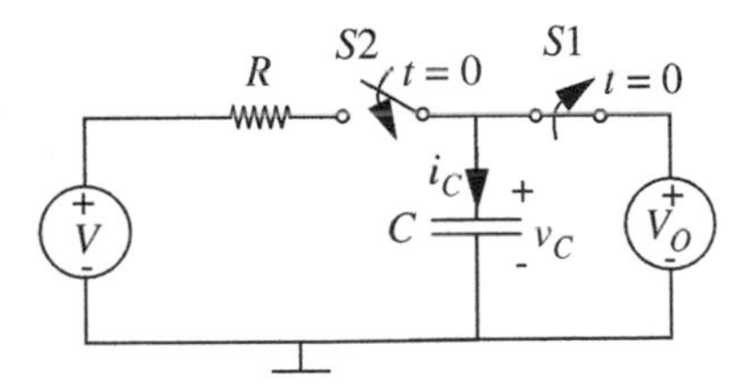


In the circuit, a DC source with value $V_O$ is applied across the capacitor using switch S1. The DC source forces the capacitor voltage to $V_O$. This DC source is switched out as shown at $t = 0$, and another DC source with voltage $V$ is switched in using switch S2. This action applies a step voltage of magnitude $V$ to the capacitor, which has an initial voltage $V_O$ at $t = 0$.

Because the drive is a step, which is constant for large $t$, we can assume a particular solution of the form

$$v_{CP} = K. \tag{10.33}$$

Substituting into Equation 10.32, we obtain

$$\frac{K}{RC} = \frac{V}{RC}.$$

which implies $K = V$. So the particular solution is

$$v_{CP} = V. \tag{10.34}$$

Summing $v_{CH}$ and $v_{CP}$, we obtain the complete solution:

$$v_C = V + Ae^{-t/RC}. \tag{10.35}$$

The initial condition can now be applied to evaluate $A$. Given that the capacitor voltage must be continuous at $t = 0$, we have

$$v_C(t = 0) = V_O.$$

Thus, at $t = 0$, Equation 10.35 yields

$$A = V_O - V.$$

The complete solution for the capacitor voltage for $t > 0$ is now

$$v_C = V + (V_O - V)e^{-t/RC} \tag{10.36}$$

where, $V$ is the input drive voltage for $t > 0$ and $V_O$ is the initial voltage on the capacitor. As a quick sanity check, substituting $t = 0$, we get $v_C(0) = V_O$, and substituting $t = \infty$, we get $v_C(\infty) = V$. Both these boundary values are what we expect, since the initial condition on the capacitor is $V_O$, and since the input voltage must appear across the capacitor after a long period of time.

By rearranging the terms, Equation 10.36 can be equivalently written as

$$v_C = V_O e^{-t/RC} + V(1 - e^{-t/RC}). \tag{10.37}$$

Finally, from Equation 9.9, the current through the capacitor is

$$i_C = C\frac{dv_C}{dt} = \frac{V - V_O}{R}e^{-t/RC}. \tag{10.38}$$

This expression for $i_C$ also matches our expectations since $i_C$ must be 0 when $t$ is large, and the since the capacitor behaves like a voltage source with voltage $V_O$ during the step transition at $t = 0$, the current at $t = 0$ must equal $(V - V_O)/R$.

These waveforms are shown in Figure 10.7b.

If we desire the voltage $v_R$ across the resistor, we can easily obtain it by applying KVL as

$$v_R = v_I - v_C$$

where we take the positive reference for $v_R$ on the input side of the resistor. Alternatively, we can obtain $v_R$ by taking the product of the current and the resistance as

$$v_R = i_C R.$$

As one final point of interest, notice that Equation 10.36 was derived assuming both an initial nonzero state ($V_O$) and a nonzero input (a step of voltage $V$).

Substituting $V = 0$ in Equation 10.36 we obtain the so called *zero input response* (*ZIR*):

$$v_C = V_O e^{-t/RC} \tag{10.39}$$

and substituting $V_O = 0$ in Equation 10.36 we obtain the *zero state response* (*ZSR*):

$$v_C = V - Ve^{-t/RC}. \tag{10.40}$$

In other words, the zero input response is the response for nonzero initial conditions, but where the input drive is zero. In contrast, the zero state response is the response of the circuit when the initial state is zero, that is, all capacitor voltages and inductor currents are initially zero.

Notice also that the total response is the sum of the ZIR and the ZSR,

as can be verified by adding the right-hand sides of Equations 10.39 and 10.40 and comparing to the right-hand side of Equation 10.36. We will have a lot more to say about the ZIR and the ZSR in Section 10.5.3.

### 10.1.4 SERIES RC CIRCUIT, SQUARE-WAVE INPUT

Examination of the waveforms in Figure 10.5a and 10.5b indicates that the presence of the capacitor has *changed the shape* of the input wave. When a square pulse is applied to the RC circuit, a decidedly non-square pulse, with slow rise and slow decay, results. The capacitor has allowed us to do a limited amount of wave shaping. This concept can be further developed by an experiment in which we drive the circuit with a square wave.

In this experiment, we will use a Thévenin source as in Figure 10.8. The source can be a standard laboratory square-wave generator. The input square wave is marked as a in Figure 10.8. Several quite distinctive wave shapes for $v_C(t)$ can be derived, depending on the relation between the *period* of the driving square wave and the *time constant* $RC$ of the network. These waveforms are all essentially variations on the solution derived in the preceding sections.

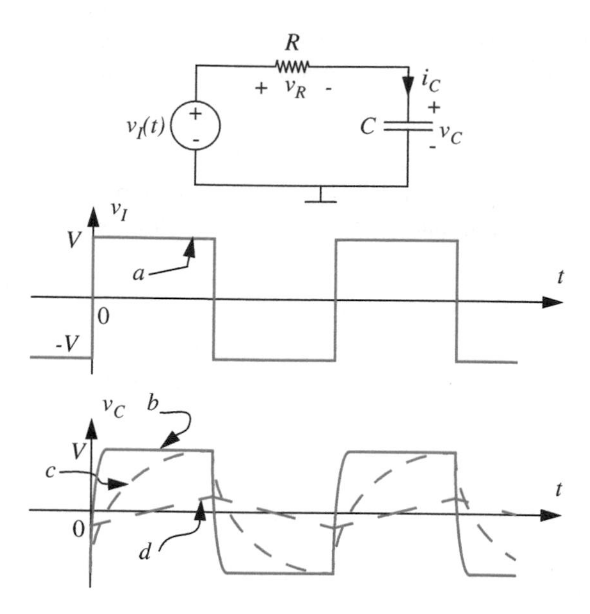


FIGURE 10.8 Response to square wave.

For the case where the circuit time constant is very short compared to the square-wave period, the exponentials go to completion relatively rapidly, as suggested by waveform b in Figure 10.8. The capacitor waveform thus closely resembles the input waveform, except for a small amount of rounding at the corners.

If the time constant is a substantial fraction of the pulse length, then the solution appears as waveform c in Figure 10.8. Note that the drawing implies that the transients still go almost to completion, so there is an upper limit on the $RC$ product for this solution to apply. Assuming, as noted here, that simple transients are complete for times greater than five time constants, the $RC$ product must be less than one-fifth of the pulse length, or one tenth the square-wave period for this solution to apply.

When the circuit time constant is much longer than the square-wave period, waveform d, shown in Figure 10.8, results. Here the transient clearly does *not* go to completion. In fact, only the first part of the exponential is ever seen. The waveform looks almost triangular, the *integral* of the input wave. This can be seen from the differential equation describing the circuit. Application of KVL gives

$$v_I = i_C R + v_C. \tag{10.41}$$

Upon substitution of the constituent relation for the capacitor, Equation 9.9, we obtain the differential equation

$$v_I = RC\frac{dv_C}{dt} + v_C. \tag{10.42}$$

It is clear from Equation 10.42 or Figure 10.8 that as the circuit time constant becomes bigger, the capacitor voltage $v_C$ must become smaller. For waveform d the time constant $RC$ is large enough that $v_C$ is much smaller than $v_I$, so in this case Equation 10.41 can be approximated by

$$v_I \simeq i_C R. \tag{10.43}$$

Physically, the current is now determined solely by the drive voltage and the resistor, because the capacitor voltage is almost zero. Integrating both sides of Equation 10.42 assuming $v_C$ is negligible, we obtain

$$v_C \simeq \frac{1}{RC} \int v_I dt + K \tag{10.44}$$

where the constant of integration $K$ is zero. Thus for large $RC$, the capacitor voltage is approximately the integral of the input voltage. This is a very useful signal-processing property. In Chapter 15 we will show that a much closer approximation to ideal integration can be obtained by adding an Op Amp to the circuit.

It is a simple matter to find the voltage across the resistor in the circuit of Figure 10.8 because we can find the current from the capacitor voltage using Equation 9.9,

$$v_R = i_C R = RC\frac{dv_C}{dt}.$$

Thus, during the charge interval, for example, from Equation 10.20, assuming the transients go to completion,

$$v_C = V(1 - e^{-t/RC}).$$

Hence

$$v_R = Ve^{-t/RC}.$$

The wave shapes in Figure 10.8 change very little if the input signal $v_I$ has zero average value, that is, if $v_I$ is changed so that it jumps back and forth from $-V/2$ to $+V/2$. Specifically, $v_C$ also has zero average value, and if the transients go to completion, as in wave forms b and c, the excursions will be $-V/2$ and $+V/2$.

## 10.2 ANALYSIS OF RL CIRCUITS

### 10.2.1 SERIES RL CIRCUIT, STEP INPUT

Figure 10.9 will serve as a simple illustration of a transient involving an inductor. (See the example discussed in Section 10.6.1 for a practical application of the analysis involving inductor transients.) The input waveform $v_S$ is assumed to

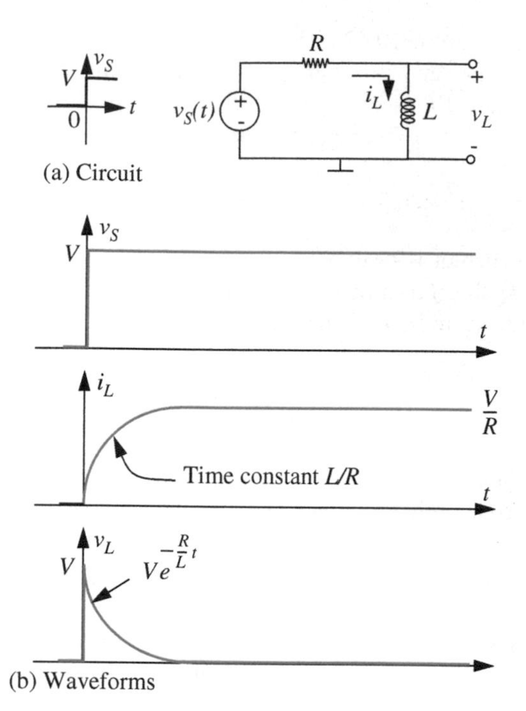


FIGURE 10.9 Inductor current buildup.

be a voltage step applied at $t = 0$ (see Figure 10.9a), and the inductor current is assumed to be zero just before the step. That is, the initial condition on the circuit is

$$i_L = 0 \qquad t < 0. \tag{10.45}$$

Suppose that we are interested in solving for the current $i_L$. As before, we can use the node method to obtain an equation involving the unknown node voltage $v_L$, and then use the constituent relation for an inductor from Equation 9.28 to substitute for $v_L$ in terms of the variable of interest to us, namely $i_L$. For variety, however, we will derive the same differential equation in $i_L$ by applying *KVL*:

$$-v_S + i_L R + L\frac{di_L}{dt} = 0. \tag{10.46}$$

The homogeneous equation is

$$L\frac{di_{LH}}{dt} + i_{LH}R = 0. \tag{10.47}$$

Assume a solution of the form

$$i_{LH} = Ae^{st}. \tag{10.48}$$

Hence

$$LsAe^{st} + RAe^{st} = 0. \quad (10.49)$$

For nonzero $A$ ($A = 0$ is a trivial solution)

$$Ls + R = 0.$$

or

$$s + \frac{R}{L} = 0 \quad (10.50)$$

$$s = -R/L. \quad (10.51)$$

Equation 10.50 is the characteristic equation for our circuit, and Equation 10.50 gives the natural frequency.

The homogeneous solution is thus

$$i_{LH} = Ae^{-(R/L)t} \quad (10.52)$$

where the time constant is in this case $L/R$.

The particular solution can be obtained by solving

$$i_{LP}R + L\frac{di_{LP}}{dt} = v_S. \quad (10.53)$$

Because the drive is a step, which is constant for large $t$, it is again appropriate to assume a particular solution of the form

$$i_{LP} = K. \quad (10.54)$$

Substituting into Equation 10.53, and noting that for large $t$, $v_S = V$, we obtain

$$KR = V.$$

or,

$$K = \frac{V}{R}. \quad (10.55)$$

So, from Equation 10.54, the particular solution is

$$i_{LP} = \frac{V}{R} \quad (10.56)$$

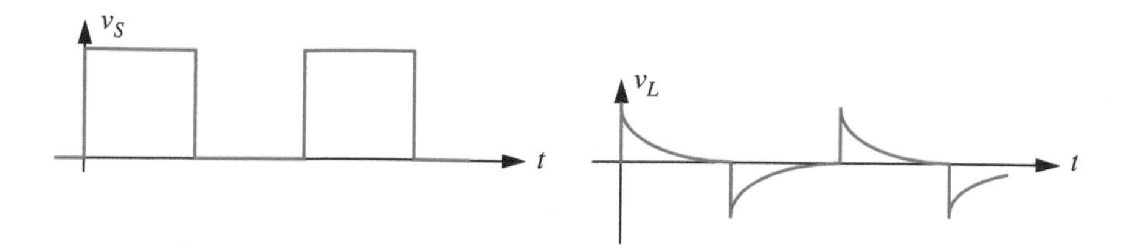


FIGURE 10.10 Response to a square-wave input.

and the complete solution is of the form

$$i_L = \frac{V}{R} + Ae^{(-R/L)t}. \tag{10.57}$$

The initial condition together with a continuity condition, can now be applied to evaluate $A$. The continuity condition for inductor current can be found from Equation 9.28. If it can be shown that the inductor voltage cannot be infinite in the circuit, then $di/dt$ must be finite, hence the *inductor current must be continuous.* For this particular circuit, with finite $v_S$, we are assured of finite $v_L$, hence $i_L$ in Equation 10.57 can be evaluated at $t = 0$, and set equal to the initial value, Equation 10.45:

$$\frac{V}{R} + A = 0. \tag{10.58}$$

The complete solution for the inductor current for $t > 0$ is now

$$i_L = \frac{V}{R}\left(1 - e^{-(R/L)t}\right) \tag{10.59}$$

and, from Equation 9.28, the voltage across the inductor is

$$v_L = L\frac{di_L}{dt} = Ve^{-(R/L)t}. \tag{10.60}$$

These waveforms are shown in Figure 10.9b. Notice that the inductor current has an initial value of 0 and a final value of $V/R$. Thus the inductor behaves like an instantaneous open circuit at $t = 0$ and a short circuit for large $t$, for the step voltage input at $t = 0$. $v_L$ is correspondingly $V$ at $t = 0$ and 0 for large $t$.

The response to a square-wave input is shown in Figure 10.10.

## 10.3 INTUITIVE ANALYSIS

The previous sections illustrated the general method of analyzing linear RC and RL circuits. The several examples with step-function drive that we worked previously suggest that such circuits have a very limited range of solutions.

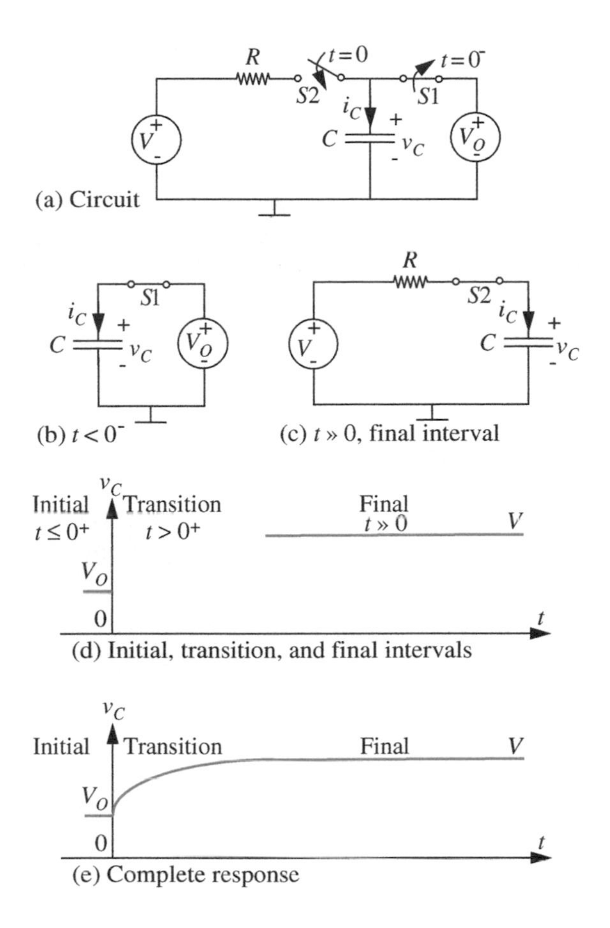


**FIGURE 10.11** Step response of series RC circuit. The arrangement of switches provides for the initial voltage $V_O$ on the capacitor, and an input step voltage of magnitude $V$ at $t = 0$.

The two basic forms that we saw are $e^{-\alpha t}$ and $(1 - e^{-\alpha t})$. Accordingly, it turns out that for simple excitations, such as the step and the impulse, the response of first-order systems can be sketched easily using some intuition.

Let us illustrate using the step response of a series RC circuit in Figure 10.11a as an example. We will address the most general case, namely one in which there is both a nonzero initial state and a nonzero input. The seemingly elaborate arrangement of switches simply provides for the initial voltage $V_O$ on the capacitor, and an input step voltage of magnitude $V$ at $t = 0$, a situation similar to that in Section 10.1.3. For the purposes of sketching our result, we will further assume that $V > V_O$. As illustrated in Figure 10.11a, switch S1 is initially closed and S2 is open, resulting in the voltage $V_O$ being applied directly across capacitor. Just before $t = 0$, that is, at $t = 0^-$, S1 is opened (S2 remains open). Then, at $t = 0$, S2 is closed (S1 remains open). The closing of S2 and opening of S1 results in an series RC circuit with a step voltage $V$ applied at $t = 0$.

Suppose we are interested in sketching the voltage $v_C$ as a function of time.[4] The form of the response can be sketched intuitively by identifying three intervals of operation as indicated in Figure 10.11d: the initial interval, which extends until $t = 0^+$ (that is, the time instant just after $t = 0$), the transition interval, which is identified as the interval after $t = 0^+$, and the final interval, where S2 has been closed and S1 has been open for a long time.

The overall response can be quickly sketched through inspection by first determining the initial and final interval values of the voltage on the capacitor.

*Initial Interval ($t \leq 0^+$)* During this initial interval, when S1 is closed ($t < 0^-$), the effective circuit is as shown in Figure 10.11b, with a DC source with voltage $V_O$ appearing across the capacitor. Thus, the capacitor voltage is $V_O$ during $t < 0^-$.

Next, notice that in the short period of time between $t = 0^-$ and $t = 0$, and still within the initial interval, the capacitor is not connected to any other circuit (recall S1 is opened at $t = 0^-$ and S2 is closed immediately thereafter at $t = 0$). Assuming the capacitor is ideal, it holds its charge and so its voltage remains at $V_O$ until the switch S1 is closed.

Then, at $t = 0$, S1 is closed, resulting in a *finite* step of magnitude $V$ being applied to a series RC circuit in which the capacitor has a voltage $V_O$ across it. Let us now determine the capacitor voltage at $t = 0^+$, just after the step. From the element law of the capacitor (Equation 9.7), we know that an instantaneous jump in capacitor voltage requires an infinite spike (that is, an impulse) in current. Since a finite step voltage applied across a resistor cannot support an infinite spike in current, we conclude that the capacitor voltage cannot change instantaneously, rather it must be continuous. Thus, the voltage across the capacitor at $t = 0^+$ must also be $V_O$. This is our initial condition on the capacitor. The voltage across the capacitor during the initial interval ($t \leq 0^+$) is sketched in Figure 10.11d.

*Final Interval ($t \gg 0$)* We next turn our attention to the final interval. To determine the capacitor voltage in the final interval, observe that our situation is identical to that of a DC source with voltage $V$ applied across the series combination of $R$ and $C$ as shown in Figure 10.11c. Since the capacitor current is proportional to the rate of change of the capacitor voltage (Equation 9.7), in a DC situation, where all transients have died out, the current flowing through the capacitor must be zero. In other words, in a DC situation, the capacitor voltage has attained some fixed value, and hence the capacitor current is zero. Effectively, the capacitor behaves like an open circuit for DC sources. Since no current is flowing, the drop across the resistor must be zero. Thus, to

4. Other branch variables in the circuit such as $i_C$ and $v_R$ share the same general form and can be derived in an analogous fashion.

satisfy KVL, the capacitor voltage must equal $V$, the voltage of the DC source. This value is sketched in the final interval in Figure 10.11d.

*Transition Interval* ($t > 0^+$) We have now sketched the initial and final values of the capacitor voltage. The transition interval for $t > 0^+$ remains to be analyzed. During this interval, observe that the capacitor voltage cannot jump instantaneously from $V_O$ to $V$ due to the continuity condition. Specifically, we know from the solution to the homogeneous equation for the RC circuit that the transient follows an exponential form, either rising $(1 - e^{-t/RC})$ or falling $(e^{-t/RC})$, with time constant RC. (For the corresponding inductor-resistor circuit the time constant will be $L/R$.) In our case, since $V > V_O$, the transient will be a rising exponential.

*Complete Response* The complete response for all of the three regions is sketched in Figure 10.11e.

The corresponding equation for the capacitor voltage that matches the initial and final values, and the exponential with time constant $RC$, for $t \geq 0$, is

$$v_C = V + (V_O - V)e^{-t/RC}.$$

In other words, for $t \geq 0$,

$$v_C = \text{final value} + (\text{initial value} - \text{final value})e^{-t/\text{time constant}} \quad (10.61)$$

or equivalently, rearranged a little bit,

$$v_C = \text{initial value}\ e^{-t/\text{time constant}} + \text{final value}(1 - e^{-t/\text{time constant}}) \quad (10.62)$$

You might want to confirm that Equation 10.61 combined with the appropriate boundary conditions results in the same solutions as obtained by solving the differential equations in the previous sections. For example, for the RC discharge transient example of Section 10.1.2, the initial capacitor voltage is given as $v_C(0)$ and the final value is zero. Substituting $V_O = v_C(0)$ and $V = 0$ into Equation 10.61, we obtain

$$v_C = v_C(0)e^{-t/RC}$$

which is the same as the expression obtained in Equation 10.26.

At this point, we take a moment to make a couple of other helpful observations. Sometimes, we desire the response related to the capacitor current. The responses related to the capacitor current can be easily determined from the voltage response and the capacitor element law. However, the current response can also be directly obtained by using the same type of insight that we used to obtain the voltage response. Here, we would seek the initial and final values of the current. In our example, the final value of the capacitor current after all transients have died out is 0. The initial value of the current (at $t = 0$) can also be determined easily. Since the capacitor voltage at $t = 0$ is $V_O$, the instantaneous current through the capacitor at $t = 0$ is given by

$$i_C(t = 0) = \frac{V - V_O}{R}$$

which is the voltage across the resistor ($V - V_O$) divided by the resistance ($R$). Thus, at the instant that the switch S1 is closed, the capacitor behaves like an *instantaneous voltage source* with voltage $V_O$. In like manner, if the initial voltage on the capacitor were zero (that is, $V_O = 0$), then the capacitor would behave like an *instantaneous short circuit*. In either case, notice that the capacitor current is not necessarily continuous, only the state variable. In our example, the capacitor current jumps from 0 to $(V - V_O)/R$ at $t = 0$. The current decays exponentially with time constant $RC$ from the initial value of $(V - V_O)/R$ at $t = 0$ to its final value of zero. The current response is plotted in Figure 10.12.

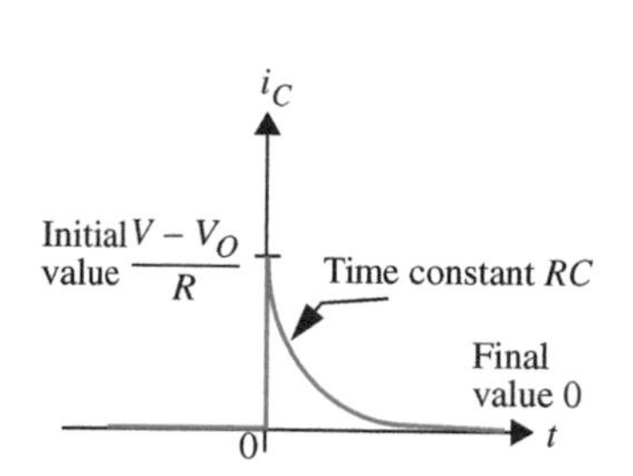


FIGURE 10.12 Current response of a series RC circuit to a step input.

Inductors can be treated in a similar manner. The key difference is that the state variable for an inductor is its current. Accordingly, the inductor current is continuous (recall, from Equation 9.26, an instantaneous jump in inductor current requires an infinite spike, that is, an impulse, in the voltage). To determine initial and final values of the inductor current, remember that the inductor behaves like a long-term short circuit for DC current sources, and like an instantaneous open circuit for abrupt transitions.[5] The time constant for circuits containing an inductor and a resistor is $L/R$. With these definitions, Equation 10.61 is equally applicable to inductor-resistor circuits.

As an inductor-resistor example, consider the current response of the series RL circuit from Figure 10.9a redrawn in Figure 10.13. As sketched in Figure 10.13, the initial current through the inductor is zero. The final current through the inductor is $V/R$, because the inductor behaves like a long-term short circuit. The time constant of the circuit is $L/R$. Substituting into Equation 10.61, we get

$$i_L = \frac{V}{R} + \left(\frac{V}{R} - 0\right) e^{-\frac{t}{L/R}}$$

or

$$i_L = \frac{V}{R}\left(1 - e^{-\frac{t}{L/R}}\right)$$

which is identical to Equation 10.59.

5. If the inductor current were nonzero, then it would behave like an instantaneous current source for abrupt transitions.

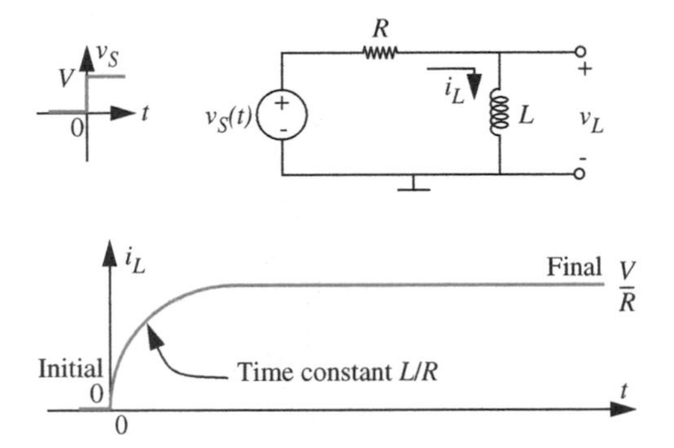


FIGURE 10.13 Series RL circuit, step response through intuitive analysis.

This section showed how we can quickly sketch the step response using intuition. A similar approach also works for impulse responses. Intuitive analysis for impulses is discussed further in Section 10.6.4.

## 10.4 PROPAGATION DELAY AND THE DIGITAL ABSTRACTION

The RC effects we have seen thus far are the source of delays in digital circuits, and are responsible for the waveforms shown in Figure 9.3 in Chapter 9, or those in Figure 10.1 in this chapter. Consider the two-inverter digital circuit shown in Figure 10.14 in which inverter A drives inverter B. Inverter A is driven by an input $v_{IN1}$ and its output is $v_{OUT1}$. Figure 10.15 replaces the inverters with their internal circuits comprising MOSFETs and resistors.

Let us begin by reviewing the basic inverter circuit. Assume that the threshold voltage for both MOSFETs is 1 volt. When $v_{IN1}$ is low ($< 1$ volt), MOSFET A is turned off, and no current flows from its drain to its source. Output voltage $v_{OUT1}$ is high. In contrast, when $v_{IN1}$ is high, MOSFET A is turned on. Its output voltage $v_{OUT1}$ is given by the voltage-divider relationship $R_{ON}/(R_{ON} + R_L)$.

Ideally, the input $v_{IN1}$ (corresponding to a sequence of 1's and 0's of the form shown in Figure 10.16) should produce the ideal output $v_{OUT1}(ideal)$.

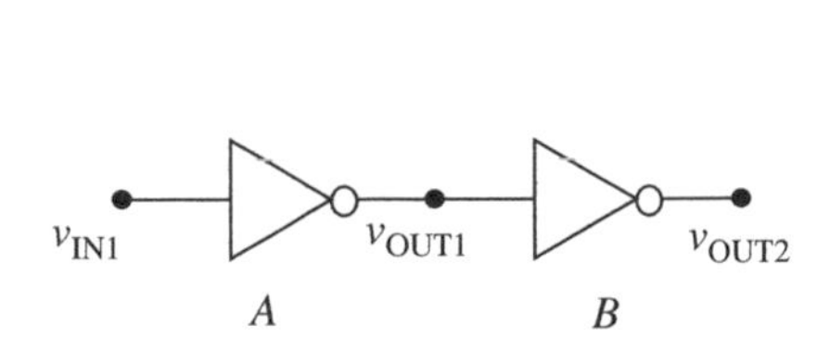


FIGURE 10.14 Inverters connected in series.

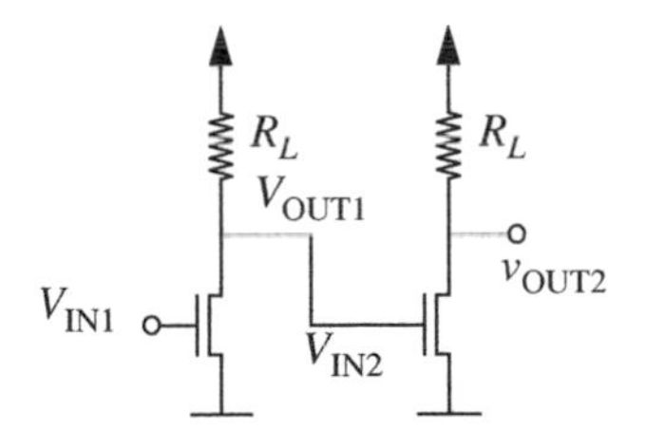


FIGURE 10.15 Internal circuits of the inverters.

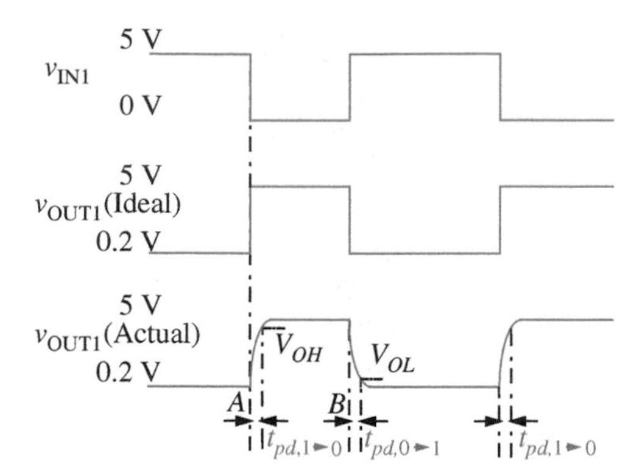


**FIGURE 10.16** Characteristics of ideal and actual inverters.

As shown in Figure 10.16, the output of an ideal inverter should show a change at the same instant as the input. Furthermore, the output should be an ideal square wave just like the input.

However, in practice, if we were to observe the output $v_{OUT1}$ on an oscilloscope, we would notice that the change in the output is not instantaneous; rather the output changes from one valid voltage level (for example, a logical 0) to another valid voltage level (for example, a logical 1) more slowly over a small period of time as suggested by the signal marked $v_{OUT1}$*(actual)* in Figure 10.16. How does this slow transition affect the behavior of the digital circuit?

Recall that the $v_{OUT1}$ signal represents a digital signal, so it must reach $V_{OH}$ so that the gate that produced it adheres to the static discipline and we obtain a nonzero noise margin. As suggested in the lowermost signal in Figure 10.16, notice that the $V_{OH}$ crossing happens at a time interval $t_{pd,1\to0}$ after the input changes from a 1 to a 0. Thus, effectively, there is a delay of $t_{pd,1\to0}$ between the moment that the input changes to a 0 to the moment that the output changes to a valid 1.

This period of time is called the *propagation delay*[6] *through inverter A for a 1 to 0 transition at the input* and is denoted as $t_{pd,1\to0}$.

As suggested in the lowermost signal in Figure 10.16,

the inverter is also characterized by a $0 \to 1$ *propagation delay*. This delay is denoted as $t_{pd,0\to1}$.

The $t_{pd,1\to0}$ and $t_{pd,0\to1}$ delays are not necessarily equal. For simplicity, we often characterize digital gates by a single delay called its *propagation delay* $t_{pd}$ and choose

$$t_{pd} = max(t_{pd,1\to0}, t_{pd,0\to1}). \tag{10.63}$$

6. The propagation delay is sometimes defined as the time interval from the 50% point of the input signal transition to the 50% point of the output signal transition.

### 10.4.1 DEFINITIONS OF PROPAGATION DELAYS

The following are more general definitions of propagation delays associated with digital gates with multiple inputs and outputs. The reader wishing to return to the computation of $t_{pd}$ for our inverter example can skip this section without loss of continuity and proceed directly to Section 10.4.2.

$t_{pd,1\rightarrow 0}$ We define $t_{pd,1\rightarrow 0}$ for a given input terminal and a given output terminal of a combinational digital circuit as the signal propagation delay from the input terminal to the output terminal for a high to low instantaneous transition at the input. More precisely, $t_{pd,1\rightarrow 0}$ for an input-output terminal pair is the time interval from the moment that the input changes from a 1 to a 0 to the moment that the output reaches a corresponding valid output voltage level ($V_{OH}$ or $V_{OL}$).

$t_{pd,0\rightarrow 1}$ Similarly, we define $t_{pd,0\rightarrow 1}$ for a given input terminal and a given output terminal of a combinational circuit as the signal propagation delay through input-output terminal pair for a low to high instantaneous transition at the input. More precisely, $t_{pd,0\rightarrow 1}$ for an input-output terminal pair is the time interval from the moment that the input changes from a 0 to a 1 to the moment that the output reaches a corresponding valid output voltage level.

$t_{pd}$ *for an Input-Output Terminal Pair:* We define the propagation delay $t_{pd}$ between an input terminal and an output terminal of a combinational circuit as

$$t_{pd} = max(t_{pd,1\rightarrow 0}, t_{pd,0\rightarrow 1})$$

where $t_{pd,1\rightarrow 0}$ and $t_{pd,0\rightarrow 1}$ are the corresponding $1 \rightarrow 0$ and $0 \rightarrow 1$ delays for the same input-output terminal pair.

*Propagation Delay* $t_{pd}$ *for a Combinational Gate:* If $t_{pd}^{i,j}$ is the propagation delay between input terminal $i$ and output terminal $j$ of a digital gate, then the propagation delay of the gate is given by

$$t_{pd} = max_{i,j} t_{pd}^{i,j},$$

which is the maximum delay of all input to output paths. The propagation delay is also called the *gate delay*.

In the simple example shown in Figure 10.16, the propagation delay through the inverter for a low to high transition at the input, $t_{pd,0\rightarrow 1}$, is also equal to the *rise time* of the output of the inverter. Similarly, $t_{pd,1\rightarrow 0}$, is also equal to the *fall time* of the inverter output. The rise and fall times are properties of output terminals of circuits, while propagation delays measure the relative signal transition times between inputs and outputs of circuits. The rise and fall times are defined as follows:[7]

*Rise Time* In general, the rise time for an output is defined as the delay in rising from its lowest value to a valid high ($V_{OH}$) at that output.
*Fall Time* The fall time for an output is defined as the delay in falling from its highest value to a valid low ($V_{OL}$) at the same output.

In general, the propagation delay and the rise/fall time are not equal. The $0 \rightarrow 1$ propagation delay for a digital circuit is the time between an input 0 to 1 transition (the input transition is assumed to happen instantaneously) and the corresponding output transition. The output transition is assumed to complete only when the output voltage crosses the appropriate output voltage threshold. The propagation delay and the rise/fall times are usually not equal when the digital circuit consists of multiple stages. When a circuit consists of multiple stages, the rise/fall time at the output is usually a function of the properties of the output circuit alone. However, the propagation delay is the sum of the delays of each of the stages.

How does the propagation delay impact our digital abstraction? Notice that the slowly rising output of the inverter now spends a nonzero amount of time in the invalid output voltage range, namely $V_{IL} \rightarrow V_{IH}$. This appears to violate the static discipline. Recall that the static discipline requires that *devices produce valid output voltages that satisfy the output thresholds when valid input voltages are supplied.* We get around this difficulty by observing that the inverter output eventually crosses the valid output threshold. Furthermore, notice that the static discipline does not take a position on time. In other words, it does not require gates to produce valid outputs instantaneously if the inputs change. Accordingly, to make the this fact explicit, we can modify the statement of the static discipline by requiring that *devices produce valid output voltages (in a finite amount of time) that satisfy the output thresholds when valid input voltages are supplied.*

*Revised statement of the static discipline* The *static discipline* is a specification for digital devices. The static discipline requires devices to interpret correctly voltages that fall within the input thresholds ($V_{IL}$ and $V_{IH}$). Provided valid

---

7. The rise and fall times are sometimes defined slightly differently. For example, the rise time of a node that transitions from a low to a high voltage might be defined as the time taken by a signal at that node to rise from 5% to 95% of the change in voltage. Alternatively, the rise time can be defined as the time taken by a signal at that node to rise from a valid low voltage $V_{OL}$ to a valid high voltage $V_{OH}$. As one more possibility, the rise time might be defined as the time taken by a signal to rise from its lowest value to 50% of the voltage difference. Corresponding definitions for the fall time also exist. The vagueness of these definitions only serves the interests of product marketeers, but for us, the important thing to learn is how to calculate the time intervals between any pair of signal values.

inputs are provided to the devices, the discipline also requires the devices to produce valid output voltages (in a finite amount of time) that satisfy the output thresholds ($V_{OL}$ and $V_{OH}$).

We can also refine our combinational gate abstraction to include the notion of a propagation delay, so that the abstraction remains valid in the presence of transitioning signals. Recall, the properties of a combinational gate as previously defined in Chapter 5.3: (1) The gate's outputs are a function of its inputs alone and (2) the gate must satisfy the static discipline. In the presence of a finite gate delay, there is a small period of time following an input transition in which the outputs do not reflect the new inputs; rather they reflect the old inputs. Thus our previously defined gate abstraction is violated. We negotiate this inconsistency by introducing a timing specification into our gate abstraction.

*Revised statement of the combinational gate abstraction* A combinational gate is an abstract representation of a circuit that satisfies these properties:

1. Its outputs will be valid no later than $t_{pd}$ after an instantaneous change in its inputs.
2. Its outputs are a function of its inputs alone (after an interval of time no greater than $t_{pd}$ following a change in its inputs).
3. It satisfies the static discipline.

Now that we have included the propagation delay of a device in its abstract specification, an additional benefit results: A gate-level circuit will now carry information on both its logic function and its speed. A rough estimate of the delay from any input to any output of a logic circuit along a path with multiple gates can be obtained by summing the propagation delays of each of the gates in that path. Thus, for example, if inverters are characterized by a $t_{pd}$ of 1 ns and OR gates with a $t_{pd}$ of 2 ns, then in the circuit in Figure 5.16 in Chapter 5, the delay from the input $A$ to output $C$ would be 2 ns, while the delay from the input $B$ to output $C$ would be 3 ns. If digital circuit designers need more accurate timing information for a circuit comprising multiple devices, or if they need to derive the $t_{pd}$ of a single device, then they must use the analysis methods discussed in the ensuing sections.

### 10.4.2 COMPUTING $t_{pd}$ FROM THE SRC MOSFET MODEL

Let us now compute the magnitude of the propagation delay. We use the switch resistor capacitor (SRC) model of the MOSFET introduced in Section 9.3.1 to determine this delay. Recall that we augmented the SR model of the MOSFET with a gate-to-source capacitor and created the SRC MOSFET model shown in Figure 10.17.

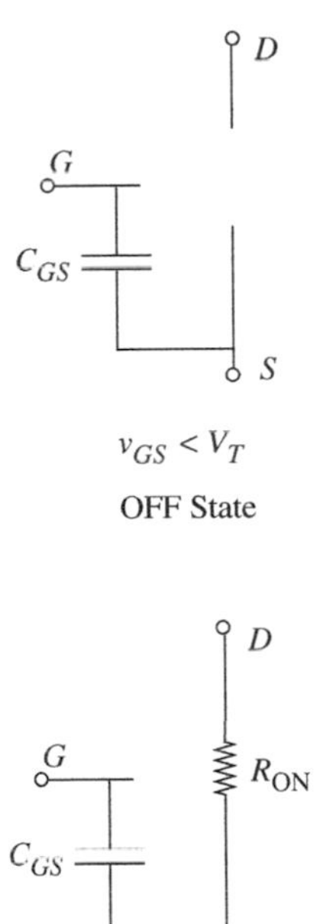


**FIGURE 10.17** The switch-resistor capacitor model of the MOSFET.

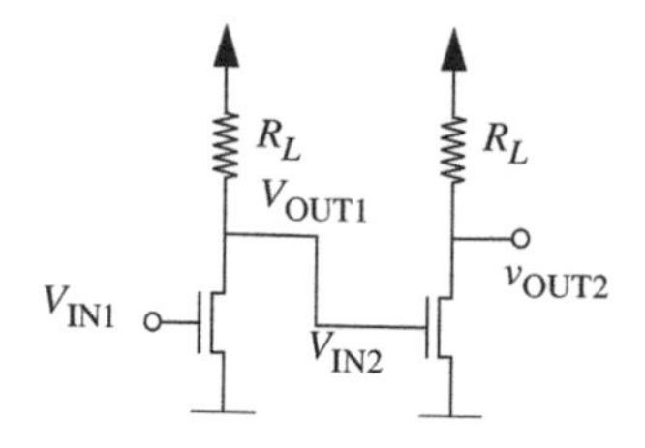


FIGURE 10.18 Internal circuits of the inverters.

Recall that the propagation delay results from the finite amount of time required for the output to transition from a given valid output voltage level to another when the input to the circuit transitions. The slower transition at the output is attributable to RC effects. Figure 10.18 replaces the inverters with their internal circuits comprising MOSFETs and resistors. Figure 10.19 further replaces the MOSFETs with their SRC circuit model when $v_{IN1}$ applied to the inverter A corresponds to a logical 1. For this $v_{IN1}$, the MOSFET in inverter A will be on, and the MOSFET in inverter B will be off. Similarly, Figure 10.20 shows the circuit model when $v_{IN1}$ applied to inverter A corresponds to a logical 0. For this $v_{IN1}$, the MOSFET in inverter A will be off, and the MOSFET in inverter B will be on. Thus, when alternating logical 1's and 0's are applied to the input to the inverter pair, and the inverters are allowed to reach steady state after each transition, the equivalent circuit model alternates between the two circuits in Figures 10.19 and 10.20.

Let us first analyze the circuit qualitatively. Consider the case where $v_{IN1}$ has been high for a long period of time and focus on the part of the circuit bounded by the dashed box in the Figure 10.19, which includes the load resistor and $R_{ON}$ of inverter A and the gate-to-source capacitor of inverter B. Since the circuit is in its steady state, the capacitor behaves as an open circuit, and so the voltage across the capacitor will be established by the voltage-divider subcircuit comprising the supply $V_S$, and the resistors $R_L$ and $R_{ON}$. Assuming $R_L \gg R_{ON}$, the capacitor voltage will have a low value (close to 0 volts).

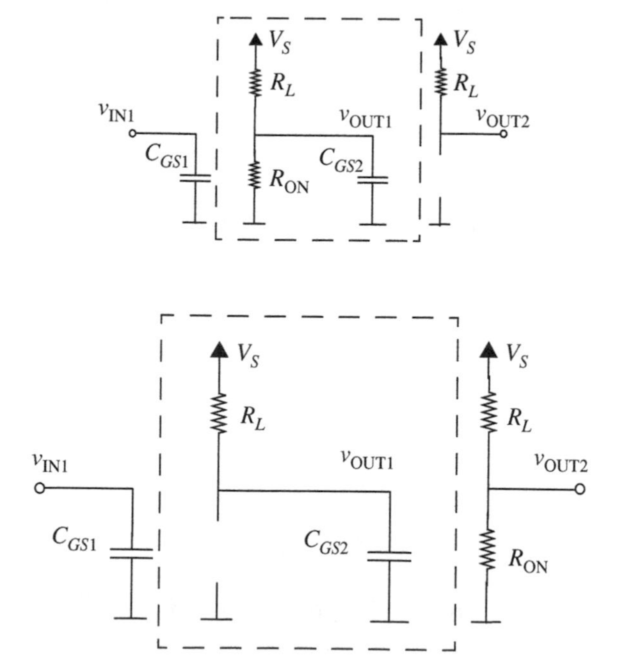


FIGURE 10.19 SRC circuit model of inverters connected in series when the input is high.

FIGURE 10.20 SRC circuit model of inverters connected in series when the input is low.

Next, focus on the time instant when the input voltage $v_{IN1}$ switches from a high to a low value (for example, 5 to 0 volts), turning the first MOSFET off. At this transition instant, the capacitor $C_{GS2}$ is almost completely discharged (assuming that $R_L \gg R_{ON}$ for the inverters). Therefore, the voltage across $C_{GS2}$, which corresponds to the voltage $v_{OUT1}$ on the output of inverter A, will be initially close to 0 V. This is depicted as the time instant A in Figure 10.16.

After the first MOSFET turns off, Figure 10.20 applies. Focus again on the part of the circuit bounded by the dashed box. It is easy to see that the circuit inside the dashed box is a first-order RC circuit. Remember, the voltage across $C_{GS2}$ is low initially. Now, $V_S$ begins to charge $C_{GS2}$ through the resistor $R_L$. The equivalent RC circuit for the devices in the box are shown in Figure 10.21. As the capacitor charges up, the output voltage of inverter A rises. This voltage must rise above the valid logical output high threshold, namely $V_{OH}$, to satisfy the static discipline. Notice that although the second MOSFET will turn on when $v_{OUT1}$ crosses its $V_T$ threshold (for example, 1 V), we require $v_{OUT1}$ to reach $V_{OH}$ to achieve a modest noise margin. Notice that the presence of the capacitor $C_{GS2}$ makes $v_{OUT1}$ take a finite amount of time to rise to the required $V_{OH}$ level. As we saw before, this interval of time is called the *propagation delay for the inverter for a high to low transition at the input* and is denoted by $t_{pd,1\rightarrow 0}$. As discussed earlier, the output capacitor charge-up time is also called the *rise time of the inverter.*

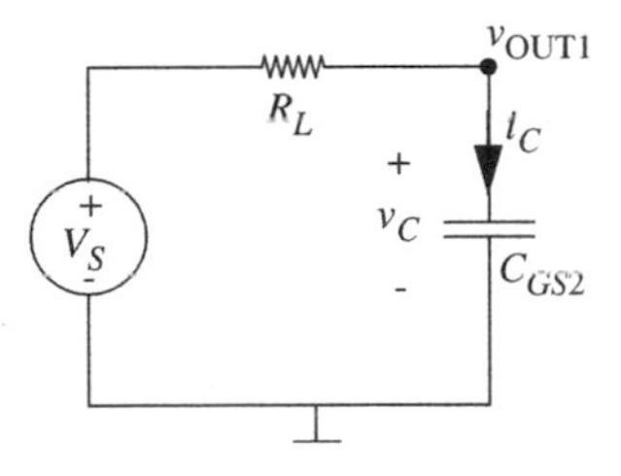

FIGURE 10.21 Equivalent circuit when $C_{GS2}$ is charging.

Next, let us consider the time instant when the input voltage switches from 0 volts to 5 volts, turning the first MOSFET on. Let us assume that this 0-V to 5-V transition happens after a sufficiently long period of time so that $C_{GS2}$ is initially charged up to its steady state value of 5 V. When the first gate is turned on, $C_{GS2}$ begins to discharge. The RC circuit and its Thévenin equivalent for the discharge is shown in Figure 10.22. For the logical 0 to logical 1 transition at the input to be reflected at the output of inverter *A*, the voltage across $C_{GS2}$ needs to go below the valid logical output low threshold, $V_{OL}$. As before, although MOSFET B will turn off when $v_{OUT1}$ drops below 1 volt, we require the output to go below $V_{OL}$ to provide for an adequate noise margin. The interval of time corresponding to the output capacitor discharge for an inverter is also called the *propagation delay for the inverter for a low to high transition at the input* and

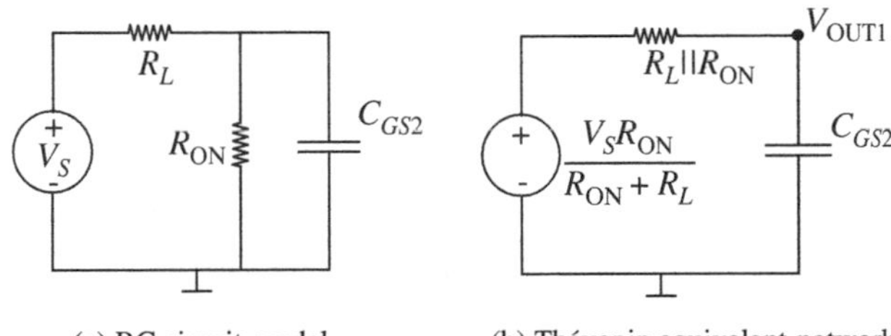

FIGURE 10.22 Equivalent circuit when $C_{GS2}$ is discharging.

is denoted by $t_{pd,0\rightarrow 1}$. Furthermore, as stated previously, the output capacitor discharge time is also called the *fall time* of inverter *A*.

The propagation delay $t_{pd}$ for inverter *A* is simply taken as the maximum of $t_{pd,0\rightarrow 1}$ and $t_{pd,1\rightarrow 0}$.

At this point, it is worth discussing a slight mismatch between the digital gate abstraction and the physical realities of computing the propagation delay. From the viewpoint of the digital gate abstraction, the propagation delay $t_{pd}$ is a property of the digital gate. Accordingly, we might say that an inverter (for example, one identical to inverter *A*) always has a propagation delay of 2 ns. However, the example discussed thus far illustrates that the propagation delay of an inverter depends not only on the characteristics of its internal components, but also on the size of the capacitance that it is driving, and therefore the propagation delay of the inverter can change depending on its environment. In particular, the propagation delay of inverter *A* in our example depends on the input capacitance of inverter *B*. Thus, strictly speaking, it makes no sense to define the propagation delay of a device in isolation. However, for convenience, we would like to characterize devices with a single $t_{pd}$ without defining their surrounding environment, so this simple device model can be used to obtain quick estimates of digital circuit delays when multiple gates are connected together. Accordingly, unless explicitly stated otherwise, device libraries or catalogs define a $t_{pd}$ for a gate assuming it is driving a "typical" load — commonly, four minimum sized inverters.[8]

## Computing $t_{pd,0\rightarrow 1}$

Let us now determine quantitatively the propagation delay for a low to high transition at the input of the inverter. Assume through the rest of this example that a valid output low voltage, $V_{OL}$, is 1 volt, and that a valid output high voltage, $V_{OH}$, is 4 volts. Also assume $R_{ON}$ is 1 kΩ, and that the threshold on-voltage for the MOSFET is 1 volt. Also assume that $R_L$ is 10 kΩ.

In this case, as discussed earlier, $C_{GS2}$ is initially charged to 5 volts. We need to determine the time taken for the capacitor voltage to drop from 5 to $V_{OL} = 1$ volts.

When the input is high, Figure 10.22 shows the equivalent circuit. Let us denote the Thévenin equivalent resistance $R_L \| R_{ON}$ as $R_{TH}$, and the Thévenin equivalent voltage $V_S R_{ON}/(R_{ON} + R_L)$ as $V_{TH}$. Let us also denote the capacitor voltage $v_{OUT1}$ by $v_C$, and the current through the circuit by $i_C$ as shown in Figure 10.23.

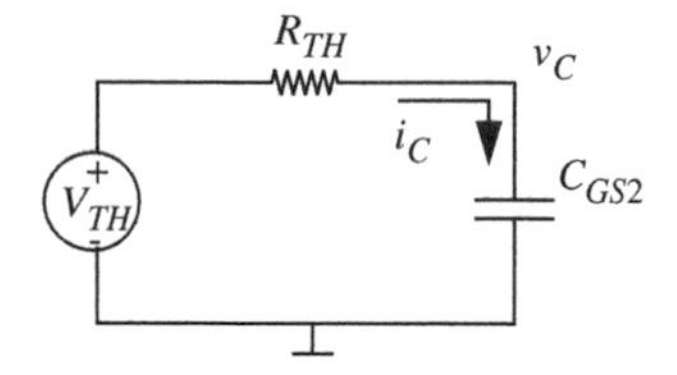


FIGURE 10.23 Equivalent circuit when $C_{GS2}$ is discharging.

8. A related metric that is sometimes used to characterize the speed of a process technology is called the fan-out-of-4 (or FO4) delay. The FO4 delay for a process technology is the propagation delay of a minimum sized inverter driving four other inverters of the same size.

Using the node method, we obtain,

$$\frac{v_C - V_{TH}}{R_{TH}} + C_{GS2}\frac{dv_C}{dt} = 0.$$

Rearranging,

$$R_{TH}C_{GS2}\frac{dv_C}{dt} + v_C = V_{TH}. \qquad (10.64)$$

Solving Equation 10.64 yields

$$v_C(t) = V_{TH} + Ae^{-t/R_{TH}C_{GS2}}. \qquad (10.65)$$

Substituting the initial condition $v_C(0) = V_S$, we obtain the final solution:

$$v_C(t) = V_{TH} + (V_S - V_{TH})e^{-t/R_{TH}C_{GS2}}. \qquad (10.66)$$

How long does it take for $v_C$ to drop below 1 volt? To obtain this duration, we must solve for the value of $t$ that satisfies

$$V_{TH} + (V_S - V_{TH})e^{-t/R_{TH}C_{GS2}} < 1.$$

In other words,

$$t > -R_{TH}C_{GS2}\ln\left(\frac{1 - V_{TH}}{V_S - V_{TH}}\right).$$

For $R_L = 10\ \text{k}\Omega$ and $R_{ON} = 1\ \text{k}\Omega$, $R_{TH} = 10000/11$, and $V_{TH} = V_S/11$. Substituting $V_S = 5$ V and $V_{TH} = 5/11$ V, the value of $t$ must satisfy

$$t > -R_{TH}C_{GS2}\ln\left(\frac{3}{25}\right).$$

Substituting for $R_{TH}$, the value of $t$ must satisfy

$$t > -\frac{10000}{11}C_{GS2}\ln\left(\frac{3}{25}\right). \qquad (10.67)$$

Suppose the gate capacitance $C_{GS2} = 100$ fF. We then have

$$t > -\frac{10}{11} \times 10^3 \times 100 \times 10^{-15} \times \ln(3/25)$$

or,

$$t > 0.1928\ \text{ns}. \qquad (10.68)$$

Thus $t_{pd,0\rightarrow 1} = 0.1928$ ns.

### Computing $t_{pd,1\rightarrow 0}$

When the input $v_{IN1}$ goes low, the circuit model that applies is shown in Figure 10.21. In this case, we know that initial voltage $V_{C0}$ on the capacitor is determined by the voltage-divider relationship:

$$V_{C0} = \frac{V_S R_{ON}}{R_{ON} + R_L} = 5/11 \text{ V}.$$

Our goal is to solve for the time it takes for the capacitor to charge up to $V_{OH} = 4$ volts.

Again, we obtain the following using the node method (writing $v_C$ in place of $v_{OUT1}$),

$$\frac{v_C - V_S}{R_L} + C_{GS2}\frac{dv_C}{dt} = 0.$$

Rearranging, we get

$$R_L C_{GS2}\frac{dv_C}{dt} + v_C = V_S. \tag{10.69}$$

Solving Equation 10.69 yields

$$v_C(t) = V_S + Ae^{-t/R_L C_{GS2}}. \tag{10.70}$$

Using the initial condition, we get

$$v_C(t) = V_S + (V_{C0} - V_S)e^{-t/R_L C_{GS2}}, \tag{10.71}$$

substituting, $V_{C0} = 5/11$ V and $V_S = 5$ V,

$$v_C(t) = 5 - (50/11)e^{-t/R_L C_{GS2}}. \tag{10.72}$$

How long does it take for $v_C$ to go above $V_{OH} = 4$ V from an initial 5/11 V? To determine the delay, we must solve for the $t$ that satisfies

$$5 - (50/11)e^{-t/R_L C_{GS2}} > 4.$$

Simplifying, we get

$$t > -R_L C_{GS2} \ln\left(\frac{11}{50}\right). \tag{10.73}$$

In other words,

$$t > -10 \times 10^3 \times 100 \times 10^{-15} \ln(11/50)$$

$$t > 1.5141 \text{ ns}. \tag{10.74}$$

The delay $t_{pd,1\rightarrow 0}$ is thus 1.5141 ns.

Notice the $R_L C_{GS2}$ factor in Equation 10.73. In typical circuits, a ballpark estimate of the delay can be obtained by simply taking the product of the capacitance and the effective resistance through which it charges. In our case, $t_{pd,1\rightarrow 0} \approx R_L C_{GS2} = 10 \times 10^3 \times 100 \times 10^{-15} = 1$ ns. Similarly, the ballpark estimate of $t_{pd,0\rightarrow 1}$ is given by $t_{pd,0\rightarrow 1} \approx R_{TH} C_{GS2} = 10/11 \times 10^3 \times 100 \times 10^{-15} = 0.09$ ns.

### Computing $t_{pd}$

By our definition, the propagation delay of the gate $t_{pd}$ is the greater of the rising and falling delays. In other words,

$$t_{pd} = max(t_{pd,0\rightarrow 1}, t_{pd,1\rightarrow 0}).$$

Therefore, $t_{pd} = 1.5141$ ns.

---

EXAMPLE 10.1 WIRE LENGTH ON A VLSI CHIP In this example, we will examine how wire length becomes an important issue in the design of VLSI chips. Consider the inverter pair circuit in Figure 10.14. Suppose the two inverters are on the opposite ends of a chip that is 1 cm on a side. The resulting long wire connecting them can no longer be treated as an ideal conductor with no resistance or capacitance. Instead we must replace the wire with an ideal wire in combination with a wire capacitance and a wire resistance. The resulting RC delays can be significantly higher than the RC delays of inverters connected to each other with short wires.

Figure 10.24 depicts graphically the wire connecting the two inverters on the VLSI chip. Assume the wire is of length $L$ and width $W$. The MOSFETs have gate lengths $L_g$ and gate widths $W_g$. Since the length of the wire is significant, we need to model it carefully. Let the wire resistance be denoted $R_{wire}$ and the wire capacitance $C_{wire}$. The circuit model for the inverter pair taking into account wire parasitics is shown in Figure 10.25.

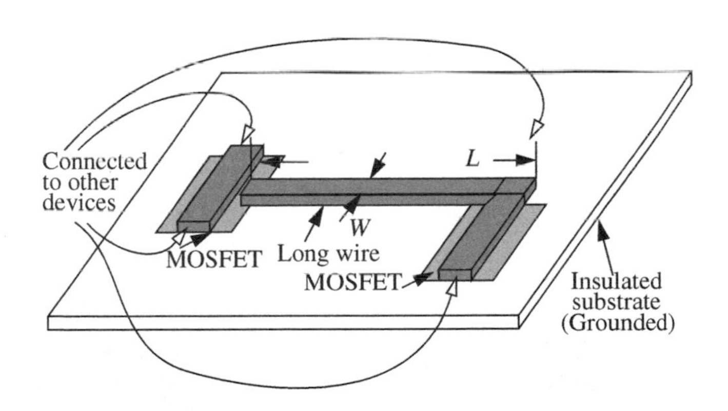


FIGURE 10.24 Long wire on a VLSI chip.

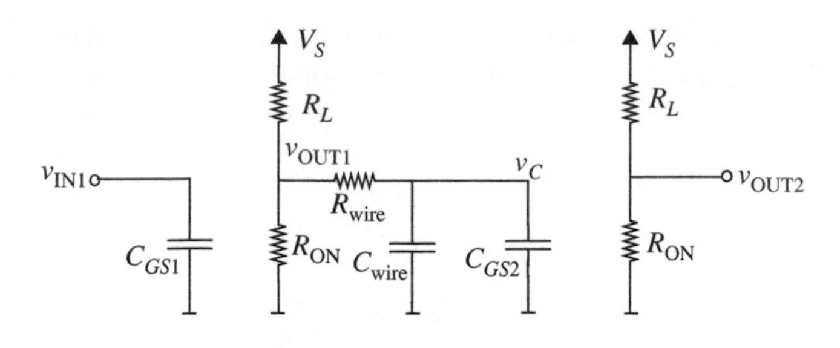


FIGURE 10.25 Circuit model of long wire on a VLSI chip.

(Here we assume that the parasitic inductance of the wire according to the model in Figure 9.7 in Chapter 9 is zero.)

If the sheet resistance of the wire is $R_\square$ (see Equations 1.10 and 1.9), we know,

$$R_{wire} = (L/W)R_\square.$$

Similarly, if $C_o$ is the capacitance per unit area of the wire, (formed between the wire, insulation, and the grounded substrate), we know

$$C_{wire} = LWC_o.$$

Clearly, the longer the wire, the larger its capacitance and resistance. Recalling that delays are related to the RC time constants, notice that the RC product for the wire is

$$R_{wire}C_{wire} = (L/W)R_\square \times LWC_o = L^2R_\square C_o.$$

The $L^2$ term in the RC product implies that wire delays grow as the square of wire lengths. Assume that the wire is 1 $\mu$m wide and 1000 $\mu$m long. Further, assume $R_\square$ is 2 $\Omega$ and $C_o$ is 2 fF/$\mu$m$^2$. Therefore,

$$R_{wire} = 1000 \times 2 = 2 \text{ k}\Omega$$

and

$$C_{wire} = 1000 \times 2\text{fF} = 2 \text{ pF}.$$

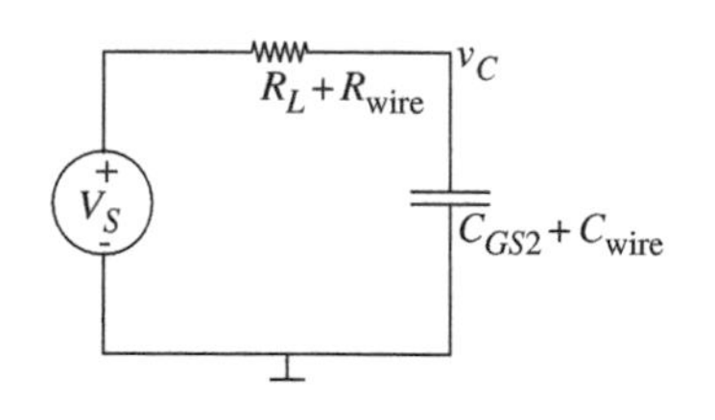


FIGURE 10.26 Charging the wire capacitor on a VLSI chip.

The RC time constant for the wire is

$$R_{wire}C_{wire} = 2 \times 10^3 \times 2 \times 10^{-12} = 4 \text{ ns}.$$

Figures 10.26 and 10.27 show the relevant circuit models for charging and discharging the wire capacitance $C_{wire}$ and gate capacitance $C_{GS2}$. Let us assume values for $V_{OL}$

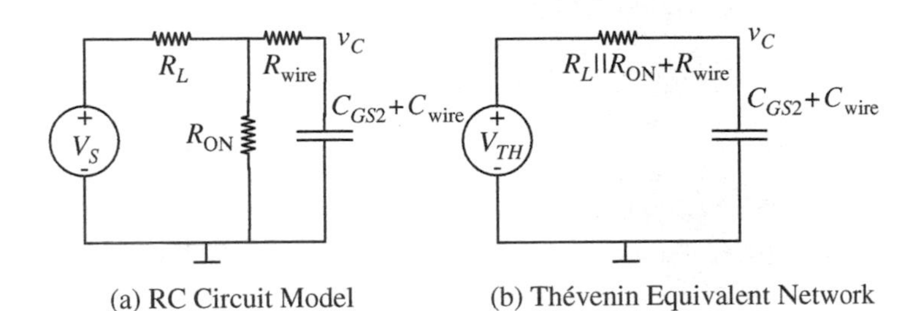


**FIGURE 10.27** Discharging the wire capacitor on a VLSI chip.

and $V_{OH}$ to be the same as those used in Section 10.4.2. In other words, $V_{OH} = 4\,\text{V}$ and $V_{OL} = 1\,\text{V}$.

When the input $V_{IN1}$ transitions from high to low, Figure 10.26 applies, and we can use the results from Section 10.4.2 to compute the propagation delay $t_{pd,1\rightarrow0}$ by using $(R_L + R_{wire})$ in place of $R_L$ and $(C_{GS2} + C_{wire})$ in place of $C_{GS2}$. Thus, the propagation delay for $R_L = 10$ k, $R_{ON} = 1\,\text{k}\Omega$, $C_{GS2} = 100$ fF is given by

$$t_{pd,1\rightarrow0} = -(R_L + R_{wire}) \times (C_{GS2} + C_{wire}) \ln(11/50)$$
$$= -(10 + 2) \times 10^3 \times (100 + 2000) \times 10^{-15} \times \ln(11/50).$$

Thus,

$$t_{pd,1\rightarrow0} = 38.15 \text{ ns}. \quad (10.75)$$

When the input $v_{IN1}$ makes a low to high transition, Figure 10.27 applies, and we can use the results from Section 10.4.2 to compute the propagation delay $t_{pd,0\rightarrow1}$ with $(R_L \| R_{ON} + R_{wire})$ in place of $R_{TH}$ and $(C_{GS2} + C_{wire})$ in place of $C_{GS2}$. For $R_L = 10\,\text{k}\Omega$, $R_{ON} = 1\,\text{k}\Omega$, $C_{GS2} = 100\,\text{fF}$, we get:

$$t_{pd,0\rightarrow1} = -(R_L \| R_{ON} + R_{wire})(C_{GS2} + C_{wire}) \ln(3/25)$$
$$= -\left(\frac{10}{11} + 2\right) \times 10^3 \times (100 + 2000) \times 10^{-15} \times \ln(3/25),$$

or,

$$t_{pd,0\rightarrow1} = 12.9 \text{ ns}. \quad (10.76)$$

Thus, we see that $t_{pd,0\rightarrow1} = 12.9$ ns, which is significantly higher than the delay when the wire effects were not included.

Choosing the larger of the rise and fall delays, we observe that $t_{pd} = 38.15$ ns. Clearly, the wire delay has increased the circuit delay by more than an order of magnitude.

## 10.5 STATE AND STATE VARIABLES

### 10.5.1 THE CONCEPT OF STATE

Capacitors and inductors can be discussed from a somewhat different point of view, one that emphasizes the *memory* aspect of the devices, as introduced in Equation 9.13 in Section 9.1.1. This section introduces an analysis of capacitor and inductor circuits based on their state, and will show that this representation facilitates computer analysis of circuits, which is particularly useful when the circuit is nonlinear or if it contains a large number of storage elements.

Let us begin by quickly reviewing the concept of state. If we apply an arbitrary current waveform to a capacitor, as in Figure 10.28, then the charge on the capacitor, and hence the capacitor voltage will be the integral of that current, as indicated in the figure.

$$q(t) = \int_{-\infty}^{t} i(t)dt. \tag{10.77}$$

It might at first appear that to perform this integral we need to know the complete current waveform from $t = -\infty$. Not so. All we need is the charge (or voltage, since $q = Cv$) at one time, and the current waveform thereafter. If the charge at $t_1$ is $q(t_1)$ then from Equation 10.77 the charge at some time $t_2$

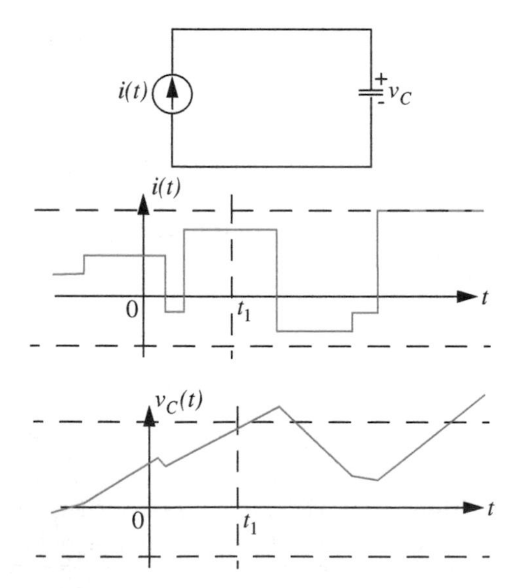


FIGURE 10.28 Voltage as a state variable.

greater than $t_1$ is

$$q(t_2) = \int_{-\infty}^{t_1} i(t) + \int_{t_1}^{t_2} i(t)dt \tag{10.78}$$

$$= q(t_1) + \int_{t_1}^{t_2} i(t)dt. \tag{10.79}$$

All of the relevant past history of the circuit prior to $t_1$ is summarized in one value, $q(t_1)$. Variables that have this property are called *state variables.* Thus Equation 10.79 indicates that if we know the value of the *state variable* at one time, and the value of the *input variable* thereafter, we can find the value of the state variable for any subsequent time.

For linear time-invariant capacitors, the capacitor voltage is also a state variable, because

$$q = Cv.$$

For an inductor, the fundamental state variable is the total flux linked by the inductor, $\lambda$. Recall from Equation 9.32, if the inductor is linear and time-invariant, the current is equally appropriate as a state variable since it is linearly related to $\lambda$ as

$$i = \frac{\lambda}{L}.$$

From this point of view, the first-order differential equations for the RC and RL circuits, Equations 10.2, 10.42, and 10.46 can all be written as *state equations*

$$\frac{d}{dt}(\text{state variable}) = f(\text{state variable, input variable}). \tag{10.80}$$

For the linear case, $f$ is a linear function, so Equation 10.80 becomes

$$\frac{d}{dt}(\text{state variable}) = K_1(\text{state variable present value}) + K_2(\text{input variable}). \tag{10.81}$$

For example, consider Equation 10.2 for the circuit in Figure 10.2a:

$$\frac{dv_C}{dt} + \frac{v_C}{RC} = \frac{i(t)}{C}$$

This equation can be written in the canonic state equation form of Equation 10.81 as

$$\frac{dv_C}{dt} = -\frac{v_C}{RC} + \frac{i(t)}{C} \tag{10.82}$$

where the only state variable is $v_C$.

### 10.5.2 COMPUTER ANALYSIS USING THE STATE EQUATION

One advantage of the state equation formulation is that even in the nonlinear case, the equations can be readily solved on a computer.[9] If the input signal and the initial value of the state variable are known, then the slope of the state variable, that is,

$$\frac{d}{dt}(\text{state variable})$$

can be found from Equation 10.81. The value of the state variable at time $t + \Delta t$ can now be estimated by standard numerical methods (Euler's method, Runge-Kutta, etc.). The process can now be repeated until the entire waveform is found.

Continuing with our example in Equation 10.82, suppose that the input signal $i(t)$ is known for all time. Also, suppose that the value of the state variable at time $t = t_0$, namely $v_C(t_0)$, is known. Then, Euler's method[10] approximates the value of the state variable at time $t = t_0 + \Delta t$ as

$$v_C(t_0 + \Delta t) = v_C(t_0) - \frac{v_C(t_0)}{RC}\Delta t + \frac{i(t_0)}{C}\Delta t. \qquad (10.83)$$

The value of $v_C$ at time $t = t_0 + 2\Delta t$ can be determined in like manner from the value of $v_C(t_0 + \Delta t)$ and $i(t_0 + \Delta t)$. Subsequent values of $v_C$ can be determined using the same process. By choosing small enough values of $\Delta t$, a computer can determine the waveform for $v_C(t)$ to an arbitrary degree of accuracy. This process illustrates the fact that the initial state contains all the information that is necessary to determine the entire future behavior of the system from the initial state and the subsequent input.

This procedure works even for circuits with many capacitors and inductors, linear or nonlinear, because these higher-order circuits can be formulated in terms of a set of first-order state equations like Equation 10.80, one for each energy-storage element (with an independent state variable) in the network. Chapter 12 discusses such an example in Section 12.10.1.

---

9. To build intuition, we will describe a simple computer method here. However, we note that other more efficient methods are employed in practice.

10. Euler's method is based on the following discrete approximation:

$$\frac{dv_C(t)}{dt} \approx \frac{v_C(t + \Delta t) - v_C(t)}{\Delta t}.$$

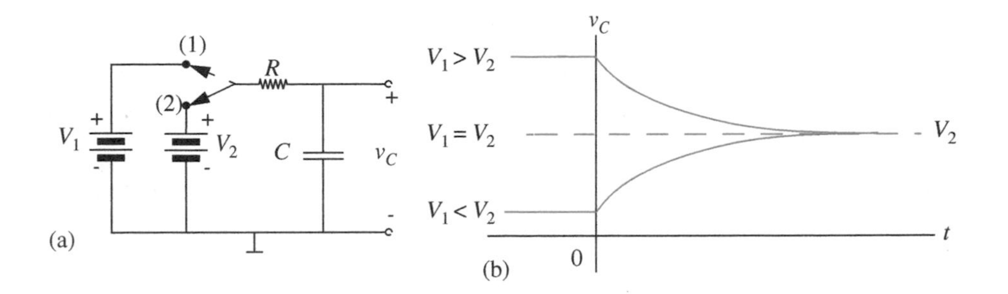


**FIGURE 10.29** Transient with initial charge on capacitor.

### 10.5.3 ZERO-INPUT AND ZERO-STATE RESPONSE

Another advantage of the state variable point of view is that it allows us to solve transient problems by superposition. Specifically, we find first the *zero-input response*, the response for the true initial conditions, with the input drive zero. Then we find the *zero-state response*, the response of the circuit when the initial state is zero, that is, all capacitor voltages and inductor currents are initially zero. The total response is the sum of the zero-input response (ZIR) and the zero-state response (ZSR).

Relating these ideas to Equation 10.81, finding the zero-input-response involves solving the equation:

$$\frac{d}{dt}(\text{state variable}) = K_1(\text{state variable present value}) \tag{10.84}$$

using the true initial conditions for the state variable. Finding the zero-state-response involves solving the equation:

$$\frac{d}{dt}(\text{state variable}) = K_1(\text{state variable present value}) + K_2(\text{input variable}) \tag{10.85}$$

with the initial value of the state variable set to 0.

Let us illustrate these ideas with an example. The circuit shown in Figure 10.29a contains a switch, which is moved from position (1) to position (2) at $t = 0$. If the switch has been in position (1) for a long time, the capacitor will be charged to the voltage $V_1$. That is, the initial condition for the circuit is

$$v_C = V_1 \qquad t < 0. \tag{10.86}$$

When the switch is moved to position (2), there will be a transient charge (or discharge) until the capacitor voltage reaches a new steady state.

The governing differential equation is the same as the previous capacitor example in Section 10.1.4, Equation 10.42:

$$v_I = RC\frac{dv_C}{dt} + v_C. \tag{10.87}$$

Writing the same equation in canonic state equation form, we get

$$\frac{dv_C}{dt} = -\frac{v_C}{RC} + \frac{v_I}{RC}. \tag{10.88}$$

First, let us first solve for the capacitor voltage directly by finding the homogeneous solution and particular solution. We will then derive the capacitor voltage by obtaining the ZIR and ZSR.

The homogeneous solution is

$$v_C = Ae^{-t/RC}. \tag{10.89}$$

By inspection from Equation 10.89, the particular solution must be

$$v_C = V_2. \tag{10.90}$$

The complete solution is the sum of these two:

$$v_C = Ae^{-t/RC} + V_2. \tag{10.91}$$

Equating Equation 10.91 at $t = 0$ to the stated initial condition, Equation 10.86,

$$v_C = V_1 = A + V_2. \tag{10.92}$$

$$A = V_1 - V_2. \tag{10.93}$$

Hence the complete solution for $t > 0$ is

$$v_C = V_2 + (V_1 - V_2)e^{-t/RC}. \tag{10.94}$$

Plots of this result are shown in Figure 10.29b. As indicated, the response depends on the relative size of $V_1$ and $V_2$. For one particular value, namely $V_1 = V_2$, there is no transient, as is obvious from physical considerations.

In Equation 10.94 the first term is the particular solution, and the second is the homogeneous solution.

Next, to obtain insight into the method involving the ZIR and ZSR, a trivial rewrite of Equation 10.94 yields

$$v_C = V_1e^{-t/RC} + V_2(1 - e^{-t/RC}). \tag{10.95}$$

Now the first term is the response to an initial *state*, in this case an initial capacitor, and no input. This we call the zero-input response (ZIR). The second

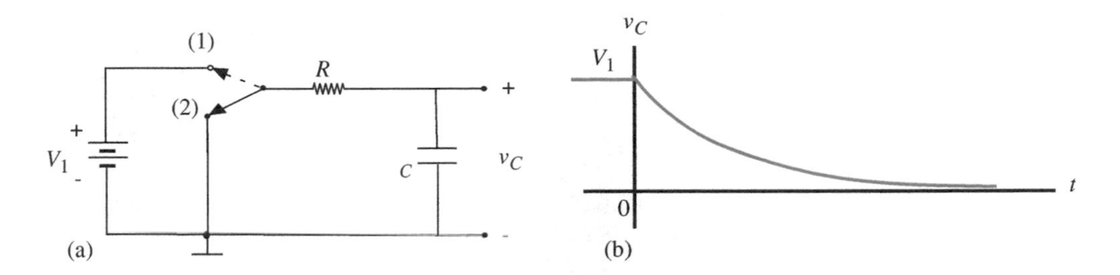


FIGURE 10.30 Zero-input subcircuit and response.

term is the response to an external input, for no initial capacitor charge: the zero-state response (ZSR).[11] To verify, let us now solve directly for the ZIR and the ZSR by superposition.

The subcircuit for finding the ZIR is shown in Figure 10.30a. As before, the capacitor is initially charged to $V_1$, but here the input for $t > 0$ is zero, so after the switch moves to position (2), the capacitor simply discharges to zero. Formally, the corresponding equation to be solved to obtain the ZIR is

$$\frac{dv_C}{dt} = -\frac{v_C}{RC} \tag{10.96}$$

with the initial condition on the capacitor voltage being $V_1$.

The homogeneous solution is

$$v_C = V_1 e^{-t/RC}. \tag{10.97}$$

This is the complete zero-input response, because the particular solution is zero.

The subcircuit for finding the zero-state response is shown in Figure 10.31a. The corresponding equation to be solved to obtain the ZSR is

$$\frac{dv_C}{dt} = -\frac{v_C}{RC} + \frac{v_I}{RC}. \tag{10.98}$$

with the initial condition on the capacitor voltage chosen as 0.

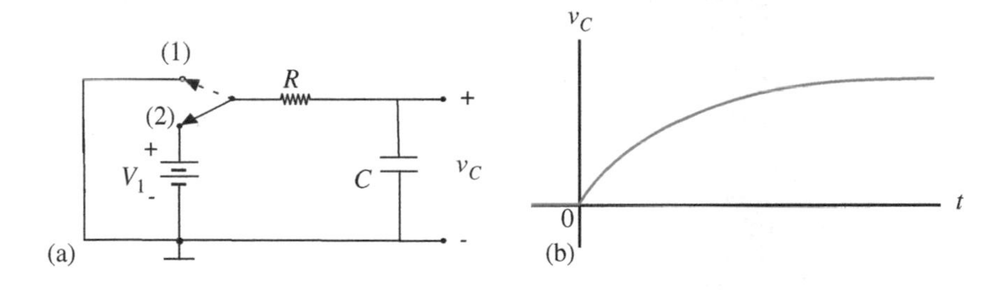


FIGURE 10.31 Zero-state subcircuit and response.

11. Notice the similarity between this and Equation 10.62.

The homogeneous solution is as in Equation 10.89, and the particular solution is again $V_2$, so the solution is of the form

$$v_C = Ae^{-t/RC} + V_2. \tag{10.99}$$

This time by definition the initial condition is zero, so evaluating right after the switch is thrown,

$$0 = A + V_2 \tag{10.100}$$

and the complete zero-state response for $t > 0$ is

$$v_C = V_2(1 - e^{-t/RC}). \tag{10.101}$$

The total response is the sum of the ZIR and ZSR from Equations 10.101 and 10.97 and is given by

$$v_C = V_1 e^{-t/RC} + V_2(1 - e^{-t/RC}). \tag{10.102}$$

Observe that Equation 10.102 agrees with the formulation of Equation 10.94.

Comments

- The particular solution and the homogeneous solution are terms which apply to a *method of solving differential equations.*
- Zero-input and zero-state responses arise from a particular way of partitioning the circuit problem into two simpler subproblems. The resulting subcircuits can be solved by finding the homogeneous solution and particular solution in each case.
- For the ZIR, the particular solution will by definition be zero. Hence *all ZIRs will be homogeneous solutions.*
- But all ZSRs are not particular solutions, because there is also a homogeneous solution associated with the ZSR. The $e^{-t/RC}$ term in Equation 10.101 is an example of this.
- A major advantage to the state variable point of view is that any arbitrary ZIR can be added to any ZSR solution we have already worked out. Thus any transient problem with zero initial conditions can be easily generalized to one with arbitrary initial conditions. This concept will be illustrated in the examples in Section 10.6.

### WWW 10.5.4 SOLUTION BY INTEGRATING FACTORS*

## 10.6 ADDITIONAL EXAMPLES

### 10.6.1 EFFECT OF WIRE INDUCTANCE IN DIGITAL CIRCUITS

Section 10.4 showed that RC effects lead to propagation delay in digital circuits. It turns out that when parasitic inductors are present, RL effects can be a similar source of propagation delay. Consider the inverter circuit shown in Figure 10.32a. Assume that a poor design has resulted in a long wire connecting the MOSFET drain to the output of the inverter. A circuit model of the inverter showing the parasitic wire inductance is depicted in Figure 10.32b.

Assume that the inverter has a 0-V input as an initial condition. The MOSFET is in its off state and the current $i_L$ through the inductor $L$ will be 0. The voltage $v_L$ across the inductor is also 0. Now suppose that a 0-V to $V_S$-step is applied to the input of the inverter as illustrated in Figure 10.32a. Assume that our goal is to determine the current $i_L$ through the inductor and the voltage $v_L$ across the inductor as a function of time.

The step input to the inverter results in a corresponding $V_S$-step applied to the $RL$ circuit at the output of the inverter as illustrated in Figure 10.32c. From the initial conditions, at $t = 0$, both $i_L$ and $v_L$ are 0. From this point on, the situation is identical to that for the RL transient discussed in Section 10.2.1 with $V$ used in place of $V_S$. Therefore the analysis presented in Section 10.2.1 applies.

It is interesting to speculate as to what will happen if the switch in Figure 10.32c is opened after being closed for a long time. When the switch is opened, the inductor current cannot go to zero instantaneously. Since a practical open switch behaves as an extremely high resistance, the current through the inductor will result in a huge voltage spike across the switch, and possibly damage it.

### 10.6.2 RAMP INPUTS AND LINEARITY

Solutions become somewhat more complicated when we move beyond simple step inputs. Consider the case of a series RC circuit with a voltage ramp drive,

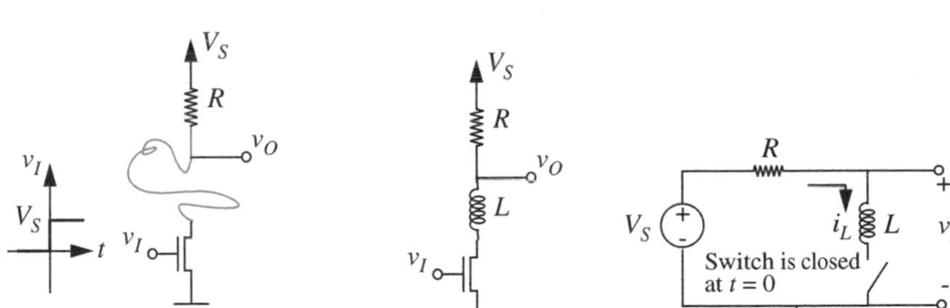


(a) Inverter with a long wire connecting the output and the MOSFET drain (b) Circuit model (c) Circuit for step input at $v_I$

**FIGURE 10.32** Inverter circuit with parasitic inductance.

that is

$$v_I = S_1 t \qquad t > 0 \tag{10.113}$$

where $S_1$ has the dimensions of volts per second. The circuit and input waveform are sketched in Figures 10.33a and b. Let us first find the zero-state response, by assuming that the capacitor is initially uncharged. The differential equation is, from before,

$$v_I = S_1 t = RC\frac{dv_C}{dt} + v_C. \tag{10.114}$$

We will solve this using our usual method of homogeneous and particular solutions. The homogeneous solution has the usual form:

$$v_C = Ae^{-t/RC}. \tag{10.115}$$

This homogeneous solution is plotted in Figure 10.33c. We must now find a particular solution that is appropriate for a ramp input. Because the drive is a ramp, a good first guess is a ramp of the same slope as the input:

$$v_C = K_2 t. \tag{10.116}$$

Substituting this into the differential equation, Equation 10.114, we obtain

$$S_1 t = RCK_2 + K_2 t. \tag{10.117}$$

Because there is no solution for $K_2$ unless $RC = 0$, our initial guess for the particular solution is not quite correct. We need another degree of freedom in the solution, so an appropriate second guess is

$$v_C = K_2 t + K_3. \tag{10.118}$$

Now from Equation 10.114:

$$S_1 t = RCK_2 + K_2 t + K_3. \tag{10.119}$$

Whence

$$S_1 = K_2 \tag{10.120}$$

$$K_3 = -S_1 RC \tag{10.121}$$

and the particular solution is

$$v_C = S_1(t - RC). \tag{10.122}$$

This is a ramp with the same slope as the input ramp, except delayed in time by one time constant, as shown in Figure 10.33d.

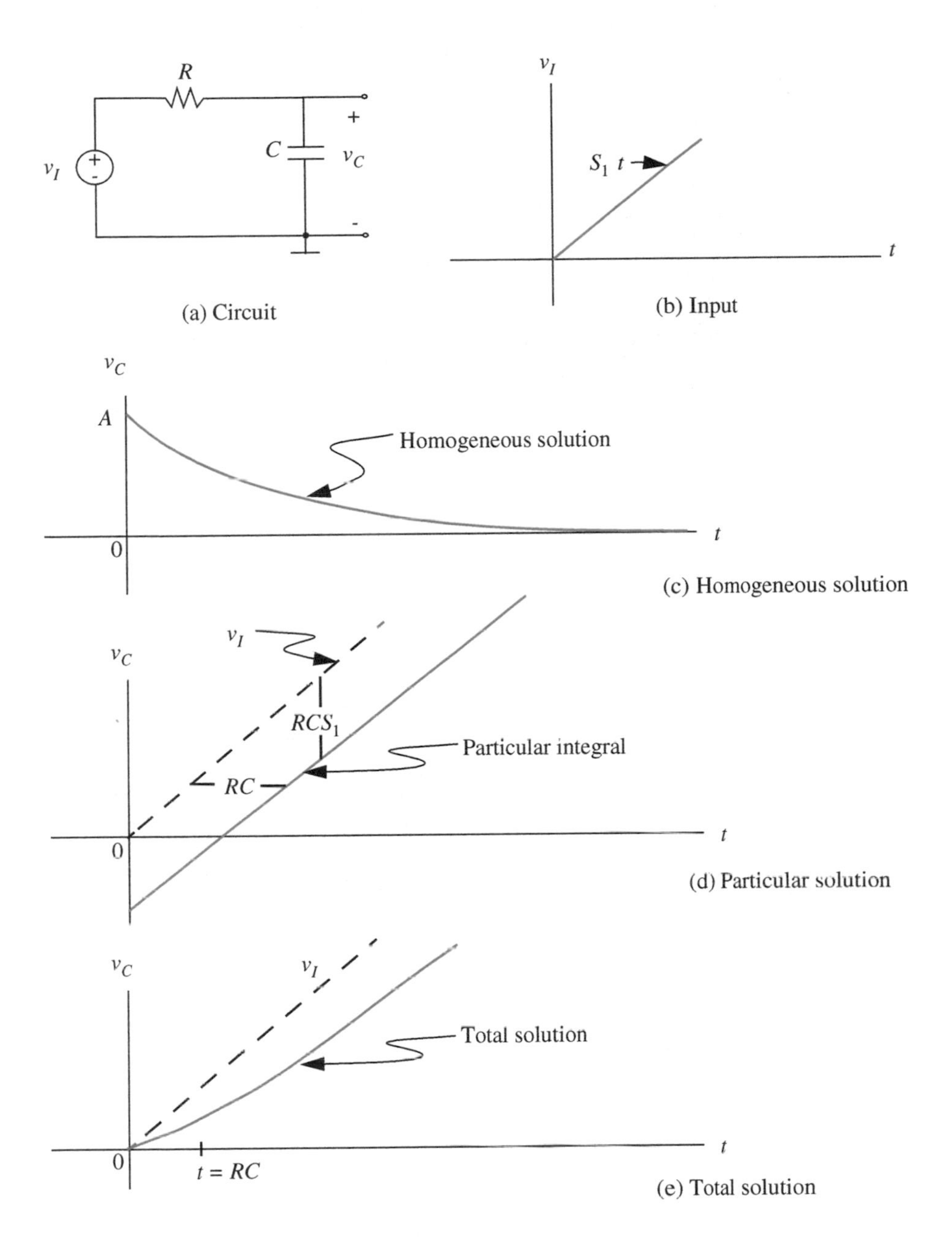


**FIGURE 10.33** Response of RC circuit to ramp.

The complete solution is of the form

$$v_C = Ae^{-t/RC} + S_1(t - RC) \quad \text{for} \quad t > 0. \tag{10.123}$$

Because we are finding the zero-state response, the initial condition is by definition zero, so evaluating Equation 10.123 at $t = 0$, we find

$$A = S_1 RC.$$

Hence the complete solution for $t > 0$ is

$$v_C = S_1(t - RC) + S_1 RCe^{-t/RC} \tag{10.124}$$

and is plotted in Figure 10.33e.

The waveforms in Figures 10.33b and 10.33e are related in a special way to those in Figures 10.2b and 10.2c. Note first that the input signal in this problem is the *integral* of the input signal in Figure 10.2 (assuming a Thévenin source). Now, from Figures 10.33e and 10.2c, or from Equations 10.124 and 10.20, the output signal here is the integral of the output signal in Figure 10.2,[12] again, assuming that we are dealing with the zero-state response.

In general, as long as one restricts integral operations to $t$ greater than zero, the *zero-state response* of the integral of some input signal is the integral of the zero-state response to that signal.

This follows from superposition if one considers integration as a summation process. In effect we are commuting two linear operators. The same is true for differentiation as well:

The response to a signal derived by differentiating an input can be obtained by differentiating the output.

Let us follow this example one step further, and consider the case where there is an initial voltage $V_0$ on the capacitor at $t = 0$, before the ramp is applied.

12. Notice that if we take the integral of Equation 10.20, namely,

$$v_C = I_0 R(1 - e^{-t/RC})$$

we obtain

$$v_C = I_0 Rt + I_0 R\, RCe^{-t/RC} + K_1$$

where, to get the zero initial condition on $v_C$, we set the constant of integration $K_1 = -I_0 R\, RC$, and obtain

$$v_C = I_0 R(t - RC) + I_0 R\, RCe^{-t/RC},$$

which is a Thévenin version of Equation 10.124.

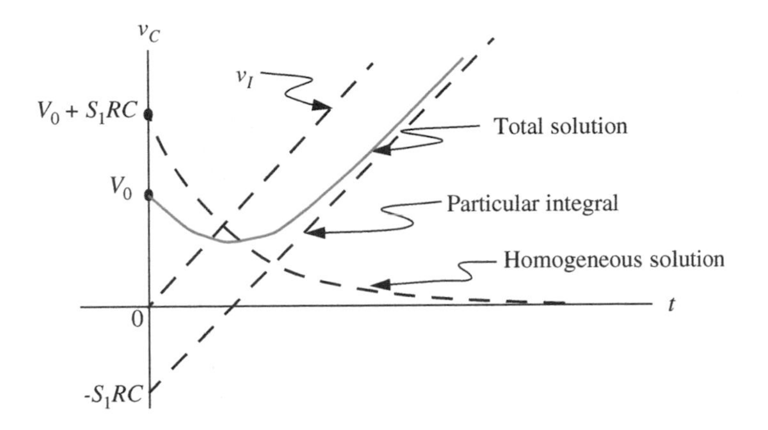


FIGURE 10.34 Ramp response for initial charge on capacitor.

Formally, the initial condition now is

$$v_C = V_0 \qquad t < 0. \tag{10.125}$$

This time around, since there is an initial condition on the capacitor, we are no longer dealing with just the zero-state response, so we cannot simply take the integral. One approach is to notice that Equation 10.124 is in effect the zero-state response, so if we find the zero-input response corresponding to the initial condition of Equation 10.125, then the complete solution is the sum of these two responses. We know from previous examples, or from Equation 10.97, that the ZIR for an RC circuit with an initial voltage $V_0$ is

$$v_C = V_0 e^{-t/RC} \tag{10.126}$$

so the total solution for $t > 0$ is

$$v_C = V_0 e^{-t/RC} + S_1 t - S_1 RC(1 - e^{-t/RC}). \tag{10.127}$$

One possible form of this solution is sketched in Figure 10.34. This example illustrates one of the advantages of the state-variable approach. Once we have found the solution for some input waveform and zero initial conditions, the solution for the same input with arbitrary initial conditions can be found by adding the appropriate ZIR solution.

---

EXAMPLE 10.2 TV DEFLECTION SYSTEM Most television sets use magnetic deflection in the cathode-ray tube. To obtain the raster scan for the picture, it is necessary to develop a ramp of current flowing through the deflection coil, as sketched in Figure 10.35. We wish to find the required waveform of $v_I$ that will

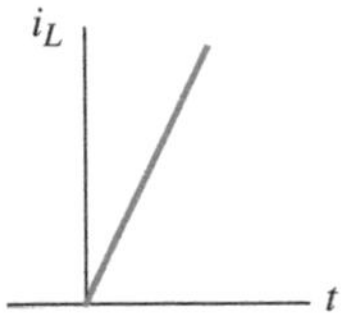

FIGURE 10.35 TV deflection coil.

generate the current ramp. The coil losses have been explicitly modeled in Figure 10.35 by the resistor $R$.

The differential equation for the circuit is, from KVL

$$v_I = iR + L\frac{di}{dt}. \tag{10.128}$$

We want a current waveform, for $t > 0$, of the form

$$i = S_1 t. \tag{10.129}$$

Hence

$$v_I = S_1 Rt + S_1 L \tag{10.130}$$

for $t > 0$. Thus to produce a ramp of current in the inductor, we need to drive with the sum of a step and a ramp.

---

WWW EXAMPLE 10.3 SOLUTION BY INTEGRATING FACTORS

---

### 10.6.3 RESPONSE OF AN RC CIRCUIT TO SHORT PULSES AND THE IMPULSE RESPONSE

It was shown in Section 10.1.4 that when the time constant of an RC circuit becomes much longer than the period of a periodic input signal, the capacitor voltage begins to approximate the *integral* of the input wave. Let us examine this property in more detail by finding the response of the RC circuit in Figure 10.36 when the input is a short pulse of amplitude $V_p$ and duration $t_p$.

We have seen several problems of this sort, so the general form of the capacitor voltage can be written by inspection. Assuming that the capacitor is initially uncharged, the response during the pulse, when the capacitor is charging is

$$v_C = V_p\left(1 - e^{-t/RC}\right) \quad 0 \le t \le t_p. \tag{10.135}$$

If $t_p$ is long enough for this transient to essentially go to completion, then at $t = t_p$, the end of the pulse, the capacitor voltage will be $V_p$. The response after the pulse has ended, when the capacitor is discharging, is thus

$$v_C = V_p e^{-(t-t_p)/RC} \quad t \ge t_p \gg RC. \tag{10.136}$$

The $t - t_p$ factor in the exponent indicates that there is a time delay in the start of the wave of an amount $t_p$. This solution is shown in Figure 10.36b.

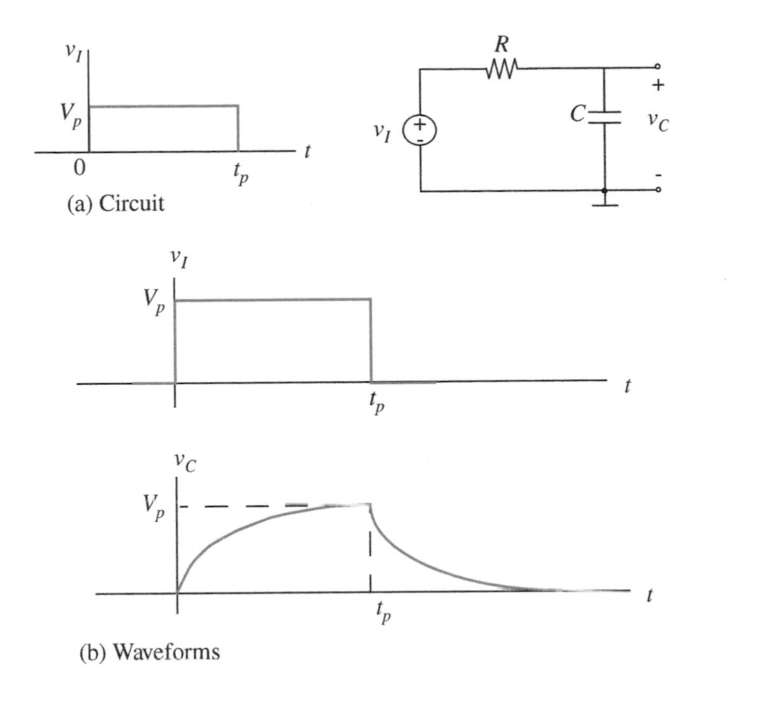


**FIGURE 10.36** Response of RC circuit to pulse.

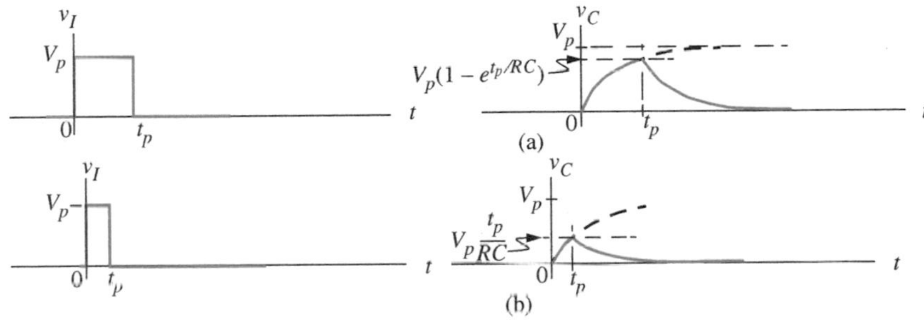


**FIGURE 10.37** Response of RC circuit to short pulse.

If the pulse is made shorter than in Figure 10.36, the charging transient can no longer reach completion. This is illustrated in Figure 10.37a. Equation 10.135 is still appropriate for the charging interval, but the response no longer reaches $V_p$. The maximum value, at $t = t_p$, is

$$v_C(t_p) = V_p \left(1 - e^{-t_p/RC}\right). \tag{10.137}$$

The discharge now has essentially the same form as before, but is smaller. That is, for $t$ greater than $t_p$, the capacitor voltage is

$$v_C = \left[V_p \left(1 - e^{-tp/RC}\right)\right] e^{-(t-t_p)/RC}. \tag{10.138}$$

If we make the pulse even shorter yet, as in Figure 10.37b, the picture actually becomes simpler. The charging part of the wave begins to look almost like a straight line. Mathematically, this can be shown by expanding the exponential in a series

$$e^{-x} = 1 - x + \frac{x^2}{2!} \cdots \tag{10.139}$$

when the charging waveform becomes, from Equation 10.135

$$v_C = V_p \left[ \frac{t}{RC} - \frac{1}{2}\left(\frac{t}{RC}\right)^2 + \cdots \right]. \tag{10.140}$$

For times much less than the time constant $RC$, that is, $t \ll RC$, we can discard all higher terms, leaving

$$v_C \simeq V_p \frac{t}{RC}, \tag{10.141}$$

which is the equation for the straight line we observe in the first part of Figure 10.37b.

Physically, when the pulse is very short the capacitor voltage is always much smaller than the pulse voltage, so during the pulse the current is roughly constant at a value

$$i_C \simeq \frac{V_p}{R}. \tag{10.142}$$

The capacitor voltage is the integral of this current, hence is a ramp:

$$v_C = \frac{1}{C} \int i_C dt \tag{10.143}$$

$$\simeq V_p \frac{t}{RC}. \tag{10.144}$$

At the end of the pulse, the capacitor voltage has reached its maximum value of

$$v_C(t_p) \simeq \frac{V_p t_p}{RC} \tag{10.145}$$

so the discharge waveform for $t$ greater than $t_p$ is

$$v_C \simeq \left[ \frac{V_p t_p}{RC} \right] e^{-(t-t_p)/RC}. \tag{10.146}$$

The important feature of this equation is that the response is now *proportional to the area ($V_p t_p$), rather than the height ($V_p$), of the input pulse.* In other words,

$$v_C \simeq \left[\frac{\text{Area of Pulse}}{RC}\right] e^{-(t-t_p)/RC}. \tag{10.147}$$

For very short pulses (that is, for $t_p \ll RC$), even the delay term in the exponent can be neglected, and the response reduces to

$$v_C \simeq \frac{\text{Area of pulse}}{RC} e^{-t/RC}. \tag{10.148}$$

Because in the limit ($t_p \ll RC$) a short pulse of large amplitude but constant area becomes an *impulse* (see Section 9.4.3) Equation 10.148 is often referred to as the *impulse response* of the circuit. In other words, if we have an impulse voltage input with area (or strength) $A$

$$v_I(t) = A\delta(t),$$

the response is given by

$$v_C \frac{A}{RC} e^{-t/RC}. \tag{10.149}$$

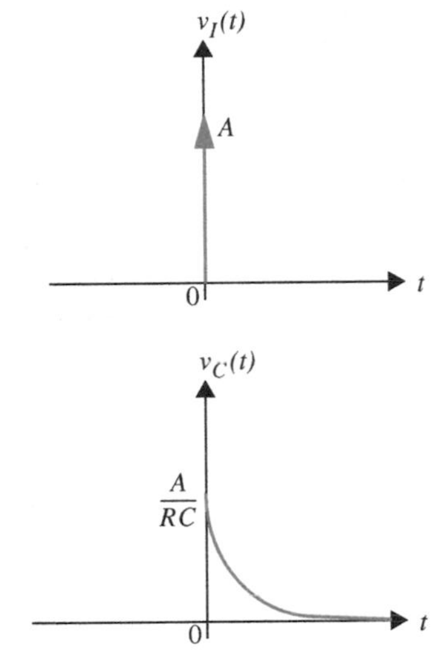


FIGURE 10.38 Impulse response of series RC circuit.

Figure 10.38 sketches the impulse voltage input and the corresponding response according to Equation 10.148.

The impulse response is a very convenient way of characterizing linear systems, because the expression contains all of the essential information about the dynamics of the system. This concept is pursued in depth in courses on signals and systems.

### 10.6.4 INTUITIVE METHOD FOR THE IMPULSE RESPONSE

The intuitive method discussed in Section 10.3 applies equally well for impulse responses. Shown in Figure 10.39a is our familiar parallel source-resistor-capacitor circuit from Figure 10.2a. Suppose that the current input is an impulse of area $Q$ that is applied at $t = 0$ as shown in Figure 10.39b. Assume we wish to find the capacitor voltage $v_C$.

As discussed in Section 10.3, we will first sketch the form of the response in the initial interval ($t < 0^+$) and in the final interval ($t \gg 0$).

Let us start by looking at the initial interval. For $t < 0$, the current source supplies zero current and hence behaves like an open circuit. Assuming that this situation has existed for a long time, the capacitor will have no charge on it, and hence $v_C$ will be 0. (If the capacitor voltage were nonzero, there would be

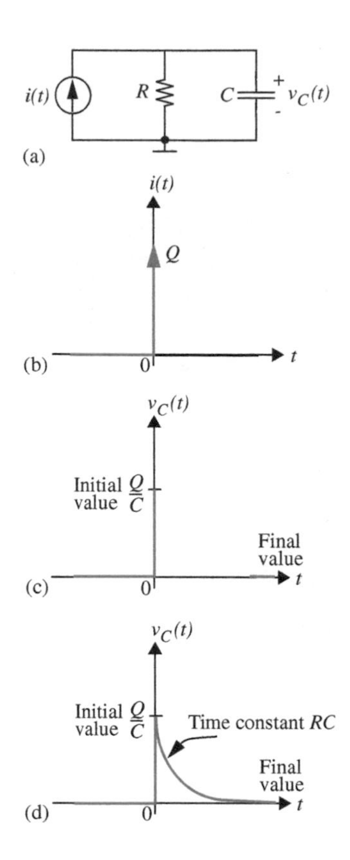

FIGURE 10.39 Intuitive method of sketching the impulse response of parallel RC circuit.

a current through $R$. This current would deplete the charge on the capacitor, till no charge exists.)

Next, the current impulse appears at $t = 0$. The capacitor behaves like an instantaneous short to the current impulse, and so the current favors the capacitor over the resistor. The entire impulse current flows through the capacitor at $t = 0$, depositing charge $Q$ on it. Corresponding to the appearance of charge $Q$ on the capacitor, from Equation 9.8, the capacitor voltage jumps instantaneously to

$$v_C(0) = \frac{Q}{C}.$$

Thus, at $t = 0^+$ we are left with the voltage $Q/C$ across the capacitor. Observe that the impulse has effectively established the initial conditions on the circuit.

This completes our intuitive analysis of the initial interval. $v_C$ during this interval is sketched in Figure 10.39c.

Next, we examine the final interval ($t \gg 0$). Since its current is zero for $t > 0$, we can again replace the current source with an open circuit. After a long period of time, a DC situation exists, and the capacitor voltage will be zero. The zero value for $v_C$ for $t \gg 0$ is also sketched in Figure 10.39c.

Finally, in the transition interval, the capacitor follows its usual exponential response with time constant $RC$. The complete curve is sketched in Figure 10.39d.

### 10.6.5 CLOCK SIGNALS AND CLOCK FANOUT

In most digital systems, a clock signal is provided to different modules of the system. A clock signal is typically a square wave between 0 volts and the supply voltage. The clock signal provides a global time base that prescribes when actions happen in systems. The use of a clock attempts to solve the following problem faced by the receiver in a pair of communicating digital systems: How to determine when a signal supplied by the sender is valid. Or conversely, how to recognize when a signal might be in the midst of transitioning to a new value. For example, we might use a stable-high clock discipline in which the sender promises to provide output signals so that they remain stable during the high part of the clock waveform. In other words, signals are allowed to transition only during the low parts of a clock. Correspondingly, the receiver promises to observe incoming signals only when the clock is high. Accordingly, the receiver circuit guarantees that its own outputs are stable when the clock is high, provided, of course, valid inputs are fed to it.

As an example of the benefits of clocked circuits, consider the digital system in Figure 10.40 in which two digital circuits are coupled to each other. Both are fed by the same timebase or clock. Inputs are fed to the first circuit in a way that input transitions happen only during the low parts of the clock signal. As shown in Figure 10.40, assume an input sequence 011 is fed to circuit 1. Similarly it produces outputs (for example 101) that are stable during the high

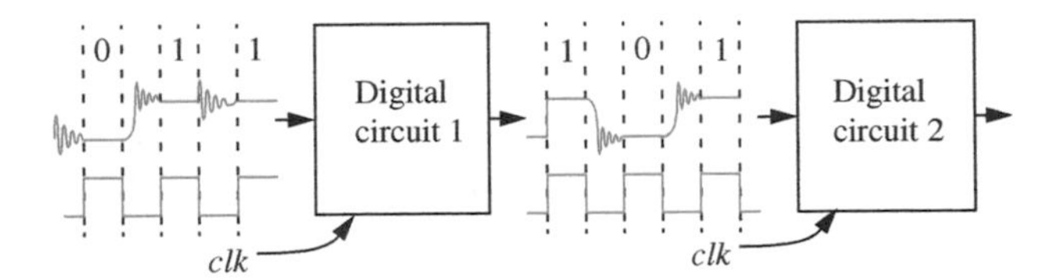


FIGURE 10.40 Clocked digital systems.

periods of the clock. Since the same clock is fed to both circuits, digital circuit 2 can observe the signal only during the periods in which the signal is valid.

Now suppose we did not use a clock. As we saw previously, RC delays cause signals to go through invalid signal levels for a finite period of time when they transition from one value to another. Without some mechanism such as a clock and an associated discipline, there would be no way in which the second digital circuit could tell when it was receiving a valid signal. As we shall see in a later chapter, telling apart a valid signal from a transitional value is particularly difficult when signals display oscillatory or ringing behavior.

The use of a clock discipline represents an instance of *time discretization.* Lumping of time into invalid periods and valid periods gives us the clocked digital abstraction and significantly simplifies the orchestrating of communication between individual circuit modules. Lumping occurs because we do not care about the precise moment when a signal is sampled, provided, of course, the signal is sampled within the valid period.

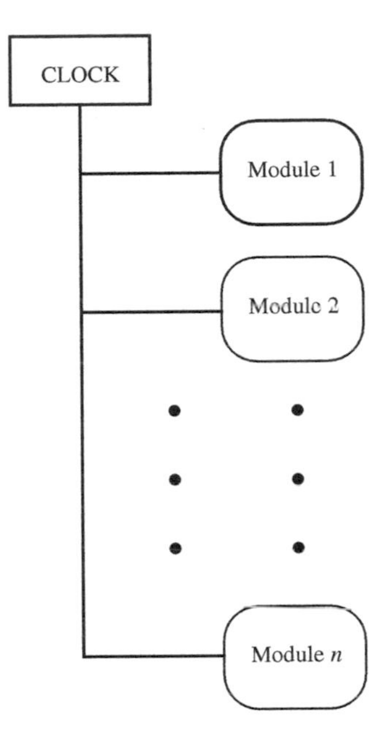


FIGURE 10.41 Clock signal for digital modules.

Figure 10.41 shows a clocked digital system in which several modules are provided a global clock timebase produced by a single clock device. One approach simply connects the clock signal generator to all the modules using one long wire. This naive approach often fails because of the RC delay associated with the long wire and the input capacitances of the driven modules. Figure 10.42 shows a circuit model for the clock distribution system. We have lumped the resistance of the wire into a single resistance $R_{wire}$. Although it is not shown in Figure 10.42, the resistance of the gate driving the clock will also appear in series with the resistance of the wire. The gate capacitors appear as parallel loads on the wire and therefore add together to yield a large equivalent capacitor:

$$C_{eq} = \sum_{i=1}^{n} C_{GSi}.$$

We know from our previous examples that slow rise and fall times at the output of a circuit result in signal delay. The rise and fall times are proportional to the *RC* time constant. A large value for *C* results in long rise and fall times, thereby limiting the clock frequency. As Figure 10.43 illustrates, notice that to achieve a valid clock signal, the clock period *T* must be larger than the sum of the rise and fall times of the clock signal. For the clock example, let us define the rise time

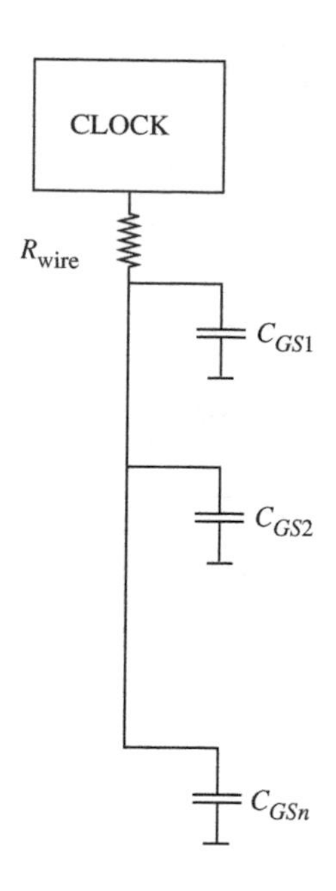


**FIGURE 10.42** Clock signal charging the gate capacitors.

$t_r$ as the time taken for the clock signal to rise from a valid output low voltage ($V_{OL}$) to a valid high output voltage ($V_{OH}$). Let us also define the fall time $t_f$ as the time taken for the clock signal to fall from a valid output high voltage ($V_{OH}$) to a valid low output voltage ($V_{OL}$). As is clear from Figure 10.43, to yield a valid digital clock signal, the clock time period must satisfy the following constraint:

$$T > t_r + t_f.$$

Figure 10.44 shows a common solution to the clock distribution problem — it limits the number of gate capacitors the signal has to drive by building a fanout buffer tree. The fanout degree of the circuit shown in Figure 10.44 is 3.

As a simple exercise, let us determine the greatest fanout degree that will support a clock frequency of 333 MHz. Let us suppose the clock signal is driven by an inverter as shown in Figure 10.45. Let us characterize the clock driver inverter by $R_L = 1\ \text{k}\Omega$, $R_{\text{ON}} = 100\ \Omega$, and $C_{GS} = 100\ \text{fF}$. Let us also assume that we desire a symmetric clock. Thus the clock period $T$ must be greater than twice the greater of the rise and fall times at the output of the inverter. Since the load resistance $R_L$ is much bigger than the ON resistance of the MOSFET, the

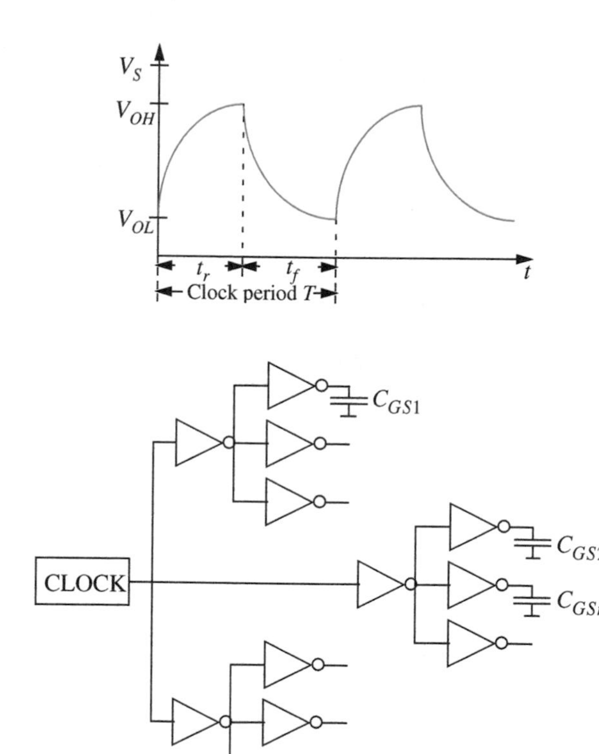


**FIGURE 10.43** Clock frequency.

**FIGURE 10.44** Fanout clock signal.

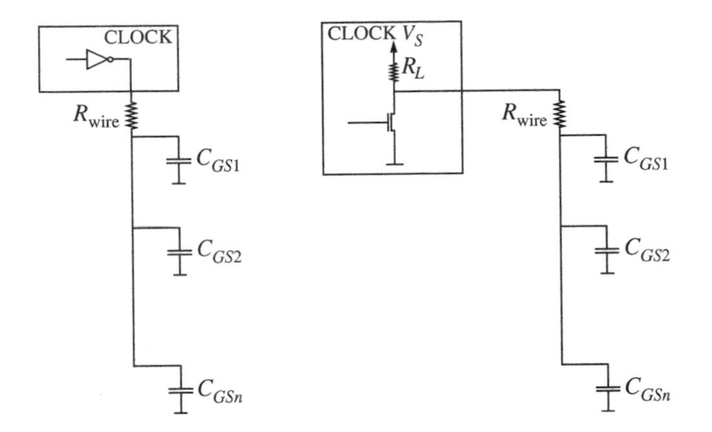


FIGURE 10.45 Clock inverter charging the gate capacitors.

rise time will be greater than the fall time. Accordingly, we focus on calculating the rise time $t_r$. As defined earlier, let $C_{eq}$ represent the total capacitance driven by the clock inverter.

Let us now compute $t_r$. As defined earlier, $t_r$ is the time taken for the clock signal to rise from $V_{OL}$ to $V_{OH}$. The equivalent circuit for computing $t_r$ is suggested in Figure 10.46. The circuit shows the equivalent capacitor being charged by the supply $V_S$ through the $R_L$ and $R_{\text{wire}}$ resistances. Let us denote $R_{eq} = R_L + R_{\text{wire}}$. The initial voltage on the capacitor $V_{C0} = V_{OL}$.

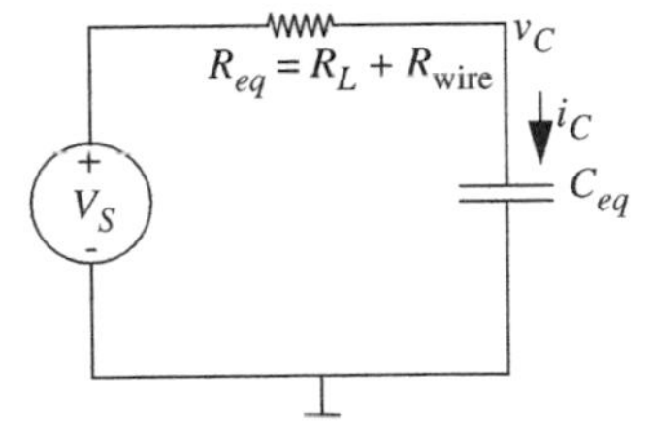


FIGURE 10.46 Equivalent circuit for determining the clock rise time.

Using the node method for the circuit shown in Figure 10.46, we get

$$\frac{v_C - V_S}{R_{eq}} + C_{eq}\frac{dv_C}{dt} = 0.$$

Rearranging, we obtain the differential equation

$$R_{eq}C_{eq}\frac{dv_C}{dt} + v_C = V_S. \tag{10.150}$$

Solving Equation 10.150, we get

$$v_C(t) = V_S + Ae^{-t/R_{eq}C_{eq}}. \tag{10.151}$$

We know that the voltage on the capacitor at $t = 0$ is $V_{OL}$. Using this initial condition, we solve for $A$ and obtain

$$v_C(t) = V_S - (V_S - V_{OL})e^{-t/R_{eq}C_{eq}}. \tag{10.152}$$

The time taken for $v_C$ to reach $V_{OH}$ from its initial value of $V_{OL}$, namely $t_r$, can be obtained from

$$V_{OH} = V_S - (V_S - V_{OL})e^{-t_r/R_{eq}C_{eq}}. \tag{10.153}$$

In other words,

$$t_r = -R_{eq}C_{eq}\ln\left(\frac{V_S - V_{OH}}{V_S - V_{OL}}\right). \tag{10.154}$$

Assuming $V_{OL} = 1$ V, $V_{OH} = 4$ V, and $V_S = 5$ V

$$t_r = -R_{eq}C_{eq}\ln\left(\frac{1}{4}\right).$$

For $R_L = 1$ kΩ and $R_{\text{wire}} \approx 0$, we get

$$t_r = 1.386 \times 10^3 C_{eq}.$$

To achieve a frequency greater than 333 MHz, the period $T$ must be less than 1/333 MHz = 3 ns. Accordingly, since $t_r < T/2 = 1.5$ ns,

$$1.5 \times 10^{-9} > 1.386 \times 10^3 C_{eq}.$$

In other words,

$$C_{eq} < 1.08 \text{ pF}.$$

Thus the total driven capacitance must be less than 1.08 pF. Suppose the inverters used in the clock buffer tree are identical to the clock driver inverters, then if the value of each gate capacitor is 100 fF, the maximum fanout degree must be less than 1080 fF / 100 fF. Thus the maximum fanout degree is 10.

### WWW 10.6.6 RC RESPONSE TO DECAYING EXPONENTIAL *

### 10.6.7 SERIES RL CIRCUIT WITH SINE-WAVE INPUT

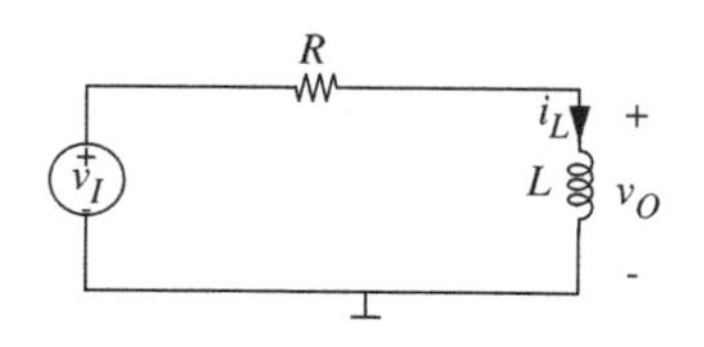

FIGURE 10.48 RL circuit with sine-wave drive.

Figure 10.48 shows a series RL circuit being driven with a sinewave voltage source suddenly applied at $t = 0$:

$$v_I = V\sin(\omega t) \qquad t > 0. \tag{10.167}$$

Let us find the voltage across the inductor, assumed to be ideal. For simplicity we assume zero initial state,

$$i_L = 0 \qquad t < 0. \tag{10.168}$$

From KVL around the loop,

$$v_I = i_L R + L\frac{di_L}{dt}. \tag{10.169}$$

The homogeneous solution, from Section 10.2.1, Equation 10.52, is

$$i_L = Ae^{-(R/L)t}. \tag{10.170}$$

Because the input is a sine wave, a reasonable first guess for the particular solution is

$$i_L = K\sin(\omega t). \tag{10.171}$$

From Equation 10.169 for $t > 0$,

$$V\sin(\omega t) = KR\sin(\omega t) + L\omega K\cos(\omega t). \tag{10.172}$$

This can't be solved for $K$ unless $L$ is zero, so our first guess is not quite right. We need another degree of freedom in the solution, so try

$$i_L = K_1 \sin(\omega t) + K_2 \cos(\omega t). \tag{10.173}$$

Now Equation 10.169 becomes

$$V\sin(\omega t) = K_1 R \sin(\omega t) + K_2 R \cos(\omega t) + K_1 \omega \cos(\omega t) - K_2 L\omega \sin(\omega t) \tag{10.174}$$

Equating sine terms, and equating cosine terms, we find

$$V = K_1 R - K_2 L\omega \tag{10.175}$$

$$0 = K_1 L\omega + K_2 R \tag{10.176}$$

which yields, via Cramer's Rule (see Appendix D),

$$K_1 = V\frac{R}{R^2 + \omega^2 L^2} \tag{10.177}$$

$$K_2 = V\frac{-\omega L}{R^2 + \omega^2 L^2}. \tag{10.178}$$

The complete solution is of the form

$$i_L = Ae^{-(R/L)t} + V\frac{R}{R^2 + \omega^2 L^2}\sin(\omega t) - V\frac{\omega L}{R^2 + \omega^2 L^2}\cos(\omega t), \quad t \geq 0. \tag{10.179}$$

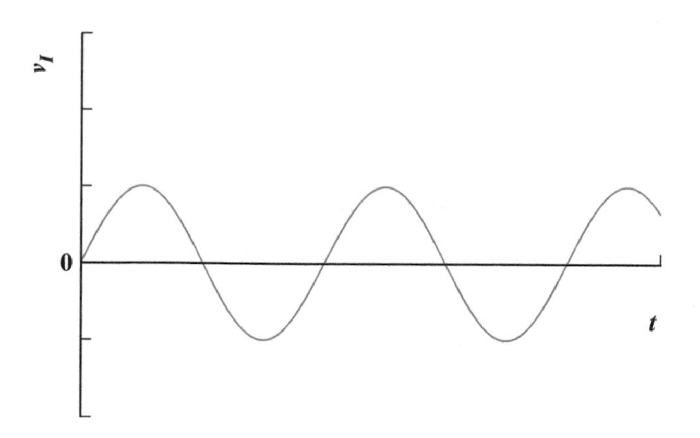


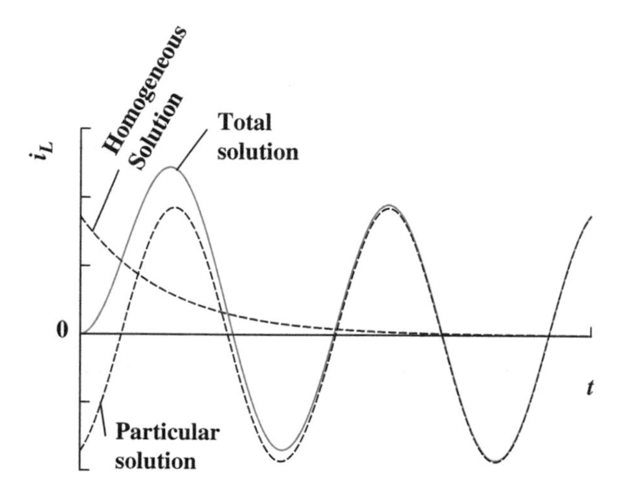


FIGURE 10.49 Waveforms for RL circuit with sinewave drive.

The value of $A$ can be found by applying the initial condition, Equation 10.168, to Equation 10.179, with $t = 0$, whence

$$A = \frac{V\omega L}{R^2 + \omega^2 L^2}. \tag{10.180}$$

The solution is shown in Figure 10.49.

Equation 10.179 is fairly easy to interpret when $t$ is large enough that the exponential term has died away. If the drive frequency is very low, such that

$$\omega \ll \frac{R}{L}, \tag{10.181}$$

then the current reduces to

$$i_L \simeq \frac{V}{R}\sin(\omega t). \tag{10.182}$$

That is, at low frequencies, the current is determined only by the resistor, and the inductor behaves like a short circuit.

At high frequencies, that is, for

$$\omega \gg \frac{R}{L} \tag{10.183}$$

Equation 10.179 reduces to

$$i_L \simeq \frac{-V}{\omega L}\cos(\omega t). \tag{10.184}$$

In this case, the current is determined almost solely by the inductor. Note that the current is still sinusoidal, but now is 90 degrees out of phase with the applied voltage. Also, the magnitude of the current becomes smaller and smaller as the frequency of the applied sinewave increases.

It is a little disappointing that such a simple circuit can lead to this level of algebra. But fortunately there is a simpler approach that can be used for linear circuits. This approach, discussed in Chapter 13, reduces all the differential equations to algebraic expressions.

## 10.7 DIGITAL MEMORY

This chapter demonstrated previously that the memory aspect of capacitors and inductors formalized using the notion of state variables provided many uses in the analog domain. The same memory property can also be utilized in the digital domain to implement digital memory using the analogous concept of digital state. Digital memory is not only an important application of capacitors, but it is of fundamental importance in its own right.

### 10.7.1 THE CONCEPT OF DIGITAL STATE

A common example of the use of memory involves the digital calculator. Suppose we wish to compute the value of the expression $(a \times b) + (c \times d)$. We might first multiply $a$ and $b$ and store the result $(a \times b)$ in memory. We might then multiple $c$ and $d$, and add the resulting value $(c \times d)$ to $(a \times b)$ by recalling the latter value from memory. Observe that the calculator contained a key to explicitly *store* a given value into the memory. Observe further that once a value was stored in memory, it could be read from memory any number of times, without affecting the value in memory. In fact, it remained valid until another value was explicitly stored in memory or it was erased. Erasure corresponds to replacing the existing value with a zero value.

Memory has many uses. In this example, memory is used as a scratchpad area to store partial results. Memory is also useful to store values input to a system from the outside world. Memory enables short-lived external inputs to be available to system circuitry for a longer period of time.

Memory is also useful in enabling better resource utilization. Suppose we wish to add three numbers $A_0$ through $A_2$. The addition can be accomplished using two adder circuits as follows: The first and second numbers are fed to the first adder. The result of the first adder and the third value are fed to the second adder. The sum $S$ is obtained as the output of the second adder.

Alternatively, we can utilize memory to accomplish the addition of three numbers with a single adder as follows: Feed the first and second numbers to the adder. Store the partial result in memory. Then feed the partial result from memory and the third number to the same adder. The adder output is the desired result.

The same concept can be generalized to add a long sequence of numbers. At any given instant, the memory stores the partial result corresponding to all the numbers that occurred till that instant. For our addition example, notice that future results depend only on the value stored in memory and future inputs. Future results do not depend on the exactly how the memory was time sequenced, just its final state. This observation stems from the concept of a "state variable" that we saw earlier. The value stored in memory is simply a *digital state variable* in a manner analogous to an analog state variable value stored on a capacitor.

The next section discusses how capacitors can be used to build digital memory.

### 10.7.2 AN ABSTRACT DIGITAL MEMORY ELEMENT

Before we discuss how to implement memory, let us first define an abstract memory element and understand how it can be used in a small system. Figure 10.50 shows an abstract memory element that can store one bit of data. It has an input $d_{IN}$, an output $d_{OUT}$, and a control input called *store*. As suggested by the waveforms in Figure 10.50, the input $d_{IN}$ is copied into memory when the store signal is high. The value stored in the memory is available to

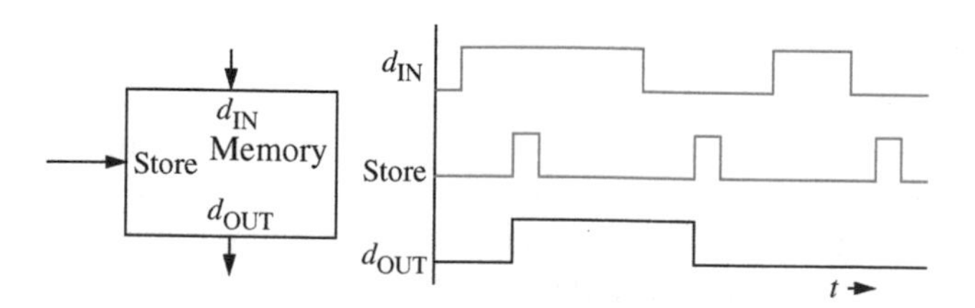


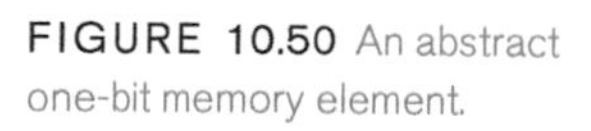
**FIGURE 10.50** An abstract one-bit memory element.

be read as the output $d_{OUT}$. If no new value is written into the memory, the last written value is stored indefinitely. If the memory is read while it is being written (that is, when the store signal is high), then the output simply reflects the value at the input.

---

EXAMPLE 10.4 MOTION DETECTOR CIRCUIT REVISITED

Let us use the memory element that we have just defined in a simple digital design. Recall the motion detector circuit from Chapter 5. The motion detector circuit was required to produce a signal $L$ to turn on a set of lights when the signal $M$ from a motion sensor was high, provided it was not daytime. We assumed that a light sensor produced a signal $D$ that was high when it was day. We had written the following logic expression for $L$:

$$L = M\overline{D}.$$

A problem with this design is that the lights that were turned on by the assertion of $M$ would go off the instant $M$ was de-asserted.[13] Let us consider a more useful design in which we require the lights to stay on even after the motion signal $M$ goes away. To make this happen, we need some form of memory to remember the occurrence of $M$, even after the $M$ signal goes away. The desired circuit uses a memory element and is shown in Figure 10.51. In this circuit, the signal $M$ is connected to the store input of the memory, and the signal $\overline{D}$ is connected to the $d_{IN}$ input of the memory. As depicted in the signal waveforms in Figure 10.51, the memory output remains high if motion is detected even when $D$ is false.[14]

---

### 10.7.3 DESIGN OF THE DIGITAL MEMORY ELEMENT

How do we implement a memory element? The memory element must be designed so that it stores indefinitely any value that has been written into it. Recall that a capacitor has the same property. Provided its discharge path has a high time constant, a capacitor can store a charge for a long period of time. Furthermore, we can use a switch to enable charging the capacitor from a given input.

---

13. To *'assert'* is to set the value to a logical 1, while to *'de-assert'* is to set the value to a logical 0.

14. Our circuit shown in Figure 10.51 has one other problem. How do the lights turn off? It should be apparent from the circuit that $L$ will go back to 0 when motion is detected during the daytime. However, relying on the appearance of signal $M$ during the daytime to turn off the lights is unappealing. One solution is to modify our memory abstraction to include a reset signal as follows: The value in memory is set to 0 when the reset input is high. (Our memory abstraction can include the additional property that the reset signal overrides the store signal if both are on at the same time.) Our motion detector circuit output $L$ can now be turned off by asserting the reset signal of the memory. The reset terminal of memory can be connected to the signal $D$ so that the memory contents and therefore the lights go off whenever it is day.

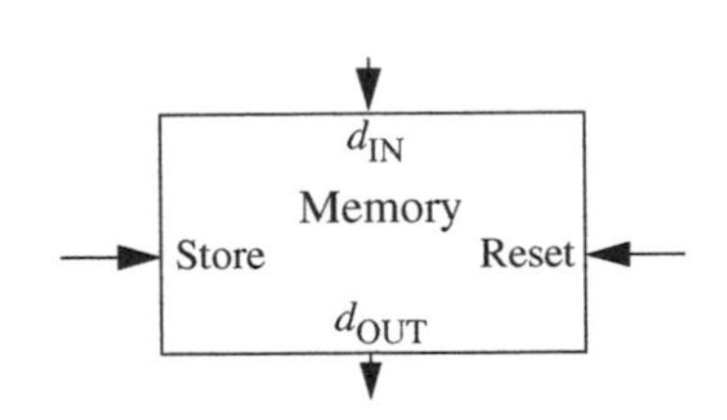

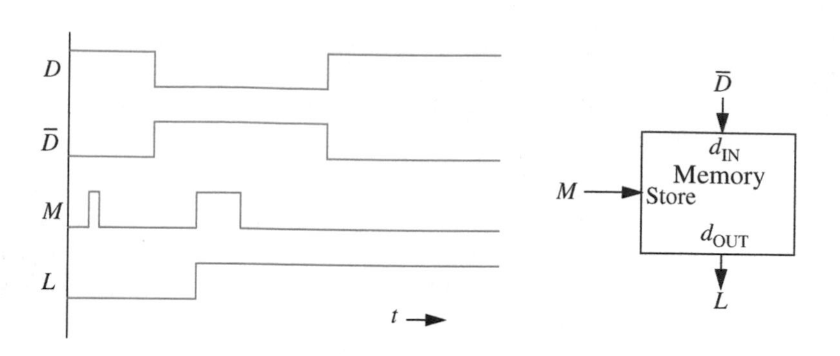


FIGURE 10.51 A motion detector circuit using memory.

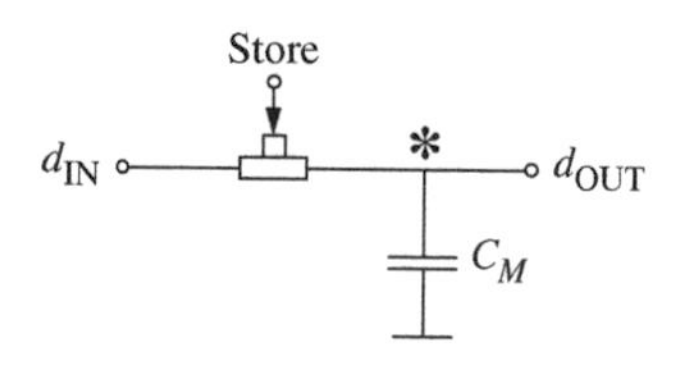


FIGURE 10.52 Circuit implementation of a memory element.

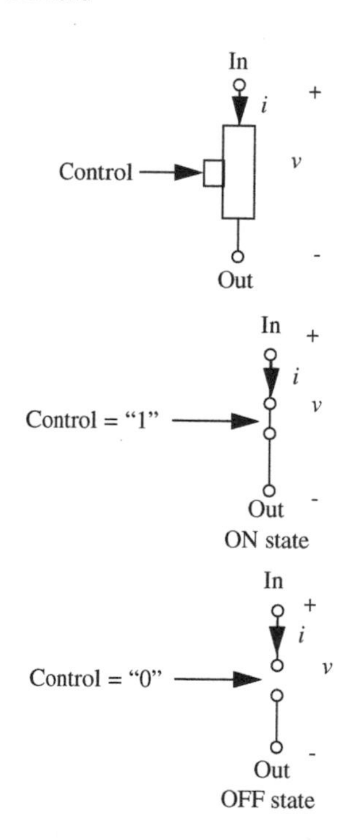


FIGURE 10.53 Three-terminal switch model.

Based on this intuition, consider the simple memory element circuit comprising a capacitor and an ideal switch shown in Figure 10.52. The switch is controlled by the store input and has the circuit model shown in Figure 10.53. When connected as shown in Figure 10.52, assume that a logical high value on the store input turns the switch into its ON state, while a logical low value on the store input turns the switch into its OFF state.

As discussed in Section 6.1, a circuit containing a switch can be analyzed by considering two linear subcircuits: one for the switch in its ON state (see Figure 10.54) and one for the switch in its OFF state (see Figure 10.56). A high on the store terminal turns the switch on, and results in the circuit illustrated in Figure 10.54. The capacitor then charges up (or discharges) to the value of the input voltage at the $d_{IN}$ terminal when the switch is turned on. Remember, the switch is symmetric about its input and output terminals. Thus, for example, if the $d_{IN}$ terminal had a high voltage corresponding to a logical 1 (produced, for example, by a voltage source, as illustrated in Figure 10.55), the capacitor will offer a high voltage at the node marked with an asterisk when the store signal is asserted. In this situation, the ideal external voltage source charges up the capacitor instantly through the ideal switch (assuming the capacitor had a low voltage initially). Alternatively, if the $d_{IN}$ terminal had a low voltage corresponding to a logical 0, the capacitor will offer a low voltage at the node marked with an asterisk when the store signal is asserted. In this latter situation, the capacitor discharges instantly through the ideal switch and the ideal external voltage source (assuming that the capacitor had a high voltage initially) and attains the same voltage as the voltage source.

Conversely, when the store signal goes low, the switch turns off (see Figure 10.56). Consequently, the $d_{OUT}$ terminal of the capacitor begins to float and the charge previously deposited on the capacitor is held in place. Thus, for example, if a high voltage had been previously stored on the capacitor, a high voltage will appear at the capacitor terminal $d_{OUT}$ even after the store signal goes low. In the ideal case, if the resistance between the $d_{OUT}$ terminal and ground is infinite, the capacitor will hold the charge forever.

The waveforms shown in Figure 10.50 will apply to the memory element circuit shown in Figure 10.52 under the following idealized assumptions: When the store signal is high, the RC time constant associated with the capacitor circuit is negligible, and when the store signal is low, the RC time constant associated with the capacitor circuit is infinite. The RC time constant of the circuit when the store signal is high is given by the product of $C_M$ and the sum of the on resistances of the switch and the driving element. Similarly, the RC time constant of the circuit when the store signal is low is given by the product of $C_M$ and the resistance seen by the capacitor.[15]

There is, however, one remaining problem with our memory element circuit. Recall that the static discipline required that our digital circuit elements such as gates be restoring. In other words, in order to obtain positive noise margins, the voltage threshold requirements on the outputs of gates was more stringent than those on the inputs. For example, the static discipline required that a $V_{IH}$ input to a gate be restored to $V_{OH}$ at the output, where $V_{OH} > V_{IH}$ for a positive noise margin. In order to inter-operate digital memory elements with our digital gates, we require that our digital memory elements satisfy the same set of voltage thresholds.

Unfortunately, our digital memory circuit as described in Figure 10.52 is non-restoring. In other words, if a voltage $V_{IH}$ corresponding to a valid 1 was applied to its input, the output of the memory element would not be restored to $V_{OH}$, rather it would be at $V_{IH}$ as well.

As suggested in Figure 10.57, a simple modification of our memory circuit can make it restoring. This design adds a pair of series connected inverters (or a buffer) to the output of our previous memory element circuit. The buffer will restore a $V_{IH}$ voltage on the capacitor terminal to a $V_{OH}$ voltage at the $d_{OUT}$ output. Interestingly, when a buffer is included in the memory element circuit, we do not need to implement a special capacitor to hold charge. Rather, the gate capacitance of the buffer $C_{GS}$ forms the memory capacitance $C_M$.

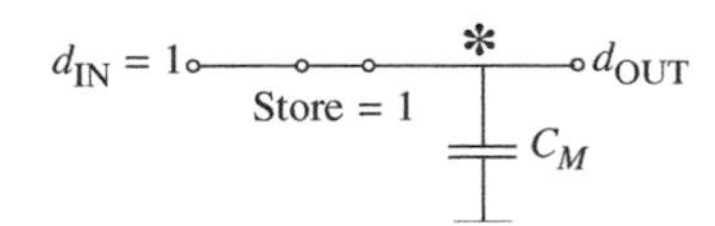


FIGURE 10.54 Charging up the memory capacitor, when the store signal is high.

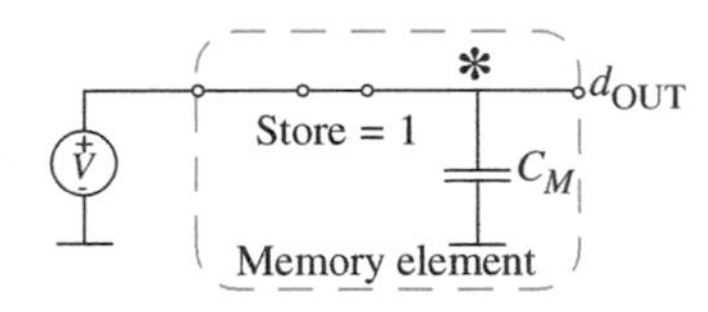


FIGURE 10.55 The memory element circuit model including the driving external source, when the store signal is high.

$d_{IN}$ Store = 0 * $d_{OUT}$ $C_M$

FIGURE 10.56 Charge storage in the memory capacitor, when the store signal is low.

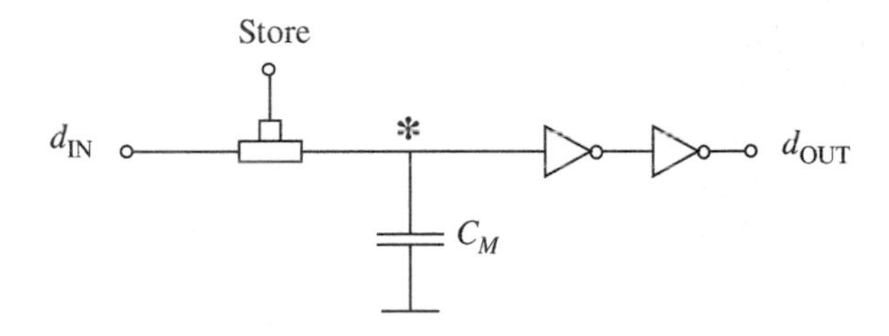


FIGURE 10.57 Circuit implementation of a signal restoring memory element.

15. We can also modify the memory element circuit to include a reset signal as follows: In this circuit, we use a second switch to discharge the capacitor to ground when the reset signal is high. Additionally, to make this circuit work, we must use non-ideal switches that are designed such that the ON resistance of the reset switch is much lower than that of the store switch. By doing so, we can ensure that reset will override the store when the store and reset are on at the same time.

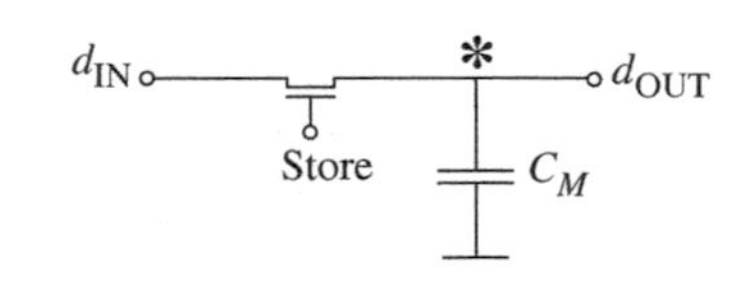

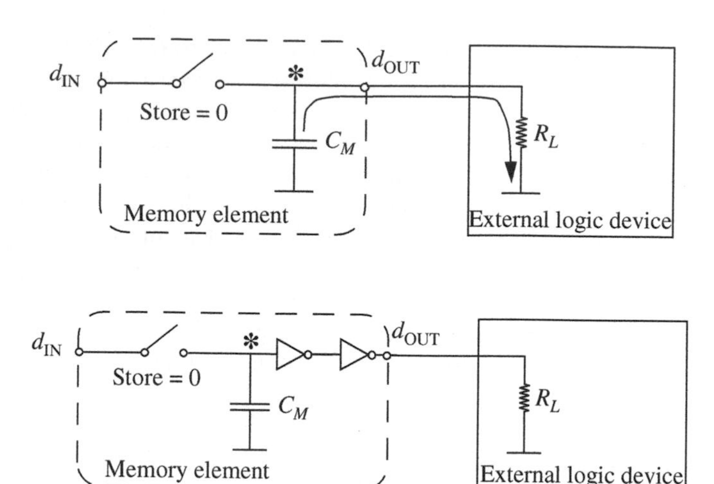


**FIGURE 10.58** Memory capacitor discharge due to load resistances for the unbuffered memory element.

**FIGURE 10.59** Memory capacitor's charge is protected in the buffered memory element.

By isolating the capacitor from the circuit that reads the stored value, the buffer offers added advantages. As shown in Figure 10.58, devices that read the value stored on the capacitor might have relatively low resistances associated with them, thereby discharging the capacitor in our original unbuffered memory circuit. In contrast, the buffered design of the memory element circuit shown in Figure 10.59 protects the capacitor's charge from the external circuit. By careful design of the memory element, the input resistance of the buffer can be made to be very large, thereby ensuring a large discharge time constant.

In practice, capacitors will leak their charge over time due to parasitic resistances. Let us suppose the capacitor gradually discharges through a parasitic resistance $R_P$ (see Figure 10.60). In this situation, for how long will the value stored in the capacitor remain valid after the store signal is de-asserted?

There are two cases to consider. First, if a 0 is stored on the capacitor, then the 0 value will be held indefinitely even with a low parasitic resistance. Notice that as the capacitor discharges to ground, the 0 stored on it will remain a 0.

The second case is more interesting. In this case, a 1 is written on the capacitor. Assume that the voltage corresponding to a 1 is $V_S$. The value stored on the capacitor will be read as a valid 1 by the buffer until it reaches the $V_{IH}$ voltage threshold. Thus the period over which the memory element will store a valid 1 is the interval over which the voltage drops from $V_S$ to $V_{IH}$. We can compute this duration from capacitor discharge dynamics (for example,

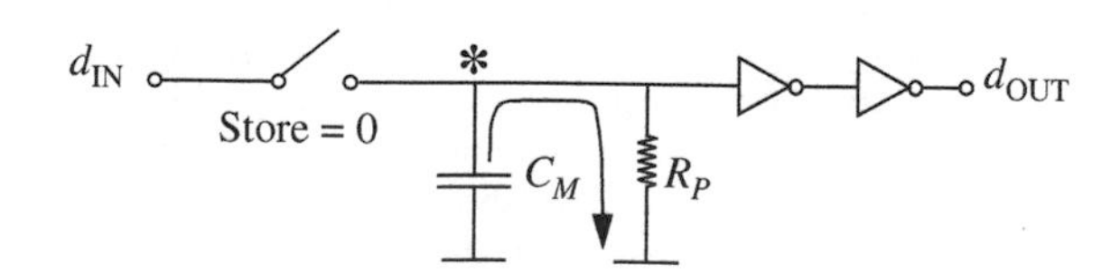


**FIGURE 10.60** Charge leakage from the memory capacitor for the buffered memory element.

see Equation 10.26). When a capacitor $C_M$ charged to an initial voltage $V_S$ discharges through a resistor $R_P$, its voltage $v_C$ as a function of time is given by the following equation:

$$v_C = V_S e^{-t/R_P C_M}.$$

The time taken for $v_C$ to drop from $V_S$ to $V_{IH}$ is given by

$$t_{V_S \rightarrow V_{IH}} = -R_P C_M \ln \frac{v_{IH}}{V_S}.$$

As an example, suppose that $C_M = 1$ pF, $R_P = 10^9$ Ω, $V_S = 5$ V, and $V_{IH} = 4$ V. Then $t_{V_S \rightarrow V_{IH}} = 0.22$ milliseconds.

### 10.7.4 A STATIC MEMORY ELEMENT

The one-bit memory element that we have discussed thus far is called a *dynamic one-bit memory element* or a *dynamic D-latch*. It is dynamic in the sense that it stores a value written into it only for a finite amount of time (due to nonzero parasitic resistances in practical implementations). The *static one-bit memory element* or a *static D-latch* is another type of memory element that has the same logic properties as the dynamic D-latch, but can store a value written into it indefinitely.

Figure 10.61 shows one possible circuit for a static memory element. In this circuit, a non-ideal switch with a very high ON resistance is connected between the power supply and the storage node of the memory element. When the output of the memory element is a logical 1, this switch is turned on and introduces a small stream of charge into the storage node to offset any leakage. Because it trickles charges into the node, this switch is called a *trickle switch*. The ON resistance of the trickle switch is made very large compared to the ON resistance of the store switch, so that the trickle input can be overridden easily by the input $d_{IN}$. A detailed circuit design of the static latch is beyond the scope of this book. The interested reader is referred to 'Principles of CMOS VLSI Design,' by Weste and Eshraghian.

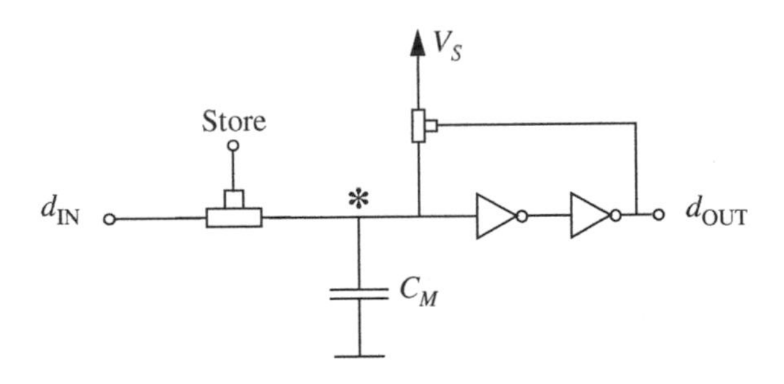


**FIGURE 10.61** Circuit implementation of a static memory element using a trickle switch.

## 10.8 SUMMARY

- The first-order differential equations that result from applying KVL and KCL to networks containing sources, resistors, and one energy-storage element can be derived using the node method or the other approaches described in Chapter 3. These differential equations can be solved by finding the homogeneous solution and the particular solution.
- The response of RC circuits resembles rising or decaying exponentials with the time constant $RC$. As an example, for a series RC circuit driven by a voltage step of amplitude $V_I$ at $t = 0$, the capacitor voltage for $t > 0$ is given by

$$v_C(t) = V_I + (V_O - V_I)e^{-t/RC}$$

  where $V_O$ is the initial voltage on the capacitor.
- In general, the response of a first order circuit (RC or RL) will be of the form

$$v_C = \text{ final value } + \text{ (initial value } - \text{ final value)}e^{-t/\text{time constant}}$$

  where the time constant is $RC$ for a resistor-capacitor circuit and $L/R$ for an resistor-inductor circuit. This form of the response in RC and RL circuits is shared by other branch variables such as the capacitor or inductor current and resistor voltage.
- Capacitors behave like open circuits when a circuit containing capacitors is driven by a DC voltage source. Conversely, a capacitor behaves like an instantaneous short circuit when inputs make an abrupt transition (for example, a step). (If the capacitor voltage were nonzero, then the capacitor would behave like a voltage source for abrupt transitions.)
- Inductors behave like short circuits when a circuit containing inductors is driven by a DC current source. Conversely, an inductor behaves like an instantaneous open circuit for inputs that make an abrupt transition (for example, a step). (If the inductor current were nonzero, then it would behave like a current source for abrupt transitions.)
- The zero-input response is the response of the system to the initial stored energy, assuming no drive.
- The zero-state response is the the response to an applied drive signal, for no initial stored energy.
- When the input signal is a short pulse (short compared to the time constant of the circuit), the response is proportional to the area of the applied pulse rather than to its height or shape.
- It is often convenient to break down a problem involving energy-storage elements into two parts. First, calculate the zero-input response, the response

of the system to the initial stored energy, assuming no drive. Then calculate the zero-state response, the response to the applied drive signal, for no initial stored energy.

- If we restrict integral operations to $t$ greater than zero, the zero-state response of the integral of some input signal is the integral of the zero-state response to that signal. The same is true for differentiation: The response to a signal derived by differentiating an input can be obtained by differentiating the output.
- The rise time for an output node is defined as the delay in rising from its lowest value to a valid high ($V_{OH}$) at that output.
- The fall time for an output node is defined as the delay in falling from its highest value to a valid low ($V_{OL}$) at the same output.
- The delay $t_{pd,1\to 0}$ for an input-output terminal pair of a gate is the time interval between a 1 to a 0 transition at the input to the moment that the output reaches a corresponding valid output voltage level ($V_{OH}$ or $V_{OL}$).
- The delay $t_{pd,0\to 1}$ for an input-output terminal pair of a gate is the time interval between a 0 to a 1 transition at the input to the moment that the output reaches a corresponding valid output voltage level ($V_{OL}$ or $V_{OH}$).

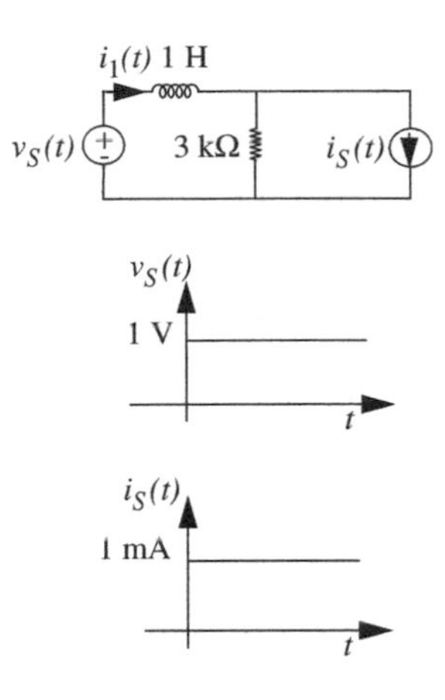

FIGURE 10.62

## EXERCISES

EXERCISE 10.1 Using superposition, determine the current $i_1(t)$ for the network shown in Figure 10.62. The network is at rest for $t < 0$.

EXERCISE 10.2 Find and sketch the zero state response for $t > 0$ in Figure 10.63. $i_S$ is a 10-mA step at $t = 0$.

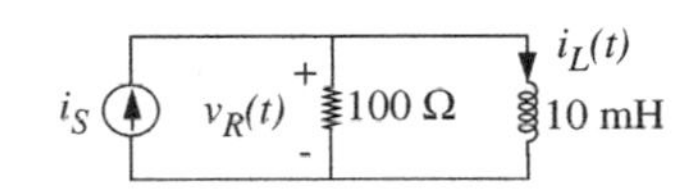

FIGURE 10.63

EXERCISE 10.3 In the circuit in Figure 10.64, $i(t) = 100\ \mu A, 0 < t < 1$ s, zero otherwise. At time $t = 2$, the voltage $v_C = 5$ V. What is $v_C$ at time $t = -1$ s?

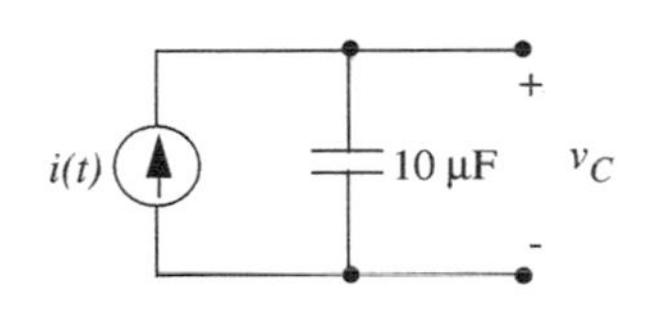

FIGURE 10.64

EXERCISE 10.4 In the circuit in Figure 10.65, the switch is closed at time $t = 0$ and opened at $t = 1$ second. Sketch $v_C(t)$ for all times.

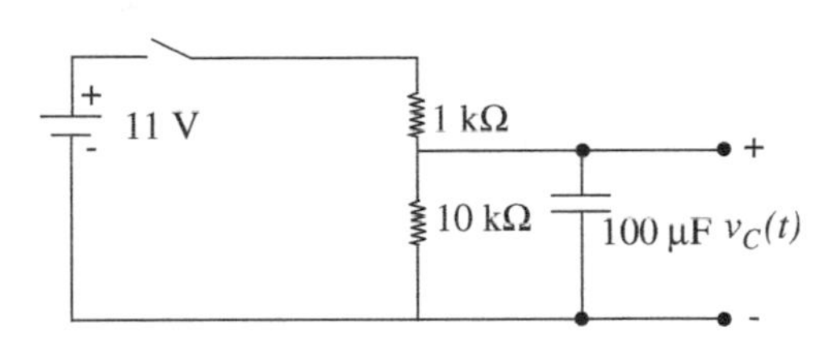

FIGURE 10.65

EXERCISE 10.5 Find and sketch the zero-input response for $t > 0$ in each network in Figure 10.66 for the given initial conditions.

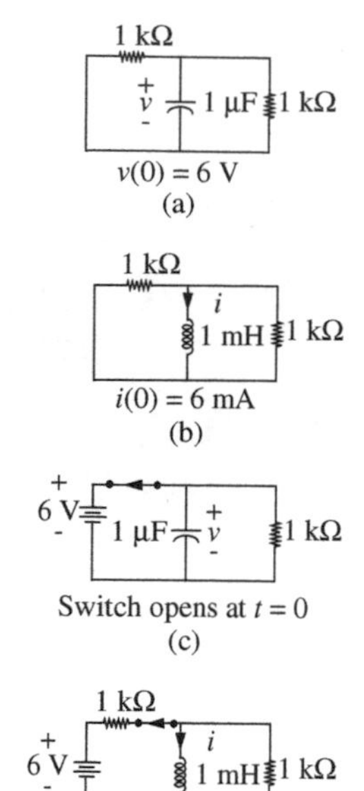


FIGURE 10.66

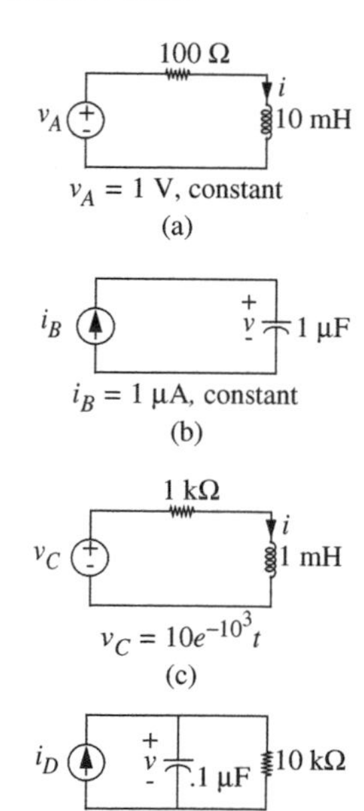


FIGURE 10.67

EXERCISE 10.6 Find and sketch the response for $t > 0$ in each network in Figure 10.67. Assume that the input is as shown for $t > 0$, and assume an initial zero state (in other words, show the zero state response).

EXERCISE 10.7 For the current source shown in Figure 10.68, assume $i_S$ consists of a single rectangular current pulse of amplitude $I_0$ amps and duration $t_0$ seconds.

a) Find the zero-state response to $i_S$.

b) Sketch the zero-state response for the cases:

i) $t_0 \gg RC$

ii) $t_0 = RC$

iii) $t_0 \ll RC$

c) Show that for $t_0 \ll RC$, (the case of a short pulse), the response for $t > t_0$ depends only on the area of the pulse ($I_0 t_0$), and not on $i_0$ or $t_0$ separately.

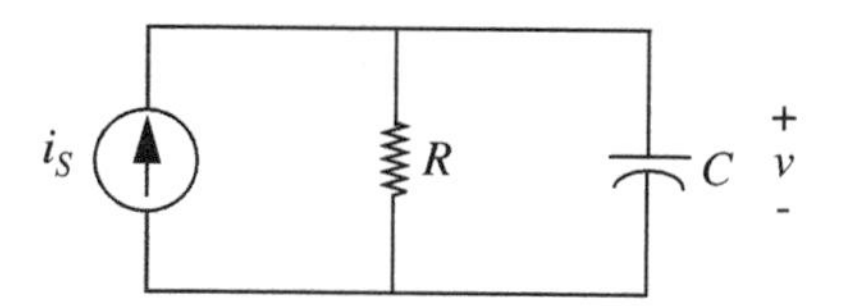


FIGURE 10.68

EXERCISE 10.8 Identify the state variable in each network in Figure 10.69. Write the corresponding state equation and find the time constants.

EXERCISE 10.9 In the circuit in Figure 10.70, $v(t) = 5$ mV for $0 < t < 1$ s, and zero otherwise. At time $t = 4$ s, $i(t) = 7$ A. What is $i(t)$ at time $t = -1$ s?

EXERCISE 10.10 Identify appropriate state variables for the network in Figure 10.71 and write the state equations.

EXERCISE 10.11 In Figure 10.72, $R_1 = 1$ kΩ, $R_2 = 2$ kΩ, and $C = 10\ \mu F$. The driving voltage $v_S = 0$ for $t < 0$. Assume $v_S$ is a 3-volt step at $t = 0$. Make a sketch of $v_C(t)$ for $t > 0$. Be sure to label the dimensions of the voltage and time axes and identify characteristic waveform shapes with suitable expressions.

EXERCISE 10.12 Identify state variables and write appropriate state equations for the circuit in Figure 10.73.

EXERCISE 10.13 Referring to Figure 10.74, before the switch is closed, the capacitor is charged to a voltage $v_S = 2$ V. The switch is closed at $t = 0$. Find an expression for $v_C(t)$ for $t > 0$. Sketch $v_C(t)$.

EXERCISE 10.14 Find the time constant of the circuit shown in Figure 10.75.

EXERCISE 10.15 A two-input RC circuit is shown in Figure 10.76. (Parts a, b, and c are independent questions.)

a) You should realize that the "bridge" of capacitors can be replaced by a single capacitor in this problem. What is the value of the single equivalent capacitor?

b) Consider operation with $i_I(t) = 0$ and $v_I(t) = 0$ for $t \geq 0$. The voltage $v_O(t)$ is known to be 1 volt at a time $t = 0$. Determine $v_O(t)$ for all $t > 0$.

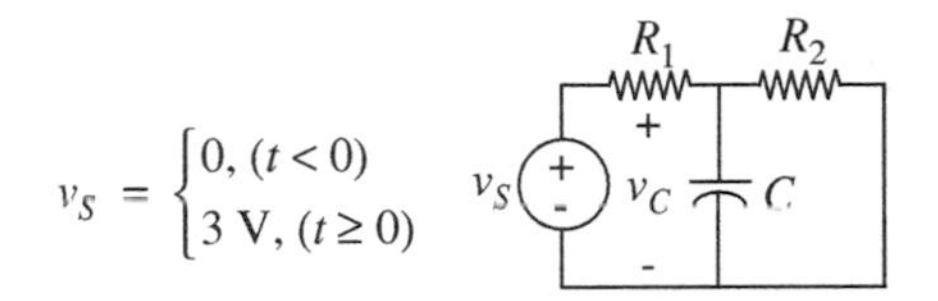


FIGURE 10.72

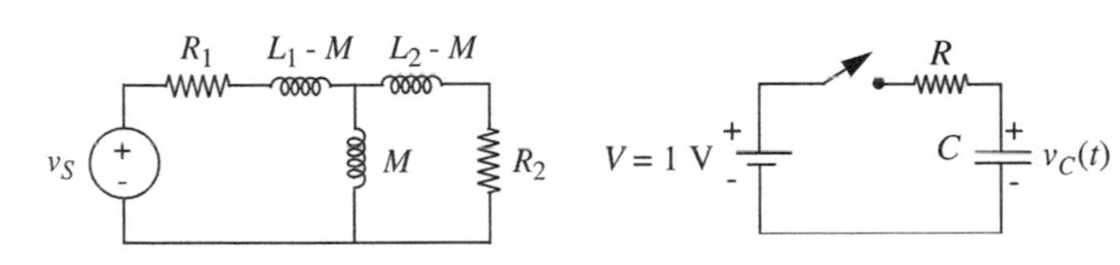


FIGURE 10.73

FIGURE 10.74

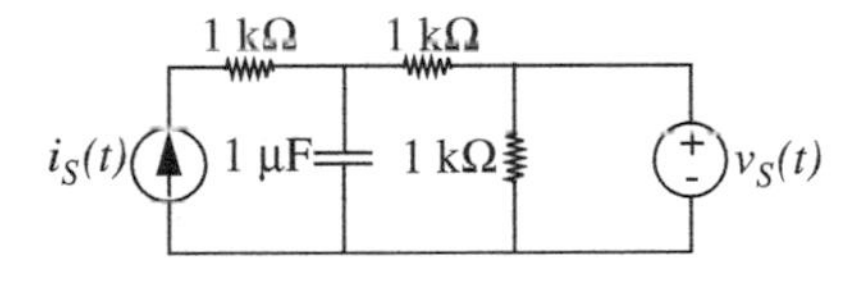


FIGURE 10.75

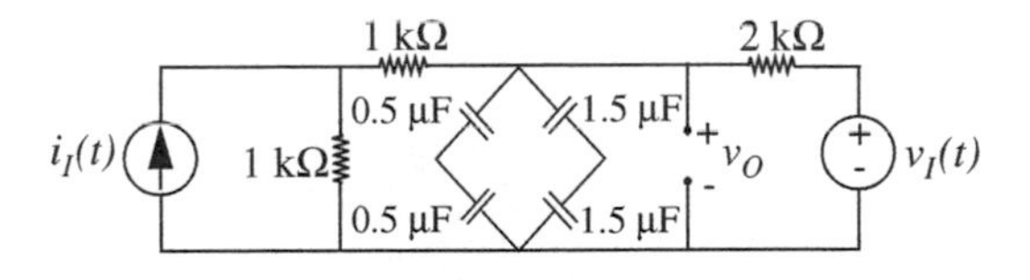


FIGURE 10.76

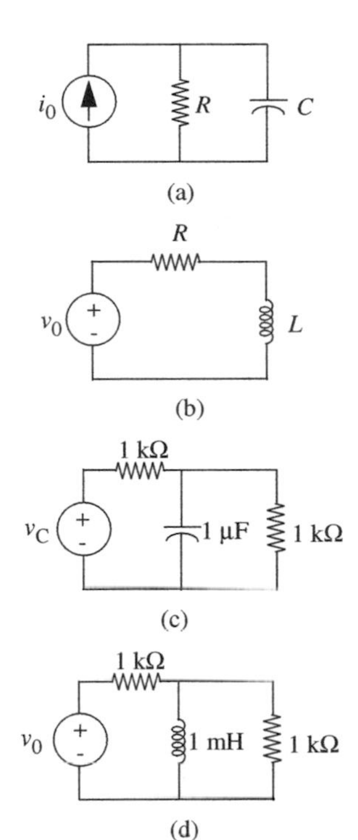


FIGURE 10.69

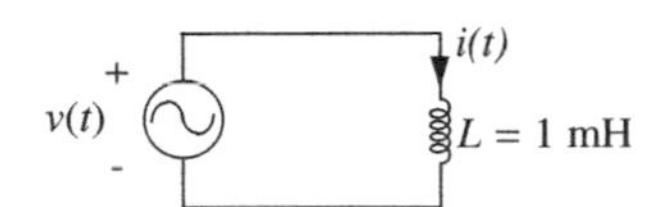


FIGURE 10.70

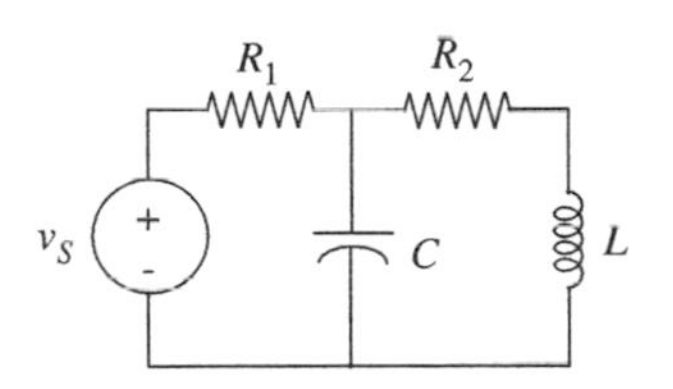


FIGURE 10.71

c) A different constraint is that sources $i_I(t)$ and $v_I(t)$ are zero for $t < 0$ and that $v_O(0) = 0$. Sources $i_I(t)$ and $v_I(t)$ undergo step transitions of +1 mA and +1 volt, respectively, at time $t = 0$. Determine $v_O(t)$ for all time.

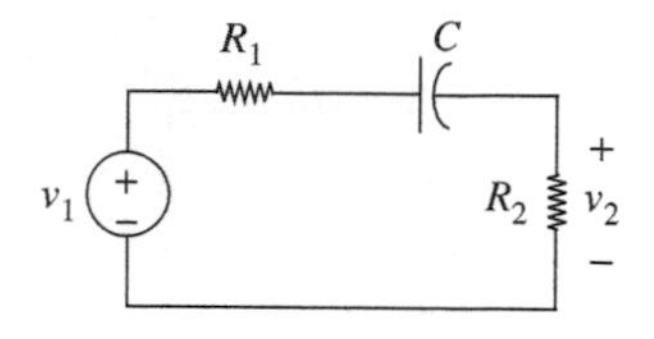


FIGURE 10.77

EXERCISE 10.16 In the circuit in Figure 10.77, $R_1 = 1\ \text{k}\Omega$, $R_2 = 2\ \text{k}\Omega$, and $C = 3\ \mu\text{F}$. Assume initial rest conditions (zero initial state), and assume that $v_1$ has a 6-volt step at $t = 0$. Find $v_2(t)$ for $t > 0$. Sketch and label.

EXERCISE 10.17 Consider the circuit shown in Figure 10.78. Sketch and label $v_O(t)$ for $i_1(t)$ a step as shown in Figure 10.79. Assume $v_O = 0$ for $t < 0$.

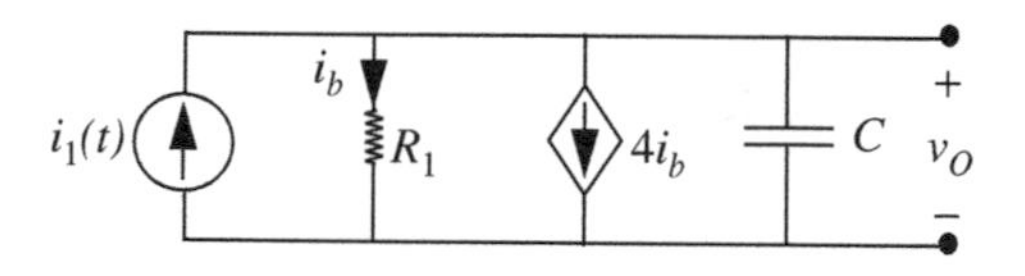


FIGURE 10.78

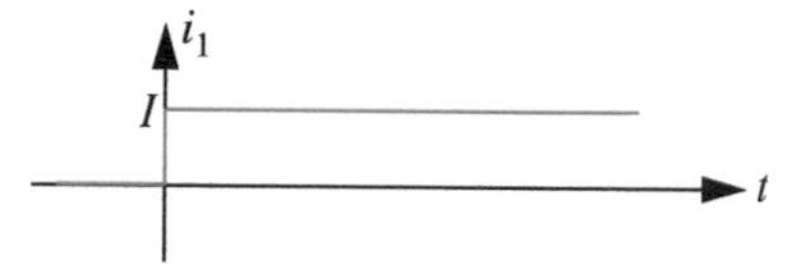


FIGURE 10.79

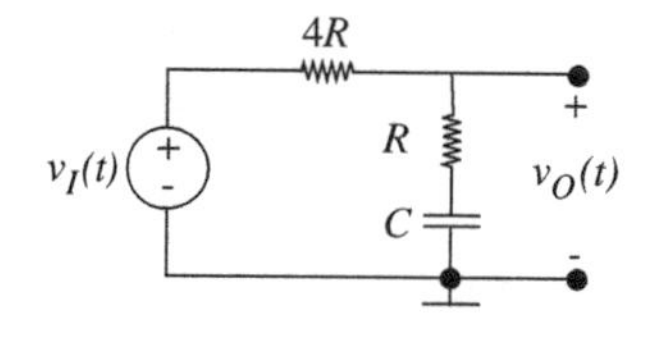


FIGURE 10.80

EXERCISE 10.18 For the circuit shown in Figure 10.80, find the characteristic equation and the zeroinput response assuming that the capacitor was initially charged to 1 volt. Label your graph.

EXERCISE 10.19 The excitation function for all four of the circuits shown in Figure 10.81 is

$$v_S(t) = 0, \qquad t < 0$$

$$v_S(t) = 10\ \text{V}, \quad t \geq 0.$$

For each of the circuits, select the time function on the right that corresponds in magnitude and shape to the output, $v_O(t)$. Assume that all capacitors and inductors have zero initial states, (the appropriate state variable is zero for $t$ less than zero). If no matching response exists, say so and explain briefly. All responses are made up of "straight lines" and "exponentials." You may choose a time function more than once. (Note that part (d) shows an op-amp circuit. Op-amps will be covered in later chapters.)

EXERCISE 10.20 An RC network is shown in Figure 10.82. The voltage $v$ and the current $i$ are constant for all time. Prior to $t = 0$, the circuit is in equilibrium with the switch closed. At time $t = 0$, the switch is opened, and it is then closed some time later. The waveform in Figure 10.83 is observed for $v_C(t)$.

What are the value of $\tau_1$, $\tau_2$, and the final value $V_1$? (Note: The figure may not be to scale.)

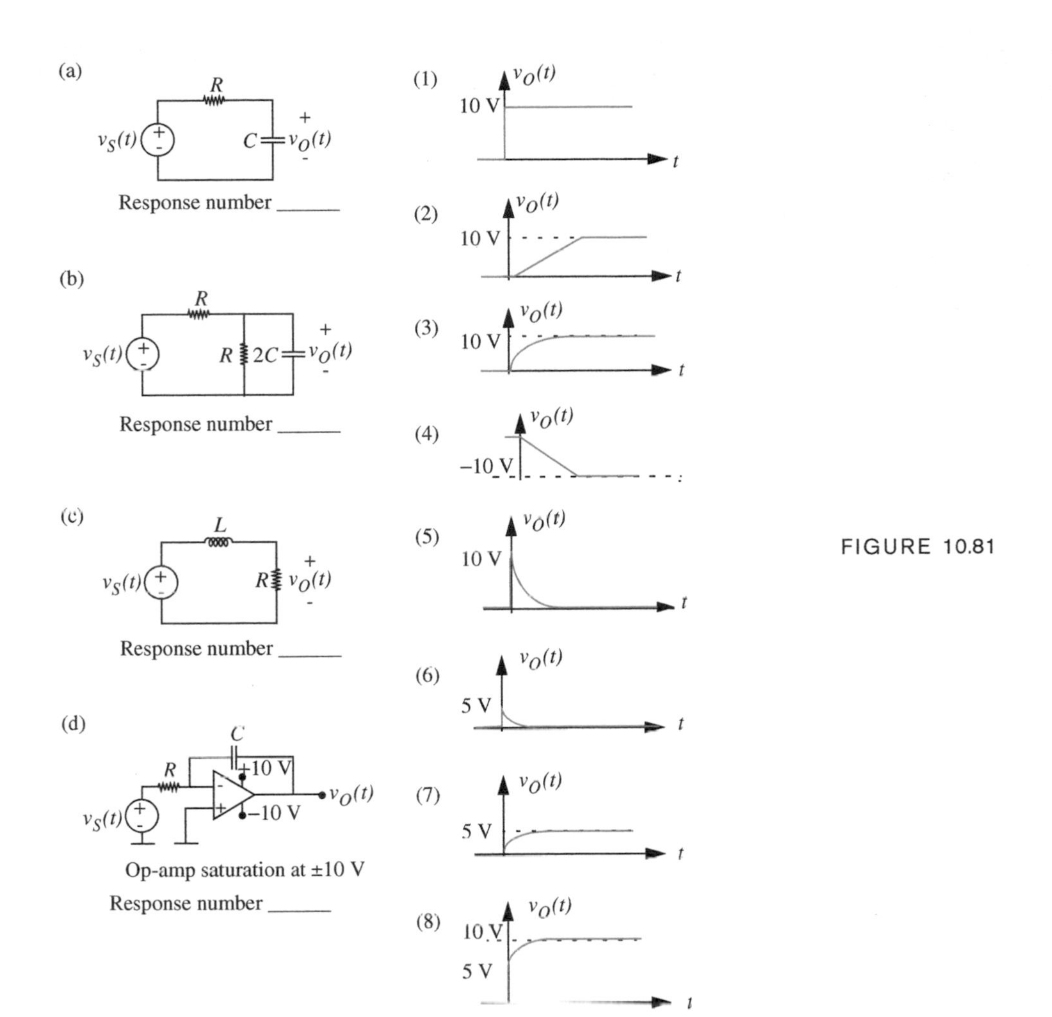


FIGURE 10.81

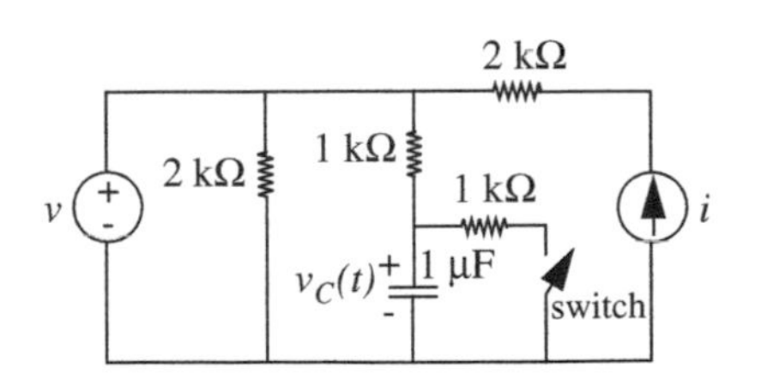


FIGURE 10.82

$v_C(t)$
2 V
Time constant $\tau_1$
Time constant $\tau_2$
$V_1$ (Final value)
t
Switch open
Switch closed

FIGURE 10.83

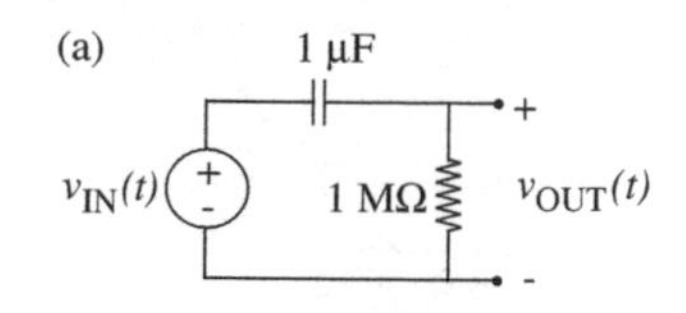


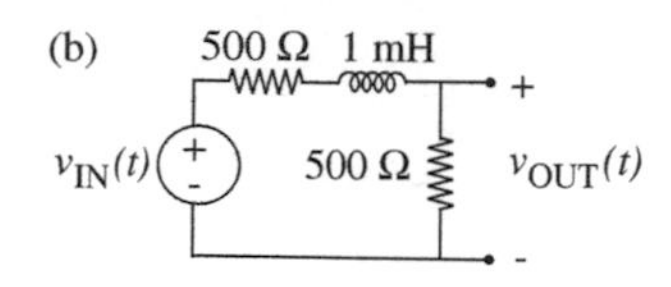


FIGURE 10.84

EXERCISE 10.21 In the two following cases in Figure 10.84 the input $v_{IN}(t) = 10u_{-1}(t)$, a 10-V step[16] starting at time $t = 0$. Give for each case:

a) The time constant of the circuit.

b) An analytic expression for the signal $v_{OUT}(t)$ as a function of time.

c) A labeled sketch of the output signal $v_{OUT}(t)$ as a function of time. Be sure to label the time and voltage scales.

EXERCISE 10.22 In each of the following cases, find by inspection and give

i) an expression for the time constant $\tau$,

ii) a sketch of the signal versus time,

iii) an analytic expression for the signal in terms of $\tau$ and any other necessary parameters.

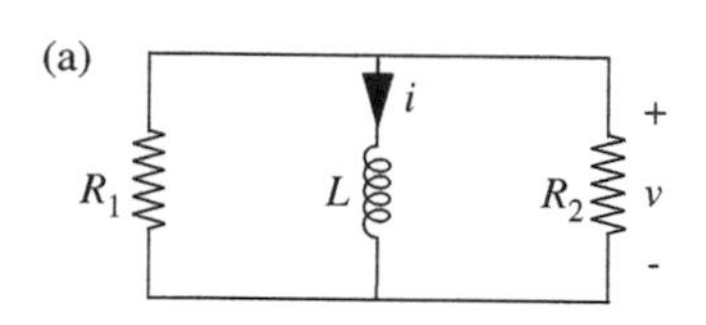


FIGURE 10.85

a) Referring to Figure 10.85, find $v(t)$ for $t > 0$ given $i(t = 0) = I_0$.

b) Referring to Figure 10.86, find $i_2(t)$ given $i_1(t = 0) = I_0/2$.

c) Referring to Figure 10.87, find $v(t)$ for $t > 0$ given that the switch is moved from 1 to 2 at $t = 0$.

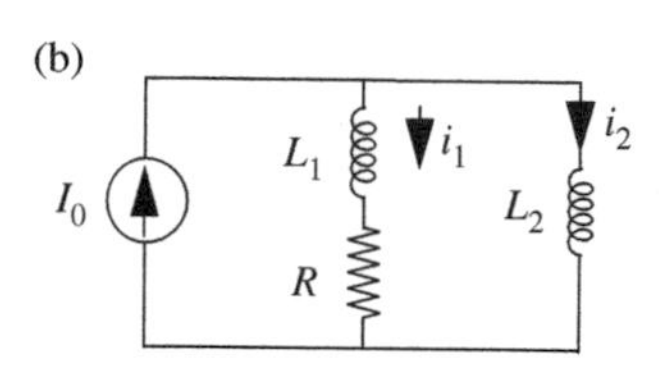


FIGURE 10.86

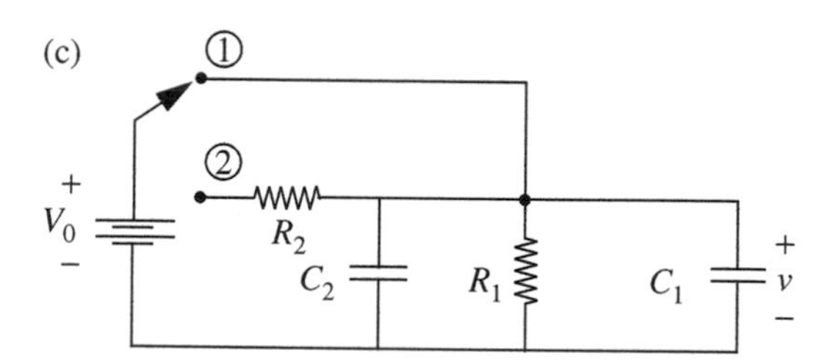


FIGURE 10.87

EXERCISE 10.23 For the circuit in Figure 10.88, with no charge on the capacitor at $t = 0$, given that if $v_I = Atu_{-1}(t)$ then $v_C = [A(t - \tau) + A\tau e^{-t/\tau}]u_{-1}(t)$. Note that $u_{-1}(t)$ represents a unit step at $t = 0$.

16. Recall that the notation $u_0(t)$ represents an impulse at time $t$ . The notation $u_n(t)$ represents the function that results from differentiating the impulse times, and the notation $u_{-n}(t)$ represents the function that results from integrating the impulse times. Thus $u_{-1}(t)$ represents the unit step at time $t$, $u_{-2}(t)$ the ramp, and $u_1(t)$ the doublet at time $t$. The unit step $u_{-1}(t)$ is also commonly represented as $u(t)$, and the unit impulse $u_0(t)$ as $\delta(t)$.

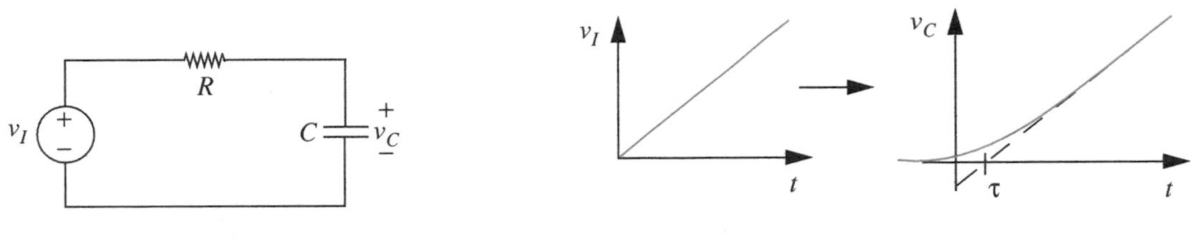


FIGURE 10.88

FIGURE 10.89

Find the following:

a) $v_C(t)$ when the input is the same as previously given but $v_C(t = 0) = V_0$.

b) $v_C(t)$ when $v_C(0) = 0$ and $v_I(t) = Bu_{-1}(t)$. Note that $u_{-1}(t)$ represents a unit step at $t = 0$.

c) $v_C(t)$ for $t \geq T$ when $v_C(0) = 0$ and

$$v_I(t) = \begin{cases} 0 & t \leq 0 \\ At & 0 \leq t \leq T \\ AT & T \leq t. \end{cases}$$

EXERCISE 10.24 A digital memory element is implemented as illustrated in Figure 10.90. Sketch the waveform at the output of the memory element for the input signals shown in Figure 10.91. Assume that the switch is ideal and that the memory element has a 0 stored in it initially.

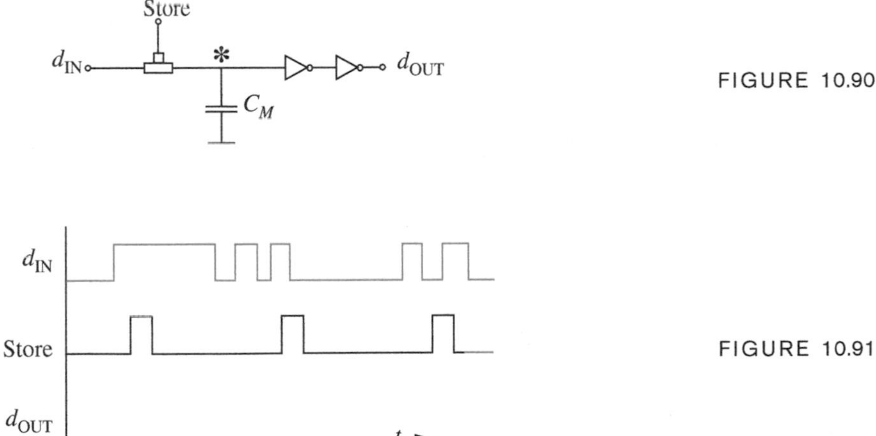


FIGURE 10.90

FIGURE 10.91

## PROBLEMS

PROBLEM 10.1 Figure 10.92a illustrates an inverter *INV1* driving another inverter *INV2*. The corresponding equivalent circuit for the inverter pair is illustrated in Figure 10.92b. *A*, *B*, and *C* represent logical values, and $v_A$, $v_B$, and $v_C$ represent voltage levels. The equivalent circuit model for an inverter based on the SRC model of the MOSFET is depicted in Figure 10.93.

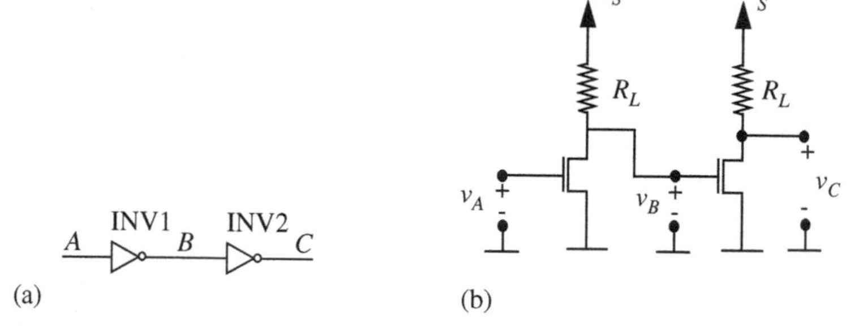


FIGURE 10.92

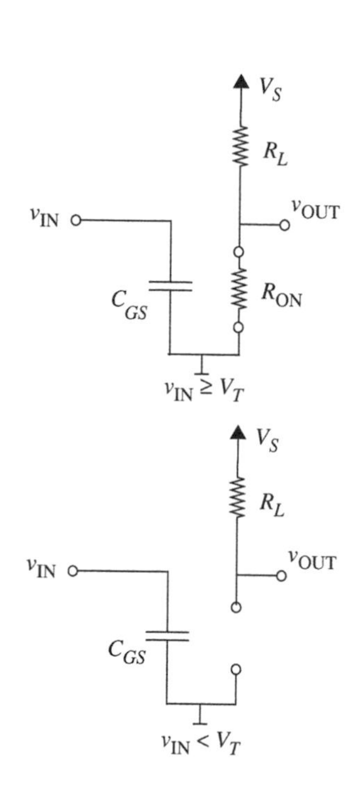


FIGURE 10.93

a) Write expressions for the rise and fall times of *INV1* for the circuit configuration shown in Figure 10.92. Assume that the inverters satisfy the static discipline with voltage thresholds $V_{IL} = V_{OL} = V_L$ and $V_{IH} = V_{OH} = V_H$.
(Hint: The rise time of *INV1* is the time $v_B$ requires to transition from the lowest voltage reached by $v_B$ (given by the voltage divider action of $R_L$ and $R_{ON}$) to $V_H$ for a $V_S$ to 0-V step transition at the input $v_A$. Similarly, the fall time of *INV1* is the time $v_B$ requires to transition from the highest voltage reached by $v_B$ (that is, $V_S$) to $V_L$ for a 0-V to $V_S$ step transition at the input $v_A$.)

b) What is the propagation delay $t_{pd}$ of *INV1* in the circuit configuration shown in Figure 10.92, for $R_{ON} = 1\ \text{k}\Omega$, $R_L = 10R_{ON}$, $C_{GS} = 1$ nF, $V_S = 5$ V, $V_L = 1$ V, and $V_H = 3$ V?

PROBLEM 10.2 The inverter-pair comprising *INV1* and *INV2* studied in Problem 10.1 (see Figure 10.92) drives another inverter *INV3* as illustrated in Figure 10.94a. Logically, the series connected pair of inverters *INV1* and *INV2* function as a buffer, as depicted in Figure 10.94b. The equivalent circuit of the buffer circuit driving *INV3* is illustrated in Figure 10.94c. For this problem, use the equivalent circuit model for an inverter based on the SRC model of the MOSFET as depicted in Figure 10.93. Assume further that each of the inverters satisfies the static discipline with voltage thresholds $V_{IL} = V_{OL} = V_L$ and $V_{IH} = V_{OH} = V_H$. Assume further that the MOSFET threshold voltage is $V_T$. (Note that to satisfy the static discipline, the following is true: $V_L < V_T < V_H$.)

a) Referring to Figure 10.94c, assume that the input to the buffer $v_A$ undergoes a step transition from 0 V to $V_S$ at time $t = 0$. Write an expression for $v_B(t)$ for $t \geq 0$

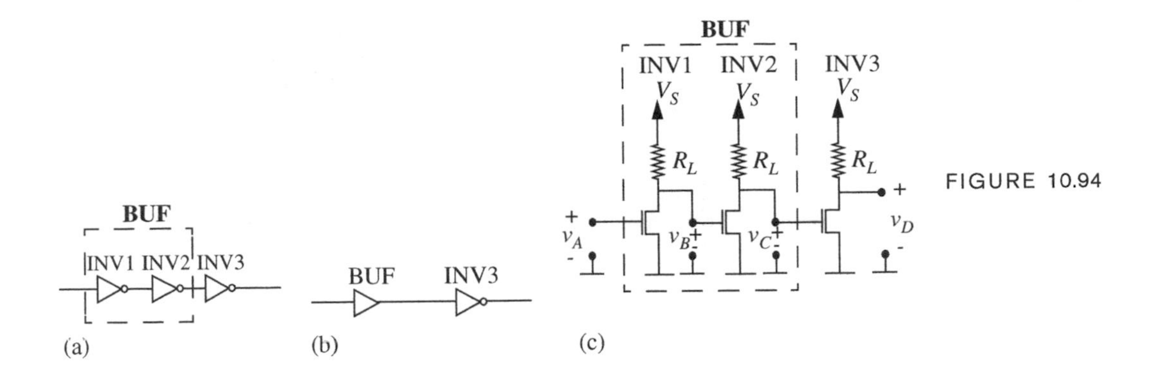


FIGURE 10.94

for the step transition in $v_A$. (Hint: See the fall time calculation in Problem 10.1a.) Sketch the form of $v_B$ for $t \geq 0$.

b) Referring to Figure 10.94c, assume that the input to the buffer $v_A$ undergoes a step transition from 0 V to $V_S$ at time $t = 0$. Write an expression for $v_C(t)$ for $t \geq 0$ for the step transition in $v_A$. (Hint: Refer to the sketch of $v_B$ drawn in part (a.) The MOSFET in *INV2* stays on for $v_B \geq V_T$, and turns off when $v_B < V_T$.) Sketch the form of $v_C(t)$ for $t \geq 0$.

c) Write an expression for the rise time of the buffer for the circuit configuration shown in Figure 10.94c. (Hint: Refer to the sketch of $v_C$ from part (b.) The rise time of the buffer is the time $v_C$ requires to transition from the lowest voltage reached by $v_C$ to $V_H$ from the time the input $v_A$ makes a step transition from 0 V to $V_S$. Note that the rise time of the buffer includes the internal buffer fall delay, which is the time $v_B$ takes to transition from $V_S$ to $V_T$, and the additional time $v_C$ takes to transition from its lowest voltage to $V_H$.)

d) Referring to Figure 10.94c, assume that the input to the buffer $v_A$ undergoes a step transition from $V_S$ to 0 V at time $t = 0$. Write an expression for $v_B(t)$ for $t \geq 0$ for the step transition in $v_A$. Sketch the form of $v_B$ for $t \geq 0$.

e) Referring to Figure 10.94c, assume that the input to the buffer $v_A$ undergoes a step transition from $V_S$ to 0 V at time $t = 0$. Write an expression for $v_C(t)$ for $t \geq 0$ for the step transition in $v_A$. (Hint: Refer to the sketch of $v_B$ drawn in part (d.) The MOSFET in *INV2* stays off for $v_B < V_T$, and turns on when $v_B \geq V_T$.) Sketch the form of $v_C(t)$ for $t \geq 0$.

f) Write an expression for the fall time of the buffer for the circuit configuration shown in Figure 10.94c. (Hint: Refer to the sketch of $v_C$ from part (e.) The fall time of the buffer is the time $v_C$ requires to transition from $V_S$ to $V_L$ from the time the input $v_A$ makes a step transition from $V_S$ to 0 V. Note that the fall time of the buffer is the sum of two components: (1) the internal buffer rise delay, or the time $v_B$ takes to transition from its lowest voltage to $V_T$ and (2) the additional time $v_C$ takes to transition from $V_S$ to $V_L$.)

g) Compute the rise time and the fall time for the buffer assuming that $R_{ON} = 1$ k, $R_L = 10R_{ON}$, $C_{GS} = 1$ nF, $V_S = 5$ V, $V_L = 1$ V, $V_T = 2$ V, and $V_H = 3$ V.

h) What is the propagation delay $t_{pd}$ of the buffer when the buffer output is connected to a single inverter using an ideal wire as shown in Figure 10.94c?

i) Notice that unlike the delay calculation in Problem 10.1, we needed the value of $V_T$ to obtain the buffer delay. Why was it necessary in the case of the buffer?

j) An approximate value for the buffer delay can be obtained by doubling the individual inverter delay. Estimate the buffer delay by using the inverter delay computed in Problem 10.1b. What is the percentage error in the value of this estimated delay as compared to the accurate buffer delay computed in part (i) of this problem?

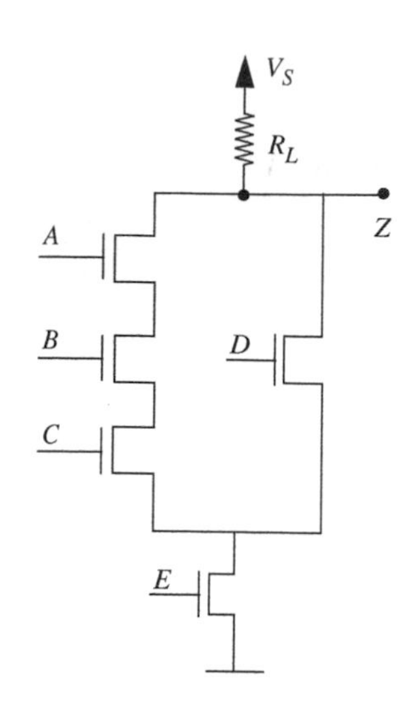

FIGURE 10.95

PROBLEM 10.3 The circuit depicted in Figure 10.95 implements the logic function $Z = \overline{(ABC + D)E}$. Suppose the output of this circuit drives an inverter with a gate capacitance of $C_{GS}$. Assume that the MOSFETs in the circuit have on resistance $R_{ON}$, and that the high and low voltage thresholds are $V_{IH} = V_{OH} = V_H$ and $V_{IL} = V_{OL} = V_L$, respectively.

a) What combination of logical inputs will result in the worst-case fall time for the circuit?

b) Derive an expression for the worst-case fall time in terms of $V_S$, $R_L$, $R_{ON}$, $V_L$, and $V_H$. Not all variables need appear in your answer.

c) Derive an expression for the worst-case rise time.

PROBLEM 10.4 Figure 10.96 illustrates an inverter *INV A* connected to another inverter *INV B* by a wire of length $l$ on a VLSI chip.

FIGURE 10.96

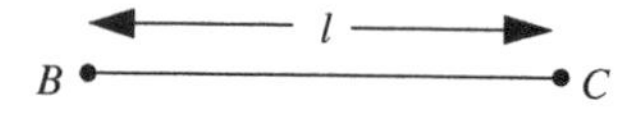


$lR_0$

$lC_0$

FIGURE 10.97

Figure 10.97 shows a lumped circuit model for the (nonideal) wire of length $l$ in a VLSI chip, and Figure 10.98 shows the equivalent circuit model for the inverter pair connected by the nonideal wire based on the SRC model for the MOSFET. Assume that the logic devices satisfy a static discipline with voltage thresholds given by $V_{IL} = V_{OL} = V_L$ and $V_{IH} = V_{OH} = V_H$, and that the supply voltage is $V_S$.

Suppose *INV A* is driven by a 0 to 1 transition at its input (denoted $v_{INA}$) at time $t = 0$. Determine $t_{pd,0\to1}$, the propagation delay through *INV A* for a 0 to 1 transition at its input. Recall that by our definition $t_{pd,0\to1}$ is the time taken by the input to *INV B*,

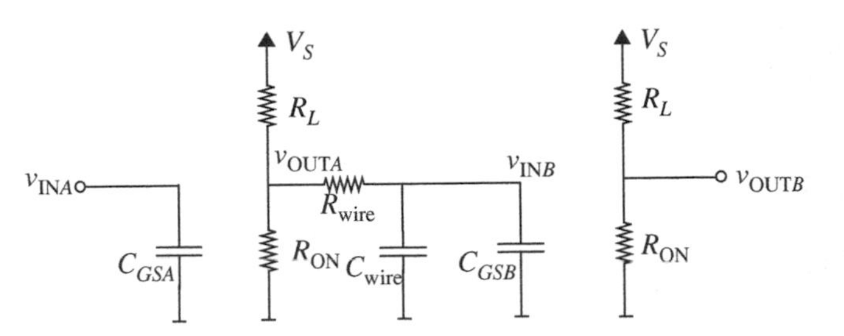


FIGURE 10.98

namely $v_{INB}$, to fall from $V_S$ to $V_L$ following the 0 to 1 transition at the input of *INV A*. Express your answer in terms of $V_S$, $V_L$, $R_{ON}$, $C_{GS}$, the wire length $l$, and the wire model parameters. By what factor does the delay increase for a $2\times$ increase in the wire length $l$?

PROBLEM 10.5 Figure 10.99 illustrates an inverter *INV A* driving $n$ other inverters *INV*1 through *INVn*. As in Problem 10.1, each of the inverters is constructed using a MOSFET and a resistor $R_L$, and the inverters satisfy the static discipline with voltage thresholds $V_{IL} = V_{OL} = V_L$ and $V_{IH} = V_{OH} = V_H$. Model the MOSFETs using the SRC model with MOSFET on resistance $R_{ON}$ and gate capacitance $C_{GS}$ as in Problem 10.1 (see Figure 10.93.)

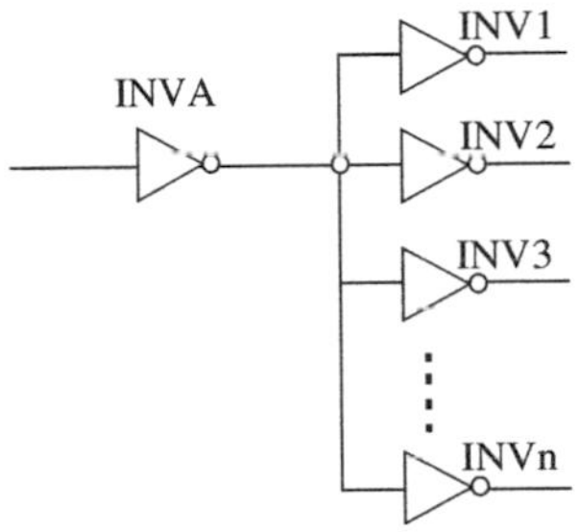

FIGURE 10.99

a) What are the rise and fall times for *INV A*? (Hint: Sum the input capacitances of each of the inverters into a single lumped value, and use your answer from Problem 10.1 to solve this part.) How does the rise time increase as the number of driven inverters $n$ increases?

b) What is the propagation delay $t_{pd}$ of *INV A* in the circuit configuration shown in Figure 10.99, for $R_{ON} = 1$ k$\Omega$, $R_L = 10R_{ON}$, $C_{GS} = 1$ nF, $V_S = 5$ V, $V_L = 1$ V, and $V_H = 3$ V.

c) Now, assume that each of the wires connecting the output of *INV A* to each of the inverters *INV*1 through *INVn* is nonideal as depicted in Figure 10.100. Model each of the wires using the model shown in Figure 10.101. Assuming that the input of *INV A* makes a step transition from 1 to 0, find the rise time at the input of any one of the inverters *INVi* driven by *INV A*.

FIGURE 10.100

d) Compute the value of the rise time determined in part (c) for the following parameters: $R_{ON} = 1$ k$\Omega$, $R_L = 10R_{ON}$, $C_{GS} = 1$ nF, $R_W = 100$ $\Omega$, $C_W = 10$ nF, $V_S = 5$ V, $V_L = 1$ V, and $V_H = 3$ V.

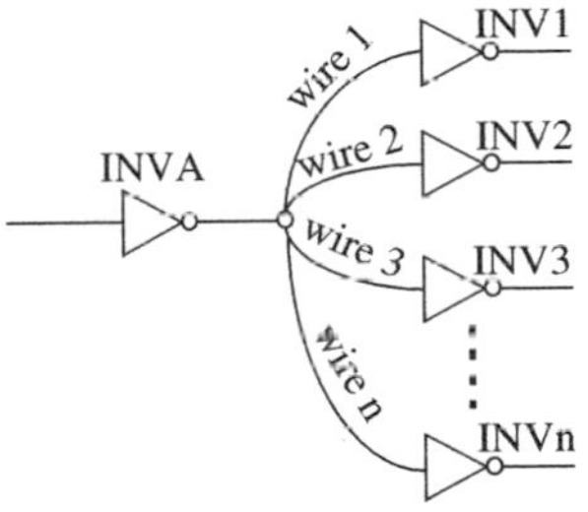

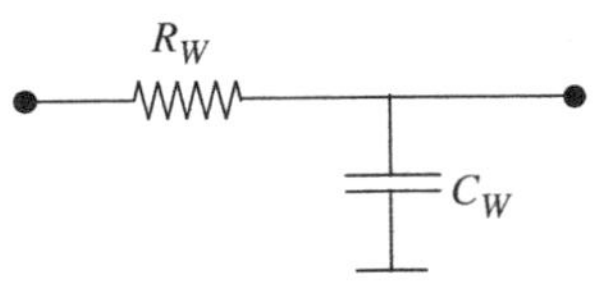

FIGURE 10.101

PROBLEM 10.6 As can be seen from the answer to Problem 10.4, long wires have a serious negative impact on the delay. One way to alleviate the wire delay problem is to introduce buffers when driving long wires, as illustrated in Figure 10.102. Assume that the buffer is constructed as depicted in Figure 10.94c using a pair of inverters identical to the inverters in this problem. In other words, the input of a buffer has a capacitance $C_{GS}$ to ground, and the output of a buffer have the same drive characteristics as an inverter output. For this problem, you will ignore the internal delay of the buffer. (See Problem 10.2c and f for a definition of the internal buffer delay.) In other words, assume that a buffer driving zero output capacitance has zero delay.

By introducing a buffer, the effective length of wire driven by either the inverter *INV A* or the buffer is $l/2$. For large $l$, given the nonlinear relationship between wire length and delay, the sum of the delays in driving the two $l/2$ wire segments is smaller than driving a single wire segment of length $l$.

INVA INVB

A B $\frac{l}{2}$ $\frac{l}{2}$ C D

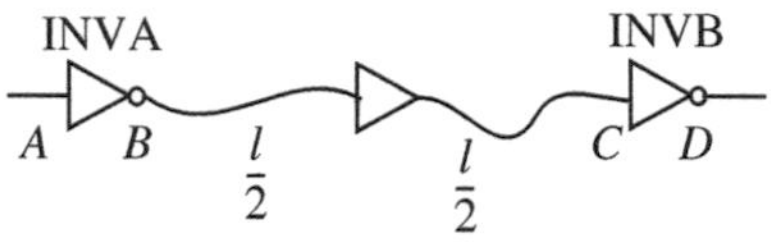

FIGURE 10.102

a) Compute the propagation delay between the input of *INV A* and the input of *INV B* for the circuit in Figure 10.102. Assume that rising transitions are longer than falling transitions at the output of either the inverters or the buffers.

   (Hint: The total delay from the input of *INV A* to the output of *INV B* is the sum of the following two quantities: (1) the propagation delay of *INV A* driving the wire segment of length $l/2$ and a capacitance $C_{GS}$ corresponding to the gate capacitance of the buffer, and (2) the propagation delay of the buffer driving the second wire segment of length $l/2$ and a capacitance $C_{GS}$ corresponding to the gate capacitance of *INV B*. Remember, the buffer has zero delay when it is driving zero output capacitance.)

b) Figure 10.103 shows a circuit in which $n-1$ buffers are introduced between *INV A* and *INV B*. *INV A* and each of the buffers drives a segment of wire of length $l/n$. Compute the propagation delay between the input of *INV A* and the input of *INV B* for this case.

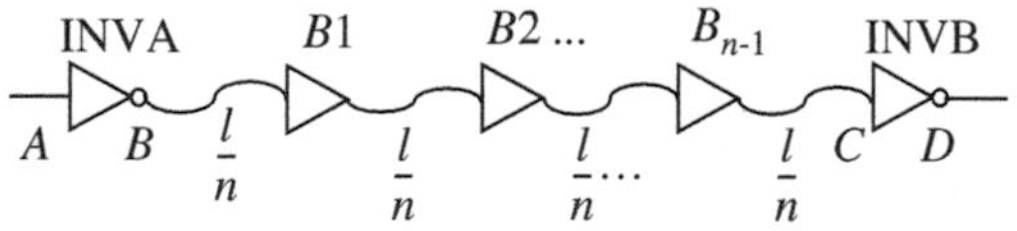

INVA $B1$ $B2$ ... $B_{n\text{-}1}$ INVB

A B $\frac{l}{n}$ $\frac{l}{n}$ $\frac{l}{n}$ ... $\frac{l}{n}$ C D

FIGURE 10.103

c) Determine the number of buffers for which the propagation delay for the circuit in Figure 10.103 is minimized.

PROBLEM 10.7 Figure 10.104 shows a buffer *BUF*1 driving a large load capacitor $C_L$. The buffer is built using an inverter pair as in Figure 10.94c. The width-to-length ratio of each NMOS transistor in the buffer is $W/L$ and the resistors have a value $R_L$. Accordingly, the gate capacitance seen at the input of the buffer is given by $(W/L)C_{GS}$. The buffer satisfies a static discipline with voltage thresholds given by $V_{IL} = V_{OL} = V_L$ and $V_{IH} = V_{OH} = V_H$. The supply voltage is $V_S$. Assume that the internal buffer delay (as defined in Problem 10.2c) is zero. Assume that there is a 0 to 1 transition at the input $A$ at time $t = 0$.

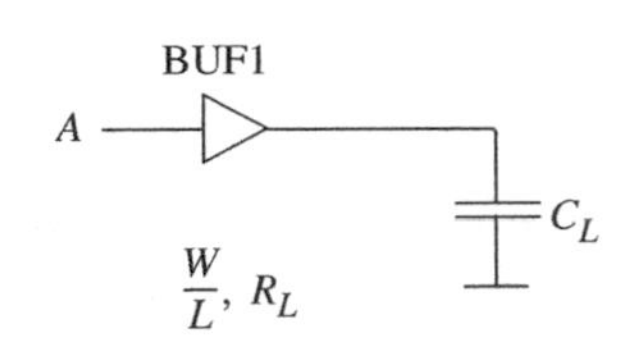

FIGURE 10.104

a) Compute the propagation delay for the buffer *BUF*1 driving the load $C_L$ for the rising transition at the input $A$.

b) Now consider Figure 10.105. This figure shows the use of a second buffer with larger transistors and smaller valued load resistors ($x > 1$) interposed between the first buffer and the load capacitor. Compute the propagation delay for the buffer *BUF*1 in series with *BUF*2 driving the load $C_L$ for the rising transition at the input $A$. Assuming that $C_L$ is much larger than the input gate capacitances of the two buffers, and that $x > 1$, is the delay computed in part (b) greater than or less than the delay computed in part (a)?

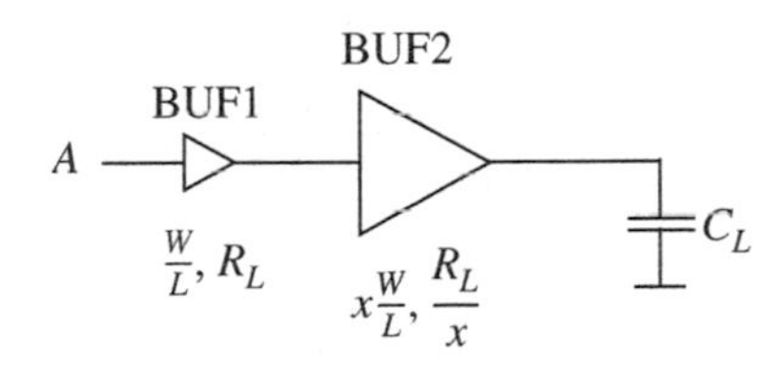


FIGURE 10.105

c) Consider Figure 10.106. This figure shows the use of a series of $n$ buffers in which *BUFi* has transistors that have a width $x$ times that of *BUFi* $-$ 1 and resistors that are a factor $x$ smaller than that of *BUFi* $-$ 1. $n$ is chosen such that $C_L$ is $x$ times the gate capacitance of *BUFn*. In other words, $n$ satisfies the equation:

$$C_L = x^n \frac{W}{L} C_{GS}.$$

Compute the propagation delay for the sequence of $n$ buffers driving the load $C_L$ for the rising transition at the input $A$. As before, assume that $C_L$ is larger than the input gate capacitances of each of the buffers and that $x > 1$.

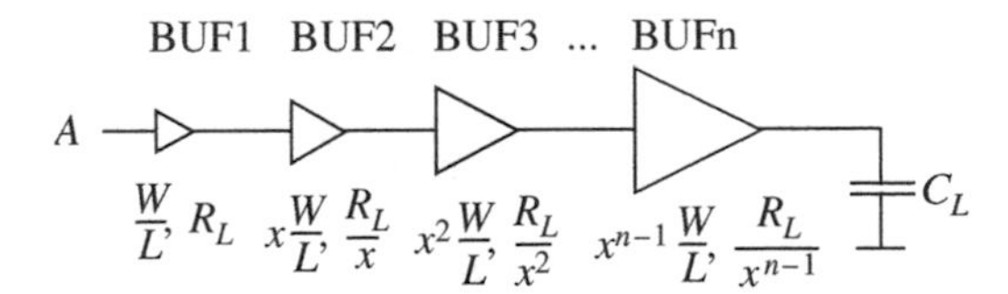


FIGURE 10.106

c) Determine the value of $x$ for which the propagation delay computed in part (b) is minimized.

PROBLEM 10.8 In this problem, you will study the effect of parasitic inductances in VLSI packages. VLSI chips are sealed inside plastic or ceramic packages and connections to certain nodes of their internal circuitry (for example, power supply, ground, input nodes, and output nodes) need to be extended outside the package. These extensions are commonly accomplished by first connecting the internal node to a metallic "pad" on the VLSI chip. In turn, the pad is connected to one end of a package "pin" using a wire that is bonded to the pad at one end and the pin at the other. The package pin, which extends outside the package, is commonly connected to external connections using a PC board.

Together the package pin, the bond wire, and the internal chip wire are associated with a nonzero parasitic inductance. In this problem, we will study the effect of the parasitic inductance associated with power supply connections. Figure 10.107 shows a model of our situation. Two inverters with load resistors $R_1$ and $R_2$ and MOSFETs with width-to-length ratios $W_1/L_1$ and $W_2/L_2$, respectively, are connected to the same power supply node on the chip that is labeled with a voltage $v_P$. Ideally this chip-level power supply node would be extended with an ideal wire outside the chip to the external power supply $V_S$ shown in Figure 10.107. However, notice the parasitic inductance $L_P$ interposed between the power supply node on the chip (marked with voltage $v_P$) and the external power supply node (marked with voltage $V_S$.)

FIGURE 10.107

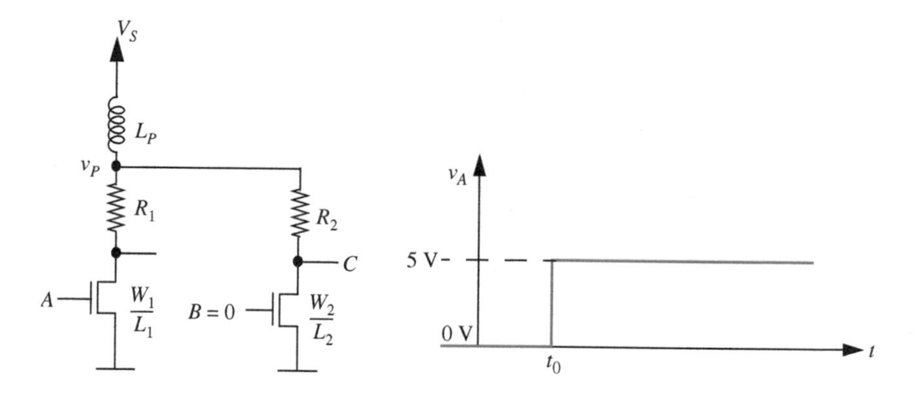


Assume that the input $B$ is 0 V at all times. Assume further that the input $A$ has 0 V applied to it initially. At time $t = t_0$, a 5-V step is applied at the input $A$. Plot the form of $v_P$ as a function of time. Clearly show the value of $v_P$ just prior to $t_0$ and just after $t_0$. Assume that the ON resistance of a MOSFET is given by the relation $(W/L)R_n$ and that MOSFET's threshold voltage is $V_T < V_S$. Also assume that $V_T < 5$ V.

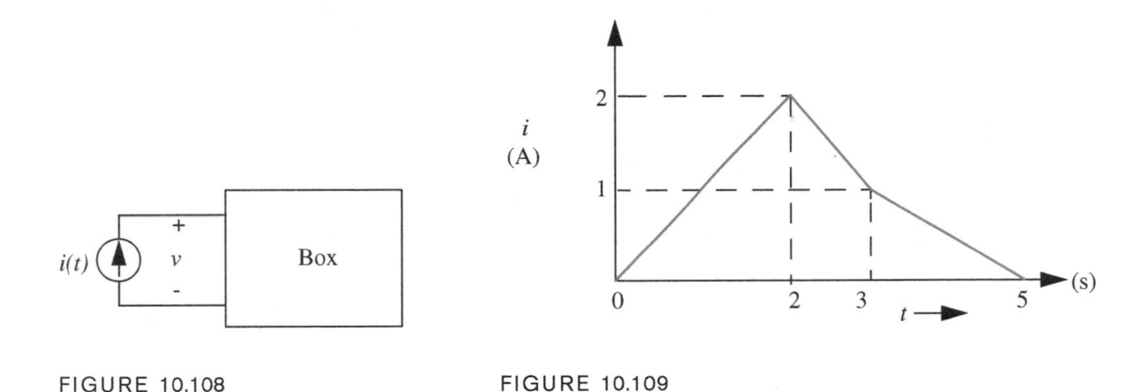


FIGURE 10.108

FIGURE 10.109

PROBLEM 10.9 A certain box, known to contain only linear elements (and no independent sources), is connected as shown in Figure 10.108.

The current waveform $i(t)$ has the form shown in Figure 10.109.

The voltage $v$ is zero for all $t < 0$, and is 1 volt for $0 < t < 2$. What is $v$ during the interval from $t = 2$ to $t = 5$? Show one simple possibility for the circuit in the box.

PROBLEM 10.10 As illustrated in Figure 10.110, a capacitor and resistor can be used to filter or smooth the waveforms we derived from a half-wave rectifier, to get something closer to a DC voltage at the output, for use in a power supply, for example.

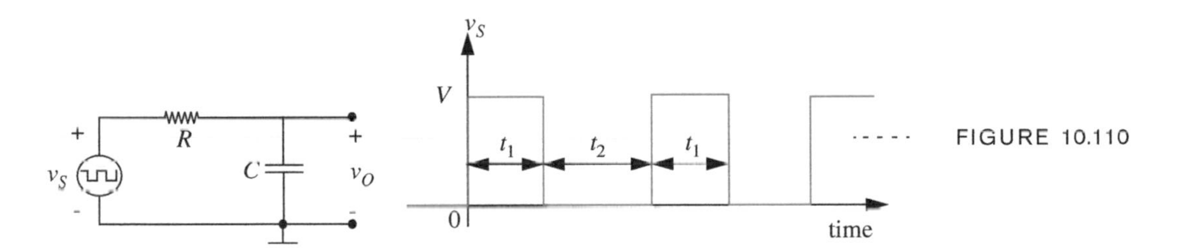


FIGURE 10.110

For simplicity, assume the voltage from source $v_S$ is a square wave. Assume that at $t = 0$, $v_O = 0$, that is, the circuit is at rest. Now assuming that $R$ is small enough to make the circuit time constant much smaller than $t_1$ or $t_2$, calculate the voltage waveforms for each half cycle of the input wave. Find the average value of the output voltage $v_O$ for $t_1 = t_2$. Sketch the waveforms carefully. For this choice of $R$, it should be clear that no useful smoothing has been accomplished.

PROBLEM 10.11 For $R$ much larger than the value used in Problem 10.10, so that the circuit time constant is much larger than $t_1$ or $t_2$, (so that the exponentials can be approximated by straight lines) calculate $v_O$ for the *first* half cycle of $v_S$, and the *second* half cycle. Sketch the result. Note that the solution does not return to the initial point of $v_O = 0$ after one cycle, so is not in the "steady state" yet.

PROBLEM 10.12 You can see from Problem 10.10 that for circuit time constant $\tau \gg t_1$ and $t_2$ the capacitor voltage starts from some value $V_{min}$ and increases when $v_S$ is positive; then when $v_S$ is zero, $v_O$ starts at some value $V_{max}$ and decreases. By definition, the "steady state" of the circuit is when $v_O$ charges from $V_{min}$ to $V_{max}$, then discharges from $V_{max}$ to the same $V_{min}$. Assuming $t_1 = t_2$, sketch the $v_O$ waveform in the steady state.

Find the average value of the voltage $v_O$. Problem 10.11 may give you a hint. Explain your answer. It may help to consider the waveform $v_S$ to be made up of a DC voltage $V/2$ and a symmetrical square wave whose values alternate between $+V/2$ and $-V/2$.

PROBLEM 10.13 This problem (see Figure 10.111) involves a capacitor and two switches. The switches are periodically driven by external clock controls at frequency $f_0$ such that first $S_1$ is closed and $S_2$ is open for time $\frac{1}{2f_0}$, and then $S_2$ is closed and $S_1$ open for time $\frac{1}{2f_0}$.

FIGURE 10.111

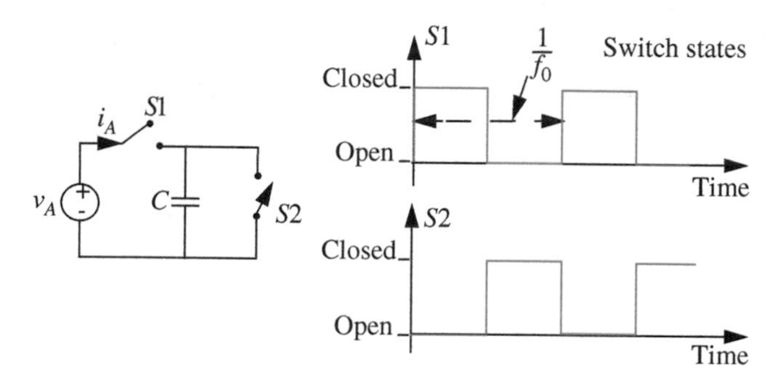


You can assume that the clock drives are *non-overlapping*; that is, $S_1$ and $S_2$ are never both closed at the same instant. $S_1$ opens just before $S_2$ closes, and $S_2$ opens just before $S_1$ closes.

a) Find an effective average current $i_A$ by determining the average rate of charge transfer over several clock cycles. Suppose $v_A = A\cos(\omega t)$ where $\omega \ll 2\pi f_0$. Sketch $i_A$ and $v_A$ on the same axes.

b) Examine your results for $i_A$ and $v_A$ from part (a). They should be in phase, and the amplitude of $i_A$ should be proportional to the amplitude of $v_A$. This is a funny form of "resistor." What is the "resistor" value? Where does the energy supplied by $v_A$ actually go?

(Comment: Circuits of this type are now commonly used in a type of MOS integrated circuit to make elements that simulate resistors with precisely controlled values. The value of such elements is that precise control of capacitor sizes and clock frequencies is easy in MOS integrated circuits, but precise control of resistor values is hard.)

PROBLEM 10.14 State variables can be used to describe the behavior of a wide range of physical systems. For each of the examples below, try to determine the following:

i) The *number* of state variables that are needed to describe the system, that is, how many state variables.

ii) Which physical variables can serve as state variables.

iii) The form of the state equations, including the identification of inputs.

iv) A simple circuit that can represent the system (an electrical analog.)

Here are the examples:

a) A hockey puck leaves a hockey player's stick with velocity $v_0$ and slides along the ice until it comes to rest (assume a very large hockey rink, or a very weak shot.)

b) Halfway through your shower each morning, the water temperature suddenly plunges toward freezing, presumably because your roommates were up earlier and showered first.

c) A simple pendulum starts from rest with an initial angular displacement $\Delta_0$, and rocks back and forth until it eventually comes to rest.

(COMMENT: Part (a) is easy if you concentrate only on the velocity, and is more difficult in terms of the circuit analogy if you include the position as well. Parts (b) and (c) lend themselves to excellent descriptions with circuit analogs.)

PROBLEM 10.15 Figure 10.112 shows the use of a filter choke.

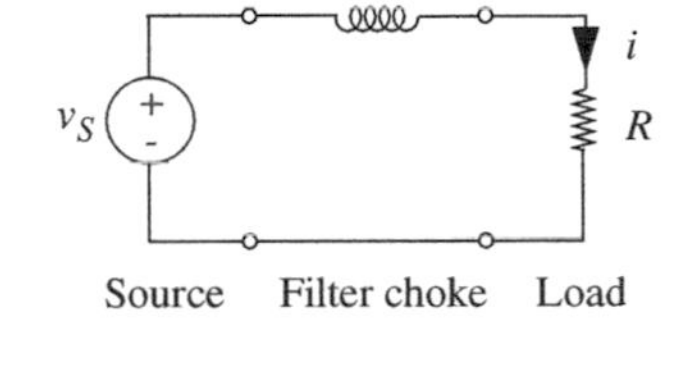


FIGURE 10.112

Assume that the waveform for $v_S$ for parts (a) and (b) is a series of square pulses starting at $t = 0$ as shown in Figure 10.113.

Assume that the waveform for $v_S$ for parts (c) and (d) is a half-rectified sine wave as shown in Figure 10.114.

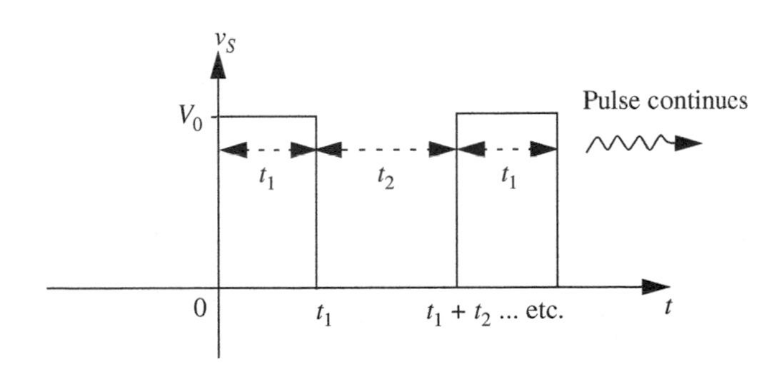


FIGURE 10.113

FIGURE 10.114

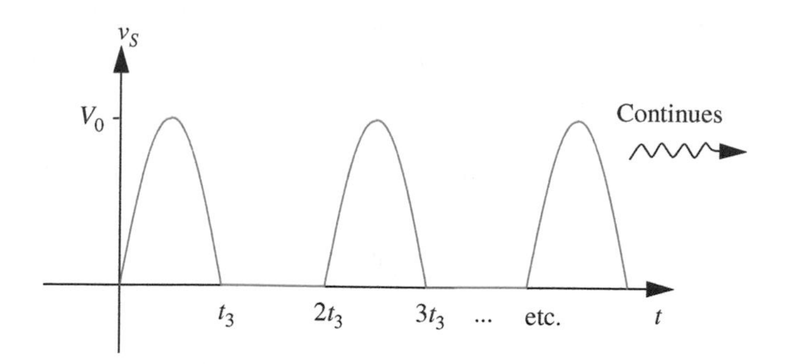


a) Assume initial rest conditions at $t = 0^-$, and assume that both $t_1$ and $t_2$ are long compared to the time constant of the network. Determine each of the following:

   i) Calculate the current waveform for the first cycle ($0 \leq t < t_1 + t_2$), the second cycle [$(t_1 + t_2) \leq t < 2(t_1 + t_2)$], and a typical cycle after stead-state periodic conditions have been reached.

   ii) How many cycles are required to go from initial rest to steady-state conditions?

   iii) In steady state, determine the average load current, the amplitude of the variations in load current through one cycle, the average energy stored in the inductor, and the ratio of this stored energy to the energy dissipated in the load during one complete cycle.

b) Repeat part (a) for the case where both $t_1$ and $t_2$ are short compared to the time constant of the network.

c) Now assume that as a filter designer, you are faced with the problem of selecting the inductor value to produce relatively smooth, ripple-free current in a load from a voltage source with a strongly pulsating value, such as the half-wave rectified sine wave shown in Figure 10.114. What method would you use to specify the inductor value with which to achieve a specified maximum variation in load current? Why might the specifications of a huge $L$ value, much larger than might be needed, be a poor design?

d) Try your hand at a design: Assume that the source waveform is half-wave rectified 60 hz 115 V AC, the load resistor is 16.2 Ω, and it is desired to have a load current ripple of 5% of the average load current. Make reasonable approximations.

PROBLEM 10.16 Consider the circuit shown in Figure 10.115.

a) Plot $v_R$ and $v_C$ for several cycles of the indicated input waveform. Assume the RC time constant is $10t_1$.

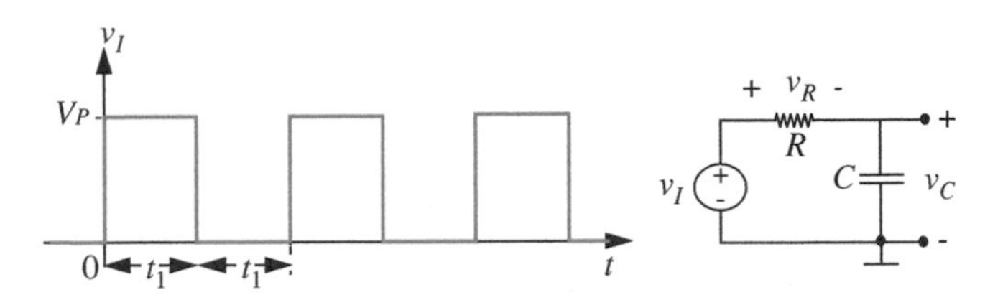


FIGURE 10.115

b) During the first several cycles, the $v_C$ waveform does not repeat, but after some time, $v_C$ is cyclic. Find and sketch this cyclic waveform. Dimension key values.

PROBLEM 10.17 Referring to Figure 10.116, for $v_I = Kt$, a ramp starting at $t = 0$, find expressions for $v_R$ and $v_L$. Plot the waveforms.

PROBLEM 10.18 Referring to Figure 10.117, given an initial inductor current $i_L(0) = 1$ mA, find the expression for $v_R$ and $v_L$. Plot the waveforms.

PROBLEM 10.19 The purpose of this problem is to illustrate the important fact that although the *zero-state response* of a linear circuit is a linear function of its input, the complete response is not. Consider the linear circuit shown in Figure 10.118.

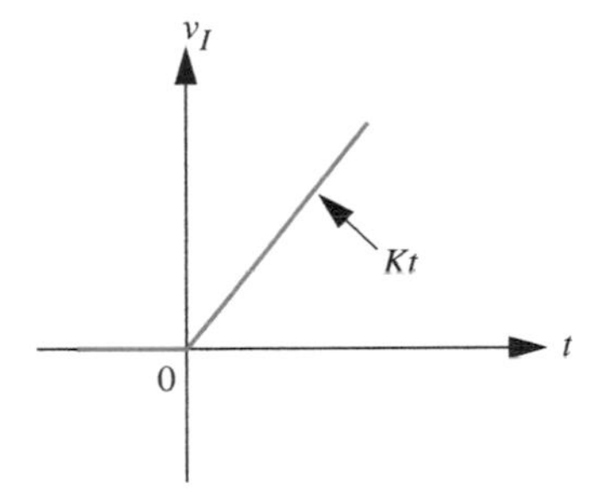


FIGURE 10.116

a) Let $i(0) = 2$ mA. Let $i_1$ and $i_2$ be the responses resulting from voltages $e_1$ and $e_2$ applied one at a time, where

$$e_1 = \begin{cases} 0, & t < 0 \\ 10 \text{ volts}, & t \geq 0 \end{cases} \tag{10.185}$$

$$e_2 = \begin{cases} 0, & t < 0 \\ 20 \text{ volts}, & t \geq 0. \end{cases} \tag{10.186}$$

Plot $i_1$ and $i_2$ as functions of $t$. Is it true that $i_2(t) = 2i_1(t)$ for all $t \geq 0$?

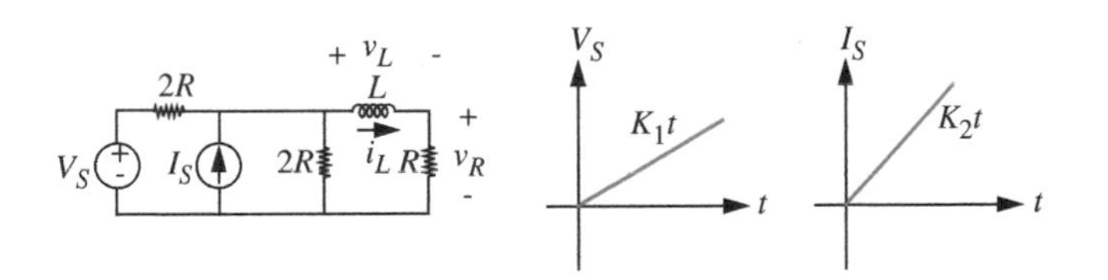


FIGURE 10.117

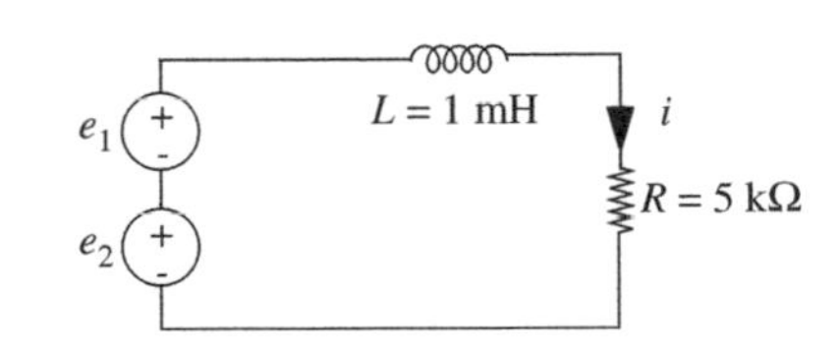


FIGURE 10.118

b) Consider now the zero-state responses due to $e_1$ and $e_2$; call them $i_1'(t)$ and $i_2'(t)$. Plot $i_1'$ and $i_2'$ as functions of $t$. Is it true that $i_2'(t) = 2i_1'(t)$ for all $t \geq 0$?

PROBLEM 10.20 In the circuit shown in Figure 10.119, the switch opens at $t = 0$. Sketch and label $i_L(t)$ and $v_L(t)$.

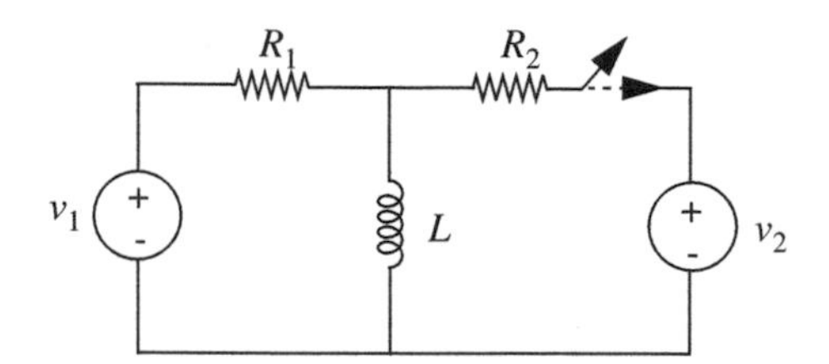


FIGURE 10.119

$v_1 = 5$ V $\quad v_2 = 3$ V, $\quad R_1 = 2$ kΩ, $\quad R_2 = 3$ kΩ, $\quad L = 4$ mH

PROBLEM 10.21 A two-input RC circuit is shown in Figure 10.120.

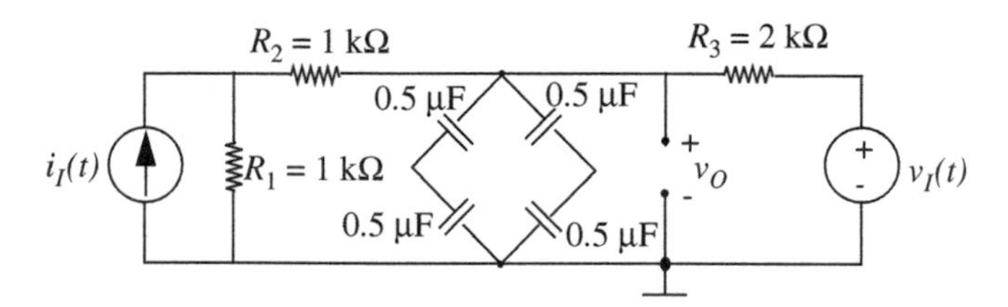


FIGURE 10.120

Consider operation with $i_I(t) = 0$, $v_I(t) = 0$ for $t \geq 0$. The voltage $v_O(t)$ is known to be 1 volt at time $t = 0$. Determine $v_O(t)$ for all $t > 0$.

A different constraint is that sources $i_I(t)$ and $v_I(t)$ are zero for $t < 0$ and that $v_O(0) = 0$. Sources $i_I(t)$ and $v_I(t)$ undergo step transitions of +1 mA and +1 volt respectively, at time $t = 0$. Determine $v_O(t)$ for all time.

PROBLEM 10.22 The neon bulb in the circuit shown in Figure 10.121 has the following behavior: The bulb remains off and acts as an open circuit until the bulb voltage $v$ reaches a threshold voltage $V_T = 65$ V. Once $v$ reaches $V_T$, a discharge occurs and the bulb acts like a simple resistor of value $R_N = 1$ kΩ; the discharge is maintained as long as the bulb current $i$ remains above the value $I_S = 10$ mA needed to sustain the discharge (even if the voltage $v$ drops below $V_T$.) As soon as $i$ drops below 10 mA, the bulb again becomes an open circuit.

a) Sketch and dimension $v(t)$ and $i(t)$, showing the first and second charging intervals.

b) Estimate the flashing rate.

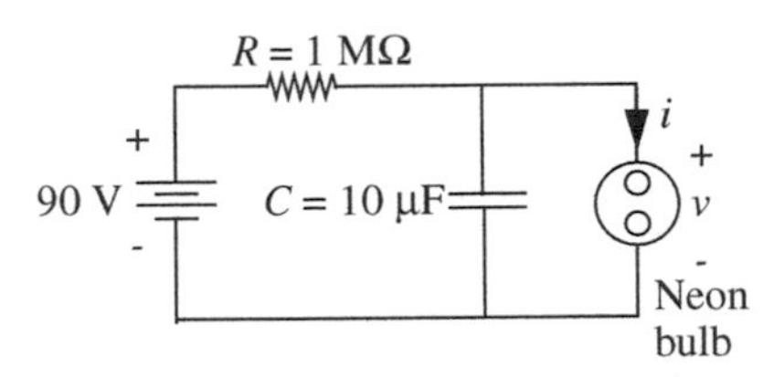


FIGURE 10.121

PROBLEM 10.23 Because of the input resistance and capacitance of an oscilloscope, laboratory observations of transients, such as the step response of the $R_1 - C_1$ circuit in Figure 10.122 may have errors in them.

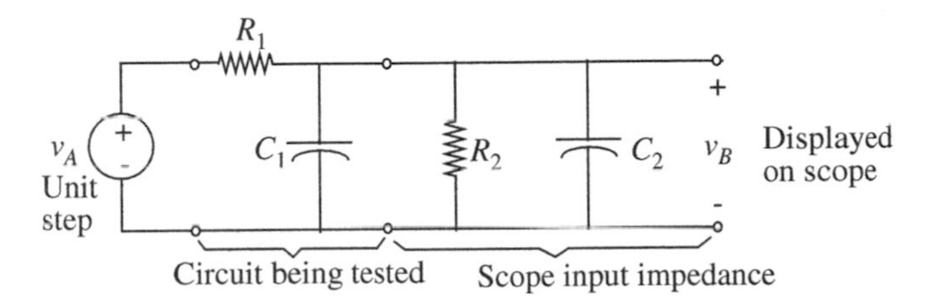


FIGURE 10.122

a) Assuming that the effect of connecting the oscilloscope to the circuit under test is to add $R_2$ and $C_2$ as shown in Figure 10.122, find and sketch the step response that will be observed at $v_B$ in this circuit. Discuss the errors introduced by the scope by comparing your result to what would be observed if the scope were ideal ($R_2 \to \infty$, $C_2 \to 0$.) Assume zero initial state.

b) A common method of coping with the errors of part (a) is to use a compensated attenuator in series with the scope (see in Figure 10.123.) For simplicity, we examine what the compensated scope displays when it is connected directly to the unit step *without* the $R_1 - C_1$ circuit of part (a). Assume zero initial state before the step is applied.

   i) What is $v_B$ immediately after the step is applied, that is, at $t = 0^+$?

   ii) What is $v_B$ as $t \to \infty$?

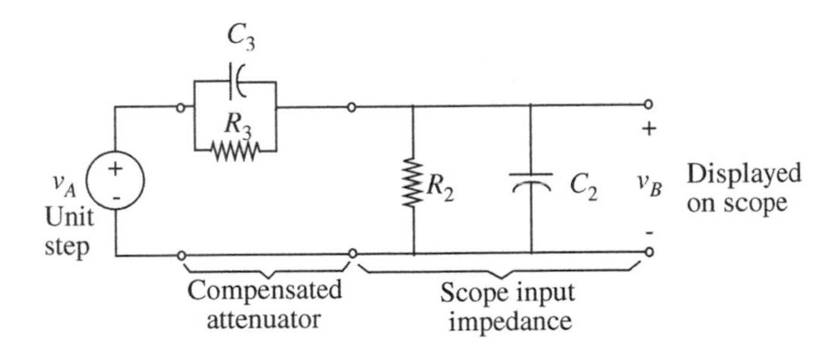


FIGURE 10.123

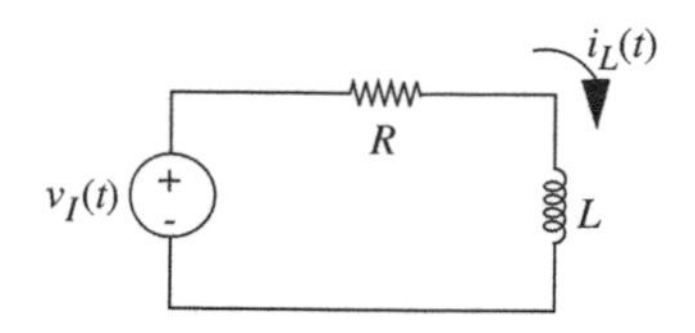


iii) Using your results, find $v_B(t)$ for all $t$.

iv) What conditions on $R_2, C_2, R_3$, and $C_3$ must be satisfied in order that there be no natural response component, that is, no transient, in $v_B(t)$? What is $v_B(t)$ in this case?

PROBLEM 10.24 The RL circuit shown in Figure 10.124 is driven with the ramp $v_I(t) = K_1 t$, for $t$ greater than zero, and $v_I(t) = 0, t < 0$.

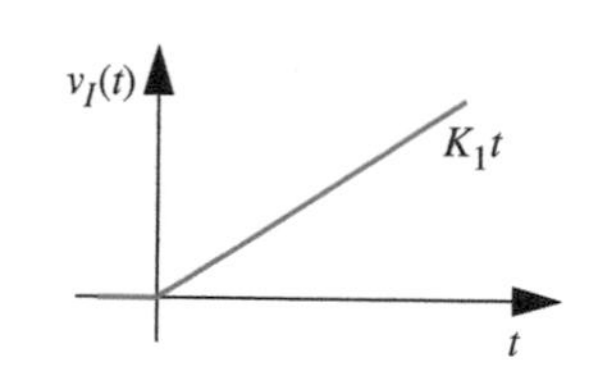


FIGURE 10.124

a) Assuming $i_L(0^-) = 0$, sketch the current $i_L(t)$. Also find an analytic expression for $i_L(t)$.

b) In some applications, such as generating a linear sweep for a magnetically deflected cathode-ray tube, we want to make $i_L(t)$ a linear ramp as shown in Figure 10.125. Find a new input waveform $v_I(t)$ such that $i_L(t) = K_2 t, t > 0$. Plot $v_I(t)$. Label all values and slopes.

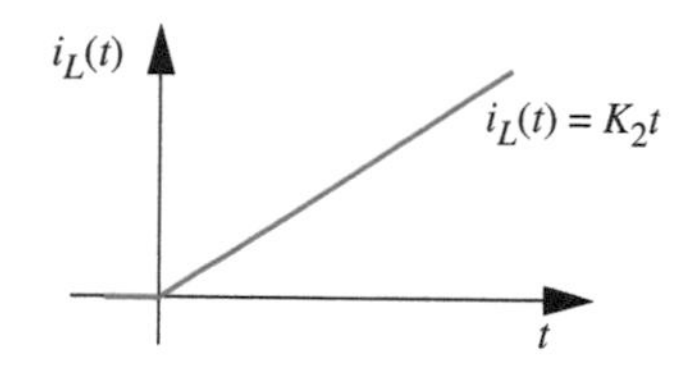


FIGURE 10.125

PROBLEM 10.25 For the RL circuit shown in Figure 10.126, sketch and label $v_R$ versus time for $t > 0$. Assume $i_L(t < 0) = 0$, and that $T_1$ is five times as long as the circuit time constant.

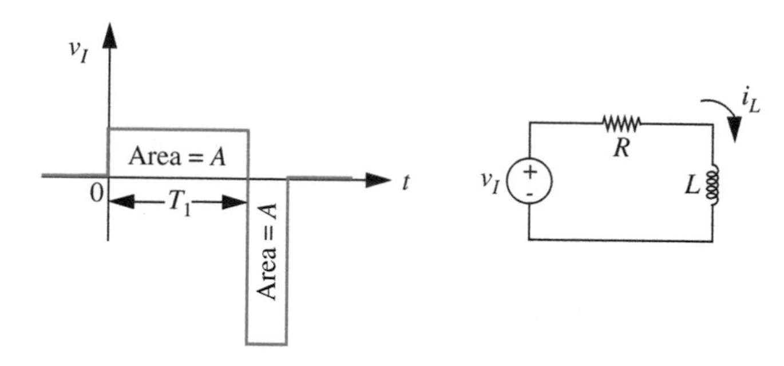


FIGURE 10.126

PROBLEM 10.26 With the capacitor initially at rest ($v_C(0) = 0$) and disconnected, the switch is closed to position (1) at time $t = 0$ in Figure 10.127.

a) Sketch the waveform $v_C(t)$ for $t > 0$. Label all relevant points on Figure 10.127 and calculate the time constant.

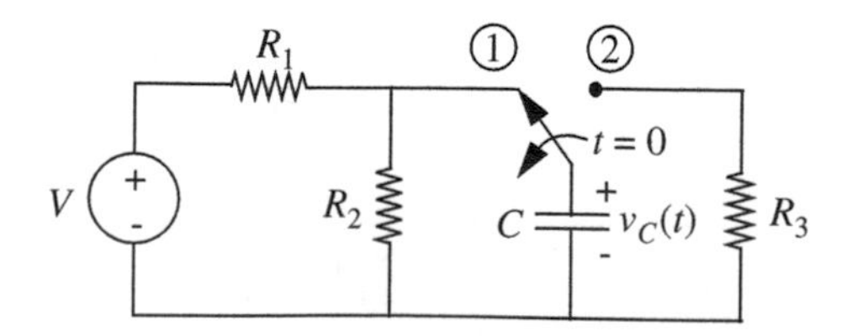


FIGURE 10.127

b) At a time $T > 0$ (at least five time constants later), the switch is thrown (instantaneously) to position (2). Sketch $v_C(t)$ for $t > T$ and label all relevant points on Figure 10.127

c) With $R_1 = R_2 = R_3$, is the time constant in part (a) greater than, less than, or equal to the time constant in part (b)?

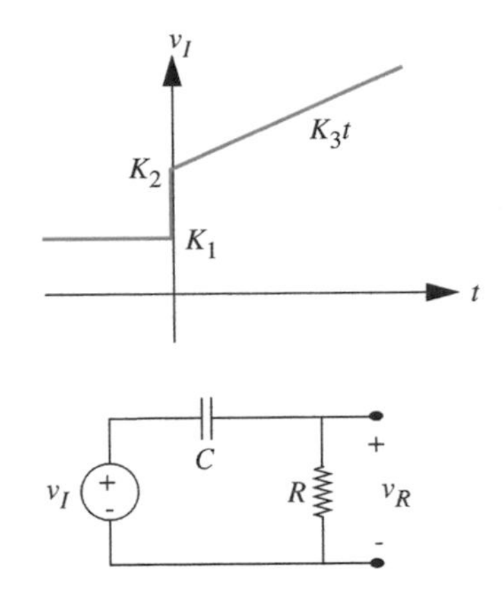


PROBLEM 10.27 For the circuit shown in Figure 10.128, sketch and label $v_R$ versus time. Assume that $v_I = K_1$ for a long time prior to $t = 0$ as illustrated in the figure.

Note that this problem can be solved in a number of simple steps by breaking the problem down into parts and solving each part. There are several ways to do this breakdown, all of roughly equal ease.

FIGURE 10.128

PROBLEM 10.28 You are given the RC circuit shown in Figure 10.129.

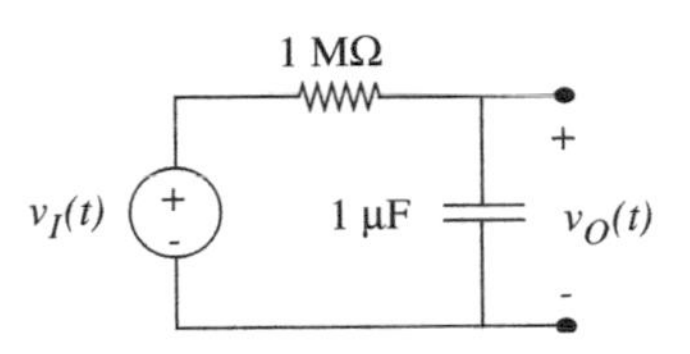


FIGURE 10.129

a) Suppose you observe that $v_O(t)$ is a triangular pulse, as shown in the sketch in Figure 10.130. Find and draw the waveform $v_I(t)$ that must be applied to produce this output signal. Label times and magnitudes, and significant parameters of the function.

b) Now the input signal is changed. You apply a ramp starting at $t = 0$, $v_I(t) = tu_{-1}(t)$, as the input signal $v_I(t)$. (Note that $u_{-1}(t)$ represents a unit step at $t = 0$.) Sketch and label the output signal $v_O(t)$ for $0 < t < 5$.

c) Give an analytic expression for the output signal $v_O(t)$ you sketched in (b).

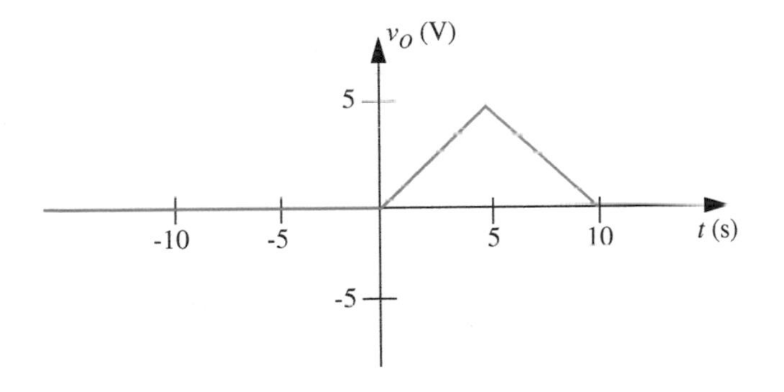


FIGURE 10.130

PROBLEM 10.29 Consider the digital memory element shown in Figure 10.131. The voltage at the storage node with respect to ground is denoted $v_M$. The figure also shows a parasitic resistance $R_P$ from the storage node to ground. This resistance will cause a leakage of the charge stored in the memory.

The signal $A$ is fed to an inverter and the inverter drives the input $d_{IN}$ of the memory element. All inverters shown in Figure 10.131 have a load resistor $R_L$ and the on

FIGURE 10.131

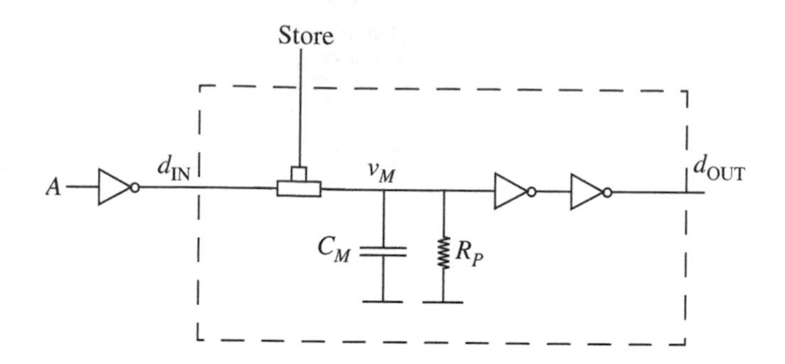


resistance of the pulldown MOSFETs in each of the inverters is $R_{ON}$. Assume that the on resistance of the switch driven by the *Store* signal is also $R_{ON}$. The supply voltage is $V_S$ and the threshold voltage for the MOSFETs is $V_T$. In doing this problem, assume that $R_P$ is much larger than either $R_{ON}$ or $R_L$.

a) Suppose that a 0-V to $V_S$-step is applied at the *Store* input of the memory element at $t = 0$. Sketch $v_M(t)$ for $t \geq 0$, assuming that $v_M(t = 0) = 0$, and that $A$ is at 0 V throughout. Assuming that $R_{ON} \ll R_P$, what is the maximum value attained by $v_M$?

b) Suppose, now, that a rectangular *pulse* of height $V_S$ is applied at the *Store* input of the memory element, and that $A$ is at 0 V throughout. The rising transition of the pulse occurs at $t = 0$ and the falling transition at $t = T$. Determine the minimum value of the pulse width $T$ so that $v_M$ can charge up to $V_H$, where $V_H = V_{IH} = V_{OH}$, the high voltage threshold of the static discipline. Assume the following: $v_M(t = 0) = 0$; $V_H < V_S$; $V_H > V_T$.

c) Let us now consider the case in which $A$ is at $V_S$ throughout, and $v_M(t = 0) = V_S$. Sketch $v_M(t)$ for $t \geq 0$, when a 0-V to $V_S$-step is applied at the *Store* input of the memory element at $t = 0$. What is the minimum value attained by $v_M$?

d) Suppose, now, that a rectangular *pulse* of height $V_S$ is applied at the *Store* input of the memory element. The rising transition of the pulse occurs at $t = 0$ and the falling transition at $t = T$. Determine the minimum value of the pulse width $T$ so that $v_M$ can discharge from $V_S$ to $V_L$, where $V_L = V_{IL} = V_{OL}$, the low voltage threshold of the static discipline. Assume as in (c) that $A$ is at $V_S$ throughout and that $v_M(t = 0) = V_S$. Assume further that $V_L < V_T$ and that $V_L$ is greater than the minimum value attainable by $v_M$.

e) Suppose the memory element is storing a 1 (assume $v_M = V_S$) at $t = 0$ and that *Store* = 0. Assuming that no further *Store* signals occur, determine the period of time for which the output ($d_{OUT}$) of the memory element will be valid. (Hint: the output becomes invalid when $d_{OUT}$ switches from 1 to 0.)

# CHAPTER 11

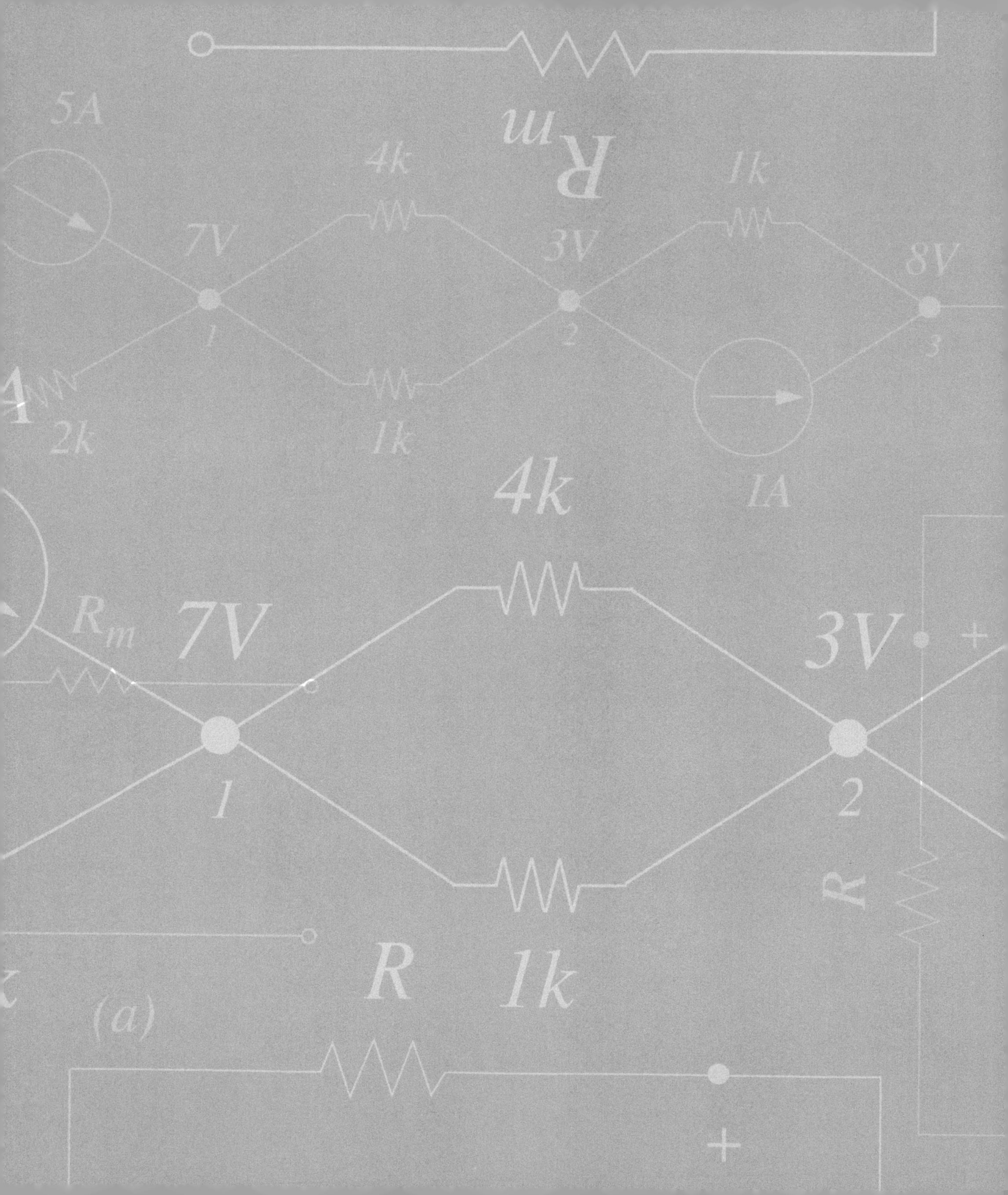
5A
4k
1k
7V
3V
8V
1
2
3
2k
1k
1A
4k
Rm
7V
3V
+
1
2
R
R
1k
(a)
+

# ENERGY AND POWER IN DIGITAL CIRCUITS

# 11

Digital circuits form the basis of a large number of battery-powered appliances used in our day to day life, including cell phones, beepers, digital watches, calculators, and laptop computers. A battery of a given weight and size stores a fixed amount of energy. The amount of time that the battery will last before requiring a replacement or a recharge depends on the amount of energy consumed by the device. Similarly, the heat generated by a device depends on its energy consumption rate, or power dissipation. Thus, both the amount of energy consumed by a device and the rate of energy consumption are critical issues in the design of circuits.

## 11.1 POWER AND ENERGY RELATIONS FOR A SIMPLE RC CIRCUIT

Let us first develop the power and energy relations for the simple RC circuit shown in Figure 11.1. Assume that the switch is closed at time $t = 0$ connecting the voltage source to the RC network. Further, assume that the capacitor has zero charge on it initially.

We know that the power delivered to a two-terminal element at any given instant of time $t$ with a voltage $v(t)$ across it and a current $i(t)$ through it is given by

$$p(t) = v(t)i(t) \tag{11.1}$$

where the current $i(t)$ is defined to be positive if it enters the element at the terminal with the positive voltage. Power is delivered to the element if $i(t)$ is positive, for example, in a resistor or a charging capacitor. A resistor dissipates

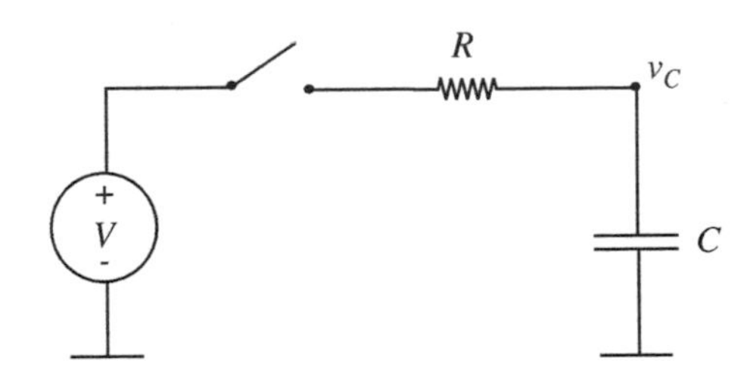


FIGURE 11.1 Energy drawn from a power source by a simple RC circuit.

energy, while a capacitor stores it. Power is delivered by an element if $i(t)$ is negative, for example, in a battery or in a discharging capacitor.

Referring to Figure 11.1, when the switch is closed, the voltage source begins to charge the capacitor $C$ through the resistor $R$. What is the instantaneous power drawn from the voltage source? The instantaneous power drawn from the voltage source is given by

$$p(t) = Vi(t) \tag{11.2}$$

where

$$i(t) = \frac{V - v_C(t)}{R}. \tag{11.3}$$

Using the equations derived for capacitor charging dynamics in Section 10.4, we can write

$$v_C = V(1 - e^{\frac{-t}{RC}}) \tag{11.4}$$

Substituting the expression for $v_C$ from Equation 11.4 into Equation 11.3, and that for $i(t)$ from Equation 11.3 into Equation 11.2, we get

$$p(t) = \frac{V^2}{R} e^{\frac{-t}{RC}}. \tag{11.5}$$

What is the total amount of energy supplied by the voltage source if the switch is closed for a long period of time? Since power is the rate at which energy is supplied, the energy $w$ supplied over an interval of time $0 \rightarrow T$, is given by

$$w = \int_0^T p(t)dt. \tag{11.6}$$

The total amount of energy supplied by the voltage source can be obtained by taking the limit as $T$ goes to $\infty$.

Thus, the total amount of energy supplied by the voltage source as $T$ goes to $\infty$ is given by

$$\begin{aligned} w &= \int_{t=0}^{t=\infty} \frac{V^2}{R} e^{\frac{-t}{RC}} dt \\ &= -\frac{V^2}{R} RC e^{\frac{-t}{RC}} \Big|_{t=0}^{t=\infty} \\ &= CV^2. \end{aligned}$$

What is the amount of energy stored in the capacitor when $T$ goes to $\infty$? After a long period of time, the capacitor will charge up to the voltage $V$. From Equation 9.18, the amount of energy stored on a capacitor with a voltage $V$ across it is given by $1/2CV^2$.

What is the amount of energy dissipated by the resistor? Since the voltage source supplies an amount of energy equal to $CV^2$, and the amount of energy stored in the capacitor is $1/2CV^2$, the remaining half of the energy supplied by the voltage source must have been dissipated in the resistor. We can verify this by explicitly computing the energy dissipated in the resistor as follows. The instantaneous power for the resistor is given by

$$\begin{aligned} p(t) &= i(t)^2 R \\ &= \left(\frac{V}{R} e^{\frac{-t}{RC}}\right)^2 R \\ &= \frac{V^2}{R} e^{\frac{-2t}{RC}}. \end{aligned}$$

The energy dissipated in the resistor over the interval $t = 0$ to $t = \infty$ is given by

$$\begin{aligned} w &= \int_{t=0}^{t=\infty} \frac{V^2}{R} e^{\frac{-2t}{RC}} dt \\ &= -\frac{V^2}{R}\frac{RC}{2} e^{\frac{-2t}{RC}} \Bigg|_{t=}^{t=\infty} \\ &= -\frac{V^2 C}{2} e^{\frac{-2t}{RC}} \Bigg|_{t=0}^{t=\infty} \\ &= \frac{1}{2} CV^2. \end{aligned}$$

Interestingly, notice that if the transients are allowed to settle, the total energy dissipated in the resistor when charging a capacitor is independent of its resistance value. The same is true when discharging a capacitor through a resistor.

## 11.2 AVERAGE POWER IN AN RC CIRCUIT

Let us now derive the average power dissipated by the slightly more complicated circuit depicted in Figure 11.2 comprising two resistors and a capacitor connected to a voltage source through a switch. Assume that a square wave

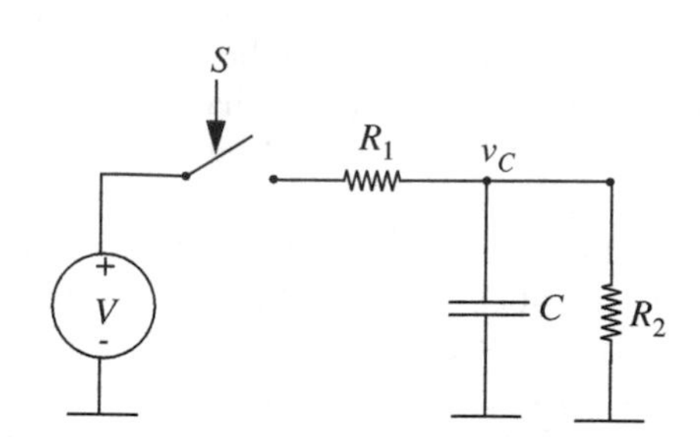


FIGURE 11.2 An RC circuit with a switch.

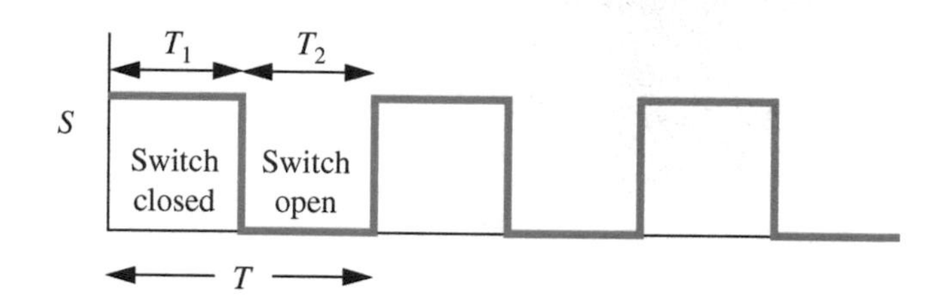


signal $S$ with a cycle time of $T$ such as that shown in Figure 11.2 controls the switch, so that the switch is closed for an interval $T_1$, then open for an interval $T_2$, and so on. When the switch is closed, the voltage source charges up the capacitor. The capacitor discharges through resistor $R_2$ when the switch is open. We will be particularly interested in the special case where the $T_1$ and $T_2$ intervals are large enough that the capacitor voltages during each of the intervals $T_1$ and $T_2$ reach their respective steady state values.

As current flows through the resistors, energy is dissipated. However, the capacitor does not dissipate energy. It simply stores energy when the switch is closed (during $T_1$) and supplies this stored energy when the switch is open (during $T_2$). We can derive the instantaneous power $p(t)$ dissipated in the circuit as a function of time. We can also derive the average power $\overline{p}$ dissipated by the circuit.

The average power is defined as the total amount of energy $w$ dissipated during some time interval, divided by the length of the time interval $T$.

In other words,

$$\overline{p} = \frac{w}{T}.$$

More specifically, if $w_1$ is the energy dissipated during the interval $T_1$ and $w_2$ is the energy dissipated during the interval $T_2$, then the average power dissipated in the circuit is given by

$$\overline{p} = \frac{w_1 + w_2}{T}. \tag{11.7}$$

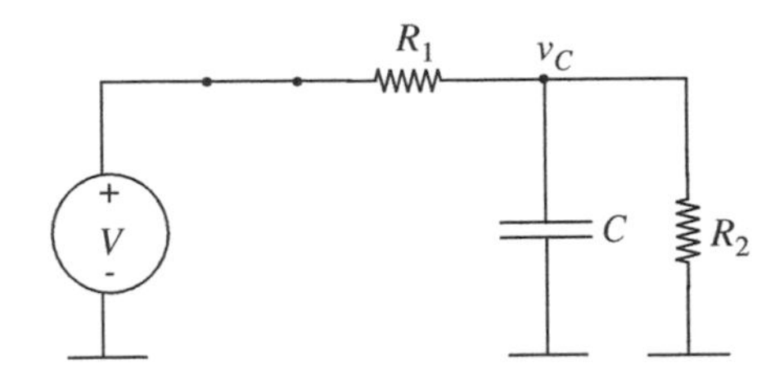


FIGURE 11.3 Equivalent circuit with the switch closed.

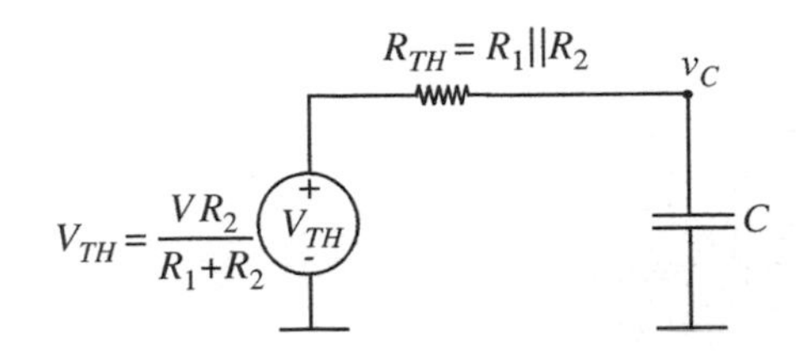


FIGURE 11.4 Thévenin equivalent circuit with the switch closed.

We will also use the fact that the energy dissipated within some interval of time is the time integral of the instantaneous power within that interval. For example, the energy dissipated within the time interval $T$ is given by

$$w_T = \int_{t=0}^{t=T} p(t)dt. \tag{11.8}$$

### 11.2.1 ENERGY DISSIPATED DURING INTERVAL $T_1$

Let us first consider the case when the switch is closed and derive the value of $w_1$. When the switch is closed, the circuit shown in Figure 11.3 applies.

To compute the energy dissipated by the circuit, we first need to determine the currents through the resistors $R_1$ and $R_2$. To facilitate the computation of the currents, let us first determine $v_C$. To simplify the calculation of $v_C$, we transform the circuit shown in Figure 11.3 to its Thévenin equivalent shown in Figure 11.4. We assume that time $t$ starts from 0 at the moment that the signal $S$ transitions from low to high. From the circuit in Figure 11.4, we can write an expression for the voltage $v_C$ as

$$v_C = V_{TH}\left(1 - e^{\frac{-t}{R_{TH}C}}\right).$$

We will be particularly interested in the special case in which $t \to \infty$. When $t \to \infty$, $v_C \to V_{TH}$.

We are now ready to determine the instantaneous power dissipated in the circuit when the switch is closed. In general, we cannot use the Thévenin equivalent circuit to determine the power consumed in the original circuit because the power computation is a nonlinear process. Therefore, returning to the circuit in Figure 11.3, the instantaneous power dissipated by the circuit when the switch is closed is given by the sum of the power dissipated in resistors $R_1$ and $R_2$.

Recalling that the power dissipated in a resistance $R$ with a voltage $v$ across it is $v^2/R$, we can write

$$\begin{aligned} p(t) &= \text{Power in } R_1 + \text{Power in } R_2 \\ &= \frac{(V - v_C)^2}{R_1} + \frac{v_C^2}{R_2} \\ &= \frac{[V - V_{TH}(1 - e^{\frac{-t}{R_{TH}C}})]^2}{R_1} + \frac{[V_{TH}(1 - e^{\frac{-t}{R_{TH}C}})]^2}{R_2}. \end{aligned}$$

We are now in a position to derive the energy dissipated in the circuit using the relationship from Equation 11.8.

$$w_1 = \int_{t=0}^{t=T_1} \left( \frac{\left[V - V_{TH}(1 - e^{\frac{-t}{R_{TH}C}})\right]^2}{R_1} + \frac{\left[V_{TH}(1 - e^{\frac{-t}{R_{TH}C}})\right]^2}{R_2} \right) dt,$$

which yields:

$$\begin{aligned} w_1 &= \frac{V_{TH}^2}{R_2}\left[t - \frac{R_{TH}C}{2} e^{\frac{-2t}{R_{TH}C}} + 2R_{TH}C e^{\frac{-t}{R_{TH}C}}\right]_{t=0}^{t=T_1} \\ &+ \frac{1}{R_1}\left[(V - V_{TH})^2 t - V_{TH}^2 \frac{R_{TH}C}{2} e^{\frac{-2t}{R_{TH}C}} - 2(V - V_{TH})V_{TH}R_{TH}C e^{\frac{-t}{R_{TH}C}}\right]_{t=0}^{t=T_1} \\ &= \frac{V_{TH}^2}{R_2}\left[T_1 - \frac{R_{TH}C}{2} e^{\frac{-2T_1}{R_{TH}C}} + 2R_{TH}C e^{\frac{-T_1}{R_{TH}C}}\right] \\ &- \frac{V_{TH}^2}{R_2}\left[-\frac{R_{TH}C}{2} + 2R_{TH}C\right] \\ &+ \frac{1}{R_1}\left[(V - V_{TH})^2 T_1 - V_{TH}^2 \frac{R_{TH}C}{2} e^{\frac{-2T_1}{R_{TH}C}} - 2(V - V_{TH})V_{TH}R_{TH}C e^{\frac{-T_1}{R_{TH}C}}\right] \\ &- \frac{1}{R_1}\left[-V_{TH}^2 \frac{R_{TH}C}{2} - 2(V - V_{TH})V_{TH}R_{TH}C\right]. \end{aligned}$$

Separating the terms containing a $T_1$ factor, we get

$$w_1 = \frac{V^2 T_1}{R_1 + R_2} + \frac{V_{TH}^2}{R_2}\left[-\frac{R_{TH}C}{2}e^{\frac{-2T_1}{R_{TH}C}} + 2R_{TH}Ce^{\frac{-T_1}{R_{TH}C}}\right]$$

$$-\frac{V_{TH}^2}{R_2}\left[-\frac{R_{TH}C}{2} + 2R_{TH}C\right]$$

$$+\frac{1}{R_1}\left[-V_{TH}^2\frac{R_{TH}C}{2}e^{\frac{-2T_1}{R_{TH}C}} - 2(V - V_{TH})V_{TH}R_{TH}Ce^{\frac{-T_1}{R_{TH}C}}\right]$$

$$-\frac{1}{R_1}\left[-V_{TH}^2\frac{R_{TH}C}{2} - 2(V - V_{TH})V_{TH}R_{TH}C\right].$$

Rearranging and simplifying,

$$w_1 = \frac{V^2 T_1}{R_1 + R_2} + \frac{CV_{TH}^2}{2}\left(1 - e^{\frac{-2T_1}{R_{TH}C}}\right). \tag{11.9}$$

When $T_1 \gg R_{TH}C$, the capacitor can be assumed to charge up to its steady-state value of $V_{TH}$, and $e^{-2T_1/R_{TH}C} \rightarrow 0$. In this situation, the above expression for $w_1$ simplifies to:

$$w_1 = \frac{V^2}{R_1 + R_2}T_1 + \frac{V_{TH}^2 C}{2} \tag{11.10}$$

where

$$V_{TH} = \frac{VR_2}{R_1 + R_2}.$$

## 11.2.2 ENERGY DISSIPATED DURING INTERVAL $T_2$

Now, let us consider the second interval $T_2$ in which the switch is off. During $T_2$, the capacitor discharges through the resistor $R_2$. For simplicity, let us assume the special case considered in the previous section in which $T_1 \gg R_{TH}C$. In this case, at the start of the second interval, the capacitor will have an initial voltage of $V_{TH}$ across it.

As in the previous section, let us first determine $v_C$. When the switch is off, the circuit shown in Figure 11.5 applies. For this derivation we assume

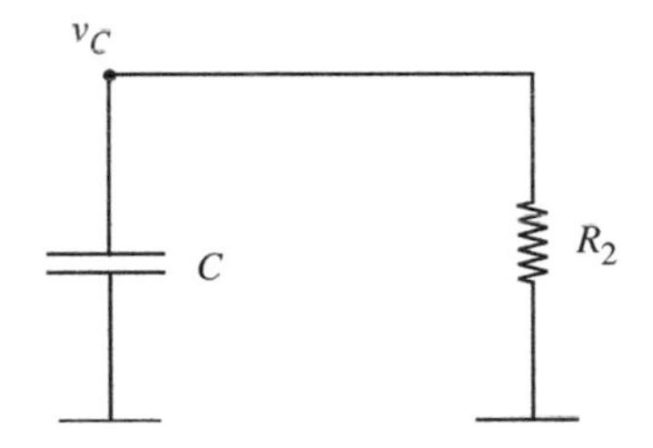


FIGURE 11.5 Equivalent circuit with the switch open.

that time $t$ starts from 0 at the moment that the signal $S$ transitions from high to low. Since the initial voltage on the capacitor is $V_{TH}$, the voltage $v_C$ is given by

$$v_C = V_{TH} e^{\frac{-t}{R_2 C}}.$$

Notice that as $t \rightarrow \infty$, the capacitor voltage $v_C \rightarrow 0$.

We are now ready to determine the instantaneous power dissipated in the circuit when the switch is open. The instantaneous power dissipated in resistor $R_2$ is given by

$$\begin{aligned} p(t) &= \frac{v_C^2}{R_2} \\ &= \frac{1}{R_2}\left(V_{TH} e^{\frac{-t}{R_2 C}}\right)^2. \end{aligned}$$

The corresponding energy consumed during $T_2$ is given by

$$w_2 = \int_{t=0}^{t=T_2} p(t)dt \tag{11.11}$$

$$= \int_{t=0}^{t=T_2} \frac{v_C^2}{R_2} dt \tag{11.12}$$

$$= \int_{t=0}^{t=T_2} \frac{1}{R_2}\left(V_{TH} e^{\frac{-t}{R_2 C}}\right)^2 \tag{11.13}$$

$$= \frac{-1}{2R_2} V_{TH}^2 R_2 C e^{\frac{-2t}{R_2 C}} \Big|_{t=0}^{t=T_2} \tag{11.14}$$

$$= \frac{V_{TH}^2 C}{2}\left(1 - e^{\frac{-2T_2}{R_2 C}}\right). \tag{11.15}$$

When $T_2 \gg R_2 C$, we can ignore the second term in the above equation and write:

$$w_2 = \frac{V_{TH}^2 C}{2}. \tag{11.16}$$

### 11.2.3 TOTAL ENERGY DISSIPATED

Combining Equations 11.10 and 11.16, for the case of $T_1 \gg R_{TH}C$ and $T_2 \gg R_2C$ we obtain the total energy dissipated in a cycle $T$:

$$w = w_1 + w_2 = \frac{V^2}{R_1 + R_2} T_1 + \frac{V_{TH}^2 C}{2} + \frac{V_{TH}^2 C}{2}.$$

Combining terms, we get

$$w = \frac{V^2}{R_1 + R_2} T_1 + V_{TH}^2 C.$$

Dividing by $T$ we get the average power $\overline{p}$:

$$\overline{p} = \frac{V^2}{(R_1 + R_2)} \frac{T_1}{T} + \frac{V_{TH}^2 C}{T}. \tag{11.17}$$

For a symmetric square wave, $T_1 = T/2$, so Equation 11.17 simplifies to:

$$\overline{p} = \frac{V^2}{2(R_1 + R_2)} + \frac{V_{TH}^2 C}{T}. \tag{11.18}$$

Equation 11.18 shows that the average power is the sum of two terms. The first term is independent of the time period of the square wave and is called the *static power* $p_{\text{static}}$. It can be computed independently by removing all capacitors and inductors from the circuit (in other words, replace the capacitors with open circuits and inductors with short circuits). The second term is related to the charging and discharging of the capacitor, and depends on the time period of the square wave. This term is called the *dynamic power* $p_{\text{dynamic}}$. In other words,

$$p_{\text{static}} = \frac{V^2}{2(R_1 + R_2)} \tag{11.19}$$

$$p_{\text{dynamic}} = \frac{V_{TH}^2 C}{T}. \tag{11.20}$$

Notice that if the switch were kept closed for a long period of time, no dynamic power is dissipated in the steady state. Notice further that the dynamic power dissipation is proportional to the capacitor value, the switching frequency, and the square of the voltage, but independent of the resistance value.

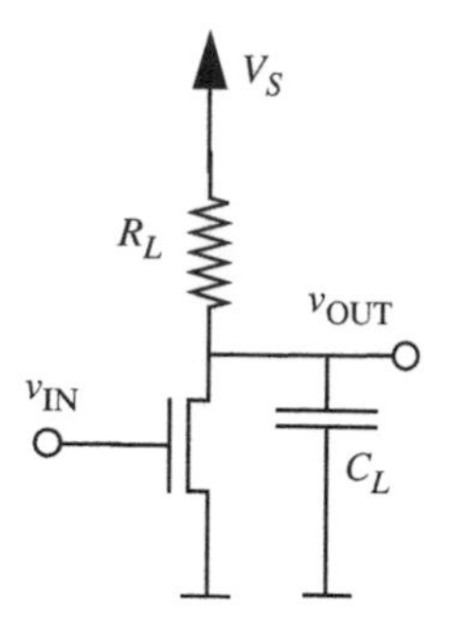


**FIGURE 11.6** Inverter with load capacitor.

## 11.3 POWER DISSIPATION IN LOGIC GATES

Let us now compute the power dissipated by our logic gates using the inverter as an example. The inverter is shown in Figure 11.6 driving a load capacitor. The load capacitor $C_L$ is the sum of the wire capacitance and the gate capacitances of the devices driven by the inverter.

As mentioned earlier, there are two different forms of power dissipated by a MOSFET inverter of the type shown in Figure 11.6 — static power and dynamic power.

- The static power dissipation $p_{\text{static}}$ is the power loss due to the static or continuous current drawn from the power supply. It is independent of the rate at which signals transition. (It can depend, however, on the state of the input signals.)
- The dynamic power dissipation $p_{\text{dynamic}}$ is the power loss due to the switching currents required to charge and discharge capacitors. As we saw previously, this component of power depends on the rate at which signals transition.

### 11.3.1 STATIC POWER DISSIPATION

Let us first compute the static power dissipated in an inverter. The static power can be determined by removing all capacitive and inductive elements from the circuit (remember, this means that we replace capacitors with open circuits and inductors with short circuits). Accordingly, we will assume that $C_L = 0$ for our static power calculation. When the MOSFET in the inverter is turned on, a resistive path exists between power and ground. So current flows through $R_L$ and the on resistance of the MOSFET, $R_{ON}$, causing static power dissipation:

$$p_{\text{static}} = \frac{V_S^2}{R_L + R_{ON}}. \tag{11.21}$$

Notice that static power is dissipated only in gates that are on.[1] When the MOSFET in the inverter is off, the static power dissipation is zero. Thus, Equation 11.21 reflects the worst-case static power dissipation.

The static power dissipation of a circuit depends on the particular set of applied inputs. When the inputs are not known, there are several ways of estimating static power dissipation. One estimate attempts to compute the worst-case power dissipation of a circuit. In this method, choose the set of

1. In practice, there are other sources of static power loss, such as that due to leakage currents, but for simplicity we will ignore these.

inputs that results in the worst-case power dissipation for the circuit. For the simple inverter, this estimate is the power dissipated when the input is a logical 1 and the MOSFET is on, for example, as computed in Equation 11.21. Another estimate is statistical, and is based on determining the expected power over all possible input sets. Each input set is assigned an occurrence probability and the power for that input set is determined. Then the expected power is computed by averaging the power for each set of inputs weighted by the occurrence probability for that input set. Yet another estimate assumes that each input to the circuit is a square wave comprising an alternating sequence of logical 1's and 0's.

---

EXAMPLE 11.1 STATIC POWER DISSIPATION Let us compute the worst-case static power dissipation for the logic gate in Figure 11.7. The worst case dissipation occurs when all the inputs are high. Suppose $R_L = 100$ kΩ and $R_{ON} = 10$ kΩ for each of the MOSFETs. Also assume that $V_S = 5$ V.

When all the inputs are high, the effective resistance $R_{eff}$ between power and ground is

$$R_{eff} = R_L + (2R_{ON} \| R_{ON} \| R_{ON}).$$

In other words, $R_{eff} = 104$ kΩ. The maximum static power dissipation is

$$p_{static} = \frac{V_S^2}{R_{eff}} = 25/104 = 0.24 \text{ mW} = 240\ \mu\text{W}. \tag{11.22}$$

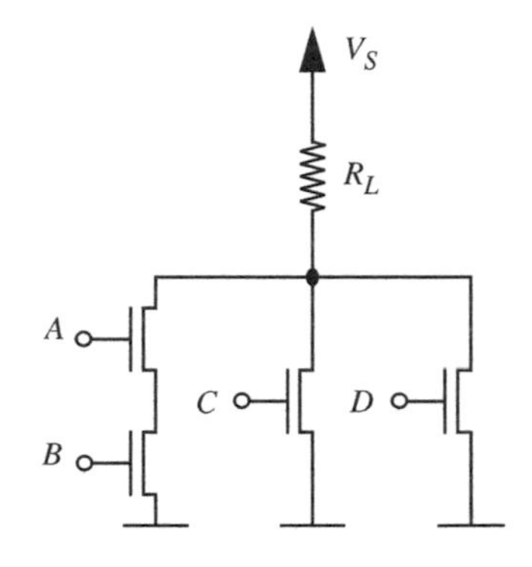

FIGURE 11.7 Worst-case static power dissipation in a logic gate.

---

## 11.3.2 TOTAL POWER DISSIPATION

Let us now compute the total power dissipated in the inverter when a time-varying input signal is applied. The total power will include both the static power and the dynamic power. Dynamic power dissipation results from the transient currents that flow through the resistors to charge and discharge the capacitor, as depicted in Figures 11.8 and 11.9.

Suppose we have a square-wave signal input to the inverter representing a sequence of alternating 1's and 0's. Let the time period of the square wave be $T$. Therefore, the frequency of the square wave is $f = 1/T$. As depicted in Figure 11.8, when the input voltage is low, the MOSFET is off, and the load capacitor $C_L$ is charged up to $V_S$ by the power supply through the resistor $R_L$. As suggested in Figure 11.9, when the input signal is high, the MOSFET is on, and the capacitor discharges through the on resistance of the MOSFET. After a long time, the voltage on the load capacitor will reach the steady state value of $V_S R_{ON}/(R_{ON} + R_L)$. Assume that the period of the input square wave is long enough for the capacitor to charge and discharge completely.

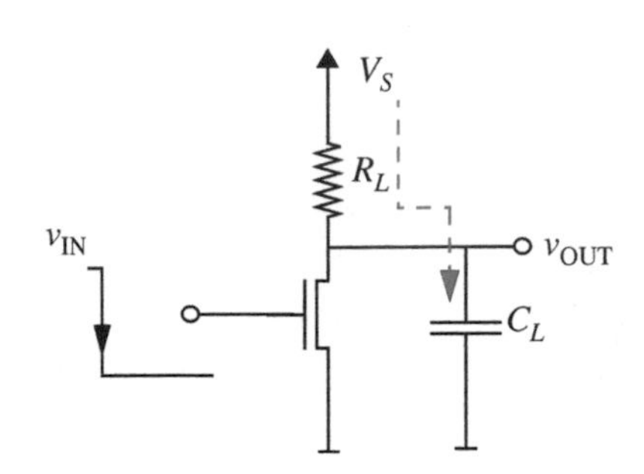


FIGURE 11.8 Charging the load capacitor.

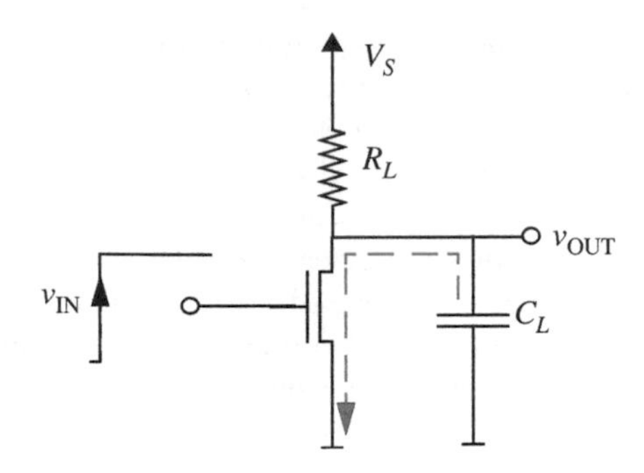


FIGURE 11.9 Discharging the load capacitor.

We will derive the average total power consumed by the inverter as in the example in Section 11.2. Let us denote the high part of the input signal as $T_1$ and the low part of the signal as $T_2$. Similarly, let $w_1$ be the energy dissipated during the interval $T_1$ and $w_2$ the energy dissipated during the interval $T_2$. Then the average power dissipated in the circuit is given by

$$\overline{p} = \frac{w_1 + w_2}{T}.$$

## Energy Dissipated During Interval $T_1$

Let us first consider the case when the input to the MOSFET switch is high and the switch is closed, and derive the value of $w_1$. When the switch is closed, the situation corresponds to that shown in Figure 11.9 and the circuit shown in Figure 11.10 applies. Figure 11.11 shows the Thévenin equivalent for this circuit.

For the circuit shown in Figure 11.11, $v_C$ is given by the following expression (assuming that time $t$ starts from 0 at the moment that the input signal

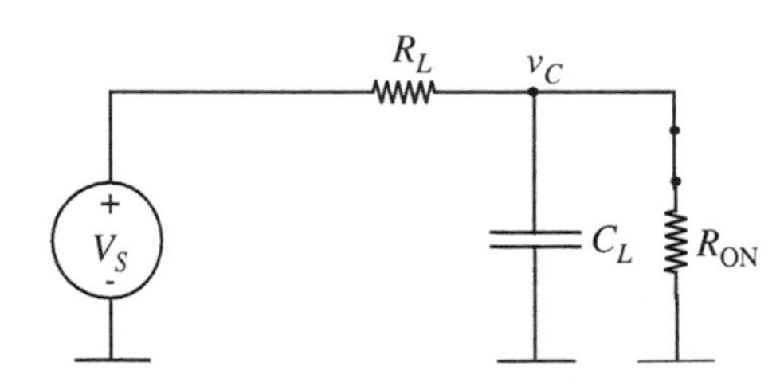


FIGURE 11.10 Equivalent circuit for the inverter with the MOSFET switch closed.

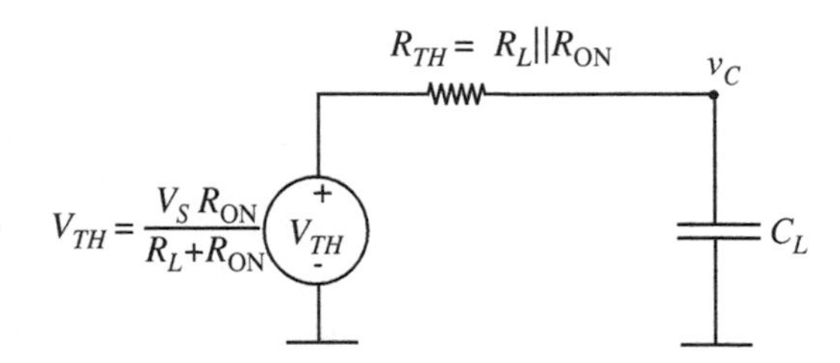


FIGURE 11.11 Thévenin equivalent circuit for the inverter with the switch closed.

transitions from low to high):

$$v_C = V_{TH} + (V_S - V_{TH})e^{\frac{-t}{R_{TH}C_L}}.$$

By substituting $t = 0$, we can verify that the capacitor is initially charged to $V_S$ when the MOSFET just turns on. Similarly, by substituting $t = \infty$, we can confirm that the final voltage on the capacitor is $V_{TH}$.

The rest of the derivation for $w_1$ follows the steps in Section 11.2. When $T_1 \gg R_{TH}C_L$, we obtain the following simplified expression for $w_1$:

$$w_1 = \frac{V_S^2}{R_L + R_{ON}}T_1 + \frac{V_S^2 R_L^2 C_L}{2(R_L + R_{ON})^2}. \quad (11.23)$$

### Energy Dissipated During Interval $T_2$

Now, let us consider the second interval $T_2$ in which the input signal is low and the switch is off. During $T_2$, the capacitor charges through the resistor $R_L$. The initial voltage on the capacitor is $V_{TH}$.

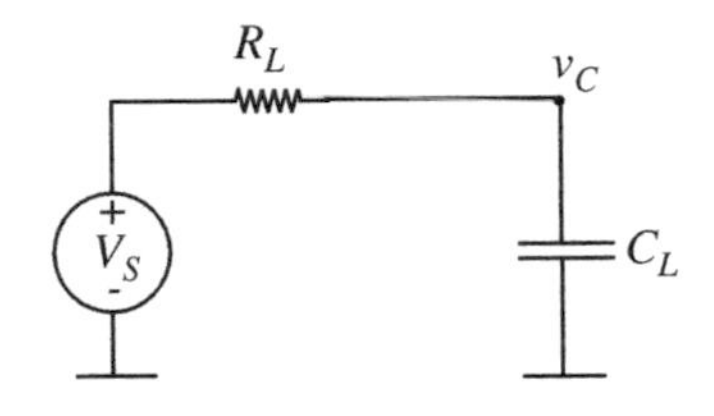


FIGURE 11.12 Equivalent circuit with the switch open for the inverter.

As in the previous section, let us first determine $v_C$. When the switch is off, the circuit shown in Figure 11.12 applies. Since the initial voltage on the capacitor is $V_{TH}$ and the final voltage is $V_S$, we can write the following expression for $v_C$:

$$v_C = V_{TH} + (V_S - V_{TH})\left(1 - e^{\frac{-t}{R_L C_L}}\right).$$

Notice that as $t \to \infty$, the capacitor voltage $v_C \to V_S$. Similarly, for $t = 0$, the capacitor voltage is $V_{TH}$.

Following the derivation in Section 11.2, we can derive the following expression for $w_2$ when $T_2 \gg R_L C_L$:

$$w_2 = \frac{V_S^2 R_L^2 C_L}{2(R_L + R_{ON})^2}.$$

### Total Energy Dissipated

Combining the expressions for $w_1$ and $w_2$, we obtain total energy dissipated by the inverter in a cycle:

$$w = w_1 + w_2 = \frac{V_S^2}{R_L + R_{ON}}T_1 + \frac{V_S^2 R_L^2 C_L}{2(R_L + R_{ON})^2} + \frac{V_S^2 R_L^2 C_L}{2(R_L + R_{ON})^2}.$$

In other words,

$$w = \frac{V_S^2}{R_L + R_{ON}} T_1 + \frac{V_S^2 R_L^2 C_L}{(R_L + R_{ON})^2}.$$

Dividing by $T$ we get the average power $\overline{p}$:

$$\overline{p} = \frac{V_S^2}{(R_L + R_{ON})} \frac{T_1}{T} + \frac{V_S^2 R_L^2 C_L}{(R_L + R_{ON})^2 T}. \quad (11.24)$$

For asymmetric square wave, $T_1 = T/2$, so Equation 11.24 simplifies to:

$$\overline{p} = \frac{V_S^2}{2(R_L + R_{ON})} + \frac{V_S^2 R_L^2 C_L}{(R_L + R_{ON})^2 T}. \quad (11.25)$$

As expected, Equation 11.25 shows that the average power is the sum of a static component[2] and a dynamic component as indicated by the following equations:

$$p_{\text{static}} = \frac{V_S^2}{2(R_L + R_{ON})} \quad (11.26)$$

$$p_{\text{dynamic}} = \frac{V_S^2 R_L^2 C_L}{(R_L + R_{ON})^2 T}. \quad (11.27)$$

Notice that the dynamic power dissipated is proportional to the frequency with which the input signal transitions. Not surprisingly, high-performance chips that clock at high frequencies dissipate a lot of power. Also notice that the power is related to the square of the supply voltage. As clock speeds of VLSI chips increase, power considerations are causing manufacturers to continually reduce the supply voltages. Whereas 5-volt power supplies were the norm in the 80s, voltages closer to 3 volts have been the norm in the 90s, and supplies closer to 1.5 volts have been commonplace after the year 2000.

It is instructive to compare the relative values of the static and dynamic power. To do so, we take the ratio of the static and dynamic power as follows:

$$\frac{p_{\text{static}}}{p_{\text{dynamic}}} = \frac{V_S^2}{2(R_L + R_{ON})} \times \frac{(R_L + R_{ON})^2 T}{V_S^2 R_L^2 C_L}.$$

2. The static power for the inverter computed here (Equation 11.26) is half that in Equation 11.21 because here we are assuming that the input is a symmetric square wave, while Equation 11.21 presented the worst case.

Simplifying and rearranging, we get

$$\frac{p_{\text{static}}}{p_{\text{dynamic}}} = \frac{R_L + R_{ON}}{R_L} \times \frac{T}{2R_L C_L}. \tag{11.28}$$

Since for normal operation of the digital gate, $T \gg R_L C_L$, we see that $p_{\text{static}} \gg p_{\text{dynamic}}$. Thus, it becomes imperative to minimize the static power.

---

EXAMPLE 11.2 DYNAMIC POWER DISSIPATION Let us compute the worst-case dynamic power dissipation for the logic gate in Figure 11.7 using Equation 11.27 under the following conditions:

- The load capacitor driven by the output of the gate has a value of $C_L = 0.01$ pF.
- The clock frequency $f = 1/T$ at which signals transition is 10 MHz. In other words, an input cannot change at a rate greater then 10 MHz.
- The power supply voltage $V_S$ is 5 V.
- $R_L = 100$ kΩ and $R_{ON} = 10$ kΩ for each of the MOSFETs.

Equation 11.27 applies when $T/2 \gg R_{TH}C_L$ and $T/2 \gg R_L C_L$. First, let us confirm that these relationships hold. Since $R_L$ is greater than $R_{TH}$, it is sufficient to verify that $T/2 \gg R_L C_L$. For the parameters supplied, $T/2 = 1/(20 \times 10^6) = 50$ ns and $R_L C_L = 100 \times 10^3 \times 0.01 \times 10^{-12} = 1$ ns. Clearly, the circuit time constants are much smaller than the signal intervals.

To obtain the worst-case dynamic power dissipation, we assume that the load capacitor is charged and discharged every cycle. Under these conditions, Equation 11.27 gives the formula for the worst-case dynamic power (with $R_{ON}$ replaced by $R_{ONpd}$, where $R_{ONpd}$ is the resistance of the pulldown network). We rewrite this equation here by replacing $R_{ON}$ in the equation with the resistance of the pulldown network $R_{ONpd}$:

$$p_{\text{dynamic}} = \frac{V_S^2 R_L^2 C_L}{(R_L + R_{ONpd})^2 T}.$$

Since it is clear from the formula that the dynamic power is maximized when $R_{ONpd}$ is minimized, we assume that all the pulldown MOSFETs are turned on when the clock signal goes high. Accordingly, if the on resistance for a MOSFET is $R_{ON}$, we must use the following value for $R_{ONpd}$:

$$R_{ONpd} = (2R_{ON} \| R_{ON} \| R_{ON}).$$

We also assume that all the input signals switch at the clock frequency. Thus the worst-case power dissipation is given by

$$p_{\text{dynamic}} = \frac{5^2 \times (100 \times 10^3)^2 \times 0.01 \times 10^{-12} \times 10 \times 10^6}{(100 \times 10^3 + 4 \times 10^3)^2}$$

$$= 2.3\ \mu\text{W}.$$

Observe that for our example, the static power dissipation (from Equation 11.22) is nearly 100 times larger than the dynamic power dissipation.

---

EXAMPLE 11.3 TOTAL POWER DISSIPATION IN A MOS INVERTER The inverter shown in Figure 11.13 drives a load capacitor $C_L$ that models the gate-to-source and interconnect capacitances of the immediate downstream circuitry. We wish to approximate the average power dissipated in the inverter given that the input voltage $v_{IN}$ is a 100-MHz square wave. In doing so, we assume that the MOSFET on-state resistance $R_{ON}$ satisfies $R_{ON} \ll R_{PU}$.

Since $R_{ON} \ll R_{PU}$, the output voltage rises much more slowly than it falls. The time constant for the rising transient is $R_{PU}C_L = 1$ ns, which is much smaller than a half period of the input square wave. Therefore, all inverter transients fully settle, and the power dissipated in the inverter may be approximated as the sum of a static dissipation plus a dynamic dissipation.

The static dissipation occurs when the MOSFET is on. The instantaneous value of this dissipation is approximately $V_S^2/R_{PU} = 22.5\ \mu$W because $R_{ON} \ll R_{PU}$. However, since the MOSFET is on only half the time, the average static power dissipation is 11.25 $\mu$W.

The dynamic dissipation is caused by the repeated charging and discharging of $C_L$. The energy $C_L V_S^2/2$ is lost during both the charging and discharging of the capacitor. Thus, the dynamic loss is $C_L V_S^2 f = 1.125\ \mu$W, where $f = 100$ MHz is the switching frequency of the input voltage.

Finally, the total average power dissipated is 12.375 $\mu$W, with the static dissipation being the dominant component of this loss.

---

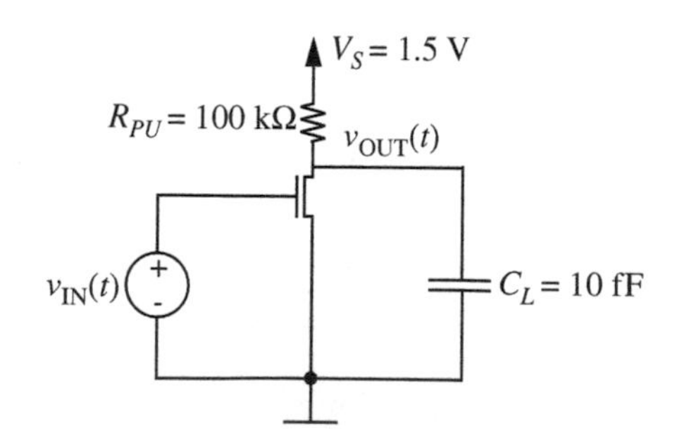


FIGURE 11.13 An inverter driving a load capacitor.

## 11.4 NMOS LOGIC

The pullup device that we have been using thus far (and seen previously in Section 6.11) in our digital gates is a resistor (for example, see the inverter in Figure 11.6). In practice, we do not really use resistors as we know them — they would take up too much area. Rather, as displayed in Figure 11.14), we might use another MOSFET with its gate connected to a second supply voltage $V_A$, where $V_A$ is at least one threshold voltage higher that the supply voltage $V_S$. This way, the pullup MOSFET remains in its ON state for any voltage between 0 and $V_S$ applied at its source.

This style of building logic gates using n-channel MOSFETs for both the pullups and the pulldowns is called *NMOS logic.*[3] The $R_{ON}$ of the pullup serves as the load resistor. The gate length of the pullup MOSFET is sized to be larger relative to the pulldown MOSFET, so the static discipline is satisfied (see Section 6.11). The power and energy calculations of Section 11.3 apply to NMOS devices with $R_{ON}$ of the pullup replacing the load resistor $R_L$ in the analyses.

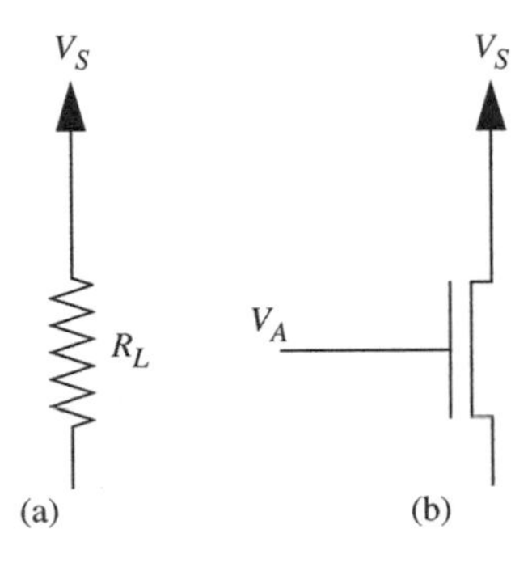


FIGURE 11.14 A MOS pullup device.

## 11.5 CMOS LOGIC

Logic gates in the NMOS logic family dissipate static power even when the circuit is idle. In the example discussed earlier, the static power was nearly 100 times larger than the power associated with signal activity. Because the pullup MOSFET is always on, the cause of the static power dissipation is the resistive path from the power supply to ground when the pulldown MOSFET is on. In this section, we introduce another type of logic called *CMOS* or *Complementary MOS*, which has no static power dissipation. Because of its low static power dissipation,[4] CMOS logic has all but replaced NMOS in modern VLSI chips.

CMOS logic makes use of a complementary MOSFET called the *p-channel MOSFET* or the *PFET*.[5] The n-channel MOSFET that we have dealt with thus far is called an *NFET*. The symbols and the SR circuit models for the NFET and PFET are shown in Figures 11.15 and 11.16, respectively. As we saw previously in Section 6.6, the NFET turns on when its $v_{GS} \geq V_{Tn}$.

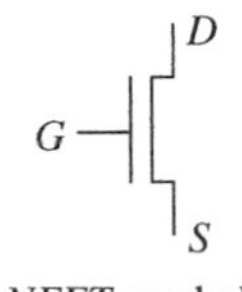


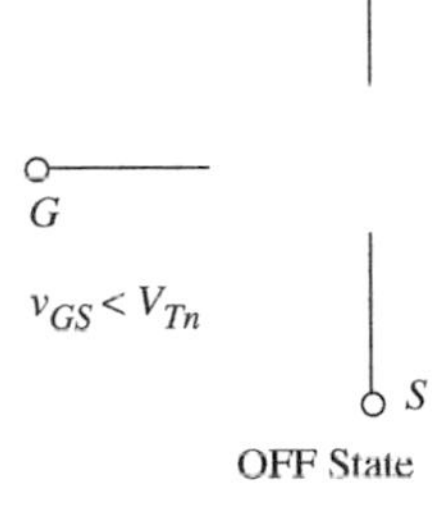


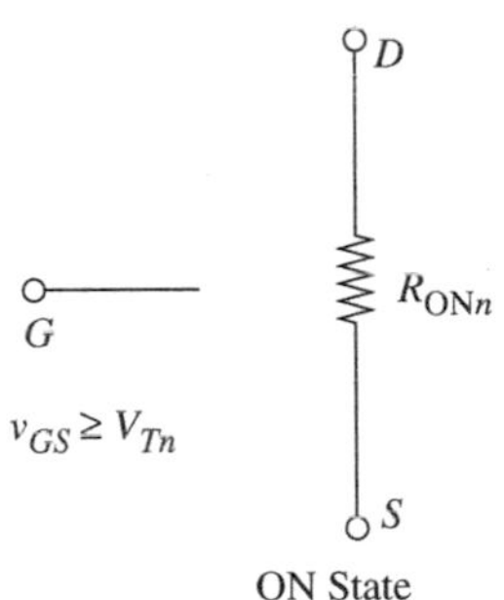


FIGURE 11.15 The switch-resistor model of the n-channel MOSFET or the NFET.

3. NMOS logic families actually use a special kind of pullup called a *depletion-mode MOSFET*, which has a negative threshold voltage. The MOSFETs that we have been dealing with thus far are called *enhancement-mode MOSFETs*. The depletion-mode MOSFET is used as a pullup with its gate connected to its source instead of the power supply. One of its advantages is that it does not require a second supply voltage.

4. Although the static power for CMOS is significantly lower than NMOS, sources of static power loss still remain, such as leakage currents.

5. We saw the p-channel MOSFET briefly in the context of amplifiers in Chapter 7 in Example 7.7. Here we will focus on its use in digital circuits.

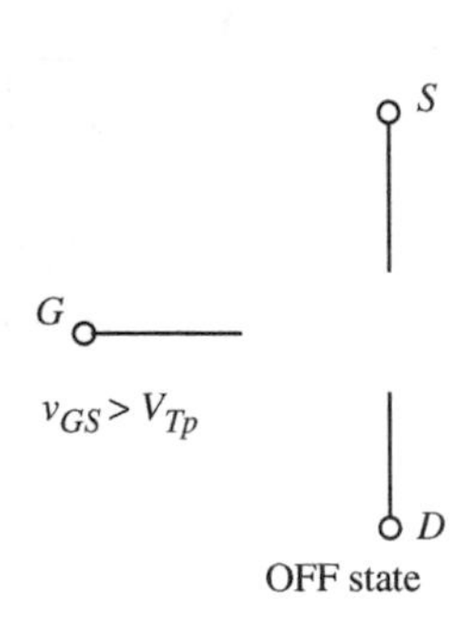


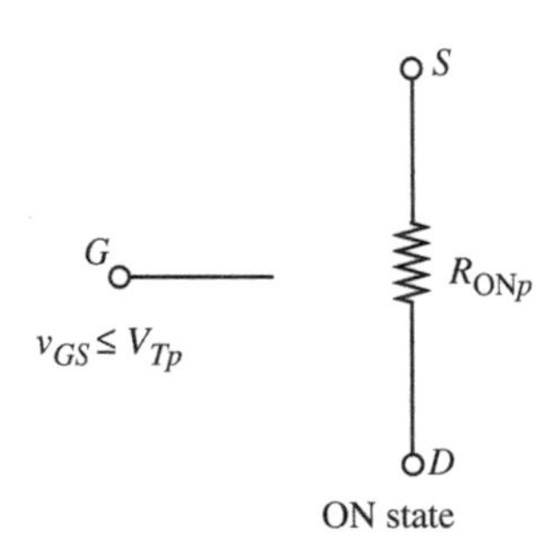


FIGURE 11.16 The switch-resistor model of the p-channel MOSFET or the PFET.

When it is on, a resistance $R_{ONn}$ appears between its drain and source. In contrast, the PFET turns on when its $v_{GS} \leq V_{Tp}$. $V_{Tp}$ is usually negative (for example, −1 volt). When it is on, a resistance $R_{ONp}$ appears between its drain and source. The drain terminal of the PFET is chosen as the terminal with the lower voltage. In contrast, the channel terminal of the NFET with the higher voltage is labeled as the drain.

As an example, an NFET with a threshold voltage $V_{Tn} = 1$ V turns on when the voltage between its gate and source is raised above 1 V. A PFET with a threshold voltage $V_{Tp} = -1$ V turns on when the voltage between its gate and source is lowered below −1 V. In other words, the PFET turns on when the voltage between its source and gate is raised above 1 V.

A CMOS inverter is shown in Figure 11.17. When the input voltage is high ($v_{IN} = V_S$), the NFET is ON and the PFET is off, resulting in a low output voltage. When the input voltage is low ($v_{IN} = 0$), the NFET turns off and the PFET turns on, resulting in a high output voltage.

CMOS logic does not suffer static power dissipation. Referring to Figure 11.17, provided the input $v_{IN}$ is at $V_S$ or 0, notice that the two complementary MOSFETs are never on at the same time. Thus, there is never any direct resistive path from the power supply to ground. Hence there is no static power dissipation.

---

EXAMPLE 11.4 POWER DISSIPATION IN A CMOS INVERTER Let us compute the dynamic power dissipated in the CMOS inverter. The CMOS inverter does not suffer from static power loss, since at any instant of time, either the pullup or the pulldown device is off, thereby precluding any continuous current flow.

The circuit model of the inverter using the SR MOSFET models for the PFET and the NFET is shown in Figure 11.18. As the model shows, no current flows directly from $V_S$ to ground for either a high or a low input signal.

Let us assume that a square-wave signal such as a clock with time period $T$ is fed to the input of the inverter as shown in Figure 11.19. Further assume that the inverter drives a load capacitor $C_L$. We shall compute the power dissipated by the inverter for this signal. A signal that transitions every cycle will result in the worst-case dynamic power dissipation.

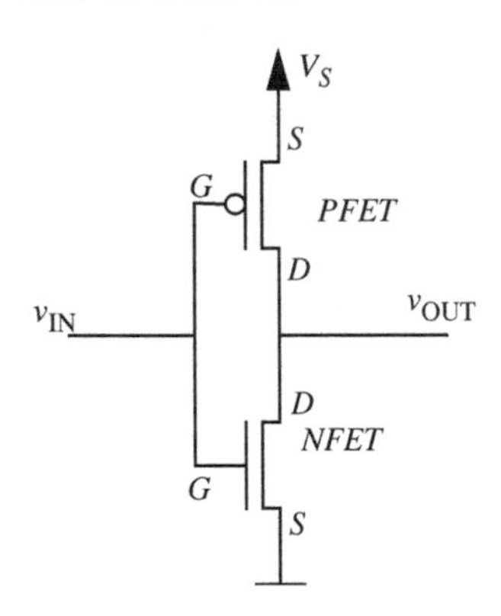


FIGURE 11.17 CMOS inverter circuit.

During the first half of the cycle, the load capacitor $C_L$ is discharged, and during the second half of the cycle, it is charged up to the supply voltage. Since static and dynamic currents don't flow simultaneously in CMOS devices, we can compute the average dynamic power dissipated using the following very simple method.

Recall that the average power dissipated is defined as the energy dissipated in a cycle divided by the cycle time $T$. During the low half cycle of the input signal, a quantity of charge equal to $Q_L$ is transferred from the power source to the capacitor. Assuming that the cycle time of the input signal is large enough for the capacitor to charge up to

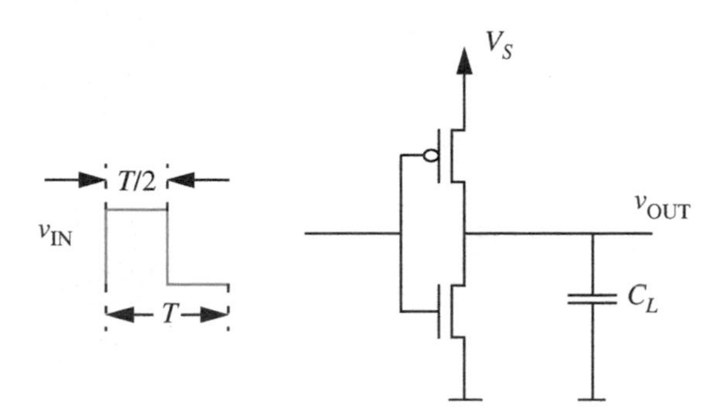


FIGURE 11.19 CMOS inverter power dissipation.

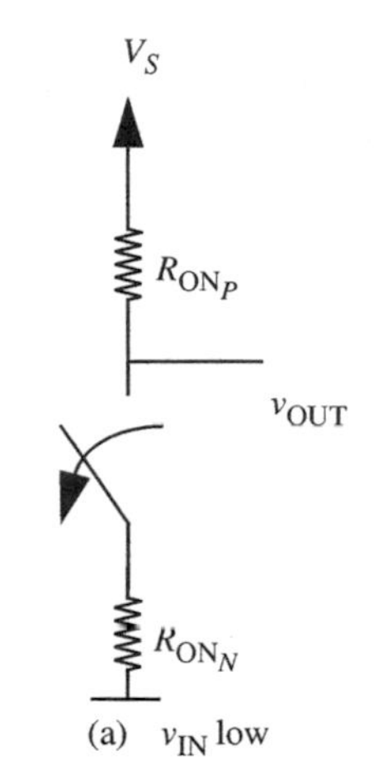


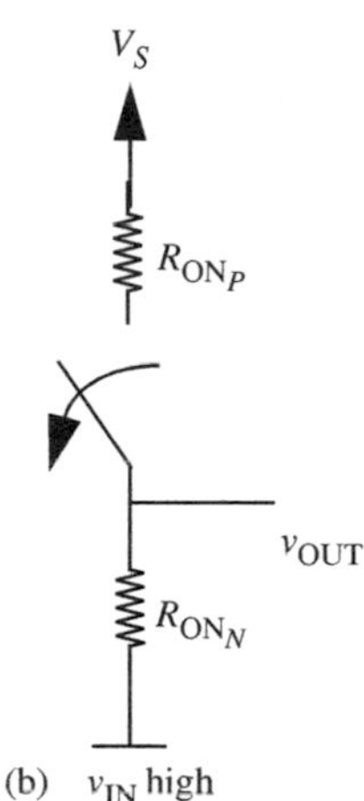


FIGURE 11.18 CMOS inverter model.

the supply voltage,

$$Q_L = C_L V_S.$$

Then, during the second half cycle, the same charge is transferred from the capacitor to ground. Effectively, a charge $Q_L$ is transferred from the power source to ground during each cycle. The amount of energy lost during this transfer is given by $V_S Q_L$. We divide this quantity by $T$ to obtain the average power. Thus,

$$p_{\text{dynamic}} = V_S Q_L / T \quad (11.29)$$

$$= V_S C_L V_S / T \quad (11.30)$$

$$= f C_L V_S^2 \quad (11.31)$$

where $f$ is the frequency of the square wave signal.

We can also derive the same answer from first principles as follows. The time-average dynamic power dissipated is the product of the power supply voltage and the average current supplied by the power supply. In other words,

$$p_{\text{dynamic}} = V_S \frac{1}{T} \int_0^T i(t)dt$$

where $V_S$ is the DC power supply voltage and $i(t)$ is the current supplied by the power source as a function of time.

Figure 11.20 shows the situation where the input signal is low (second half cycle) and the load capacitor is charging up through the on resistance of the PFET. Figure 11.21 shows the situation where the input signal is high (first half cycle) and the load capacitor discharges to ground through the on resistance of the NFET.

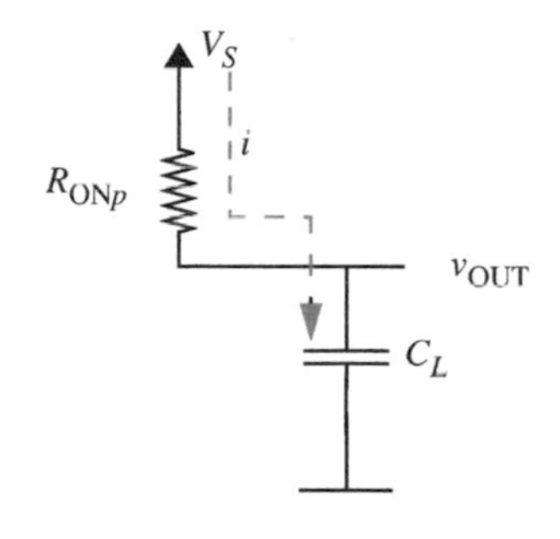


FIGURE 11.20 Load capacitor charging.

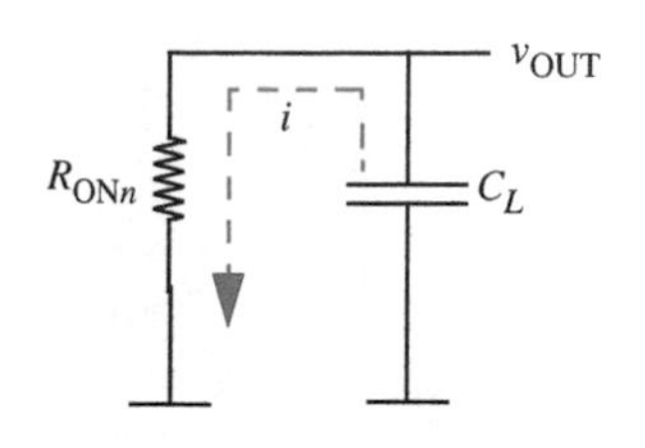


**FIGURE 11.21** Load capacitor discharging.

Because the pullup switch is off in the first half cycle (when the input is high), notice that the power supply directly provides a current only during the second half cycle (when the input is low). Therefore, power is delivered by the power supply only during the interval $T/2 \rightarrow T$. Thus,

$$p_{\text{dynamic}} = V_S \frac{1}{T} \int_{T/2}^{T} i(t)dt. \tag{11.32}$$

If $Q(t)$ is the charge on the capacitor as a function of time, we have

$$i(t) = \frac{dQ(t)}{dt}.$$

Therefore

$$p_{\text{dynamic}} = V_S \frac{1}{T} \int_{T/2}^{T} \frac{dQ(t)}{dt} dt. \tag{11.33}$$

Furthermore, we know that

$$Q(t) = C_L v_{\text{OUT}}(t)$$

where $v_{\text{OUT}}(t)$ is the voltage across the load capacitor as a function of time. Differentiating both sides with respect to $t$, we get

$$\frac{dQ(t)}{dt} = C_L \frac{dv_{\text{OUT}}(t)}{dt}.$$

Substituting for $dQ(t)/dt$ in Equation 11.33 and observing that $v_{\text{OUT}}$ rises from 0 to $V_S$ during the second half cycle, we obtain:

$$p_{\text{dynamic}} = V_S C_L \frac{1}{T} \int_{0}^{V_S} dv_{\text{OUT}} \tag{11.34}$$

$$= V_S C_L \frac{1}{T} V_S \tag{11.35}$$

$$= \frac{V_S^2 C_L}{T}. \tag{11.36}$$

The dynamic power dissipated by the CMOS inverter is therefore $V_S^2 C_L / T$. If $T = 1/f$,

the dynamic power dissipated by the CMOS inverter is $fV_S^2 C_L$.

In reality, it is difficult to obtain instantaneous input rise and fall times. When the rise and fall times are finite, there will a short period of time in the middle of an input transition

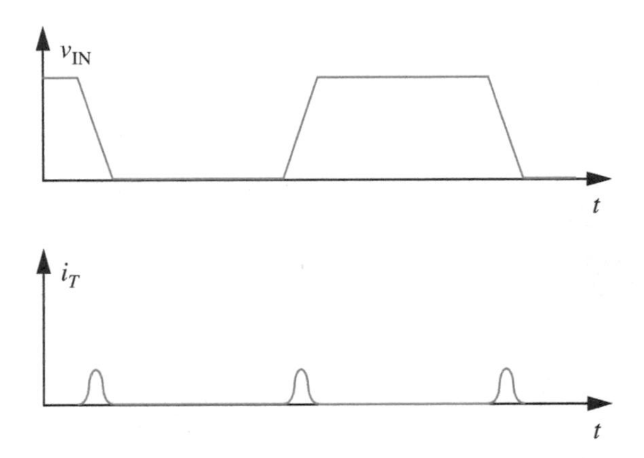


**FIGURE 11.22** Static power loss due to nonzero signal rise and fall times.

interval (for example, when $v_{IN} = 2.5$ V) in which both MOSFETs will be turned on, resulting in a current path from the power supply to ground. This transient switching current, $i_T$, depicted in Figure 11.22 is another source of dynamic power loss, but we ignore it in our analyses.

---

EXAMPLE 11.5 POWER DISSIPATION IN ANOTHER CMOS INVERTER The circuit shown in Figure 11.23 is the same as that shown in Figure 11.13 except that the inverter in Figure 11.23 is a CMOS inverter. As in Example 11.3, we wish to approximate the average power dissipated in the inverter given that $v_{IN}$ is a 100-MHz square wave. Again, we will assume that the on-state resistance of both MOSFETs is so small that the switching transients of the inverter fully settle.

Because the two MOSFETs in Figure 11.23 are never on at the same time, the CMOS inverter does not exhibit static dissipation. This a major advantage of CMOS logic over NMOS logic. The only dissipation is dynamic dissipation, which is the same as for the inverter in Example 11.3 because $V_S$, $C_L$, and $f$ are all the same. Thus the average power dissipated in the CMOS inverter is 1.125 $\mu$W.

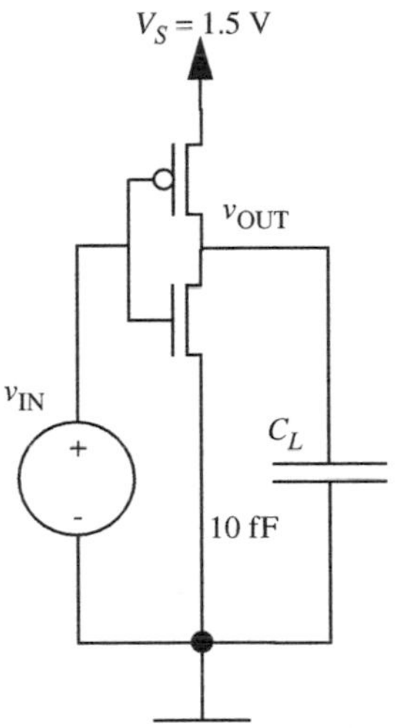

$V_S = 1.5$ V, $v_{OUT}$, $v_{IN}$, $C_L$, 10 fF

**FIGURE 11.23** CMOS inverter driving a load capacitor.

---

EXAMPLE 11.6 POWER CONSUMED BY A MICROPROCESSOR In this example, we will estimate the average power consumed by a microprocessor based on the simple formula developed in Equation 11.31. The Raw microprocessor designed at MIT using IBM's 180-nm, SA27E CMOS technology process, had about 3 million gates (assume each gate is equivalent to a 2-input NAND gate) and clocked at 425 MHz. Assume each gate offered a load capacitance of approximately 30-nF, and the nominal supply voltage is 1.5 V. Assume further that approximately 25% of the gates switch values in a given cycle.

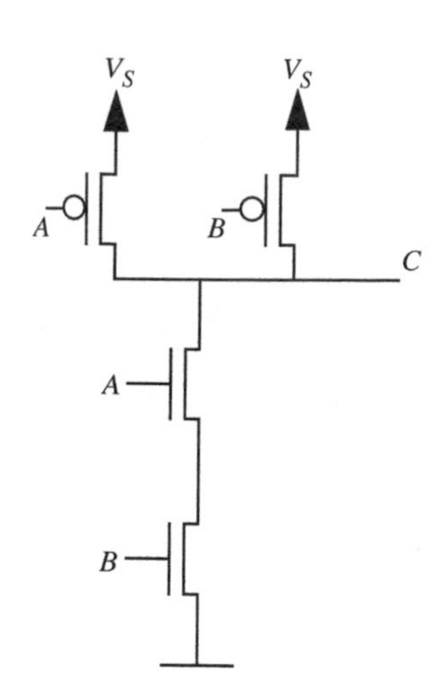


FIGURE 11.24 CMOS NAND gate.

| A | B | C |
|---|---|---|
| 0 | 0 | 1 |
| 0 | 1 | 1 |
| 1 | 0 | 1 |
| 1 | 1 | 0 |

TABLE 11.1 Truth table.

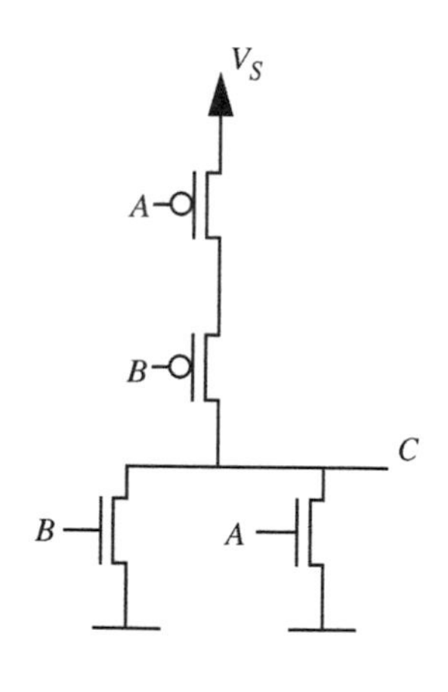


FIGURE 11.25 CMOS NOR gate.

Substituting the values from the Raw microprocessor into Equation 11.31, we obtain the dynamic power consumed by the entire chip as

$$\begin{aligned}&(\text{Fraction Switching}) \times (\#\text{Gates}) \times fC_L V_S^2\\ &\quad = 0.25 \times (3 \times 10^6) \times (425 \times 10^6) \times (30 \times 10^{-15}) \times 1.5^2\\ &\quad = 21.5\ \text{W}.\end{aligned}$$

---

### 11.5.1 CMOS LOGIC GATE DESIGN

How do we build logic gates such as NANDs and NORs using CMOS technology? Let us look at a few examples and then generalize to arbitrary logic functions. We have already seen one example — the inverter. The CMOS inverter comprised a pulldown NFET and a pullup PFET.

#### CMOS NAND Gate

As you can verify, the circuit in Figure 11.24 implements a logic function with the truth table shown in Table 11.1. This is the truth table for a *NAND* gate. Notice that pulldown circuit comprising two series-connected NFETs is the same as in the NMOS logic implementation. The pullup circuit performs a complementary function and comprises parallel-connected PFETs. In other words, the pullup is off when the pulldown circuit is on, and vice versa. There is no static power dissipation in this gate.

#### CMOS NOR Gate

Similarly, we can verify that the circuit in Figure 11.25 implements the truth table for a *NOR* gate as shown in Table 11.2.

The two pulldown NFETs are connected in parallel as in the corresponding NMOS implementation. The PFET pullups are series connected to form the complementary network.

#### Other Logic Functions

As evident from the previous examples, CMOS logic gates can be visualized as comprising two complementary modules: the familiar pulldown circuit comprising NFETs, and a complementary pullup module using PFETs. If we are interested in implementing the logic function $f$, the NFET pulldown network is designed so it offers a short circuit when $f$ is FALSE and an open circuit when $f$ is TRUE. Similarly, the PFET pullup network is designed so it offers a short circuit when $f$ is TRUE and an open circuit when $f$ is FALSE. Thus, the CMOS implementation of logic function $f$ will assume the form shown in Figure 11.26. In the figure, $\bar{f}$ is the complement of $f$. In other words,

$$\bar{f} = \text{NOT} f.$$

Let us construct a CMOS circuit for the function

$$f(A, B, C) = (\overline{A} + \overline{B})C.$$

Assume that the inputs are available in both their TRUE and complement forms. In the CMOS circuit, the pulldown network must be on when $f$ is FALSE. Similarly, we must construct a complementary pullup circuit $\overline{f}$ that is off when $f$ is on. Let us derive an expression for $\overline{f}$:

$$\overline{f(A, B, C)} = \overline{(\overline{A} + \overline{B})C} \tag{11.37}$$

$$= \overline{\overline{(AB)}\;\overline{\overline{\overline{C}}}} \tag{11.38}$$

$$= \overline{\overline{AB + \overline{C}}} \tag{11.39}$$

$$= AB + \overline{C}. \tag{11.40}$$

| A | B | C |
|---|---|---|
| 0 | 0 | 1 |
| 0 | 1 | 0 |
| 1 | 0 | 0 |
| 1 | 1 | 0 |

TABLE 11.2 Truth table.

An application of the ideas in Figure 11.26 leads to the circuit in Figure 11.27. We can verify that the circuit indeed correctly implements the logic by developing its truth table and comparing it to that of $f$.

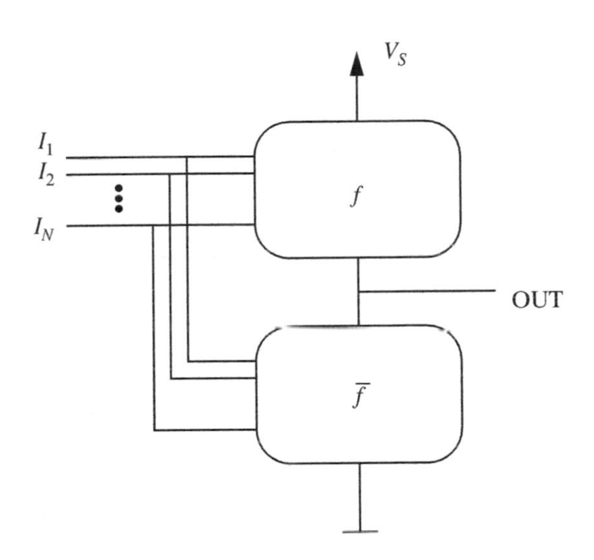


FIGURE 11.26 CMOS configuration to implement the logic function $f$.

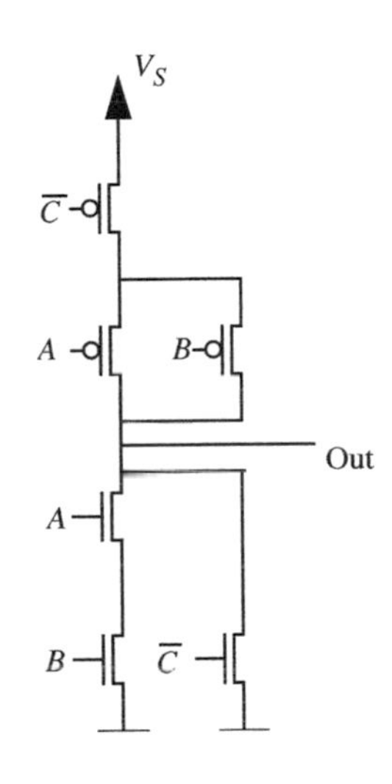


FIGURE 11.27 CMOS implementation of $f(A, B, C) = (\overline{A} + \overline{B})C$.

## 11.6 SUMMARY

- We began this chapter by analyzing the power and energy dissipated in first-order resistor-capacitor networks driven by voltage-step inputs. The results of this analysis were then used to determine the losses in NMOS logic gates. Most importantly, we observed that, if the transients in NMOS logic gates are allowed to settle, then the average power dissipated in the gates could be decomposed into two parts: static losses and dynamic losses. Static loss results when a pull-up resistor is connected across the power supply by one or more closed MOSFETs. Dynamic loss results from the repeated charging and discharging of the MOSFET gate-to-source capacitances, and hence increases linearly with switching speed. We further observed that in NMOS logic gates in which the switching transients settle, the static losses are always much larger than the dynamic losses. Finally, this observation motivated the development of CMOS logic gates, which do not exhibit static losses. CMOS logic is therefore much more energy efficient that NMOS logic.
- The dynamic power loss in a CMOS gate driving other CMOS gates was found to be

$$CV^2f,$$

where $C$ is the total driven downstream capacitance, $V$ is the power supply voltage, and $f$ is the switching frequency. As digital circuits become ever faster, this loss increases linearly with $f$. Therefore, in order to reduce dynamic losses, and hence reduce the associated thermal management problems, CMOS logic circuits are built from MOSFETs having ever decreasing gate-to-source capacitances, and are built to operate from ever decreasing power supply voltages.[6] Also, circuits that are not in use are now commonly shut down to avoid dynamic losses.

## EXERCISES

EXERCISE 11.1 An inverter built using an NMOS transistor and a resistor $R_L$ drives a capacitance $C_L$. The power supply voltage is $V_S$ and the on resistance of the MOSFET is $R_{ON}$. The threshold voltage for the MOSFET is $V_T$. Assume that logical 0's are represented using 0 V and logical 1's using $V_S$ volts.

a) Determine the steady-state power consumed by the inverter when a 0 is applied to its input.

b) Determine the steady-state power consumed by the inverter when a 1 is applied to its input.

6. As of 2004, for example, commonly available technology processes use power supply voltages of 1 V to 1.5 V.

c) Determine the static power and the dynamic power consumed by the inverter when a sequence of the form 01010101 ... is applied to its input. Assume that signal transitions (0 to 1, or 1 to 0) happen every $T$ seconds. Assume further than $T$ is much greater than the circuit time constant.

d) Assuming the input in part(c), by what factor does the dynamic power decrease if (i) $T$ is increased by a factor of 2, (ii) $V_S$ is decreased by a factor 2, and (iii) $C_L$ is decreased by a factor 2?

e) Suppose that the inverter must satisfy a static discipline with high and low voltage thresholds $V_{IH} = V_{OH} = V_H = V_{OL} = V_L$, respectively. You are given a MOSFET with on resistance $R_{ON}$ and threshold $V_T$. Assume that $V_L < V_T < V_H < V_S$ choose a value for $R_L$ in terms of the other circuit parameters such that the power consumed by the inverter is minimized.

EXERCISE 11.2 Determine $\bar{f}$ for the following functions. Express your answer in a simplified sum of products form (Hint: use De Morgan's laws.)

a) $f = \overline{A - B}$

b) $f = \overline{A + B}$

c) $f = A + B$

EXERCISE 11.3 Give a CMOS implementation (using NMOS and PMOS transistors only) of the following logic functions. In doing these exercises, is the value of the on resistance of the MOSFETs needed? Why or why not?

a) $f = \overline{A - B}$

b) $f = \overline{A + B}$

c) $f = A + B$

EXERCISE 11.4 Write a truth table and a boolean expression that describes the operation of each of the digital circuits in Figure 11.28.

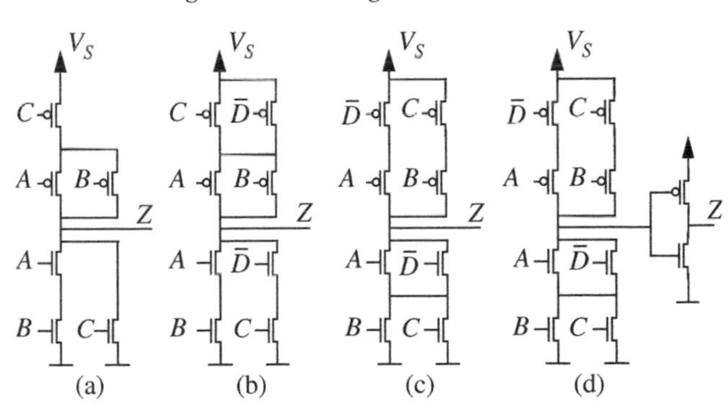


FIGURE 11.28

# PROBLEMS

PROBLEM 11.1 This problem examines the power dissipated by a small digital logic circuit. The circuit comprises a series-connected inverter and NOR gate as shown in Figure 11.29. The circuit has two inputs, A and B, and one output, Z. The inputs are assumed to be periodic with period $T_4$ as shown in Figure 11.29. Assume that $R_{ON}$ for each MOSFET is zero.

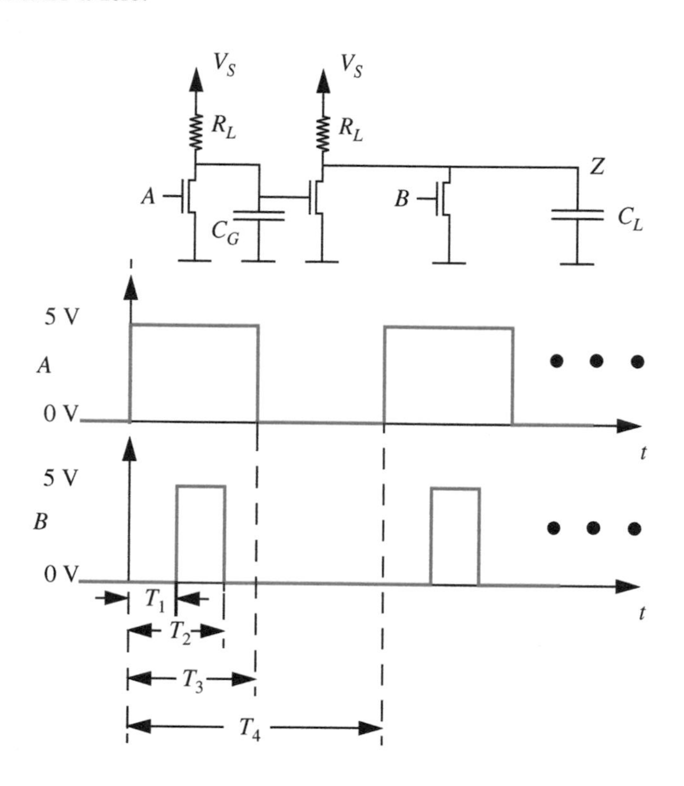


FIGURE 11.29

a) Sketch and clearly label the waveform for the output Z for $0 \leq t \leq T_4$. In doing so, assume that $C_G$ and $C_L$ are both zero.

b) Derive the time-average static power consumed by the circuit in terms of $V_S$, $R_L$, $T_1$, $T_2$, $T_3$, and $T_4$. Here, time-average power is defined as the total energy dissipated by the gate during the period $0 \leq t \leq T_4$ divided by $T_4$.

c) Now assume that $C_G$ and $C_L$ are nonzero. Derive the time-average dynamic power consumed by the circuit in terms of $V_S$, $R_L$, $C_G$, $C_L$, $T_1$, $T_2$, $T_3$, and $T_4$. In doing so, assume that the circuit-time constants are all much smaller than $T_1$, $T_2 - T_1$, $T_3 - T_2$, and $T_4 - T_3$.

d) Evaluate the time-average static and dynamic powers for $V_S = 5$ V, $R_L = 10$ kΩ, $C_G = 100$ fF, $C_L = 1$ pF, $T_1 = 100$ ns, $T_2 = 200$ ns, $T_3 = 300$ ns, and $T_4 = 600$ ns.

e) What is the amount of energy consumed by the circuit in 1 minute for the parameters in part (d)?

f) By what percentage does the total time-average power consumption drop if the power supply voltage $V_S$ drops by 30%?

PROBLEM 11.2 Implement the logic function $Z = \overline{A+B+CD}$ using NMOS transistors alone. In other words, use an NMOS transistor in place of the pull-up resistor. Your implementation must satisfy a static discipline with low and high voltage thresholds given by $V_{IL} = V_{OL} = V_L$ and $V_{IH} = V_{OH} = V_H$, where $0 < V_L < V_T < V_H < V_S$. $V_S$ is the power supply voltage. As your answer, specify the $W/L$ values for the pullup and the pulldown transistors.

For what combination of inputs does the circuit dissipate the greatest amount of static power? Determine the static power dissipation for this combination of inputs.

PROBLEM 11.3 A circuit consists of $N$ inverters, where $N \gg 1$. Each inverter is built using a NMOS transistor and a resistor $R_L$. The power supply voltage is $V_S$ and the on resistance of the MOSFETs is $R_{ON}$. The threshold voltage for the MOSFETs is $V_T$.

a) Suppose we do not know how the inverters are connected to each other or to the inputs and outputs of the circuit. How might you estimate the amount of static power that the circuit is likely to consume?

b) Suppose it is known that the inverters are connected in series as one long chain. Estimate the amount of static power dissipated by the circuit.

PROBLEM 11.4 Consider the digital memory element illustrated in Figure 11.30. Assume that the inverters are implemented using a pulldown NMOS transistor with on resistance $R_{ON}$, and a pullup resistor $R_L$. The power supply voltage is $V_S$. What is the instantaneous power dissipated by the memory element when it stores a logical 1? What is the instantaneous power dissipated by the memory element when it stores a logical 0?

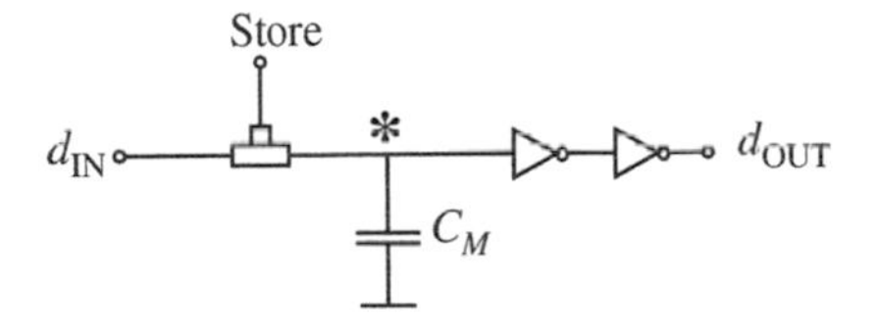


FIGURE 11.30

PROBLEM 11.5 Give a CMOS implementation (using NMOS and PMOS transistors only) of the following logic functions:

1. $(A+B) \cdot (C+D)$

2. $\overline{(A+B) \cdot (C+D)}$

3. $\overline{A} \cdot \overline{B} \cdot C \cdot D$

4. $(\overline{Y \cdot W})(\overline{X \cdot W})(\overline{\overline{X} \cdot Y \cdot \overline{W}})$

PROBLEM 11.6

a) Express $\overline{F}$ in a simplified sum-of-products form given that $F = A\overline{B} + C\overline{D}$.

b) Implement the logic function $F = A\overline{B} + C\overline{D}$ with an NMOS digital logic circuit that obeys the static discipline defined by the low-level and high-level logic thresholds $V_{IL} = V_{OL} = V_L$ and $V_{IH} = V_{OH} = V_H$, respectively. Assume the supply voltage is $V_S$, and that the on-state resistance of the NMOS transistors is $R_{ON}$. Determine the lowest value of the pull-up resistor $R_{PU}$ for which the circuit will obey the static discipline in terms of $R_{ON}$, $V_S$, $V_L$, and $V_H$; not all variables need appear in your answer.

c) Implement the logic function $F = A\overline{B} + C\overline{D}$ with a CMOS digital logic circuit. (Hint: make use of the result from part (a).)

d) Suppose that the NMOS and CMOS circuits above drive a capacitance $C_L$. Assume that the on-state resistance of both the PMOS and NMOS transistors is $R_{ON}$. For both the NMOS and CMOS circuits determine the worst-case output rise time. For the purpose of this problem, assume that the worst-case output rise time is the time the output takes to go from 0 V to $V_H$. Sketch the form of the output for both the NMOS and the CMOS circuit.

e) Suppose that the inputs are arranged such that $B = 1$, $C = 0$, and $D = 1$, and that a 0-V to 5-V square-wave signal is applied to the input $A$. Assume the square wave cycle time is $T$, and that $T$ is large enough so that the output comes close to its steady state value for both falling and rising transitions. Under these conditions, compute the power consumed by the CMOS and NMOS circuits when driving the capacitance $C_L$ load.

# CHAPTER 12

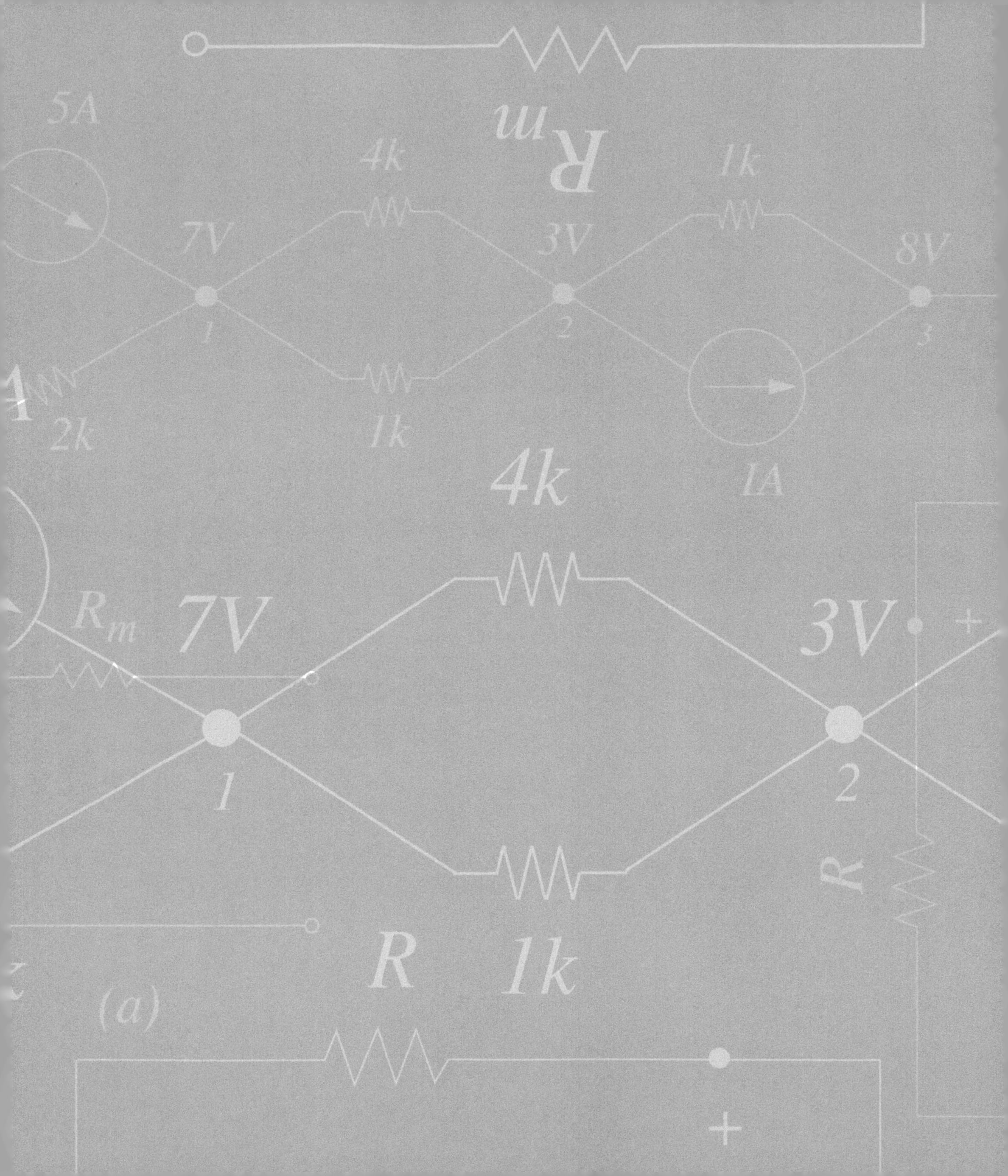

5A
4k
1k
7V
3V
8V
1
2
3
2k
1k
1A
4k
$R_m$
7V
3V
1
2
+
R
R
1k
(a)
+

# TRANSIENTS IN SECOND-ORDER CIRCUITS

# 12

Many familiar physical systems that exhibit oscillatory behavior, such as clock pendulums, automobile suspensions, tuned filters in radios, and inter-chip digital interconnections are predominantly second-order systems. That is, their dynamics are well described by second-order differential equations. Second-order systems contain two energy storage elements with independent states. For example, a second-order circuit could contain two independent capacitors, two independent inductors, or one capacitor and one inductor. In contrast, the circuits studied in Chapter 10 contained only one energy storage element that is, one capacitor or one inductor. Therefore, their dynamics were described by first-order differential equations. As we shall see in this chapter, the dynamics of first-order circuits and second-order circuits can be very different.

To illustrate the typical behavior of second-order circuits, and to motivate their study, consider the behavior of the two cascaded inverters shown in Figure 12.1. Given a square-wave input to the first inverter at $v_{IN1}$, we expect to see a square-wave output at $v_{OUT1}$ and $v_{IN2}$, as illustrated in Figure 12.2. However, in Chapter 9 we saw that the presence of the parasitic MOSFET gate capacitance in the second inverter results in the slower output waveform

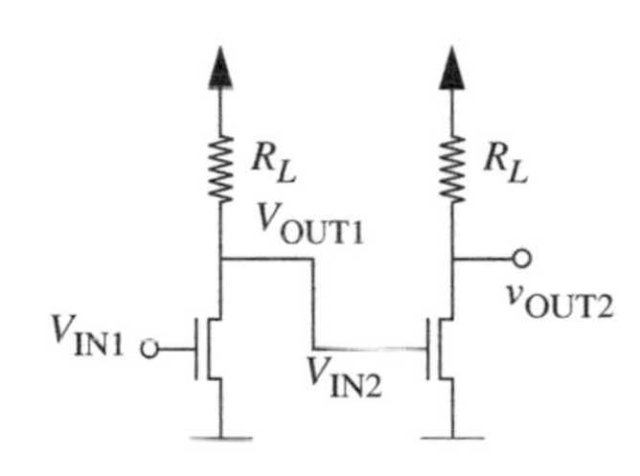


FIGURE 12.1 Two cascaded inverters.

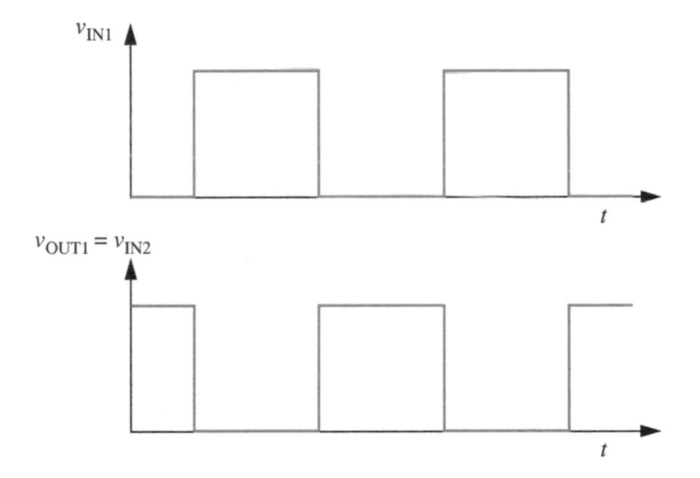


FIGURE 12.2 Ideal response of the first inverter to a square-wave input.

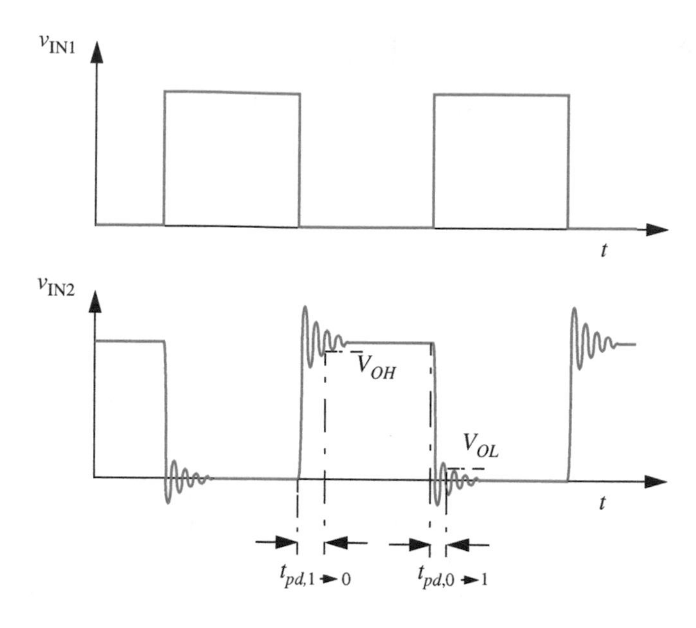


**FIGURE 12.3** Second-order response of the cascaded inverters to a square-wave input.

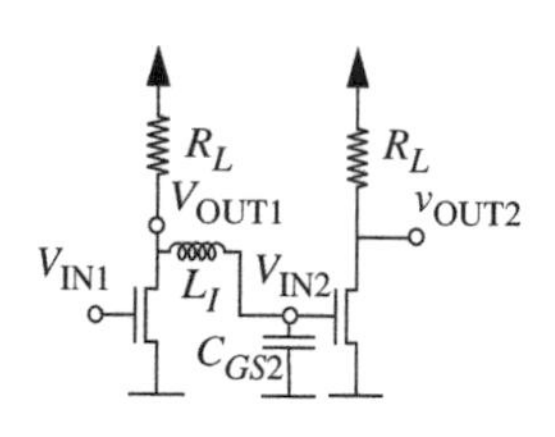


**FIGURE 12.4** Cascaded inverters with parasitic wiring inductance and gate capacitance.

shown in Figure 9.3. In Section 10.4 we analyzed this waveform in detail and found that it contains a decaying exponential. This decaying exponential is the homogeneous response of the first-order circuit formed by the Thévenin equivalent of the first inverter and the MOSFET gate capacitance in the second inverter. Now suppose that the interconnect between the two inverters is long, so that its parasitic inductance also becomes important. In this case, the dynamics of the cascaded inverters change considerably, and the waveform of $v_{IN2}$ takes on the second-order character shown in Figure 12.3.

With the inclusion of the gate capacitance $C_{GS2}$ of the MOSFET in the second inverter, and the interconnect inductance $L_I$ between then two inverters, the circuit shown in Figure 12.1 becomes the circuit shown in Figure 12.4. The latter circuit contains a capacitor and an inductor, and so it is a second-order circuit. Note that $v_{OUT1}$ and $v_{IN2}$ are now no longer equal. The second-order nature of the new circuit is more readily seen in Figure 12.5, which is extracted

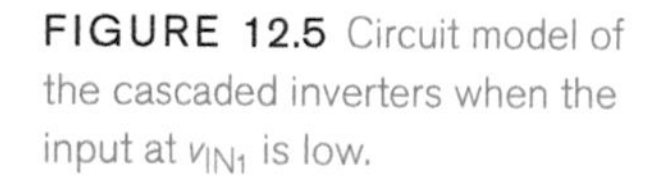
**FIGURE 12.5** Circuit model of the cascaded inverters when the input at $v_{IN1}$ is low.

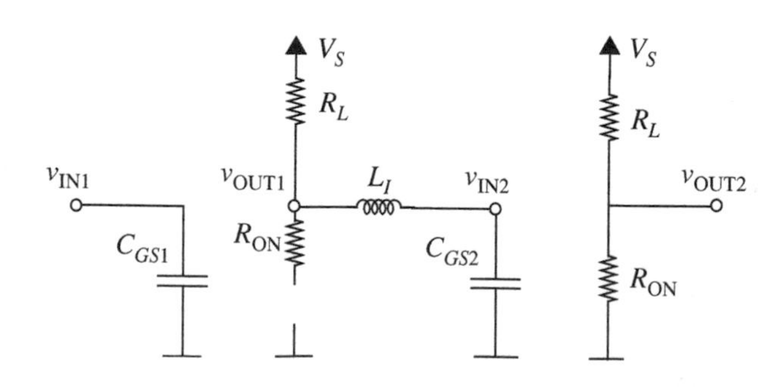

from Figure 12.4 for the case of a low-level input at $v_{IN1}$. The circuit shown in Figure 12.5 contains the series combination of the resistor $R_L$, the capacitor $C_{GS2}$, and the inductor $L_I$. It is the interaction of these three elements that leads to the oscillatory waveform observed in Figure 12.3. We shall study this second-order circuit in more detail in Section 12.5.

Close inspection of the waveform of $v_{IN2}$ shown in Figure 12.3 reveals that both the correct operation and the speed of the cascaded inverters depends on the second-order behavior of the capacitor-inductor-resistor circuit formed between the two inverters. The oscillatory character of the waveform of $v_{IN2}$ following each rising and falling transition is referred to as *ringing*. As can be seen in Figure 12.3, the output of the first inverter as received by the second inverter is valid only after the ringing transitions settle above $V_{OH}$ for a rising transition, and below $V_{OL}$ for a falling transition. Waiting for the ringing to settle within the output thresholds results in the signal propagation delays $t_{pd,1\rightarrow 0}$ and $t_{pd,0\rightarrow 1}$. This too is analyzed in more detail in Section 12.5.

In the remainder of this chapter, we will study the behavior of a variety of second-order circuits similar to the one shown in Figure 12.5. We begin with the simplest circuit involving only one capacitor and one inductor. Following that, we study more complex circuits by including resistors and then sources. As we shall see, there are many such second-order circuits. Fortunately, their behavior is much the same, and so we need examine only several of them in detail. We will also study the behavior of two-capacitor and two-inductor second-order circuits. Finally, we close by addressing several issues of general importance to the analysis of second-order and higher-order circuits.

## 12.1 UNDRIVEN LC CIRCUIT

The simplest second-order circuit is the undriven circuit having one capacitor and one inductor, shown in Figure 12.6. It is lossless because it contains no elements that dissipate energy. That is, it contains no resistors. It is undriven because it contains no independent sources to provide external stimuli. Therefore, it offers us the opportunity to focus on the internal, or homogeneous, behavior of the circuit itself. The undriven response is also the Zero Input Response (ZIR) of the circuit. Of course, without a source one should ask how did the operation of the circuit begin. We will temporarily defer the answer to this question and simply assume for the moment that its branch voltages and currents are not all zero.

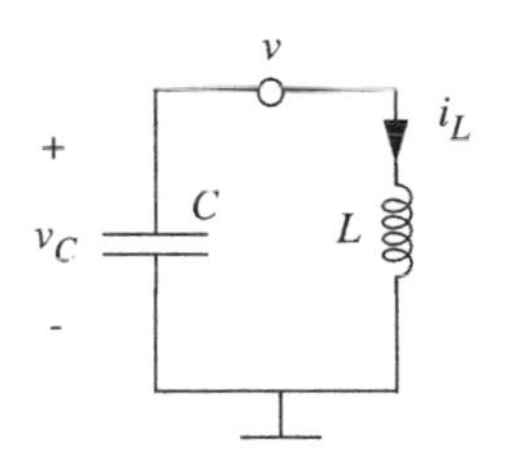


FIGURE 12.6 A simple second-order circuit.

To examine the behavior of the circuit shown in Figure 12.6, we can employ the node analysis method outlined in Chapter 3. Since a ground node is already selected in Figure 12.6, and since the unknown voltage at the other node is already labeled with the node voltage $v$, Steps 1 and 2 of the node analysis are already complete. Note that $v$ will be the primary unknown for the node analysis.

In addition to the node voltage $v$, the states of the two circuit elements are of interest. These are the capacitor voltage $v_C$ and the inductor current $i_L$, and they too are labeled in Figure 12.6. The states are related to $v$ according to

$$v_C(t) = v(t) \tag{12.1}$$

$$i_L(t) = \frac{1}{L}\int_{-\infty}^{t} v(\tilde{t})d\tilde{t} \tag{12.2}$$

where $\tilde{t}$ is a dummy variable of integration. Equation 12.1 follows from the fact that $v$ and $v_C$ represent the same voltage, and Equation 12.2 follows from the constitutive law for the inductor given in Equation 9.30. Thus, once the node voltage $v$ is determined, both $v_C$ and $i_L$ are easily determined from Equations 12.1 and 12.2, respectively.

As observed in Equation 12.1 we have chosen to use two separate symbols for the same voltage. Those symbols are $v$ and $v_C$. We have done this to distinguish between the node voltage $v$ and the branch voltage $v_C$, which happen to be the same. In the future, for simplicity in such cases we will use only one symbol.

Returning now to the node analysis, we complete Step 3 by writing KCL in terms of $v$ for the node at which $v$ is defined. This yields

$$C\frac{dv(t)}{dt} + \frac{1}{L}\int_{-\infty}^{t} v(\tilde{t})d\tilde{t} = 0. \tag{12.3}$$

The first term in Equation 12.3 is the capacitor current, and follows from the constitutive law for the capacitor given in Equation 9.9. The second term in Equation 12.3 is the inductor current. Because the circuit contains an inductor, Equation 12.3 contains a time integral. To remove this integral, we differentiate Equation 12.3 with respect to time, and also divide by C, to obtain

$$\frac{d^2v(t)}{dt^2} + \frac{1}{LC}v(t) = 0 \tag{12.4}$$

which is easier to work with.

To complete the node analysis, we complete Steps 4 and 5 by solving Equation 12.4 for $v$, and using it to determine $i_L$ and $v_C$, for example. Equation 12.4 is an ordinary second-order linear differential equation with constant coefficients. As we did with first-order systems in Section 10.11, we can obtain a general solution to our second-order differential equation using the following steps:

1. Find the homogeneous solution. To find the homogeneous solution, the drive is set to zero.

2. Find the particular solution.
3. The total solution is then the sum of the homogeneous solution and the particular solution. Use the initial conditions to solve for the remaining constants.

In our Equation 12.4, there is no drive to begin with, so Equation 12.4 is also the homogeneous equation. Thus, the homogeneous solution is also the total solution. To obtain the homogeneous solution, we proceed as in Section 10.1.1 for solving first-order homogeneous equations. We expect the homogeneous solution to our linear, constant-coefficient, ordinary, second-order differential equation (Equation 12.4) to be a superposition of two terms of the form

$$Ae^{st}$$

where $A$ is a coefficient and $s$ is a frequency. The substitution of this candidate term into Equation 12.4 yields

$$As^2e^{st} + A\frac{1}{LC}e^{st} = 0. \tag{12.5}$$

After factoring out $A$ and $e^{st}$, this becomes

$$A\left(s^2 + \frac{1}{LC}\right)e^{st} = 0. \tag{12.6}$$

Since $e^{st}$ is never zero for finite $st$, and since $A = 0$ is a trivial solution that leads to $v = 0$, it follows from Equation 12.6 that

$$s^2 + \frac{1}{LC} = 0. \tag{12.7}$$

The two roots are

$$s_1 = +j\omega_o$$
$$s_2 = -j\omega_o \tag{12.8}$$

where

$$\omega_o \equiv \sqrt{\frac{1}{LC}} \tag{12.9}$$

and where $j$ denotes $\sqrt{-1}$. Therefore, the solution for $v$ is a linear combination of the two functions

$$e^{s_1 t} \text{ and } e^{s_2 t}$$

and takes the form

$$v(t) = A_1 e^{s_1 t} + A_2 e^{s_2 t} \tag{12.10}$$

where $A_1$ and $A_2$ are as yet unknown constants that are equivalent to the two constants of integration encountered when integrating Equation 12.4 twice to find $v$. Substituting for $s_1$ and $s_2$ from Equation 12.8, the solution for $v$ becomes

$$v(t) = A_1 e^{+j\omega_o t} + A_2 e^{-j\omega_o t}.$$

However, rather than work with these complex exponential functions, it is more intuitive to work with their scaled sum and difference, namely $\cos(\omega_o t)$ and $\sin(\omega_o t)$.[1] Thus, we can take the solution for $v$ to be

$$v(t) = K_1 \cos(\omega_o t) + K_2 \sin(\omega_o t) \tag{12.11}$$

where $K_1$ and $K_2$ are as yet unknown constants.

Equation 12.7 is referred to as the *characteristic equation* of the circuit because it summarizes the internal dynamics of the circuit. The roots of the characteristic equation, $s_1$ and $s_2$, are called the *natural frequencies* of the circuit because they indicate the oscillation frequency of the natural circuit when no forcing drive is present (see Equations 12.8 and 12.11). We have seen similar equations and frequencies before in the context of first-order resistor-capacitor and resistor-inductor circuits. For example, in Chapter 10, the characteristic equations took the form

$$s + \frac{1}{RC} = 0$$

for first-order RC circuits, and

$$s + \frac{R}{L} = 0$$

for first-order LR circuits. Their roots led to the natural frequencies $-1/RC$ and $-R/L$, respectively, and the associated time constants $RC$ and $L/R$, characteristic of first-order circuits.

To complete the solution to Equation 12.4 we must determine the unknown constants $K_1$ and $K_2$. To do so, we need specific information about $v$. Mathematically, this is provided by specifying $v$ and $dv/dt$ at a particular time. That is, we provide initial conditions from which Equation 12.4 can be integrated. However, in working with electronic circuits, it is more common to know the states of a circuit at the initial time, and this information must be used

---

1. Recall from the Euler relation, $e^{j\omega_o t} = \cos(\omega_o t) + j\sin(\omega_o t)$, that $e^{+j\omega_o t} + e^{-j\omega_o t} = 2\cos(\omega_o t)$ and that $e^{+j\omega_o t} - e^{-j\omega_o t} = 2j\sin(\omega_o t)$ (See Appendices C and B for more details).

to find $v$ and $dv/dt$ at that time. Nevertheless, let us assume for the moment that we know $v$ and $dv/dt$ at an initial time, and complete the solution for $v$; we will return to the initial states shortly. Without loss of generality we will choose the initial time to be $t = 0$. In other words, we assume we know the initial conditions

$$v(0) \quad \text{and} \quad \frac{dv}{dt}(0).$$

From Equation 12.11 evaluated at $t = 0$ we see that

$$v(0) = K_1 \tag{12.12}$$

and from the derivative of Equation 12.11 evaluated at $t = 0$ we see that

$$\frac{dv}{dt}(0) = \omega_o K_2. \tag{12.13}$$

Equations 12.12 and 12.13 can be solved to yield

$$K_1 = v(0) \tag{12.14}$$

$$K_2 = \frac{1}{\omega_o}\frac{dv}{dt}(0). \tag{12.15}$$

The two unknown coefficients are now known in terms of the initial conditions. Finally, combining Equations 12.11, 12.14 and 12.15 yields

$$v(t) = v(0)\cos(\omega_o t) + \frac{1}{\omega_o}\frac{dv}{dt}(0)\sin(\omega_o t) \tag{12.16}$$

as the solution for $v$ given that we know $v$ and $dv/dt$ at $t = 0$. It both satisfies Equation 12.4 and matches the initial conditions.

While Equation 12.16 is the solution for $v$, it is not expressed in terms of the state variables $i_L$ and $v_C$ evaluated at $t = 0$, which is often more useful. To make it so, we must determine $v$ and $dv/dt$ in terms of $i_L$ and $v_C$. From Equation 12.1 we can immediately determine $v$ in terms of $v_C$. Thus evaluating $v(t)$ at $t = 0$, we get

$$v(0) = v_C(0). \tag{12.17}$$

Next, by combining Equations 12.2 and 12.3 we can determine $dv/dt$ in terms of $i_L$ according to

$$C\frac{dv(t)}{dt} = -i_L(t). \tag{12.18}$$

Thus, evaluating $dv(t)/dt$ at $t = 0$, we get

$$\frac{dv}{dt}(0) = -\frac{1}{C} i_L(0). \tag{12.19}$$

Then, combining Equation 12.16 with Equations 12.17 and 12.19 yields

$$v(t) = v_C(0)\cos(\omega_\circ t) - \sqrt{\frac{L}{C}} i_L(0)\sin(\omega_\circ t) \tag{12.20}$$

as the completed solution for $v$ given the state variables of the circuit at $t = 0$; Equation 12.9 has also been used to simplify the result.

The final step in our analysis of the circuit shown in Figure 12.6 is to determine the state variables. From Equations 12.20 and 12.1 we find that

$$\begin{aligned} v_C(t) &= v_C(0)\cos(\omega_\circ t) - \sqrt{\frac{L}{C}} i_L(0)\sin(\omega_\circ t) \\ &= \sqrt{v_C^2(0) + \frac{L}{C} i_L^2(0)} \cos\left(\omega_\circ t + \tan^{-1}\left(\sqrt{\frac{L}{C}}\frac{i_L(0)}{v_C(0)}\right)\right) \end{aligned} \tag{12.21}$$

and from Equations 12.20 and 12.18 we find that

$$\begin{aligned} i_L(t) &= \sqrt{\frac{C}{L}} v_C(0)\sin(\omega_\circ t) + i_L(0)\cos(\omega_\circ t) \\ &= \sqrt{\frac{C}{L}}\sqrt{v_C^2(0) + \frac{L}{C} i_L^2(0)} \sin\left(\omega_\circ t + \tan^{-1}\left(\sqrt{\frac{L}{C}}\frac{i_L(0)}{v_C(0)}\right)\right) \end{aligned} \tag{12.22}$$

where the second equalities in Equations 12.21 and 12.22 both result from the application of trigonometric identities.[2] Again, Equation 12.9 has been used to

---

2. The relevant identities are

$$a\cos(x) - b\sin(x) = \sqrt{a^2 + b^2}\cos(x + \tan^{-1}(b/a))$$

and

$$a\sin(x) + b\cos(x) = \sqrt{a^2 + b^2}\sin(x + \tan^{-1}(b/a)).$$

These are discussed further in Appendix B.

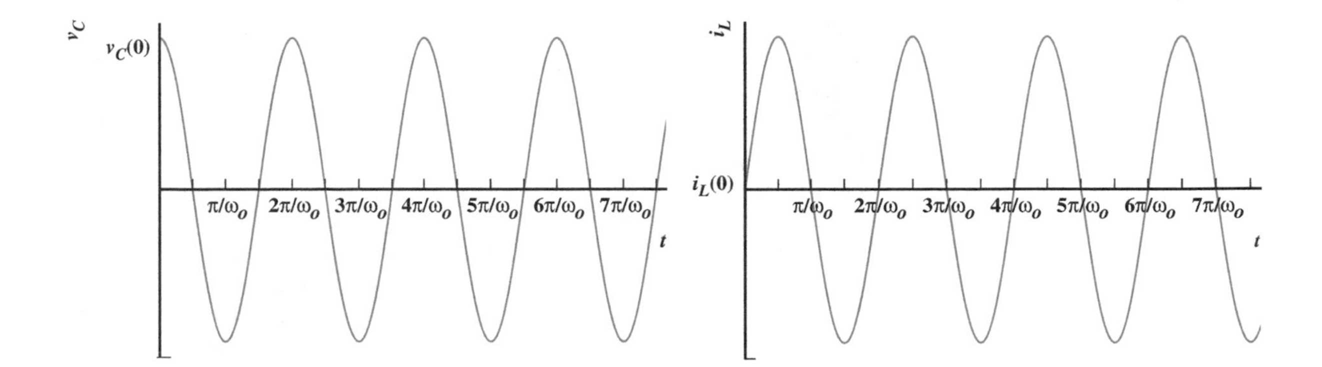


**FIGURE 12.7** $i_L$ and $v_C$ for the undriven LC circuit shown in Figure 12.6 for the special case of $i_L(0) \equiv 0$.

simplify the results. We could also have used Equation 12.2 to determine $i_L$ according to

$$i_L(t) = i_L(0) + \frac{1}{L}\int_0^t v(\tilde{t})d\tilde{t}. \tag{12.23}$$

Substitution of Equation 12.20 into Equation 12.23 does yield Equation 12.22, but using Equation 12.18 is easier. Note that Equation 12.23 displays the memory property of the inductor first observed in 9.33.

A close look at Equations 12.21 and 12.22 shows that, for all choices of initial condition, $i_L$ and $v_C$ are sinusoidal functions that are a quarter cycle out of phase with each other. The initial conditions affect only their amplitudes and their common absolute phase. Therefore, we can illustrate the behavior of the circuit without loss of understanding by considering the special case of

$$i_L(0) \equiv 0,$$

for example. The resulting expressions for $i_L$ and $v_C$ for $i_L(0) = 0$ are given as follows, and Figure 12.7 shows the corresponding evolution of both $i_L$ and $v_C$:

$$v_C(t) = v_C(0)\cos(\omega_o t) \tag{12.24}$$

$$i_L(t) = \sqrt{\frac{C}{L}}v_C(0)\sin(\omega_o t). \tag{12.25}$$

Figure 12.7 illustrates several important points. As mentioned earlier, both branch variables, or states, are sinusoidal in time. Because of this, the peaks of one state occur at the zeros of the other. This behavior underlies the ringing seen in Figure 12.3, and is identical to that of many other lossless second-order oscillators, such as a spring and mass or a linearized pendulum in which mass position and velocity are the two states. Given the definitions of positive $i_L$ and

$v_C$ in Figure 12.6, $v_C$ leads $i_L$ by a quarter cycle. Thus, the greatest positive slope in $i_L$ occurs at the positive peaks of $v_C$, and the greatest negative slope in $i_L$ occurs at the negative peaks of $v_C$, in accordance with the constitutive law for the inductor. Similarly, the greatest positive slope in $v_C$ occurs at the negative peaks of $i_L$, and the greatest negative slope in $v_C$ occurs at the positive peaks of $v_C$, in accordance with the constitutive law for the capacitor. So, each state drives the growth of the other.

There is also an important energy interpretation to Figure 12.7. It is that the oscillations in $i_L$ and $v_C$ carry out a repetitive exchange of energy between the inductor and the capacitor. Indeed, the state of each element drives the growth of the other at the expense of the energy it stores. To see this, consider the energy $w_E$ stored in the capacitor, the energy $w_M$ stored in the inductor, and the total energy $w_T$ stored between them. In the general case, the substitution of Equations 12.21 and 12.22 into Equations 9.18 and 9.36, respectively, yields

$$w_E = \left(\frac{1}{2}Cv_C^2(0) + \frac{1}{2}Li_L^2(0)\right) \cos^2\left(\omega_\circ t + \tan^{-1}\left(\sqrt{\frac{L}{C}}\frac{i_L(0)}{v_C(0)}\right)\right) \tag{12.26}$$

$$w_M = \left(\frac{1}{2}Cv_C^2(0) + \frac{1}{2}Li_L^2(0)\right) \sin^2\left(\omega_\circ t + \tan^{-1}\left(\sqrt{\frac{L}{C}}\frac{i_L(0)}{v_C(0)}\right)\right) \tag{12.27}$$

and so $w_T$ is given by

$$w_T = w_E + w_M = \frac{1}{2}Cv_C^2(0) + \frac{1}{2}Li_L^2(0). \tag{12.28}$$

Thus, the total energy $w_T$ is constant in time. This is the case because there are no resistors in the circuit that could dissipate the energy. Note too that the energy completely exchanges between the two elements since both $w_E$ and $w_M$ periodically go to zero. This behavior is also identical to that of many other lossless second-order mechanical oscillators in which kinetic and potential energies are repetitively exchanged. To illustrate this, Figure 12.8 shows the $w_E$, $w_M$, and $w_T$ for the special case of $i_L(0) \equiv 0$. Note that the energies exchange at the frequency $2\omega_\circ$ because both $i_L$ and $v_C$ go to zero twice during the period $2\pi/\omega_\circ$.

To close this section, let us summarize three important observations. The first observation is that second-order capacitor-inductor circuits are capable of oscillation. This is in contrast to first-order resistor-capacitor and resistor-inductor circuits. The second observation is that we have now seen a third time constant to go along with the $RC$ time constant associated with first-order

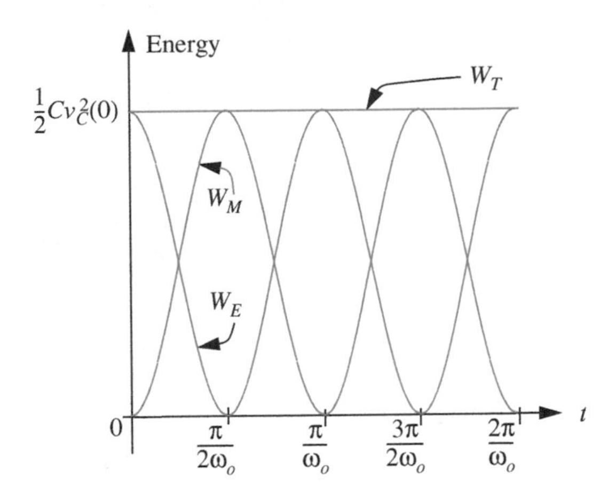


**FIGURE 12.8** $w_E$, $w_M$, and $w_T$ for the circuit shown in Figure 12.6 for the special case of $i_L(0) \equiv 0$.

resistor-capacitor circuits, and the $L/R$ time constant associated with first-order resistor-inductor circuits. That time constant is $\sqrt{LC}$, which is the time constant associated with second-order capacitor-inductor circuits. The third observation is the meaning of the ratio $\sqrt{L/C}$ when associated with second-order capacitor-inductor circuits. This ratio is a characteristic of energy storage. Because $w_E$ and $w_M$ exchange completely, it follows from Equations 12.21 and 12.22 that

$$\frac{C}{2}v_{C_{\text{Peak}}}^2 = \frac{L}{2}i_{L_{\text{Peak}}}^2 \quad \Rightarrow \quad \frac{v_{C_{\text{Peak}}}}{i_{L_{\text{Peak}}}} = \sqrt{\frac{L}{C}} \tag{12.29}$$

where $v_{C_{\text{Peak}}}$ and $i_{L_{\text{Peak}}}$ are the peak values of $v_C$ and $i_L$, respectively. Thus, the ratio $\sqrt{L/C}$, which has the units of resistance, originates from energy considerations and is the ratio of the peak values of the two states. The ratio $\sqrt{L/C}$ is called the *characteristic impedance*. In the following section, we will see that the parameter $\sqrt{L/C}$ also helps characterize the dynamics of damped second-order circuits.

---

EXAMPLE 12.1 AN UNDRIVEN LC CIRCUIT For the circuit shown in Figure 12.6, suppose that $C = 1\ \mu$F and $L = 100\ \mu$H. What is the oscillation frequency of $i_L$ and $v_C$? Further, suppose that at some time, $i_L = 0.5$ A and $v_C = 10$ V. What will be the peak values of $i_L$ and $v_C$?

From Equation 12.9 with $C = 1\ \mu$F and $L = 100\ \mu$H,

$$\omega_o = \frac{1}{\sqrt{LC}} = 10^5 \text{ rad/s}$$

or approximately 15.9 kHz. At the time of the measurement of $i_L$ and $v_C$,

$$w_E = 50\ \mu\text{J} \quad \text{and} \quad w_M = 12.5\ \mu\text{J}.$$

Thus

$$w_T = 62.5\ \mu\text{J}.$$

The peak value of $i_L$ occurs when this energy is stored entirely in the inductor. In other words,

$$\frac{1}{2} L i_{L_{\text{Peak}}}^2 = w_T = 62.5\ \mu\text{J}.$$

So,

$$i_{L_{\text{Peak}}} \approx 1.12\ \text{A}.$$

Similarly, the peak value of $v_C$ occurs when the energy is stored entirely in the capacitor. In other words,

$$\frac{1}{2} C v_{C_{\text{Peak}}}^2 = w_T = 62.5\ \mu\text{J}.$$

So,

$$v_{C_{\text{Peak}}} \approx 11.2\ \text{V}.$$

Also, note that $C$, $L$, $v_{C_{\text{Peak}}}$, and $i_{L_{\text{Peak}}}$ do satisfy Equation 12.29. This corresponds to

$$\sqrt{L/C} = 10\ \Omega.$$

---

EXAMPLE 12.2 ANOTHER UNDRIVEN LC CIRCUIT For the circuit shown in Figure 12.6, as in the previous example, suppose that $C = 1\ \mu\text{F}$ and $L = 100\ \mu\text{H}$. Further, suppose that at $t = 0$ the inductor current $i_L = 0$ and the capacitor voltage $v_C = 1$ V. Plot the waveforms for $t > 0$ for $i_L$ and $v_C$.

Since

$$\omega_o = \frac{1}{\sqrt{LC}} = 10^5\ \text{rad/s}$$

we know that the waveforms for the voltage and the current will be sinusoids of frequency $\omega_o = 10^5$ rad/s.

Further, since the initial value of the current is given to be 0, we have from Equations 12.24 and 12.25

$$v_C(t) = v_C(0)\cos(\omega_o t) = 1.0\cos(10^5 t)$$

$$i_L(t) = \sqrt{\frac{C}{L}}\, v_C(0)\sin(\omega_o t) = 0.1 \times \sin(10^5 t).$$

The waveforms are plotted in Figure 12.9.

---

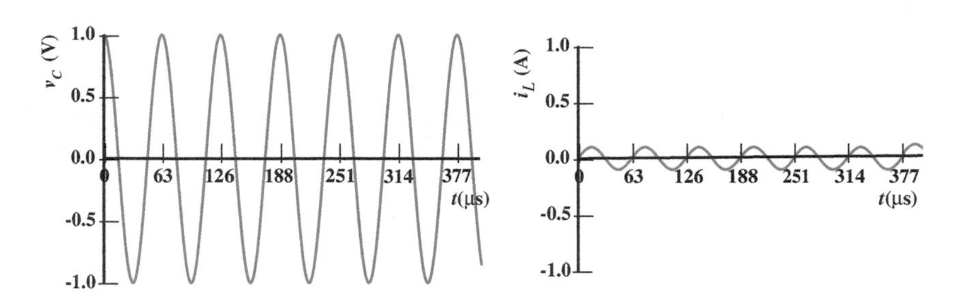


FIGURE 12.9 $i_L$ and $v_C$ for the undriven LC circuit shown in Figure 12.6 for $i_L(0) = 0$ and $v_C(0) = 1.0$.

---

EXAMPLE 12.3 SPRING-MASS OSCILLATOR The spring-mass oscillator shown in Figure 12.10 is also a lossless second-order system. Its motion is described by

$$M\frac{d^2x(t)}{dt^2} + K\,x(t) = 0. \tag{12.30}$$

What is its oscillation frequency?

The equation of motion for the spring-mass oscillator is the same as Equations 12.3 and 12.4, but with $M$ replacing $C$, $1/K$ replacing $L$, and $x$ replacing $v$. So, by analogy to Equation 12.9, its oscillation frequency is

$$\sqrt{K/M}.$$

Similar analogies can be made to every aspect of our analysis of the capacitor-inductor circuit.

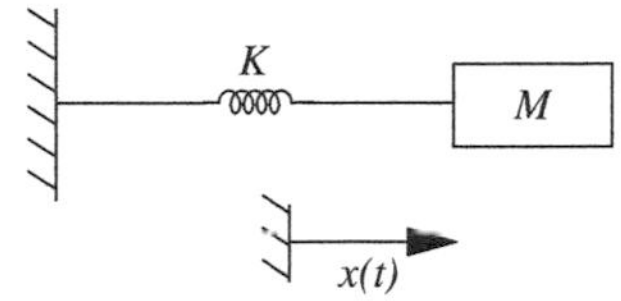

FIGURE 12.10 A second-order spring-mass oscillator. Note that the coiled object in this figure is a spring, not an inductor.

---

EXAMPLE 12.4 AN IDEALIZED SWITCHED POWER SUPPLY In this example, we will analyze the idealized switching charge pump shown in Figure 12.11. Such charge pumps are used in voltage converters to translate one DC voltage to another DC voltage (for example, from a 1.5-volt battery to an electronic amplifier circuit that needs a 3-volt DC source). The purpose of the charge pump is to transfer energy losslessly from the voltage source through the inductor and into the capacitor. In doing so, it charges the capacitor and builds up the voltage across it.

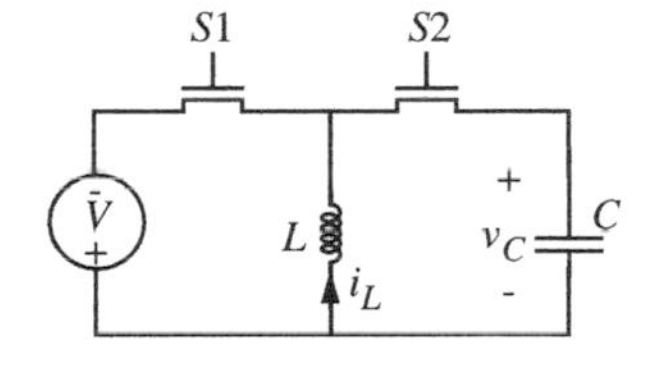


FIGURE 12.11 An idealized switching charge pump.

The charge pump operates cyclically as shown in Figure 12.12. To begin a cycle, switch S1 closes for the duration $T$ causing the current in the inductor to ramp up. During this time switch S2 remains open, and so the capacitor charge and voltage remain constant. Next, switch S1 opens and switch S2 closes. This switch action disconnects the source from the inductor, and since the inductor current does not change instantaneously, it redirects the inductor current into the capacitor. The inductor current now rings down as the capacitor voltage rings up. Finally, switch S2 opens when the inductor current first goes to zero, and both switches remain open until the start of the next cycle.

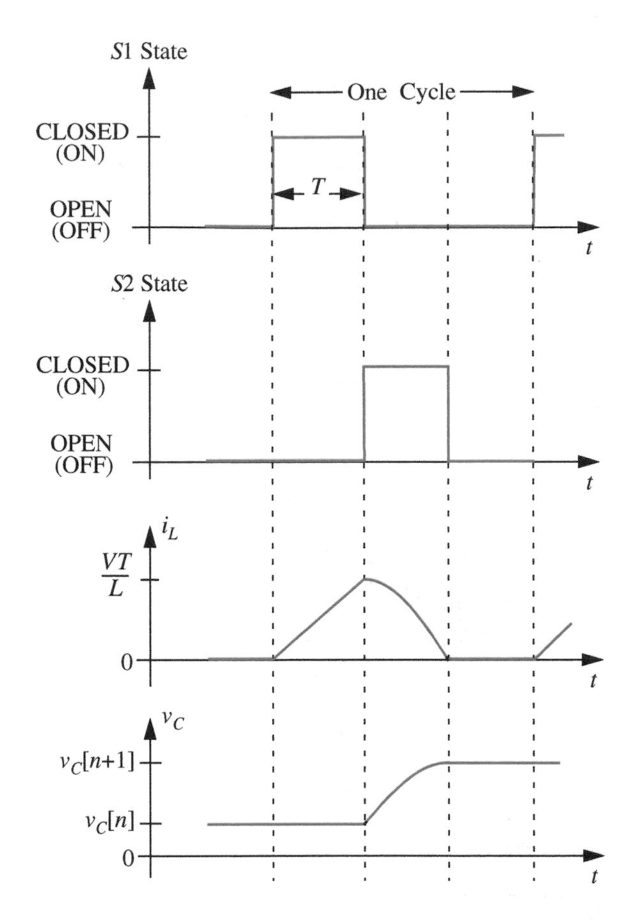


FIGURE 12.12 Charge pump operation.

Our goal is to determine the value of the capacitor voltage as a function of time. To do so, we will analyze the behavior of the circuit during each interval of operation in sequence. To begin, let us find the inductor current $i_L$ as it ramps up over the beginning of each cycle.

*S1 Closed, S2 Open* During this initial interval, because S1 is closed and S2 is open, the DC voltage $V$ appears directly across the inductor, and the inductor current $i_L$ ramps up. This aspect of charge pump operation has already been analyzed in Subsection 9.4.2. In particular, if we define $t = 0$ to occur at the beginning of a cycle, then $i_L$ is given by Equation 9.78 for $0 \leq t \leq T$. Thus, $i_L$ builds up as a ramp.

Now suppose that we wish to find the capacitor voltage $v_C$ at the end of the $n$th cycle. This is most easily done through energy considerations. From Equation 9.36 we see that

$$i_L = \frac{VT}{L} \tag{12.31}$$

at the end of each current ramp. Therefore, from Equation 9.36, the energy $w_M$ stored in the inductor at the end of each current ramp is given by

$$w_M = \frac{L}{2}\left(\frac{VT}{L}\right)^2 = \frac{V^2T^2}{2L}. \tag{12.32}$$

*S1 Open, S2 Closed* This energy is completely transferred to the capacitor during the next interval of the cycle over which switch S2 is closed and S1 is open. Indeed, switch S2 opens to end the second interval just as $i_L$ goes to zero and the energy transfer is completed.

*S1 Open, S2 Open* The cycle ends with both S1 and S2 open. All the energy is now stored in the capacitor. This sequence of actions is repeated during each cycle.

Therefore, the energy $w_E$ stored in the capacitor at the end of the $n$th cycle grows as

$$w_E[n] = w_E[n-1] + \frac{V^2T^2}{2L} \tag{12.33}$$

beginning from

$$w_E[0] = 0 \tag{12.34}$$

where the notation $[n]$ is used to show that $n$ is a cycle index rather than continuous time. Equations 12.33 and 12.34 can be solved to yield

$$w_E[n] = n\frac{V^2T^2}{2L} \tag{12.35}$$

as an explicit statement of the energy stored in the capacitor at the end of the $n$th cycle. Finally, by combining Equations 9.18 and 12.35, we find that

$$v_C[n] = V\sqrt{n\frac{T^2}{LC}}. \tag{12.36}$$

Thus, the voltage across the capacitor is related to the voltage $V$ of the input source, and grows with the square root of $n$. This example will be revisited later on in this chapter and its operation as a DC-DC converter will be established more explicitly.

Finally, suppose that we wish to find the details of $i_L$ and $v_C$ during the $n$th ringing period in which switch S1 is off and switch S2 is on. In this case, we use Equations 12.21 and 12.22 with one modification and several substitutions. The modification is that the sign of $i_L$ in both equations must be reversed since $i_L$ is defined in opposite directions in Figures 12.6 and 12.11. The substitutions are $-VT/L$ for $i_L(0)$, $v_C[n]$ for $v_C(0)$, and $t-T$ for $t$ since $t=0$ occurs at the beginning of a cycle when switch S1 closes.

WWW EXAMPLE 12.5 GRAPHICAL INTERPRETATION

## 12.2 UNDRIVEN, SERIES RLC CIRCUIT

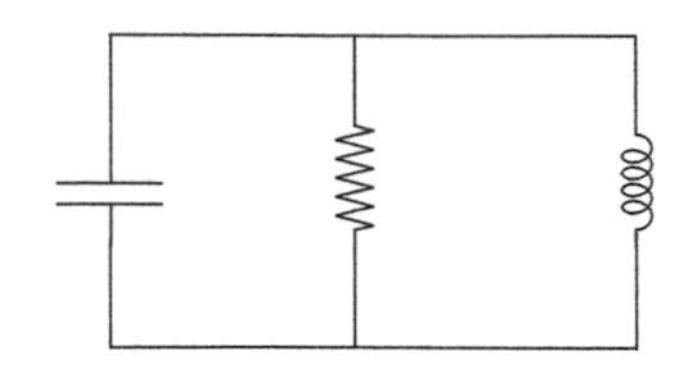

(a) Parallel RLC circuit

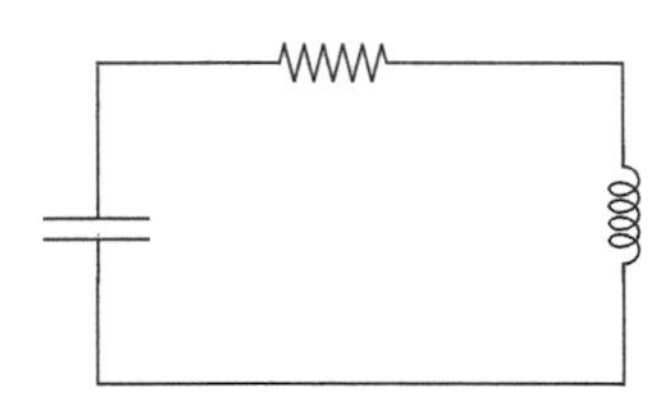

(b) Series RLC circuit

**FIGURE 12.14** Two second-order circuits with one resistor each.

The next step in our study of second-order circuits is to include a loss mechanism through which energy stored in the circuit will dissipate. Second-order circuits can exhibit loss for two reasons. First, real capacitors and real inductors are lossy. A common loss mechanism in a capacitor is dielectric leakage, which can be modeled with a parallel resistor. A common loss mechanism in an inductor is the resistive loss in its winding, which can be modeled with a series resistor. Second, we may purposefully introduce loss into a circuit in order to modify its behavior. For example, we might wish to suppress the oscillations seen in Figure 12.3. Again, this is accomplished by including one or more resistors.

In this section, we will focus on undriven second-order capacitor-inductor circuits to which we have added a single resistor. There are two ways in which we can add a resistor to the circuit shown in Figure 12.6. We can either place it in parallel or in series with the two original elements. The two resulting circuits are shown in Figure 12.14. We will study the series RLC circuit in more detail here, and in Section 12.4 will discuss the corresponding parallel circuit. As we shall see shortly, the presence of the resistors in these circuits, and their associated losses, changes the behavior of the original circuit significantly. Most importantly, the energy stored in the circuit is no longer constant; rather, it decays in time. As a consequence, the circuit states also decay in time.

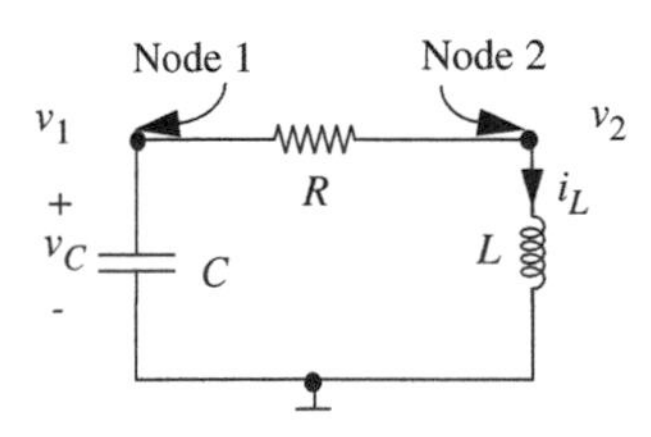


**FIGURE 12.15** The series second-order circuit shown in Figure 12.14b.

Let us now examine the behavior of the series circuit shown in Figure 12.14b, which is redrawn in Figure 12.15. To analyze the behavior of this circuit we can again employ the node method, and this analysis closely parallels that of Section 12.1. Since a ground node is already selected in Figure 12.15, and since the unknown node voltages are already labeled as $v_1$ and $v_2$, we may proceed immediately to Step 3 of node analysis. Here, we write KCL at Nodes #1 and #2 in terms of $v_1$ and $v_2$. This yields

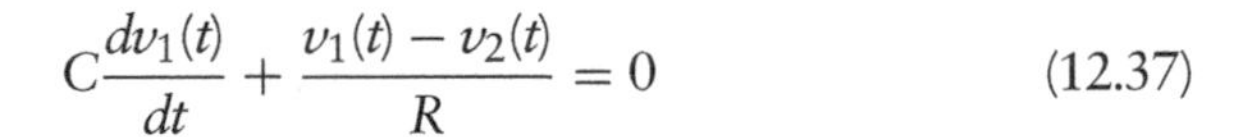

$$C\frac{dv_1(t)}{dt} + \frac{v_1(t) - v_2(t)}{R} = 0 \qquad (12.37)$$

for Node #1, and

$$\frac{v_2(t) - v_1(t)}{R} + \frac{1}{L}\int_{-\infty}^{t} v_2(\tilde{t})d\tilde{t} = 0 \qquad (12.38)$$

for Node #2. To treat these two equations simultaneously, we first use Equation 12.37 to determine $v_2$ in terms of $v_1$, and then substitute the result into Equation 12.38 to obtain a second-order differential equation in $v_1$. This yields

$$v_2(t) = RC\frac{dv_1(t)}{dt} + v_1(t) \qquad (12.39)$$

$$\frac{d^2v_1(t)}{dt^2} + \frac{R}{L}\frac{dv_1(t)}{dt} + \frac{1}{LC}v_1(t) = 0. \tag{12.40}$$

Note that to arrive at Equation 12.40 we have divided Equation 12.38 by $C$, and differentiated it with respect to time.

To complete the node analysis, we complete Steps 4 and 5 by solving Equation 12.40 for $v_1$, and using it to determine $v_2$ and other branch variables of interest. Equation 12.40 is an ordinary second-order homogeneous linear differential equation with constant coefficients. Since the circuit does not have a drive, its homogeneous solution is also its total solution. Thus, as with Equation 12.4, we expect its solution also to be a superposition of two terms of the form

$$Ae^{st}.$$

The substitution of this candidate term into Equation 12.40 yields

$$A\left(s^2 + \frac{R}{L}s + \frac{1}{LC}\right)e^{st} = 0 \tag{12.41}$$

from which it follows that

$$s^2 + \frac{R}{L}s + \frac{1}{LC} = 0. \tag{12.42}$$

Equation 12.42 is the *characteristic equation* of the circuit. It is slightly more complex than Equation 12.7 because of the term proportional to $s$. We will see shortly that this term is responsible for damping and energy loss. To simplify Equation 12.42, and to put it in a form that is more standard for the characteristic equation in second-order circuits, we write it as

$$s^2 + 2\alpha s + \omega_\circ^2 = 0 \tag{12.43}$$

where

$$\alpha \equiv \frac{R}{2L} \tag{12.44}$$

$$\omega_\circ \equiv \frac{1}{\sqrt{LC}}; \tag{12.45}$$

note that Equation 12.45 is the same as Equation 12.9. Equation 12.43 is a quadratic equation having two roots. Those roots are

$$s_1 = -\alpha + \sqrt{\alpha^2 - \omega_\circ^2} \tag{12.46}$$

$$s_2 = -\alpha - \sqrt{\alpha^2 - \omega_\circ^2}. \tag{12.47}$$

Therefore, the solution for $v_1$ is a linear combination of the two functions $e^{s_1 t}$ and $e^{s_2 t}$, and takes the form

$$v_1(t) = A_1 e^{s_1 t} + A_2 e^{s_2 t} \quad (12.48)$$

where $A_1$ and $A_2$ are as yet unknown constants that are equivalent to the two constants of integration encountered when integrating Equation 12.40 twice to find $v_1$. Note that $s_1$ and $s_2$ are the two natural frequencies of the circuit.[3]

To complete the solution to Equation 12.40 we must again determine $A_1$ and $A_2$. To do so, we need specific information about $v_1$, which we will again be provided by specifying $v_1$ and $dv_1/dt$ at an initial time, again chosen to be $t = 0$. As mentioned earlier, it is actually more common to know $i_L$ and $v_C$ at that initial time, and so we must use this information to first determine $v_1$ and $dv_1/dt$ at the initial time, and then $A_1$ and $A_2$. Since $v_1$ and $v_C$ represent the same voltage,

$$v_C(t) = v_1(t) \quad (12.49)$$

so that

$$v_1(0) = v_C(0). \quad (12.50)$$

Next, the constitutive law for the capacitor yields

$$i_L(t) = -C\frac{dv_1}{dt}(t) \quad (12.51)$$

so that

$$\frac{dv_1}{dt}(0) = -\frac{1}{C} i_L(0). \quad (12.52)$$

3. At this point, it is worth dwelling for a moment on the two natural frequencies $s_1$ and $s_2$, and writing a few useful equalities related to them. Adding Equations 12.46 and 12.47 gives us

$$s_1 + s_2 = -2\alpha,$$

subtracting them yields

$$s_1 - s_2 = 2(\sqrt{\alpha^2 - \omega_o^2}),$$

and multiplying them yields

$$s_1 s_2 = \omega_o^2.$$

Because they are the roots, both $s_1$ and $s_2$ satisfy the characteristic equation given in Equation 12.42, or in its more general form, Equation 12.43.

Next, we evaluate Equation 12.48 and its derivative at $t = 0$, and equate the results to Equations 12.50 and 12.52 to obtain

$$v_1(0) = A_1 + A_2 = v_C(0) \tag{12.53}$$

$$\frac{dv_1}{dt}(0) = s_1A_1 + s_2A_2 = -\frac{1}{C}i_L(0). \tag{12.54}$$

Equations 12.53 and 12.54 can be jointly solved for $A_1$ and $A_2$ to obtain

$$A_1 = \frac{Cs_2v_C(0) + i_L(0)}{C(s_2 - s_1)} \tag{12.55}$$

$$A_2 = \frac{Cs_1v_C(0) + i_L(0)}{C(s_1 - s_2)}, \tag{12.56}$$

which can be substituted into Equation 12.48 to yield

$$v_1(t) = \frac{Cs_2v_C(0) + i_L(0)}{C(s_2 - s_1)}e^{s_1t} + \frac{Cs_1v_C(0) + i_L(0)}{C(s_1 - s_2)}e^{s_2t}. \tag{12.57}$$

Finally, substitution of Equation 12.57 in Equations 12.49 and 12.51 yields

$$v_C(t) = \frac{Cs_2v_C(0) + i_L(0)}{C(s_2 - s_1)}e^{s_1t} + \frac{Cs_1v_C(0) + i_L(0)}{C(s_1 - s_2)}e^{s_2t} \tag{12.58}$$

$$i_L(t) = -s_1\frac{Cs_2v_C(0) + i_L(0)}{(s_2 - s_1)}e^{s_1t} - s_2\frac{Cs_1v_C(0) + i_L(0)}{(s_1 - s_2)}e^{s_2t} \tag{12.59}$$

as the states of the series circuit. This completes the formal analysis of the circuit shown in Figure 12.15[4].

---

4. The two circuits shown in Figure 12.14 are duals of one another, and so the response of one can be directly constructed from the response of the other. To see this, note that KVL applied to the single loop in the circuit shown in Figure 12.15 results in

$$\frac{d^2i_L}{dt^2} + \frac{R}{L}\frac{di_L}{dt} + \frac{i_L}{LC} = 0,$$

and the element law for the capacitor results in

$$i_L = -C\frac{dv_C}{dt}.$$

Similarly, KCL applied to the upper node in the circuit shown in Figure 12.14a results in

$$\frac{d^2v_C}{dt^2} + \frac{G}{C}\frac{dv_C}{dt} + \frac{v_C}{LC} = 0,$$

Let us now examine the dynamic behavior of $v_C$ and $i_L$ as expressed by Equations 12.58 and 12.59. To do so, it is convenient to consider three separate cases defined by the relative sizes of $\alpha$ and $\omega_o$. These cases are as follows:

$$\begin{aligned} \alpha < \omega_o &\Rightarrow \text{under-damped dynamics;} \\ \alpha = \omega_o &\Rightarrow \text{critically-damped dynamics;} \\ \alpha > \omega_o &\Rightarrow \text{over-damped dynamics.} \end{aligned}$$

As we shall see in the following three sections, the dynamic behavior of the series RLC circuit shown in Figure 12.15 is quite different for these three cases.[5]

### 12.2.1 UNDER-DAMPED DYNAMICS

The case of under-damped dynamics is characterized by

$$\alpha < \omega_o$$

or, after substitution of Equations 12.44 and 12.45 , by

$$R/2 < \sqrt{L/C}.$$

As $R$ becomes small, the corresponding resistor approaches a short circuit, and so the circuit shown in Figure 12.15 approaches that shown in Figure 12.6. Therefore, we should expect the under-damped dynamics to be oscillatory in nature. As we shall see shortly, this is indeed the case.

With $\alpha < \omega_o$, the quantity inside the radicals in Equations 12.46 and 12.47 is negative, and so $s_1$ and $s_2$ are again complex numbers. To simplify matters, and to make clear the complex nature of $s_1$ and $s_2$, we define $\omega_d$

---

and the element law for the inductor results in

$$v_C = L\frac{di_L}{dt},$$

where $v_C$ is the capacitor voltage defined positively at the upper node, $i_L$ is the inductor current defined positively in the downwards direction, and $G$, $C$ and $L$ are the conductance, capacitance and inductance of the resistor, capacitor and inductor respectively. A comparison of the two sets of equations shows that the second set can be constructed from the first with $i_L$ replaced by $v_C$, $-v_C$ replaced by $i_L$, $R$ replaced by $G$, $L$ replaced by $C$ and $C$ replaced by $L$. Therefore, the homogeneous response of the circuit shown in Figure 12.14a can be constructed directly from Equations 12.58 and 12.59 in the same way. See, for example, Equations 12.100 and 12.101. Indeed, all results in Section 12.4 can be derived from those in Section 12.2 in the same way. Further, duality continues to hold when the circuit shown in Figure 12.14a is extended to include a parallel current source while the circuit shown in Figure 12.14b is extended to include a series voltage source. Thus, all results in Section 12.6 can be derived directly from the results in Section 12.5.

5. Interestingly, as we will see in Section 12.4, the dynamics of the parallel RLC circuit will be essentially identical to that of the series circuit for all three cases, except for a reversal in the role of $R$. The identical dynamics for the two circuits arises because they both have the same characteristic equation when it is represented in the standard form shown in Equation 12.43.

according to

$$\omega_d \equiv \sqrt{\omega_o^2 - \alpha^2} \tag{12.60}$$

so that Equations 12.46 and 12.47 become

$$s_1 = -\alpha + j\omega_d \tag{12.61}$$

$$s_2 = -\alpha - j\omega_d. \tag{12.62}$$

The real and imaginary parts of $s_1$ and $s_2$ are now more apparent.

Since $s_1$ and $s_2$ are now complex, the exponentials in Equations 12.58 and 12.59 are also complex. Thus, $v_C$ and $i_L$ will exhibit both oscillatory and decaying behavior. To see this, we substitute Equations 12.61 and 12.62 into Equations 12.58 and 12.59, and use the Euler relation given by

$$e^{j\omega_d t} = \cos(\omega_d t) + j\sin(\omega_d t)$$

and the fact that $LCs_1s_2 = 1$, to obtain

$$\begin{aligned} v_C(t) &= v_C(0)e^{-\alpha t}\cos(\omega_d t) + \left(\frac{\alpha C v_C(0) - i_L(0)}{C\omega_d}\right) e^{-\alpha t}\sin(\omega_d t) \\ &= \sqrt{v_C^2(0) + \left(\frac{\alpha C v_C(0) - i_L(0)}{C\omega_d}\right)^2}\, e^{-\alpha t} \\ &\quad \cos\left(\omega_d t - \tan^{-1}\left(\frac{\alpha C v_C(0) - i_L(0)}{C\omega_d v_C(0)}\right)\right) \end{aligned} \tag{12.63}$$

$$\begin{aligned} i_L(t) &= i_L(0)e^{-\alpha t}\cos(\omega_d t) + \left(\frac{v_C(0) - \alpha L i_L(0)}{L\omega_d}\right) e^{-\alpha t}\sin(\omega_d t) \\ &= \sqrt{i_L^2(0) + \left(\frac{v_C(0) - \alpha L i_L(0)}{L\omega_d}\right)^2}\, e^{-\alpha t} \\ &\quad \sin\left(\omega_d t + \tan^{-1}\left(\frac{L\omega_d i_L(0)}{v_C(0) - \alpha L i_L(0)}\right)\right). \end{aligned} \tag{12.64}$$

These expressions for $v_C$ and $i_L$ more clearly expose the oscillatory and decaying behavior of the circuit states. Sketches of $v_C$ and $i_L$ are shown in Figure 12.16 for the special case of

$$i_L(0) = 0.$$

From Figure 12.16, we see that the capacitor voltage is nearly maximum when the inductor current is zero, and vice versa. A more careful examination of Equations 12.63 and 12.64, however, reveals that the circuit states are not exactly in quadrature as they were for the circuit shown in Figure 12.6. In fact,

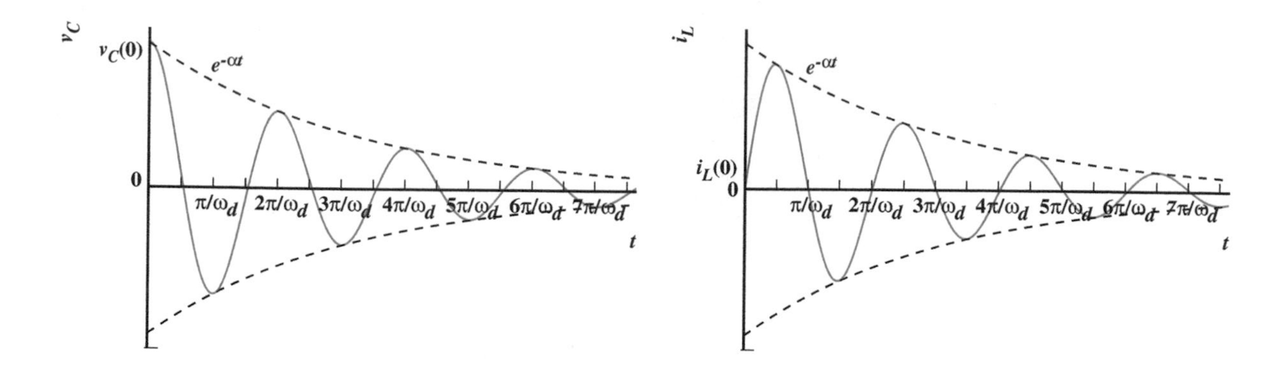


**FIGURE 12.16** Waveforms of $v_C$ and $i_L$ in series RLC circuit for the case of $i_L(o) = o$.

examination of Equations 12.63 and 12.64, for example, with $i_L(0) = 0$, shows that the quadrature lead of $v_C$ with respect to $i_L$ is reduced by $\phi = \tan^{-1}(\alpha/\omega_d)$. Nonetheless, the peaks of $v_C$ occur when $i_L$ is nearly zero, and vice versa. This indicates that the energy is sloshing back and forth, stored first in the electric field of the capacitor, and then in the magnetic field of the inductor. All underdamped second-order systems have this property. A simple pendulum is an obvious example. Here the exchange is between kinetic energy and potential energy: The kinetic energy is maximum when the potential energy is zero, and vice versa. We defer a more detailed analysis of the stored energy to Section 12.3.

As $R \to 0$, that is, as the corresponding resistor approaches a short circuit, it is apparent from Equations 12.44 and 12.60 that $\alpha \to 0$ and $\omega_d \to \omega_o$, respectively. Therefore, as $R \to 0$, remembering that $\omega_o = 1/\sqrt{LC}$, Equations 12.63 and 12.64 reduce to Equations 12.21 and 12.22, respectively. This is expected because Figure 12.15 reduces to Figure 12.6 as $R \to 0$. Limiting behavior such as this can often be used to check the validity of an analysis.

Using Equations 12.63 and 12.64 it is now possible to interpret physically the parameters $\alpha$, $\omega_o$, and $\omega_d$ in the context of the circuit shown in Figure 12.15. The factor $e^{-\alpha t}$ produces the decay or damping in $v_C$ and $i_L$, hence $\alpha$ is referred to as the *damping factor* of the circuit. Larger values of $\alpha$ cause the circuit states to decay more rapidly.

In the absence of damping and the associated energy dissipation in the circuit, that is for $R = 0$ and hence $\alpha = 0$, the circuit states would oscillate at the frequency

$$\omega_o = \frac{1}{\sqrt{LC}}$$

as they did in the case of the circuit shown in Figure 12.6. Thus, $\omega_o$ is referred to as the *undamped natural frequency* or the *undamped resonance frequency*.

In the presence of damping, the circuit states oscillate at the lower frequency $\omega_d$, hence $\omega_d$ is referred to as the *damped natural frequency*. With sufficiently large damping, $\omega_d$ goes to zero and the circuit ceases to oscillate. This case is studied in the next two subsections.

Since $\omega_o$, $\omega_d$, and $\alpha$ are directly related to the roots of the characteristic equation, $s_1$ and $s_2$ (see Equations 12.60, 12.61, and 12.62), it should also be clear at this point why $s_1$ and $s_2$ are called the natural frequencies of the system.

Equation 12.60 indicates that $\alpha$, $\omega_d$, and $\omega_o$ respectively, form the two sides and hypotenuse of a right triangle. In fact, this is part of a more comprehensive picture of the location in the complex plane of the roots of the characteristic equation. These locations, given by Equations 12.61 and 12.62, are shown in Figure 12.17. Note that as $R$ varies so do $\alpha$ and $\omega_d$; $\omega_o$, however, remains constant so that $s_1$ and $s_2$ remain the constant distance $\omega_o$ from the complex-plane origin for the case of under-damped dynamics.

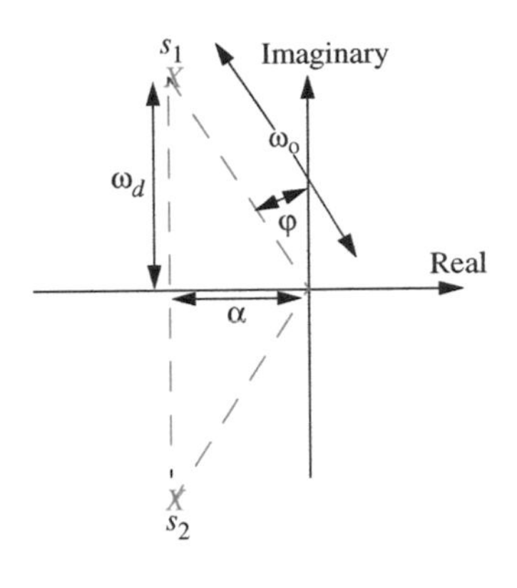


FIGURE 12.17 The location of $s_1$ and $s_2$ in the complex plane.

In contrast to the circuit behavior studied in Section 12.1, Equations 12.63 and 12.64 are characterized by two important rates, or frequencies. The first frequency is $\omega_d$, which determines the rate at which the states oscillate. The second frequency is $\alpha$, which determines the rate at which the states decay. As a consequence, another important characteristic of the circuit behavior described by Equations 12.63 and 12.64 is the relative size of $\alpha$ with respect to $\omega_o$. This is usually expressed in terms of the *Quality Factor* Q of the circuit defined by

$$Q \equiv \frac{\omega_o}{2\alpha}. \tag{12.65}$$

For the series circuit shown in Figure 12.15, $Q$ is evaluated by substituting Equations 12.44 and 12.45 into Equation 12.65. This yields

$$Q = \frac{1}{R}\sqrt{\frac{L}{C}}. \tag{12.66}$$

If the damping factor $\alpha$ is small compared to $\omega_o$, as is characteristic of an under-damped circuit, then $Q$ will be large, and the circuit will oscillate for a long time near the frequency $\omega_o$. For the series circuit, this is achieved with a small $R$, that is, with the corresponding resistor near a short circuit. Alternatively, to purposefully damp any oscillations and make them slower, one would make $R$ large.

The preceding discussion suggests an interesting interpretation of $Q$. From Equations 12.63 and 12.64 we see that the period of oscillation of the circuit states is $2\pi/\omega_d$. Thus, the period of $Q$ oscillations is $2\pi Q/\omega_d$. In the latter period of time the same equations show that the amplitude of the circuit states will decay by

$$e^{-2\pi Q\alpha/\omega_d} \approx e^{-\pi}$$

for $\omega_d \approx \omega_o$. Thus, as illustrated in Figure 12.18, the state amplitudes of an under-damped circuit will decay to approximately $e^{-\pi}$, or 4%, of their original values in $Q$ cycles of oscillation.[6]

6. Or to approximately 20% of their original values in $Q/2$ cycles.

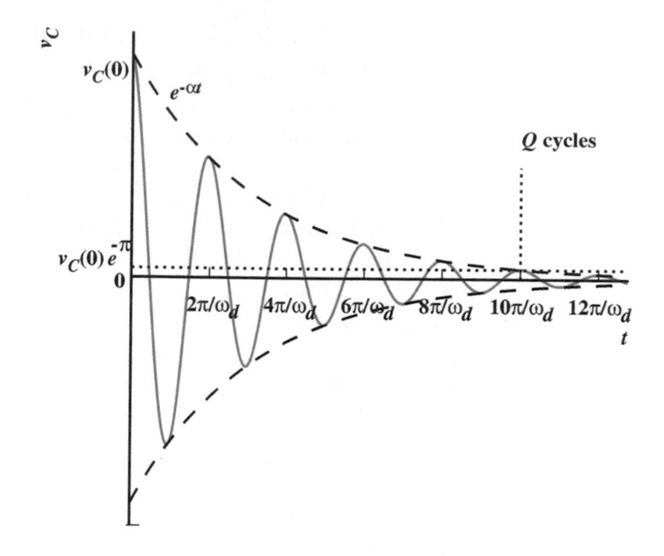


**FIGURE 12.18** Waveform of $v_C$ in under-damped, undriven, series RLC circuit for the case of $i_L(0) = 0$, with a $Q$ of 5.

## 12.2.2 OVER-DAMPED DYNAMICS

The case of over-damped dynamics is characterized by

$$\alpha > \omega_\circ$$

or, after substitution of Equations 12.44 and 12.45, by

$$R/2 > \sqrt{L/C}.$$

In this case, the quantity inside the radicals in Equations 12.46 and 12.47 is positive, and so both $s_1$ and $s_2$ are real. For this reason, the dynamic behavior of $v_C$ and $i_L$, as expressed by Equations 12.58 and 12.59, does not exhibit oscillation. Rather, it involves two real exponential functions that decay at different rates, as the two equations show.

The expressions for $v_C$ and $i_L$ for the case of $i_L(0) = 0$ with over-damping are obtained from Equations 12.58 and 12.59, and are shown here:

$$v_C(t) = \frac{s_2 v_C(0)}{(s_2 - s_1)} e^{s_1 t} + \frac{s_1 v_C(0)}{(s_1 - s_2)} e^{s_2 t} \tag{12.67}$$

$$i_L(t) = -s_1 \frac{C s_2 v_C(0)}{(s_2 - s_1)} e^{s_1 t} - s_2 \frac{C s_1 v_C(0)}{(s_1 - s_2)} e^{s_2 t}. \tag{12.68}$$

Since $\alpha > \omega_\circ$ for over-damped circuits, note that $s_1$ and $s_2$ are both real in the preceding two equations.

Figure 12.19 compares the waveforms of $v_C$ and $i_L$ for the case of $i_L(0) = 0$ with under-, over-, and critical-damping. We will address the critically-damped

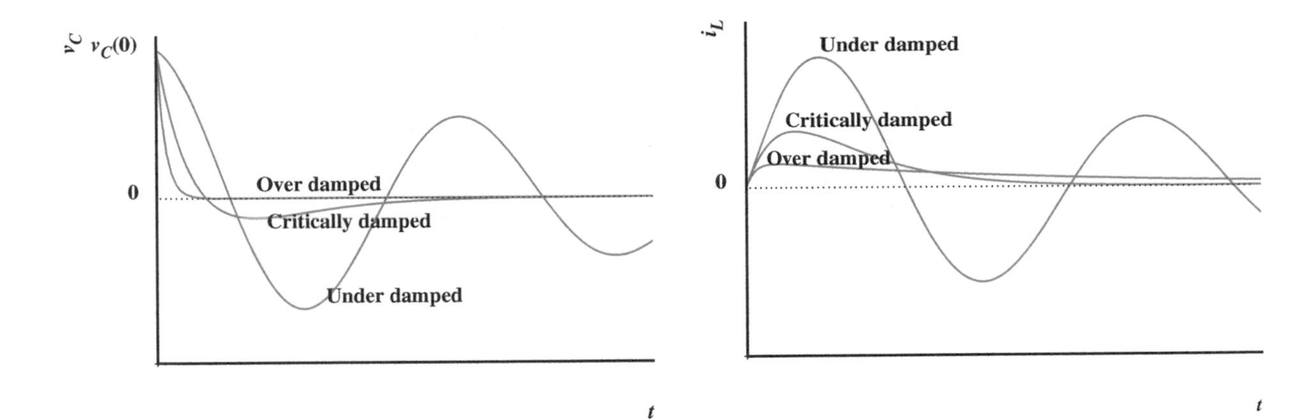


**FIGURE 12.19** Waveforms of $v_C$ and $i_L$ in undriven, series RLC circuit for the case of $i_L(0) = 0$ with under-, over-, and critical-damping.

circuit in the next subsection. Notice that the circuit displays ringing behavior only for the under-damped case.

As $R$ becomes large, in particular larger than $2\sqrt{L/C}$, it becomes a significant open circuit between the capacitor and inductor. In this way it absorbs the oscillating voltage that the capacitor and inductor share for smaller values of $R$. As a consequence, the energy exchange between the capacitor and inductor is interrupted, and the circuit ceases to oscillate. Instead, its behavior is more like that of an independent capacitor and an independent inductor discharging through the resistor. To see this, let us determine the asymptotic values of $s_1$ and $s_2$ as $R$ becomes large and hence as $\alpha$ becomes large. They are

$$s_1 = \alpha + \sqrt{\alpha^2 - \omega_\circ^2} = \alpha\left(1 + \sqrt{1 - \left(\frac{\omega_\circ}{\alpha}\right)^2}\right) \approx 2\alpha = \frac{R}{L} \tag{12.69}$$

$$s_2 = \alpha - \sqrt{\alpha^2 - \omega_\circ^2} = \alpha\left(1 - \sqrt{1 - \left(\frac{\omega_\circ}{\alpha}\right)^2}\right) \approx \alpha\frac{\omega_\circ^2}{2\alpha^2} = \frac{1}{RC}. \tag{12.70}$$

As expected, the corresponding time constants approach $L/R$ and $RC$, the time constants of an independent inductor-resistor circuit and an independent capacitor-resistor circuit. Note that, for over-damped dynamics, $\alpha > \omega_\circ$ from which it follows that $L/R$ is the faster time constant and $RC$ is the slower time constant.

### 12.2.3 CRITICALLY-DAMPED DYNAMICS

The case of critically-damped dynamics is characterized by

$$\alpha = \omega_\circ.$$

In this case, it again follows from Equations 12.46 and 12.47 that

$$s_1 = s_2 = -\alpha$$

and that the characteristic equation, Equation 12.43, has a repeated root. Because of this, $e^{s_1 t}$ and $e^{s_2 t}$ are no longer independent functions, and so the general solution for $v_1$ is no longer the superposition of these two functions as given by Equation 12.48. Rather, it is the superposition of the repeated exponential function

$$e^{s_1 t} = e^{s_2 t} = e^{-\alpha t} \text{ and } te^{-\alpha t}.$$

From this observation, and Equations 12.49 and 12.51, it follows that $v_C$ and $i_L$ will exhibit similar behavior.

Perhaps the easiest way to determine $v_C$ and $i_L$ for the case of critical-damping is to evaluate Equations 12.63 and 12.64 under the conditions of that case. To do so, observe from Equation 12.60 that, for critical-damping $\omega_\circ = \alpha$, and so $\omega_d = 0$. Therefore, we can obtain $v_C$ and $i_L$ for the case of critical-damping by evaluating Equations 12.63 and 12.64 in the limit $\omega_d \to 0$. To do so, substitute the approximations $\cos(\omega_d t) \approx 1$ and $\sin(\omega_d t) \approx \omega_d t$, as $\omega_d t \to 0$, into the first equalities in Equations 12.63 and 12.64, and cancel the resulting terms involving $\omega_d$. This results in

$$v_C(t) = v_C(0)e^{-\alpha t} + \frac{\alpha C v_C(0) - i_L(0)}{C} te^{-\alpha t} \tag{12.71}$$

$$i_L(t) = i_L(0)e^{-\alpha t} + \frac{v_C(0) - \alpha L i_L(0)}{L} te^{-\alpha t}. \tag{12.72}$$

From Equations 12.71 and 12.72 we see that $v_C$ and $i_L$ contain both the decaying exponential function $e^{-\alpha t}$ and the function $te^{-\alpha t}$, as expected.

---

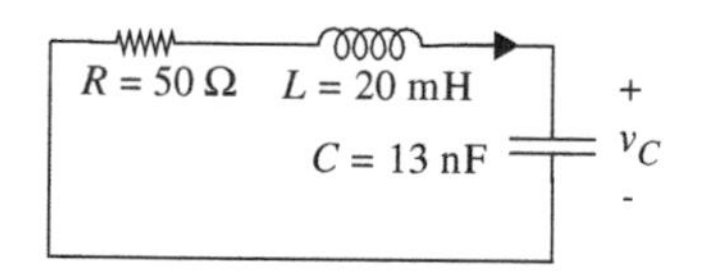


FIGURE 12.20 Undriven series RLC circuit.

EXAMPLE 12.6 ZERO INPUT RESPONSE OF A SERIES RLC CIRCUIT What is the general shape of the transient response of the undriven, series RLC circuit in Figure 12.20, assuming that the capacitor and inductor have some nonzero initial state?

The circuit in Figure 12.20 is an undriven series RLC circuit, the same as discussed in Section 12.2. Its damping factor is given by

$$\alpha = \frac{R}{2L} = 1,250 \text{ rad/s},$$

and its undamped resonance frequency

$$\omega_o = \sqrt{\frac{1}{LC}} = 62,017 \text{ rad/s},$$

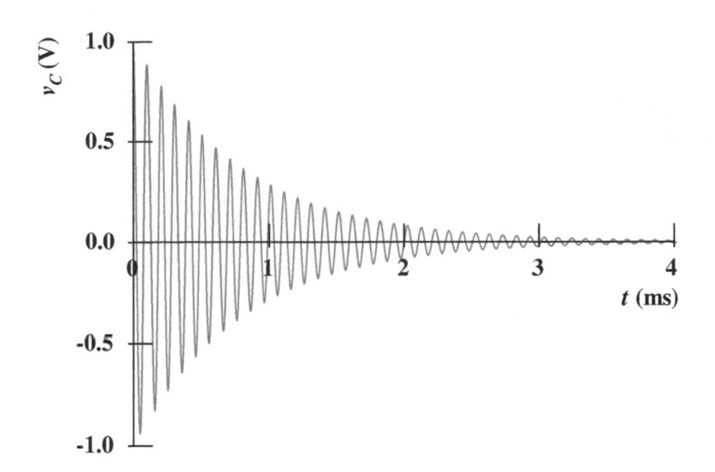


FIGURE 12.21 The response of the undriven series RLC circuit.

or, 9.8704 kHz. Since

$$\alpha < \omega_o$$

we conclude that the circuit is under-damped and will therefore produce a sinusoidal response.

As given in Equation 12.66, its $Q$ is $\sqrt{L/C}/R$, which evaluates to approximately 25.

Following the interpretation of $Q$ discussed in Section 12.2.1, the transient response of any branch variable in the circuit will be an underdamped decaying sinusoid that decays to 4% of its amplitude in 25 cycles of oscillation.

The oscillation frequency is given by Equation 12.60, which is

$$\omega_d = \sqrt{\omega_o^2 - \alpha^2} = 62{,}005 \text{ rad/s},$$

or, 9.8684 kHz.

The waveform for $v_C$ assuming $v_C(0) = 1$ V and $i_L(0) = 0$ is plotted in Figure 12.21.

## 12.3 STORED ENERGY IN TRANSIENT, SERIES RLC CIRCUIT

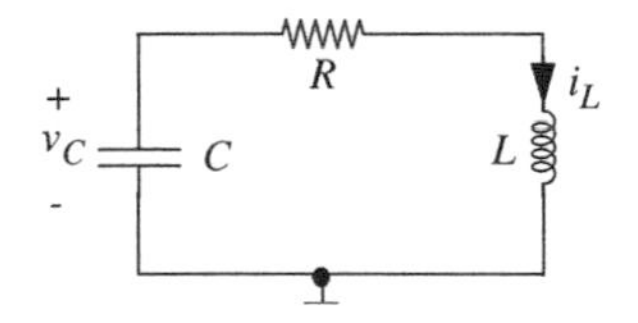


FIGURE 12.22 Analysis of the stored energy in a series RLC circuit.

Let us now calculate the stored energy in the series RLC circuit. Specifically, we calculate the decay of energy stored in the under-damped series RLC circuit (see Figure 12.22) previously analyzed in Section 12.2.1. Recall, the under-damped case applies when

$$\alpha < \omega_o.$$

To simplify matters, we will further assume that $Q \gg 1$. For $Q \gg 1$, $\alpha \ll \omega_o$, so $\omega_d \approx \omega_o$ and $\alpha \ll \omega_d$.

The voltage and current expressions for this special case are derived from Equations 12.63 and 12.64, and simplify to

$$v_C(t) \approx \sqrt{v_C^2(0) + \frac{L}{C} i_L^2(0)}\; e^{-\alpha t} \cos\left(\omega_d t + \tan^{-1}\left(\sqrt{\frac{L}{C}} \frac{i_L(0)}{v_C(0)}\right)\right) \tag{12.73}$$

$$i_L(t) \approx \sqrt{\frac{C}{L}} \sqrt{v_C^2(0) + \frac{L}{C} i_L^2(0)}\; e^{-\alpha t} \sin\left(\omega_d t + \tan^{-1}\left(\sqrt{\frac{L}{C}} \frac{i_L(0)}{v_C(0)}\right)\right). \tag{12.74}$$

The energy stored in the capacitor is given by

$$w_E(t) = \frac{1}{2} C v_C(t)^2$$

and that in the inductor is given by

$$w_M(t) = \frac{1}{2} L i_L(t)^2.$$

Substituting for $v_C$ and $i_L$, we get

$$w_E(t) \approx \left(\frac{1}{2} C v_C^2(0) + \frac{1}{2} L i_L^2(0)\right) e^{-2\alpha t} \cos^2\left(\omega_d t + \tan^{-1}\left(\sqrt{\frac{L}{C}} \frac{i_L(0)}{v_C(0)}\right)\right) \tag{12.75}$$

$$w_M(t) \approx \left(\frac{1}{2} C v_C^2(0) + \frac{1}{2} L i_L^2(0)\right) e^{-2\alpha t} \sin^2\left(\omega_d t + \tan^{-1}\left(\sqrt{\frac{L}{C}} \frac{i_L(0)}{v_C(0)}\right)\right). \tag{12.76}$$

The total energy stored in the circuit is the sum of the energy in the capacitor and that in the inductor and is given by

$$w_T(t) = w_E(t) + w_M(t) \approx \left(\frac{1}{2} C v_C^2(0) + \frac{1}{2} L i_L^2(0)\right) e^{-2\alpha t}. \tag{12.77}$$

Let us examine the expression for the energy stored in the capacitor (Equation 12.75). The expression is made up of three factors. The first

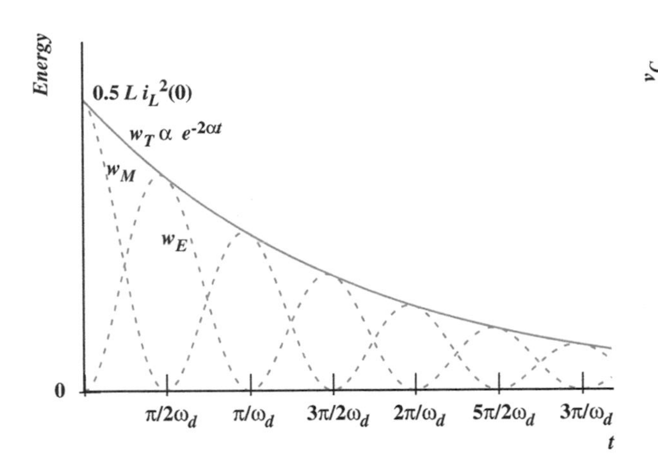


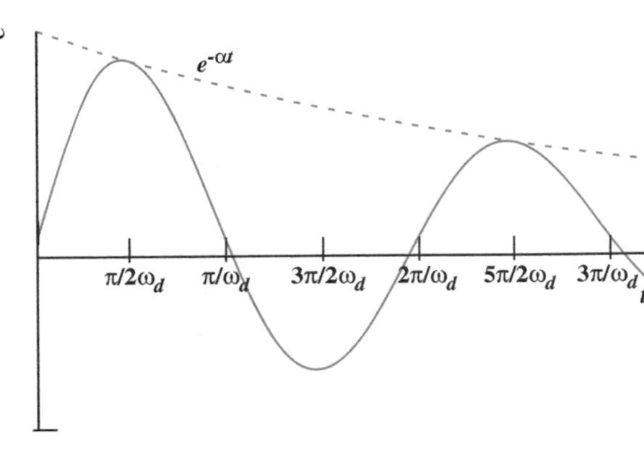


**FIGURE 12.23** Energy in RLC transient.

factor term,

$$\left(\frac{1}{2}Cv_C^2(0) + \frac{1}{2}Li_L^2(0)\right),$$

is the initial stored energy ($w_T(0)$) in the system. The second factor represents the decay of energy with time. Finally, by rewriting the third factor as

$$\cos^2\left(\omega_d t + \tan^{-1}\left(\sqrt{\frac{L}{C}}\frac{i_L(0)}{v_C(0)}\right)\right) = \frac{1 + \cos 2\left(\omega_d t + \tan^{-1}\left(\sqrt{\frac{L}{C}}\frac{i_L(0)}{v_C(0)}\right)\right)}{2}$$

we can see that the energy is sloshing back and forth between the capacitor and inductor, twice per cycle of the transient ring.

Through a comparison with the results of Section 12.1, we also see that for large $Q$, that is, for a relatively short-circuited $R$ and hence light damping, the introduction of a resistor into the circuit causes an exponential decay of the states and the stored energy. A sketch of the energy versus time is shown in Figure 12.23 for the case

$$v_C(0) = 0.$$

The length of time for the stored energy to dissipate can now be readily calculated. Obviously the controlling function in Equation 12.77 is the exponential term $e^{-2\alpha t}$. This can be rewritten using Equation 12.65 as

$$\text{decay} = e^{-\omega_o t/Q}. \tag{12.78}$$

Since $\omega_d \simeq \omega_o$ for large $Q$, then in $Q$ cycles

$$\omega_o t \simeq \omega_d t = 2\pi Q.$$

Hence in $Q$ cycles the decay term is

$$\text{decay} = e^{-2\pi}, \tag{12.79}$$

which is much smaller than one. Hence we conclude the following:

The energy in the transient decays to a very small value (approximately 0.2%) in about $Q$ cycles.

Notice again the significance of $Q$.

The preceding discussion suggests yet another interpretation for $Q$. From Equation 12.77 it is apparent that the energy stored in the circuit during an oscillation cycle is approximately $w_T(0)e^{-2\alpha t}$. From Equations 12.73 and 12.77 it is also apparent that the average value of $v_C^2$ is approximately $w_T(0)e^{-2\alpha t}/C$ during that same cycle. Therefore the energy dissipated in the resistor during the cycle is approximately $2\pi w_T(0)e^{-2\alpha t}/RC\omega_d$. Division of the stored energy by the dissipated energy leads to the conclusion that

$$Q \approx 2\pi \frac{\text{Energy stored during an oscillation cycle}}{\text{Energy dissipated during an oscillation cycle}} \tag{12.80}$$

as an energy interpretation for $Q$. This interpretation is common to all under-damped second-order systems.

## WWW 12.4 UNDRIVEN, PARALLEL RLC CIRCUIT*

### WWW 12.4.1 UNDER-DAMPED DYNAMICS

### WWW 12.4.2 OVER-DAMPED DYNAMICS

### WWW 12.4.3 CRITICALLY-DAMPED DYNAMICS

## 12.5 DRIVEN, SERIES RLC CIRCUIT

In the next two sections, we combine the results of the previous sections and study the behavior of second-order circuits that have both damping and an external input. In this section, we study the series circuit obtained by including a series voltage source in Figure 12.15. The next section studies the parallel circuit obtained by including a parallel current source in WWW Figure 12.24. Through the use of Thévenin and Norton equivalence, the results actually apply to many other circuits as well.

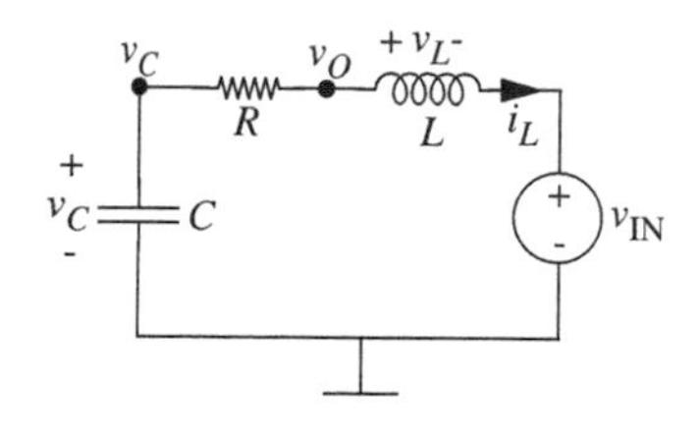


FIGURE 12.26 The series second-order circuit with a resistor, capacitor, inductor, and voltage source.

Consider now the circuit shown in Figure 12.26. As in previous subsections of this chapter, we will analyze the behavior of this circuit using the node method beginning at Step 3. In doing so, we will follow the analyses presented in Subsection 12.2 very closely, using $v_C$ and $v_O$ as the unknown node voltages.

We begin by completing Step 3 of the node method. To do so, we write KCL in terms of $v_C$ and $v_O$ for the node at which they are defined to obtain

$$C\frac{dv_C(t)}{dt} + \frac{v_C(t) - v_O(t)}{R} = 0 \tag{12.116}$$

for the node at which $v_C$ is defined, and

$$\frac{v_O(t) - v_C(t)}{R} + \frac{1}{L}\int_{-\infty}^{t}(v_O(\tilde{t}) - v_{IN}(\tilde{t}))d\tilde{t} = 0 \tag{12.117}$$

for the node at which $v_O$ is defined. To treat these two equations simultaneously, we first use Equation 12.116 to determine $v_O$ in terms of $v_C$, and then substitute the result into Equation 12.117 to obtain a second-order differential equation in $v_C$. This yields

$$v_O(t) = RC\frac{dv_C(t)}{dt} + v_C(t) \tag{12.118}$$

$$\frac{d^2v_C(t)}{dt^2} + \frac{R}{L}\frac{dv_C(t)}{dt} + \frac{1}{LC}v_C(t) = \frac{1}{LC}v_{IN}(t). \tag{12.119}$$

Note that to arrive at Equation 12.119 we have divided Equation 12.117 by C, and differentiated it with respect to time. Unlike Equation 12.4, Equation 12.119 is an inhomogeneous differential equation because it is driven by the external signal $v_{IN}$.

To complete the node analysis, we complete Step 4 and 5 by solving Equation 12.119 for $v_C$, and using it to determine $i_L$ and any other variables of interest. Since the capacitor and inductor share the same current, $i_L$ in particular can be obtained from

$$i_L(t) = -C\frac{dv_C(t)}{dt}. \tag{12.120}$$

Here, the negative sign follows from the opposing definitions of positive capacitor and inductor currents.

To solve Equation 12.119, we employ our usual method of solving differential equations:

1. Find the homogeneous solution $v_{CH}(t)$.
2. Find the particular solution $v_{CP}(t)$.
3. The total solution is then the sum of the homogeneous solution and the particular solution as follows:

$$v_C(t) = v_{CH}(t) + v_{CP}(t).$$

Then, use the initial conditions to solve for the remaining constants.

The homogeneous solution $v_{CH}(t)$ to Equation 12.119 is obtained by solving this differential equation with $v_{IN} \equiv 0$. With $v_{IN} \equiv 0$, the circuit shown in Figure 12.26 is identical to that shown in Figure 12.15, and so the two circuits have the same homogeneous equation. Thus, borrowing the homogeneous solution from Equation 12.48, we can write

$$v_{CH}(t) = K_1 e^{s_1 t} + K_2 e^{s_2 t} \tag{12.121}$$

where $K_1$ and $K_2$ are as yet unknown constants, that will be determined from the initial conditions after the total solution has been formed. $s_1$ and $s_2$ are the roots of the characteristic equation

$$s^2 + 2\alpha s + \omega_\circ^2 = 0 \tag{12.122}$$

where $\alpha$ and $\omega_\circ$ are given by

$$\alpha \equiv \frac{R}{2L} \tag{12.123}$$

$$\omega_\circ \equiv \sqrt{\frac{1}{LC}}; \tag{12.124}$$

The roots are given by

$$s_1 = -\alpha + \sqrt{\alpha^2 - \omega_\circ^2} \tag{12.125}$$

$$s_2 = -\alpha - \sqrt{\alpha^2 - \omega_\circ^2}. \tag{12.126}$$

As with the undriven, series RLC circuit in Section 12.2, the circuit exhibits under-damped, over-damped, or critically-damped behavior depending on the relative values of $\alpha$ and $\omega_\circ$:

$$\begin{aligned} \alpha < \omega_\circ \quad &\Rightarrow \quad \text{under-damped dynamics;} \\ \alpha = \omega_\circ \quad &\Rightarrow \quad \text{critically-damped dynamics;} \\ \alpha > \omega_\circ \quad &\Rightarrow \quad \text{over-damped dynamics.} \end{aligned}$$

For brevity, the rest of the section will assume that

$$\alpha < \omega_\circ$$

so that the circuit displays under-damped dynamics. For the under-damped case, $s_1$ and $s_2$ are complex and can be written as

$$s_1 = -\alpha + j\omega_d$$

$$s_2 = -\alpha - j\omega_d$$

where, as in Section 12.2.1,

$$\omega_d \equiv \sqrt{\omega_\circ^2 - \alpha^2}. \tag{12.127}$$

Since $s_1$ and $s_2$ are complex, the exponentials in Equation 12.121 are also complex. To expose the resulting oscillatory and decaying behavior, we can rewrite into a more intuitive form the homogeneous solution in Equation 12.121 using the complex notation for $s_1$ and $s_2$ as in Equation 12.127 and the Euler relation:

$$v_{CH}(t) = A_1 e^{-\alpha t}\cos(\omega_d t) + A_2 e^{-\alpha t}\sin(\omega_d t) \tag{12.128}$$

where $\omega_d$ is defined as in Equation 12.127, and where $A_1$ and $A_2$ are unknown constants we will evaluate later depending on the initial conditions of the circuit.

Next, we need to find $v_{CP}(t)$. Knowing it, we can write the total solution as

$$v_C(t) = v_{CP}(t) + v_{CH}(t) = v_{CP}(t) + A_1 e^{-\alpha t}\cos(\omega_d t) + A_2 e^{-\alpha t}\sin(\omega_d t). \tag{12.129}$$

At this point, only $v_{CP}$, and $A_1$ and $A_2$, remain as unknowns.

We will now proceed to find $v_{CP}$, and then $A_1$ and $A_2$, for two cases of $v_{IN}$, namely a step and an impulse. That is, we will proceed to find the step response and the impulse response of the circuit. To simplify matters, we will continue to assume that the circuit is under-damped, that both the step and the impulse occur at $t = 0$, and that the circuit is initially at rest prior to that time. The latter assumption suggests that we are interested in the zero-state response[7] and provides these initial conditions

$$v_C(0) = 0$$

$$i_L(0) = 0$$

for the solution of Equation 12.119 after the step and impulse occur, that is, for $t > 0$. Arbitrary initial conditions will modify only $A_1$ and $A_2$.

## 12.5.1 STEP RESPONSE

Let $v_{IN}$ be the voltage step given by

$$v_{IN}(t) = V_\circ u(t) \tag{12.130}$$

and shown in Figure 12.27.

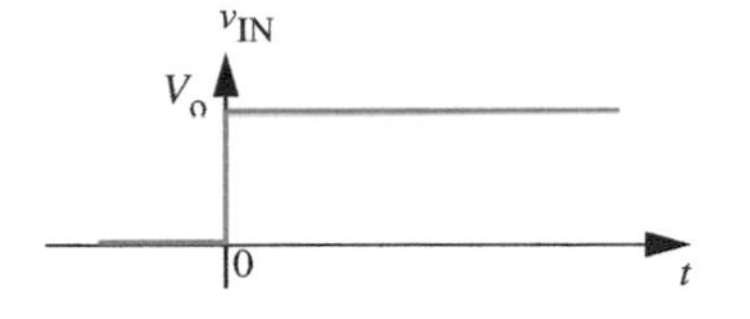


FIGURE 12.27 A voltage step input.

7. Recall, the zero-state response is the response of the circuit for zero initial state.

To find $v_{CP}$, we substitute Equation 12.130 into Equation 12.119 and obtain

$$\frac{d^2 v_C(t)}{dt^2} + \frac{R}{L}\frac{dv_C(t)}{dt} + \frac{1}{LC} v_C(t) = \frac{1}{LC} V_\circ \tag{12.131}$$

for $t > 0$. Any function that satisfies Equation 12.131 for $t > 0$ is an acceptable $v_{CP}$. It does not matter whether that function satisfies the initial conditions or not. One such function is

$$v_{CP}(t) = V_\circ. \tag{12.132}$$

Thus, we have the particular solution for a step input. The total solution is given by summing the homogeneous solution (Equation 12.128) and the particular solution (Equation 12.132) as

$$v_C(t) = V_\circ + A_1 e^{-\alpha t}\cos(\omega_d t) + A_2 e^{-\alpha t}\sin(\omega_d t), \tag{12.133}$$

again for $t > 0$. Additionally, the substitution of Equation 12.133 into Equation 12.120 yields

$$i_L(t) = (\alpha C A_1 - \omega_d C A_2)e^{-\alpha t}\cos(\omega_d t) + (\omega_d C A_1 + \alpha C A_2)e^{-\alpha t}\sin(\omega_d t), \tag{12.134}$$

also for $t > 0$. Now only $A_1$ and $A_2$ remain as unknowns.

In Chapter 9, we saw that the voltage across a capacitor is continuous unless the current through it contains an impulse. We also saw that the current through an inductor is continuous unless the voltage across it contains an impulse. Since $v_{IN}$ contains no impulses, we can therefore assume that both $v_C$ and $i_L$ are continuous across the step at $t = 0$. Consequently, since both states are zero for $t \leq 0$, Equations 12.133 and 12.134 must both evaluate to zero as $t \to 0$. This observation allows us to use the initial conditions to determine $A_1$ and $A_2$. Evaluation of both equations as $t \to 0$, followed by the substitution of the initial conditions, yields

$$v_C(0) = V_\circ + A_1 \tag{12.135}$$

$$i_L(0) = \alpha C A_1 - \omega_d C A_2. \tag{12.136}$$

Equations 12.135 and 12.136 can be solved to yield

$$A_1 = v_C(0) - V_\circ \tag{12.137}$$

$$A_2 = \frac{\alpha v_C(0) - \alpha V_\circ - i_L(0)/C}{\omega_d}. \tag{12.138}$$

Since we are given that $v_C(0) = 0$ and $i_L(0) = 0$,

$$A_1 = -V_o \tag{12.139}$$

$$A_2 = -\frac{\alpha}{\omega_d}V_o. \tag{12.140}$$

Finally, the substitution of Equations 12.139 and 12.140 into Equations 12.133 and 12.134 yields[8]

$$v_C(t) = V_o\left(1 - \frac{\omega_o}{\omega_d}e^{-\alpha t}\cos\left(\omega_d t - \tan^{-1}\left(\frac{\alpha}{\omega_d}\right)\right)\right)u(t) \tag{12.141}$$

$$i_L(t) = -\frac{V_o}{\omega_d L}e^{-\alpha t}\sin(\omega_d t)u(t); \tag{12.142}$$

Equations 12.124 and 12.127 have also been used to simplify the results. Note that the unit step function $u$ has been introduced into Equations 12.141 and 12.142 so that they are valid for all time. The validity of Equations 12.141 and 12.142 can be demonstrated by observing that they satisfy the initial conditions, and Equations 12.119 and 12.120, respectively, for all time. Because they do, our assumption that the states are continuous at $t = 0$ is justified.

Figure 12.28 shows $v_C$ and $i_L$ as given by Equations 12.141 and 12.142. Notice the overshoot of $v_C$ above the input voltage $V_o$ during the initial transient. Although the average value of $v_C$ is close to $V_o$ during the transient, the peak value is closer to $2V_o$.

As expected, the ringing in both states now decays as $t \to \infty$. This decay is well characterized by the quality factor $Q$, as defined in WWW Equation 12.108 and discussed shortly thereafter. In fact, because the circuits shown in Figures 12.15 and 12.26 have the same homogeneous response, the entire discussion of $\alpha$, $\omega_d$, and $\omega_o$ given in Subsection 12.2 applies here as well.

Another observation concerns the short-time behavior of the circuit. We have seen in Chapter 10 that the transient behavior of an uncharged capacitor is to act as a short circuit during the early part of a transient, while the corresponding transient behavior of an uncharged inductor is to act as an open circuit. This

---

8. We can also substitute the expressions for $A_1$ and $A_2$ with nonzero initial conditions as given by Equations 12.137 and 12.138 into Equation 12.133, for example, and obtain a more general form of the solution

$$v_C(t) = V_o + (v_C(0) - V_o)e^{-\alpha t}\cos\omega_d t + \frac{(\alpha v_C(0) - \alpha V_o - i_L(0)/C)}{\omega_d}e^{-\alpha t}\sin\omega_d t.$$

We can obtain the ZSR from this general solution by substituting $v_C(0) = 0$ and $i_L(0) = 0$. Alternatively, we can obtain the ZIR by substituting $V_o = 0$ and using the appropriate initial conditions for the state variables.

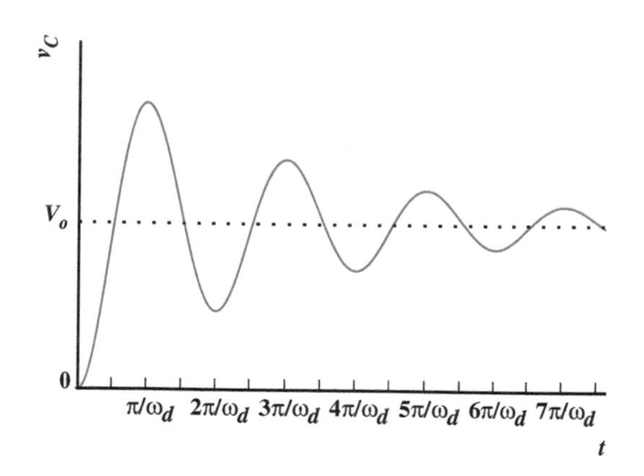


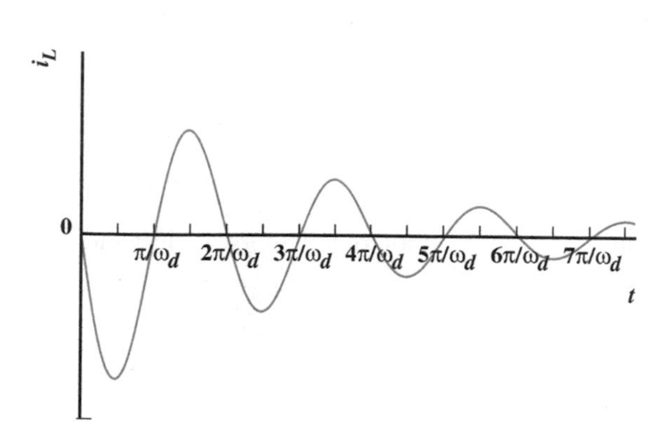


**FIGURE 12.28** $v_C$ and $i_L$ for the series RLC circuit circuit shown in Figure 12.26 for the case of a step input through $v_{IN}$.

behavior is observed in Figure 12.28 since $v_{IN}$ drops entirely across the inductor (and $v_C$ is 0) at the start of the transient, and $i_L$ ramps up correspondingly.

A related observation concerns the long-time behavior of the circuit. We have also seen in Chapter 10 that the transient behavior of a capacitor is to act as an open circuit as $t \to \infty$, while the corresponding transient behavior of an inductor is to act as a short circuit. This behavior is also observed in Figure 12.28, since $v_{IN}$ appears entirely across the capacitor as $t \to \infty$, since $v_C \to V_\circ$.

Figure 12.28 also explains the ringing transients seen in Figure 12.3. This is because the circuit shown in Figure 12.26, from which the response in Figure 12.28 is derived, is essentially the same as the subcircuit containing the inductor $L_I$ shown in Figure 12.5.

---

## EXAMPLE 12.7 STEP RESPONSE OF SERIES RLC CIRCUIT

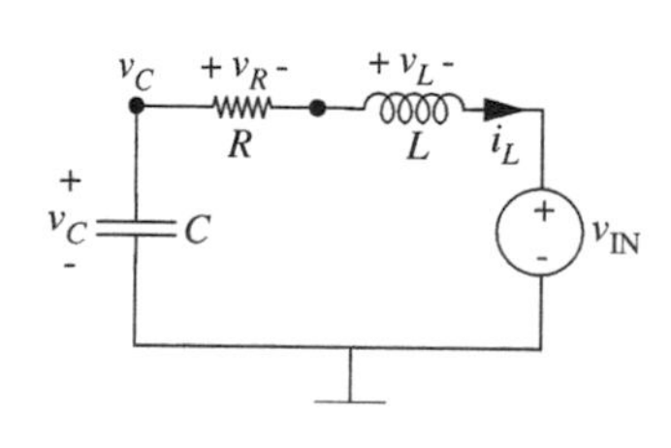


**FIGURE 12.29** A driven, series RLC circuit.

In this example we evaluate the step response of the circuit shown in Figure 12.29 with $R = 50\ \Omega$, $L = 20$ mH, and $C = 13$ pF. Assume that $v_{IN}$ steps from 0 to $V_\circ$ at $t = 0$, where $V_\circ = 1$ V. In other words, $v_{IN} = V_\circ u(t) = u(t)$ V, where $u(t)$ is the unit step function. Let us suppose that we are interested in obtaining $i_L$, $v_R$, $v_C$, and $v_L$.

Following Equations 12.123, 12.124, and 12.127, respectively, this circuit is characterized by $\alpha = 1.25$ krad/s, $\omega_\circ = 62.017$ krad/s, and $\omega_d = 62.005$ krad/s.

From Equation 12.142

$$i_L(t) = -0.8064 \text{ mA } e^{-(1250\ s^{-1}\ t)} \sin\left(62005\ \frac{\text{rad}}{\text{s}}\ t\right) u(t).$$

Multiplying this result by 50 Ω yields

$$v_R(t) = -40.32 \text{ mV } e^{-(1250\ s^{-1}\ t)} \sin\left(62005\ \frac{\text{rad}}{\text{s}}\ t\right) u(t).$$

From Equation 12.141

$$v_C(t) = 1\text{ V}\left(1 - 1.0002\, e^{-(1250\ s^{-1}\ t)} \cos\left(62005\ \frac{\text{rad}}{\text{s}}\ t - 20.14\ \text{mrad}\right)\right) u(t).$$

Finally, using $v_L = L\, di_L/dt$,

$$v_L(t) = -1.0002\text{ V}\, e^{-(1250\ s^{-1}\ t)} \cos\left(62005\ \frac{\text{rad}}{\text{s}}\ t + 20.14\ \text{mrad}\right) u(t).$$

Since $i_L$ is common to all four circuit elements in Figure 12.29, all branch variables are now known. Finally, as a quick check on our work, note from the preceding that $v_C - v_R - v_L = v_{IN} = 1\text{ V}\, u(t)$ as is required by applying KVL around the one loop in the circuit.

### 12.5.2 IMPULSE RESPONSE

Let $v_{IN}$ be the impulse given by

$$v_{IN} = \Lambda_o \delta(t) \tag{12.143}$$

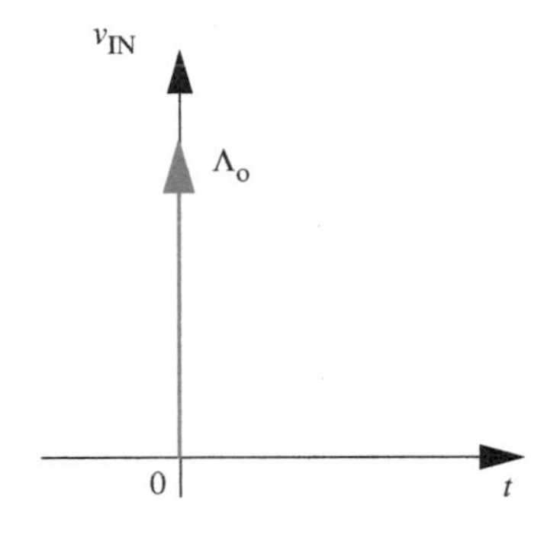


FIGURE 12.30 The voltage impulse $v_{IN}$.

as shown in Figure 12.30. Because $v_{IN}$ is an impulse, it vanishes for $t > 0$. Therefore, Equation 12.119 reduces to a homogeneous equation for $t > 0$. Thus the impulse response of the circuit is essentially a homogeneous response, this response is identical to that studied in Subsection 12.2. In fact, as observed in Section 10.6.4 for a first-order circuit, the role of the impulse is to establish initial conditions for the subsequent homogeneous response.

As previously observed on several occasions, the transient behavior of an uncharged capacitor is to act as a short circuit during the early part of a transient, while the corresponding transient behavior of an uncharged inductor is to act as an open circuit. Since the inductor acts as an open circuit during a sudden transition, the impulse in $v_{IN}$ falls entirely across the inductor while $v_C$ remains zero at $t = 0$. An important consequence of this is that the flux linkage $\Lambda_o$ delivered by $v_{IN}$ is delivered entirely to the inductor, and so $i_L$ steps to $-\Lambda_o/L$ at $t = 0$. This establishes the initial conditions immediately after the impulse. In other words, these will be our initial conditions for the rest of our analysis:

$$i_L(0) = -\frac{\Lambda_o}{L}$$

$$v_C(0) = 0.$$

Since Equation 12.119 reduces to a homogeneous equation, the simplest acceptable particular solution is

$$v_{CP}(t) = 0. \tag{12.144}$$

We obtain the total solution by adding this particular solution to the homogeneous solution described by Equation 12.128:

$$v_C(t) = v_{CH}(t) + v_{CP}(t) = A_1 e^{-\alpha t} \cos(\omega_d t) + A_2 e^{-\alpha t} \sin(\omega_d t), \qquad (12.145)$$

again for $t > 0$. Additionally, since the capacitor and inductor share the same current, $i_L$ can be obtained by substituting Equation 12.145 into

$$i_L(t) = -C \frac{dv_C(t)}{dt} \qquad (12.146)$$

as follows:

$$i_L(t) = (\alpha C A_1 - \omega_d C A_2) e^{-\alpha t} \cos(\omega_d t) + (\omega_d C A_1 + \alpha C A_2) e^{-\alpha t} \sin(\omega_d t), \qquad (12.147)$$

also for $t > 0$.

We will now determine the unknowns $A_1$ and $A_2$ from the initial conditions.

$$v_C(0) = A_1 = 0 \qquad (12.148)$$

$$i_L(0) = -\omega_d C A_2 - \alpha C A_1 = -\frac{\Lambda_\circ}{L}. \qquad (12.149)$$

Equations 12.148 and 12.149 can be rearranged to yield

$$A_1 = 0 \qquad (12.150)$$

$$A_2 = \frac{\Lambda_\circ}{LC\omega_d} - \frac{\alpha v_C(0)}{\omega_d} = \frac{\Lambda_\circ}{LC\omega_d}. \qquad (12.151)$$

Finally, the substitution of Equations 12.150 and 12.151 into Equations 12.145 and 12.147 yields

$$v_C(t) = \left(\frac{\Lambda_\circ}{LC\omega_d}\right) e^{-\alpha t} \sin(\omega_d t) u(t) \qquad (12.152)$$

$$i_L(t) = -\left(\frac{\Lambda_\circ}{L}\right) e^{-\alpha t} \cos(\omega_d t) - \left(\frac{\alpha \Lambda_\circ}{L\omega_d}\right) e^{-\alpha t} \sin(\omega_d t) u(t), \qquad (12.153)$$

where the unit step function $u$ has been introduced into Equations 12.152 and 12.153 so they are valid for all time.

Finally, we may also determine $v_L$ from KVL according to

$$v_L(t) = v_C(t) - v_{IN}(t) - i_L(t) R. \qquad (12.154)$$

Note that Equations 12.152 and 12.153 satisfy the initial conditions established by the impulse, and that they respectively satisfy Equations 12.119 and 12.120 for all time. Because they do, they justify our interpretation of the circuit behavior at $t = 0$.

Because the impulse response of the circuit is essentially a homogeneous response, this response is identical to that studied in Section 12.2. In fact, Equations 12.63 and 12.64 are identical to Equations 12.152 and 12.153 with the substitution of zero for $v_C(0)$ and $-\Lambda_\circ/L$ for $i_L(0)$ in the former equations.

Alternatively, we can determine the circuit response to the impulse input by simply differentiating the step response given in Equations 12.224 and 12.225. As discussed in Section 12.6, we do this by applying the operator $(\Lambda_\circ/V_\circ)d/dt$ because the impulse input can be derived by applying the same operator to the step input as

$$(\Lambda_\circ/V_\circ)\frac{d}{dt}V_\circ u(t) = \Lambda_\circ\delta(t).$$

This results in

$$\begin{aligned}
v_C(t) &= \frac{\Lambda_\circ}{V_\circ}\frac{d}{dt}\left(V_\circ\left(1 - \frac{\omega_\circ}{\omega_d}e^{-\alpha t}\cos\left(\omega_d t - \tan^{-1}\left(\frac{\alpha}{\omega_d}\right)\right)\right)u(t)\right) \\
&= \omega_\circ\Lambda_\circ e^{-\alpha t}\sin\left(\omega_d t - \tan^{-1}\left(\frac{\alpha}{\omega_d}\right)\right)u(t) \\
&\quad + \frac{\alpha\omega_\circ\Lambda_\circ}{\omega_d}e^{-\alpha t}\cos\left(\omega_d t - \tan^{-1}\left(\frac{\alpha}{\omega_d}\right)\right)u(t) \\
&\quad + \Lambda_\circ\left(1 - \frac{\omega_\circ}{\omega_d}e^{-\alpha t}\cos\left(\omega_d t - \tan^{-1}\left(\frac{\alpha}{\omega_d}\right)\right)\right)\delta(t) \\
&= \Lambda_\circ\frac{\omega_\circ^2}{\omega_d}e^{-\alpha t}\sin(\omega_d t)u(t)
\end{aligned} \tag{12.155}$$

$$\begin{aligned}
i_L(t) &= \frac{\Lambda_\circ}{V_\circ}\frac{d}{dt}\left(-\frac{V_\circ}{\omega_d L}e^{-\alpha t}\sin(\omega_d t)u(t)\right) \\
&= -\frac{\Lambda_\circ}{L}e^{-\alpha t}\cos(\omega_d t)u(t) + \frac{\alpha\Lambda_\circ}{\omega_d L}e^{-\alpha t}\sin(\omega_d t)u(t) \\
&\quad - \frac{\Lambda_\circ}{\omega_d L}e^{-\alpha t}\sin(\omega_d t)\delta(t) \\
&= -\frac{\Lambda_\circ}{L}\frac{\omega_\circ}{\omega_d}e^{-\alpha t}\cos\left(\omega_d t + \tan^{-1}\left(\frac{\alpha}{\omega_d}\right)\right)u(t).
\end{aligned} \tag{12.156}$$

Note that terms involving the impulse $\delta$ vanish in Equations 12.155 and 12.156 because $\delta$ is itself zero everywhere except $t = 0$, and the coefficients of the impulse in both the equations are zero at $t = 0$.[9]

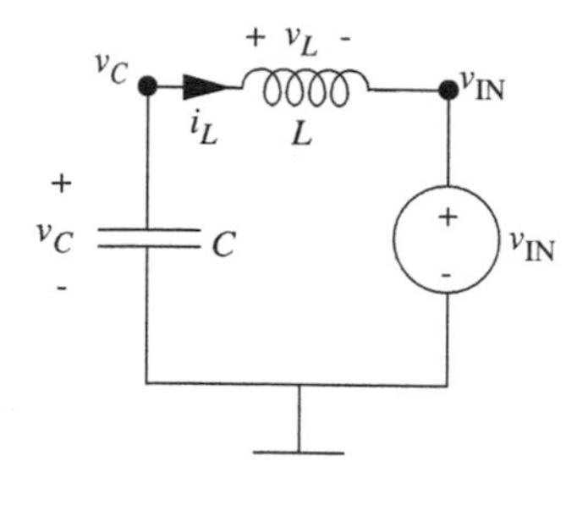


**FIGURE 12.31** The driven, series LC circuit.

EXAMPLE 12.8 SERIES LC CIRCUIT DRIVEN BY A STEP
Consider the driven, series LC circuit in Figure 12.31. This circuit differs from the series RLC circuit in Figure 12.26 in that the series resistance is zero. Suppose that we are interested in the ZSR for the circuit for a voltage step drive given by

$$v_{IN}(t) = V_\circ u(t).$$

The zero-state response of this circuit to the step input can be obtained from the ZSR of the series RLC circuit (Equations 12.141 and 12.142) by assuming $R = 0$. When the resistance is zero,

$$\alpha = 0$$

and

$$\omega_d = \omega_\circ$$

(see Equations 12.123 and 12.210). Thus Equations 12.141 and 12.142 simplify to

$$v_C(t) = V_\circ(1 - \cos(\omega_\circ t))u(t) \qquad (12.157)$$

$$i_L(t) = \frac{V_\circ}{\omega_\circ L}\sin(\omega_\circ t)u(t). \qquad (12.158)$$

Figure 12.32 shows $v_C$ and $i_L$ as given by Equations 12.157 and 12.158. Notice that the oscillations in both states do not decay and continue indefinitely because there are no resistors in the circuit to damp their response. Similarly, here, $v_C$ undergoes a two-fold overshoot. However, the average value of $v_C$ is $V_\circ$.

EXAMPLE 12.9 CASCADED INVERTERS We now apply the results of this section to a practical problem, namely the study of the cascaded inverters shown in Figure 12.1, and modeled in Figure 12.4. What is new in the latter figure is the parasitic wiring inductance between the two inverters, which makes the circuit a second-order circuit. As discussed in Subsections 9.3.2 and 9.3.3, parasitic inductance is present in all wiring. In some cases it is particularly important, and these cases are generally characterized by a large $Q$ when they are second-order in nature.

9. Note that Figure 12.17 is helpful in simplifying trigonometric expressions involving $\alpha$, $\omega_d$, and $\omega_\circ$.

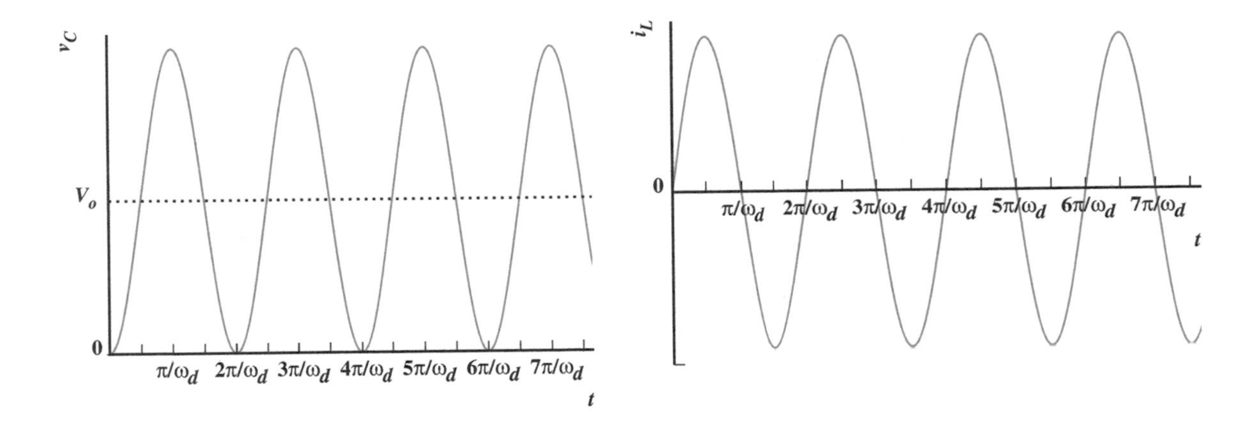


**FIGURE 12.32** $v_C$ and $i_L$ for the series LC circuit circuit for the case of a step input through $v_{IN}$.

Long clock and data lines between gates within computer chips can have significant parasitic inductance, perhaps as much as tens of pico Henries per millimeter. This inductance becomes increasingly important as clock rates increase. Similarly, long power supply lines within computer chips and on printed-circuit boards (see Figure 12.33) can have significant parasitic inductance. This inductance becomes increasingly important as circuits are switched on and off dynamically in an effort to save power and reduce heat dissipation when the circuits are functionally inactive. Finally, as data lines pass across chip boundaries they acquire parasitic inductance arising from the bonding wires within the chips (see Figure 12.34) and the interconnection between the chips. This inductance can be as high as tens of nano Henries or more.

For the sake of discussion, consider Figure 12.4 in which one inverter, acting as a pad buffer, drives a signal off one chip to the input of a second inverter on another chip.

**FIGURE 12.33** An example printed-circuit board showing wiring traces. (Photograph Courtesy of Maxim Integrated Products.)

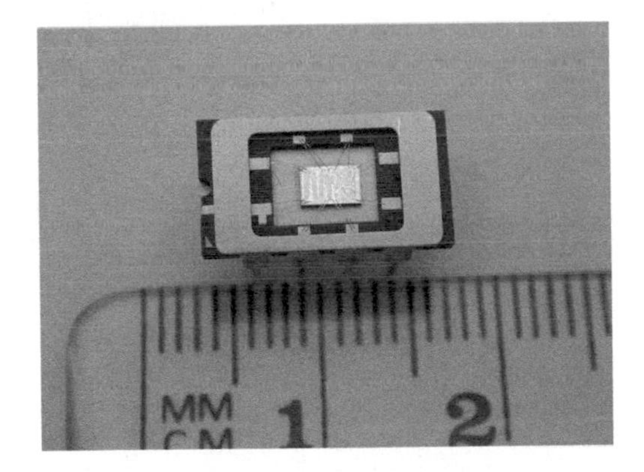


**FIGURE 12.34** Bond wires connecting integrated circuit to the posts of a package head frame. (Photograph Courtesy of Maxim Integrated Products.)

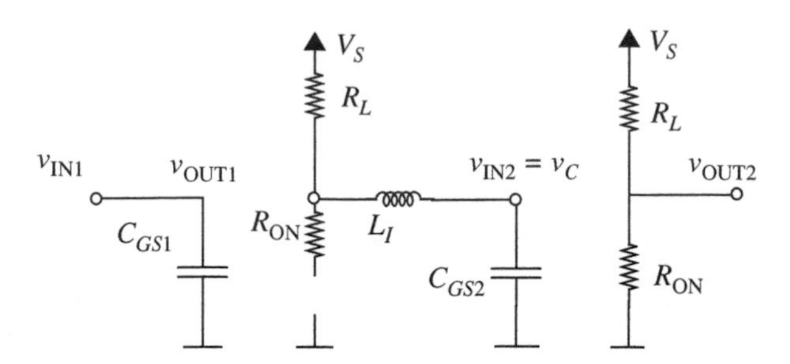


**FIGURE 12.35** Circuit model of the cascaded inverters when the input at $v_{IN1}$ is low.

The corresponding circuit model shown in Figure 12.5 is repeated here in Figure 12.35 for convenience. Consistent with this example, we assume that the driving inverter is characterized by

$$R_L = 900\ \Omega$$
$$R_{ON} = 100\ \Omega$$
$$V_S = 5\ \text{V}.$$

We also assume that the receiving inverter is characterized by

$$C_{GS2} = 0.1\ \text{pF}$$

and that the parasitic wiring inductance is well modeled by

$$L_I = 100\ \text{nH}.$$

Given this system description, let us examine the transient voltage that appears at $v_C$, where

$$v_C = v_{IN2}$$

is the gate voltage of the receiving inverter, and determine its impact on the propagation delay. Here there are two separate cases, namely a rising transient and a falling transient at the node with voltage $v_C$.

The complexity of handling successive rising and falling transients is most easily managed by modeling the driving inverter with its Thévenin equivalent, which is summarized in Figure 12.36. This in turn leads to a simpler model for the interconnection as shown in Figure 12.37. Figure 12.36 shows that

$$v_{TH} = 5\ \text{V}$$
$$R_{TH} = 900\ \Omega$$

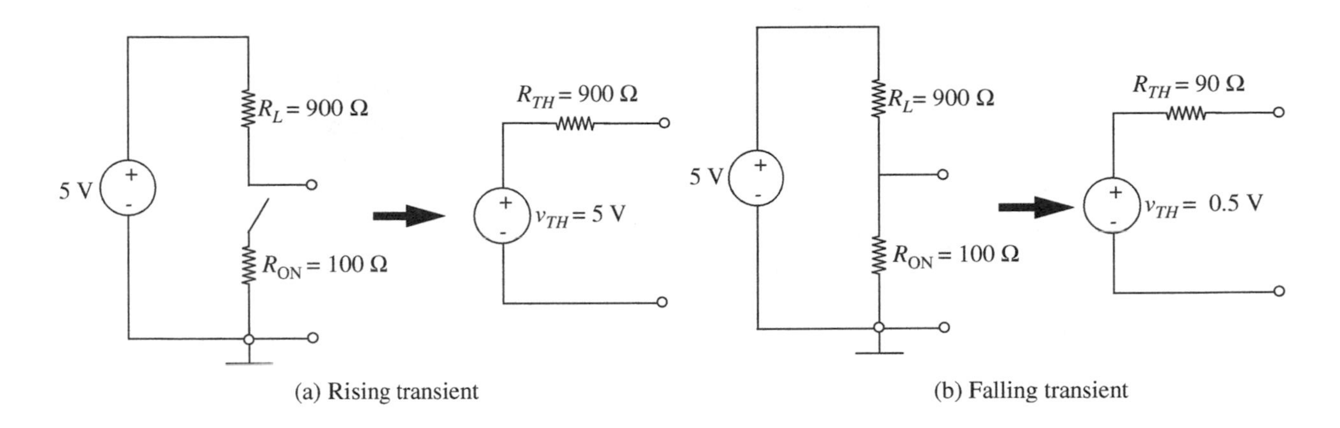


FIGURE 12.36 Thévenin equivalent of the driving inverter during a rising and falling transition.

when the driving inverter drives a high output during a rising transient. Similarly, it shows that

$$v_{TH} = 0.5 \text{ V}$$

$$R_{TH} = 90\ \Omega$$

when the driving inverter drives a low output during a falling transient.

There are two complications to Figure 12.36 that we have not studied so far. The first complication is that $R_{TH}$ is different for the rising and falling transitions. However, this complication is inconsequential because $R_{TH}$ is piecewise constant. Thus, we can separately analyze the rising transient during which $R_{TH} = 900\ \Omega$, and the falling transient during which $R_{TH} = 90\ \Omega$. Note too that the circuit remains linear even though $R_{TH}$ is time varying.

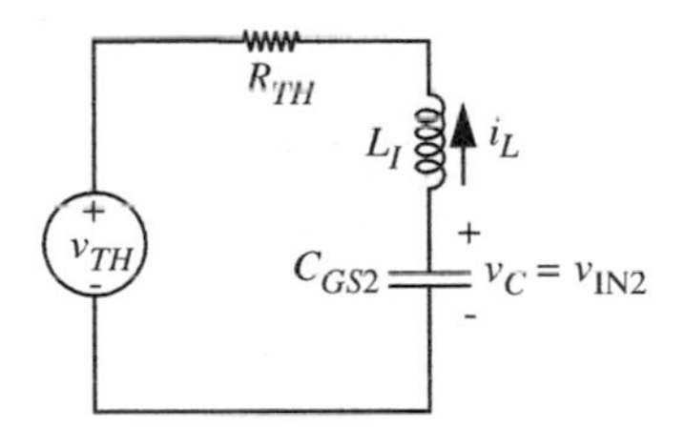


FIGURE 12.37 Equivalent circuit of the two-inverter circuit.

The second complication is that $v_{TH}$ does not step to or from 0 V. Rather it falls from 5 V to 0.5 V for a falling transition, and rises from 0.5 V to 5 V for a rising transition. Because the circuit in Figure 12.37 is linear, we can use the method of superposition to decompose this problem into two simpler problems. Accordingly, we will break $v_{TH}$ into two component voltage sources connected in series as shown in Figure 12.38. The first component, $\bar{v}_{TH}$, is constant at 0.5 V. The second component, $\tilde{v}_{TH}$, steps from 0 V to 4.5 V for a rising transition, and falls from 4.5 V to 0 V for a falling transition. The circuit response to these two components may be superimposed to find the total response. We will denote the $v_C$ response to $\bar{v}_{TH}$ by $\bar{v}_C$, and we will denote the corresponding response to $\tilde{v}_{TH}$ by $\tilde{v}_C$. Similarly, we will denote the corresponding $i_L$ responses as $\bar{i}_L$ and $\tilde{i}_L$.

The circuit response to the constant $\bar{v}_{TH}$ is most easily determined by observing that the capacitor and inductor in Figure 12.37, respectively, behave as an open circuit and a

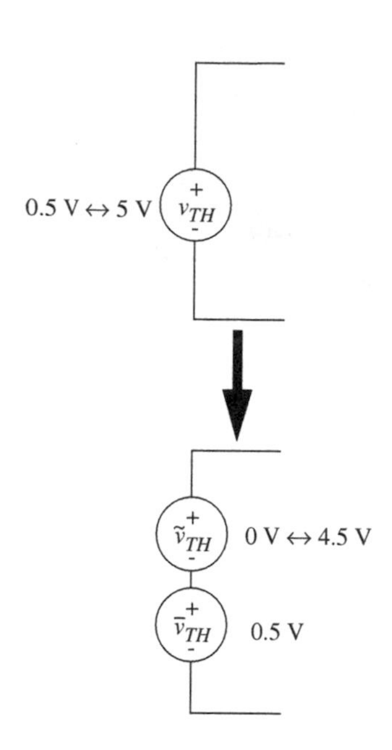


FIGURE 12.38 Breaking up $v_{TH}$ into two additive components, each of whose responses can be added together to obtain the total response through superposition.

short circuit after long periods of time. Therefore

$$\bar{v}_C(t) = 0.5 \text{ V} \tag{12.159}$$

$$\bar{i}_L(t) = 0 \text{ A}. \tag{12.160}$$

We will now determine $\tilde{v}_C$ and $\tilde{i}_L$, and then the total $v_C$ and $i_L$, first for the falling transient and then for the rising transient. During the analysis of each transient we will assume that the previous transient has fully settled.

## Falling Transient

Prior to a falling transient, $v_{TH}$ is assumed to equal 5.0 V for a long time. Since the capacitor and inductor, respectively, behave as an open circuit and a short circuit after long periods of time, the circuit states will have settled to $v_C = 5.0$ V and $i_L = 0$ prior to the transient. These are then the initial states for the falling transient. For simplicity, we will assume that the falling transient begins at $t = 0$. Therefore, the varying component of the circuit states begin a falling transient from

$$\tilde{v}_C(0) = v_C(0) - \bar{v}_C = 5.0 \text{ V} - 0.5 \text{ V} = 4.5 \text{ V} \tag{12.161}$$

$$\tilde{i}_L(0) = i_L(0) - \bar{i}_L = 0 \text{ A} - 0 \text{ A} = 0 \text{ A}. \tag{12.162}$$

This transient resembles that of an undriven, series RLC circuit with an initial voltage on the capacitor, and is studied in Subsection 12.2. To determine whether the circuit is under-, over-, or critically-damped, we first compute the parameters that describe the response, namely

$$\alpha = \frac{R_{TH}}{2L_I} = 4.5 \times 10^8 \text{ rad/s} \tag{12.163}$$

$$\omega_\circ = \frac{1}{\sqrt{L_I C_{GS2}}} = 1.0 \times 10^{10} \text{ rad/s} \tag{12.164}$$

$$\omega_d = \sqrt{\omega_\circ^2 - \alpha^2} = 0.999 \times 10^{10} \text{ rad/s} \tag{12.165}$$

$$Q = \frac{\omega_\circ}{2\alpha} = 11. \tag{12.166}$$

Since $\alpha < \omega_\circ$, the circuit is under-damped, and so we may use Equations 12.63 and 12.64 to determine the falling transient response.

With the substitution of Equations 12.161, 12.162, 12.163, and 12.165 into Equations 12.63 and 12.64, the varying component of the falling transient is found to be

$$\tilde{v}_C(t) = 4.5\, e^{-(4.5\times 10^8\, t)} \cos(0.999 \times 10^{10}\, t - 0.045) \text{ V} \tag{12.167}$$

$$\tilde{i}_L(t) = 4.5 \times 10^{-3}\, e^{-(4.5\times 10^8\, t)} \sin(0.999 \times 10^{10}\, t) \text{ A}. \tag{12.168}$$

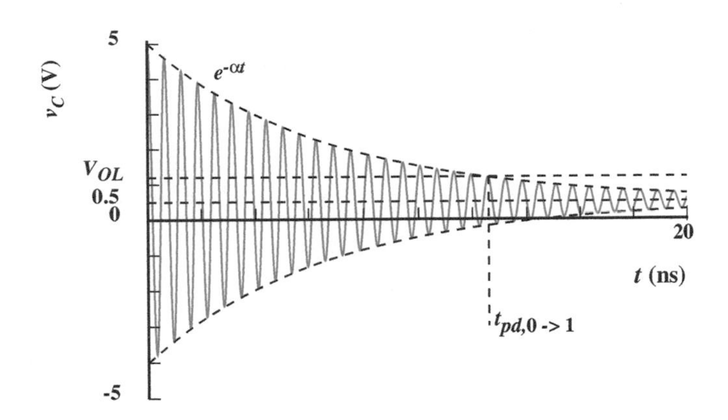


FIGURE 12.39 $v_C$ transient during a falling transition.

The total transient is then obtained by summing the individual responses as follows:

$$v_C(t) = 0.5 \text{ V} + 4.5\, e^{-(4.5\times10^8\, t)} \cos(0.999 \times 10^{10}\, t - 0.045) \text{ V} \tag{12.169}$$

$$i_L(t) = 4.5 \times 10^{-3}\, e^{-(4.5\times10^8\, t)} \sin(0.999 \times 10^{10}\, t) \text{ A.} \tag{12.170}$$

The transient in $v_C$ is shown in Figure 12.39.

Interestingly, when ringing occurs during the transient, calculating the propagation delay is not as straightforward as in the RC case. In the RC case, the propagation delay of the driving inverter was simply the time required for the output voltage to fall below the valid output low threshold, $V_{OL}$ (see Section 10.4.2). For our discussion, we will assume that when the inductance of the connecting wire is relevant, the delay associated with the driving inverter includes the effect of both the inductor $L_I$ and the gate capacitance $C_{GS2}$. Accordingly, we can see from the circuit model in Figure 12.37 that the relevant output node for computing the propagation delay of the driving inverter is $v_{IN2} = v_C$. With ringing behavior during a falling transition, this output is a valid low only when all oscillations remain below the $V_{OL}$ threshold. The corresponding $t_{pd,0\rightarrow1}$ propagation delay is depicted in Figure 12.39; recall that for an inverter $t_{pd,0\rightarrow1}$ is the delay of the inverter for a low to high transition at the input, and hence a high to low transition at the output.

Finally, notice that if the ringing is sufficiently under damped then $v_C$ can fall below $V_{OL}$ and then rise again above $V_{OH}$ one or more times. However, this behavior does not violate the combinational gate abstraction because the definition of the combinational gate in the presence of delay (see Section 10.4.1) does not constrain the gate's operation following an input transition during a time interval whose length is equal to the propagation delay.

### Rising Transient

Prior to a rising transient, $v_{TH}$ is assumed to equal 0.5 V for a long time. Since the capacitor and inductor, respectively, behave as an open circuit and a short circuit after long periods of time, the circuit states will have settled to $v_C = 0.5$ V and $i_L = 0$ prior to the transient. These are then the initial states for the rising transient. Again for simplicity, we will assume that the rising transient begins at $t = 0$. Therefore, the varying component of the circuit states begin a rising transient from

$$\tilde{v}_C(0) = v_C(0) - \bar{v}_C = 0.5 \text{ V} - 0.5 \text{ V} = 0 \text{ V} \tag{12.171}$$

$$\tilde{i}_L(0) = i_L(0) - \bar{i}_L = 0 \text{ A} - 0 \text{ A} = 0 \text{ A}. \tag{12.172}$$

This transient resembles that of a series RLC circuit that is initially at rest, driven by a voltage step of 4.5 volts, and is studied in Section 12.5. We first compute the parameters that describe the response, namely

$$V_\circ = \tilde{v}_{TH} = 4.5 \text{ V} \tag{12.173}$$

$$\alpha = \frac{R_{TH}}{2L_I} = 4.5 \times 10^9 \text{ rad/s} \tag{12.174}$$

$$\omega_\circ = \frac{1}{\sqrt{L_I C_{GS2}}} = 1.0 \times 10^{10} \text{ rad/s} \tag{12.175}$$

$$\omega_d = \sqrt{\omega_\circ^2 - \alpha^2} = 8.9 \times 10^9 \text{ rad/s} \tag{12.176}$$

$$Q = \frac{\omega_\circ}{2\alpha} = 1.1. \tag{12.177}$$

Since $\alpha < \omega_\circ$, the circuit is under-damped, and so we may use Equations 12.141 and 12.142 to determine the rising transient response. With the substitution of Equations 12.173 through 12.176 into Equations 12.141 and 12.142, the varying component of the rising transient is found to be

$$\tilde{v}_C(t) = 4.5(1 - 1.1\, e^{-(4.5\times 10^9\, t)} \cos(8.9 \times 10^9\, t - 0.47)) \text{ V} \tag{12.178}$$

$$\tilde{i}_L(t) = 5.1 \times 10^{-3}\, e^{-(4.5\times 10^9\, t)} \sin(8.9 \times 10^9\, t) \text{ A}. \tag{12.179}$$

The total transient is then

$$v_C(t) = 0.5 \text{ V} + 4.5(1 - 1.1\, e^{-(4.5\times 10^9\, t)} \cos(8.9 \times 10^9\, t - 0.47)) \text{ V} \tag{12.180}$$

$$i_L(t) = 5.1 \times 10^{-3}\, e^{-(4.5\times 10^9\, t)} \sin(8.9 \times 10^9\, t) \text{ A}. \tag{12.181}$$

This transient is shown in Figure 12.40. Again we see that the transient rings, although with greater damping due to an increase in $R_{TH}$. With this ringing behavior, the output is not valid until all oscillations remain above the $V_{OH}$ threshold. The corresponding $t_{pd,1\to 0}$ is depicted in Figure 12.40.

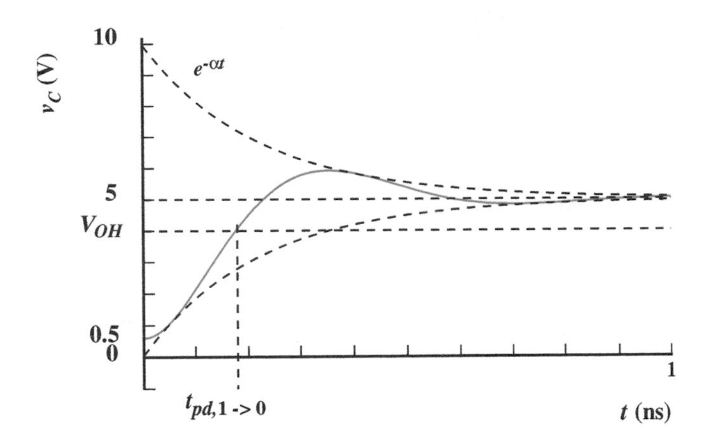


FIGURE 12.40 $v_C$ transient during a rising transition.

EXAMPLE 12.10 ANOTHER SWITCHED POWER SUPPLY In this example, we will analyze the behavior of the switched power supply circuit shown in Figure 12.41.The purpose of the circuit is to convert the DC input voltage $V$ to a different output voltage $v_{OUT}$. The MOSFETs in the circuit operate as switches, and the square-wave inputs to the MOSFET's are shown in Figure 12.42. For variety, the switched power supply circuit in this example has a slightly different arrangement of switches than the one previously discussed in Figure 12.11. We will further assume that the switches have some resistance associated with them. Specifically, assume that the MOSFET's have $R_{ON} = R$, the inductor has inductance $L$, and the capacitor has

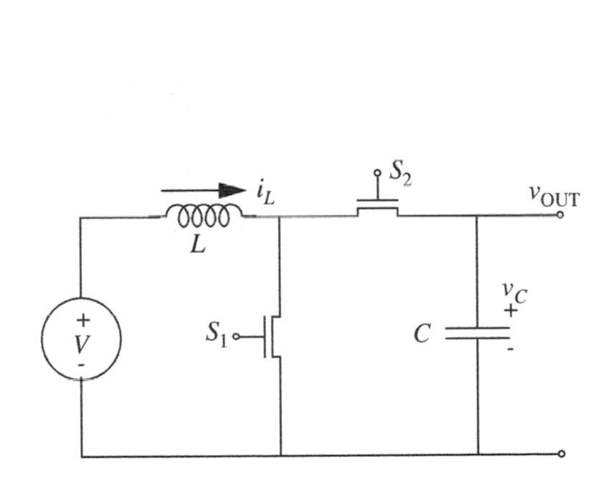


FIGURE 12.41 RLC circuit with switches.

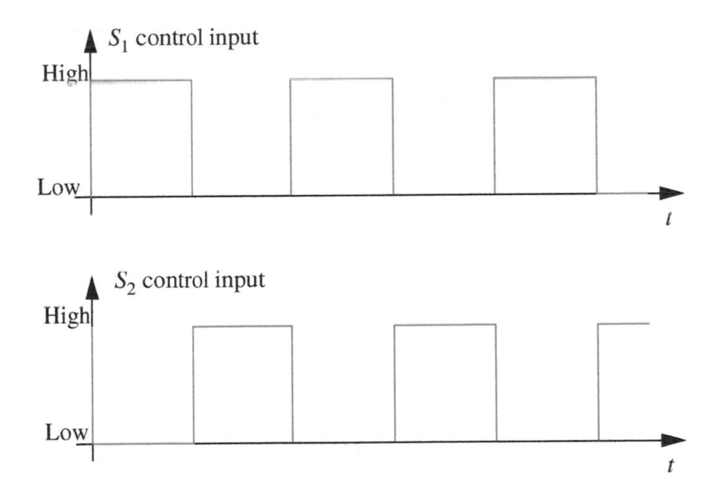


FIGURE 12.42 Input to the switches.

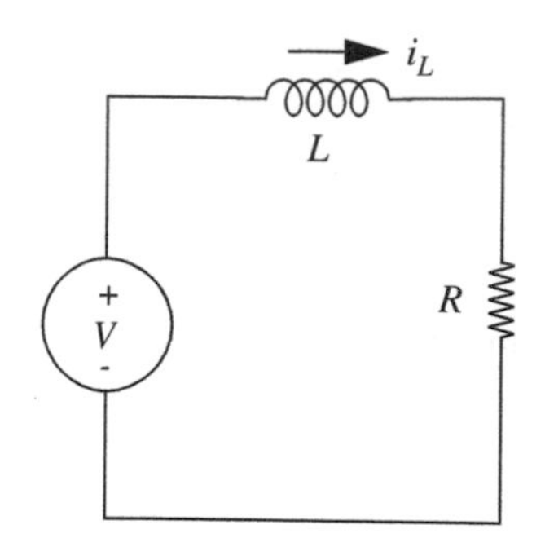


FIGURE 12.43 The equivalent RL circuit when $S_1$ is closed and $S_2$ is open. $R$ is the on resistance of the MOSFET switch. After a long period of time, the final value of the current through the inductor will be $V/R$.

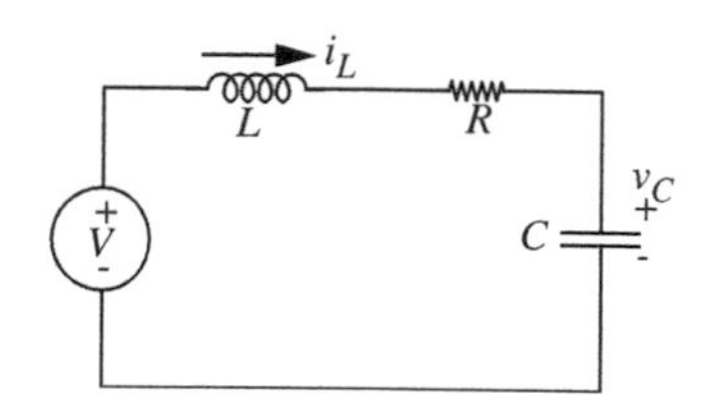


FIGURE 12.44 The equivalent RLC circuit when $S_2$ is closed and $S_1$ is open.

capacitance $C$. We also assume that the MOSFET's have no capacitances associated with them. What is the behavior of this circuit, and in particular, how does $v_{OUT}$ change over time?

*Intuition* We first provide an intuitive explanation of the circuit behavior. When the switch $S_1$ is closed and $S_2$ is open, we have a series RL circuit as shown in Figure 12.43. Over time, the current through the inductor in this circuit will build up and eventually reach $V/R$ (if given enough time).

Next, when $S_1$ is open and $S_2$ is closed, we have a series resonant RLC circuit as shown in Figure 12.44. When all the transients have died out, the capacitor will have the same voltage ($V$) as the voltage source and no current flows through the inductor (again, if given enough time).

Let us first consider the case where $S_1$ is opened and $S_2$ is kept closed for a *long period of time.* At the instant that $S_1$ is opened and $S_2$ is closed, the inductor has a current equal to $V/R$ flowing through it. Since the current flowing through the inductor cannot change instantaneously, the same current will flow into the capacitor through the closed switch $S_2$. As the capacitor charges, its voltage rises and the current decreases. After some amount of time, the current reaches zero. If the capacitor is small enough, or if there was a large enough initial voltage on it, its voltage might be at a higher value than the voltage of the source $V$. Let us assume this is the case. Because the capacitor voltage reaches a higher value than $V$, it begins to supply energy and a current begins to flow through the inductor in the reverse direction. This process continues as the energy oscillates between the inductor and the capacitor. If the given switch settings exist for a long enough period of time, the oscillations will die out as the resistor dissipates the energy.

Let us now consider the case where $S_1$ is opened and $S_2$ is kept closed only for a *short amount of time.* This case starts out just like the previous case. In other words, at the instant that $S_1$ is opened and $S_2$ is closed, the inductor has a current equal to $V/R$ flowing through it. Since the current flowing through the inductor cannot change instantaneously, the same current will flow into the capacitor through the closed switch $S_2$. As the capacitor charges, its voltage rises past $V$ (again, assuming a small enough capacitor, or a large enough initial voltage on the capacitor).

Now, an interesting scenario arises if we close $S_1$ and open $S_2$ after the capacitor voltage has risen past $V$, but before the current reverses direction. Since $S_2$ is now open, the capacitor has no path to discharge, and so it holds the final value of its voltage. Then when $S_1$ is opened and $S_2$ is closed again, the inductor current charges up the capacitor further, thereby further increasing its voltage. If this process is repeated, notice that the capacitor voltage will keep rising indefinitely over time.

If, however, a resistive load is applied to the capacitor output as illustrated in Figure 12.48, its output will not keep rising indefinitely; rather the capacitor gets discharged (fully or partially depending on the relative charging and discharging time constants). By adjusting the switching intervals over which the capacitor is allowed to charge and discharge we can achieve a range of average voltage values (including values that are higher or lower than the input voltage) at the output of the circuit. This property forms the basis of DC/DC converter power supplies.

The following discussion provides a more detailed analysis of the circuit behavior, assuming there is no output load resistor. We will consider the two cases: (a) $S_1$ closed, $S_2$ open, which forms a series RL circuit, and (b) $S_1$ open, $S_2$ closed, which forms a series RLC circuit.

*$S_1$ Closed, $S_2$ Open: Series RL Circuit* When $S_1$ is closed and $S_2$ is open, we have the series RL circuit shown in Figure 12.43. We will assume that $S_1$ is closed long enough for the transient to die out. Therefore, the current through the inductor will reach $V/R$ before $S_1$ is opened.

*$S_1$ Open, $S_2$ Closed: Series RLC Circuit* When $S_1$ is open and $S_2$ is closed, we have the driven, series RLC circuit shown in Figure 12.44, which is identical to the circuit in Figure 12.26. This time around, we will analyze this circuit from first principles following the method shown in Section 12.2. First, let us collect our information on the drive voltage and the initial conditions. We know that the drive voltage for the circuit is $V$.

Let us now determine the initial conditions. For convenience, we will take $t = 0$ as the instant that $S_1$ is opened and $S_2$ is closed. The state of the circuit at $t = 0$ is given by

$$i_L(0) = \frac{V}{R} \tag{12.182}$$

$$v_C(0) = v_0 \tag{12.183}$$

where $v_0$ is the voltage on the capacitor at $t = 0$. Just before $S_1$ is opened and $S_2$ is closed, there was a current of $V/R$ flowing in the inductor and a voltage $v_0$ on the capacitor. Therefore, these are our initial conditions. The initial voltage on the capacitor $v_0$ will be 0 when the circuit starts from rest. If there is no load resistor connected to the capacitor, then this voltage will simply be the final value on the capacitor from the previous charging cycle.

We know that the total solution for $v_C$ and $i_L$ is given by the sum of the homogeneous solution ($v_{CH}$ and $i_{LH}$) and particular solution ($v_{CP}$ and $i_{LP}$). We know from Equation 12.121 that the homogeneous solution is given by

$$v_{CH} = K_1 e^{s_1 t} + K_2 e^{s_2 t}. \tag{12.184}$$

Further, using

$$i_{LH} = C\frac{dv_{CH}}{dt}$$

we obtain the homogeneous solution for the current as

$$i_{LH} = K_1 C s_1 e^{s_1 t} + K_2 C s_2 e^{s_2 t}. \tag{12.185}$$

In the preceding equations, $s_1$ and $s_2$ are given by

$$s_1 = -\alpha + \sqrt{\alpha^2 - \omega_\circ^2}$$

$$s_2 = -\alpha - \sqrt{\alpha^2 - \omega_\circ^2}$$

where

$$\alpha \equiv \frac{R}{2L}$$

$$\omega_\circ \equiv \sqrt{\frac{1}{LC}}.$$

If we wait long enough for all the transients to die out, there will be no current flowing in the inductor and the capacitor voltage will be the same as the voltage source. Thus, we have the following particular solution:

$$i_{LP} = 0 \tag{12.186}$$

$$v_{CP} = V. \tag{12.187}$$

Therefore the total solution is given by

$$v_C = K_1 e^{s_1 t} + K_2 e^{s_2 t} + V \tag{12.188}$$

$$i_L = K_1 C s_1 e^{s_1 t} + K_2 C s_2 e^{s_2 t}. \tag{12.189}$$

Now, we are in a position to solve for $K_1$ and $K_2$ using the initial conditions as follows:

$$v_0 = K_1 + K_2 + V \tag{12.190}$$

$$\frac{V}{R} = K_1 C s_1 + K_2 C s_2. \tag{12.191}$$

Solving for $K_1$ and $K_2$ we get

$$K_1 = \frac{(v_0 - V)C s_2 - \frac{V}{R}}{C(s_2 - s_1)} \tag{12.192}$$

$$K_2 = \frac{(v_0 - V)C s_1 - \frac{V}{R}}{C(s_1 - s_2)}. \tag{12.193}$$

Substituting the preceding expressions for $K_1$ and $K_2$ into Equations 12.188 and 12.189, we obtain the complete solutions:

$$v_C = \frac{(v_0 - V)Cs_2 - \frac{V}{R}}{C(s_2 - s_1)} e^{s_1 t} + \frac{(v_0 - V)Cs_1 - \frac{V}{R}}{C(s_1 - s_2)} e^{s_2 t} + V \tag{12.194}$$

$$i_L = \frac{(v_0 - V)Cs_2 - \frac{V}{R}}{C(s_2 - s_1)} Cs_1 e^{s_1 t} + \frac{(v_0 - V)Cs_1 - \frac{V}{R}}{C(s_1 - s_2)} Cs_2 e^{s_2 t}. \tag{12.195}$$

Suppose we choose the element values in our switched power supply circuit such that

$$\alpha < \omega_\circ,$$

then the circuit will be under-damped, and $s_1$ and $s_2$ will be complex. $s_1$ and $s_2$ can now be written as

$$s_1 = -\alpha + j\omega_d$$

$$s_2 = -\alpha - j\omega_d$$

where

$$\omega_d = \sqrt{\omega_\circ^2 - \alpha^2}.$$

As discussed in Section 12.5, for complex $s_1$ and $s_2$, our solution in Equation 12.195 can be rewritten using the Euler relation into the following form:

$$v_C = V + A_1 e^{-\alpha t} \sin(\omega_d t) + A_2 e^{-\alpha t} \cos(\omega_d t) \tag{12.196}$$

$$i_L = A_3 e^{-\alpha t} \sin(\omega_d t) + A_4 e^{-\alpha t} \cos(\omega_d t). \tag{12.197}$$

The form of the waveforms for $v_C$ and $i_L$ are shown in Figures 12.45 and 12.46. The waveforms in the figures assume that $S_1$ is left open and $S_2$ is left closed for a long period of time.

Notice that the voltage waveform in Figure 12.45 first increases and the current waveform first decreases. This corresponds to the observation we made earlier that the current can't change direction within an inductor instantaneously. Thus, if we close $S_1$ and open $S_2$ quickly, (for example, before $t_1$, which is the instant that the inductor current goes to zero as seen in Figure 12.46), we can accumulate more charge, and hence more voltage, on the capacitor during each cycle. In this situation, the waveform of the capacitor voltage $v_{OUT}$ (where $v_{OUT} = v_C$) will look like that shown in Figure 12.47.

The value of the voltage $v_{OUT}$ at the end of the $n$th cycle can be calculated iteratively. Let $v_{OUT}[n]$ denote the value of $v_{OUT}$ at the end of the $n$th cycle. In the first iteration, we compute $v_{OUT}[1] = v_C$ by substituting $v_0 = 0$ and $t = t_1$ in Equation 12.197. In the second iteration, we compute $v_{OUT}[2]$ by substituting $v_0 = v_{OUT}[1]$ and $t = t_1$ in

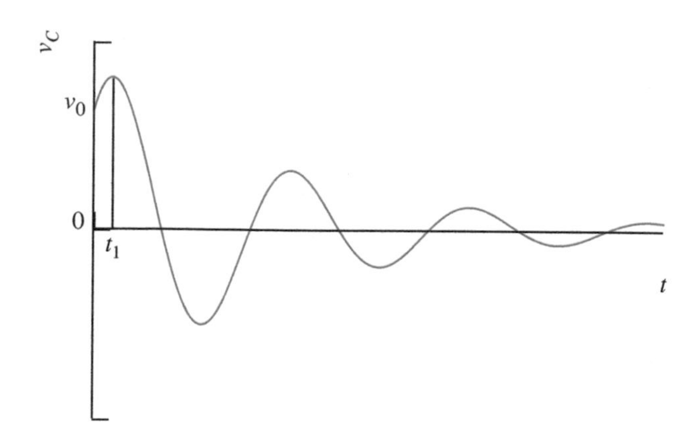


FIGURE 12.45 Components of the capacitor voltage waveform.

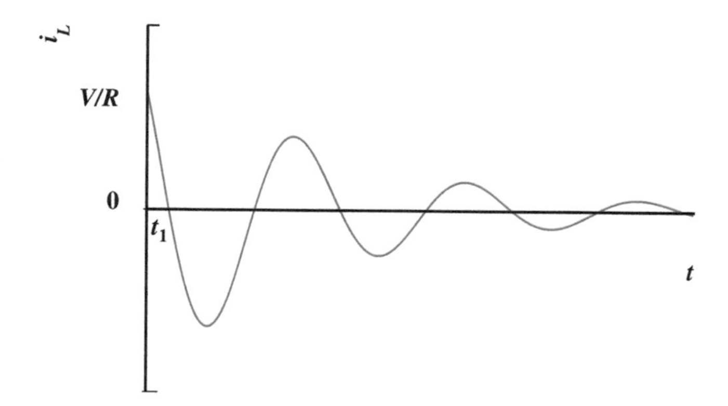


FIGURE 12.46 Components of the inductor current waveform.

Equation 12.197. Using this iterative process, we can determine the value of the output voltage after $n$ cycles.

Alternatively, if are interested only in the maximum value of the output voltage at the end of each cycle, and we do not care about the exact waveforms, we can use the much simpler energy method discussed in Example 12.4, with one difference. In our example, since the switches have a finite resistance, and since we have assumed that $S_1$ is closed and $s_2$ is open for a long period of time, $i_L$ in Equation 12.31 is computed differently as

$$i_L = \frac{V}{R}.$$

*Charging and Discharging the Capacitor through a Load Resistor* Let us make the circuit a little more interesting by adding a load resistor $R_L$ at the output port as shown in Figure 12.48. How does this new circuit behave?

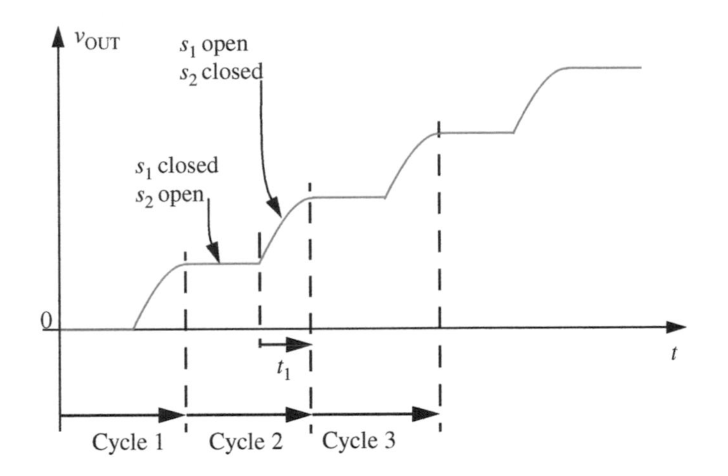


FIGURE 12.47 Behavior of $v_{OUT}$ over time.

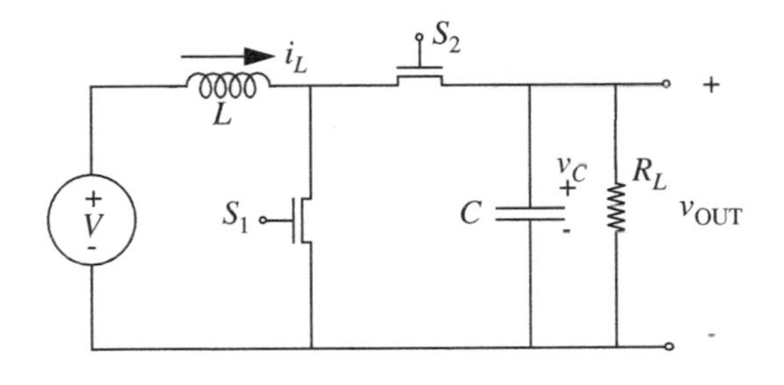


FIGURE 12.48 The modified circuit.

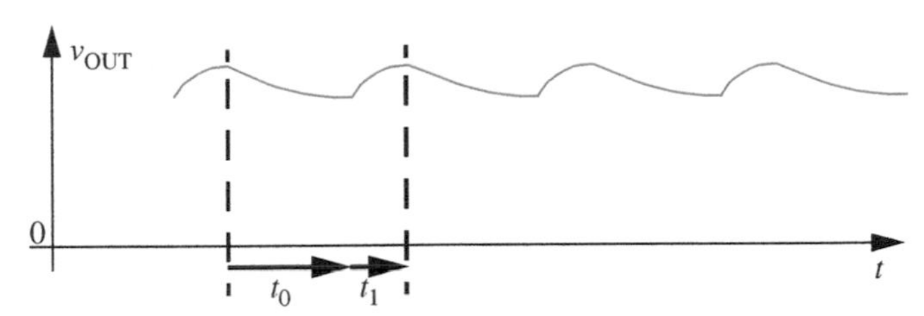


FIGURE 12.49 Behavior of the output port.

The mathematical computation of the form of the output voltage is complicated, so we do not present it here. Instead, we present a qualitative discussion of the circuit, behavior. First, when $S_2$ is closed and $S_1$ is open, we have a slightly more complicated RLC circuit, which behaves in a manner similar to the original circuit. Essentially, the inductor current charges the capacitor to a higher voltage.

Now, when $S_1$ is closed and $S_2$ is opened, we have two subcircuits operating. The first circuit is a series RL circuit identical to the original circuit. The second circuit is an RC circuit in which the capacitor discharges through the load resistor $R_L$. Depending on the exact values of the charging and discharging capacitor time constants, we might obtain the waveform shown in Figure 12.49 at the output port. The waveform in the figure assumes that $R_L C \gg t_0$, where $t_0$ is the time interval for which $S_2$ is open.

## 12.6 DRIVEN, PARALLEL RLC CIRCUIT*

### 12.6.1 STEP RESPONSE

### 12.6.2 IMPULSE RESPONSE

## 12.7 INTUITIVE ANALYSIS OF SECOND-ORDER CIRCUITS

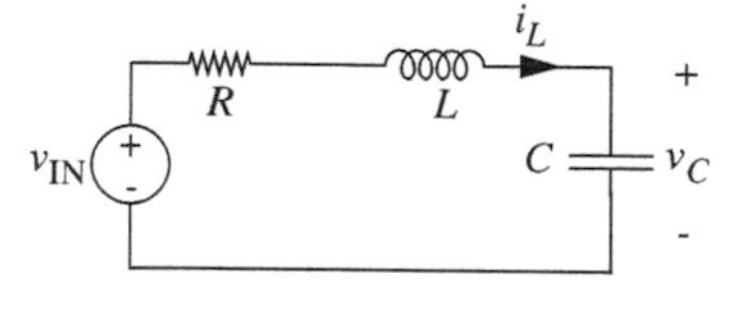

FIGURE 12.55 A driven, series RLC circuit.

Second-order circuits are amenable to a quick, intuitive analysis when they are driven by simple inputs such a step or an impulse, much like first-order circuits (see Section 10.3). To illustrate, we will show how $v_C$, the voltage across the capacitor, in the series RLC circuit in Figure 12.55 can be plotted by inspection. We will assume the following element values:

$$L = 100\ \mu\text{H}$$
$$C = 100\ \mu\text{F}$$
$$R = 0.2\ \Omega.$$

The initial state of the circuit at $t = 0$ is given by

$$v_C(0) = 0.5\ \text{V}$$
$$i_L(0) = -0.5\ \text{A}.$$

The circuit is driven by a DC voltage source, with

$$v_{IN} = 1\ \text{V}.$$

Based on the initial conditions and the drive, we can immediately determine the initial value and the final value of $v_C$. From the initial conditions that we are given, we know that

$$v_C(0) = 0.5\ \text{V}$$
$$i_L(0) = -0.5\ \text{A}.$$

In the steady state, the capacitor behaves like an open circuit. Therefore, the inductor current vanishes and the input drive appears across the capacitor. Thus,

$$v_C(\infty) = v_{IN} = 1\ \text{V}$$
$$i_L(\infty) = 0.$$

The initial and final values of $v_C$ are plotted in Figure 12.56a.

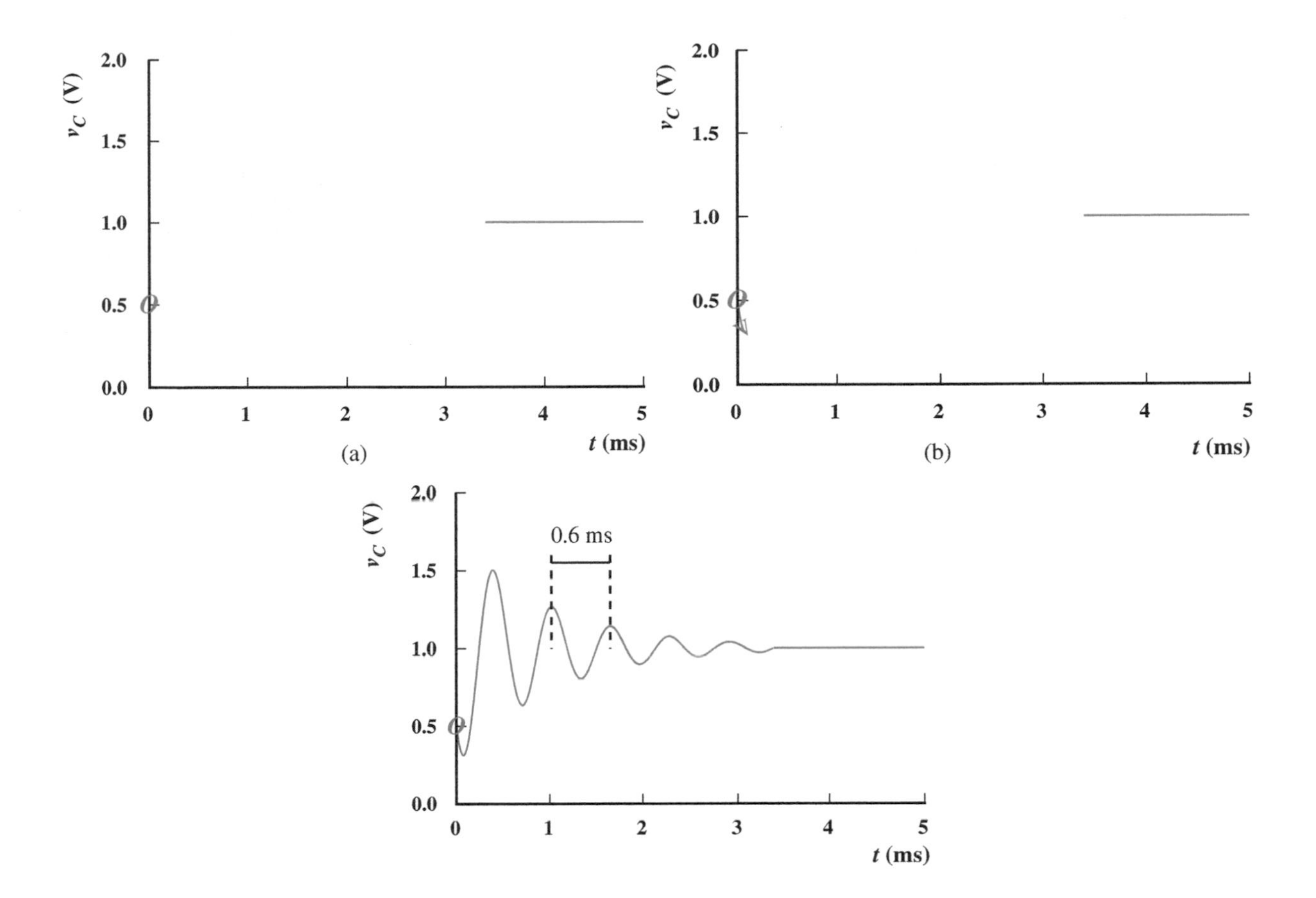


FIGURE 12.56 Sketching the form of $v_C$.

Next, we obtain additional information by writing the characteristic equation for the circuit. A simple method for writing the characteristic equation by inspection will be discussed in Sections 14.1.2 and Section 14.2 in Chapter 14. That method will be based on the impedance approach, which will be discussed in Chapters 13 and 14. For now, proceeding based on what we know thus far, we can obtain the characteristic equation by writing the differential equation for the system with the drive set to zero,

$$\frac{d^2v_C(t)}{dt^2} + \frac{R}{L}\frac{dv_C(t)}{dt} + \frac{1}{LC}v_C(t) = 0$$

and then substituting the candidate solution $Ae^{st}$ and dividing throughout by $Ae^{st}$

$$s^2 + \frac{R}{L}s + \frac{1}{LC} = 0.$$

Comparing to the standard form of the characteristic equation

$$s^2 + 2\alpha s + \omega_o^2$$

we obtain a few more parameters that describe the behavior of our second-order system

$$\alpha = \frac{R}{2L} = 10^3 \text{ rads/s}$$

$$\omega_o = \sqrt{\frac{1}{LC}} = 10^4 \text{ rads/s}$$

Since

$$\alpha < \omega_o,$$

we conclude that the system is under-damped. The oscillation frequency is given by

$$\omega_d = \sqrt{\omega_o^2 - \alpha^2} \approx 9950 \text{ rads/s} \approx 1584 \text{ cycles/s}$$

and the quality factor

$$Q = \frac{\omega_o}{2\alpha} = 5.$$

Since $Q = 5$, we also know that the system will ring for approximately 5 cycles.

To complete the picture, we must now combine the boundary values shown in Figure 12.56a with a sinusoid of frequency 1584 Hz (or cycle time 0.6 ms) that decays over about 5 cycles. To do so, it helps to determine the initial trajectory of the capacitor voltage (increasing or decreasing) starting from its initial value of 0.5 V. It turns out that we can obtain this information by looking at the initial state on the other memory element, namely the inductor. The initial inductor current is given as $-0.5$ *A*. In the absence of a driving impulse, since the magnitude of this current cannot change instantaneously, and since the given direction of the initial current tends to discharge the capacitor, we can conclude that the capacitor voltage will tend to decrease. This decreasing initial trajectory of the capacitor voltage is illustrated in Figure 12.56b.

Knowing the initial trajectory, we can stitch in a sinusoid that decays over about 5 cycles with the correct initial trajectory. The resulting approximate sketch is illustrated in Figure 12.56c. Notice that a simple intuitive analysis allowed us to guess the values of the following parameters of the curve in Figure 12.56c:

1. The initial value.
2. The final value.

3. The initial trajectory of the curve.
4. The frequency of ringing.
5. The approximate length of the time interval over which the ringing lasts.

Unfortunately, our list does not include the maximum amplitude of the envelope that governs the decay of the sinusoid, a parameter that when combined with the rate of decay $\alpha$ would add even more accuracy to our sketch. Although we can determine this value, it adds significant complexity of our analysis, and so we will not attempt to solve for the general form. It turns out, however, that the maximum amplitude can be calculated much more easily when an initial state and a driving voltage are not both present. We will work out the following example to illustrate this fact.

---

EXAMPLE 12.11 INTUITIVE ANALYSIS EXAMPLE This example shows how the maximum initial amplitude of the decaying sinusoid can be calculated with ease when an initial state and a driving voltage are not present simultaneously. Suppose we set the voltage drive to zero in our circuit in Figure 12.55, but we let all other conditions remain the same. In other words, suppose that

$$L = 100\ \mu\text{H}$$
$$C = 100\ \mu\text{F}$$
$$R = 0.2\ \Omega.$$

Further, suppose that the initial state of the circuit at $t = 0$ is given by

$$v_C(0) = 0.5\ \text{V}$$
$$i_L(0) = -0.5\ \text{A}.$$

However, the input drive is given by

$$v_{IN} = 0.$$

As we did in Section 12.7, we can quickly guess the following:

1. The initial value of the capacitor voltage is given by

$$v_C(0) = 0.5\ \text{V}.$$

2. Since there is no drive, and since there is a dissipative element in the circuit, the final value of the capacitor voltage is given by

$$v_C(\infty) = 0\ \text{V}.$$

These initial and final values are plotted in Figure 12.57a.

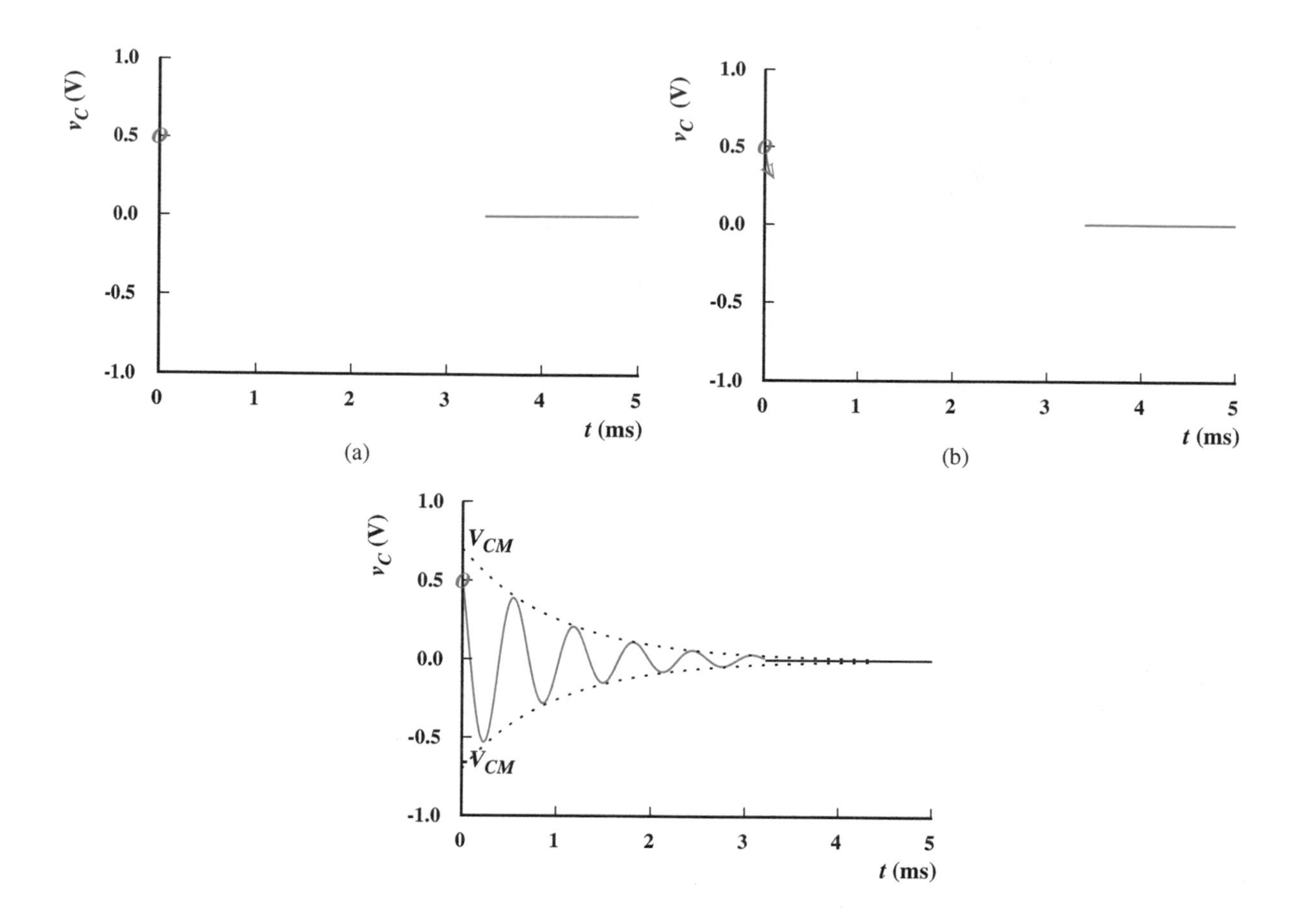


FIGURE 12.57 Sketching the form of $v_C$ when the drive is set to zero.

3. The capacitor voltage will tend to decrease initially, since the initial inductor current tends to discharge the capacitor. This decreasing initial trajectory of the capacitor voltage is illustrated in Figure 12.57b.
4. The ringing frequency is $\omega_d \approx 9950$ rads/s.
5. Since $Q = 5$, the ringing will last for approximately 5 cycles.

Additionally, we can use energy arguments to determine the envelope bounding the decaying capacitor voltage. Recall that the magnitude of the capacitor voltage peaks whenever the system energy resides entirely in the capacitor, and the inductor current is zero. At $t = 0$, had all the energy been stored in the capacitor, the capacitor voltage would have attained an absolute maximum value that we denote by $V_{CM}$. Accordingly, the decay in the capacitor voltage will be governed by a pair of exponential curves (positive and negative) with initial values $+V_{CM}$ and $-V_{CM}$ at $t = 0$, decaying to zero in about 5 cycles, as illustrated by the dotted curves in Figure 12.57c.

The value of $V_{CM}$ can be computed from the total energy in the system at $t = 0$ as follows:

$$\frac{1}{2}Cv_C(0)^2 + \frac{1}{2}Li_L(0)^2 = \frac{1}{2}CV_{CM}^2.$$

Substituting for $v_C(0)$ and $i_L(0)$ we obtain

$$V_{CM} \approx 0.7 \text{ V}.$$

The form of the decaying sinusoid can now be completed as also shown in Figure 12.57c.

---

EXAMPLE 12.12 INTUITIVE ANALYSIS EXAMPLE: IMPULSE RESPONSE In this example, we will use intuition to sketch the form of $v_C$ in the parallel RLC circuit shown in Figure 12.58 when it is driven by an input impulse current given by

$$i_{IN}(t) = Q_o\delta(t).$$

The strength of the impulse is given as

$$Q_o = 10^{-4} \text{ C}.$$

Assume that

$$L = 100\ \mu\text{H}$$

$$C = 100\ \mu\text{F}$$

$$R = 5\ \Omega$$

and that the circuit is initially at rest (that is, both the capacitor voltage and the inductor current are zero before the impulse).

Recalling that the impulse serves to establish initial conditions on the circuit, let us first determine the values of the state variables immediately following the impulse, namely,

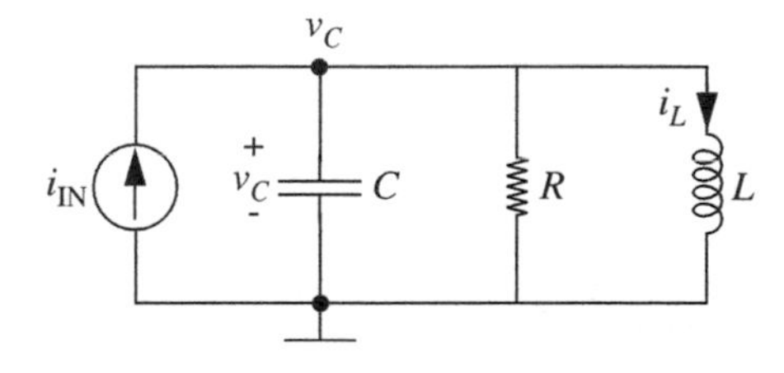


FIGURE 12.58 A parallel second-order circuit with a resistor, capacitor, inductor, and current source.

$v_C(0^+)$ and $i_L(0^+)$. The capacitor looks like an instantaneous short to the current pulse, and so the impulse passes entirely through the capacitor depositing a charge of $Q_o = 1$ C on it. Thus,

$$v_C(0^+) = \frac{Q_o}{C} = 1 \text{ V}$$

and $i_L(0^+) = 0$.

As we did in Section 12.7, we can now determine the following:

1. The initial value of the capacitor voltage is 1 V.
2. Since there is no drive, and since there is a dissipative element in the circuit, the final value of the capacitor voltage is given by

$$v_C(\infty) = 0 \text{ V}.$$

3. The capacitor voltage will tend to decrease initially, since all the energy starts out in the capacitor. Since there is no drive, the maximum voltage across the capacitor is also 1 V.
4. Since we have a parallel RLC circuit,

$$\alpha = \frac{1}{2RC} = 10^3 \text{ rads/s}$$

$$\omega_o = \frac{1}{\sqrt{LC}} = 10^4 \text{ rads/s}$$

and the ringing frequency is

$$\omega_d = \sqrt{\omega_o^2 - \alpha^2} \approx= 10^4 \text{ rads/s}.$$

5. The quality factor

$$Q = \frac{\omega_o}{2\alpha} = 5,$$

so the ringing will last for approximately 5 cycles.

The form of $v_C(t)$ can now be plotted as shown in Figure 12.59.

---

## 12.8 TWO-CAPACITOR OR TWO-INDUCTOR CIRCUITS

In the previous sections of this chapter, we focused on second-order circuits containing one capacitor and one inductor. It is also possible to construct a

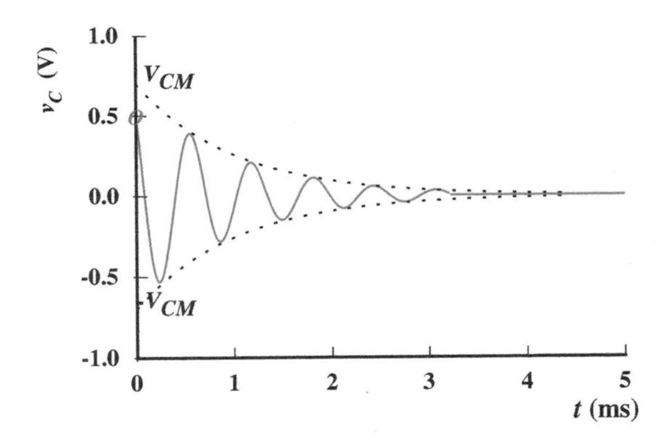

FIGURE 12.59 Sketching the form of $v_C$ for a parallel RLC circuit driven by an impulse current input.

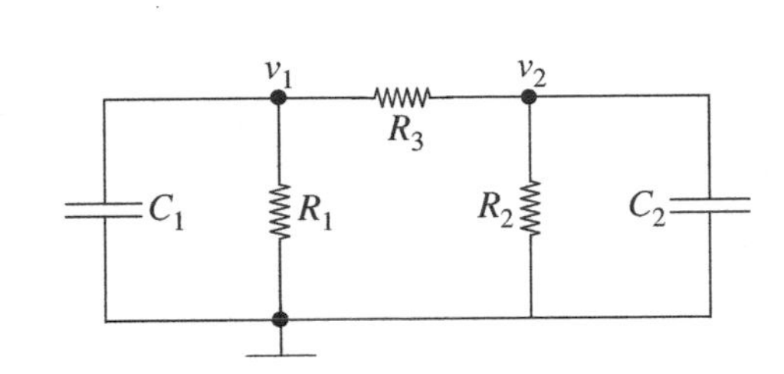

FIGURE 12.60 A second-order circuit containing two independent capacitors.

second-order circuit from two capacitors or two inductors. In this section, we will briefly examine the analysis and behavior of such circuits.

As an example of a two-capacitor circuit, consider the circuit shown in Figure 12.60. To analyze this circuit we again employ the node method beginning with Step 3, just as we have in previous sections. To do so, we write KCL at Nodes #1 and #2 in terms of $v_1$ and $v_2$. This yields

$$C_1 \frac{dv_1(t)}{dt} + \frac{1}{R_1} v_1(t) + \frac{1}{R_3}(v_1(t) - v_2(t)) = 0 \qquad (12.238)$$

for Node #1, and

$$C_2 \frac{dv_2(t)}{dt} + \frac{1}{R_2} v_2(t) + \frac{1}{R_3}(v_2(t) - v_1(t)) = 0 \qquad (12.239)$$

for Node #2. To treat these two equations simultaneously, we first use Equation 12.238 to determine $v_2$ in terms of $v_1$, and then substitute the result into Equation 12.239 to obtain a second-order differential equation in $v_1$. This yields

$$v_2(t) = R_3 C_1 \frac{dv_1(t)}{dt} + \left(1 + \frac{R_3}{R_1}\right) v_1(t) \qquad (12.240)$$

$$\frac{d^2 v_1(t)}{dt^2} + \left(\frac{1}{R_1 C_1} + \frac{1}{R_2 C_2} + \frac{1}{R_3 C_1} + \frac{1}{R_3 C_2}\right) \frac{dv_1(t)}{dt} + \left(\frac{1}{R_1 R_2 C_1 C_2} + \frac{1}{R_1 R_3 C_1 C_2} + \frac{1}{R_2 R_3 C_1 C_2}\right) v_1(t) = 0. \qquad (12.241)$$

Equation 12.241 is an ordinary second-order linear homogeneous differential equation with constant coefficients similar to those derived in earlier sections of this chapter.

To complete the node analysis, we complete Steps 4 and 5 by solving Equation 12.241 for $v_1$, and using it to determine $v_2$ and any other branch variables of interest. Given the form of Equation 12.241, we expect its solution to be a superposition of two terms of the form $Ae^{st}$. The substitution of this candidate term into Equation 12.241 yields

$$s^2 + 2\alpha s + \omega_\circ^2 = 0 \tag{12.242}$$

as the characteristic equation of the circuit where

$$\alpha \equiv \frac{1}{2}\left(\frac{1}{R_1C_1} + \frac{1}{R_2C_2} + \frac{1}{R_3C_1} + \frac{1}{R_3C_2}\right) \tag{12.243}$$

$$\omega_\circ^2 \equiv \frac{1}{R_1R_2C_1C_2} + \frac{1}{R_1R_3C_1C_2} + \frac{1}{R_2R_3C_1C_2}. \tag{12.244}$$

Except for the details of $\alpha$ and $\omega_\circ$, Equation 12.242 is the same as every other characteristic equation seen so far in this chapter. Because Equation 12.242 is a quadratic equation it has two roots. Those roots are

$$s_1 = -\alpha + \sqrt{\alpha^2 - \omega_\circ^2} \tag{12.245}$$

$$s_2 = -\alpha - \sqrt{\alpha^2 - \omega_\circ^2}. \tag{12.246}$$

Therefore, the solution for $v_1$ is a linear combination of the two functions $e^{s_1t}$ and $e^{s_2t}$, and takes the form

$$v_1(t) = A_1e^{s_1t} + A_2e^{s_2t} \tag{12.247}$$

where $A_1$ and $A_2$ are unknown constants that depend on the initial states of the two capacitors. To find $A_1$ and $A_2$, two equations are required. The first equation comes from the evaluation of $v_1$ in Equation 12.247 at the initial time. The second equation comes from the substitution of Equation 12.247 into Equation 12.240 to determine $v_2$ followed by the evaluation of $v_2$ at the initial time. The solution of the two equations yields $A_1$ and $A_2$; we will not carry out the details here.

On the surface, it appears that the analysis and behavior of a second-order circuit containing two independent capacitors is identical to that of a second-order circuit containing one capacitor and one inductor. The same conclusion is reached by examining a circuit containing two independent inductors. While this conclusion is largely true, there is one important difference between second-order circuits containing two capacitors or two inductors, and second-order circuits containing one capacitor and one inductor. That difference is that the latter circuits can exhibit under-damped oscillatory behavior, while the former circuits cannot. That is, $s_1$ and $s_2$ are always real and non-positive, for second-order circuits containing two independent capacitors or two independent inductors. In fact, we can extend this statement, without proof, to higher-order circuits as well.[10] Circuits containing only resistors and capacitors, or only resistors and inductors, will have characteristic equations with only real non-positive roots. For this reason, such circuits can not oscillate. Rather, their states may have only as many as $N-1$ zero crossings where $N$ is the order of the circuit. The exact number of zero crossings actually depends upon the initial conditions of the circuit.

To see that $s_1$ and $s_2$ as given in Equations 12.245 and 12.246 are always real and negative, we examine the term inside the radicals. With the substitution of Equations 12.243 and 12.244, this term becomes

$$\alpha^2 - \omega_\circ^2 = \frac{1}{4}\left(\frac{1}{R_1C_1} + \frac{1}{R_3C_1} - \frac{1}{R_2C_2} - \frac{1}{R_3C_2}\right)^2 + \frac{1}{R_3^2C_1C_2}, \quad (12.248)$$

which is always positive. Therefore, $s_1$ and $s_2$ are always real and negative, and the circuit exhibits only over-damped dynamics.

---

EXAMPLE 12.13 A NUMERICAL EXAMPLE Plot $v_1$ and $v_2$ for the circuit in Figure 12.60 given that

$$R_1 = R_2 = R_3 = 1\ M\Omega$$

$$C_1 = C_2 = C_3 = 1\ \mu F$$

and initial states on the capacitors given by

$$v_1(0) = 0$$

and

$$v_2(0) = 1\ V.$$

10. For an outline of a proof, see Problem 4.6 in W. M. Siebert, *Circuits, Signals, and Systems*, MIT Press, 1986.

Substituting these values into Equation 12.241, we obtain the second-order differential equation that must be solved to obtain $v_1$:

$$\frac{d^2 v_1(t)}{dt^2} + 4\frac{dv_1(t)}{dt} + 3v_1(t) = 0.$$

Substituting the candidate term $Ae^{st}$ into the preceding differential equation we get the characteristic equation

$$s^2 + 4s + 3 = 0.$$

Comparing corresponding terms with the standard form of the characteristic equation for the following second-order circuits:

$$s^2 + 2\alpha s + \omega_\circ^2 = 0$$

we can write

$$\alpha = 2 \text{ rads/s}$$

$$\omega_\circ = \sqrt{3} \text{ rads/s.}$$

The two roots of the characteristic equation are

$$s_1 = -1$$

$$s_2 = -3.$$

Therefore, the solution for $v_1$ is a linear combination of the two functions $e^{-t}$ and $e^{-3t}$, and takes the form

$$v_1(t) = A_1 e^{-t} + A_2 e^{-3t}.$$

The corresponding solution for $v_2$ is related to that for $v_1$ by Equation 12.240 and is given by

$$v_2(t) = \frac{dv_1(t)}{dt} + 2v_1(t) = A_1 e^{-t} - A_2 e^{-3t}.$$

Substituting the initial conditions ($v_1(0) = 0$ and $v_2(0) = 1$) into the expressions for $v_1$ and $v_2$, we get

$$A_1 + A_2 = 0$$

$$A_1 - A_2 = 1.$$

These two equations can be solved to yield

$$A_1 = \frac{1}{2}$$

$$A_2 = -\frac{1}{2}.$$

Thus, the solutions for $v_1$ and $v_2$ are

$$v_1(t) = \frac{1}{2}e^{-t} - \frac{1}{2}e^{-3t}$$

$$v_2(t) = \frac{1}{2}e^{-t} + \frac{1}{2}e^{-3t}.$$

Plots of $v_1$ and $v_2$ are given in Figure 12.61.

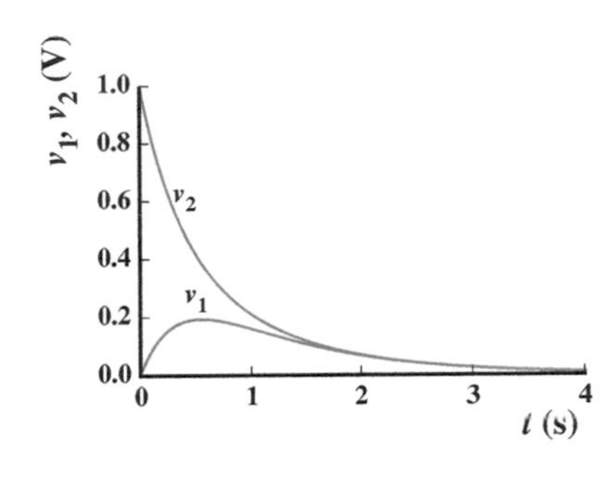


FIGURE 12.61 $v_1$ and $v_2$ for the two-capacitor circuit.

## 12.9 STATE-VARIABLE METHOD

In the preceding sections of this chapter, we used the node method to analyze the behavior of various second-order circuits. This form of analysis directly yields the node voltages within a circuit expressed as functions of time. However, as noted earlier, we are often more interested in the circuit states than we are in the node voltages. In this case, we must use the node voltages to determine the states, and this takes additional effort. As we shall see shortly, there exists an alternative method of analysis that is particularly useful when the circuit states are of primary interest. We will refer to this method as *state-variable analysis.* Its principal advantage is that it offers a more direct way to obtain the equations which govern state evolution, and hence a more direct way to determine the states themselves. Of course, to determine the node voltages from the results of a state-variable analysis also requires additional effort. Thus, when both the states and node voltages are of interest, there may be no best choice of analysis.

The first step of a state-variable analysis is to derive the differential equations that explicitly govern the evolution of the circuit states. The second step is to solve these *state equations* for the states as functions of time. The states may then be used to determine any other branch variables of interest. Here, we will focus only on the first step since the solution of the differential equations that are the state equations may be executed using the same method employed earlier in this chapter, or the method developed in Section 12.10.

The desired state equations express the derivative of each state as functions of the states themselves, and any external signals applied through independent sources. As we shall see now, there is a relatively simple method for deriving these equations. To motivate this method, consider the constitutive laws for the capacitor and inductor, namely Equations 9.9 and 9.28. They can be written as

$$\frac{dv_C(t)}{dt} = \frac{1}{C} i_C(t) \tag{12.249}$$

$$\frac{di_L(t)}{dt} = \frac{1}{L} v_L(t), \tag{12.250}$$

respectively. From these equations we see that the derivative of the capacitor voltage is proportional to the capacitor current, and that the derivative of the inductor current is proportional to the inductor voltage. Therefore, to determine expressions for the state derivatives, we can equivalently find expressions for the capacitor currents and inductor voltages. These expressions should be derived in terms of the states and any independent external signals. This suggests a method for deriving the state equations.

First, we replace each capacitor by an independent voltage source having a voltage equal to the corresponding capacitor state. Additionally, we replace each inductor by an independent current source having a current equal to the corresponding inductor state. Second, we analyze the new circuit, which now contains only sources and resistors, to find the capacitor currents and inductor voltages. This analysis may be carried out using the node method developed in Chapter 3. It results in expressions that depend on the independent sources within the new circuit, namely the original independent sources and the sources representing the circuit states. Finally, we substitute the expressions for the capacitor currents and inductor voltages into equations of the form of Equations 12.249 and 12.250. This yields the desired state equations.

To illustrate the state-variable analysis of a circuit, consider the analysis of the circuit shown in WWW Figure 12.50. To analyze this circuit we replace the capacitor by a voltage source and the inductor by a current source as shown in Figure 12.62. Next we analyze this circuit to determine $i_C$ and $v_L$. This results in

$$i_C(t) = i_{IN}(t) - i_L(t) - \frac{v_C(t)}{R} \tag{12.251}$$

$$v_L(t) = v_C(t). \tag{12.252}$$

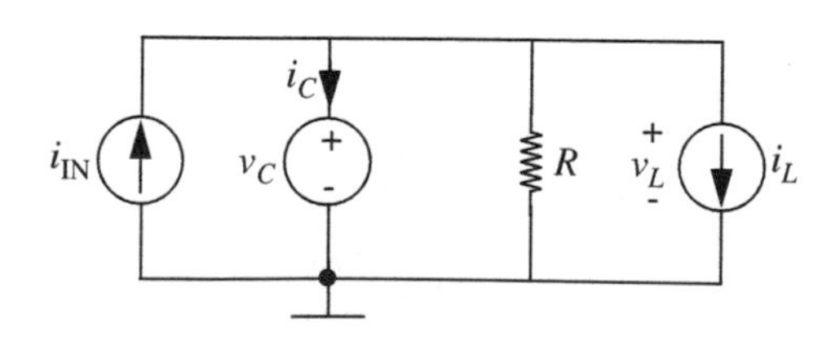


FIGURE 12.62 The equivalent circuit used for the state-variable analysis of the circuit shown in WWW Figure 12.50.

Finally, the substitution of Equations 12.251 and 12.252 into Equations 12.249 and 12.250, respectively, yields

$$\frac{dv_C(t)}{dt} = \frac{1}{C}i_{IN}(t) - \frac{1}{C}i_L(t) - \frac{1}{RC}v_C(t) \quad (12.253)$$

$$\frac{di_L(t)}{dt} = \frac{1}{L}v_C(t). \quad (12.254)$$

Equations 12.253 and 12.254 are the desired state equations.

It is important to emphasize that a state-variable analysis introduces no new physics. Rather it offers only an alternative analysis of the same circuit. To see this for the previous example, note that the substitution of Equation 12.254 into Equation 12.253 to eliminate $v_C$ yields WWW Equation 12.201. Thus, the state equations predicts the same circuit behavior as found previously in Section 12.6.

To close this section, it is valuable to identify the similarities and differences of the node and state-variable analyses. To begin, it is again important to emphasize that both analyses predict the same behavior for any given circuit. The main difference is that they do so in terms of different sets of variables, and through different mathematical mechanics. A node analysis does so in terms of the node voltages, and often results in a single high-order differential equation. A state-variable analysis does so in terms of the states, and results in a set of coupled first-order differential equations. For this reason, if the initial conditions for a circuit analysis are expressed in terms of the states, then the state-variable analysis will make easier use of that information. Further, a coupled set of first-order differential equations is more likely to be compatible with typical numerical analysis packages. On the other hand, it is often the node voltages, defined with respect to a common ground, that are most easily measured in an experiment. Thus, both sets of variables are useful.

## WWW 12.10 STATE-SPACE ANALYSIS*

### WWW 12.10.1 NUMERICAL SOLUTION*

## WWW 12.11 HIGHER-ORDER CIRCUITS*

## 12.12 SUMMARY

- The primary goal of this chapter was to examine the behavior of second-order circuits, primarily circuits containing at least one capacitor and one inductor. To analyze these and other second-order circuits we again relied on the node method of analysis. The mechanics of this analysis were essentially unchanged from Chapter 10, except for the details of the solution of second-order, as opposed to first-order, differential equations.

- Through our analysis of second-order circuits we observed that their behavior can be very different than the behavior of first-order circuits. Most importantly, a circuit containing one capacitor and one inductor can exhibit oscillations that correspond to an exchange of energy between the capacitor and inductor. Not surprisingly, these oscillations were found to decay in the presence of energy loss (for example, when a resistor is introduced in the circuit). To characterize this oscillatory behavior we introduced several key parameters: the undamped natural frequency (or undamped resonance frequency, or simply resonance frequency)

$$\omega_o \equiv 1/\sqrt{LC}$$

the damping factor $\alpha$, the damped natural frequency

$$\omega_d \equiv \sqrt{\omega_o^2 - \alpha^2}$$

and the quality factor

$$Q \equiv \omega_o/2\alpha.$$

The details of $\alpha$ depend on the circuit topology. For a parallel resonant circuit,

$$\alpha = \frac{1}{2RC}$$

and for a series resonant circuit,

$$\alpha = \frac{R}{2L}.$$

- The response of second-order systems can be classified as under-damped, critically-damped, or over-damped according to

$$\begin{aligned} \alpha < \omega_o &\quad \Rightarrow \quad \text{under-damped;} \\ \alpha = \omega_o &\quad \Rightarrow \quad \text{critically-damped;} \\ \alpha > \omega_o &\quad \Rightarrow \quad \text{over-damped.} \end{aligned}$$

- When the system is under-damped, the parameter $\omega_d$ determines the rate at which the states oscillate, and $\alpha$ determines the rate at which the states decay. $\omega_o$ is the oscillation frequency in the absence of damping. Intuitively, $Q$ determines the amount of ringing exhibited by the circuit, and is the approximate number of cycles after which the energy in an RLC circuit can be considered to have decayed to zero.
- The zero-input response (ZIR) is the response of the system to the initial stored energy, assuming no drive.
- The zero-state response (ZSR) is the the response to an applied drive signal, for no initial stored energy.
- The zero-input response of under-damped second-order systems resemble sinusoids with amplitudes that decay with time. As an example, the ZIR for an under-damped parallel resonant circuit, with $V_o$ as the initial voltage on the capacitor and zero initial current through the inductor, is given by

$$v_C(t) = V_o \frac{\omega_o}{\omega_d} e^{-\alpha t} \cos(\omega_d t + \phi)$$

for the capacitor voltage, and

$$i_L(t) = \frac{V_o}{\omega_d L} e^{-\alpha t} \sin(\omega_d t)$$

for the inductor current, for $t > 0$. In the previous equation

$$\phi = tan^{-1}\left(\frac{\alpha}{\omega_d}\right).$$

- To facilitate the analysis of second-order circuits, we also introduced two new methods of circuit analysis, namely state-variable analysis and state-space analysis. State-variable analysis was introduced as an alternative to node analysis, which is particularly useful when the state variables of a circuit, as opposed to the node voltages, are of primary interest. State-space analysis was introduced as a means of solving the coupled first-order differential equations that commonly result from a state-variable analysis. However, we also saw that it could be used as a means of solving the differential equations that naturally result from a node analysis. Finally, we briefly examined the analysis of higher-order circuits. The important finding from that examination was that the analysis of higher-order circuits can be carried out in a manner identical to the analysis of second-order circuits.

## EXERCISES

EXERCISE 12.1

a) Is the zero input response of the circuit shown in Figure 12.64 under-damped, over-damped, or critically-damped?

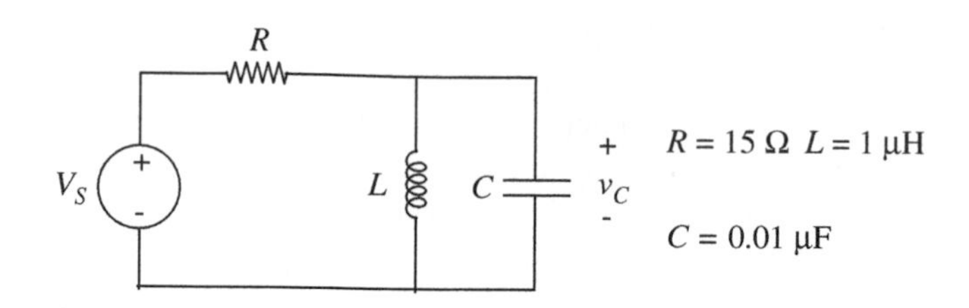


FIGURE 12.64

b) What is the form of the zero input response ($v_C$) for the same circuit? Make a rough sketch.

c) Compare the envelope of the zero input response with the rate of decay of the zero input response of the RC circuit in Figure 12.65.

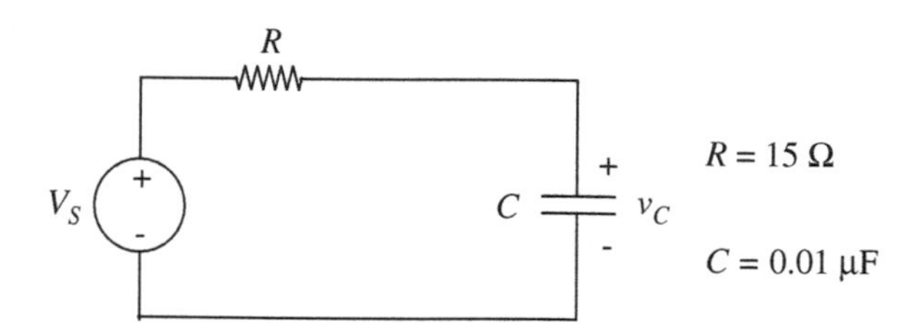


FIGURE 12.65

How do they differ?

EXERCISE 12.2 For each of the circuits in Figure 12.66, find and sketch the indicated zero-input response corresponding to the indicated initial conditions

a) In Figure 12.66, find $v_2$, assuming $v_1(0) = 1$ V, $v_2(0) = 0$

b) In Figure 12.67, find $v$, assuming $i(0) = 0$, $v(0) = 1$ V

c) Repeat (b), but with the resistor changed to 5 Ω.

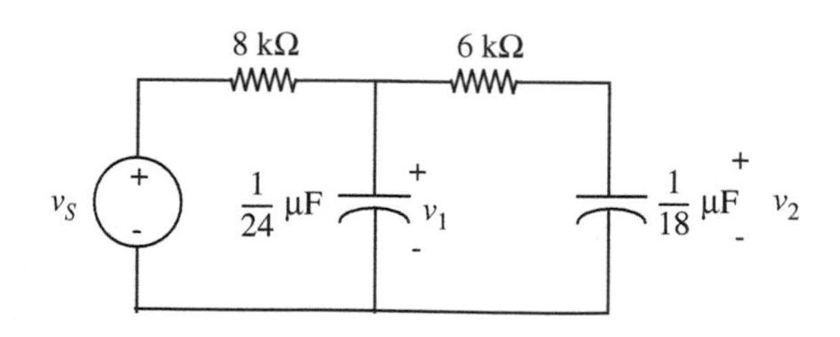


FIGURE 12.66

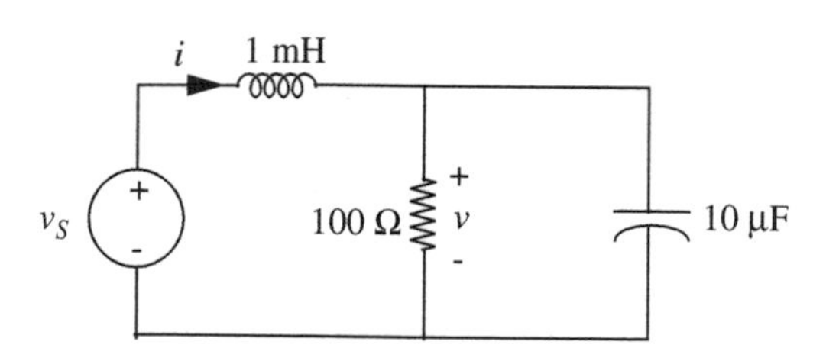


FIGURE 12.67

EXERCISE 12.3 In the circuit in Figure 12.68, a constant voltage source of 10 V is applied at $t = 0$. Find all branch voltages and all branch currents at $t = 0^+$ and at $t = \infty$ given $i_l(0^-) = 2$ A and $v_4(0^-) = 4$ V.

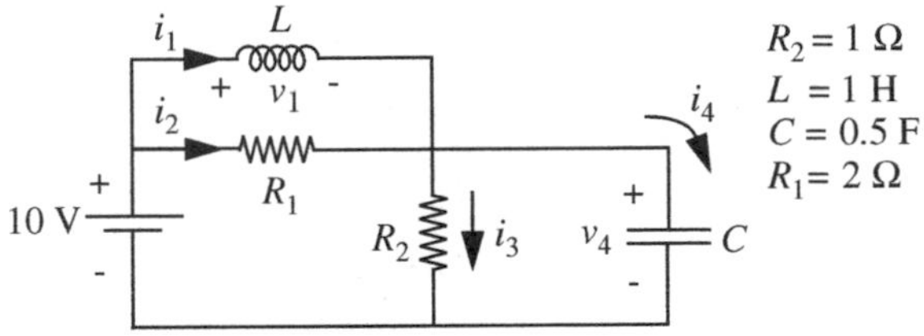

$R_2 = 1\ \Omega$
$L = 1$ H
$C = 0.5$ F
$R_1 = 2\ \Omega$

FIGURE 12.68

EXERCISE 12.4 Is the zero-input response of the circuit in Figure 12.69 under-damped, over-damped, or critically-damped? (Provide some kind of justification for your answer, either a calculation or a sentence of explanation.)

$L - 1\ \mu$H $C = 0.01\ \mu$F and $R_1 = R_2 = 15\ \Omega$

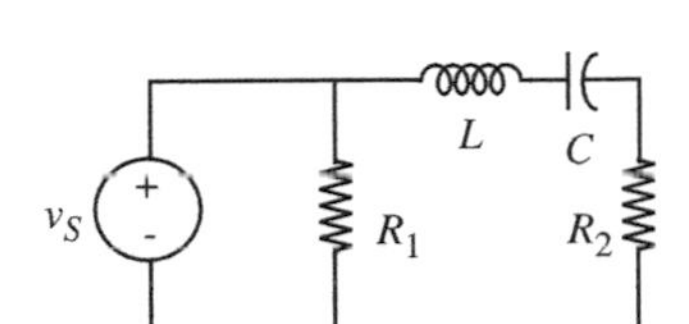


FIGURE 12.69

EXERCISE 12.5 In the circuit in Figure 12.70, the inductor current and capacitor voltage have been constrained by some external magic to be $i_L = 5$ A, $v_C = -6$ volts. At $t = 0$, the external restraints are removed, and the natural response of the circuit is allowed to evolve. Find the initial slopes of the state variables.

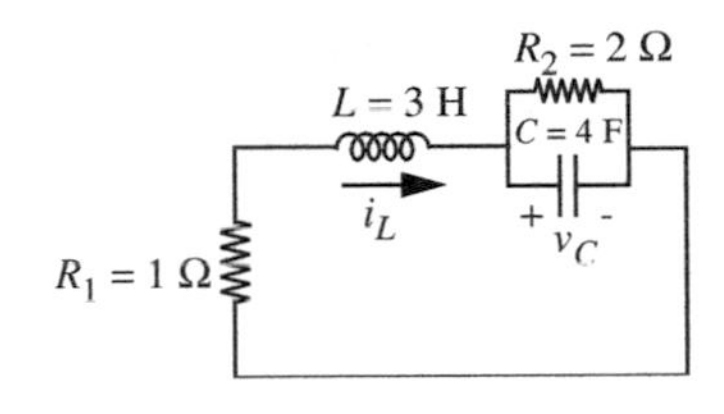


FIGURE 12.70

EXERCISE 12.6

a) Write the differential equations for the circuit in Figure 12.71 in state-variable form.

b) Assuming $v_C(0) = 0$, sketch $v_C(t)$ for a very short pulse of height $v_i$. Don't work it out: just show the *form*.

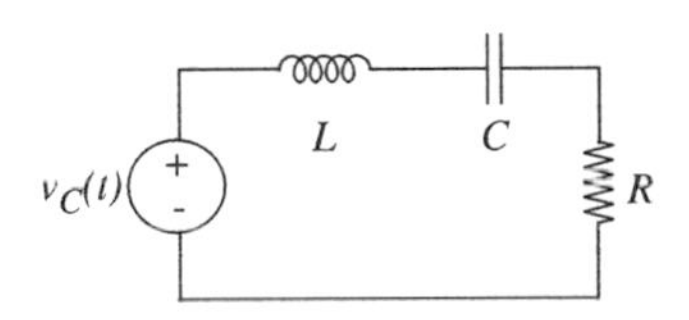


FIGURE 12.71

EXERCISE 12.7 Solve the following sets of coupled first-order state equations for $t > 0$ with the indicated inputs and initial values. Plot the positions of the natural frequencies in the complex plane. Sketch the state trajectories.

a)

$$\frac{dx_1}{dt} = -3x_1 + x_2$$

$$\frac{dx_2}{dt} = x_1 - 3x_2$$

$$x_1(0) = 2$$

$$x_2(0) = 0$$

b)

$$\frac{dx_1}{dt} = -4x_2$$

$$\frac{dx_2}{dt} = 4x_1$$

$$x_1(0) = 2$$

$$x_2(0) = 0$$

EXERCISE 12.8 Find the roots of the characteristic polynomial (often called the network natural frequencies) in each of the networks in Figure 12.72.

Numerical values: $R_1 = 10\ \Omega$, $L = 10\ \mu\text{H}$, $C = 10\ \mu\text{F}$, $R_2 = 2\ \Omega$.

## PROBLEMS

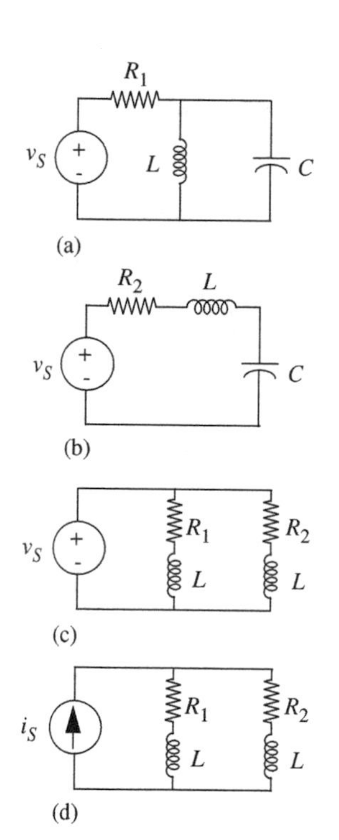


FIGURE 12.72

PROBLEM 12.1 Electrical networks are used to model physical systems governed by linear differential equations. The most important problems that arise in such modeling concern the interplay of accuracy and simplicity. It is usually very important to know when certain effects can safely be ignored in order to simplify the model and subsequent analysis. Such knowledge can be obtained by understanding the consequences of making the simplifying assumptions.

Two networks that could be used to model an acoustic system are shown in Figure 12.73. It is known that the inductance $L$ is small (specifically $L << (R^2C)/4$) but it is not known whether a circuit model with no inductances will be adequate. You are to help answer this problem by determining the difference in the responses of the capacitor voltage $v_C$ for the two circuits. Specifically, assume:

$$i_S(t) = Iu_{-1}(t) \qquad \text{(a step of amplitude I)}$$

$$v_C(0^-) = 0$$

$$i_L(0^-) = 0.$$

Determine $v_C(t)$ for $t > 0$ for both circuits. You should identify the effects of the inductance on such characteristics of the response as the natural frequencies, approximate behavior for small $t$, and asymptotic behavior.

You can greatly simplify the form of your results by making use of some assumptions derived from Taylor's theorem. For $x << 1$,

$$\sqrt{1-x} \simeq 1 - 1/2x \tag{12.297}$$

and

$$e^{-x} \simeq 1 - x. \tag{12.298}$$

PROBLEM 12.2 Capacitor $C_1$ has an initial voltage $v_1(0) = V$. Capacitor $C_2$ is initially uncharged, $v_2(0) = 0$. The voltage across element $A$ tends to zero as time tends to infinity. At time $t = 0$, the switch is closed. Refer to Figure 12.74.

a) Compute the initial charge of the system.

b) Find the voltage across both capacitors a long time after the switch has been closed. Remember that the total charge of the system must be conserved.

c) Find the energy stored in the system after a long time.

d) Find the ratio of final stored energy to initial energy. Where did the rest of the energy go?

e) Assume element $A$ is a resistor $R$. Find its voltage or current, and from that, find out the energy lost in it.

f) Find the ratio of lost energy to initial energy. Is it what you expected? Does it depend on $R$?

g) What would happen if an inductor was placed in series with $R$? Sketch the behavior of the current. (No calculations are needed.)

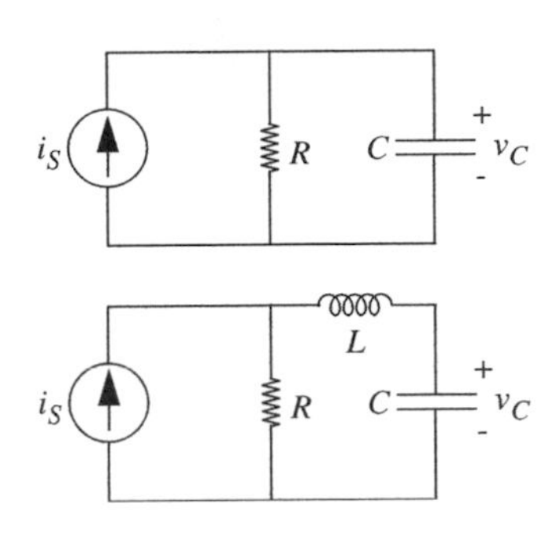

FIGURE 12.73

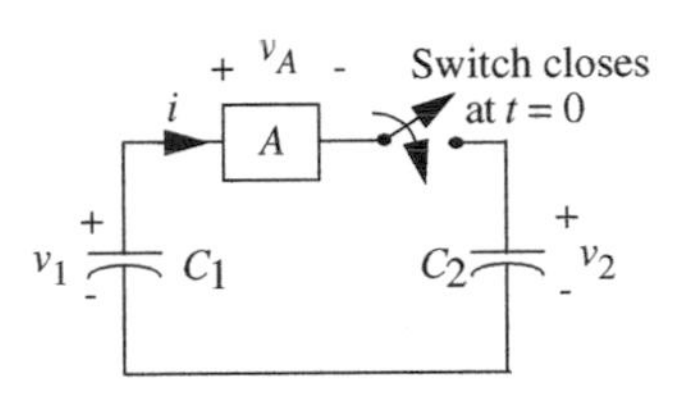

FIGURE 12.74

PROBLEM 12.3 Shown in Figure 12.75 is one possible circuit model for a transformer, for use where there can be a common ground between primary and secondary. Assume: $L_1 = 2.5$ H, $L_2 = 0.025$ H, $M = k\sqrt{L_1 L_2}$, where $k < 1$, $R_1 = 1$ kΩ, $R_2 = 10$ Ω.

a) Write the state equations for this network using $i_1$ and $i_2$ as state variables, and using the given circuit model to represent the transformer.

b) Determine the behavior of the natural frequencies of the network as a function of the coupling constant $k$. In particular, what are the natural frequencies in the limit of small $k$, and in the so-called tight-coupling limit, where $k$ approaches unity?

c) Assume that $v_S$ is a 1-volt square pulse of length 5 msec. Find $v_2(t)$ for the case $k = .98$. Is the output a good replica of a square pulse, or are there obvious departures from the square pulse shape?

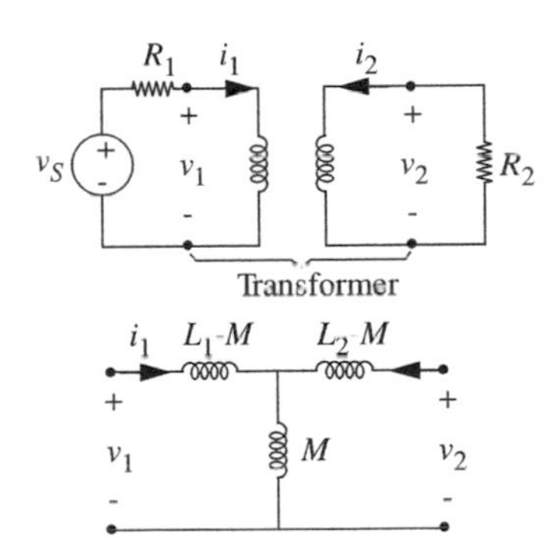

FIGURE 12.75

PROBLEM 12.4 Assuming $y(t) = Be^{st}$, for each differential equation, find the particular solution and the general form of the homogeneous solution. Plot the natural frequencies in the complex plane.

Assume $\tau$, $\alpha$, $\omega_0$ are constants. Do not worry about the dimensions of the right-hand side. Assume $B$ always has the appropriate dimension.

1) $\frac{dx}{dt} + \frac{x}{\tau} = y$

2) $\frac{dx}{dt} + \frac{x}{\tau} = \frac{dy}{dt}$

3) $\frac{x}{\tau} = \frac{y}{\tau} + \frac{dy}{dt}$

4) $\frac{d^2x}{dt^2} + \omega_0^2 x = y$

For 5) and 6), assume $\alpha$ and $\omega_0$ are both positive numbers.

5) $\frac{d^2x}{dt^2} + 2\alpha\frac{dx}{dt} + \omega_0^2 x = y$ Assume $\alpha > \omega_0$.

6) $\frac{d^2x}{dt^2} + 2\alpha\frac{dx}{dt} + \omega_0^2 x = \frac{dy}{dt}$ Assume $\alpha < \omega_0$.

PROBLEM 12.5 The circuit in in Figure 12.76 is the electrical analogue of a temperature control system.

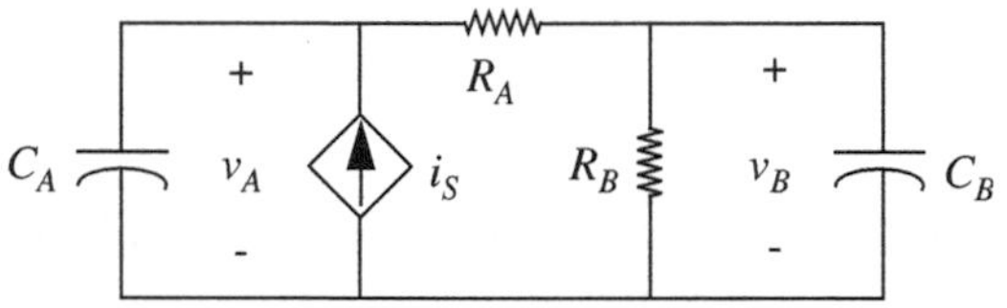

FIGURE 12.76

Assuming $C_A = 1$ F, $C_B = 4$ F, $R_A = 1\ \Omega$, $R_B = 4\ \Omega$.

$i_S = K(V_0 - v_B)^2$ where $K = 25$ A/V$^2$, $V_0 = 1.1$ V.

a) Write dynamical equations for this network in state form. Use $v_A$ and $v_B$ as state variables.

(As a check on your state equations, the stable steady-state value of $v_B$ is 1 V. That is, you should have $dv_A/dt = dv_B/dt = 0$ for $v_B = 1$ V.)

b) Now assume $v_A = V_A + v_0$ and $v_B = V_B + v_b$, where $V_A$ and $V_B$ are the steady-state values and $v_a$ and $v_b$ are small variations. Determine a small-signal *linear* circuit model in which $v_a$ and $b_b$ are the state variables.

c) Is the zero-input response of the small-signal circuit under-damped, over-damped, or critically-damped?

PROBLEM 12.6 In the circuit in Figure 12.77, the switch has been in position 1 for all $t < 0$. At $t = 0$, the switch is moved to position 2 (and remains there for $t > 0$). Find and sketch $v_C(t)$ and $i_L(t)$ for $t > 0$.

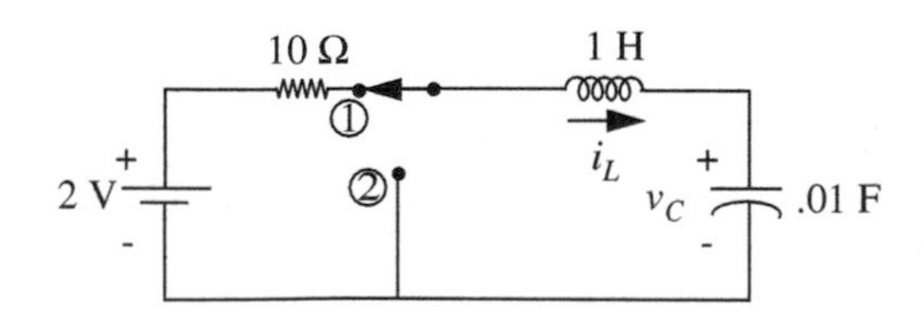


FIGURE 12.77

PROBLEM 12.7 Figure 10.107 (Problem 10.8 in the chapter on first-order transients) illustrated a parasitic inductance associated with VLSI package pins. Figure 12.78 is a modification of Figure 10.107 and shows a lumped parasitic capacitor $C_P$ associated with the power node within the VLSI chip. In this problem, we will study the combined effect of the parasitic inductance $L_P$ and capacitance $C_P$.

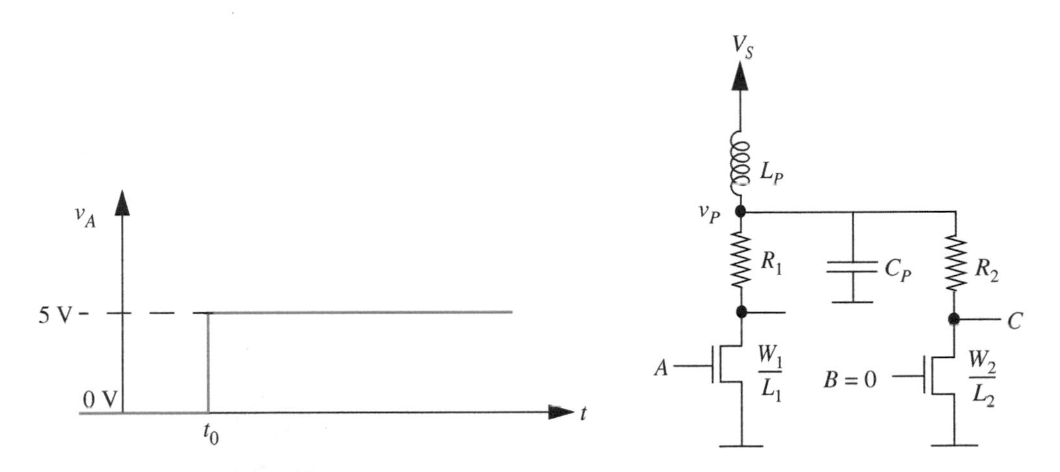


FIGURE 12.78

Assume that the input $B$ is 0 V at all times. Assume further that the input $A$ has 0 V applied to it initially. At time $t = t_0$, a 5-V step is applied at the input $A$. Plot the form of $v_P$ as a function of time for the under-damped and over-damped cases, assuming that $v_P = V_S$ for $t < t_0$. Clearly show the value of $v_P$ just prior to $t_0$ and just after $t_0$. Assume that the on resistance of a MOSFET is given by the relation $(L/W)R_n$ and that the MOSFET's threshold voltage is $V_T < V_S$. Also assume that $V_T < 5$ V. Compare this result with that for the inductor acting alone as computed in Problem 10.8 (Figure 10.107) in the chapter on first-order transients.

# CHAPTER 13

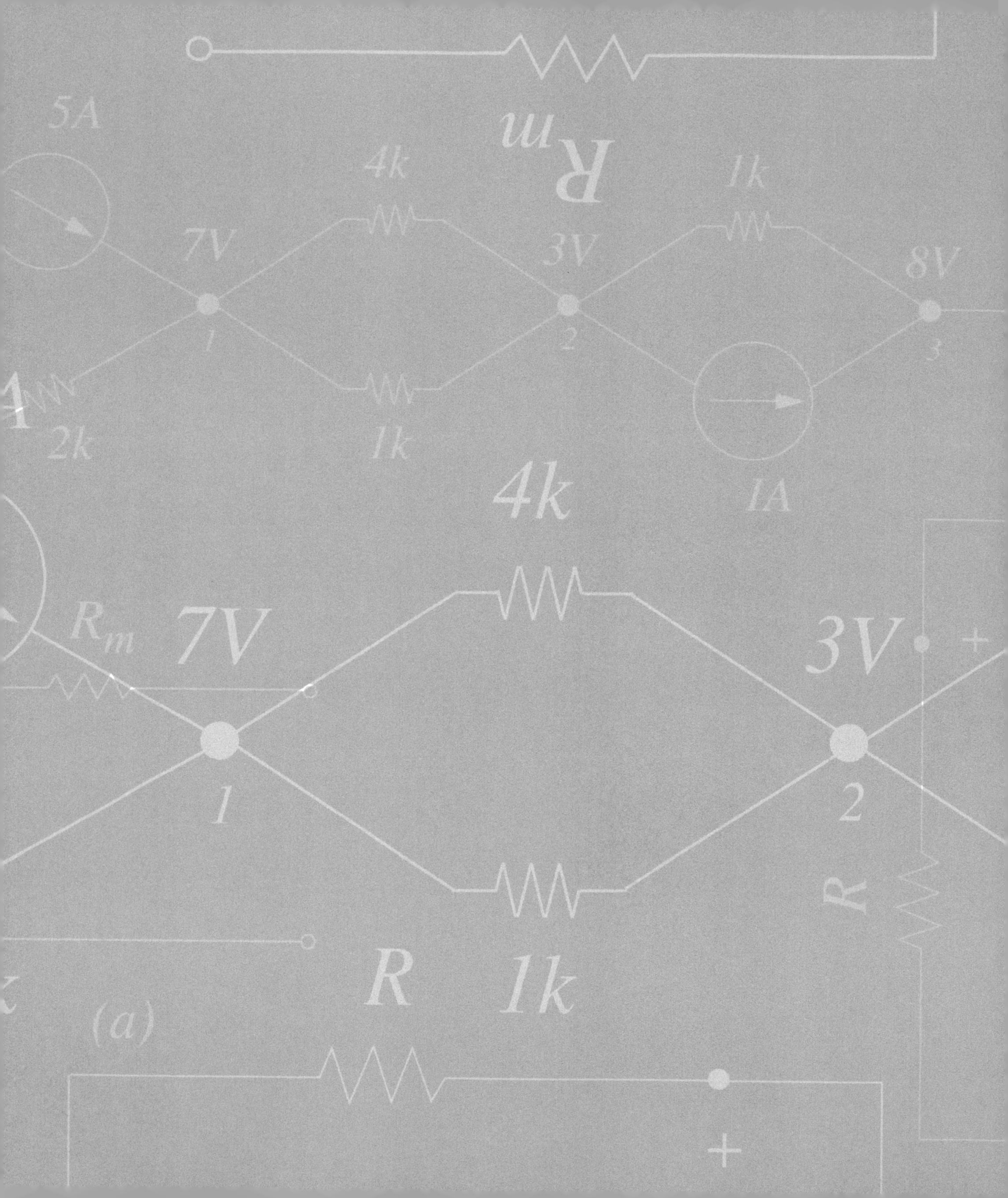
5A
4k
1k
7V
3V
8V
1
2
3
2k
1k
1A
4k
R_m
7V
3V
+
1
2
R
R
1k
(a)
+

# SINUSOIDAL STEADY STATE: IMPEDANCE AND FREQUENCY RESPONSE

# 13

## 13.1 INTRODUCTION

This chapter represents a major change in point of view for circuit analysis, hence it is important to review where we have been and where we are going. The analysis method discussed in preceding chapters has four basic steps:

1. Draw a circuit model of the problem.
2. Formulate the differential equations.
3. Solve these equations. If the equations are linear, then find the homogeneous solution and the particular solution. If the equations are nonlinear, then numerical methods often are required.
4. Use the initial conditions to evaluate the constants in the homogeneous solution.

This approach, diagrammed in the top of Figure 13.1, is basic and powerful, in that it can handle both linear and nonlinear problems, but often it involves substantial mathematical manipulations if the drive signals are other than simple impulses, steps, or ramps. Thus there is considerable incentive to look for easier methods of solution, even if these methods are more restricted in application. Simplified methods are indeed possible if the system is linear and time invariant, and we assume *sinusoidal* drive and focus on the steady-state behavior. Because in many design applications such as audio amplifiers, oscilloscope vertical amplifiers, and Op Amps linearity is a basic design constraint, systems that are linear or at least incrementally linear represent a large and important class, hence are worthy of special attention. Further, more complicated input signals such as square waves can be considered to be the sum of many sinusoids; hence the problem can be solved by superposition.

Equally important, we often characterize systems by their frequency response (that is, sinusoidal response). Examples include our hearing, audio equipment, ultrasonic pest deterrents, and wireless network receivers. The frequency related behavior of such systems is as important as their time-domain behavior. Therefore, the sinusoidal steady state response is useful because it is a natural and convenient way to describe the behavior of linear systems.

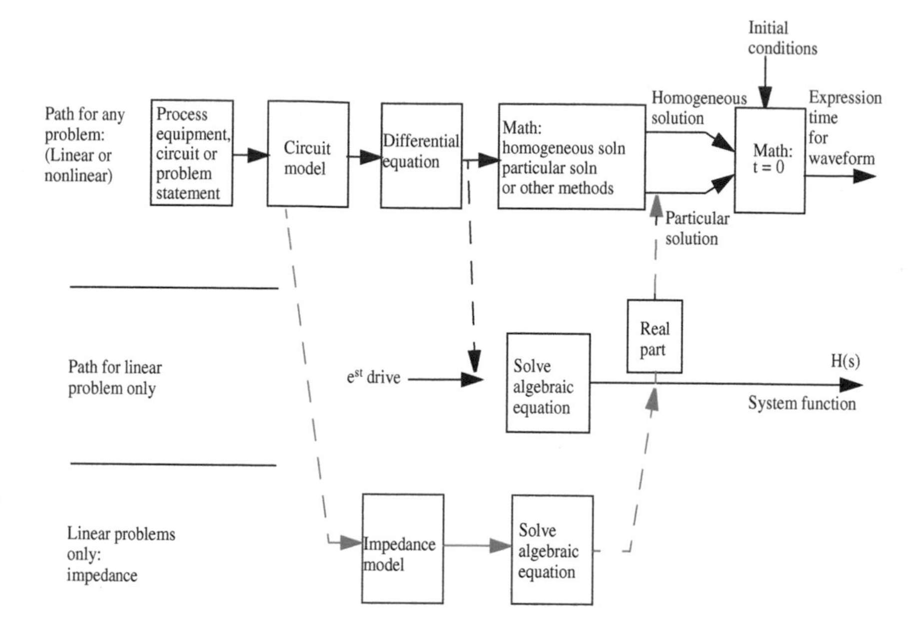


FIGURE 13.1 Analysis methods.

We wish to show in this chapter that the solution to linear circuit problems is greatly simplified by assuming a drive of the form $e^{st}$ as illustrated in the center panel of Figure 13.1, primarily because under this assumption the differential equation is transformed into an algebraic equation, and because the response to a sinusoidal drive can be directly obtained from the response to the $e^{st}$ drive. This leads further to a shorthand solution method involving the concept of *impedance*, whereby the algebraic equation can be found directly from the circuit model, without writing the differential equation at all, as diagrammed in the lowermost panel of the Figure 13.1.

The insight behind the employment of a drive of the form $e^{st}$, where $s = j\omega$, is the following: Recall, we wish to find the system response in the steady state[1] to a sinusoidal input of the form $\cos(\omega t)$. We will show that directly solving system differential equations with a sinusoidal input leads to a tangle of trigonometry and is very complicated. (You have already seen an example of a direct solution of an RL circuit for a sinusoidal drive in Section 10.6.7.) Instead, we employ the following mathematical trick: Realizing that

$$e^{j\omega t} = \cos(\omega t) + j\sin(\omega t) \qquad (13.1)$$

(the Euler relation), we first obtain with relative ease the circuit response to an unrealizable drive of the form $e^{j\omega t}$. The resulting response will contain a real

1. Interestingly, the substitution of $s = j\omega$ will give us the response of the circuit in sinusoidal steady state. Although not covered in this book, the use of Laplace Transforms where we substitute $s = \sigma + j\omega$ will yield the total response.

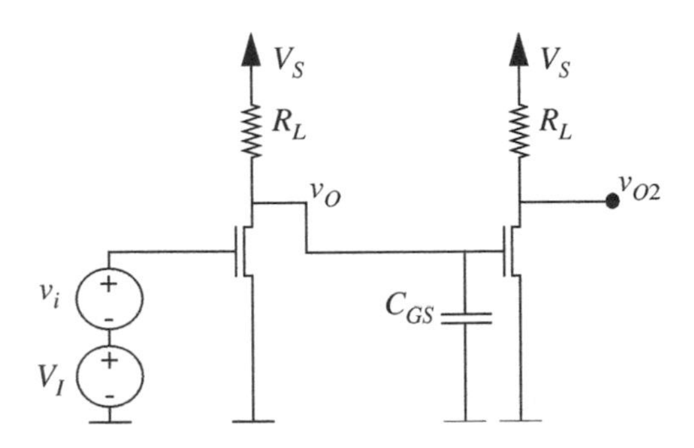


**FIGURE 13.2** A two-stage MOSFET amplifier showing the MOSFET gate capacitor.

part and an imaginary part. For real linear systems, by superposition, the real part of the response is due to the real part of the input (namely, $\cos(\omega t)$) and the imaginary part of the response is due to the imaginary part of the input (namely, $j\sin(\omega t)$). Accordingly, by taking the real part of the response to $e^{j\omega t}$, we obtain the response to a real sinusoidal input of the form $\cos(\omega t)$. (Similarly, by taking the imaginary part of the response to $e^{j\omega t}$, we obtain the response to an input of the form $\sin(\omega t)$.)

To motivate the study of methods based on the sinusoidal steady state, let us present an example of the type of problem that can be solved with ease using these methods. Suppose we construct the linear small-signal amplifier shown in Figure 13.2 by concatenating two single stage MOSFET amplifiers of the type studied in Chapter 8. The DC voltage $V_I$ is chosen to bias the first stage appropriately, and the DC value of the first stage output voltage $V_O$ provides the bias for the second stage. Figure 13.2 further shows the presence of a capacitor $C_{GS}$ at the input node of the second stage (for example, reflecting the gate capacitance of the MOSFET in the second stage).

Suppose, now, that we wish to find the first-stage output voltage $v_o$ in response to $v_i$, a small sinusoidal signal applied to the input of the amplifier. In particular, we are interested in determining how the presence of the capacitor $C_{GS}$ affects the amplification afforded by the first amplifier stage. Suppose, further, that we do not care about initial transients, rather, we are interested in the steady-state behavior when all transients have died out. Experimental application of a sinusoid to the input and measurement of the response $v_o$ will show very different behavior as the frequency of the input is swept from a low to a high value. We will observe that for low-frequency signals the gain of the first stage is no different from our earlier calculations in Chapter 8 in the absence of the capacitor $C_{GS}$. However, we will also observe that the presence of the capacitor makes the gain of the amplifier fall off rapidly at high frequencies.

Analytical analysis based on the methods we learned in the previous chapters would suggest writing the differential equation for the circuit comprising the resistor $R_L$ and the capacitor $C_{GS}$ and finding the forced response to an applied sinusoid. As demonstrated by the example in Section 10.6.7, this type of analysis is very cumbersome. In contrast, the analysis methods that we will learn

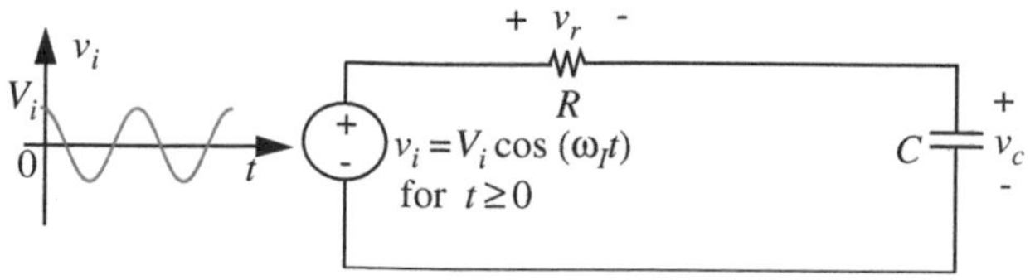

**FIGURE 13.3** RC circuit with tone burst in. The amplitude of the input waveform is $V_i$, where $V_i$ is real.

in this chapter will make this a trivial exercise. In particular, Section 13.3.4 will analyze the circuit of Figure 13.2 in detail and explain the observed behavior.

## 13.2 ANALYSIS USING COMPLEX EXPONENTIAL DRIVE

To illustrate this new approach, let us analyze the simple linear first-order RC circuit shown in Figure 13.3, and presume that we wish to find the capacitor voltage $v_C$, in response to a cosine wave suddenly applied at $t = 0$, often called a *tone burst*. The tone burst is mathematically represented as

$$v_i = V_i \cos(\omega_1 t) \quad \text{for } t \geq 0,$$

where $V_i$ is the amplitude of the cosine, and $\omega_1$ its frequency. (Note that we do not use $\omega_o$ in the input signal to avoid confusion with the $\omega_o$ used to represent the undamped natural frequency in a second order systems.)

The differential equation for the circuit is

$$v_i = v_C + RC\frac{dv_C}{dt}. \tag{13.2}$$

Let us attempt to solve this differential equation by summing its homogeneous and particular solutions. Recall, when dealing with circuit responses, the homogeneous solution is also called the *natural response*, and the particular solution is also called the *forced response*. Recall further that the forced response depends on the external inputs to the circuit. Let us denote the homogeneous solution as $v_{CH}$ and the particular or forced solution as $v_{CP}$. Then, we know that the total solution is given by

$$v_C = v_{CH} + v_{CP}.$$

### 13.2.1 HOMOGENEOUS SOLUTION

From Equation 13.2, the homogeneous solution can be derived by solving

$$RC\frac{dv_{CH}}{dt} + v_{CH} = 0. \tag{13.3}$$

As we have seen in Chapter 10, the homogeneous solution for this equation is

$$v_{ch} = K_1 e^{-t/RC} \tag{13.4}$$

where $K_1$ is a constant to be determined from the initial conditions.

### 13.2.2 PARTICULAR SOLUTION

The straightforward approach to finding the particular or forced solution $v_{cp}$ involves finding any solution to the differential equation

$$v_i = v_{cp} + RC\frac{dv_{cp}}{dt}. \tag{13.5}$$

Since the input $v_i$ is given by

$$v_i = V_i \cos(\omega_1 t)$$

(where $V_i$ is real), this amounts to finding any solution to

$$V_i \cos(\omega_1 t) = v_{cp} + RC\frac{dv_{cp}}{dt}. \tag{13.6}$$

Obviously the forced response $v_{cp}$ must be some combination of sines and cosines, so we assume

$$v_{cp} = K_2 \sin(\omega t) + K_3 \cos(\omega t) \tag{13.7}$$

or, equivalently,

$$v_{cp} = K_4 \cos(\omega t + \Phi). \tag{13.8}$$

There is nothing wrong with this approach, except that it leads to a tangle of trigonometry. So, we will abandon this path.

Instead, let us launch out in a slightly different direction. The Euler relation

$$e^{j\omega t} = \cos(\omega t) + j\sin(\omega t) \tag{13.9}$$

shows that $e^{j\omega t}$ contains the cosine term we want, in addition to an unwanted sine term. Hence, by a sort of inverted superposition argument, we replace the actual source $v_i$ with a source of the form

$$\tilde{v_i} = V_i e^{s_1 t} \tag{13.10}$$

and return later to unscramble the cosine and sine parts. In this equation we have used

$s_1$ as a shorthand for $j\omega_1$,

and have included a "~" above $v_i$ to indicate that this is not the true drive voltage. For consistency, we will use the same notation for all variables related to this fake drive voltage. The differential equation to find the particular solution to $\tilde{v_i}$ now becomes

$$\tilde{v_i} = V_i e^{s_1 t} = \tilde{v_{cp}} + RC\frac{d\tilde{v_{cp}}}{dt}. \tag{13.11}$$

It is clear that a reasonable assumption for the particular solution is

$$\tilde{v_{cp}} = V_c e^{st} \tag{13.12}$$

in which we must somehow find $V_c$ and $s$. On substitution of the assumed particular solution into Equation 13.11, we obtain

$$V_i e^{s_1 t} = V_c e^{st} + RCsV_c e^{st}. \tag{13.13}$$

We note first that $s$ must equal $s_1$, otherwise Equation 13.13 cannot be satisfied for all time. Now, on the basis that $e^{st}$ can never be zero for finite values of $t$, the $e^{s_1 t}$ terms can be divided out, to yield an *algebraic equation relating the complex amplitudes of the voltages* rather than a differential equation relating the voltages as time functions:

$$V_i = V_c + V_c RCs_1 \tag{13.14}$$

which can be solved to yield

$$V_c = \frac{V_i}{1 + RCs_1} \quad \text{for} \quad s_1 \neq -\frac{1}{RC} \tag{13.15}$$

a restriction clearly satisfied in this case because $s_1 = j\omega_1$ where $\omega_1$ is a real number. Thus Equation 13.15 becomes

$$V_c = \frac{V_i}{1 + j\omega_1 RC} \tag{13.16}$$

or, from Equation 13.12, the particular solution for the fake input $\tilde{v_i}$ is

$$\tilde{v_{cp}} = \frac{V_i}{1 + j\omega_1 RC} e^{j\omega_1 t}. \tag{13.17}$$

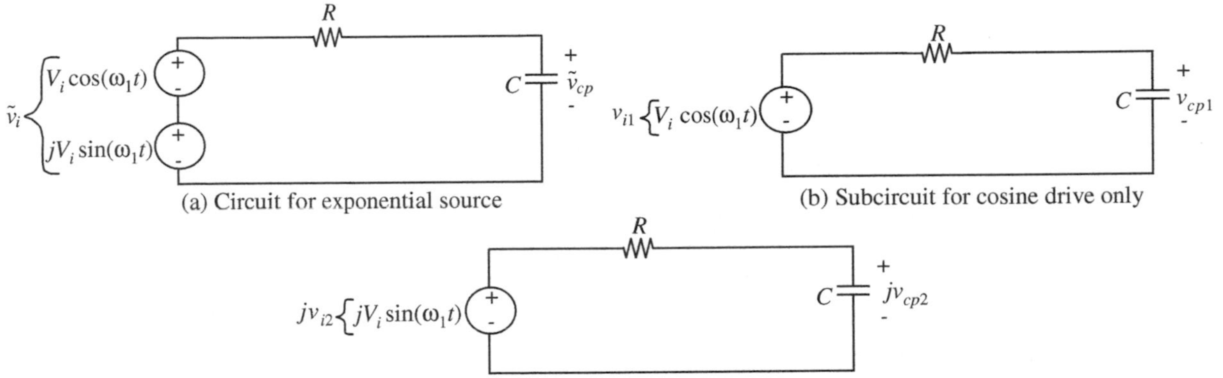

(a) Circuit for exponential source

(b) Subcircuit for cosine drive only

(c) Subcircuit for sine drive only

FIGURE 13.4 RC circuit with exponential drive $e^{st}$.

At this point you should protest. No waveform measured in the laboratory will have a "$j$" associated with it. The problem arises because we have used a complex rather than a real drive. That is, we have analyzed the circuit shown in Figure 13.4a, rather than Figure 13.3. The complex exponential drive $\tilde{v}_i$ can be represented by the Euler relation as the sum of two sources as depicted in Figure 13.4a. *If the circuit is linear,* the two-source circuit can be analyzed by superposition, as suggested in Figure 13.4b and 13.4c. Specifically, the voltage $\tilde{v_{cp}}$ can be found by summing the response to $V_i \cos(\omega t)$, as obtained from Figure 13.4b with $j$ times the response to $V_i \sin(\omega t)$, as found from Figure 13.4c:

$$\tilde{v_{cp}} = v_{cp1} + jv_{cp2}. \tag{13.18}$$

From the perspective of Figure 13.3, we have calculated in Equation 13.18 the response $\tilde{v_{cp}}$, and what we really want is $v_{cp1}$. Notice that $v_{cp1}$ is none other than the $v_{cp}$ that we had originally set out to find, namely, the solution to Equation 13.6. So we want to "de-superimpose" the two sources in Figure 13.4a. This is a simple matter because of the $j$ flag: $v_{cp1}$ is the real part of $\tilde{v_{cp}}$.

The next task, then, is to find an easy way of calculating the real part of a complex expression. (Those readers who are a little hazy about manipulation of complex numbers, and in particular the conversion between rectangular and polar form, should review at this point Appendix C on complex numbers or a suitable math text.) In this specific problem, we must find the real part of the $\tilde{v_{cp}}$ expression, Equation 13.17. The difficulty is that the expression has two factors, one in Cartesian or rectangular form, and the other in polar, whereas multiplication is simpler if both factors are in polar form. Hence we rewrite Equation 13.17 in polar form as

$$\tilde{v_{cp}} = \frac{V_i}{\sqrt{1 + (\omega_1 RC)^2}} e^{j\Phi} e^{j\omega_1 t} \tag{13.19}$$

where

$$\Phi = \tan^{-1}\left(\frac{-\omega RC}{1}\right). \tag{13.20}$$

Now, to find the real part of $\tilde{v_{cp}}$, we use the Euler relation to write Equation 13.19 as

$$\tilde{v_{cp}} = \frac{V_i}{\sqrt{1+(\omega_1 RC)^2}}\cos(\omega_1 t + \Phi) + j\frac{V_i}{\sqrt{1+(\omega_1 RC)^2}}\sin(\omega_1 t + \Phi)$$

from which the real part of $\tilde{v_{cp}}$ is available by inspection,

$$v_{cp1} = \frac{V_i}{\sqrt{1+(\omega_1 RC)^2}}\cos(\omega_1 t + \Phi). \tag{13.21}$$

This, finally, is the particular solution of Equation 13.6.

### 13.2.3 COMPLETE SOLUTION

The complete expression for the capacitor voltage in response to a cosine tone burst is the sum of this particular solution ($v_{cp1}$) and the homogeneous solution ($v_{ch}$) previously found in Equation 13.4:

$$v_c = K_1 e^{-t/RC} + \frac{V_i}{\sqrt{1+(\omega_1 RC)^2}}\cos\left(\omega_1 t + \Phi\right). \tag{13.22}$$

The one remaining unknown constant, $K_1$, can be found from the initial conditions by setting $t$ to zero in the usual manner. However, as we will see shortly, we usually do not care about the first term.

### 13.2.4 SINUSOIDAL STEADY-STATE RESPONSE

Under sinusoidal drive, we are almost always interested in the *steady-state* value of the capacitor voltage, which can be readily obtained from Equation 13.22 by assuming $t$ is very large. When $t \to \infty$, Equation 13.22 reduces to

$$v_c = \frac{V_i}{\sqrt{1+(\omega_1 RC)^2}}\cos\left(\omega_1 t + \Phi\right), \tag{13.23}$$

which is simply the particular solution to a cosine input (compare with Equation 13.21). For a cosine input, the steady-state response is often termed the *response to a cosine*. The corresponding complete response is termed the *response to a cosine burst*, and includes both the homogeneous and particular terms. In Equation 13.23, the $V_i/\sqrt{1+(\omega_1 RC)^2}$ factor gives the amplitude (or magnitude) of the response, and $\Phi$ is the phase. The phase is the angular difference between the output and input sinusoids. Notice that both the magnitude and phase (see Equation 13.20) of the response are frequency dependent.

Equation 13.22 is really quite general, in that it gives the capacitor voltage for any amplitude and any frequency of cosine tone burst. For example, it is obvious that at low frequencies, (that is, for $\omega_1$ small), and after the transient has died away:

$$v_c \simeq V_i \cos(\omega_1 t). \tag{13.24}$$

Thus, after the transient has died away, the output looks almost like the input. We conclude that for $\omega_1$ small, the capacitor behaves like an open circuit. Further, for $\omega_1$ large, that is, at high frequencies, after the transient has died away

$$v_c \simeq \frac{V_i}{\omega_1 RC} \cos(\omega_1 t - 90°) \tag{13.25}$$

so the output will be sinusoidal, but 90 degrees out of phase with the input, and much smaller. At high frequencies then, the magnitude of the capacitor voltage will get very small, so we can say that the capacitor begins to behave like a short circuit.

There are four general conclusions to be drawn from this specific example:

1. The use of an $e^{st}$ drive reduces a differential equation to an algebraic equation, thereby simplifying the solution. This solution process replaces trigonometry with complex algebra, which is a wise trade.

2. The last couple of pages from Equation 13.17 to Equation 13.22, although necessary for completeness, did not add any new insight about the circuit behavior. For example, the same information about the form of $v_c$ in the steady state, or its value at low frequencies and high frequencies could have been found from Equation 13.17, or even from the *complex amplitude* $V_c$, Equation 13.16, just as easily as from Equation 13.21, without the intervening "real part" calculation.[2]
   For example, the steady state value of $v_c$ (or the particular or forced response) can be determined from the value of $V_c$ as

$$v_c = Re\left[V_c e^{j\omega_1 t}\right] \tag{13.26}$$

2. Recall from Equation 13.16 that $V_c$ is the complex amplitude of the forced response to our fake input $\tilde{v}_i = V_i e^{j\omega_1 t}$.

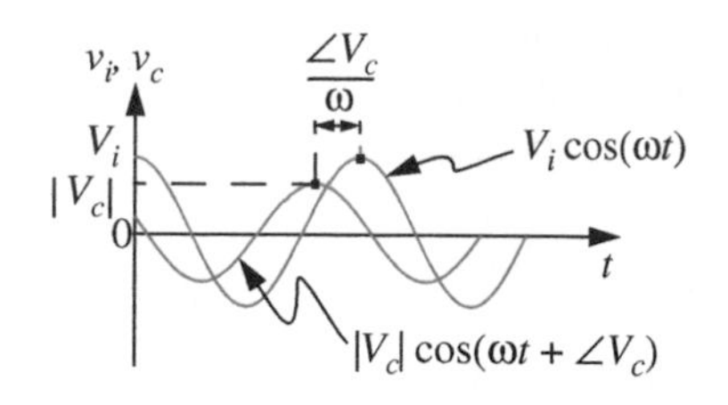


FIGURE 13.5 The amplitude and phase of the response $v_C$ compared to the input sinusoid $v_i$.

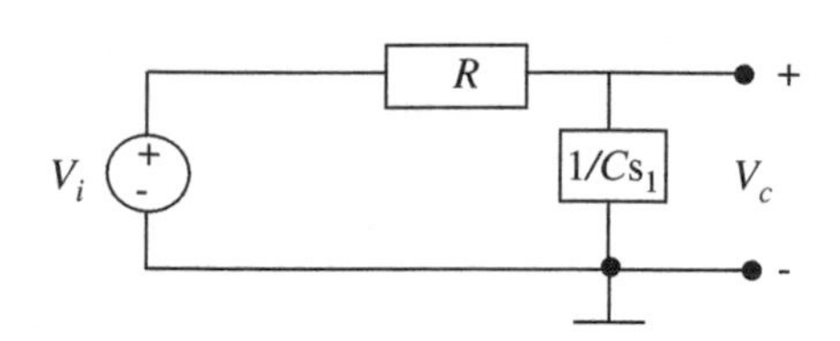


FIGURE 13.6 A circuit interpretation of Equation 13.15.

or, equivalently,

$$v_c = |V_c| \cos(\omega_1 t + \angle V_c). \tag{13.27}$$

Figure 13.5 shows a sketch of the input cosine and the output response with the various magnitudes and phases marked. Notice that the complex amplitude $V_c$ carries both the amplitude and phase information of the response ($|V_c|$ and $\angle V_c$ respectively) in an easily accessible manner. Thus, our analysis can stop at Equation 13.16.

3. The denominator of the $V_c$ expression, Equation 13.15, has the same form as the characteristic polynomial in the homogeneous solution, (see Chapter 10, Equation 10.9 for example) so the value of $s$ in the homogeneous solution could have been found from this denominator without any formal solution of the homogeneous equation. That this is a general result can be shown by examining the two derivations.

4. The $V_c$ expression, Equation 13.15, looks very much like a voltage divider expression, especially if we divide through by $Cs_1$.

$$V_c = \frac{1/Cs_1}{R + 1/Cs_1} V_i. \tag{13.28}$$

This suggests a very simple method for finding the complex amplitude $V_c$ directly from the circuit: Redraw the circuit, replacing resistors with $R$ boxes, capacitors with $1/Cs_1$ boxes, and cosine sources by their amplitudes, in this case $V_i$, as shown in Figure 13.6. Now $V_c$ can be found in one line. But what are these boxes? And what is $V_c$? The next section will provide the answers.

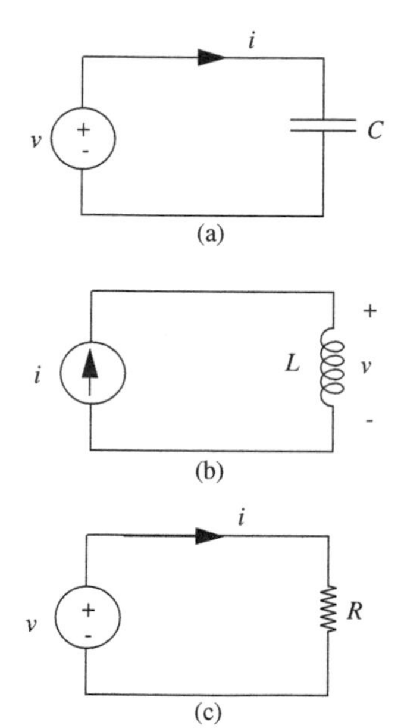


FIGURE 13.7 Impedance calculations.

## 13.3 THE BOXES: IMPEDANCE

To get a better idea of the meaning of the boxes in Figure 13.6, let us examine some trivial cases, as sketched in Figure 13.7. In Figure 13.7a, a voltage source

$V_i \cos(\omega_1 t)$ is connected across a capacitor, hence

$$i = C\frac{dv}{dt}. \tag{13.29}$$

On the basis of Section 13.1, assume the voltage and current are of the form

$$v = Ve^{st} \tag{13.30}$$

$$i = Ie^{st} \tag{13.31}$$

where, as before, we use $s$ as a shorthand notation for $j\omega$.

On substituting these relations in Equation 13.29, and dividing by $e^{st}$ (never zero for finite $s$ and $t$) we find

$$I = CsV \tag{13.32}$$

or

$$V = \frac{1}{Cs}I. \tag{13.33}$$

Similar calculations on the inductor and the resistor yield

$$V = LsI \tag{13.34}$$

$$V = IR. \tag{13.35}$$

These equations indicate that for linear $R$, $L$, or $C$, in each case the complex amplitude of the voltage is related to the complex amplitude of the current by very simple algebraic expressions which are generalizations of Ohm's Law. The constants relating $V$ to $I$ in Equations 13.33, 13.34, and 13.35 are called *impedances*, and these equations are the constituent relations for $C$, $L$, and $R$ expressed in impedance form. The constituent relations for these elements and for voltage and current sources are summarized in Figure 13.8.

Just as we used $R$ to denote resistances, we commonly use the letter $Z$ to denote impedances.

Thus, the impedances of an inductor, a capacitor, and a resistor are given by

$$Z_L = sL = j\omega L \tag{13.36}$$

$$Z_C = \frac{1}{sC} = \frac{1}{j\omega C} \tag{13.37}$$

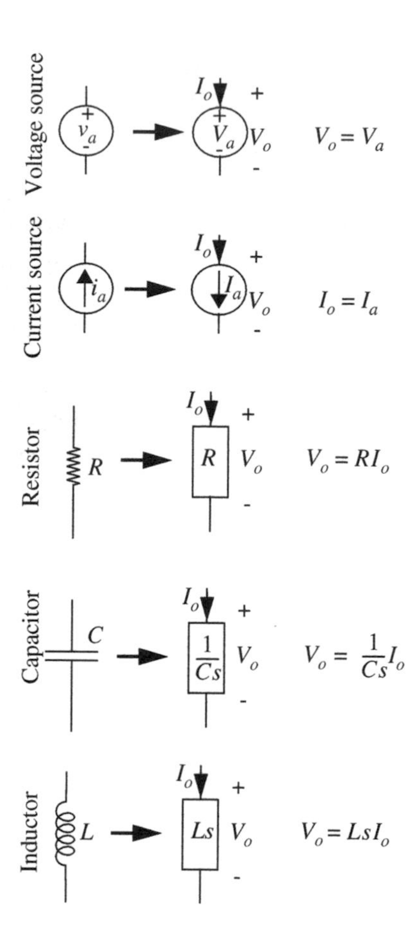


**FIGURE 13.8** Constituent relations for a voltage source, a current source, and *R*, *L*, and *C* in impedance form. Note that $V_o$ and $I_o$ are terminal variables, while $V_a$ and $I_a$ are element parameters. Note also that $s = j\omega$.

and

$$Z_R = R \tag{13.38}$$

respectively.

Furthermore, just as the conductance was defined as the reciprocal of resistance, we define admittance as the reciprocal of impedance.

Impedances are complex numbers in general. They are also frequency dependent. Figure 13.9 plots the magnitude of the impedances of an inductor, a capacitor, and a resistor as a function of frequency. The curves in the figure reinforce the following intuition developed in Chapter 10 and summarized in Section 10.8:

Inductors behave like short circuits for DC (or very low frequencies) and like open circuits for very high frequencies. Capacitors behave like open circuits for DC (or very low frequencies) and like short circuits for very high frequencies.

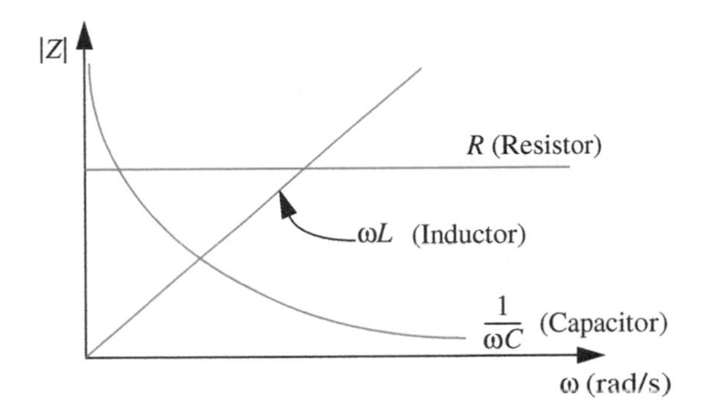


**FIGURE 13.9** Frequency dependence of the impedances of inductors, capacitors, and resistors.

Now, generalizing from these results, the relations among complex amplitudes of voltages and currents for any linear RLC network can be found by replacing the (sinusoidal) sources by their complex (or real) amplitudes, and replacing resistors by $R$ boxes, capacitors by $1/Cs$ boxes, and inductors by $Ls$ boxes. The resultant diagram is called the *impedance model* of the circuit. The complex voltages and currents in circuits can now be found by standard linear circuit analysis: Node Equations, Thévenin's theorem, etc.

The impedances follow the same combination rules as resistors, for example, impedances in series add, although here the addition involves complex numbers.

Therefore, the intuitive method based on series and parallel simplifications also applies.

We note that the impedance representation does not change the topology of the circuit — devices are simply replaced by their corresponding impedance models drawn as boxes. The reason is that KVL and KCL apply to a given circuit irrespective of the form of the drive. In other words, KVL and KCL apply irrespective of whether the voltages and currents are sinusoids, DC values, or any other form, for that matter. The impedance form simply assumes sinusoidal drive and response and captures in a convenient form individual device behavior when the drives are sinusoids. Thus, because the KVL and KCL equations are unchanged for sinusoidal drive, the circuit topology remains the same because it captures the same information as expressed by KVL and KCL.

If desired, the expressions for the actual voltages and currents, the particular solutions or forced responses in Chapter 10 parlance, can be found by multiplying the corresponding complex variable by $e^{j\omega t}$ and taking the real part.

For example, to obtain the actual voltage $v_x(t)$ from the corresponding complex variable $V_x(j\omega)$, we use

$$v_x(t) = Re\left[V_x(j\omega)\, e^{j\omega t}\right] \tag{13.39}$$

or equivalently,

$$v_x(t) = |V_x| \cos(\omega t + \angle V_x). \tag{13.40}$$

We emphasize again, however, that this step is usually not necessary, because the complex amplitude expression contains all the key information about circuit behavior.

At this point it is necessary to explicitly introduce a notation for voltages and currents to clearly differentiate complex amplitudes from time functions. We abide by the international standard in this matter:

- DC or operating-point variables: uppercase symbols with uppercase subscripts (for example, $V_A$)
- Total instantaneous variables: lowercase symbols with uppercase subscripts (for example, $v_A$)
- Incremental instantaneous variables: lowercase symbols with lowercase subscripts (for example, $v_a$)
- Complex amplitudes or complex amplitudes of incremental components, and real amplitudes of sinusoidal input sources: uppercase symbols with lowercase subscripts (for example, $V_a$)

Summarizing, the impedance method allows us to determine with ease the steady-state response of any linear RLC network for a sinusoidal input. The method works with complex amplitudes of voltages and currents at its variables and has the following general steps:

1. First, replace the (sinusoidal) sources by their complex (or real) amplitudes. For example, the input voltage $v_A = V_a \cos(\omega t)$ is replaced by its real amplitude $V_a$.
2. Replace circuit elements by their impedances, namely, resistors by $R$ boxes, inductors by $Ls$ boxes, and capacitors by $1/Cs$ boxes. Here $s = j\omega$. The resulting diagram is called the *impedance model* of the network.
3. Now, determine the complex amplitudes of the voltages and currents in the circuit by any standard linear circuit analysis technique — Node method, Thévenin method, intuitive method based on series and parallel simplifications, etc.
4. Although this step is not usually not necessary, we can then obtain the time variables from the complex amplitudes. For example, the time variable corresponding to node variable $V_o$ is given by

$$v_O(t) = |V_o| \cos(\omega t + \angle V_o). \tag{13.41}$$

EXAMPLE 13.1 REVISITING THE RC EXAMPLE To illustrate the power of the method, let us revisit the RC circuit from Figure 13.3 (redrawn here as Figure 13.10a for convenience) and analyze it using the impedance method just described. As before, suppose that we wish to find the steady-state capacitor voltage $v_C$ in response to an input of the form $v_i = V_i \cos \omega_1 t$.

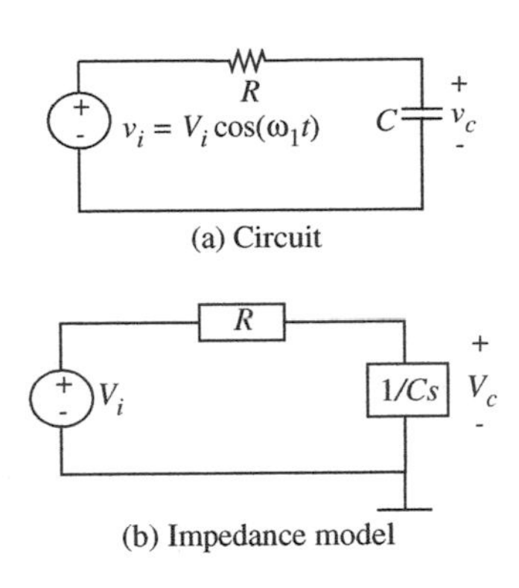


FIGURE 13.10 Impedance model of RC circuit with sinusoidal input.

Figure 13.10b shows the corresponding impedance model. In the model, notice that we have replaced the input voltage $v_i$ with the real amplitude $V_i$, and the capacitor voltage $v_C$ with the complex amplitude $V_c$, according to the first step of the impedance method. Further, according to the second step of the method, we have replaced the resistor with an $R$ box and the capacitor with a box with impedance $1/Cs$. As before, $s$ is a shorthand notation for $j\omega$.

We can derive an expression for $V_c$ by applying the generalized voltage divider relation in the impedance model in Figure 13.10b as

$$V_c = \frac{Z_C}{Z_R + Z_C} V_i \tag{13.42}$$

where $Z_R$ and $Z_C$ are the impedances of the resistor and the capacitor, respectively. Substituting the actual impedance values, we obtain

$$V_c = \frac{1/Cs}{R + 1/Cs} V_i = \frac{1}{RCs + 1} V_i. \tag{13.43}$$

Since $s$ is a shorthand for $j\omega$, at a specific frequency $\omega_1$,

$$V_c = \frac{1}{1 + j\omega_1 RC} V_i. \tag{13.44}$$

Having obtained the expression for $V_c$, the complex amplitude of the desired voltage, we have completed the third step of the impedance method. Amazingly, notice that we have arrived at the same result as in Equation 13.16 in a few easy steps.

Although not always necessary, we will proceed with the fourth step of the impedance method and obtain the actual voltage $v_C$ as a function of time. We can do so by substituting the magnitude and phase of $V_c$, and the frequency of our input into Equation 13.41 as follows:

$$\begin{aligned} v_C(t) &= |V_c| \cos(\omega_1 t + \angle V_c) \\ &= \frac{V_i}{\sqrt{1 + (\omega_1 RC)^2}} \cos\left(\omega_1 t + \tan^{-1} \frac{-\omega_1 RC}{1}\right). \end{aligned}$$

Not surprisingly, this expression for $v_C(t)$ is the same as that in Equation 13.23, but derived with significantly less effort.

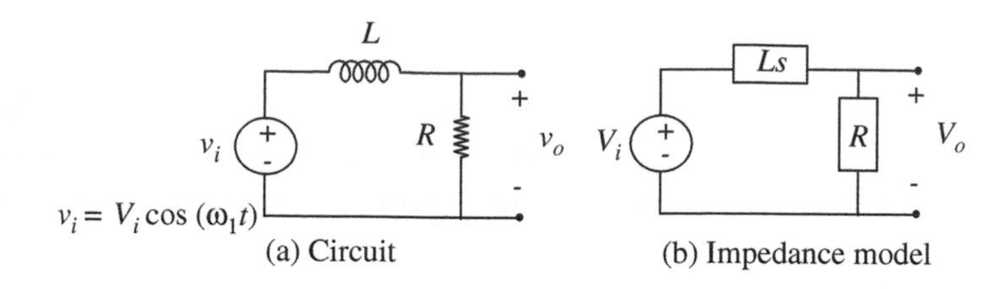


FIGURE 13.11 Impedance model of LR circuit.

As a final note, this is the forced response for a cosine wave drive. If the excitation is a tone burst, then the homogeneous solution must be added to obtain the complete solution.

---

### 13.3.1 EXAMPLE: SERIES RL CIRCUIT

Next, as a further illustration of the concept of impedance, let us find the voltage across the resistor in the RL circuit of Figure 13.11, assuming that $v_i = V_i \cos(\omega_1 t)$. Figure 13.11b shows the corresponding impedance model.

The generalized voltage divider relation for $V_o$ in the impedance model, Figure 13.11b, is

$$V_o = \frac{Z_R}{Z_R + Z_L} V_i \tag{13.45}$$

where $Z_R$ and $Z_L$ are the impedances of the resistor and the inductor, respectively. Substituting the actual impedance values, we obtain

$$V_o = \frac{R}{R + Ls} V_i. \tag{13.46}$$

Recall from Equation 13.10 that we have been using $s$ as a shorthand for $j\omega$. So at any frequency, $\omega_1$,

$$V_o = \frac{R}{R + j\omega_1 L} V_i. \tag{13.47}$$

The denominator on the right-hand side of Equation 13.47 is the impedance $Z$ seen by the voltage source at the frequency $\omega_1$. In other words,

$$Z(j\omega_1) = R + j\omega_1 L.$$

To find $v_o$, the actual output voltage as a time function, convert Equation 13.47 to polar form, then substitute into Equation 13.39

$$V_o = \frac{R}{\sqrt{R^2 + \omega_1^2 L^2}} e^{j\Phi} V_i \tag{13.48}$$

where

$$\Phi = \tan^{-1} -\omega_1 L/R. \tag{13.49}$$

From Equation 13.39, the time function is

$$v_o(t) = \frac{R}{\sqrt{R^2 + \omega_1^2 L^2}} V_i \cos(\omega_1 t + \Phi). \tag{13.50}$$

This is the forced response for a cosine wave drive. If the excitation is a tone burst, then the homogeneous solution must be added to obtain the complete solution. By comparing Equation 13.50 to Equation 13.48, we conclude that the complex amplitude $V_o$ in Equation 13.48 is a complex number containing information about both the amplitude and the phase of the sinusoidal output waveform $v_o(t)$ at any frequency.

Again it is easy to find $v_o(t)$ when the drive frequency $\omega_1$ is either low or high. At low frequencies, ($\omega_1$ small) we note from Equations 13.50 and 13.49 that $\Phi = 0$ and

$$V_o = V_i \tag{13.51}$$

so after the transient has died out, $v_o(t) \simeq v_i(t)$. That is, the resistor voltage looks just like the drive voltage. The inductor at low frequencies must behave like a short circuit, because its impedance approaches zero for $\omega$ small.

At high frequencies, specifically, where $\omega_1$ is such that $(\omega_1 L)^2 \gg R^2$,

$$|V_o| \simeq \frac{R}{\omega_1 L} V_i \tag{13.52}$$

and $\Phi$ approaches $-90°$. Thus in this frequency range $v_o(t)$ becomes smaller and smaller in amplitude with increasing frequency, and lags behind $v_i$ by nearly 90°.

At this point, we can also look at the impedance of each component ($Z_L$ and $Z_R$) in Figure 13.11 and develop the same qualitative intuition about the behavior of the circuit from the voltage-divider relationship. For instance, when $\omega_1$ is small, the impedance of the inductor is small, and so $Vo = V_i$. Similarly, for high frequencies, the impedance of the inductor becomes much greater than that of the resistor and so $V_o$ becomes very small.

---

EXAMPLE 13.2 RL EXAMPLE WITH NUMBERS Let us rework the example of Figure 13.11 and obtain the amplitude of $v_o$ using numbers this time around. Suppose that

$$L = 1 \text{ mH}$$

$$R = 1 \text{ k}\Omega$$

$$v_i = V_i \cos(2\pi ft), \quad \text{where} \quad V_i = 10$$

where we will look at three values of the frequency $f$: 100 kHz, 1 MHz, and 10 MHz.

Using impedances and the voltage divider relation

$$V_o = \frac{Z_R}{Z_R + Z_L} V_i \tag{13.53}$$

$$= \frac{1000}{1000 + 0.001s} V_i \tag{13.54}$$

$$= \frac{1000}{1000 + 0.001s} 10 \tag{13.55}$$

$$= \frac{10}{1 + 0.000001s} \tag{13.56}$$

where $s = j2\pi f$.

We can also write Equation 13.54 in the form of a transfer function $H(s)$ relating the complex output voltage to the complex input voltage:

$$H(s) = \frac{V_o}{V_i} = \frac{1000}{1000 + 0.001s}. \tag{13.57}$$

Since

$$v_o = |V_o| \cos(2\pi f + \angle V_o)$$

using Equation 13.56, we get

$$\text{Amplitude of } v_o = |V_o|$$

$$= \left| \frac{10}{1 + j\,0.000001 \times 2\pi f} \right|$$

$$= \frac{10}{\sqrt{1 + 3.9 \times 10^{-11} f^2}}.$$

Let us now plug in the three values of $f$ and obtain the amplitudes of the response at those frequencies

$f = 100$ kHz:

$$\text{Amplitude of } v_O = \frac{10}{\sqrt{1 + 3.9 \times 10^{-11}(100000)^2}}$$
$$= 8.5 \text{ V}$$

$f = 1$ MHz:

$$\text{Amplitude of } v_O = \frac{10}{\sqrt{1 + 3.9 \times 10^{-11}(1000000)^2}}$$
$$= 1.6 \text{ V}$$

$f = 10$ MHz:

$$\text{Amplitude of } v_O = \frac{10}{\sqrt{1 + 3.9 \times 10^{-11}(10000000)^2}}$$
$$= 0.16 \text{ V}.$$

The numbers clearly show the increasing impedance of the inductor as the frequency is increased. At the relatively low frequency of 100 kHz, the low impedance of the inductor causes the amplitude of the response to be similar to that of the input signal (8.5 V versus 10 V). Conversely, at the high frequency of 10 MHz, the amplitude of the response is much smaller than that of the input (0.16 V versus 10 V).

Thus far, in this example, we computed the amplitude of $v_O$ at three specific frequencies. In general, we can also graph any parameter of interest as a function of frequency. Commonly, we graph the form of the amplitude and the phase of the transfer function $V_o/V_i$ as a function of frequency $\omega$ (Equation 13.57). Although this seems difficult to do by inspection, we will learn a technique for doing so in Section 13.4. A computer generated plot of the magnitude and phase of $V_o/V_i$ versus frequency is shown in Figure 13.12. The same magnitude graph plotted on a log-log scale in Figure 13.13

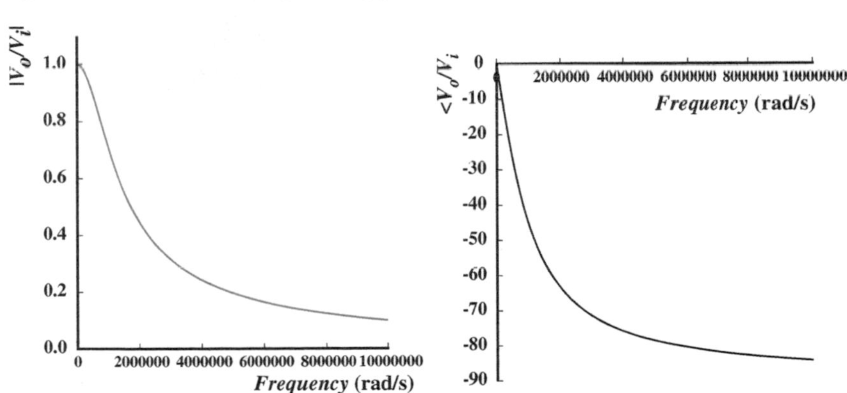


FIGURE 13.12 Magnitude and phase of $V_o/V_i$ versus frequency $\omega$.

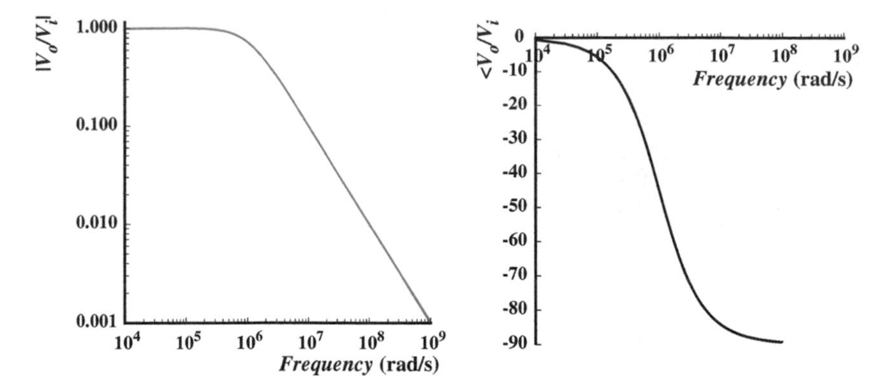


**FIGURE 13.13** Magnitude and phase of $V_o/V_i$ versus frequency $\omega$ on log scales.

is much more revealing. The corresponding phase graph is also plotted on a log scale. Notice the interesting frequency related behavior of the magnitude and phase plots in Figure 13.13. The log plot instantly reveals that high frequencies are severely attenuated while low frequencies are passed through unattenuated. We will have much more to say about this variable frequency view of responses in Section 13.4.

### 13.3.2 EXAMPLE: ANOTHER RC CIRCUIT

Let us now work another example with numerical quantities. Figure 13.14a shows an RC circuit driven by a sinusoidal voltage $v_i = 10\cos(1000\,t)$ V. Find the impedance $Z$ seen by the voltage source at the frequency $\omega = 1000$ rad/s. Also, find the voltage $V_r$ across the resistor.

Figure 13.14b shows the impedance model for the circuit. Since the capacitor and the resistor are in series, the impedance $Z$ seen by the source is given by the sum of the impedances of the capacitor and the resistor as

$$Z = 500 \times 10^3 + \frac{1}{1 \times 10^{-9} s} \ \Omega$$

where, we have used the shorthand $s = j\omega$. Since $\omega = 1000$ rad/s, we get

$$Z = 500 \times 10^3 + \frac{1}{1 \times 10^{-9} j1000} \ \Omega$$

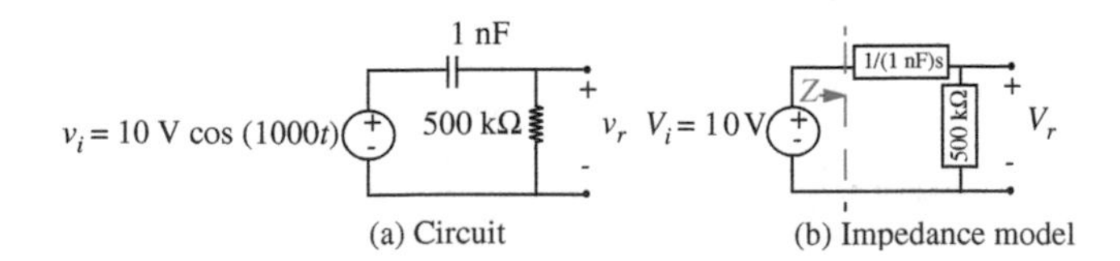


**FIGURE 13.14** Impedance model of RC circuit.

or

$$Z = 0.5 - j\,\text{M}\Omega.$$

Next, let us determine $V_r$. Applying generalized voltage divider relation for $V_r$ in the impedance model,

$$V_r = \frac{500 \times 10^3}{500 \times 10^3 + \frac{1}{1\times 10^{-9}s}} V_i. \tag{13.58}$$

Substituting $s = j1000$ and $V_i = 10$ we get

$$V_r = \frac{500 \times 10^3}{500 \times 10^3 + \frac{1}{1\times 10^{\;9}j1000}} 10. \tag{13.59}$$

Simplifying,

$$V_r = \frac{0.5\ \text{M}\Omega}{0.5\ \text{M}\Omega - j\ \text{M}\Omega} 10$$

or,

$$V_r = \frac{5}{0.5 - j}.$$

Impedance analysis can end here, since the expression for $V_r$ contains all the information about the amplitude and the phase of the time function $v_r$. However, let us take the extra step and determine $v_r$ by writing

$$v_r - Re\left[\frac{5}{0.5 - j} e^{j1000t}\right].$$

Simplifying, and writing the previous expression in polar form

$$v_r = Re\left[4.47 e^{j1.1} e^{j1000t}\right].$$

Taking the real part,

$$v_r = 4.47 \cos(1000t + 1.1)\ \text{V}.$$

Notice from Equation 13.59 that if the frequency of the input were increased from 1000 rad/s to, say, $10^6$ rad/s, $V_r \approx 10$ V. This says that

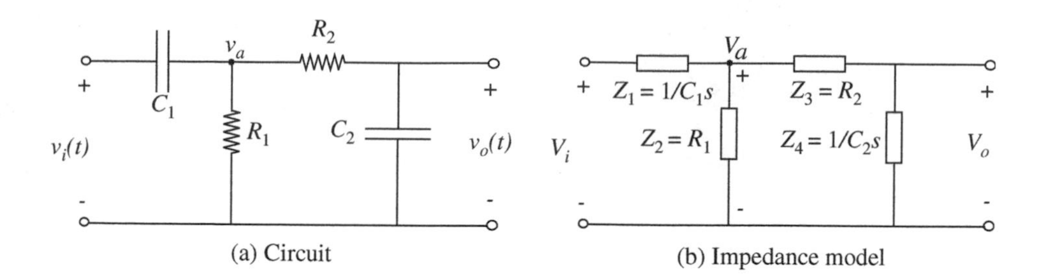


FIGURE 13.15 A second-order circuit example.

the capacitor offers very low impedance compared to that of the resistor at high frequencies, and the entire input voltage falls across the resistor.

On the other hand, if the frequency of the input is decreased from 1000 rad/s to, say, 1 rad/s, $V_r \approx 0$ V. This says that the capacitor has a very high impedance compared to that of the resistor at low frequencies. Therefore, nearly all of the input voltage falls across the capacitor, resulting in $V_r \approx 0$.

### 13.3.3 EXAMPLE: RC CIRCUIT WITH TWO CAPACITORS

Consider the second-order circuit containing resistors and capacitors shown in Figure 13.15a. The impedance model of the circuit is shown in Figure 13.15b.

Suppose we are interested in deriving $v_o(t)$ for an input of the form $v_i(t) = V_i \cos(\omega t)$. Observing that

$$v_i(t) = Re\left[V_i e^{j\omega t}\right]$$

our usual method will be to find the output response $V_o e^{j\omega t}$ to the input $V_i e^{j\omega t}$, and then to determine the actual output voltage $v_o(t)$ using

$$v_o(t) = Re\left[V_o e^{j\omega t}\right].$$

For convenience, as in the past, we will use the variable $s$ as a substitute for $j\omega$.

Let us first use the impedance method to derive the complex output voltage amplitude $V_o$ as a function of the input voltage amplitude $V_i$. We will use the method of series and parallel simplification from Section 2.4 to do so. Notice that the voltage $V_a$ can be obtained using a voltage-divider relationship between $1/C_1s$ and $R_1\|(R_2 + 1/C_2s)$. Then, observe that $V_o$ can be obtained from yet another voltage-divider relationship between $R_2$ and $1/C_2s$. Accordingly, we have

$$V_a = \frac{\left(R_2 + \frac{1}{C_2s}\right)\|R_1}{\left(R_2 + \frac{1}{C_2s}\right)\|R_1 + \frac{1}{C_1s}} V_i \tag{13.60}$$

$$V_o = \frac{\frac{1}{C_2 s}}{R_2 + \frac{1}{C_2 s}} V_a \tag{13.61}$$

$$= \left( \frac{\left(R_2 + \frac{1}{C_2 s}\right) \| R_1}{\left(R_2 + \frac{1}{C_2 s}\right) \| R_1 + \frac{1}{C_1 s}} \right) \left( \frac{\frac{1}{C_2 s}}{R_2 + \frac{1}{C_2 s}} \right) V_i. \tag{13.62}$$

Simplifying, we get

$$V_o = \frac{R_1 C_1 s}{R_1 R_2 C_1 C_2 s^2 + (R_1 C_1 + R_1 C_2 + R_2 C_2)s + 1} V_i. \tag{13.63}$$

Equation 13.63 written in the form of a transfer function $H(s)$ relating the complex output voltage to the complex input voltage is

$$H(s) = \frac{V_o}{V_i} = \frac{R_1 C_1 s}{R_1 R_2 C_1 C_2 s^2 + (R_1 C_1 + R_1 C_2 + R_2 C_2)s + 1}. \tag{13.64}$$

Let us assume the following set of parameters: $R_1 = 1\ \text{k}\Omega$, $R_2 = 1\ \text{k}\Omega$, $C_1 = 1\ \text{mF}$, and $C_2 = 1\ \text{mF}$. For these parameters, we get:

$$V_o = \frac{s}{s^2 + 3s + 1} V_i. \tag{13.65}$$

Factoring the denominator polynomial, we get

$$V_o = \frac{s}{\left(s - \frac{-3-\sqrt{5}}{2}\right)\left(s - \frac{-3+\sqrt{5}}{2}\right)} V_i. \tag{13.66}$$

Observe that the denominator polynomial has real roots at $(-3 - \sqrt{5})/2$ and $(-3 + \sqrt{5})/2$.

Substituting $s = j\omega$, we obtain the complex amplitude $V_o$ as a function of $V_i$ for a given frequency $\omega$:

$$V_o = \frac{j\omega}{\left(j\omega - \frac{-3-\sqrt{5}}{2}\right)\left(j\omega - \frac{-3+\sqrt{5}}{2}\right)} V_i. \tag{13.67}$$

In general, $V_i$ and $V_o$ are complex amplitudes.[3] We can determine the actual time-varying voltage by multiplying the complex amplitude by $e^{j\omega t}$

---

3. In this instance, we know that $V_i$ is real since we chose

$$v_i(t) = Re\left[V_i e^{j\omega t}\right] = V_i \cos(\omega t).$$

and taking the real part of the resulting expression. Therefore the complex amplitude $V_o$ given by Equation 13.67 contains all the information about the amplitude and phase of the actual sinusoidal output $v_o(t)$. Therefore we could very well stop at this point. However, as an exercise, let us go ahead and determine the actual output voltage.

The actual output voltage $v_o(t)$ is given by

$$v_o(t) = Re\left[V_o e^{j\omega t}\right] = Re\left[\frac{j\omega}{\left(j\omega - \frac{-3-\sqrt{5}}{2}\right)\left(j\omega - \frac{-3+\sqrt{5}}{2}\right)} V_i e^{j\omega t}\right].$$

Simplifying,

$$v_o(t) = Re\left[A_1 A_2 A_3 V_i e^{j(\phi_1 - \phi_2 - \phi_3)} e^{j\omega t}\right] \tag{13.68}$$

where

$$A_1 = \omega, \quad A_2 = \frac{1}{\sqrt{\omega^2 + \frac{7+3\sqrt{5}}{2}}}, \quad A_3 = \frac{1}{\sqrt{\omega^2 + \frac{7-3\sqrt{5}}{2}}}$$

and

$$\phi_1 = \frac{\pi}{2}, \quad \phi_2 = \tan^{-1}\frac{2\omega}{3+\sqrt{5}}, \quad \phi_3 = \tan^{-1}\frac{2\omega}{3-\sqrt{5}}.$$

In the preceding equation, notice that the expression within the brackets excluding the $e^{j\omega t}$ term is simply a polar representation of the complex amplitude $V_o$. In other words,

$$V_o = A_1 A_2 A_3 V_i e^{j(\phi_1 - \phi_2 - \phi_3)}. \tag{13.69}$$

In Equation 13.69, $A_1 A_2 A_3 V_i$ is the magnitude of $V_o$ and $(\phi_1 - \phi_2 - \phi_3)$ is the phase.

Now, to obtain $v_o(t)$, we can simplify Equation 13.68 to get

$$v_o(t) = A_1 A_2 A_3 V_i \cos(\omega t + \phi_1 - \phi_2 - \phi_3). \tag{13.70}$$

The preceding expression for $v_o(t)$ is actually the particular solution for an excitation given by $v_i(t) = V_i \cos(\omega t)$, and is the steady-state response of the circuit for a cosine wave excitation. If we care to obtain the complete solution, we must add in the homogeneous solution as illustrated in Equation 13.22.

A computer generated plot of the magnitude and phase of $V_o/V_i$ (Equation 13.69) versus frequency is shown in Figure 13.16. The same magnitude

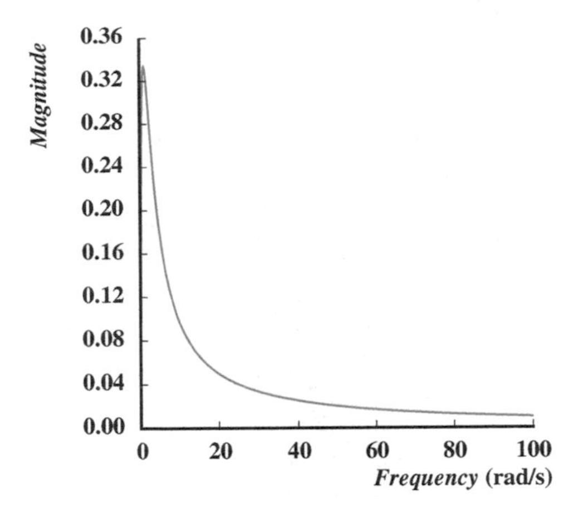


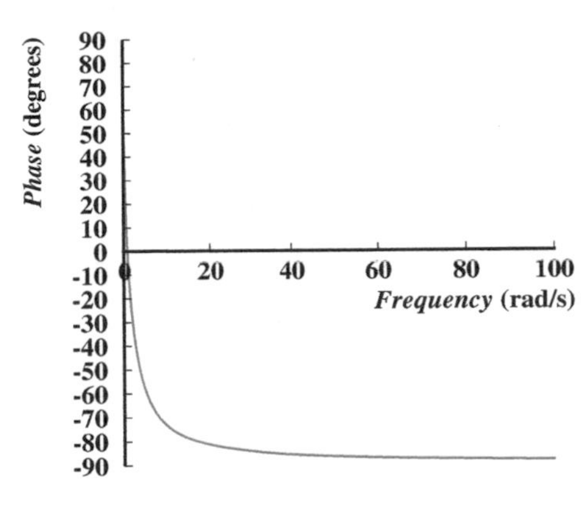


FIGURE 13.16 Magnitude and phase of $V_O$ (assuming $V_i$ is unity) versus frequency $\omega$.

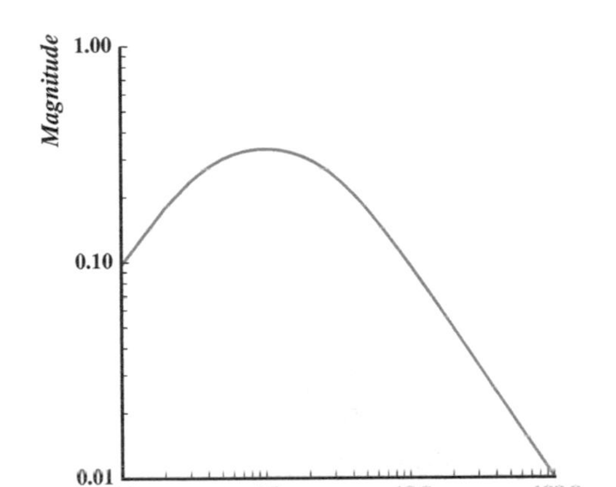


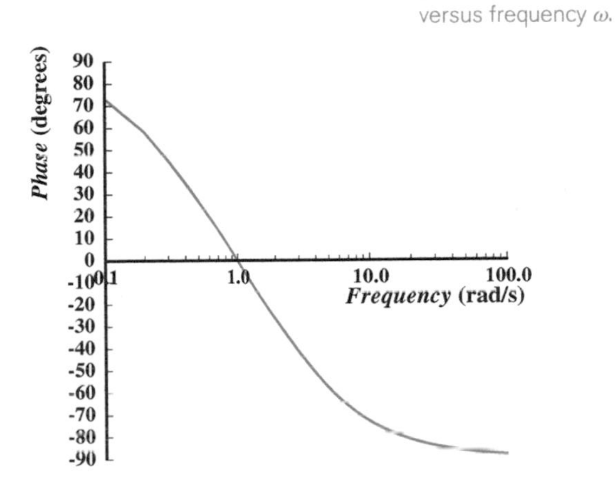


FIGURE 13.17 Magnitude and phase of $V_O$ (assuming $V_i$ is unity) versus frequency $\omega$ on log scales.

graph plotted on a log-log scale in Figure 13.17 is much more revealing. The corresponding phase graph is also plotted on a log scale. Notice the interesting frequency related behavior of the magnitude and phase plots in Figure 13.17. The log plot reveals that both low and high frequencies are severely attenuated.

---

EXAMPLE 13.3 NODE METHOD ANALYSIS WITH IMPEDANCES Consider again the circuit shown in Figure 13.15a and its impedance model shown in Figure 13.15b. The node method applies equally well when impedances are used in place of resistances and complex amplitudes are used in place of time functions

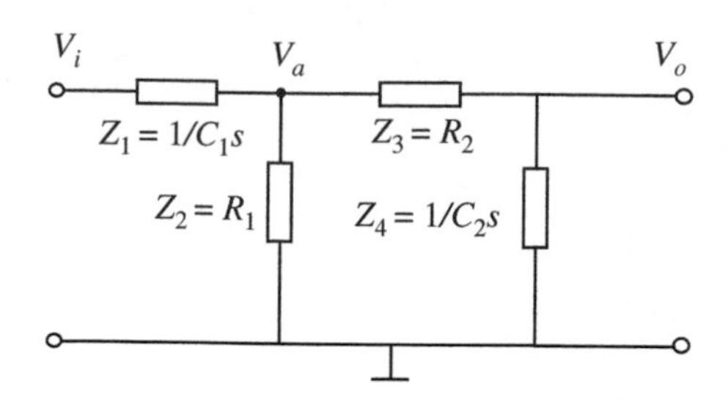


**FIGURE 13.18** Using node analysis with impedances.

for sources. Let's use the node method to determine $V_o$ as a function of the input voltage amplitude $V_i$.

As the first and second steps of node analysis, Figure 13.18 shows our choice of ground node and the node voltages labeled with respect to this selection of a ground node.

As the third step of node analysis, let us write KCL for the nodes with unknown node voltages. For the node with voltage $V_a$ we have,

$$\frac{V_i - V_a}{Z_1} - \frac{V_a}{Z_2} - \frac{V_a - V_o}{Z_3} = 0.$$

Substituting for the impedance values,

$$\frac{V_i - V_a}{\frac{1}{C_1 s}} - \frac{V_a}{R_1} - \frac{V_a - V_o}{R_2} = 0$$

or, simplifying,

$$V_i - \left(1 + \frac{1}{R_1 C_1 s} + \frac{1}{R_2 C_1 s}\right) V_a + \frac{1}{R_2 C_1 s} V_o = 0 \tag{13.71}$$

and, for the node with voltage $V_o$ we have,

$$\frac{V_a - V_o}{R_2} - \frac{V_o}{1/C_2 s} = 0.$$

Simplifying, we get

$$\frac{1}{R_2 C_2 s} V_a - \left(1 + \frac{1}{R_2 C_2 s}\right) V_o = 0. \tag{13.72}$$

Eliminating $V_a$ from Equations 13.71 and 13.72 and solving for $V_o$ in terms of $V_i$ we get

$$\frac{V_o}{V_i} = \frac{R_1 C_1 s}{R_1 R_2 C_1 C_2 s^2 + (R_1 C_1 + R_1 C_2 + R_2 C_2)s + 1}. \tag{13.73}$$

It should come as no surprise that the solutions for $V_o$ in Equations 13.73 and 13.64 are the same.

### 13.3.4 EXAMPLE: ANALYSIS OF SMALL-SIGNAL AMPLIFIER WITH CAPACITIVE LOAD

Consider the two-stage MOSFET amplifier shown in Figure 13.19. When biased properly, this circuit behaves as a linear amplifier for small signals. This example was used in Section 13.1 (Figure 13.2) to motivate the study of methods based on the sinusoidal steady state. There we made the experimental observation that the presence of the capacitor $C_{GS}$ makes the gain of the amplifier fall off rapidly at high frequencies. We now explain why.

Let us analyze the steady-state response of the first stage of the amplifier to a small sinusoid signal applied at its input when the output of the first stage is loaded with the gate capacitance $C_{GS}$ of the second stage. In other words, our goal is to find the relationship between $v_o$ and $v_i$ in the presence of the capacitor $C_{GS}$ when $v_i = V_i \cos(\omega t)$.

We first construct the small-signal circuit model for the first stage including the load capacitor. We can ignore the loading of the second stage because the MOSFET has an infinite input resistance. It is easy to see that the small-signal circuit model can be developed from the circuit in Figure 8.16 by adding the load capacitor $C_{GS}$ as illustrated in Figure 13.20.

If $C_{GS}$ were absent, we would get the usual amplifier response relation

$$\begin{aligned} v_o &= -K(V_I - V_T)R_L v_i \\ &= -g_m R_L v_i. \end{aligned}$$

When $v_i = V_i \cos(\omega t)$, and $C_{GS} = 0$, the output is simply an amplified version of the input, and is given by

$$v_o = -g_m R_L V_i \cos(\omega t). \tag{13.74}$$

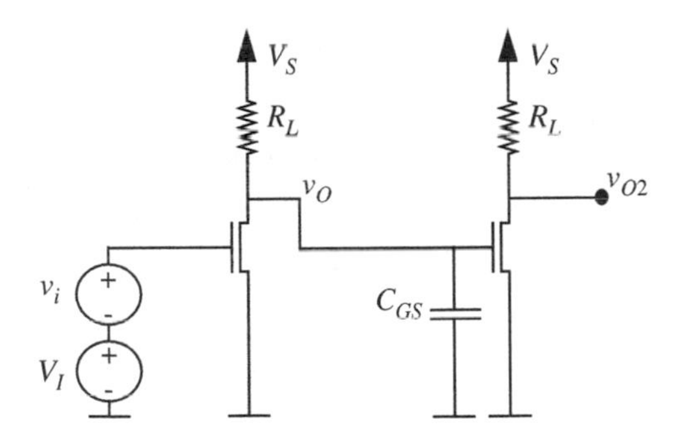


FIGURE 13.19 A two-stage MOSFET amplifier showing the MOSFET gate capacitor.

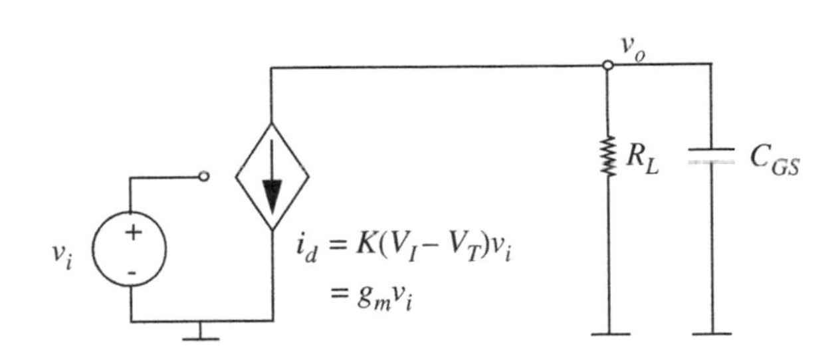


FIGURE 13.20 Small-signal circuit model for a MOSFET amplifier including a load capacitor.

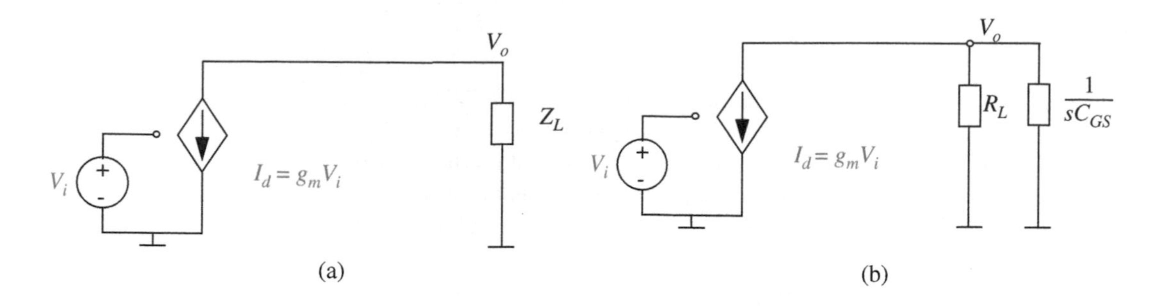


**FIGURE 13.21** Impedance model.

To find the relation between $v_o$ and $v_i$ in the presence of the capacitor, we first draw the impedance model of the circuit as shown in Figure 13.21a, and replace the load resistance $R_L$ with the effective load impedance $Z_L$ in the computation of the amplifier response. As illustrated in Figure 13.21b, the load impedance $Z_L$ is given by

$$Z_L = R_L \parallel \frac{1}{sC_{GS}}$$
$$= \frac{R_L}{1 + sR_L C_{GS}}.$$

From the circuit in Figure 13.21a, we can write the following expression for the complex amplitude of the output:

$$V_o = -g_m Z_L V_i = -g_m \frac{R_L}{1 + sR_L C_{GS}} V_i.$$

Substituting $s = j\omega$, the response becomes

$$V_o = -g_m \frac{R_L}{1 + j\omega R_L C_{GS}} V_i. \tag{13.75}$$

As usual, we can derive the time domain value $v_o$ of the output voltage by multiplying the complex amplitude $V_o$ by $e^{j\omega t}$ and taking the real part. In other words,

$$v_o = Re\left[V_o(j\omega)e^{j\omega t}\right]. \tag{13.76}$$

To simplify the analysis, let us first convert $V_o$ to polar form:

$$V_o = -g_m \frac{R_L}{\sqrt{1 + (\omega R_L C_{GS})^2}} V_i e^{j\phi} \quad \text{where} \quad \phi = \tan^{-1}(-\omega R_L C_{GS}).$$

Substituting the polar form of $V_o(j\omega)$ into Equation 13.76 we get

$$v_o = Re\left[-g_m \frac{R_L}{\sqrt{1 + (\omega R_L C_{GS})^2}} V_i e^{j(\omega t + \phi)}\right] \tag{13.77}$$

$$= -g_m \frac{R_L}{\sqrt{1 + (\omega R_L C_{GS})^2}} V_i \cos(\omega t + \phi) \tag{13.78}$$

where $\phi = \tan^{-1}(-\omega R_L C_{GS})$.

Let us analyze the response at various frequencies. Equation 13.78 shows that at low frequencies ($\omega \to 0$) the expression for $v_o$ is no different from that in Equation 13.74. Thus the response to a DC signal or a very low frequency sinusoid is similar to the response when the capacitor is absent. This is not surprising because the capacitor behaves as an open circuit at very low frequencies.

However, notice that as the frequency of the input sinusoid increases, the amplitude of the output sinusoid decreases. In fact, for very high frequencies ($\omega \gg 1/R_L C_{GS}$), the amplitude of $v_o$ tends to 0. The insight behind the high frequency result is that the capacitor behaves as a short for high frequencies. The resulting zero impedance of the load reduces the amplifier gain to 0.

## 13.4 FREQUENCY RESPONSE: MAGNITUDE AND PHASE VERSUS FREQUENCY

We characterize the behavior of a network by its *frequency response.*

*Frequency response* A plot of the magnitude and the phase of the network's transfer function as a function of frequency.

*Transfer function* Also known as a *system function*, is the ratio of the complex amplitude of the network output to the complex amplitude of the input.

Equation 13.57 is one example of a system function, and Figure 13.13, a plot of the magnitude and phase of the system function versus frequency, is the frequency response.

The frequency response contains a lot of information about the system. It includes a magnitude plot and a phase plot, both as a function of frequency. The magnitude of the network's transfer function is the ratio of the amplitudes of the output and the input, and indicates the *gain* of the system as a function of frequency. The phase is the angular difference between the output and the input sinusoids.

Observing the frequency response of networks represents a major difference in perspective from the preceding chapters. The earlier chapters presented time-domain analyses in which our focus was on finding an output signal value as a function of time for a given input signal also specified as a function of time. For example, as shown in Figure 10.2, the step response of an RC network plotted the output voltage of an RC network as a function of time in response to a step input. In contrast, the frequency response represents a *frequency domain analysis* in which output behavior is presented as a function of input frequency. In a frequency domain analysis our goal is to determine the magnitude and phase of the output in response to an input sinusoidal signal of a given frequency in steady state.

Plotting the frequency response of a network with the aid of a computer for arbitrary system transfer functions as illustrated in Figure 13.13 is quite easy. Nevertheless, it is still useful to be able to make approximate sketches of the frequency response by inspection. We've already seen that obtaining insight into the frequency response at low frequencies and high frequencies is a relatively simple matter, as demonstrated by Equations 13.51 and 13.52 in Section 13.3.1. At intermediate frequencies, however, the relation between output and input is somewhat more complicated, especially for a network with several inductors or capacitors. This section will show how the general shape of the frequency response for first-order circuits can be sketched by inspection. Chapter 14 will do the same for an important class of second-order circuits. We will resort to computer analysis for other circuits.[4]

Section 13.4.1 begins by reviewing the frequency response of resistors, capacitors, and inductors. This simple process will help us build some intuition, which will then be used in Section 13.4.2 to develop a more general method of sketching by inspection the frequency response for circuits containing a single storage element and a resistor.

### 13.4.1 FREQUENCY RESPONSE OF CAPACITORS, INDUCTORS, AND RESISTORS

Resistors, inductors, and capacitors result in transfer functions of the form $s$, $1/s$, or a constant. Recall from Section 13.3 that the element laws for resistors, inductors, and capacitors in terms of complex amplitudes of voltages and currents are given by

$$\text{Resistor:} \quad V_o = R I_o$$

$$\text{Inductor:} \quad V_o = sLI_o = j\omega LI_o$$

$$\text{Capacitor:} \quad V_o = \frac{1}{sC} I_o = \frac{1}{j\omega C} I_o$$

4. There do exist methods for sketching frequency response plots for arbitrary circuits without the use of a computer. One of these, the Bode plot method, is discussed in Section 13.4.2. The popularity of these methods, however, has waned in recent times due to the widespread use of computers.

respectively. We can rewrite the preceding expressions in the form of $V_o/I_o$ transfer functions that relate the complex voltage amplitudes to the complex current amplitudes as a function of frequency as follows:

$$\text{Resistor:} \quad \frac{V_o}{I_o} = H(j\omega) = R$$

$$\text{Inductor:} \quad \frac{V_o}{I_o} = H(j\omega) = j\omega L$$

$$\text{Capacitor:} \quad \frac{V_o}{I_o} = H(j\omega) = \frac{1}{j\omega C}.$$

Here each of the transfer functions is an impedance. These transfer functions are complex numbers. They are also frequency dependent. The corresponding magnitudes and phases are given by

$$\text{Resistor:} \quad \left|\frac{V_o}{I_o}\right| = R \quad \text{and} \quad \angle\frac{V_o}{I_o} = 0$$

$$\text{Inductor:} \quad \left|\frac{V_o}{I_o}\right| = \omega L \quad \text{and} \quad \angle\frac{V_o}{I_o} = 90^\circ$$

$$\text{Capacitor:} \quad \left|\frac{V_o}{I_o}\right| = \frac{1}{\omega C} \quad \text{and} \quad \angle\frac{V_o}{I_o} = -90^\circ.$$

Figure 13.22 plots the frequency response for the resistor, inductor, and the capacitor using the following values for the elements:

$$R = 1\ \Omega$$

$$L - 1\ \mu\text{H}$$

$$C = 1\ \mu\text{C}.$$

As shown in Figure 13.22, the frequency response is a plot of the magnitude and phase of the transfer function versus frequency.

Frequency response plots are commonly plotted using log scales. As we will see shortly, log scales make the magnitudes of the responses due to capacitors and inductors appear as straight lines in the graph. Log scales also allow us to observe the response over many orders of magnitude variation in the frequency without necessarily compressing the low frequency behavior of the response (near zero rad/s) into the magnitude and phase axes. The frequency responses for the resistor, inductor, and the capacitor can be sketched on log scales by

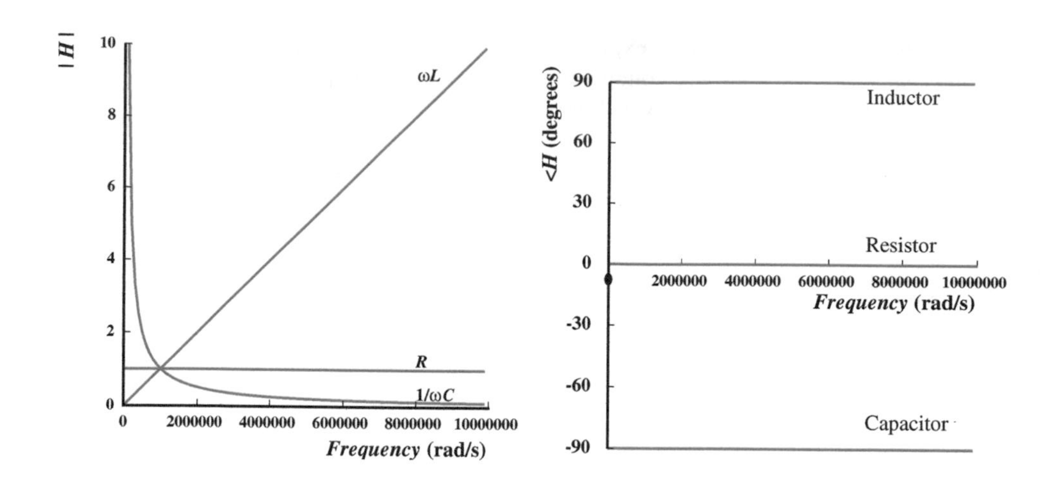


**FIGURE 13.22** Frequency response of inductors, capacitors, and resistors plotted on linear scales.

using the following relations:[5]

$$\text{Resistor:} \quad \log\left|\frac{V_o}{I_o}\right| = \log R$$

$$\text{Inductor:} \quad \log\left|\frac{V_o}{I_o}\right| = \log \omega L = \log L + \log \omega$$

$$\text{Capacitor:} \quad \log\left|\frac{V_o}{I_o}\right| = \log \frac{1}{\omega C} = -\log C - \log \omega.$$

Figure 13.23 shows the corresponding plots. We can make several observations about the plots. The first set of observations relate to the nature of logarithmic plots in general.

- If $x$ is the variable being plotted using a log scale, $x$ is *multiplied* by a fixed factor for each fixed length increment along the axis. In contrast, on a linear scale, $x$ is incremented by a fixed amount for each fixed length increment along the axis. For example, equal length increments along the frequency axis on a linear scale might correspond to the values $0$, $2 \times 10^6$, $4 \times 10^6$, $6 \times 10^6$, $8 \times 10^6$, and so on. On the other hand, equal length

5. In deriving the log relations here and in the future, we assume that both the left- and right-hand sides of the equations are divided by appropriate unit constants before taking the logs so that the arguments to the log functions are unitless.

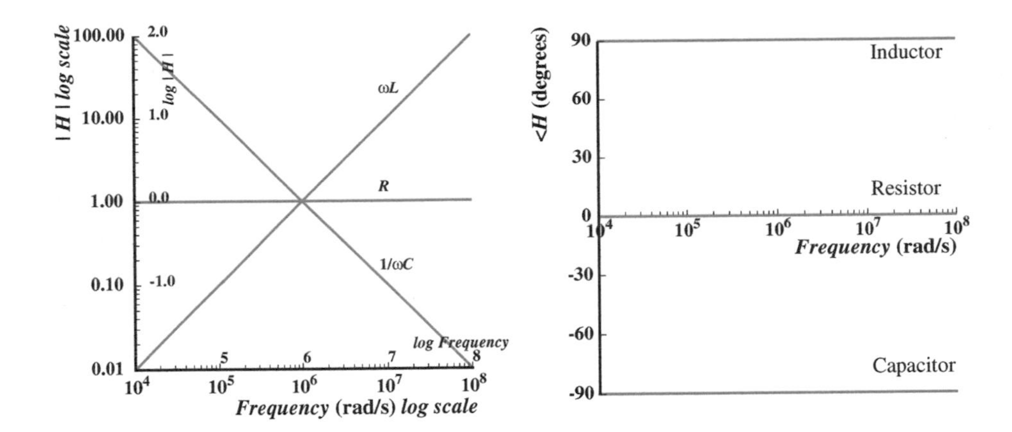


FIGURE 13.23 Frequency response of inductors, capacitors, and resistors plotted on log scales.

increments along the frequency axis on a log scale might correspond to the values $10^4$, $10^5$, $10^6$, $10^7$, and so on.[6]

- There are two equivalent ways of plotting log functions:

1. Plot $\log x$ on a linear scale.

2. Plot $x$ on a logarithmic scale.

In Figure 13.23, we have plotted the magnitude functions both ways. In other words, for the abscissa, we plot both $\log \omega$ on a linear scale and $\omega$ on a logarithmic scale. For the ordinate, $R$, $\omega L$, and $1/\omega C$ are also plotted both ways. In the future, we will choose to plot $x$ on a logarithmic scale.[7]

On the phase plot, the horizontal scale is logarithmic, and the vertical scale is linear.

- The function

$$|H| = \omega$$

6. Historically, an octave is used to indicate a 2 times change in frequency. For example, 2 kHz is a 1-octave increase in frequency from 1 kHz. Similarly, 4 kHz is a 2-octave increase from 1 kHz, and 8 kHz is a 3-octave increase from 1 kHz. In like manner, 500 Hz is 1 octave below 1 kHz.

Another useful term is a *decade*. A decade is a range of frequencies in which the highest frequency is 10 times the lowest frequency. For example, the range from 1 kHz to 10 kHz is 1 decade, and that from 1 kHz to 100 kHz is 2 decades.

7. In the literature it is also common to plot the response magnitude in decibels (dB), defined as

$$\text{Response ratio in dB} = 20 \log_{10} |H(j\omega)|. \qquad (13.79)$$

plots as a straight line with slope of +1 in log space, given consistent horizontal and vertical scales, because it changes by a factor of 10 for a factor of 10 increase in $\omega$.

- Correspondingly,

$$|H| = 1/\omega$$

plots as a straight line with slope of −1 in log space. (Notice that $log|H| = \log 1/\omega = -\log\omega$.)

- The value of $L$ in $|H| = \omega L$ establishes the offset. Writing

$$\log|H| = \log L + \log\omega.$$

Thus, $\log|H| = 0$ when $\log L = -\log\omega$, or when $\omega = 1/L$. Put another way, $|H| = 1$, when $\omega = 1/L$.

The second set of observations relate to the specific magnitude and phase curves for our three transfer functions.

- Notice in Figure 13.23 that the magnitude curve for the inductor appears as a straight line with +1 slope on log-log scales. This implies that the magnitude of the transfer function (or the impedance of the inductor) increases with increasing frequency.

  Because $L = 1\ \mu$H, the magnitude curve for the inductor passes through unity for $\omega = 1/L = 10^6$ rad/s.

- In contrast, the magnitude curve for the capacitor appears as a straight line with −1 slope on log-log scales. This implies that the magnitude of the transfer function decreases with increasing frequency.

  Because $C = 1\ \mu$F, the magnitude curve for the capacitor passes through unity for $\omega = 1/C = 10^6$ rad/s.

- The magnitude plot for the resistor appears as a horizontal line.

- The phase plot for the inductor shows that the inductor causes a fixed phase shift of 90°, while that for the capacitor indicates a phase shift of −90°. The resistor does not introduce any phase shift.

As we will see next, these simple plots for resistors, capacitors, and inductors provide much of the necessary insight to plot the frequency responses for circuits containing a resistor and a storage element.

### 13.4.2 INTUITIVELY SKETCHING THE FREQUENCY RESPONSE OF RC AND RL CIRCUITS

Let us now examine the frequency response of circuits with a single storage element and a single resistor, an important class of first-order circuits, and see how their responses can be sketched by inspection. Such circuits result in transfer functions of the form $1/(s+a)$, $(s+a)$, $s/(s+a)$, $(s+a)/s$, where $a$ is some constant. We will illustrate the approach using as an example the series RL circuit from Figure 13.11 in Section 13.3.1.

The input-output relationship as a function of the input drive frequency $\omega$ for the series RL circuit of Figure 13.11 is given by

$$V_o = \frac{R}{R + sL} V_i \tag{13.80}$$

(see Equation 13.47). Dividing by $V_i$, we obtain its transfer function:

$$H(j\omega) = \frac{V_o}{V_i} = \frac{R}{R + sL}. \tag{13.81}$$

For reasons that will be clear shortly, let's rewrite this in a more standard form as

$$H(j\omega) = \frac{V_o}{V_i} = \frac{R/L}{R/L + s}. \tag{13.82}$$

The magnitude of the transfer function is

$$|H(j\omega)| = \left| \frac{R/L}{R/L + j\omega} \right| \tag{13.83}$$

and its phase is

$$\angle H(j\omega) = \tan^{-1} \frac{-\omega L}{R}. \tag{13.84}$$

This frequency response using logarithmic scales for the horizontal and vertical axes was previously shown in Figure 13.13. We repeat here in Figure 13.24 a computer-generated plot of the same response plot (assuming, as before, $L = 1$ mH, $R = 1$ k$\Omega$, so that $R/L = 10^6$ rad/s).

The same frequency response can also be sketched easily by making the following observations about the magnitude and phase plots. Let us first deal with the magnitude plot. Observe from Figure 13.24 that the magnitude plot is asymptotic to two straight lines. At low frequencies ($\omega \to 0$), the magnitude becomes

$$|H(j\omega)| \simeq 1. \tag{13.85}$$

Thus, at low frequencies the magnitude is unity, and hence appears as a horizontal line.

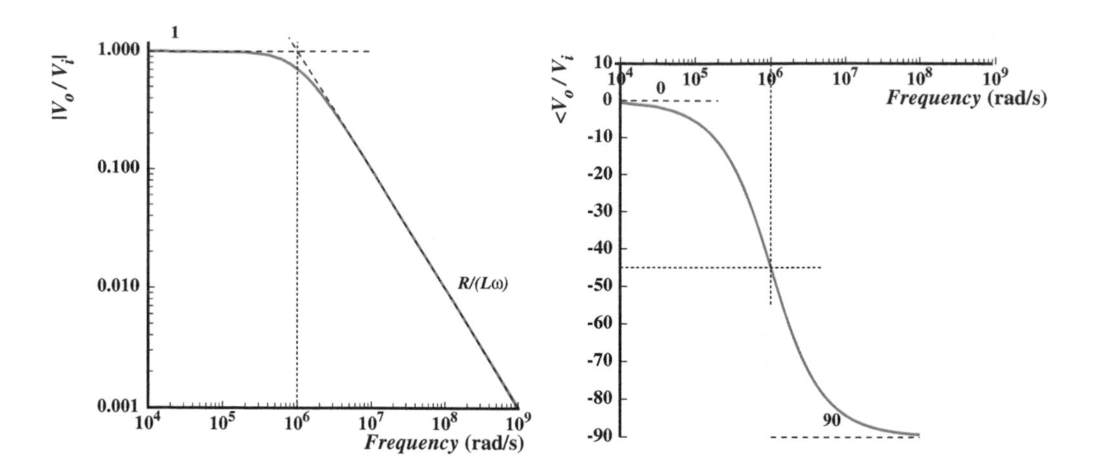


FIGURE 13.24 Magnitude and phase of $V_o/V_i$ versus frequency $\omega$ on log scales.

At high frequencies ($\omega \to \infty$), the $\omega$ term dominates the expression in the denominator and the magnitude becomes

$$|H(j\omega)| = \frac{R/L}{\omega}. \tag{13.86}$$

From our observations on log plots in Section 13.4.1, we know that the log plot for the expression in Equation 13.86 will appear as a straight line with a slope of $-1$ for consistent horizontal and vertical scales, and passes through the point $|H| = 1$ when $\omega = R/L$.

Clearly the two asymptotes intersect at

$$\omega = \frac{R}{L} = 10^6 \text{ rad/s}$$

called the *break frequency* or corner frequency. At the break frequency, the true magnitude of $H(j\omega)$ is

$$|H(j\omega)| = |10^6/(10^6 + j10^6)| = \frac{1}{\sqrt{2}} = 0.707.$$

Thus the break frequency is also called the 0.707 frequency. At this frequency, the real and imaginary parts of the function are equal.[8]

8. Since 0.707 in decibels is $20 \log 0.707 = -3$ dB, the break frequency is also called a $-3$ dB frequency. Since $0.707^2 = 0.5$, the break frequency is also called the half-power point. When the magnitude curve begins to dip after the break frequency, the break frequency is also called the cutoff frequency.

As is clear from Figure 13.24, the high- and low-frequency asymptotes approximate the frequency response curve pretty well.

Let us now address the phase plot. Like the magnitude plot, the phase curve can also be approximated by the low and high frequency asymptotes. At low frequencies the phase becomes

$$\angle H(j\omega) \simeq 0 \tag{13.87}$$

while at high frequencies the phase is

$$\angle H(j\omega) = -90°. \tag{13.88}$$

Notice that the phase curve goes smoothly from $0°$ at $\omega = 0$ to $-90°$ at $\omega = \infty$. At the break point, the real and imaginary parts of Equation 13.83 are equal, hence the angle of $H(j\omega)$ is $-45°$ at this frequency.

Examination of Equations 13.16 and 13.47 (or Equations 13.21 and 13.50 for the corresponding time functions) shows that the RC circuit of Figure 13.3 and the RL circuit of Figure 13.11 will have frequency response plots of the same form: unit magnitude and zero phase at low frequencies, and magnitude falling as $1/\omega$ (slope of $-1$ in log-log plot) with $-90°$ phase shift at high frequencies. These functions are called *low-pass filters*, because in signal-processing terms they pass low frequencies and reject high frequencies. In other words, they do not affect low frequencies, while they provide a low gain for high frequencies. However, in many circuit applications the opposite effect is desired. We wish to get rid of low frequencies, and pass high frequencies. Decoupling amplifier stages so that the DC offset from one stage does not affect the next stage is a case in point. We will study these and other filters in more detail in Section 13.5.

To summarize, the frequency response of circuits containing a single storage element and a single (Thévenin ) resistor is of the form $1/(s+a)$, $(s+a)$, $s/(s+a)$, $(s+a)/s$, where $a$ is some constant, and can be sketched intuitively as follows:

- Magnitude Plot
  1. Sketch the low frequency asymptote.
  2. Sketch the high frequency asymptote. The two asymptotes intersect at the break frequency $a$.
- Phase Plot
  1. Sketch the low frequency asymptote.
  2. Sketch the high frequency asymptote.

3. At the break frequency, the phase will be 45° or −45° as appropriate. Draw a smooth line starting with the low frequency asymptote, passing through 45° or −45° as appropriate at the break frequency, and finishing off at the high frequency asymptote.

---

**EXAMPLE 13.4 INTUITIVE SKETCH OF THE FREQUENCY RESPONSE OF RC CIRCUIT** Let us sketch the frequency response for the transfer function relating $V_r$ to $V_i$ for the RC circuit shown in Figure 13.14.

From Equation 13.58, we can immediately write the transfer function as

$$\frac{V_r}{V_i} = H(s) = \frac{500 \times 10^3}{500 \times 10^3 + \frac{1}{1\times10^{-9}s}}. \tag{13.89}$$

Simplifying,

$$H(s) = \frac{s}{s + 2000}.$$

Substituting, $s = j\omega$,

$$H(j\omega) = \frac{j\omega}{j\omega + 2000}. \tag{13.90}$$

Since the transfer function is of the form $s/(s+a)$, we can apply our intuitive method for sketching the frequency response:

- Magnitude Plot

1. Sketch the low-frequency asymptote.
   The low-frequency asymptote ($\omega \ll 2000$) is given by:

   $$|H| = \frac{\omega}{2000}.$$

2. Sketch the high-frequency asymptote. The two asymptotes intersect at the break frequency.
   The high-frequency asymptote ($\omega \gg 2000$) is given by:

   $$|H| = 1.$$

   The two asymptotes intersect at the break frequency:

   $$\omega = 2000 \quad \text{rad/s}.$$

   The dashed lines in the magnitude plot in Figure 13.25 show these two asymptotes. They intersect at $\omega = 2000$ rad/s. A computer-generated plot of

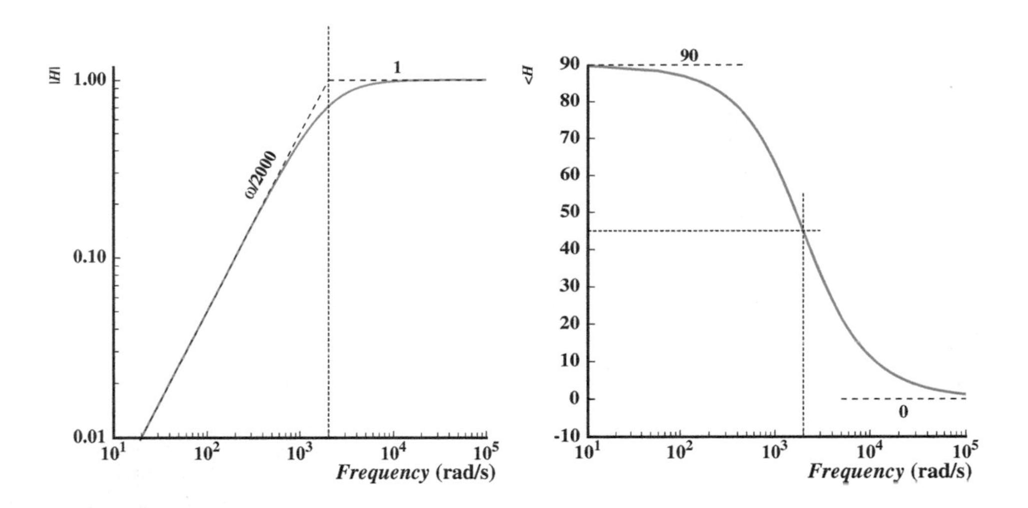


FIGURE 13.25 Frequency response.

the magnitude versus frequency is also shown. Together, the asymptotes are a fairly good approximation of the magnitude curve.

- Phase Plot

1. Sketch the low-frequency asymptote.

   The low-frequency asymptote for the phase is given by:

$$\angle H = 90^\circ.$$

2. Sketch the high-frequency asymptote.

   The high-frequency asymptote for the phase is given by:

$$\angle H - 0^\circ.$$

   The dashed lines in the phase plot in Figure 13.25 show these two asymptotes.

3. The point $(2000, 45^\circ)$ denoting the phase at the break frequency is also marked. The true phase curve is also shown, starting with the low-frequency asymptote, passing through $45^\circ$ at the break frequency, and finishing off at the high-frequency asymptote.

---

### WWW 13.4.3 THE BODE PLOT: SKETCHING THE FREQUENCY RESPONSE OF GENERAL FUNCTIONS *

WWW EXAMPLE 13.5 BODE PLOT FOR SERIES RL CIRCUIT

WWW EXAMPLE 13.6 ANOTHER BODE PLOT EXAMPLE

## 13.5 FILTERS

The frequency response of several of the circuits considered in the previous sections indicated their frequency selective behavior (for example, the RL circuit in Figure 13.11 or the RC circuit in Figure 13.15). We can use such circuits to process signals according to their frequency. Circuits used in this manner are called *filters*. Filters are a major application of frequency domain analysis. The signal-processing property of filtering is fundamental to the operation of all television, radio, and cellular phone receivers, which must select one transmitted signal from among many present at the receiver antenna.

The frequency response plots (see Figure 13.24) of the RL circuit in Figure 13.11 show that it rejects (that is, attenuates) high frequencies and passes (that is, does not affect) low frequencies, and therefore behaves like a *low-pass filter*. The RC circuit in Figure 13.15 behaves like a *band-pass filter* because it passes frequencies that fall within a certain band and rejects both very low frequencies and high frequencies (see Figure 13.17). In general, we can build many other types of filters as well. Figure 13.30 shows in abstract form the magnitude curves of the frequency response for several types of filters.

This section analyzes circuits from a filter point of view. Let us begin with a detailed analysis of a simple RC circuit with a view towards using it as a filter. To illustrate the concept of filtering, let us find the voltage across the capacitor in the simple RC circuit of Figure 13.31a, assuming that $v_i = V_i \cos(\omega t)$. In particular, we wish to find the amplitude of the output signal as the frequency $\omega$ of the input signal is changed.

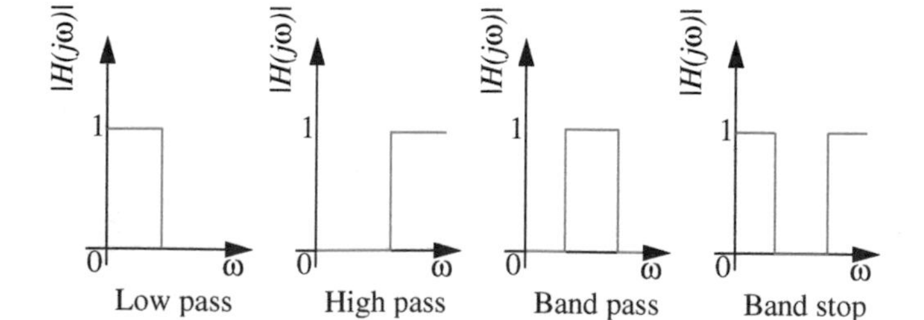


FIGURE 13.30 Various forms of filters.

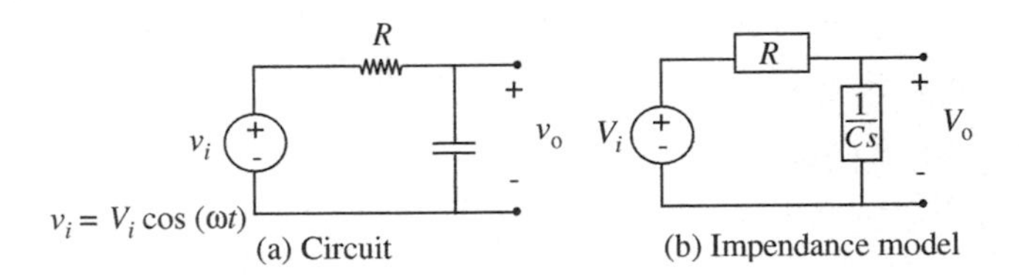


FIGURE 13.31 A simple RC filter circuit and its impedance model.

The generalized voltage divider relation for $V_o$ in the impedance model, Figure 13.31b, is

$$V_o = \frac{\frac{1}{Cs}}{R + \frac{1}{Cs}} V_i = \frac{1/RC}{s + 1/RC} V_i. \tag{13.101}$$

Recall from Equation 13.10 that we have been using $s$ as a shorthand for $j\omega$. So at any frequency, $\omega$, the complex output voltage is given by:

$$V_o = \frac{1/RC}{j\omega + 1/RC} V_i. \tag{13.102}$$

The corresponding system function is

$$H(j\omega) = \frac{V_o}{V_i} = \frac{1/RC}{j\omega + 1/RC}. \tag{13.103}$$

It is easy to see that the magnitude of the system function at low frequencies is close to unity. At high frequencies, on the other hand, the magnitude of the system function approaches 0. Because low frequencies are passed and high frequencies are rejected or attenuated, this circuit acts as a low pass filter.

Figure 13.32 plots the frequency response of the RC circuit assuming that the RC time constant of the circuit is $RC = 1/20$ seconds. The shape of the

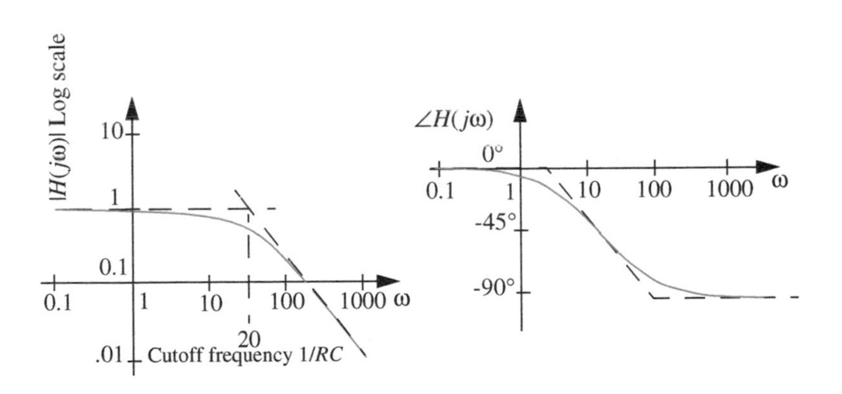


FIGURE 13.32 Frequency response of the simple RC filter circuit.

magnitude plot is indicative of a low-pass filter. The break frequency is 20 rad/s. This says that the filter begins to reject input signals whose frequencies are in the vicinity of 20 rad/s. The level of attenuation increases as the frequency increases beyond the break frequency.

Frequency cutoff begins in the vicinity of the break frequency. Hence the break frequency is also called the *cutoff frequency*. Thus we can design our RC low-pass filter to have any cutoff frequency by an appropriate choice of the RC time constant. As pictured in Figure 13.33, the higher the value of $RC$, the lower the cutoff frequency of the filter.

Finally, noting the similar forms of Equations 13.75 and 13.103, we conclude that Figure 13.32 reflects the frequency response of the circuit in Figure 13.21 as well. In fact, the circuit in Figure 13.21 is a Norton version of the circuit in Figure 13.31.

### 13.5.1 FILTER DESIGN EXAMPLE: CROSSOVER NETWORK

Jeb is building a stereo amplifier system and needs some help. As illustrated in Figure 13.34, Jeb needs to take the output of his CD player and somehow split it into high and low frequencies, then pass each of the signals through MOSFET amplifiers, and then send the amplified high and low signals to a

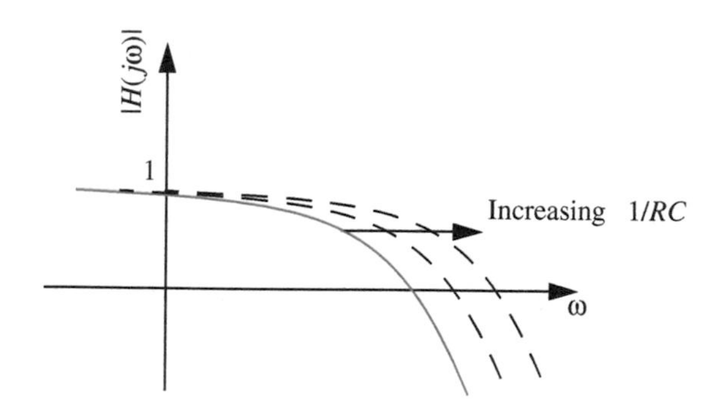


FIGURE 13.33 Designing the cutoff frequency (or break frequency) of a filter.

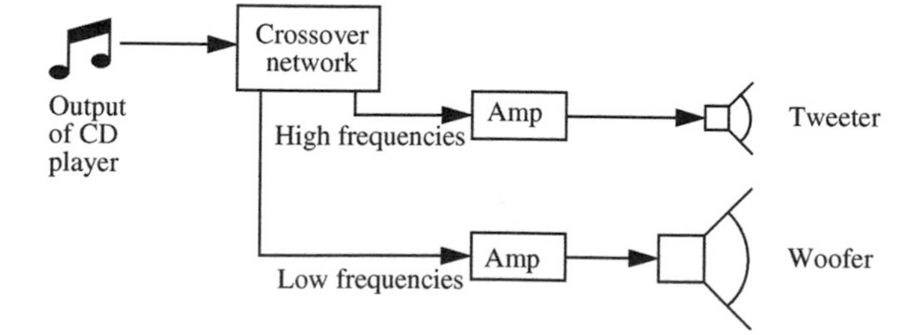


FIGURE 13.34 Crossover system for an amplifier.

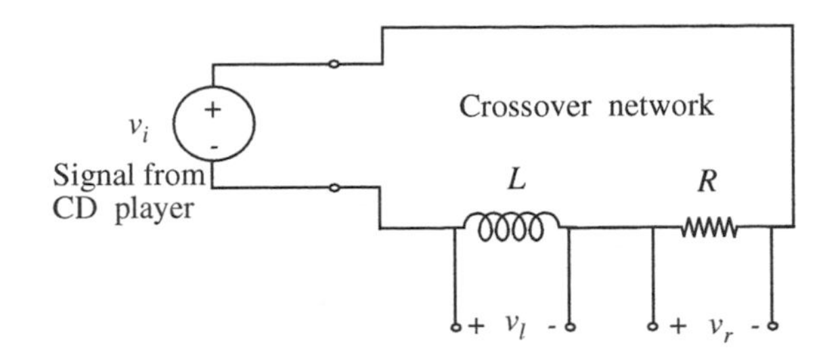


FIGURE 13.35 Crossover network circuit.

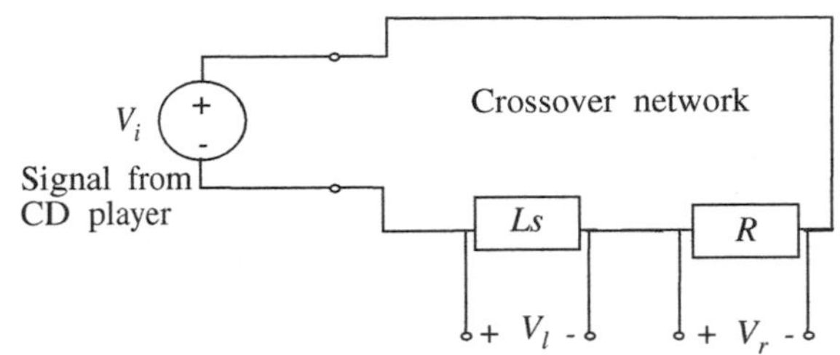


FIGURE 13.36 Impedance model of the crossover network.

tweeter and a woofer, respectively. Together, the tweeter and woofer form the speaker system.

Jeb asks his science teacher for help. The science teacher tells him, "Take a resistor and an inductor and connect them in series. Then take the high frequencies off one element and the low frequencies of the other," and then rushes off to teach a class. Unfortunately, Jeb forgot to ask which element to connect to the low-frequency woofer and which to connect to the high-frequency tweeter.

Jeb goes to one of his friends Nina for help, and sketches the network suggested by the science teacher (see Figure 13.35). Nina has just read about filters in her electronics class and is confident she can figure this out quickly.

As a first step, Nina draws the impedance model of the circuit as shown in Figure 13.36.

She then writes down the transfer functions corresponding to the inductor and resistor outputs as follows:

$$\frac{V_l}{V_i} = -\frac{Ls}{Ls + R}$$

$$\frac{V_r}{V_i} = -\frac{R}{Ls + R}.$$

Substituting, $s = j\omega$,

$$\frac{V_l}{V_i} = -\frac{j\omega L}{j\omega L + R}$$

$$\frac{V_r}{V_i} = -\frac{R}{j\omega L + R}.$$

By taking the limit as $\omega \to 0$, Nina observes that signal across the inductor, $V_l$, goes to zero. Similarly, $V_l$ tends to $V_i$ as $\omega \to \infty$. Thus Nina concludes

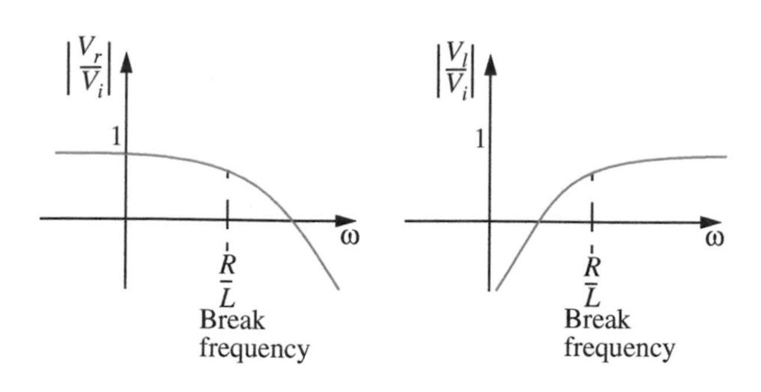


FIGURE 13.37 The magnitude of the transfer functions for the two signals from the crossover network.

that the transfer function for the signal across the inductor resembles that of a high-pass filter.

In contrast the signal across the resistor, $V_r$, tends to $V_i$ as $\omega \rightarrow 0$. It tends to zero as $\omega \rightarrow \infty$. Thus the transfer function for the signal across the resistor resembles that of a low-pass filter.

Nina plots the two transfer functions in Figure 13.37, and based on her analysis, recommends that Jeb connect the signal from the inductor to the tweeter and the signal from the resistor to the woofer.

Jeb then asks Nina to walk down with him to an electronics store, eShack Inc., so they can select an appropriate resistor and inductor. Since the human ear responds to frequencies up to 20 kHz,[9] Nina decides that an appropriate break frequency for the filters is $f_{\text{break}} = 5$ kHz. In radians, the break frequency is given by:

$$\omega_{\text{break}} = 2\pi 5000$$

or 31,416 radians. Thus, Nina begins to look for an inductor and a resistor with values $L$ and $R$ such that $R/L = 31416$. She is able to find a resistor of value 100 Ω and an inductor of value 3.2 mH, thus solving Jeb's design problem.

## 13.5.2 DECOUPLING AMPLIFIER STAGES

In this section, we will discuss an application of an RC filter. Recall the MOSFET amplifier circuit discussed in Chapter 8. The same circuit is shown in Figure 13.38a. The circuit has an input port and an output port. Figure 13.38b

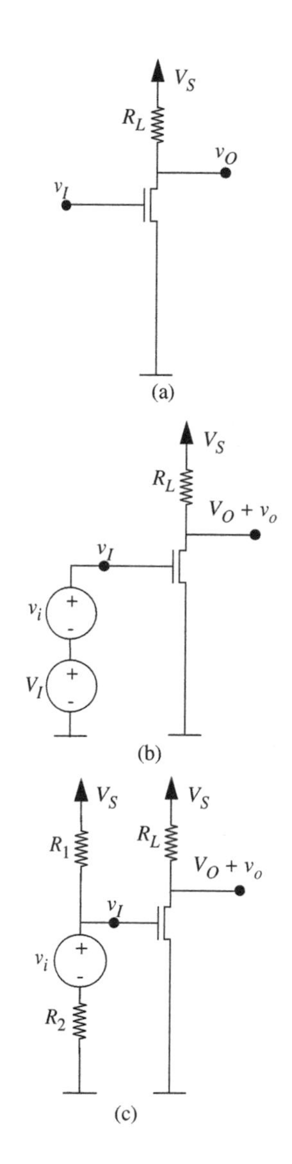


FIGURE 13.38 Amplifier circuit discussed in Chapter 8.

9. In reality, only a young human ear responds to frequencies up to 20 kHz. Humans lose their sensitivity to high frequencies as they grow older at the rate of roughly 1 Hz per day starting in their teenage years. Loud music accelerates the sensitivity loss!

shows the same amplifier circuit supplied with an input DC bias voltage $V_I$ and a small signal input $v_i$. The bias voltage $V_I$ might be provided by a resistor divider circuit as in Figure 13.38c. Unlike the DC bias, the small signal input $v_i$ is commonly a time-varying voltage, for example a sine wave. Correspondingly, the output voltage $v_O$ is the sum of a DC output bias $V_O$ and a small-signal voltage $v_o$.

A problem with the amplifier circuit shown in Figure 13.38a is that it cannot be *cascaded* easily. In other words, the output port of an amplifier A cannot easily be coupled to the input of another amplifier B because the biasing of one stage affects the biasing of all cascaded stages. This form of coupling might be useful if a designer wants to build an amplifier whose gain is greater than that of a single amplifier stage. Notice in Figure 13.38 that the output of the amplifier is the sum of a bias output voltage $V_O$ and a small-signal output $v_o$. Notice further that the output of our amplifier in Figure 13.38 (call it amplifier A) cannot be fed directly to the input of another amplifier B because of the biasing requirement for saturation operation of B. If the output of amplifier A is fed directly to the input of amplifier B, then the output bias voltage of A will become the input bias for B, and amplifier design requirements often result in conflicting values for the output and input bias voltages.

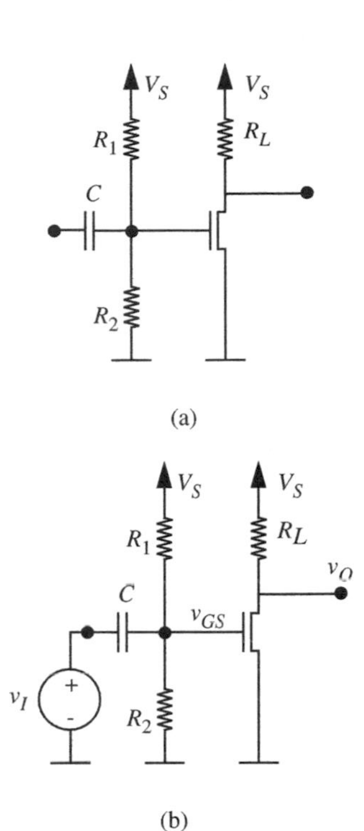


FIGURE 13.39 New amplifier circuit with decoupling capacitor.

We will now discuss another implementation of the amplifier circuit based on an input coupling capacitor as shown in Figure 13.39a. This circuit makes use of the fact that most small-signal inputs of interest are time-varying signals with a reasonably high-frequency content. As illustrated in Figure 13.39b, a small-signal input can be fed directly to the input port of the amplifier in Figure 13.39a, even if the small signal is superimposed over some DC bias voltage. As we will show shortly, the capacitor $C$ and the resistors $R_1$ and $R_2$ filter out any DC bias voltage over which the small signal is superimposed, and pass through the time-varying small signal. Furthermore, we will show that the resistive divider formed by $R_1$ and $R_2$ provides the necessary input bias voltage.

Let us determine the voltage $v_{GS}$ at the gate of MOSFET as a function of the input voltage $v_I$ for the amplifier in Figure 13.39b. We will use the impedance method to do so. Our goal is to determine $v_{GS}$ as a function of frequency so we can see how the circuit behaves for DC for and time-varying signals. Since the gate input of the MOSFET is an open circuit, the subcircuit relevant to the computation of $v_{GS}$ is shown in Figure 13.40. The corresponding impedance model is shown in Figure 13.41.

Figure 13.41 shows that $V_{gs}$ is a function of two inputs: $V_i$ and $V_S$. We will use superposition to compute their combined effect. Let $V_{gsi}$ be the component due to $V_i$ and $V_{gss}$ be the component due to $V_S$.

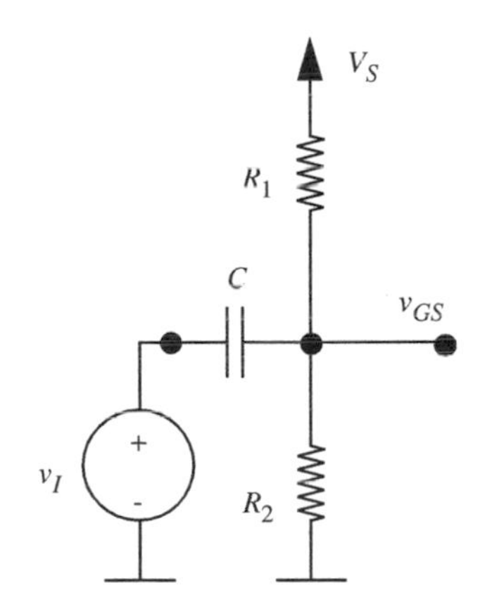


FIGURE 13.40 Subcircuit to compute $v_{GS}$.

To find the effect of $V_S$ acting alone, namely $V_{gss}$, we short out $V_i$, and obtain the equivalent circuit shown in Figure 13.42a. The MOSFET has infinite input impedance, so it does not figure in the calculation of $V_{gss}$. (We ignore the MOSFET's gate capacitance for simplicity.) $V_{gss}$ can now be found through a simple voltage divider relationship:

$$V_{gss} = \frac{1/Cs \parallel R_2}{R_1 + 1/Cs \parallel R_2} V_S \quad (13.104)$$

$$= \frac{R_2}{R_1 R_2 Cs + (R_1 + R_2)} \quad (13.105)$$

$$= \frac{R_2}{R_1 R_2 Cj\omega + (R_1 + R_2)}. \quad (13.106)$$

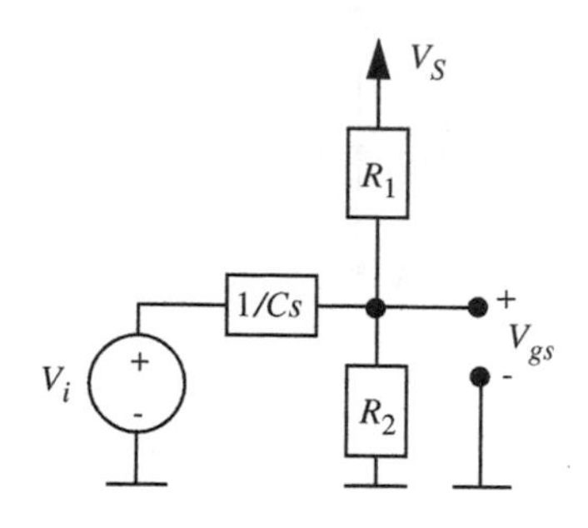


FIGURE 13.41 Impedance model relevant to the computation of $V_{gs}$.

Since $V_S$ is a DC voltage, we know that its frequency $\omega$ is 0. Therefore,

$$V_{gss} = \frac{R_2}{R_1 + R_2} V_S. \quad (13.107)$$

Notice that $V_{gss}$ does not depend on $C$.

Now let us short out $V_S$ and compute $V_{gsi}$. The relevant equivalent circuit is shown in Figure 13.42b. The voltage $V_{gsi}$ can again can be found through a simple voltage-divider relationship:

$$V_{gsi} = \frac{R_1 \parallel R_2}{1/Cs + R_1 \parallel R_2} V_i \quad (13.108)$$

$$= \frac{R_{eq}}{1/Cs + R_{eq}} V_i \quad (13.109)$$

$$= \frac{R_{eq}Cs}{1 + R_{eq}Cs} V_i \quad (13.110)$$

where $R_{eq} = R_1 \parallel R_2$. Substituting $s = j\omega$

$$V_{gsi} = \frac{R_{eq}Cj\omega}{1 + R_{eq}Cj\omega} V_i. \quad (13.111)$$

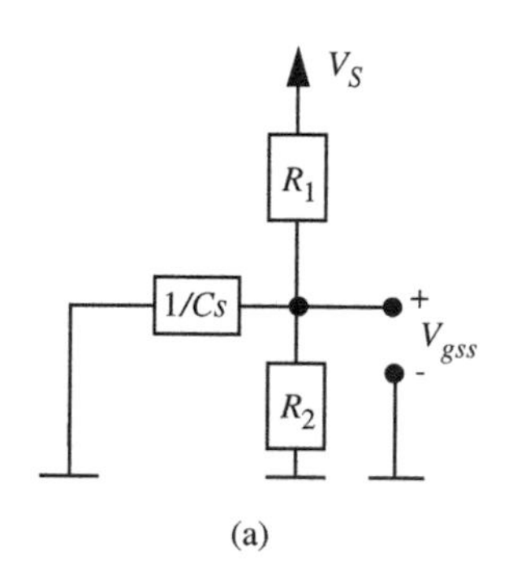


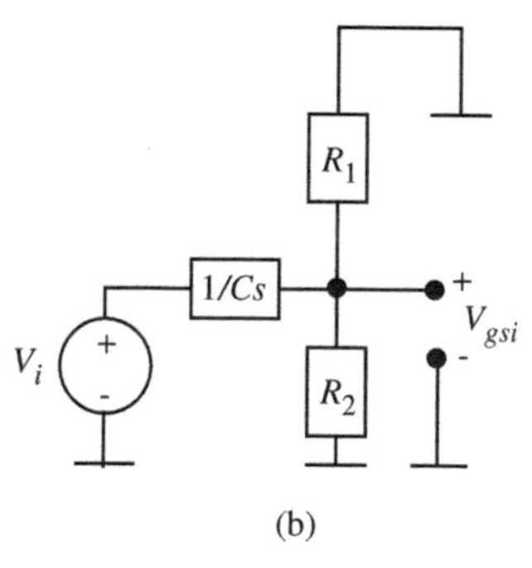


FIGURE 13.42 (a) Equivalent circuit to compute $V_{gss}$, the component due to $V_S$; (b) equivalent circuit to compute $V_{gsi}$, the component due to $V_i$.

We can also write a transfer function $H(j\omega)$ relating the complex input voltage to $V_{gsi}$ as follows:

$$H(j\omega) = \frac{V_{gsi}}{V_i} = \frac{R_{eq}Cj\omega}{1 + R_{eq}Cj\omega}. \quad (13.112)$$

Let us analyze Equation 13.112 at low and high frequencies:

- When $\omega$ is small, $H(j\omega)$ is close to 0. In particular, $\omega = 0$ for DC signals, and $H(j\omega) = 0$. This implies that the circuit will filter out any DC bias over which the input signal is superimposed. Thus, any DC bias voltage

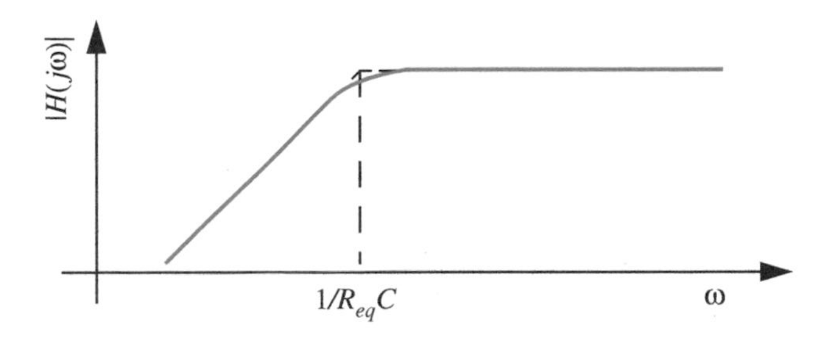


**FIGURE 13.43** Frequency response plot (magnitude only) for the high-pass filter formed by $C$, $R_1$, and $R_2$.

carried by the input will not affect the node voltage $v_{GS}$, which means that the biasing of the MOSFET in Figure 13.39 is independent of the applied input $v_I$. We will discuss the issue of biasing in more detail shortly.

- When $\omega$ is large, $H(j\omega)$ is close to unity. Therefore high frequencies are passed through without any attenuation.

The magnitude portion of the frequency response plot for the preceding system function is shown in Figure 13.43. It is easy to see that the plot resembles that for a high-pass filter. Signals with frequencies significantly greater than the break frequency $1/R_{eq}C$ will pass through, while signals with frequencies significantly lower than $1/R_{eq}C$ will be attenuated.

Next, to study the impact of cascading, suppose the input $V_i$ is composed of a DC component $V_{i0}$ and a time-varying component $V_{i\omega}$, as might be the case if the output of a previous amplifier stage is used as an input to the next stage. In other words, suppose

$$V_i = V_{i0} + V_{i\omega}$$

then, from Equation 13.111,

$$V_{gsi} = \frac{R_{eq}Cj\omega}{1 + R_{eq}Cj\omega} V_i \tag{13.113}$$

$$= \frac{R_{eq}Cj\omega}{1 + R_{eq}Cj\omega} V_{i0} + \frac{R_{eq}Cj\omega}{1 + R_{eq}Cj\omega} V_{i\omega}. \tag{13.114}$$

Since $V_{i0}$ is a DC signal, the first term in the preceding equation vanishes, so

$$V_{gsi} = \frac{R_{eq}Cj\omega}{1 + R_{eq}Cj\omega} V_{i\omega}. \tag{13.115}$$

Therefore, if we choose $R_{eq}C$ to be large compared to the lowest frequency component of interest in the input signal, ($\omega \gg 1/R_{eq}C$), we get

$$V_{gsi} \simeq V_{i\omega}. \tag{13.116}$$

This completes our derivation of $V_{gsi}$. The expression for $V_{gsi}$ demonstrates that the DC component of the input signal does not influence $V_{gsi}$.

To complete the analysis, we obtain $V_{gs}$ by adding the contributions due to $V_S$ and $V_i$:

$$V_{gs} = V_{gss} + V_{gsi} = \frac{R_2}{R_1 + R_2} V_S + \frac{R_{eq}Cj\omega}{1 + R_{eq}Cj\omega} V_{i\omega}. \tag{13.117}$$

Further, choosing a large value for $R_{eq}C$ so that the lowest frequency component of the input signal ($\omega$) is much greater than $1/R_{eq}C$, we get

$$V_{gs} = \frac{R_2}{R_1 + R_2} V_S + V_{i\omega}. \tag{13.118}$$

Equation 13.118 indicates that the voltage at the gate of the MOSFET is the sum of two components: a DC bias voltage, $R_2 V_S/(R_1 + R_2)$, and $V_{i\omega}$, the time-varying component of the input. The values of $R_1$ and $R_2$ can be chosen to achieve a desired input DC bias. Thus, from the viewpoint of biasing the amplifiers, as shown in Figure 13.44, multiple amplifier stages can be designed independently, and cascaded without the output bias of one stage impacting the input bias of the next stage.[10] The decoupling capacitor is instrumental in

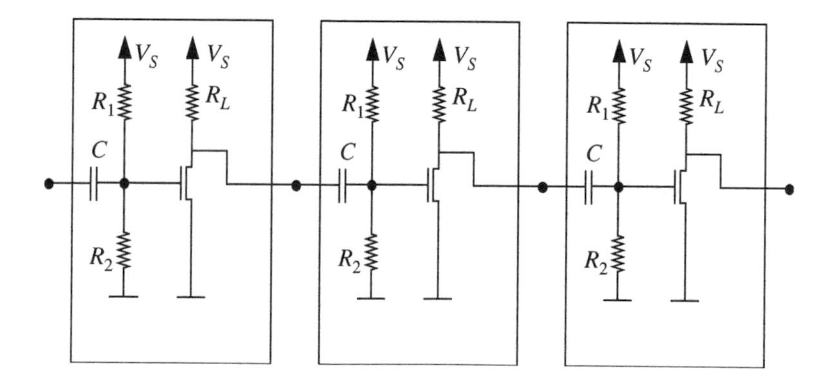


FIGURE 13.44 Cascading multiple amplifiers together.

10. Note, however, that the small-signal gain of each amplifier stage depends on the load offered by the subsequent stage. Specifically, when computing the small-signal gain, assuming the capacitor $C$ behaves as a short for the frequencies of interest, the load seen by each of the first and second stages will not by $R_L$ alone, rather it will be $R_L$ in parallel with $R_1$ and $R_2$. We saw an example of a loading effect, specifically a capacitive loading effect, in Section 13.3.4.

filtering out the DC bias component of the input, thereby allowing multiple amplifiers to be coupled directly.

## 13.6 TIME DOMAIN VERSUS FREQUENCY DOMAIN ANALYSIS USING VOLTAGE-DIVIDER EXAMPLE

In this section, we compare frequency domain versus time domain analysis using the compensated attenuator (or compensated voltage divider) as an example. The simple voltage dividers discussed in Section 2.3.4 do not work very well at high frequencies. In most circuits, some parasitic shunt capacitance is present, represented in Figure 13.45a by capacitor $C_2$. If the drive is a low-frequency sine wave, the attenuation of the voltage divider is $R_2/(R_1 + R_2)$ as expected, but at high frequencies the capacitance causes the attenuation to increase beyond this desired value. This effect can be readily shown by a frequency domain analysis of the voltage divider.

### 13.6.1 FREQUENCY DOMAIN ANALYSIS

The frequency domain analysis begins by constructing the impedance model for the circuit, Figure 13.45b. The generalized voltage divider relation is

$$V_o = \frac{R_2 \parallel (1/C_2 s)}{R_1 + R_2 \parallel (1/C_2 s)} V_i, \tag{13.119}$$

which simplifies to

$$V_o = \frac{R_2}{R_1 + R_2 + R_1 R_2 C_2 s} V_i. \tag{13.120}$$

At high frequencies, ($\omega$ large, and $s = j\omega$), this expression reduces to

$$V_o \simeq \frac{1}{j\omega R_1 C_2} V_i \tag{13.121}$$

so $v_o(t)$ becomes an increasingly small sinusoid, lagging 90° behind the input sine wave. Thus the attenuation of the voltage divider is not fixed as desired,

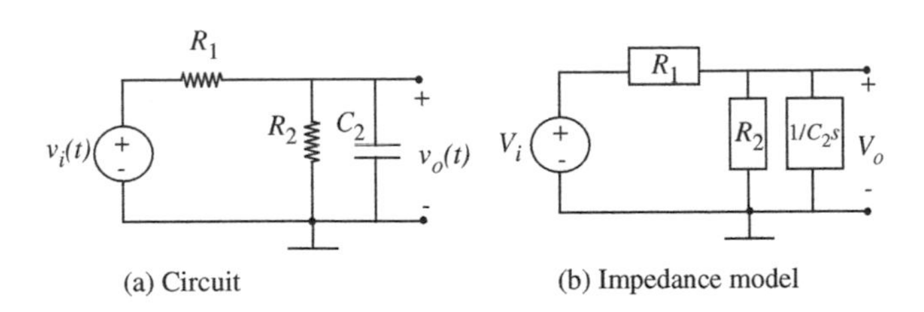


FIGURE 13.45 Voltage divider with parasitic capacitance.

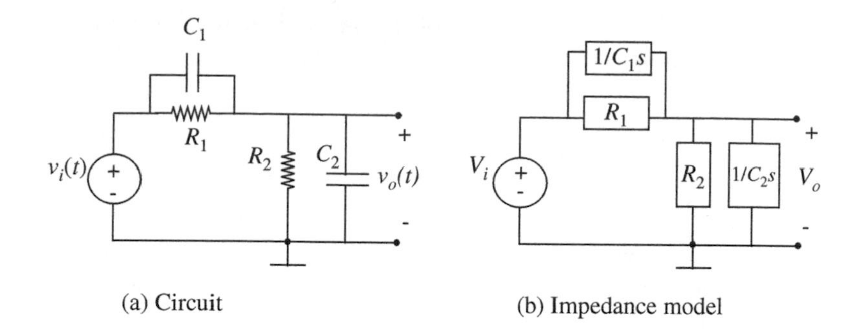


FIGURE 13.46 Compensated voltage divider.

but increases with increasing frequency. This effect is readily observable if one tries to build a 2:1 divider at the input of an oscilloscope by using two one-megohm resistors for $R_1$ and $R_2$. (Such a divider circuit might be used in a scope probe, for example.) Two problems will arise. The *input resistance* of the scope will change the desired attenuation at low frequencies, and the scope input capacitance will cause the attenuation to change with frequency, as can be seen from Equation 13.120.

To remedy the capacitance problem, it is necessary to add a small capacitor in parallel with the series resistor, as shown in Figure 13.46. This design is used in all good oscilloscope probes. Analysis of the impedance model yields

$$V_o = \frac{R_2 \parallel (1/C_2s)}{R_1 \parallel (1/C_1s) + R_2 \parallel (1/C_2s)} V_i, \tag{13.122}$$

which expands to

$$V_o = \frac{R_2(R_1C_1s + 1)}{R_1(R_2C_2s + 1) + R_2(R_1C_1s + 1)} V_i. \tag{13.123}$$

If we choose one particular value of $C_1$ such that

$$R_1C_1 = R_2C_2, \tag{13.124}$$

then Equation 13.123 reduces to

$$V_o = \frac{R_2}{R_1 + R_2} V_i \tag{13.125}$$

so the real output time function is

$$v_o(t) = \frac{R_2}{R_1 + R_2} v_i(t) \tag{13.126}$$

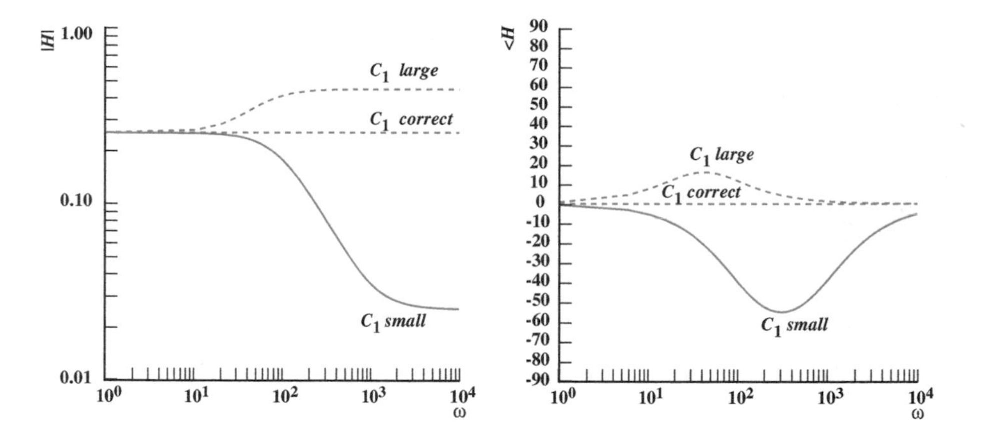


FIGURE 13.47 Magnitude and phase for compensated attenuator

independent of frequency! The voltage divider has now been *compensated* for the effect of $C_2$. Let us examine Equation 13.123 in more detail, and form the frequency response plots for under-compensation, correct compensation, and over-compensation. For this purpose it is helpful to rearrange Equation 13.123, to obtain the system function in our standard form,

$$H(s) = \frac{V_o}{V_i} = \left(\frac{C_1}{C_1 + C_2}\right)\left(\frac{s + \frac{1}{R_1 C_1}}{s + \frac{(R_1+R_2)}{R_1 R_2 (C_1+C_2)}}\right). \tag{13.127}$$

Let us assume that

$$\frac{C_1}{C_1 + C_2} = 0.025$$

$$R_1 C_1 = 1 \text{ ms}$$

and

$$\frac{R_1 + R_2}{R_1 R_2 (C_1 + C_2)} = 100$$

and so that Equation 13.127 in our standard form becomes

$$H(j\omega) = \frac{V_o}{V_i} = \frac{0.025(1000 + j\omega)}{100 + j\omega}. \tag{13.128}$$

The frequency response for the under compensated case ($C_1$ too small) is plotted with solid lines in Figure 13.47. The plots appropriate for

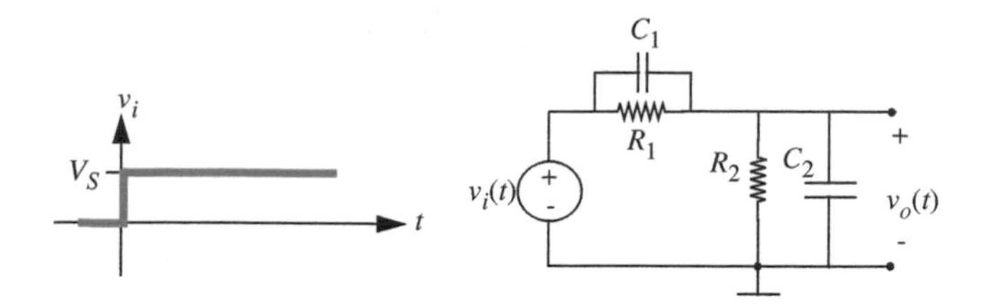


FIGURE 13.48 Compensated attenuator, step drive.

under- and over-compensation are also shown in dashed lines. The circuit is properly compensated when the attenuation is constant, independent of frequency.

### 13.6.2 TIME DOMAIN ANALYSIS

To relate back to the analysis methods used in Chapter 10, let us now calculate the response of the attenuator to a voltage step, as shown in Figure 13.48. Note first that the system is still first order, in spite of the two capacitors. We know this because it is not possible to specify two independent initial conditions, only one.

Let us assume that the capacitor $C_2$ is initially uncharged, so the initial condition is

$$v_o(t = 0) = 0. \tag{13.129}$$

It follows that the initial voltage on $C_1$ must also be zero, if we assume $v_i = 0$ for $t < 0$. A homogeneous solution of the form

$$v_o = Ke^{st} \tag{13.130}$$

is appropriate. (Note that here $s$ is not a shorthand notation for $j\omega$.) But as noted in Section 13.2, we have already solved this problem: The characteristic polynomial is the denominator of the system function, in this case the denominator of Equation 13.127. Setting the characteristic polynomial to zero, we find

$$s = \frac{-1}{(R_1 \parallel R_2)(C_1 + C_2)}. \tag{13.131}$$

By inspection, the particular solution, which will satisfy the differential equation after the transient has died away, is

$$v_o = \frac{R_2}{R_1 + R_2} V_S \tag{13.132}$$

where $V_S$ is the height of the applied step. Thus the complete solution is of the form:

$$v_o = Ke^{-t/(R_1 \| R_2)(C_1+C_2)} + \frac{R_2}{R_1 + R_2} V_S \quad \text{for} \quad t > 0. \tag{13.133}$$

The constant $K$ cannot be evaluated in the usual way (that is, setting $v_o = 0$ for $t = 0$ and solving for $K$), because the two capacitors form a loop facing the voltage source. Thus when the step is applied, in theory infinite current flows through the capacitors for an instant, and the capacitor voltages change instantaneously. Thus immediately after the step, at $t = 0^+$, the initial condition on $v_o$ will not be 0.

Instead, we can proceed as follows to find the initial condition at $t = 0^+$. Because the same current flows through each capacitor, both capacitors receive equal charge:

$$q = C_1 v_{c1} = C_2 v_o. \tag{13.134}$$

Also

$$v_i = v_{c1} + v_o; \tag{13.135}$$

hence right after the step, at $t = 0^+$,

$$v_o = \frac{C_1}{C_1 + C_2} V_S. \tag{13.136}$$

This, then, is the initial condition to use.

Now $K$ can be evaluated by setting $t$ to zero in Equation 13.133, and equating the voltage to that given in Equation 13.136

$$\frac{C_1}{C_1 + C_2} V_S = K + \frac{R_2}{R_1 + R_2} V_S \tag{13.137}$$

whence

$$K = \left( \frac{C_1}{C_1 + C_2} - \frac{R_2}{R_1 + R_2} \right) V_S. \tag{13.138}$$

The complete solution is sketched in Figure 13.49 for $C_1$ too small, correctly chosen, and too large. The correct choice is obviously the one that makes $K$ equal to zero in Equation 13.138, so that there is no transient. This is equivalent to making the capacitor voltage at $t = 0^+$, Equation 13.136, equal to the final

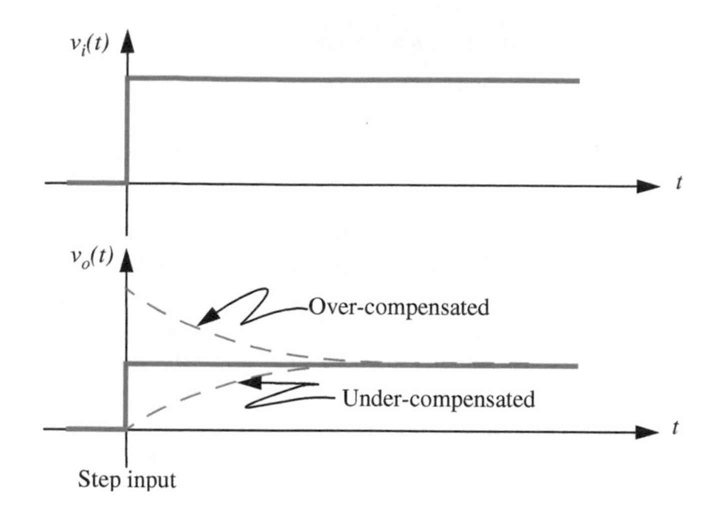


**FIGURE 13.49** Time response of compensated attenuator to a step input.

capacitor voltage derived from the particular solution, Equation 13.132. These conditions both require, as before:

$$R_1C_1 = R_2C_2. \tag{13.139}$$

### 13.6.3 COMPARING TIME DOMAIN AND FREQUENCY DOMAIN ANALYSES

It is possible to tie together the *time-domain* solution with the previous *frequency-domain* solution by assuming the driving waveform to be a square wave. The time-domain solution for the over-compensated case ($C_1$ is large) now looks like Figure 13.50. In particular, make a note of the relatively higher values of the output signal at the transition points.

In terms of the frequency domain, the square-wave drive can be considered as a sum of the sine waves:[11] A sine wave with the same period as the square wave, a sine wave at three times this frequency and one-third the amplitude, another at five times the frequency and one-fifth the amplitude, etc. The higher frequency sine waves are called *harmonics*. It is relatively easy to visualize how each of these components will be altered as they pass through the attenuator by examining the appropriate frequency response plot. Again looking at the over-compensated case ($C_1$ is large), the magnitude plot of Figure 13.47 shows

11. The theory behind such a decomposition of waveforms into sinusoids can be found in any text that deals with Fourier Series, for example, *Signals and Systems*, Alan V. Oppenheim and Alan S. Willsky, 1996, Prentice Hall Publishers.

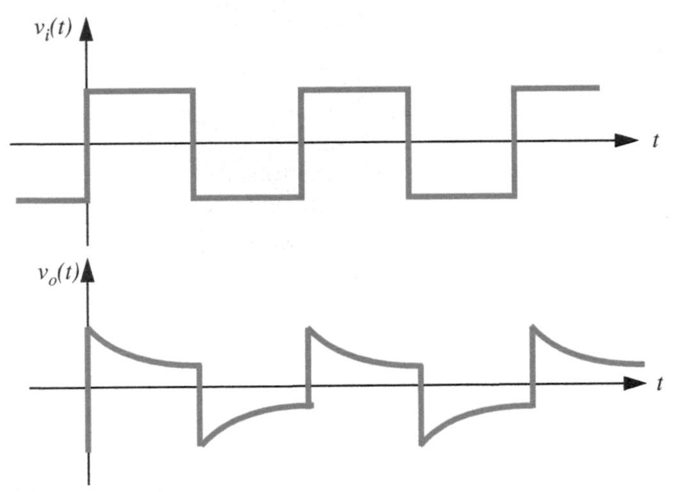


FIGURE 13.50 Time response of compensated attenuator to square-wave input.

that the high-frequency harmonics will be larger than they are supposed to be compared to the low frequencies, and the phase plot of the same figure shows that they will be shifted in phase.

We can correlate the larger amplitude of the high-frequency harmonics in the output as observed in the frequency domain with the relatively larger values of the square wave immediately following a signal transition in the time domain.

The important conclusions from these two quite different analysis methods are as follows:

- The two analyses provide complementary views of the same circuit.
- Often in an experimental situation it is easier to adjust $C_1$ for perfect compensation by the time-domain approach of looking at the response to a square wave. The alternative technique more consistent with the frequency-domain view is to apply first a low-frequency sine wave, and then a high-frequency sine wave, and check that the response amplitude is the same in both cases.

## 13.7 POWER AND ENERGY IN AN IMPEDANCE

In this section, we wish to address some of the issues of power and energy flow in RLC circuits. As discussed in Chapter 11, power and energy are critical issues in the design of circuits. The size of the battery required by a device so it will function for a desired amount of time is related to the energy efficiency

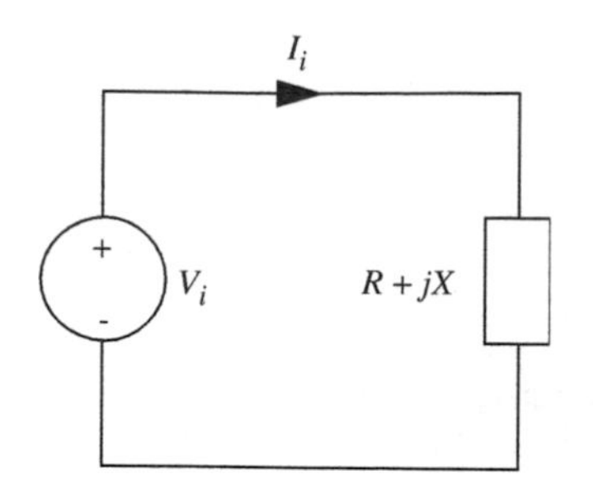


FIGURE 13.51 Power calculations in an arbitrary impedance.

of the device. Similarly, the cooling requirements for a device depend on the power dissipated by the device.

### 13.7.1 ARBITRARY IMPEDANCE

Let us examine first the power delivered to some arbitrary impedance $Z = R + jX$ by a sinusoidal source, as depicted in Figure 13.51. The quantity $X$ is usually referred to as the *reactance* of the circuit.

A general sinusoidal drive can be written as:

$$v_i(t) = |V_i| \cos(\omega t + \phi). \tag{13.140}$$

Hence the complex amplitudes of the voltage and current are

$$V_i = |V_i| e^{j\phi} \tag{13.141}$$

$$I_i = \frac{V_i}{Z} = \frac{|V_i| e^{j\phi}}{R + jX} \tag{13.142}$$

$$= \frac{|V_i| e^{j(\phi - \theta)}}{\sqrt{R^2 + X^2}} \tag{13.143}$$

$$= |I_i| e^{j(\phi - \theta)}. \tag{13.144}$$

where

$$\theta = \tan^{-1} \frac{X}{R}. \tag{13.145}$$

The power delivered to the impedance is by definition the product of $v(t)$ and $i(t)$.

Because power is *not a linear function of $v$* and $i$, we must be cautious about using impedance concepts in power calculations.

Thus we start with *time expressions* such as Equation 13.140 rather than complex amplitudes. From Equation 13.143, the current as a function of time is

$$i(t) = Re\left[I_i e^{j\omega t}\right] \tag{13.146}$$

$$= \frac{|V_i|}{\sqrt{R^2 + X^2}} \cos(\omega t + \phi - \theta). \tag{13.147}$$

Hence, from Equations 13.140 and 13.147 the instantaneous power is given by:

$$p(t) = vi = \frac{|Vi|^2}{\sqrt{R^2 + X^2}}[\cos(\omega t + \phi)][\cos(\omega t + \phi - \theta)] \tag{13.148}$$

$$= \frac{1}{2}\frac{|V_i|^2}{\sqrt{R^2 + X^2}}[\cos(2\omega t + 2\phi - \theta) + \cos\theta]. \tag{13.149}$$

Thus, in general, the instantaneous power for sinusoidal drive has a sinusoidal component at twice the frequency of the input signal, and the DC component. We will examine this expression for some simple cases shortly, but first let us complete the general derivation by calculating the *average power*, because this is the quantity that determines your monthly bill from the power company.

Because the average value of cos $(\omega t)$ is zero, the average power flowing into an arbitrary impedance is just the DC term in Equation 13.149:

$$\bar{p} = \frac{1}{2}\frac{|V_i|^2}{\sqrt{R^2 + X^2}}\cos\theta. \tag{13.150}$$

From Equations 13.141 and 13.143, this can be written as

$$\bar{p} = \frac{1}{2}|V_i \parallel I_i|\cos\theta \tag{13.151}$$

where $V_i$ and $I_i$ are the complex amplitudes of the voltage and current respectively, and $\theta$ is the angle between them. The term $\cos\theta$ is often called the *power factor.*

The average power in terms of complex amplitudes of voltages and currents is one half the product of the two magnitudes multiplied by the cosine of the angle between them.

The average power can also be written directly in terms of complex voltage and complex current. Again from Equations 13.141 and 13.143

$$\bar{p} = \frac{1}{2}Re[V_i I_i^*] \tag{13.152}$$

$$= \frac{1}{2}Re[V_i^* I_i] \tag{13.153}$$

where $I_i^*$ is the complex conjugate of $I_i$, and $V^*$ is the complex conjugate of $V$. Using this notation, 1/2 $VI^*$ is often called *complex power*, whence the real part of the complex power is the average power, the "real" power, as per Equation 13.151, and the imaginary part is called *reactive power.*

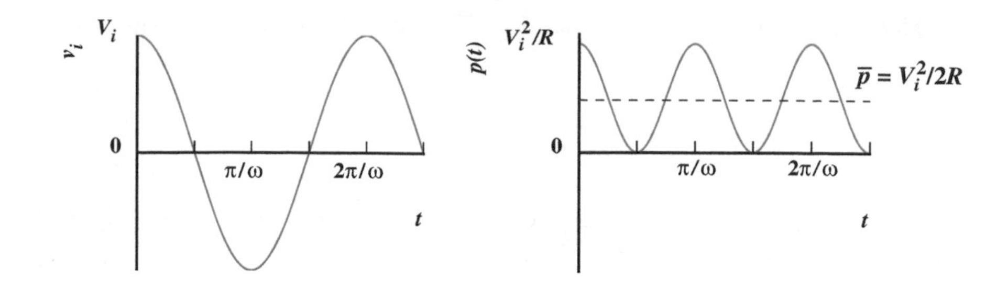


FIGURE 13.52 Power flow in a pure resistance.

## 13.7.2 PURE RESISTANCE

To develop some insight concerning power and energy flow, let us examine a few special cases. First, assume that the impedance in Figure 13.51 is a pure resistance $R$. That is, $X = 0$. Further, we assume for simplicity that the time origin is selected to make the voltage drive a cosine wave, that is, $\phi = 0$ in Equation 13.140. Then 13.149 is reduced to:

$$p(t) = \frac{V_i^2}{2R}(1 + \cos 2(\omega t)). \tag{13.154}$$

Again we have a double-frequency term and a DC term. A plot of power as a function of time for the resistive case is shown in Figure 13.52. From the figure or from Equation 13.154, the average power dissipated in the resistor for sinusoidal drive is

$$\overline{p} = \frac{V_i^2}{2R}. \tag{13.155}$$

Remember from Section 1.8.1, this is exactly one half of the power delivered by the DC voltage of the same amplitude. Recall, also from Section 1.8.1, the voltage unit called the *root-mean-square* voltage, abbreviated *rms*, which is related to the peak amplitude of the sinewave by the square root of two

$$V_{\text{rms}} = \frac{V_i}{\sqrt{2}}. \tag{13.156}$$

In terms of the rms unit, average power is

$$\overline{p} = \frac{(V_{\text{rms}})^2}{R} \tag{13.157}$$

just as for DC power. For non-sinusoidal voltages, the general definition of rms voltage is, as the name implies,

$$V_{\text{rms}} = \sqrt{\overline{(v(t))^2}}. \tag{13.158}$$

Most voltages related to the AC power line are quoted in terms of rms values, unless specifically designated as peak values. Thus the 115-V AC power from a wall socket is 115 volts rms, or $115 \times \sqrt{2} = 162.6$ volts peak.

### 13.7.3 PURE REACTANCE

Next, we examine the case where the impedance in Figure 13.51 consists only of inductors and/or capacitors, that is, $R = 0$. Regardless of the circuit configuration, at any given frequency the impedance must look like either a pure inductor or a pure capacitor (although it will change from one to the other as frequency changes, when we go through a resonant frequency). If the circuit is inductive, then from Equation 13.145, $\theta = \pi/2$. Again assuming a cosine drive voltage ($\phi = 0$), Equation 13.149 reduces to:

$$p(t) = \frac{V_i^2}{2X} \cos(2\omega t - \pi/2) \tag{13.159}$$

$$= \frac{V_i^2}{2X} \sin 2(\omega t). \tag{13.160}$$

If the circuit is capacitive at the frequency of interest, then $X$ in Equation 13.143 must equal $-1/\omega C$, hence from Equation 13.145 with $R = 0$, $\theta = -\pi/2$, and

$$p(t) = -\frac{V_i^2}{2X} \sin(2\omega t). \tag{13.161}$$

Power flow as a function of time for both of these cases is shown in Figure 13.53. Note that in both cases the *average power is zero*. Thus, circuits with inductors and capacitors but no resistors do not dissipate any power. The L's and C's absorb power for two quarters of each cycle, and deliver the power back to the source during the other two quarter cycles. Power companies are not happy about this state of affairs, because they still must supply the power depicted in Figure 13.53, even though they get it all back a few milliseconds later. The problem is that the current associated with this instantaneous power causes $i^2R$ power losses in the transmission lines, and the power company must pay for this power loss, even though the customer is consuming zero power on the average.[12]

Although there is no average power supplied to this lossless circuit in the sinusoidal steady state, there is energy stored on the average. For example, for

12. Not surprisingly, it turns out that the customers pay for it eventually, because the rates assume a loss factor.

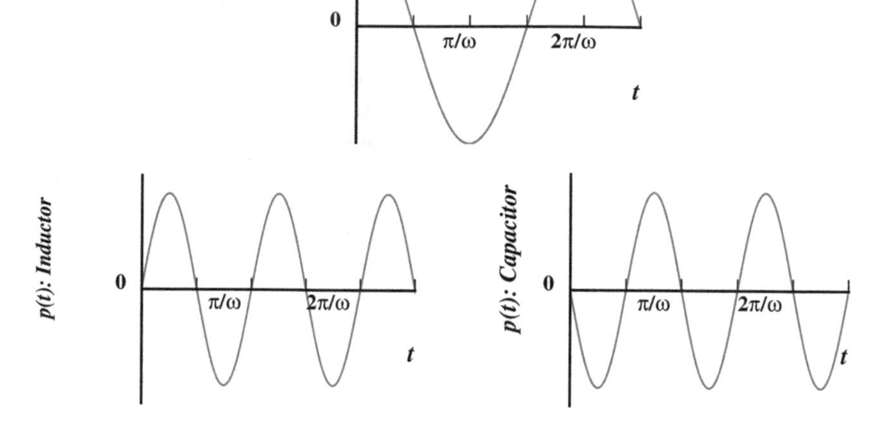


**FIGURE 13.53** Power flow in inductor and capacitor. The maximum value of $p(t)$ for both the inductor and capacitor is $1/2(V_i^2/X)$.

a capacitor, the stored energy is, from Equation 9.18,

$$W_C = \frac{1}{2}Cv(t)^2. \tag{13.162}$$

For sinusoidal $v(t)$,

$$W_C = \frac{1}{2}C(V_i \cos(\omega t))^2 \tag{13.163}$$

$$= \frac{1}{2}CV_i^2\left(\frac{1}{2} + \frac{1}{2}\cos(2\omega t)\right). \tag{13.164}$$

Again a DC term and a double-frequency term. Hence the average stored energy is

$$\overline{W_C} = \frac{1}{4}CV_i^2. \tag{13.165}$$

A similar derivation for an inductor yields

$$\overline{W_L} = \frac{1}{4}LI_i^2 \tag{13.166}$$

for an inductor current of the form

$$i_L(t) = I_i \cos(\omega t). \tag{13.167}$$

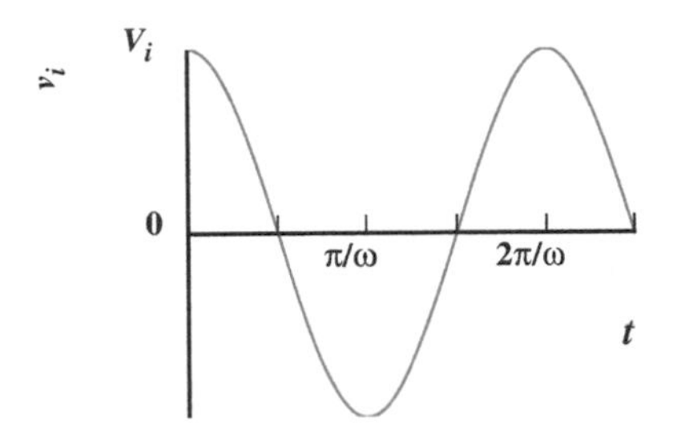

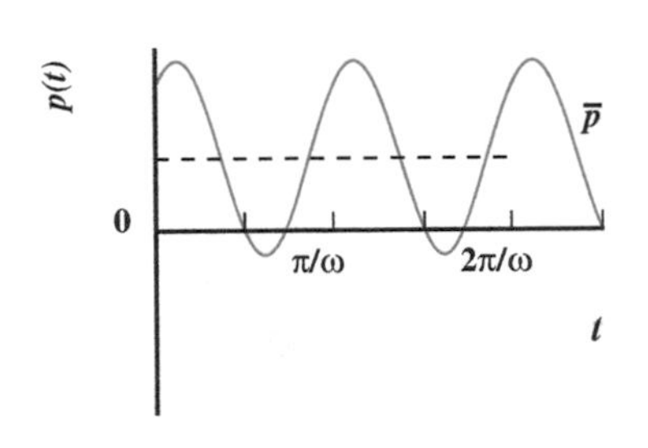


FIGURE 13.54 Power flow in an inductive circuit. The average power is given by $\overline{p} = 1/2 V_i^2/\sqrt{R^2+X^2}\cos\theta$. The maximum value of $p(t)$ is $1/2 V_i^2/\sqrt{R^2+X^2}(\cos\theta + 1)$, and the minimum value is $1/2 V_i^2/\sqrt{R^2+X^2}(\cos\theta - 1)$.

For the general case when the network contains resistors, capacitors, and inductors, the power flow will have some intermediate form between Figure 13.52 and Figure 13.53. Assuming that the circuit is net inductive at the frequency of interest, then $\theta$ is positive but less than $\pi/2$, and Equation 13.149 with $\phi = 0$ becomes

$$p(t) = \frac{1}{2}\frac{V_i^2}{\sqrt{R^2+X^2}}[\cos(2\omega t - \theta) + \cos\theta].$$

The power waveform is as depicted in Figure 13.54.

### 13.7.4 EXAMPLE: POWER IN AN RC CIRCUIT

Let us examine a specific circuit with both resistive and reactive components, the RC circuit of Figure 13.55. To calculate the average power from either Equation 13.151 or 13.152, we must find the complex amplitude of the current. By inspection of Figure 13.55:

$$I_i = \frac{V_i}{Z} = \frac{V_i}{R + 1/j\omega C} \tag{13.168}$$

$$= \frac{V_i}{\sqrt{R^2 + (1/\omega C)^2}} e^{-j\theta} \tag{13.169}$$

where

$$\theta = \tan^{-1}\left(\frac{1}{\omega RC}\right) \tag{13.170}$$

Now the average power dissipated in the circuit is, from Equation 13.151:

$$\overline{p} = \frac{1}{2}\frac{V_i^2}{\sqrt{R^2 + (1/\omega C)^2}}\cos(\theta) \tag{13.171}$$

$$= \frac{1}{2}\frac{V_i^2}{|Z|}\cos(\theta). \tag{13.172}$$

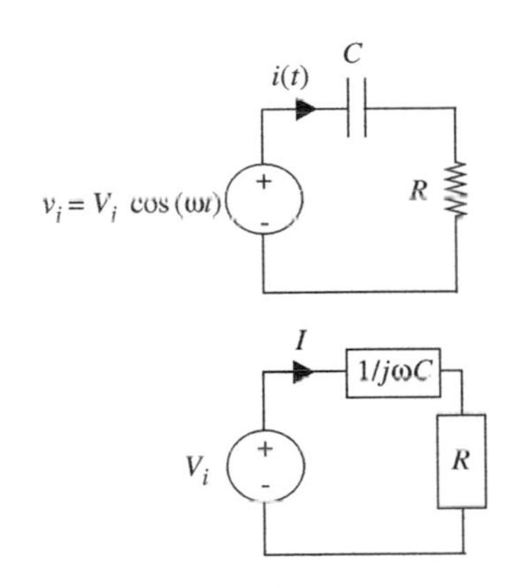


FIGURE 13.55 Series RC circuit.

Note from Equation 13.171 that if we choose $\omega$ such that

$$R = \frac{1}{\omega C} \tag{13.173}$$

that is, at the break frequency or corner frequency of the circuit, then

$$\bar{p} = \frac{1}{2}\frac{V_i^2}{2R}. \tag{13.174}$$

Hence the frequency $\omega = 1/RC$ is also called the *half-power frequency* of the impedance, because the average power is one half of the value found in Equation 13.151 for the capacitor being a short circuit. Note also that because there is no average power dissipated in the capacitor, the average power dissipated in the resistor must be identical to the average power supplied by the source to the impedance.

## WWW EXAMPLE 13.7 MAXIMIZING POWER TRANSFER USING A TRANSFORMER

## WWW EXAMPLE 13.8 NON-IDEAL TRANSFORMERS

## 13.8 SUMMARY

- Sinusoidal steady state is an important characterization of a linear system. It comprises a frequency response, which includes a gain plot and a phase plot as a function of frequency.
- The impedance approach provides an analysis of circuits for sinusoidal inputs, which complements the time-domain calculations of Chapter 12, by showing the behavior of the circuit as a function of frequency.
- By assuming complex exponential drives instead of sinusoidal drives for linear time-invariant circuits, the differential equations describing circuit behavior reduce to algebraic equations.
- These algebraic equations can be found directly by using impedance. The constituent relations for $R$, $L$, and $C$, relating complex amplitudes are

$$V = IR$$

$$V = Ls\,I$$

and

$$V = (1/Cs)\,I$$

where $s$ is a shorthand notation for $j\omega$. Accordingly, the impedance of an inductor is $sL$, that for a capacitor is $1/sC$ and that for a resistor is $R$.

- We extended our variable notation to distinguish between total variables, DC operating values, small-signal variables, and complex amplitudes.
  - We denote total variables with lowercase letters and uppercase subscripts, for example, $v_D$.
  - DC operating-point variables using all uppercase, for example, $V_D$.
  - Incremental values using all lowercase letters, for example, $v_d$.
  - Complex amplitudes use uppercase letters and lowercase subscripts, for example, $V_d$.
- Inductors behave like short circuits for DC (or very low frequencies) and like open circuits for very high frequencies. Capacitors behave like open circuits for DC (or very low frequencies) and like short circuits for very high frequencies.
- The steady state values of the real voltages or currents (functions of time) can be found from their corresponding complex amplitudes by multiplying the complex amplitude by $e^{j\omega t}$ and taking the real part. For example,

the steady state value of $v_c$ can be determined from the value of $V_c$ as

$$v_c = Re\left[V_c e^{j\omega t}\right]$$

or, equivalently,

$$v_c = |V_c|\cos(\omega t + \angle V_c).$$

- The impedance method allows us to determine with ease the steady-state response of any linear RLC network for a sinusoidal input. The method works with complex amplitudes of voltages and currents at its variables and has the following steps:

  1. First, replace the (sinusoidal) sources by their complex (or real) amplitudes. For example, the input voltage $v_A = V_a\cos(\omega t)$ is replaced by its amplitude $V_a$.
  2. Replace resistors by $R$ boxes, inductors by $Ls$ boxes, and capacitors by $1/Cs$ boxes. The resulting diagram is called the impedance model of the network.
  3. Now, determine the complex amplitudes of the voltages and currents in the circuit by any standard linear circuit analysis technique — Node method, Thévenin method, etc.
  4. Although this step is not usually not necessary, we can then obtain the time variables from the complex amplitudes.

- The frequency response characterizes the behavior of a network as a function of frequency. Frequency domain analysis of a network is carried out by examining the network's system function, which is the ratio of the complex amplitude of the network output to the complex amplitude of the input.

- A frequency response plot is a convenient way of summarizing how a network behaves as function of frequency. A frequency response plot has two graphics:

  - the log magnitude of the system function plotted against log frequency, and
  - the angle of the system function plotted against log frequency.

- $|H(j\omega)| = \omega$ plots as a straight line with slope of $+1$ in log space, given consistent horizontal and vertical scales.

  Correspondingly, $|H(j\omega)| = 1/\omega$ plots as a straight line with slope of $-1$ in log space.

- The frequency response for system functions arising from circuits containing a single storage element and a single (Thévenin) resistor is of the form

$1/(s+a)$, $(s+a)$, $s/(s+a)$, $(s+a)/s$, where $a$ is some constant. Such responses can be intuitively graphed as follows:

- The magnitude plot is sketched by drawing the low-frequency and the high-frequency asymptotes. The two asymptotes intersect at the break frequency $a$.
- The phase plot can also be graphed by sketching the low-frequency and the high-frequency asymptotes. At the break frequency, the phase will be $45°$ or $-45°$ as appropriate.

- The average power in terms of complex amplitudes of voltages and currents is one-half the product of the two magnitudes multiplied by the cosine of the angle between them.

## EXERCISES

EXERCISE 13.1 Find the magnitude and phase of each of the following expressions:

a) $(8 + j\,7)(5e^{j30°})(e^{-j39°})(0.3 - j\,0.1)$

b) $$\frac{(8.5 + j\,34)(20e^{-j25°})(60)(\cos(10°) + j\sin(10°))}{(25e^{j20°})(37e^{j23°})}$$

c) $(25e^{j30°})(10e^{j27°})(14 - j\,13)/(1 - j\,2)$

d) $(13e^{j30^{(15° + j1.5)}})(6e^{(1-j30°)})$

EXERCISE 13.2 Find the real and imaginary parts of the following expressions:

a) $(3 + j\,5)(4e^{j50°})(7e^{-j20°})$

b) $(10e^{j50°})(e^{j20°})$

c) $(10e^{j50°})(e^{j\omega t})$

d) $Ee^{j\omega t}$ where $E = |E|e^{j\Theta}$

EXERCISE 13.3 Find the system function $V_L/I$ for the network shown in Figure 13.62. Then find the response $v_l(t)$ for $i(t) = I\cos(\omega t)$ under steady-state conditions.

EXERCISE 13.4 Referring to Figure 13.63, given $i(t) = I_0 \cos(\omega t)$, where $I_0 = 3$ mA and $\omega = 10^6$ rad/s, determine $v(t)$ in the sinusoidal steady state. Assume $R = 1$ kΩ and $L = 1$ mH.

EXERCISE 13.5 The two-terminal linear network in Figure 13.64 is known to contain exactly two elements. The magnitude of the impedance function is as shown (log-log coordinates).

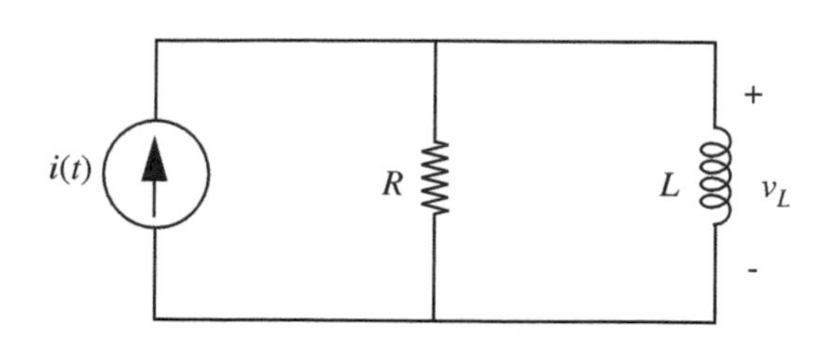


FIGURE 13.62

FIGURE 13.63

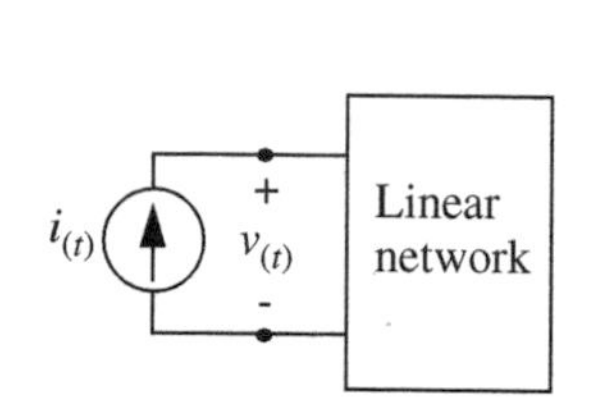


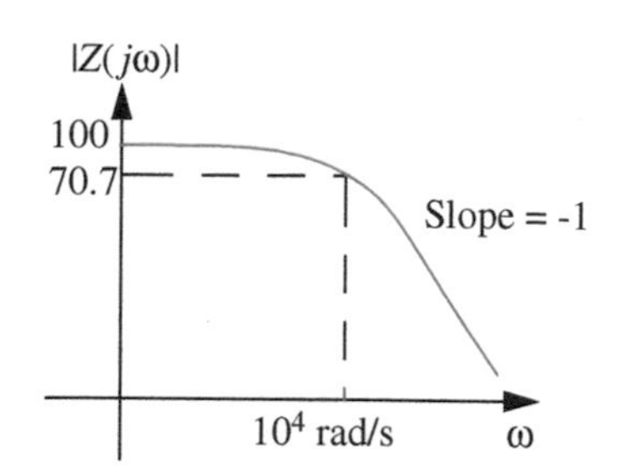


FIGURE 13.64

Draw a two-element circuit that has the impedance magnitude function indicated in your sketch. Specify the numerical value of each element.

EXERCISE 13.6 For each of the circuits shown in Figure 13.65, select the magnitude of the frequency response for the system function (that is, impedance, admittance,

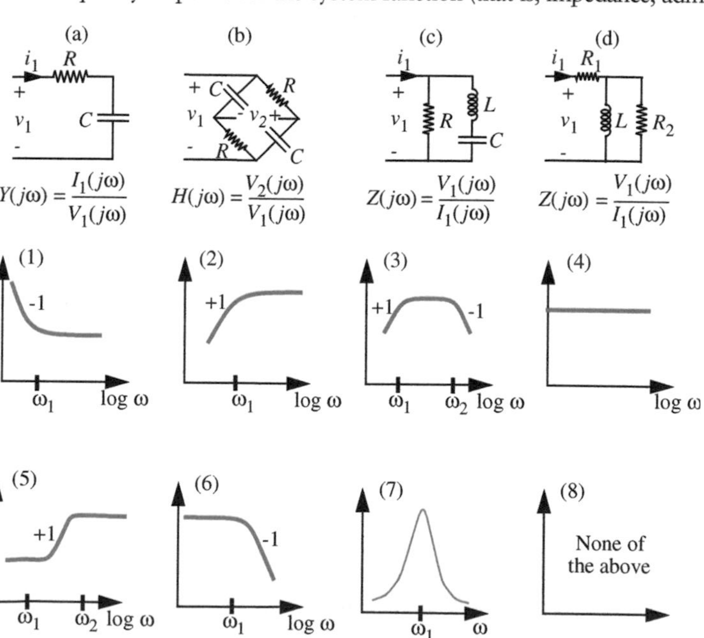


FIGURE 13.65

or transfer function) from those given. It is not necessary to relate the critical frequencies to the circuit parameters, and you may choose a magnitude response more than once.

Please note that the magnitude responses, except (7), are sketched on a log-log scale, with slopes labeled.

EXERCISE 13.7 A linear network is excited with a sinusoidal voltage $v_I(t) = \cos\left(t - 5\pi/8\right)$ for all time, as shown in Figure 13.66.

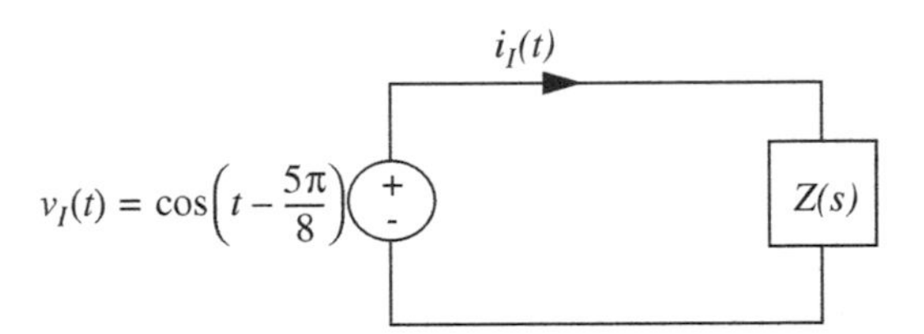


FIGURE 13.66

The current observed under the sinusoidal steady-state conditions is $i_I(t) = \sqrt{2}\sin(t + \pi/8)$.

What is $Z(s = j1)$, the impedance of the network at an excitation frequency of one radian per second?

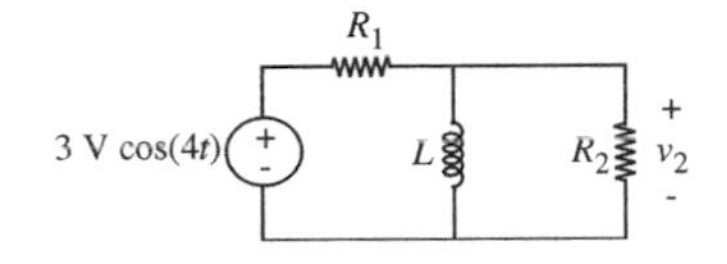


FIGURE 13.67

EXERCISE 13.8 Find $v_2(t)$ in the sinusoidal steady state in Figure 13.67. Assume $L = 10$ H, $R_1 = 120\ \Omega$, and $R_2 = 60\ \Omega$.

EXERCISE 13.9 A sinusoidal test signal is applied to a linear network that is constructed from exactly two circuit elements as shown in Figure 13.68.

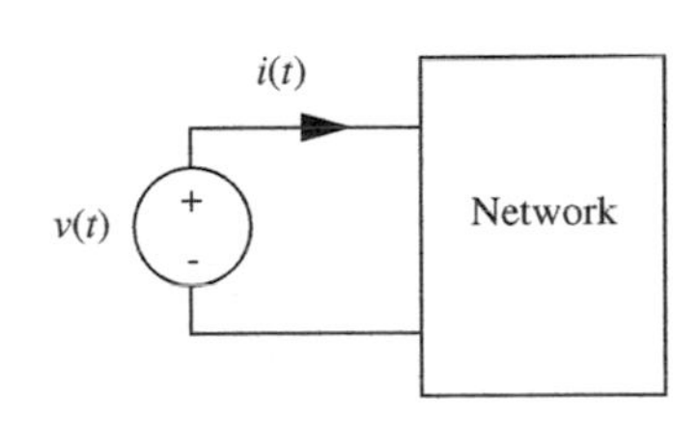


FIGURE 13.68

The magnitude portion of the Bode plot for the impedance $Z(j\omega) = V(j\omega)/I(j\omega)$ is shown in Figure 13.69.

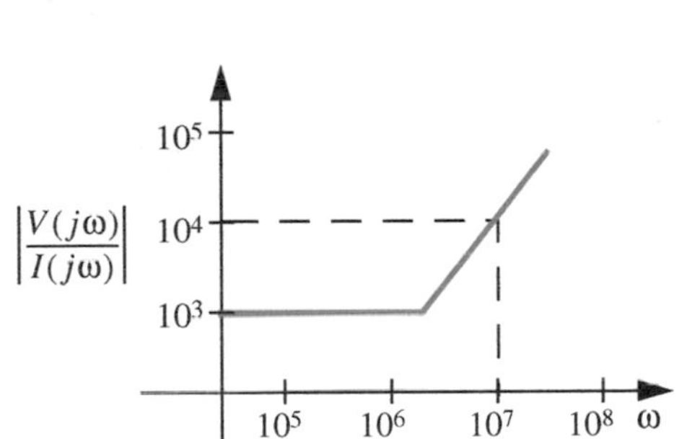


FIGURE 13.69

Draw the network and find the element values.

EXERCISE 13.10 The circuit shown in Figure 13.70 is a highly simplified model of a power transmission system.

FIGURE 13.70

$v_1(t)$ and $v_2(t)$ are the voltages of two power generators:

$v_1 = V\cos(\omega t)$ $\;v_2 = V\cos(\omega t + |\Phi)$

Find the Thévenin equivalent of this circuit at the terminals 1–2 in terms of a complex amplitude $V_{oc}$ and a complex Thévenin impedance $Z_{th}$.

EXERCISE 13.11 Write expressions for $H(j\omega) = V_o/V_i$, its magnitude $|H(j\omega)|$, and its phase angle $\angle H(j\omega)$, as a function of $\omega$ in the four cases shown in Figure 13.71.

(a)

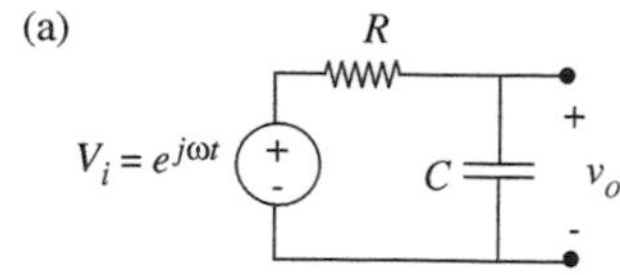


(b)



(c)

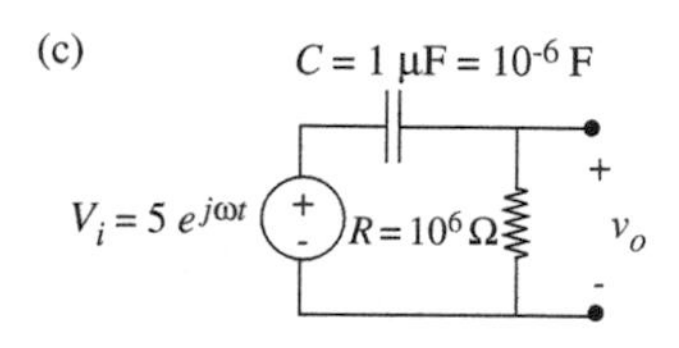


(d)

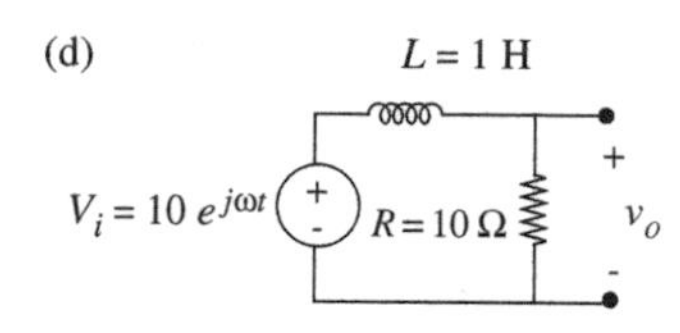


FIGURE 13.71

EXERCISE 13.12 Plot the log magnitude and the phase angle, both as functions of frequency (on a logarithmic scale), of the following complex quantity:

$$H(j\omega) = \frac{1 - j\omega}{1 + j\omega}.$$

Label all significant asymptotes, slopes, and break points.

EXERCISE 13.13 In the network shown in Figure 13.72,

$R = 1\ \text{k}\Omega$ $\;C_1 = 20\ \mu\text{F}$ $\;C_2 = 20\ \mu\text{F}$.

a) Determine the magnitude and phase of $H(j\omega)$, the transfer function relating $V_O/V_i$.

b) Given $v_i(t) = \cos(100t) + \cos(10000t)$, determine the sinusoidal steady state output voltage, $v_o(t)$.

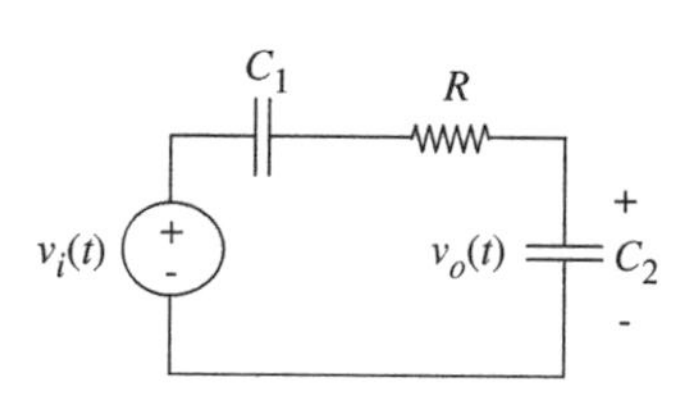


FIGURE 13.72

EXERCISE 13.14 Find $v_2(t)$ in the sinusoidal steady state for the circuit in Figure 13.73.

$L = 10$ H $\;R_1 = 120\ \Omega$ $\;R_2 = 60\ \Omega$

EXERCISE 13.15

a) Write the transfer function $V_o(s)/V_i(s)$ for the circuit in Figure 13.74.

b) Write the transfer function $I_a(s)/V_i(s)$.

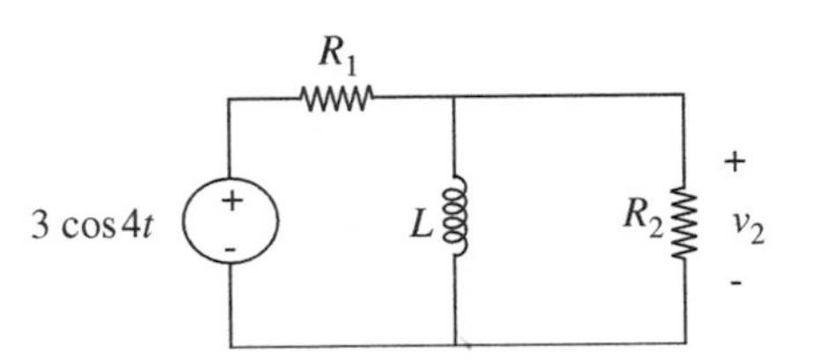


FIGURE 13.73

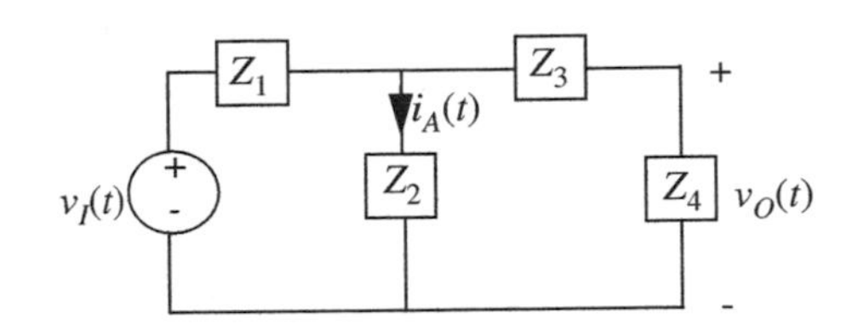


FIGURE 13.74

EXERCISE 13.16 Write the transfer functions $V_O(s)/V_i(s)$, $I_a(s)/V_i(s)$ in the circuit in Figure 13.75.

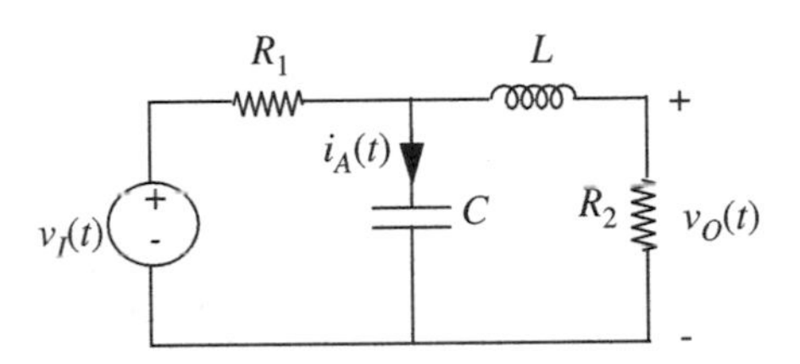


FIGURE 13.75

EXERCISE 13.17 Write the transfer function $I_a(s)/I_s(s)$ for the circuit in Figure 13.76.

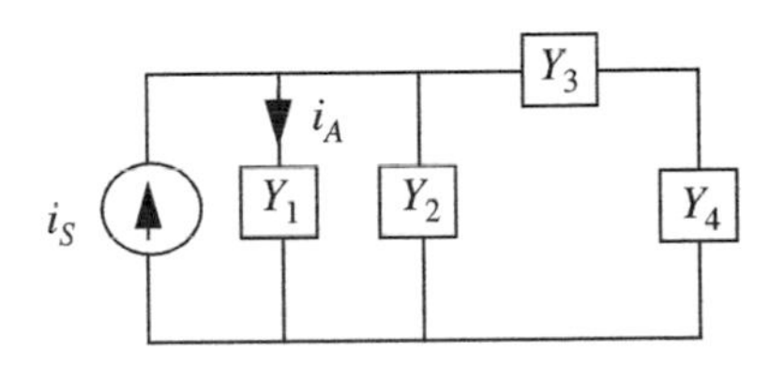


FIGURE 13.76

EXERCISE 13.18 Find $I_a/I_s$ in the circuit in Figure 13.77.

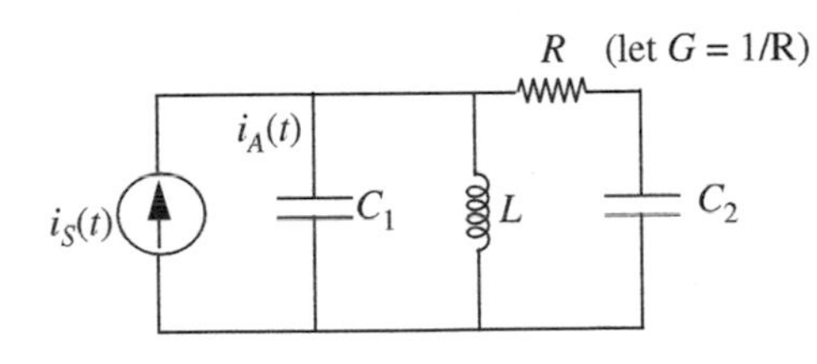


FIGURE 13.77

## PROBLEMS

PROBLEM 13.1 For each of the networks shown in Figure 13.78:

a) Determine an expression for the indicated complex impedance or transfer function.

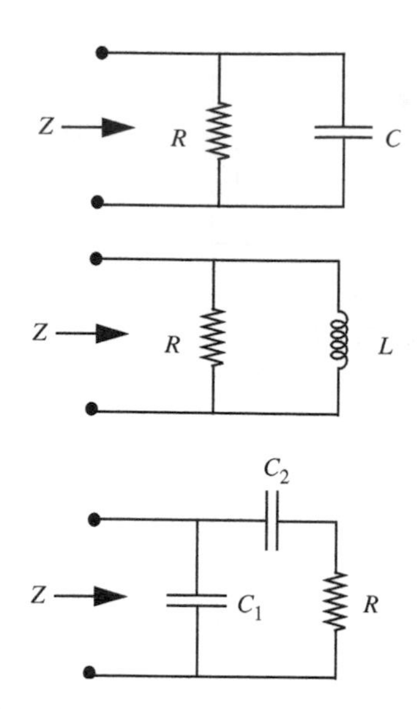


FIGURE 13.78

b) Sketch the magnitude and angle of the indicated quantity as a function of frequency. You may use either linear or log-log coordinates, but it is recommended that you learn to use both kinds of axes.

PROBLEM 13.2 Shown in Figure 13.79 is one possible circuit model for a transformer, for use where there can be a common ground between primary and secondary.

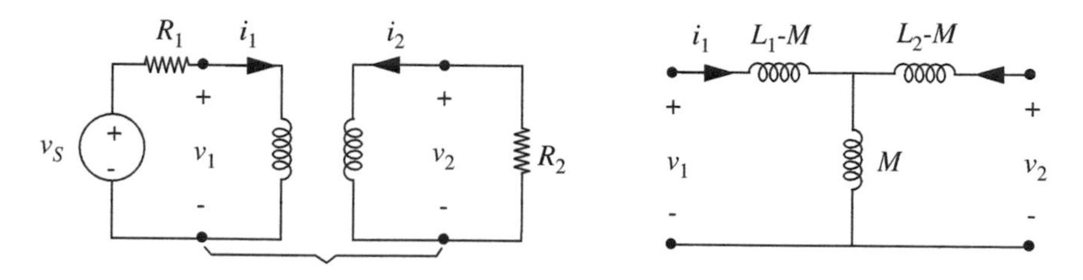


FIGURE 13.79

Assume:

$L_1 = 2.5$ H, $L_2 = 0.025$ H, $M = k\sqrt{L_1 L_2}$ where $k < 1$, $R_1 = 1$ kΩ, $R_2 = 10$ Ω.

a) Determine an expression for the sinusoidal steady-state transfer function $V_2/V_s$.

b) In the tight-coupling limit, $k \to 1$, the two natural frequencies are far apart. (See Problem 12.3 in the previous chapter.) For this specific case, sketch the magnitude and angle of the transfer function on log-log scales.

PROBLEM 13.3 An electrical system has the following transfer function:

$$H(j\omega) = \frac{Y(j\omega)}{X(j\omega)} = \frac{10^5(10 + j\omega)(1000 + j\omega)}{(1 + j\omega)(100 + j\omega)(10000 + j\omega)}.$$

a) Plot the magnitude of $H(j\omega)$ in decibels versus the logarithm of frequency, labeling all 3dB points.

b) Sketch the phase of $H(j\omega)$ versus the logarithm of frequency.

c) For what values of $\omega$ does the magnitude of $H(j\omega)$ equal $0db$? What is the relationship between the magnitudes of $X(j\omega)$ and $Y(j\omega)$ at these frequencies?

d) List the frequencies at which the phase of $H(j\omega)$ equals $-45$ degrees.

PROBLEM 13.4 Refer to Figure 13.80 for this problem. Assume $R_1 = 1$ kΩ and $L_1 = 10$ mH.

a) Find the transfer function $H(j\omega) = V_1/V_o$.

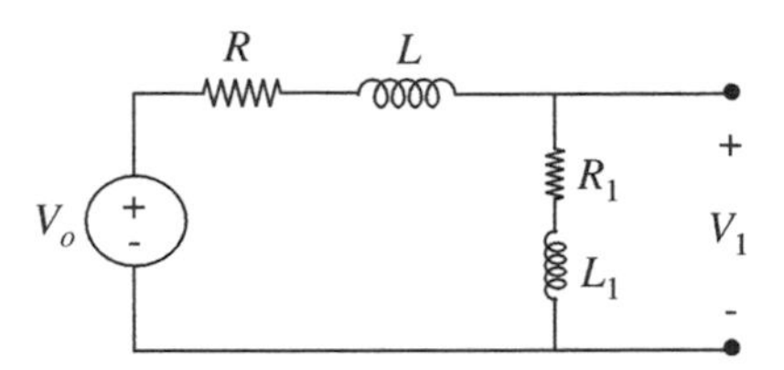


FIGURE 13.80

b) Find $R$ so that the $DC$ gain is 1/10.

c) Find a value of $L$ so that the response at high frequencies is equal to response at $DC$.

d) Plot $H(j\omega)$ (magnitude and phase) versus. log $\omega$ for the values of $R$ and $L$ found previously.

PROBLEM 13.5 This problem examines the simple doorbell circuit commonly used in homes (see Figure 13.81).

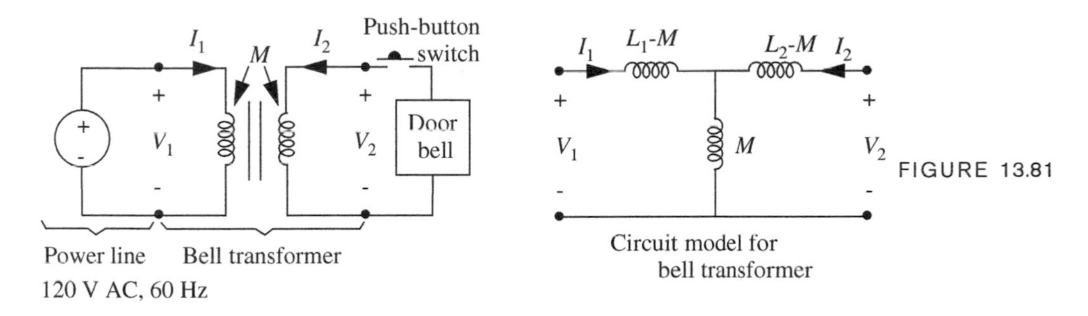


FIGURE 13.81

Data for the transformer in Figure 13.81 is:

$L_1 = 2.5$ H, $L_2 = .025$ H, $M = k\sqrt{L_1 L_2}$, where $k < 1$.

a) In the limit $k \simeq 1$, what is the voltage $V_2$ with the push-button switch not pressed (open)? You should use root-mean-square amplitudes for all quantities. The voltage source is given as 120-$V$ root-mean-square $M = 0.25$ H.

b) The doorbell operates by repetitive making and breaking of a contact and can normally be modeled as a 10 Ω resistance at 60 Hz. Determine the magnitude of the root-mean-square primary current $I_1$ under normal doorbell operation (push button closed, doorbell = 10 Ω) in the limit of $k \simeq 1$.

c) An important safety issue in such circuits is the prevention of fire in the event that the doorbell should accidently stick with its contact closed, thus becoming equal to a short circuit. This can be accomplished by adjusting the value of $k$. Find the value of $k$ that will limit the root-mean-square primary current to 500 mA for the case where the push button is pressed and the doorbell acts like a short circuit.

PROBLEM 13.6 In the circuit in Figure 13.82, the switch has been in Position (1) for a long time. At $t = 0$, the switch is moved instantly to Position (2). For the particular parameter values of this circuit, the complete output waveform for *all* time greater than zero is

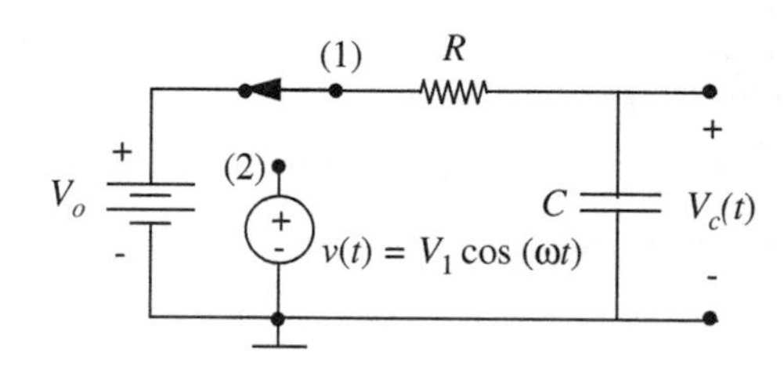


FIGURE 13.82

$$v_c(t) = |V_c| \cos(\omega t + \Phi)$$

a) Find $|V_c|$ and $\Phi$ in terms of $V_1, \omega, R$, and $C$.

b) Find $V_o$ in terms of $|V_c|, \omega, R$, and $C$ required to produce the $v_c(t)$ waveform.

# CHAPTER 14

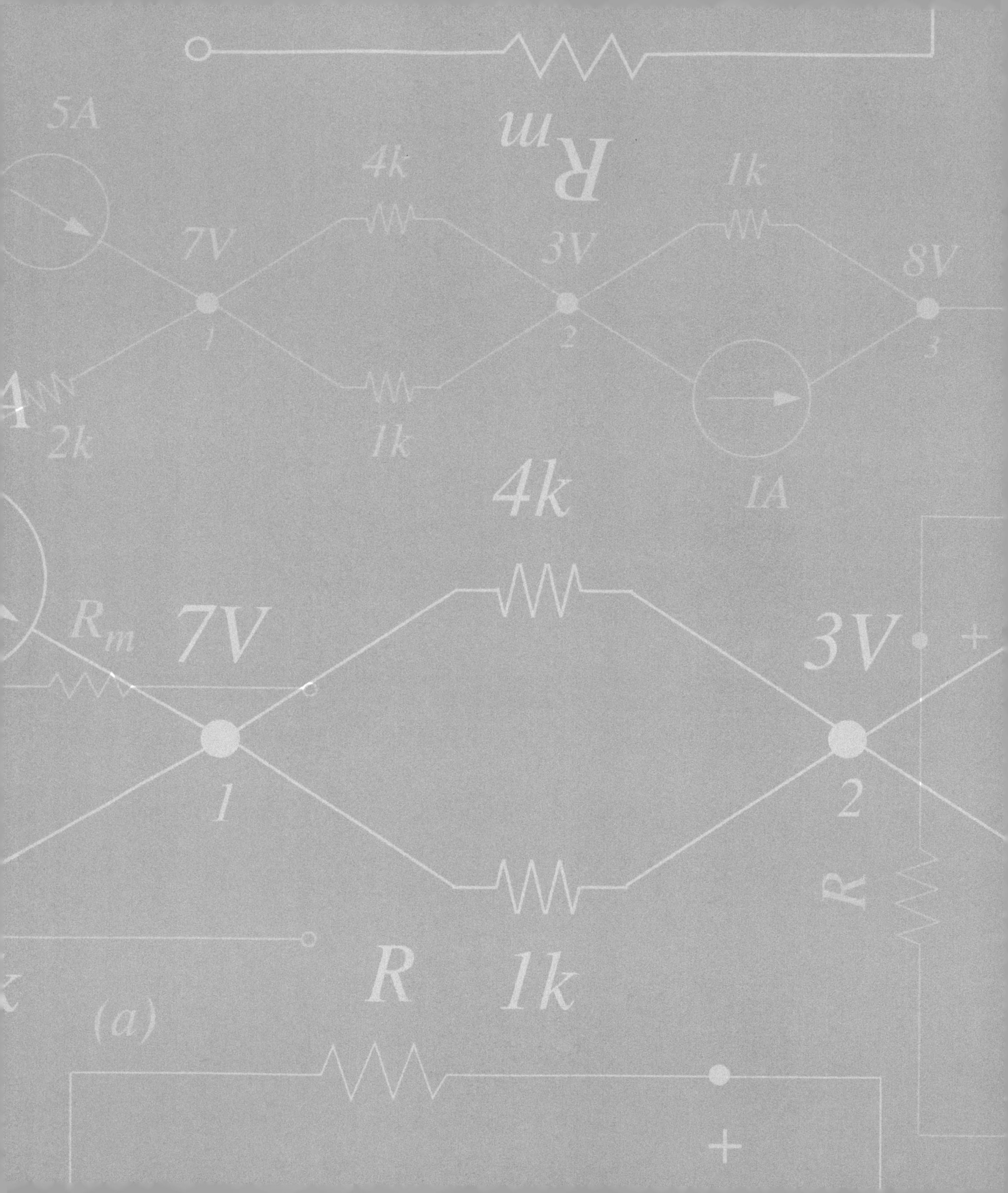

5A
4k
1k
7V
3V
8V
1
2
3
2k
1k
1A
4k
$R_m$
7V
3V
+
1
2
R
R
1k
(a)
+

# 14 SINUSOIDAL STEADY STATE: RESONANCE

Chapter 12 showed that circuits containing an inductor and a capacitor could display oscillatory behavior when the circuit was under-damped. This chapter will show that oscillatory behavior occurs when the system function has *complex roots.* Such systems are called *resonant systems.* The behavior of resonant circuits was characterized by parameters such as their quality factor and resonant frequency. This chapter revisits resonant circuits from an impedance and frequency point of view.

Resonant circuits are useful in analog design to build filters with high selectivity such as radio tuners and channel selectors in cell phones and wireless networks. Resonant circuits are also used to build oscillators to produce sinusoids of a given frequency. The same oscillators also form the basis of clock generators in digital design.

Many physical systems can also be modeled as second-order resonant circuits. Because the injection of even miniscule amounts of energy at or near the resonant frequency of a second-order resonant circuit can cause a massive and sustained response, physical structures such as buildings and bridges are carefully modeled and designed to avoid such responses. In fact, resonators excited at their resonance can be viewed as energy accumulators in the sense that they continuously extract and store energy from their excitation. In this case, the only limit to the amplitude of their response is internal dissipation or nonlinear behavior resulting from the stresses of large amplitude responses. The most notorious case of such a response is the Tacoma Narrows Bridge disaster. Alternating winds injected enough energy into the bridge structure at its resonant frequency that the entire bridge entered into resonance and began to sway back and forth and finally collapsed. Sections 14.5.2 and 14.6 will provide more insight into this type of resonant behavior.

## 14.1 PARALLEL RLC, SINUSOIDAL RESPONSE

The response of a second-order system, specifically a parallel RLC system as in Figure 14.1, to a brief pulse and to a step was calculated in Chapter 12. We now wish to examine this same system from the impedance point of view, to show how such circuits can be used as filters. In particular, we will discuss the factors that affect the selectivity of the filter. To tie back to the calculations in Chapter 12, let us first calculate the total time-domain response $v(t)$, when the

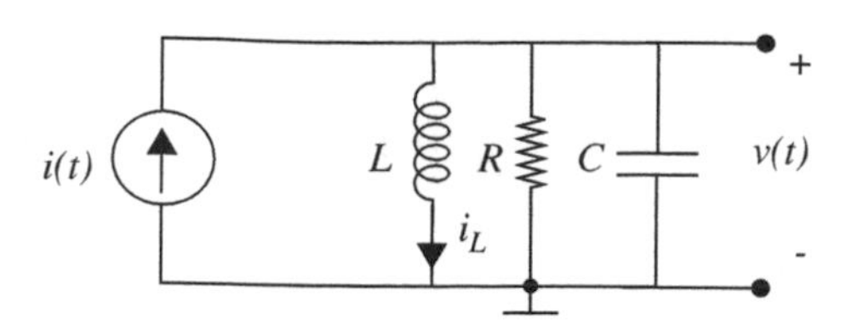


FIGURE 14.1 Parallel RLC circuit.

circuit is driven with a tone burst at some frequency $\omega_1$:

$$i(t) = I_o \cos(\omega_1 t) \quad \text{for} \quad t > 0. \tag{14.1}$$

We acknowledge that the total response to a sinusoid is rarely necessary; we are more often interested in just the forced (or particular) response to sinusoids. Accordingly, you may skip directly to Section 14.1.2 and then to Section 14.2 without loss of continuity. We, however, will plow through the total response for completeness.

Application of KCL to the top node gives

$$i(t) = C\frac{dv}{dt} + \frac{v}{R} + i_L \tag{14.2}$$

and the constituent relation for the inductor is

$$v = L\frac{di_L}{dt}. \tag{14.3}$$

Differentiating Equation 14.2, and substituting Equation 14.3, we obtain a second-order differential equation describing the system:

$$\frac{1}{C}\frac{di}{dt} = \frac{d^2v}{dt^2} + \frac{1}{RC}\frac{dv}{dt} + \frac{1}{LC}v. \tag{14.4}$$

As in the past, we will solve this differential equation by summing together the homogeneous and particular solutions, $v_h$ and $v_p$, respectively.

### 14.1.1 HOMOGENEOUS SOLUTION

The homogeneous solution for this equation was worked out in detail in Chapter 12, so will be only briefly reviewed here. The homogeneous equation is

$$\frac{d^2v}{dt^2} + \frac{1}{RC}\frac{dv}{dt} + \frac{1}{LC}v = 0. \tag{14.5}$$

Assuming a homogeneous solution of the form:

$$v_h = Ke^{st} \tag{14.6}$$

the characteristic equation is

$$s^2 + \frac{1}{RC}s + \frac{1}{LC} = 0 \tag{14.7}$$

which, to simplify notation, is written in canonic form as

$$s^2 + 2\alpha s + \omega_o^2 = 0 \tag{14.8}$$

where

$$\omega_o^2 = \frac{1}{LC}$$

and

$$\alpha = \frac{1}{2RC}.$$

We saw in Chapter 12 that the system is resonant, that is, displays oscillatory behavior, when it is under-damped. As further discussed in Section 12.2.1, under-damped systems are characterized by the condition:

$$\omega_o > \alpha. \tag{14.9}$$

Since we are focusing on resonant systems in this chapter, we will assume that the system is under-damped, that is, $\omega_o > \alpha$. Under this assumption, the characteristic equation has these two roots:

$$s_a = -\alpha + j\omega_d \tag{14.10}$$

$$s_b = -\alpha - j\omega_d \tag{14.11}$$

where

$$\omega_d^2 = \omega_o^2 - \alpha^2. \tag{14.12}$$

It follows from our assumption ($\omega_o > \alpha$) that the two roots identified in Equations 14.10 and 14.11 are complex. Furthermore, the roots form a complex conjugate pair. In other words, resonant systems are characterized by a pair of complex conjugate roots.

Hence the homogeneous solution is

$$v_h(t) = e^{-\alpha t}\left[K_a e^{j\omega_d t} + K_b e^{-j\omega_d t}\right] \tag{14.13}$$

$$= Ke^{-\alpha t}\cos\left(\omega_d t + \theta\right) \tag{14.14}$$

where $K$ and $\theta$ are constants to be determined from the two initial conditions, the inductor current and the capacitor voltage before the tone burst, after the expression for the total solution has been written.

### 14.1.2 PARTICULAR SOLUTION

Now let us find the particular solution for this system using the impedance approach. One possible particular solution is the steady-state response of the system $v_p(t)$ to the cosine signal $I_o \cos(\omega_1 t)$. The impedance model derived from the original circuit, Figure 14.1, is shown in Figure 14.2. The complex constants $I_o$ and $V_p$ in Figure 14.2 are related to the original time variables by the expressions:

$$i(t) = Re\left[I_o e^{s_1 t}\right] = I_o \cos \omega_1 t \tag{14.15}$$

$$v_p = Re\left[V_p e^{s_1 t}\right]. \tag{14.16}$$

Direct application of KCL to the top node in the impedance model yields

$$I_o = \frac{V_p}{Ls_1} + \frac{V_p}{R} + \frac{V_p}{1/Cs_1}. \tag{14.17}$$

Solving for $V_p$ we find

$$V_p = \frac{I_o}{1/Ls_1 + 1/R + Cs_1} \tag{14.18}$$

$$= \frac{I_o s_1/C}{s_1^2 + s_1/RC + 1/LC}. \tag{14.19}$$

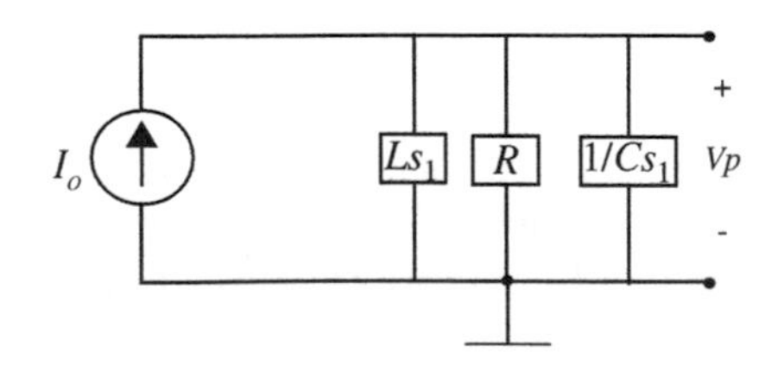


FIGURE 14.2 Impedance model, parallel RLC.

Note that we could have obtained the characteristic equation, Equation 14.7, without writing the differential equation by using the denominator of this system function.

The particular solution for this system can now be found from Equations 14.16 and 14.18:

$$v_p(t) = |V_p| \cos\left(\omega_1 t + \angle V_p\right). \tag{14.20}$$

This example is continued in Section 14.1.3 where the homogeneous and particular solutions are added to derive the total solution.

### 14.1.3 TOTAL SOLUTION FOR THE PARALLEL RLC CIRCUIT

Now we are in a position to calculate the total solution, or the complete time function $v(t)$, which is the capacitor voltage in response to the cosine tone burst. The complete solution for the cosine tone burst drive is the sum of the homogeneous solution $v_h$ (Equation 14.14), and the particular solution $v_p$ (Equation 14.20):

$$v(t) = Ke^{-\alpha t} \cos\left(\omega_d t + \theta\right) + |V_p| \cos\left(\omega_1 t + \angle V_p\right) \tag{14.21}$$

where the constants $K$ and $\theta$ are chosen to match the initial conditions for $i_L$ and $v_C$.

Equation 14.21 gives the complete response for our parallel circuit for a cosine tone burst of frequency $\omega_1$. As $t$ becomes large, the first term dies away, and only the second cosine term with frequency $\omega_1$ remains. Accordingly, the second term is the steady-state response to a cosine of frequency $\omega_1$.

Equation 14.21 further shows that, in general, the two cosine terms in this expression are not at the same frequency. The first term is the natural response, at a frequency $\omega_d$, the damped natural frequency of the system. The second term is the forced or driven response, at the frequency $(\omega_1)$ of the input signal. Thus one would expect interference or beating to occur between these two components. Computer generated plots of $v(t)$ (see Figure 14.3) from Equation 14.21 show the interference effect clearly for $\omega_1 \approx \omega_d$. Observation of the response of any high Q resonant circuit[1] to a step cosine slightly off the resonant frequency will reveal such interference.

1. Those of you taken aback by the sudden reappearance of Q (introduced in Section 12.4.1, Equation 12.65) take heart from the fact that we will have a lot more to say about Q from a frequency point of view in Section 14.2. For now, it suffices to understand that for a high Q resonant circuit the value of the damping factor $\alpha$ is small, and therefore the natural response of the circuit will last for a long time.

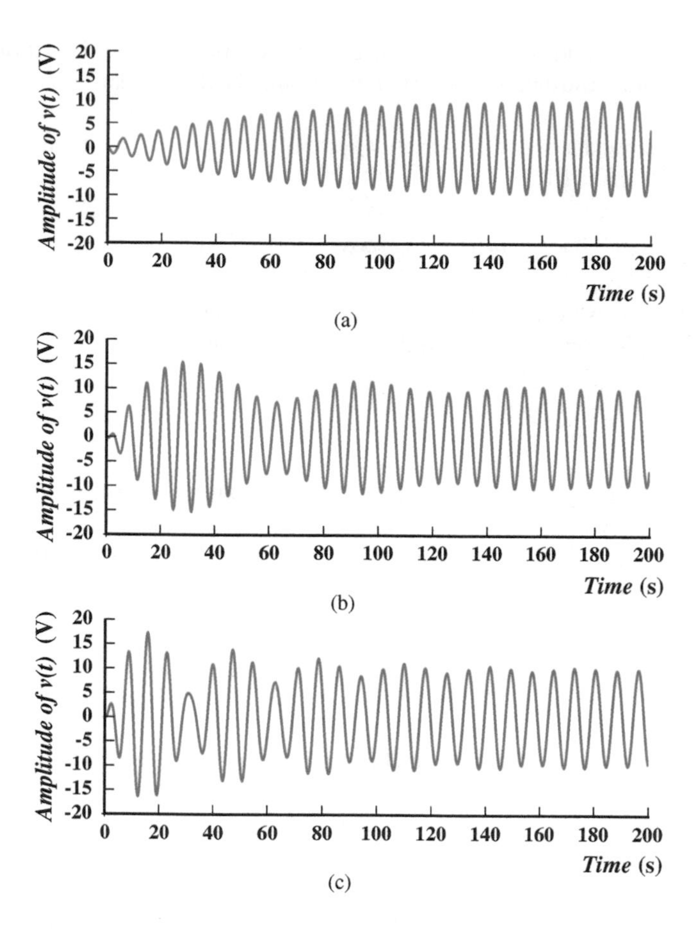


FIGURE 14.3 Computer calculation of the response $v(t)$ to a step cosine: (a) drive at $\omega_d$, ($\omega_1 = \omega_d = 1$ rad/s); (b) drive a little below $\omega_d$, ($\omega_d = 1$ rad/s, $\omega_1 = 0.9$ rad/s); (c) drive frequency lower still, ($\omega_d = 1$ rad/s, $\omega_1 = 0.8$ rad/s).

Next, Figure 14.4 shows plots of $v(t)$ when $\omega_1 \ll \omega_d$ and $\omega_1 \gg \omega_d$. When the drive frequency is very low (for example, Figure 14.4b), the response looks almost like that for a step.

This section analyzed the total response of the resonant circuit. However, as mentioned earlier, we tend to be less interested in the total response, and more concerned with the particular solution or the *steady-state response.* Accordingly, Section 14.2 will analyze the steady state response of Equation 14.19 in more detail. It will also show how to draw the frequency response plot for quadratic roots, and revisit the parameters $Q$, $\alpha$, $\omega_o$, and $\omega_d$, the stalwarts of second-order systems, from a frequency point of view. The frequency response plot of our parallel resonant circuit will show vividly the filtering that occurs when a signal is passed through the resonant circuit.

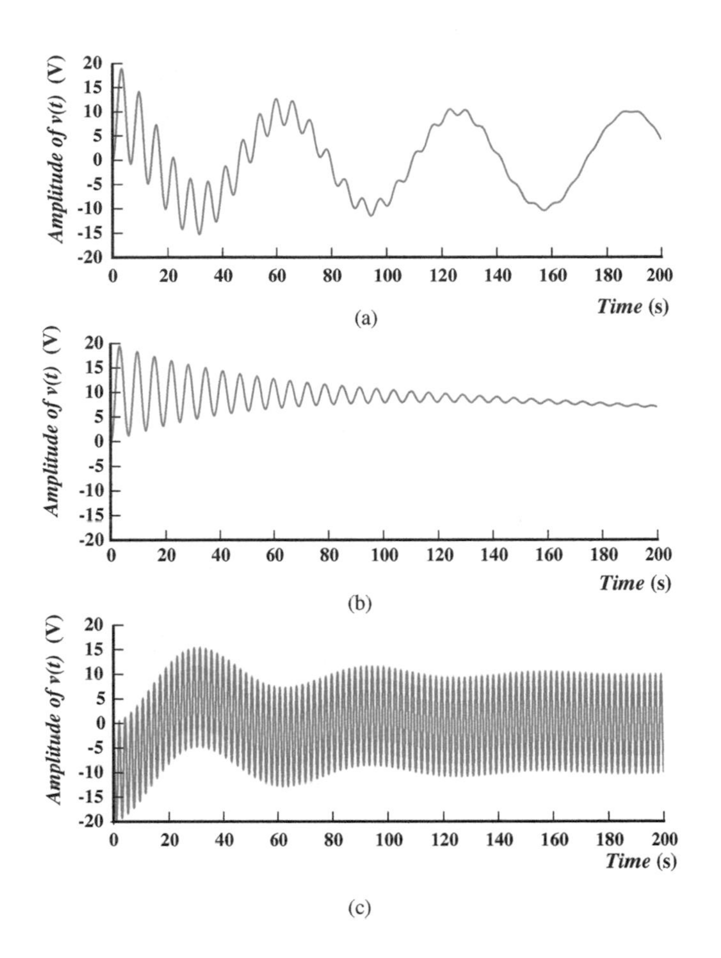


**FIGURE 14.4** Computer calculation of the response $v(t)$ to a step cosine: (a) drive at a much lower frequency than $\omega_d$, that is, $\omega_1 \ll \omega_d$, ($\omega_d = 1$ rad/s, $\omega_1 = 0.1$ rad/s); (b) drive near DC, ($\omega_d = 1$ rad/s, $\omega_1 = 0.004$ rad/s); (c) drive frequency much greater than $\omega_d$, ($\omega_d = 0.1$ rad/s, $\omega_1 = 3$ rad/s).

## 14.2 FREQUENCY RESPONSE FOR RESONANT SYSTEMS

The previous section determined the total response of a parallel RLC circuit for a sinusoidal input by solving its differential equation, which tended to be a fairly grungy calculation. But as noted previously, we tend to be more interested in the steady-state response of circuits to sinusoidal inputs. As introduced in Chapter 13, plotting the frequency response is a convenient way of visualizing how the circuit responds in the steady state to sinusoids of various frequencies. The frequency response of a network is examined by plotting two graphics:

- the log magnitude of the system function of the network plotted against log frequency, and
- the angle of the system function plotted against log frequency.

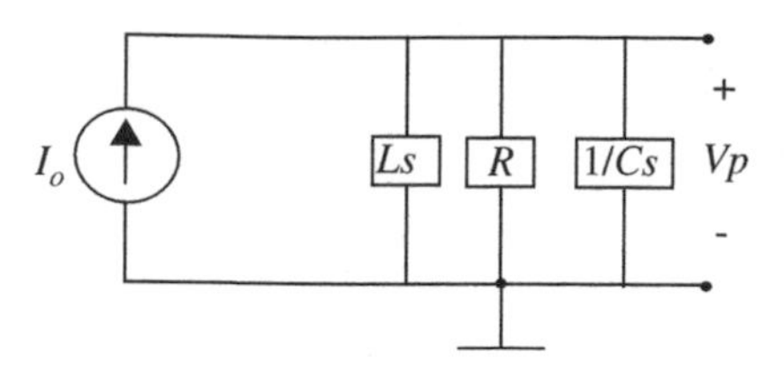


FIGURE 14.5 Impedance model, parallel RLC.

A network's system function is the ratio of the complex amplitude of the output to the complex amplitude of the input.

Let us study the frequency response of our parallel RLC resonant circuit. Its impedance model is shown in Figure 14.5. Its system function can be written by inspection using the impedance method as illustrated in Section 14.1.2. Referring to Figure 14.5, we can write the following expression relating the output $V_p$ to the input $I_o$:

$$V_p = \frac{I_o}{1/Ls + 1/R + Cs}.$$

Thus, the system function is given by

$$H(s) = \frac{V_p}{I_o} = \frac{1}{1/Ls + 1/R + Cs} \tag{14.22}$$

$$= \frac{s/C}{s^2 + s/RC + 1/LC.} \tag{14.23}$$

Notice that the denominator of Equation 14.23, which computes the steady-state response of our parallel RLC circuit, and is the characteristic equation, yields a pair of complex roots under certain conditions. From Equation 14.9, we know that the roots are complex when

$$\omega_o > \alpha$$

or specifically, when

$$\sqrt{\frac{1}{LC}} > \frac{1}{2RC}$$

and we expect the frequency response to look substantially different from those we have seen thus far. The rest of this section will focus on the frequency response of system functions with complex roots. This discussion will expand the repertoire of system functions we have seen thus far, which included both first- and second-order system functions with real roots (Chapter 13).

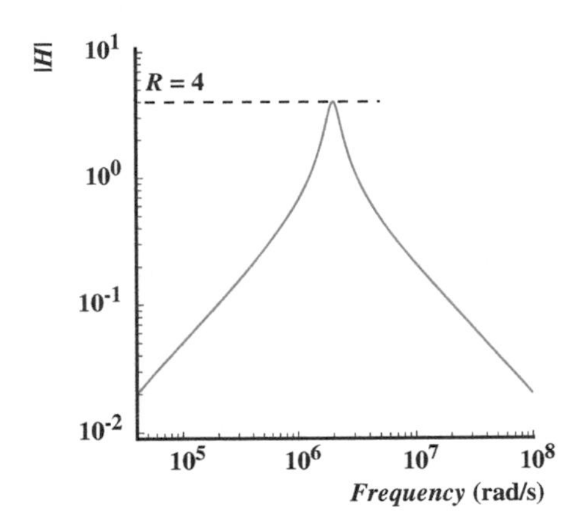


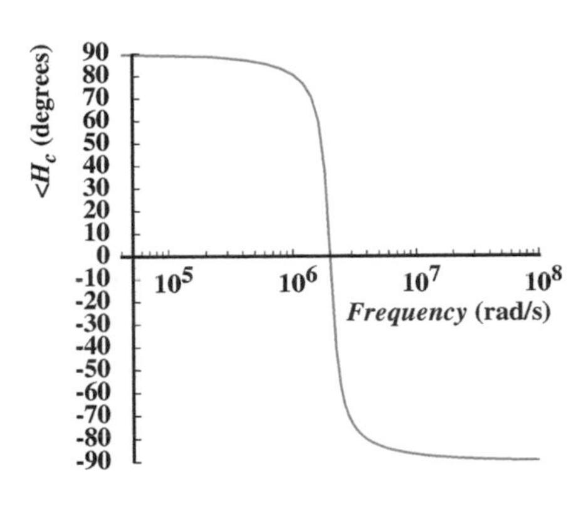


**FIGURE 14.6** Computer-generated plot of the magnitude and phase of the frequency response.

For concreteness, let us examine the frequency response of the circuit for the following element values:

$$L = 0.5\ \mu\text{H}$$
$$C = 0.5\ \mu\text{F}$$
$$R = 4\ \Omega.$$

For these element values, the system function becomes

$$H(s) = \frac{2 \times 10^6 s}{s^2 + 0.5 \times 10^6 s + 4 \times 10^{12}}. \tag{14.24}$$

For the element values that we have chosen, the denominator of Equation 14.24 indeed yields a pair of complex roots, and hence the system is resonant. Figure 14.6 shows a computer-generated plot of the corresponding frequency response. The magnitude plot clearly displays the frequency sensitive behavior of the circuit: Both low and high frequencies are attenuated, giving this response the characteristics of a bandpass filter.

More interestingly, observe the behavior of both the magnitude and phase plots at the frequency $2 \times 10^6$ rads/sec. At this frequency, the magnitude plot displays a sharp peak and the phase plot shows an abrupt phase transition. The rest of this section will provide more insight into this response, and will discuss how we can quickly guess the form of the frequency response for system functions with complex roots. Section 14.2.1 will further show that the sharp transitions of the magnitude and phase at the resonant frequency are directly related to the complex roots. Section 14.4 goes on to show how we can sketch the complete form of the response of resonant system functions without the use of a computer.

To obtain insight into resonant system functions, let us start by examining the system function from Equation 14.23, repeated here for convenience:

$$H(s) = \frac{V_p}{I_o} = \frac{s/C}{s^2 + s/RC + 1/LC}.$$

Observe that the denominator of the system function can be written in the following standard form for second-order systems:

$$s^2 + 2\alpha s + \omega_o^2 \tag{14.25}$$

where

$$\omega_o = \sqrt{\frac{1}{LC}} \tag{14.26}$$

and

$$\alpha = \frac{1}{2RC} \tag{14.27}$$

which is the same as the characteristic polynomial for second-order systems (see Equation 12.85 from Chapter 12). Depending on the relative values of $\alpha$ and $\omega_o$, the roots of this second-order polynomial will be real or complex. Shortly, we will show that the behavior of the system response depends heavily on the nature of these roots, and that the values of $\omega_o$ and $\alpha$ provide substantial insight into the form of the frequency response plot. We will also study the correspondence between the frequency domain interpretation of $\alpha$ and $\omega_o$ and our previous time-domain interpretation of Chapter 12.

As a first step towards obtaining some insight into the relationship between the system function and the shape of the frequency response, we divide Equation 14.25 throughout by $s/C$ and rewrite as

$$H(s) = \frac{1}{1/Ls + 1/R + Cs}. \tag{14.28}$$

To simplify further, we set $G = 1/R$ in Equation 14.28 and write $s$ as $j\omega$ and $1/j\omega$ as $-j/\omega$, which yields

$$H(j\omega) = \frac{1}{G + j\left(\omega C - 1/\omega L\right)}. \tag{14.29}$$

Certain features are already obvious without recourse to any complicated calculations. At one particular frequency, the L and C terms in the denominator will cancel, and $|H|$ will be maximum. This cancellation occurs where

$$\omega C = 1/\omega L. \tag{14.30}$$

At this frequency,

$$\omega = \omega_o = \frac{1}{\sqrt{LC}}. \tag{14.31}$$

This frequency is called the *resonant frequency* $\omega_o$ of the system. In the context of the homogeneous solution in the time domain developed in Chapter 12, this was called the undamped resonant frequency. Notice also that this resonant frequency is none other than the $\omega_o$ obtained by writing the system function in standard form as in Equation 14.25.

In our example, $L = 0.5\ \mu\text{H}$ and $C = 0.5\ \mu\text{C}$, so

$$\omega_o = \sqrt{\frac{1}{LC}} = 2 \times 10^6 \text{ rad/s.}$$

The peak in the value of $H(s)$ occurs at this frequency, as can be verified by looking at the magnitude plot in Figure 14.6.

Next, let us focus on the behavior of the circuit at this resonant frequency. At the resonant frequency,

$$|H(j\omega_o)| = R \tag{14.32}$$

so the complex amplitude of the capacitor voltage simplifies to

$$V_p = I_o/G = I_o R. \tag{14.33}$$

The capacitor voltage (time function) at this particular frequency is thus

$$v_p(t) = I_o R \cos\left(\omega_o t\right). \tag{14.34}$$

Thus at resonance the effect of the inductor cancels out the effect of the capacitor, and the *circuit behaves like a pure resistor.*

Put another way, the parallel connection of the inductor and capacitor offers infinite impedance to an input signal whose frequency is $\omega = \omega_o$.

Next, let us study the behavior of the circuit for very small and very large value frequencies. For $\omega$ very small, Equation 14.29 indicates that

$$H(j\omega) \simeq j\omega L \tag{14.35}$$

or that

$$V_p \simeq j\omega L I_o \tag{14.36}$$

and hence

$$v_p(t) \simeq wLI_o \cos\left(\omega t + \pi/2\right). \tag{14.37}$$

That is, the *circuit behaves like an inductor.* This is not surprising since the effect of the smallest impedance in a parallel circuit dominates. Relating back to the frequency response, Equation 14.35 further implies that the low-frequency asymptote of the magnitude plot will resemble that of an inductor (see Figure 13.9).

Similarly, for $\omega$ very large, we find that

$$H(j\omega) \simeq \frac{1}{j\omega C} \tag{14.38}$$

or that

$$V_p \simeq \frac{I_o}{j\omega C} \tag{14.39}$$

and

$$v(t) \simeq \frac{I_o}{\omega C} \cos\left(\omega t - \pi/2\right) \tag{14.40}$$

and *the circuit appears to contain only a capacitor.* In terms of the frequency response, Equation 14.38 implies that the high frequency asymptote of the magnitude plot will resemble that of a capacitor (see Figure 13.9).

At this point, although we do not as yet have a complete understanding of the frequency response (for example, the cause of the peakiness of certain second-order system functions), we know enough to develop the form of the response plot by identifying a few constraints. Specifically, Equations 14.32, 14.35 and 14.38 establish the basic structure of the frequency response, Figure 14.7. As illustrated in the figure, the system function magnitude has low-frequency and high-frequency asymptotes of $\omega L$ and $1/\omega C$, respectively. From Equations 14.29 and 14.30, these asymptotes intersect at $\omega_o$, the resonant frequency. Further, we know at frequency $\omega = \omega_o$, the system function $H(j\omega) = R$. These three constraints are shown in Figure 14.7. Together, the three constraints point to the form of the actual magnitude curve as shown in Figure 14.8.

The phase of $H(j\omega)$ is also relatively easy to guess. From Equations 14.35 and 14.38, it is easy to see that the low- and high-frequency asymptotes of the

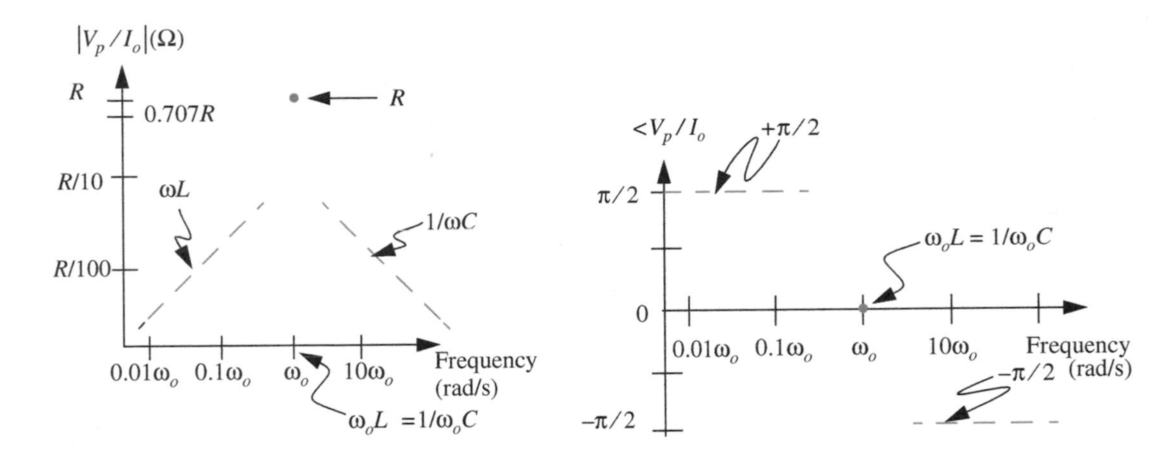


FIGURE 14.7 Frequency response asymptotes for parallel RLC circuit.

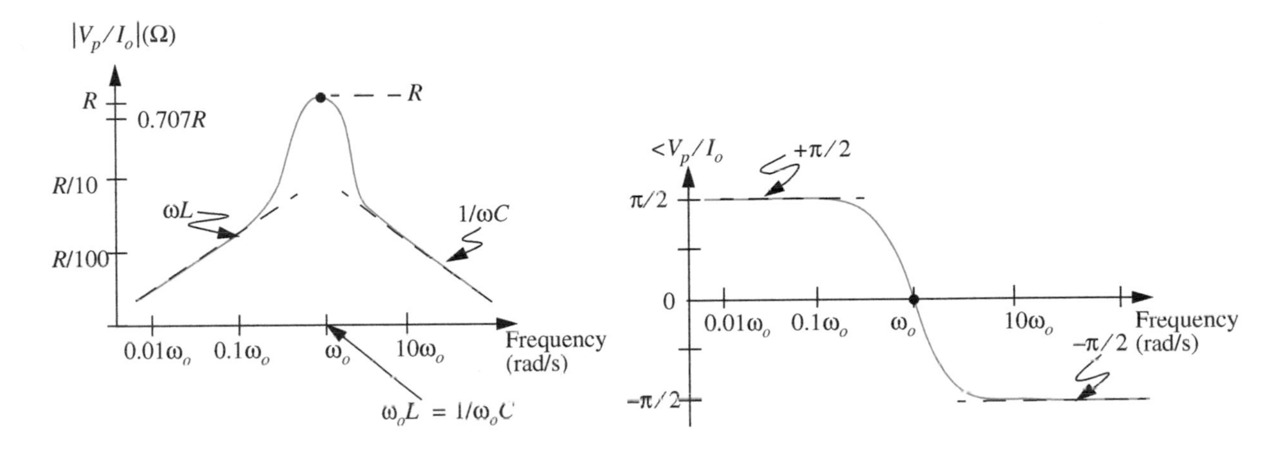


FIGURE 14.8 Form of the frequency response for parallel RLC circuit.

phase are +90° and −90°, respectively. Furthermore, at resonance, since

$$H(j\omega_o) = R$$

the angle is zero. These three phase constraints are shown in Figure 14.7. Compare the three phase constraints in Figure 14.7 with the actual curve drawn in Figure 14.8.

The process discussed here for guessing the form of the frequency response for Equation 14.23 generalizes to other resonant systems and can be summarized as follows:

- Magnitude Plot Constraints
  1. Mark the low-frequency asymptote.
  2. Mark the high-frequency asymptote.
  3. Mark $|H(j\omega_o)|$, the magnitude of the system function at the frequency $\omega_o$. The frequency $\omega_o$ can be determined by writing the system function in standard form (Equation 14.25).
- Phase Plot Constraints
  1. Mark the low-frequency asymptote.
  2. Mark the high-frequency asymptote.
  3. Mark $\angle H(j\omega_o)$, the angle of the system function at the frequency $\omega_o$.

---

EXAMPLE 14.1 TRANSFER FUNCTION FOR PARALLEL RLC CIRCUIT Determine the transfer function $H_c = V_c/I$ for the parallel RLC circuit shown in Figure 14.9 given that

$$i(t) = 0.1 \text{ A} \cos(2\pi f t)$$

$$L = 0.1 \text{ mH}$$

$$C = 1\ \mu\text{F}$$

$$R = 10\ \Omega$$

Sketch the asymptotes of the magnitude and phase of the frequency response. Determine the values of $Q$, $\omega_o$, $\alpha$, the two $\omega.707$ frequencies, and the bandwidth. Write the time-domain expression for the steady state value of $v_c$ for $f = 1$ MHz.

We can obtain $V_c$ by multiplying the current $I$ by the impedance of the parallel $R$, $L$, $C$ combination as:

$$V_c = I \frac{1}{1/Ls + 1/R + sC}.$$

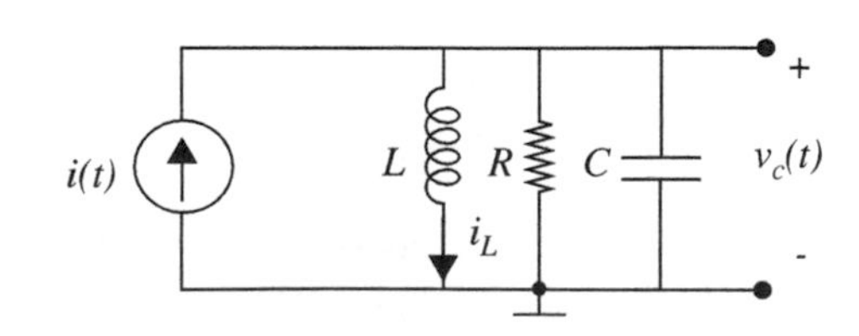


FIGURE 14.9 Parallel resonant circuit example.

The transfer function is given by

$$H_c = \frac{V_c}{I} = \frac{1}{1/Ls + 1/R + sC}$$

or in standard form,

$$H_c = \frac{s/C}{1/LC + s/RC + s^2}.$$

Substituting the actual element values,

$$H_c = \frac{10^6 s}{10^{10} + 10^5 s + s^2}. \tag{14.41}$$

The frequency response obtained by substituting $s = j\omega$ is

$$H_c(j\omega) = \frac{j10^6\omega}{(10^{10} - \omega^2) + j10^5\omega}.$$

To determine the form of the magnitude plot of the frequency response, we must find the low- and high-frequency asymptotes and the value of the response at $\omega_o$. The low-frequency asymptote is given by

$$H_c(j\omega) = \frac{j\omega}{10^4}$$

and the high-frequency asymptote is

$$H_c(j\omega) = \frac{10^6}{j\omega}.$$

Comparing the denominator of Equation 14.41 to the canonic form $s^2 + 2\alpha s + \omega_o^2$, we get

$$\omega_o = 10^5$$

and

$$|H_c(j\omega_o)| = 10.$$

The corresponding low- and high-frequency phase asymptotes are 90°, and −90°, respectively. The phase at $\omega_o$ is 0°.

The dashed lines in the plots in Figure 14.10 show the low- and high-frequency asymptotes, and the $X$ symbols mark the values at $\omega_o$. The solid lines show the actual plots.

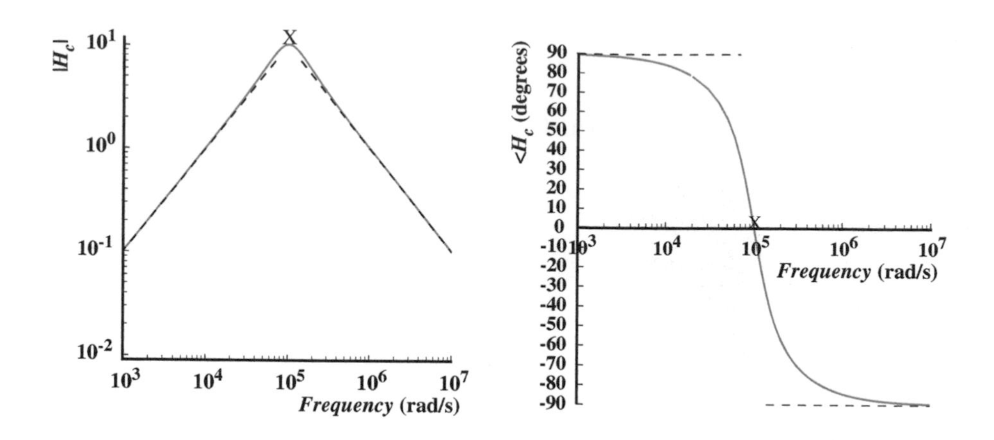


FIGURE 14.10 Magnitude and phase of the frequency response.

Finally, to determine the time domain expression for $v_c$, we know that

$$V_c = H_c I$$

where, since $i(t) = 0.1 \cos(2\pi ft)$,

$$I = 0.1 \text{ A}$$

Thus, the time-domain expression for $v_c$ in the steady state is given by:

$$v_c(t) = |0.1H_c(\omega)| \cos(\omega t + \angle H_c(\omega)).$$

At $f = 1$ MHz, or $\omega = 2\pi 10^6$ rad/s, this expression becomes

$$v_c(t) = 0.016 \cos(2\pi 10^6 t - 89°).$$

### 14.2.1 THE RESONANT REGION OF THE FREQUENCY RESPONSE

We will now take a closer look at the resonant region of the frequency response in Figure 14.6 where there are sharp transitions in the magnitude and phase. In particular, we would like to determine the width of the resonant peak and the factors that affect its sharpness.

To obtain some indication of the width of the resonance in Figure 14.6, two points that are easy to calculate are the frequencies where $|H(j\omega)|$ is down to 0.707 (or $1/\sqrt{2}$) of its maximum value.[2] These frequencies $\omega_{0.707}$ can be readily

2. A frequency at which $|H(j\omega)|$ falls to $1/\sqrt{2}$ or 0.707 of its maximum value is called a 0.707 frequency or $\omega_{0.707}$. As defined for first-order circuits in Chapter 13, such a frequency is also called a *half power frequency*. Since 0.707 in decibels is $20\ log(0.707) = -3$dB, it is also called a $-3$dB frequency.

calculated from Equation 14.29 because $|H|$ will be at 0.707 of its maximum when the denominator becomes $G(1 \pm j1)$. (Notice that when the denominator of $H$ becomes $G(1 \pm j1)$, the magnitude $|H|$ becomes $1/\sqrt{2}G$, which is equivalent to $0.707/G$.) In other words, for $\omega_1 = \omega_{0.707}$, $|H|$ will be at 0.707 of its maximum, and

$$G + j\left(\omega_{0.707}C - 1/\omega_{0.707}L\right) = G(1 \pm j1).$$

Simplifying,

$$G = \pm\left(\omega_{0.707}C - 1/\omega_{0.707}L\right). \tag{14.42}$$

This is a quadratic in $\omega_{0.707}$:

$$\omega_{0.707}^2 \pm \frac{G}{C}\omega_{0.707} - \frac{1}{LC} = 0. \tag{14.43}$$

Solving for $\omega_{0.707}$

$$\omega_{0.707} = \pm\frac{G}{2C} \pm \sqrt{\left(\frac{G}{2C}\right)^2 + \frac{1}{LC}}. \tag{14.44}$$

The two positive roots, namely,

$$\omega_{0.707} = +\frac{G}{2C} + \sqrt{\left(\frac{G}{2C}\right)^2 + \frac{1}{LC}} \tag{14.45}$$

and

$$\omega_{0.707} = -\frac{G}{2C} + \sqrt{\left(\frac{G}{2C}\right)^2 + \frac{1}{LC}} \tag{14.46}$$

diagramed as in Figure 14.11a on a linear frequency scale, indicate that the width of the curve between the two 0.707 frequencies, usually called the *bandwidth*, is $G/C$. In other words,

$$\text{Bandwidth} = \frac{G}{C} = \frac{1}{RC}.$$

Comparing Equation 14.25 with the standard form for second order circuits, Equation 14.25, we can write

$$\frac{1}{RC} = 2\alpha = \text{Bandwidth}.$$

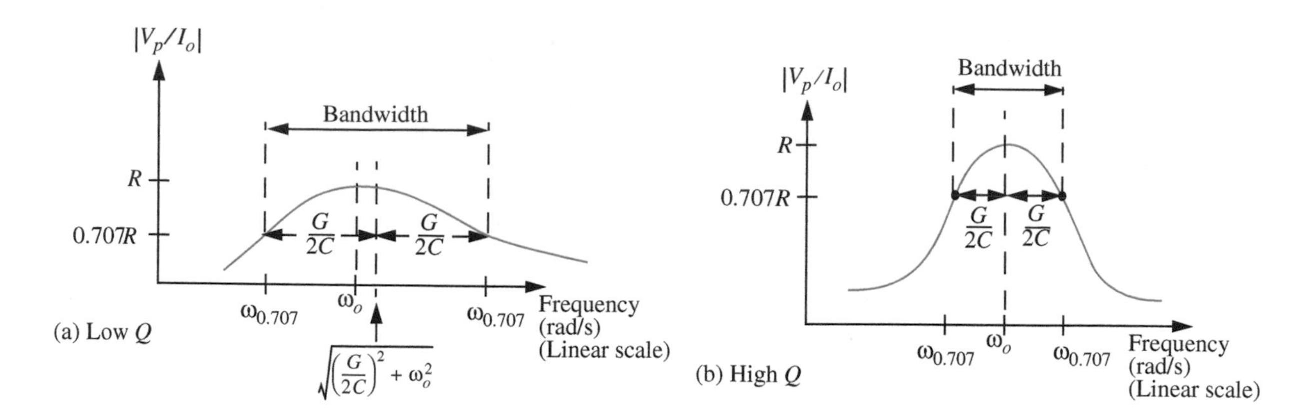


FIGURE 14.11 Bandwidth calculation; in the figure, $G = 1/R$.

Recall, we came across this same *damping factor* in Equation 12.85 in Chapter 12 from a time-domain point of view. There, in the time domain view, the damping factor was an indication of how quickly the natural response died out. Notice the presence of the decaying exponential term containing $\alpha$ in Equation 14.21, the time domain response for our parallel RLC circuit.

Notice also from Figure 14.11a that the 0.707 frequencies are not usually symmetric about $\omega_o$, rather they are symmetric about

$$\sqrt{\left(\frac{G}{2C}\right)^2 + \frac{1}{LC}}$$

or

$$\sqrt{\alpha^2 + \omega_o^2}.$$

However, for low values of G, this value is close to the resonance frequency. Therefore, the terms *center frequency* and *resonance frequency* are used interchangeably.

A useful gauge of the sharpness of the resonance is the ratio of the resonance frequency to the bandwidth:

$$\frac{\text{Resonance frequency}}{\text{Bandwidth}} = \frac{\omega_o}{G/C} = Q = \omega_o RC = \frac{R}{\omega_o L}. \quad (14.47)$$

Again, this very *Quality Factor* was introduced in Chapter 12 (Equations 12.65 and 12.66) from a quite different point of view. There, in the time domain view, Q indicated the length of time for which the circuit would "ring" when excited by an input such as a step. If we know the quality factor Q and the frequency $\omega_o$, we can derive the bandwidth as:

$$\text{Bandwidth} = \frac{\omega_o}{Q}. \quad (14.48)$$

The relative importance of the two terms under the radical in Equations 14.45 and 14.46 can be assessed by noting from Equation 14.31 that

$$\frac{1}{LC} = \omega_o^2. \tag{14.49}$$

By dividing out the $\omega_o^2$ from under the radical, and substituting from Equation 14.47, we obtain

$$\omega_{0.707} = +\frac{G}{2C} + \omega_o\sqrt{1 + \frac{1}{4Q^2}} \tag{14.50}$$

and

$$\omega_{0.707} = -\frac{G}{2C} + \omega_o\sqrt{1 + \frac{1}{4Q^2}}. \tag{14.51}$$

For $Q$ greater than five, the radicals in the expressions for the two 0.707 frequencies are within 1/2 percent of unity. In such a case, it is reasonable to neglect the small offset, and assume that the resonant curve is symmetric about $\omega_o$, as illustrated Figure 14.11b, and as summarized in the following equations:

$$\omega_{0.707} \approx \omega_o + \frac{G}{2C} \tag{14.52}$$

and

$$\omega_{0.707} \approx \omega_o - \frac{G}{2C}. \tag{14.53}$$

Knowing the five constraints, namely the two 0.707 frequencies in addition to the resonance frequency $\omega_o$ and the low- and high-frequency asymptotes, we can guess the form of the frequency response more accurately than that in Figure 14.8. The five constraints, along with the actual curve, are shown in Figure 14.12.

The phase of $H(j\omega)$ is also relatively easy to sketch. At the 0.707 frequency, the real and imaginary parts of the denominator in Equation 14.29 are equal (see Equation 14.42), so the phase must be +45° below resonance, and −45° above. These two constraints, along with the low- and high-frequency asymptotes at +90° and −90° respectively, and the phase at resonance of zero, form five constraints that allow us to draw a fairly accurate phase plot as also shown in Figure 14.12.

At this point, a few observations are in order. Recall from Equation 14.47 that $Q = R/\omega_o L$. Thus, it should be clear from this expression for $Q$ and

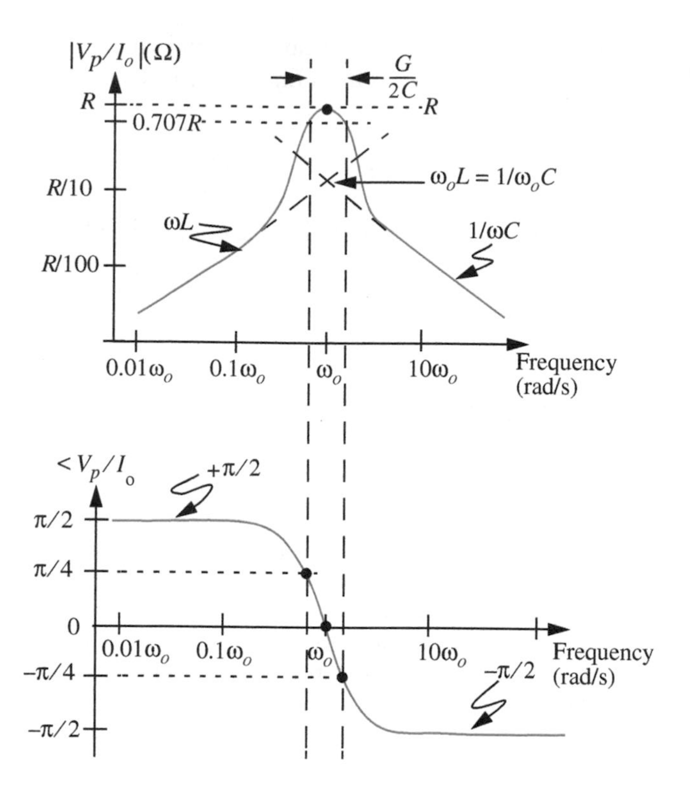


FIGURE 14.12 Sketch of the frequency plot for parallel RLC circuit along with the five constraints, namely the two 0.707 frequencies, the resonance frequency $\omega_O$, and the low- and high-frequency asymptotes.

Figure 14.12 that $Q$ is the ratio of the peak height of the curve to the height of the intersection point of the asymptotes. Thus, $Q$ is an indication of the "peakiness" of the frequency response curve. Several magnitude and phase plots illustrating the relationship between the peakiness and $Q$ are shown in Figures 14.13 and 14.14.

Next, relating to the time domain (see the discussion surrounding Equation 12.65 in Chapter 12), since

$$Q = \frac{\omega_O}{2\alpha} \tag{14.54}$$

a high value of $Q$ means that the damping factor $\alpha$ is small compared to $\omega_O$, and the circuit will ring for a long time when excited by a step or an impulse.[3]

3. In fact, it can be shown that $Q$ itself is an approximate measure of the number of oscillations.

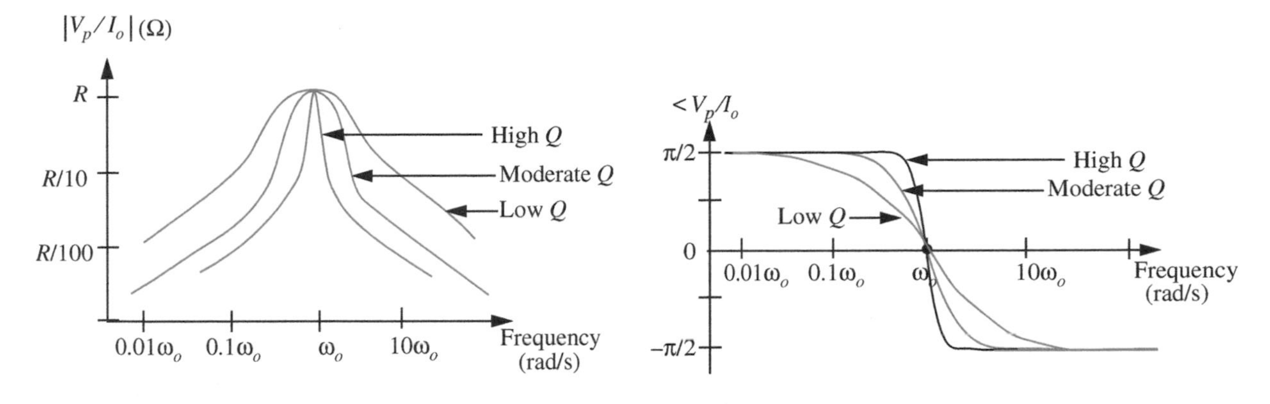


FIGURE 14.13 High Q versus low Q circuits.

FIGURE 14.14 Phase of high Q and low Q circuits.

A high value of $Q$, which implies that $\omega_o$ is large compared to $\alpha$, also means that the roots of the characteristic polynomial will be complex. Notice that the roots of the characteristic polynomial (Equation 14.25) are given by

$$-\alpha + \sqrt{\alpha^2 - \omega_o^2} \quad \text{and} \quad -\alpha - \sqrt{\alpha^2 - \omega_o^2}.$$

We can also see from Equation 14.54 that for $Q > 0.5$, the roots will be complex because

$$\alpha < \omega_o.$$

Notice, further, from our observations in Section 12.4.1, that when $\alpha < \omega_o$ the circuit is under-damped. We now see the explicit correlation between complex roots, resonant circuits, under-damping, and peakiness in the frequency response.

Finally, using the frequency response plot in Figure 14.11 or Figure 14.12, on the basis of Equation 14.20, it is relatively simple to visualize the *filtering* that occurs when a signal is passed through the resonant circuit. Any frequency components near the resonant frequency will pass through the system relatively unattenuated, but all other frequency components will be substantially attenuated and shifted in phase. If, for example the filter input is a 990-Hz square wave, and the filter resonant frequency is 3000 Hz, the output will be nearly sinusoidal at 2970 Hz, because the filter will pass this third-harmonic component of the square wave and reject the fundamental and other harmonics. As mentioned previously, this signal-processing property of *filtering* is fundamental to the operation of all television, radio, and cellular phone receivers, which must select one transmitted signal from among many present at the receiver antenna.

EXAMPLE 14.2 DETERMINING CRITICAL PARAMETERS

Determine the values of $Q$, $\omega_o$, $\alpha$, the two $\omega_{0.707}$ frequencies, and the bandwidth, for the circuit shown in Figure 14.9 in Example 14.1 given that

$$i(t) = 0.1\cos(2\pi ft)$$
$$L = 0.1 \text{ mH}$$
$$C = 1\ \mu\text{F}$$
$$R = 10\ \Omega.$$

Plot the magnitude and phase curves on both log and linear scales for those element values. Also, keeping the resonance frequency and the peak magnitude constant, show frequency response plots for $Q = 0.5, 0.75, 1, 2, 4, 8,$ and $16$.

As worked in Example 14.1, the transfer function for the circuit in Figure 14.9 written in standard form is

$$H_c = \frac{s/C}{1/LC + s/RC + s^2} = \frac{10^6 s}{10^{10} + 10^5 s + s^2}. \tag{14.55}$$

Comparing the denominator of Equation 14.55 to the canonic form $s^2 + 2\alpha s + \omega_o^2$, we get

$$\omega_o = \sqrt{\frac{1}{LC}} = 10^5 \text{ rad/s}$$

$$\alpha = \frac{1}{2RC} = \frac{10^5}{2} \text{ rad/s}$$

$$Q = \frac{\omega_o}{2\alpha} = R\sqrt{\frac{C}{L}} = 1. \tag{14.56}$$

The two $\omega_{0.707}$ frequencies are given by Equations 14.45 and 14.46 as[4]

$$\omega_{0.707} = +\frac{G}{2C} + \sqrt{\left(\frac{G}{2C}\right)^2 + \frac{1}{LC}} = 1.618 \times 10^5 \text{ rad/s}$$

and

$$\omega_{0.707} = -\frac{G}{2C} + \sqrt{\left(\frac{G}{2C}\right)^2 + \frac{1}{LC}} = 0.618 \times 10^5 \text{ rad/s}.$$

4. where $G = 1/R$

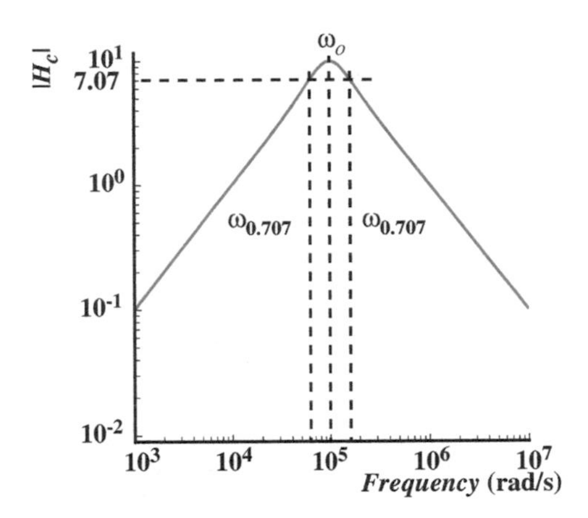


FIGURE 14.15 Magnitude of the frequency response on a log scale.

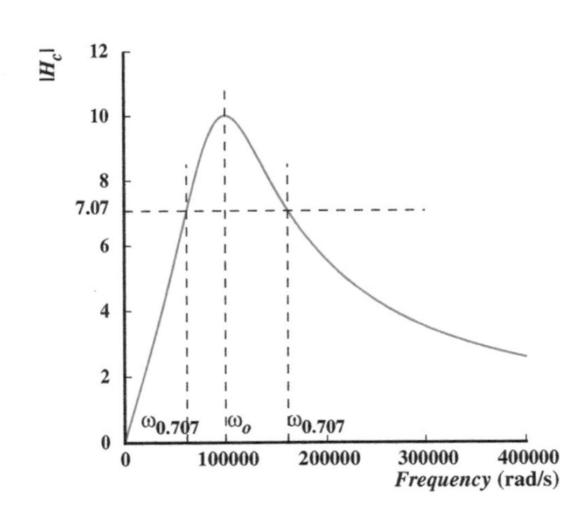


FIGURE 14.16 Magnitude of the frequency response on a linear scale.

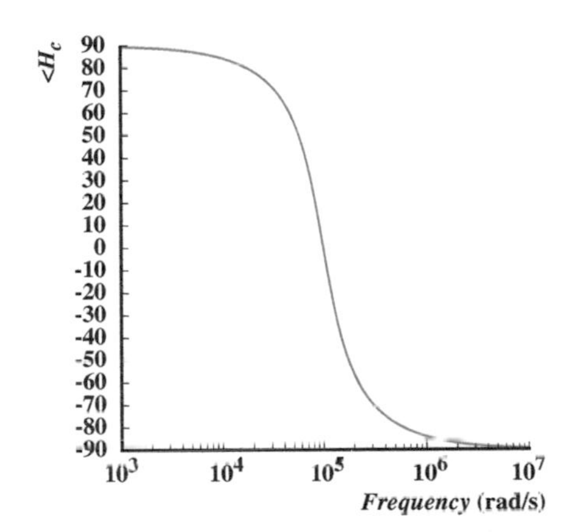


FIGURE 14.17 Phase of the frequency response on a log scale.

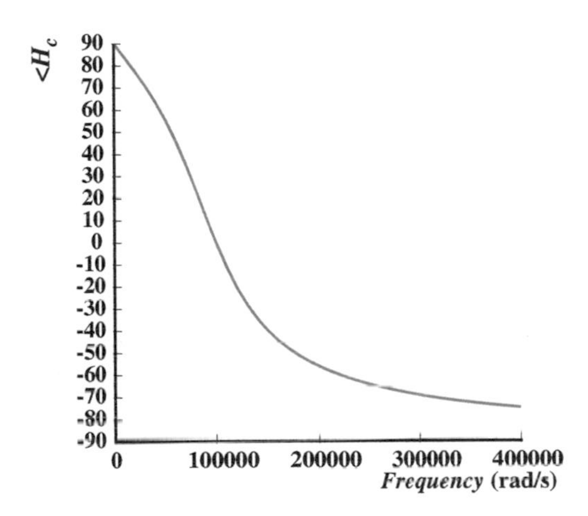


FIGURE 14.18 Phase of the frequency response on a linear scale.

The bandwidth is

$$1.618 \times 10^5 - 0.618 \times 10^5 = 10^5 \text{ rad/s}.$$

The bandwidth and the resonance frequency $\omega_o$ are marked in Figure 14.15, which shows the magnitude plot on a log scale. The bandwidth and resonance frequency are also marked on the magnitude plot drawn on a linear scale in Figure 14.16. The phase is plotted in Figures 14.17 and 14.18 on log and linear scales, respectively.

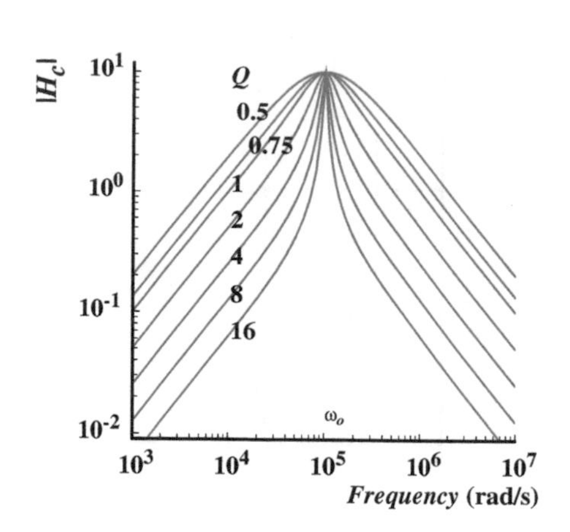


FIGURE 14.19 Magnitude for different values of Q.

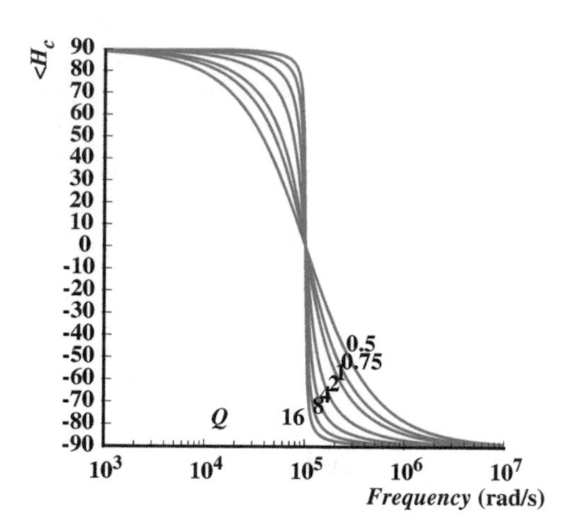


FIGURE 14.20 Phase for different values of Q.

Figures 14.19 and 14.20 show the frequency response for different values of $Q$, keeping the resonance frequency and the peak magnitude constant. To keep the peak magnitude the same, we keep $R$ constant at 10 Ω. Similarly, to keep the resonance frequency the same we keep $\sqrt{1/LC}$ a constant at $10^5$ rad/s. We obtain different values of $Q$ by choosing different $C/L$ ratios (Equation 14.56), while keeping both $R$ and $LC$ constant. Thus, for example,

$$L = 0.1 \text{ mH}$$
$$C = 1\ \mu\text{F}$$
$$R = 10\ \Omega$$

yield $Q = 1$, while

$$L = 0.05 \text{ mH}$$
$$C = 2\ \mu\text{F}$$
$$R = 10\ \Omega$$

yield $Q = 2$, and

$$L = 0.2 \text{ mH}$$
$$C = 0.5\ \mu\text{F}$$
$$R = 10\ \Omega$$

yield $Q = 0.5$.

## 14.3 SERIES RLC

A second topology for RLC circuits is shown in Figure 14.21, the *series resonant circuit.* Direct analysis of the impedance model, Figure 14.21b, yields

$$I = \frac{V_i}{R_s + Ls + 1/Cs} \tag{14.57}$$

$$= \frac{(s/L)\, V_i}{s^2 + sR_s/L + 1/LC}. \tag{14.58}$$

Thus,

$$H(s) = \frac{I}{V_i} = \frac{(s/L)\, V_i}{s^2 + sR_s/L + 1/LC}. \tag{14.59}$$

Again we obtain an expression identical in form to Equation 14.23, so this too is a resonant circuit. The difference this time is that the *current* is the output variable, so here the current is maximum at resonance, whereas for the parallel circuit of Figure 14.1, the voltage was maximum. By comparing corresponding terms in the two derivations, for this series circuit,

$$\text{Resonant frequency} = \omega_o = \frac{1}{\sqrt{LC}} \tag{14.60}$$

$$\text{Bandwidth} = \frac{R_s}{L}. \tag{14.61}$$

Comparing this expression for the bandwidth with the expression for the damping factor $\alpha$ developed in our time-domain analysis for a series RLC circuit (see Section 12.12), we can write

$$\text{Bandwidth} = 2\alpha.$$

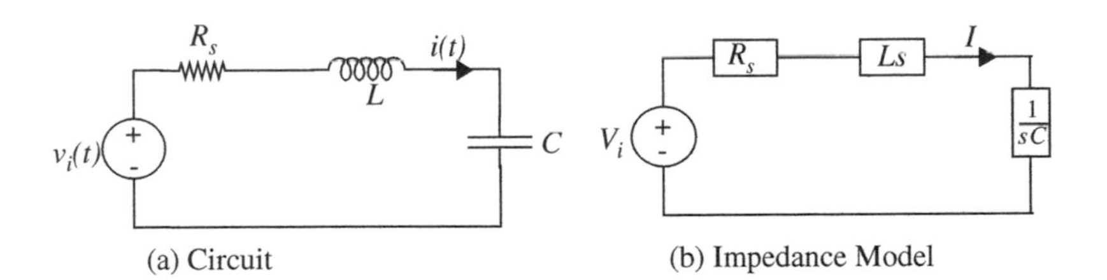


FIGURE 14.21 Series resonant circuit.

Furthermore, because Q is defined in this chapter as the ratio of resonance frequency to bandwidth, for the series circuit,

$$Q = \frac{\omega_o L}{R_s} \tag{14.62}$$

the same relation we found in Chapter 12 (Equation 12.109) by examining the response to a short pulse. Comparison with the corresponding expression for the parallel RLC circuit indicates that for high Q in a parallel resonant circuit, R should be large, whereas in the series case, $R_s$ should be small. Confusing, but correct.

The plots of the magnitude and phase of $H(j\omega)$ versus $\omega$, Figure 14.12, again apply to this circuit, except now the system function is defined as

$$H(j\omega) = \frac{I}{V_i}. \tag{14.63}$$

---

EXAMPLE 14.3 TRANSFER FUNCTION OF SERIES RLC CIRCUIT Determine the transfer function $H_r = V_r/V_i$ for the series RLC circuit shown in Figure 14.22, given that

$$v_i(t) = 0.1\cos(2\pi f t)$$

$$L = 0.1 \text{ mH}$$

$$C = 1\ \mu\text{F}$$

$$R = 5\ \Omega.$$

Sketch the asymptotes of magnitude and phase of the frequency response. Determine the values of Q, $\omega_o$, and $\alpha$. Write the time-domain expression for the steady state value of $v_r$ for $f = 1$ MHz.

From the impedance model, we get the transfer function

$$H_r = \frac{V_r}{V_i} = \frac{R}{Ls + R + \frac{1}{sC}}. \tag{14.64}$$

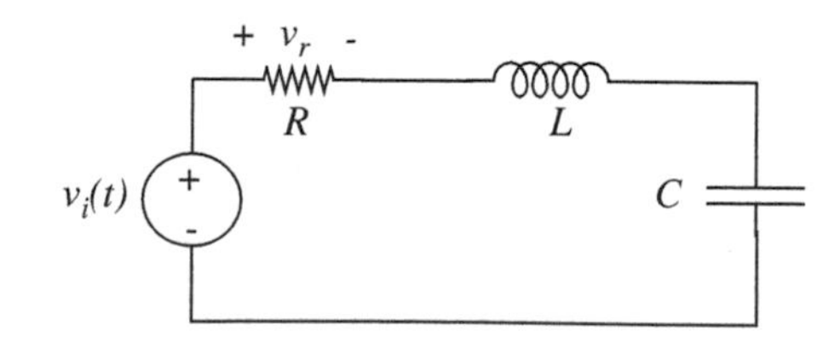


FIGURE 14.22 Series resonant circuit example.

Multiplying the numerator and denominator by $s/L$ we get

$$H_r = \frac{\frac{sR}{L}}{s^2 + s\frac{R}{L} + \frac{1}{LC}}.$$

Substituting the actual element values,

$$H_r = \frac{5 \times 10^4 s}{s^2 + 5 \times 10^4 s + 10^{10}}. \tag{14.65}$$

The frequency response obtained by substituting $s = j\omega$ is

$$H_r(j\omega) = \frac{j5 \times 10^4 \omega}{(10^{10} - \omega^2) + j5 \times 10^4 \omega}.$$

To determine the form of the magnitude plot of the frequency response, we must find the low- and high-frequency asymptotes and the value of the response at $\omega_o$. The low-frequency asymptote is given by

$$H_r(j\omega) = \frac{j\omega}{2 \times 10^5}$$

and the high-frequency asymptote is

$$H_r(j\omega) = \frac{5 \times 10^4}{j\omega}.$$

Comparing the denominator of Equation 14.65 to the canonic form $s^2 + 2\alpha s + \omega_o^2$, we get

$$\omega_o = 10^5 \text{ rad/s}$$

$$\alpha = 2.5 \times 10^4 \text{ rad/s}$$

and

$$Q = \frac{\omega_o}{2\alpha} = 2.$$

At $\omega = \omega_o$,

$$|H_r(j\omega_o)| = 1.$$

The corresponding low- and high-frequency phase asymptotes are 90° and −90°, respectively. The phase at $\omega_o$ is 0°.

The dashed lines in the plots in Figure 14.23 show the low- and high-frequency asymptotes, and the $X$ symbols mark the values at $\omega_o$. The solid lines show the actual plots.

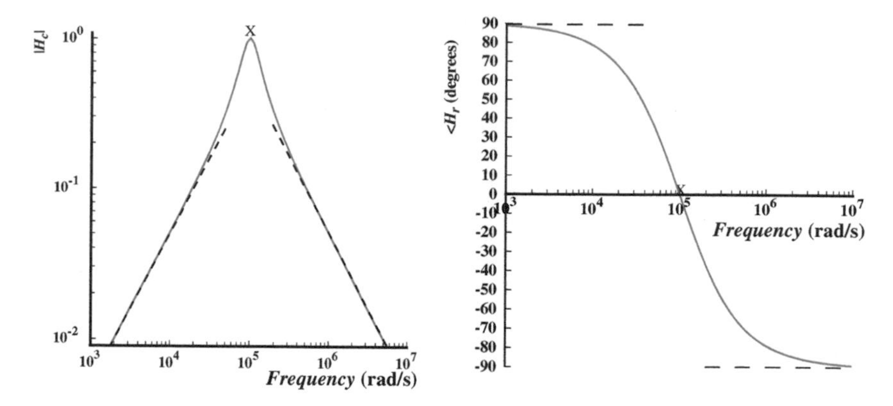


FIGURE 14.23 Magnitude and phase of the frequency response.

Finally, from

$$V_r = H_r V_i$$

where $V_i = 0.1$ V, we can obtain time domain expression for $v_r$ in the steady state as

$$v_r(t) = |0.1H_r(\omega)| \cos(\omega t + \angle H_r(\omega)).$$

At $f = 1$ MHz, or $\omega = 2\pi 10^6$ rad/s, this expression becomes

$$v_r(t) = 0.0008 \cos(2\pi 10^6 t - 89.5°).$$

---

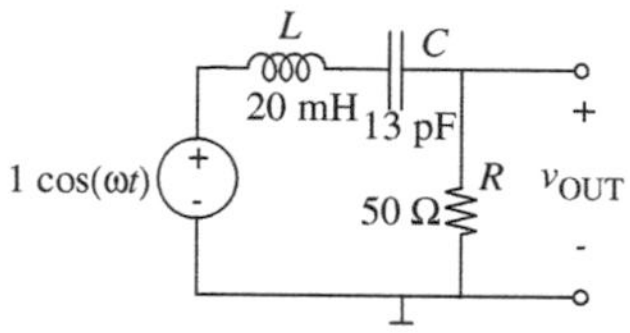

EXAMPLE 14.4 METAL DETECTOR USING A RESONANT CIRCUIT The circuit shown in Figure 14.24 can be used as a metal detector. To do so, the inductor is wound as a large flat coil. In the presence of nearby metal, the inductance of the coil changes, and so $v_{OUT}$ changes as well. Suppose that we can detect a 0.1-mV change in the amplitude of $v_{OUT}$ in the sinusoidal steady state. What is the corresponding minimum detectable change in inductance, and how should the frequency $\omega$ be chosen to maximize the sensitivity of the metal detector?

L 20 mH, C 13 pF, R 50 Ω, 1 cos(ωt), $v_{OUT}$

FIGURE 14.24 A metal detector circuit.

To analyze the metal detector, we first compute $V_{out}$, the complex amplitude of $v_{OUT}$ in the sinusoidal steady state. The magnitude of $V_{out}$ is then the amplitude of $v_{OUT}$ in the sinusoidal steady state. Thus, using impedances

$$|V_{out}| = \frac{\omega RC}{\sqrt{(1 - \omega^2 LC)^2 + (\omega RC)^2}} \text{ 1 V},$$

which is also the amplitude of $v_{OUT}$. Next, we differentiate the amplitude of $v_{OUT}$ with respect to $L$ to determine the sensitivity. This yields

$$\frac{d|V_{out}|}{dL} = \frac{\omega^3 RC^2(1 - \omega^2 LC)}{((1 - \omega^2 LC)^2 + (\omega RC)^2)^{3/2}} \text{ 1 V}.$$

For the parameters given in Figure 14.24, the absolute value of the sensitivity is maximized near 62.920 krad/s, or 10.014 kHz. This is slightly above the resonance frequency:

$$\omega_o = \sqrt{1/LC} = 62.017 \text{ krad/s}.$$

At this frequency, the peak sensitivity is approximately −484.3 V/H. Therefore, given a minimum measurable change in voltage amplitude of 0.1 mV, the minimum measurable change in inductance is approximately 0.2 $\mu$H, or about 0.001% of the coil inductance.

---

EXAMPLE 14.5 ANOTHER RLC CIRCUIT EXAMPLE The input $v_z$ in the second-order circuit in Figure 14.25 is a sinusoid. Determine the impedance $Z$. Determine also $\omega_o$, $\alpha$, $\omega_d$, and $Q$ for the circuit. Show that the circuit is resonant for the element values indicated in Figure 14.25. Plot the magnitude and phase of the system function $I_z/V_z$ as a function of frequency, and sketch the low- and high-frequency asymptotes.

Using the impedance model, the impedance $Z$ is given by:

$$Z = \frac{1}{\frac{1}{R+Ls} + sC}$$

$$= \frac{\frac{s}{C} + \frac{R}{LC}}{s^2 + s\frac{R}{L} + \frac{1}{LC}}.$$

The desired system function is the admittance $I_z/V_z = 1/Z$ and is given by:

$$H(s) = \frac{I_z}{V_z} = \frac{1}{Z} = \frac{s^2 + s\frac{R}{L} + \frac{1}{LC}}{\frac{s}{C} + \frac{R}{LC}} \tag{14.66}$$

Comparing

$$s^2 + s\frac{R}{L} + \frac{1}{LC}$$

to the canonic form

$$s^2 + 2\alpha s + \omega_o^2$$

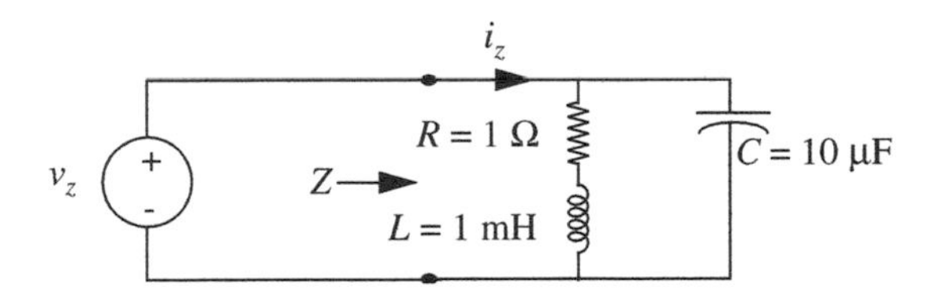


FIGURE 14.25 A second-order circuit.

we get

$$\omega_o = \sqrt{\frac{1}{LC}}$$

and

$$\alpha = \frac{R}{2L}.$$

Thus, we can compute

$$\omega_d = \sqrt{\omega_o^2 - \alpha^2}$$

and

$$Q = \frac{\omega_o}{2\alpha}.$$

For

$$L = 1 \text{ mH}$$

$$C = 10\ \mu\text{F}$$

$$R = 1\ \Omega$$

we get

$$\omega_o = 10^4 \text{ rad/s},$$

$$\alpha = 500 \text{ rad/s},$$

$$\omega_d = 9988 \text{ rad/s},$$

and

$$Q = 10.$$

Since $Q > 0.5$, or equivalently, since $\omega_o > \alpha$, the roots of the characteristic equation are complex and therefore the circuit is resonant.

Substituting the numerical quantities into our system function, we get

$$H(s) = \frac{s^2 + 1000s + 10^8}{10^5 s + 10^8}. \tag{14.67}$$

Let us now determine the the low- and high-frequency asymptotes, and the magnitude and phase of the response at the resonant frequency. For low frequencies, the system function in Equation 14.67 reduces to

$$H(s) = 1.$$

This directly yields the low-frequency asymptote for the magnitude. The corresponding low-frequency phase asymptote is 0°. This asymptote implies that the admittance is similar to that of a 1-Ω resistor.

For high frequencies, the system function becomes

$$H(s) = 10^{-5} s,$$

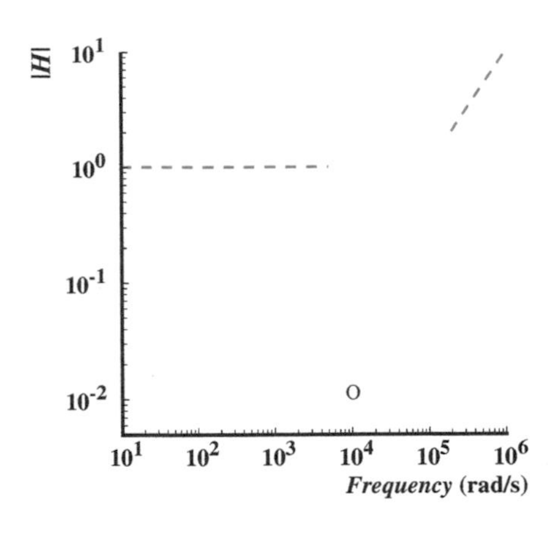


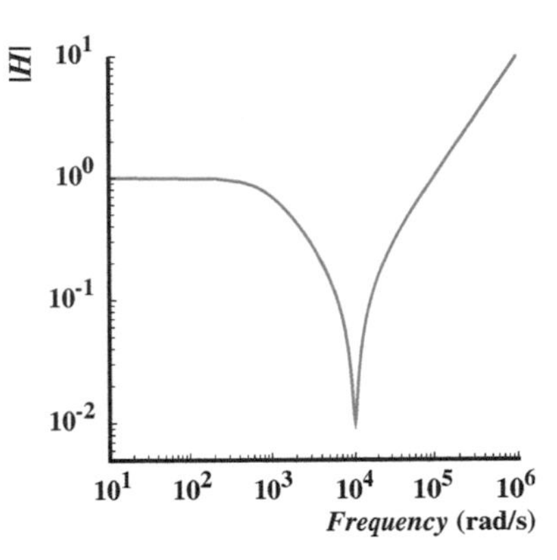


FIGURE 14.26 Magnitude versus frequency.

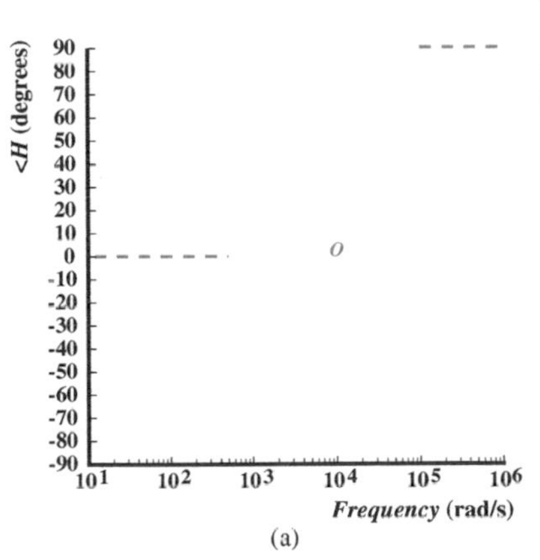


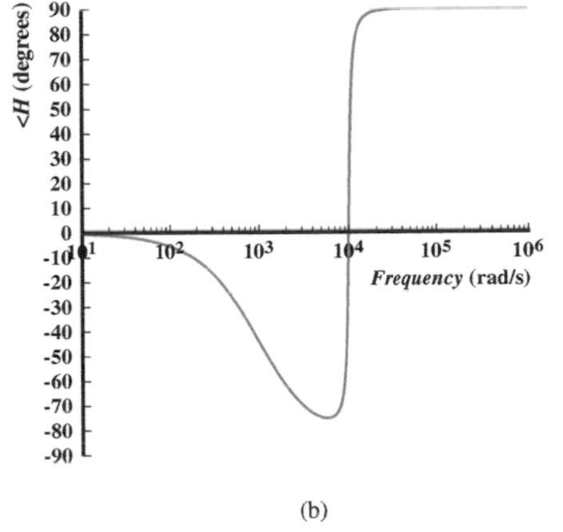


FIGURE 14.27 Phase versus frequency.

which gives us our high-frequency asymptote. This asymptote implies that at high frequencies the admittance is similar to that of a 10-$\mu$F capacitor.

The magnitude and phase of the response at the resonance frequency $\omega_o = 10^4$ rad/s is obtained by substituting $\omega = 10^4$ in Equation 14.67 as

$$|H(j\omega_o)| \approx 0.01$$

and

$$\angle H(j\omega_o) \approx 6^\circ.$$

The three constraints for the magnitude and phase are sketched in Figures 14.26a and 14.27a, respectively. Figures 14.26b and 14.27b contain the corresponding computer generated plots for the magnitude and phase of the system function versus frequency.

## www 14.4 THE BODE PLOT FOR RESONANT FUNCTIONS *

### www EXAMPLE 14.6 BODE PLOT EXAMPLE

## 14.5 FILTER EXAMPLES

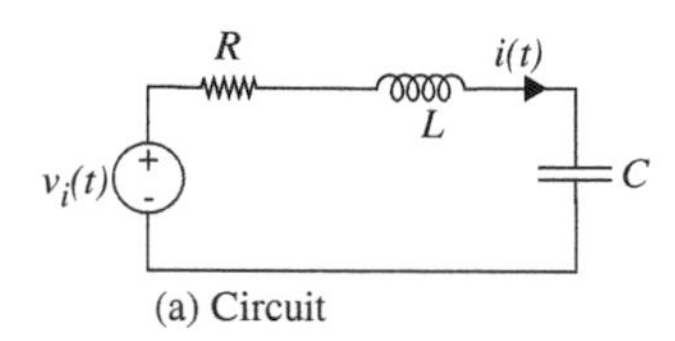

(a) Circuit

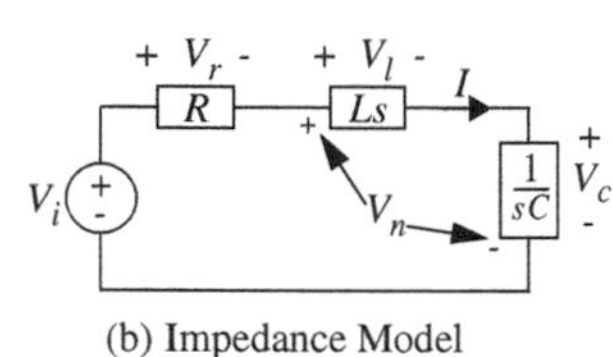

(b) Impedance Model

**FIGURE 14.32** Resonant series RLC circuit.

Depending on where the output is taken, the series and parallel RLC resonant circuits can be used as highly selective filters of various types. The higher the $Q$, the higher the selectivity. Here, we revisit the series resonance circuit (see Figure 14.32a) and demonstrate how various types of filters can be derived from the same basic circuit.

Figure 14.32b shows the impedance model of the series RLC circuit. Applying the impedance method, we obtain the following relation between the complex amplitude of the current $I$ and the input voltage $V_i$:

$$I = \frac{V_i}{R + Ls + 1/Cs} \tag{14.68}$$

$$= \frac{(s/L)V_i}{s^2 + sR/L + 1/LC}. \tag{14.69}$$

We can also rewrite the denominator of the expression for $I$ in our general form as

$$I = \frac{(s/L)V_i}{s^2 + 2\alpha s + \omega_o^2} \tag{14.70}$$

where $\omega_o$ and $\alpha$ for the series resonance circuit are given by

$$\omega_o = \sqrt{\frac{1}{LC}}$$

$$\alpha = \frac{R}{2L}.$$

The corresponding system function relating $I$ and $V_i$ is

$$H(s) = \frac{I}{V_i} \tag{14.71}$$

$$= \frac{s/L}{s^2 + sR/L + 1/LC} \tag{14.72}$$

$$= \frac{s/L}{s^2 + 2\alpha s + \omega_o^2}. \tag{14.73}$$

We will now show that the system function relating the *voltage* across each of the elements (see Figure 14.32b) to the input voltage represents different kinds of filters. For concreteness, we will plot our results using the following element values:

$$L = 1\ \mu\text{H}$$

$$C = 1\ \mu\text{F}$$

$$R = 1\ \Omega.$$

These element values result in

$$\omega_o = 10^6 \text{ rad/s}.$$

In Hertz, the resonant frequency is $10^6/2\pi = 159,154$ Hz.

The damping factor is

$$\alpha = 5 \times 10^5 \text{ s}$$

and the quality factor is

$$Q = \frac{\omega_o}{2\alpha} = 1.$$

### 14.5.1 BAND-PASS FILTER

First, let us look at the behavior of the voltage across the resistor, $V_r$. Multiplying the expression for the current in Equation 14.70 by the impedance $R$, we get

$$V_r = IR = \frac{\frac{sR}{L} V_i}{s^2 + 2\alpha s + \omega_o^2},$$

which leads to the following system function relating $V_r$ to $V_i$:

$$H_r(s) = \frac{V_r(s)}{V_i(s)} = \frac{sR/L}{s^2 + 2\alpha s + \omega_o^2}.$$

Since $\alpha = R/2L$, we can write the system function for the voltage across the resistor as

$$H_r(s) = \frac{2\alpha s}{s^2 + 2\alpha s + \omega_o^2}.$$

We can plot the frequency response for this system function by substituting $s = j\omega$ and taking the magnitude and phase of $H_r(s)$. The computer-generated

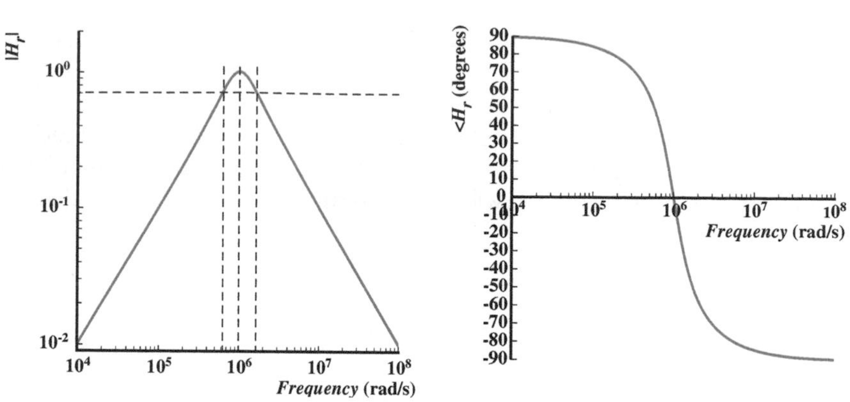

FIGURE 14.33 Frequency response of the bandpass filter.

frequency response corresponding to the preceding system function is shown in Figure 14.33.

It is clear from Figure 14.33 that $H_r$ represents a bandpass filter as we discussed in Section 14.2. As also discussed in Section 14.2, notice that the magnitude of the bandpass system function at $\omega = \omega_o = 10^6$ rad/s is unity.

The bandwidth is

$$\text{Bandwidth} = \frac{\omega_o}{Q} = \frac{10^6}{1} = 10^6 \text{ rad}$$

as can also be verified from Figure 14.33 by taking the difference between the high- and low-frequencies at which the magnitude falls to $1/\sqrt{2}$ of the peak value.

It is instructive to plot the frequency response for a range of values of $Q$ by choosing different values for $R$ (see Figure 14.34). For the series resonant circuit, since

$$Q = \frac{\omega_o L}{R}$$

we can hold the values of $L$ and $C$ constant, and choose resistance values of 2 Ω, 1 Ω, 0.1 Ω, and 0.01 Ω, to obtain Q's of 0.5, 1, 10, and 100, respectively. It is easy to see from Figure 14.34 that higher the value of $Q$, the greater the "peakiness" or selectivity of the curve. Notice also that the phase curves show a corresponding sharper transition as $Q$ increases.

### 14.5.2 LOW-PASS FILTER

Let us now look at the voltage across the capacitor, $V_c$. Multiplying the expression for the current in Equation 14.70 by the impedance of the capacitor $1/sC$,

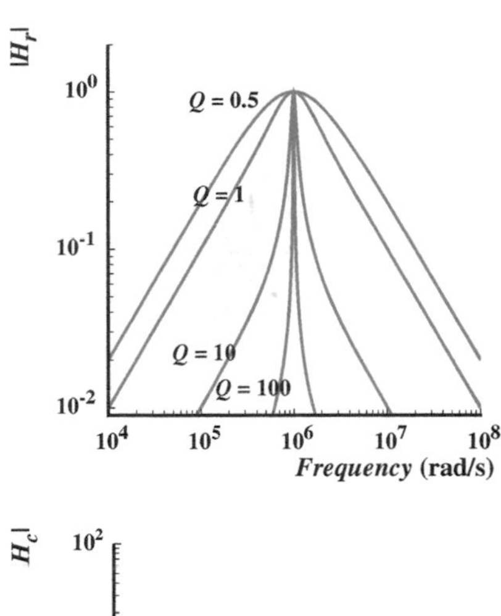


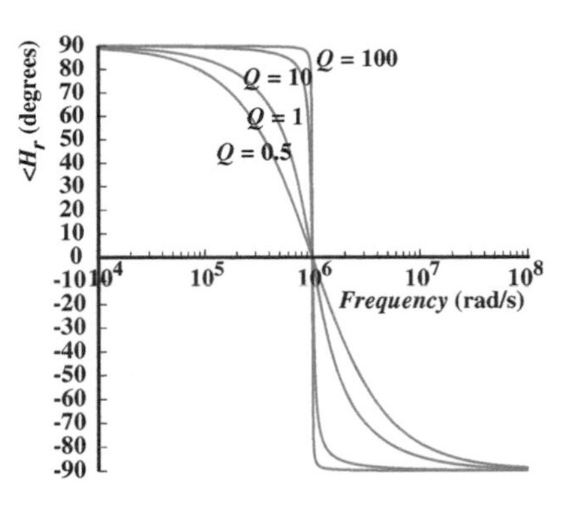


**FIGURE 14.34** Frequency response of the bandpass filter for several values of Q.

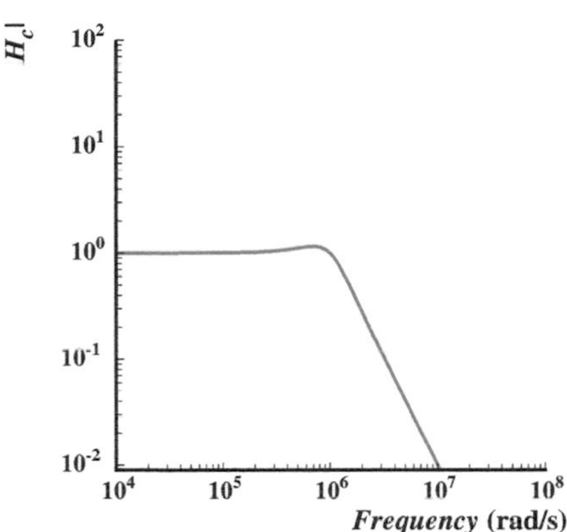


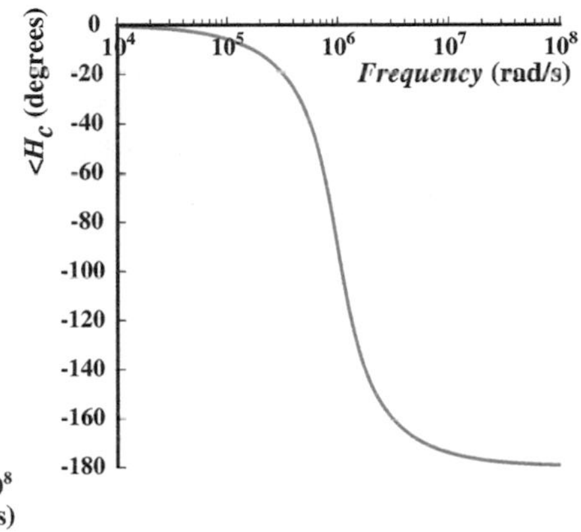


**FIGURE 14.35** Frequency response of the lowpass filter for Q= 1.

we get

$$V_c = \frac{I}{sC} = \frac{\frac{1}{LC}V_i}{s^2 + 2\alpha s + \omega_o^2},$$

which leads to the following system function relating $V_c$ to $V_i$:

$$H_c(s) = \frac{V_c(s)}{V_i(s)} = \frac{\frac{1}{LC}}{s^2 + 2\alpha s + \omega_o^2}.$$

Since $\omega_o^2 = 1/LC$, we can write the system function for the voltage across the capacitor as

$$H_c(s) = \frac{\omega_o^2}{s^2 + 2\alpha s + \omega_o^2}.$$

The frequency response plot corresponding to $H_c$ is shown in Figure 14.35. Because it passes through low-frequency signals unattenuated, $H_c$ represents a low-pass filter.

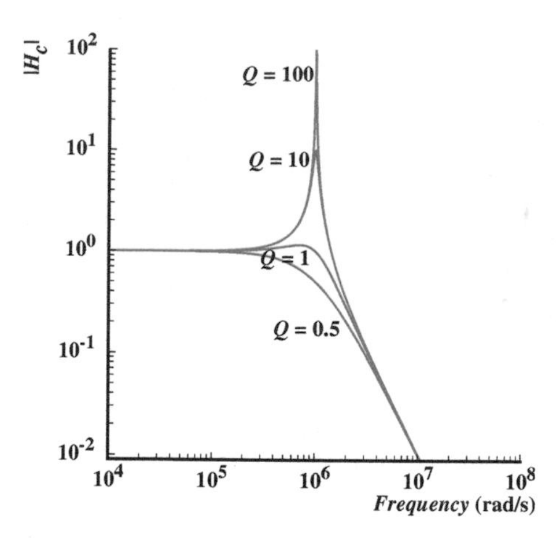


FIGURE 14.36 The magnitude curves of the frequency response for the low-pass filter for several values of Q.

The magnitude curves of the frequency response for a range of values of $Q$ are shown in Figure 14.36. As in our bandpass example, $Q$ can be increased by decreasing the value of the resistance $R$. Notice that the response curve displays some very interesting behavior for values of the drive frequency $\omega$ that are close to $\omega_o$. Specifically, the magnitude of the capacitor voltage for drive frequency $\omega = \omega_o$ can far exceed the input drive voltage for large values of $Q$. This is in stark contrast to the magnitude of the voltage across the resistor, which never exceeds the input drive voltage (see Figure 14.34). Furthermore, as with the bandpass filter, the higher the value of $Q$, the greater the "peakiness" of the curve near $\omega_o$.

Let us now derive the relationship between $Q$ and the magnitude of the response at $\omega_o$. We know that

$$V_c(s) = \frac{\omega_o^2 V_i}{s^2 + 2\alpha s + \omega_o^2}.$$

To obtain the response for any frequency $\omega$, we substitute $s = j\omega$ as follows:

$$V_c = \frac{\omega_o^2 V_i}{(j\omega)^2 + 2\alpha j\omega + \omega_o^2}$$

Substituting $\omega = \omega_o$ and simplifying we obtain the following response at resonance:

$$V_c = \frac{j\omega_o V_i}{2\alpha}.$$

Taking the magnitude and substituting $Q = \omega_o/2\alpha$ we get at resonance:

$$|V_c| = QV_i. \tag{14.74}$$

This tells us that the magnitude of the capacitor voltage in a series RLC circuit that is driven at its resonance frequency is $Q$ times the input voltage! If, for example, a series resonant circuit with a Q of 100 is connected to a 10-V sinusoidal source, then at resonance the capacitor voltage will be 1000-V! Put another way, even small excitations can cause massive responses in second-order circuits when the excitation frequency is close to that of the circuit's resonance frequency. This now sheds some insight into the Tacoma Narrows Bridge disaster. Because the frequency of alternating winds was close to the bridge's resonant frequency, the bridge began to sway back and forth and finally collapsed. We will have a lot more to say about $Q$ and the response of resonant circuits in Section 14.6.

---

EXAMPLE 14.7 RESONANT RESPONSE OF A DRIVEN HIGH-Q CIRCUIT The circuit shown in Figure 14.37 is a resonant circuit driven by a 1-Volt cosinusoidal voltage source. We wish to find: (a) the frequency at which the capacitor voltage $v_C$ has its largest amplitude, and the value of that amplitude; (b) the undamped resonance frequency (or, simply, resonance frequency) of the circuit, and the amplitude of $v_C$ at that frequency; (c) the damped resonance frequency of the circuit, and the amplitude of $v_C$ at that frequency; (d) the amplitude of $v_C$ at the frequency of 1 kHz; and (e) the amplitude of $v_C$ at the frequency of 100 kHz. To carry out the analysis, we first determine $V_c$, the complex amplitude of $v_C$, and then take its magnitude. The magnitude of $V_c$ is the amplitude of $v_C$ in sinusoidal steady state. Using impedances, the magnitude of the complex amplitude $V_c$ is given by:

$$|V_c(\omega)| = \frac{1\text{ V}}{\sqrt{(1-\omega^2 LC)^2 + (\omega RC)^2}}.$$

Again, this is also the amplitude of $v_C$ in sinusoidal steady state. The phase of $v_C$ is the angle of $V_c$.

(a) To find the maximum amplitude, we take the derivative and find $\omega$ for which the derivative goes to zero. Doing so, we find that the amplitude of $v_C$ is maximized when

$$\omega = \sqrt{\frac{1}{LC} + \frac{R^2}{2L^2}},$$

or when $\omega = 61.992$ krad/s, or 9.8664 kHz, given the parameters shown in Figure 14.37. At this frequency, the amplitude of $v_C$ is 24.8120 V, considerably higher than the magnitude of the 1-Volt drive!

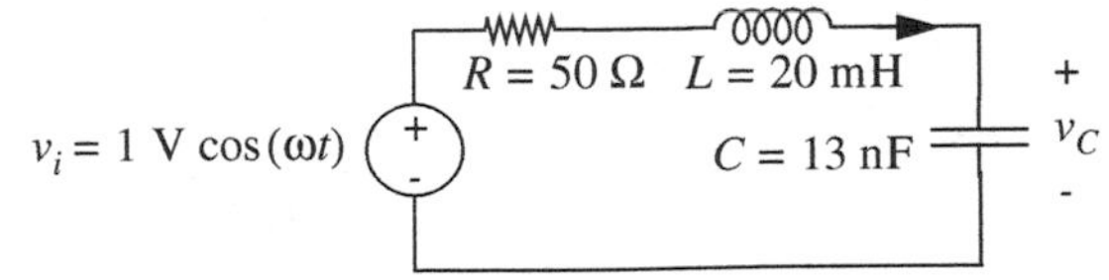

$R = 50\ \Omega$ $L = 20$ mH

$v_i = 1\text{ V}\cos(\omega t)$ $C = 13$ nF $v_C$

FIGURE 14.37 Series resonant circuit.

(b) Substituting the resonance frequency $\omega_o = \sqrt{1/LC}$ for $\omega$ in the preceding expression for $|V_c|$, we obtain

$$|V_c(\omega_o)| = \frac{1\text{ V}}{\omega_o RC} = \frac{\sqrt{L/C}\,1\text{ V}}{R} = Q \times 1\text{ V}.$$

Recall that for a series resonant circuit $Q = \omega_o L/R = (\sqrt{L/C})/R$. Note from the previous equation that at the resonance frequency, the amplitude of the output is $Q$ times the amplitude of the input. For the parameters in Figure 14.37, the resonance frequency $\omega_o = 62.017$ krad/s, or 9.8074 kHz. At the resonance frequency, the amplitude of $v_C$ is 24.8069 V, which is not the maximum amplitude, but is very close it. Note too that $Q = 24.8069$.

(c) The damped resonant frequency is $\omega_d = \sqrt{\omega_o^2 - \alpha^2} = \sqrt{(1/LC) - (R/2L)^2}$. This is the oscillation frequency of the homogeneous response, and thus the oscillation frequency of the response to initial conditions. For the parameters in Figure 14.37, $\omega_d = 62.005$ krad/s, or 9.8684 kHz. At this frequency, the amplitude of $v_C$ is 24.8107 V, which is again very close to the maximum amplitude.

(d) At 1 kHz the amplitude of $v_C$ is 1.01 V. It is clear that at this frequency, the input is being passed through without degradation to the output. However, the amplitude is significantly lower than those near the resonance frequency, as derived previously.

(e) At the higher frequency of 100 kHz the amplitude of $v_C$ is 0.01 V, which is significantly lower than that of the input.

Thus, this circuit is behaving like a low-pass filter as it passes low frequencies without degradation, while it significantly attenuates high frequencies. However, since this circuit has a very high $Q$, it probably will not serve as a useful low-pass filter, as it can produce voltages that are significantly higher than the input when driven near its resonance frequency. If a more or less flat response is desired at low frequencies (from DC to approximately the resonance frequency) then the circuit designer must change the circuit parameters to obtain a lower value of $Q$, for example, $Q = 1$.

One final point is worth noting: The circuit parameters shown in Figure 14.37 make for a high-Q circuit, with $Q \approx 25$. In this case, the undamped resonance frequency, the damped resonance frequency, and the frequency at which the circuit yields the maximum amplitude of $v_C$ are all very nearly equal. Further, the amplitudes at all three frequencies are all very nearly equal to $Q$ times the input amplitude.

---

### 14.5.3 HIGH-PASS FILTER

The voltage across the inductor is obtained by multiplying the current $I$ with the impedance of the inductor $sL$:

$$V_l = IsL = \frac{s^2 V_i}{s^2 + 2\alpha s + \omega_o^2},$$

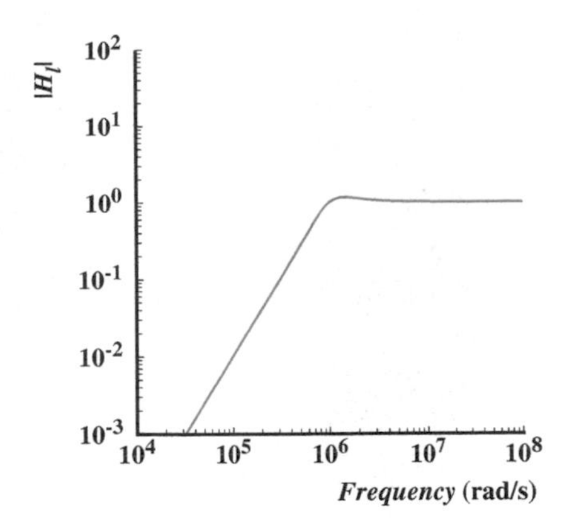


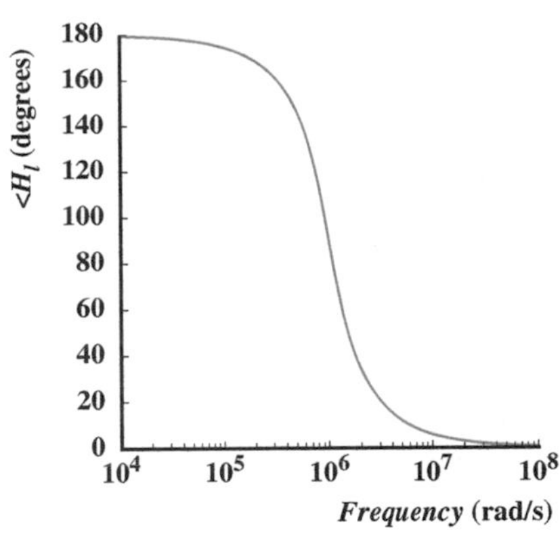


FIGURE 14.38 Frequency response of the high-pass filter.

which leads to the following system function relating $V_l$ to $V_i$:

$$H_l(s) = \frac{V_l(s)}{V_i(s)} = \frac{s^2}{s^2 + 2\alpha s + \omega_o^2}.$$

The frequency response plot for $H_l$ is shown in Figure 14.38. Because high-frequency signals pass through unattenuated, $H_l$ represents a high-pass filter.

## 14.5.4 NOTCH FILTER

A notch filter is also called a *bandstop filter*. It eliminates a range of frequencies about a *notch frequency*. As shown here, the system function corresponding to the voltage $V_n$ forms a notch filter. Multiplying $I$ with the combined series impedance of the inductor and the capacitor we have

$$V_n = I(sL + \frac{1}{sC}) = \frac{(s^2 + \frac{1}{LC})V_i}{s^2 + 2\alpha s + \omega_o^2},$$

which leads to the following system function relating $V_n$ to $V_i$:

$$H_n(s) = \frac{V_n(s)}{V_i(s)} = \frac{(s^2 + \frac{1}{LC})}{s^2 + 2\alpha s + \omega_o^2}.$$

The frequency response plot for $H_n$ as shown in Figure 14.39 clearly demonstrates that $H_n$ behaves as a notch filter. In fact, at $\omega_o$, $H_n$ goes to 0.

The four types of filters constructed using a resonant RLC circuit are summarized in Figure 14.40. The general behavior of each of the filters can be deduced quickly by observing the behavior of each circuit element within the filter for low, moderate, and high frequencies. For example, because a capacitor behaves like an open circuit for low frequencies, and a short circuit for high

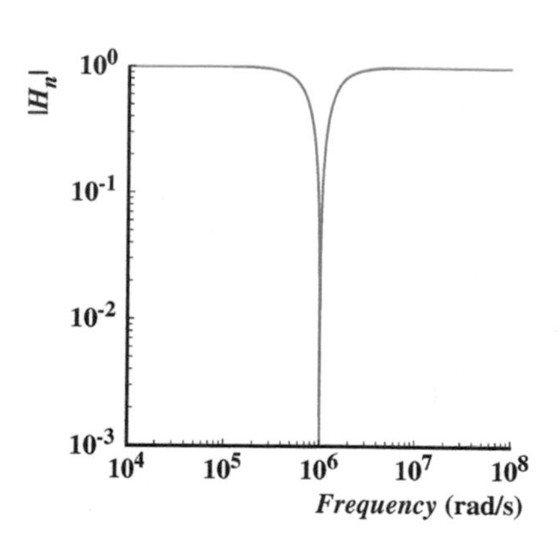


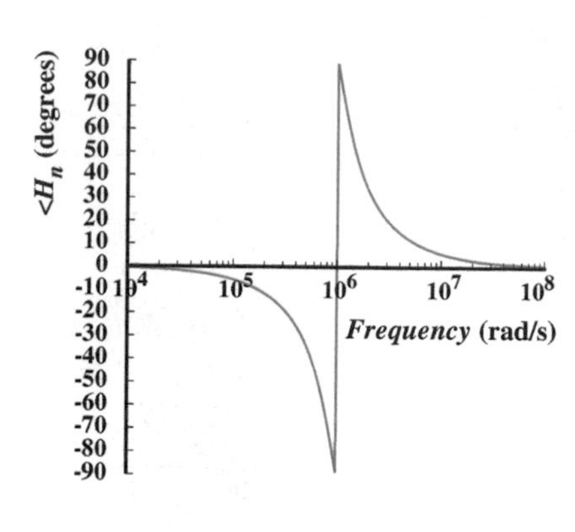


FIGURE 14.39 Frequency response of the notch filter.

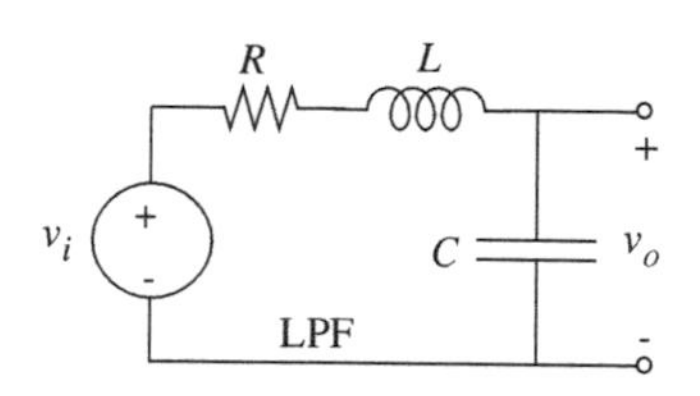


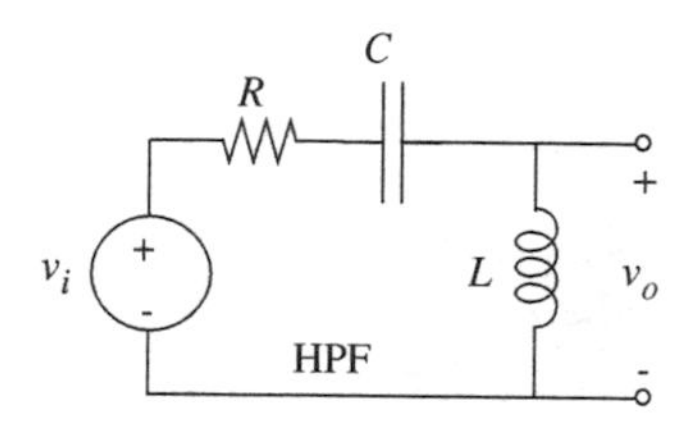


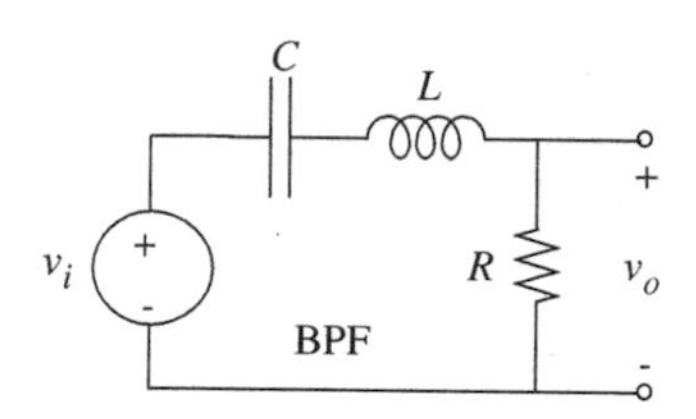


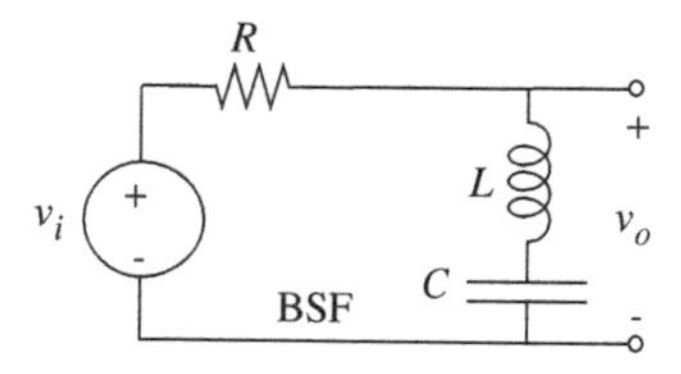


FIGURE 14.40 Filters built using resonant RLC circuit. The filters include a low-pass filter (LPF), a high-pass filter (HPF), a bandpass filter (BPF), and a bandstop filter (BSF).

frequencies, it is easy to see that the circuit built by taking the output across the capacitor is a low-pass filter.

## 14.6 STORED ENERGY IN A RESONANT CIRCUIT

Let us examine the energy flow in a series resonant circuit such as Figure 14.32, and obtain further insight into its highly sensitive behavior for input frequencies close to its resonance, as suggested by Equation 14.74. Assuming a cosine wave for the voltage source:

$$v_i = V_i \cos(\omega t)$$

(where $V_i$ is real), the complex amplitude of the current is

$$I = \frac{V_i}{R + j(\omega L - 1/\omega C)}. \tag{14.75}$$

If we assume that the circuit is being driven at its resonant frequency,

$$\omega = \omega_o = \frac{1}{\sqrt{LC}}, \tag{14.76}$$

then

$$I = \frac{V_i}{R}. \tag{14.77}$$

Therefore, if the series RLC circuit is driven at its resonant frequency, that is at $\omega = \omega_o$, we see that

$$V_c = \frac{I}{j\omega_o C} = \frac{V_i}{j\omega_o RC} = -jV_i \left(\frac{\omega_o L}{R}\right) \tag{14.78}$$

$$V_l = j\omega_o LI = jV_i\left(\frac{\omega_o L}{R}\right). \tag{14.79}$$

Using the expression for the $Q$ of a series resonant circuit, Equation 14.62, we find

$$|V_c| = |V_l| = QV_i. \tag{14.80}$$

That is, the voltage across either the capacitor or the inductor in a series resonant circuit is $Q$ times the input voltage when the circuit is driven at its resonant frequency.

However, notice from Equations 14.78 and 14.79 that the *sum* of the inductor and capacitor voltages is zero. Thus, the combination of the two elements appears as a short. Any solace one obtains from this fact, however, should be short lived because for high $Q$ resonant circuits, the capacitor or inductor voltage can still be massive, and can damage an element if its voltage rating is exceeded.[5] On a more positive note, this principle is used in instruments for measuring the $Q$ of inductors.

Next, to better understand the goings on within the capacitor and inductor, let us examine the stored energy at resonance. From Equation 14.78

$$v_C = Re\left[V_c e^{j\omega_o t}\right] \tag{14.81}$$

$$= Re\left[-jV_i Q e^{j\omega_o t}\right] \tag{14.82}$$

$$= V_i Q \sin \omega t. \tag{14.83}$$

Hence, from Equation 13.162, the stored energy is

$$w_C = \frac{1}{2} C V_i^2 Q^2 \sin^2(\omega_o t) \tag{14.84}$$

$$= \frac{1}{4} C V_i^2 Q^2 \left(1 - \cos(2\omega_o t)\right). \tag{14.85}$$

For the inductor,

$$i_L = Re\left[I e^{j\omega_o t}\right] \tag{14.86}$$

$$= \frac{V_i}{R}\cos(\omega_o t) \tag{14.87}$$

---

5. Note that it is not strictly necessary for the circuit to be driven with a sinusoid at its resonant frequency for a potentially harmful response, rather any signal that has even a miniscule *component* at the resonant frequency can cause a huge and sustained response.

$$w_L = \frac{1}{4}\frac{L}{R^2}V_i^2\left(1 + \cos(2\omega_o t)\right). \tag{14.88}$$

By substituting both Equation 14.62 and 14.76 into Equation 14.85, the stored energy in the capacitor can be written in a form closer to Equation 14.88:

$$w_C = \frac{1}{4}\frac{L}{R^2}V_i^2\left(1 - \cos(2\omega_o t)\right). \tag{14.89}$$

Now it is obvious that at resonance, the total stored energy in the system is *constant*:

$$w_{\text{total}} = w_L + w_C = \frac{1}{2}\frac{L}{R^2}V_i^2. \tag{14.90}$$

The energy is first stored in the inductor, then in the capacitor, shifting back and forth at twice the input frequency.

If the circuit is not driven at its resonant frequency, the stored energy will no longer be constant. $w_{\text{total}}$ will have a time dependence, requiring reactive power from the source.

It is possible to define a quality factor Q based on stored and dissipated energy at resonance:

$$Q = \frac{\text{Stored energy}}{\text{Average energy dissipated per radian}}. \tag{14.91}$$

Because at resonance $I = V_i/R$, the average power dissipated in the resistor is

$$\overline{p} = \frac{V_i^2}{2R}. \tag{14.92}$$

The average energy dissipated per radian is this quantity divided by the frequency expressed in radians per second,

$$w_{\text{diss}} = \frac{V_i^2}{2R\omega_o}. \tag{14.93}$$

Substituting into Equation 14.91 from Equations 14.90 and 14.93, we obtain

$$Q = \frac{LV_i^2/2R^2}{V_i^2/2R\omega_o} = \frac{\omega_o L}{R} \tag{14.94}$$

as before.

We have now seen three definitions for the quality factor Q of a resonant circuit. The first, encountered in Chapter 12, was based on the ratio of the undamped resonant frequency to the damping factor for transient excitation (Equation 12.65):

$$Q = \frac{\omega_o}{2\alpha}. \tag{14.95}$$

The second, derived in this chapter (Equation 14.47), was based on the width of the resonant peak in the frequency response for sinusoidal excitation:

$$Q = \frac{\omega_o}{\omega_2 - \omega_1} \tag{14.96}$$

where $\omega_2 - \omega_1$ is the bandwidth, and $\omega_1$ and $\omega_2$ are the frequencies where the response magnitude is down to 0.707 of its peak value.

The third is the relation in terms of stored and dissipated energy at resonance, Equation 14.91:

$$Q = \frac{\text{Stored energy}}{\text{Average energy dissipated per radian}}. \tag{14.97}$$

These definitions all reduce to the same value for second-order circuits, but they yield slightly different values in higher order circuits.

---

EXAMPLE 14.8 TIME-DOMAIN VERSUS FREQUENCY-DOMAIN BEHAVIOR FOR A HIGH-Q RLC CIRCUIT This example uses the quality factor $Q$ for the circuit in Figure 14.37 to deduce the general form of the frequency domain response and the time domain response (specifically, the zero input response) of the circuit. We will focus on $v_C$ and consider two cases: $R = 50\ \Omega$ and $R = 500\ \Omega$.

Using the impedance method, the magnitude of the system function relating $V_c$ to $V_i$ is given by

$$|H_c(\omega)| = \frac{1}{\sqrt{(1 - \omega^2 LC)^2 + (\omega RC)^2}}.$$

$Q$ for the series resonant circuit is given by

$$Q = \frac{1}{R}\sqrt{\frac{L}{C}}.$$

For $R = 50\ \Omega$, $Q = 25$. For this high value of $Q$, following the discussion of $Q$ in Section 14.5.2, we expect to see a peaky frequency response, as can be confirmed by observing the magnitude plot of the frequency response of $H_c$ in Figure 14.41a. The plot in Figure 14.41a also shows that the peak value is 25.

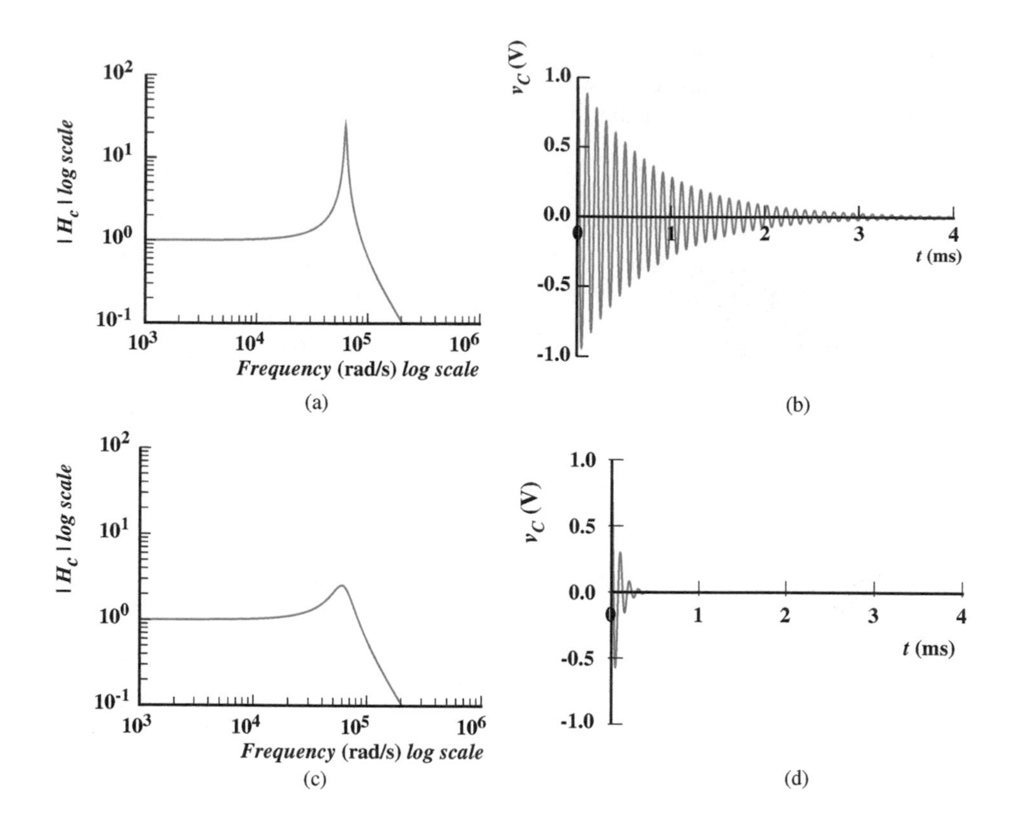


FIGURE 14.41 Time-domain versus frequency-domain behavior of an RLC circuit: (a) frequency response for $Q = 25$, ($R = 50\ \Omega$); (b) transient response $Q = 25$, ($R = 50\ \Omega$); (c) frequency response for $Q = 2.5$, ($R = 500\ \Omega$); (d) transient response for $Q = 2.5$, ($R = 500\ \Omega$).

In the time domain, according to the interpretation of $Q$ discussed in Section 12.2.1, the high value of $Q$ implies that the circuit will ring for many cycles if the input is set to 0 and an initial voltage is present on the capacitor. Figure 14.41b shows the zero input response for this circuit for $v_C(0) = 1$ V. The time domain plot in Figure 14.41b also shows that the circuit oscillates for approximately 25 cycles before it decays to an unobservable level in the graph.

For $R = 500\ \Omega$, $Q = 2.5$, a rather modest value. Thus, we do not expect to see significant peakiness in the frequency response, as can be confirmed by observing Figure 14.41c.

Similarly, from a time-domain viewpoint, we do not expect the circuit to ring for many cycles. Rather, we expect the transients due to an initial voltage on the capacitor to die out quickly. Figure 14.41d confirms this observation.

## 14.7 SUMMARY

- Resonant systems are characterized by a quadratic expression of the form $s^2 + 2\alpha s + \omega_o^2$, with complex roots, in its system function. Systems with complex roots in their system functions display oscillatory behavior.
- The impedance approach provides an analysis of resonant circuits which complements the time-domain calculations of Chapter 12, by showing the behavior of the circuit as a highly selective filter. The selectivity of the filter is related to the quality factor $Q$ of the circuit.
- The performance of a resonant circuit is summarized by its frequency response. The frequency response comprises plots of magnitude and phase angle versus frequency.
- The following constraints provide intuition into the shape of the frequency response (including magnitude and phase) for resonant second-order systems:

    1. the low-frequency asymptote,
    2. the high-frequency asymptote, and
    3. the magnitude and phase of the response at the resonant frequency.

- The quality factor $Q$, the resonant frequency $\omega_o$, and the damping factor $\alpha$ are three key parameters that characterize the behavior of resonant systems. These three parameters can be determined by inspection by writing the resonant system function in standard form, such that a quadratic expression of the form $s^2+2\alpha s+\omega_o^2$ is identifiable. The parameters $\alpha$ and $\omega_o$ are directly identified from the quadratic term, while $Q$ is obtained from:

$$Q = \frac{\omega_o}{2\alpha}.$$

The bandwidth and damped resonant frequency $\omega_d$ are two other important parameters in resonant systems and are given by:

$$\text{Bandwidth} = 2\alpha$$

and

$$\omega_d = \sqrt{\omega_o^2 - \alpha^2},$$

respectively. $\omega_d$ is the frequency at which a resonant circuit actually oscillates. The value of $\omega_d$ is close to that of $\omega_o$ for high-$Q$ circuits.

- For the parallel RLC resonant structure, the voltage across the parallel combination reaches a maximum at

$$\omega_o = 1/\sqrt{LC}, \tag{14.98}$$

the resonant frequency. The damping factor is given by:

$$\alpha = \frac{1}{2RC}$$

and the quality factor $Q$ is

$$Q = \frac{\omega_o}{2\alpha} = R\sqrt{\frac{C}{L}}.$$

The bandwidth for the parallel resonant structure is

$$\text{Bandwidth} = 2\alpha = \frac{1}{RC}.$$

- In the series RLC resonant structure, the current through the elements is maximum at the resonance frequency:

$$\omega_o = 1/\sqrt{LC}.$$

The damping factor is given by:

$$\alpha = \frac{R}{2L}$$

and the quality factor $Q$ is

$$Q = \frac{\omega_o}{2\alpha} = \frac{1}{R}\sqrt{\frac{L}{C}}.$$

The bandwidth for the series resonant structure is

$$\text{Bandwidth} = 2\alpha = \frac{R}{L}.$$

- The bandwidth is related to the resonant frequency by the quality factor:

$$Q = \frac{\text{Resonant frequency}}{\text{Bandwidth}} \tag{14.99}$$

so high $Q$ means narrow bandwidth (or high selectivity).

- Other equivalent definitions for $Q$ are

$$Q = \frac{\omega_o}{2\alpha} \tag{14.100}$$

and

$$Q = \frac{\text{Stored energy}}{\text{Average energy dissipated per radian}} \tag{14.101}$$

at resonance.

- Using $Q$ as the common parameter, the time-domain step response of a circuit can be visualized from the circuit's frequency response, and vice versa. For example, a "peaky" gain versus frequency plot implies ringing in the step response, while the absence of peakiness implies a quick decay of the step response.

- Resonant systems are the basis of second-order filters including the LPF, HPF, BPF, and BSF.

- In later chapters, we will see many RC active filter topologies that exhibit resonance without requiring inductors.[6]

EXERCISE 14.1

a) For the circuit in Figure 14.42, assume a sinusoidal steady state at a fixed frequency $\omega_0$. Determine an equivalent circuit for the $R - L$ parallel combination ($Z_1$) in terms of a resistor $R'$ in series with a suitable inductance $L'$.

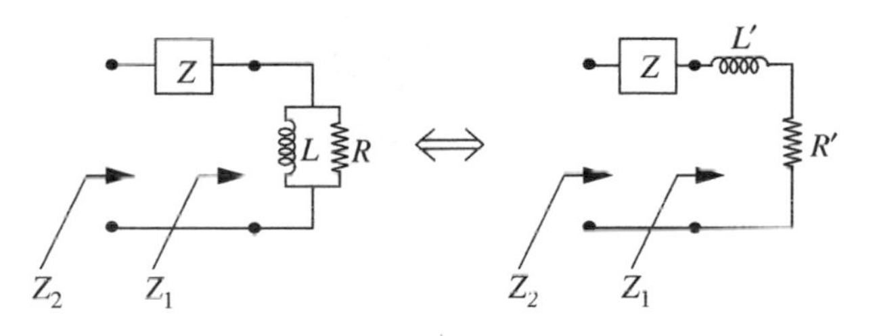


FIGURE 14.42

b) Determine the impedance $Z$ that must be added in series with $Z_1$ such that the total impedance $Z_2$ is equivalent to a pure resistance at frequency $\omega_0$. What is this value of this resistance?

EXERCISE 14.2 For a parallel RLC network with $R = 1$ k$\Omega$, $L = 1/12$ H, $C = 1/3$ $\mu$F, find $\omega_0, f_0, \alpha, Q_0, \omega_d, \omega_1, \omega_2$, and $\beta = \omega_2 - \omega_1$. ($\omega_1$ and $\omega_2$ are the half-power frequencies.)

---

6. We care about this because inductors are hard to fabricate in integrated circuits.

EXERCISE 14.3 A parallel resonant RLC circuit (see Figure 14.43) driven by a current source, 0.2 ($\cos \omega t$), (units of amperes) shows a maximum voltage response amplitude of 80 V at $\omega = 2500$ rad/s. and 40 V at 2200 rad/s. Find $R$, $L$, and $C$.

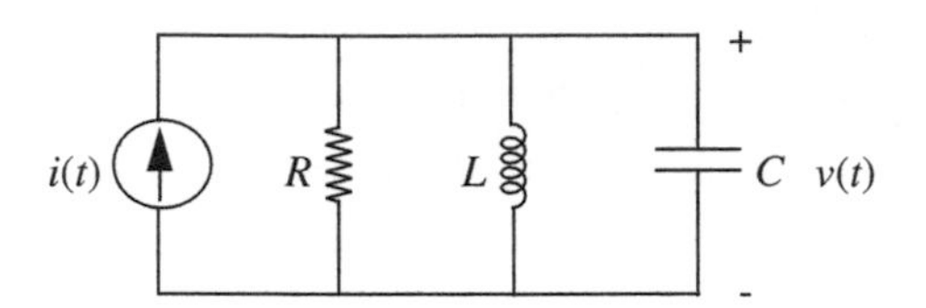


FIGURE 14.43

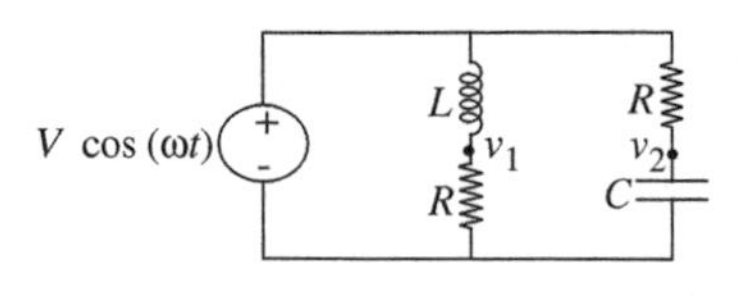


FIGURE 14.44

EXERCISE 14.4 Find an expression for the value of $L$ that will balance the bridge (see Figure 14.44) to make $v_1 - v_2 = 0$, for an input voltage $V \cos(\omega t)$.

EXERCISE 14.5 One or two of the following statements made about the second-order RLC network in Figure 14.45 is/are inconsistent with the rest. Circle the inconsistent statement(s):

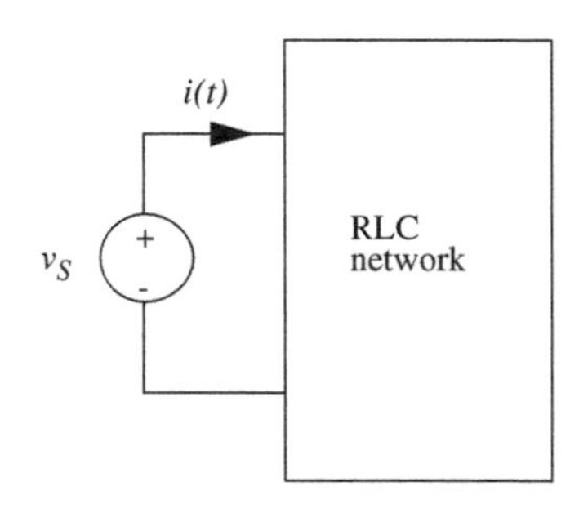


FIGURE 14.45

a) The natural frequencies $s_1$ and $s_2$ of this circuit are as shown in the complex plane (see Figure 14.46).

b) $Q = 1.2$.

c) The admittance function $Y(j\omega) = I(j\omega)/V_s(j\omega) = j2\omega/[(169 - \omega^2) + j10\omega]$.

d) The step response for $t > 0$ is of the form:

$$i(t) = Ae^{-5t} \cos(12t + \phi). \tag{14.102}$$

e) The steady-state response to $v_s(t) = B \cos(25t)$ is of the form:

$$i(t) = C \cos(25t + \Phi). \tag{14.103}$$

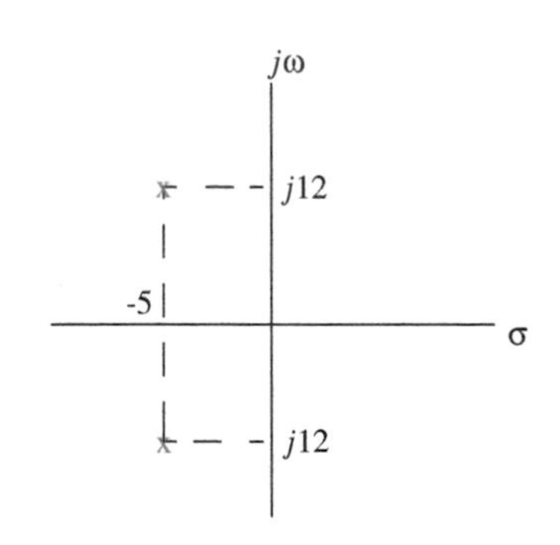


FIGURE 14.46

EXERCISE 14.6 Consider the network shown in Figure 14.47.

a) Show that by proper choice of the value of $L$, the impedance $V_i(s)/I_i(s) = Z_i(s)$ can be made independent of $s$. What value of $L$ satisfies this condition?

b) With $L$ as determined in part (a), what is the value of $Z_i$?

c) Assume that the capacitor voltage and the inductor current are both zero for $t < 0$. Determine $i_C(t)$ for $t > 0$ when $v_I(t)$ is a unit step.

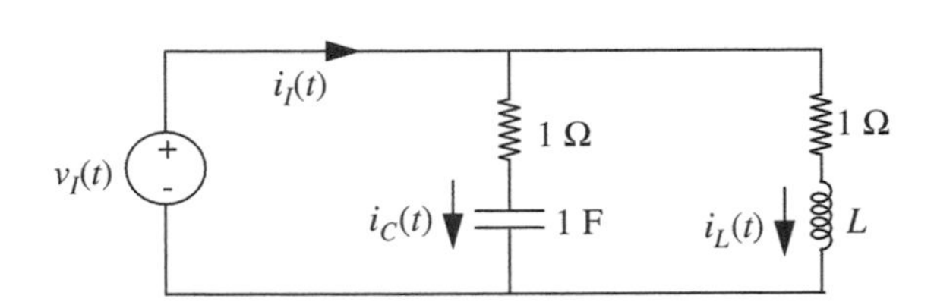


FIGURE 14.47

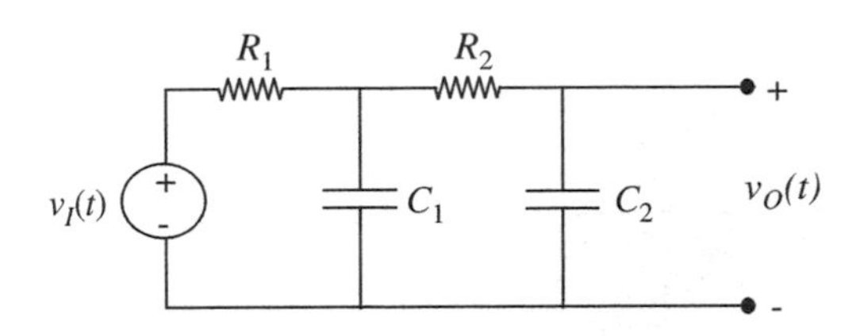


FIGURE 14.48

EXERCISE 14.7 Each of the following parts makes a statement about a second-order system. Indicate whether the statement is true or false.

a) The network shown in Figure 14.48 (with both $R$'s and $C$'s positive) can exhibit natural responses of the form $e^{-\alpha t} \sin \omega t$.

b) The natural response of an RLC network is given by: $v_O(t) = 25e^{-5t} \cos(12t + \pi/7)$. The $Q$ of the network is 1.2.

c) For the circuit shown in Figure 14.49, the output voltage under sinusoidal steady state conditions is zero.

d) The circuit shown in Figure 14.50 contains 3 energy storage elements and thus has 3 natural frequencies.

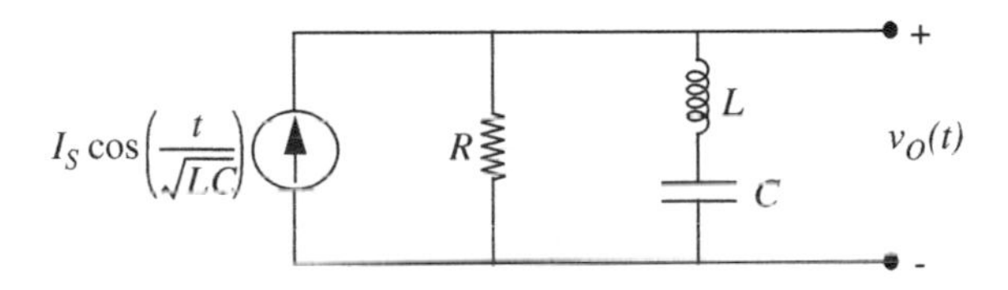


FIGURE 14.49

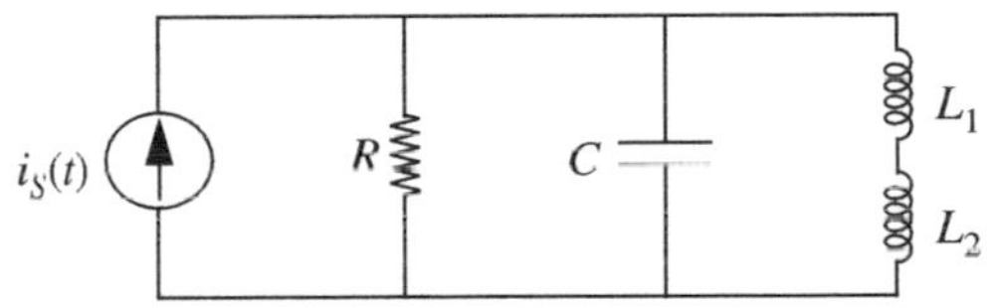

FIGURE 14.50

EXERCISE 14.8 The voltage-transfer ratio of a certain network is shown in Figure 14.51 in Bode-plot form.

This transfer ratio can be expressed in the form:

$$\frac{V_o(s)}{V_i(s)} = \frac{Ks}{(s^2 + s\omega_0/Q + \omega_0^2)(\tau s + 1)}. \tag{14.104}$$

Determine the parameters $K$, $Q$, $\omega_0$, and $\tau$.

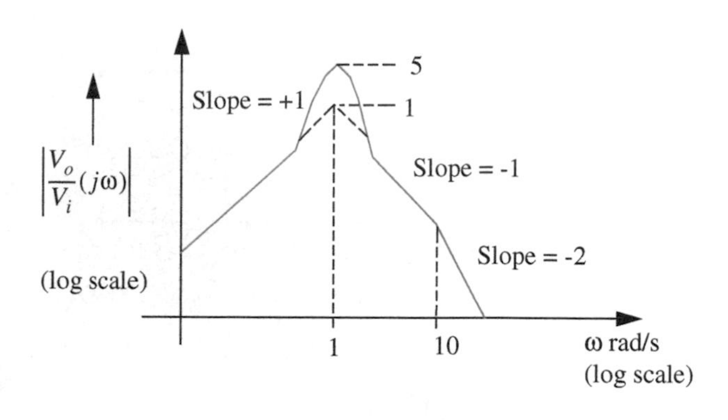


FIGURE 14.51

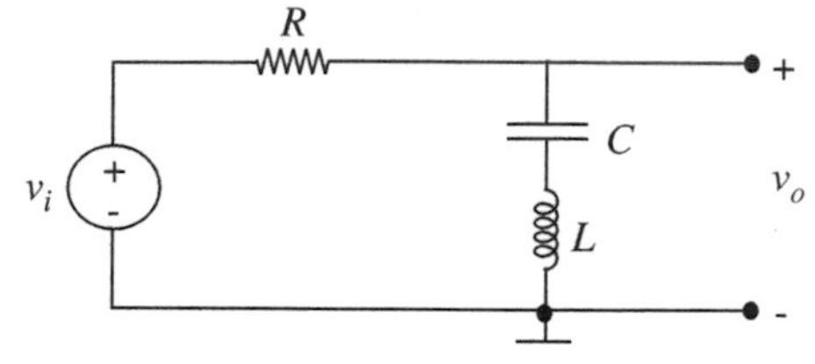

FIGURE 14.52

EXERCISE 14.9

a) In the circuit in Figure 14.52, find an expression for the complex amplitude $V_o$ as a function of $V_i$ after transients have died out, assuming $v_i$ is a sinusoid: $v_i = V_i \cos(\omega t)$.

b) Find $v_o(t)$ at the frequency $\omega_0 = 1/\sqrt{LC}$.

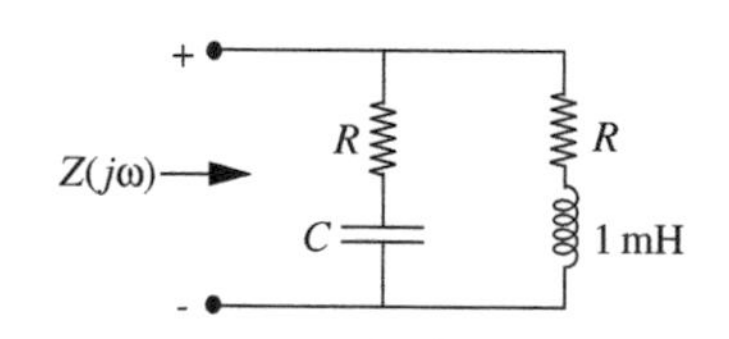


FIGURE 14.53

EXERCISE 14.10 The impedance of the network shown in Figure 14.53 is found to be 2 kΩ and is purely real at all frequencies. The value of the inductor is one mH as shown. What are the values of $R$ and $C$?

## PROBLEMS

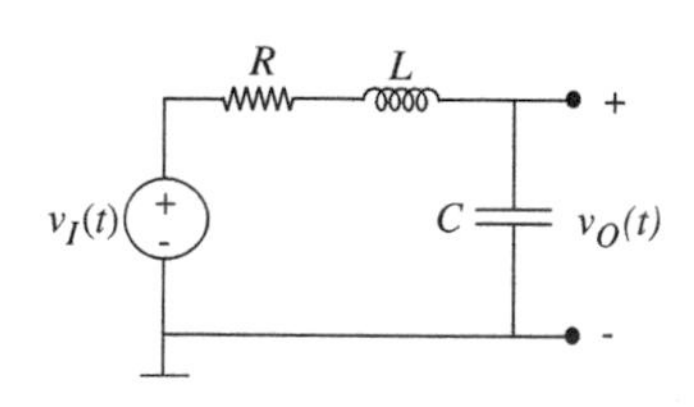


FIGURE 14.54

PROBLEM 14.1 For the series-resonant circuit in Figure 14.54, draw the impedance model, and find the transfer function $V_o/V_i$. Sketch the Bode plot of log magnitude and phase of this function versus log frequency by sketching the asymptotes, then sketching the function. This is a second-order low-pass filter.

For this topology, the maximum amplitude does not occur at the resonant frequency $\omega_0$ (prove this, but don't work out all the math). However, this is a small effect for all but very low $Q$. Find expressions for the resonant frequency (defined as the frequency where the $s^2$ and the $s^0$ terms cancel in the denominator) and the $Q$.

PROBLEM 14.2 Consider the circuit in Figure 14.55.

a) Draw the Bode plot of $|Z(w)|$ for $R = L = C = 1$. What is the resonant frequency?

b) Draw the Bode plot of $|Z(w)|$ for $R = 1, L = C = 2$. What is the resonant frequency?

c) Comment on the results of part (a) and part (b).

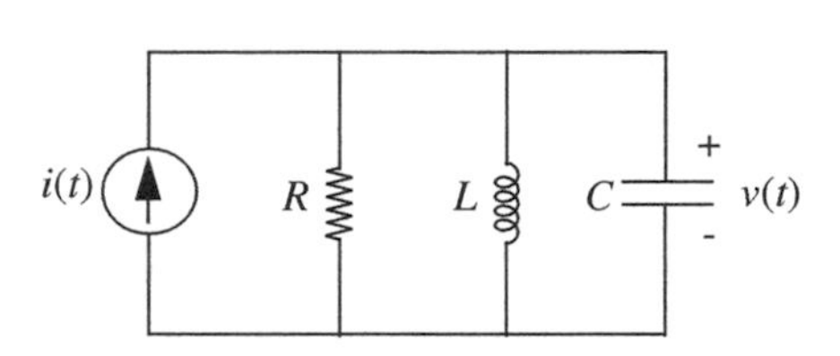


FIGURE 14.55

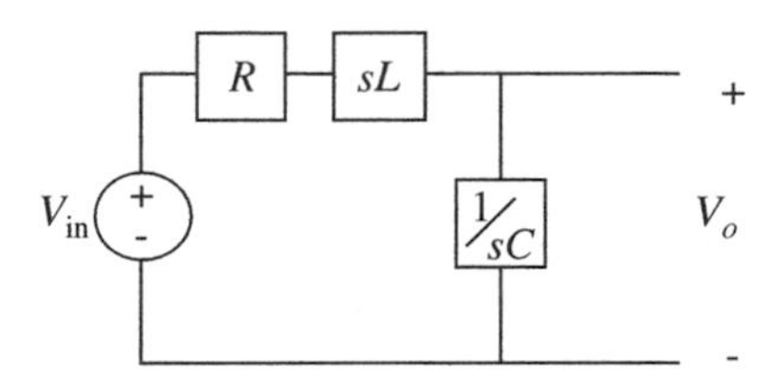


FIGURE 14.56

PROBLEM 14.3 The circuit shown in Figure 14.56 has an input voltage $v_{in1}(t) = V_1 \cos(120\pi t)$, and $L = 500$ mH, $C = 80\ \mu$F, and $R = 50\ \Omega$.

a) Compute the transfer function $H(s) = V_o(s)/V_{in1}(s)$.

b) Set $v_{in1}(t) = 0$. What is the equivalent complex impedance of the circuit evaluated between $V_o$ and ground?

c) Parts (a) and (b) might lead you to believe that Thévenin's Theorem also applies to complex impedances. If this is true, then we can replace the circuit between $V_o$ and ground by a complex Thévenin impedance ($Z_{th}$) and a complex open circuit voltage ($V_{oc}$). Taking $v_{in1}(t) = 10 \cos(120\pi t)$ compute $Z_{th}$ and $V_{oc}$.

d) Having represented the circuit by its Thévenin's equivalent, we wish to connect it to another circuit having $v_{in2}(t) = 10 \cos(200t)$ as shown in Figure 14.57.

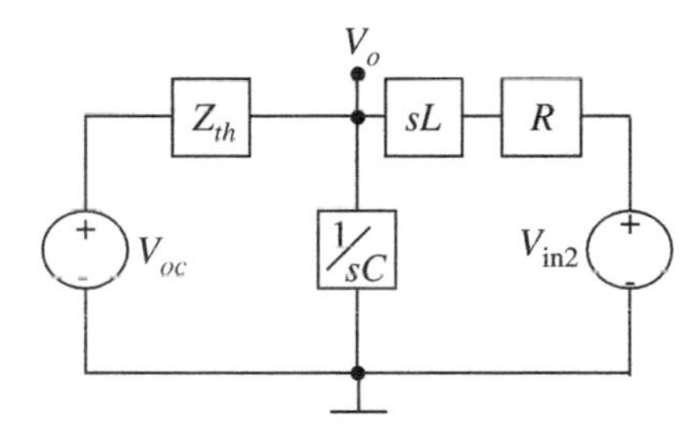


FIGURE 14.57

1) Are there any problems with this approach? If so state them explicitly.

2) Compute the complex $V_o$ for this circuit.

3) Now let $v_{in1} = v_{in2} = 10 \cos(120\pi t)$. Evaluate $V_o$ for this case.

4) If $v_{in1}(t) = v_{in2}(t) = 10 \cos(120\pi t)$ compute the real output voltage $v_o(t)$.

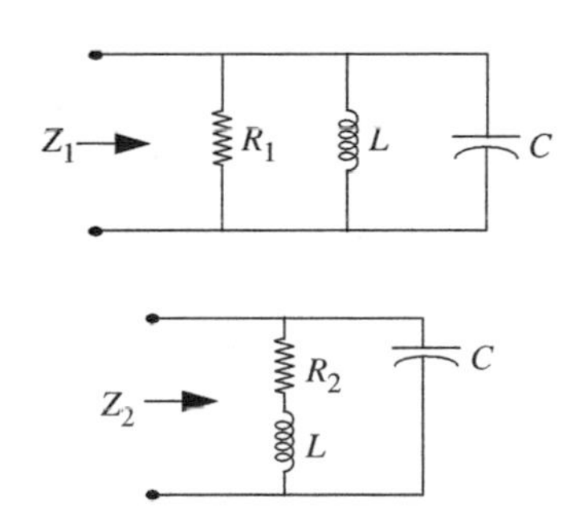


FIGURE 14.58

PROBLEM 14.4

a) Determine $\omega_o, \alpha, \omega_d$, and $Q_1$ for each of the circuits in Figure 14.58 ($Q_1 = \omega_o/2\alpha$).

b) Assume $L = 1$ mH and $C = 10\ \mu$F. Find values of $R_1$ and $R_2$ that will yield $Q_1 = 10$. What is the ratio of $R_1$ to $R_2$?

c) Make a parallel $L' - R'$ equivalent circuit for the $L - R_2$ series combination (as in Exercise 14.1) and use this equivalent circuit to calculate what the ratio of $R_1$ and $R_2$ in part (b) should be for $Q_1 = 10$ in both circuits. How large is the discrepancy, if any?

d) Using the values for $R_1$ and $R_2$ found in part (b), make plots of $|Z_1|$ and $|Z_2|$ versus frequency and $\angle Z_1$ and $\angle Z_2$ versus frequency. Identify the following features of your plot:

  i) The maximum impedance, the frequency $\omega_r$ at which this occurs, and the phase angle at $\omega_r$.

  ii) The frequencies $\omega_1$ and $\omega_2$ at which $|Z|$ is $1/\sqrt{2}$ smaller than the maximum, and the phase angles at $\omega_1$ and $\omega_2$. Calculate the quantity $Q_2 = \omega_r/(\omega_2 - \omega_1)$.

e) Now suppose that you have just been given a "parallel resonant" circuit $Z$, but you don't know whether it is of the $Z_1$ form or the $Z_2$ form. Suggest a step-by-step experimental procedure based on measurements of $|Z|$ and perhaps $\angle Z$ as a function of frequency to determine the following:

  i) which of the two forms of parallel resonant circuit is the best model, and

  ii) specific values for the three elements, $R, L$, and $C$.

PROBLEM 14.5

a) Write down the differential equation describing the circuit in Figure 14.59.

b) Write the transfer function $V_o(s)/V_i(s)$.

c) Solve for $i_I(t)$ assuming $v_I(t) = \cos(\omega t)$ (let $\omega = 1$).

d) Plot the roots of the characteristic polynomial from part (b) on the complex s-plane (assume $R^2C^2 < 4CL$).

PROBLEM 14.6

a) In the circuit in Figure 14.60, given that $v_S = V_S \cos(\omega t)$, where $\omega = 10^6$ rad/s. Design a lossless coupling network containing one inductor and one capacitor that will maximize the power transferred to the antenna at frequency $\omega$.

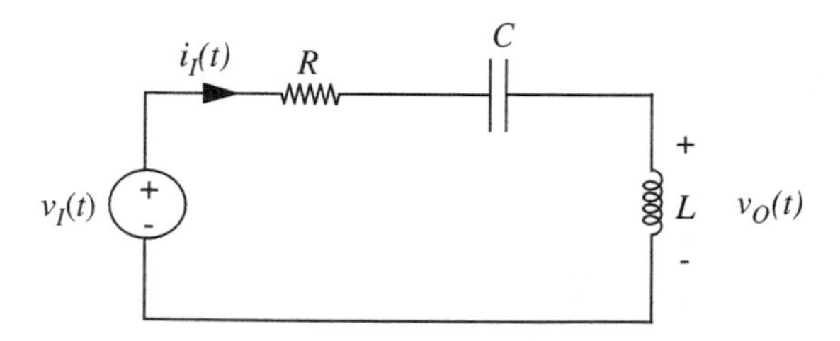


FIGURE 14.59

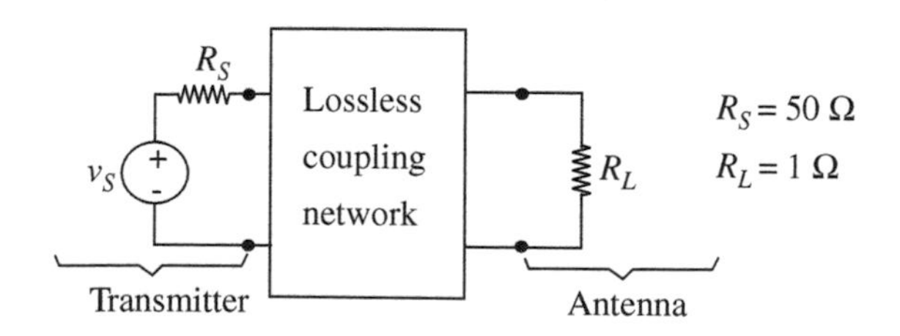


FIGURE 14.60

b) Now suppose that $v_S = V_S \cos(\omega t) + \epsilon \cos(3\omega t)$, where $\epsilon$ represents a small amount of third-harmonic distortion introduced by nonlinearities somewhere in the transmitter. Since the FCC forbids the broadcast of harmonics, it is important to check that coupling networks do not inadvertently favor the coupling of harmonics to the transmitter. For your design in (a), calculate how much third harmonic reaches the antenna.

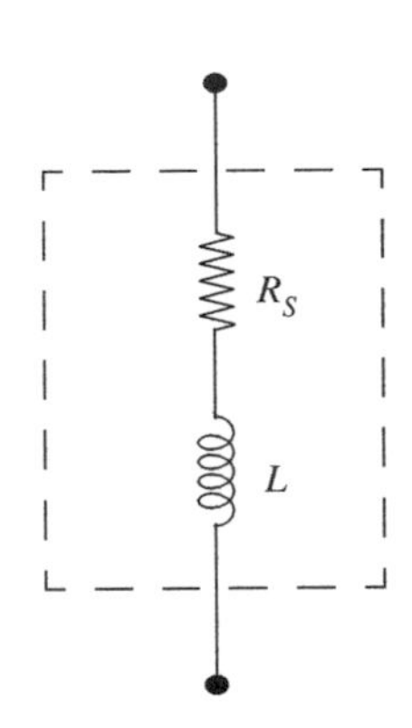


Simple model of a physical inductor

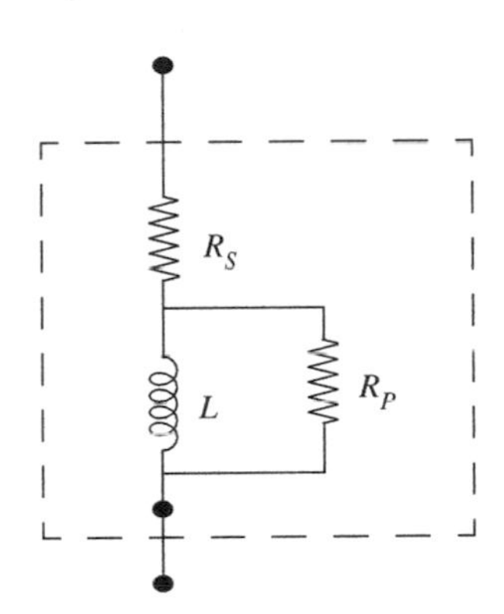


More complex model

FIGURE 14.61

PROBLEM 14.7 Refer to Figure 14.61 for this problem.

The $Q$ of a physical energy storage element may be defined as

$$Q_1 = \frac{Im(Z)}{Re(Z)} \tag{14.105}$$

where $Z$ is the terminal impedance of the element. The $Q$ may also be defined in terms of energy as

$$Q_2 = \frac{2\pi < W >}{E_{\text{diss/cycle}}} \tag{14.106}$$

where $< W >$ is the average stored energy and $E_{\text{diss/cycle}}$ is the energy dissipated per cycle.

a) For the simple inductor model, calculate and compare $Q_1$ and $Q_2$ as functions of frequency.

b) For the more complex model, and assuming $R_P \gg R_S$, sketch $Q_1$ as a function of $\omega$ making reasonable approximations.

c) Suppose two inductors with the same $Q_1$ and $Q_{10}$ are connected in series. Express $Q_1$ for the series combination in terms of $Q_{10}$.

PROBLEM 14.8 Communications receivers require high-Q circuits to separate signals broadcast on adjacent channels. Due to losses, modeled by the parallel resistance $r$, there is a limit to the $Q$ that can be achieved with passive components. In the amplifier circuit in Figure 14.62, a variable resistor $R_F$ has been added which has the effect of increasing the $Q$ of the passive tuned circuit.

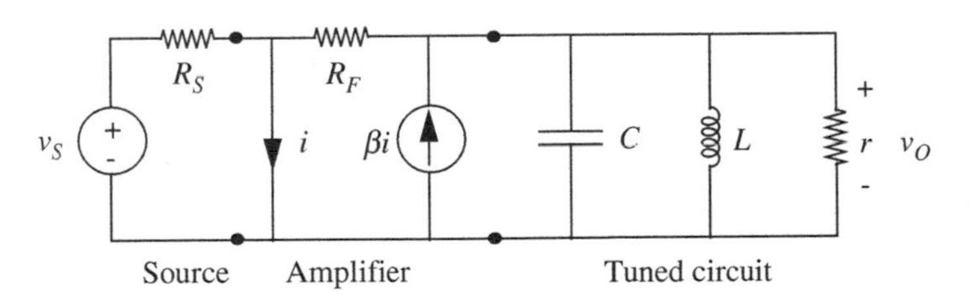


FIGURE 14.62

$$R_S = 1\ \text{k}\Omega, r = 10000\ \Omega, \quad L = \frac{100}{\pi}\mu\text{H}, \quad \beta = 11, \quad R_F \text{ and } C \text{ variable.}$$

a) Consider first the tuned circuit by itself, disconnected from the amplifier. If $C$ is chosen so that the circuit has a 1-MHz resonant frequency, what is its $Q$?

b) Determine the overall transfer function $H(s) = V_o/V_s$.

c) Select values for $C$ and $R_F$ so that the overall frequency response is peaked at a frequency 1 MHz and has a half-power band width of 2 kHz. (Note, the half-power bandwidth $= 2\alpha$.) What is the $Q$ in this case?

PROBLEM 14.9

a) Consider the two circuits in Figure 14.63.

FIGURE 14.63

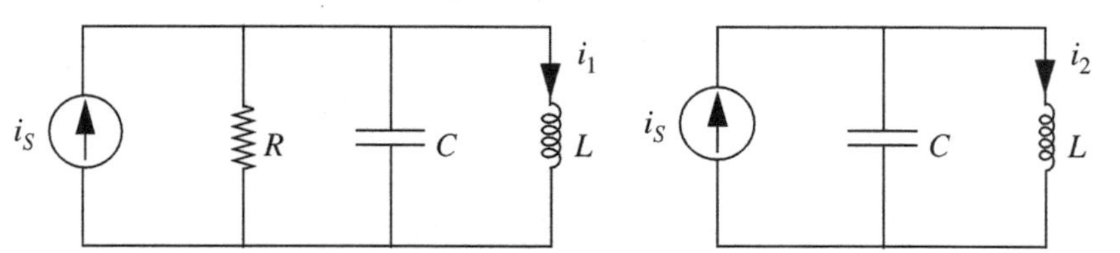


Determine the following transfer functions:

$$H_1(s) = I_1/I_s \text{ and } H_2(s) = I_2/I_s.$$

b) Given $i_s(t) = u_{-1}(t)$, draw the circuits as they would appear in steady state. (Recall that $u_{-1}(t)$ represents a unit step at time $t = 0$.) What are the "forced responses" $i_1^F$ and $i_2^F$?

c) Calculate the "natural responses" $i_1^N$ and $i_2^N$. Assume
$i_L(0) = 0, \quad v_C(0) = 0, \quad R \gg \sqrt{L/4C}$.
Why is $i_2^F$ not the complete steady-state response of the second circuit?

d) Write the step response $i_1 = i_1^F + i_1^N$ and $i_2 = i_2^F + i_2^N$ in terms of $\omega_0$ and $Q$.
Answer:

$$i_1(t) = 1 - e^{-\omega_o t/2Q}\left(\frac{1}{2Q}\sin(\omega_o T) + (\cos\omega_o t)\right)$$

$$i_2(t) = 1 - \cos\omega_o t.$$

e) $i_2(t)$ reaches maxima/minima at $t = \frac{n\pi}{\omega_o}, n = 0, 1, 2, \ldots$. For what value of $n$ does $i_1^N(\frac{n\pi}{\omega_o}) = \frac{1}{5}i_2^N(\frac{n\pi}{\omega_o})$?
For $Q = 5, 50$, and 500 calculate:

$$\frac{i_1^N(\frac{2\pi}{\omega_o})}{i_2^N(\frac{2\pi}{\omega_o})}. \tag{14.107}$$

Sketch $i_1(t)$ for $Q = 50$.

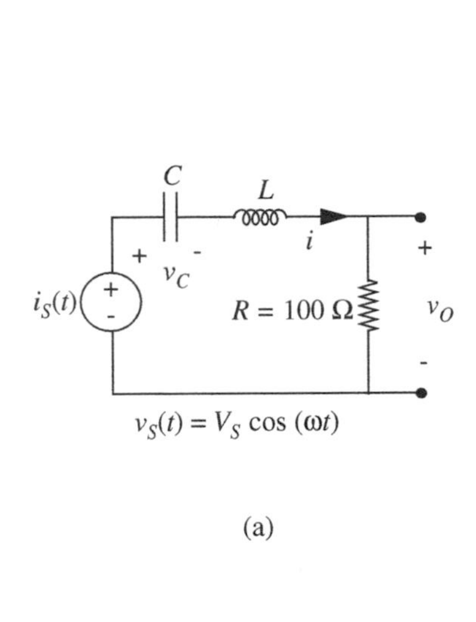


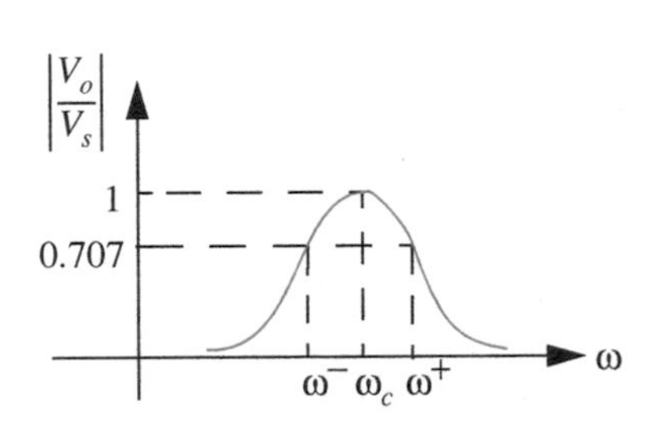


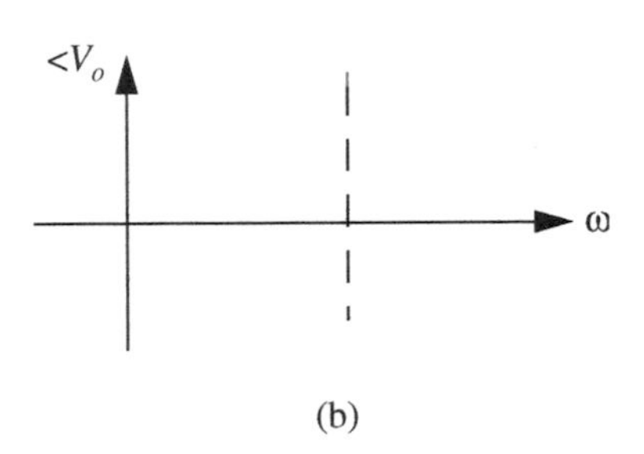


FIGURE 14.64

PROBLEM 14.10 The circuit in Figure 14.64a is to be used as a bandpass filter having the magnitude-frequency curve shown in Figure 14.64b (linear coordinates). The input voltage is

$$v_s(t) = V_s \cos(\omega t)$$

and

$$w_c = 1 \times 10^6 \text{ rad/s}$$

$$w^+ = 1.05 \times 10^6$$

$$w^- = 0.95 \times 10^6 \qquad (14.108)$$

a) Find the appropriate values of $L$ and $C$. Using these values:

   i) Sketch $\angle V_o$ vs. $\omega$.

   ii) Let $v_S = 10 \cos 10^6 t$. Calculate $v_C(t)$, $i(t)$, and $v_O(t)$.

   iii) For $v_S = 10 \cos 10^6 t$, determine the total stored energy $W_s$ and the time-averaged power dissipated.

PROBLEM 14.11 An RLC circuit is shown in Figure 14.65.

The magnitude of $I_i/V_i(j\omega)$ is measured and is as plotted in Figure 14.66 (on log-log coordinates).

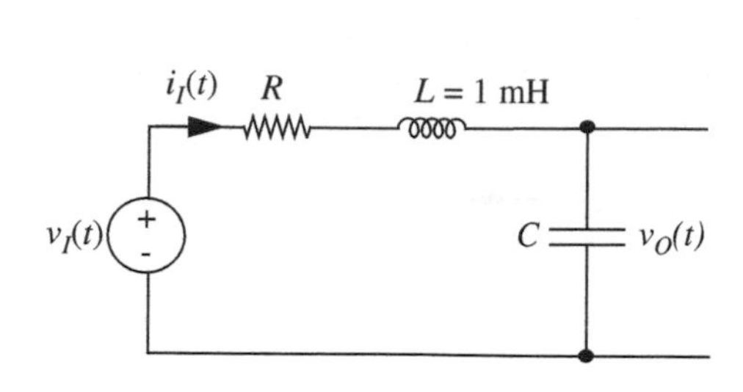


FIGURE 14.65

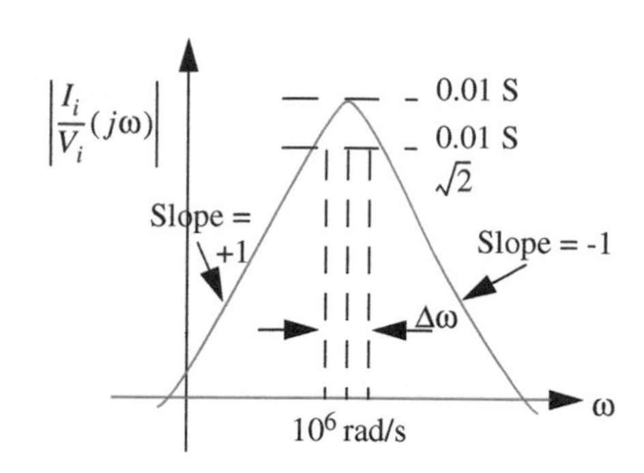


FIGURE 14.66

a) What is the value of $C$?

b) What is the value of $R$?

c) What is the value of $\Delta\omega$?

d) The circuit is now excited with a unit step of voltage. The values of $i_I(t)$ and $v_O(t)$ are zero prior to time $t = 0$.
Sketch the signal $v_O(t)$ for $t$ greater than zero, labeling important features.

PROBLEM 14.12 Refer to Figure 14.67 for this problem.

$v_A = A\cos(400t) \quad A = 141 \text{ kV}, \quad L = 0.25 \text{ H}$

This problem examines a simple model of an electric power system. The source $v_A$ represents the generator in the power plant. The inductance L represents the net effect of all power lines and transformers. The customer's load is represented by resistance $R_L$ to which the capacitor $C$ is added in parts (b) and (c).

a) No capacitor. $R_L = 100\ \Omega$. Find the magnitude of $v_B$ and the average power dissipated in $R_L$.

b) In an attempt to improve on the situation in part (a), the customer adds a capacitor in parallel with his load. He finds that a 25-$\mu$F capacitor works well. Find the magnitude of $v_B$ and the power dissipated in $R_L$ for $R_L = 100\ \Omega$ and $C = 25\ \mu$F.

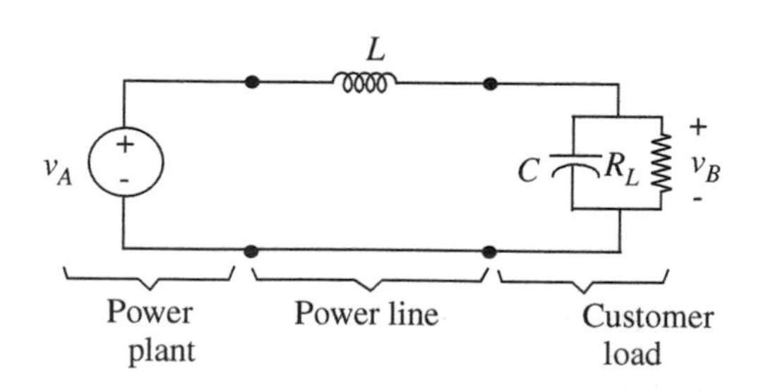


FIGURE 14.67

c) The customer is now very happy. However, before going home for the night, he turns off 90% of his load (making $R_L = 1\ \text{k}\Omega$), at which point sparks and smoke begin to appear in the equipment still connected to the power line. The customer calls you in as a consultant to straighten things out:

   i) Why did sparks appear when the customer tried to turn off 90% of the load?

   ii Assuming a variable $R_L$ in the range $100 \le R_L \le 1000\ \Omega$ provide the customer with a simple formula he can use to calculate the right value of $C$ so that the magnitude of $v_B$ is always equal to 141 kV.

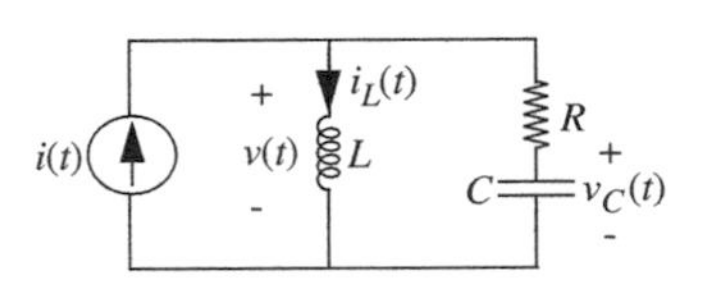

FIGURE 14.68

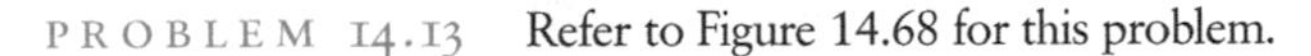
PROBLEM 14.13 Refer to Figure 14.68 for this problem.

$\frac{R}{2L} = 5 \quad \frac{1}{LC} = 16 \quad R = 25 \quad \frac{1}{RC} = 1.6$

a) Assume that $i(t) = 0$ for $t > 0$, and that $i_L(0) = 0, v_C(0) = V_o$. Find $v_C(t)$ for $t > 0$. Simplify your answer, and make a rough sketch of $v_C(t)$ showing its behavior.

b) Find the transfer function (system function) relating $V(s)$ to $I(s)$.

c) When $i(t) = 2e^{-3t}$, it is known that the voltage $v(t)$ can be expressed as:

$$v(t) = Ae^{s_1 t} + Be^{s_2 t} + De^{-3t}. \tag{14.109}$$

Find $s_1, s_2$, and $D$. (You need not find $A$ and $B$.)

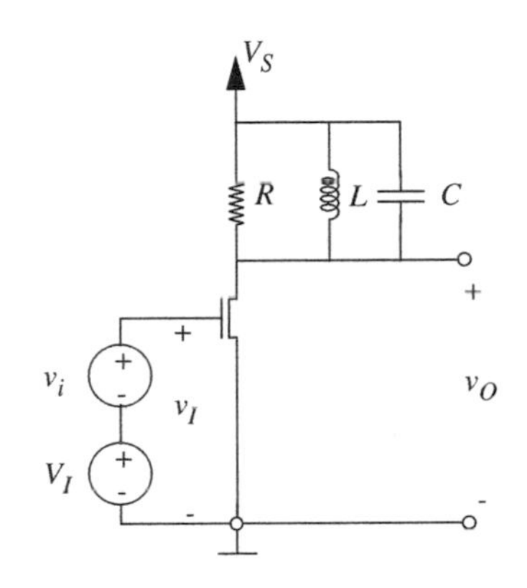

FIGURE 14.69

PROBLEM 14.14 Refer to Figure 14.69 for this problem.

$V_T = 1\ \text{V} \quad K = 1\ \text{mA/V}^2$

For $v_i(t)$ a small sinusoidal voltage, choose $V_I, R, L$, and $C$ to give a resonance at $\omega = 10^5$ rad/s, $Q = 10$, and an incremental gain $v_o/v_i$ at resonance of $-2$. Use the incremental model.

PROBLEM 14.15 The two networks shown in Figure 14.70 are driven in sinusoidal steady state by the voltage $v_I(t) = V_I \cos(\omega t)$. Their outputs take the form $v_O(t) = V_O \cos(\omega t + \phi)$.

a) For both networks, find $V_O$ and $\phi$ as functions of $V_I$ and $\omega$ using impedance methods.

b) For both networks, let $R = 1000\ \Omega$, $L = 47$ mH, and $C = 4.7$ nF. Plot and clearly label $V_O/V_I$ for $2\pi \times 10^3$ rad/s $\le \omega \le 2\pi \times 10^5$ rad/s; use a linear axis for $V_O/V_I$, and a logarithmic axis for $\omega$. You need only plot enough points to outline the dependence of $V_O/V_I$ on $\omega$.

c) Describe the filtering function of each network, and how each network acts to perform its function.

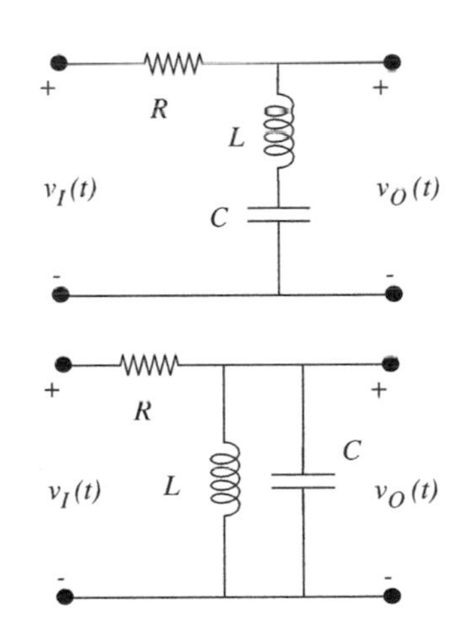

FIGURE 14.70

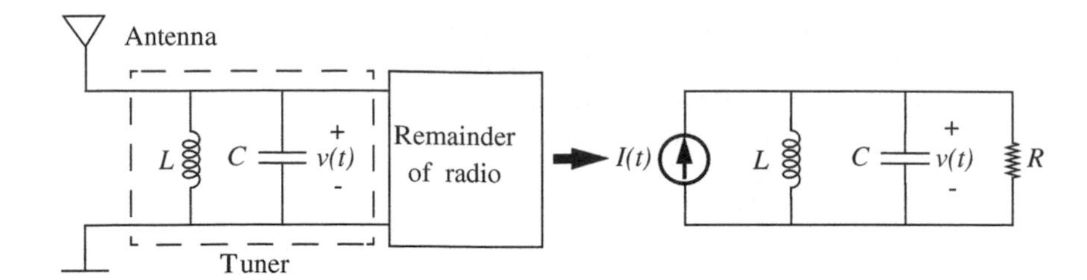


FIGURE 14.71

PROBLEM 14.16 This problem examines the very simple tuner for an AM radio shown in Figure 14.71. Here, the tuner is the parallel inductor and capacitor. The injection of radio signals into the tuner by the antenna is modeled by a current source, while the Norton resistance of the antenna in parallel with the remainder of the radio is modeled by a resistor. (You can learn more about antenna modeling in follow-on courses in Electromagnetic Waves.) The AM radio band extends from 540 kHz through 1600 kHz. The information transmitted by each radio station is constrained to be within ±5 kHz of its center frequency. (You can learn more about AM radio transmission in courses in signals and systems.) To prevent frequency overlap of neighboring stations, the center frequency of each station is constrained to be a multiple of 10 kHz. Therefore, the purpose of the tuner is to pass all frequencies within 5 kHz of the center frequency of the selected station, while attenuating all other frequencies.

a) Assume that $I(t) = I\cos(\omega t)$. Find $v(t)$ where $v(t) = V\cos(\omega t + \phi)$, and both $V$ and $\phi$ are functions of $\omega$. Note that $v(t)$ is the output of the tuner, namely the signal that is passed on to the remainder of the radio.

b) For a given combination of $I$, $C$, $L$, and $R$, at what frequency is V maximized?

c) Assume that $L = 365\ \mu$H. Over what range of capacitance must $C$ vary so that the frequency of maximum $V/I$ may be tuned over the entire AM band? (Note that tuning the frequency of maximum $V/I$ to the center frequency of a particular station tunes in that station.)

d) As a compromise between passing all frequencies within 5 kHz of a center frequency and rejecting all frequencies outside that band, let the design of $R$ be such that $V(1\text{ MHz} \pm 5\text{ kHz})/V(1\text{ MHz}) \approx 0.25$ when the tuner is tuned to 1 MHz. Given this design criterion, determine R.

e) Given your design for R, determine $V(1\text{ MHz} \pm 10\text{ kHz})/V(1\text{ MHz})$. Also, determine $Q$ for the tuner and its load resistor when the tuner is tuned to 1 MHz.

# CHAPTER 15

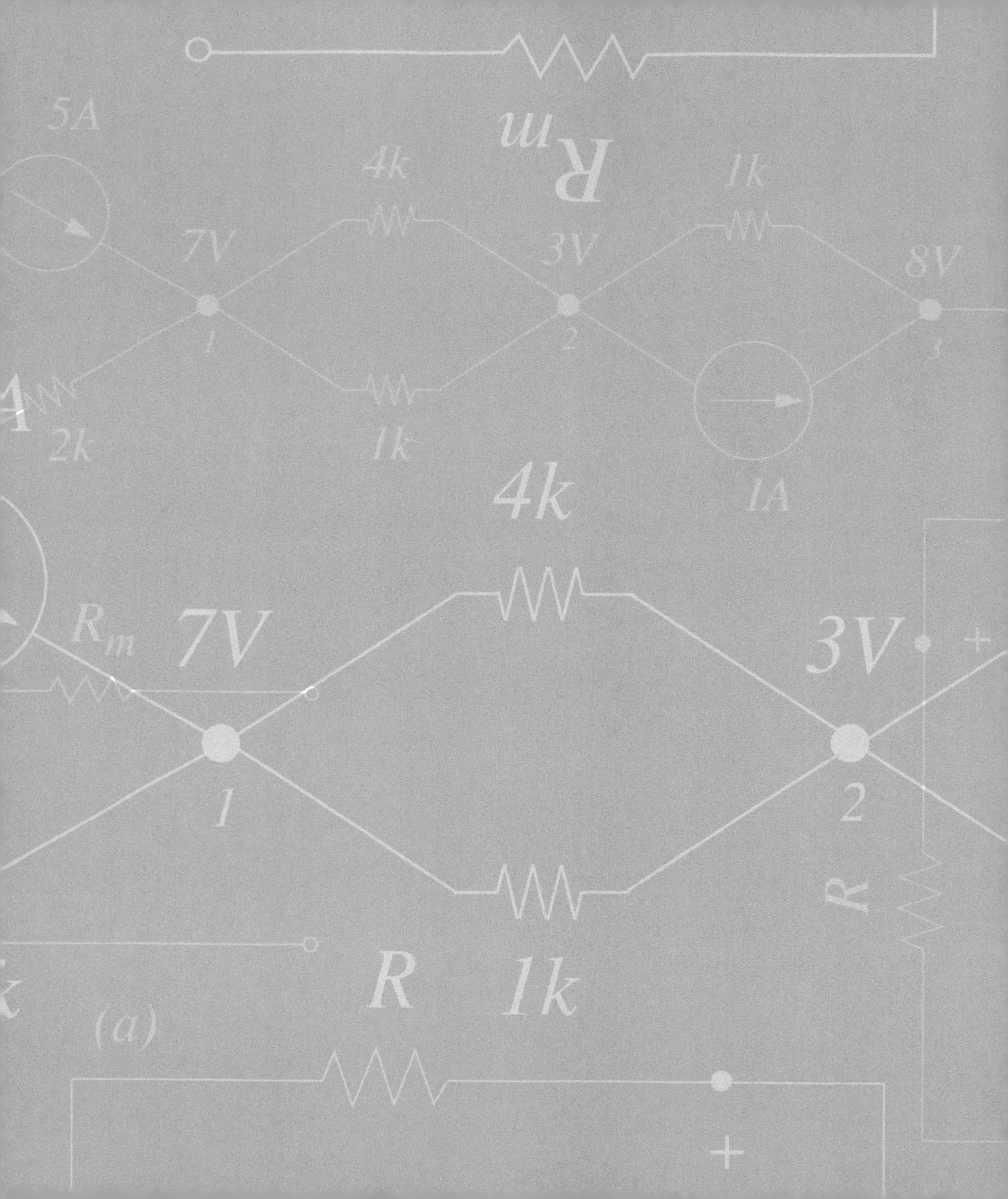

5A
4k
1k
7V
3V
8V
1
2
3
2k
1k
1A
4k
$R_m$
7V
3V
+
1
2
R
R
1k
(a)
+

# THE OPERATIONAL AMPLIFIER ABSTRACTION

# 15

## 15.1 INTRODUCTION

This chapter introduces a very powerful amplifier abstraction called the *operational amplifier* or *Op Amp*. Much as the gate abstraction forms the foundation of most of digital electronics, the operational amplifier forms the basis for much of electronic circuit design.

The Op Amp is a multistage two-input differential amplifier that is designed to be an almost ideal control device, specifically, a voltage-controlled voltage source. An abstract representation of the operational amplifier shown in Figure 15.1 suggests it is a four-port device. The four ports are an input port, an output port, and a pair of power ports. A $+V_S$-voltage (for example, 15 volts) is applied at the plus power port and a $-V_S$-voltage (for example, $-15$ volts) is applied at the minus power port. An input voltage (the control) applied across the non-inverting and inverting input terminals of the Op Amp is amplified by a large amount and appears at the output port. In the operational amplifier abstraction, the input impedance across the input port is infinity, and the output impedance is zero. The gain, or the factor by which the input voltage is amplified, is also infinity.

This chapter uses the Op Amp to construct more complex circuits using its simple, abstract model. Internally, the Op Amp itself is a moderately complicated circuit (see, for example, Figure 15.2) and its design is beyond the scope of this book. Briefly, it contains an input stage not unlike the differential amplifier

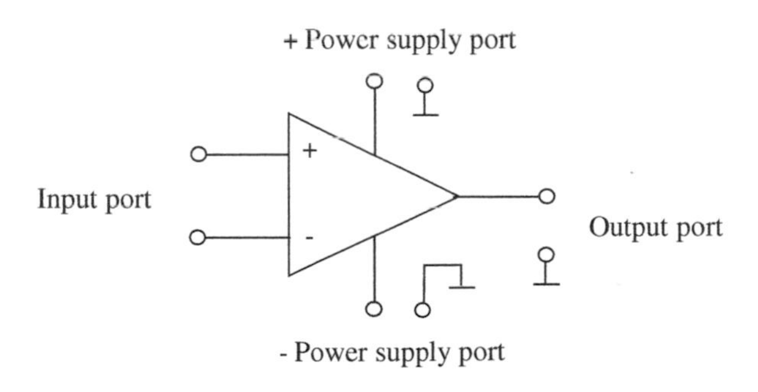


FIGURE 15.1 The operational amplifier abstraction.

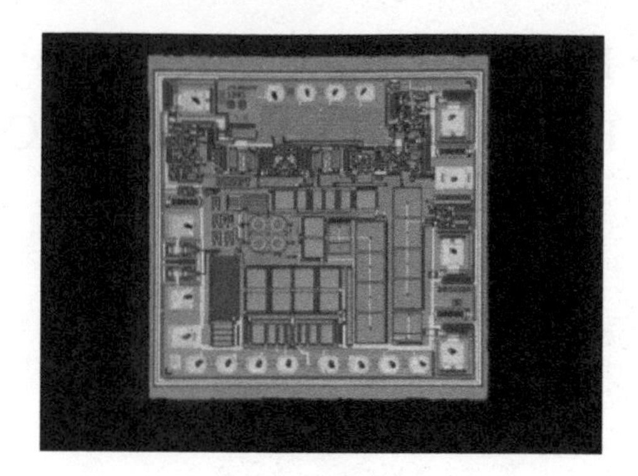

**FIGURE 15.2** A chip photograph of the MAX406 Op Amp from Maxim Integrated Products. The chip is roughly 2mm on a side. (Photograph Courtesy of Maxim Integrated Products)

discussed in Example 7.21 in Chapter 7, or Example 8.3 in Chapter 8. This differential input stage gives the Op Amp its high input resistance, and a high gain. It also converts the differential input voltage to a single-ended output.[1] Typical Op Amps also have a second stage similar to the second stage in Example 7.21 in Chapter 7, which provides additional amplification and level shifts the output voltage to zero when both inputs are equal. Op Amps may also have an output stage similar to the buffer illustrated in Figure 8.40 in Chapter 8, which gives the Op Amp its low output impedance.

In this chapter, initially, our discussion will be in terms of circuits containing Op Amps and resistors. After the basic ideas of Op Amps used as dependent sources and negative feedback have become familiar, circuits with both capacitors and resistors will be introduced.

### 15.1.1 HISTORICAL PERSPECTIVE

The name *operational amplifier* originates from the bygone days of the analog computer (1940–1960), in which the constants in differential equations were represented by the gains of amplifiers. Thus these amplifiers, constructed from balanced pairs of specially manufactured vacuum tubes, had to have reliable, known, fixed gains. Because transistors are inherently more temperature-dependent than vacuum tubes, it was at first thought that satisfactory transistor Op Amps could not be built. But in 1964, it was discovered that by fabricating balanced transistor pairs close together on a single silicon chip to minimize thermal gradients, the temperature problems could be overcome. And thus were born in rapid succession the 703, the 709, and then the ubiquitous 741. Op Amps are rarely used for analog computers now, but instead have become universal building blocks in all aspects of analog circuitry.

---

1. Op Amps with single-ended inputs are also useful. Example 15.1 discusses one such circuit.

## 15.2 DEVICE PROPERTIES OF THE OPERATIONAL AMPLIFIER

The symbol and standard labeling for the operational amplifier are shown in Figure 15.3a. The two required external power supplies have been explicitly shown in the diagram, although showing them is not the usual practice. All five currents have been labeled, in addition to appropriate node voltages, referred to the indicated common ground terminal. In this primitive circuit, the voltage $v_i$ is used to *control* the output voltage $v_o$. Let us examine this control function in detail to find out both the extent of the control, and the cost of the control; that is, how much power must be applied from source $v_i$ to control a given amount of power at the $v_o$ terminal. To address the first problem, we set up the circuit exactly as in Figure 15.3, and measure the output voltage $v_o$, both as a function of time and as a function of $v_i$, assuming $v_i$ is some low-frequency sinusoid. The results are shown in Figures 15.3b and 15.3c. Note the difference in scale of the voltage axes, indicating that the output voltage is perhaps 300,000 times as large as the input voltage. The plot of $v_o$ versus $v_i$ shows a region around the origin where $v_o$ is fairly linearly related to $v_i$, but much beyond this range the control becomes ineffective, and $v_o$ stays at a fixed voltage, or *saturates*, at roughly either $+12$ volts or $-12$ volts, depending on the polarity of $v_i$. The curves will also differ for different samples of the same Op Amp type.

Separate measurements on the device, not illustrated in Figure 15.3, would indicate that the maximum output current $i_o$ is about 10 mA for the 741, and that the input currents $i^-$ and $i^+$ are extremely small, of the order of $10^{-7}$ amps. Thus it is obvious without any formal calculation that the amount of input power required for the control function is orders of magnitude smaller than the power that can be controlled at the output.

The curve of output voltage versus input voltage, Figure 15.3c, is nonlinear. But we also observe that the device has very large voltage gain, defined as $\Delta v_o/\Delta v_i$. We certainly would be willing to sacrifice substantial amounts of gain in return for a corresponding improvement in linearity. Fortunately, the addition of two resistors to the circuit results in precisely this trade-off. Figure 15.4 shows one possible circuit configuration, and the resulting relation between $v_o$ and $v_i$. We will have more to say about this circuit in Section 15.3.1.

### 15.2.1 THE OP AMP MODEL

To gain some insight about how the circuit in Figure 15.4 is working, we first need a circuit model that approximates the Op Amp behavior illustrated in the data in Figure 15.3. On the basis of the preceding chapters, we are led to assign node voltages as in Figure 15.3a, and apply KCL to the circuit. The current law equation turns out to be not very helpful, but it is important to understand why, so we proceed. From Figure 15.3a,

$$i^+ + i^- + i_{p1} + i_{p2} + i_o = 0. \tag{15.1}$$

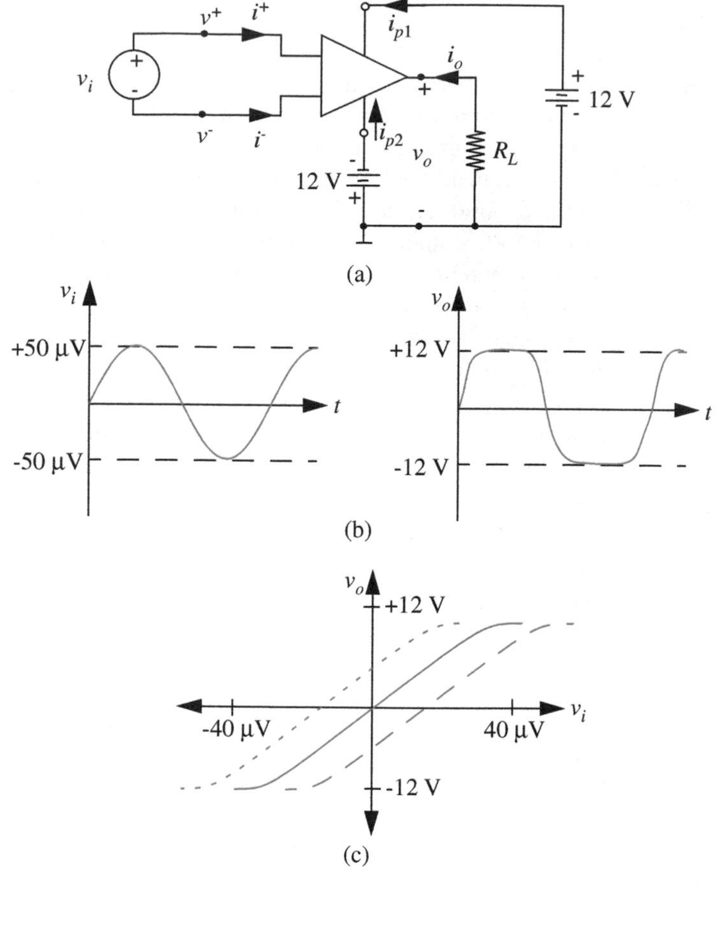


FIGURE 15.3 Operational amplifier characteristic. As illustrated by the dashed lines in (c), different devices of the same type might have different characteristics. The characteristics might also depend on temperature.

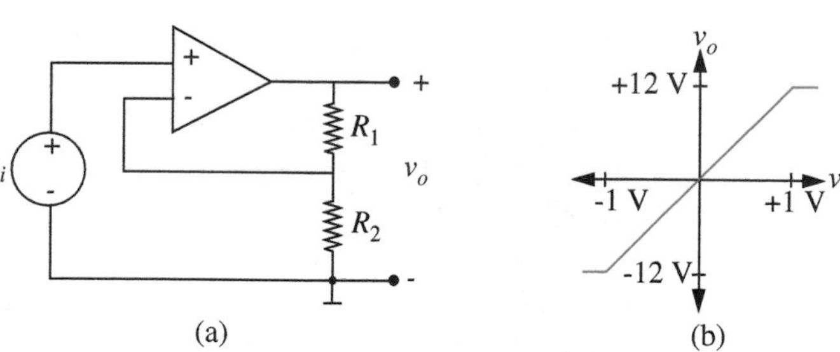


FIGURE 15.4 Non-inverting Op Amp amplifier. The characteristic in (b) assumes $R_1/R_2 = 11$.

As noted in Equation 15.1, $i^+$ and $i^-$ are about four orders of magnitude smaller than $i_o$ hence

$$i_o \simeq -i_{p1} - i_{p2}. \tag{15.2}$$

But $i_{p1}$ and $i_{p2}$ are both power supply currents, so Equation 15.2 merely states that the output current comes from the power supplies. Important, but not very useful (except possibly for the calculation of power dissipation).

Figure 15.3c offers more insight. We see that in the center of the characteristic, the output voltage is approximately proportional to the input voltage, or, more precisely, to the difference between $v^+$ and $v^-$. (Note that $v^+$ and $v^-$ are *labels* for voltages, and hence each can be positive or negative, depending on the circuit.) If we idealize this relationship by making it linear, then the curve of Figure 15.5 results. The curve can now be expressed mathematically as

$$v_o = A(v^+ - v^-). \tag{15.3}$$

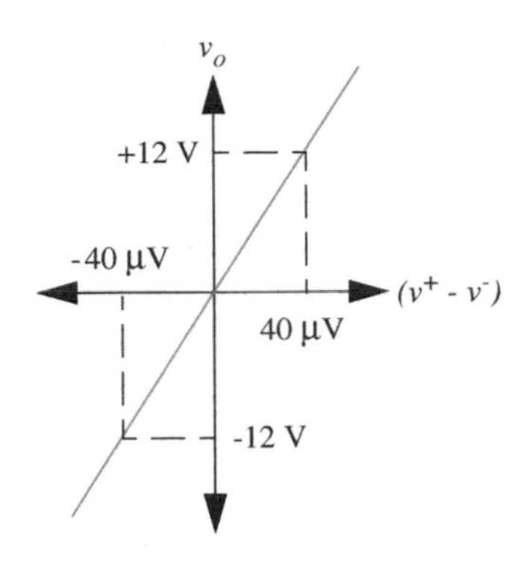


FIGURE 15.5 Linearized characteristic.

This is the mathematical representation of a *voltage-dependent voltage source*, controlled by $(v^+ - v^-)$. For this particular device the constant A, the voltage gain, is 300,000.

The model in Figure 15.6 represents Equation 15.3 in circuit terms. To clearly distinguish the dependent source from an independent source, as before, all dependent sources are represented by diamond-shaped symbols. The disembodied wires on the left of the diagram are distressing at first sight, but merely indicate that the input current to this ideal voltage-controlled voltage source is zero by definition; that is, $i^+ = i^- = 0$.

The dependent source of Figure 15.6 by itself is clearly an imperfect model of an Op Amp. The saturation so clearly present in Figure 15.3c is missing from Figure 15.5 and from the model of Figure 15.6, as is the temperature dependence. To simplify the initial discussion, we shall ignore saturation effects in Op Amps when discussing linear circuits by assuming that we always operate in the central linear part of the amplifier characteristic. We will specifically examine saturation behavior of Op Amps in Section 15.7.

As a summary, the idealized Op Amp model shown in Figure 15.6 has the following properties:

- The output voltage

  $$v_o = A(v^+ - v^-)$$

  where the gain $A \to \infty$. The output resistance is 0.
- The input currents $i^+ = 0$ and $i^- = 0$. Accordingly, the input resistance is infinite.

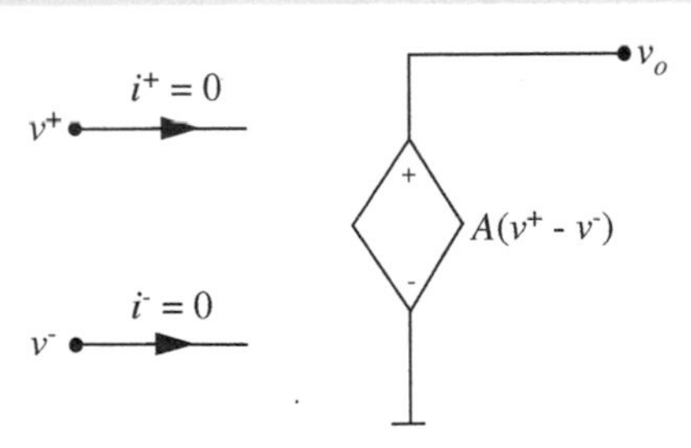


FIGURE 15.6 Voltage-controlled voltage source.

## 15.3 SIMPLE OP AMP CIRCUITS

### 15.3.1 THE NON-INVERTING OP AMP

Now we are in a position to find an analytical relation between $v_O$ and $v_i$ for the circuit in Figure 15.4. We replace the Op Amp by the linear model in Figure 15.6, as shown in Figure 15.7, then analyze this linear circuit by the methods of Chapter 3. The voltage variables defined in Figure 15.7 are in fact the node variables for the circuit, so we can use the node method to derive three independent expressions relating the three unknown voltages.

First, notice that

$$v^+ = v_i \tag{15.4}$$

since $v_i$ is the branch voltage between $v^+$ and the ground node.

Next, recalling that the model specifically assumes no input current, that is, $i^- \simeq 0$, we write the node equation at the node with voltage $v^-$ as

$$\frac{v^-}{R_2} + \frac{v^- - v_O}{R_1} = 0.$$

or,

$$v^- = \frac{R_2}{R_1 + R_2} v_O. \tag{15.5}$$

The dependent-source relation yields our third equation:

$$v_O = A(v^+ - v^-). \tag{15.6}$$

Substituting and solving, we obtain

$$v_O = \frac{Av_i}{1 + A\frac{R_2}{R_1+R_2}}. \tag{15.7}$$

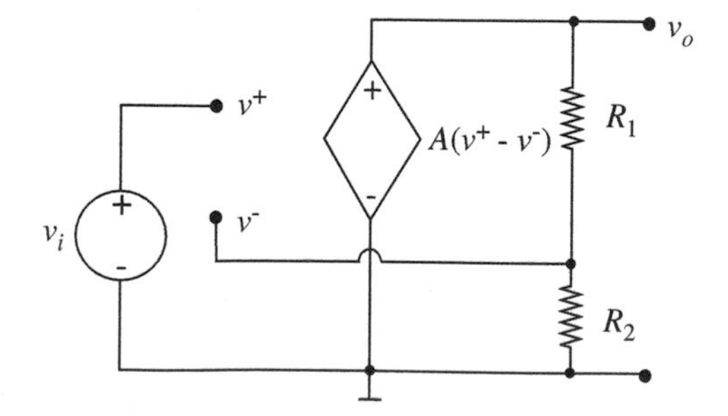


FIGURE 15.7 Model of a non-inverting Op Amp amplifier.

Recall that $A$ is very large, 300,000 in our case, so if the voltage divider does not introduce too much attenuation,

$$\frac{AR_2}{R_1 + R_2} \gg 1 \tag{15.8}$$

hence we can neglect the "1" term in the denominator of Equation 15.7 to obtain the approximate result:

$$v_o \simeq \frac{R_1 + R_2}{R_2} v_i. \tag{15.9}$$

This is an important result. It says that the relation between $v_o$ and $v_i$, is *almost independent of the somewhat unreliable gain constant A of the original Op Amp.*

In other words, because resistor values are stable, reliable and very insensitive to temperature, we expect $v_o$ in this circuit to be a stable reliable function of $v_i$. But this reliability has come at a price: The gain is now much less than for the Op Amp alone—somewhere between 1 and 1,000 depending on the choice of $R_1$ and $R_2$ (but not more than 1000, or the inequality, Equation 15.8, will no longer be valid).

Several important conclusions can be drawn from this simple example:

- It is possible to construct from a high-gain Op Amp and a pair of resistors a reliable amplifier with a known fixed gain. This particular configuration is called the *non-inverting connection.*

- *Negative feedback.*

The basic structure of this circuit, in which some of the output signal is brought back to the input of the circuit, and compared with the input signal, is called *negative feedback.*

For the feedback to be negative in simple Op Amp circuits, the attenuated output signal must be fed back to the $v^-$ terminal. If the output signal is fed back only to the $v^+$ input, very different behavior results, as we shall see. The first-order consequences of these connections will be explored in this chapter, but more complex issues of stability and oscillations are dealt with in books on Signals and Systems.

- We have chosen to model the Op Amp by the dependent source of Figure 15.6, which is a voltage-controlled voltage source, for obvious reasons.

- Although the +12 volt and −12 volt DC power supplies are obviously necessary for Op Amp operation, (they power the voltage-controlled voltage source), their inclusion in the circuit model we use for analysis is not very helpful, because the KCL calculation does not yield a useful relation. Calculating the current through a voltage source rarely provides useful insight, because a voltage source can support any current.

The use of feedback as a way of building stable reliable systems is so intertwined with our daily lives that we are totally unaware of it. Familiar examples are household furnace controls, and cruise controls and anti-lock brakes on automobiles.

### 15.3.2 A SECOND EXAMPLE: THE INVERTING CONNECTION

Another very common Op Amp circuit, the "inverting connection", is shown in Figure 15.8a. For negative feedback, the signal from the output must find its way to the *negative* terminal of the Op Amp, as shown.

If we use the Op Amp model in Figure 15.6, hereafter referred to as the *ideal Op Amp model*, then the circuit model for the inverting amplifier is as shown in Figure 15.8b. Following the same analysis method as before, we will derive three independent equations relating the three unknown node voltages $v^+$, $v^-$, and $v_O$. Accordingly, by inspection:

$$v^+ = 0.$$

Summing the currents at the $v^-$ node, we find, assuming the $v^-$ terminal of the Op Amp draws no current:

$$\frac{(v_i - v^-)}{R_a} + \frac{(v_O - v^-)}{R_b} = 0. \tag{15.10}$$

Hence

$$v^- = \frac{R_b}{R_a + R_b} v_i + \frac{R_a}{R_a + R_b} v_O.$$

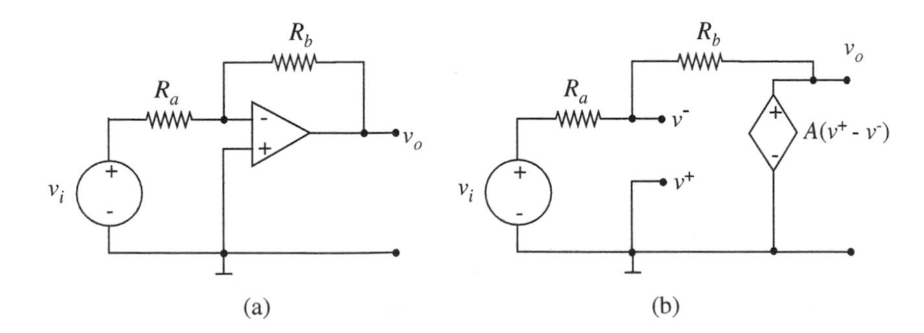


FIGURE 15.8 Inverting Op Amp.

The Op Amp output relation yields

$$v_o = A(v^+ - v^-). \tag{15.11}$$

Substituting and solving, we obtain

$$v_o = \frac{-AR_b/(R_a + R_b)}{1 + AR_a/(R_a + R_b)} v_i. \tag{15.12}$$

As before, if we assume $A$ is of the order of $10^5$, and the resistor ratio $R_a/(R_a + R_b)$ is not less than 0.001, then

$$A\frac{R_a}{R_a + R_b} \gg 1 \tag{15.13}$$

and Equation 15.12 can be approximated as:

$$v_o \simeq -\frac{R_b}{R_a} v_i. \tag{15.14}$$

Again we have a relation between the input and the output voltage that is almost independent of the unreliable gain $A$, and dependent only on resistor ratios. But this time the output signal is *inverted* compared to the input signal, as indicated by the minus sign. Equation 15.14 for the inverting connection and the corresponding equation for the non-inverting case, Equation 15.9, are encountered so frequently that they rapidly become primitives in our circuit analysis repertoire, as with the voltage-divider and current-divider relations.

One might be tempted to use superposition on $v_i$ and the dependent source $A(v^+ - v^-)$ in Figure 15.8b to find $v^-$, but as discussed in Section 3.5.1 this is a hazardous approach. The problem is that the value of the dependent source is controlled by some other variable in the circuit, so we are not free to simply set the source to zero.

The safest rule to follow is: Do *not* set dependent sources to zero in superposition calculations.

---

EXAMPLE 15.1 SINGLE-ENDED AMPLIFIER Circuits containing single-input amplifiers can be analyzed in much the same way as circuits containing Op Amps, as this example shows. Consider the circuit shown in Figure 15.9, which contains a single-input inverting amplifier having gain $-A$. Except for its finite gain, the amplifier is assumed to be ideal. Thus, its input current is zero, it drives $v_{OUT} = -Av_{MID}$, and its negative feedback makes the circuit stable.

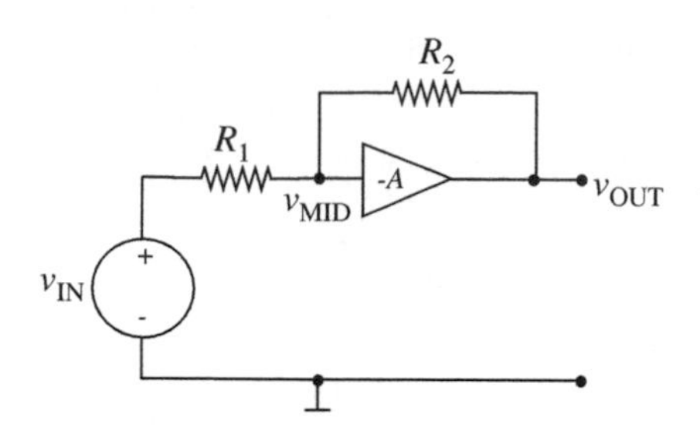


FIGURE 15.9 A single-ended amplifier in a feedback loop. Note that power supply and ground connections to the Op Amp are not shown.

Following the node method,

$$0 = \frac{v_{MID} - v_{IN}}{R_1} + \frac{v_{MID} - v_{OUT}}{R_2}$$

with

$$v_{OUT} = -Av_{MID}.$$

Combining these two equations yields

$$v_{OUT} = \frac{-A(R_2/R_1)}{A + 1 + (R_2/R_1)} v_{IN},$$

which is identical to the result obtained for the inverting amplifier constructed with an Op Amp. For example, with $R_1 = 1\ \text{k}\Omega$, $R_2 = 100\ \text{k}\Omega$, and $A = 10^5$, $v_{OUT} = 99.9v_{IN}$.

Further, in the limit $A \rightarrow \infty$,

$$v_{OUT} = -\frac{R_2}{R_1} v_{IN}.$$

---

### 15.3.3 SENSITIVITY

It is helpful at this point to be more precise about just how "independent" $v_o$ really is to changes in the Op Amp gain $A$. Let $G$ be the gain $v_o/v_i$ of the Op Amp circuit. Then for the non-inverting connection, for example, we find from Equation 15.7:

$$G = \frac{v_o}{v_i} = \frac{A}{1 + A\frac{R_2}{R_1+R_2}}. \tag{15.15}$$

Taking the differential, assuming small changes in A and constant $R_1$ and $R_2$, we obtain

$$dG = \frac{1}{(1 + A\frac{R_2}{R_1+R_2})^2} dA. \tag{15.16}$$

The fractional change in circuit gain is then, from Equation 15.15,

$$\frac{dG}{G} = \left( \frac{1}{1 + A\frac{R_2}{R_1+R_2}} \right) \frac{dA}{A}. \tag{15.17}$$

Thus with *negative feedback* a given percentage change in the Op Amp gain $A$ results in a much smaller percentage change in the overall circuit gain G, smaller by a factor $1 + AR_2/(R_1 + R_2)$. Note from Equation 15.7 that this is exactly the factor by which the gain is reduced as a result of applying the feedback. By inspection of Figure 15.7, the gain term $AR_2/(R_1 + R_2)$ represents the gain for a signal traveling all the way around the feedback loop: through the Op Amp with its gain of $A$, then through the feedback resistor network with a "gain" of $R_2/(R_1 + R_2)$, (hence called the loop gain). In general, for negative feedback, gain changes are suppressed by a factor 1 + (loop gain), and the overall gain is reduced by this same factor.

### 15.3.4 A SPECIAL CASE: THE VOLTAGE FOLLOWER

A useful circuit for isolating one electrical system from another is the *voltage follower* shown in Figure 15.10. Comparison with Figure 15.4a indicates that this circuit is a degenerate case of the non-inverting connection, in which $R_1 = 0$ and $R_2 = \infty$. Hence, from Equation 15.9, the input-output relation for the follower is

$$v_o \simeq v_i. \tag{15.18}$$

That is, within a part of $10^{-5}$ or so, the output voltage is equal to the input voltage. An obvious question: Why not just use a piece of copper wire to get the gain of one in Equation 15.18? To answer, we need only look at the currents. The current that must be supplied by the input source is $i^+$, hence is a few nanoamps. The maximum current that can be supplied to some load by the Op Amp output circuit is a few milliamperes. Thus for a one-volt signal level, the circuit is drawing perhaps $10^{-8}$ watts from the signal source, but can deliver $10^{-3}$ watts to the load resistor $R_L$. A piece of wire obviously produces no such power gain. Said in another way, the Op Amp is providing isolation

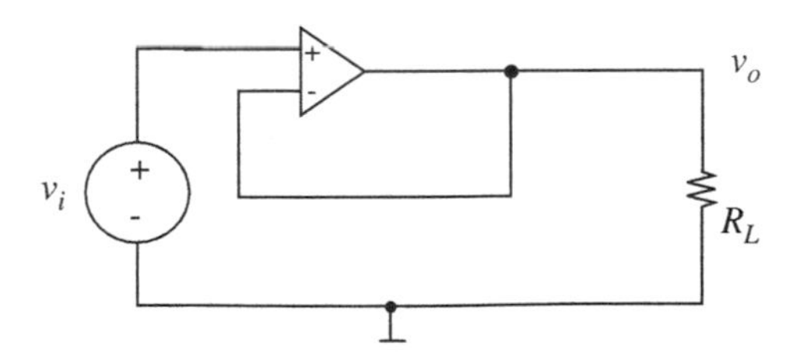


FIGURE 15.10 Voltage follower.

between the input and the output parts of the circuit, in the sense that the output resistor $R_L$ can be changed by many orders of magnitude, with a corresponding orders-of-magnitude change in output current, but the output voltage and the input current will be virtually unchanged. This isolation is referred to as *buffering*.

### 15.3.5 AN ADDITIONAL CONSTRAINT: $v^+ - v^- \simeq 0$

In all preceding Op Amp calculations, we have made an approximation that because the so-called loop gain in the denominator is much bigger than one, the "one" term can be neglected. This approximation is almost always valid in Op Amp calculations. It is the factor 1+ (loop gain) that determines how insensitive the circuit is to changes in the Op Amp gain constant $A$ (see Equation 15.17, for example), hence large loop gain is clearly a desirable design goal. If the loop gain is almost always going to be large, it seems a bit clumsy (although clearly correct) to make the circuit calculations without taking this fact into account until the last line. One would hope that with some hindsight, it might be possible to make the "large loop gain" assumption at the start of the circuit calculation, thereby simplifying the math. Let us re-examine the circuit of Figure 15.8b with this in mind.

We know that for most Op Amps, A will be 100,000 or larger, and the maximum allowed $v_o$ will be about 12 V (see Figure 15.3c). Hence the largest value of $(v^+ - v^-)$ for linear operation will be around 120 mV, a voltage orders-of-magnitude smaller than either the input or the output voltage. On this basis it is reasonable to assume, as before, $i^+ \simeq 0$, and $i^- \simeq 0$, but *include an additional constraint*:

$$v^+ - v^- \simeq 0. \tag{15.19}$$

Not equal to zero, just small compared to other circuit voltages. When these three constraints are applied to the circuit in Fig 15.8b, we find

$$v^+ = 0$$

$$v^- \simeq 0.$$

Hence KCL at the $v^-$ node yields

$$\frac{v_i}{R_a} + \frac{v_o}{R_b} \simeq 0 \tag{15.20}$$

(compare with Equation 15.10). Solving for $v_o$, we find

$$v_o \simeq -\frac{R_b}{R_a} v_i \tag{15.21}$$

as before, except this time the calculation is much simpler, because the combined constraints of approximately zero voltage and approximately zero current are quite powerful. For the non-inverting circuit of Figure 15.7, for example, we can write, using the voltage-divider relation,

$$v_i = v^+ \simeq v^- = \frac{R_2}{R_1 + R_2} v_O. \tag{15.22}$$

Hence

$$v_O \simeq \frac{R_1 + R_2}{R_2} v_i \tag{15.23}$$

as before. The voltage constraint of Equation 15.19 is also called the *virtual ground constraint*,[2] and can be interpreted in physical terms by noting that the output of a circuit with negative feedback must adjust itself to *force* $(v^+ - v^-)$ to be nearly zero, because that nearly-zero voltage is in turn multiplied by 100,000 to become the output voltage.

The $v^+ - v^- \simeq 0$ constraint can be applied only if the Op Amp is not saturated and the feedback is negative; that is, the net feedback signal comes from the output back to the negative input terminal.

## 15.4 INPUT AND OUTPUT RESISTANCES

### 15.4.1 OUTPUT RESISTANCE, INVERTING OP AMP

Negative feedback has a profound effect on the Thévenin-equivalent input and output resistances of circuits. To illustrate, we calculate first the Thévenin output resistance of the simple inverting Op Amp assumed to be operating in the active (non-saturated) region, that is, the circuit in Figure 15.8b. Obviously if we model the Op Amp by the ideal Op Amp model, the Thévenin output resistance is by definition zero, with or without feedback. So to show any effect, we must use a more accurate device model that includes some finite resistance in series with the dependent source, as in Figure 15.11. One way of calculating the Thévenin output resistance is to apply a test current $i_t$, at the output terminals, as shown in Figure 15.11, and calculate the resulting voltage $v_t$, when all other *independent sources*, in this case $v_i$, are set to zero.

2. Or more accurately, the *virtual short constraint*, or the *virtual node constraint*, since the inverting and non-inverting inputs need not always be at ground potential.

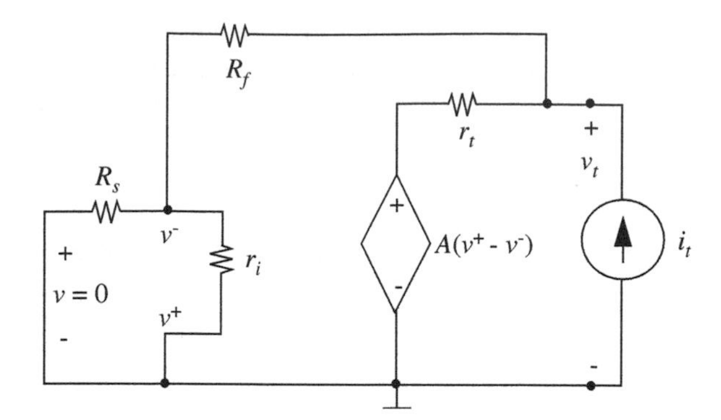


**FIGURE 15.11** Calculation of output resistance.

In calculating the Thévenin resistance do *not* casually set dependent sources to zero, as their value is dictated by some other variable in the circuit which may or may not be zero.

The calculation of $v_t$ is straightforward. We use the node method with conductances in place of resistances for convenience. In other words, we use $g_i = 1/r_i$, $G_s = 1/R_s$, $G_f = 1/R_f$, and $g_t = 1/r_t$. Applying KCL at the nodes with unknown node voltages, we get the following three independent equations:

$$v^+ = 0 \tag{15.24}$$

$$v^- = \frac{G_f}{G_f + G_s + g_i} v_t \tag{15.25}$$

$$i_t + \left[A\left(v^+ - v^-\right) - v_t\right] g_t + \left(v^- - v_t\right) G_f = 0. \tag{15.26}$$

$$\tag{15.27}$$

To simplify the mathematics, we now assume for this calculation that $r_i$ is infinite ($g_i = 0$), because it is always much larger than $R_s$ or $R_f$. Now, eliminating $v^+$ and $v^-$ from Equation 15.26,

$$\frac{i_t}{v_t} = G_o = \frac{AG_f g_t}{G_f + G_s} + g_t + \frac{G_f G_s}{G_f + G_s}. \tag{15.28}$$

Thus the output conductance is the sum of three conductances. The first term is the effect of the feedback, the second term is the output conductance of the Op Amp alone, and the third term in resistance notation is $R_f + R_s$, hence is the effect of the feedback resistors in the absence of the Op Amp. For large $A$, this last term is not important, so the Thévenin output conductance with feedback is

$$\frac{i_t}{v_t} = G_o \simeq g_t \left[ 1 + \frac{AG_f}{G_f + G_s} \right] \tag{15.29}$$

or, in more familiar terms

$$G_o \simeq g_t \left[ 1 + A \frac{R_s}{R_s + R_f} \right]. \tag{15.30}$$

Hence the Thévenin output resistance of the circuit is

$$R_o \simeq \frac{r_t}{1 + A\frac{R_s}{R_s + R_f}}. \tag{15.31}$$

For large loop gain

$$R_o \simeq \frac{r_t}{A\frac{R_s}{R_s + R_f}}. \tag{15.32}$$

The Thévenin output resistance $r_t$ for the Op Amp alone, without feedback is typically of the order of 1000 ohms, so for large $A$ and reasonable $R_s$ and $R_f$, the overall Thévenin output resistance $R_o$ for this topology circuit is a fraction of an ohm.

Equation 15.31 is in fact a general result. For any linear circuit in which the feedback resistor is sampling the output *node voltage* (rather than the output current), the Thévenin equivalent output resistance with feedback is equal to the output resistance without feedback, divided by a factor 1+ (loop gain), the same factor involved in gain calculations and calculation of sensitivity to changes in the gain constant $A$.

### 15.4.2 INPUT RESISTANCE, INVERTING CONNECTION

To calculate the Thévenin-equivalent input resistance of the inverting Op Amp circuit, we apply a test source at the input, and measure the resulting response. (There are no internal independent sources to be set to zero.) In Figure 15.12 we have chosen to drive with a test voltage $v_t$, and calculate the resulting current $i_t$. As before, it is equally valid to apply a test current source, and calculate the resulting voltage. The calculations are greatly simplified if the circuit topology is taken into account. The input consists of two elements in series: the resistor $R_s$, and a complicated circuit that will reduce to the Thévenin-equivalent input resistance of the rest of the Op Amp circuit. Recognizing this, we can first calculate the resistance *to the right of* the $R_s$ (just set $R_s$ to zero in Figure 15.12)

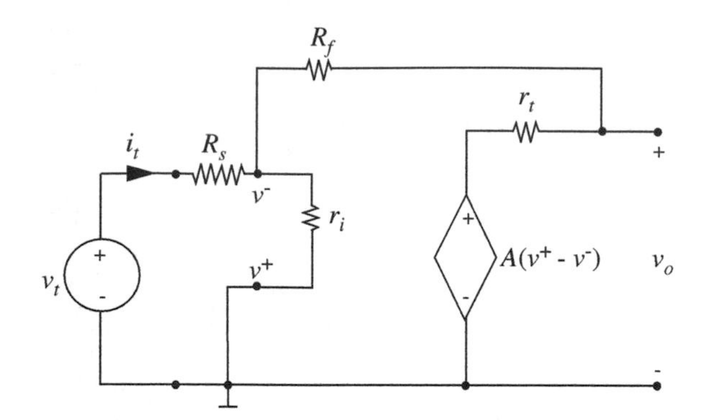


FIGURE 15.12 Input resistance calculation.

and then add $R_s$ to this calculated value to get the complete answer. We will denote the resistance of the Op Amp circuit to the right of $R_s$ as $R_i$, and the complete input resistance, including resistor $R_s$, as $R_i'$.

Because we chose a test voltage, and $R_s$ is zero for now, the control variable is directly constrained:

$$v^+ = 0 \tag{15.33}$$

$$v^- = v_t. \tag{15.34}$$

Now apply KCL at the input node:

$$i_t = \frac{v_t}{r_i} + \frac{v_t - A(v^+ - v^-)}{R_f + r_t}. \tag{15.35}$$

hence

$$\frac{i_t}{v_t} = G_i = \frac{1}{r_i} + \frac{1}{R_f + r_t} + \frac{A}{R_f + r_t}. \tag{15.36}$$

Again we have the sum of three conductances. So the corresponding resistance expression, the Thévenin input resistance for the circuit, is the parallel combination of three terms:

$$R_i = r_i \parallel (R_f + r_t) \parallel \left(\frac{R_f + r_t}{A}\right) \tag{15.37}$$

the Op Amp input resistance, the feedback resistor plus Op Amp output resistor, and an effective resistance generated by the feedback. For large $A$,

$$R_i \simeq \frac{R_f + r_t}{A} \tag{15.38}$$

that is, we expect the input resistance to be *very low*. For example, for a typical case of $R_f = 10\ \text{k}\Omega$, $r_t = 1000\ \Omega$, $A = 10^5$, the input resistance measured at the $v^-$ terminal will be 0.1 ohm. Simple physical reasoning serves to support this result. If we imagine applying a small voltage to the input, say 0.1 mV, then the Op Amp will immediately drive $v_o$ to -A times 0.1 mV, or -10 volts. So resistor $R_f$ has a large voltage across it, hence a large current will flow. This large current must come from the input source, and is $10^5$ times as large as one might expect for such a small input voltage. Large current for small voltage means the effective input resistance will be very small, in fact roughly the feedback resistor $R_f$ divided by $A$.

In accordance with our initial assumptions, the complete input resistance of the inverting Op Amp, including resistor $R_s$, is

$$R_i' = R_i + R_s \tag{15.39}$$

as can be verified by calculating the input resistance directly from Figure 15.12 including $R_s$. Because $R_i$ is so small,

$$R_i' \simeq R_s. \tag{15.40}$$

### 15.4.3 INPUT AND OUTPUT R FOR NON-INVERTING OP AMP

The active-region output resistance of the non-inverting Op Amp circuit can be calculated in much the same way as for the inverting circuit. We set the *independent* source to zero and apply a test current source to the output terminals, as shown in Figure 15.13. Now calculate $v_t$. As usual, we apply the node method to find three independent equations. First find expressions for $v^+$ and $v^-$, and then write KCL at the output node. Again we assume $r_i$ is much larger than $R_2$ to simplify the math:

$$v^+ = 0 \tag{15.41}$$

$$v^- = v_t \frac{R_2}{R_1 + R_2} \tag{15.42}$$

$$i_t - \frac{v_t}{R_1 + R_2} - \frac{v_t - A(v^+ - v^-)}{r_t} = 0. \tag{15.43}$$

Hence

$$\frac{i_t}{v_t} = G_o = \frac{1}{R_1 + R_2} + \frac{1}{r_t} + \frac{AR_2/(R_1 + R_2)}{r_t}. \tag{15.44}$$

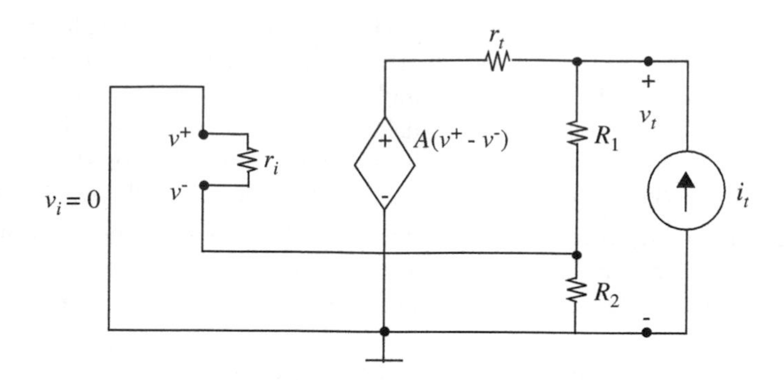


FIGURE 15.13 Output resistance calculation, non-inverting circuit.

For large $A$ and reasonable $R_1$ and $R_2$,

$$R_o \simeq \frac{r_t}{AR_2/(R_1 + R_2)}. \tag{15.45}$$

This is the Thévenin output resistance $r_t$ of the Op Amp alone, divided by the loop gain, or, more accurately, from Equation 15.44, 1 + (loop gain). As before, the output resistance is *very low*.

The input resistance for the active (nonsaturated) region can be found from the circuit in Figure 15.14. As before, we need expressions for $v^+$ and $v^-$, and a KCL equation involving $i_t$:

$$v^+ = v_t \tag{15.46}$$

$$v^- = v_t - i_t r_i. \tag{15.47}$$

KCL at Node 1 yields

$$i_t + \frac{A(v^+ - v^-) - v^-}{R_1 + r_t} - \frac{v^-}{R_2} = 0. \tag{15.48}$$

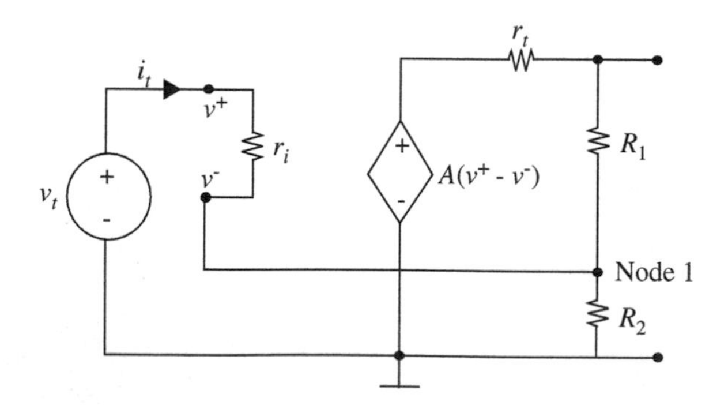


FIGURE 15.14 Input resistance calculation, non-inverting Op Amp.

Substituting and solving, assuming A is large, we find

$$R_i = \frac{v_t}{i_t} \simeq r_i \left[ \frac{AR_2}{R_1 + r_t + R_2} \right]. \tag{15.49}$$

This expression shows that for the non-inverting connection, the effective input resistance in the active region is *very high*, (roughly the Op Amp input resistance $r_i$ multiplied by the loop gain) in contrast to the result for the inverting case, Equation 15.38. Reasoning physically, if we apply a voltage $v_t$ at the input, the output voltage adjusts itself so that $v^-$ is very nearly equal to $v_t$, so there is very little voltage across $r_i$, hence much less current flowing in it than we might expect. Hence the circuit input resistance is large. This property enables the non-inverting connection to be particularly useful in buffering applications.

This point of view suggests an alternative approach to the calculation. If we assume at the outset that $v^+ - v^- \simeq 0$, then

$$v_t \simeq A(v^+ - v^-) \frac{R_2}{R_2 + R_1 + r_t}. \tag{15.50}$$

But $v^+ - v^-$, although small, must not be zero for finite $r_t$:

$$v^+ - v^- = i_t r_i. \tag{15.51}$$

When Equation 15.51 is substituted into Equation 15.50, we find $R_i$ as before (Equation 15.49).

### WWW 15.4.4 GENERALIZATION ON INPUT RESISTANCE *

### 15.4.5 EXAMPLE: OP AMP CURRENT SOURCE

We have shown that both the inverting and non-inverting Op Amp connections have very low output resistance, that is, they approximate ideal voltage sources. But in some circuit applications, we may want the Op Amp to look like a current source, that is, we want a very *high* output resistance. It follows from the discussion at the end of Section 15.4.1 that such a design can be realized by a change in the *topology of the output circuit.*

In the two circuits already discussed, the feedback network sends a signal back to the negative input terminal that is proportional to the *output voltage* $v_O$. Thus the circuit tends to stabilize this variable, thereby creating a voltage source. By analogy, to make a current source, we must arrange to feed back a signal proportional to the *output current* flowing in the circuit being driven by the Op Amp. One possible topology is shown in Figure 15.15a. The circuit looks, at first glance, like the non-inverting connection shown in Figure 15.15b, but there is an important difference. In the new topology, the resistor $R_L$ we are trying to drive is now part of the voltage divider feedback network. Thus in Figure 15.15a we are using the resistor $R_s$ to *sample the current through* $R_L$, whereas in Figure 15.15b $R_1$ and $R_2$ sample the *voltage across* $R_L$. The distinction seems trivial until we think in terms of $R_L$ varying in value, or even being nonlinear. Then it is clear that there is a fundamental difference in the two topologies.

Once the topological issues are understood, the circuit analysis is trivial. Assuming $v^+ \simeq v^-$, we note from Figures 15.15a or 15.15c

$$v^+ = v_i \tag{15.52}$$

$$v^- = i_L R_s \tag{15.53}$$

$$v^+ \simeq v^-. \tag{15.54}$$

Therefore

$$i_L \simeq \frac{v_i}{R_s} \tag{15.55}$$

*independent of the value of* $R_L$.

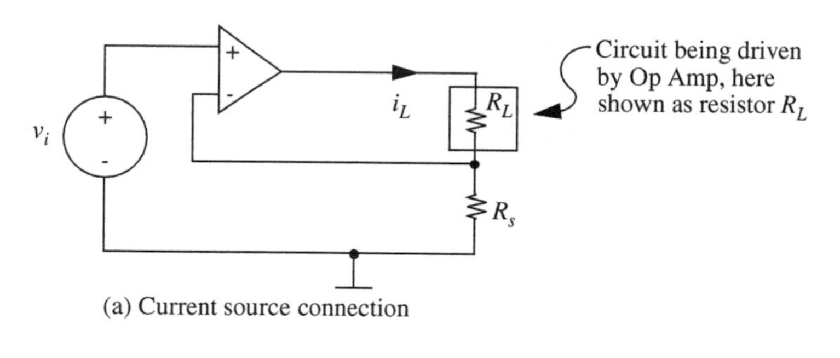


(a) Current source connection

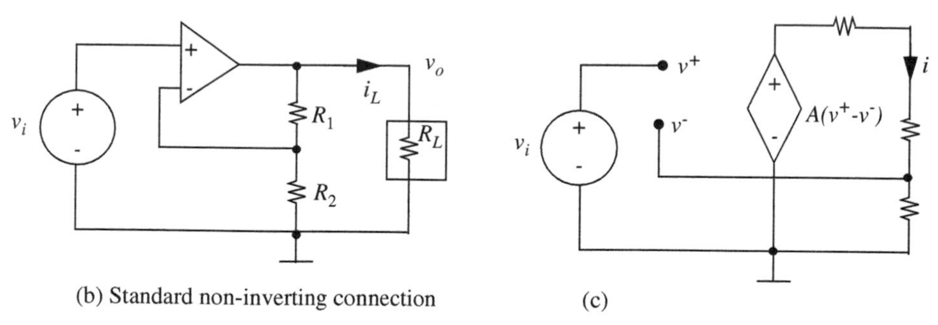


(b) Standard non-inverting connection

(c)

FIGURE 15.15 Op Amp current source.

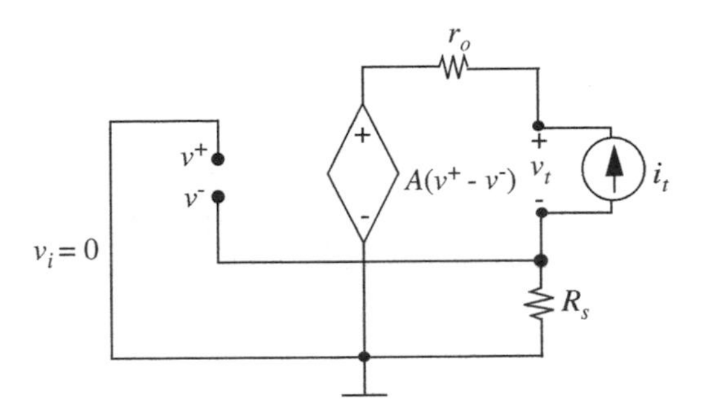


FIGURE 15.16 Output resistance of current source.

The fact that the current through $R_L$ is independent of the value of $R_L$ suggests that the Op Amp circuit looks like a current source. It is a simple matter to verify this more formally: Replace $R_L$ by a test source, and find the Thévenin output resistance of the circuit. In this case we choose a test current source $i_t$, as in Figure 15.16:

$$v^- = -i_t R_s \tag{15.56}$$

$$v^+ = 0 \tag{15.57}$$

$$v_t = A(v^+ - v^-) + i_t r_o - v^- \tag{15.58}$$

$$= (1 + A) i_t R_s + i_t r_o \tag{15.59}$$

$$R_o = \frac{v_t}{i_t} = (1 + A) R_s + r_o. \tag{15.60}$$

For reasonable circuit parameters, $R_o$ could well be many megohms.

Again these results can be generalized to summarize the effect of negative feedback on the effective output resistance of a circuit. If the Op Amp, the load resistor $R_L$ and the feedback network appear to be connected in series, in a loop, hence sharing a common current, then the output resistance will be high. If the Op Amp, $R_L$, and the feedback circuit all appear to be in parallel, tied to a common node, sharing a common voltage, then the output resistance will be low.

## 15.5 ADDITIONAL EXAMPLES

This section contains a number of examples of Op Amp circuits. They are intended both to illustrate the versatility of the Op Amp as a circuit design building block and to serve as a review and extension of analysis techniques introduced earlier in this chapter.

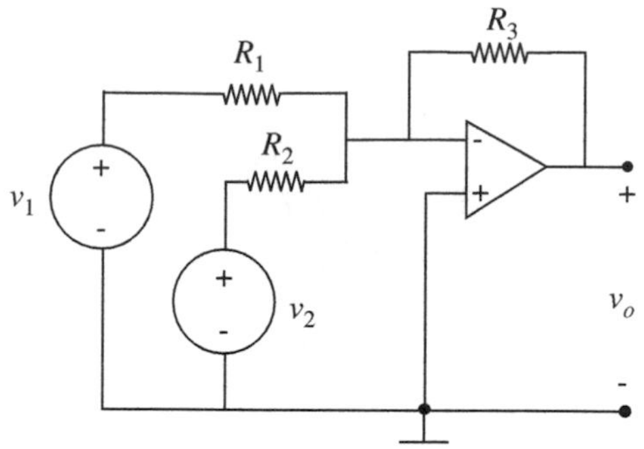

FIGURE 15.17 Adder.

### 15.5.1 ADDER

An Op Amp circuit for adding two signals together is shown in Figure 15.17. If we assume $v^+ \simeq v^-$, then application of KCL to the $v^-$ node yields

$$\frac{v_1}{R_1} + \frac{v_2}{R_2} + \frac{v_o}{R_3} \simeq 0. \tag{15.61}$$

Therefore

$$v_o \simeq -\left(\frac{R_3}{R_1}v_1 + \frac{R_3}{R_2}v_2\right), \tag{15.62}$$

which represents the weighted sum of the two input signals.[3] Note that within the accuracy of the voltage constraint $v^+ - v^- \simeq 0$, the two input signals do not cross-couple; that is, no current from $v_2$ flows in $R_1$, and vice versa. Thus the circuit is an *ideal adder*.

### 15.5.2 SUBTRACTER

If we wish to take the difference between two signals, then the circuit of Figure 15.18 is appropriate. Direct application of superposition to the *independent sources* yields the two subcircuits shown in Figures 15.18b and 15.18c. In Figure 15.18b, source $v_2$ has been set to zero. On the assumption of $i^+ \simeq 0$, there will be no current through $R_3$ and $R_4$, so $v^+ \simeq 0$, and the topology is seen to be that of an inverting amplifier. Hence

$$v_{oa} = -\frac{R_2}{R_1}v_1. \tag{15.63}$$

3. Because the Op Amp model is linear, the same result can be derived using superposition.

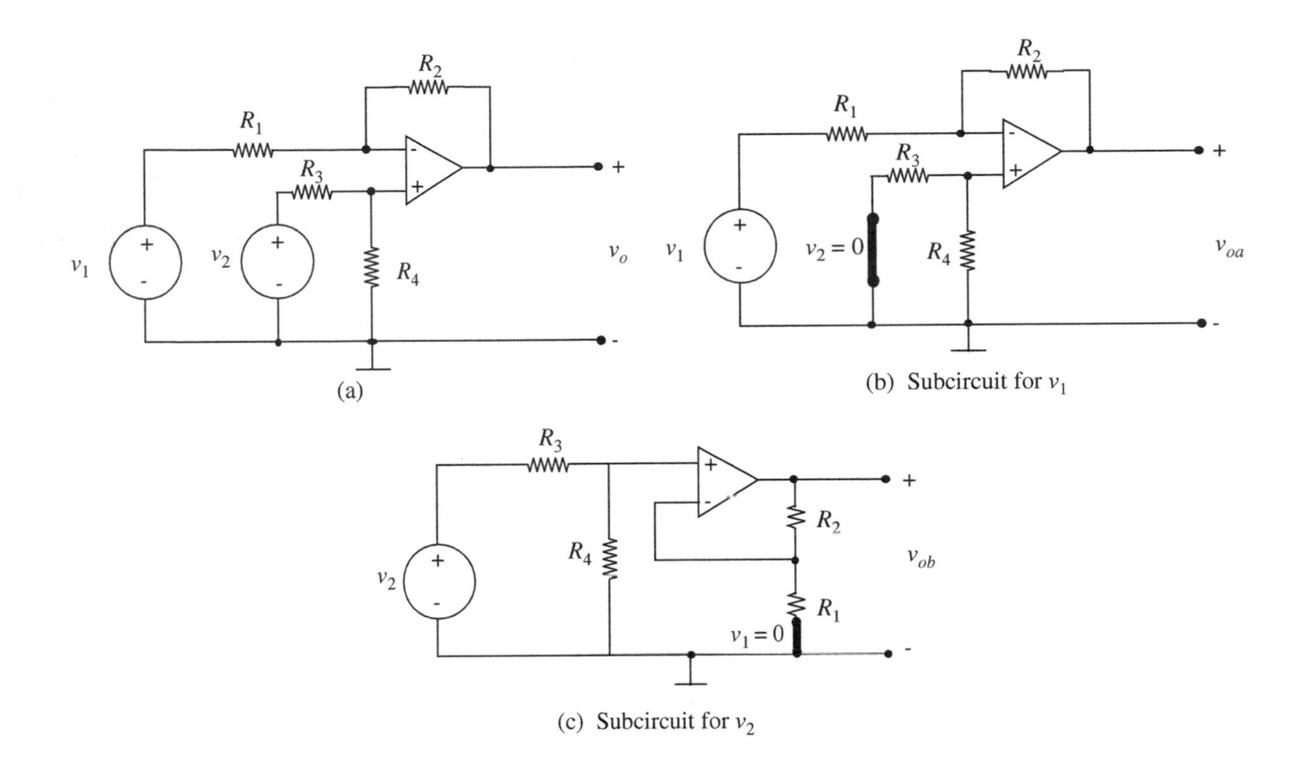


FIGURE 15.18 Subtracter.

When source $v_1$ is set to zero, and the circuit slightly rearranged, the non-inverting topology emerges, with a voltage divider at the input, as indicated in Figure 15.18c. Hence

$$v_{ob} = \left(\frac{R_1 + R_2}{R_1}\right)\left(\frac{R_4}{R_3 + R_4}\right) v_2. \tag{15.64}$$

The total output voltage is the sum of the two voltages $v_{oa}$ and $v_{ob}$. To make a subtracter, the resistor ratios in Equations 15.63 and 15.64 should be equal. This can be achieved by setting $R_3 = R_1$ and $R_4 = R_2$. Then

$$v_o = \frac{R_2}{R_1}(v_2 - v_1). \tag{15.65}$$

Now $v_o$ is proportional to the difference between the two input voltages.

## 15.6 OP AMP RC CIRCUITS

### 15.6.1 OP AMP INTEGRATOR

The circuit in Figure 15.19 gives a much closer approximation to ideal integration than the simple RC circuits discussed in Chapter 10. The analysis to

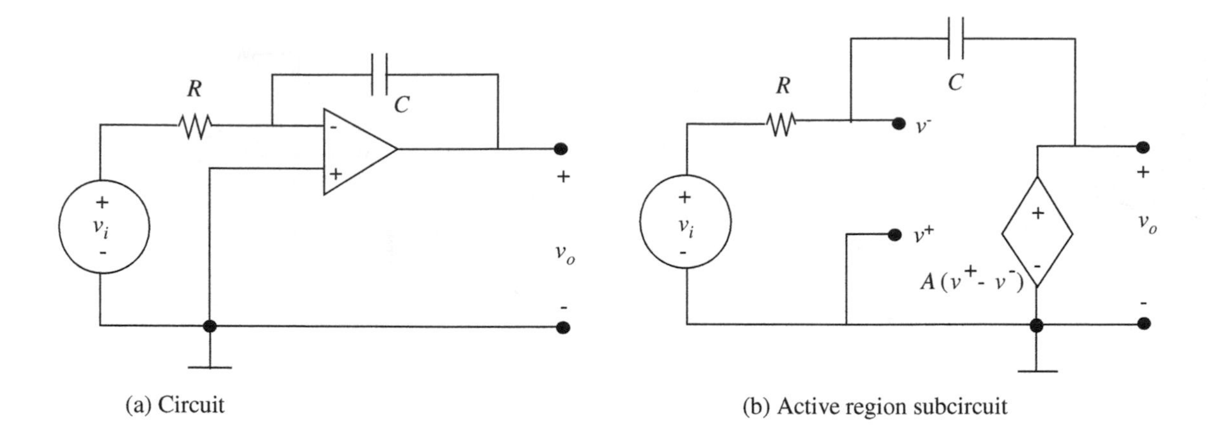


FIGURE 15.19 Op Amp Integrator.

show this is quite straightforward. Assuming linear-region operation, we replace the Op Amp by the dependent-source model, as in Figure 15.19b, and analyze the resulting linear circuit using the node method. KCL at the $v^-$ node yields

$$\frac{v_i - v^-}{R} + \frac{Cd(v_O - v^-)}{dt} = 0. \tag{15.66}$$

If we assume at the outset that the Op Amp gain $A$ is large enough to ensure that

$$v^+ \simeq v^- \tag{15.67}$$

then because $v^+ = 0$, Equation 15.66 reduces to

$$\frac{v_i}{R} + \frac{Cdv_O}{dt} \simeq 0 \tag{15.68}$$

or,

$$v_O \simeq -\frac{1}{RC}\int v_i dt. \tag{15.69}$$

That is, the circuit calculates the (negative) integral of the input voltage.

A more exact calculation involves substituting the Op Amp equation:

$$v_O = A(v^+ - v^-) \tag{15.70}$$

into Equation 15.66, again noting $v^+ = 0$:

$$\frac{v_i}{R} - \frac{v^-}{R} - CA\frac{dv^-}{dt} - C\frac{dv^-}{dt} = 0. \tag{15.71}$$

Hence

$$RC(1 + A)\frac{dv^-}{dt} + v^- = v_i. \tag{15.72}$$

The effective time constant of the circuit (by analogy with Equation 10.150, for example) is

$$\tau = (1 + A)RC. \tag{15.73}$$

Thus the time constant associated with the passive elements alone is multiplied by the gain of the Op Amp. This is often referred to as the Miller Effect, originally in reference to the fact that a small input to output capacitance in early vacuum tubes seriously limited the frequency response of amplifier circuits. The time constant can be made very large for modest component values. For example, if the RC time constant is 1 second, and $A$ is $10^5$ or greater, the effective circuit time constant in measured in days. On this time scale almost any waveform lasting for less than a minute or so will seem like a "short pulse." Thus the analysis of Section 10.6.3 is applicable, and on the time scale of minutes, the circuit acts like an integrator.

The ultimate test of an integrator is to apply a small voltage step, $V$, and see how closely the integrator output conforms to a ramp. From Equations 15.70 and 15.72,

$$(1 + A)RC\frac{dv_O}{dt} + v_O = -AV. \tag{15.74}$$

For $v_i$ a small fixed value $V$ after $t = 0$, $v_O$ will follow the usual exponential charging curve toward $(-AV)$, (see Equation 10.101, for example). That is,

$$v_O = -AV\left(1 - e^{-t/(1+A)RC}\right). \tag{15.75}$$

This curve is plotted in Figure 15.20 on the basis that the RC time constant (without the Op Amp) is roughly one second. Obviously on the time scale of minutes, the circuit looks like an almost-perfect integrator, provided, of course, the Op Amp is always operating in the active region.

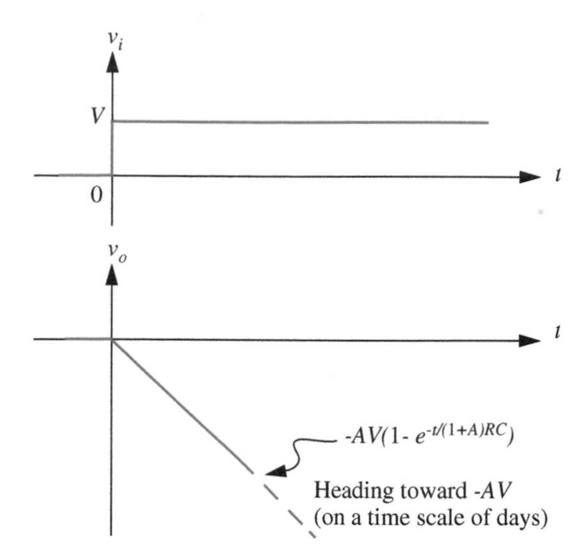


FIGURE 15.20 Waveforms of integrator.

## 15.6.2 OP AMP DIFFERENTIATOR

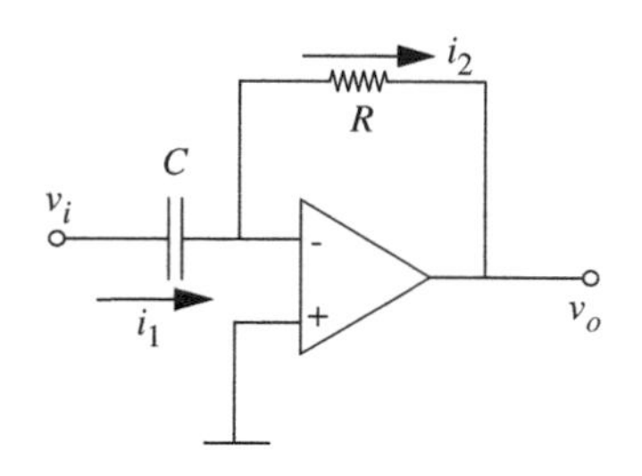


FIGURE 15.21 Differentiator circuit.

The Op Amp differentiator shown in Figure 15.21 complements the integrator. Because $v^- \simeq v^+$ and $v^+ = 0$, we know that the current $i_1$ through the capacitor is given by:

$$i_1 = C\frac{dv_i}{dt}.$$

Since virtually no current flows into the Op Amp, $i_1 = i_2$, and therefore

$$v_o = -Ri_1.$$

Eliminating $i_1$ from the preceding two equations, we obtain

$$v_o = -RC\frac{dv_i}{dt}. \tag{15.76}$$

That is, this circuit calculates the (negative) time derivative of the input voltage.

Sample input and output waveforms for the differentiator are shown in Figure 15.22. For the square-pulse input shown, the outputs are a pair of spikes each at the time instant the input makes a transition. As illustrated in the example, the differentiator circuit is often used in detecting shape transitions in waveforms.

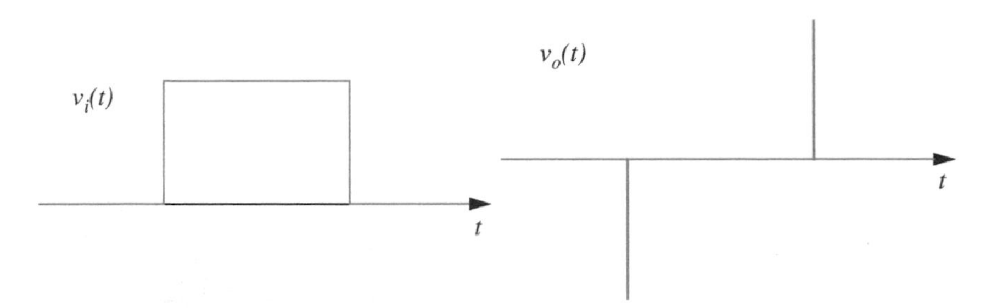


FIGURE 15.22 Differentiator waveforms.

### 15.6.3 AN RC ACTIVE FILTER

An Op Amp embedded in a more complicated RC circuit is shown in Figure 15.23a. This is an *RC active filter*, with all of the useful resonance properties of a capacitor-inductor circuit. To show this, we calculate the output voltage $v_o$ in terms of $v_i$. First draw the linear-region circuit model with the dependent source, Figure 15.23b. Then write Node equations, taking current entering the node as positive. We assume at the outset $v^+ - v^- \simeq 0$. because $v^+$ is zero in this circuit, the appropriate constraint is $v^- \simeq 0$. For Node $v_1$,

$$(v_i - v_1)g_1 - C_1\frac{dv_1}{dt} + C_2\frac{d(v_o - v_1)}{dt} = 0 \tag{15.77}$$

and for Node $v^-$

$$C_1\frac{dv_1}{dt} + v_o g_2 = 0. \tag{15.78}$$

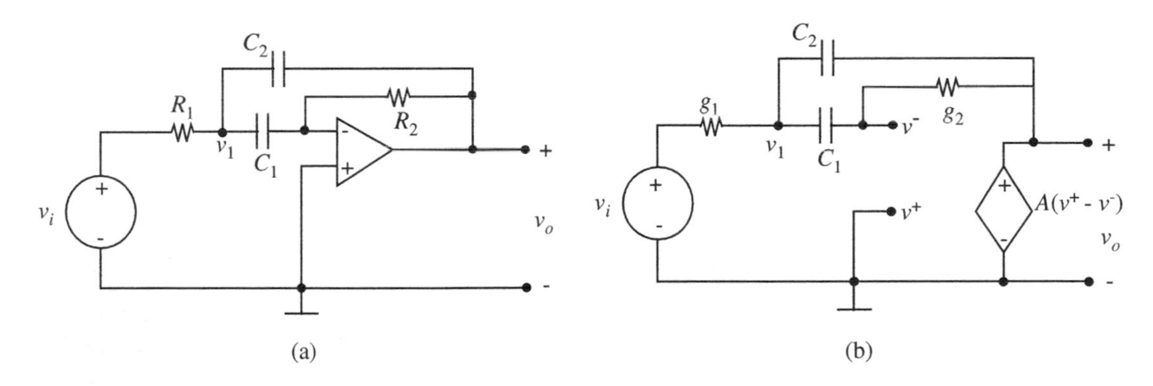


FIGURE 15.23 Op Amp RC active filter.

Hence

$$v_i g_1 = g_1 v_1 + (C_1 + C_2)\frac{dv_1}{dt} - C_2\frac{dv_o}{dt} \tag{15.79}$$

$$0 = C_1\frac{dv_1}{dt} + v_o g_2. \tag{15.80}$$

These equations can be solved by taking the derivative of both sides of both equations, and eliminating terms in $v_1$ and its derivative by substitution from Equation 15.80 and the derivative of Equation 15.80. By so doing we obtain a second order differential equation for $v_o$:

$$\frac{d^2 v_o}{dt^2} + g_2\frac{C_1 + C_2}{C_1 C_2}\frac{dv_o}{dt} + \frac{g_1 g_2}{C_1 C_2}v_o = -\frac{g_1}{C_2}\frac{dv_i}{dt}. \tag{15.81}$$

This equation is identical in form to that of an RLC resonator (see Equation 12.119), but this circuit contains no inductors. The effect of an inductor is created by an active element, in this case the Op Amp, and the capacitors, hence the name *RC active filter*. The advantages of an RC active filter (this is only one realization; there are many others) are that it can provide a power gain unlike an RLC network and that it does not require inductors. Because inductors are difficult to fabricate in VLSI technology, this is an important design advantage for integrated circuits. Furthermore, inductors are not very ideal elements, especially for low-frequency applications (for example, for frequencies below perhaps 100 kHz). Thus in this frequency range, resonant circuits are often built out of Op Amps, resistors, and capacitors.

The properties of filter circuits were explored previously in Chapters 10 and 13. As we did in Chapter 13, the circuit of Figure 15.23 can also be analyzed using the impedance method by using impedance values $1/sC_1$ and $1/sC_2$ for the capacitors (see Section 15.6.4). We will also see other examples of impedance based analysis for Op Amps later in this chapter.

Since Equation 15.81 is identical in form to Equation 12.119 for the RLC circuit, we can readily determine the behavior of our RC active filter. Notice that the output response $v_o$ of the Op Amp RC active filter corresponds to the capacitor voltage $v_C$ in Section 12.5. The equation corresponding to the series RLC circuit in Section 12.5 was

$$\frac{d^2 v_C}{dt^2} + \frac{R}{L}\frac{dv_C}{dt} + \frac{1}{LC}v_C = \frac{1}{LC}v_{IN} \tag{15.82}$$

with the damping factor $\alpha = R/2L$ and the undamped resonant frequency $\omega_o = 1/\sqrt{LC}$.

Thus, the corresponding damping factor in our Op Amp circuit is

$$\alpha = g_2 \frac{C_1 + C_2}{2C_1C_2} \tag{15.83}$$

and the undamped resonant frequency is

$$\omega_o = \sqrt{\frac{g_1 g_2}{C_1 C_2}}. \tag{15.84}$$

## 15.6.4 THE RC ACTIVE FILTER— IMPEDANCE ANALYSIS

Let us analyze the Op Amp active filter circuit of Section 15.6.3 for a sinusoidal drive. Since the Op Amp is a linear device (namely, a VCVS) we can use the impedance method for the analysis.

The circuit configuration is repeated in Figure 15.24a. The impedance model for the circuit is shown in Figure 15.24b. The circuit is sufficiently complicated that node analysis is advisable. At node $V_1$ assuming $v^+ \simeq v^-$,

$$(V_i - V_1)\, g_1 + (V_o - V_1)\, sC_2 - V_1 sC_1 = 0 \tag{15.85}$$

where $g_1 = 1/R_1$. At the $V^-$ node,

$$V_1 sC_1 + V_o g_2 = 0 \tag{15.86}$$

where $g_2 = 1/R_2$.

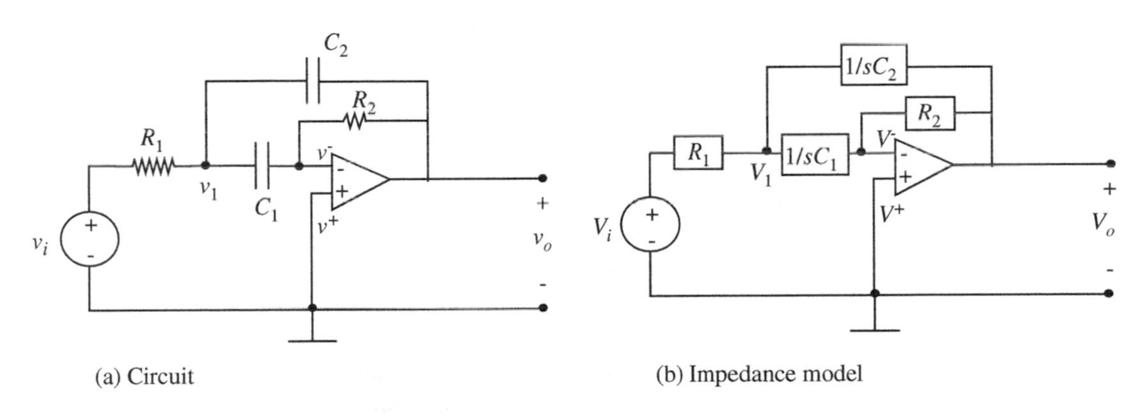


FIGURE 15.24 RC active filter analysis using the impedance method.

Now $V_o$ can be found by Cramer's Rule. First, rewrite with source terms on the left

$$V_i g_1 = V_1 \left[g_1 + s(C_1 + C_2)\right] - V_o s C_2 \qquad (15.87)$$

$$0 = V_1 s C_1 + V_o g_2 \qquad (15.88)$$

(These equations should be compared to the corresponding differential equations, Equations 15.79 and 15.80.) Solving for the complex amplitude $V_o$,

$$V_o = \frac{-g_1 s C_1 V_i}{\left[g_1 + s(C_1 + C_2)\right] g_2 + s^2 C_1 C_2} \qquad (15.89)$$

$$= \frac{-g_1 s C_1 V_i}{g_1 g_2 + s(C_1 + C_2) g_2 + s^2 C_1 C_2} \qquad (15.90)$$

$$= \frac{-s\left(g_1/C_2\right) V_i}{s^2 + s\frac{C_1+C_2}{C_1 C_2} g_2 + \frac{g_1 g_2}{C_1 C_2}}. \qquad (15.91)$$

Equation 15.91 has exactly the form of Equation 14.19 (except for the minus sign), hence the circuit is equivalent to a parallel RLC filter. By comparing corresponding terms we find, as in Chapter 12,

$$\text{Resonant frequency} = \omega_o = \sqrt{\frac{g_1 g_2}{C_1 C_2}} \qquad (15.92)$$

$$\text{Bandwidth} = g_2 \frac{C_1 + C_2}{C_1 C_2}. \qquad (15.93)$$

With these scaling factors, the frequency response plot of Figure 14.12 directly applies to this circuit (except for the additional 180° in the phase), along with all other properties discussed in Section 14.1.

### WWW 15.6.5 SALLEN-KEY FILTER

## 15.7 OP AMP IN SATURATION

Thus far we have used the Op Amp in its active region. In the active region, the voltage-controlled voltage source model of the Op Amp shown in Figure 15.6 applies. Furthermore, when negative feedback is applied, and the Op Amp is operated in the active region, we can use the input voltage constraint given by $v^- \simeq v^+$. However, the voltage-controlled voltage source model and the input voltage constraint no longer apply when the Op Amp output reaches saturation. In saturation, the Op Amp output will be close to one of the power supply voltages, +12 or −12 V. In positive saturation, the output will be close to +12 V and in negative saturation the output will be close to −12 V.

The Op Amp exits the active region and enters the saturation region when the external inputs are such that the Op Amp output is required to go above +12 or below −12 V. As an example, suppose the Op Amp has power supply voltages of +12 and −12, then if two volts are applied as the input to a non-inverting Op Amp amplifier circuit with a gain of 10, the Op Amp will be driven into positive saturation. Similarly, if minus two volts are applied to the same amplifier, the Op Amp will be driven into negative saturation.

How do we model the Op Amp when it is in saturation? When an Op Amp enters positive saturation, a near-short circuit forms between the Op Amp output and the positive power supply. Similarly, when the Op Amp enters negative saturation, a near-short circuit forms between the Op Amp output and the negative power supply. Accordingly, simple positive and negative saturation models for the Op Amp are illustrated in Figure 15.28. The normal dependent voltage source is not shown in the saturation models because its output gets limited by the power supply voltages and so it turns into a simple voltage source.

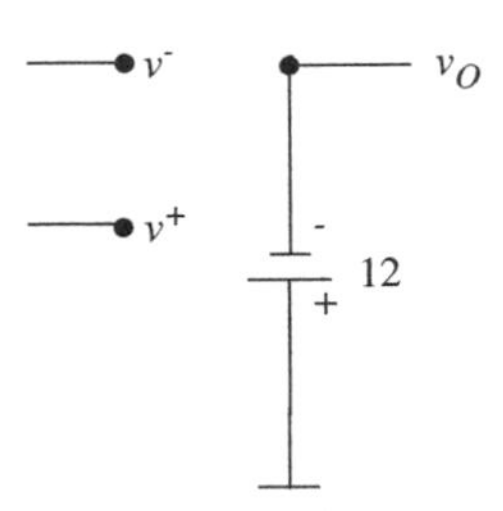


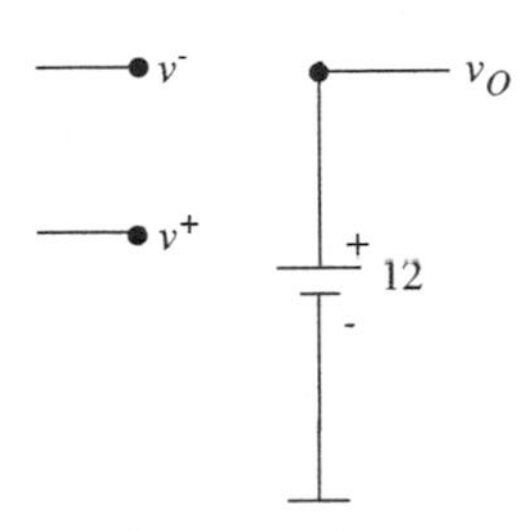


FIGURE 15.28 Op Amp model in saturation.

## 15.7.1 OP AMP INTEGRATOR IN SATURATION

If the Op Amp integrator in Figure 15.19 is driven into negative saturation, then the appropriate subcircuit is shown in Figure 15.29. Our negative saturation model for the Op Amp says that $v_O$ is fixed at some voltage close to the negative power supply voltage, say −12 V. Because of the 12-V battery assumed at the output when the Op Amp is in saturation, the dependent source is no longer involved in the calculations, so the circuit reduces to a simple series RC configuration, as indicated in Figure 15.29b.

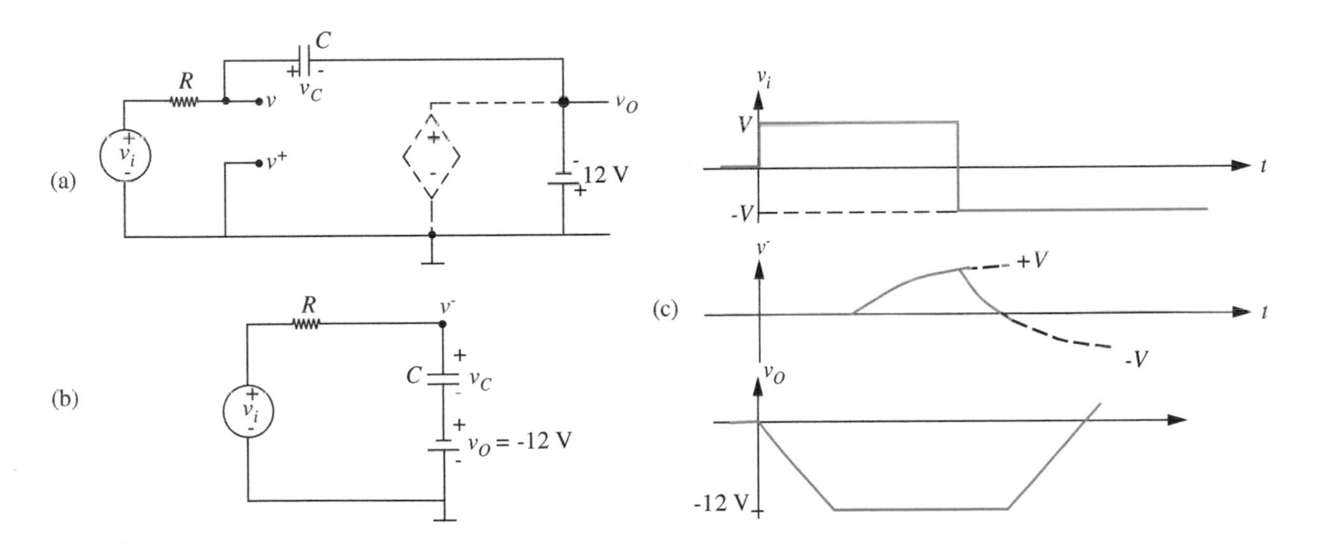


FIGURE 15.29 Integrator Op Amp in saturation.

Assuming a step input of $v_i$ as in Figure 15.29c, the solution for $v_C$ and $v^-$ can be found by inspection using the methods of Section 10.5.3. The voltage across $v_C$ at the instant before entering saturation is

$$v_{\text{init}} = +12\,\text{V} \tag{15.104}$$

because $v^-$ at this point is almost zero due to the Op Amp constraint. If the transient went to completion, the final capacitor voltage would be

$$v_{\text{final}} = V + 12\,\text{V}. \tag{15.105}$$

Hence from Equation 10.62, assuming a time origin at the instant of entering saturation,

$$v_C = 12e^{-t/RC} + (V + 12)(1 - e^{-t/RC}) \tag{15.106}$$

$$= V(1 - e^{-t/RC}) + 12. \tag{15.107}$$

(This is a general result for fixed voltages in series with capacitors: By superposition such problems can be solved without the fixed voltage, then the fixed voltage can be added back in. In other words, the transient part is unaffected by the fixed voltage.) It follows that

$$v^- = V(1 - e^{-t/RC}) \tag{15.108}$$

(a result that an experienced analyst of RC circuits would have written down directly). Waveforms appropriate to these equations are shown in Figure 15.29c.

An important issue of circuit performance is how long it takes the circuit to recover from the effects of saturation. To this end, assume that the input step is now reversed in polarity. The Op Amp will still be held in saturation by the large positive voltage on the $v^-$ terminal, and will remain in saturation until $v^-$ has decayed virtually to zero. Figure 15.29b remains the appropriate circuit representation for this interval. Thus with $v_i = -V$,

$$v^- = -V + 2Ve^{-t/RC} \tag{15.109}$$

$$v_O = -12 \tag{15.110}$$

where the time origin is now defined to be at the instant of negative transition of $v_i$. Appropriate waveforms are shown in Figure 15.29c. As noted earlier, this saturation state persists until $v^-$ has decayed almost to zero. Only then will the Op Amp come out of saturation, and integrate back toward zero, as shown in Figure 15.29c.

In summary, driving the Op Amp into saturation has two serious consequences on the performance of the integrator. First, the integration is truncated

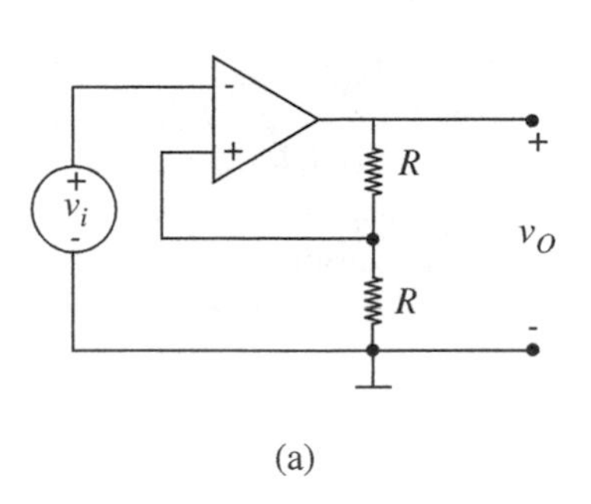


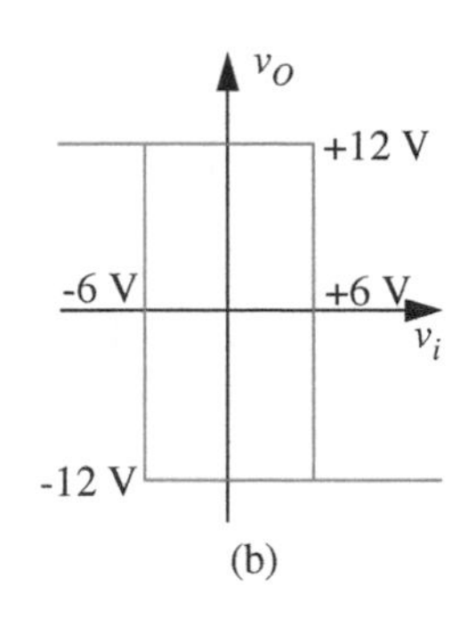


FIGURE 15.30 Positive feedback.

when the Op Amp saturates. Second, when the input wave goes negative, there is a substantial *delay* before the circuit recovers, and again acts as an integrator.

## 15.8 POSITIVE FEEDBACK

In every Op Amp circuit discussed thus far, the feedback network has been connected from the Op Amp output to the *negative* input terminal of the Op Amp. Such a connection provides *negative feedback*, which tends to make the circuit more linear, more temperature independent, more reliable. An obvious question: What happens if the feedback is connected to the *positive* input terminal, as in Figure 15.30a?

The complete relation between $v_O$ and $v_i$, Figure 15.30b, shows both saturation and hysteresis. To understand the circuit action, assume that $v_O$ is initially positive and $v_i$ is negative. Then, because of the feedback, $v^+$ is still at $+6$.

To get the Op Amp out of saturation, $v_i$ must be made positive enough to bring $(v^+ - v^-)$ approximately to zero, hence $v_i$ must be approximately $+6$ volts. If $v_i$ is slightly more positive than $+6$, $v_O$ will be driven negative, whereupon $v^+$ will be driven negative, driving $v_0$ even more negative. Hence, $v_O$ undergoes a regenerative negative transition to $-12$ V. Now $v_i$ must be made more negative than $-6$ volts to initiate a regenerative transition to $+12$ V. the width of the hysteresis region can be controlled by the ratio of the feedback resistors.

The circuit is obviously no longer a linear amplifier: The positive feedback has enhanced rather than suppressed the basic nonlinearity of the unadorned Op Amp. One application of this circuit is as a digital comparator, to convert a continuous analog signal to a two-state signal.

### 15.8.1 RC OSCILLATOR

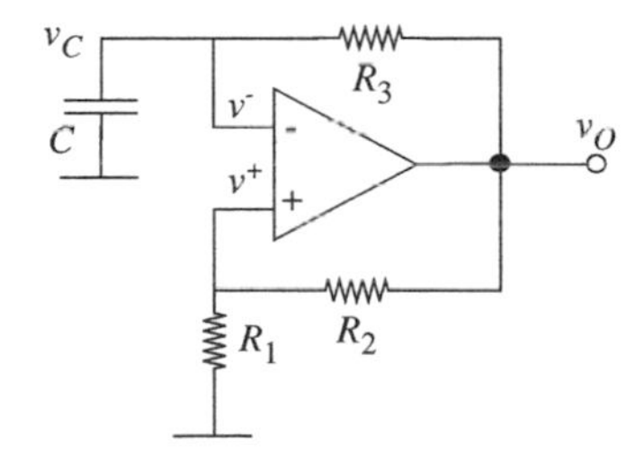


FIGURE 15.31 RC oscillator circuit.

Shown in Figure 15.31 is another Op Amp circuit that uses positive feedback. Assume that the power supply voltages are $V_S$ and $-V_S$. This circuit behaves as an oscillator, and uses positive feedback to saturate the Op Amp at both positive and negative values of $V_S$. Let's examine how this oscillator works.

Let us first analyze the circuit qualitatively, referring to the waveforms for $v_C$ and $v_O$ in Figure 15.33. As depicted in Figure 15.33, let us assume that the system starts from rest so the capacitor voltage $v_C = 0$. Thus the inverting terminal $v^-$ of the Op Amp is at 0 V. Let us also assume that the output is in positive saturation initially, in other words at the positive power supply voltage, $V_S$. Since the output is fed back to the positive input, we observe that

$$v^+ = \frac{V_S R_1}{R_1 + R_2}.$$

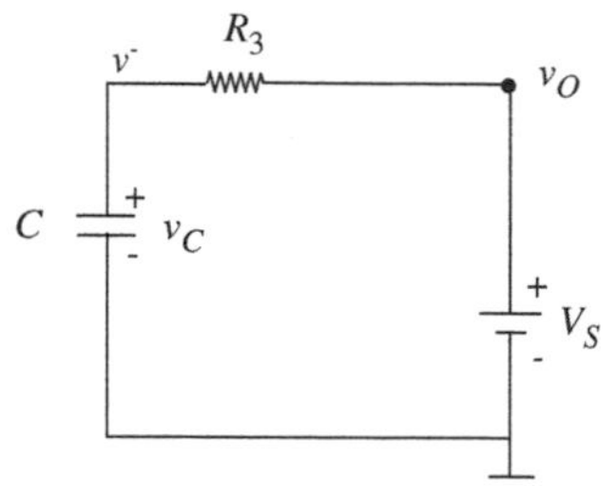

FIGURE 15.32 Equivalent circuit for the RC oscillator when the Op Amp is in positive saturation.

This positive voltage at the non-inverting input terminal will result in a positive voltage difference at the Op Amp input port (between $v^+$ and $v^-$), and consequently, the output will continue to be driven to the positive saturation voltage, namely $V_S$. The equivalent circuit is as shown in Figure 15.32. The capacitor $C$ begins to charge up towards $V_S$ through the resistor $R_3$. Since no current flows into the $v^-$ terminal, the charging dynamics are that of a simple RC circuit.

As the capacitor charges up, eventually its voltage $v_C$ crosses $v^+ = V_S R_1/(R_1 + R_2)$, resulting in an effective negative voltage across the Op Amp input port, namely across the $v^+$ and $v^-$ terminals. The Op Amp amplifies this negative voltage difference at its input to a large negative voltage at its output. Since the negative voltage at the output is fed back to the non-inverting terminal by the voltage divider formed by $R_1$ and $R_2$, the non-inverting terminal voltage becomes negative, which makes the voltage difference at the Op Amp input even more negative, and which in turn makes the output voltage fall even more. This positive feedback process continues until the output reaches the negative saturation voltage $-V_S$. At this point, we have

$$v^+ = \frac{-V_S R_1}{R_1 + R_2}.$$

Notice that the output voltage transitions from $V_S$ to $-V_S$ very quickly at the moment that the capacitor voltage $v_C$ crosses $V_S R_1/(R_1 + R_2)$.

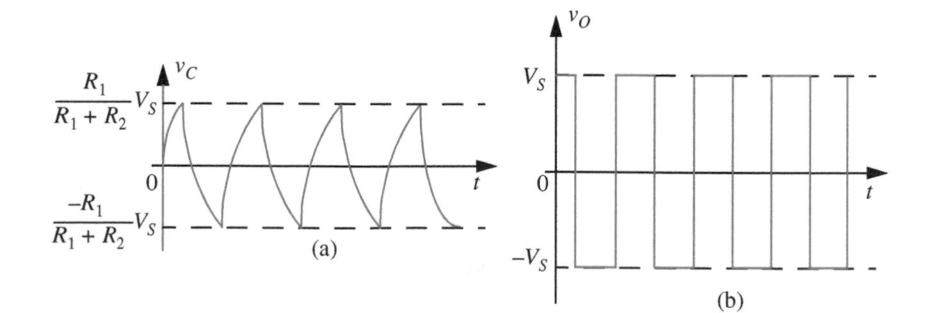


FIGURE 15.33 Oscillator behavior.

Therefore, at the instant the output reaches $-V_S$ and $v^+$ transitions to $-V_S R_1/(R_1 + R_2)$, we can assume that the capacitor voltage is still at approximately $V_S R_1/R_1 + R_2$, since the voltage across the capacitor changes much more slowly.

Now, since the capacitor voltage $v_C$ is higher than the output voltage, the capacitor begins to discharge through $R_3$. Figure 15.34 shows that the equivalent circuit that applies. When the capacitor voltage falls below $-V_S R_1/R_1 + R_2$, the voltage $v^-$ will be lower than $v^+$ resulting in a positive voltage difference at the Op Amp input. The Op Amp amplifies this positive difference to a positive voltage at its output, which when fed back to the non-inverting terminal causes a larger positive voltage to appear across the Op Amp input. The resulting positive feedback causes the Op Amp output to go into positive saturation. Thus, the output voltage reaches $V_S$ and that at $v^+$ will again be

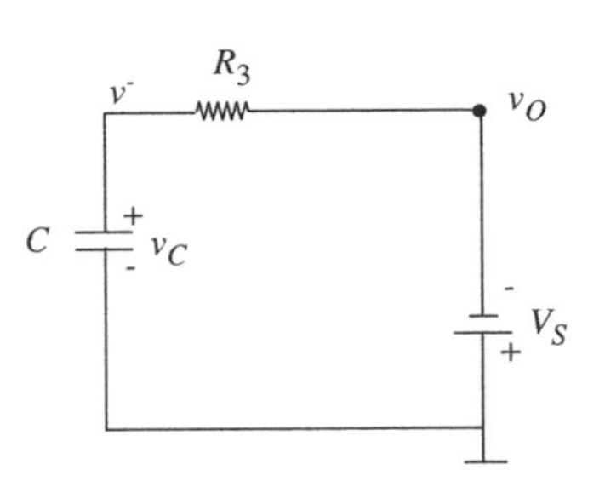

FIGURE 15.34 Equivalent circuit for the RC oscillator when the Op Amp is in negative saturation.

$$v^+ = \frac{V_S R_1}{(R_1 + R_2)}.$$

As in the beginning, the capacitor voltage is now lower than the output and therefore the capacitor begins to charge up. This cycle repeats and results in a square wave at the output of the Op Amp.

Let us derive the time period of the oscillator in Figure 15.31. Assume that at time $T_1$, $v_O$ transitions from $V_S$ to $-V_S$ as illustrated in Figure 15.35. We know that at $T_1^-$, $v_O = V_S$ and from the voltage-divider relationship, we know that $v^+ = R_1 V_S/(R_1 + R_2)$. We also know that $v^-$ is lower than $v^+$ at $T_1^-$, and that the capacitor voltage is increasing.

At time $T_1^+$, $v^-$ becomes slightly greater than $v^+$. In other words, $v^- \approx V_S R_1/(R_1 + R_2)$. Output $v_O$ transitions virtually instantaneously to $-V_S$ and $v^+$ becomes $-R_1 V_S/(R_1 + R_2)$. The capacitor now begins to discharge and $v^-$ begins to decrease. We know that $v_O$ will transition from low to high when $v^-$ falls below $v^+ = -R_1 V_S/(R_1 + R_2)$.

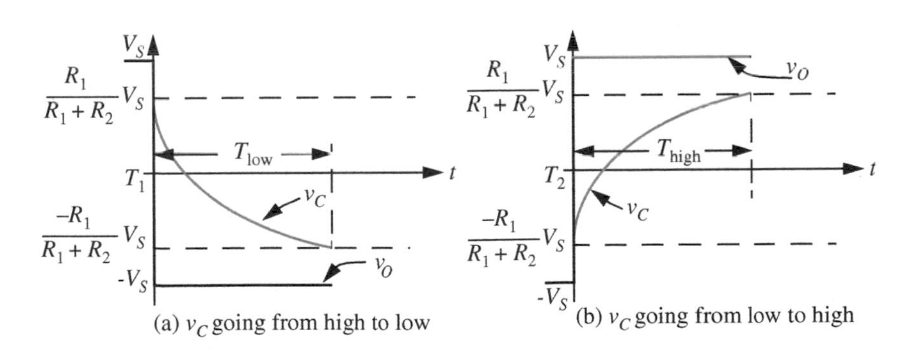

FIGURE 15.35 Computing the time period of the oscillator.

Thus the interval $T_{low}$ is the time taken for the capacitor to discharge from its initial value of $V_S R_1/(R_1 + R_2)$ to its final value of $-V_S R_1/(R_1 + R_2)$. The capacitor discharge dynamics are governed by a simple first-order differential equation whose solution is given by

$$v_C = -V_S + \left(\frac{R_1}{R_1 + R_2} + 1\right) V_S e^{-t/R_3 C}. \tag{15.111}$$

We need to find $T_{low}$, the time taken for $v_C$ to drop below $-R_1 V_S/(R_1 + R_2)$ from Equation 15.111. In other words, we need to solve for the time that satisfies

$$v_C = -V_S + \left(\frac{R_1}{R_1 + R_2} + 1\right) V_S e^{-t/R_3 C} < -\frac{R_1}{R_1 + R_2} V_S. \tag{15.112}$$

Thus,

$$-V_S + \left(\frac{R_1}{R_1 + R_2} + 1\right) V_S e^{-T_{low}/R_3 C} = -\frac{R_1}{R_1 + R_2} V_S,$$

which yields

$$T_{low} = R_3 C \ln\left(1 + \frac{2R_1}{R_2}\right). \tag{15.113}$$

It is easy to verify that the duration of the high period $T_{high}$ is exactly the same as the low period. Thus, the period $T$ of the oscillator is simply

$$T = 2R_3 C \ln\left(1 + \frac{2R_1}{R_2}\right).$$

## WWW 15.9 TWO-PORTS*

## 15.10 SUMMARY

- The Op Amp is a widely used amplifier abstraction that forms the foundations of much of electronic circuit design. Op Amp devices are constructed using primitive elements such as transistors and resistors.
- The Op Amp is a four-ported device. The ports include an input port with terminals usually labeled $v^+$ and $v^-$, an output port with one terminal labeled $v_o$, and the other being ground, a positive power supply port with a $+V_S$ voltage applied with respect to ground, and a negative power supply port with a $-V_S$ voltage applied with respect to ground. Although the ground terminal is not explicitly shown in the Op Amp symbol, it is very much a part of all Op Amp circuits.
- The Op Amp behaves like a voltage-dependent voltage source. Its input-output relationship can can be expressed mathematically as

$$v_o = A(v^+ - v^-)$$

  where $A$ is a large number called the open loop gain of the amplifier. In most practical Op Amp applications, $A$ is treated as infinity.
- Most useful Op Amp circuits are built using the negative feedback connection, in which a portion of the output signal of the Op Amp is fed back to the $v^-$ input of the Op Amp. Examples of Op Amp circuits built this way include inverting and non-inverting amplifiers, buffers, adders, integrators, and differentiators.
- We commonly apply the constraint:

$$v^+ \approx v^-$$

  in analyzing Op Amp circuits, if the Op Amp is not saturated and the feedback is negative.
- Op Amp circuits are sometimes built using the positive feedback connection, in which a portion of the output signal of the Op Amp is fed back to the $v^+$ input of the Op Amp. Examples of such Op Amp circuits include oscillators and comparators.

## EXERCISES

EXERCISE 15.1 Find the Thévenin equivalent for the circuit in Figure 15.41. The circuit contains two resistors and a dependent current source.

EXERCISE 15.2 Calculate $v_O$ in terms of $I_1$, $V_1$, and $V_2$ in Figure 15.42. You may assume the operational amplifier has ideal characteristics.

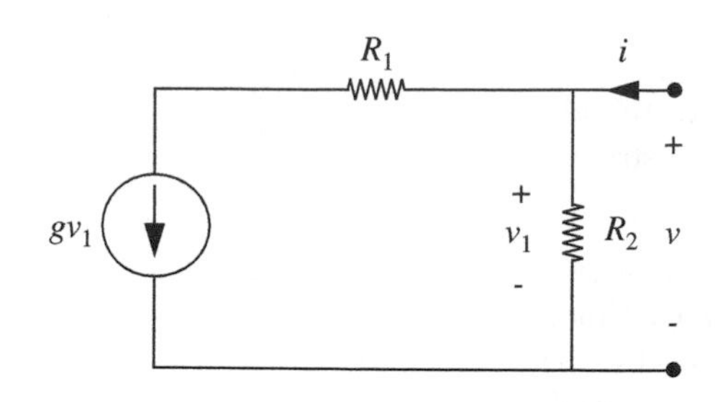


FIGURE 15.41

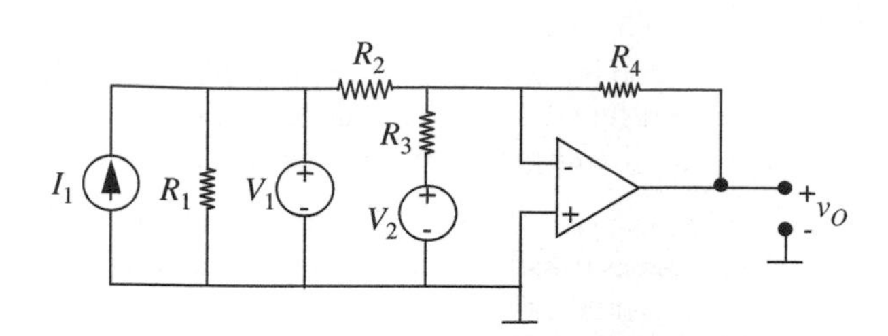


FIGURE 15.42

EXERCISE 15.3 Calculate the sensitivity of the gain, dG/G, as a function of fractional change in Op Amp gain, dA/A for the inverting Op Amp connection shown in Figure 15.43.

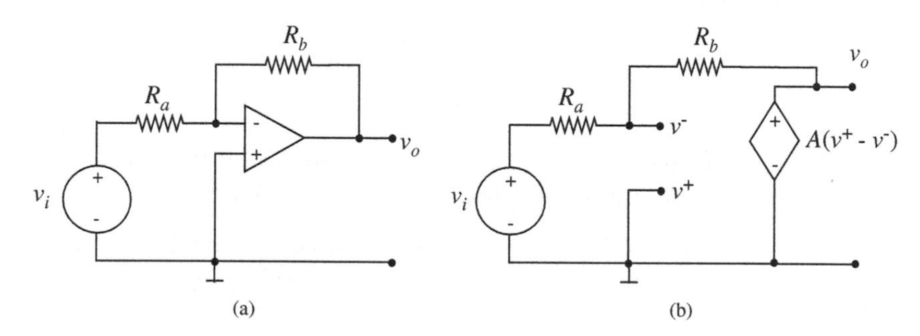


FIGURE 15.43 Inverting Op Amp.

EXERCISE 15.4 The circuit in Figure 15.44 is called a differential amplifier.

a) Using the ideal Op Amp model, derive an expression for the output voltage $v_O$ in terms of $v_1, v_2, R_1, R_2, R_3$, and $R_4$.

b) Does connecting a load resistor $R_L$ between the output and ground change the previous expression for $v_O$? Why?

c) Let $v_1 = v_2$ and $R_1 = 1\ \text{k}\Omega$, $R_2 = 30\ \text{k}\Omega$, and $R_3 = 1.5\ \text{k}\Omega$. Find $R_4$ so that $v_O = 0$.

d) Let $v_2 = 0$ and $v_1 = 1$ V. Using the preceding resistor values (including that computed for $R_4$), find $v_O$.

EXERCISE 15.5 For the circuit shown in Figure 15.45, D is a silicon diode, where $i = I_S\left(e^{qv/nkT} - 1\right)$, $kT/q = 26$ mV, and $n$ is between 1 and 2.

a) Find $v_O$ in terms of $v_1$ and $R_1$.

b) Make a quick sketch of the answer to (a).

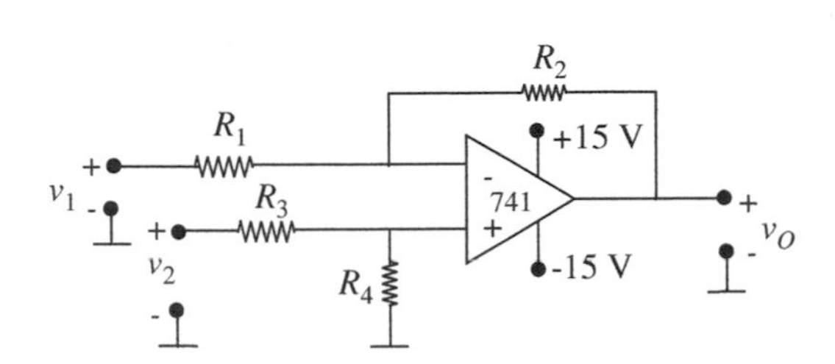


FIGURE 15.44

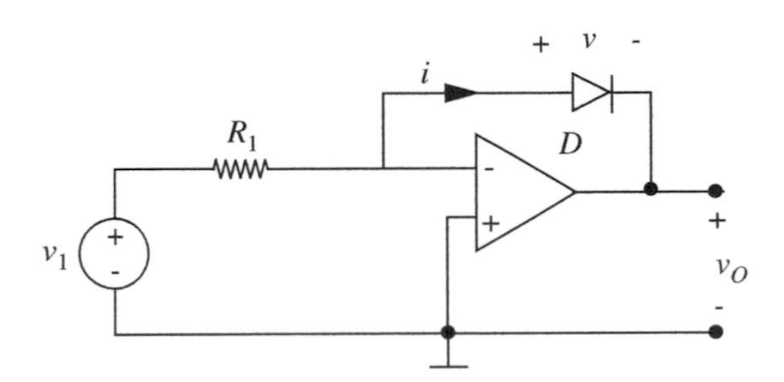


FIGURE 15.45

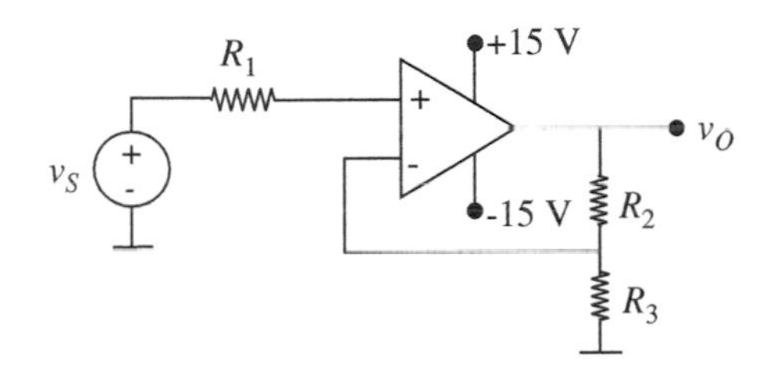


FIGURE 15.46

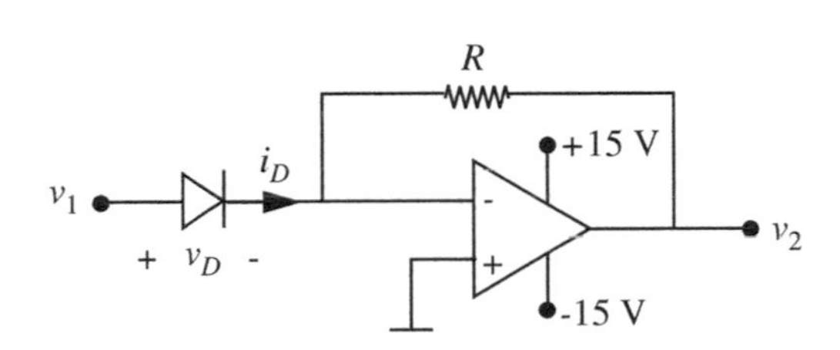


FIGURE 15.47

EXERCISE 15.6 Refer to the figure in Figure 15.46 for this exercise.

Given that $v_S = 2\ \cos(\omega t)$ (in volts), make a sketch of $v_O(t)$ through one complete cycle. Be sure to label the dimensions of the voltage and time axes and identify characteristic waveform shapes with suitable expressions. (Make reasonable assumptions based on your lab experience.)

EXERCISE 15.7 Refer to Figure 15.47 for this exercise.

Diode data $i_D = I_S\left(e^{qv_D/kt} - 1\right)$

where $I_S = 10^{-12}$ A

and $kT/q = 25$ mV.

For $v_1$ in the range $|v_1| < 575$ V, how should the value of $R$ be chosen to keep the Op Amp in the linear region? Make reasonable approximations.

EXERCISE 15.8 Find the Norton equivalent circuit to the left of terminal pair a–a′ in Figure 15.48.

EXERCISE 15.9 In the circuits (a) and (b) shown in Figure 15.49 the operational amplifiers are ideal and have infinite gain. If the input to each amplifier is $v_1 = I$ V, what is the output voltage $v_O$ for (a) and for (b)?

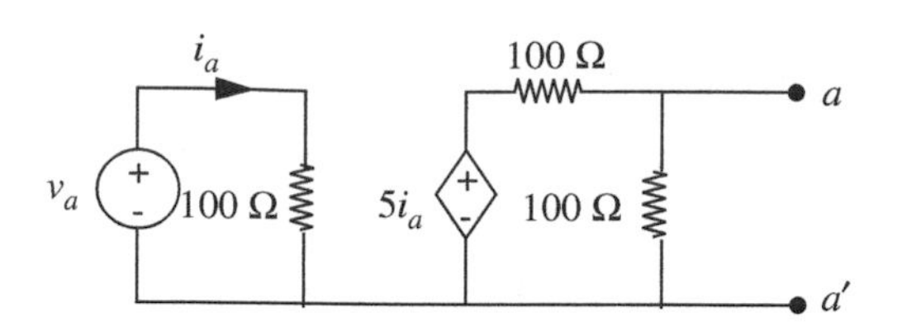


FIGURE 15.48

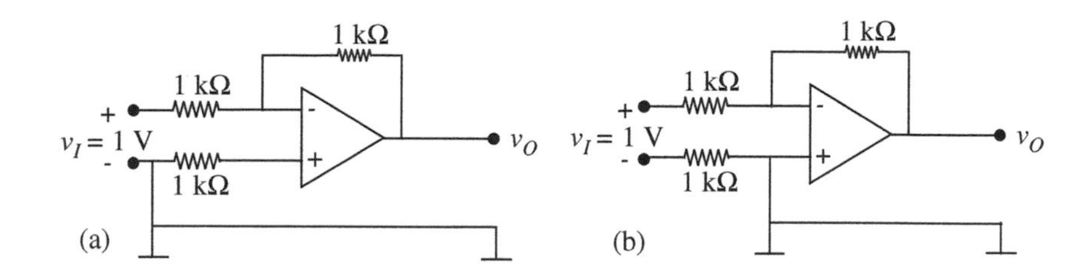


FIGURE 15.49

EXERCISE 15.10 You may assume that the operational amplifiers used in the connections shown in Figure 15.50 have very high gain and input resistance, and low output resistance when operating in the linear region.

The input signals have the form shown in Figure 15.51.

a) Plot the output voltage $v_O$ for the circuit of Figure 15.50a for $A = 1$ V. Note: In all of your plots, be sure to clearly indicate peak values and times when signals change character abruptly.

b) Plot the output voltage $v_O$ for the circuit of Figure 15.50a for $A = 10$ V.

c) Plot the output voltage $v_O$ for the circuit of Figure 15.50b for $A = 10$ V.

EXERCISE 15.11 For the circuit shown in Figure 15.52 (which includes a voltage-controlled voltage source) determine:

a) The input resistance $v_I/i_I$.

b) The Thévenin equivalent resistance at the terminals $a$ and $b$.

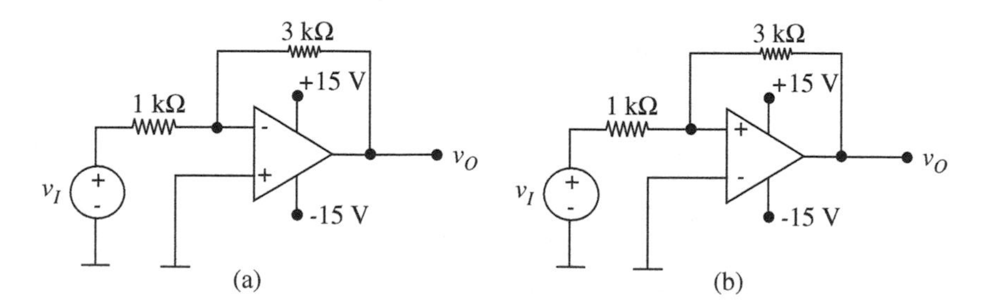


FIGURE 15.50

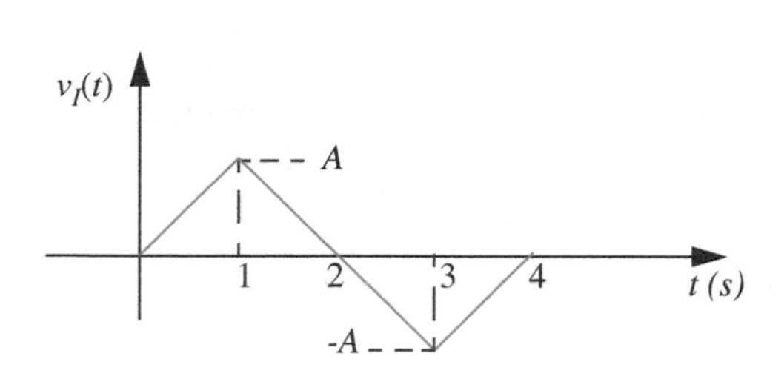


FIGURE 15.51

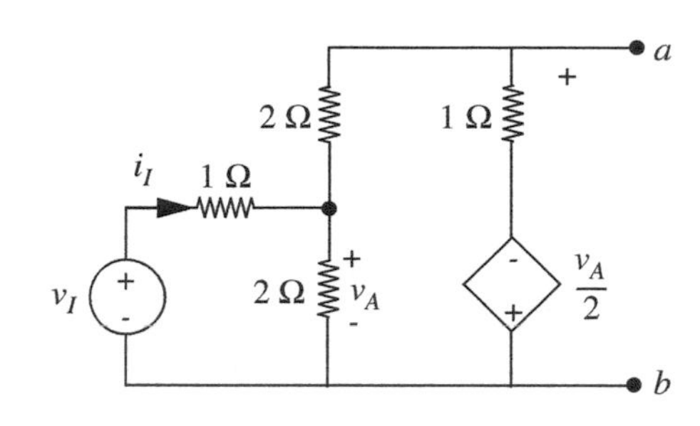


FIGURE 15.52

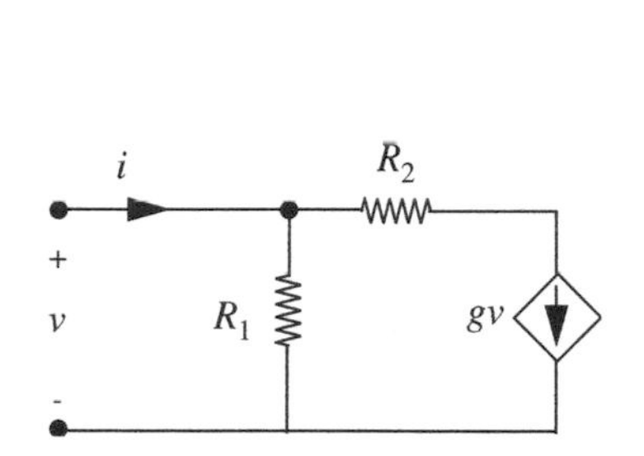


FIGURE 15.53

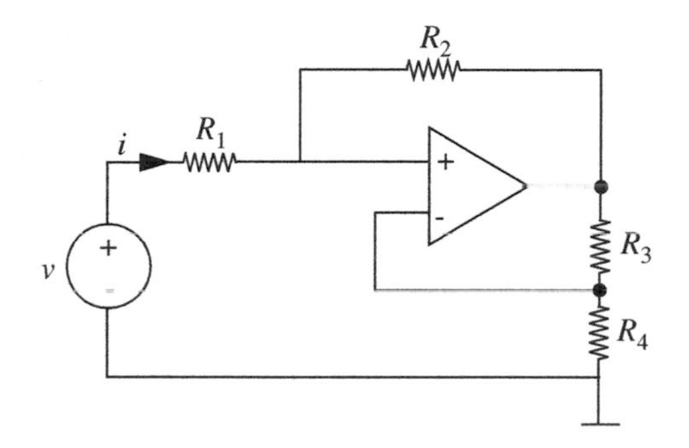


FIGURE 15.54

EXERCISE 15.12 Find and label clearly the Thévenin equivalent for the network in Figure 15.53.

EXERCISE 15.13 Find $i$ in terms of $v$ for the linear network in Figure 15.54. Assume an idealized operational amplifier.

EXERCISE 15.14 Determine the Thévenin equivalent for the circuit shown in Figure 15.55, to the left of terminal pair a–a′. The circuit contains a current-controlled voltage source.

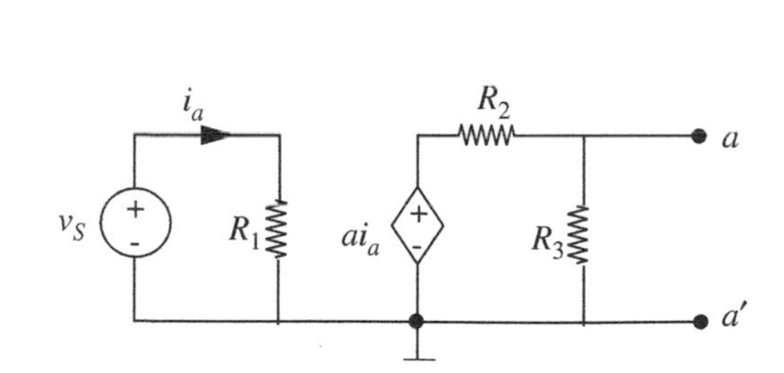


FIGURE 15.55

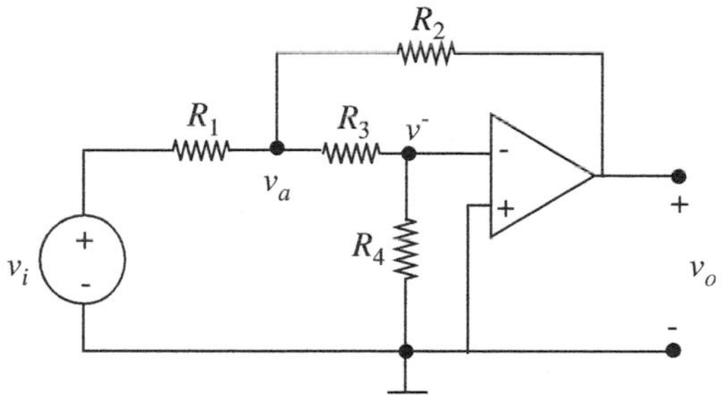

FIGURE 15.56

EXERCISE 15.15

a) Draw a circuit model for the Op Amp circuit in Figure 15.56.

b) Write the node equations for the $v_a$ and the $v^-$ nodes, and enough more independent relations to specify $v_o$ in terms of $v_i$. Do not solve.

EXERCISE 15.16 For the circuit in Figure 15.57 find $v_{\text{out}}$ as a function of $v_1, v_2, R_a$, and $R_b$ in the *limit of very high Op Amp gain*. Assume input resistance $r_i = \infty$, output resistance $r_t = 0$, and non-saturated operation.

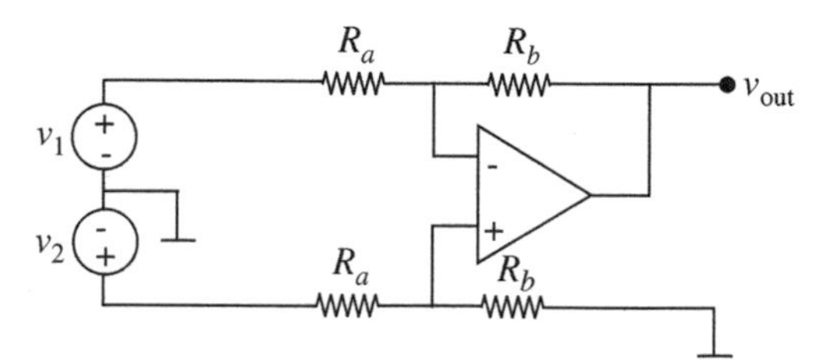


FIGURE 15.57

EXERCISE 15.17 For the circuit in Figure 15.58 find $i_1$ as a function of $v_i$, $R_1$, $R_2$, and the Op Amp gain $A$. Assume input resistance $r_i = \infty$ output resistance $r_t = 0$, and non-saturated operations.

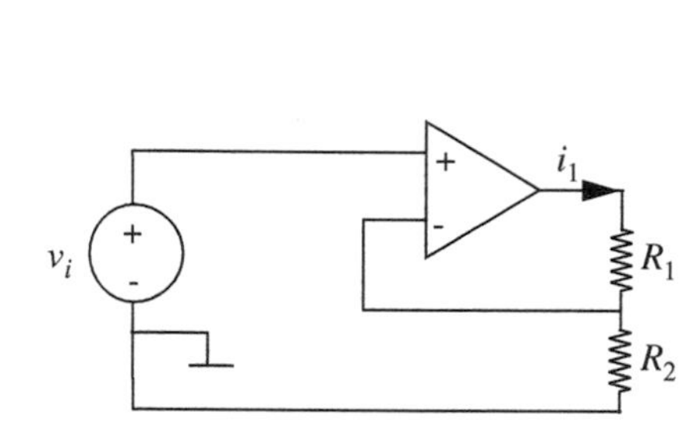


FIGURE 15.58

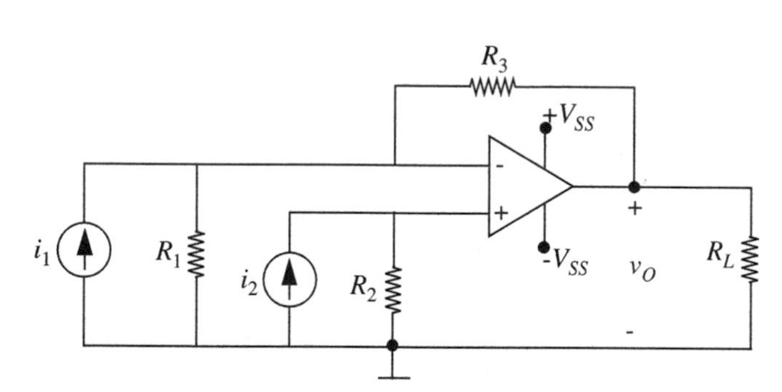


FIGURE 15.59

EXERCISE 15.18 Consider the circuit illustrated in Figure 15.59.

Assume that the operational amplifier is ideal with input resistance $r_i$ very large and output resistance $r_t$ negligibly small, so that $i^+ \simeq, i^- \simeq 0$, and $v_O = A\left(v^+ - v^-\right)$, with $A$ very large. Assume it is operating in its linear range.

a) Draw a linear equivalent circuit for this circuit valid for operation with the Op Amp in its linear range.

b) Derive an expression for $v_O$ as a function of $i_1$,$i_2$, and the resistors in the circuit.

EXERCISE 15.19 In the circuit in Figure 15.60 determine the voltage gain $G = v_o/v_i$:

a) when terminal x is connected to terminal a.

b) when terminal x is connected to terminal b. Assume the Op Amp is ideal.

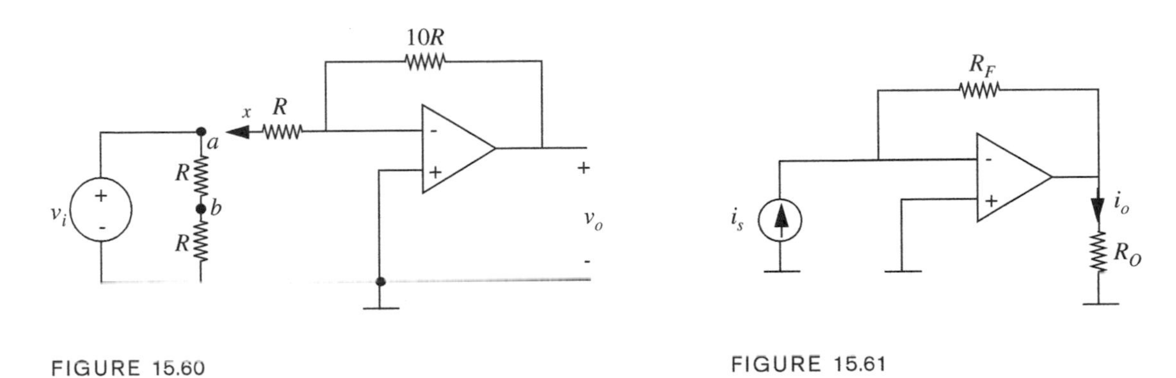


FIGURE 15.60

FIGURE 15.61

EXERCISE 15.20 For the amplifier shown in Figure 15.61, find the current transfer ratio $i_o/i_s$. Assume that the Op Amp is ideal.

EXERCISE 15.21 Find the Thévenin output resistance of the circuit shown in Figure 15.62. That is, find the resistance seen looking in at the terminals X X, the terminals that drive the load resistance $R_L$. (Resistor $R_L$ should not be included when you make this calculation.) *Do not* assume $v^+ \simeq v^-$, as it leads to trouble here. Now, state a condition on the value of $R_S$ to ensure that the circuit acts as a current source driving $R_L$.

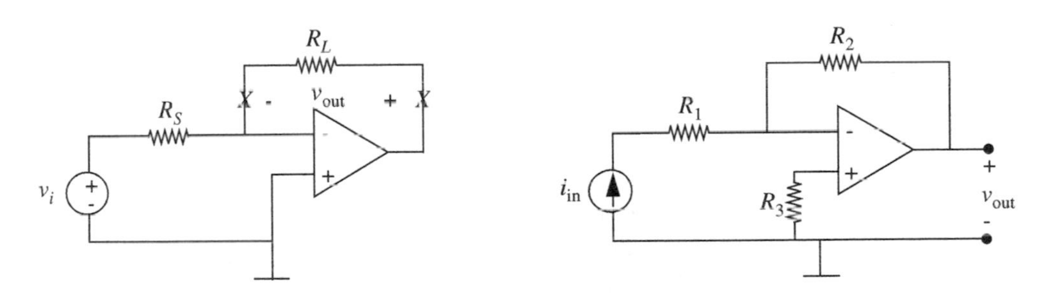


FIGURE 15.62

FIGURE 15.63

EXERCISE 15.22 For the Op Amp circuit in Figure 15.63:

a) Assume that the Op Amp is ideal (very large gain $A$, zero output resistance, infinite input resistance, operating in the linear region) and find $v_{out}$ as a function of $i_{in}$, $R_1$, $R_2$, and $R_3$.

b) Draw the circuit model, assuming the Op Amp has finite $A$, keeping the other assumptions from (a).

c) Analyze the circuit and find an expression for $v_{out}$ as a function of $i_{in}$, $R_1$, $R_2$, and $R_3$ and (finite) $A$.

EXERCISE 15.23 The operational amplifier circuit shown in Figure 15.64 is driven with a ramp.

You may assume that the operational amplifier has infinite open-loop gain, zero output resistance, and infinite input resistance, and that the capacitor voltage is zero for $t < 0$. What are the value of $v_O(t)$ at $t = 0^+$ and $t = 2$ ms?

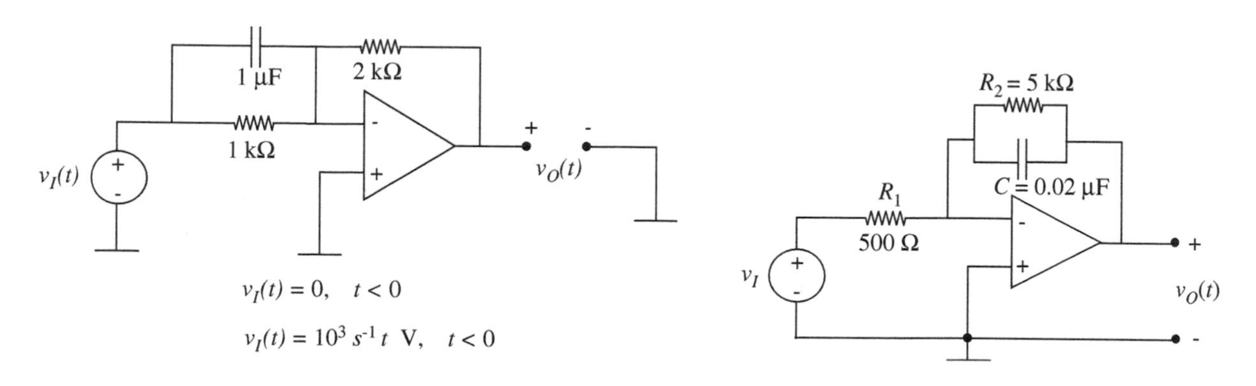


FIGURE 15.64

FIGURE 15.65

EXERCISE 15.24 An operational amplifier is connected as shown in Figure 15.65.

a) What is the gain of the amplifier for $\omega = 0$?.

b) Find the expression for $V_o(j\omega)/V_i(j\omega)$.

c) At what frequency does $|V_o|$ fall to 0.707 of its low-frequency value?

EXERCISE 15.25 For the circuit shown in Figure 15.66, determine $V_{out}(s)$ in terms of $V_{in}(s)$.

EXERCISE 15.26 $R_1 = R_2 = 20\ \Omega$ $C = 2.4\mu$F $L = 0.25$ mH

Find the system function $H(s) = V_b/V_a$ for the circuit in Figure 15.67.

EXERCISE 15.27 For the circuit shown in Figure 15.68, select the magnitude of the frequency response for the system function given. It is not necessary to relate the critical frequencies to the circuit parameters.

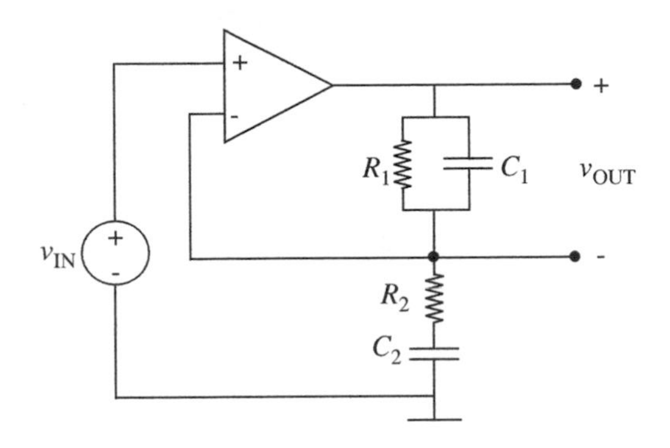


FIGURE 15.66

FIGURE 15.67

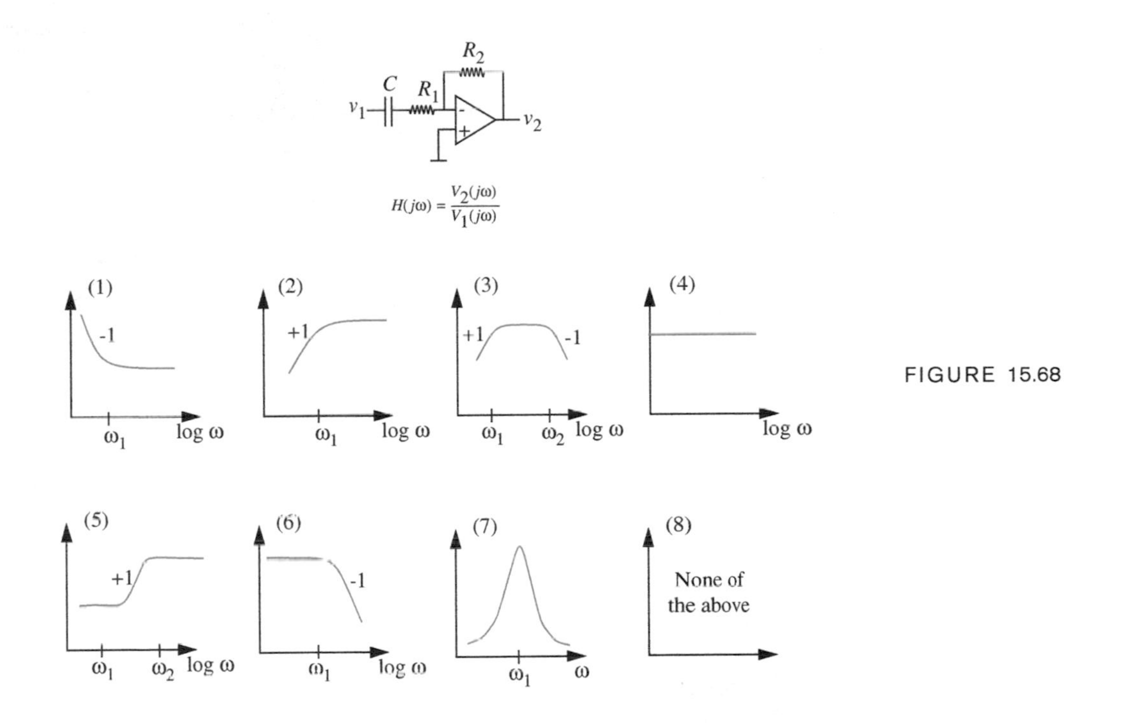


FIGURE 15.68

Please note that the magnitude responses, except (7), are sketched on a log-log scale, with slopes labeled.

## PROBLEMS

PROBLEM 15.1 The circuit shown in Figure 15.69 is very similar to the standard non-inverting Op Amp except that $R_L$ is some external resistor, and we are interested in showing that the current through $R_L$ is nearly constant, regardless of the value of $R_L$, that is, the circuit acts like a *current source* for driving $R_L$.

FIGURE 15.69

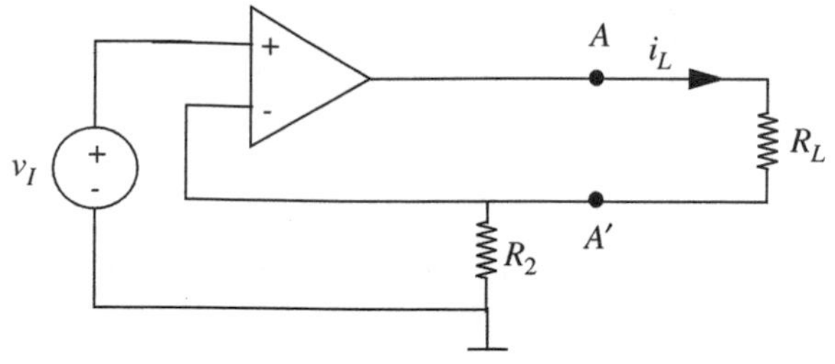

a) Using the ideal Op Amp assumption of large gain, zero output resistance, infinite input resistance, show that the expression for $i_L$ as a function of $v_I$ is independent (or weakly dependent) on $R_L$.

b) To verify the "current source" action more directly, find the Thévenin equivalent resistance looking to the left of terminals AA′, with $R_L$ an open circuit.

PROBLEM 15.2 Zener diodes are most often used to establish stable reference voltages, independent of power supply variations, and independent of any lingering AC signals that may be present in the power supply.

a) For the characteristics shown in Figure 15.70, find $v_O$ assuming $v_A$ is a clean DC voltage of value 15 V.

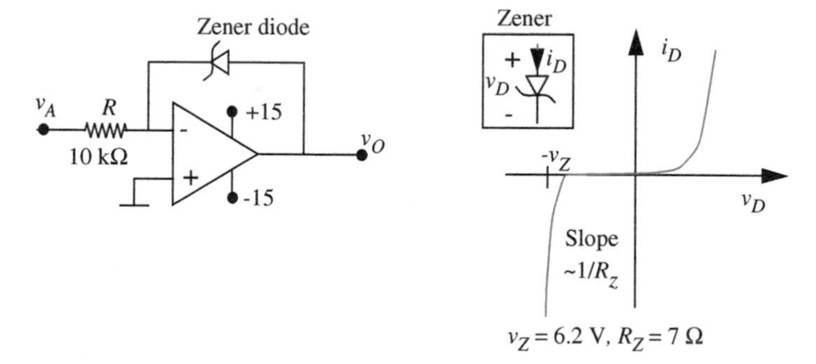


FIGURE 15.70

b) Determine the sensitivity of $v_O$ to changes in $v_A$. That is, find $dv_O/dv_A$. If $v_A$ has 0.1 V of DC drift or so of 120-*Hz* AC ripple, how much drift or ripple shows up on $v_O$?

PROBLEM 15.3 Consider the circuit in Figure 15.71.

Find $v_O$ assuming that all Op Amps are ideal and operating in the linear region.

PROBLEM 15.4 You are faced with the problem of constructing a current transmitter, a circuit that forces a load current $i_L$ into a load under accurate control of a

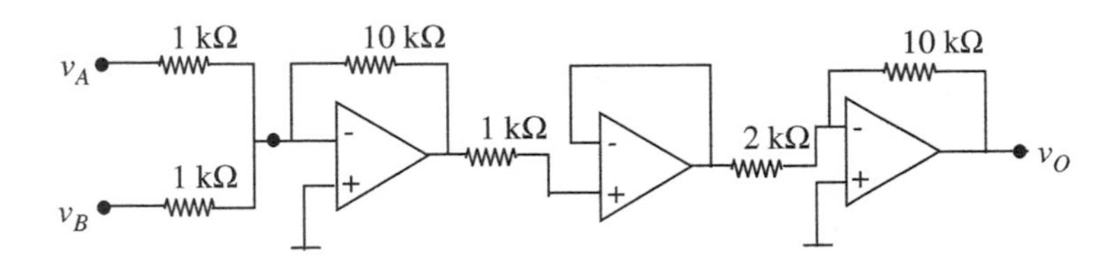


FIGURE 15.71 $v_A = 0.1$ V and $v_B = 0.2$ V.

source voltage $v_S$, independent of variations in load resistance. That is, you need a voltage-controlled current source.

The design requirements for your problem are to achieve

$$i_L = -Kv_S$$

where $K = 10$ mA/V for the ranges $|v_S| < 1$ V, $R_L < 1$ kΩ.

While looking through a handbook of practical circuits, you come across the schematic in Figure 15.72 as a proposed solution to your problem. The question is, will it work?

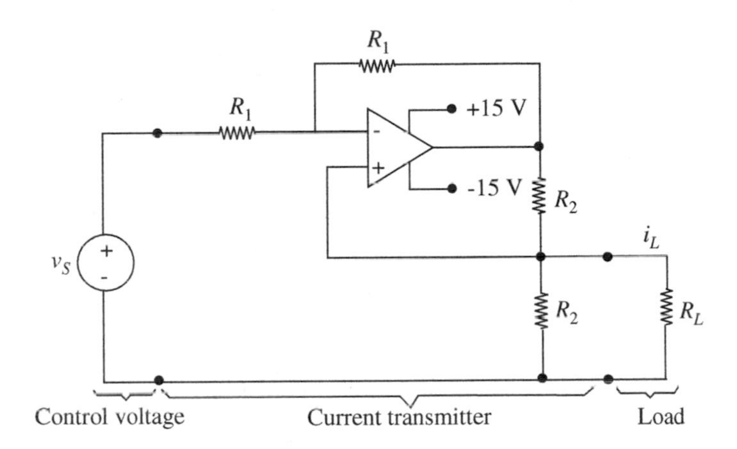


FIGURE 15.72

a) As a first step, analyze the basic principle of operation of the above circuit. Show explicitly whether it is capable of performing the desired function.

b) Next, determine whether there will be any problems in selecting resistor values $R_1$ and $R_2$ to meet the specifications for your particular application. You should draw on experience with Op Amp limitations. Can you meet the specs?
(Note: Part (a) is easy. Part (b) is endless, so look only for the *larger* issues, that is, major sources of error or failure.

PROBLEM 15.5 Find the Norton equivalent of the circuit in Figure 15.73 looking into terminals $A$ and $A'$.

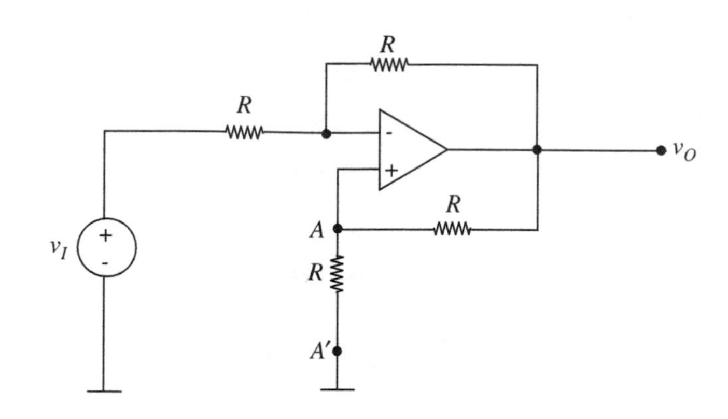


FIGURE 15.73

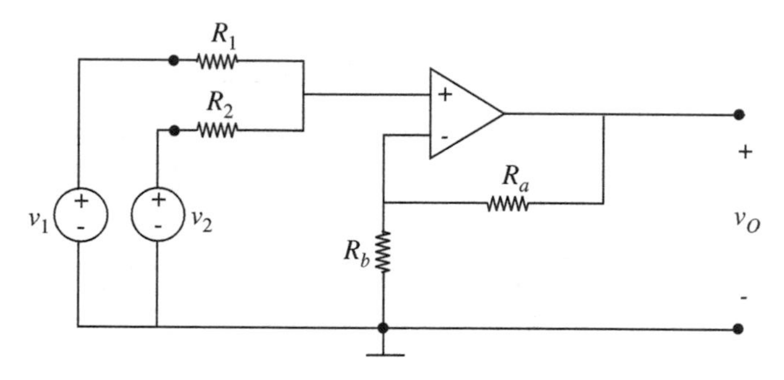


FIGURE 15.74

PROBLEM 15.6 You are asked to design the circuit shown in Figure 15.74 so that the output voltage $v_O$ is the weighted sum of $v_1$ and $v_2$; specifically:

$$v_O = 3v_1 + 5v_2$$

It is known that the magnitudes of $v_1$ and $v_2$ are never larger than 1 volt.

a) Determine the values for $R_1, R_2, R_a,$ and $R_b$ that will make the circuit perform that sum.

b) Given that the op amp is powered from +15 and −15 V, and has output current limits of +1 mA and −1 mA, redesign if necessary to meet these additional design constraints.

c) How would you change the design to perform the sum:

$$v_O = -3v_1 - 5v_2$$

using only one Op Amp (given Figure 15.74, a two-op amp design is obviously trivial, but unnecessarily complicated)?

PROBLEM 15.7 For the circuit in Figure 15.75, assuming an ideal Op Amp with large $A$:

a) Calculate $v_O$ in terms of $v_I$ and the resistor values.

b) Find $i$ in terms of $v_I$ and the resistor values.

c) For what resistor values in (a) will the voltage gain become infinite? Explain why this occurs (one sentence).

d) Find the limits on the solutions in the (a) and (b) imposed by using a real Op Amp.

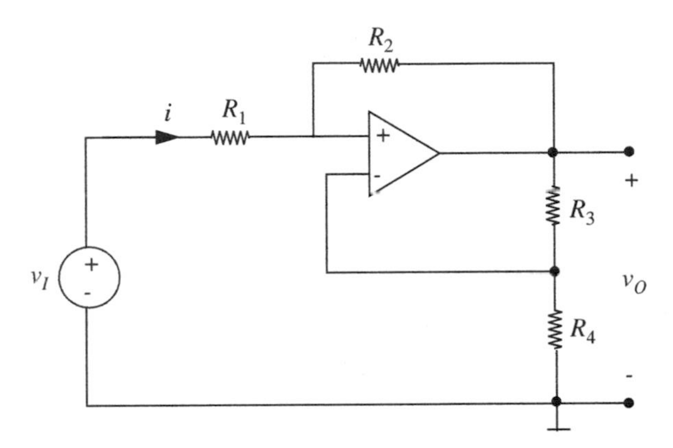


FIGURE 15.75

PROBLEM 15.8 Choose values for $R_1$ through $R_5$ in Figure 15.76 so that

$$v_O = +2v_1 - 5v_2 - v_3 - 3v_4.$$

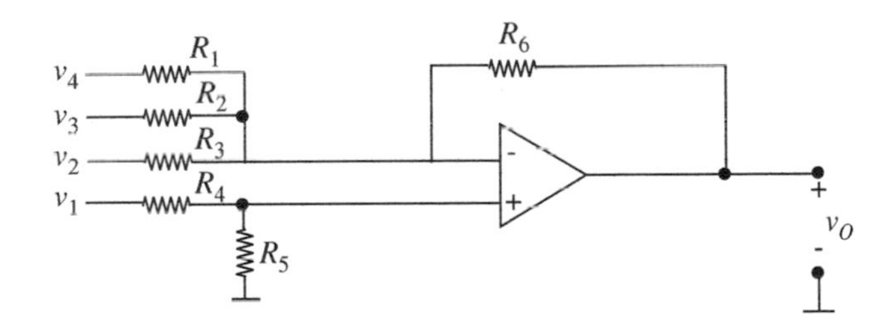


FIGURE 15.76

You may assume the operational amplifier has ideal characteristics.

PROBLEM 15.9 For the circuit in Figure 15.77, find $v_O$ in terms of $v_I$. Analyze with literal resistor values, then substitute numbers: $R_1 = R_2 = R_3 = 10\ \text{k}\Omega$, $R_4 = 100\ \Omega$.

PROBLEM 15.10 This question concerns the circuit illustrated in Figure 15.78:

The operational amplifier is a high gain unit ($A = 10^5$) with high input resistance, $r_i$, and negligibly low output resistance, $r_t$. Assume that it is operating in its linear region.

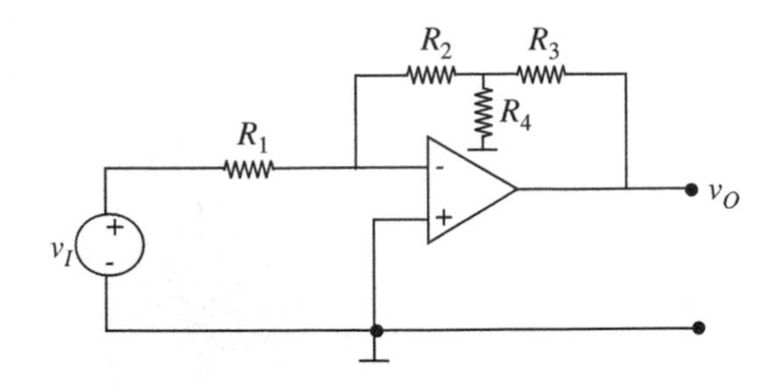


FIGURE 15.77

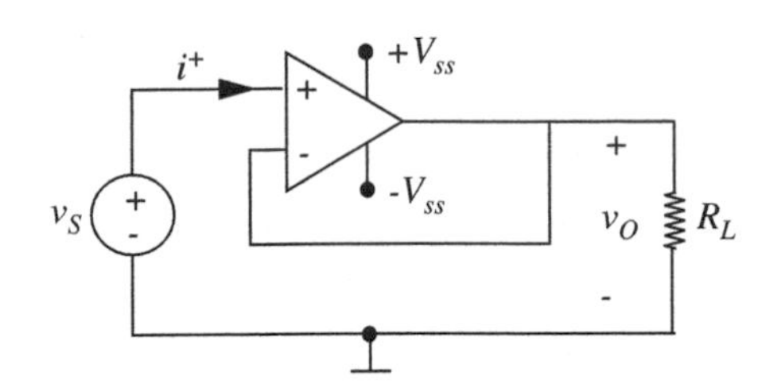


FIGURE 15.78

The following data is given:

$$v_S = 1\text{ V}$$
$$i^+ = 10\text{ pA} = 10^{-11}\text{ A}$$
$$R_L = 1\text{ k}\Omega.$$

a) What is $v_O$? (Accurate to within 1%.)

b) What is the power delivered by the source $v_S$? What is the power dissipated in the load resistor, $R_L$?

c) The power dissipated in the load resistor, $R_L$, is much larger than the power supplied by the source, $v_S$. Where does this additional power come from?

PROBLEM 15.11 The equivalent circuit of an amplifier is shown in Figure 15.79.

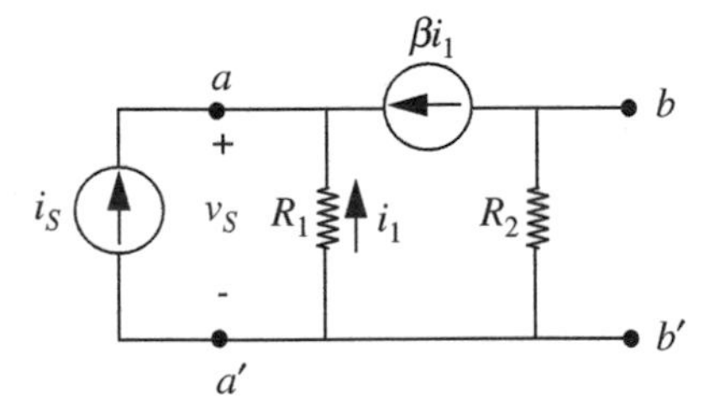


FIGURE 15.79

a) Find the input resistance seen by the current source $i_S$ at the input terminals $a$–$a'$.

b) Find the output resistance seen at the output terminals $b$–$b'$ (with the current source shut off).

PROBLEM 15.12 For the circuit in Figure 15.80 find $v_O$ in terms of $v_1$ and $v_2$. You can use in your analysis the ideal Op Amp model.

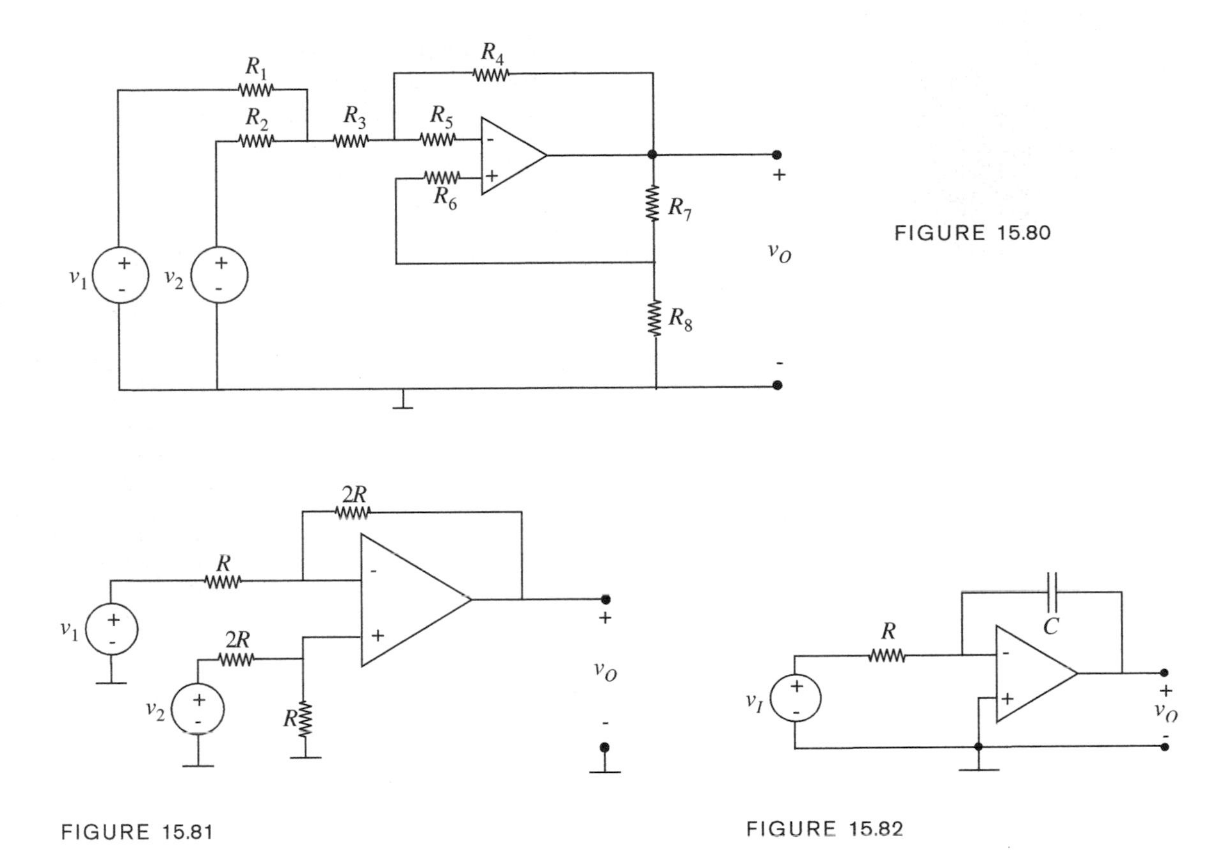


FIGURE 15.80

FIGURE 15.81

FIGURE 15.82

PROBLEM 15.13 An operational amplifier circuit is shown in Figure 15.81.

You may assume that the operational amplifier has ideal characteristics, including zero input current and output resistance and further make the simplifying assumption that its open-loop gain is infinite. Also, assume that the amplifier does not saturate.

a) With $v_2 = 0$, what is the value of the gain $v_O/v_1$?

b) Voltage $v_2$ is now made 3 volts. Plot the $v_O$ vs. $v_1$ characteristics. Be sure to show important values and slopes.

PROBLEM 15.14 By combining Op Amps with RC circuits, we can make circuits that perform elementary mathematical operations, such as integration and differentiation. The circuit in Figure 15.82 is, over some range, an integrator.

a) Use the ideal Op Amp model to determiner the ideal function performed by this circuit.

b) Based on your knowledge of Op Amp limitations, indicate the constraints that must be placed on the component values R and C to achieve satisfactory operation, assuming that the input is a sine wave with angular frequency $\omega$ and peak amplitude $A$. Express your answer as a constraint on the RC product imposed by the voltage limit, and a separate constraint imposed by the current limit.

c) For practical reasons, $R$ usually should not be greater than 1 megohm. Calculate the value of $C$ required to meet the voltage constraint listed above for operation at 20 Hz and above, and $A = 1$ V.

PROBLEM 15.15 The capacitor you calculated in Problem 15.14c is (or should be) much larger than the maximum capacitor that can be included on a VLSI chip. For this reason, the circuit in Figure 15.82 must usually be built of Op Amps, discrete $R$'s and $C$'s. To allow the circuit to be built on a chip, the resistor is replaced by a *switched capacitor*, which can produce a very large "effective resistor" with reasonable capacitor values. This circuit is shown in Figure 15.83.

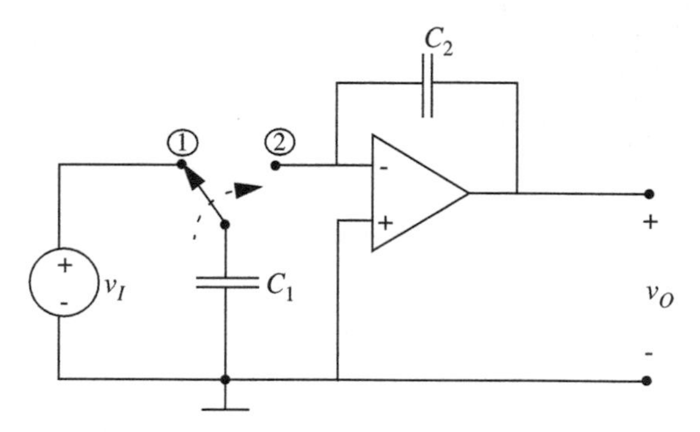


FIGURE 15.83

At time $t = t_1$, the switch moves to position (1), and $C_1$ charges (instantly) to voltage $v_1(t_1)$. Then at time $t_2$, the switch moves to position (2), and $C_1$ discharges into $C_2$. Assuming that the usual Op Amp approximation of $(v^+ - v^-) \simeq 0$ can still be used, calculate the charge that is "dumped" at each cycle, hence the average current (a function of both $v_I$ and the switching frequency $f_c$), and hence the effective resistance of the switched capacitor. Also, show that the overall system equation relating $v_O$ to $v_I$ is the same as in Problem 5.14.

PROBLEM 15.16 In Figure 15.83, what are the constraints on $C_1$ and $C_2$ set by the Op Amp voltage and current limits? Calculate the appropriate values of $C_1$, $C_2$, and $f_c$ for operation at 20 Hz and above. Can the circuit now be built on an IC chip if we replace the switch by MOS transistors, and $C_{max} = 100$ pF?

PROBLEM 15.17 Design a differentiator circuit out of RC circuits and Op Amps.

Calculate the constraints as in Problem 5.14b.

PROBLEM 15.18 This problem deals with switched-capacitor circuits introduced in Problem 15.15. Referring to Figure 15.84, assume both $S_1$ switches are closed for time $1/2f_0$ with $S_2$ open, and $S_2$ closed for $1/2f_0$ with $s_1$ open. Assume no overlap, that is, and $S_1$ and $S_2$ switches are never both closed at the same time.

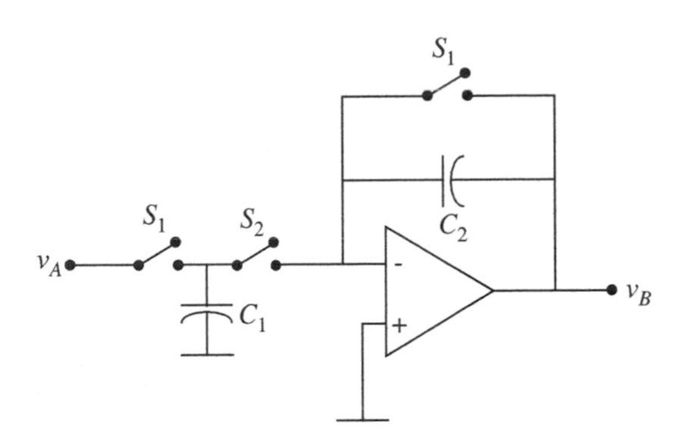


FIGURE 15.84

a) For $v_A = A$ volts (constant), go through one complete clock cycle, identifying the charge on each capacitor and the voltage at each node.

b) Now assume $v_A = A\cos(\omega t)$ where $\omega \ll 2\pi f_0$. Sketch $v_B$. In the circuit as constructed, $v_B$ is zero half the time. During the *other* half cycles, $v_B$ and $v_A$ are related by a simple gain expression, just as in a normal inverting amplifier. What is the "gain"?

PROBLEM 15.19 Figure 15.85 is a *practical* implementation of a switched capacitor circuit (see Problem 15.15). As in the previous problem, it is useful to examine the behavior of an "average $v_B$" over a clock cycle.

a) Show that if $v_A = A$ volts (constant), the cycle-average of $v_B$ has a steady-state value equal to $-(C_1/C_2)A$. In other words, for low-frequency signals, the circuit behaves like a non-inverting amplifier with gain $-(C_1/C_2)$.

b) Show, for $v_A$ a step of amplitude $A$ volts, and assuming $v_B$ is initially zero, that the cycle average of $v_B$ "charges up" to its steady-state value with time constant $\tau = c_3/f_0C_2$. That is, show that the cycle-average of $v_B$ obeys a first-order linear differential equation with time constant $C_3/f_0C_2$.

PROBLEM 15.20

a) Use the ideal Op Amp model to determine the ideal function performed by the circuit in Figure 15.86.

b) Based on your knowledge of Op Amp limitations, discuss the accuracy with which the circuit will perform the intended function, or indicate any constraints that must

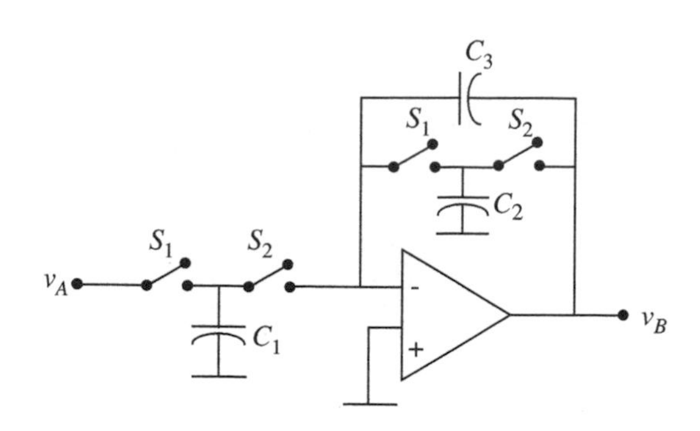


FIGURE 15.85

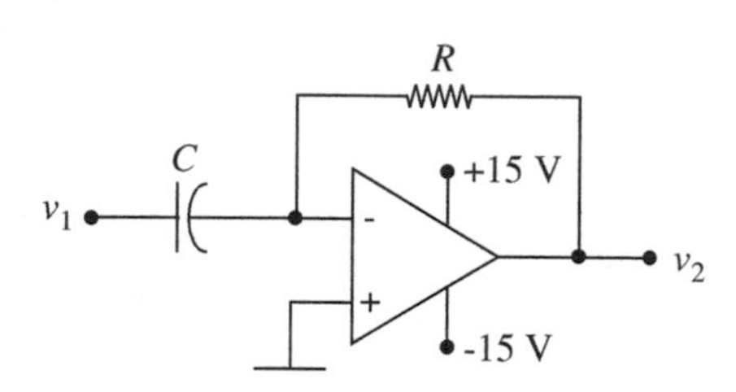


FIGURE 15.86

be placed on the component values R and C to achieve satisfactory operation, assuming that the input is as follows:

i: A sine wave with angular frequency $\omega$ and peak amplitude $A$.

ii: A triangle wave with period $T$ and peak amplitude $A$.

iii: A square wave with period $T$ and peak amplitude $A$.

c) The leakage of an actual capacitor can often be modeled by a large resistor in parallel with an ideal capacitor. What effects on circuit performance would capacitor leakage have?

PROBLEM 15.21

a) Using the "ideal operational amplifier" assumption, that is, infinite gain, infinite input resistance, and zero output resistance, determine the relationship between $v_O(t)$ and $v_I(t)$ in Figure 15.87.

b) If the signal $v_I(t)$ is the rectangular pulse in Figure 15.88, sketch $v_O(t)$ for $t > 0$, assuming that $v_O(0) = 0$.

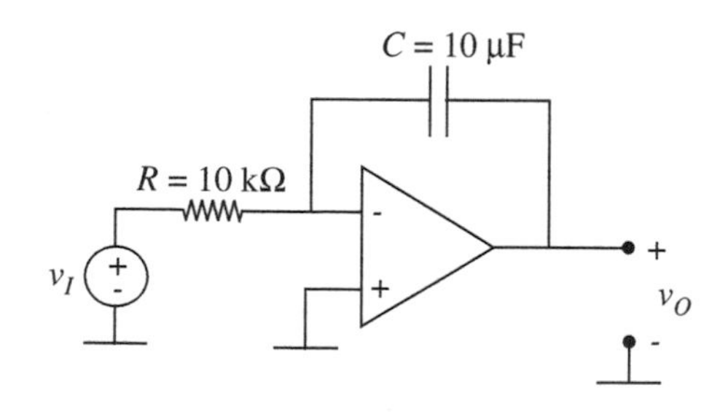


FIGURE 15.87

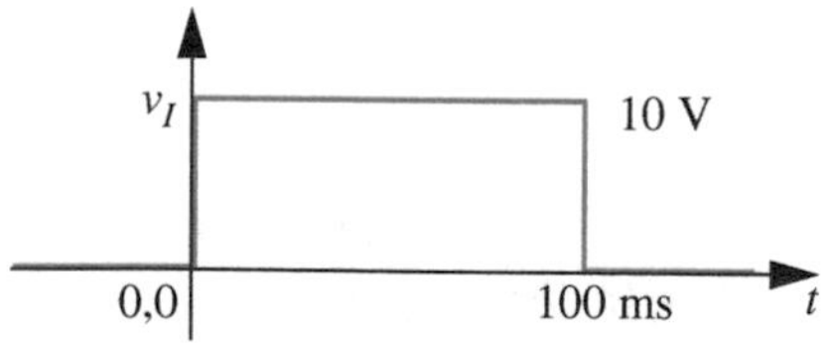

FIGURE 15.88

PROBLEM 15.22 An operational amplifier is connected as shown in Figure 15.89.

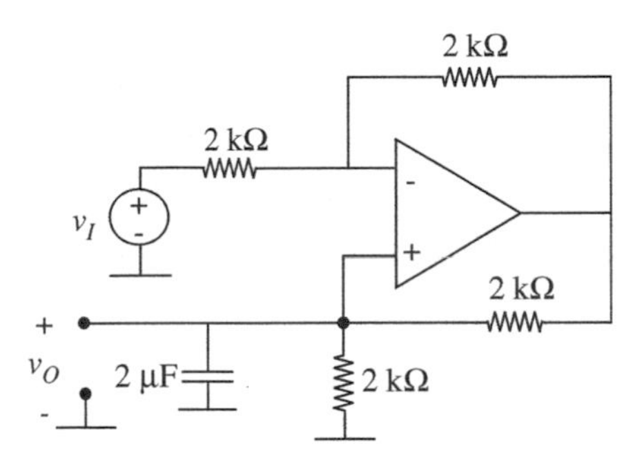


FIGURE 15.89

The voltage $v_I$ is 2 volts for $0 < t < 1$ ms, and 0 otherwise. Assuming that $v_O = 0$ for $t < 0$, sketch $v_O$ for $t > 0$.

PROBLEM 15.23 Consider the two circuits in Figure 15.90.

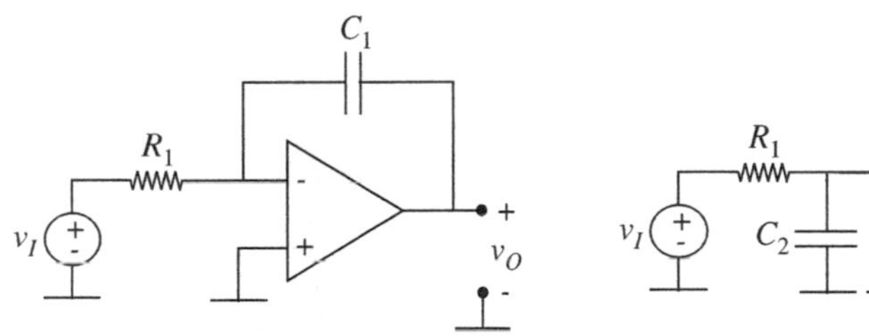


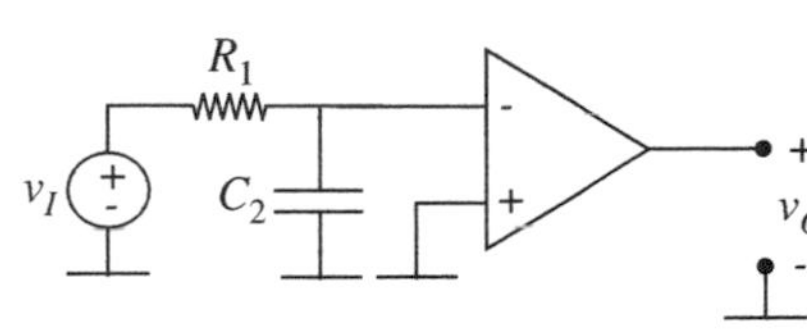


FIGURE 15.90

Use the Op Amp model to find the transfer function $v_O/v_I$ for the two circuits.

Assume only *moderate* gain (say 100) for the Op Amp so you cannot assume $v^+ = v^-$. How large does $C_2$ have to be compared to $C_1$ in order for the two circuits to behave the same? The increase in the effective size of $C_1$ because of the gain of the amplifier is called the Miller Effect, and is used in Op Amp design.

PROBLEM 15.24 Assuming an ideal Op Amp (large gain, $v^+ \simeq v^-$, $r_{in}$ infinite, $r_{out}$ zero, but including amplifier saturation effects).

a) Plot a curve of $i_{IN}$ versus $v_{IN}$ between $-20$ and $+20$ V for the circuit in Figure 15.91, assuming $R_2 = R_3$. Dimension your plot.

b) A capacitor is initially charged to 1 volt (switch in position (1)) in Figure 15.92, then connected to the circuit at $t = 0$ (switch in position (2)). Sketch and dimension the waveform $v_C(t)$ for $t$ greater than zero.

PROBLEM 15.25 An operational amplifier is connected as shown in Figure 15.93.

a) Assuming that the amplifier has infinite gain and infinite input resistance and zero output resistance, determine the relationship between $v_O(t)$ and $v_I(t)$.

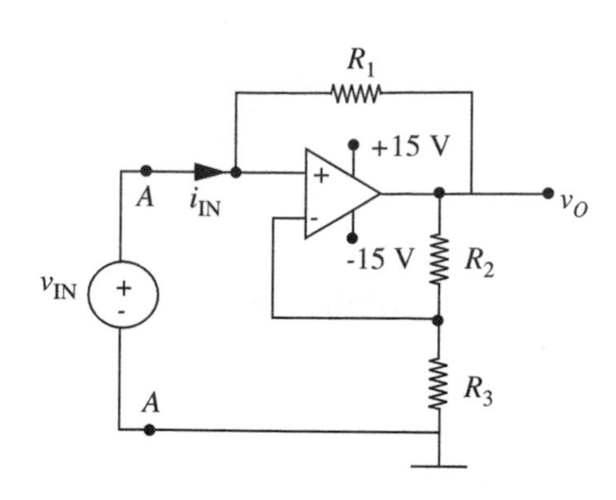


FIGURE 15.91

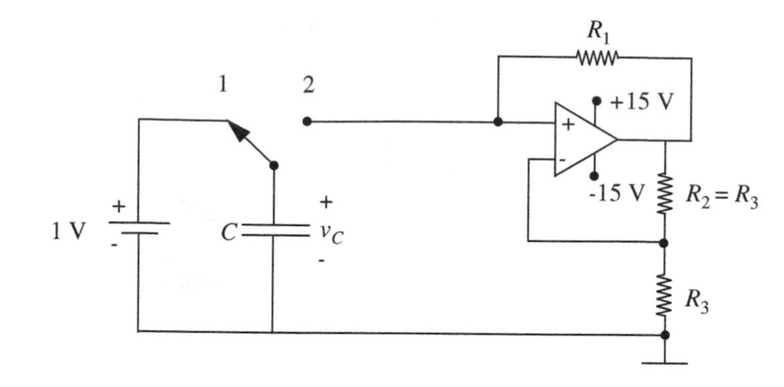


FIGURE 15.92

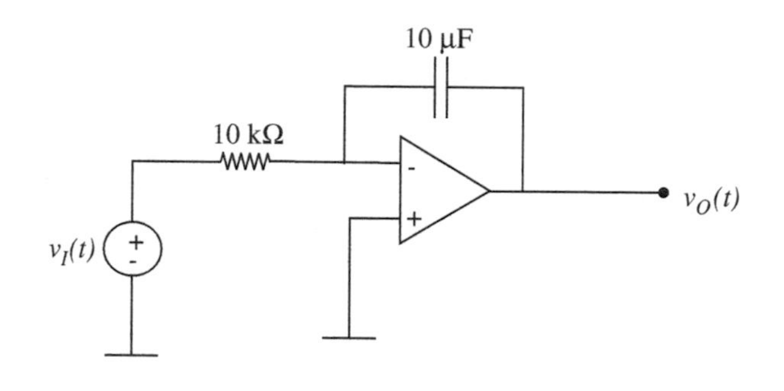


FIGURE 15.93

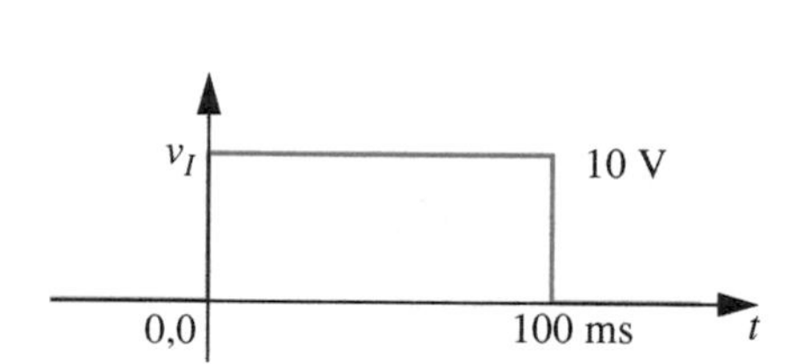


FIGURE 15.94

b) The signal $v_I(t)$ is a rectangular pulse as Figure 15.94. Assuming that $v_O(0) = 0$, draw $v_O(t)$, for $t > 0$.

c) The operational amplifier is now connected as in Figure 15.95. The voltage $v_O(t)$ is held at zero (by some means not shown) for $t < 0$. The switch is initially in the up position, connecting the 10 kΩ resistor to a fixed voltage $V_F$. At time $t = 100$ ms, the switch is thrown to the down position. The observed voltage $v_O(t)$ is shown in Figure 15.96.

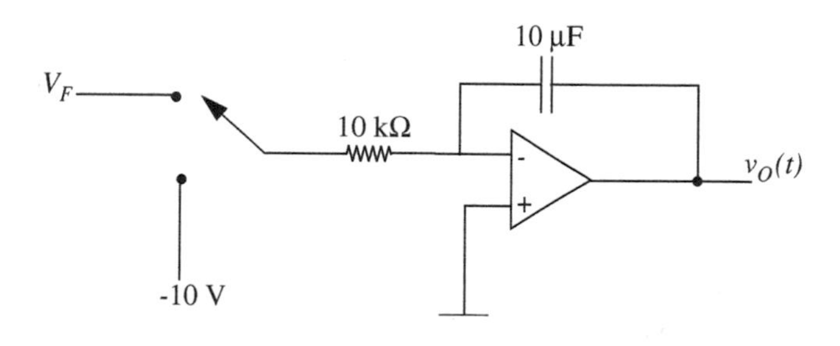


FIGURE 15.95

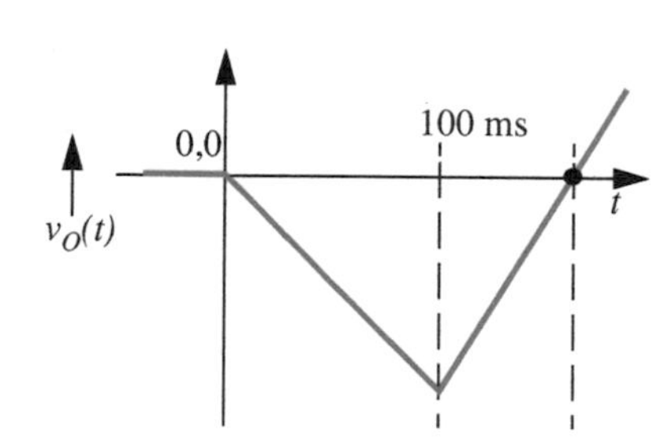


FIGURE 15.96

Determine the relationship between $V_F$ and $\tau$, the time required for $v_O(t)$ to return to zero volts.

PROBLEM 15.26 We wish to show that the circuit shown in Figure 15.97 behaves in a manner very similar to an RLC circuit.

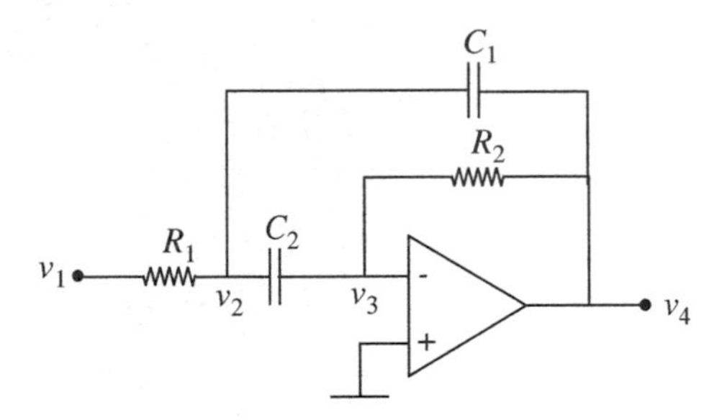


FIGURE 15.97

a) Write the node equations for $v_2$ and $v_3$.

b) Simplify these equations by using the Op Amp assumption, that is, $v^- \simeq v^+$. This allows you to neglect $v_3$ terms compared to $v_4$ terms, and $dv_3/dt$ terms compared to $dv_2/dt$ and $dv_4/dt$ terms, provided $C_1$ and $C_2$ are comparable. (You must later check on this last assumption.)

c) Find the characteristic equation. Compare with the RLC case.

d) For the following numerical values, is the circuit under-, over-, or critically-damped? What is the $Q$ of the circuit, in *literal* form?

$C_1 = C_2 = .01\ \mu\text{F}$

$R_1 = 10\ \Omega$

$R_2 = 1\ \text{k}\Omega$

PROBLEM 15.27 What is the differential equation relating to $v_O$ to $v_I$ in the network in Figure 15.98? Assume that the Op Amps are ideal.

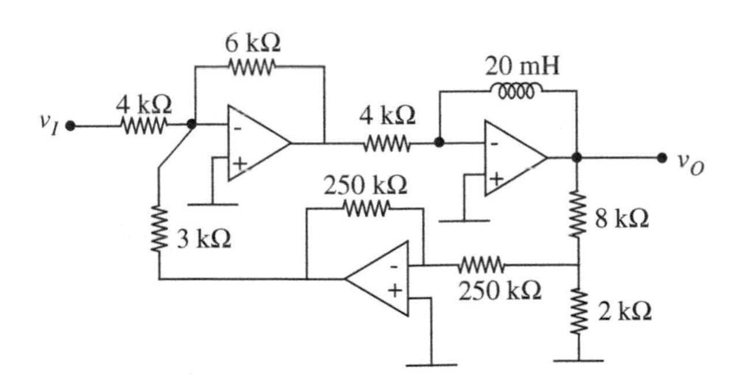


FIGURE 15.98

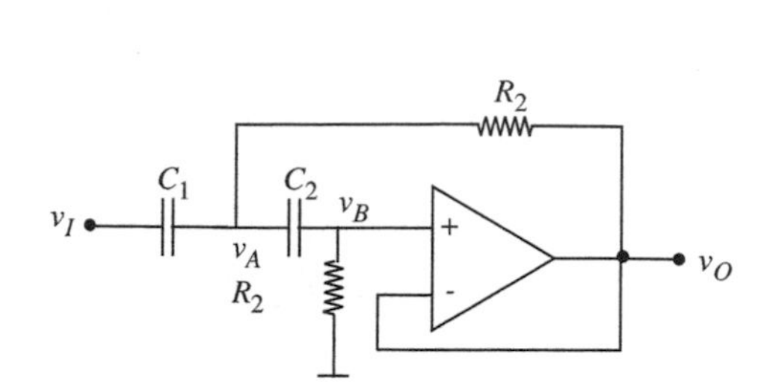


FIGURE 15.99

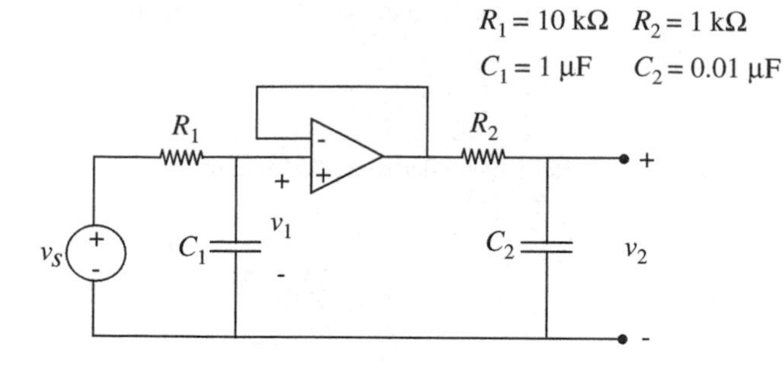


FIGURE 15.100

PROBLEM 15.28 The circuit in Figure 15.99 behaves in a manner very similar to an RLC circuit.

a) Write the node equations.

b) Assume $v_A = V_a e^{st}, v_B = V_b e^{st}$, and find the characteristic equation.

c) Find $\alpha$ and $\omega_o$ in terms of $C_1, C_2, G_1$, and $G_2$.

PROBLEM 15.29

a) Find $H_1(S) = V_1/V_s$ in Figure 15.100. Plot and dimension log $|H_1|$ and $\angle H_1$ vs. log $\omega$.

b) Find $H_2(S) = V_2/V_1$. Plot and dimension log $|H_2|$ and $\angle H_2$ vs. log $\omega$.

c) Find $H_t(S) = V_2/V_s = H_1(S)H_2(S)$. Plot and dimension log $|H_t|$ and $\angle H_t$ vs. log $\omega$. Compare with the plots you obtained in parts (a) and (b).

PROBLEM 15.30

a) Find the transfer function for the network in Figure 15.101.

FIGURE 15.101

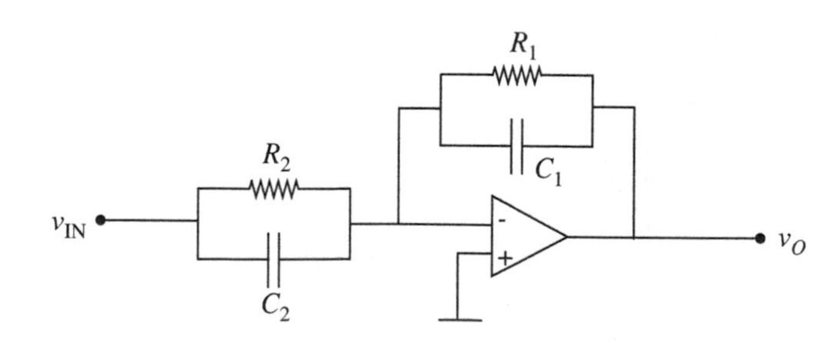

b) Synthesize the function $V_o/V_{in} = -(s+4)/(s+6)$ using the circuit in Figure 15.101. That is, find values of $R_1, R_2, C_1$, and $C_2$ that satisfy $V_o/V_{in}$. You may use capacitors of 1 $\mu F$.

PROBLEM 15.31 The circuit shown in Figure 15.102 is a capacitance multiplier. It may be incorporated into circuits that might otherwise require unrealistically large physical capacitors. You may assume that the operational amplifier has ideal characteristics.

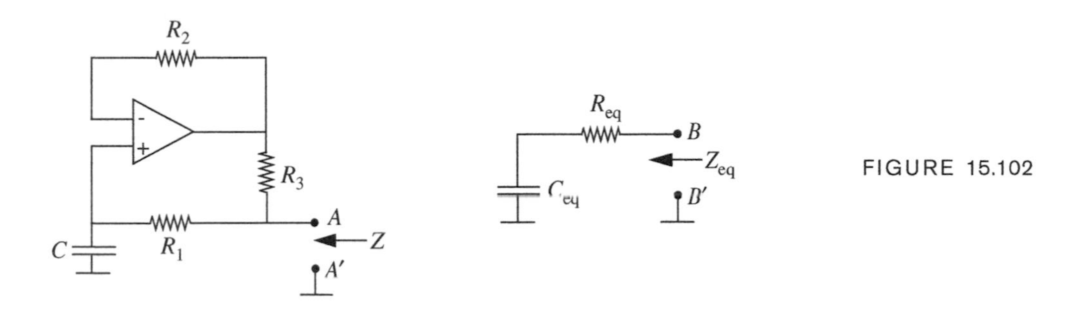


FIGURE 15.102

a) Find the impedance $Z$ looking into terminal A–A′ for the circuit.

b) Show that the model on the right corresponds to an impedance equivalent to the result obtained in part (a).

c) For $R_1 = R_2 = 10$ MΩ, and $R_3 = 1$ kΩ, what is $C_{eq}$ in terms of $C$?

PROBLEM 15.32 Show that the Op Amp circuit in Figure 15.103 has the same form of transfer function as the circuit in Problem 14.1 (shown on the left-hand side of Figure 15.103). Find expressions for the resonant frequency and the $Q$.

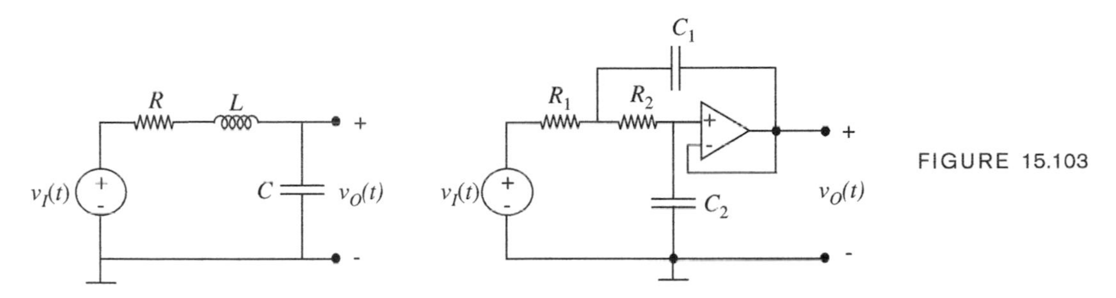


FIGURE 15.103

PROBLEM 15.33 The circuit in Figure 15.104 is a switched capacitor filter. The switches $S_1$ and $S_2$ are driven by nonoverlapping clocks as in Problem 15.15. Both $S_1$ switches are closed for time $1/2f_c$ with $S_2$ open, and $S_2$ closed for $1/2f_c$ with $S_1$ open. $V_{in} = A\cos(\omega t)$, $\omega \ll 2\pi f_0$.

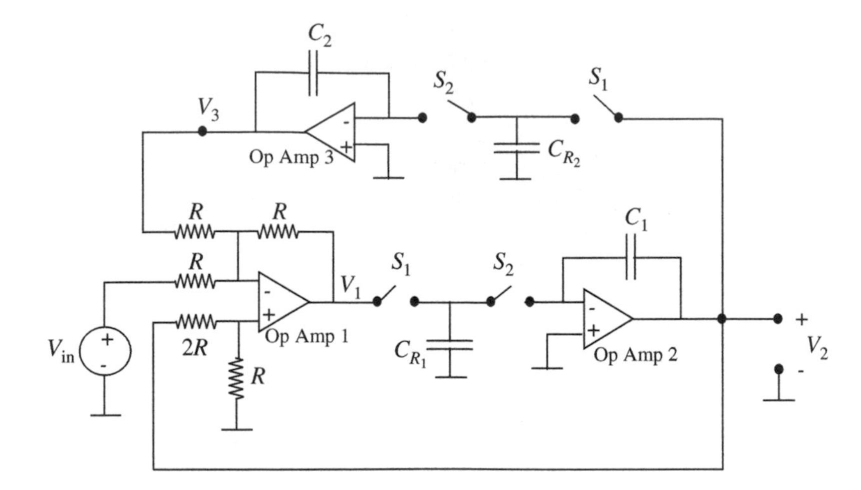


FIGURE 15.104

a) Find (in the sinusoidal steady state) the transfer functions $V_3/V_2$ and $V_2/V_1$. Refer to Problem 15.15 to see how to handle the switches. Note that there are no switches across $C_1$ and $C_2$.

b) Now find a simple equation to describe the operation of Op Amp 1, that is, find an expression for $V_1$ in terms of $V_2$, $V_{in}$, and $V_3$. (Note that in all of our impedance calculations, we have been implicitly assuming that the relation among $V$'s for such a circuit is the same as the relation among the time variables $v(t)$.)

c) Now substitute from (a) into (b) to find the overall transfer function $V_2/V_{in}$. Find expressions for the resonant frequency $\omega_0$ and the bandwidth $\Delta\omega$ in terms of the circuit constants. The easiest way to do this is to get the transfer function into the form:

$$V_o = \frac{K_s V_{in}}{s^2 + 2\alpha s + \omega_0^2} \tag{15.122}$$

and work by analogy to the parallel RLC case. How does the resonant frequency $\omega_0$ depend on the clock frequency $f_c$?

PROBLEM 15.34 The circuit shown in Figure 15.105 behaves like an RLC circuit.

a) Find the transfer function $V_4/V_1$. (You may assume that the Op Amp is ideal, that is $V^+ = V^-$ to simplify your calculations.)

b) Sketch the magnitude of the transfer function $|V_4/V_1|$ versus frequency. Indicate the frequency at which the peak occurs, the magnitude of the transfer function at the peak, and the $Q$ of the resonance. Use the following numerical values:

$C_1 = C_2 = 0.01\ \mu\text{F}\quad R_1 = 10\ \Omega\quad R_2 = 1\ \text{k}\Omega$

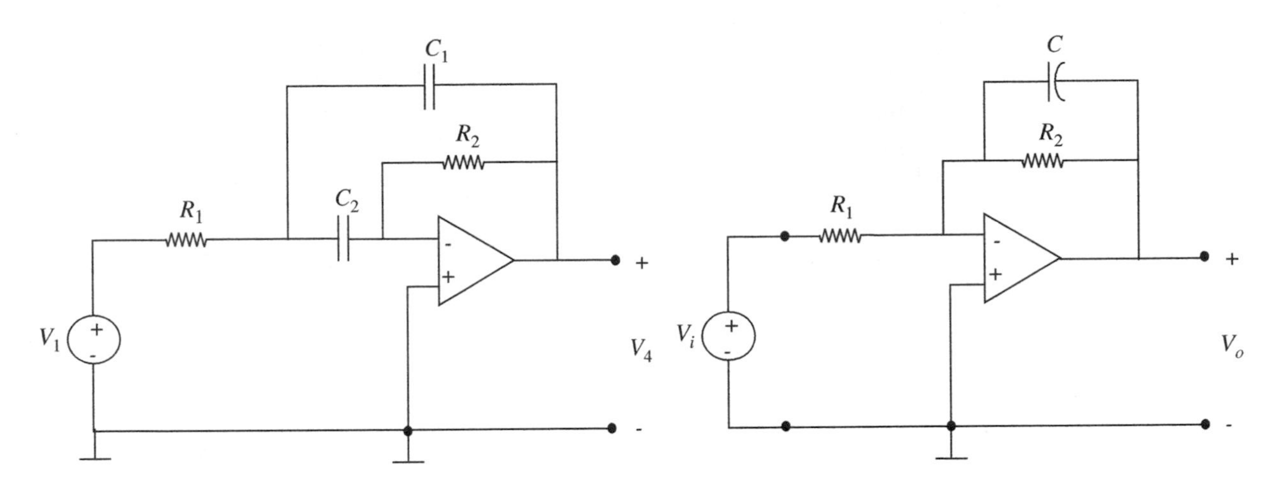


FIGURE 15.105

FIGURE 15.106

c) This circuit is known as an RC active filter. Is it a low-pass, high-pass, or band-pass filter? What is the expression for bandwidth in terms at $R_1, C_1$, etc.? That is, $B = \omega_2 - \omega_1$ where $\omega_1$ and $\omega_2$ are the half power frequencies.

PROBLEM 15.35

a) Find an expression for the complex amplitude ratio $V_o/V_i$ for the active filter circuit in Figure 15.106, given that $R_2 = 10R_1$. Sketch the Bode plot, $|V_o/V_i|$ versus $\omega$ and $V_o/V_i$ versus $\omega$.

b) An equivalent filter can be made with the circuit shown in Figure 15.107. Find the value of $C_2$ needed to make a filter equivalent to that in part (a), assuming that $R_1$ and $R_2$ are the same here as for part (a). How does the value of $C_x$ here compare to that of $C$ in the filter of part (a)?

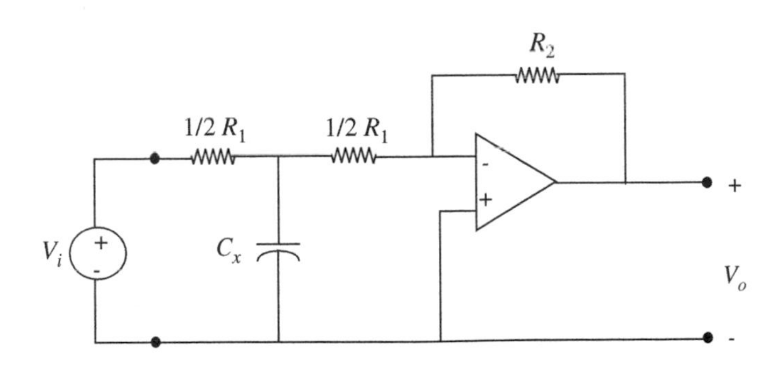


FIGURE 15.107

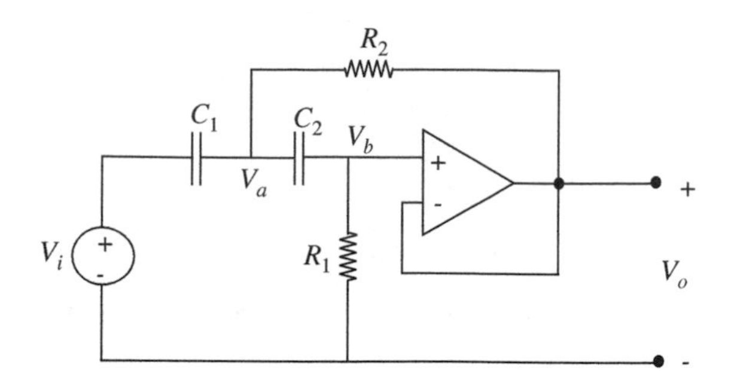


FIGURE 15.108

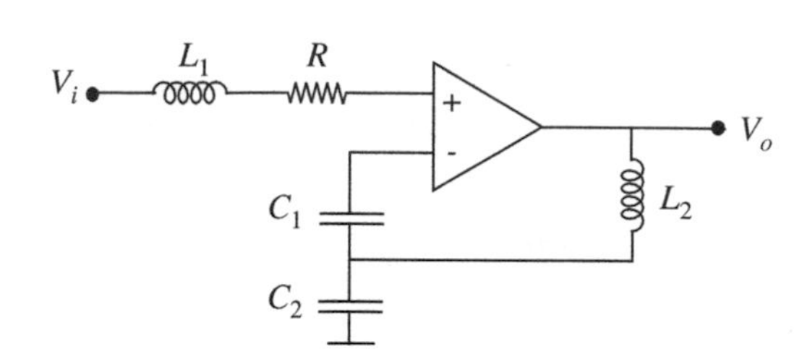


FIGURE 15.109

PROBLEM 15.36 The circuit shown in Figure 15.108 behaves in a way very similar to an RLC circuit.

a) Write the sinusoidal steady state node equations for the complex amplitudes $V_a$ and $V_b$.

b) Solve for $V_o/V_i$ using the results in (a), and noting that $V_o = V_b$.

c) Assuming the circuit is under-damped, sketch the magnitude of the transfer function $|V_o/V_i|$ versus frequency. Indicate the frequency at which the peak occurs, the magnitude of the transfer function at the peak, and the $Q$ of the resonance.

PROBLEM 15.37 Plot the frequency response (magnitude and phase) of the active filter shown in Figure 15.109. Assume the Op Amp is ideal.

PROBLEM 15.38 The circuit shown in Figure 15.110 has a resonance very similar to an RLC circuit.

a) Write the sinusoidal steady-state equations for $V_2$ and $V_3$.

b) Solve for $V_4/V_1$ using the results in (a), and noting that $V_3 = -V_4/A$, where the Op Amp gain $A$ can be assumed to be very large.

FIGURE 15.110

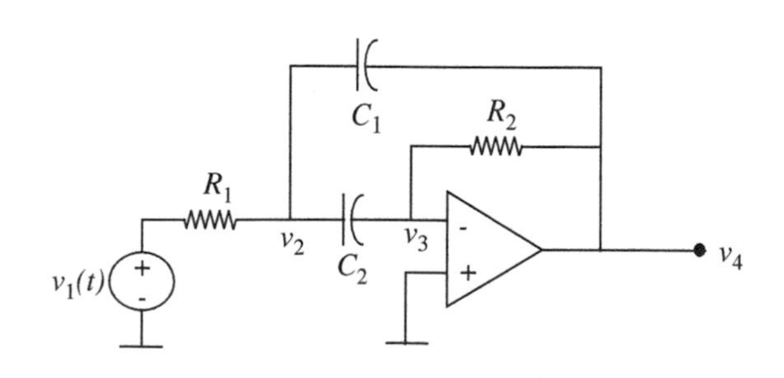

c) Assuming now that $C_1 = C_2 = 0.1\ \mu F$, $R_1 = 10\ \Omega$, $R_2 = 1\ k\Omega$, sketch the magnitude of the transfer function $|V_4/V_1|$ versus frequency. Indicate the frequency at which the peak occurs, the magnitude of the transfer function at the peak, and the $Q$ of the resonance.

PROBLEM 15.39 For the circuit in the figure in Figure 15.111

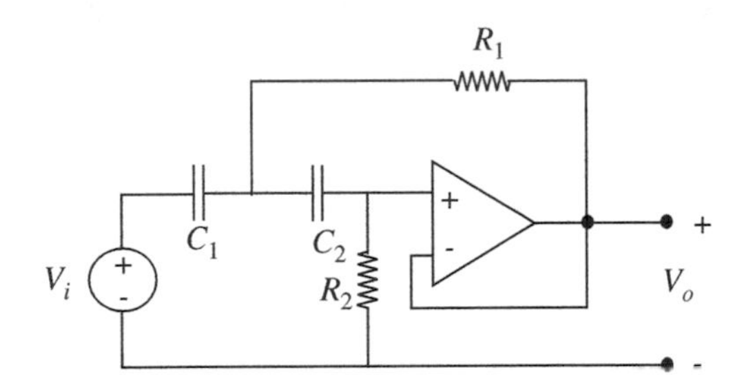


FIGURE 15.111

a) Find a set of equations which, if solved, would give $V_o/V_i$.

b) Assuming that these equations, when solved, yield

$$V_o/V_i = \frac{(j\omega C_1)(j\omega C_2)}{G_1G_2 + j\omega(C_1 + C_2)G_2 + (j\omega)^2 C_1C_2}. \tag{15.123}$$

Find the expression for the undamped resonant frequency ($\omega_0$) of the circuit.

c) Find an expression for the low-frequency asymptote of $V_o/V_i$. (Zero is not an acceptable answer.)

d) Find an expression for the high-frequency asymptote of $V_o/V_i$. (Zero is not an acceptable answer.)

e) Assuming $Q = 1/2$, sketch the magnitude and phase of $V_o/V_i$ versus $\omega$. Specify coordinates and dimension key features.

PROBLEM 15.40 Tech Hi-Fi advertises a car stereo system that can deliver 10 watts average power into a 4-$\Omega$ speaker. Given your demonstrated proficiency in electronics, you decide to build one using an (hefty) Op Amp. To save yourself the problems associated with designing the receiver, you plan to use a small transistor AM-FM radio as the signal source.

You try the circuit shown in Figure 15.112.

In the following parts, you may assume that the hefty Op Amp has very high open-loop gain, zero output resistance, infinite input resistance, and other good features.

a) What is the operating-point value of the voltage at the output of the operational amplifier?

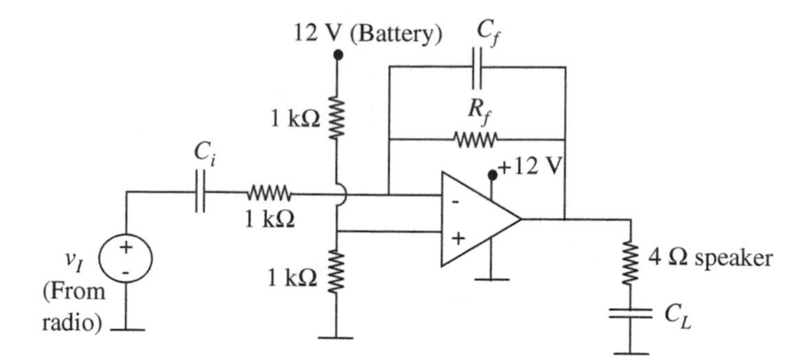


FIGURE 15.112

b) Why is capacitor $C_L$ included?

c) Assume that the maximum signal from your radio is 1 volt peak to peak. What is the maximum value of $R_f$ that ensures the operational amplifier will remain in the linear region?

d) What is the maximum average power that can be delivered to the 4-Ω speaker with $v_I$ as a constant amplitude sinusoid?

e) In spite of your answers to parts (b) and (c), assume that you choose $R_f = 10$ kΩ and that capacitor $C_L$ is very large. In order to reduce low frequency noise, you decide that you should make the lower half-power frequency 100 radians per second. What value of $C_i$ should be selected? You also want to filter high-frequency noise by making the upper half-power frequency $10^5$ radians per second. What value of $C_f$ should be selected?

PROBLEM 15.41

a) Using the ideal Op Amp assumptions, write the node equations for the complex voltage for the circuit in Figure 15.113. Solve for $V_o$.

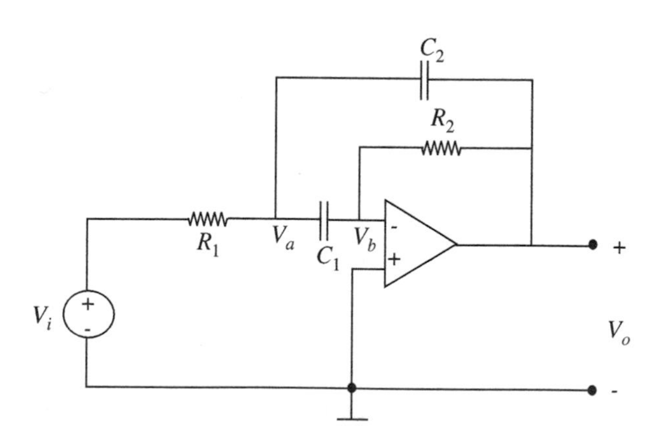


FIGURE 15.113

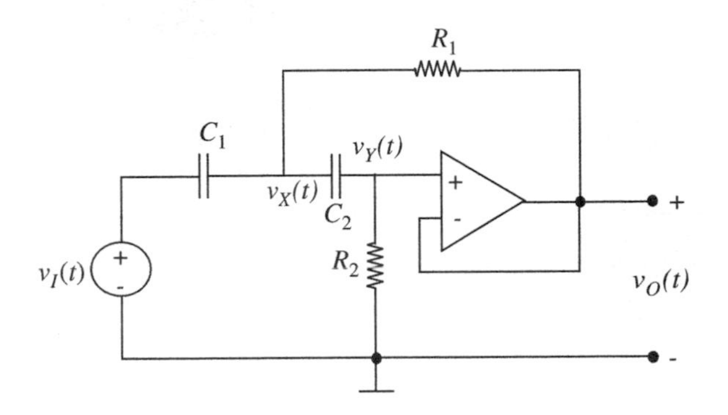


FIGURE 15.114

b) Assume $V_o$ is of the form:

$$V_o = \frac{sKV_i}{s^2 + 2\alpha s + \omega_0^2}. \tag{15.124}$$

If a short pulse is now applied to this circuit, the output voltage after the pulse is

$$v_O(t) = 3e^{-100t}\sin(1000t + 20^o). \tag{15.125}$$

For $K = 400(sec^{-1})$ find the response $v_O(t)$ in the steady state to a one-volt cosine wave at the resonant frequency:

$$v_I(t) = 1\text{ V}\cos\left(\omega_0 t\right). \tag{15.126}$$

(Provide *numbers* for $\omega_0$, etc.)

c) Repeat (b), for a one-volt cosine wave at the lower 0.707 frequency $\omega_1$.

PROBLEM 15.42

a) For the circuit in Figure 15.114 write the node equations needed to find $V_o(s)$ in terms of $V_i(s)$. Your answer *must* be arranged with the source terms on the left, the unknown variables on the right, and must use *conductances* $g(= 1/R)$.

b) Solving these equations, you should obtain for $C_1 = C_2$,

$$V_o(s) = \frac{s^2 V_i}{s^2 + s\frac{2}{R_2 C} + \frac{1}{R_1 R_2 C^2}}. \tag{15.127}$$

For $R_1 = 1\text{ k}\Omega$, find the values of $R_2$ and $C$ that give a $Q$ of 10 and a resonant frequency, defined as the frequency where the $s^2$ term and the $s^0$ term cancel in the denominator of Equation 15.127, of $\omega_o = 1000$ rad/s.

PROBLEM 15.43 For the network shown in Figure 15.115:

FIGURE 15.115

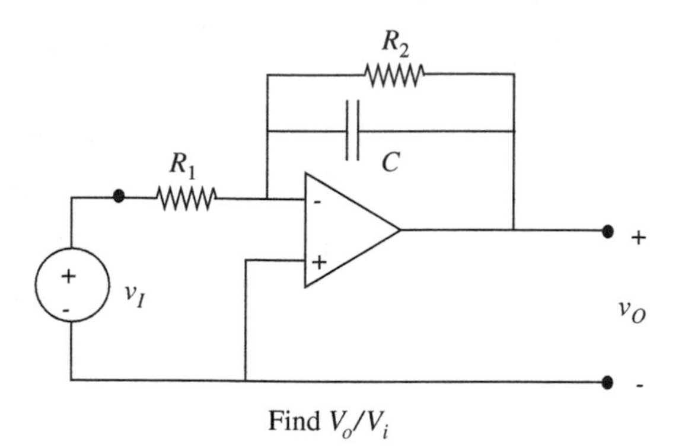


Find $V_o/V_i$

a) Determine an expression for the indicated transfer function.

b) Sketch the magnitude and angle of the indicated quantity as a function of frequency. You may use either linear or log-log coordinates, but it is recommended that you learn to use both kinds of axes.

# CHAPTER 16

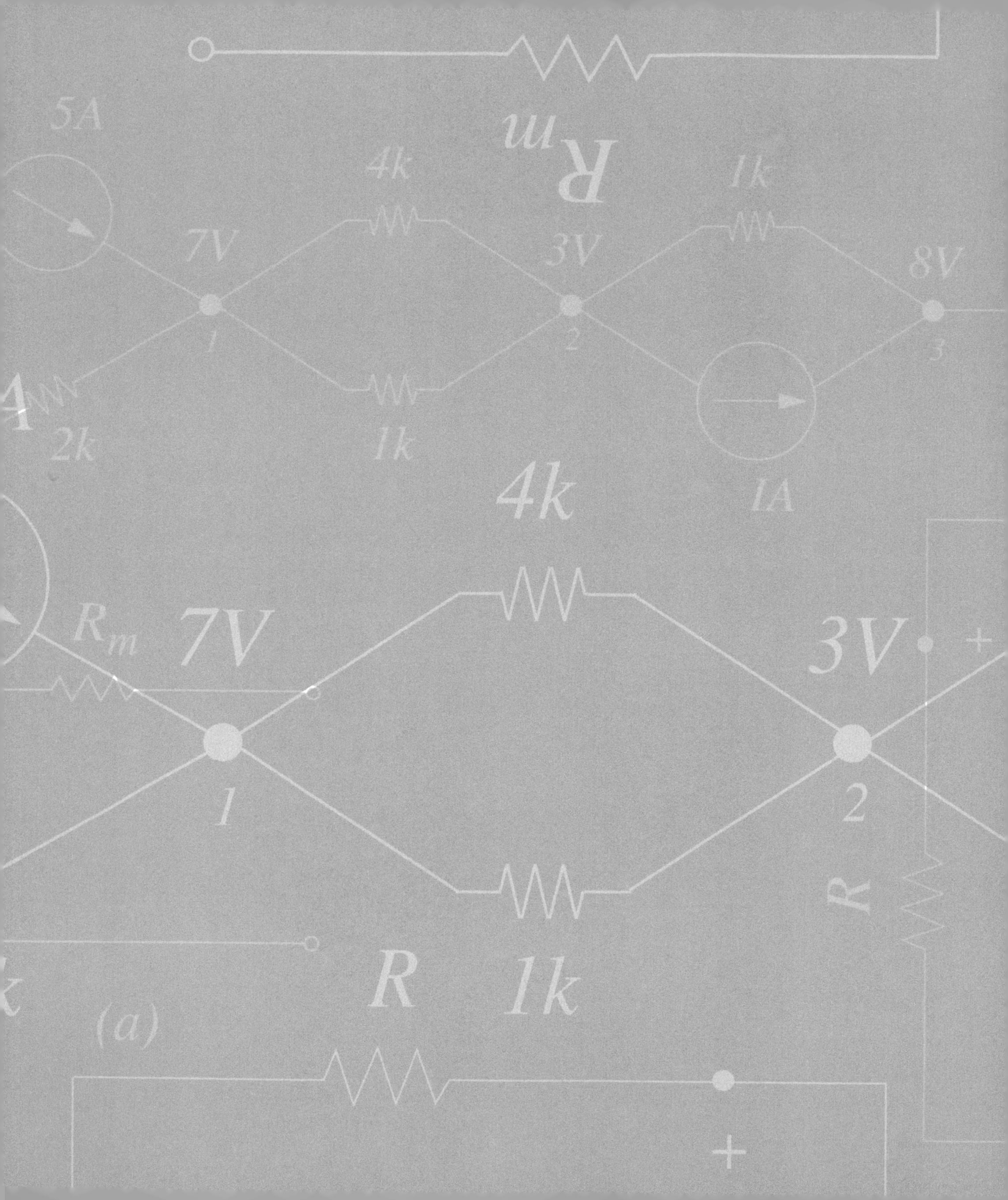

5A
4k
1k
7V
3V
8V
1
2
3
2k
1k
1A
4k
$R_m$
7V
3V
+
1
2
R
R
1k
(a)
+

# DIODES

# 16

## 16.1 INTRODUCTION

The diode was introduced in Chapter 4 as an example of a nonlinear device. We used its nonlinear $v$–$i$ characteristic to develop several methods of analyzing nonlinear circuits. It turns out that the diode is a particularly useful nonlinear device, and merits closer examination. Figure 16.1 shows a few discrete diodes. This chapter explores several useful diode circuits, and develops additional analysis methods that are specific to diodes.

## 16.2 SEMICONDUCTOR DIODE CHARACTERISTICS

We will consider semiconductor diodes made out of silicon. Recall, we have previously seen an example of a silicon-based device, namely the MOSFET. Let us first review briefly the properties of silicon and semiconductors. Silicon is an element in the cubic crystal class of Group IV in the periodic table (along with germanium). In pure crystalline silicon, each atom forms covalent bonds with its nearest neighbors, so that at room temperature almost all valence electrons are involved in the structural bonding, and very few are free to move about the crystal. Hence pure silicon at room temperature is a very poor conductor of electricity.

However, if minute amounts of impurities are added to the silicon by high temperature diffusion, or by ion implantation, for example, then the electrical

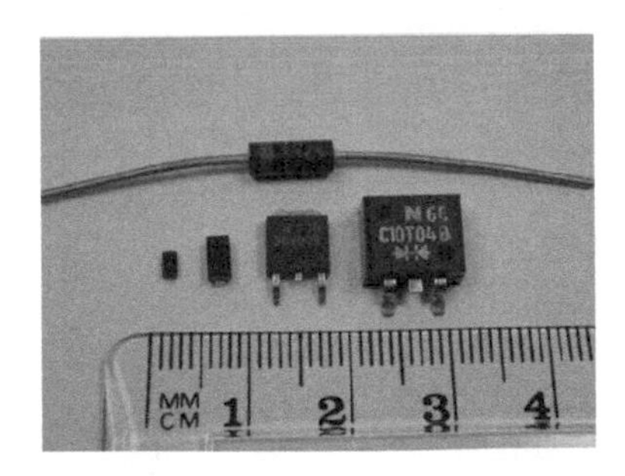


FIGURE 16.1 Discrete diodes. ( Photograph Courtesy of Maxim Integrated Products.)

properties change dramatically. Add a part per million of a Group V element, such as phosphorus, to the Group IV element silicon, and we obtain a crystal that has many "mobile" electrons not involved in covalent bonds. Hence the material is now a good conductor. We call this material n-type silicon (*n* standing for negative, indicating mobile negative charge carriers, that is, electrons). Similarly, if we add small amounts of a Group III element such as aluminum to the pure silicon, the resulting crystal will have a large deficit of electrons in the bonding structure. One useful way of visualizing the effect of this deficit is to imagine that we have created not a deficit of negative charges, but a *surplus of positive charges*, which we call *holes.* (A hole is thus a convenient way of representing the absence of an electron.) This is called p-type silicon, signifying mobile positive charge carriers. Both n-type and p-type silicon are electrically neutral, because the constituents were electrically neutral. But unlike pure silicon, both are relatively good conductors of electricity. You may recall the use of both p-type and n-type silicon in the fabrication of MOSFETs.

One way to make a semiconductor diode is to metallurgically create a wafer of silicon containing n-type material adjoining p-type material. In an n-channel MOSFET, for example, the n-type drain juxtaposed to the p-type channel region forms a diode. The circuit symbol for the diode, shown in Figure 16.2, emphasizes this asymmetric structure by denoting the p region with an arrow, and the n region as a line. If a battery and a resistor are connected to the diode to make the p region positive with respect to the n region, as in Figure 16.3a, large currents will flow. This is called *forward bias.* But if the battery is connected in the opposite way (Figure 16.3b), to make the n region more positive than the p region to *reverse bias* the diode, almost no current will flow. This gross asymmetry in electrical behavior is the essence of the semiconductor diode.

An analytical expression for the relation between the voltage $v_D$ and the current $i_D$ for the diode can be derived from semiconductor physics:

$$i_D = I_s(e^{v_D/V_{TH}} - 1). \tag{16.1}$$

The parameter, $V_{TH} = kT/q$, is called the thermal voltage, and the constant $I_s$ is the saturation current. For silicon $I_s$ is typically $10^{-12}$ amps. $q$ is the charge

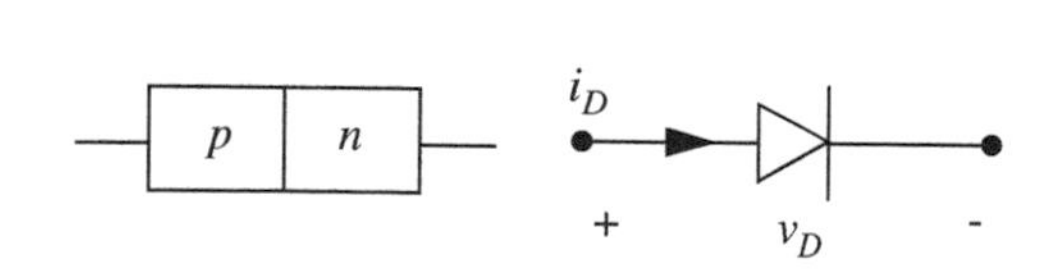


FIGURE 16.2 A semiconductor diode.

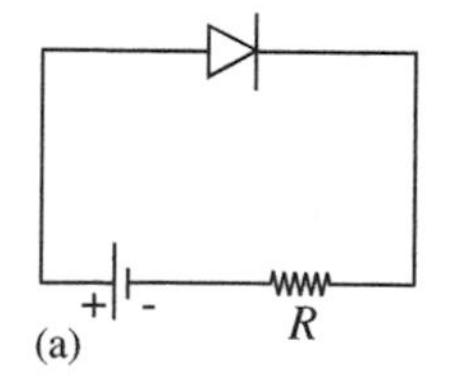


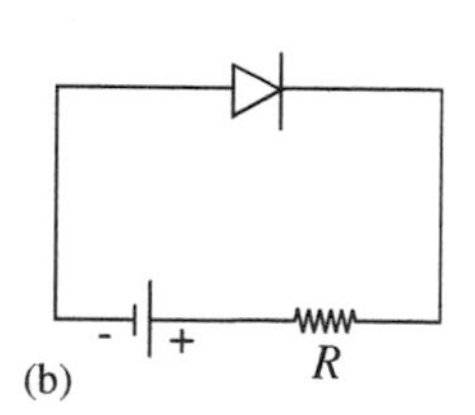


FIGURE 16.3 (a) Forward bias; (b) reverse bias.

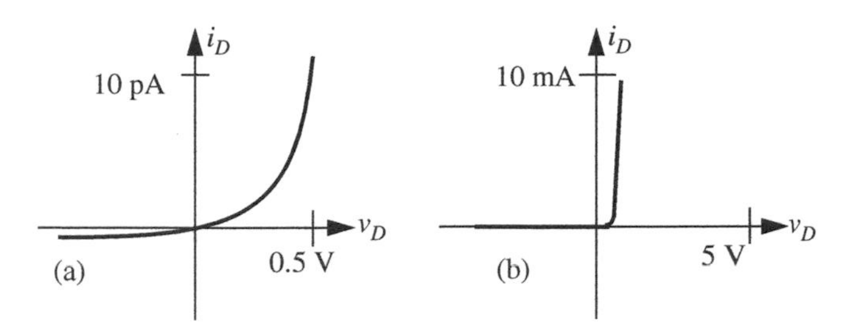


FIGURE 16.4 $v$–$i$ characteristics of a silicon diode.

of an electron,[1] $k$ is the Boltzmann's constant,[2] and $T$ is the temperature in kelvins.[3] At room temperature, $kT/q$ is approximately 0.025 volts.

Typical measured $v$–$i$ characteristics for a silicon diode are shown in Figure 16.4. If we plot on a current scale of pico-amps ($10^{-12}$ amps), as in Figure 16.4a, then the expected exponential shape appears. But if we plot on a more typical scale of milliamps (Figure 16.4b), then the curve looks quite different. The current appears to be zero until the diode voltage is almost 0.6 volts, at which point the characteristic rises very sharply. This apparent knee is entirely due to the mathematical behavior of exponentials, rather than some physically-related threshold in the device. Nonetheless, on the scale of milliamperes, silicon diodes appear to have a voltage threshold of 0.6 to 0.7 V, (0.2 V for germanium diodes). This threshold has important consequences for semiconductor circuit design, some detrimental but others of great value. For example, recall that digital logic depends critically on the presence of such a threshold, as we saw in Chapter 6.

---

EXAMPLE 16.1 ANALYSIS OF A DIODE-BASED TEMPERATURE MEASUREMENT CIRCUIT To achieve the greatest possible computational performance, microprocessors in notebook computers and servers operate with a variable frequency clock. The faster the clock, the more operations per second a microprocessor can perform. However, as its clock frequency increases, a microprocessor becomes hotter for reasons that are discussed in Chapter 11. Generally, the temperature of a microprocessor should be limited to about 110 °C. To increase its performance, the clock frequency of a microprocessor is increased until the thermal limit is reached. How does a microprocessor determine its temperature?

Diodes in a microprocessor can be used to sense temperature. For example, the MAXIM MAX1617 device measures temperature by forcing two different currents through a

---

1. The electron charge $q = 1.602 \times 10^{-19}$ C.

2. The Boltzmann's constant $k = 1.380 \times 10^{-23}$ J/°K.

3. The temperature in kelvins can be obtained from the temperature in degrees Celsius as follows:

$$T\ [°K] = T\ [°C] + 273.15\ [°C].$$

diode and comparing the resulting voltages. For sufficiently large voltages, the diode equation can be approximated by:

$$i_D \approx I_s e^{qv_D/kT}.$$

Therefore, the voltage $v_D$ across the diode is given approximately by:

$$v_D = \frac{kT}{q} \ln\left(\frac{i_D}{I_s}\right).$$

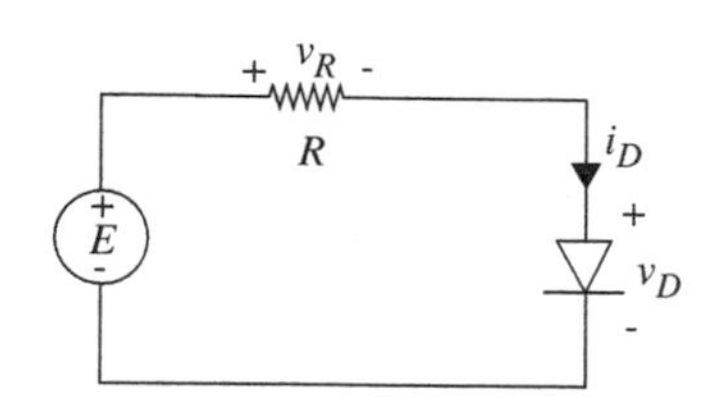


FIGURE 16.5 Circuit with diode.

To measure temperature, the MAX1617 first forces the current $i_{D1}$ through the diode, and next forces the current $i_{D2}$ through the diode. The resulting voltages, $v_{D1}$ and $v_{D2}$, are then differenced to obtain

$$v_{D1} - v_{D2} = \frac{kT}{q} \ln\left(\frac{i_{D1}}{i_{D2}}\right).$$

This voltage difference is proportional only to absolute temperature in kelvins if the ratio between $i_{D1}$ and $i_{D2}$ is fixed.

Suppose $i_{D1} = 100\ \mu$A and $i_{D2} = 10\ \mu$A. Then, for $T = 300$ °K, or 27 °C, the voltage difference is 59.5 mV. If the temperature rises to $T = 383$ °K, or 110 °C, the voltage difference rises to 76.0 mV.

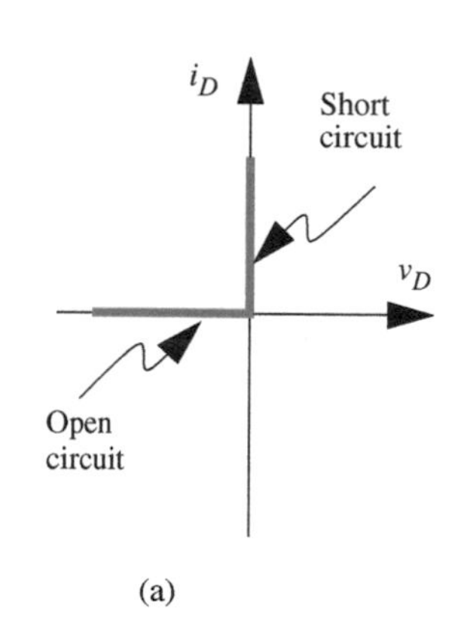


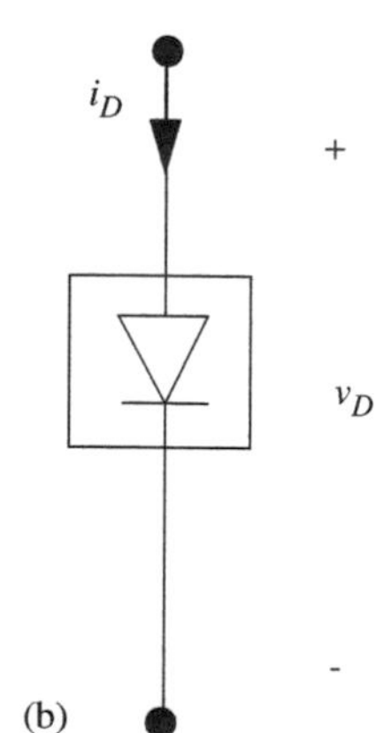


FIGURE 16.6 Ideal diode model.

## 16.3 ANALYSIS OF DIODE CIRCUITS

Given the analytical expression for the diode characteristic, Equation 16.1, how can we calculate the voltages and currents in a simple circuit such as Figure 16.5? Depending on our requirements, as discussed in Chapter 4, we can use one of the four methods of analyzing nonlinear circuits developed previously: (1) analytical solutions, (2) graphical analysis, (3) piecewise-linear analysis, and (4) incremental analysis. However, circuits with multiple diodes and other elements get algebraically complex quickly and become virtually impossible to analyze directly. Fortunately, it turns out that the dichotomous behavior of the diode under forward bias and reverse bias allows us to decompose more complex diode circuits into simple subcircuits, each of which can be independently analyzed using one of the four methods. This decomposition method is called *the method of assumed diode states.*

### 16.3.1 METHOD OF ASSUMED STATES

Recall from the graphical construction illustrated in Figure 16.6 (as well as from the original definition of the ideal diode outlined in Equations 16.2 and 16.3), that the ideal diode has two mutually exclusive states: *the ON state, for which the diode voltage $v_D$ is zero (the diode is a short circuit)*, and *the OFF state,* *where the diode current $i_D$ is zero (the diode is an open circuit)*. This suggests a very simple analysis technique: Draw two subcircuits corresponding to the two diode states, and analyze each subcircuit. Because in each subcircuit the diode is either a short or an open circuit, the subcircuits are linear. Hence linear analysis methods can again be used to find the output voltage. Some insight is then required to piece together a complete solution from these two parts.

$$\text{Diode ON:} \quad v = 0 \quad \text{for all positive } i. \tag{16.2}$$

$$\text{Diode OFF:} \quad i = 0 \quad \text{for all negative } v. \tag{16.3}$$

To illustrate, consider the half-wave rectifier circuit discussed in Chapter 4 (redrawn in Figure 16.7). Let us analyze this circuit using the method of assumed states and the ideal diode piecewise linear model depicted in Figure 16.6.

The two subcircuits for the half-wave rectifier corresponding to the two diode states are shown in Figure 16.8a and 16.8b. One circuit applies whenever

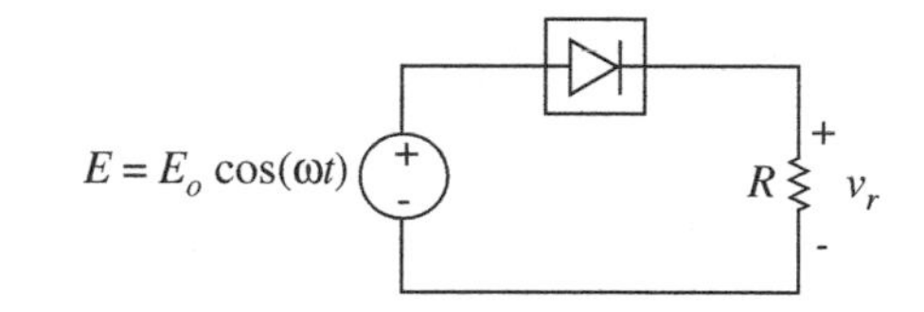


FIGURE 16.7 Half-wave rectifier circuit using the ideal diode model.

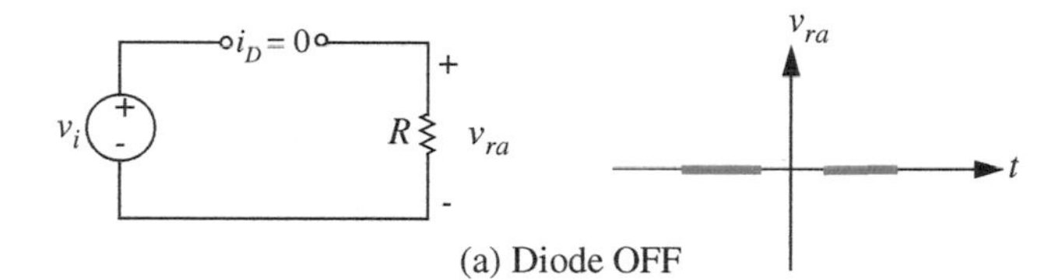


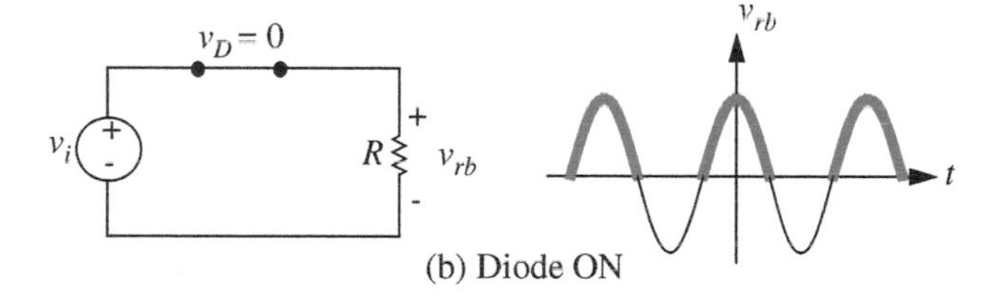


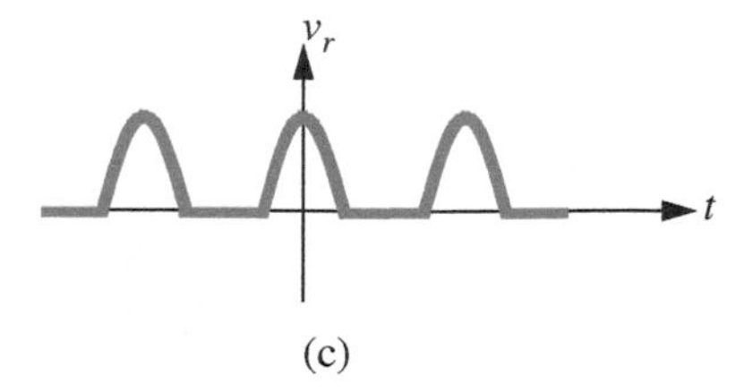


FIGURE 16.8 Analysis by assumed diode states.

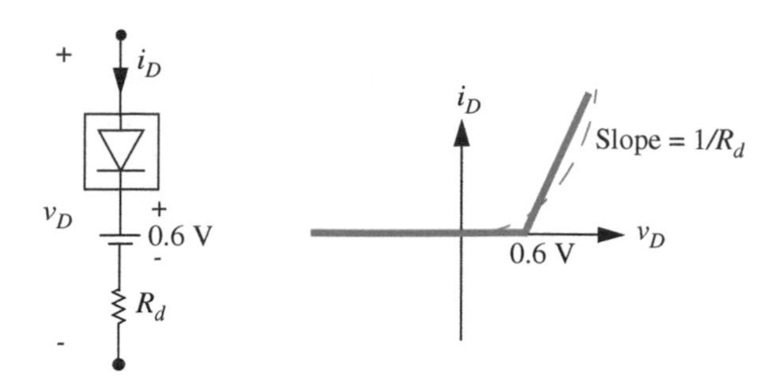


FIGURE 16.9 Diode model comprising an ideal diode, the 0.6-volt source, and a resistor.

the diode is OFF, the other when the diode is ON. Analysis of these two circuits is trivial in this case: For the diode OFF, Figure 16.8a, $v_{ra} = 0$, and for the diode ON, Figure 16.8b, $v_{rb} = v_i$. Now we must apply the diode constraints, Equations 4.34 and 4.35, to find which portions of these waveforms are valid. The ON circuit applies only for $i_D$ positive, hence only for $v_i$ positive, as indicated by the darker line segments. The OFF circuit applies only for $v_D$ negative, hence $v_i$ negative, so the valid parts of the $v_r$ waveform here are again darkened. The two parts of the solution can now be combined to yield the complete solution for the output wave, Figure 16.8c.

Generalizing from this simple example, to analyze a diode circuit by the method of assumed states:

1. Draw a subcircuit for each possible state (ON or OFF) of the diodes. For one diode there are two subcircuits. For $n$ diodes there are $2^n$ such states, and hence $2^n$ subcircuits.
2. Analyze each resulting linear circuit to find an expression for the desired output variable. Because in each subcircuit the diode is either a short or an open circuit, the subcircuits are linear. Hence linear analysis methods can be used.
3. Establish the range of validity of each of the expressions in (2); then assemble the appropriate segments to form the complete output waveform.

---

EXAMPLE 16.2 METHOD OF ASSUMED STATES WITH IMPROVED PIECEWISE LINEAR DIODE MODEL The method of assumed states can also be applied with other diode models and analysis methods. To illustrate, let us use the diode model comprising the ideal diode and both the 0.6-volt source and the resistor $R_d$ shown in Figure 16.9 (see Chapter 4.4.1 for details) to re-analyze the half-wave rectifier circuit of Figure 16.7.

An appropriate circuit model is shown in Figure 16.10a. The subcircuits for the ON and the OFF states of the diode are shown in Figures 16.10b and 16.10c. As both subcircuits are linear (by definition), analysis is easy.

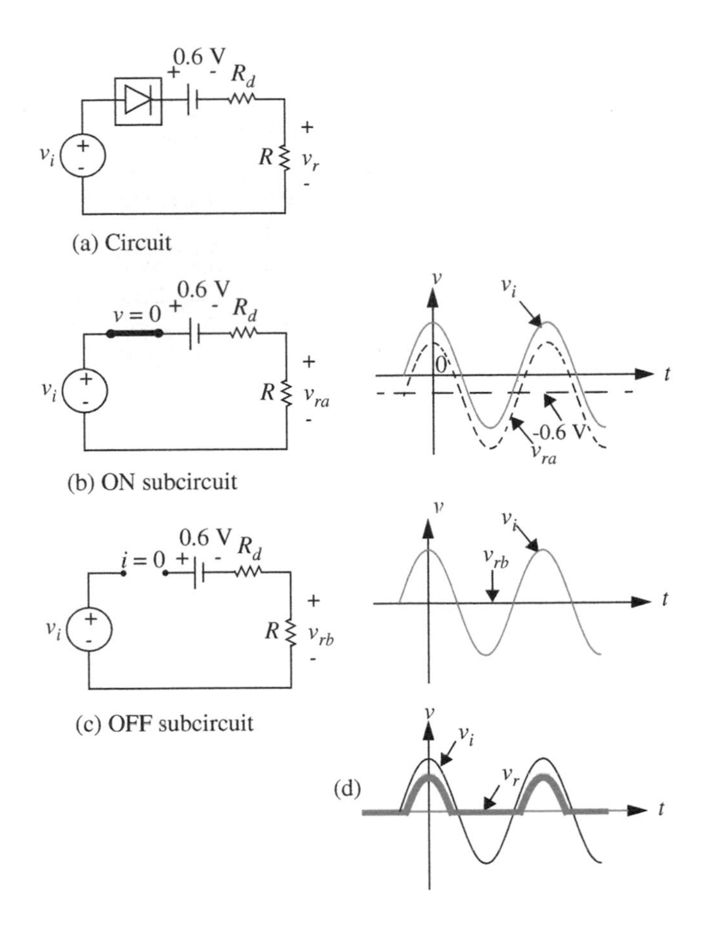


FIGURE 16.10 Analysis with more accurate model.

For the ON state,

$$v_{ra} = (v_i - 0.6)\frac{R}{R + R_d} \tag{16.4}$$

and for the OFF state, $v_r = 0$.

The regions of validity for these waveforms can be found from the constraint equations for the ideal diode, Equations 16.2 and 16.3. The ON subcircuit only applies for $i_D$ positive, hence the valid regions of $v_{ra}$ are the positive segments. The OFF subcircuit must therefore fill in the gaps. More formally, the OFF circuit applies *when the voltage across the ideal diode is negative*, hence for $v_i$ less than 0.6 V. The composite solution is shown in Figure 16.10d.

## 16.4 NONLINEAR ANALYSIS WITH RL AND RC

Circuits that contain one energy-storage element (capacitor or inductor) and resistive nonlinearities such as diodes, Op Amps, and MOSFETs are very common in electronic systems: sweep circuits in oscilloscopes or television sets, rectifier circuits in all types of equipment. Fortunately, such circuits can be analyzed and designed quite readily using a combination of two techniques we have already discussed. If we represent each nonlinear element by a piecewise linear model, then the circuit can be represented by two or more subcircuits, each representing one diode state. By definition each of these subcircuits is linear, and contains one L or one C, so can be solved by methods of Chapter 10.

### 16.4.1 PEAK DETECTOR

A simple example is shown in Figure 16.11, a circuit identical to the half-wave rectifier already discussed, except for an added capacitor. The output waveform of this circuit will follow the positive peaks of the input wave, so is more efficient for converting AC to DC. The node equation at the $v_C$ node is

$$i_D = \frac{v_C}{R} + C\frac{dv_C}{dt}. \tag{16.5}$$

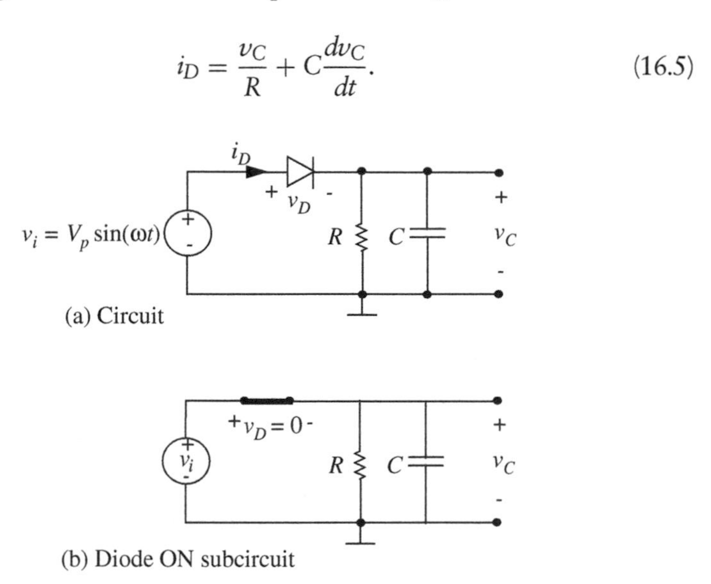


FIGURE 16.11 Peak detector.

The $v$–$i$ relation for the semiconductor diode can be written in terms of $v_C$ by noting that

$$v_D = v_i - v_C. \tag{16.6}$$

Hence

$$i_D = I_s\left(e^{q(v_i - v_C)/kT} - 1\right). \tag{16.7}$$

Substituting this into Equation 16.5 and recasting as a state equation we obtain

$$\frac{dv_C}{dt} = -\frac{v_C}{RC} + \frac{I_s}{C}\left[e^{q(v_i - v_C)/kT} - 1\right]. \tag{16.8}$$

This equation could be solved by standard numerical methods, but much insight can be obtained from a piecewise-linear solution.

If we model the diode as an ideal diode, then two linear RC subcircuits result, one for the diode ON, and the other for the OFF state, as shown in Figures 16.11b and 16.11c. For the diode ON,

$$v_C = v_i. \tag{16.9}$$

For the diode OFF, the drive is disconnected from the capacitor, so $v_C$ is a zero-input response, of the form:

$$v_C = Ke^{-t/RC}. \tag{16.10}$$

The constraints which determine the range of validity of each of these solutions are derived from the conditions on the diode states, repeated from Section 4.4:

$$\text{Diode ON:}\quad i_D \quad \text{positive,} \quad v_D = 0 \tag{16.11}$$

$$\text{Diode OFF:}\quad v_D \quad \text{negative,} \quad i_D = 0. \tag{16.12}$$

Applying Equation 16.12 to the OFF circuit, we find that to keep $v_D$ negative:

$$v_i < v_C. \tag{16.13}$$

In the ON state, Equation 16.11 requires that $i_D$ be positive. From Equation 16.5,

$$i_D = \left[\frac{v_C}{R} + C\frac{dv_C}{dt}\right] > 0. \tag{16.14}$$

The solution for a sine wave $v_i$ can now be sketched. As shown in Figure 16.12, during the first quarter cycle, the source is charging the capacitor, so $i_D$ is positive, and the diode is ON. Now as the sine wave comes down from its

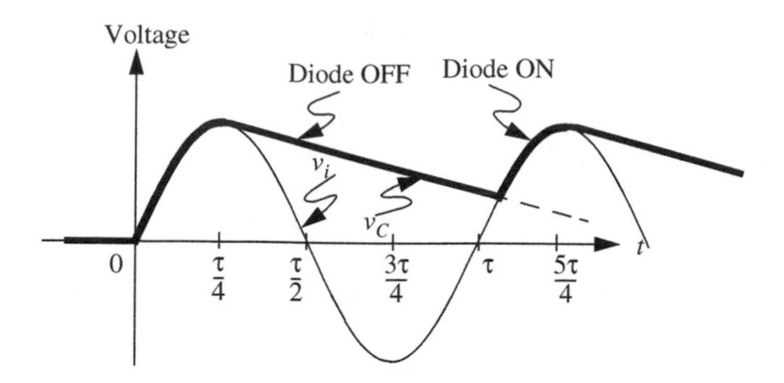


**FIGURE 16.12** Peak detector waveforms.

peak, the capacitor starts to discharge, but from Equation 16.14 this forces $i_D$ to zero, and the diode changes to the OFF state. A simple exponential discharge follows of the form given in Equation 16.10. At some later point in the cycle, the input voltage rises up to the capacitor voltage, the constraint of Equation 16.13 is no longer met, and the diode switches back to the ON state.

It is somewhat messy to calculate the exact voltages or times at which the diode switches state. But frequently in nonlinear circuit design the fundamental design constraints of the circuit make exact calculations of this sort unnecessary. For example, for rectifier applications, it is desirable to have the RC time constant much longer than the sine wave period, as we shall see. In this case the diode switches to the OFF state only a few degrees beyond the peak of the wave. Hence the starting amplitude for $v_C$ in the diode OFF state is roughly the peak of the input sine wave, and the OFF state waveform is

$$v_C \simeq V_p e^{-\frac{(t-\tau/4)}{RC}}. \tag{16.15}$$

Note that the peak of the size wave occurs at $t = \tau/4$, where $\tau$ is the cycle time. For the very long time-constant case, this discharge will be approximately linear:

$$v_C \simeq V_p \left(1 - \frac{t - \tau/4}{RC}\right). \tag{16.16}$$

The return to the ON state occurs when $v_i = v_C$, hence when

$$V_p \sin(\omega t) = V_p \left(1 - \frac{t - \tau/4}{RC}\right), \quad 2\pi < \omega t \leq \frac{5\pi}{2}. \tag{16.17}$$

This is still a transcendental equation, but it can be solved readily on a calculator.

Often we are interested only in an upper bound on the size of the ripple on the $v_C$ wave when the circuit is in the "steady state," that is, when the waveforms become repetitive. In this case the first quarter cycle in Figure 16.12 is ignored, and attention is focused on the repeating waveform thereafter. This would be the case if we were using the circuit as a rectifier to convert a 60-Hertz AC signal to DC to supply DC power to Op Amps and MOSFETs. Rather than solving Equation 16.17 for the turn-on time, and then calculating the capacitor voltage at turn-on, we just assume the transient continues for the complete period of the sine wave. This gives a slightly larger ripple than the actual value, so designs based on this approximation will be conservative. We are in effect assuming that Equation 16.16 applies for the entire period $\tau$ of the input sine wave, and at the end of one cycle, $v_C$ instantly jumps to $V_p$, then decays again. Under this assumption the peak-to-peak value of the ripple will be, from Equation 16.16,

$$\text{Peak-to-peak ripple} \simeq V_p \frac{\tau}{RC}. \tag{16.18}$$

If, for example, the RC time constant is chosen to be ten times the period of the sine wave, the peak-to-peak ripple will be 10% of $V_p$, the peak of the input sine wave. The DC voltage from the rectifier is the average value of $v_C$. This can be found by inspection of the $v_C$ waveform in Figure 16.12, again, assuming that the transient continues for the complete period of the sine wave:

$$v_C = \frac{1}{\tau}\int_0^{\tau} V_p\left(1 - \frac{t}{RC}\right) dt \tag{16.19}$$

$$= V_P\left(1 - \tau/2RC\right). \tag{16.20}$$

For $RC = 10\tau$, the DC is 0.95 $V_p$.

### 16.4.2 EXAMPLE: CLAMPING CIRCUIT

A simple *diode clamping* circuit is shown in Figure 16.13. Assuming that the diode can be modeled by an ideal diode, then the two subcircuits for the ON and OFF states have a very simple form, as shown in Figure 16.13b and c. In the ON state,

$$v_C = v_i \tag{16.21}$$

$$v_o = 0. \tag{16.22}$$

Any time $v_i$ is less than $v_C$, the circuit will be forced into the ON state. The degenerate nature of the circuit, which is modeled here as having zero resistance in this state, may be troublesome. Think of adding a small resistance associated with the source, or with the forward-biased diode. Now to keep the diode ON, the diode current $i_D$ must be positive, so

$$v_i < v_C. \tag{16.23}$$

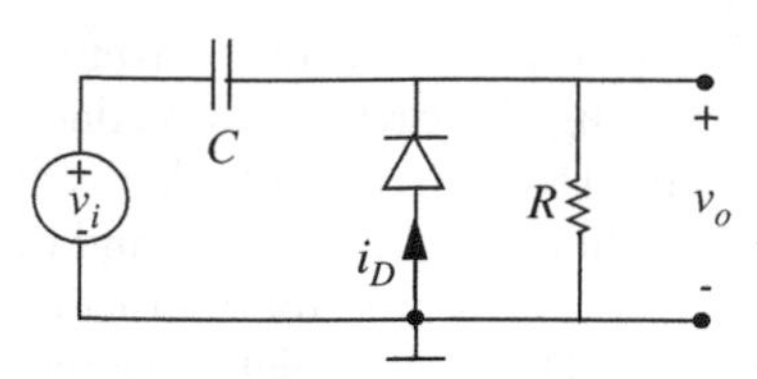


(a) Complete circuit

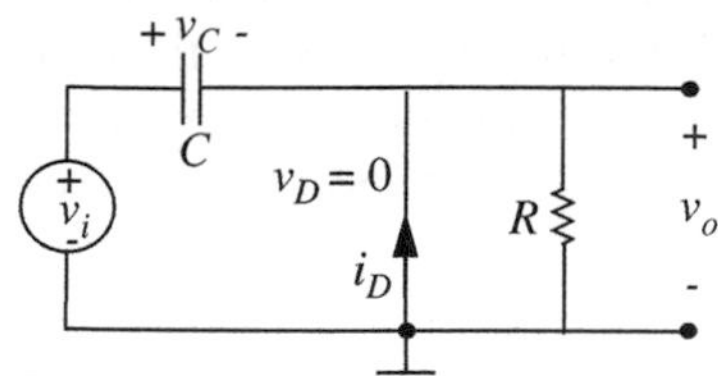


(b) Diode ON subcircuit

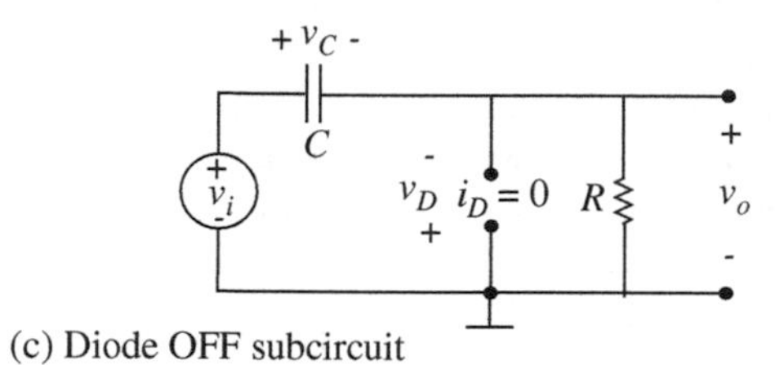


(c) Diode OFF subcircuit

**FIGURE 16.13** Diode clamping circuit.

In the OFF state, the circuit reduces to the linear RC circuit discussed in Chapter 10. If we assume $v_i$ is a square wave to simplify the problem, then in this state $v_C$ will be

$$v_C = V_{\text{init}} e^{-t/RC} + V_{\text{final}}(1 - e^{-t/RC}). \tag{16.24}$$

The resistor voltage can be found from the capacitor voltage:

$$v_O = i_C R \tag{16.25}$$

$$= RC \frac{dv_C}{dt}. \tag{16.26}$$

For the square-wave input case, this can be evaluated from Equation 16.24:

$$v_O = \left[V_{\text{final}} - V_{\text{init}}\right] e^{-t/RC}. \tag{16.27}$$

To keep the circuit in the OFF state, $v_D$ must be negative, hence from KVL

$$v_C - v_i = v_D < 0. \tag{16.28}$$

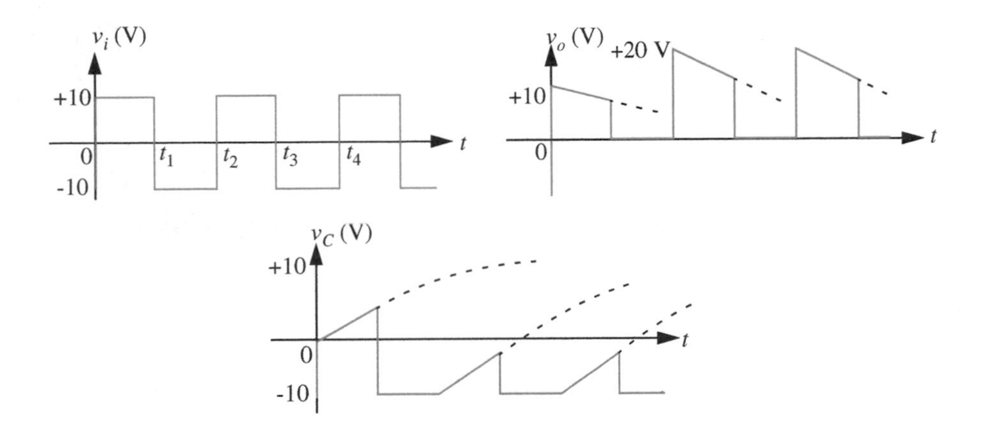


**FIGURE 16.14** Diode clamp waveforms.

Hence the circuit constraint is

$$v_C < v_i. \tag{16.29}$$

Now the waveforms can be sketched. Assuming that the capacitor is initially uncharged and the input voltage goes positive to +10 V on the first half cycle, as suggested in Figure 16.14, then the initial circuit state is the OFF state, from Equation 16.29. Hence the capacitor will charge toward +10 volts. Here $V_{\text{init}} = 0$, and $V_{\text{final}} = 10$, so

$$v_C = 10\left(1 - e^{-t/RC}\right). \tag{16.30}$$

At the same time

$$v_o = 10e^{-t/RC}. \tag{16.31}$$

The Figure 16.14 show these waveforms for the case where the transients only go part-way to completion, that is, for the RC time constant long compared to the period of the square-wave.

The input waveform undergoes an abrupt transition at $t_1$, at the end of the first half-cycle, dropping to $-10$ volts. Because $v_C$ had been positive, the constraint of Equation 16.23 now applies, and the circuit is forced into the ON state. Physically, the diode is ON, so the capacitor is connected directly across $v_i$, and $v_o$ is zero. Note that for this over-idealized circuit the capacitor voltage is forced to change instantaneously from some positive voltage to $-10$ volts in response to the input transition. A more realistic model which included either diode forward resistance or a resistance associated with the source would eliminate this anomaly, and produce a rapid transition in $v_C$ rather than an instantaneous jump.

As noted in the waveforms, the ON state persists until the input wave undergoes a transition back to +10 volts. As soon as $v_i$ moves away from

$-10$ volts, the constraint of Equation 16.29 will be satisfied, and the circuit will be forced to the OFF state. The capacitor starts at $-10$ volts, and would charge to $+10$ volts if the input remained at $+10$ for long enough, so from Equation 16.24

$$v_C = 10 - 20e^{-t/RC}. \qquad (16.32)$$

Also, from Equation 16.27, the output voltage must be

$$v_o = 20e^{-t/RC}. \qquad (16.33)$$

Reasoning physically from the circuit, right after the transition the capacitor is still charged to $-10$ volts, and the source is at $+10$ volts, so the voltage across the resistor at this point is 20 volts. If it were not for the intervening transition, the resistor voltage would decay to zero; hence Equation 16.33.

At $t_3$, $v_i$ moves abruptly to $-10$ V, and forces the diode to the ON state, as before. From here on, the waveform patterns repeat.

The waveforms in Figure 16.14 have been drawn for the case where the capacitor discharges appreciably during each half-cycle. This case was chosen for clarity rather than functionality. When the circuit is used as a *DC restorer* in a television set, the RC time constant would normally be chosen to be much longer than the period of the input wave. In such designs, the output wave has the same shape as the input wave, except that the waveform is shifted so as to be always positive. The most negative value of the input voltage is now *clamped* to 0 V.

### WWW 16.4.3 A SWITCHED POWER SUPPLY USING A DIODE

## WWW 16.5 ADDITIONAL EXAMPLES

### WWW 16.5.1 PIECEWISE LINEAR EXAMPLE: CLIPPING CIRCUIT

### WWW 16.5.2 EXPONENTIATION CIRCUIT

### WWW 16.5.3 PIECEWISE LINEAR EXAMPLE: LIMITER

### WWW 16.5.4 EXAMPLE: FULL-WAVE DIODE BRIDGE

### WWW 16.5.5 INCREMENTAL EXAMPLE: ZENER-DIODE REGULATOR

### WWW 16.5.6 INCREMENTAL EXAMPLE: DIODE ATTENUATOR

## 16.6 SUMMARY

- The following is an analytical expression for the relation between the voltage $v_D$ and the current $i_D$ for the diode:

$$i_D = I_s(e^{v_D/V_{TH}} - 1).$$

The parameter, $V_{TH} = kT/q$, is called the thermal voltage. The constant $I_s$ for silicon is typically $10^{-12}$ amps. $q$ is the charge of an electron, $k$ is the Boltzmann's constant, and $T$ is the temperature in kelvins. At room temperature, $kT/q$ is approximately 0.025 V.

- The ideal diode model approximates the $v$–$i$ characteristics of a diode using two straight-line segments given by:

$$\text{Diode ON:} \quad v = 0 \quad \text{for all positive } i$$

$$\text{Diode OFF:} \quad i = 0 \quad \text{for all negative } v.$$

- A more accurate diode model comprises an ideal diode in series with a voltage source, and can be summarized as:

$$\text{Diode ON (vertical segment):} \quad v_D = 0.6 \text{ V} \quad \text{for} \quad i_D > 0$$

$$\text{Diode OFF (horizontal segment):} \quad i_D = 0 \quad \text{for} \quad v_D < 0.6 \text{ V}.$$

- An even more accurate diode model comprises an ideal diode in series with a voltage source and a resistor, and can be summarized as:

$$\text{Diode ON (vertical segment):} \quad v_D = 0.6 \text{ V} + i_D R_d \quad \text{for } i_D > 0$$

$$\text{Diode OFF (horizontal segment):} \quad i_D = 0 \quad \text{for } v_D < 0.6 \text{ V}.$$

- The method of assumed states to analyze a diode circuit has the following steps:

  1. Draw a subcircuit for each possible state (ON or OFF) of the diodes. For one diode there are two subcircuits. For $n$ diodes there are $2^n$ such states, and hence $2^n$ subcircuits.

  2. Analyze each resulting linear circuit to find an expression for the desired output variable. Because in each subcircuit the diode is either a short or an open circuit, the subcircuits are linear. Hence linear analysis methods can be used.

  3. Establish the range of validity of each of the expressions in (2), then assemble the appropriate segments to form the complete output waveform.

## EXERCISES

**EXERCISE 16.1** Determine and graph the $v$–$i$ relation imposed at its port by the network shown in Figure 16.27. Assume that the diode is ideal.

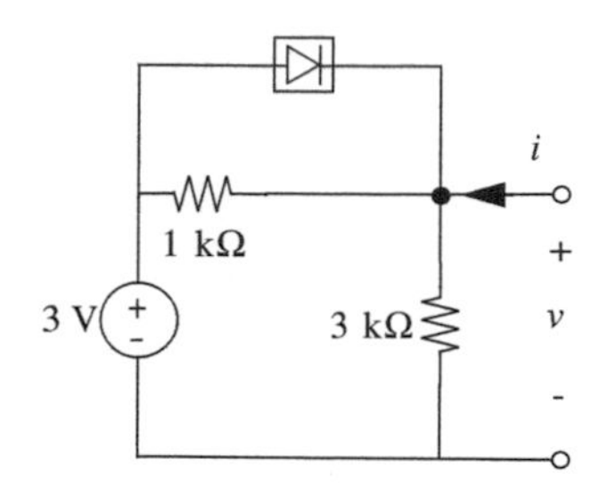


FIGURE 16.27

**EXERCISE 16.2** Consider the circuit shown in Figure 16.28. Determine and graph $v_{OUT}$ as a function of $v_{IN}$ for the following two diode models. Clearly label the breakpoints between neighboring piecewise-linear regions in the graph. In addition, indicate the regions of the graph that correspond to the different on/off state combinations of the diodes.

a) Assume that each diode is ideal.

b) Model each diode as an ideal diode in series with a 0.6-V source, as shown in Figure 16.29.

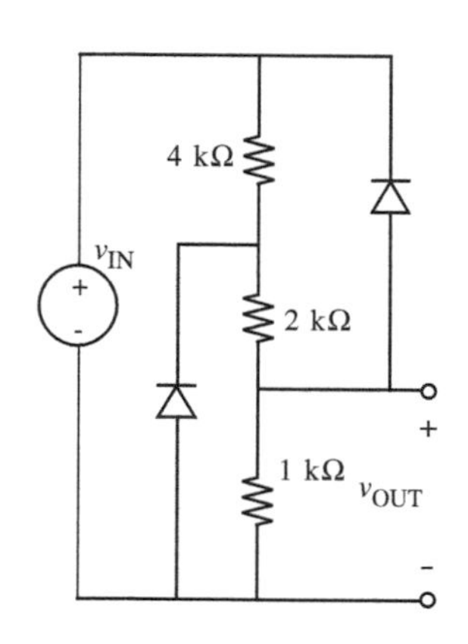


FIGURE 16.28

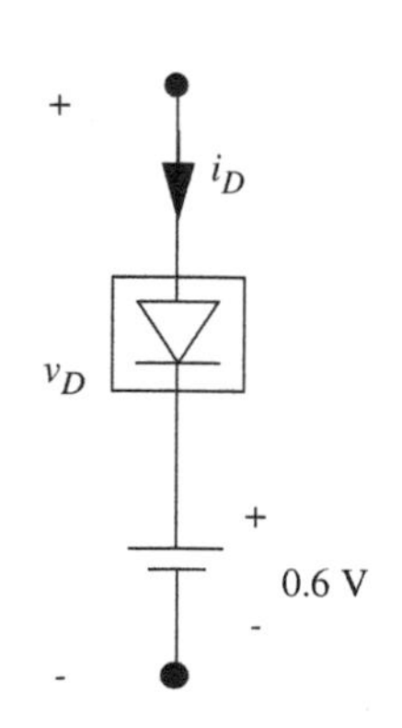


FIGURE 16.29

**EXERCISE 16.3** The diode in the circuit shown in Figure 16.30 is ideal. Assuming that $v_{IN}(t) = 1 \text{ V} \sin(2\pi 100 \text{ rad/s } t)$, determine and graph $v_{OUT}$ for $0 \leq t \leq 20$ ms.

**EXERCISE 16.4** Determine and graph $v_{OUT}$ as a function of $v_{IN}$ for the circuit shown in Figure 16.31. In doing so, model the diode as shown in Figure 16.29, and assume that the Op Amp is ideal. Also, contrast the input-output behavior of the circuit shown in Figure 16.31 with that of the half-wave rectifier studied in the chapter on nonlinear analysis.

**EXERCISE 16.5** This exercise explores the use of superposition to analyze networks containing diodes. For this purpose, assume that the diodes in Figure 16.32 are all ideal.

a) Let $v_{IN2} = 0$. Determine $v_{OUT}$ as a function of $v_{IN1}$.
b) Let $v_{IN1} = 0$. Determine $v_{OUT}$ as a function of $v_{IN2}$.
c) Finally, determine $v_{OUT}$ for the general case in which both $v_{IN1}$ and $v_{IN2}$ are nonzero.
d) Your answer to Part (c) should not be a superposition of your answers to Parts (a) and (b). Why not?

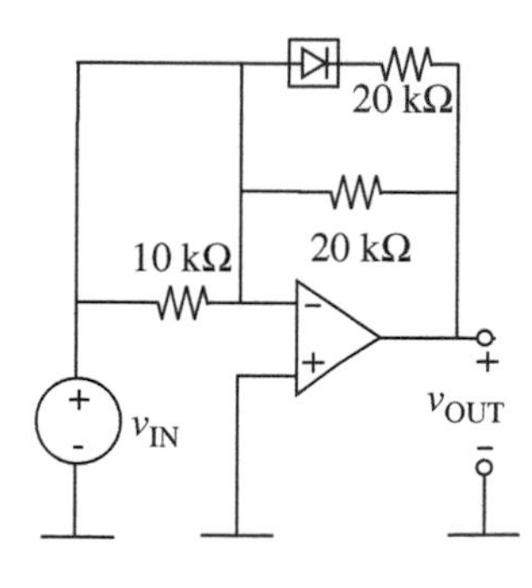


FIGURE 16.30

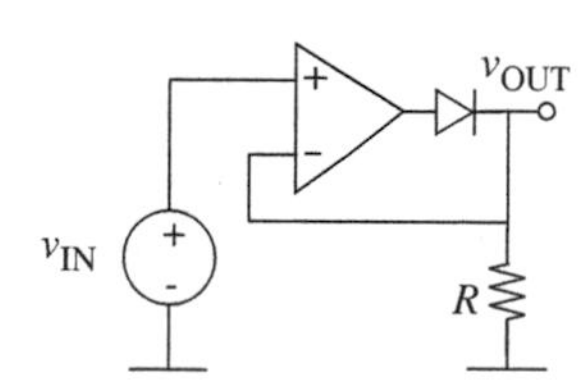


FIGURE 16.31

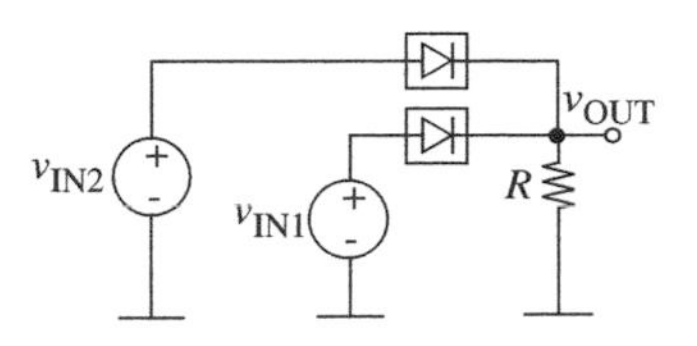


FIGURE 16.32

PROBLEM 16.1 For the two circuits shown in Figure 16.33, determine and graph $v_{OUT}$ as a function of $v_{IN}$. Assume that all diodes and Op Amps are ideal.

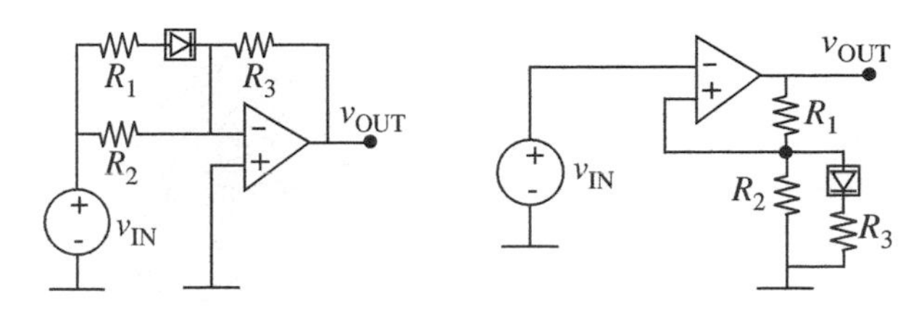


FIGURE 16.33

PROBLEM 16.2 The diodes in the networks shown in Figure 16.34 are ideal. Both networks are driven by a voltage source which produces a pulse of amplitude $V_o$ for a duration $T$. Prior to $t = 0$, both networks are at rest.

a) Find $v_C(t)$ and $i_L(t)$ for $0 \leq t \leq T$.
b) Find $v_C(t)$ and $i_L(t)$ for $T \leq t$.

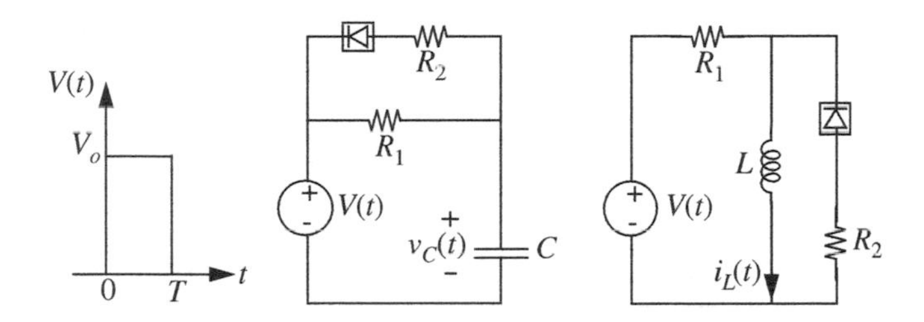


FIGURE 16.34

PROBLEM 16.3 The diodes in the networks shown in Figure 16.35 are ideal. Both networks are driven by a current source which produces a pulse of amplitude $I_o$ for a duration $T$. Prior to $t = 0$, both networks are at rest.

a) Find $v_C(t)$ and $i_L(t)$ for $0 \leq t \leq T$.
b) Find $v_C(t)$ and $i_L(t)$ for $T \leq t$.

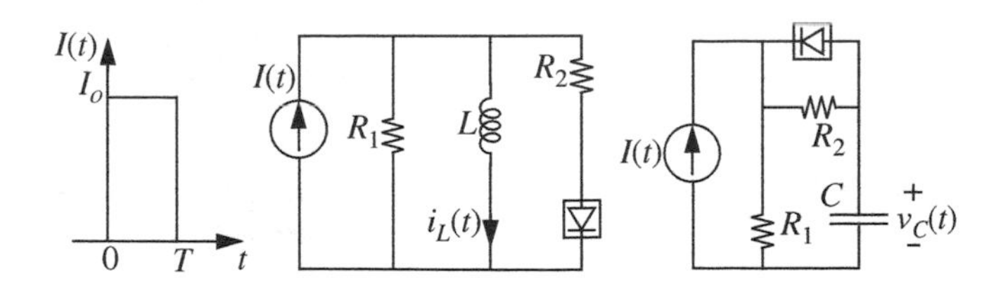


FIGURE 16.35

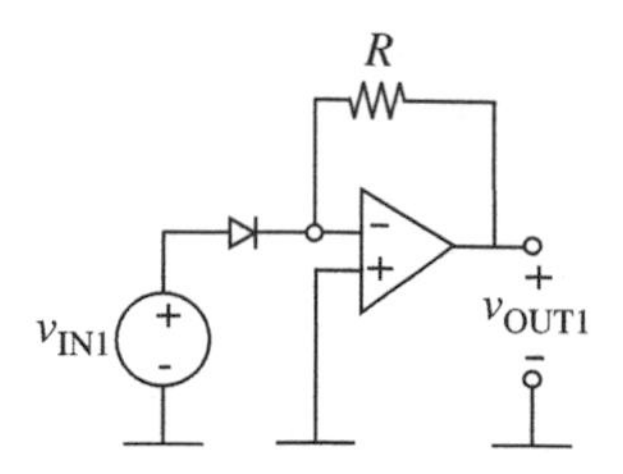


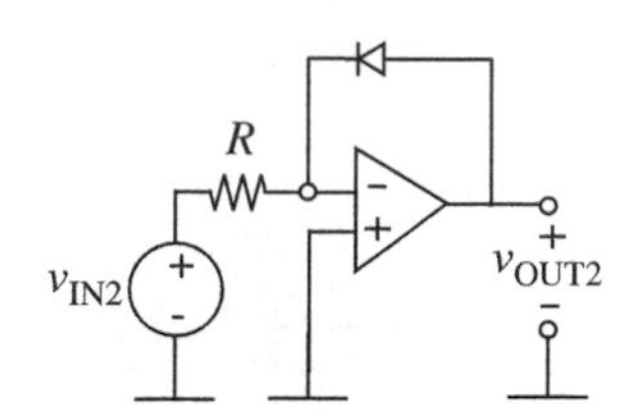


FIGURE 16.36

PROBLEM 16.4 This problem studies the construction of multipliers, dividers, and exponentiators using diodes and Op Amps. Throughout this problem assume that the diodes exhibit the relation $i_D \approx I_S e^{qv_D/kT}$, and that the Op Amps are ideal.

a) For both circuits shown in Figure 16.36, determine $v_{OUT}$ as a function of $v_{IN}$. Also, in view of the approximation used to describe the behavior of the diodes, state the range of $v_{IN}$ over which the analysis holds.
b) Multiplication can be performed by adding logarithms. Using this fact, construct a circuit that produces an output voltage that is proportional to the product of two input voltages. State the input-output relation of the circuit, and state the range of input voltages over which the circuit will act as a multiplier.
c) Division can be performed by subtracting logarithms. Using this fact, construct a circuit that produces an output voltage that is proportional to the quotient of two input voltages. State the input-output relation of the circuit, and state the range of input voltages over which the circuit will act as a divider.
d) Exponentiation can be performed by scaling logarithms. Using this fact, construct circuits that produce an output voltage that is proportional to the square and cube, respectively, of an input voltage. State the input-output relation of the circuits, and state the range of input voltages over which the circuits will act as exponentiators.

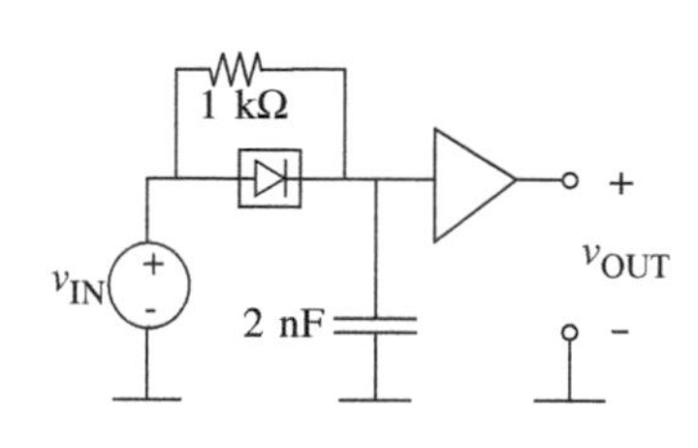


FIGURE 16.37

PROBLEM 16.5 Determine $v_{OUT}$ for the circuit shown in Figure 16.37 given that $v_{IN}$ is a 100-kHz square wave that switches between 0 V and 5 V. The buffer in the circuit produces an output of 0 V for an input of 2.5 V and below; it produces an output of 5 V for an input above 2.5 V. Assume that the diode is ideal.

PROBLEM 16.6 The circuit shown in Figure 16.38 is a very simple power supply for a resistive load. With a sufficiently large value for C, it produces a reasonably constant $v_{OUT}$ from a 60-Hz input of the form $v_{IN} = 10 \text{ V} \cos(2\pi \text{ } 60 \text{ rad/s } t)$.

a) Also shown in Figure 16.38 is a graph of $v_{IN}$ and $v_{OUT}$. Assume that $C = 10^3$ $\mu$F. Determine $T_1$ and $T_2$, the times at which $v_{OUT}$ breaks away from and returns to

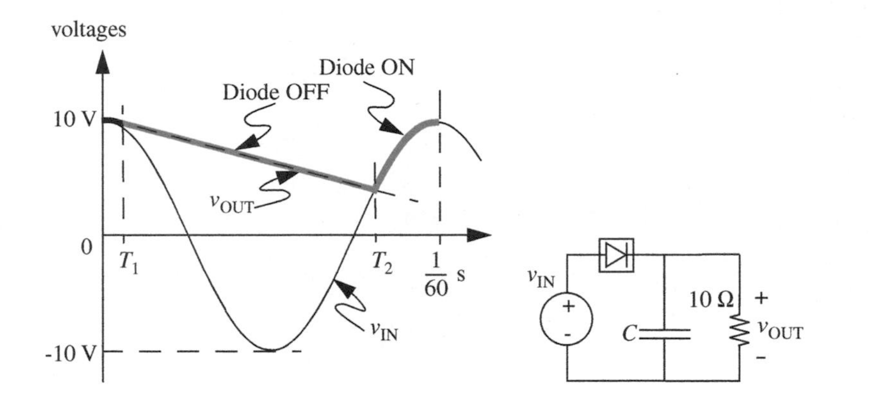


FIGURE 16.38

$v_{IN}$, respectively, as shown in the figure. Also find $v_{OUT}(T_2)$, the minimum value of $v_{OUT}$.

b) Repeat Part (a) for $C = 10^4\ \mu F$.

c) Approximately how large should $C$ be if $v_{OUT}$ is to drop no more that 0.1 V?

# APPENDIX A

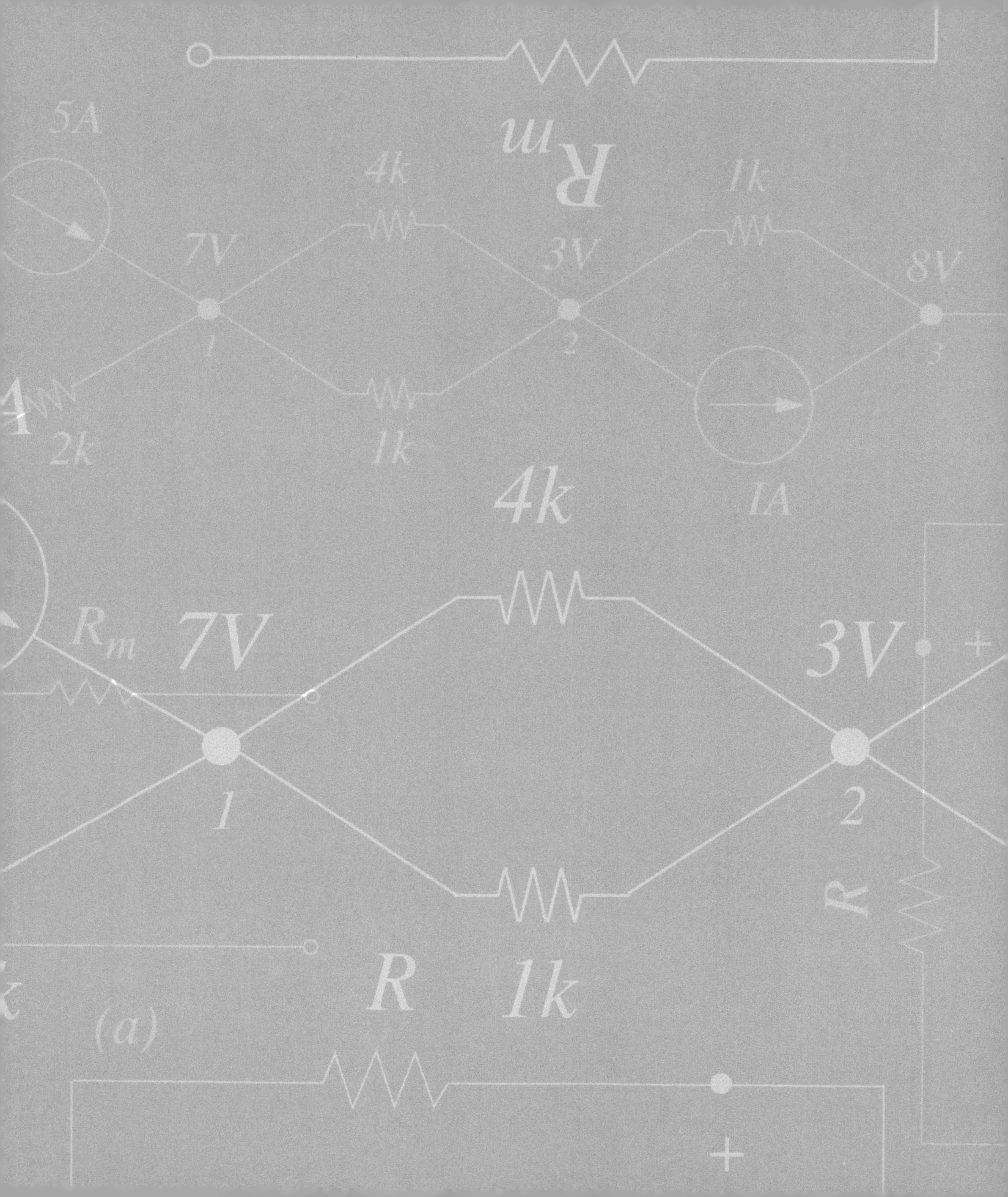
5A
4k
1k
7V
3V
8V
2k
1k
1A
4k
$R_m$
7V
3V
1
2
R
R
1k
(a)

# APPENDIX A

# MAXWELL'S EQUATIONS AND THE LUMPED MATTER DISCIPLINE

This appendix develops the constraints of the lumped matter discipline and demonstrates that the constraints result in a simplification of Maxwell's Equations into algebraic equations.

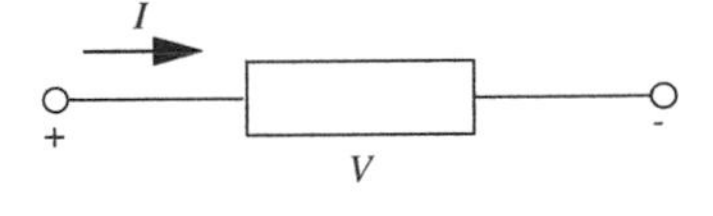


FIGURE A.1 A lumped circuit element.

## A.1 THE LUMPED MATTER DISCIPLINE

Lumped circuits comprise lumped elements connected by ideal wires. A lumped element has the property that a unique terminal voltage $V(t)$ and terminal current $I(t)$ can be defined for it. As depicted in Figure A.1, for a two-terminal element, $V$ is the voltage across the terminals of the element, and $I$ is the current through the element. As we will show shortly, the voltage and the current are defined for an element or for points within a circuit only under certain constraints that we collectively call the *lumped matter discipline.*

Let us use our familiar lightbulb as an example and derive the conditions under which we can treat pieces of matter as lumped elements for inclusion in electronic circuits. Suppose for the sake of discussion the lightbulb is made out of a cylindrical piece of filament of length $l$ and cross-sectional area $a$ as depicted in Figure A.2.

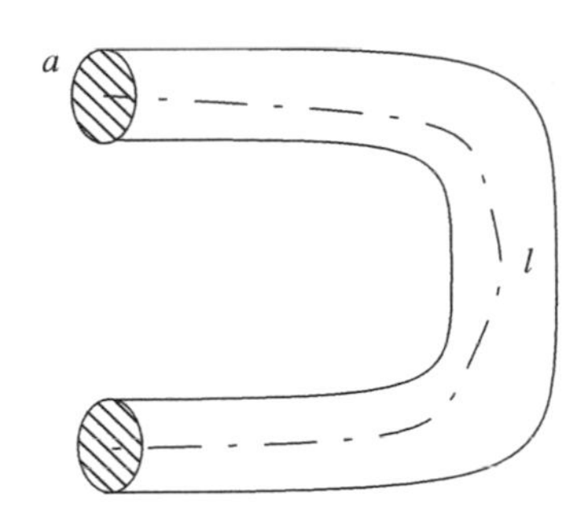


FIGURE A.2 Resistive bulb filament.

As shown in Figure A.3, let us assume that the terminals labeled $x$ and $y$ are attached to the end surfaces of the filament, and that the end faces are equipotential surfaces. Let us determine the set of conditions under which (1) a unique voltage $V$ can be defined across $x$ and $y$ and (2) a unique current $I$ can be defined through $x$ and $y$.

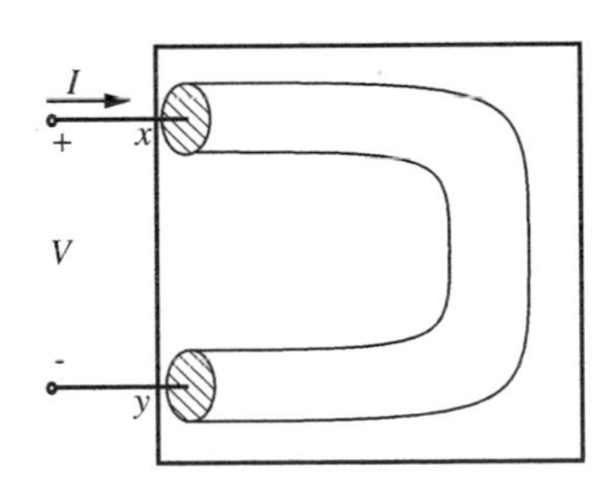


FIGURE A.3 Defining a voltage and current for the terminals of the filament.

### A.1.1 THE FIRST CONSTRAINT OF THE LUMPED MATTER DISCIPLINE

Let's begin with the voltage. We define voltage as the line integral of the electric field $\mathbf{E}$ according to[1]

$$V_{yx} = -\int_x^y \mathbf{E} \cdot \mathbf{dl}.$$

1. Alternatively, observing that

$$qV_{yx} = -\int_x^y q\mathbf{E} \cdot \mathbf{dl},$$

**TABLE A.1** Maxwell's Equations (for a vacuum). The fifth equation is the continuity equation implicit in Maxwell's Equations. It can be derived by combining the time derivative of the first equation with the divergence of the fourth equation. $\mathbf{E}$ is the electric field, $\mathbf{B}$ the magnetic flux density, $\rho$ the charge density (note that this $\rho$ is different from the resistivity used in computing the resistance of an element), $\mathbf{J}$ the current density, $\epsilon_0$ the permittivity of free space, $\mu_0$ the magnetic permeability of free space, $\Phi_E$ the electric flux, and $\Phi_B$ the magnetic flux. $\Phi_E$ is defined as the area integral of $\mathbf{E}$ and $\Phi_B$ as the area integral of $\mathbf{B}$.

| DIFFERENTIAL FORM | INTEGRAL FORM | POPULAR NAME |
|---|---|---|
| $\nabla \cdot \mathbf{E} = \frac{\rho}{\epsilon_0}$ | $\oint \mathbf{E} \cdot d\mathbf{S} = \frac{q}{\epsilon_0}$ | Gauss's law for electricity |
| $\nabla \cdot \mathbf{B} = 0$ | $\oint \mathbf{B} \cdot d\mathbf{S} = 0$ | Gauss's law for magnetism |
| $\nabla \times \mathbf{E} = -\frac{\partial \mathbf{B}}{\partial t}$ | $\oint \mathbf{E} \cdot d\mathbf{l} = -\frac{\partial \Phi_B}{\partial t}$ | Faraday's law of induction |
| $\nabla \times \mathbf{B} = \mu_0 \epsilon_0 \frac{\partial \mathbf{E}}{\partial t} + \mu_0 \mathbf{J}$ | $\oint \mathbf{B} \cdot d\mathbf{l} = \mu_0 \epsilon_0 \frac{\partial \Phi_E}{\partial t} + \mu_0 i$ | Ampere's law (extended) |
| $\nabla \cdot \mathbf{J} = -\frac{\partial \rho}{\partial t}$ | $\oint \mathbf{J} \cdot d\mathbf{S} = -\frac{\partial q}{\partial t}$ | Continuity equation |

Note that $\mathbf{E}$ is a vector. As illustrated in Figure A.4, the preceding equation indicates that the voltage depends on the path $x \to y$. However, for our lumped abstraction to apply, we require that we be able to assign a unique voltage between the $x$ and $y$ terminals. Clearly, this voltage cannot be a function of the specific path between the $x$ and $y$ points. It seems we have already hit a snag in our attempt to create the abstraction of a lumped element to which the lumped parameters $V$ and $I$ apply.

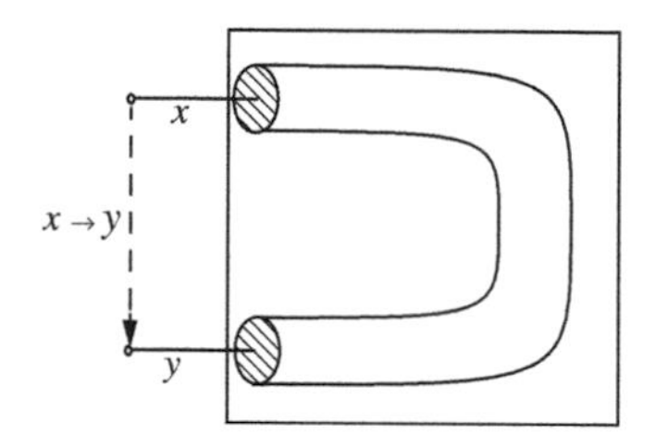


**FIGURE A.4** The voltage between the $x$ and $y$ terminals.

Put another way, we know from Maxwell's Equations (summarized in Table A.1) that

$$\oint \mathbf{E} \cdot d\mathbf{l} = -\frac{\partial \Phi_B}{\partial t}$$

where $\mathbf{E}$ is the electric field and $\Phi_B$ is the magnetic flux which passes through the surface outlined by the closed path of the integral, as depicted in Figure A.5. We also know the preceding equation as Faraday's law of induction. Notice that if we choose the two points $x$ and $y$ to be the same, then we can obtain a nonzero value for $\int_x^y \mathbf{E} \cdot d\mathbf{l} = \oint \mathbf{E} \cdot d\mathbf{l}$. Thus it appears our definition of potential difference or voltage has no useful meaning in this case. However, are there constraints that we can apply for which a unique voltage can be defined?

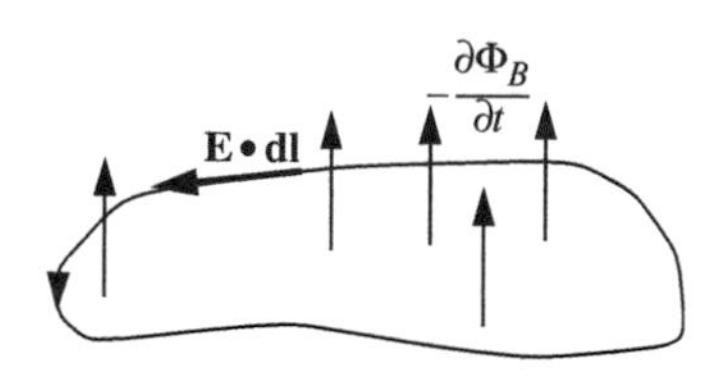


**FIGURE A.5** Pictorial depiction of $\oint \mathbf{E} \cdot d\mathbf{l}$.

In the absence of a time-varying magnetic flux, we can write

$$\oint \mathbf{E} \cdot d\mathbf{l} = 0.$$

---

the voltage $V_{yx}$ at point $y$ measured relative to point $x$ can also be defined as the energy required to move a particle with unit charge against the force due to the electrical field from $x$ to $y$.

The preceding equation says that integral of **E** over a closed circuit vanishes in the absence of a time-varying magnetic flux. Suppose we choose the closed circuit to include the points $x$ and $y$ as shown in Figure A.6. We can then write

$$\underset{\text{Path 1}}{\int_x^y \mathbf{E}\cdot d\mathbf{l}} + \underset{\text{Path 2}}{\int_y^x \mathbf{E}\cdot d\mathbf{l}} = 0$$

or

$$\underset{\text{Path 1}}{\int_x^y \mathbf{E}\cdot d\mathbf{l}} = -\underset{\text{Path 2}}{\int_y^x \mathbf{E}\cdot d\mathbf{l}}$$

or

$$\underset{\text{Path 1}}{\int_x^y \mathbf{E}\cdot d\mathbf{l}} = \underset{\text{Path 2.}}{\int_x^y \mathbf{E}\cdot d\mathbf{l}}$$

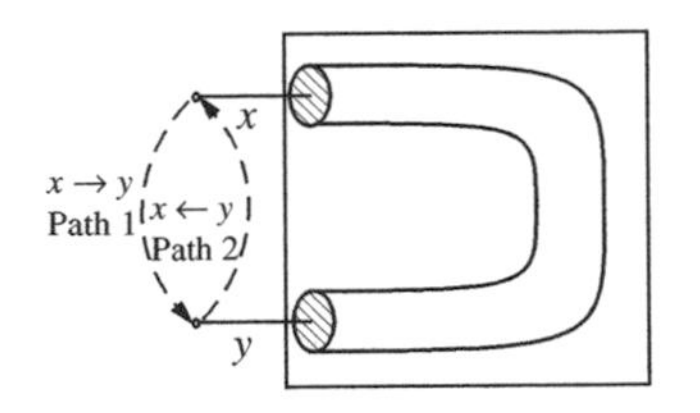


FIGURE A.6 Line integral of **E** over a closed circuit that includes the points $x$ and $y$.

Notice that for this equality to hold even when *Path 1* and *Path 2* are chosen independently, each integral must be independent of path. It follows from this that the computed voltage $V_{yx}$ between any pair of points $x$ and $y$ given by $-\int_x^y \mathbf{E}\cdot d\mathbf{l}$ is independent of the path.[2] Thus, we have our desired outcome: We can ascribe a unique potential difference or voltage across the terminals $x$ and $y$, provided

$$\frac{\partial \Phi_B}{\partial t} = 0.$$

Additionally, we assume that the rate of change of flux is 0 for all time, so that the voltage can be a uniquely defined function of time. This directly leads to the first constraint of our lumped matter discipline.

*First constraint of the lumped discipline* Choose lumped element boundaries such that

$$\frac{\partial \Phi_B}{\partial t} = 0$$

for all time, through any closed path outside the element.

Since we have assumed that the rate of change of flux is 0 for all time, and because any flux build up would require a non zero rate of change of flux, it follows that the flux must also be 0.[3]

2. Notice that the internal behavior of the element can be arbitrarily complicated, but the specific relationship between the voltage and current at the terminals will completely characterize its behavior to any circuit to which this element is connected.

3. We can arrive at the same property in a different way, as follows: A nonzero but constant flux external to an element can be the result of a current flowing internal to the element, or produced by an external source.

### A.1.2 THE SECOND CONSTRAINT OF THE LUMPED MATTER DISCIPLINE

Now, let us focus on the current $I$. The current $I$ through some cross-sectional surface ($S_z$) of the filament at some point $z$ is given by

$$I = \int_{S_z} \mathbf{J} \cdot d\mathbf{S}$$

where $\mathbf{J}$ is the current density at a given point within the filament. Note that $\mathbf{J}$ is a vector. If we choose the surface of the filament $S_x$ at the terminal $x$, then we get the amount of current entering the filament. Alternatively, if we choose the surface $S_y$ at the $y$ terminal, then we get the amount of current leaving the filament. What can we say about the current entering and leaving the filament?

It turns out that $\mathbf{J}$ can be a complicated function of position. So let us try to answer the preceding question using the continuity equation (derived from Maxwell's Equations), which gives the following relationship between the surface integral of the current out of a closed surface and the time derivative of the charge enclosed by the surface (see Figure A.7):

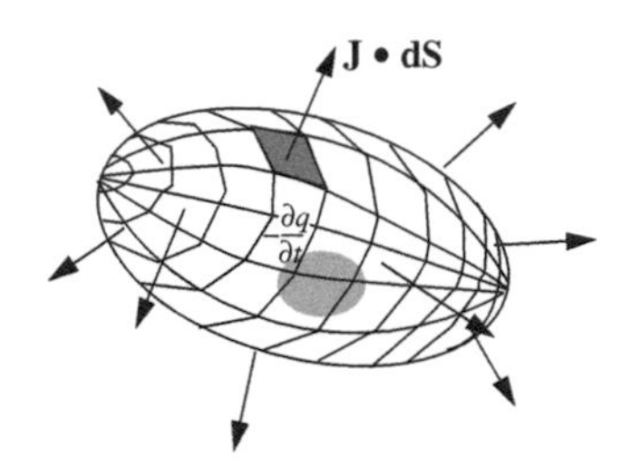


FIGURE A.7 Pictorial depiction of $\oint \mathbf{J} \cdot d\mathbf{S}$.

$$\oint \mathbf{J} \cdot d\mathbf{S} = -\frac{\partial q}{\partial t}.$$

In the preceding equation, $q$ is the total amount of charge within the closed surface. If we choose the closed surface to envelop the entire filament as depicted in Figure A.8, then $q$ will be the total charge within the enclosed volume. Let us assume that the faces of the filament at the terminals $x$ and $y$ are the only entry and exit points for the current, and that there is no charge outside the element. It is clear from the preceding equation that the current into the element will not equal the current out of the element in the presence of a time-varying total charge within the element. Thus, it makes no sense to define a current "through" the element in the presence of a time-varying total internal charge.

---

If an internal current produces a significant amount of flux, then a time-varying current will produce a nonzero time-varying flux — a situation we have explicitly disallowed. Thus, the flux resulting from an internal current must be negligible. If the flux is significant, then we will introduce a new lumped element called the inductor and capture the flux inside it, thereby adhering to the constraint.

Next, consider the case where there is an external source producing a temporally-constant amount of flux. Clearly we would like to be able to define a fixed voltage across the terminals of our element even when the element moves. However, since a moving element would create the same effect as a time-varying flux — a situation we have disallowed, the external flux must also be 0.

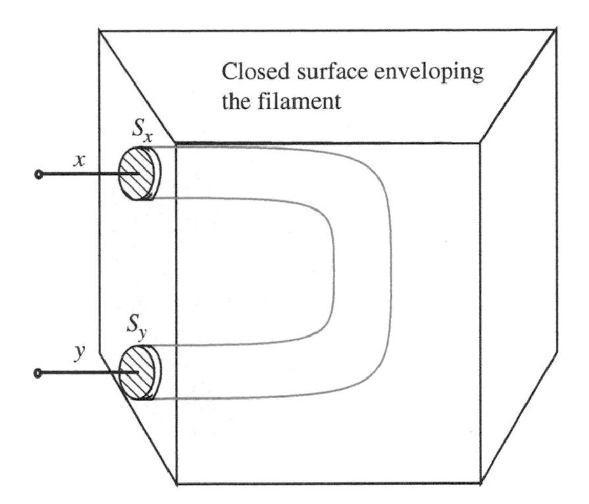


FIGURE A.8 Integral of $\mathbf{J}$ over a closed surface.

However, consider the situation in which there is no time-varying charge within the element. In other words, consider the situation in which

$$\frac{\partial q}{\partial t} = 0$$

for the element taken as a whole. With this situation, we derive the nice result that

$$\oint \mathbf{J} \cdot \mathbf{dS} = 0.$$

Simply stated, when there is no total time-varying charge within an element, the total current into the element is 0. Returning to our filament example, if there is no current flow across the curved cylindrical surface of the filament, we can rewrite the total current flowing into the element as the different between the current flowing into the end surface at $x$ and the current flowing out of the end surface at $y$:

$$\oint \mathbf{J} \cdot \mathbf{dS} = -\int_{S_x} \mathbf{J} \cdot \mathbf{dS} + \int_{S_y} \mathbf{J} \cdot \mathbf{dS} = 0.$$

Since the sum of the two components of the current must be 0, it follows they must be equal. Thus we are able to define a meaningful current flowing through the element when there is no net time varying charge within the element. This outcome directly leads to the second constraint of the lumped matter discipline.

*Second constraint of the lumped discipline* Choose the lumped element boundaries so that there is no total time varying charge within the element. In other words, choose element boundaries such that

$$\frac{\partial q}{\partial t} = 0$$

for all time, where $q$ is the total charge within the element.

Notice that we have assumed explicitly that the rate of change of charge is zero for all time so that the current can be an arbitrary function of time. Since we have assumed that the rate of change of charge is zero for all time, and because any charge build up would require a nonzero rate of change of charge, it follows that the net charge within any element must also be zero.[4]

### A.1.3 THE THIRD CONSTRAINT OF THE LUMPED MATTER DISCIPLINE

Finally, let us consider the practical matter of the propagation speeds of electromagnetic waves. The lumped element approximation requires that we be able to define a voltage $V$ between a pair of element terminals and a current through the terminal pair. Defining a current through the element means that the current in must equal the current out. Now consider the following thought experiment. Apply a current pulse at the terminal $x$ of the filament at time instant $t$ and observe both the current into terminal $x$ and the current out of terminal $y$ at a time instant $t + dt$ very close to $t$. If the filament were long enough or if $dt$ were small enough, the finite speed of electromagnetic waves might result in our measuring different values for the current in and the current out.

We fix the problem created by the finite propagation speeds of electromagnetic waves by adding a third constraint. We include the constraint that the timescale of interest in our problem be much larger than electromagnetic propagation delays through our elements. Put another way, the size of our lumped elements must be much smaller than the wavelength associated with the $V$ and $I$ signals.[5]

Under the preceding speed constraints, electromagnetic waves can be treated as if they propagated instantly through a lumped element. By neglecting propagation effects, the lumped element approximation becomes analogous to

4. If an element does store charge, then we will collect equal amounts of charges of opposite polarities inside a new lumped element called a *capacitor*, so that there is no net charge inside the element.

5. More precisely, the wavelength that we are referring to is that wavelength of the electromagnetic wave launched by the signals.

the point-mass simplification, in which we are able to ignore many physical properties of elements such as their length, shape, size, and location.

*Third constraint of the lumped discipline* Operate in the regime in which signal timescales are much larger than the propagation delay of electromagnetic waves across the lumped elements.

### A.1.4 THE LUMPED MATTER DISCIPLINE APPLIED TO CIRCUITS

Circuits are sets of lumped elements connected by ideal wires. A node is formed at the junction point at which the terminals of two or more elements are connected. We choose the wires such that they obey the lumped matter discipline, so the wires themselves are also lumped elements. For their voltages and currents to be meaningful, the constraints that apply to lumped elements apply to entire circuits as well. In other words, for voltages between any pair of points in the circuit and for currents through wires to be defined, any segment of the circuit must obey a set of constraints similar to those imposed on each of the lumped elements.

Accordingly, the lumped matter discipline for circuits can be stated as:

1. The rate of change of magnetic flux linked with any portion of the circuit must be 0 for all time.
2. The rate of change of the charge at any node in the circuit must be 0 for all time. A node is any point in the circuit at which two or more element terminals are connected using wires.
3. The signal timescales must be much larger than the propagation delay of electromagnetic waves through the circuit.

Notice that the first two constraints follow directly from the corresponding constraints applied to lumped elements. Remember that a node is simply a junction point of a set of wires, which are themselves lumped elements. So, the first two constraints do not imply any new restrictions beyond those already assumed for lumped elements.[6] The third constraint for circuits, however, imposes a stronger restriction on signal timescales than for elements, because a circuit can have a much larger physical extent than a single element. The third constraint says that the circuit must be much smaller in all its dimensions than the wavelength of light at the highest operating frequency of interest.

6. As seen in Chapter 9, it turns out that voltages and currents in circuits result in electric and magnetic fields, thus appearing to violate the set of constraints to which we promised to adhere. In most cases these are negligible. However, when their effects cannot be ignored, we explicitly model them using elements called capacitors and inductors.

## A.2 DERIVING KIRCHHOFF'S LAWS

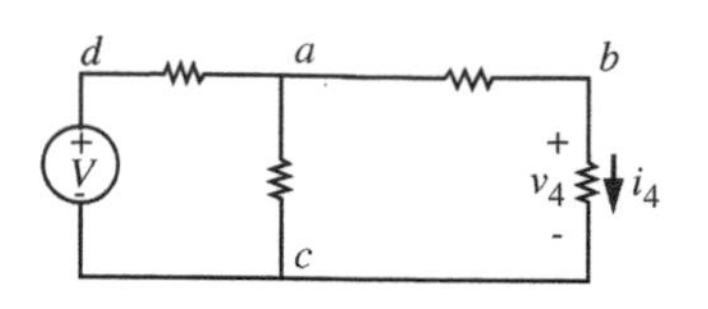


FIGURE A.9 Simple resistive network.

This section uses the lumped matter discipline to derive Kirchhoff's laws from Maxwell's Equations. To illustrate the basic ideas, let us suppose that we are interested in deriving the voltages across and the currents through each of the elements in the circuit in Figure A.9.

In general, we can resort to Maxwell's Equations and the related continuity equation to solve the circuit. The relevant equations are:

$$\oint \mathbf{E} \cdot \mathbf{dl} = -\frac{\partial \Phi_B}{\partial t}$$

and

$$\oint \mathbf{J} \cdot \mathbf{dS} = -\frac{\partial q}{\partial t}.$$

Recall that according to the lumped matter discipline we have agreed to constrain ourselves to the circuit domain in which

$$\frac{\partial \Phi_B}{\partial t} = 0$$

for closed circuit loops, and

$$\frac{\partial q}{\partial t} = 0$$

for circuit nodes. In this constrained domain, the general equations can be simplified to the following:

$$\oint \mathbf{E} \cdot \mathbf{dl} = 0 \tag{A.1}$$

and

$$\oint \mathbf{J} \cdot \mathbf{dS} = 0. \tag{A.2}$$

Equation A.1 says the line integral of the field around any closed path must equal 0. Similarly, Equation A.2 says that the surface integral of the current over any surface must be 0. Of course, Equations A.1 and A.2 are valid only under the lumped matter discipline.

Applying Equation A.1 to the closed loop defined by the three circuit edges $a \to b$, $b \to c$, and $c \to a$, as depicted in Figure A.10 we obtain

$$\oint \mathbf{E} \cdot \mathbf{dl} = \int_a^b \mathbf{E} \cdot \mathbf{dl} + \int_b^c \mathbf{E} \cdot \mathbf{dl} + \int_c^a \mathbf{E} \cdot \mathbf{dl} = 0.$$

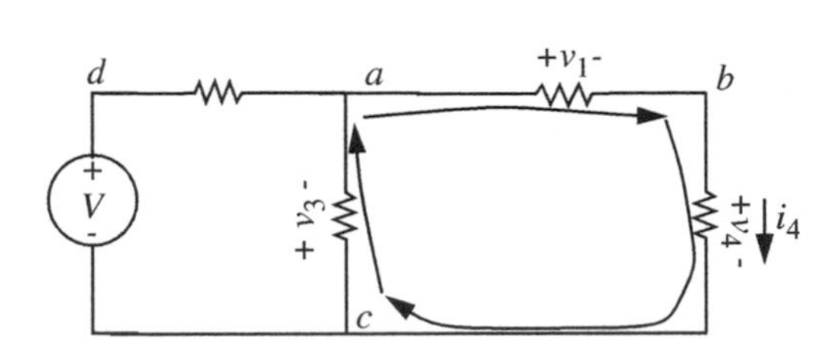


FIGURE A.10 The line integral over a closed loop in the network.

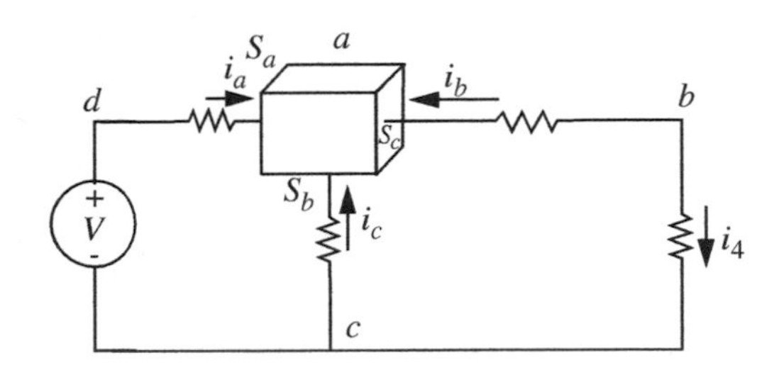


FIGURE A.11 The surface integral over a closed surface in the network.

Since we know that $\int \mathbf{E} \cdot \mathbf{dl} = 0$ along an ideal wire is 0, and since the potential difference $V_{xy}$ across the $xy$ terminals of an element is given by

$$V_{xy} = \int_x^y \mathbf{E} \cdot \mathbf{dl},$$

we can write

$$\int_a^b \mathbf{E} \cdot \mathbf{dl} + \int_b^c \mathbf{E} \cdot \mathbf{dl} + \int_c^a \mathbf{E} \cdot \mathbf{dl} = v_1 + v_2 + v_3 = 0.$$

In other words, the sum of the voltages along any closed path in the circuit must equal 0. Accordingly, we can write Kirchhoff's voltage law:

*KVL* The algebraic sum of the voltages around any closed path in a network must be zero.

We will now derive Kirchhoff's current law. Let us apply Equation A.2 to the closed box-like surface depicted in Figure A.11. We notice that there are currents flowing only through surfaces $S_a$, $S_b$, and $S_c$. Therefore,

$$\oint \mathbf{J} \cdot \mathbf{dS} = \int_{S_a} \mathbf{J} \cdot \mathbf{dS} + \int_{S_b} \mathbf{J} \cdot \mathbf{dS} + \int_{S_c} \mathbf{J} \cdot \mathbf{dS} = 0.$$

Since our currents are confined to the wires entering the three surfaces we obtain

$$\int_{S_a} \mathbf{J} \cdot \mathbf{dS} + \int_{S_b} \mathbf{J} \cdot \mathbf{dS} + \int_{S_c} \mathbf{J} \cdot \mathbf{dS} = -i_a - i_b - i_c = 0.$$

In other words, the sum of the currents flowing into any closed surface must be zero. Simply put, the preceding statement is a statement of the conservation of charge. We can now write Kirchhoff's current law:

*KCL* The current flowing out of any junction or *node* must equal the current flowing in. That is, the algebraic sum of the currents flowing into any node must be zero.

## A.3 DERIVING THE RESISTANCE OF A PIECE OF MATERIAL

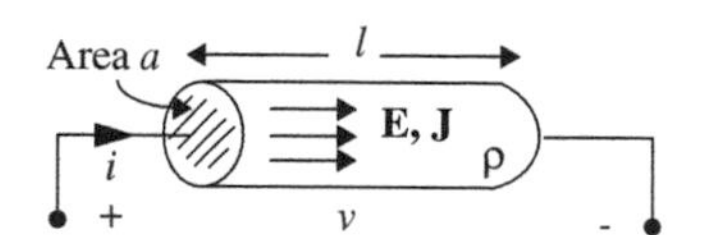


FIGURE A.12 A cylindrical-wire shaped resistor.

The resistance of a piece of material depends on its geometry. As illustrated in Figure A.12, assume the resistor has a conducting channel with cross-sectional area $a$, length $l$, and resistivity $\rho$. This channel is terminated at its extremes by two conducting plates that extend to form the two terminals of the resistor.

For lumped elements that both obey Ohm's Law and satisfy the lumped matter discipline, we can obtain a lumped resistance value from the microscopic form of Ohm's Law:

$$\mathbf{E} = \rho \mathbf{J} \tag{A.3}$$

where **J** is the current density, $\rho$ is the resistivity, and **E** is the electrical field at any point within the resistor.

As the current $i$ enters the resistor through a terminal, it spreads out to conduct uniformly through the channel. This current is given by

$$i = \int \mathbf{J} \cdot d\mathbf{S}$$

evaluated at any cross-sectional surface.[7]

The voltage across the resistor[8] is defined as

$$v = \int \mathbf{E} \cdot d\mathbf{l}.$$

We can substitute the expressions for $v$ and $i$ into Ohm's Law to get

$$R = \frac{v}{i} = \frac{\int \mathbf{E} \cdot d\mathbf{l}}{\int \mathbf{J} \cdot d\mathbf{S}}. \tag{A.4}$$

For a cylindrical-wire shaped resistor with cross-sectional area $a$ and length $l$, with the terminals taken at the circular end surfaces (see Figure A.12),

7. We are able to obtain $i$ in this fashion directly as a result of our second constraint.

8. We know this voltage is unique because of our first constraint.

the Equation A.4 reduces to the following equation through cylindrical symmetry:

$$R = \frac{El}{Ja},$$

where $E$ and $J$ are the magnitudes of the electrical field **E** and current density **J**, respectively. From Equation A.3 we know that $E/J = \rho$, so we get

$$R = \rho\frac{l}{a}. \tag{A.5}$$

Similarly, the resistance of a cuboid shaped resistor with length $l$, width $w$, and height $h$ is given by

$$R = \rho\frac{l}{wh} \tag{A.6}$$

when the terminals are taken at the pair of surfaces with area $wh$.

# APPENDIX B

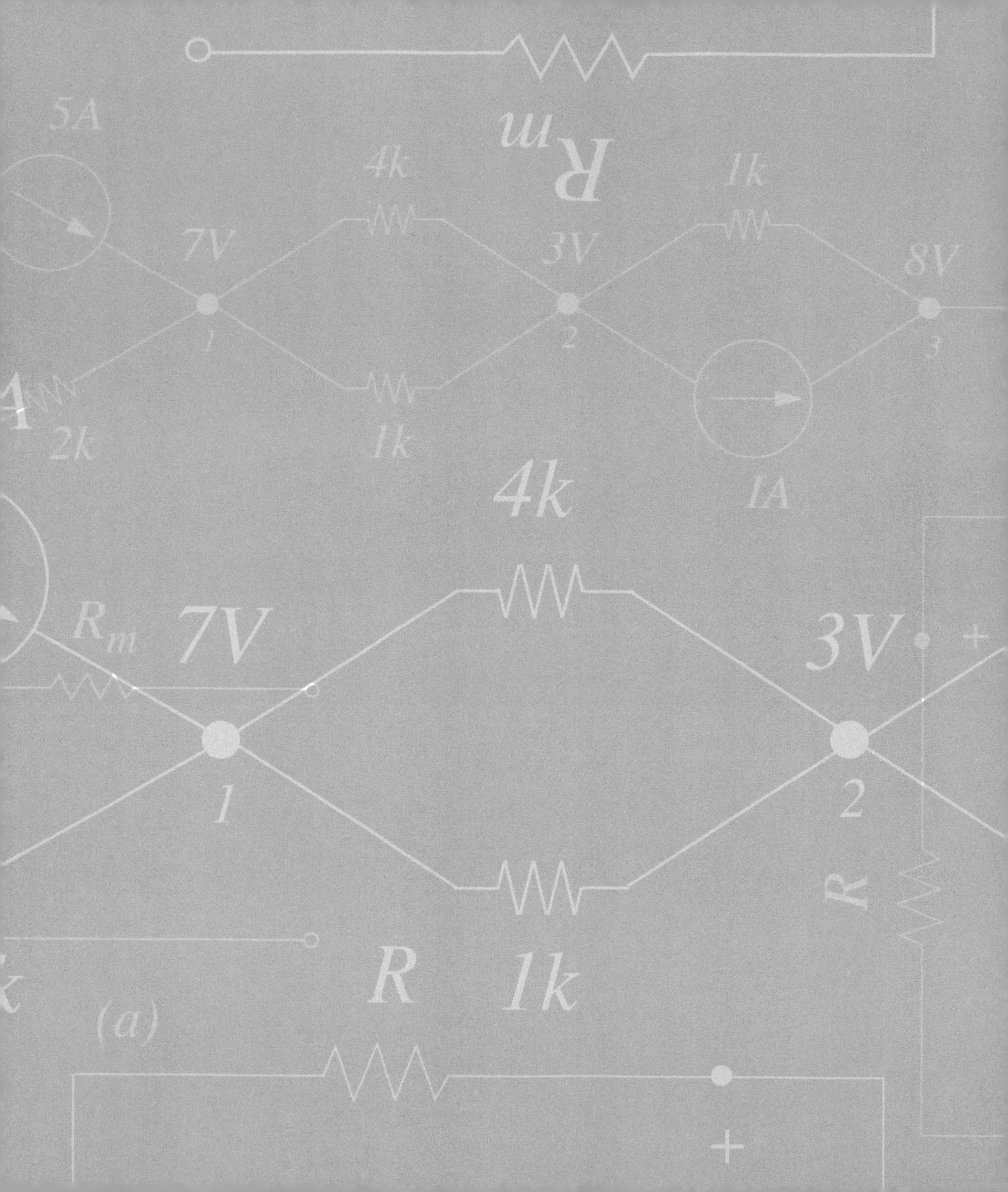
5A
4k
1k
7V
3V
8V
1
2
3
2k
1k
1A
4k
$R_m$
7V
3V
+
1
2
R
R
1k
(a)
+

# APPENDIX B

# TRIGONOMETRIC FUNCTIONS AND IDENTITIES

This appendix briefly reviews the three trigonometric functions $\cos(\theta)$, $\sin(\theta)$, and $\tan(\theta)$, and various identities involving them. These functions are often encountered during the analysis of transients in second-order linear circuits, and during the analysis of any linear circuit in sinusoidal steady state.

Consider a point located on the unit circle in the $x$–$y$ plane. If the angular position of the point around the circle is the angle $\theta$ measured from the $x$-axis, then the $x$ and $y$ coordinates of the point are $\cos(\theta)$ and $\sin(\theta)$, respectively. This defines the functions $\cos(\theta)$ and $\sin(\theta)$, as shown in Figure B.1. Additionally, we consider here the ratio of these two functions, namely $\tan(\theta) \equiv \sin(\theta)/\cos(\theta)$. All three functions are shown in Figure B.2.

In the identities that follow, $\theta$ is treated as a constant angle, specified in radians.[1] However, the identities hold whether $\theta$ is constant or not. It could just as well be a function of time or any other variable. In fact, it is often a function of time.

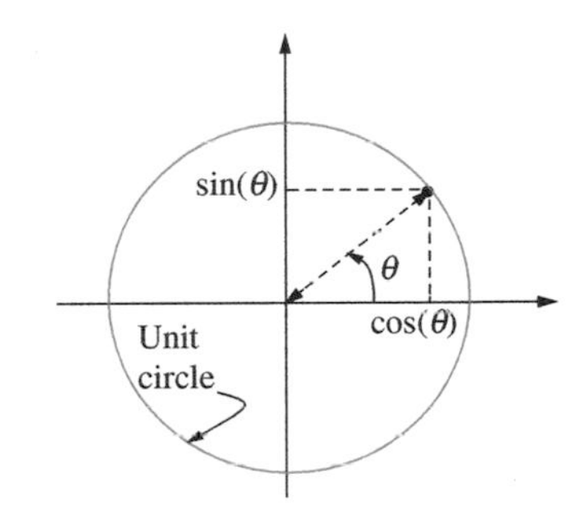


FIGURE B.1 The definitions of $\cos(\theta)$ and $\sin(\theta)$ as the $x$ and $y$ coordinates, respectively, of a point on the unit circle in the $x$–$y$ plane.

## B.1 NEGATIVE ARGUMENTS

$$\cos(-\theta) = \cos(\theta) \tag{B.1}$$

$$\sin(-\theta) = -\sin(\theta) \tag{B.2}$$

$$\tan(-\theta) = -\tan(\theta) \tag{B.3}$$

---

1. Angles measured in degrees can be converted into radians using

$$\theta_{rads} = \frac{2\pi}{360}\theta_{degs}$$

Notice, $2\pi$ radians is equivalent to 360 degrees, or one cycle.

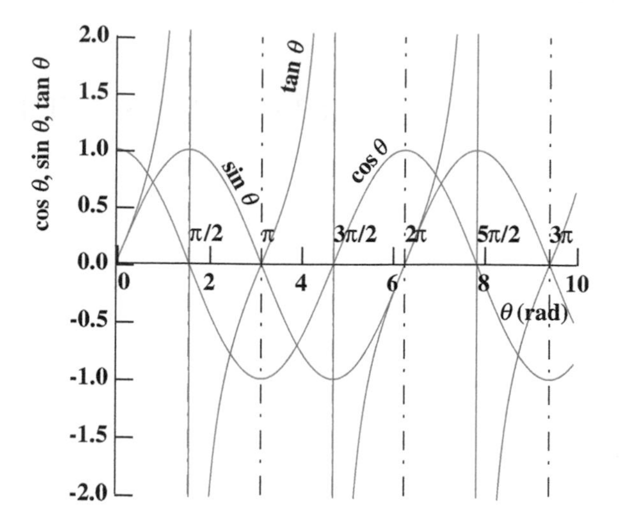


FIGURE B.2 The functions $\cos(\theta)$, $\sin(\theta)$, and $\tan(\theta)$.

## B.2 PHASE-SHIFTED ARGUMENTS

$$\cos\left(\theta \pm \frac{\pi}{2}\right) = \cos\left(\theta \mp \frac{3\pi}{2}\right) = \mp \sin(\theta) \tag{B.4}$$

$$\sin\left(\theta \pm \frac{\pi}{2}\right) = \sin\left(\theta \mp \frac{3\pi}{2}\right) = \pm \cos(\theta) \tag{B.5}$$

$$\tan\left(\theta \pm \frac{\pi}{2}\right) = \frac{-1}{\tan(\theta)} \tag{B.6}$$

$$\cos(\theta \pm \pi) = -\cos(\theta) \tag{B.7}$$

$$\sin(\theta \pm \pi) = -\sin(\theta) \tag{B.8}$$

$$\tan(\theta \pm \pi) = \tan(\theta) \tag{B.9}$$

$$\cos(\theta \pm 2\pi) = \cos(\theta) \tag{B.10}$$

$$\sin(\theta \pm 2\pi) = \sin(\theta) \tag{B.11}$$

$$\tan(\theta \pm 2\pi) = \tan(\theta) \tag{B.12}$$

## B.3 SUM AND DIFFERENCE ARGUMENTS

$$\cos(\theta_1 \pm \theta_2) = \cos(\theta_1)\cos(\theta_2) \mp \sin(\theta_1)\sin(\theta_2) \tag{B.13}$$

$$\sin(\theta_1 \pm \theta_2) = \sin(\theta_1)\cos(\theta_2) \pm \cos(\theta_1)\sin(\theta_2) \tag{B.14}$$

$$\tan(\theta_1 \pm \theta_2) = \frac{\tan(\theta_1) \pm \tan(\theta_2)}{1 \mp \tan(\theta_1)\tan(\theta_2)} \tag{B.15}$$

## B.4 PRODUCTS

$$\cos(\theta_1)\cos(\theta_2) = \frac{1}{2}(\cos(\theta_1 - \theta_2) + \cos(\theta_1 + \theta_2)) \tag{B.16}$$

$$\sin(\theta_1)\cos(\theta_2) = \frac{1}{2}(\sin(\theta_1 - \theta_2) + \sin(\theta_1 + \theta_2)) \tag{B.17}$$

$$\sin(\theta_1)\sin(\theta_2) = \frac{1}{2}(\cos(\theta_1 - \theta_2) - \cos(\theta_1 + \theta_2)) \tag{B.18}$$

## B.5 HALF-ANGLE AND TWICE-ANGLE ARGUMENTS

$$\cos(\theta/2) = \pm\sqrt{\frac{1 + \cos(\theta)}{2}} \tag{B.19}$$

$$\sin(\theta/2) = \pm\sqrt{\frac{1 - \cos(\theta)}{2}} \tag{B.20}$$

$$\tan(\theta/2) = \frac{1 - \cos(\theta)}{\sin(\theta)} = \frac{\sin(\theta)}{1 + \cos(\theta)} = S\sqrt{\frac{1 - \cos(\theta)}{1 + \cos(\theta)}};$$

$$S = \begin{cases} +1 \text{ for } \theta/2 \text{ in Q1 or Q3} \\ -1 \text{ for } \theta/2 \text{ in Q2 or Q4} \end{cases} \tag{B.21}$$

$$\cos(2\theta) = \cos^2(\theta) - \sin^2(\theta) \tag{B.22}$$

$$\sin(2\theta) = 2\sin(\theta)\cos(\theta) \tag{B.23}$$

$$\tan(2\theta) = \frac{2\tan(\theta)}{1 - \tan^2(\theta)} \tag{B.24}$$

## B.6 SQUARES

$$\cos^2(\theta) = \frac{1}{2}(1 + \cos(2\theta)) \tag{B.25}$$

$$\sin^2(\theta) = \frac{1}{2}(1 - \cos(2\theta)) \tag{B.26}$$

$$\cos^2(\theta) + \sin^2(\theta) = 1 \tag{B.27}$$

## B.7 MISCELLANEOUS

The scaled sums and differences of sinusoidal functions are among the most common identities we will use. Notice from the following equations that the scaled sums and differences of sinusoids (of the same frequency) are also sinusoids:

$$A_1\cos(\theta) + A_2\sin(\theta) = \sqrt{A_1^2 + A_2^2}\ \cos\left(\theta - \tan^{-1}\left(\frac{A_2}{A_1}\right)\right) \tag{B.28}$$

$$= \sqrt{A_1^2 + A_2^2}\ \sin\left(\theta + \tan^{-1}\left(\frac{A_1}{A_2}\right)\right) \tag{B.29}$$

$$A_1\cos(\theta) - A_2\sin(\theta) = \sqrt{A_1^2 + A_2^2}\ \cos\left(\theta + \tan^{-1}\left(\frac{A_2}{A_1}\right)\right) \tag{B.30}$$

$$= \sqrt{A_1^2 + A_2^2}\ \sin\left(\theta - \tan^{-1}\left(\frac{A_1}{A_2}\right)\right) \tag{B.31}$$

## B.8 TAYLOR SERIES EXPANSIONS

$$\cos(\theta) = 1 - \frac{\theta^2}{2!} + \frac{\theta^4}{4!} - \frac{\theta^6}{6!}\cdots \tag{B.32}$$

$$\sin(\theta) = \frac{\theta}{1!} - \frac{\theta^3}{3!} + \frac{\theta^5}{5!} - \frac{\theta^7}{7!}\cdots \tag{B.33}$$

$$\tan(\theta) = \theta + \frac{\theta^3}{3} + \frac{2\theta^5}{15} + \frac{17\theta^7}{315} + \frac{62\theta^9}{2835}\cdots \tag{B.34}$$

## B.9 RELATIONS TO $e^{j\theta}$

$$e^{\theta} = 1 + \frac{\theta}{1!} + \frac{\theta^2}{2!} + \frac{\theta^3}{3!} + \frac{\theta^4}{4!} + \frac{\theta^5}{5!}\cdots \tag{B.35}$$

$$\begin{aligned} e^{j\theta} &= 1 + \frac{j\theta}{1!} - \frac{\theta^2}{2!} - \frac{j\theta^3}{3!} + \frac{\theta^4}{4!} + \frac{j\theta^5}{5!}\cdots \\ &= \left(1 - \frac{\theta^2}{2!} + \frac{\theta^4}{4!}\cdots\right) + j\left(\frac{\theta}{1!} - \frac{\theta^3}{3!} + \frac{\theta^5}{5!}\cdots\right) \\ &= \cos(\theta) + j\sin(\theta) \end{aligned} \tag{B.36}$$

Equation B.36 is called the Euler relation.

$$\cos(\theta) = \frac{e^{j\theta} + e^{-j\theta}}{2} \tag{B.37}$$

$$\sin(\theta) = \frac{e^{j\theta} - e^{-j\theta}}{2} \tag{B.38}$$

$$\tan(\theta) = \frac{e^{j\theta} + e^{-j\theta}}{e^{j\theta} - e^{-j\theta}} \tag{B.39}$$

# APPENDIX C

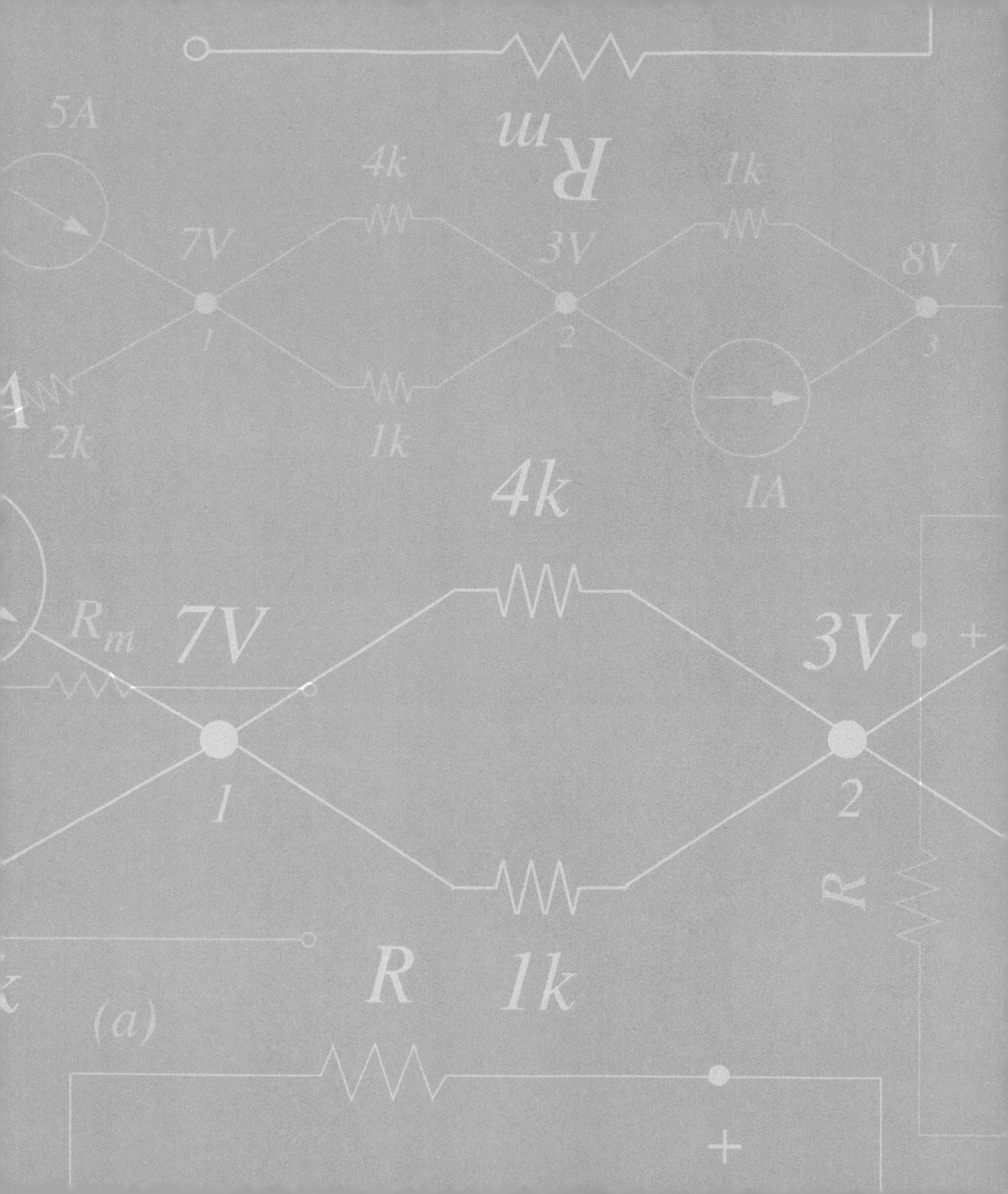

5A
4k
1k
7V
3V
8V
1
2
3
2k
1k
IA
4k
R_m
7V
3V
+
1
2
R
R
1k
(a)
+

# APPENDIX C

# COMPLEX NUMBERS

A *complex number*, $z$ for example, takes the form:

$$z = a + jb \tag{C.1}$$

where $a$ and $b$ are both real numbers, and $j$ is the *imaginary* unit defined according to:

$$j^2 = -1. \tag{C.2}$$

Here, $a$ is referred to as the *real part* of $z$, and $b$ is referred to as the *imaginary part* of $z$. These two parts can be extracted from $z$ using the real-part function $\Re(\,)$ and the imaginary-part function $\Im(\,)$, respectively. Thus, we write

$$a = \Re(z) \tag{C.3}$$

$$b = \Im(z) \tag{C.4}$$

and more generally,

$$z = \Re(z) + j\Im(z). \tag{C.5}$$

If $\Im(z) = 0$, then $z$ is a purely real number. If $\Re(z) = 0$, then $z$ is a purely imaginary number. Otherwise, $z$ is a complex number.

## C.1 MAGNITUDE AND PHASE

A complex number can be viewed as a point in the two-dimensional *complex plane*, as shown in Figure C.1 for $z$ as given in Equation C.1. The distance $r$ measured from the origin to the point is referred to as the *magnitude* of $z$, and the angle $\theta$ measured from the real axis to the radius on which the point lies is referred to as the *angle*, or *phase*, of $z$. Thus we write

$$r = |z| \tag{C.6}$$

$$\theta = \angle z. \tag{C.7}$$

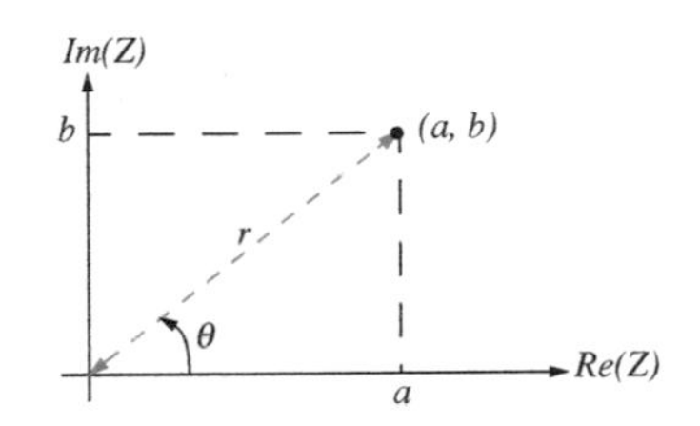


FIGURE C.1 The location of the complex number $z$ in the complex plane. The distance $r$ is the magnitude of $z$, and the angle $\theta$ are referred to as the *angle*, or *phase*, of $z$.

## C.2 POLAR REPRESENTATION

Similarly, a complex number can be viewed as a vector in the complex plane. In this case, its components are $a$ and $b$, or $\Re(z)$ and $\Im(z)$, and its magnitude and direction are $r$ and $\theta$, or $|z|$ and $\angle z$.

From the geometry of Figure C.1 it is apparent that

$$r = \sqrt{a^2 + b^2} \tag{C.8}$$

$$\theta = \tan^{-1}\left(\frac{b}{a}\right). \tag{C.9}$$

In Equation C.9, the $\tan^{-1}(\ )$ function is understood to be a two-argument inverse having full range such that $0 \leq \angle z < 2\pi$ or $-\pi \leq \angle z < \pi$; the choice of range is a matter of convenience. From Equations C.8 and C.9 we can identify the more general expressions:

$$|z| = \sqrt{\Re(z)^2 + \Im(z)^2} \tag{C.10}$$

$$\angle z = \tan^{-1}\left(\frac{\Im(z)}{\Re(z)}\right). \tag{C.11}$$

Equations C.10 and C.11 express $|z|$ and $\angle z$ in terms of $\Re(z)$ and $\Im(z)$. It is also possible to invert these expressions. Again from the geometry of Figure C.1, this yields

$$a = r\cos(\theta) \tag{C.12}$$

$$b = r\sin(\theta), \tag{C.13}$$

from which we may identify the general expressions

$$\Re(z) = |z|\cos(\angle z) \tag{C.14}$$

$$\Im(z) = |z|\sin(\angle z). \tag{C.15}$$

In summary, $\Re(z)$ and $\Im(z)$ are the Cartesian coordinates of $z$ in the complex plane while $|z|$ and $\angle z$ are the polar coordinates of $z$ in the same plane. Correspondingly, Equations C.10 and C.11 are a Cartesian-to-polar coordinate transformation, while Equations C.14 and C.15 are a polar-to-Cartesian coordinate transformation.

We can now use the polar coordinates $|z|$ and $\angle z$ to form an alternative expression for a complex number. Beginning with $z$ as presented in Equation C.1,

$$z = a + jb = r\cos(\theta) + jr\sin(\theta) = r(\cos(\theta) + j\sin(\theta)) = re^{j\theta}. \tag{C.16}$$

The first equality in Equation C.16 follows from the substitution of Equations C.12 and C.13. The last equality follows from the substitution of the Euler identity:

$$\cos(\theta) + j\sin(\theta) \equiv e^{j\theta}, \tag{C.17}$$

which is derived in Equation B.36 using Taylor Series expansions for the functions $e^{(\,)}$, cos( ) and sin( ). From Equation C.16 we can identify the more general expression:

$$z = |z|e^{j\angle z}. \tag{C.18}$$

Equation C.18 is the polar equivalent of Equation C.5. We will use these two expressions interchangeably. Which one is preferred depends on the application. For example, we shall see shortly that addition and subtraction are most easily carried out using complex numbers in Cartesian form; while multiplication, division, and determining a magnitude are most easily carried out using complex numbers in polar form.

## C.3 ADDITION AND SUBTRACTION

Mathematical operations are performed on complex numbers just as they are on purely real numbers. For example, the addition and subtraction of the two complex numbers $(a_1 + jb_1)$ and $(a_2 + jb_2)$ proceeds according to:

$$(a_1 + jb_1) + (a_2 + jb_2) = a_1 + jb_1 + a_2 + jb_2 = (a_1 + a_2) + j(b_1 + b_2) \tag{C.19}$$

$$(a_1 + jb_1) - (a_2 + jb_2) = a_1 + jb_1 - a_2 - jb_2 = (a_1 - a_2) + j(b_1 - b_2). \tag{C.20}$$

Thus, the real and imaginary parts of complex numbers add and subtract separately, just like the components of a vector. This is because the real and imaginary parts are defined along orthogonal axes of the complex plane. Because of this the addition and subtraction of complex numbers in polar form is less convenient.

## C.4 MULTIPLICATION AND DIVISION

The multiplication and division of complex numbers proceeds as directly as does their addition and subtraction. The only difference is that we commonly substitute $-1$ for each occurrence of $j^2$, as permitted by Equation C.2.

For example, the multiplication and division of the two complex numbers $(a_1 + jb_1)$ and $(a_2 + jb_2)$ proceeds according to:

$$(a_1 + jb_1)(a_2 + jb_2) = a_1a_2 + ja_1b_2 + jb_1a_2 + j^2b_1b_2$$
$$= (a_1a_2 - b_1b_2) + j(a_1b_2 + a_2b_1) \quad \text{(C.21)}$$
$$\frac{a_1 + jb_1}{a_2 + jb_2} = \frac{a_1 + jb_1}{a_2 + jb_2} \cdot \frac{a_2 - jb_2}{a_2 - jb_2} = \frac{(a_1a_2 + b_1b_2) + j(b_1a_2 - a_1b_2)}{a_2^2 + b_2^2}$$
$$= \frac{a_1a_2 + b_1b_2}{a_2^2 + b_2^2} + j\frac{b_1a_2 - a_1b_2}{a_2^2 + b_2^2}. \quad \text{(C.22)}$$

Note in particular the use of $(a_2 - jb_2)$ in Equation C.22 to clear the denominator of terms involving $j$. While multiplication and division of complex numbers in Cartesian form is not inconvenient, it is much more convenient using complex numbers in polar form. For example,

$$r_1e^{j\theta_1}r_2e^{j\theta_2} = r_1r_2e^{j(\theta_1+\theta_2)} \quad \text{(C.23)}$$
$$\frac{r_1e^{j\theta_1}}{r_2e^{j\theta_2}} = \frac{r_1}{r_2}e^{j(\theta_1-\theta_2)}. \quad \text{(C.24)}$$

Equations C.23 and C.24 demonstrate that magnitudes multiply and divide, and that angles add and subtract, during the multiplication and division of complex numbers, respectively. Taking powers of complex numbers is equally convenient when the numbers are in polar form.

## C.5 COMPLEX CONJUGATE

The *complex conjugate* of $z$, denoted here by $z^*$ is defined such that

$$\Re(z^*) = \Re(z) \quad \text{(C.25)}$$
$$\Im(z^*) = -\Im(z). \quad \text{(C.26)}$$

For $z$ as given in Equation C.1,

$$z^* = a - jb \quad \text{(C.27)}$$

while for $z$ as given in Equation C.16,

$$z^* = re^{-j\theta}. \quad \text{(C.28)}$$

In general, $z^*$ can be derived from $z$ by replacing each occurrence of $j$ in $z$ by $-j$.

By combining $z$ with $z^*$, several useful relations can be expressed. In particular,

$$zz^* = |z|^2 \tag{C.29}$$

$$\frac{z+z^*}{2} = \Re(z) \tag{C.30}$$

$$\frac{z-z^*}{2j} = \Im(z) \tag{C.31}$$

$$\frac{1}{z} = \frac{z^*}{zz^*}\frac{z^*}{|z|^2}. \tag{C.32}$$

Each of these relations can be readily proven with the substitution of Equations C.1 and C.27. Equation C.29 is a particularly useful means of computing $|z|$ when $z$ is expressed in Cartesian form. In fact, it was used in Equation C.22 to clear the denominator of terms involving $j$.

## C.6 PROPERTIES OF $e^{j\theta}$

In Equations C.16 and C.17 we introduced the complex number $e^{j\theta}$. As we shall see shortly, numbers of this form are very important for our purposes, and to this end it has several important properties. First,

$$|e^{j\theta}| = 1 \tag{C.33}$$

$$\angle e^{j\theta} = \theta. \tag{C.34}$$

This can be seen by comparing $e^{j\theta}$ with Equation C.17, or by substituting Equation C.18 into Equation C.29. Second,

$$\Re(e^{j\theta}) = \cos(\theta) \tag{C.35}$$

$$\Im(e^{j\theta}) = \sin(\theta). \tag{C.36}$$

These relations can be seen by comparing Equations C.18 and C.5.

## C.7 ROTATION

Finally, the multiplication of another complex number by $e^{j\theta}$ acts to rotate that number in the complex plane by the angle $\theta$. To see this consider that:

$$\left(r_1 e^{j\theta_1}\right)\left(e^{j\theta_2}\right) = r_1 e^{j(\theta_1+\theta_2)}. \tag{C.37}$$

Here, the multiplication of $r_1 e^{j\theta_1}$ by $e^{j\theta_2}$ preserves the magnitude $r_1$ while adding $\theta_2$ to the angle $\theta_1$. Thus, the complex number $r_1 e^{j\theta_1}$ is rotated in the complex plane by the angle $\theta_2$.

## C.8 COMPLEX FUNCTIONS OF TIME

Our discussion of complex numbers to this point has focused on constant complex numbers. However, none of the discussion has actually relied on $z$ being constant. Indeed, the entire discussion applies to complex functions of time, and in particular to the time function $e^{j\omega t}$, in which the substitution $\theta = j\omega t$ has been made. This time function is central to our study of the sinusoidal steady state operation of linear electronic circuits. Following Equations C.33 through C.36, we see that:

$$|e^{j\omega t}| = 1 \tag{C.38}$$

$$\angle e^{j\omega t} = \omega t \tag{C.39}$$

$$\Re(e^{j\omega t}) = \cos(\omega t) \tag{C.40}$$

$$\Im(e^{j\omega t}) = \sin(\omega t). \tag{C.41}$$

Equations C.38 and C.39 show that $e^{j\omega t}$ is a unit vector in the complex plane that rotates with the angular frequency $\omega$. Equations C.40 and C.41 are projections of this vector onto the real and imaginary axes, respectively. An angular frequency $\omega$ measured in radians per second is equivalent to the angular frequency $\omega/2\pi$ in cycles per second.

## C.9 NUMERICAL EXAMPLES

We close this appendix with a several numerical examples. For these examples, let

$$z_1 = -2 + j2 \tag{C.42}$$

$$z_2 = 1 + j\sqrt{3}. \tag{C.43}$$

We can find the complex conjugates of $z_1$ and $z_2$ using Equations C.25 and C.26. This yields

$$z_1^* = -2 - j2 \tag{C.44}$$

$$z_2^* = 1 - j\sqrt{3}. \tag{C.45}$$

We can find the real and imaginary parts of $z_1$ and $z_2$ by association with Equation C.5, or by using Equations C.30 and C.31. By either method,

$$\Re(z_1) = -2; \quad \Im(z_1) = 2 \tag{C.46}$$

$$\Re(z_2) = 1; \quad \Im(z_2) = \sqrt{3}. \tag{C.47}$$

We can find the magnitude and angle of $z_1$ and $z_2$ using Equations C.10 and C.11. This yields

$$|z_1| = 2\sqrt{2} \ ; \quad \angle z_1 = \frac{3\pi}{4} \tag{C.48}$$

$$|z_2| = 2 \ ; \quad \angle z_2 = \frac{\pi}{3}. \tag{C.49}$$

We can find the polar form of $z_1$ and $z_2$ using Equation C.18. This yields

$$z_1 = 2\sqrt{2}e^{j\frac{3\pi}{4}} \tag{C.50}$$

$$z_2 = 2e^{j\frac{\pi}{3}}. \tag{C.51}$$

Following Equations C.19 and C.20, the sum and difference of $z_1$ and $z_2$ are

$$z_1 + z_2 = -1 + j\left(2 + \sqrt{3}\right); \quad z_1 - z_2 = -3 + j\left(2 - \sqrt{3}\right). \tag{C.52}$$

Following Equations C.21 and C.22, the product and ratio of $z_1$ and $z_2$ are

$$z_1 z_2 = -\left(2\sqrt{3} + 2\right) - j\left(2\sqrt{3} - 2\right); \quad \frac{z_1}{z_2} = \left(\frac{\sqrt{3}}{2} - \frac{1}{2}\right) + j\left(\frac{\sqrt{3}}{2} + \frac{1}{2}\right). \tag{C.53}$$

Alternatively, following Equations C.23 and C.24, the product and ratio of $z_1$ and $z_2$ are

$$z_1 z_2 = 4\sqrt{2}e^{j\frac{13\pi}{12}}; \quad \frac{z_1}{z_2} = \sqrt{2}e^{j\frac{5\pi}{12}}. \tag{C.54}$$

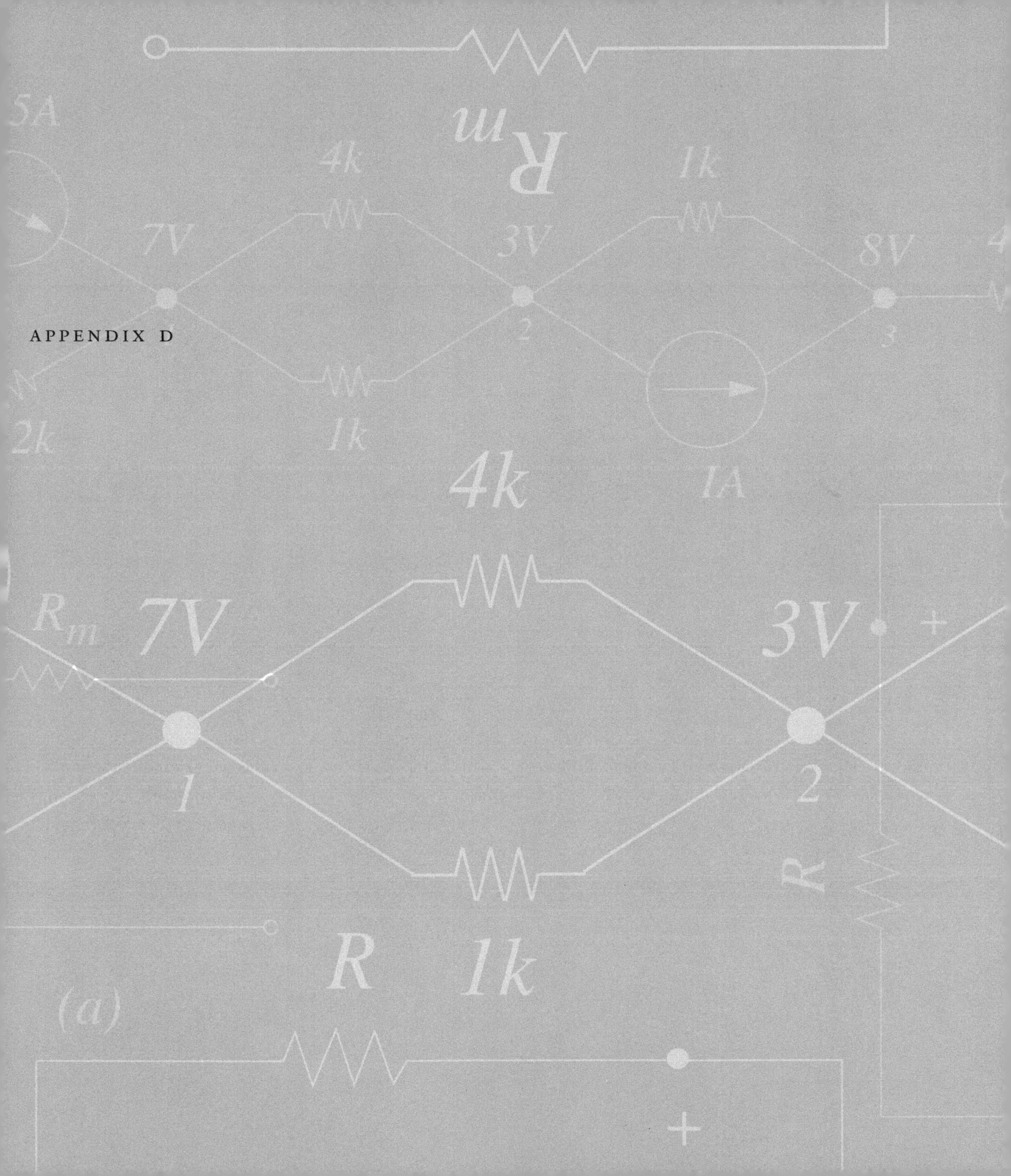

APPENDIX D

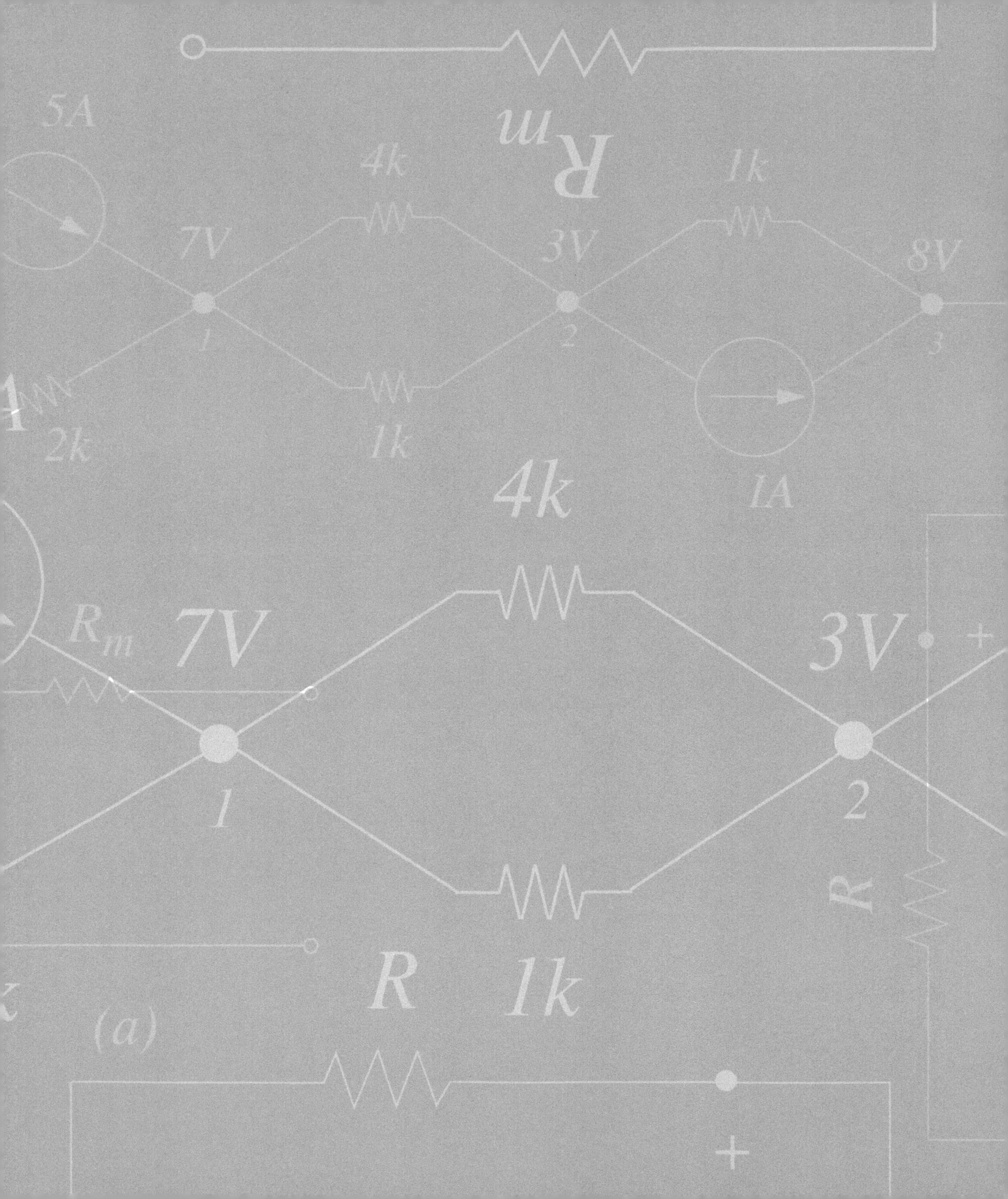

5A
4k
1k
7V
3V
8V
1
2
3
2k
1k
1A
4k
$R_m$
7V
3V
1
2
R
R
1k
(a)

# APPENDIX D

# SOLVING SIMULTANEOUS LINEAR EQUATIONS

The need to solve a simultaneous set of linear algebraic equations is a common occurrence during the analysis of electronic circuits. Of course, such equations are easily solved for special cases with numerical analysis packages, but at times the insight gained from an analytic solution is more valuable. To this end, this appendix reviews the analytic solution of equations of the form:

$$\mathcal{M}x = y$$

where $\mathcal{M}$ is a known matrix, $y$ is a known vector, and $x$ is a vector of unknowns.[1] From the start, we assume that the equations have a unique solution, and so $\mathcal{M}$ is a square matrix with $\det(\mathcal{M}) \neq 0$.

Consider the case of two equations and two unknowns. In this case,

$$\begin{bmatrix} M_{11} & M_{12} \\ M_{21} & M_{22} \end{bmatrix} \begin{bmatrix} x_1 \\ x_2 \end{bmatrix} = \begin{bmatrix} y_1 \\ y_2 \end{bmatrix}. \tag{D.1}$$

The solution to Equation D.1 is

$$\begin{bmatrix} x_1 \\ x_2 \end{bmatrix} = \frac{1}{\Delta} \begin{bmatrix} M_{22} & -M_{12} \\ -M_{21} & M_{11} \end{bmatrix} \begin{bmatrix} y_1 \\ y_2 \end{bmatrix} \tag{D.2}$$

$$\Delta = M_{11}M_{22} - M_{12}M_{21}; \tag{D.3}$$

here, $\Delta = \det(\mathcal{M})$. The validity of Equation D.2 can be verified by direct substitution into Equation D.1. Consider next the case of three equations and

---

1. For a more detailed treatment, the reader is referred to G. Strang, *Linear Algebra and its Applications*, Academic Press 1988.

three unknowns. In this case,

$$\begin{bmatrix} M_{11} & M_{12} & M_{13} \\ M_{21} & M_{22} & M_{23} \\ M_{31} & M_{32} & M_{33} \end{bmatrix} \begin{bmatrix} x_1 \\ x_2 \\ x_3 \end{bmatrix} = \begin{bmatrix} y_1 \\ y_2 \\ y_3 \end{bmatrix}. \tag{D.4}$$

The solution to Equation D.4 is

$$\begin{bmatrix} x_1 \\ x_2 \\ x_3 \end{bmatrix} = \frac{1}{\Delta} \begin{bmatrix} M_{22}M_{33} - M_{23}M_{32} & M_{32}M_{13} - M_{12}M_{33} & M_{12}M_{23} - M_{22}M_{13} \\ M_{31}M_{23} - M_{21}M_{33} & M_{11}M_{33} - M_{31}M_{13} & M_{21}M_{13} - M_{11}M_{23} \\ M_{21}M_{32} - M_{31}M_{22} & M_{31}M_{12} - M_{11}M_{32} & M_{11}M_{22} - M_{21}M_{12} \end{bmatrix} \begin{bmatrix} y_1 \\ y_2 \\ y_3 \end{bmatrix} \tag{D.5}$$

$$\begin{aligned} \Delta = M_{11}M_{22}M_{33} &+ M_{12}M_{23}M_{31} + M_{13}M_{21}M_{32} - M_{31}M_{22}M_{13} \\ &- M_{32}M_{23}M_{11} - M_{33}M_{21}M_{12}; \end{aligned} \tag{D.6}$$

again, $\Delta = \det(\mathcal{M})$. The validity of Equation D.5 can be verified by direct substitution into Equation D.4.

Finally, for higher-order cases, we can turn to an elimination process, or to Cramer's Rule,[2] although the algebra may be excessive in either case. Cramer's Rule states that:

$$x_n = \frac{\det(\mathcal{B}_n)}{\det(\mathcal{M})} \tag{D.7}$$

where $\mathcal{B}_n$ is the matrix formed by replacing the $n$th column of $\mathcal{M}$ with $y$.

2. Again, see G. Strang, *Linear Algebra and its Applications*, Academic Press 1988.

# ANSWERS TO SELECTED PROBLEMS

## CHAPTER 1

**Ex1.1** $R = 12\ \Omega$

**Ex1.3** $\frac{V_{DC}^2}{R}$

## CHAPTER 2

**Ex2.1** (a) $2.5\ \Omega$ (b) $1\ \Omega$ (c) $2R$

**Ex2.3** (a) $10\ \Omega$ (b) $1\ \Omega$ (c) $2\ \Omega$ (d) $2\ \Omega$

**Ex2.5** (a) $R_1 + R_2 + R_3$, (b) $\frac{R_1R_2 + R_3(R_1 + R_2)}{R_1 + R_2}$ (c) $\frac{R_1(R_2 + R_3)}{R_1 + R_2 + R_3}$ (d) $\frac{R_1R_2}{R_1 + R_2} + \frac{R_3R_4}{R_3 + R_4}$ (e) $\frac{(R_1 + R_2)(R_3 + R_4)}{R_1 + R_2 + R_3 + R_4}$

**Ex2.7** $R_2$ and $R_3$

**Ex2.9** (b) 2 (c) 3 (d) (Depending on your assignment of branch variables, your answer may be different.) KVL: $V_A + V_E + V_C + V_B = 0$, $V_C - V_D = 0$ KCL: $i_B - i_C - i_D = 0$, $i_A - i_B = 0$, $-i_A + i_E = 0$ (e) $i_A = i_B = i_E = 0.2A$ $i_C = 1\ A$ $i_D = -0.8\ A$ (f) $V_D = -2\ V$, $V_C = -2\ V$, $V_E = 2\ V$, $V_B = 1\ V$, $V_A = -1\ V$

**Pr2.1** $0.5\ \Omega$

**Pr2.3** $\frac{4}{5}\ \Omega$

**Pr2.5** (a) $R_T = R_1 + R_2 + R_3$ (b) $R_T = \frac{R_1R_2R_3}{R_1R_2 + R_1R_3 + R_2R_3}$ (c) $R_T = \frac{R_1R_2 + R_1R_3}{R_1 + R_2 + R_3}$ (d) $R_T = R_1 + \frac{R_2R_3}{R_2 + R_3}$

(e) $R_T = \frac{R_1R_3 + R_1R_4 + R_2R_3 + R_2R_4}{R_1 + R_2 + R_3 + R_4}$

**Pr2.7** $i_3 = -\frac{vR_2}{R_1R_2 + R_1R_3 + R_2R_3}$

**Pr2.9** Power = 2 W

**Pr2.13** (a) $i_1 = \frac{v_AR_2 + v_AR_3 - v_BR_2}{R_1R_2 + R_2R_3 + R_1R_3}$, $i_2 = \frac{v_AR_3 + v_BR_1}{R_1R_2 + R_2R_3 + R_1R_3}$, $i_3 = \frac{v_BR_2 + v_BR_1 - v_AR_2}{R_1R_2 + R_2R_3 + R_1R_3}$

**Pr2.15** $v_4 = \frac{vR_2R_4 + IR_1R_2R_4 + IR_1R_3R_4 + IR_2R_3R_4}{R_1R_2 + R_1R_3 + R_1R_4 + R_2R_3 + R_2R_4}$

**Pr2.17** $v_C = 225\ V$

## CHAPTER 3

**Ex3.1** 8/53 A

**Ex3.3** Left: $V_{OC} = I_SR_2$, $R_T = R_1 + R_2$, Right: $V_{OC} = \frac{I_SR_2R_3}{R_1 + R_2 + R_3}$, $R_T = R_3 \| (R_1 + R_2)$

**Ex3.5** 1/3 V

**Ex3.7** $I_{SC} = 1$ mA, and $R_T = 8\ \text{k}\Omega$

**Ex3.9** (a) $i(t) = \frac{1}{4}(v_1(t) + v_2(t))$ (b) Energy $= \frac{1}{16}\int_{T_1}^{T_2} (v_1(t) + v_2(t))^2 dt$ (c) $\int_{T_1}^{T_2} v_1 \cdot v_2 \cdot dt \equiv 0$

**Ex3.11** $R_T = 2\ \Omega$ and $V_{OC} = 6\ V$

**Ex3.13** $R_T = R_2$ and $v_T = I_3 \cdot R_2 + V_3$

**Ex3.15** (1) $(g_1 + g_3 + g_5)\, v_a - g_3 \cdot v_b + 0 \cdot v_c = g_1 \cdot V$, (2) $-g_3 \cdot v_a + (g_3 + g_4)\, v_b - g_4 \cdot v_c = I$, (3) $0 \cdot v_a - g_4 \cdot v_b + (g_2 + g_4 + g_6) \cdot v_c = g_2 \cdot V$

**Ex3.17** $R_T = \frac{R_1(R_2 + R_3)}{R_1 + R_2 + R_3}$, $V_{OC} = \frac{R_1 R_2 \cdot I + (R_2 + R_3)\, V}{R_1 + R_2 + R_3}$

**Ex3.19** $R_T = 100\ \Omega$, $V_{OC} = 16\frac{2}{3}$ V, $I_{SC} = 1/6$ A

**Ex3.22** (a) $R_T = R_6 + R_7 + R_8$ and $V_{OC} = I \cdot R_6$, (b) $R_T = R_4 \| (R_1 + R_3)$, and $I_{SC} = V/(R_2 + R_3)$

**Ex3.24** $R_T = 5\ \text{k}\Omega$, $V_{OC} = 49$ V

**Ex3.26** $V \cdot g_1 = v_a\,(g_1 + g_2 + g_4) - v_b \cdot g_4$, and $V \cdot g_3 - I = +v_a\,(-g_4) + v_b\,(g_3 + g_4)$

**Pr3.1** 15 A

**Pr3.3** 8.57 V

**Pr3.5** (a) $R_{eq} = R$, (b) $v_{TH} = 0.125$ V, $R_{TH} = 1\ \Omega$

**Pr3.7** (a) 0, b) i) $V\left(\frac{R}{R + R_1} - \frac{1}{2}\right)$, ii) $\frac{V(R - R_1)}{3R + 5R_1}$, c) $R_{TH} = R$, $V_{TH} = 0$.

**Pr3.9** $\frac{A_0}{2} - 4$ V

**Pr3.13** (a) $R_{TH} = 100\ \text{k}\Omega$, $v_T = -10\beta V_S$ (b) $R_L = R_T$

## CHAPTER 4

**Ex4.3** $i_D = 4.7$ mA, $v_D = 5.7$ V

**Ex4.5** (a) $i = 2 \cdot I_s\left(e^{q \cdot V_D/KT} - 1\right)$, (b) $i = I_s\left(e^{q \cdot V_D/2KT} - 1\right)$

**Ex4.7** Diode on: $i(t) = (V_1(t) + 5V)/R$; Diode off: $i(t) = 0$

**Pr4.1** (a) $i_A = \frac{2Rc_2 v_I + Rc_1 + 1 - \sqrt{(Rc_1 + 1)^2 - 4Rc_2(Rc_0 - v_I)}}{2R^2 c_2}$ for $v_I \geq Rc_0$; $i_A = 0$ otherwise,

$$v_A = \frac{\sqrt{(Rc_1 + 1)^2 - 4Rc_2(Rc_0 - v_I)} - (Rc_1 + 1)}{2Rc_2}$$ for $v_I \geq Rc_0$, $V_A = V_I$ otherwise

(b) $V_A = \frac{\sqrt{(Rc_1 + 1)^2 - 4Rc_2(Rc_0 - V_I)} - (Rc_1 + 1)}{2Rc_2}$, $I_A = \frac{2Rc_2 V_I + Rc_1 + 1 - \sqrt{(Rc_1 + 1)^2 - 4Rc_2(Rc_0 - V_I)}}{2R^2 c_2}$

(c) $\frac{\Delta i_a}{\Delta v_i} = \frac{1}{R}\left(1 - \frac{1}{\sqrt{(Rc_1 + 1)^2 + 4R^2 c_0 c_2 + 4Rc_2 V_I}}\right)$ (d) $1 - \frac{1}{\sqrt{(Rc_1 + 1)^2 + 4R^2 c_0 c_2 + 4Rc_2 V_I}}$

(e) $\Delta i_A = \frac{1}{1.02R}\left(v_I - \frac{\sqrt{(1.02Rc_1 + 1)^2 - 4.08Rc_2(1.02Rc_0 - v_I)} - (1.02Rc_1 + 1)}{2c_2(1.02R)^2}\right) - \frac{1}{R}\left(v_I - \frac{\sqrt{(Rc_1 + 1)^2 - 4Rc_2(Rc_0 - v_I)} - (Rc_1 + 1)}{2c_2 R^2}\right)$

(f) $\dfrac{di_A}{dv_A} = 2c_2V_A + c_1; \;\; V_A \geq 0$ (g) $r_N = \dfrac{1}{2c_2V_A + c_1}, \;\; i_a = \dfrac{v_0 \cos \omega t}{R + \dfrac{R}{\sqrt{(Rc_1+1)^2 - 4Rc_2(Rc_0 - V_I)} - 1}}$

(h) (i) $I_A = \dfrac{20Rc_2 + Rc_1 + 1 - \sqrt{(Rc_1+1)^2 - 4Rc_2(Rc_0 - 10)}}{2R^2c_2}$ (ii) $i_a = \dfrac{1}{R + \dfrac{R}{\sqrt{(Rc_1+1)^2 - 4Rc_2(Rc_0 - 10)} - 1}}$

(iii) $i_A = \dfrac{20Rc_2 + Rc_1 + 1 - \sqrt{(Rc_1+1)^2 - 4Rc_2(Rc_0 - 10)}}{2R^2c_2} + \dfrac{1}{R + \dfrac{R}{\sqrt{(Rc_1+1)^2 - 4Rc_2(Rc_0 - 10)} - 1}}$

(iv) $i_A = \dfrac{22Rc_2 + Rc_1 + 1 - \sqrt{(Rc_1+1)^2 - 4Rc_2(Rc_0 - 11)}}{2R^2c_2}$

(v) $i_a = \dfrac{2Rc_2 - \sqrt{(Rc_1+1)^2 - 4Rc_2(Rc_0 - 11)} + \sqrt{(Rc_1+1)^2 - 4Rc_2(Rc_0 - 10)}}{2R^2c_2}$

(vi) $\dfrac{1}{R + \dfrac{R}{\sqrt{(Rc_1+1)^2 - 4Rc_2(Rc_0 - 10)} - 1}} - \dfrac{2Rc_2 - \sqrt{(Rc_1+1)^2 - 4Rc_2(Rc_0 - 11)} + \sqrt{(Rc_1+1)^2 - 4Rc_2(Rc_0 - 10)}}{2R^2c_2}$

**Pr4.3** (a) $i \approx 1.4$ A; $v \approx 2.8$ V (b) $i \approx 1.9$ A; $v \approx 2.9$ V (d) $i \approx 1$ A; $v \approx 3$ V

**Pr4.5** (a) $v_o = 0.024\Delta v$ (b) DC: 4.5 V AC: 1.2 mV (c) 25 Ω

**Pr4.7** Assume $I_{pss} = 5$ mA and $V_p = 5$ V. (a) $i = \dfrac{2V_S - \left(\dfrac{V_P^2}{RI_{DSS}} + 2V_P\right) + \sqrt{\left(\dfrac{V_P^2}{RI_{DSS}} + 2V_P\right)^2 - \dfrac{4V_P^2V_S}{RI_{DSS}}}}{2R}$ for $V_S < V_P + I_{DSS}R$

(b) $V_S = 5$ V; $i_{average} = 3.1$ mA, $V_S = 10$ V; $i_{average} = 5$ mA, $V_S = 15$ V; $i_{average} = 5$ mA

**Pr4.9** (a) ii; if S current source, i (b) 1 A

**Pr4.11** (a) $R_{TH} = 0.5$ kΩ, $V_{OC} = \dfrac{1}{4}v_I$ (b) $v_D = 0.6$ V, $i_D = 0.8$ mA (c) $r_d = \dfrac{V_{TH}}{I_S} \exp\left(\dfrac{-V_D}{V_{TH}}\right) = 9.44 \times 10^{-4}\Omega$

(d) $v_d = 7.55 \times 10^{-9} \cos(\omega t)$

**Pr4.13** $v_{out} = \dfrac{R}{R + 500} 10^{-3} \sin(\omega t)$

## CHAPTER 5

**Ex5.1** $Z = \overline{X} + \overline{Y}$

**Ex5.3** $Z = \overline{WXY}$

**Ex5.5** 100, 0100

**Ex5.7** (c) $\overline{B}\,\overline{C}\,\overline{D}$, $B\overline{D}$, $B + \overline{D}$, $\overline{B}\,\overline{C}D$ (d) $\overline{B}\,\overline{C}\,\overline{D}$, 0, 1, $\overline{B}CD$

**Ex5.9** (c) 0.5 V (d) 4.4 V (e) 1.5 V (f) 3.5 V (g) Yes. $NM_0 = 1$ V and $NM_1 = 0.9$ V

**Pr5.1** (a) $AB + CD$ (b) $A\overline{B} + \overline{C}D$ (c) $A\overline{B} + BC$ (d) $\overline{B} + C$ (e) $AB + AC + \overline{B}C$ (f) 1

**Pr5.3** $OUT_2 = ABCD$, $OUT_1 = \overline{A}CD + B\overline{C}D + BC\overline{D} + A\overline{B}C + A\overline{B} \cdot \overline{C}D + AB\overline{C} \cdot \overline{D}$, $OUT_0 = \overline{A} \cdot \overline{B} \cdot \overline{C}D + \overline{A} \cdot \overline{B}C\overline{D} + \overline{A}B\overline{C} \cdot \overline{D} + \overline{A}BCD + A\overline{B} \cdot \overline{C} \cdot \overline{D} + A\overline{B}CD + AB\overline{C}D + ABC\overline{D}$

**Pr5.5** $OUT0 = IN \cdot \overline{S_1} \cdot \overline{S_2}$, $OUT1 = IN \cdot \overline{S_1} \cdot S_0$, $OUT2 = IN \cdot S_1 \cdot \overline{S_0}$, $OUT3 = IN \cdot S_1 \cdot S0$

**Pr5.7** $Z = \overline{A3} \cdot \overline{A2} \cdot \overline{A1} \cdot A0 + \overline{A3} \cdot \overline{A2} \cdot A1 \cdot \overline{A0} + \overline{A3} \cdot A2 \cdot \overline{A1} \cdot \overline{A0} + \overline{A3} \cdot A2 \cdot A1 \cdot A0 + A3 \cdot \overline{A2} \cdot \overline{A1} \cdot \overline{A0} + A3 \cdot \overline{A2} \cdot A1 \cdot A0 + A3 \cdot A2 \cdot \overline{A1} \cdot \overline{A0} + A3 \cdot A2 \cdot A1 \cdot \overline{A0}$

**Pr5.9** $OUT0 = IN0$, $OUT1 = \overline{IN0}IN1 + IN0\overline{IN1}$

**Pr5.11** $C_1 = \overline{A_1}A_0B_1B_0 + A_1A_0\overline{B_1}B_0 + A_1B_1 + B_1B_0C_0 + A_1A_0C_0 + A_1B_0C_0 + A_0B_1C_0$

**Pr5.13** (c) 0.5 V (d) 4.4 V (e) 1.6 V (f) 3.2 V (g) 8 (f) $NM_0 = 1.1$ V, $NM_1 = 1.2$ V, unchanged

## CHAPTER 6

**Ex6.3** (b) yes (c) no (d) 2 (e) 2

**Ex6.5** 2.27 mW

**Ex6.7** (b) 0.5 (c) 4.4 (d) 1.6 (e) 3.2 (f) 1.1 (g) 1.2 (h) 2.4

**Pr6.1** (a) $Z = \overline{A + B}$ (b) $Z = \overline{A}BC$

**Pr6.3** $N = \dfrac{100k}{(V_S{-}1)R_{ON}}$, $P_{MAX} = \dfrac{V_S^2}{100k + NR_{ON}}$

**Pr6.5** $n \leq \dfrac{V_{OL}R}{(V_S - V_{OL})R_{ON}}$, $m$:any value, $P_{MAX} = \dfrac{V_S^2}{R}$ as $m$ becomes large

**Pr6.7** $Area = \dfrac{1}{12\sqrt{2}} + \dfrac{3}{2\sqrt{2}}$

## CHAPTER 7

**Ex7.1** $v_O = V_S - (RK)^{\frac{1}{3}}$

**Ex7.3** $v_B = \dfrac{R_BV_S - K}{R_A + R_B}$

**Ex7.5** (a) $R_{ON} = \dfrac{2}{K(5 - V_T)}$

**Ex7.7** (a) $v_O = V_S - \dfrac{KR_Lv_I^2}{2}$ (b) $0 \leq i_{DS} \leq \dfrac{1 + KR_LV_S - \sqrt{1 + 2KR_LV_S}}{KR_L^2}$ (c) $V_I = \dfrac{\sqrt{1 + 2KR_LV_S - 1}}{2KR_L}$,

$V_O = \dfrac{3KR_LV_S - 1 + \sqrt{1 + 2KR_LV_S}}{4KR_L}$, $I_{DS} = \dfrac{1 + KR_LV_S - \sqrt{1 + 2KR_LV_S}}{4KR_L^2}$

**Ex7.9** (b) $v_O = V_S - i_CR_L$ (c) $i_C = \beta\dfrac{v_I - 0.6\,\text{V}}{R_I}$ (d) $i_E = i_B(\beta + 1)$ (e) $v_O = 6.2 - 2v_I$ (f) $v_O = 4.8$ V, $i_B = 0.2\ \mu$A, $i_C = 20\ \mu$A and $i_E = 20.2\ \mu$A.

**Pr7.1** $V_O = V_A - V_T - \sqrt{\dfrac{W_2L_1}{L_2W_1}(V_B - V_T)^2}$

**Pr7.3** (d) $\sqrt{\dfrac{2V_S}{KR} - \dfrac{2V_T}{KR} + \dfrac{2}{K^2R^2} - \sqrt{\dfrac{4}{K^4R^4} + \dfrac{8V_S}{K^3R^3}}} \leq v_{IN} \leq v_T + \sqrt{\dfrac{2V_S}{KR} - \dfrac{2V_T}{KR}}$

**Pr7.5** $V_T \leq v_{IN} \leq V_S + V_T$

**Pr7.7** (a) $v_S = -V_T - \sqrt{\frac{1}{K}}$, $v_O = V_S - \frac{R_L I}{2}$ (b) $\frac{W}{L} = \frac{2K}{K_n}$, $V_B = V_T + \sqrt{\frac{I}{K}} - V_S$

**Pr7.9** (b) $\frac{V_S - V_T}{R_C} \geq I \geq \frac{V_S - V_T - V_L}{R_C - R_L}$

**Pr7.11** (b) $i_D = \frac{K}{2}(V_S - v_{IN} - V_T)^2$, $v_{OUT} = V_S - \frac{KR_D}{2}(V_S - v_{IN} - V_T)^2$ (c) $-V_T \leq v_{OUT} \leq V_S$

**Pr7.15** $v_{OUT} = v_{IN} - V_T + \frac{1}{KR_S} - \sqrt{\frac{2(v_{IN} + V_S - V_T)}{KR_S} + \frac{1}{K^2 R_S^2}}$

**Pr7.19** (b) $\beta' = (\beta + 2)\beta$ (c) 1.2 V

## CHAPTER 8

**Ex8.1** (a) $V_O = V_S - \frac{KR_L}{2}(V_I - V_T)^2$ (b) $\frac{dv_O}{dv_I}|_{v_I = V_I} = -KR_L(V_I - V_T)$

**Ex8.3** current source $i_{DS} = \frac{K}{2}$ so that the small signal model is an open circuit

**Ex8.5** (a) $v_I - V_T \leq V_S - \frac{KR_L}{2}(v_I - V_T)^2$, $\frac{\sqrt{1+2KR_L V_S - 1}}{KR_I} \leq v_O \leq V_S$ (b) $V_I = V_T + \frac{\sqrt{1+2KR_L V_S - 1}}{2KR_L}$,
$V_O = \frac{3KR_L V_S + \sqrt{1+2KR_L V_S - 1}}{4KR_L}$ (c) $\frac{\sqrt{1+2KR_L V_S - 1}}{2KR_L}$ (d) $\frac{1 - \sqrt{1+2KR_L V_S}}{2}$
(e) $v_o = \frac{A}{2}(1 - \sqrt{1 + 2KR_L V_S})\sin(\omega t)$

**Ex8.7** (a) $V_O = 10$ V (c) $-50$ (d) $v_O = -0.05\sin(\omega t)$ (e) $r_i = 100\ \text{k}\Omega$ and $r_o = 50\ \text{k}\Omega$ (f) $\frac{i_o}{i_b} = -50$ and $\frac{v_o}{v_i}\frac{i_o}{i_b} = 1250$

**Pr8.1** (a) $V_{MID} = \sqrt{\frac{2(V_S - V_{OUT})}{KR}} + v_T$, $V_{IN} = \sqrt{\frac{2(V_S - V_{MID})}{KR}} + v_T$ (b) $G_m = K^2R^2\left[V_S - .5KR(V_{IN} - v_T)^2 - v_T\right](V_{IN} - v_T)$
(c) 136

**Pr8.3** $-3RKV_{IN}^2$

**Pr8.5** (a) $\frac{v_o}{v_i} = \frac{-\beta R_L}{R_B}$ (b) $\frac{v_o}{v_i} = \frac{-\beta R_L R_1}{R_B(R_1 + R_2)}$

**Pr8.7** $\sqrt{2V_S KR - 2V_T KR}$

**Pr8.9** (b) $V_{OUT} = \frac{1}{KR_S} + V_I - V_T - \sqrt{\frac{2}{KR_s}(V_{IN} + V_S - V_T) + \frac{1}{K^2R_S^2}}$ (d) $\frac{dV_{OUT}}{dV_{IN}} = 1 - (2KR_S[V_{IN} + V_S - V_T] + 1)^{-\frac{1}{2}}$
(e) $\frac{v_{test}}{i_{test}} = R_S$ (f) infinite

**Pr8.11** (b) $-\frac{R_L R_E}{R_L + R_E}K(V_{IN} - V_T)$

**Pr8.13** (a) $V_O = \frac{V_I - 0.6}{1 + \frac{R_I}{(\beta+1)R_E}}$ and $I_E = \frac{V_I - 0.6}{R_E + \frac{R_I}{(\beta+1)}}$ (c) $\frac{v_o}{v_i} = \frac{1}{1 + \frac{R_I}{(\beta+1)R_E}}$ (d) $r_o = (R_E \| R_I)/\left(1 + \beta\frac{R_E \| R_I}{R_I}\right)$
and $r_i = R_I + \beta R_E$ (f) $\frac{i_o}{i_b} = (\beta + 1)\frac{R_E}{R_E + R_O}$ and Power Gain $= (\beta+1)^2\frac{R_E^2}{(R_E + R_O)^2}\frac{1}{R_I + (\beta+1)R_E \| R_O}$

## CHAPTER 9

**Ex9.1** (a) $3/4\,\mu F$ (b) $4\,\mu F$ (c) $4/3\,\mu F$

## CHAPTER 10

**Ex10.1** $i_1(t) = \frac{4}{3}\left(1 - e^{t/\tau}\right)$ mA for $t \geq 0$; $\tau = \frac{1}{3}$ ms

**Ex10.3** $-5$ volts

**Ex10.5** (a) $v = 6e^{-t/\tau}$, $\tau = 500\,\mu s$ (b) $i = (6 \times 10^{-3})e^{-t/\tau}$, $\tau = 2\,\mu s$ (c) $v = 6e^{-t/\tau}$, $\tau = 1$ ms (e) $i = (6 \times 10^{-3})e^{-t/\tau}$, $\tau = 1\,\mu s$

**Ex10.7** (a) For $0 \leq t \leq t_0$, $v = RI_0\left(1 - e^{-t/RC}\right)$, and for $t > t_0$, $v = RI_0\left(1 - e^{-t_0/RC}\right)e^{-(t-t_0)}/RC$

**Ex10.9** 2A

**Ex10.11** $v_C = 2\left(1 - e^{-t/\tau}\right)$, for $\tau = \frac{20}{3}$ ms

**Ex10.13** $v_C = 1 + e^{-t/\tau}$

**Ex10.15** (a) $C_{EQ} = 1\,\mu F$ (b) $\tau = 1$ ms, $v_0(t) = 1 \cdot e^{-t/\tau}$ (c) $v_0(t) = \left(1 - e^{-t/\tau}\right)$; $\tau = 1$ ms for $t > 0$

**Ex10.17** $v_0(t) = \frac{I_1 R_1}{5}\left(1 - e^{-t/\tau}\right)$, $\tau \frac{R_1 C}{5}$

**Ex10.19** (A) $v_0(t) = 10\text{ V}\left(1 - e^{-t/\tau}\right)$; $\tau = R \cdot C$, (B) $v_0(t) = 10\text{V}\left(\frac{R}{R+R}\right)\left(1 - e^{-t/\tau}\right)$; $\tau = R \cdot C$, (C) $v_0(t) = 10\left(1 - e^{-t/\tau}\right)$; $\tau = L/R$, (D) $v_0 = \frac{-10}{RC}t$

**Ex10.21** (a) (i) $\tau = 1$ s (ii) $v_0 = 10e^{-t/\tau}$; $\tau = 1$ s (b) (i) $\tau = 1\,\mu$ s (ii) $v_0(t) = 5\left(1 - e^{-t/\tau}\right)$; $\tau = 1\,\mu s$

**Ex10.23** (a) $v_c = \left[A(t - RC) + (V_0 + ARC)e^{-t/RC}\right]u_{-1}(t)$ (b) $v_c = B\left(1 - e^{-t/RC}\right)$

(c) $v_c(t) = AT + \left[ARC\left(e^{-T/RC} - 1\right)\right]e^{-(t-T)/RC}$

**Pr10.1** (a) $t_{rise} = -\tau \ln\left(\dfrac{V_S - V_H}{V_S - V_S\dfrac{R_{ON}}{R_{ON}+R_L}}\right)$ $\tau = R_L C_{GS}, t_{fall} = -\tau \ln\left(\dfrac{V_L - V_S\dfrac{R_{ON}}{R_{ON}+R_L}}{V_S - V_S\dfrac{R_{ON}}{R_{ON}+R_L}}\right)$ $\tau = C_{GS}\dfrac{R_{ON}R_L}{R_{ON}+R_L}$

(b) $t_{pd} = 8.2\,\mu s$

**Pr10.3** (a) $A$, $B$, $C$, and $E$ must all be high and $D$ must be low (b) $t_{fall} = -\tau_{fall}\ln\left(\dfrac{V_L - V_S\dfrac{4R_{ON}}{4R_{ON}+R_L}}{V_S - V_S\dfrac{4R_{ON}}{4R_{ON}+R_L}}\right)$

$\tau_{fall} = C_{GS}\dfrac{4R_{ON}R_L}{4R_{ON}+R_L}$ (c) $t_{rise} = -\tau_{rise}\ln\left(\dfrac{V_S - V_H}{V_S - V_S\dfrac{2R_{ON}}{2R_{ON}+R_L}}\right)$

**Pr10.5** (a) $t_{rise} = -\tau \ln\left(\dfrac{V_S - V_H}{V_S - V_S\dfrac{R_{ON}}{R_{ON}+R_L}}\right)$ $\tau = nC_{GS}R_L$ (b) $t_{rise} = n8.2\,\mu s$ (c) $t_{rise} = -\tau \ln\left(\dfrac{V_S - V_H}{V_S - V_S\dfrac{R_{ON}}{R_{ON}+R_L}}\right)$

$\tau = (C_W + CGS)(nR_L + R_W)$ (d) $t_{rise} = (0.9 + n90.3)\,\mu s$

**Pr10.9** $v=-1$ V for $2<t<3$ and $v=-1/2$ V for $3<t<5$

**Pr10.11** $0<t<t_1: v_O(t)=\frac{V}{RC}t, t_1<t<t_1+t_2: \quad v_O(t)=\frac{Vt_1}{RC}-\frac{Vt_1}{(RC)^2}(t-t_1)$

**Pr10.13** (a) $i_{AVG}=CV_Af_0$ (b) $R=\frac{v_A}{i_A}=\frac{1}{Cf_0}$

**Pr10.17** $v_L=\tau K(1-e^{-t/\tau}), v_R=Kt-\tau K\left(1-e^{-t/\tau}\right) \quad \tau=L/R$

**Pr10.19** (a) not true (b) true

**Pr10.21** First: $v_O=e^{-t/\tau}$, Second: $v_O=1-e^{-t/\tau}$, $\tau=0.5$ ms

**Pr10.23** (a) $v_B=v_A\frac{R_2}{R_1+R_2}\left(1-e^{-t/\tau}\right), \tau=\frac{R_1R_2(C_1+C_2)}{R_1+R_2}$ (b) (i) $v_B(0^-)=0$ (ii) $v_B(t\to\infty)=v_A\frac{R_2}{R_3+R_2}$

**Pr10.27** $v_R=(K_2-K_1)e^{-t/\tau}+K_3\tau\left(1-e^{-t/\tau}\right)$

**Pr10.29** (a) $V_S$ (b) $T_{min}=-C_M(R_L+R_{ON})\ln\left(1-\frac{V_H}{V_S}\right)$ (c) $\frac{R_{ON}}{R_{ON}+R_L}V_S$ (d) $T_{min}=-C_M\left(R_{ON}+\frac{R_{ON}R_L}{R_{ON}+R_L}\right)$ $\ln\left(\frac{V_L-\frac{R_{ON}}{R_{ON}+R_L}V_S}{\frac{R_L}{R_{ON}+R_L}V_S}\right)$ (e) $-C_MR_P\ln\left(\frac{V_\tau}{V_S}\right)$

## CHAPTER 11

**Ex11.1** (a) $P_{steady-state,0}=0$, (b) $P_{steady-state,1}=\frac{V_S^2}{R_{ON}+R_L}$, (c) $P_{static}=\frac{V_S^2}{2\,(R_L+R_{ON})}$, $P_{dynamic}=\frac{V_S^2R_L^2C_L}{(R_L+R_{ON})^2\,T}$, (d)(i) halved, (ii) quartered, (iii) halved, (e) Maximize $R_L$ while looking out for dynamic constraints

**Pr11.1** (b) $\frac{V_S^2}{R_L}\left(\frac{-T_1+T_2+T_4}{T_4}\right)$ (c) $\frac{V_S^2}{T_4}(C_G+2C_L)$ (d) $P_{static}=2.9$ mW, $P_{dynamic}=87.5\ \mu$W, (e) 0.18 J (f) 51%

**Pr11.3** (b) $P_{\text{static}}=\frac{N}{2}\cdot\frac{V_S^2}{R_L+R_{ON}}$

## CHAPTER 12

**Ex12.1** (a) $2\alpha=\frac{1}{RC}, \omega_o^2=\frac{1}{LC}$, since $\alpha<\omega_o$, underdamped, (b) $v_C=Ke^{-\alpha t}\cos\left(\omega_dt+\phi\right), \omega_d=\sqrt{\omega_o^2-\alpha^2}$, $\phi=\tan^{-1}\left(\frac{\alpha}{\omega_d}\right)$, $\omega_o=10\times10^6, \alpha=3.33\times10^6$, (c) $v_C$ in $RC$ circuit decays as $e^{-t/RC}$, while $v_C$ in RLC circuit decays with "envelope" $e^{-t/2RC}$

**Ex12.3** $t=0^+: i_1=2\text{A}, \ v_1=6\text{V}, \ i_2=3\text{A}, \ v_2=6\text{V}, \ i_3=4\text{A}, \ v_3=4\text{V}, \ i_4=1\text{A}, \ v_4=4\text{V}$. At $t=\infty$: $i_1=10\text{A}, \ v_1=0, \ i_2=0, \ v_2=0, \ i_3=10\text{A}, \ v_3=10\text{V}, \ i_4=0, \ v_4=10\text{V}$

**Ex12.5** $\frac{dv_C}{dt}\big|_{t=0^+}=2\text{V/s}, \frac{di_L}{dt}=\frac{1}{3}\text{A/s}$

**Ex12.7** (a) $x_1=e^{-2t}+e^{-4t}, \ x_2=e^{-2t}-e^{-4t}$, (b) $x_1=2\cos(4t), \ x_2=2\sin(4t)$

**Pr12.1** with small inductor: $v_C(t) = IR - \frac{LIR}{L - R^2C}e^{\frac{-Rt}{L}} + \frac{IR^3C}{L - R^2C}e^{\frac{-t}{RC}}$, without inductor: $v_C(t) = IR\left(1 - e^{\frac{-t}{RC}}\right)$.

**Pr12.3** (a) $i_1' = \frac{L_2}{M^2 - L_1L_2}R_1i_1 - \left(\frac{R_2}{M} + \frac{R_2L_1L_2}{M(M^2 - L_1L_2)}\right)i_2 - \frac{L_2}{M^2 - L_1L_2}v_S$, $i_2' = \frac{-M}{M^2 - L_1L_2}R_1i_1 + \frac{R_2L_1}{M^2 - L_1L_2}i_2 + \frac{M}{M^2 - L_1L_2}v_S$, (c) $v_2(t) = \left(0.05e^{-20202t} - 0.05e^{-20000t}\right)u(t) - \left(0.05e^{-20202(t-0.005)} - 0.05e^{-20000(t-0.005)}\right) \times u(t - 0.005)$

**Pr12.5** (a) $C_A\frac{dv_A}{dt} + \frac{v_A - v_B}{R_A} = K(V_0 - v_B)^2, C_B\frac{dv_B}{dt} + \frac{v_B}{R_B} = \frac{v_A - v_B}{R_A}$, (b) $i_s = -2K(V_0 - V_B)v_b$, (c) Overdamped

## CHAPTER 13

**Ex13.1** (a) $MAG = 16.8$, $PHASE = 13.75$ deg, (b) $MAG = 45.47$, $PHASE = 18°$, (c) $MAG = 2136$, $PHASE = 78°$, (d) $MAG = 47.3$, $PHASE = -15°$

**Ex13.3** $\frac{V_L}{I} = \frac{RLs}{Ls + R}, v_L(t) = \frac{RLI\omega}{\sqrt{(L\omega)^2 + R^2}} \cdot \cos(\omega t + \phi), \phi = \tan^{-1}\left(\frac{R}{\omega L}\right)$

**Ex13.5** $Z = \frac{R}{j\omega RC + 1}, \frac{1}{RC} = 10^4\text{rad/s}, R = 100\ \Omega, C = 1\ \mu\text{F}$

**Ex13.7** $Z_{s=j} = \frac{1}{\sqrt{2}}e^{-(\pi/4)j}$

**Ex13.9** $\frac{R}{L} = 2 \times 10^6\text{rad/s}, \frac{v(j\omega)}{I(j\omega)} = R + L\omega j$

**Ex13.11** (a) $\frac{V_0}{V_i} = \frac{1}{\sqrt{(\omega RC)^2 + 1}}e^{j\phi}$, $\phi = \tan^{-1}(-RC\omega)$, (b) $\frac{V_0}{V_i} = \frac{\omega L}{\sqrt{(\omega L)^2 + R^2}}e^{j\phi}$, $\phi = \tan^{-1}\left(\frac{R}{\omega L}\right)$, (c) $\frac{V_0}{V_i} = \frac{RC\omega}{\sqrt{(RC\omega)^2 + 1}}e^{j\phi}$, $\phi = \tan^{-1}\left(\frac{1}{RC\omega}\right)$, (d) $\frac{V_0}{V_i} = \frac{R}{\sqrt{(\omega L)^2 + R^2}}e^{j\phi}, \phi = \tan^{-1}\left(-\frac{\omega L}{R}\right)$

**Ex13.13** (a) $\frac{V_0}{V_i} = \frac{1}{\sqrt{1 + \frac{\omega^2}{100^2}}}\left(\frac{1}{2}\right)e^{j\phi}, \phi = \tan^{-1}\left(-\frac{\omega}{100}\right)$, (b) $v_0(t) = \frac{1}{2\sqrt{2}}\cos(100t - 45°) + \frac{1}{200.01}\cos(10{,}000t - 89.4°)$

**Ex13.15** (a) $\frac{V_0}{V_i} = \frac{Z_2 \cdot Z_4}{(Z_2 + Z_3 + Z_4) \cdot Z_1 + (Z_3 + Z_4) \cdot Z_2}$,
(b) $\frac{I_a(s)}{V_i(s)} = \frac{Z_3 + Z_4}{(Z_3 + Z_4)Z_2 + Z_1(Z_2 + Z_3 + Z_4)}$

**Ex13.17** $\frac{I_a(s)}{I_s(s)} = \frac{Y_\parallel}{Y_\parallel + Y_1}\ ; Y_\parallel = \frac{Y_2(Y_3 + Y_4)}{Y_2 + Y_3 + Y_4}$

**Pr13.1** (a) (i) $Z = \frac{R}{1 + j\omega RC}$ (ii) $Z = \frac{j\omega RL}{R + j\omega L}$ (iii) $Z = \frac{j\omega RC_2 + 1}{j\omega C_1 - \omega^2C_1C_2R + j\omega C_2}$

**Pr13.3** (c) $H(j\omega) = 0$ db at $\omega = 10^5$ (d) 1; 10; 100; 1,000; 10,000

**Pr13.5** (a) 12 (b) $I = \frac{\sqrt{409}}{25}$ (c) $k = \sqrt{\frac{13}{5}}$

## CHAPTER 14

**Ex14.1** (a) $L' = \dfrac{R^2L}{(L\omega_o)^2 + R^2}$, $R' = \dfrac{\omega_o^2 L^2 R}{(L\omega_o)^2 + R^2}$, (b) $C = \dfrac{R^2 + (\omega_o L)^2}{R^2\,\omega_o{}^2 L}$

**Ex14.3** $R = 400\,\Omega,\ L = 23.7\,\text{mH},\ C = 6.7\mu\text{F}$

**Ex14.5** (b) is inconsistent with the other statements, $Q = 1.3$ actually

**Ex14.7** (a) False (roots are real and negative), (b) False ($Q = 1.3$), (c) True (at $s = \dfrac{j}{\sqrt{LC}}, |H(j\omega)| = 0$), (d) False (system is second order)

**Ex14.9** (a) $V_O(s) = \dfrac{1 + LCs^2}{1 + RCs + LCs^2} V_i(s)$, (b) $v_O(t) = 0$

**Pr14.1** $\dfrac{V_O}{V_L} = \dfrac{1}{(1 - \omega^2 LC) + j\omega RC}$, $\omega_0 = \dfrac{1}{\sqrt{LC}}$, $Q = \dfrac{1}{R}\sqrt{\dfrac{L}{C}}$

**Pr14.3** (a) $\dfrac{1}{(1 - \omega^2 LC) + j\omega RC}$ (b) $Z_{eq} = \dfrac{R + j\omega L}{(1 - \omega^2 LC) + j\omega RC}$ (c) $V_{oc} = 2.03e^{j(120\pi t + 0.311)}$, $Z_{th} = 39.6e^{j(1.622)}$

**Pr14.5** (a) $\dfrac{1}{L}v'_I(t) = i'' + \dfrac{R}{L}i' + \dfrac{1}{LC}i$ (b) $\dfrac{LCs^2}{LCs^2 + RCs + 1}$ (c) $i(t) = \dfrac{C}{\sqrt{(1 - LC)^2 + R^2C^2}}\cos\left[t + tan^{-1}\left(\dfrac{1 - LC}{RC}\right)\right]$

(d) $-\dfrac{R}{2L} \pm i\sqrt{\dfrac{1}{LC} - \dfrac{R^2}{4L^2}}$

**Pr14.7** (a) $Q_1 = \dfrac{L\omega}{R_S}$, $Q_2 = \dfrac{L\omega}{2R_S}$ (b) $Q_1 \approx \dfrac{R_P L_\omega}{R_S R_P + L^2\omega^2}$

**Pr14.9** (a) $H_1(s) = \dfrac{R}{R + RLs + RLCs^2}$, $H_2(s) = \dfrac{1}{1 + LCs^2}$ (b) $i_1^F = 1, i_2^F = 1$ (c) $i_1^N = -e^{-\alpha t}\left(\dfrac{1}{2Q}\sin(\omega_0 t)\right) + \cos(\omega_0 t), i_2^N = -\cos(\omega_0 t)$ (d) $i_1(t) = 1 - e^{-\omega_o t/2Q}\left(\dfrac{1}{2Q}\sin(\omega_o t)\right) + \cos(\omega_o t), i_2(t) = 1 - \cos(\omega_o t)$

**Pr14.11** (a) $C = 10^{-9}$ F (b) $R = 100\,\Omega$ (c) $\Delta\omega = 100{,}000\dfrac{rad}{s}$ (d) $v_O(t) = 1 - e^{-5000t}[0.005\sin(988{,}749t) + \cos(988{,}749t)]$

**Pr14.13** (a) $v_C = \dfrac{4V_0}{3}e^{-2t} - \dfrac{V_0}{3}e^{-8t}$ (b) $\dfrac{RLCs^2 + Ls}{LCs^2 + RCs + 1}$ (c) $s_1 = -2, s_2 = -8, D = -42$

**Pr14.15** (a) (i) $V_O = \left(\dfrac{1 - LC\omega^2}{\sqrt{(1 - LC\omega^2)^2 + (\omega RC)^2}}\right)V_I$, $\phi = -\tan^{-1}\left(\dfrac{\omega RC}{1 - LC\omega^2}\right)$ (ii) $V_O = |V_O(j\omega)| = \left(\dfrac{\omega L}{\sqrt{R^2(1 - LC\omega^2)^2 + (\omega L)^2}}\right)V_I$, $\phi = \angle V_O(j\omega) = \dfrac{\pi}{2} - \tan^{-1}\left(\dfrac{\omega L}{R(1 - LC\omega^2)}\right) = \tan^{-1}\left(\dfrac{R(1 - LC\omega^2)}{\omega L}\right)$

(c) (i) notch (ii) band-pass

## CHAPTER 15

**Ex15.1** $R_{th} = \dfrac{R_2}{gR_2 + 1}, v_{th} = 0$

**Ex15.3** $\dfrac{dG}{G} = \dfrac{1}{1 + AR_A/(R_A + R_B)}$

**Ex15.5** $v_O = \dfrac{-nkT}{q}\ln\left(\dfrac{v_1}{I_S R_1} + 1\right)$

**Ex15.7** $R \leq 1539\,\Omega$

**Ex15.9** (a) $v_O = -v_I$, (b) $v_O = -\frac{1}{2}v_I$

**Ex15.11** (a) $2\,\Omega$, (b) $2/3\,\Omega$

**Ex15.13** $i = V\left(\frac{R_3}{R_1R_3 - R_2R_4}\right)$

**Ex15.15** $(v_i - v_a)g_1 + (v^- - v_a)g_3 + (v_O - v_a)g_2 = 0$ and $(v_a - v^-)g_3 + (0 - v^-)g_4 = 0$, and either $v_O = A(v^+ - v^-)$ and $v^+ = 0$, or $v^+ \approx v^-$ and $v^+ = 0$

**Ex15.17** $i_1 = \frac{A}{R_1 + (1 + A)R_2}v_i$

**Ex15.19** (a) $v_o = -10v_i$, $(b)v_o = -\frac{10}{3}v_i$

**Ex15.21** $R_{TH} = (1 + A)R_S$, $R_S(1 + A) \gg R_L$

**Ex15.23** $v_0(t = 0^+) = -2$ V and $v_0(t = 1\,\text{ms}) = -4$ V

**Ex15.25** $V_{\text{out}} = \frac{R_1C_2s}{(R_1C_1s + 1)(R_2C_2s + 1)} \cdot V_{\text{in}}$

**Ex15.27** (2)

**Pr15.1** $i_L = \frac{Av_I}{AR_2 + R_2 + R_L} \approx \frac{v_I}{R_2}$

**Pr15.1** $AR_2$

**Pr15.3** 1.5 V

**Pr15.5** $i_N = \frac{v_I}{R}, R_{TH} = R$

**Pr15.7** (a) $v_{OUT} = v_{IN}\frac{(R_2)(R_3 + R_4)}{-R_1(R_3 + R_4) + R_4(R_1 + R_2)}$, b) $i = v_{IN}\frac{R_3}{-R_1(R_3 + R_4) + R_4(R_1 + R_2)}$, c) $R_1R_3 = R_2R_4$

**Pr15.9** $\frac{V_{OUT}}{v_{IN}} = -\frac{(R2\|R4) + R3}{R1} = -1.9091.$

**Pr15.11** (a) $R_{IN} = \frac{R_1}{1 + \beta}$, (b) $R_{OUT} = R_2$

**Pr15.13** (a) $\frac{v_O}{v_1} = -2$

**Pr15.17** $C < \frac{I_L}{A\omega}; RC < \frac{V_L}{A\omega}$

**Pr15.21** (a) $v_O = -10 \int v_I dt$

**Pr15.23** $C_2 = C_1(A + 1).$

**Pr15.25** (c) $\tau = \frac{v_F}{100}$

**Pr15.27** $v_O - 2 * 10^{-6}\frac{dv_O}{dt} = 7.5 * 10^{-6}\frac{dv_I}{dt}.$

**Pr15.33** (a) $\frac{V_2}{V_1} = \frac{-1}{R_1C_1s}$, $\frac{V_3}{V_2} = \frac{-1}{R_2C_2s}$, (b) $V_1 = V_2 - V_3 - V_{IN}$, (c) $\frac{V_2}{V_{IN}} = \frac{R_2C_2s}{R_1R_2C_1s^2 - R_2C_2s + 1}$, $\Delta\omega = \frac{1}{R_1C_1}$, $\omega_0 = \sqrt{\frac{1}{R_1R_2C_1C_2}}$.

**Pr15.35** (a) $\frac{V_O}{V_I} = \frac{-10}{10R_1Cs+1}$, (b) $C_x = 40C$.

**Pr15.37** $V_o = V_i\frac{C_2L_2s^2+1}{C_2L_2s^2}$.

**Pr15.39** (a) $\frac{V_O}{V_I} = \frac{R_1R_2C_1C_2s^2}{R_1R_2C_1C_2s^2+R_1(C_1+C_2)s+1}$, (b) $\omega_o = \sqrt{\frac{G_1G_2}{C_1C_2}}$, (c) $\left|\frac{V_o}{V_i}\right| = 2(\omega-\omega_o)+1$, (d) $\left|\frac{V_o}{V_i}\right| = 1$.

**Pr15.41** (a) $\left|\frac{V_o}{V_I}\right| = \frac{-R_2C_1s}{R_1R_2C_1(C_1+C_2)s^2+R_1C_2s+1}$, (b) $v_O(t) = 2\cos(1005t)$,

(c) $v_O(t) = 1.3758\cos(1005t - 47.5 \text{ degrees})$

**Pr15.43** (a) $\frac{V_o}{v_i} = \frac{-R_2}{j\omega R_1R_2C+R_1}$

# FIGURE ACKNOWLEDGEMENTS

**Figure 1.18** courtesy of Anant Agarwal, the Raw Group.

**Figure 1.17** courtesy of Intel Corporation.

**Figure 6.38**, "Intel 0.13um generation logic transistor", courtesy of Intel Corporation.

**Figures 1.2, 1.12, 1.16, 6.7, 6.32, 9.9, 9.10, 12.33, 12.34, 15.2, 16.1** courtesy of Maxim Integrated Products.

# INDEX

NOTE: Web-based material is listed by chapter and page range (WWW Chapter Number:Page Range).

## Special characters



## Numbers



## A



## B



## C

## D

## E

## F



## G



## H

## M



## N

## O

## S